# Preparation of Compounds Labeled with Tritium and Carbon-14

# Preparation of Compounds Labeled with Tritium and Carbon-14

ROLF VOGES
J. RICHARD HEYS
THOMAS MOENIUS

WILEY
A John Wiley and Sons, Ltd, Publication

This edition first published 2009


*Registered office*
John Wiley & Sons Ltd, The Atrium, Southern Gate, Chichester, West Sussex, PO19 8SQ, United Kingdom

For details of our global editorial offices, for customer services and for information about how to apply for permission to reuse the copyright material in this book please see our website at www.wiley.com.



***Library of Congress Cataloging-in-Publication Data***

Voges, R. (Rolf)
Preparation of compounds labeled with tritium and carbon-14 / Rolf Voges,
J. Richard Heys, Thomas Moenius.
p. cm.
Includes bibliographical references and index.
ISBN 978-0-470-51607-2
1. Organic compounds–Synthesis. 2. Radiolabeling. 3. Tritium. 4. Carbon–Isotopes.
I. Heys, J. R. (J. Richard) II. Moenius, Thomas. III. Title.
QD262.V54 2009
572'.36–dc22 2008046982

A catalogue record for this book is available from the British Library.

ISBN 9780470516072 (Hbk)

Typeset in 10/12pt Times by Thomson Digital, Noida, India.

MIX
Paper from responsible sources
FSC® C013604

# Contents

# Preface

The field of organic radiochemical synthesis, like any scientific discipline, grows over time with the development of new knowledge, while remaining rooted in the old from which it is continually nourished. Despite the recent appearance of substantive accounts of newer, related fields such as synthesis with short-lived isotopes, no comprehensive account of the synthesis of tritium- and carbon-14-labeled compounds has appeared for a long time. Yet tritium and carbon-14 isotopes remain the cornerstone isotopes for research in the life sciences, broadly defined. Previous texts on synthesis with carbon-14 or tritium, or both, achieved their different goals well, but the newest of these is 25 years old. Many specialist reviews have been written since then, but by their nature they are narrowly focused or lack depth and are therefore of limited utility.

Despite the lack of an up-to-date comprehensive text in this area, one may question the value a new book may have today, given the increasing extent to which organic chemists rely on electronic databases for synthetic information. In the present case, there are at least two affirmative answers to such a question. First, it is intended that this book will be available from the publisher electronically in several forms, some searchable. Second and more important, in contrast to one-question-one-answer information sources, this book is intended to provide thorough discussions of strategies and methods, to show connections between different elements, exemplifying, comparing and contrasting them. Thereby, we hope to provide the reader with not only a source of information, but also opportunities to acquire deeper understanding, perspective and a stimulus to more creative thinking in the field. This ambition is reflected in the inclusion of chapters devoted to some of the ways that the generation of synthetic strategies for tritium and carbon-14-labeled compounds are different from those of standard (nonisotopic) organic synthesis.

A second important organizing principle of this book is its focus on labeled reagents and building blocks, in contrast to the more reaction-based orientation that characterizes most sources of standard organic synthetic chemistry information (although reaction-based information is thoroughly addressed here as well). Accordingly, this book is intended to complement, not to replace, these other resources. In standard organic synthesis, the conception and 'desktop' evaluation of candidate synthetic routes are carried out through consideration of possible retrosynthetic disconnections or forward synthetic transformations relevant to the target molecule. Generally, the reagents and small building blocks required for these transformations tend to be regarded as incidental, because they are usually available from commercial sources at reasonable costs. However, in organic *radiochemical* synthesis it is precisely the choice of the most appropriate isotopically labeled reagent or small building block that is critical to optimal route planning. Therefore, it is through consideration of both the available nonisotopic reactions and routes, *and* the available isotopic reagents and building blocks and their reactions, that good radiochemical synthetic planning can be accomplished.

Many times throughout the book, reaction conditions described in the original publications are mentioned either in the text or beneath the reaction schemes. We have included this information to assist the reader's understanding of the practical dimensions of the chemistry, and as an aid to comparison of different reactions. *This information should not be used as directions for actual work in the laboratory.* Instead, the reader should refer to the original cited literature for the relevant details.

Writing this book has only been possible because of the efforts of the many, many colleagues who devised, carried out and reported on the chemistry of which this text is built. We thank them. The authors are further indebted to many colleagues for their kindness in answering questions and offering suggestions and advice. We wish to give special thanks to those who generously devoted their time and expertise to review parts of the manuscript and offer helpful comments and corrections. They include Hendrik Andres, A. Jon Bloom, Karl Cable, Grazyna Ciszewska, Chad Elmore, Crist Filer, Albrecht Glänzel, Thomas Hartung, John Herbert, Barry Kent and his colleagues at GE Healthcare, Scott Landvatter, Bill Lockley, Franz Maier, Ulrich Pleiss, Tapan Ray, Ines Rodriguez, Sebastien Roy, Rhys Salter, Piet Swart, Alain Schweitzer, Bill Wheeler, Chris Willis and Markus Zollinger. Finally, our deepest gratitude goes individually to Anna, Elisabeth, Ulrike, Hanna and Moritz who have been generous, loving, accommodating, and as patient as reasonable people can be, during our work on this book.

# Glossary

| | |
|---|---|
| ADME | absorption, distribution, metabolism, excretion |
| AFA | acetic formic anhydride |
| AIBN | $\alpha,\alpha'$-azo-bis(isobutyronitrile); $\alpha,\alpha'$-azobis (2-methyl-propionitrile) |
| Alpine-Borane | lithium *B*-isopinocampheyl-9-borabicyclo[3.3.1]nonane |
| BABS | *N*-bromoacetylbornane-2,10-sultam |
| 9-BBN | 9-borabicyclo[3.3.1]nonane |
| BINAP | 2,2′-bis-(diphenylphosphino)-1,1′-binaphthyl |
| Bn | benzyl |
| Boc | *tert*-butyloxycarbonyl |
| Boc-BMI | 1-(*tert*-butoxycarbonyl)-2-*tert*-butyl-3-methyl-4-imidazolidinone |
| BOP-Cl | bis(2-oxo-3-oxzolidinyl)phosphinic chloride |
| BSA | bis(trimethylsilyl)acetamide |
| Burgess reagent | (methoxycarbonylsulfamoyl)triethylammonium hydroxide |
| Bz | benzoyl |
| CBS catalysts | substituted 1,3,2-oxazaborolidine (Corey-Bakshi-Shibata) catalysts |
| CCD | charge-coupled device |
| CDI | 1,1′-cabonyldiimidazole |
| CDT | 1,2′-carbonyldi-1,2,4-triazole |
| Ci | Curie; 1 Ci = $3.7 \times 10^{10}$ Bq |
| CNC | 1-chloro-1-nitrosocyclohexane |
| COD, cod | 1,5-cyclooctadiene |
| Cp* | pentamethylcyclopentadiene |
| Cy, cy | cyclohexyl |
| CyTMG | *N*-cyclohexyl-*N*′, *N*′, *N*″, *N*″-tetramethyl guanidine |
| d.e. | diastereomeric excess |
| d.p. | diastereomeric purity |
| DABCO | 1,4-diazabicyclo[2.2.2]octane |
| DAST | (diethylamino)sulfur trifluoride |
| Davies reagent | 10-camphorsulfonyloxaziridine |
| dba | dibenzylideneacetone |
| DBAC | di-*tert*-butyl azodicarboxylate |
| DBH | 1,3-dibromo-5,5-dimethylhydantoin |
| DBN | 1,5-diazabicyclo[4,3,0]non-5-ene |

| | |
|---|---|
| DBU | 1,8-diazabicyclo[5,4,0]undec-7-ene |
| DCC, DCCI | 1,3-dicyclohexylcarbodiimide |
| DDQ | 2,3-dichloro-5,6-dicyano-1,4-benzoquinone |
| DEAD | diethyl azodicarboxylate |
| (+)-or (−)-DET | diethyl (L)- or (D)-tartrate |
| DIAD | diisopropyl azodicarboxylate |
| DIBAL | diisobutylaluminum hydride |
| DIEA | diisopropylethylamine |
| DIH | 1,3-diiodo-5,5-dimethylhydantoin |
| DIP-Cl | *B*-chlorodiisopinocampheylborane |
| DMA, DMAC | *N,N*-dimethylacetamide |
| DMAP | 4-(dimethylamino)pyridine |
| DMPU | 1,3-dimethyl-3,4,5,6-tetrahydro-2(1*H*)-pyrimidinone |
| DMS | dimethylsulfide |
| DPMGBS | *N*-(*N′*-diphenylmethylideneglycyl)bornane-10,2-sultam |
| DPPA | diphenylphosphoryl azide |
| dppe | 1,2-bis(diphenylphosphino)ethane |
| dppf | 1,1′-bis(diphenylphosphino)ferrocene |
| EC | Enzyme Commission (enzyme nomenclature) |
| e.e. | enantiomeric excess |
| e.p. | enantiomeric purity |
| EDC, EDCI | 1-[3-(dimethylamino]propyl]-3-ethylcarbodiimide |
| Et-DUPHOS | 1,2-bis(2,5-diethylphosphino)ethane |
| Evans reagent | 4-substituted-1,3-oxazolidin-2-one |
| galvinoxyl | [2,6-di-*tert*-butyl-α-(3,5-di-*tert*-butyl-4-oxo-2,5-cyclohexadien-1-ylidene)*p*-tolyloxy |
| GPE | *N*-glycylpseudoephedrine |
| hfpd | 1,1,1,5,5,5-hexafluoropentane-2,4-dionate |
| HMDS | hexamethyldisilazane |
| HMPA | hexamethylphosphoamide |
| HOBt | 1-hydroxybenztriazole |
| homochiral | enantiomerically pure, enantiopure |
| HOSu | *N*-hydroxysuccinimide |
| HSCIE | high-temperature solid-state catalytic isotope exchange |
| HYTRA | 2-hydroxy-1,2,2-triphenylacetate |
| $Ipc_2BCl$ | *B*-chlorodiisopinocampheylborane (DIP-Cl) |
| isotopomer | the isotopically labeled version of a compound; one of a set of compounds differing from one another only in the number and/or distribution of isotopic atoms |
| KIE | kinetic isotope effect |
| Lawesson's reagent | 2,4-bis(4-methoxyphenyl)-1,3-dithia-2,4-diphosphetane-2,4-disulfide |
| Lindlar catalyst | 5% palladium on calcium carbonate poisoned with lead |
| MCPBA | 3-chloroperbenzoic acid |
| MOM | methoxymethyl |
| MPEG | *mono*-methyl polyethylene glycol |

| | |
|---|---|
| MsCl | methanesulfonyl chloride |
| NBS | *N*-bromosuccinimide |
| NCS | *N*-chlorosuccinimide |
| NMM | 4-methylmorpholine |
| NMO | 4-methylmorpholine *N*-oxide |
| NMP | 1-methyl-2-pyrrolidinone |
| Oppolzer reagent | (2*R*)- or (2*S*)-bornane-10,2-sultam |
| Pa, kPa | pascal, kilopascal; 1 atm = 101,325 Pa |
| PABS | *N*-phosphonoacetylbornane-2,10-sultam |
| $PCy_3$ | tricyclohexylphosphine |
| Periodinane | 1,1,1-tris(acetyloxy)-1,1-dihydro-1,2-benziodoxol-3(1H)-one (Dess-Martin reagent) |
| PET | positron emission tomography |
| PPTS | pyridinium *p*-toluenesulfonate |
| proton sponge | *N,N,N′,N′*-tetramethyl-1,8-naphthalenediamine |
| py | pyridine |
| QWBA | quantitative whole body autoradiography |
| RAMP | (*R*)-1-amino-2-(methoxymethyl)pyrrolidine |
| salen | *N,N′*-bis(3,5-di-*tert*-butylsalicylidine)-1,2-cyclohexanediamine |
| SAMP | (*S*)-1-amino-2-(methoxymethyl)pyrrolidine |
| Seebach reagent | 1-(*tert*-butoxycarbonyl)-3-methyl-4-imidazolidinone |
| TBAF | tetrabutylammonium fluoride |
| TBDMS | *tert*-butyldimethylsilyl |
| TBDPS | *tert*-butyldiphenylsilyl |
| TBTU | *O*-benzotriazol-1-yl-*N,N,N′,N′*-bis(tetramethylene) uronium salt |
| TFA | trifluoroacetic acid |
| TFAA | trifluoroacetic anhydride |
| TIPS | triisopropylsilyl |
| TMEDA | *N,N,N′*N′-tetramethylethylenediamine |
| TMS | trimethylsilyl |
| TPAP | tetrapropylammonium perruthenate |
| Trioxane | *c*-$(HCHO)_3$ |
| trisyl | 2,4,6-triisopropylbenzenesulforyl |
| triton B | benzyltrimethylammonium hydroxide |
| UDP | uridine 5′-diphosphate |
| **Xc** | chiral auxiliary |
| VOC | vinyloxycarbonyl |
| Z | benzyloxycarbonyl |

Isotopic designators in chemical names: Naming conventions generally follow IUPAC rules (*Pure and Applied Chemistry*, ***1979***, *51*, 353–380; www.iupac.org/publications/pac/) ("brackets preceding name of labeled unit")

Isotopic designators in chemical formulas: These generally follow common usage, which is to place a superscript number before the atom to indicate the mass of the isotope that replaces the natural isotope at that position, regardless whether it replaces the natural isotope in all molecules (isotopic substitution) or only some molecules (isotope labeling).

Isotopic designators in chemical structures:

| | |
|---|---|
| * | preferred symbol to indicate that site is labeled or substituted with tritium or carbon-14 ($^3$H may be used instead to indicate stereochemistry) |
| o | indicated site is labeled or substituted with deuterium, carbon-11 or carbon-13 ($^2$H may be used instead to indicate stereochemistry) |
| Δ # ˆ | indicated site is an alternative/optional/conditional position for a label (identity of isotope will be noted in Figure) |
| ( ) | when used with an isotopic designator, indicates minor isotopic content |

Examples of the use of isotopic names, designators and symbols:

[$^{14}$C]Methyl cyanoacetate   $N{\equiv}CCH_2COO^{14}CH_3$ 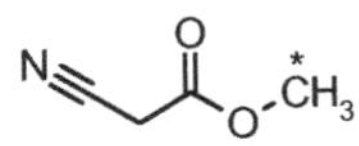

Methyl [$^{14}$C]cyanoacetate   $N{\equiv}{}^{14}CCH_2COOCH_3$ 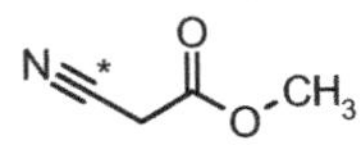

Methyl cyano[1-$^{13}$C]acetate   $N{\equiv}CCH_2{}^{13}COOCH_3$ 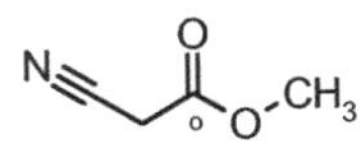

Methyl cyano[2-$^{13}$C]acetate   $N{\equiv}C^{13}CH_2COOH_3$ 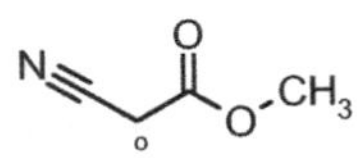

Sodium [$^3$H]borohydride   $NaB^3H_4$ (regardless of specific activity)

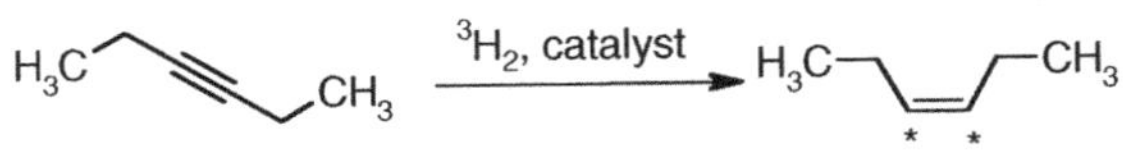


*cis*-[3,4-$^3$H]hex-3-ene

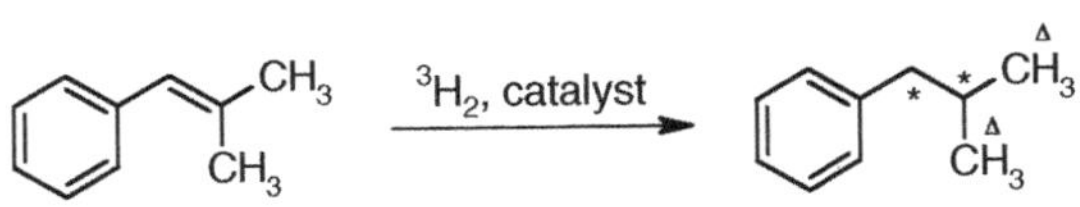


* = positions labeled in reduction without isotope scrambling (2-methyl-1-phenyl[1,2-$^3$H]propane)

*/Δ = positions labeled in reduction with isotope scrambling (2-[$^3$H]methyl-1-phenyl[1,2,3-$^3$H]propane)

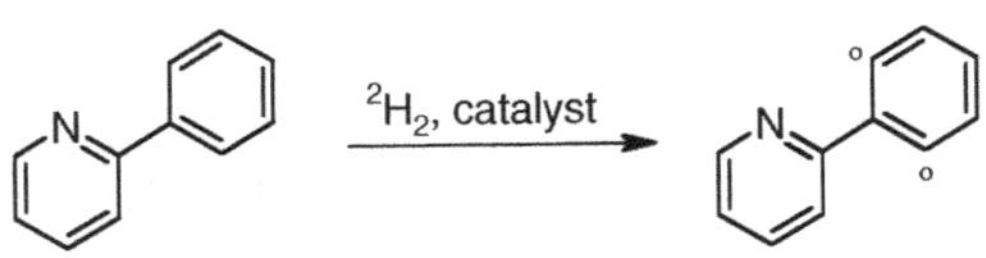


2-[2,6-$^2$H]phenylpyridine

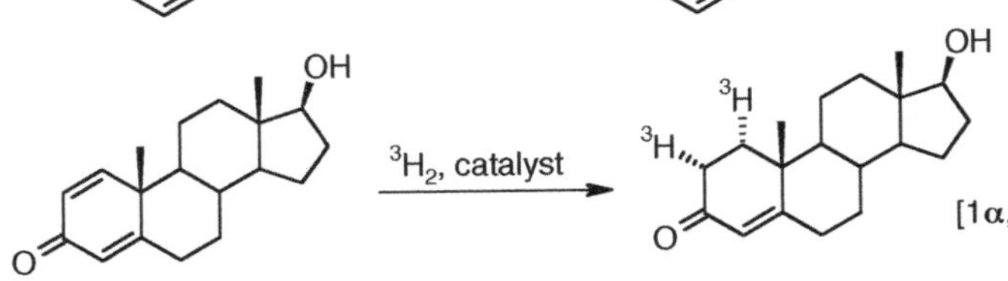


[1α,2α-$^3$H]testosterone

# Author Biographies

**Rolf Voges** studied chemistry at the Universities of Marburg and Freiburg, where he received his Ph.D. in organic chemistry on investigations into steric isotope effects, for which he received the Gödeke Award. After postdoctoral research on stereoselective syntheses he joined the isotope group of Sandoz Pharma AG (now Novartis AG) being involved for thirty years in organic radiochemical synthesis as Head of the Isotope Laboratories and then Head of Isotope Section. He is author or coauthor of about 40 publications, one patent, and coeditor of two previous conference proceeding volumes in the field, co-organizer of two international symposia on the synthesis and application of isotopically labeled compounds, founder and co-organizer of eleven Bad Soden meetings of the Central European Division of the International Isotope Society (CED-IIS), and co-editor of the proceedings. For three years he held a leadership position in the IIS, serving as its 2001 president. In recognition of his scientific achievements and his service to the isotope society he received the IIS-CED Award in 1995 and in 2003 the IIS-Award. He is now retired and lives in southwestern Germany, near the Swiss and French borders.

Contact: Oberer Birkenweg 18, 79540 Lörrach, Germany; red.voges@t-online.de

**Richard Heys** received his Ph.D. in organic chemistry from Stanford University in 1976 and conducted postdoctoral research in the chemistry department at Yale; both involved the synthesis of radiolabeled compounds and their use in elucidation of biosynthetic pathways. His subsequent 29-year career in organic radiochemical synthesis both in the laboratory and as a manager took him to the Radiochemistry Department of Midwest Research Institute (now part of Aptuit, Inc.), Smith Kline & French Laboratories/SmithKline Beecham Pharmaceuticals (now GlaxoSmithKline) and AstraZeneca Pharmaceuticals. Author or coauthor of over 85 publications, 8 patents and a previous conference proceedings volume in the field, organizer of an international symposium on the synthesis of isotopically labeled compound and holder of leadership positions (including president and CFO) in the International Isotope Society for 9 years, he is retired and lives in northwestern Connecticut.

Contact: P.O. Box 576, Litchfield, CT 06759; rheys@optonline.net

**Thomas Moenius** received his Ph.D. in organic chemistry from the University of Erlangen-Nürnberg in 1986. He is a member of the isotope group of Novartis Pharma AG, working in the field of carbon-14 and tritium labeling. Since 2007 he is also European Editor for Journal for Labelled Compounds and Radiopharmaceuticals.

Contact: Novartis Pharma AG, WSJ.507.951, Novartis Campus CH-4056, Basel, Switzerland; thomas.moenius@novartis.com

# 1

# Introduction

Compounds labeled with carbon-14 or tritium have for decades been used in a vast number and wide range of applications, especially in the life sciences[1,2], including research and development of human and animal pharmaceuticals and crop protection agents. Notwithstanding new technological developments, and in some cases because of them, the value of these isotopes in various research areas continues to be great.

In studies of the interactions of small molecules (both synthetic and natural) with receptors, enzymes and other complex biological molecules and systems, compounds labeled with tritium are indispensable because they can be detected and quantified at the nanomolar level of concentration. The tritium label facilitates measurements of the affinity of labeled ligands to their cognate receptors, the densities of receptors in tissue preparations, and the development of high-throughput assays for assessment of the interactions of test compounds with receptors. Analogously, the activity of enzymes can be studied, and potential nonnative substrates screened, by use of a test substrate labeled in such a way as to signal (e.g., by release of radioactivity or change in chromatographic mobility of the substrate) the chemical transformation catalyzed by the enzyme.

Similarly, carbon-14-labeled compounds have no equal for assessment of their metabolism *in vitro* (such as with hepatocytes, cytochrome P450 subtypes or other enzyme or subcellular tissue preparations), or for *in vivo* characterization of their absorption, distribution, metabolism and excretion (ADME) in animals and humans[3], as they can be detected by several different methods and accurately quantified in complex biological matrices. One of the newer of these methods is accelerator mass spectrometry (AMS)[4], whose exquisite sensitivity allows the use of far smaller quantities of carbon-14 than standard ADME studies, therefore providing increased safety margins with regard to radiation exposure to human volunteers.

Compounds labeled with isotopes such as carbon-14 or tritium have also contributed to numerous advances in studies of biochemistry[5], biosynthetic pathways[6], enzyme mechanisms[7], elucidation of organic reaction mechanisms[8] and environmental sciences.

*Preparation of Compounds Labeled with Tritium and Carbon-14* Rolf Voges, J. Richard Heys and Thomas Moenius

Clearly, the value of carbon-14 and tritium isotopes in research is dependent upon their incorporation into compounds of interest. This is made possible by the availability of a wide variety of preparative methods capable of furnishing study compounds possessing the desired isotope(s) in specific locations within the chemical structure and in suitable levels of enrichment.

The successful practitioner in standard (nonisotopic) synthetic organic chemistry needs to possess a broad knowledge of reactions and reagents, the ability to plan a practicable sequence of reactions starting from readily available starting materials and ending with the synthetic target, a facility in executing chemical laboratory operations efficiently and safely, and a working knowledge of analytical methods sufficient to ensure that the progress of a synthesis can be adequately assessed and to obtain information helpful in improving reaction parameters.

The synthesis of compounds labeled with isotopes requires the synthetic chemist to have additional expertise because the synthetic target must be assembled so as to contain one or more isotopic atoms. The preparation of compounds labeled with carbon-14 and tritium requires the ability to deal with a far smaller selection of starting materials compared with standard synthetic chemistry, the ability to plan reaction sequences that generate the correct chemical structures required isotopes in the appropriate positions, knowledge of the circumstances under which these isotopes' $\beta^-$ emissions may lend additional instability to compounds and of ways to avoid or mitigate these effects. Moreover, work with tritium requires mindfulness of how this isotope's vulnerability to loss by exchange processes can be affected by its position in the chemical structure and by the conditions to which the compound is subjected. These are properties relevant not only to predictions about the stability and utility of the tritiated products, but also to the practicability of preparing them by tritium-for-hydrogen exchange when it would be advantageous to do so. Lastly, the fact that carbon-14 and tritium are unstable nuclei means that the practitioner must be well trained and familiar with the proper handling of radioactive materials.

Given all this, it is fair to say that organic synthesis with isotopes is a demanding specialty field within organic synthesis. This book is intended to be both a learning tool for scientists new to the field, and a continuing resource for radiochemical synthesis chemists throughout their careers. It is assumed that the reader has, at a minimum, a practical knowledge of synthetic organic chemistry and a good working knowledge of the chemistry laboratory.

*The authors emphasize the importance of safe working practices and expect that readers make themselves familiar with, and take care to work at all times in accordance with, their national, local and institutional radiation safety protocols regarding carbon-14 and tritium, to maintain good practices of contamination monitoring, and are competent in the control and remediation of radioactive contamination.* Some general guidelines have been published[9].

The organization of this book is as follows. The remainder of this chapter provides short accounts of purification, analysis and storage and stability of compounds labeled with carbon-14 and tritium, and descriptions of some common techniques and technologies unique to work with these isotopes. Chapter 2 discusses some strategies particularly appropriate for planning syntheses of compounds labeled with carbon-14 and tritium, an appropriate topic for inclusion because there are distinct differences vis-à-vis the ways nonisotopic synthetic problems are approached, and an appreciation of these differences is key to effective work in the field. The discussion of one strategy unique to organic radiochemical synthesis, *reconstitution*, is considered worthy of its own chapter, and is elaborated in Chapter 10.

The main parts of the book are devoted to presentations and critical discussion of the use of building blocks, reactions and reagents. These sections are arranged in ways appropriate for each isotope: preparation of tritium-labeled compounds is in large part organized by methodological approach, while preparation of carbon-14-labeled compounds is organized by the various isotopically labeled building blocks. Though most labeling reactions with tritium involve incorporation of the isotope from tritium gas or tritiated water sources into the intact carbon frameworks of final products or late stage synthetic intermediates, sometimes the use of tritiated building blocks is more appropriate. The aim of planning is therefore to identify appropriate substrates and methods for introducing the label. Chapter 3 discusses methods of exchange labeling with tritium gas or tritiated water, and Chapter 4 presents methods of synthesis utilizing tritiated reagents and the relatively small number of readily available tritiated building blocks.

On the other hand, the preparation of compounds labeled with carbon-14 usually involves some amount of carbon framework construction, and a number of carbon-14-labeled building blocks are available for this purpose. Therefore the planning process for syntheses of carbon-14-labeled compounds involves evaluation of synthetic pathways and selection of building blocks, including one or more containing the carbon-14 label. Chapters 5–9 present the most frequently used carbon-14-labeled building blocks and discuss their use.

Finally, two chapters cover, in methodologically oriented fashion, the chemical synthesis of enantiomerically pure $^{3}$H- and $^{14}$C-labeled compounds (Chapter 11) and biological methods of preparation (Chapter 12).

This book is intended to be useful for the researcher in any of several ways. It can be used as a text, which by study in the entirety can bring a newcomer in this field up to a reasonable level of competence. It can be used by scientists faced with specific labeling tasks as a source of reference for comprehensive and critical information on the utility of particular methods, reagents and building blocks. And finally it is hoped that scientists working in the field will find that browsing the book will stimulate new ideas for labeling, provide reminders of methods that can be productively employed in future projects, or spark creative thinking for problem solving in the field.

At times throughout the book the authors have included examples using deuterium or carbon-13 (or even carbon-11) when, in their opinion, the methods are likely to be applicable to tritium and carbon-14, respectively. In particular we note the growing importance of synthesis of carbon-11-labeled compounds as the utility of positron emission tomography grows rapidly[10]; however, synthesis with short-lived isotopes such as carbon-11 and fluorine-18 is a subfield in itself and is covered elsewhere[11].

## 1.1 Physical Properties of Tritium and Carbon-14

The properties of tritium and carbon-14 are well suited for use as tracers in many life sciences and chemistry applications. Table 1.1 lists the important physical properties of the isotopes.

**Table 1.1** *Physical properties of tritium and carbon-14*

| | Tritium | Carbon-14 |
|---|---|---|
| Half-life | 12.3 years | 5730 years |
| Specific activity | 29.2 Ci/milliatom | 62.4 mCi/milliatom |
| Maximum energy of radiation ($\beta^-$) | 18.6 keV | 156 keV |
| Mean energy of radiation | 5.7 keV | 56 keV |
| Decay product | $^3He^+$ (stable) | $^{14}N^+$ (stable) |
| Maximum penetration of radiation | | |
| Air | ca 6 mm | ca 20 cm |
| Water | ca 6 $\mu$m | ca 250 $\mu$m |
| Glass/concrete | ca 2 $\mu$m | ca 170 $\mu$m |

Tritium is prepared in a $^6Li(n,\alpha)^3H$-reaction by irradiation of appropriate lithium-6-enriched compounds (e.g. LiF) or alloys (Li–Al, Li–Mg) with a high flux of neutrons in a nuclear reactor. Some of the tritium evolves as $^3H_2$ gas from the target through recoil during the generation process, and the rest is retained in the solid from which it is liberated by chemical methods.

Carbon-14 is produced in a $^{14}N(n,p)^{14}C$ reaction, also in a nuclear reactor, by irradiation of solid beryllium or aluminum nitride or a saturated solution of ammonium nitrate for periods ranging from 1 to 3 years. Afterwards the target is dissolved in half-concentrated sulfuric acid and the effluent gases are oxidized over an appropriate catalyst. [$^{14}$C]Carbon dioxide resulting from this procedure is absorbed by an aqueous solution of sodium hydroxide and $Ba^{14}CO_3$ is precipitated by addition of aqueous barium hydroxide. Barium [$^{14}$C]carbonate is the standard chemical form for storage and commerce, and it is the universal starting material from which all other carbon-14-labeled compounds are prepared. Because of the omnipresence of environmental carbon, the isotopic purity of $Ba^{14}CO_3$ is normally in the range of 80–90%, corresponding to specific activities of 50–56 mCi/mmol. Material of higher specific activities up to 62 mCi/mmol is commercially available, but it is considerably more expensive and only needed in exceptional cases.

These isotopes emit low-energy $\beta^-$ particle (electron) radiation that does not require shielding for worker safety, as the radiation cannot penetrate the skin. Only with large amounts of carbon-14 can detectable secondary X-radiation occur. This radiation, Bremsstrahlung[12], is produced when electrons are decelerated in the Coulomb fields of atomic nuclei. As the energy of Bremsstrahlung is proportional to both the energy of the electron and the atomic number of the matter through which it passes, it is very low for carbon-14 used or stored in normal laboratory vessels. Routine precautions must be taken, however, to avoid internal exposure to these isotopes through ingestion, inhalation, contact with open wounds, or topical contact with compounds that may be absorbed transdermally. This is easily accomplished by working in fume hoods or glove boxes when there is any possibility of airborne radioactivity, by wearing suitable gloves at all times and by refraining from eating, drinking or smoking in the laboratory. Monitoring of laboratory spaces, equipment and personnel for contamination is easily accomplished in the case of carbon-14 using thin-window Geiger counters; analogous monitoring for tritium can only be accomplished by using windowless gas proportional counting devices. Usually the most expedient method for monitoring of surfaces for removable tritium or carbon-14 contamination is by wiping the surface with a moist cotton swab or filter paper disk and measuring the radioactivity on the wiper material by liquid scintillation counting. Internal exposure of personnel is most easily monitored by regular urine radioanalysis.

As carbon and hydrogen are fundamental components of every organic compound, they can be replaced with carbon-14 and tritium without changing compounds' chemical makeup. Therefore, the chemical and physical properties of compounds labeled with tritium or carbon-14 are very similar to those of their unlabeled counterparts. Metabolically, they behave the same with one exception: if a metabolic transformation involves oxidation at a carbon atom whose hydrogen has been replaced by tritium, that metabolic pathway may be slowed because of the greater energy that is required to break a carbon–tritium bond compared with a carbon–hydrogen bond (primary isotope effect), and in rare cases this can cause significant alterations in the ratio of two or more different metabolites ('metabolic switching'[13], see also Chapter 2). Also, in rare cases the small differences in polarity and/or $pK_a$ caused by tritium isotopic substitution can become apparent during the use of especially sensitive separation methods, such as high performance liquid chromatography (HPLC), in which the retention times of labeled and unlabeled congeners may be different (see Section 1.3.1 below) (secondary isotope effect). Because of the small difference in mass between carbon-14 and carbon-12 these effects are very small, and in life sciences experiments they can usually be neglected.

The long half-lives of these isotopes rarely make it necessary to correct for natural decay; an exception is the long-term ($>1$ year) storage of tritium-labeled compounds.

The range of specific activities available in compounds labeled with these isotopes is suitable for tracer applications extending from mass balance studies in drug metabolism research to detailed investigations of the interactions of small and medium-sized molecules with biological macromolecules such as receptors and enzymes, and from tracing of biosynthetic pathways to the elucidation of chemical reaction mechanisms. Both isotopes can be detected with high sensitivity by a variety of instruments readily available in life science laboratories (liquid scintillation counters, $^3H$ nuclear magnetic resonance (NMR) instruments, mass spectrometers, phosphorimagers[14]), allowing discrimination and measurement of labeled compounds in complex biological samples.

## 1.2 Purification

The methods suitable for purifying compounds labeled with tritium or carbon-14 are fundamentally the same as those for similar nonlabeled compounds on the same mass scale, which is typically in the tens to hundreds of milligrams for carbon-14-labeled compounds, and micrograms to a few milligrams for tritiated compounds. Books such as Microscale Manipulations in Chemistry title as appropriate[15] are useful guides to techniques and devices for manipulating small quantities of compounds. Besides small mass scales, the most important source of constraint on laboratory methods is the need to control the material so as to minimize exposure of the worker and contamination of the laboratory. Approaches to purifying any material should take into consideration the possibility of radiation damage to compounds, which can produce impurities different from those encountered in a corresponding unlabeled compound.

Chromatographic methods are by far the most useful ones for the purification of compounds labeled with carbon-14 and tritium. There is a variety of methods having medium to high resolving power, the most common of which are flash chromatography[16] and its automated cousins, closed column methods such as HPLC or medium pressure liquid chromatography (MPLC), the more recently emerging supercritical fluid chromatography (SFC), and the more classical planar techniques of preparative radial flow chromatography and thin layer chromatography (TLC). The choice of method depends on the equipment available, the mass scale, and the ease of separation of the impurities[17].

Another common purification method is recrystallization. It is operationally simple and can be done on quantities down to the milligram scale using conventional micro-scale techniques and apparatus. It is relatively easy to conduct the required manipulations so as to avoid the inadvertent dispersal of particulates. Compounds sensitive to radiation-generated oxygen radicals in solution can be protected by working under an inert gas atmosphere[18].

## 1.3 Analysis

Analytical characterization of tritium and carbon-14-labeled compounds used in life sciences usually includes the following aims:

(a) To provide evidence of chemical identity;
(b) To measure chemical, and if appropriate enantiomeric, purity;
(c) To measure radiochemical purity (and, rarely, radionuclidic purity);
(d) To determine the specific activity;
(e) To determine or confirm the site(s) of labeling within the molecule.

The analyses most pertinent to each compound are determined by its intended use and the method of its synthesis. The specifications or acceptable numerical limits will depend upon the intended use and the requirements of applicable local procedures, institutional standards or government regulations. The level of detail with which analytical procedures are prescribed, the skill and care with which they are conducted and the quality of data interpretation all vary significantly, according to the expertise of the analyst and local standards of practice. Such differences can be expected to result in correspondingly higher or lower risk to the success of the studies in which the compounds are used. The goal for all analytical measurements should be to minimize subjectivity.

The standards for thoroughness of analysis, degree of procedural rigor, etc. are generally flexible for compounds used in research and early drug discovery studies (where the setting of specifications may be relatively informal), somewhat more formal for studies such as ADME in animals (where some institutional specifications or standards of practice usually exist). The standards are highest for compounds intended for human radiolabel studies (where extensive and detailed prescriptions must be adhered to and formal oversight of procedures and independent review of written records and data are common).

### 1.3.1 Chemical Identity

Analyses pertaining to chemical identity are intended to provide evidence that the structure and, if appropriate, stereochemistry, of a compound are in accordance with that claimed.

NMR is a useful analytical method because it can provide quite detailed information about chemical structure; current widely available instruments are sensitive enough so that $^{1}H$ NMR analysis is feasible for all carbon-14-labeled compounds and all tritium-labeled compounds except the cases where very limited quantities of high specific activity samples are available. $^{13}C$ NMR should also be routinely run when possible, especially for carbon-14-labeled compounds (see below). 2D-NMR methods, which are within reach of most laboratories, are very powerful for assessment of structural details, and should be considered whenever 1D methods leave ambiguity. Mass spectrometric analysis is universally recommended: it can provide not only confirmation of molecular weight, but also data suitable for calculation of specific activity (see Section 1.3.4 below). Classical methods should not be discounted; for example, infrared analysis provides a detailed fingerprint when a reference standard is available for comparison. The potential for isotope-induced changes in NMR and IR spectra should also be recognized, but these changes are quantitatively predictable and need not detract from the quality of the analysis.

Matching of chromatographic retention times is an unreliable indicator of chemical identity, for two reasons. Firstly, it is not uncommon for closely related compounds to have indistinguishable retention characteristics, even in chromatographic systems of high resolving power. Neither is it uncommon that a byproduct or analog of the intended synthesis product is closely related to it, and therefore to the reference standard. Secondly, the higher the resolving power of the method, the more likely it is that the presence of isotopes can alter the retention characteristics of a compound, causing the labeled compound and its unlabeled reference standard to appear chromatographically nonidentical. This phenomenon, called isotopic fractionation[19] has been recognized for decades and has more recently been the subject of a review[20]. Furthermore, the $pK_a$s of amines may be altered by substitution of tritium (or deuterium) for hydrogen on adjacent carbon atoms, resulting in a significant change in retention time if the HPLC mobile phase has a pH value near the $pK_a$ of the amine[21].

### 1.3.2 Chemical (and Enantiomeric) Purity

Chemical purity is defined as the weight of the compound of interest contained in a sample of given weight, usually expressed in percent. For radiolabeled compounds, it is usually calculated arithmetically from measurements of the respective HPLC/UV peak areas obtained after multiple injections of known weights of a sample and of a reference standard of known chemical purity. In the absence of a suitable reference standard, it is only possible to establish a chemical purity *relative to* an available sample of the authentic compound. The ratio of the HPLC/UV peak area of the analyte to the total area of all peaks in an HPLC chromatogram is most definitely *not* a measure of the chemical purity, because such a measurement fails to take into account either the absorbance characteristics of the observed impurities or any impurities that do not absorb light at the wavelength of detection, nor do they account for any impurities not eluting from the column, or any solvents or inorganic salts that may be present in the sample.

Chemical purity measurements are often not performed on tritiated compounds of high specific activity. Such materials are usually prepared in submilligram quantities, and, even if it were feasible to weigh samples accurately enough to prepare solutions of known mass concentration, the required manipulations would increase the risk of decomposition of the compound (see Section 1.5). Fortunately, high specific activity tritiated compounds are used in such small (mass) quantities that unlabeled impurities are unlikely to cause problems. The absence of major impurities can be assessed to a modest degree of certainty by HPLC analysis with UV detection.

Measurements of enantiomeric purity are most conveniently accomplished using chiral HPLC analysis against authentic samples of enantiomerically pure and racemic materials. Alternative methods include NMR with a chiral shift reagent and optical rotation measurements.

### 1.3.3 Radiochemical (and Radionuclidic) Purity

Radiochemical purity, in analogy to chemical purity, is the ratio of radioactivity contained in the compound of interest to the total radioactivity of the sample. Radiochemical purity is usually measured chromatographically in order to exploit its separation power, and HPLC with online radioactivity detection is the most preferred method because of its superior resolution and detail-rich radioactivity profiling. Prior to the advent of modern HPLC radiodetectors, the eluate stream was collected in fractions that were counted by liquid scintillation counting (LSC) in order to construct a histogram from which the radiochemical purity could be extracted. This method is sometimes used with extremely low specific activity samples; however, the newer technique of automated fraction collection in multiwell microplates followed by high-throughput solid scintillation counting allows a degree of fractionation high enough to approach the resolution of on-line detection[22].

Thin layer chromatography continues to be used in spite of its relatively lower resolving power compared with HPLC, and potential problems in quantitation, such as self-absorption of radiation within the layer, especially with tritium (ca 2 $\mu$m path length of $\beta^-$ in solid materials, see Section 1.1 above). One advantage compared with HPLC radio-detection is that impurities not eluting through HPLC columns may be detected on the TLC plate. In this technique, radiometry is conducted after development and drying of the TLC plate; available methods are of two types, linear and two-dimensional. Linear methods are based on single- or multi-wire gas proportional counters sensitive to a narrow band parallel to the direction of plate development. Two-dimensional methods include film-based autoradiography, which is of relatively low quantitative power, and the more modern techniques based on phosphor imaging screens or high-resolution crossed-wire proportional counters and high sensitivity CCD cameras, both of which are supported by computerized measurement systems.

Since it is relatively common that compounds coelute or nearly coelute in even high-resolution chromatographic systems, it is always recommended that radiochemical purity analyses be carried out in two different chromatographic systems, as unlike one another as possible. For example, the combination of one reverse-phase HPLC or TLC method and one normal-phase HPLC or TLC method is usually recommended, but two reverse-phase HPLC analyses using different column types and mobile phases may also be acceptable.

It should be noted that these chromatographic–radiometric methods have inherent shortcomings that must be understood and taken into account both in the chromatography component and the radiodetection component of the assay.

The chromatographic part: in HPLC, there is the possibility that one or more radioactive components are not detected because they fail to exit the column by the end of the monitoring period. An effective check for this possibility is to measure the quantity of radioactivity exiting the column and to compare it with the amount injected. In TLC, there is the possibility that one or more components may be lost or diminished through volatilization before scanning can be completed. Before running a TLC analysis, the absence of volatile components may be confirmed by measuring the radioactivity in a sample before and after drying *in vacuo*.

The radiometric part: there are two phenomena involved. First, unlike UV detectors for HPLC, which have virtually no baseline noise, in HPLC radiodetection and TLC radioscanning there is always a significant baseline noise, resulting from a combination of environmental radiation, detector noise and the decay statistics of low-level radioactivity. The presence of this noise makes it difficult to distinguish between baseline and minor radioactive impurities, and even more difficult to measure them accurately. The second is that radiodetector peaks of eluting components tend to be broader and to have more pronounced tails than the UV detector peaks for the same components. This reduction in resolution is caused by the additional internal volume of the in-line radiodetector cell and associated plumbing that the sample must travel through after it exits the UV detector. Resolution can be further degraded by suboptimal scintillant flow rates (for liquid scintillant radiodetector cells) and peak tailing characteristic of many solid scintillant cells. There is an unavoidable tradeoff between maximization of radiodetector resolution and sensitivity (signal-to-noise ratio). These characteristics of radiodetector performance may make it more difficult to recognize and accurately quantify impurities running close to the compound of interest, and more difficult to judge the point at which the tail of the peak of interest returns to baseline. The uncertainties involved make accurate and consistent interpretation difficult, and make the data reduction process vulnerable to subjective judgments and therefore person-to-person variability. It is not unusual for measurements by two scientists using the same instrumentation for the analysis of the same sample to vary by 2%, and by 1% between successive injections by the same person. It is therefore not surprising that it is sometimes difficult to be certain whether a 98% radiochemical purity level (typically specified for ADME studies) has been met. Most radiometric instruments allow the operator to establish settings for control of a variety of parameters for data processing, such as baseline correction, peak detection and peak deconvolution, but correctly establishing these parameters requires skill, and the same parameter settings may not be optimal for all analyses.

*Radionuclidic* purity is only of concern in the context of dual-isotope labeling, or if cross-contamination from a laboratory mishap is suspected. Radionuclidic purity is best measured by liquid scintillation counting; modern LSC instruments have detectors and analysis software designed to discriminate quantitatively between the different isotopes used in the life sciences, except at very low counting levels. Radionuclidic purity is entirely distinct from isotopic purity, or content, of compounds labeled with stable isotopes, such as deuterium or carbon-13. Such information may be very important to the utility of stable-labeled compounds such as internal standards for mass spectrometric quantitation assays[23].

### 1.3.4 Specific Activity

There are two ways in which specific activity is expressed, radioactivity per unit mass (e.g., mCi/mg) and radioactivity per molar unit (e.g., Ci/mmole or mCi/mmole).

The former, used more frequently for carbon-14-labeled compounds and low specific activity tritiated compounds, is most simply determined by preparing a solution of known mass concentration and measuring the radioactivity of defined volumes by LSC. Solutions already made up for chemical purity determinations can be used for this assay.

The latter expression of specific activity is used almost always for high specific activity tritiated compounds, and often for carbon-14-labeled compounds. Measures of radioactivity per molar unit can be calculated from mass spectrometry data[24]. In this analysis, the distribution of isotopic species (e.g., $^3H_0$, $^3H_1$, $^3H_2$, ...) is determined by measuring the relevant peak intensities in the molecular ion envelope and correcting them for naturally occurring isotopes (e.g., $^{13}C$, $^{34}S$) present in the molecule; this can be done manually or by use of readily available computer algorithms. The contribution of each isotopic species to the total can be used to calculate the average number of isotopic atoms per molecule and thence, from the molar specific activity of the pure isotope, the molar specific activity of the compound.

Interconversion between expressions of radioactivity per unit mass and radioactivity per molar unit must take into account the fact that the former has to be corrected for the chemical purity of the sample, whereas the latter does not.

### 1.3.5 Position of Label

The importance of knowing the location of carbon-14 or tritium atoms within compounds depends primarily upon their intended use, but may be of interest also in studies of reaction mechanisms, molecular rearrangements and mechanisms of isotope exchange.

In most syntheses the location of carbon-14 atoms follows logically from the route of synthesis. However, there are cases in which the position of a label was altered by previously unrecognized or incompletely understood reaction pathways, as illustrated by the reactions of **1** to **2**[25] and **3** to **4**[26] in Figure 1.1. In these cases, the carbon-13 labels and $^{13}C$ NMR were

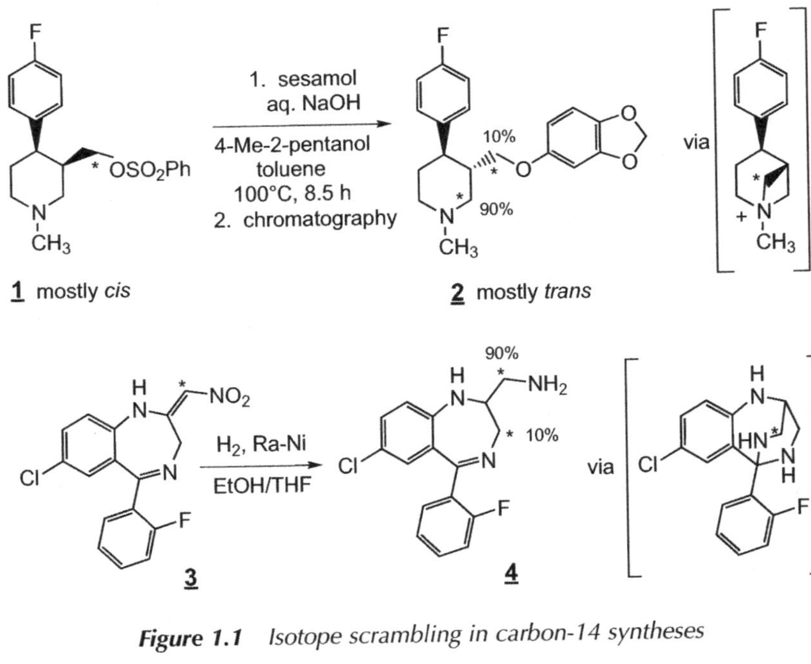


**Figure 1.1** *Isotope scrambling in carbon-14 syntheses*

used to elucidate the phenomena. When it is of interest to confirm independently the location of carbon-14 atoms themselves within a structure, $^{13}$C NMR spectroscopy can also be used for this purpose. As carbon-14 contains none of the 1.1% natural abundance carbon-13 isotope present in nonlabeled sites, $^{13}$C NMR spectra measured under conditions where all atoms are relaxed between pulses (and all natural carbon sites therefore give signals of equal area or intensity), those carbon sites substituted with carbon-14 can be recognized by the diminution of the corresponding signals. As the extent of diminution is proportional to the carbon-14 isotopic enrichment at each site, the information can also be used in specific activity calculations.

Compounds labeled with carbon-14 biosynthetically may have complex patterns of isotope incorporation and/or low carbon-14 enrichment; in such cases compound prepared in carbon-13 pilot studies may be used as a surrogate for analysis.

For tritiated compounds, $^{3}$H NMR is universally the method of choice for identification and quantitation of label distribution[27]. Tritium probes for NMR instruments have become widely available, the sensitivity of tritium detection is greater atom-for-atom than that for protium, and tritium chemical shifts are almost identical to those of protium. These factors make $^{3}$H NMR an extremely powerful technique for quantitative determination of the distribution of tritium in samples of tritiated molecules down to the low mCi range in routine instruments, and as low as the tens of $\mu$Ci range using high-field cryoprobe-equipped instruments[28]. Spectra recorded with and without suppression of protium–tritium coupling, and application of various 2D-NMR techniques[29] have been used to interpret complex $^{3}$H NMR spectra. Even the measurement of $^{3}H_1$, $^{3}H_2$ and $^{3}H_3$ isotopomers of methylene and methyl groups can be accomplished through generally predictable peak multiplicities and geminal and vicinal isotope effects on chemical shifts[30].

## 1.4 Stability and Storage of Compounds Labeled with Tritium or Carbon-14

Because radiolabeled compounds are continuously exposed to their own radiation, they are often observed[31] to decompose much faster than their unlabeled counterparts. Therefore, the purity of radiolabeled compounds should either be measured by analysis shortly before use or by applying knowledge of stability data already accumulated for the same compound under the same storage conditions. However, in the latter case, the large batch-to-batch variability in stability often noted with radiolabeled compounds makes it risky to estimate purity levels of individual batches over time. Occasional observations of radioracemization[32] underline the need for appropriate purity controls to ensure the quality of compounds. It is common for isotope laboratories maintaining inventories of tracers to devote significant resources to purity monitoring and repurifications. One strategy sometimes used to mitigate these efforts is to store late synthetic intermediates rather than final products. This is most effective with intermediates that are more stable than final products and the transformation from one to the other is simple, and in cases where portions of the final product do not require frequent dispensing.

In any case, a good understanding of the nature of radiolytic decomposition and knowledge of ways to minimize its impact can be of great benefit[33,34]. Modes of decomposition have been summarized[35] as shown in Table 1.2.

*Table 1.2* Modes of decomposition of radiolabeled compounds

| Mode of decomposition | Cause | Mitigation |
|---|---|---|
| Primary (internal) | Isotopic decay | None at a given specific activity |
| Primary (external) | Interaction with radioactive emission | Dispersal of labeled molecules |
| Secondary | Interaction with molecules excited by the radiation | Dispersal of reactive molecules, Free radical scavenging, Cooling to low temperatures |
| Chemical | Chemical instability | Cooling to low temperatures |

**Primary (internal) decomposition** is caused by disintegration of unstable nuclei within compounds and is of concern mainly for those labeled with short-lived isotopes such as sulfur-35, phosphorus-32 and phosphorus-33. Primary (internal) decomposition can be neglected for carbon-14-labeled compounds, as the half-life is very long (5730 years). In the case of tritium ($T_{1/2} = 12.3$ years), primary (internal) decomposition of approximately 5%/year can usually be ignored for mono-labeled compounds because the remaining molecular fragments are nonradioactive. However, compounds containing multiple tritium atoms or dual $^3H/^{14}C$ labels will produce radiolabeled impurities upon disintegration of one of the tritium atoms, at a rate proportional to the number of tritium atoms remaining in the molecule. For example, a compound containing four tritium atoms per molecule would be expected to generate impurities amounting to about $(4 - 1) \times 5\% = 15\%$ after 1 year of storage, from primary (internal) decomposition alone.

**Primary (external) decomposition** arises from direct interaction of the emitted ($\beta^-$) particles with nearby labeled molecules. The emission energy of even low-energy emitters such as tritium and carbon-14 is two to four orders of magnitude greater than the average energy of an organic bond (mean $\beta^-$ energies for these isotopes = 5.7–56 keV or $8.3 \times 10^4 - 8.2 \times 10^5$ kcal/mol). Therefore, this interaction can easily lead to decomposition of the molecules impacted, which may number in the hundreds per $\beta^-$ particle. Primary (external) decomposition is likely to be significant only for tritiated and carbon-14-labeled compounds in a neat state, i.e., a pure solid, bulk liquid or thin film. It is much more consequential for tritiated compounds for two reasons. First, the path length of tritium $\beta^-$ particles is so short (~6 $\mu$m) that practically all their energy[36] is deposited within the bulk of the compound mass. Second, the radiation density within the bulk of tritiated compounds is usually very high, owing to the specific activities common for such compounds. In contrast, the longer path length of carbon-14 $\beta^-$ emissions permits much of their energy to escape the mass of neat carbon-14-labeled compounds, and radiation densities are lower, in proportion to these compounds' lower specific activities. The combination of these two factors reduces the impact of primary (external) decomposition in carbon-14-labeled compounds. Moreover, crystal lattice energies can additionally contribute significantly to compound stability, so that compounds in crystalline form are often more stable than in their amorphous, liquid or glassy form.

For labeled compounds dissolved in a solvent or in contact with air, it is mostly solvent or air molecules that are activated, typically forming radicals. **Secondary decomposition** can result from the interaction of these radicals with labeled molecules. Usually this is by far the largest contributor to the decomposition of labeled compounds, and the one that most needs to be controlled in order to extend the stability of compounds in storage.

Radiolytic decomposition is a multi-dimensional process, depending on a variety of parameters such as chemical structure, specific activity, impurities and storage conditions (formulation, solvent and storage temperature). Attempts have been made to calculate '*G*(-M)'values*[38], semiempirical measures that are meant to characterize the vulnerability of systems to damage by radiation, based on chemical structure. Unfortunately, as *G*(-M) values strongly depend on experimental data, and are importantly influenced by the impurity profile of individual batches, their practical value appears to be questionable[34]. General strategies for stabilizing radiolabeled compounds are usually effective and are discussed below, with the acknowledgment that it is difficult to predict in individual cases[32] what the effect of any particular parameter will be.

**General aspects:** One source of information on storage conditions for particular compounds, or compounds related to the compound of interest, is the catalogs of commercial radiochemical vendors. These companies have strong incentives to maximize the shelf-lives of their compounds, and they have had extensive experience with a variety of chemicals. As a rule, the use of highly pure solvents, tight container closures, controlled temperature and avoidance of unnecessary freezing–thaw cycles is recommended. Some discussion is provided below on three areas of consideration for stabilizing radiolabeled compounds.

(a) **Dispersal of radiolabeled molecules**

As discussed above, primary (external) decomposition is of concern mostly for tritiated compounds; it can be reduced by dispersal of the molecules, and dissolution in a solvent is the most commonly practiced method. Criteria for solvent selection to minimize reactive species are discussed in the next section. However, from a practical point of view the solvent should be chemically compatible with the compound, able to keep it in solution at the selected concentration and storage temperature, and either be removable with appropriate ease or be compatible with the intended use of the solute. The optimum concentration for a particular compound is a compromise between maximization of dispersal on the one hand and disadvantages of excessive dilution on the other. The disadvantages include increased vulnerability of the solute to chemical decomposition caused by dissolved oxygen or trace organic, acidic or metal contaminants in the solvent, and the impracticality of storing large volumes of solution.

Our own practical experience recommends concentrations of 0.5–10 mCi per mL of highly pure, deoxygenated solvent, and storage in several small quantities instead of one large one to avoid frequent opening. Radicals are readily formed upon interaction of radiation with oxygen, and compounds sensitive to these radicals are more prone to decomposition. Compounds containing divalent sulfur are highly sensitive; oxidation of amines has been documented; and data collected on the stability of MK0677[39] indicate the susceptibility of its benzyloxy moiety.

If the radiolabeled compound is to be stored in its natural physical state, it might be stabilized by dilution with unlabeled carrier to the lowest specific activity needed. While blending of a liquid radiolabeled compound and carrier can be simply

*$G$(-M) is defined as the number of molecules irreversibly changed per 100 eV of energy absorbed; see also Ref. 37.

accomplished by mixing, blending of labeled and unlabeled solids should be carried out by complete dissolution together and subsequent recrystallization of the blend, in order to avoid particle-scale inhomogeneity.

Traces of heavy metals can significantly catalyze the rate of decomposition. Therefore, for example, it is recommended to remove metal catalysts from crude products as soon as practicable after the reactions in which they have been used.

(b) **Minimizing reactive species**

For many years benzene either pure or in combination with methanol[40] was the solvent of choice for the storage of less polar radiochemicals. By virtue of its $\pi$-orbitals, benzene's ability to absorb energy and to stabilize excited states rationalized this choice. Nowadays toluene, which has equivalent stabilizing properties, is usually used instead of benzene because of the latter's high toxicity. Alcohols such as methanol or ethanol are commonly used for dissolution of more polar compounds. Both act as radical scavengers[41] and are widely observed to extend the shelf-life of a variety of compound types when used either as solvent or as a cosolvent. Other solvents such as pentane and ethyl acetate, being neither radical inhibitors nor promoters, are sometimes used either as solvents or cosolvents. In contrast, chloroform, dichloromethane[42], water and ether-type solvents tend to form radicals and therefore can destabilize radiolabeled compounds. Water tends to form highly reactive hydroxyl radicals in the presence of radiation, so water alone is a poor choice of solvent. However, addition of at least 2% ethanol to aqueous solutions effectively traps radicals, so effectively so that 95 : 5 aqueous ethanol is quite commonly used for the storage and shipment of polar compounds such as amino acids, sugars and nucleotides. Similarly, dichloromethane–methanol or water–methanol solvent mixtures may be suitable alternatives[43].

A significant advantage in using toluene or ethanol in storage solutions is their dual role as solvents and stabilizers. Other effective radical scavengers, such as dimethyl sulfide, 2-mercaptoethanol[38], and benzyl alcohol, are occasionally used as stabilizing additives to solutions of labeled compounds in various solvents, but usually only in cases where they are compatible with the intended studies as they are difficult to separate from the labeled solute.

(c) **Lowering the storage temperature**

As the decomposition of radioactive compounds by interaction with radicals or other activated molecules proceeds by chemical reactions with perceptible activation energies, the reaction rates are temperature dependent. Therefore, the lower the storage temperature the lower the rate of decomposition should be, and the optimum storage temperature is the lowest temperature practicable. This rule is well documented not only for solids and solutions, but also for frozen solutions[44].

If the storage temperature is lower than the freezing point of a solution, the solution should be frozen rapidly, for example by immersion in liquid nitrogen, rather than being placed in the freezer in the liquid state. Slow freezing of solutions can result in crystallization or precipitation of the solute, or cause migration of the solute towards the center of the volume (the last to freeze), thereby greatly increasing the local concentration of the compound and offsetting the benefit of molecular dispersion.

If liquid nitrogen storage is used, stringent precautions must be taken in packaging samples. The vial closures routinely used in organic chemistry laboratories do not prevent the slow insinuation of nitrogen into vials, whether immersed in liquid nitrogen or stored in the $-140\,^{\circ}C$ zone above it. They can then explode as they warm up after being removed from the storage facility. Plastic cryovials with silicone-ring seals are suitable for storage of samples in liquid nitrogen, but they are incompatible with many low-polarity organic solvents. Flame-sealed glass ampules are the most secure but they are fragile and troublesome to prepare. The use of 'cryo' class labeling materials or markers is recommended.

Freezers designed to maintain temperatures down to $-140\,^{\circ}C$ are commercially available, but they are costly; therefore many radiochemistry laboratories compromise by storing stocks in more readily available $-80\,^{\circ}C$ or $-60\,^{\circ}C$ freezers.

## 1.5 Specialist Techniques and Equipment

Synthetic work with tritium- or carbon-14-labeled compounds differs in several ways from synthetic work with unlabeled materials, and this reality influences the way isotope scientists plan and conduct their work (see Chapter 2). Differences include:

- The need for more careful containment of materials, owing to the requirements of radiation safety and the desire to avoid the spread of contamination (subject to the specifics of national and institutional regulations);
- The use of smaller scales, especially with tritium;
- The higher likelihood that radioactive compounds will be unstable (discussed in general in Section 1.4 above, and mentioned for specific compounds where they are discussed throughout this book).

These differences impose greater constraints upon radiochemical synthesis work compared with those of 'regular' organic synthesis, constraints that can limit the feasibility of certain operations or significantly increase the difficulty of conducting them. Over the decades, various practices have been developed in efforts to manage or mitigate these constraints, usually through individual ingenuity and inventiveness. Regrettably, no thorough compilation of these "tricks of the trade" exists. While it is not the intention here to cover this subject comprehensively, some important aspects will be discussed. Other sources of information are available[45].

Methods and technologies for manipulation of reactants and reaction mixtures should take account of the potential volatility of radioactive components, including possible byproducts. A practical distinction can be made among components that are (a) gases at or near room temperature; (b) liquids whose boiling points (up to about 180–200 °C) render them suitable for manipulation by static vacuum transfer; (c) materials of low but significant volatility (liquids b.p. ~200 to ~280 °C and solids with a tendency to sublimate); and (d) nonvolatiles.

Fundamentally, the handling of nonvolatile materials need not differ significantly from that of unlabeled materials, except that special care should be taken against scattering of less dense solids by fume hood drafts or static buildup, which is a much more severe problem because of ionization induced by emitted $\beta^-$ particles.

However, substantial practical efficiencies can be achieved by simplification of procedures. For example, many small-scale reactions can be conducted in inexpensive screw-cap vials with 'flea'-sized magnetic stirring bars. Inert atmospheres and exclusion of moisture can be maintained by using septum stoppers and inert gas supplied through syringe needles. Heating is conveniently provided by an oil bath. Workups can be performed in the same or analogous vials, by separating layers using Pasteur pipets, removing solvents with gentle streams of nitrogen, etc. Utilizing such disposable supplies reduces the amount of glass washing required.

Materials that are gases at or near room temperature must unavoidably be handled in a closed vacuum line. A large number of glass vacuum line systems have been described for the manipulation of such materials[46], and extensive descriptions of vacuum line techniques are available[18,47]. As liquid nitrogen is generally used to condense materials in order to manipulate and transfer them on a vacuum manifold, tritium gas, [$^{14}$C]carbon monoxide and methane isotopomers, which have boiling points below 77 K, require different approaches. Commercial stainless steel vacuum systems have been developed for manipulation of tritium gas and its storage by reversible uptake on depleted uranium[48], supplanting the older Toepler pump technology[†]. The newer systems take advantage of the reversible, exothermic reaction of tritium gas with uranium metal to give $U^3H_3$[49]. Tritium gas can be expelled from the $U^3H_3$ storage bed with excellent control (equilibrium $^3H_2$ pressure = 1 atm (760 Torr) at 436 °C and $\sim 1.3 \times 10^{-3}$ Pa ($\sim 10^{-6}$ Torr) at 25 °C), and any excess of tritium gas is spontaneously reabsorbed by the bed as it cools.

There is as yet no analogous manifold available for handling $^{14}CO$; however, a glass setup for its generation and use has been described[50]. Among the so-called 'condensible' gases (b.p. >77 K), which are routinely manipulable using liquid nitrogen cooling, [$^{14}$C]carbon dioxide is usually prepared and used as needed on vacuum manifolds or simple glass vacuum assemblies, but a commercial steel $^{14}CO_2$ storage and handling manifold has recently become available[51]. Carbon-14-labeled gases purchased commercially are usually packaged in breakseal vessels with ground glass fittings that can be attached to the user's vacuum line and opened under controlled conditions.

Vacuum-transferrable volatile materials (b.p. up to about 180–200 °C) often encountered include (aside from those prepared as intermediates in the radiochemical laboratory) commercial building blocks such as [$^{14}$C]methyl iodide and other low-molecular-weight carbon-14-labeled alkyl halides, methanol, ethanol, benzene, acetic and haloacetic acids, acetyl and haloacetyl chlorides and dimethylformamide. These compounds are most appropriately handled on vacuum manifolds in the same way as gases, but some may, with proper experimental design, be used without such systems. In the latter case, it is strongly recommended that safety measures be taken against the possibility of the release of volatile radioactivity.

The most problematic materials can be those with low but significant volatilities. The main reason is that the potential for volatilization under particular conditions may not be evident or easily evaluated beforehand. For example, [$^{14}$C]benzoic acid can evaporate at a significant rate by sublimation on a rotary evaporator with a little warming, and some nonvolatile [$^{14}$C]aryl compounds, safely handled in the open, can gain significant volatility when substituted with fluoro groups. Experimental plans should be examined for possible

[†] The 'mercury air pump', August Toepler (1836–1912): lecturer, Academy Poppelsdorf, Bonn (1859–1864); chair of chemistry and chemical technology, Polytechnic Institute, Riga (1864–1868); professor, University of Graz (1868–1876); chair, Experimental Physics and director, Physical Institute, Dresden Technical University (1876–1900).

cases such as these, and if appropriate, modification of reaction conditions should be seriously considered. Tracer runs may be useful in developing appropriate methods.

A separate track in terms of intelligent handling of potentially problematic materials or reactions is the redesign of syntheses to make them more efficient—even if they do not involve volatile materials at all. There are two key drivers for synthesis redesign: reducing the need for manipulations of materials, and reducing the volatility of intermediates while preserving their relevant reactivity. Achievement of these goals has the added benefit of reducing the likelihood of radiation-induced decomposition (especially critical for polymerizable intermediates such as radiolabeled acrylates).

The reduction of manipulations can be accomplished through modification of experimental procedures so as to run as many reaction steps as possible in one pot, and/or to perform transfers of labeled intermediates from one reaction vessel into another without opening up the system, whether it be a complex vacuum manifold or a simple glassware assembly. Such tactics have been called 'telescoping' reactions[45a]. An example of a simple procedure is the preparation and use of $^{14}CO$ already mentioned. A more complex one, is illustrated for [$^{14}$C]methyl 3,5-dinitrobenzoate (7) in Figure 1.2[52]. Reduction of $^{14}CO_2$ with $LiAlH_4$ produces lithium tetramethoxyaluminate (5), which requires release of [$^{14}$C]methanol by addition of a proton source. Tetrahydrofurfuryl alcohol was selected for this purpose because of its relatively low volatility (b.p. 178 °C). The reduction itself was conducted in the solvent 2(3)-(tetrahydrofurfuryloxy) tetrahydropyran (b.p. 106–108 °C/6 mmHg) for the same reason. The [$^{14}$C]methanol (6) thus produced (94% yield) was expelled by purging the solution with a gentle stream of helium then trapped in a solution of 3,5-dinitrobenzoyl chloride and pyridine in ethyl acetate to generate the easily handled carbon-14-labeled ester, from which [$^{14}$C]methyl iodide can easily be prepared. This procedure is generally applicable for the synthesis of C1–C5 [$^{14}$C]alcohols that have reasonable volatility; higher alcohols are better separated from the solvent by fractional distillation (see Chapter 5, Section 5.1.2).

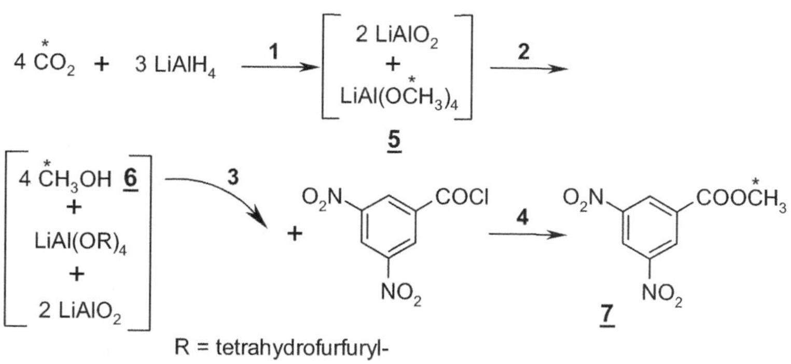


**Reaction conditions**: **1**. 2(3)-(tetrahydrofurfuryloxy)tetrahydropyran (T-solvent), $LiAlH_4$, 0 °C, 10 min; **2**. tetrahydrofurfuryl alcohol; **3**. He (50 mL/min), 110 °C; **4**. py, EtOAc; r.t., 16 h.

***Figure 1.2*** *Synthesis of [$^{14}$C]methyl 3,5-dinitrobenzoate*

A virtuosic example of telescoping[53], depicted in Figure 1.3, began with barium [$^{14}$C]carbonate and ended after six synthetic steps with [1-$^{14}$C]glycerol (**9**). It was carried out without any workups or isolations of intermediates; only the continuous extraction of [1-$^{14}$C]acrylic acid (**8**) might have been avoided. The overall yield of purified [1-$^{14}$C]glycerol at a specific activity of 10 mCi/mmol was 35%, based on $Ba^{14}CO_3$.

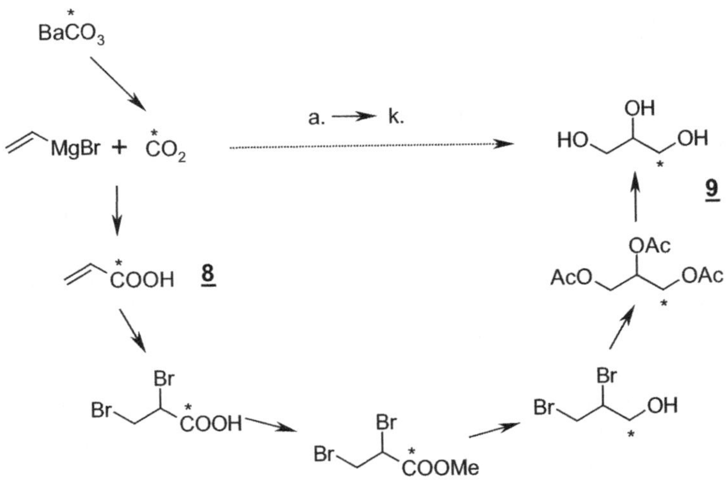


**Reaction sequence:** **a.** $Et_2O$; −20 °C, 20 min, **b.** add aq. $H_2SO_4$, **c.** continuous extraction with $Et_2O$ (tr. hydroquinone), **d.** concentrate to 10 mL with $N_2$ stream, **e.** add $Br_2$; r.t., 1 h, **f.** add $CH_2N_2/Et_2O$; 0 °C, 3 h, **g.** add $AlCl_3/Et_2O$, $LiAlH_4$; −80 °C, 3 h, **h.** add wet $Et_2O$, **i.** add HOAc, NaOAc; reflux, 15 h, **j.** evap volatiles, **k.** add aq. NaOH; reflux, 6 h.

***Figure 1.3*** *'Telescoped' synthesis of [1-$^{14}$C]glycerol*

An alternative tactic for managing volatile intermediates is to 'devolatilize' them by covalently combining them with molecular entities that render the resulting compounds nonvolatile, preserves the reactivities of the radiolabeled moieties, and are removable after having served their purpose.

This principle can be illustrated by the synthesis of the carbon-14-labeled azulene derivative [$^{14}$C]**11**, an advanced intermediate in the preparation of an anti-inflammatory and anti-ulcerative drug candidate[54] (Figure 1.4). The route developed for the unlabeled compound utilized the condensation of butyraldehyde with the lactone **10**; substitution of [1-$^{14}$C]butyraldehyde in the reaction would lead to [$^{14}$C]**11** possessing the label at the desired position. However, rather than prepare, handle and utilize radiolabeled butyraldehyde (b.p. 75 °C), especially since the reaction is conducted at 78 °C, the investigators chose to utilize the synthetically equivalent but nonvolatile building block **12**. Condensation of **12** with the lactone partner proceeded similarly to that with butyraldehyde, then the oxygen function was removed in three high-yield steps as shown.

Another method of devolatilization is to use solid-phase synthesis methodologies. The preparation of 2-propyl[1-$^{14}$C]octanoic acid (**16**)[55] might serve as an example, even though the labeled substrate is not highly volatile. (Figure 1.5) [1-$^{14}$C]Octanoic acid (**14**) was

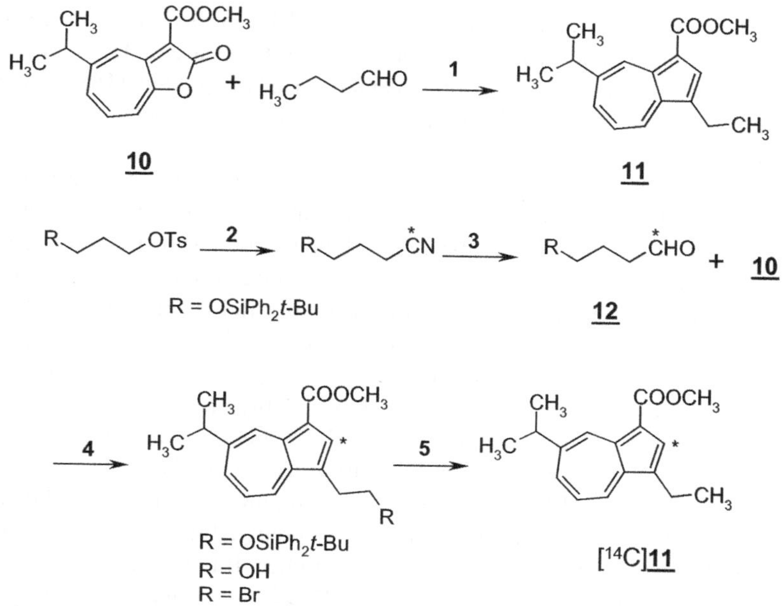


**Reaction conditions:** **1**. morpholine, EtOH; reflux, 7 h; **2**. $K^{14}CN$, 18-crown-6, MeCN; 90–100 °C, 9 h; **3a**. DIBAH, hexane/$Et_2O$; −25 °C, 2 h, **b**. 5% aq. $H_2SO_4$; 0 °C, 30 min; **4**. morpholine, EtOH; reflux, 8 h; **5a**. TBAF, THF; 20 °C, 1 h, **b**. $PPh_3$, $CBr_4$; 0 °C, 1.5 h then 20 °C, 30 min, **c**. $Bu_3SnH$, toluene; 100 °C, 2.5 h.

***Figure 1.4*** *Synthesis of methyl 3-ethyl-7-isopropyl[2-$^{14}$C]azulene-1-carboxylate.*

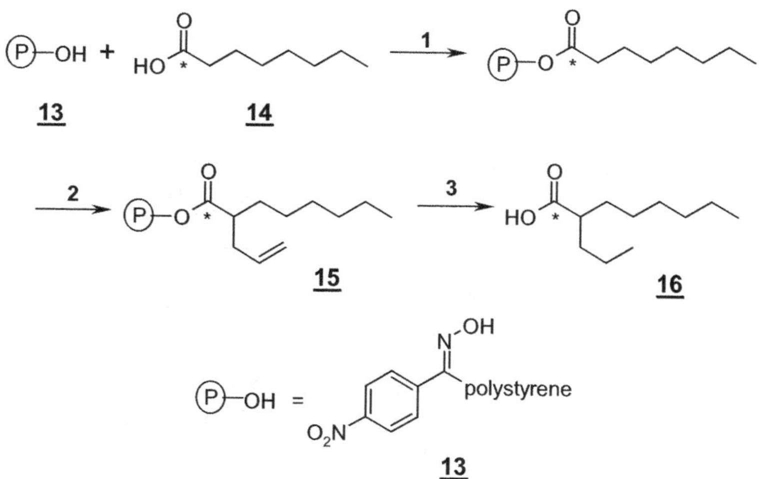


**Reaction conditions**: **1.** EDC, $CH_2Cl_2$; **2.** LDA/THF, −40 °C; allyl bromide; **3.** $H_2$, Pd/C, MeOH.

***Figure 1.5*** *Solid-phase synthesis of 2-propyl[1-$^{14}$C]octanoic acid*

attached to the oxime-functionalized polystyrene resin **13** then alkylated with allyl bromide to give the solid-phase supported 2-allyl derivative **15**. Exposure of **15** to mild catalytic hydrogenation reduced the allyl double bond and released the product (**16**) from the resin. Other examples, using carbon-14[56] and developed for use with labeled materials[57], have been published.

The devolatilization concept has been extended to certain commonly used isotope sources with the additional advantage of making them more stable or storable. $[^3H/^{14}C]$Methyl nosylate (**17**) has been developed[58] as a substitute for tritiated or carbon-14-labeled methyl iodide, both of which are relatively unstable and difficult to handle (Figure 1.6). Ester **17** is a nonvolatile, easily purifiable solid that is substantially less radiolytically sensitive than the corresponding methyl halides ($[^3H]$methyl nosylate at a specific activity of >80 Ci/mmol suffered no appreciable decomposition after storage at 39 mCi/mL for 14 weeks at 4 °C in hexane/ethyl acetate[58]). This derivative is reported[58] to possess similar reactivity to $[^3H/^{14}C]$methyl iodide in a variety of reactions, and to provide greater flexibility during use in synthesis.

9-Phenyl[9-$^2$H]fluorine (**18**), a nonvolatile solid with p$K_a \sim 18$, has been shown[59] to be a source of deuterium to quench carbanions; it was prepared in high isotopic content by Pd-catalyzed deuterio-debromination of 9-bromo-9-phenylfluorene using deuterium gas. A number of model compounds (exemplified by **19**–**22**) were labeled, in good chemical yields and excellent isotope incorporation, by metallation (*n*-BuLi, THF, −78 °C) and subsequent addition of **18** to the lithio species. The deuterium in **18** is acidic enough to undergo transmetallation with a wide variety of carbanions, but not so acidic as to be air- or moisture-sensitive. Although the analogous tritiated reagent has not been reported in the literature, it could be prepared analogously, appears likely to be stable on storage and could prove to be a good alternative to high specific activity tritiated water.

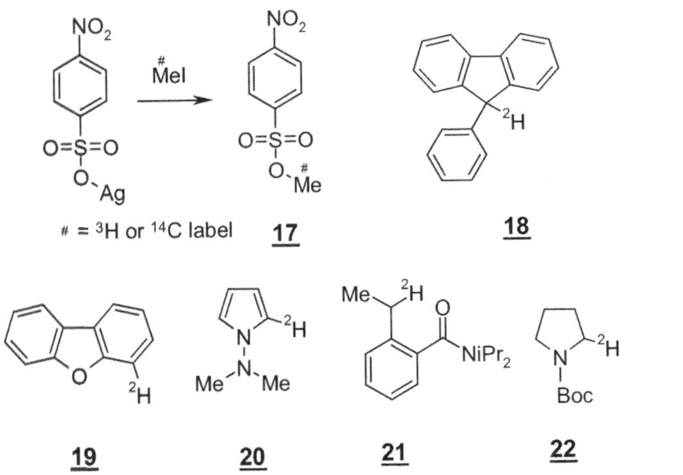


*Figure 1.6* Devolatilized sources of tritium/carbon-14 and deuterium

## References


1. Important classic texts include: a. Kamen, M.D., *Radioactive Tracers in Biology;* Academic Press: New York, ***1948*** and 2nd Edn., ***1951***; b. Wang, C.H. and Willis, D.L., *Radiotracer Methodology in Biological Science;* Prentice-Hall: Englewood Cliffs, NJ, ***1965***.
2. Recent texts and reviews include: a. *Radiotracers in Drug Development*, Lappin, G. and Temple, S., Eds.; CRC Press: Boca Raton, FL, ***2006***; b. Marathe, P.H.; Shyu, W.C. and Humphries, W.G., *Curr. Pharm. Design*, ***2004***, *10*, 2991–3008; c. Dalvie, D., *Curr. Pharm. Design*, ***2000***, *6*, 1009–1028.
3. Roffey, S.J.; Obach, R.S.; Gedge, J.I. and Smith, D.A., *Drug Metab. Rev.*, ***2007***, *39*, 17–43.
4. Brown, K.; Dingley, K.T. and Turteltaub, K.W., *Methods Enzymol.*, ***2005***, *402*, 423–443; Vogel, J. S. and Love, A.H., *Methods Enzymol.*, ***2005***, *402*, 402–422. For a recent example of the application of AMS to ADME studies see Beumer, J.H.; Garner, R.C.; Cohen, M.B.; Galbraith, S.; Duncan, G.F.; Griffin, T.; Beijnen, J.H. and Schellens, F.H.M., *Investig. New Drugs*, ***2007***, *25*, 327–334. For a recent example of AMS in microdosing and experimental INDs see McLean, M. A.; Tam, C.-Y. J.; Baratta, M.T.; Hollimen, C.L.; Ings, R.M. and Galluppi, G.R., *Drug Devel. Res.*, ***2007***, *68*, 14–22.
5. Hansen, P.E., *J. Label. Compd. Radiopharm.*, ***2007***, *50*, 967–981.
6. Mahmud, T., *J. Label. Compd. Radiopharm.*, ***2007***, *50*, 1039–1051.
7. a. Cleland, W.W., *J. Label. Compd. Radiopharm.*, ***2007***, *50*, 1006–1015; b. Fitzpartick, P.F., *J. Label. Compd. Radiopharm.*, ***2007***, *50*, 1016–1025.
8. a. Hoff, R.H. and Hengge, A.C., *J. Label. Compd. Radiopharm.*, ***2007***, *50*, 1026–1038; b. Lloyd-Jones, G.C. and Munoz, M.P., *J. Label. Compd. Radiopharm.*, ***2007***, *50*, 1072–1087; c. Parkin, G., *J. Label. Compd. Radiopharm.*, ***2007***, *50*, 1088–1114; d. Westaway, K.C., *J. Label. Compd. Radiopharm.*, ***2007***, *50*, 989–1005; e. Happel, J., *Isotopic Assessment of Heterogeneous Catalysis;* Academic Press: Orlando, FL, ***1986***.
9. *Guide to Working Safely with Radiolabeled Compounds;* Amersham Life Sciences: Amersham International plc, UK, ***1992.***
10. Welch, M.J. and Redvanly, C.S.; *Handbook of Radiopharmaceuticals;* John Wiley & Sons, Ltd: Chichester, ***2003.***
11. Antoni, G.; Kihlberg, T and Långstrom, B. in Welch, M.J. and Redvanly, C.S. *Handbook of Radiopharmaceuticals;* John Wiley & Sons, Ltd: Chichester, ***2003***, pp. 141–194; see also the proceedings of the *17th International Symposium on Radiopharmaceutical Science,* Aachen, 30 Apr–4 May ***2007*** published in, *J. Label. Compd. Radiopharm.* ***2007***, *50*, 255–641.
12. Friedlander, G.; Kennedy, J.W.; Macias, E.S. and Miller, J.M., *Nuclear and Radiochemistry,* 3rd Edn; John Wiley & Sons, Inc.: New York, ***1981***, pp. 222, 579.
13. a. Horning, M.G.; Thenot, J.P.; Bouwsma, O.; Nowlin, J. and Lertratanangkoon, K.,in *Adv. Pharmacol. Ther., Proc. 7th Intl. Cong. Pharmacol. 1978*, Tillement, J.P., Ed.; ***1979***, *7*, 245–256; b. Miwa, G.T. and Lu, A.Y., *BioEssays*, ***1987***, *7*, 215–219.
14. Schweitzer, A., *Spatial Imaging of Radioactivity in Animal Tissues and Organs in Pharmacokinetics*, 2nd Edn; Marcel Dekker: New York, ***1995.***
15. Ma, T.S. and Horak, V., *Microscale Manipulations in Chemistry;* John Wiley & Sons, Inc.: New York, ***1976.***
16. Still, W.C.; Kahn, M. and Mitra, A., *J. Org. Chem.*, ***1978***, *43*, 2923.
17. Sheppard, G., *The Radiochromatography of Labeled Compounds, Review 14;* The Radiochemical Centre Ltd.: Amersham, UK, ***1972.***
18. Shriver, D.F. and Drezdzon, M.A., *The Manipulation of Air-Sensitive Compounds*, 2nd Edn; John Wiley & Sons, Inc.: New York, ***1986.***
19. Klein, P.D., *Adv. Chromatogr.*, ***1966***, *3*, 3–65.
20. Filer, C.N., *J. Label. Compd. Radiopharm.*, ***1999***, *42*, 169–197.
21. See, for example, Heys, J.R., *J. Chromatogr.*, ***1987***, *407*, 37–47.
22. Bruin, G.J.; Waldmeier, F.; Boernsen, K.O.; Pfaar, U.; Gross, G. and Zollinger, M., *J. Chromatogr. A*, ***2006***, *1133*, 184–194.
23. Richards, D.P.; Sojo, L.E. and Keller, B.O., *J. Label. Compd. Radiopharm.*, ***2007***, *50*, 1124–1136.

24. Moenius, Th.; Richmond, R.; Masero, R.; Jean, C. and Nufer, R., in *Synthesis and Applications of Isotopically Labeled Compounds 1994*, Allen, J. and Voges, R., Eds.; John Wiley & Sons, Ltd: Chichester, ***1995***, pp. 475–479.
25. Willcocks, K.; Barnes, R.D.; Rustidge, D.C. and Tidy, D.J.D., *J. Label. Compd. Radiopharm.*, ***1993***, *33*, 783–794.
26. Zhang, Y.; Woo, P.W.K.; Hartman, J.; Colbry, N.; Huang, Y. and Huang, C.C., *Tetrahedron Lett.*, ***2005***, *46*, 2087–2091.
27. A recent brief review is available: Filer, C.N., *J. Radioanal. Nucl. Chem.*, ***2006***, *268*, 663–669.
28. a. Alexakis, E.; Bloxsidge, J.P.; Jones, J.R. and Lockley, W.J.S., *J. Label. Compd. Radiopharm.*, ***2007***, *50*, 300–302; b. Alexakis, E.; Bloxsidge, J.P.; Jones, J.R. and Lockley, W.J.S., *J. Label. Compd. Radiopharm.*, ***2006***, *49*, 406–407; c. Bloxsidge, J.P.; Garman, R.N.; Gillies, D.G.; Jones, J.R. and Lu, S.-Y., in *Synthesis and Applications of Isotopically Labeled Compounds*, Vol. 8, Dean, D.C.; Filer, C.N. and McCarthy, K.E., Eds.; John Wiley & Sons, Ltd: Chichester, ***2004***, pp. 381–384.
29. a. Vogt, F.G.; Freyer, A.J.; Levinson, S.H.; Shu, A.Y.L. and Heys, J.R., *Magn. Resonance Chem.*, ***2005***, *43*, 147–155; b. Williams, P.G.; Morimoto, H. and Wemmer, D.E., *J. Am. Chem. Soc.*, ***1988***, *110*, 8038–8044.
30. Evans, E.A.; Warrell, D.C.; Elvidge, J.A. and Jones, J.R., *Handbook of Tritium NMR Spectroscopy and Applications;* John Wiley & Sons, Ltd: Chichester, ***1985***.
31. Tolbert, B.M.; Adams, P.T.; Bennett, E.L.; Hughes, A.M.; Kirk, M.R.; Lemmon, R.M.; Noller, E. M.; Ostwald, R. and Calvin, M., *J. Am. Chem. Soc.*, ***1953***, *75*, 1867–1868.
32. Considerable variability in reported stability data underline the complexity of the situation (compare ref. 34).
33. a. Bayly, R.J. and Evans, E.A., *J. Label. Compd.*, ***1966***, *2*, 1–34; b. Bayly, R.J. and Evans, E.A., *J. Label. Compd.*, ***1967***, *3* (Suppl. 1), 349–379; c. *Guide to the Self-Decomposition of Radiochemicals*, Amersham Life Sciences, UK, ***1992***; d. Suzzoni-Dezard, S. and Pichat, L., *LÁtualite Chim.*, ***1997***, *7*, 14–17.
34. Evans, E.A., *Tritium and Its Compounds*, 2nd Edn.; John Wiley & Sons: New York, ***1974***, pp. 663–672.
35. Bayly, R.J and Weigel, H., *Nature*, ***1960***, *188*, 384–387.
36. Friedlander, G.; Kennedy, J.W.; Macias, E.S. and Miller, J.M., *Nuclear and Radiochemistry*, 3rd Edn; John Wiley & Sons, Inc.: New York, ***1981***, pp. 206–242.
37. Rochlin, P., *Chem. Rev.*, ***1965***, *65*, 685–696.
38. Bonner, W.A., *Radiation Res.*, ***1999***, *152*, 83–87.
39. Jones, A.N.; Dean, D.C.; Jenkins, H.J.; Melillo, D.G.; Nargund, R.P. and Wallace, M.A., *J. Label. Compd. Radiopharm.*, ***1996***, *38*, 561–565.
40. Ekström, A. and Garnett, J.L., *J. Label. Compd. Radiopharm.*, ***1967***, *3*, 167–183.
41. Sheppard, G.; Sheppard, H.C. and Stivala, J.F., *J. Label. Compd.*, ***1974***, *10*, 557–567.
42. Roller, P.R.; Keefer, L.K.; Bradford, W.W. and Reist, E.J., *J. Label. Compd. Radiopharm.*, ***1981***, *18*, 1261–1272.
43. Fredenhagen, A., *J. Label. Compd. Radiopharm.*, ***2001***, *44*, 971–972.
44. Bayly, R.J. and Evans, E.A., *Storage and Stability of Compounds Labeled with Radioisotopes, Review No. 7*, Amersham Life Sciences, UK, ***1968***, p. 80.
45. a. McCarthy, K.E., *Current Pharmaceutical Design*, ***2000***, *6*, 1057–1083; b. Calvin, M.; Heidelberger, C.; Reid, J.C.; Tolbert, B.M. and Yankwich, P.E., *Isotopic Carbon;* John Wiley & Sons, Inc.: New York, ***1949***; c. Evans, E.A., *Tritium and its Compounds*, 2nd Edn; John Wiley & Sons, Inc.: New York, ***1974***, pp. 190–237.
46. a. Oekonomopulos, R., in *Synthesis and Applications of Isotopically Labeled Compounds 1988*, Baillie, T.A. and Jones, J.R., Eds.; Elsevier: Amsterdam, ***1989***, pp. 53–54; b. Eberle, A.N. and Zeller, A., *Helv. Chim. Acta*, ***1985***, *68*, 1880–1892; c. Mezo, I.; Seprodi, J.A.; Teplan, I.; Morgat, J.-L.; Fromageot, P.; Toth, G. and Sirokman, F., *J. Label. Compd. Radiopharm.*, ***1978***, *14*, 557–567; d. Devillers, J.; Winand, M. and Bettens, B., *J. Label. Compd. Radiopharm.*, ***1976***, *12*, 219–229; e. Marche, P.; Girma, J.-P.; Morgat, J.-L.; Fromageot, P.; Ghelis, C.; Dubrasquet, M. and Bonfils, S., *Eur. J. Biochem.*, ***1975***, *50*, 375–382; f. Morgat, J.-L.; Desmares, J. and Cornu, M., *J.

*Label. Compd.*, ***1975***, *11*, 257–264; g. Muramatsu, M. and Tajima, K., *J. Label. Compd.*, ***1966***, *2*, 304–313; h. Gosztonyi, T. and Walde, N., *J. Label. Compd.*, ***1966***, *2*, 155–173; i. Otto, P.Ph.H.L. and Juppe, G., *J. Label. Compd.*, ***1965***, *1*, 115–127; j. Birkofer, L. and Hempel, K., *Chem. Ber.*, ***1963***, *96*, 1373–1381; k. Glascock, R.F. and Pope, G.S., *Biochem. J.*, ***1960***, *75*, 328–335.

47. Plesch, P.H., Ed., *High Vacuum Techniques for Chemical Syntheses and Measurements;* Cambridge University Press: Cambridge, ***2005***; c. Bubner, M. and Schmidt, J., *Laboratorien, Apparature und Analysen bei chemischen Synthesen Kohlenstoff-14 markierter Verbindungen;* Georg Thieme Verlag: Leipzig, ***1966***, pp. 39–89.
48. RC TRITEC, Teufen, Switzerland (rctritec.com); IN/US Systems, Inc., Fairfield, NJ (www.inus.com).
49. a. Rapkin, E.; Steele, G. and Schavey, R., *Am. Lab.*, October ***1995***, pp. 31–38; b. Spedding, F.H.; Newton, A.S.; Warf, J.C.; Johnson, O.; Nottorf, R.W.; Johns, I.B. and Daane, A.H., *Nucleonics*, January ***1949***, pp. 4–15.
50. Elmore, C.S.; Dean, D.C. and Melillo, D.G., *J. Label. Compd. Radiopharm.*, ***2000***, *43*, 1135–1144
51. RC TRITEC, Teufen, Switzerland (www.rctritec.com).
52. Voges, R., Internal Procedure, Novartis Switzerland.
53. Pichat, L.; Herbert, M. and Aubert, F., *J. Label. Compd.*, ***1965***, *1*, 66–78.
54. Suzaka, H.; Tomiyama, T. and Ikegami, S., *J. Label. Compd. Radiopharm.*, ***1990***, *28*, 901–910.
55. Ho, J.Z.; Tang, C. and Braun, M.P., *J. Label. Compd. Radiopharm.*, ***2007***, *50*, 496–497.
56. Tan, W.; Iyer, R.P.; Jiang, Z.; Yu, D. and Agrawal, S., *Tetrahedron Lett.*, ***1995***, *36*, 5323–5326.
57. Austin, C.A.; Smith, D. and Hartley, R.C., *J. Label. Compd. Radiopharm.*, ***2007***, *50*, 502–503.
58. Pounds, S., in *Synthesis and Applications of Isotopically Labeled Compounds*, Vol. 8, Dean, D. C.; Filer, C.N. and McCarthy, K.E., Eds.; John Wiley & Sons, Ltd: Chichester, ***2004***, pp. 469–472.
59. a. Cintrat, J.-C. and Rousseau, B., *J. Label. Compd. Radiopharm.*, ***1998***, *41*, 1017; b. Cintrat, J.-C.; Pillon, F. and Rousseau, B., *Tetrahedron Lett.*, ***2001***, *42*, 5001–5003.

# 2

# Strategies for Target Preparation

'Target' is the term used to refer to the isotopically labeled compound intended as the goal of an isotopic synthesis or preparation. Any plan or strategy for preparation of a target must take into account a number of factors, the most important of which are its specifications and the resources available to the isotope chemist. The specifications of a target include the identity of the isotope(s), quantity of radioactivity, specific activity, position(s) of the label(s) within the structure, limit(s) of purity, dose, dosage form. It is the intended use of a target that determines, or places limits upon, the specifications; the optimum properties of a target are defined by the type and details of the study(ies) in which it will be used.

The resource limits the isotope chemist must deal with include the timeline, budget, laboratory facilities and equipment, regulatory requirements, the availability of labeled and unlabeled chemical intermediates and standards, unlabeled target and relevant synthetic experimental procedures.

Often, the ideal specifications for a target may be impossible to attain within the limitations of available resources. This will necessitate a compromise that needs to be consistent with both the constraints on the scientist preparing the target and the requirements of the study. Moreover, as organic radiochemical synthesis is rarely routine, unexpected results may occur during target preparation, forcing a change in plan. Therefore, it is very important that there is good communication between isotope chemist and customer during the entire course of target synthesis.

One example of such a compromise has been driven by the incompatibility between the time required for traditional syntheses of carbon-14-labeled compounds for ADME studies and the increasingly aggressive timelines of modern drug development. The solution for some organizations has been to effect 'fast labeling approaches' (one- or two-step transformations)[1] capable of delivering targets adequate for use in early phase studies, leaving the more resource-intensive preparations of targets for later phase studies until when (if ever) they are needed. These fast approaches often involve preparation of tritium-labeled targets that, although not optimum because of this isotope's greater liability toward metabolic loss, meet the short timelines and are able to provide data of a level of quality sufficient

*Preparation of Compounds Labeled with Tritium and Carbon-14* Rolf Voges, J. Richard Heys and Thomas Moenius

to support the preliminary nature of early studies. More advanced studies on selected candidates may additionally be supported by syntheses of carbon-14-labeled targets.

Given the resource limitations listed above and the capacity constraints on radiochemists, outsourcing of part or all of target preparation to commercial custom synthesis companies is increasingly included in the range of options considered for managing project loads. Depending on capacity, financial resources and confidentiality issues, the preparation of early or advanced intermediates or entire syntheses (especially resyntheses) might be outsourced. For many radiochemistry groups, selective outsourcing allows them to offset limitations in in-house technical capabilities, enhance their capacity and/or accelerate their delivery timelines.

The first part of this chapter discusses the links between the types of studies most frequently conducted with isotopically labeled compounds (targets), and the range of target specifications that are usually most desirable for each. The second part of this chapter outlines different conceptual approaches to the preparation of targets and how they can guide the chemist in the planning of target syntheses or preparations.

## 2.1 Formulating Target Specifications

The first step requires a determination of the quantity of target needed, identity(ies) of the isotope(s), specific activity, chemical and radiochemical purity, and position(s) of label(s) within the molecule. Each factor is strongly dependent on the type and design of the envisaged study(ies).

General information on the specifications for studies in drug development is given in Table 2.1 as a guide to the synthetic chemist on the quantities involved. The optimum requirements for each study will depend upon a number of variables specific to the chemical and biological properties of the study compound, dose range, animal species, specific study aims, analytical parameters, etc. In contrast, the specifications for target compounds used in environmental studies (such as those conducted for registration of crop sciences chemicals), and studies of biosynthetic pathways, enzyme or organic reaction mechanisms, for example, are entirely dependent on the needs of the particular experiment and no general guidelines are possible.

In studies such as receptor binding and assays using internal standards, no structural alteration of the target is anticipated, so that the position of label is less important. In mechanistic studies in enzymology or organic chemistry, the nature of the chemical transformations is usually known or anticipated, and the label position can be selected accordingly. However, in the several kinds of studies where metabolic transformations may occur, there is more uncertainty about the nature and extent of chemical changes to be anticipated. Therefore it is particularly important that target specifications be established in close cooperation with the ADME experts. Some basic considerations are discussed below[2].

As it is the radioactivity that is detected, quantified and/or traced, the isotopic label(s) must remain a part of the molecule, or of the molecular fragment(s), which is of most interest. In the absence of specific information about the metabolism of a particular target, usually that part is labeled which is of higher structural complexity, shows a higher likelihood of toxicological relevance and/or is characterized by less data. Once the behavior of this isotopomer in an initial study is determined, the information gained may illuminate

**Table 2.1** *Typical quantities and characteristics of labeled compounds for common types of studies**

| Study | Sensitivity required | Specific activity | Isotope | Label location | RCP | CP | Quantity |
|---|---|---|---|---|---|---|---|
| Receptor characterization, receptor binding assays | > 1000 dpm per tissue sample for LSC or > 1000 dpm per $mm^2$ in ROI of tissue slice for phosphorimaging | 20–100 Ci/mmole | Tritium | Anywhere except positions exchangeable under physiological conditions | > 98% | Free of receptor-binding contaminants | 1–10 μCi/experiment |
| Metabolism *in vitro* (hepatocytes or enzyme preparations) | Radio-HPLC requires at least 200–1000 dpm for detection of a metabolite (LumaPlate/TopCount) | > 1 mCi/mmole | Tritium or carbon-14 | Metabolically stable; remains in metabolite(s) of interest | > 98% | > 95% | 10 uCi/experiment |
| Tissue distribution by means of QWBA | > 1000 dpm $^3H$/ > 100 dpm $^{14}C$ per tissue sample for LSC or > 1000 dpm $^3H$/ > 100 dpm $^{14}C$ per $mm^2$ in ROI of tissue slice for phosphorimaging | Determined by molecular weight, animal and dose | carbon-14 preferred | Any but sites that may be metabolized to $^{14}CO_2$ | > 98% | > 95% | Mouse/rat dose (per kg): 0.2–1 mCi $^{14}C$ or 2–4 mCi $^3H$. Dog/monkey dose (per kg): 20–50 μCi $^{14}C$ or 100–500 μCi $^3H$ |
| ADME *in vivo* | Radio-HPLC requires at least 200–1000 dpm for detection of a metabolite (LumaPlate/TopCount) | Determined by molecular weight, animal and dose | carbon-14 preferred | Metabolically stable; remains in metabolite(s) of interest | > 98% | > 95% | Mouse/rat dose (per kg): 0.2–1 mCi $^{14}C$ or 2–4 mCi $^3H$. Dog/monkey dose (per kg): 20–50 μCi $^{14}C$ or 100–500 μCi $^3H$ |

(*continued*)

**Table 2.1** *(Continued)*

| Study | Sensitivity required | Specific activity | Isotope | Label location | RCP | CP | Quantity |
|---|---|---|---|---|---|---|---|
| Human radiolabel studies | Radio-HPLC requires at least 200–1000 dpm for detection of a metabolite (LumaPlate/TopCount); sensitivity may be vastly increased by use of AMS** | Determined by molecular weight and dose | carbon-14 preferred | Metabolically stable; remains in metabolite(s) of interest | >98% | >98% | Determined by regulatory and ethical limits on radiation dose to subjects |
| Quantification by MS-spectroscopy (Internal standard) | Less than 1% overlap of molecular cluster of labeled and unlabeled materials | n.a. | deuterium, carbon-13 or nitrogen-15 | Chemically stable | n.a. | >95% | Milligram range |
| Biosynthetic pathways | Same sensitivities as above for LSC, radio-chromatography or 2D imaging | Per study design | carbon-14 preferred | In part of precursor expected to be incorporated into compound of interest | >98% | >95% | 1–100 mCi for fermentations or intact organisms |
| Enzyme mechanisms; organic reaction mechanisms | Same sensitivities as above for LSC, radio-chromatography or 2D imaging | Per study design | Per study objective | As required to address specific hypothesis | >98% | >95% | μCi range per experiment |

*based on current technologies and general practices.
**see Reference 4, Chapter 4.

the need for further studies with other target isotopomers. An example[3] that required labeling of two parts of a molecule is the $GABA_A$ receptor partial agonist **1a**. As suggested by initial *in vitro* biotransformation studies, **1a** underwent significant metabolic scission to the benzimidazole alcohol **2a** and the fluoropyridine imidazole **3a**.

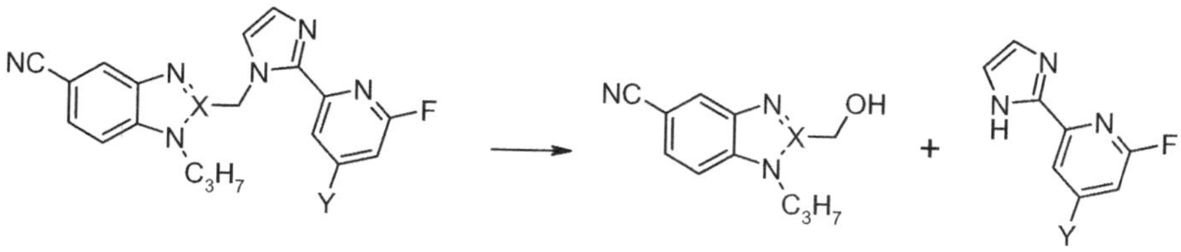

**1a**: X = $^{12}$C, Y = $^{1}$H
**b**: X = $^{14}$C; Y = $^{3}$H

**2a**: X = $^{12}$C
**b**: X = $^{14}$C

**3a**: Y = $^{1}$H
**b**: Y = $^{3}$H

***Figure 2.1*** *NADPH-dependent cleavage of **1a** by hepatic microsomes (rat, dog, human)*

Dual labeling to **1b** allowed a comprehensive qualitative profiling of its metabolism. The application of different isotopes permitted unequivocal attribution of the individual metabolites to the moieties **2** and **3**. However, experimental limitations did not allow their exact quantification at very low levels.

The decision on the type of isotope (tritium or carbon-14) is mainly influenced by the specific activity required and the ease of its incorporation into the target. In cases of low-dosed drugs, however, there might be no alternative to tritium[4].

Owing to the intrinsic properties of the hydrogen-carbon bond, tritium labels may be at risk of loss from [$^3$H]targets, either in aqueous solution of varying pH ('chemical instability') or by enzymatic oxidation ('metabolic instability'). Both processes form tritiated water, whose presence may result in misleading or ambiguous data[5]. An illustrative example[6] is the comparison of pharmacokinetic results from studies of the two isotopomers **4a** and **4b** of the $\alpha_7$ nicotinic acetylcholinergic agonist.

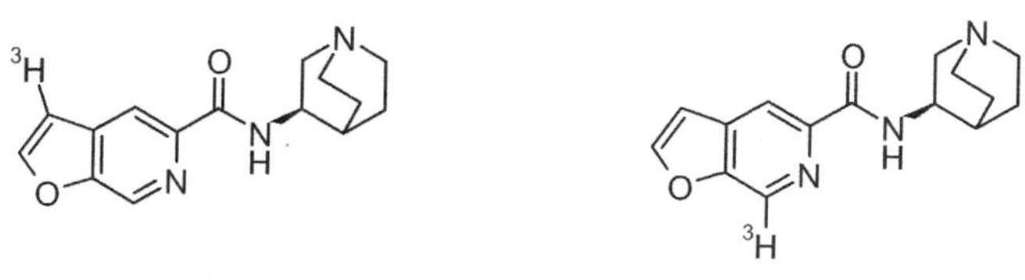

**4a** **4b**

***Figure 2.2*** *Structures of the $^3$H-isotopomers **4a** and **4b***

In Sprague–Dawley rats the more easily labeled derivative [3-$^3$H]isotopomer **4a** showed a terminal half-life (radioactivity) of 133 h; while the [7-$^3$H]isotopomer **4b** had a terminal half-life (radioactivity) of only 31 h. It turns out that the label in **4a** is readily lost in metabolism and the measured half-life is most likely that of $^3$HHO, and only the **4b** data actually reflect the pharmacokinetics of the target molecule itself, the label location in this isotopomer being metabolically stable.

As demonstrated in the case of SDZ HDC912 (**5**)[7], identification of metabolically stable positions for placement of tritium is not always straightforward. Since the dog was identified to be a predictive species all isotopomers were tested in this species. Each of three isotopomers proved to undergo some metabolic loss of label, so that a large experimental effort had to be expended to find a label position of acceptable metabolic stability. In this case [5-$^3$H]SDZ HDC912 (**5**) was identified as the most suitable target for ADME studies in man.

**5**

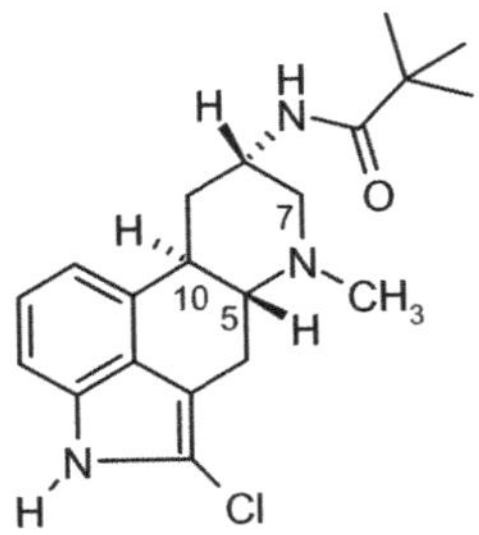


| Position | C5 | C7 | C10 |
|---|---|---|---|
| % $^3$HHO in dog | 3 | 6 | 10 |

***Figure 2.3*** *Metabolic stabilities of different SDZ HDC912 isotopomers*

Computer-assisted metabolism programs[8] designed to predict the relative site lability of test molecules, as well as quantum chemical calculations, have been suggested[1] to be of value in identifying optimal labeling positions. Even though the use of such resources may significantly reduce the risk in the choice of label location, initial pilot studies are recommended for actual measurements of a target's metabolic stability, before investing in full *in vitro* or ADME studies. One good measure[9a] of *in vivo* metabolic (in)stability of tritium labels is the amount of radioactivity in the lyophilizate of the urine. Even with carbon-14-labeled compounds, metabolism may be extensive enough to produce $^{14}CO_2$. Since this results in a deficit in the recovery of radioactivity as measured by the standard assays (excreta, residue in tissues, etc.), such a result may indicate the need for quantification of radioactivity in exhaled air in order to completely characterize the compound's metabolic fate[9b].

Replacement of C–H bonds subject to cytochrome P-450-catalyzed oxidation with C–$^2$H or C–$^3$H bonds may increase substrate flux through alternative pathways, a well-known phenomenon termed metabolic switching[10]. This phenomenon occurs when the $^1H/^2H$ or $^1H/^3H$ kinetic isotope effect (KIE) is large enough to influence the turnover of the primary pathway. The effect is to alter not only the ratio of the metabolites, but also plasma levels and drug clearance. An example is (+/–) tramadol (**6**)[11] (Figure 2.4). In this case the $OC^2H_3$ isotopomer is *O*-demethylated at only one-fourth the rate of nonlabeled tramadol, resulting in an increase in the extent of *N*-demethylation so that it becomes a major source of metabolite.

Rules[12a] described in the literature data provide some guidance for assessing the likelihood for consequences of labeling at different positions; the most important factors being the rate-determining metabolic steps and the relative importance of the affected

6

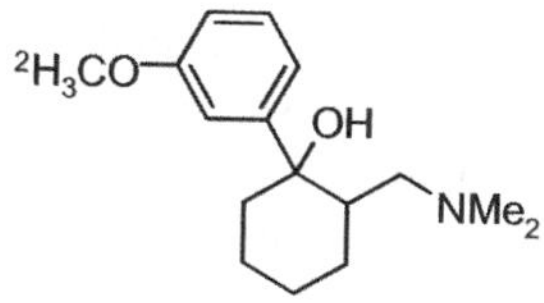


*Figure 2.4 Structure of (+/–)-[O-methyl-$^2H_3$]tramadol*

pathway among others. The findings with 6 are consistent with the notion that oxidation at the *O*-methyl carbon is rate determining and that there is a significant $^1H/^2H$ KIE in that reaction. This is not uncommon in the metabolic oxidations of aliphatic C–H bonds. In contrast, enzymatic hydroxylation at [$^{2/3}$H]labeled sites of aromatic rings often does not show a KIE because the rate-determining step is actually an epoxidation[12b].

Although there are distinct difficulties with tritium, this isotope is an indispensable tool in modern drug development. However, because of the risk of metabolic instability of tritium labels, tritiated targets require careful design and testing to ensure their suitability.

## 2.2 Planning Radiotracer Preparations

Standard, or nonisotopic, organic synthesis involves building up molecules from simpler, more readily available starting materials. In nontrivial cases planning of synthetic routes might make use of retrosynthetic analysis (Figure 2.5)[13] thinking backwards step by step to identify bonds that can be formed to build up the final product most efficiently. The stepwise reduction of the molecular complexity (B, C, D) continues until readily available starting materials are identified. This planning approach is often supported by information from reaction-based computer databases (i.e. SCIFINDER). Having been planned in a backwards direction (A → D), the actual synthesis is then conducted in the forward direction (D → A), which will be called *construction* strategy.

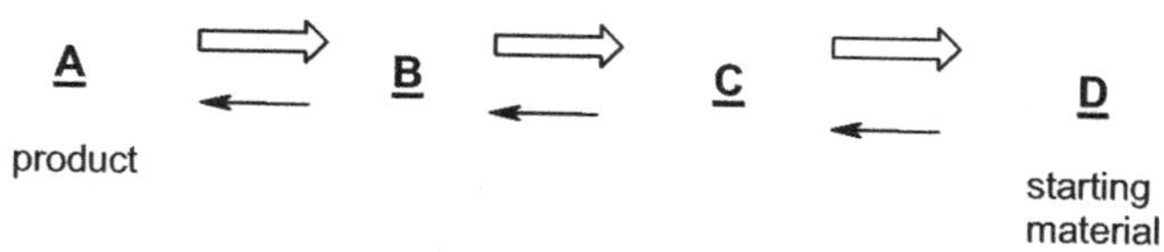


*Figure 2.5 Strategy for organic synthesis design*

Planning the preparations of isotopically labeled organic compounds differs in significant respects. In addition to particular conditions for construction strategy within isotope labeling, the reconstitution and the derivatization strategies are additional options for isotope labeling.

### 2.2.1 The Construction Strategy

The construction strategy is closely related to that of organic chemistry. Since radiolabeling in pharmaceutical and crop sciences is part of an integrated development process, experimental information on the synthesis process as well as unlabeled material(s) are often available. Therefore, making use of the available project information while keeping in mind the given labeling constraints assists development of the optimal construction plan. The main constraints to isotope labeling are the limited availability of isotopically labeled starting materials and the technical demands of their handling.

As illustrated in Figure 2.6, the labeled building block may enter into the synthetic pathway at any point; however, the later it enters the better, in order to minimize the handling and purification of radioactive intermediates and maximize the overall radiochemical yield.

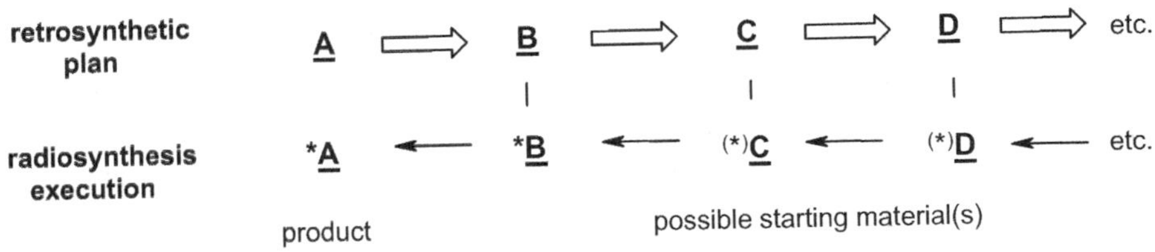


***Figure 2.6*** *Illustration of the construction strategy*

As detailed elsewhere in this book, all tritiated and carbon-14-labeled building blocks derive from a very few sources, ultimately tritium gas or barium [$^{14}$C]carbonate. Although a reasonable number of labeled building blocks are available commercially, usually the greater number of synthetic steps they are away from these isotopic sources, the greater their cost. Therefore, consideration should be given during synthesis planning to utilizing simpler labeled building blocks or preparing them in house from simpler compounds. Chapter 4, Section 4 and Chapters 5–9 provide extensive information, both on the preparation of a large number of commonly used labeled building blocks and on their various uses in organic radiochemical synthesis. Figure 2.7 below summarizes schematically for carbon-14 the transformation of some building blocks, and shows their interrelations in the 'trees' – $^{14}CO_2$, $M^{14}CN$, $H^{14}C{\equiv}^{14}CH$ and $BaN^{14}CN$ – according to which Chapters 5–9 are organized.

### 2.2.2 Reconstitution Strategies

Reconstitution strategies are discussed in detail in Chapter 10. As illustrated in Figure 2.8, they formally encompass three kinds of approaches. The first, 'exchange' (or 'direct replacement'), involves direct introduction of the label into the unlabeled target molecule ($A \rightarrow A^*$). The second, 'indirect replacement' introduces the label via a functionalization of the unlabeled target molecule without disrupting its structural framework ($A \rightarrow A' \rightarrow A^*$). The label is then introduced during the process of restoring the original structure. The third type, 'disconnection/reconnection' involves changing the structural framework of the unlabeled target, usually by excision of part of the structure, followed by reconstruction of the molecule in labeled form ($A \rightarrow\rightarrow B \rightarrow\rightarrow A^*$).

Introduction of isotopes by exchange (direct replacement) is the most straightforward method of preparation of labeled targets from their unlabeled analogs. This approach is used mainly in tritium chemistry, where its advantage of simplicity is, however, offset by wide variation in (and sometimes by the unpredictability of) the extent and position of labeling. Therefore substrate-specific optimization might be

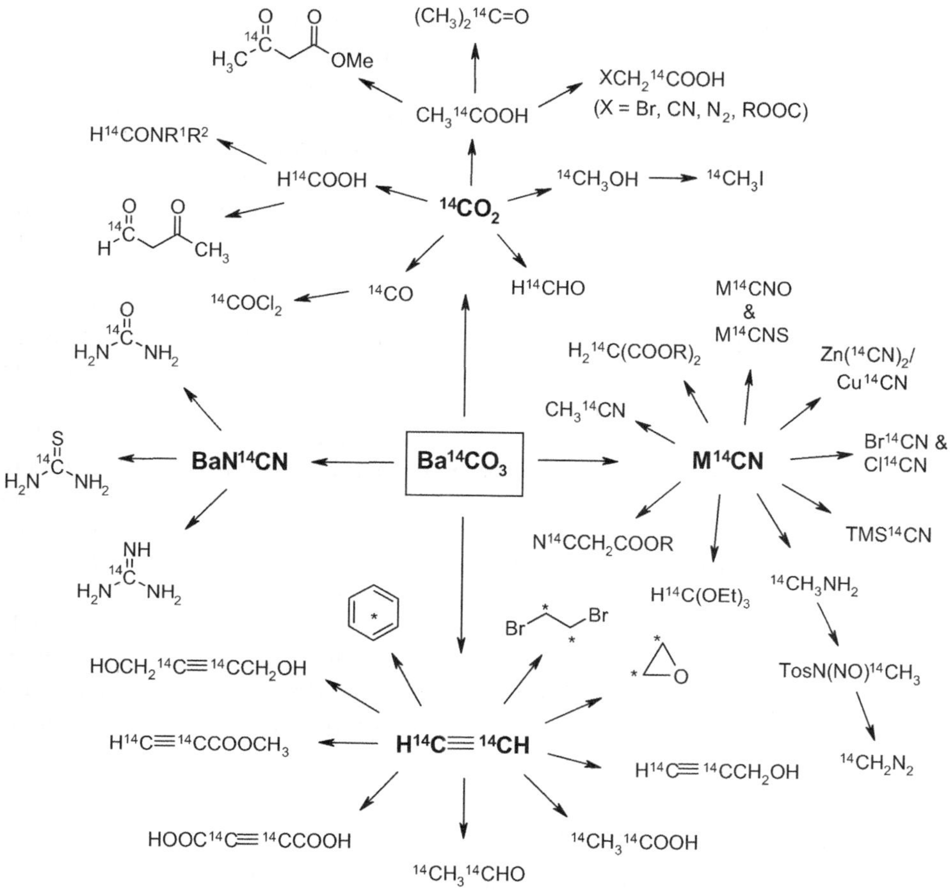


***Figure 2.7*** *The four 'trees' of building blocks derived from Ba$^{14}$CO$_3$*

required to test and optimize the reaction conditions. The great variety of methods for exchange-labeling of compounds with tritium are the subject of Chapter 3 and Chapter 10, Section 10.1.1. In addition, a few methods for exchange labeling with carbon-14, usually involving small, one-carbon moieties, have been described; these are discussed in Chapter 10, Section 10.1.2.

***Figure 2.8*** *Three approaches to labeling by reconstitution*

There is a much wider variety of indirect replacement approaches. In most cases these approaches introduce the labels at structurally predefined positions generated by (formally) oxidative processes, which are then followed in a second step by reductive operations. Examples of such approaches preferentially designed for the introduction of tritium include halogenation followed by tritiodehalogenation, the introduction of carbon–carbon multiple bonds followed by catalytic tritiation, the oxidation of carbon–heteroatom bonds followed by reduction using tritide reagents, etc. These approaches, which do not alter the skeleton of the target in the process, are discussed in Chapter 4, Sections 1–3 and Chapter 10, Sections 10.1.1.2–4.

'Disconnection/reconnection' strategies use cleavage or excision of a small segment of the molecule, which is then replaced with its labeled analog. Alternatively the labeled moiety is attached first, then the analogous original fragment is cleaved off; the labeling of methyl amines is a structural class in which this strategy has been frequently used. The simplest disconnection/reconnection sequences, such as replacement of *N*-, or *S*-alkyl groups with their isotopic analogs have been conducted in both manners, while more complex replacement operations are almost always carried out by excision first and replacement second. The entire range of such variations and examples of each strategy are discussed in Chapter 10, and they make use of tritiated and carbon-14-labeled building blocks discussed in Chapter 4, Section 4, and Chapters 5–9.

In general, reconstitution strategies are most valuable in the labeling of complex, highly functionalized molecules that would present formidable challenges for multistep synthesis (construction strategy). Reconstitution approaches have also been used for the preparation of target analogs, wherein the labeled moiety used in the reconstitution is a structural variant of the one excised.

### 2.2.3 The Derivatization Strategy

Derivatization is simply the attachment of a labeled building block to the unlabeled target. Clearly, this produces not the structurally identical isotopomer, but a labeled derivative. Nevertheless, this approach can be useful if the performance of the derivative in the intended studies mimics to an acceptable degree that of the original structure; i.e., the changes in the physical or chemical properties incurred in the derivatization do not alter the study parameters that are measured. Derivatization can be advantageous for complex compounds such as proteins, peptides or complex natural products that would otherwise be difficult to label. Preliminary evaluation of the derivative in the intended assay is essential to ensure its suitability.

$$\underline{A} + {}^*\underline{R} \longrightarrow \underline{A}-{}^*\underline{R}$$

***Figure 2.9*** *Illustration of the derivatization strategy*

### 2.2.4 Selection of an Appropriate Strategy

In practice, more than one approach might be applicable to the labeling of a given target. In order to identify the most effective strategy, a number of issues, peculiar to the preparation of compounds labeled with tritium or carbon-14, warrant consideration.

- **Availability of isotopic materials**
  The viability of any labeling strategy depends on the availability or accessibility of the appropriate isotope source, reagent or labeled building block. Even though the option of custom synthesis provides a considerable arsenal, cost and timely delivery still remain relevant factors. On the other hand for in-house preparation the magnitude of the required effort has to be taken into consideration. Extremely costly or difficult-to-handle intermediates might occasionally be avoided by adaptation of the strategy design. An example of this is the restricted delivery for [$^{14}$C]phosgene, which might require a different labeling approach, such as outsourcing the [$^{14}$C]phosgene reaction or modification of the synthesis to accommodate an alternative labeled reagent.
- **Availability of practical/technical capabilities**
  Handling of radiolabeled materials requires dedicated facilities and techniques and sometimes special equipment or instrumentation. The availability of these resources and experience with these techniques therefore influence the options for synthesis design.
- **Radiation**
  Regulatory standards may discourage the use of some types of compounds, such as volatile or chemically reactive radioactive materials. Examples of ways to avoid volatile intermediates are given in Chapter 1. As also discussed in Chapter 1, radiation is not only a risk for the operator, but it may also have an impact on the stability of the target or a particular intermediate. Since the magnitude of these risks depends significantly on the structure and physical properties of the compounds, good synthesis planning should either take these risks into consideration and address them in their handling or completely avoid them by adapted synthesis design.
- **Number of radiolabeled steps**
  In order to minimize expenditures for radiolabeled starting materials, the risk of contamination and radioactive waste, and to make optimal use of laboratory capacity, it is a general goal to keep the number of radiolabeled steps to a minimum. Keys to achieving these ends are synthesis plans that allow introduction of the label at a late stage, and the use of convergent syntheses. An example is the labeling of NVP MTH958 (**9**)[14] (Figure 2.10). The key intermediate in both approaches is (*R*)-*N*-Boc-3-trimethylsilyl [1-$^{14}$C]alanine (**7**). While the linear synthesis requires six hot steps from **7** to obtain the final target **9**, the optimized convergent synthesis involves reaction of **7** with the azido-protected building block **8** and reduction of the product to afford the target **9** in only two radiolabeled steps from **7**. An advantage of the convergent alternative was not only the reduced number of radioactive steps, but also the higher radiochemical yield (9% versus 1%).

  Nevertheless, there are also examples of long linear syntheses, such as the 13-step synthesis of [$^{14}$C]JNZ092[15], that are viable because they take advantage of the availability of well-elaborated, robust research/development procedures. The decision on a final strategy therefore needs to balance the value of already existing information against the gain and the effort required to develop alternative syntheses.
- **Availability of unlabeled intermediates/targets**
  As already outlined above, the development process may make available unlabeled intermediates/target that can support the radiosynthesis, either as reference materials for the analytical proof of identity, the development of an appropriate chromatographic

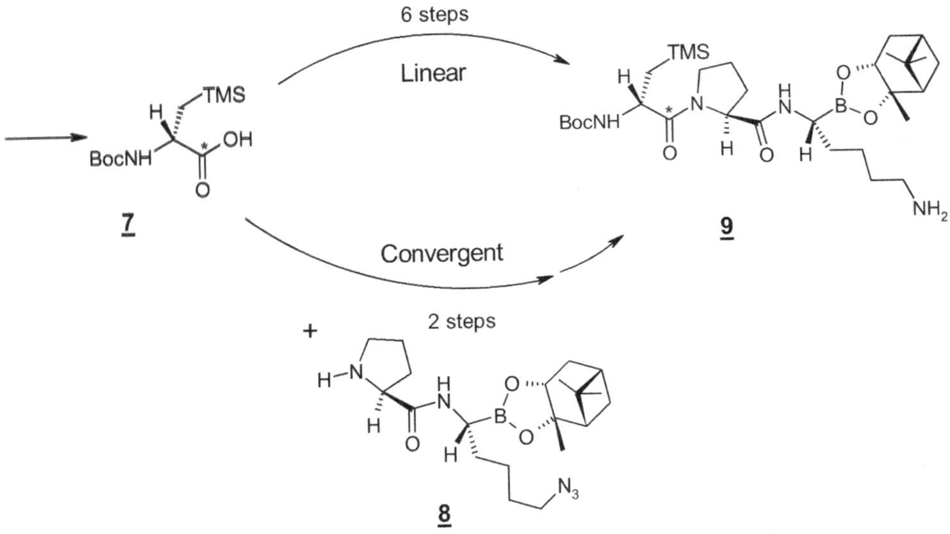


*Figure 2.10* *Linear versus convergent synthesis of [$^{14}$C]NVP MTH958* **9**

separation system (e.g. HPLC) or for testing and optimization of individual synthesis steps. In addition, the availability of adequate quantities of unlabeled late intermediates and/or unlabeled target compound may open the opportunity for reconstitution approaches, either at the level of late intermediates or the target.

- **Common intermediates**
  Modern drug development requires radiolabeled targets for early profiling of compounds. Therefore, it is not unusual for several labeled compounds of closely related structure to be needed. One way to address such situations in an efficient way is to prepare an adequate quantity of a common synthetic intermediate in isotopically labeled form. Portions of it can subsequently be carried on to the individual targets as required. Ethyl 4-benzyloxy-[3-$^{14}$C]indole-2-carboxylate (**10**) (Figure 2.11) is an example of an intermediate that gave access to a series of ß-blocking agents and their metabolites[16]. Decisive for the suitability of **10** in such a role was the differential reactivity of the 4-benzyloxy and the 2-ethyl carboxylate substituents, which permitted the synthetic options necessary to access the individual targets.

A similar strategy might be followed for particularly unstable targets. Storage of a suitable, more stable intermediate and synthesis on demand from portions of this intermediate would reduce the need for resyntheses.

### 2.2.5 Case Studies of Strategy Development

The design of a radiosynthesis is a multi-dimensional process open to a variety of strategies and methodologies. There are usually a number of different approaches to the same target and often there is no 'right' solution, but only an 'individual' solution determined by the particular conditions. The following two examples illustrate the different labeling strategies for [$^{3}$H]paclitaxel (**11**) (Figure 2.12) and [$^{14}$C]imatinib (**25**) (Figure 2.19).

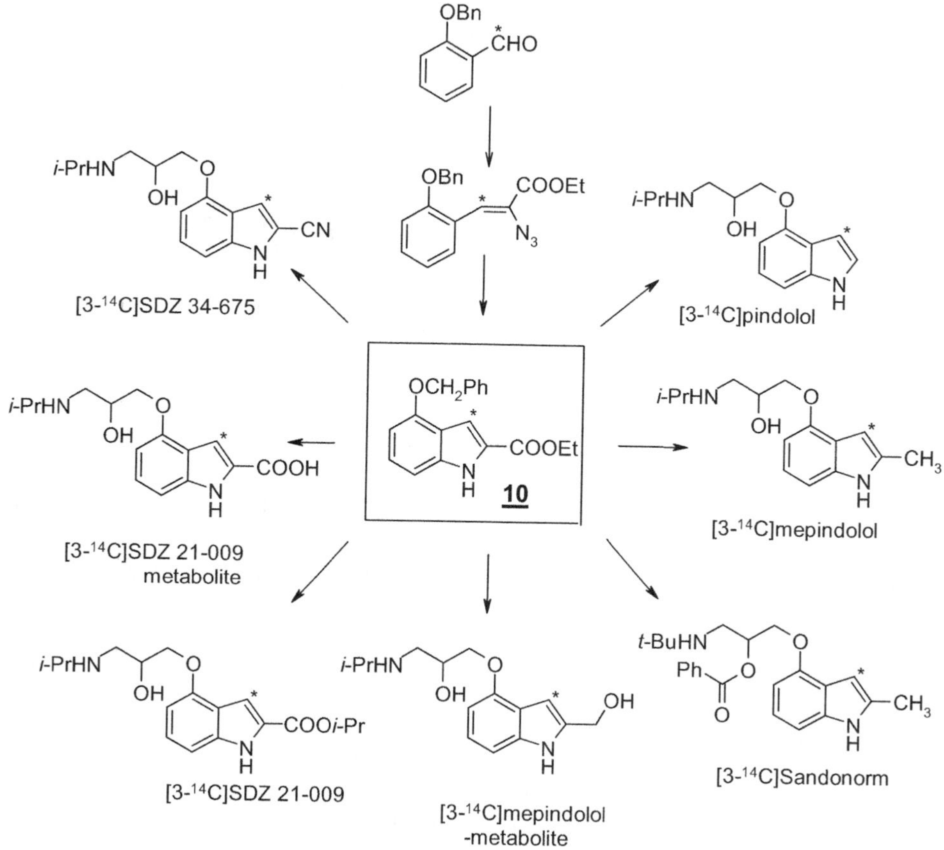


***Figure 2.11*** *Ethyl 4-benzyloxy[3-$^{14}$C]indole-2-carboxylate, a common precursor for the synthesis of a series of ß-blocking agents*

**Example Paclitaxel**

Paclitaxel (**11**) is an ester-type molecule consisting of (2*R*,3*S*)-*N*-benzoyl-3-phenylisoserine and the highly complex baccatin III moiety (Figure 2.12). For decades, this formidable structure has presented challenges to isotope labeling scientists whose interests were to make available labeled versions appropriate for the various biological studies stimulated by paclitaxel's unique pharmacological properties. In early days, synthetic knowledge to operate on the structure was scarce and radiochemists had to get by with methods of limited effectiveness. As more information on paclitaxel's chemical reactivity and synthetic transformations was developed, and new radiochemical methodology became available, both could be applied to the preparation of isotopomers labeled with greater selectivity and to higher specific activity. This has resulted in unusually large and widespread efforts for the labeling of a single molecule, which makes palclitaxel a useful case study in which to highlight the interplay of various factors involved in the selection and execution of labeling strategies.

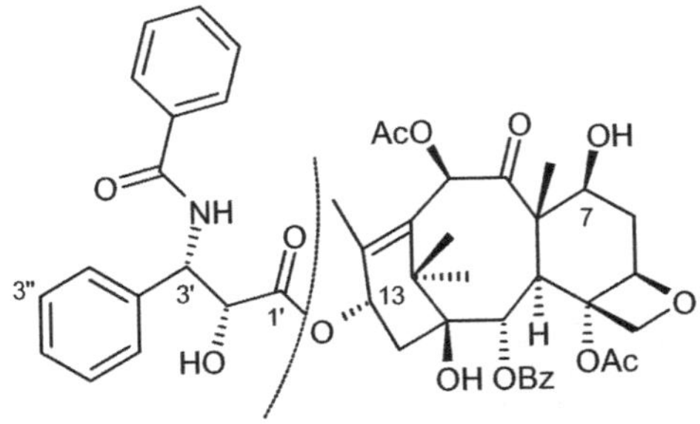


11

***Figure 2.12*** *Structure of paclitaxel*

The earliest reported radiolabeling of paclitaxel[17], at a time when very little was known about the chemistry of the molecule, was accomplished by a single-step direct isotope exchange strategy. The compound remained largely intact through the process (25 mg of paclitaxel, 25 Ci of >90 atom % $^3H_2O$, $Rh/Al_2O_3$, dioxane; 80 °C, 5 h), and the isolated product, only a few mCi at the relatively low specific activity of 446 mCi/mmol and with undetermined label location(s), was nevertheless adequate for measurement of its binding constant to polymerized microtubule protein.

Later, as the different reactivities of paclitaxel's hydroxyl groups became known, a derivatization approach was employed[18] in order to obtain a product of higher specific activity. This three-step sequence (Figure 2.13) afforded 7-[$^3$H]acetylpaclitaxel (**12**) of specific activity 1.4 Ci/mmol in about 15% radiochemical yield. The biological activity of **12** towards preventing microtubule disassembly was determined to be about one-half that of the parent compound, but the compound was useful for comparing the tubulin binding of derivatives. Later, a slightly modified protocol[19] was used to synthesize 7-[1-$^{14}$C]acetylpaclitaxel at a specific activity of 48 mCi/mmol, for use in cytotoxicity assays.

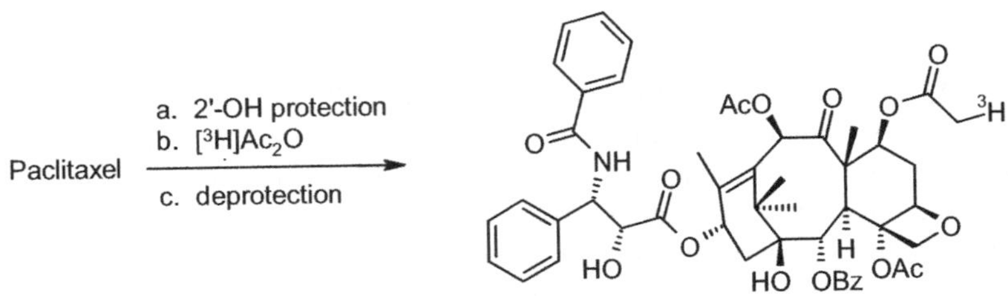


12

***Figure 2.13*** *Derivatization strategy for labeling paclitaxel*

By the early 1990s, *de novo* syntheses of the paclitaxel side-chain and methods for coupling it ($\beta$-lactam synthon method) with the baccatin III system had been published[20]. This synthetic methodology was exploited to prepare[21] (via a disconnection/reconnection strategy) side-chain-labeled [$^3$H]paclitaxel at a much higher specific activity (19.3 Ci/mmol) than previously possible, finally providing a tool for high sensitivity assays. This was accomplished through preparation of a protected baccatin III from paclitaxel, total synthesis of the racemic bromine-substituted side-chain (**13**), and the coupling of them together to give a mixture of epimers from which 3″-bromopaclitaxel (**14**) was isolated after deprotection. Its catalytic tritiodebromination and deprotection gave **[3″-$^3$H]11**. Although the synthesis was lengthy, only the last step involved radioactive materials and the product was labeled site specifically. (At that time, baccatin III was only available by reductive cleavage of paclitaxel[22]. As baccatin III has since become commercially available, this synthesis today might be considered a construction approach since it would not need to use paclitaxel as starting material.)

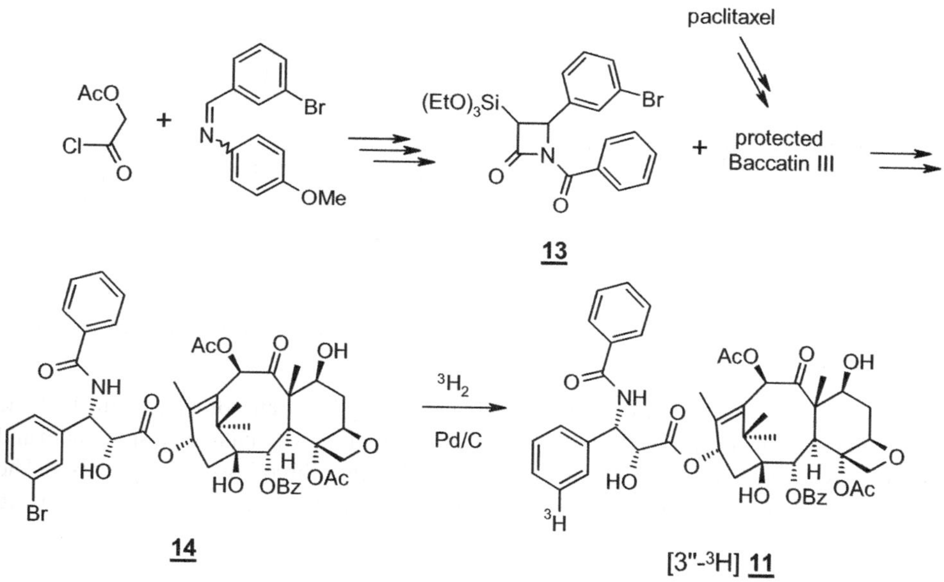


***Figure 2.14*** *Use of a multistep disconnection/reconnection strategy for preparing tritium-labeled paclitaxel*

Shortly thereafter, stereoselective procedures were used to prepare two carbon-14 isotopomers labeled in the side-chain, as illustrated in Figure 2.15. [3′-*N*-carbonyl-$^{14}$C]-paclitaxel was prepared[22] in 36% overall radiochemical yield starting with [carbonyl-$^{14}$C]-benzoylation of a preformed nonracemic azetidinone derivative and proceeding through intermediate **15**. [3′-$^{14}$C]paclitaxel was prepared[23] by a longer route in 11% overall radiochemical yield starting from benz[$^{14}$C]aldehyde and an appropriately protected (*S*)-threonine, and proceeding through **16**. These [$^{14}$C]paclitaxel isotopomers, being labeled site specifically with carbon-14, were appropriate for use in classical pharmacological and drug metabolism studies in which tracing of the fate of the side-chain was required.

In the first specific site labeling of the baccatin III part of the paclitaxel molecule, a reconstitution strategy was used[21]. Oxidation of baccatin III (**17**) (Figure 2.16), obtained by hydrolytic cleavage of paclitaxel, to its 13-keto analog, protection of its 7-OH group (**18**)

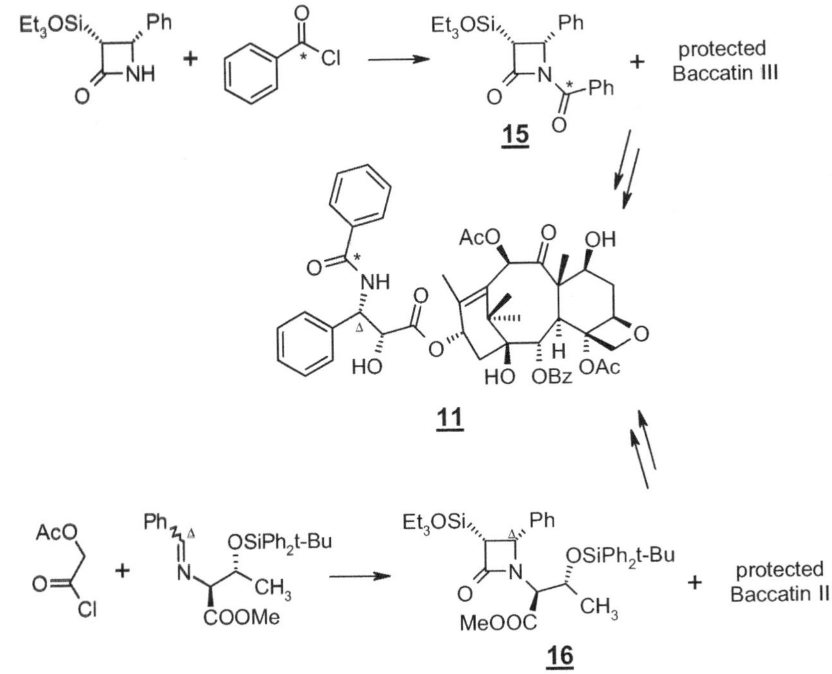


*Figure 2.15* *Use of multistep construction strategies for preparing carbon-14-labeled paclitaxel*

and reduction of the carbonyl with [$^3$H]borane–tetrahydrofuran complex furnished the protected [13-$^3$H]baccatin III ([$^3$H]**19**). Subsequent coupling with unlabeled side-chain and deprotection afforded [13-$^3$H]paclitaxel ([13-$^2$H] **11**) in a three-hot-steps synthesis. The specific activity of 1.6 Ci/mmol reported might have been significantly increased by using carrier-free tritide reagent (Chapter 4, Section 4.3.8).

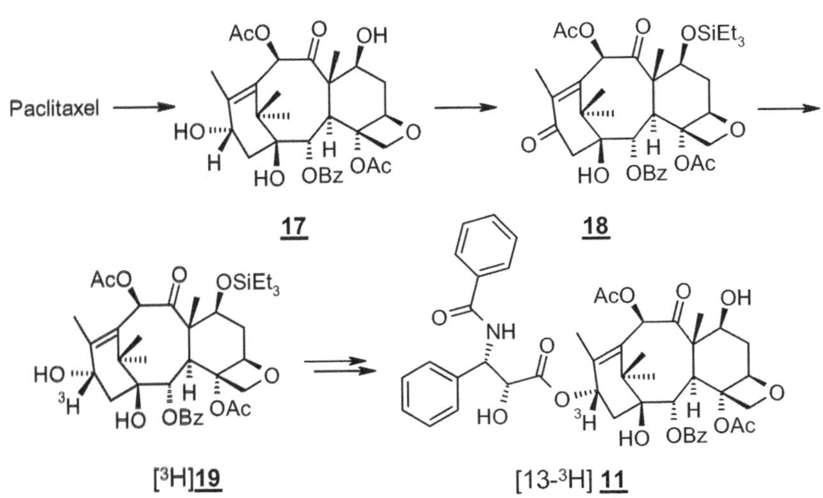


*Figure 2.16* *Use of a reconstitution strategy for baccatin III-labeling of paclitaxel*

More recently, improved procedures for preparing radiolabeled paclitaxel have appeared. The development of a new construction approach type synthesis of [1′,2′-$^{14}C_2$]-paclitaxel (**[1′,2′-$^{14}C_2$]11**)[24] to give material of a higher specific activity (104 mCi/mmol, 34% overall yield) was enabled by the availability of Oppolzer chemistry[25], as shown in Figure 2.17. Secondly, an exchange approach used the organoiridium-mediated tritium exchange methodology[26] developed during the 1990s (see Chapter 3) to label paclitaxel in a single step to higher specific activities than ever before (53–61 Ci/mmol).

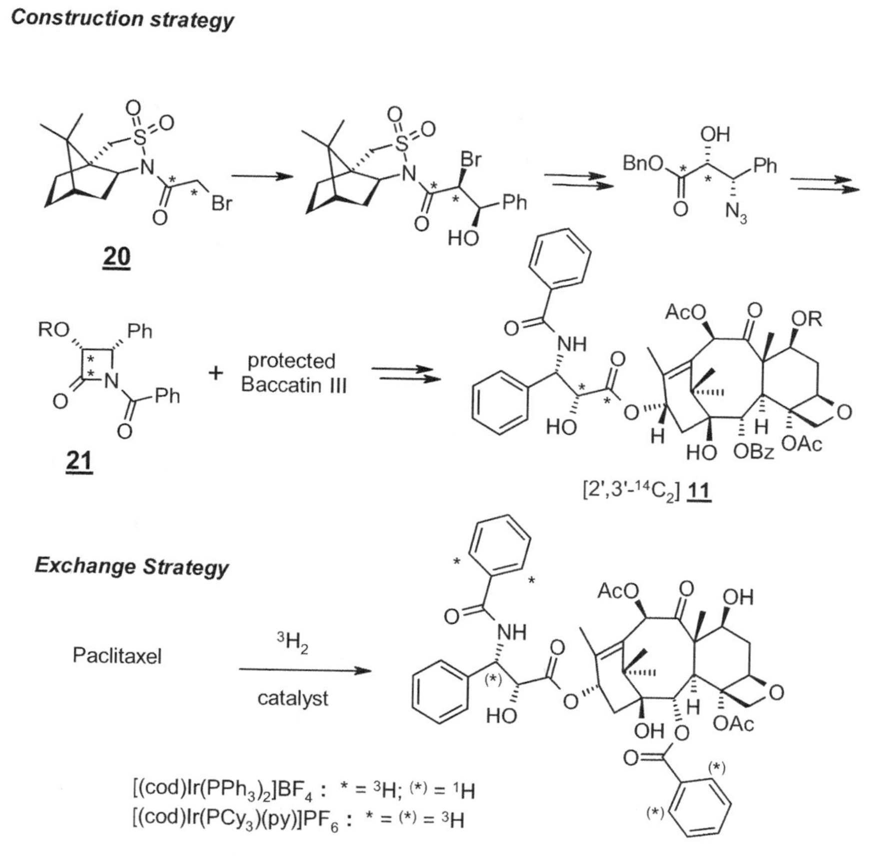


***Figure 2.17*** *More recent strategies for preparation of radiolabeled paclitaxel (11)*

Finally, in a demonstration that a reconstitution strategy can also be used to prepare a labeled analog of the unlabeled starting material, [$^{13}C_6$]BMS-275183 (24) was synthesized[27] as shown in Figure 2.18.

## Example [$^{14}$C]Imatinib

Imatinib (Gleevec®, 25) (Figure 2.19) is a highly effective drug for the treatment of chronic myelogenous leukemia and is also indicated for the treatment of metastatic and unresectable malignant gastrointestinal stromal tumors (GIST)[28]. In contrast to paclitaxel, imatinib is an entirely synthetic molecule. As with any drug candidate with strong prospects for

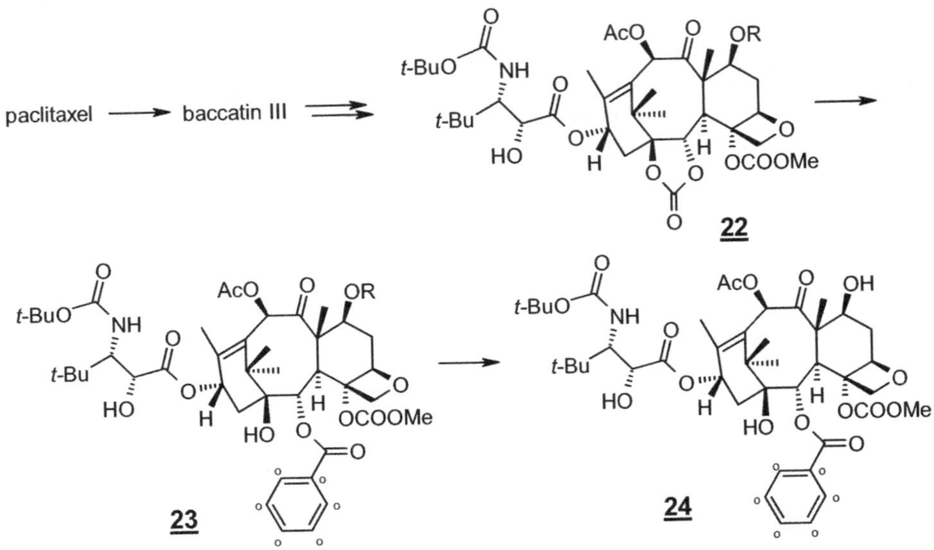


*Figure 2.18* Application of a reconstitution strategy to the synthesis of paclitaxel analog [$^{13}C_6$] BMS-275183 (**24**)

entering clinical development, a large chemical development effort was undertaken to elaborate a total synthesis having a high level of efficiency and dependability, in order to provide for clinical supplies and eventually commercial production. This is the situation commonly encountered by radiochemistry scientists: both a good synthesis route and various synthetic intermediates are already available, thereby offering the radiochemist much to choose from.

This synthetic route[29] is shown in Figure 2.19 in a manner indicating the retrosynthetic plan that, lacking such a route in advance, the radiochemist might have formulated. In this case the problem was reduced to formulating strategies appropriate for preparations via $^{2}$H, $^{13}$C, $^{14}$C and $^{15}$N-labeled building blocks[30]. Five opportunities that provide different positions of label, allowing different degrees of labeling and requiring different sources of label, are given in the Figure 2.19.

**Opportunity 1:** an indirect replacement approach used demethylation of the *N*-methyl piperazine moiety to obtain nor-imatinib (**29**), which was subsequently remethylated with [$^{11}$C]methyl iodide (**30**) to provide a target for PET studies[31]. This labeling position is metabolically unstable and was therefore not suitable for use in ADME studies.

**Opportunity 2:** a first construction strategy made use of the available intermediate **26**. Reaction of benzyl chloride derivative **26** with commercially available 1-methyl [$^{2}H_8$]piperazine (**31**) afforded the [$^{2}H_8$]imatinib (**25**). In a one-step synthesis a sufficiently labeled target (M + 8) was accessible to provide an internal standard for MS-quantification.

**Opportunity 3:** an additional opportunity provided the intermediate **27**. Since acylation of **27** with 4-(chloromethyl)benzoyl chloride gave immediate access to **26**, the carboxamide moiety appeared to be a favorable position for a carbon-14 label. This identified the carboxylic acid **32** as a labeled precursor, easily accessible by [$^{14}$C]carboxylation of the

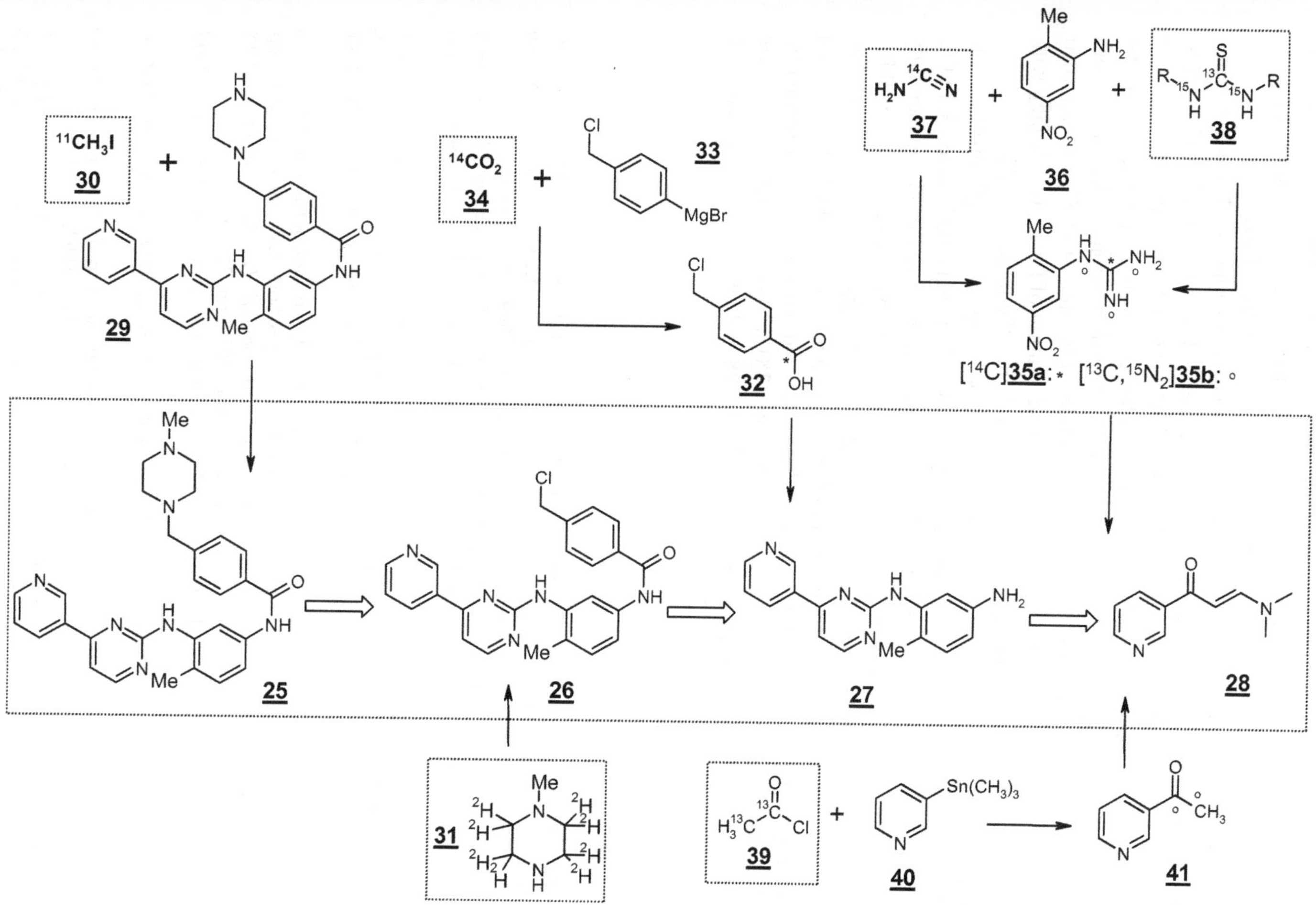


*Figure 2.19* Application of different strategies for the syntheses of various isotopomers of Imatinib (25)

Grignard reagent (**33**). The required isotope source of this approach is therefore $^{14}CO_2$ (**34**) (Chapter 5, Section 5.1). Unfortunately, preliminary ADME investigations indicated that the intermediate **27** was a metabolite of interest, therefore indicating a requirement for incorporation of the carbon-14 and stable label into the skeleton of **27**.

**Opportunity 4:** a suitable precursor for the carbon-14 labeling of **27** was the correspondingly labeled guanidine derivative **35**, suggesting [$^{14}$C]cyanamide (**37**) (Chapter 9, Section 9.1) as the appropriate labeled building block. Because of the deactivating influence of the *meta*-nitro group of **36**, the previously developed reaction conditions had to be slightly modified to accomplish the radiosynthesis. Transformation of the aniline into its hydrochloride and reaction with [$^{14}$C]cyanamide (*n*-butanol, 100 °C, 6 h) afforded [guanidyl-$^{14}$C]**35a** in 65% radiochemical yield. Condensation of [guanidyl-$^{14}$C]**35a** with intermediate **28** gave the labeled metabolite **27** and thence the target [$^{14}$C]**25**.

**Opportunity 5:** preparation of stable isotope-labeled metabolite **27** involved the labeling of two intermediates, the guanidine **35b** and the propenone derivative **28**. In this approach [$^{13}$C, $^{15}N_2$]guanidine (Chapter 9, Section 9.2) was prepared from commercially available [$^{13}$C,$^{15}N_2$]urea. Bis-bocylation, followed by mercury(II)-assisted nucleophilic displacement with the deactivated aniline **36** gave [$^{13}$C, $^{15}N_2$]**35b** after final deprotection.

The labeling of the second building block used commercially available [$^{13}C_2$]acetyl chloride **39** (Chapter 6, Section 6.1.2) in a Pd(II)-catalyzed reaction with 3-trimethyltin-pyridine **40** to obtain 1-pyridine-3-yl[$^{13}C_2$]ethanone (**28**) in 60–70% yield. Subsequent coupling with DMF-dimethylacetal gave [1,2-$^{13}C_2$]**28**, which when combined with [$^{13}$C, $^{15}N_2$]**35b** gave [$^{13}C_3$, $^{15}N_2$]**27**.

## References


1. Stolle, W.T.; Easter, J.A.; Jian, Z.; Mahindroo, V.K.; Maxwell, B.D.; Skaddam, M.B., in *Synthesis and Application of Isotopically Labelled Compounds*, Vol. 8, Dean, D.C.; Filer, C.N. and McCarthy, K.E., Eds; John Wiley & Sons, Ltd: Chichester, ***2004***, pp. 447–450.
2. See also: Zhang, D.; Comezoglu, S.N.; in *Drug Metabolism in Drug Design and Development 2008*, Zhang, D.; Zhu, M. and Humphreys, G. Eds; John Wiley & Sons, Ltd: Chichester, ***2008***, pp. 573–604.
3. Shaffer, C.L. and Langer, C.S., *J. Pharm. Biomed. Anal.*, ***2007***, *43*, 1195–1205; see also: Prakash, C.; Kamel, A. and Miao, Z., in *Synthesis and Application of Isotopically Labelled Compounds*, Vol. 8, Dean, D.C.; Filer, C.N. and McCarthy, K.E., Eds; John Wiley & Sons, Ltd: Chichester, ***2004***, pp. 115–120.
4. a. Gray, A.; Wilkinson, D.J. and Seddon, H., in *Synthesis and Application of Isotopically Labelled Compounds*, Vol. 7, Pleiss, U. and Voges, R., Eds; John Wiley & Sons, Ltd: Chichester, ***2001***, pp. 496–499. b. Voges, R., in *Synthesis and Application of Isotopically Labelled Compounds 1988*, Baillie, T.A. and Jones, J.R., Eds; Elsevier, Amsterdam, ***1989***, pp. 33–40.
5. Kim, H.; Prelusky, D.; Wang, L.; Palamanda, J. and Nomeir, A.A., *American Pharm. Rev.*, ***2004***, 7, 44–48.
6. Shaffer, C.L.; Gunduz, M.; Thornburgh, B.A. and Fate, G.D., *Drug Metab. Disposition*, ***2006***, *34* (9), 1615–1623.
7. a. Voges, R. and Kohler, B. in *Synthesis and Application of Isotopically Labelled Compounds 1991*, Buncel, E. and Kabalka, G.W., Eds.; Elsevier: Amsterdam, ***1992***, pp. 74–81.
8. a. Voges, R.; von Wartburg, B.R. and Loosli, H.R. in *Synthesis and Application of Isotopically Labelled Compounds 1985*, Muccino, R.R., Ed.; Elsevier: Amsterdam, ***1985***, pp. 371–376; b. Afezlius, L.; Arnby, H.C.; Broo, A.; Carlsson, L.; Isaksson, C.; Jurva, U.; Kjellander, B.; Kolmodin, K.; Nilsson, K.; Raubacher, F. and Weidolf, L., *Drug Metabolism Reviews*, ***2007***, *39*, 61–86.

9. a. Tse, F.L.S. and Jaffe, J.M., *Preclinical Drug Disposition – A Laboratory Handbook*, Vol. 46, Chap. 2, Marcel Dekker: New York, ***1991***, pp. 17–29; b. Tse, F.L.S. and Jaffe, J.M., *Preclinical Drug Disposition – A Laboratory Handbook*, Vol. 46, Chap. 3, Marcel Dekker: New York, ***1991***, pp. 66–68.
10. Miwa, G.T. and Lu, A.Y., *BioEssyas*, ***1987***, *7* (5), 215.
11. Hamm, S.; von Unruh, G.E.; Paar, W.D. and Dengler, H.J., *Isotopenpraxis Environ. Health Stud.*, ***1994***, *30*, 99–110.
12. a. Browne, T.R.; in *Stable Isotopes in Pharmaceutical Research*, Elsevier: Amsterdam, ***1997***, pp. 13–18; b. Guroff, G.; Daly, J.; Jerina, D.M.; Rensen, J. and Udenfirend, S., *Science*, ***1967***, *157*, 1524–1530.
13. Corey, E.J. and Cheng, X.-M., *The Logic of Chemical Synthesis*, John Wiley & Sons, Inc: New York, ***1989***.
14. Voges, R.; Kohler, B.; Metz, Y. and Riss, B., in *Synthesis and Application of Isotopically Labelled Compounds*, Vol. 7, Pleiss, U. and Voges, R., Eds.; John Wiley & Sons, Ltd: Chichester, ***2001***, pp. 295–299.
15. Moenius, Th.; Andres, H.; Nozulak, J.; Salter, R.; Ray, T.; Burtscher, P.; Zueger, C.; Schnelli, P. and Voges, T., *J. Label. Compd. Radiopharm.*, ***2007***, *50*, 616–619.
16. Voges, R.; Griesser, R. and Schreier, E., in *Synthesis and Application of Isotopically Labelled Compounds 1982*, Duncan, W.P. and Susan, A.B., Eds; Elsevier: Amsterdam, ***1983***, pp. 209–214.
17. Parness, J. and Horwitz, S.B., *J. Cell Biol.*, ***1981***, *91* (2), 479–487.
18. Chenu, J.; Takoudiu, M.; Wright, M.; Senihl, V. and Guenard, D., *J. Label. Compd. Radiopharm.*, ***1987***, *24*, 1245–1255.
19. Rao, C.S.; Chu, J.-J.; Liu, R.-S. and Lai, Y.-K., *Bioorg. Med. Chem.*, ***1998***, *6*, 2193–2204.
20. a. Holton, R.A., US Patent 5229526, July 20, ***1993***; b. Ojima, I.; Kuduk, S.D.; Slater, J.C.; Gimi, R.H. and Sun, C.M., *Tetrahedron*, ***1996***, *52*, 209–224.
21. Taylor, G.F.; Thornton, S.S.; Tallent, C.R. and Kepler, J.A., *J. Label. Compd. Radiopharm.*, ***1993***, *33*, 501–515.
22. Walker, D.G.; Swigor, J.E.; Kant, J. and Schroeder, D.R., *J. Label. Compd. Radiopharm.*, ***1994***, *36*, 973–980
23. Walker, D.G.; Standridge, R.D. and Swigor, J.E., *J. Label. Compd. Radiopharm.*, ***1995***, *34*, 479–492.
24. Rodriguez, I. and Voges, R., *J. Label. Compd. Radiopharm.*, ***2000***, *43*, 169–176.
25. Voges, R., in *Synthesis and Application of Isotopically Labelled Compounds (1994)*, Allen, J. and Voges, R., Eds., John Wiley & Sons, Ltd: Chichester, ***1995***, pp. 1–26.
26. Shu, A.Y.L. and Heys, J.R., *Tetrahedron Lett.*, ***2000***, *41*, 9015–9019.
27. Ogan, M.D.; Kadow, J. and Rinehart, J.K., in *Synthesis and Application of Isotopically Labelled Compounds*, Vol. *8*, Dean, D.C.; Filer, C.N. and McCarthy, K.E., Eds.; John Wiley & Sons, Ltd: Chichester, ***2004***, pp. 463–466.
28. a. Gschwind, H.-P.; Pfaar, U.; Waldmeier, F.; Zollinger, M.; Sayer, C.; Zbinden, P.; Hayes, M.; Pokorny, R.; Seiberling, M.; Ben-Am, M.; Peng, B. and Gross, G., *Drug Metab. Disposition*, ***2005***; doi:10.1124/dmd.105.00483; b. Radford, I.R., *Curr. Opinion Investigational Drugs*, ***2002***, *3*(3), 492–499.
29. Meisenbach, M., Abstracts, *37th Middle Atlantic Regional Meeting of the American Chemical Society, May 22–25, 2005* (***2005***).
30. Salter, R.; Bordeaux, K.; Burtscher, P.; Metz, Y.; Moenius, Th.; Rodriguez, I.; Ruetsch, R.; Voges, R. and Zueger, C., *J. Label. Compd. Radiopharm.*, ***2006***, *46*, 208–210.
31. Kil, K.-E.; Ding, Y.-S.; Lin, K.-S.; Alexoff, D.; Kim, S.W.; Shea, C.; Xu, Y.; Muench, L. and Fowler, J., *Nucl. Med. Biol.*, ***2007***, *34*(2), 153–163.

# 3

# Preparation of Tritium-Labeled Compounds by Isotope Exchange Reactions

Although tritium-labeled compounds can be prepared by chemical synthesis, as are carbon-14-labeled compounds, they may also be prepared by directly replacing one or more hydrogen atoms with tritium, without any change in the chemical structure of the substrate (*direct replacement*, Chapter 2). Depending on a number of variables, isotopic exchange may be superior to any synthetic approach, or be a valuable component of an efficient synthetic strategy. The structure of the target compound is an important variable, some compounds being impractical to label by any exchange method, or amenable to labeling only in positions that are inappropriate for the compounds' intended uses. Another parameter is the vulnerability of the incorporated label to loss by back-exchange during purification, storage or use, or by other chemical processes or biological transformations encountered in use. Other factors include the level of specific activity (efficiency of label incorporation) achievable, distribution of label in the molecule, ease of purification. In appropriate cases, exchange labeling can be advantageous in that the unlabeled target compound can be tritiated in one step, requiring only a few milligrams of material. Alternatively, a synthetic intermediate may be labeled by exchange, reducing the number of labeled steps compared with synthesis from other available tritiated building blocks.

This chapter discusses each of six categories of exchange labeling processes, together comprising a variety of methods worthy of consideration as simple alternatives to more complex synthetic strategies.

Procedures using deuterium are presented only in cases where tritium examples are lacking and when (in the authors' opinion) they will work with the analogous tritium source. Any adjustments of procedure deemed necessary for making them work with

*Preparation of Compounds Labeled with Tritium and Carbon-14* Rolf Voges, J. Richard Heys and Thomas Moenius

tritium are described in place. As a specific activity level appropriate for a tritiated compound may be achieved by less (or *much* less) than complete replacement of hydrogen by tritium at the relevant site(s), these procedures are not recommended (although some may be suitable) for preparing deuterium-labeled compounds, which usually require high levels of isotopic content. A comprehensive review of hydrogen–deuterium exchange has recently appeared[1].

Notwithstanding the above, a *cautionary note* is warranted. In practical experience, it is not uncommon that model studies with deuterium do not reliably predict results for tritium. There are many reasons why the results may differ. Hydrogen isotope effects can play a part, but as kinetic analyses are highly dependent upon the nature of the exchange reaction and in most cases are not straightforward, the reader interested in more rigorous analysis is referred elsewhere[2]. Consideration of some influential variables discussed below may guide the investigator towards more successful experimental design and evaluation of results.

**Isotope effects:** Unlike equilibrium isotope effects, kinetic isotope effects usually play some role in hydrogen isotope exchange processes. The nature of the kinetic isotope effect in a particular exchange reaction depends upon the mechanism of the exchange and the nature of the rate-determining step. Although generalizations should be applied only with great caution to specific cases, a few may be illustrative. For example, in electrophilic exchange under acidic aqueous conditions the protonation step is often rate determining, and in such cases it would be expected that tritium incorporation into substrate would be slower than a deuterium model reaction by a factor proportional to $k^{2}\mathrm{H}/k^{3}\mathrm{H}$ for that step, everything else being equal (see below). For reactions proceeding by base-mediated deprotonation, that step is often rate determining; therefore, the rate of tritium incorporation would not be expected to differ from that of a deuterium model study, everything else being equal. Exchange reactions involving catalysts or other (isotopic) hydrogen carriers are more complex and defy such straightforward generalizations.

**Isotope concentration:** While deuterium model reactions are often conducted with pure deuterium oxide as the isotope source, tritium oxide is rarely used at anything close to nuclidic purity (note that tritiated water at 50 Ci/mL, the highest specific activity normally available commercially, has a tritium/hydrogen ratio of only about 1.6/98.4). Therefore, the concentration of tritium in $^{3}\mathrm{HHO}$ is usually *much* lower than that of deuterium in $^{2}\mathrm{H}_2\mathrm{O}$, and this difference will be important if the source concentration is a factor in the rate equation. Analogously, model exchange reactions with deuterium gas are often done at one atmosphere of pressure, whereas in most cases tritium gas is used at lower pressures. This can result in substantially slower tritium exchange rates.

**Extent of progress towards equilibration:** If an exchange reaction progresses to complete equilibration, then the isotopic abundance of all exchangeable sites in the substrate molecule will be equal to that of the exchangeable pool (see below). Rarely is it determined experimentally whether an exchange reaction has progressed to equilibrium, and rarely in practice is complete equilibration achieved. Reactions are usually terminated at arbitrary times, and in any case some catalysts lose their activity during the course of the reaction. Therefore, in many reactions the specific activity of the product is lower, sometimes much lower, than that of the isotope source, or the degree of labeling may differ at different substrate sites. Given these factors, results of model reactions should be interpreted with great caution.

**Exchangeable pool:** Complete substitution of tritium for hydrogen in a substrate site is only possible with an infinite excess of tritium source. This is a feature that is commonly ignored in extrapolating the utility of a catalyst or exchange method from deuterium to tritium. In running deuterium trials, investigators often use a vast excess of deuterium (gas, for example); this is not the way most tritiation reactions are conducted. If only a modest excess of tritium gas or tritiated water is used, the limit of product specific activity may be significantly lower than that of the isotope source. This limit will be determined by the ratio of tritium to hydrogen in the entire exchangeable pool. Included in the exchangeable pool are the source tritium and any hydrogen diluent, the hydrogens of the substrate subject to catalyzed exchange, any easily exchangeable substrate hydrogens (OH, NH, etc.), hydrogen initially bound to the catalyst, and adventitious water in the solvent or attached to glass surfaces.

## 3.1 Homogeneous Acid- or Base-Catalyzed Exchange

Compounds tritiated from protic tritium sources (e.g., tritiated water) by means of acid- or base-catalyzed exchange may be more likely to lose their label by back-exchange in protic solvents, compared with those labeled by means of a catalyst which is separated after the labeling is accomplished. However, the key measure of utility for these compounds is the stability of the label under the actual conditions of use. Tritium incorporated into a substrate under the influence of strong acid or strong base or high temperature into low-acidity carbon sites may be indefinitely stable through any conceivable biological study. Tritium installed into acidic sites under milder conditions may be much more labile, though it might still prove suitable for a particular experiment. Knowledge of p$K_a$ values, mechanisms of exchange, exchange rates of related chemical systems or information from testing of the compound prior to use are among the means that can be used to determine whether the compound will be useful for the intended application.

### 3.1.1 Exchange without Added Acid or Base

It is only by a practical distinction that exchange labeling without added catalysts is discussed separately from exchange in acidic or basic media, for in most cases the exchange can reasonably be understood to occur via protonation/deprotonation sequences, perhaps mediated by acidic or basic ancillary functional groups within the substrate itself. Rarely, radical processes resulting from tritium decay are invoked. This is not a commonly used preparative method for tritium-labeled compounds, owing to the usually slow rate of isotope incorporation, and the higher likelihood of loss of label by back-exchange during use. However, for compounds that are thermally stable but are unstable at low or high pHs, tritium incorporation may proceed at a practical rate with heating, whilst loss of label by back-exchange may be slow enough under the conditions of its use as to render it quite adequate for the intended use.

Among simple molecules, pyridine undergoes exchange of the C2-H and C6-H on heating in $^2H_2O$ to 225 °C[3], and a variety of mononuclear and polynuclear nitrogen heteroaromatics and their methyl derivatives do likewise, both adjacent to nitrogen and in the methyl groups[4], yet the labels in all these compounds are quite stable at ambient temperature. 2-Methyl-substituted derivatives of benzimidazole, benzothiazole, indole and benzofuran were found[5] to undergo selective exchange at the methyl groups when heated in neutral $^2H_2O$ solution at 145–185 °C, at rates decreasing in the order listed, and only at above 225 °C did the aryl sites begin to undergo exchange. However, even though such procedures could be applied with tritiated water there are more practical ways to tritiate such simple compounds. These include acid-catalyzed exchange (Section 3.1.2), exchange catalyzed by metal salts and metal complexes (Section 3.3) and labeling from tritium gas (Section 3.5).

There are examples of the uncatalyzed exchange with $^3$HHO at the $\alpha$-positions of steroidal ketones[6]. The ketone functions were immediately subjected to synthetic transformations in order to prevent tritium loss by back-exchange. The C2 positions of imidazole-containing pharmaceuticals have been similarly exchanged[7]; in this case, the half-time for loss of the tritium label was determined to be 2.4 days at 37 °C and pH 7, adequate for the intended study. This figure is similar to that measured[8] for [imidazole-2-$^2$H]histidine, namely 2.8 days at 37 °C and pH 8.2.

It is with more complex compounds, which are more difficult to label by other means, that exchange in neutral tritiated media can have real value.

It has long been known[9] that the C8 hydrogen of purines is exchangeable with relative ease, even under neutral conditions[10]. The C2-H of adenine exchanges about $10^3$ slower than C8-H[11]; mechanisms and pH profiles of exchange rates of purines and related compounds have been determined[12]. Purine nucleosides and nucleotides, and purines in oligonucleotides, tRNAs and DNA exchange readily[13] with heating at pH 7–8, but depurination can occur[14] at acidic pHs. *E. coli* B DNA, tritiated by heating in $^3$HHO for 20 min at 100 °C, was labeled to 62% of the theoretical level based on known rates of purine C8 exchange, and the label was stable enough for the material to be used in DNA–DNA hybridization experiments[15]. In contrast, hydrogen isotope exchange in pyrimidines requires the presence of platinum or palladium catalysts[16]. The half-time of exchange of purine C8 hydrogens at neutral pH at 90 °C is measured in hours, and at −20 °C in years[16,17], so purine-containing compounds can be exchange labeled quickly under heating then stored almost indefinitely in the cold without significant back-exchange. Two oligonucleotides (a 21-mer containing ten purines and a 20-mer containing three purines) were heated for 6 h at 90 °C in a solution of tritiated water (46 mCi/milliatom H) buffered to pH 7.8. The products, stripped of labile tritium and purified by chromatography, had specific activities of better than 75% of the calculated equilibrium values based on the number of purines in each. Incubated in aqueous buffer at 37 °C for 72 h, these [$^3$H]oligonucleotides lost only 10% of their radioactivity to back-exchange[17], providing a measure of their utility in biological experiments. The relative rates of tritium exchange into purines at various positions in nucleic acid units and tRNAs has been used to deduce information about local conformations and tertiary structure[18]. The deuterium labeling of bleomycin $A_2$ was accomplished[19] by microwave induced heating at 165 °C for 6 min in $^2H_2O$; the product isolated in moderate yield had an isotope distribution of one to seven deuterium atoms per molecule. By analogy with previous labeling trials with a model imidazole and a model bis-thiazole, the major sites of labeling in bleomycin $A_2$ (**1**) were inferred to be as indicated in Figure 3.1. Whether such a molecule would withstand analogous treatment with tritiated water without undue radiolysis is uncertain.

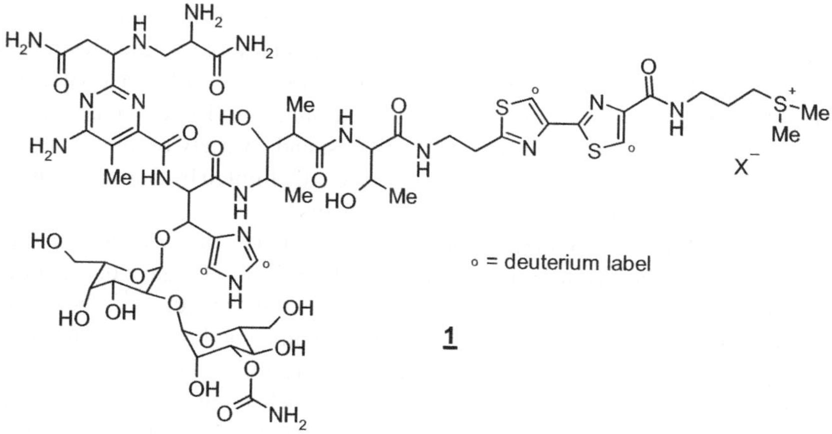


***Figure 3.1*** *Structure of [$^2$H]bleomycin $A_2$*

### 3.1.2 Exchange under Acidic Conditions

A great many acids have been used to catalyze tritium-for-hydrogen exchange, including common mineral acids, organic acids such as trifluoroacetic acid, Lewis acids such as boron tribromide, and complex acids such as $^3HHO/AlCl_3$ and $^3HHO/H_3PO_4/BF_3$. Most of the strongest are not commonly used in pharmaceutical or biological research, as their acidity is incompatible with many complex, multifunctional molecules. Many early examples of their use, mostly with simple molecules, have been collected[20]. However, as discussed in this section, a number of acid-catalyzed procedures have been shown to have good utility in particular applications. One example[21] is the preparation of the pharmaceutical-related compound [$^3$H]2 (220 mCi/mmol), by tritiation of 2,6-dichlorophenol with $^3HHO/AlCl_3$ followed by its coupling with the right-hand fragment of the molecule.

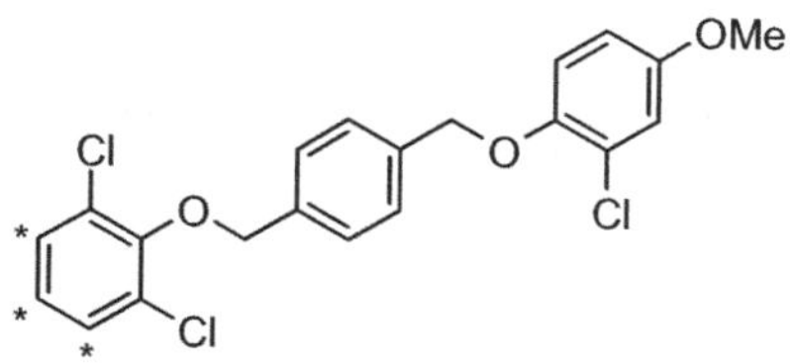


***Figure 3.2*** *Structure of tritiated 1-(2,6-dichlorophenoxy)methyl-4-(2-chloro-4-methoxy) phenoxymethylbenzene*

Generally, strongly acidic media can introduce tritium into aryl positions, and those activated for electrophilic substitution are more easily exchanged than those possessing electron-withdrawing (deactivating) substituents. The order of reactivity for reagent systems for exchange of benzene hydrogens[22] is $^3HHO/AlCl_3 > {}^3HHO/H_3PO_4/BF_3 > {}^3HHSO_4$. Only the strongest acid systems can mediate exchange at alkyl C–H bonds. Many nitrogen heterocycles can be exchanged under mildly Brønsted-acidic conditions but less readily in concentrated acids, consistent with an ylide mechanism[23], which requires the intermediacy of an N-unprotonated species. The use of dilute acids at elevated temperatures has been extensively studied and reported[4]. Typical conditions of 0.1–0.5 M HCl/$^3$HHO at 150–250 °C for hours to days are sufficient to equilibrate aryl hydrogens with solvent deuterium/tritium in alkylbenzenes, phenols, anilines, and a variety of mono- and polycyclic heteroaromatics. However, the method is significantly limited by side reactions such as ester/amide hydrolysis and alkyl group rearrangements.

Anilines are well known to undergo rapid exchange under mildly acidic conditions[24]. Labeling of desipramine (3) in dilute acid (1 M HCl/$^2H_2O$, 80 °C) gave deuterium incorporation into the aryl positions *ortho* and *para* to the amino substituent[25]; the labels were sufficiently stable for use of the compound under nonacidic conditions, and the equivalent procedure utilizing tritiated water should give similar results, taking isotope effects into account. In many cases mild acid-catalyzed labeling has been used as part of a synthetic sequence ending in a product less prone to back-exchange. For example, *m*-amino[$^3$H] levamisole (4) was prepared[26] at a specific activity of 39 Ci/mmol from the unlabeled compound by heating the HCl salt with a small amount (2 μL, 5 Ci) of $^3$HHO; it was then deaminated via treatment of the crude salt with $NaNO_2$ at 4 °C followed by reduction with hypophosphorous acid at room temperature without significant loss of label. Tritium labels in the end product levamisole (5) would be chemically quite stable. Whereas benomyl (6) and related compounds could not be deuterated in the aryl ring by acid- or base-catalyzed exchange, phenylenediamine was readily perdeuterated (2 N $^2$HCl/$^2H_2O$, reflux, 18 h), then converted to [$^2H_4$]benomyl and its derivative (7) via 2-amino[$^2H_4$]benzimidazole[27].

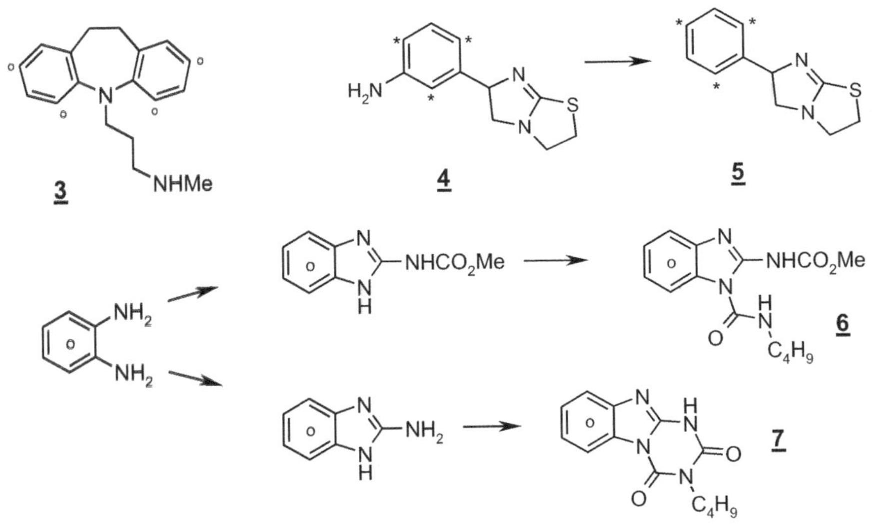


**Figure 3.3** *Deuterium/tritium labeling of desipramine (3), levamisole (5) and benomyl (6)*

Another strategic approach to labeling a compound is through *in situ* formation of a derivative more easily exchangeable than the parent[28]. Several related methods for isotopic labeling of C2 of α-amino acids utilize this strategy, most of them in acidic conditions. One method[29] generally involves heating amino acids or *N*-acetylamino acids in the presence of

$^2H_2O$ or $^3HHO$ and an excess of acetic anhydride. These conditions very likely result in transient formation of the azlactone (**8**), whose C4-H undergoes exchange as indicated. Ring opening then releases the *N*-acetylamino acid labeled at C2. Racemization occurs during this process; enzymatic deacetylation has been used to resolve the product when (*R*)- or (*S*)-$\alpha$-amino [2-$^n$H]acids are desired. This procedure has been shown to label selectively the carboxy-terminal amino acid of peptides[30]. The azlactones (**8**) of some amino acids have been prepared separately and shown to give $\alpha$-amino [2-$^2$H]acids upon treatment in deuterium oxide solution with a catalytic amount of pyridine[30]. A second method involves the heating of *N*-unsubstituted amino acids or esters in acetic acid solution in the presence of salicylaldehyde[31]. In this case it is the intermediacy of an imine/iminium species (e.g. **9**) that undergoes exchange at the position corresponding to the amino acid C2. Both C2 and C3 hydrogens were replaced with deuterium when *N*-unsubstituted amino acids were heated under slightly acidic conditions in the presence of pyridoxal and Al(III) ion[32]. This reactivity was attributed to the formation of an aluminum complex with the amine nitrogen and carboxylate of the amino acid and the C3 hydroxyl of pyridoxal.

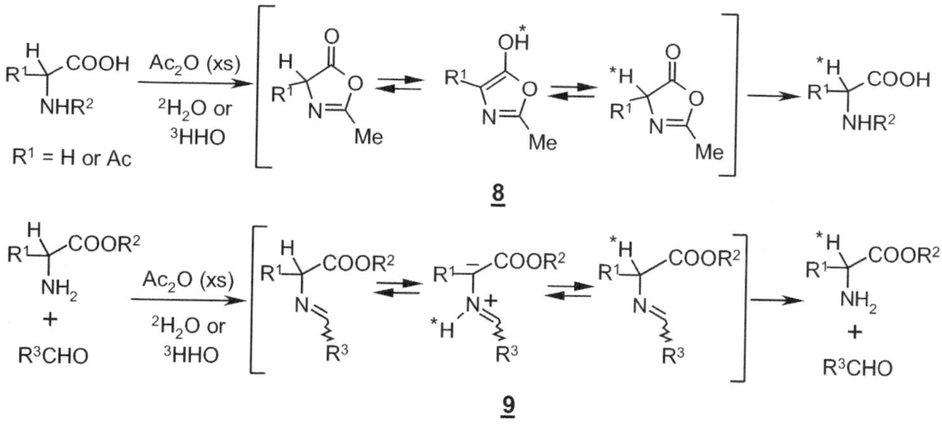


***Figure 3.4*** *Deuterium/tritium exchange-labeling at C2 of amino acids via imine/iminium species*

The stronger trifluoroacetic [$^3$H]acid has been found useful[33] in tritiating the open ring positions of the plant terpenoid gossypol (**10**) with good mass recovery (Figure 3.5). Treatment of **10** in THF/TF[$^3$H]A/$^3$HHO (from $^3$HHO and trifluoroacetic anhydride in a molar ratio of 7.7:1) at 75 °C for 45 min gave [4,4′-$^3$H]gossypol with a specific activity of about 45% of the theoretical. A similar method was found to be an improvement over other acids and basic systems for the deuteration of ferrocene and some substituted ferrocenes[34]. Duplication of such conditions using tritiated water would provide [$^3$H]ferrocene at a specific activity of approximately 3.75 times greater than the molar specific activity of the $^3$HHO. Both of these procedures are practicable for use of tritiated water of very high specific activity.

An alternative to the use of tritiated aqueous acid or acid/organic solvent mixtures is tritiated heptafluorobutyric acid, which combines good solvating power for organic substrates, low volatility and simple preparation, from heptafluorobutyric anhydride and tritiated water, without dilution of tritium by ancillary hydrogens. The method was shown

**10**

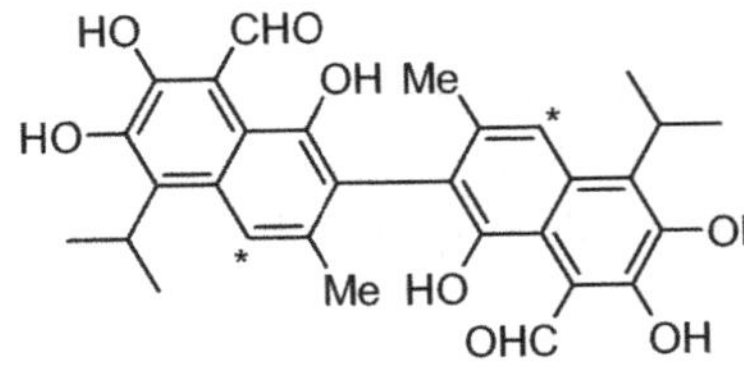

*Figure 3.5* *Trifluoroacetic acid-catalyzed exchange labeling of gossypol*

to be effective in several model compounds[35], and has been used for tritiation of **11**[36] (to 330 mCi/mmol from $^3$HHO at 555 mCi/milliatom H) and of **12**[37] (to 13.5–17.6 Ci/mmol from carrier-free $^3H_2O$) (Figure 3.6).

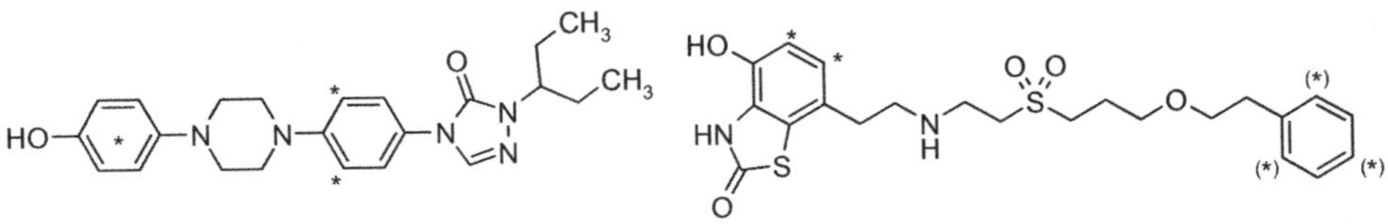

**11** **12**

* = tritium label
(*) = minor/uncertain tritium label

*Figure 3.6* *Compounds tritiated by heptafluorobutyric acid-catalyzed exchange*

Many Lewis acids have been used to catalyze exchange labeling of compounds with tritium. They, like Brønsted acid systems, show the general site selectivity of electrophilic aromatic substitution, but factors such as substrate stability and rate of exchange vary from reagent to reagent. For example, addition of $^3$HHO to solutions of **13** and **14** in $BF_3/Et_2O$ and keeping the mixture at room temperature for periods of days to weeks gave the compounds labeled in the rings to levels of 57% and 41% of the specific activity of the tritiated water, respectively[38] (Figure 3.7).

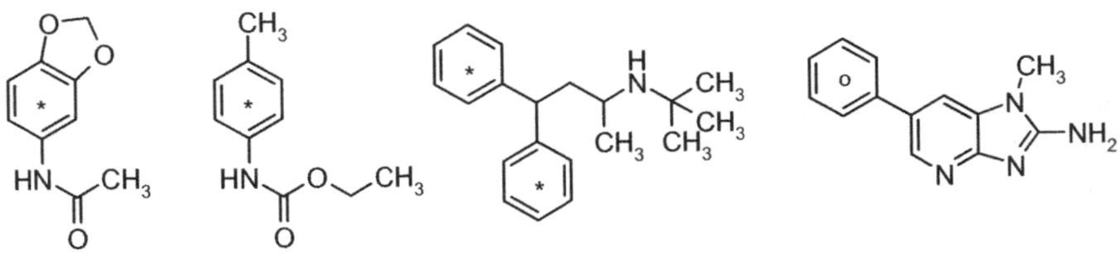

**13** **14** **15** **16**

*Figure 3.7* *Compounds tritiated or deuterated by boron trifluoride etherate-catalyzed exchange*

The complex $BF_3/^2H_3PO_4$ has been used extensively since its introduction[39], when it was shown to deuterate a number of simple aromatics. It often requires less heating or shorter reaction times for maximum labeling than do Brønsted acids; it is prepared by adding at least three equivalents of deuterium oxide to $P_2O_5$ then saturating the resulting liquid with gaseous $BF_3$. Treatment of terodiline (**15**) with this reagent, prepared[40] with $^3HHO$ of specific activity 45 mCi/milliatom H, at room temperature for four days, gave a 31% yield of [$^3H$]terodiline at 347 mCi/mmol, equivalent to complete equilibration of label at all aromatic positions. Several reports provide further insight into the regioselectivity achievable, although the work was conducted with deuterium and not tritium. For example, treatment of the food mutagen PHIP with $BF_3/^2H_3PO_4$ in $SO_2$ in a closed vessel at 90 °C gave deuterated PHIP (**16**)[41] with high isotope incorporation into the phenyl ring only (85% $^2H_5$, 14% $^2H_4$, 1% $^2H_3$); labeling of the heterocyclic part of the molecule was prevented by irreversible N-protonation by the strongly acidic medium.

Labeling of enterolactone[42] (**17**) with a $^2H_3PO_4/^2H_2O/BF_3$ reagent ($^2H_2O$ and $P_2O_5$ in 8:1 molar ratio) at room temperature for 20 h resulted in deuteration at all eight free aryl sites (Figure 3.8); by comparison only the six sites *ortho* and *para* to the hydroxyl groups were exchanged[43] by treatment with $PBr_3/^2H_2O$. A further example[44] of differential labeling is genistein (**18**); refluxing of this flavonoid in $CF_3COO^2H$ for 2 days gave (6,8,3′,5′-$^2H_4$)genistein. The deuterium at C6 and C8 are relatively labile, so refluxing the tetradeutero product for 30 min in 1% AcCl/MeOH removed them completely, giving (3′,5′-$^2H_2$)genistein, whose labels were stable. Alternatively, aryl-hexadeuterated genistein was formed under $^2H_3PO_4/BF_3$ treatment at 55 °C for 3 days, and acidic back-exchange of the C6- and C8-deuterium atoms as above provided (2′,3′,5′,6′-$^2H_4$)genistein. Therefore, if tritium were to be used instead of deuterium, tactics could be selected either to maximize specific activity by labeling the largest number of sites, or to restrict labeling to specific parts of the molecule. A number of other flavonoids and isoflavonoids labeled at selected sites have been prepared[45,46] by similar procedures, and the possible utility of the method with other acid-stable phenolic compounds is apparent.

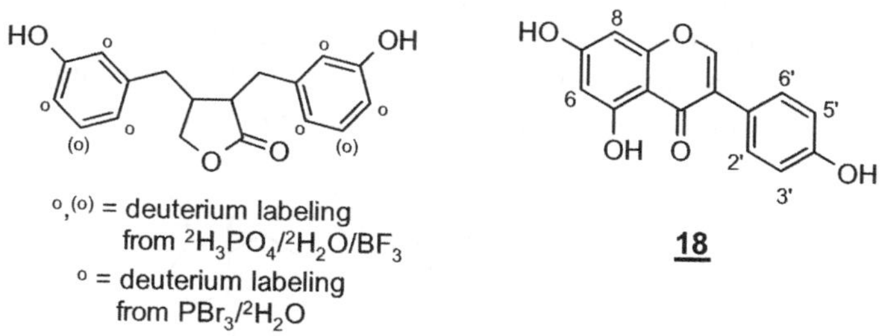


**17** **18**

***Figure 3.8*** *Lignanolactones labeled by acid-catalyzed deuterium exchange*

Although their reactivity is incompatible with many multifunctional compounds of biological interest, ethylaluminum dichloride (EADC) and boron tribromide[47] are worthy of note because they can be very effective in mediating the labeling of simpler compounds or synthetic intermediates. Aromatic[48] and polycyclic aromatic[49] and some heteroaromatic[50] substrates treated with EADC at room temperature followed by addition of a small amount of tritiated water have been labeled to maximal levels in just a few minutes; various halo and alkyl substituents are tolerated. Unfortunately, rearrangements of aryl substituents can occur during the process[51]. Several simple steroids have been labeled with EADC, or $BBr_3$ using a similar procedure[52], in either cyclohexane or chloroform solution; however, phenolic and keto steroids have perhaps been labeled more often using base-catalyzed exchange procedures (Section 3.1.3). Generally, EADC was observed to be more reactive than $BBr_3$ and chloroform a more active medium than cyclohexane; for some substrates, decomposition competed with labeling.

### 3.1.3 Exchange under Basic Conditions

Many of the same limitations of acid-catalyzed hydrogen isotope exchange, such as substrate stability, specific activity and potential vulnerability of the product to lose label by back-exchange, apply also to base-catalyzed exchange. However, some basic procedures and reagents can direct isotopic hydrogen into significantly more stable positions than can acidic ones. Base-catalyzed exchange procedures can be classified as thermodynamically controlled, in which reversible deprotonation–reprotonation occurs, and kinetically controlled, in which a substrate is deprotonated irreversibly and the resulting anion quenched with isotopic hydrogen.

*Thermodynamic Control*

A wide variety of carbon-acidic compounds are subject to hydrogen isotope exchange in alkaline aqueous media, including aldehydes, ketones, esters, nitriles, nitro compounds and phenols. A thorough discussion of deuterium labeling of such classes is available in A. F. Thomas's excellent book[53]; many of the procedures described therein have been modified for tritium labeling. Base-catalyzed exchange labeling of heterocyclic compounds has been reviewed[54]. Compounds labeled at C-H acidic positions may be subject to loss of label and are therefore not always useful in biological experiments. However, these labeling procedures are frequently used to label intermediates whose activating function is subsequently modified or removed during conversion into target compounds. This approach was common in steroid and terpene biosynthesis research[55]. A recent example[56] is provided in a preparation of a tritium-labeled vitamin D analog **19** (Figure 3.9).

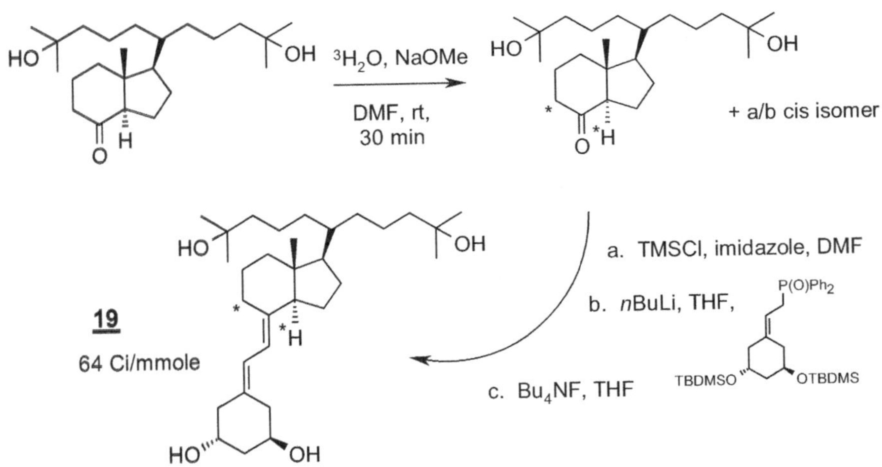


*Figure 3.9 Tritium introduced into stable positions of a target via an activating function in a synthetic intermediate*

Nevertheless, many compounds labeled by basic aqueous exchange are useful without further modification. Some of the more recent applications of this approach, involving useful chemoselectivity, include the tritiation[57] of 2-oxoglutaric acid at C3 (**20**) (disodium salt, $^3$HHO, 80 °C, 3 h) for study of enzyme mechanisms, the tritiation[58] of several investigational drug compounds possessing methylsulfonyl groups (e.g. **21**) ($^3$HHO, THF, DBU, rt, 18 h) for metabolism studies and labeling[59] of the phytohormone **22** ($^3$HHO, DMF, $Et_3N$, 80 °C) for biosynthetic studies (Figure 3.10).

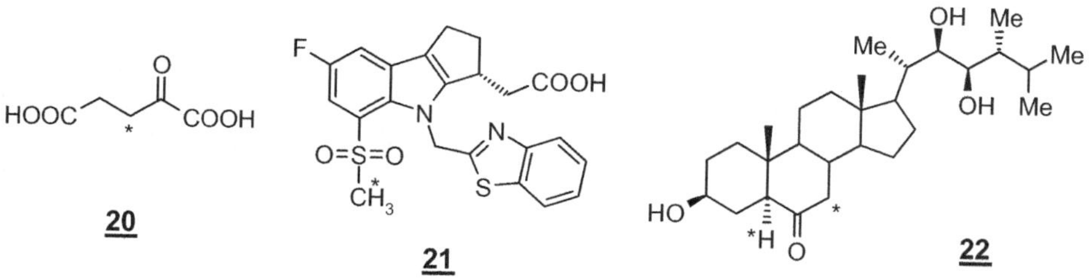


*Figure 3.10* *Example compounds labeled in chemically labile positions useful for certain studies*

Phase transfer catalysis has been exploited with beneficial effect in isotope exchange reactions. Chemically sensitive compounds that would not survive under the conventional base catalysis conditions can be labeled successfully. For example **23** was tritiated[60] in benzene solution using $^3$HHO/NaOH/*n*-$Bu_4$NBr, with a recovery of 80%, and the base-sensitive steroids **24** and **25** were tritiated without double bond isomerization (location of tritium not reported) (Figure 3.11)[61].

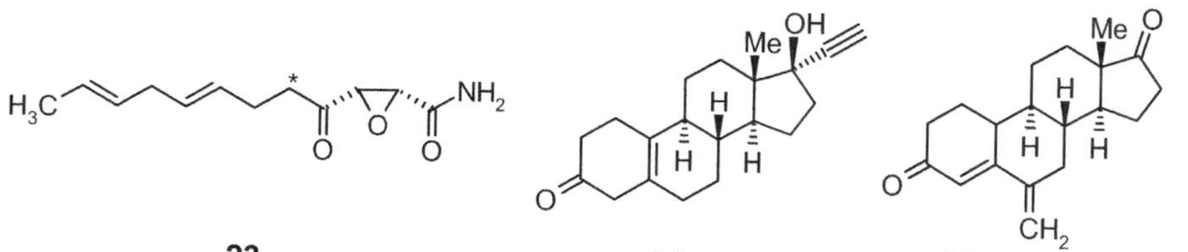


*Figure 3.11* *Chemically sensitive compounds labeled via phase-transfer catalysis*

Advantage has been taken of the enhanced basicity of reagents transported from the aqueous into the organic phase owing to their reduced solvation. This effect was demonstrated[62] in room temperature reactions using $^2H_2O$/NaOD/*n*-$Bu_4NHSO_4$ and hexane, benzene or mesitylene solutions of test compounds, exemplified by their deuterated products **26–28** (Figure 3.12). Candidates for exchange under these conditions are hydrogens of benzylic methylene groups, $\alpha$- and $\beta$-hydrogens of furans and thiophenes and their benzo derivatives, and hydrogens of dihalobenzenes, all of which have p$K_a$s of 38 or less (DMSO scale). The method could be used with $^3$HHO with little or no modification and little dilution by protons.

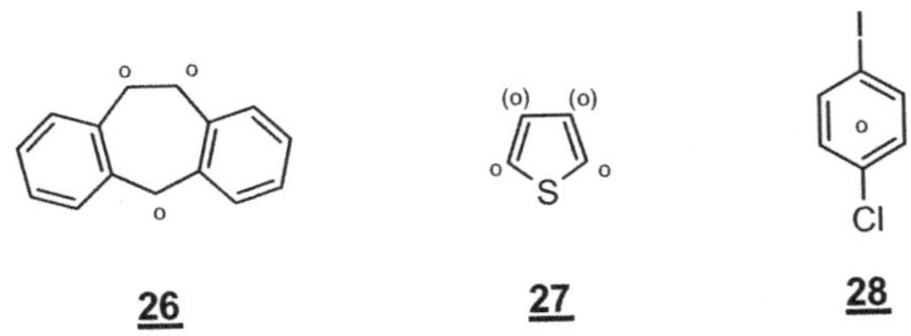


*Figure 3.12* Compounds labeled by exploitation of basicity enhanced by phase-transfer catalysis

$\alpha$-Amino acids possessing unsubstituted amine functions have been exchanged-labeled with deuterium at C2 (with racemization) by treatment under reflux in $KO^2H/^2H_2O$ in the presence of a catalytic amount of pyridoxal[29d]. The mechanism is analogous to that described for the acid-catalyzed C2 exchange of amino acids with salicylaldehyde (Section 3.1.2).

*Kinetic Control*

Stoichiometric C-deprotonation of a compound results in a carbanion which can be quenched by addition of tritiated water or another tritium source, either *in situ* or after some manipulation of the deprotonated substrate. This method can have advantages over thermodynamic exchange procedures, including greater chemo- or regioselectivity, the achievement of higher specific activity or economy in the use of tritium.

Simple examples include the preparation[63] of monomethyl [2-$^3$H]succinate (**29**) for enzyme mechanism studies (Figure 3.13). The carboxyl, C2-dianion was prepared with LDA, freed of all volatiles, then treated with $CF_3COO^3H$ prepared from 0.44 molar equivalents (vs **29**) of tritiated water, thus utilizing the entire quantity of tritium without dilution. Likewise, the *N*-acetylazetidinone **30** (previously deuterated at C3 by a different exchange procedure) was treated with LiHMDS to selectively form the anion at the acetate C2 position, which was then quenched by addition of tritiated water[64]. Compound **30** was used as an intermediate for the preparation of a substrate for studies of the enzyme clavaminate synthase.

*Figure 3.13* Compounds labeled by tritonation of ester enolates

The chemically sensitive biocompatible biodegradable polymers, poly($\varepsilon$-caprolactone)[65] and poly-rac.-lactic acid[66], of interest in controlled drug delivery, were tritiated adjacent to the carboxyl groups by treatment with 0.25 equivalent of LDA followed by $^3$HHO. Although the degree of labeling was sufficient to provide good sensitivity as tracers in biological systems, it was significantly below theoretical levels.

When LDA (or to a lesser extent LiHMDS) is used as the base to deprotonate a substrate, isotope incorporation from added tritium or deuterium can be considerably less than theoretical because the identical proton removed from the substrate can remain in the lithium enolate–amine complex, and reprotonate the substrate in preference to added (isotopic) protons – a phenomenon termed 'internal proton return'[67]. This can be prevented by removing the offending proton; two common ways are to add an equivalent of alkyllithium to the enolate–amine complex, or to silylate the enolate to form the silyl enol ether then, in a separate step, regenerate the enolate ion using an alkyllithium[68]. Studies illustrating the effectiveness of both strategies have been reported using deuterium[69]; tritiated water could be used in place of deuterium oxide without significant modification of these procedures.

Procedures involving direct metalation of a nonenolizing C–H bond followed by quenching of the anion with a tritium source usually are effective in emplacing the label in positions less prone to back-exchange than those discussed above. In practice, the utility of metallations is limited by the ability of a substrate's functionality to survive exposure to a strong base, but this approach has wide utility. Organolithium reagents for example have had widespread use, and their compatibility with various functional groups is well-delineated. In a simple example, dithienobenzene **31** was tritiated[70] at C2 by sequential treatment with *n*-BuLi followed by $^3$HHO (Figure 3.14). The special value of the metallation approach to tritium labeling, however, arises from two characteristics: the wide variety of readily available organometallic bases[71] which allows for selection of a base appropriate to the substrate at hand, and the phenomenon of 'directed *ortho* metallation'[72], which provides a control and predictability over the site of metalation (usually an aryl C–H bond). For example[73], compound **32** was deprotonated (*s*-BuLi, TMEDA, THF, −78 °C) then quenched with tritiated trifluoroacetic acid prepared from trifluoroacetic anhydride and $^3$HHO; the specific activity of the product was 78% of the theoretical level. Ferrocene **33** was selectively monodeuterated[74] in 85% isotopic purity (*t*-BuLi, $Et_2O$, r.t.; $^2H_2O$), and (+)-pisatin[75] (**34**) to 23.9 Ci/mmol (*n*-BuLi, $^3H_2O$).

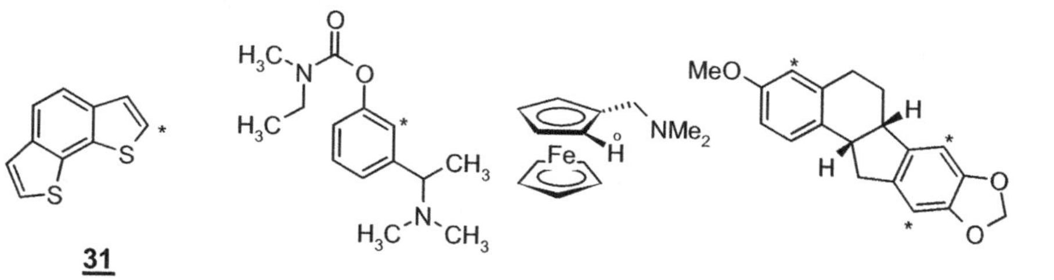


**31** **32** **33** **34**

***Figure 3.14*** *Compounds labeled by direct metalation and $^3$HHO or $^2H_2O$ quenching*

Considering the difficulty of obtaining and using protic tritium sources at high isotopic abundance, the ability to use tritium gas, routinely available at near 100% isotopic purity, is very attractive for preparing compounds at high specific activity. Therefore, the quenching of a metalated arene with tritium gas to provide the corresponding tritiated arene, as in the case of ibogaine[76] (**35**) is of interest (Figure 3.15). A specific activity of 13.8 Ci/mmol was achieved, and the label was shown by $^3$H NMR spectroscopy to be a mixture of species monotritiated at one of the two sites indicated, indicating that two isomeric metalated species were formed under the reaction conditions. As (isotopic) hydrogen gas acted as the acid in the labeling step, anions from aryl C–H bonds of acidity no greater than about p$K_a$ = 37 would be expected[77] to react. Unfortunately this excludes many compounds that might seem to be good candidates, as illustrated by the failure of metalated rimonabant (SR141716, **36**) to react efficiently with deuterium gas[78].

**35**

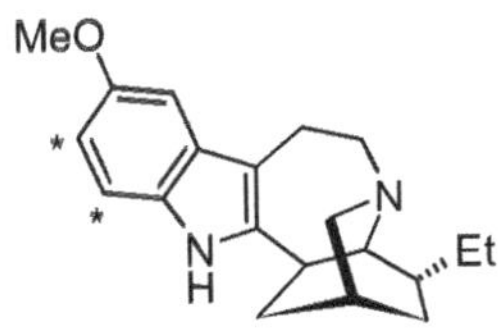


**36**

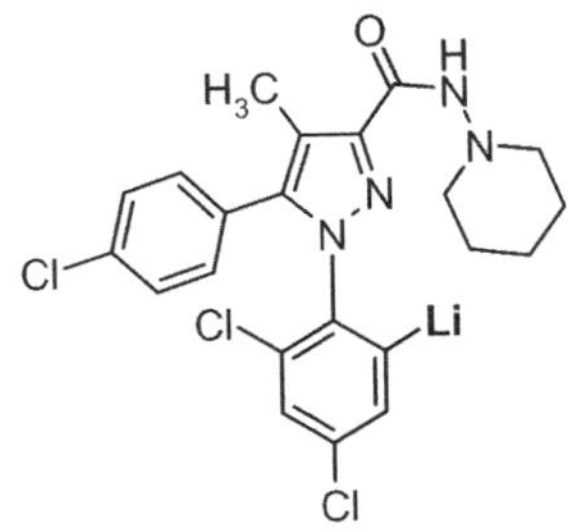


*Figure 3.15* *Use of tritium or deuterium gas in quenching metalated organics*

9-Phenyl[9-$^2$H]fluorene (**37**), a nonvolatile solid with p$K_a$ ~18, has been shown[79] to be a source of protic deuterium to quench carbanions (Figure 3.16); it was prepared in high isotopic content by Pd-catalyzed deuterium–bromine replacement of 9-bromo-9-phenylfluorene using deuterium gas. A number of model compounds (exemplified by **38**–**41**) were labeled, in good chemical yields and with very good isotope incorporation, by metallation (*n*-BuLi, THF, −78 °C) and subsequent addition of **37**. The deuterium in **37** is acidic enough to undergo transmetallation with a wide variety of carbanions, but not so acidic as to be air- or moisture-sensitive. Although the analogous tritiated reagent has not been published it could be prepared analogously, may be stable on storage (dissolved in a radical-trapping solvent) and could prove to be a good alternative to high specific activity tritiated water.

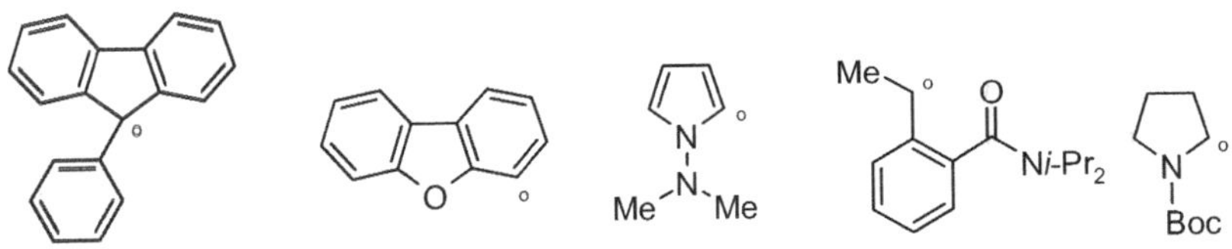


**37** **38** **39** **40** **41**

*Figure 3.16* *Use of 9-phenyl[9-$^2$H]fluorene (**37**) as label source for substrates **38**–**41***

## 3.2 Heterogeneous Catalysis with Tritium in Solvent

By far the most widely utilized catalysts for exchanging solvent-based tritium (usually water) with substrates are metals. However, a variety of other catalytic materials have been investigated for this purpose; in addition there are catalytic species which, although themselves normally soluble, have been 'heterogenized' to modify their properties or improve the utility of the method.

### 3.2.1 Metals

The use of supported and unsupported metals as catalysts for exchange labeling of substrates in solution from solvent tritium was thoroughly reviewed in 1974[80,81] and again in 1987[82]. By the time these reviews were written the utility of the method had been well established for labeling of a wide variety of compounds. The reaction systems utilized group 8, 9 and 10 metals in tritiated water or mildly acidic solutions (e.g., aqueous acetic acid). Of these metals platinum and palladium were found to be generally the most active, and nickel slightly less so. Frequently the most active catalysts are prepared by *in situ* reduction of metal oxides with hydrogen or tritium gas[83] or of metal halides by hydride reagents such as sodium borohydride. Exchange in substrates does not usually occur at useful rates below about 50 °C, and most reactions require heating at 100–180 °C for periods from a few hours to 24 h or more, so this method is not often used for heat-sensitive substrates. In most cases, substrate-specific optimization is required to identify optimal conditions.

Although the distribution of label in complex substrates was not usually well defined in early work, studies on simple model compounds[84,85] provided data for formulation of useful mechanistic models for metal-catalyzed exchange, thereby rationalizing the observed regioselectivity. A later experimental survey[86] utilizing $^3$H NMR to full advantage more accurately quantified data for platinum. More recently, careful studies[87] have led to the development of a model of catalyst action that accounts for the effects of some experimental variables on the efficiency of metal-mediated catalytic hydrogen isotope exchange from isotopic water sources. In this model the key isotope species for exchange is $[^3H^+(^3H_2O)_n]$ clusters adsorbed on metal active sites. These clusters are most highly enriched with tritium when two conditions are met. First, the tritium oxide component is produced by reduction of the oxide of the metal catalyst-to-be rather than from an exogenous source. Second, the just-reduced metal catalyst is thoroughly vacuum degassed immediately prior to use, thus ensuring maximal catalytic formation of the $^3H^+$ component, which arises through reverse spillover[88]. Fulfilling these conditions requires that the freshly produced $^3H_2O$ must be removed by vacuum transfer, the newly reduced metal activated under vacuum, then the $^3H_2O$ transferred back in. Although the original procedure, and the one still commonly used, involves combining portions of catalyst from a bottle, $^3H_2O$ or $^3$HHO, solvent and substrate; more recent studies have clearly shown the superiority of one or another variation of the procedure involving initial reduction of supported PdO or $PtO_2$, which produces both $^3H_2O$ and highly active catalyst.

Compound classes labeled by heterogeneous metal-catalyzed tritium exchange from water include amino acids, peptides and proteins, sugars, polycyclic aromatic hydrocarbons, purines, pyrimidines and nucleosides, alkyl-substituted aromatics, alicyclic compounds and steroids, *N*-heterocycles and anilines and some *O*- and *S*-heterocycles. The method is firmly fixed in the repertoire of methods of the commercial custom labeling industry.

In general, platinum displays a preference for aryl over alkyl exchange; palladium displays the opposite preference, and nickel a *strong* preference for alkyl over aryl exchange. In aryl exchange, all three metals are relatively insensitive to electronic factors but are sensitive to steric factors, *meta*- and *para*-labeling being much faster than *ortho*. In alkyl labeling all three metals are selective for benzyl positions over those more distant from phenyl groups, but no general rule is apparent for methyl vs methylene vs methine site preferences. In *N*-heterocycles and anilines positions next to nitrogen are usually labeled more rapidly than other ring locations. Metals on supports often have slightly different levels of activity and regioselectivities compared with the free metals. When metal/tritiated water systems were directly compared with metal/tritium gas systems, the latter usually resulted in higher levels of isotope incorporation.

A recent comparison[89] of hydrogen-activated forms of 5% Pt/C with 10% Pd/C in deuterium exchange in $^2H_2O$ demonstrated some of the differences in their site selectivity, as exemplified by **42–44**.

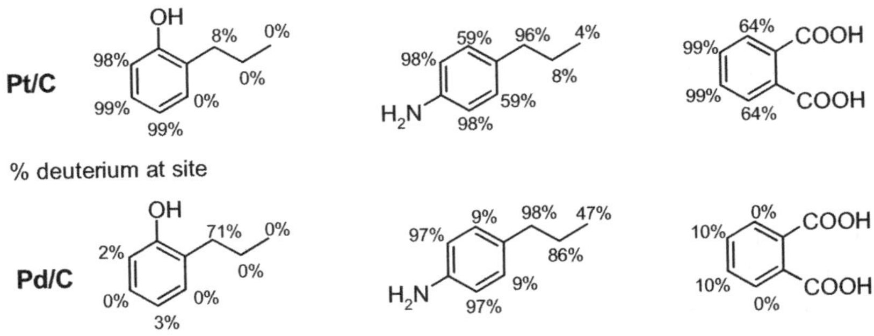


**42** (r.t.) **43** (80°C) **44** (180°C)

***Figure 3.17*** *Site selectivities of Pt/C and Pd/C in catalysis of exchange from $^2H_2O$*

Further, an apparently synergistic effect was observed[90] in the labeling of hindered aryl sites in some model compounds similar to **42–44**. In these cases a mixture of Pt/C and Pd/C gave much higher levels of deuterium incorporation than either catalyst alone.

Advantages of the method using the combination of heterogeneous metal catalyst and protic isotope source include the ability to use neutral, or much less strongly acidic media than most homogeneous metal salt exchange reactions, and the avoidance of reductive side reactions characteristic of metal-catalyzed exchange with tritium gas. Disadvantages include the limitation of specific activity compared with tritium gas as the isotope source (although this can increasingly be overcome by adoption of new procedures), the need to safely contain the pressure generated by reactions heated above the boiling point of the solvent, and the potential for substrate decomposition because of high temperatures and high radiation densities within the reaction. Racemization usually accompanies labeling at homochiral methines[22].

Organic cosolvents are frequently used, a primary reason being solubilization of compounds with low water solubility. Another is to allow for achievement of higher specific activity levels in the product. If the cosolvent is aprotic, smaller amounts of higher specific activity water can be used than would otherwise be possible; moreover, the increase in volume per unit radioactivity may reduce the risk of radiolytic effects on the substrate.

*Platinum*

Platinum is frequently the most effective metal for labeling complex compounds possessing sterically unencumbered aryl sites. Two examples[91] are **45** and **46**, which were labeled at the indicated positions using platinum prepared by $NaBH_4$ reduction of $PtO_2$, and $^3HHO$ at 50 Ci/mL (ca 450 mCi/milliatom H). In a comparison of microwave heating with conventional heating in a similar reaction system (Pt/$^3$HHO/diglyme), a substantial enhancement of the $^3H/H$ exchange rate in **47** was reported[92]. The specific activity of tritiated **47** (approximately 5.2 mCi/mmol using $^3HHO$ of 8.9 mCi/milliatom H) obtained by conventional heating at 100–120 °C for 48 h was equalled in only 1 h under microwave heating at the same temperature.

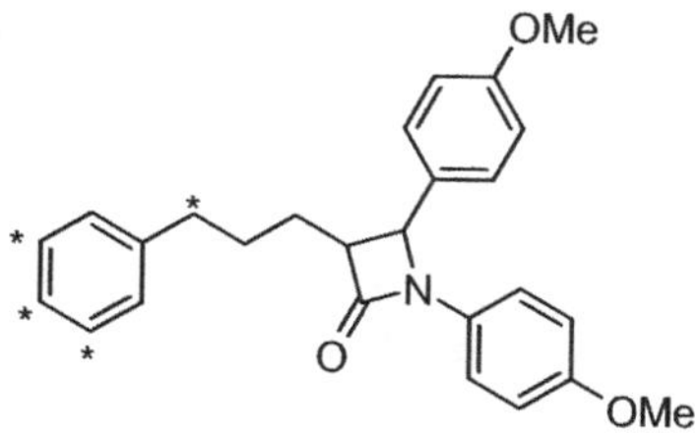


**45**

Diglyme/160°C/72h
57 mCi/mmol

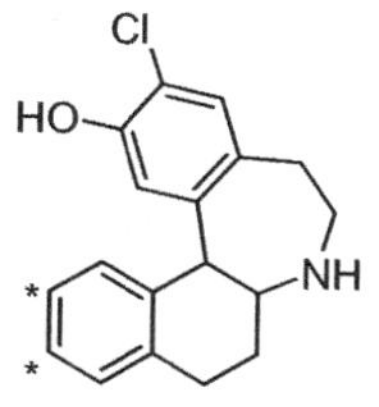


**46**

Dioxane/110°C/18h
216 mCi/mmol

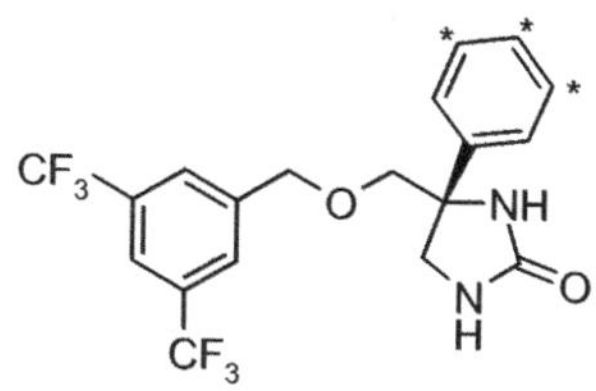


**47**

***Figure 3.18*** *Labeling from tritiated water catalyzed by* in situ *reduced platinum(IV) oxide*

Some exchange reactions utilizing platinum gave results that were not so clearly platinum catalyzed. For example, compound **48** was obtained[93] at a specific activity of 19.8 Ci/mmol after heating for 48 h at 140 °C in dioxane/$^3H_2O$ with $NaBH_4$-activated platinum. The location of label (assessed by $^3$H NMR) is more indicative of base-catalyzed exchange than platinum-catalyzed exchange. Similarly, the distribution of label in apigenin (**49**) after undergoing exchange in tritiated acetic acid at 130 °C for 4 h in the presence of platinum black[94] is the same as that resulting from acid-catalyzed exchange in similar flavonoids (see Section 3.1.2).

**48**

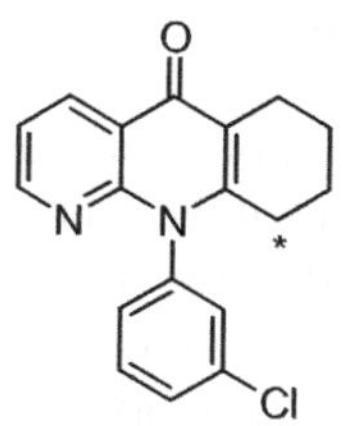


**49**

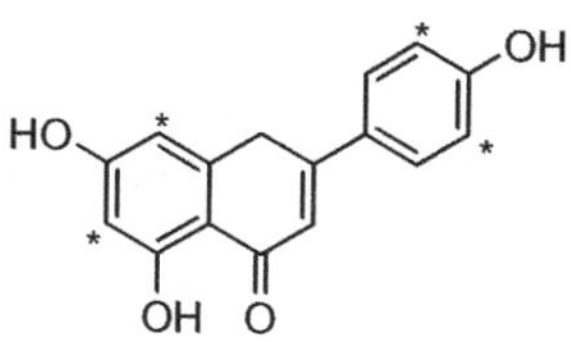


***Figure 3.19*** *Labeling by undetermined catalytic mechanisms*

*Palladium*

Supported and unsupported forms of palladium have been used with tritiated water to label a wide variety of complex compounds, many of which do not survive the reductive conditions of catalyzed labeling with tritium gas. Generally, the recovery of substrate and the level of label incorporation are strongly dependent upon several reaction parameters, so that experimentation is usually required to optimize conditions for each new compound.

As a rough generalization, many of the best results are obtained by forming the tritium oxide and active metal catalyst *in situ* by stirring the oxide form of the catalyst with tritium gas in the presence of an organic cosolvent, then adding the substrate, sealing the reaction vial and heating. Temperatures exceeding 150 °C are often required. Ciprofloxacin (**50**)[95], alprazolam (**51**)[96], zaleplon (**52**)[97] are successful examples; however, the specific activities shown are the highest obtained in each case out of a large number of experimental trials. The locations of labels in the molecules were not reported. It is apparent that the effects of various experimental variables were being explored. As described above, more recent studies[87] have clarified matters to some degree; however, as a rule substrate-specific optimization is still required.

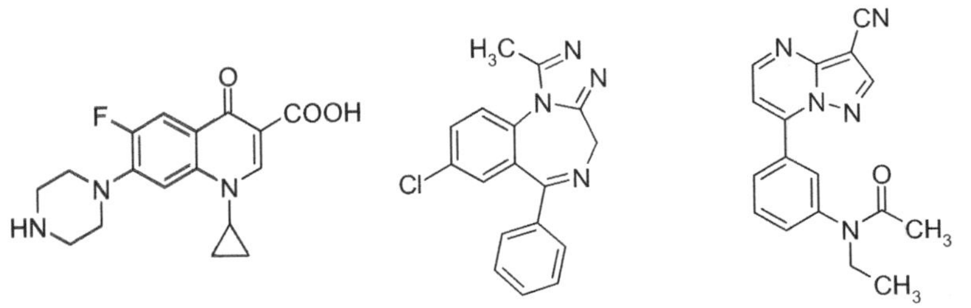


**50** (35 Ci/mmol) **51** (27 Ci/mmol) **52** (18.5 Ci/mmol)

***Figure 3.20*** *Compounds labeled from tritiated water by palladium(0) catalysis*

Substrates that are more heat-sensitive, such as cyclosporin A and FK-506, could be labeled[98] by using shorter heating times, achieving specific activities of 4 Ci/mmol and 16 Ci/mmol, respectively. Product purifications were more difficult, and the yield of the latter was only 10%. Thiols such as coenzyme A have been labeled[99] (~3.8 Ci/mmol) by similar procedures. In contrast, aliphatic compounds such as unsaturated fatty acids, phospholipids and eicosanoids are labeled only to 0.54–3.8 Ci/mmol[100]; nevertheless, considering the simplicity of the procedure compared with the alternatives, it may be the method of choice.

Conducting Pd/C–$^2H_2O$ labeling reactions in the presence of a small amount of hydrogen gas results in a more active catalyst[101], as can pretreatment of the catalyst with $NaBH_4$[102]. Nevertheless, elevated temperatures are still required (110 °C for nucleosides, 110–180 °C for N-heterocycles, ≥140 °C for all but the benzylic positions of aralkylic compounds)[103]. The 'catalytic' hydrogen gas rapidly becomes equilibrated with solvent deuterium[104]. Very dilute tritiated water was used as proof of principle[105], but this method would not be practical in its current form for the preparation of high specific activity compounds because it requires neat water as solvent/labeling source.

*Nickel*

Nickel is also an effective catalyst[82] for the exchange labeling of alkylaromatics; compared with platinum and palladium, it is more active in exchanging aliphatic hydrogens and much less active toward aromatic sites. To be effective at all, it must be prepared in an active form, such as Raney nickel, but this catalyst can cause unwanted side reactions such as deoxygenation. Nevertheless, Raney nickel has been used with success in the deuteration of aromatic acids[106], phenols, anilines and pyridines[107] under conventional heating; and for microwave-assisted deuteration of ionic liquids[108]. Regioselective deuterium labeling of certain carbohydrates has been accomplished without epimerization, sometimes with organic cosolvents, either with conventional heating[109] or with ultrasound enhancement[110] or microwave irradiation[111]. Organic cosolvents are compatible with this process, and a study of the effects of reaction temperature and cosolvents on the regioselectivity of labeling of indole and *N*-methylindole has been reported[112], as has a follow-up study comparing the effects of microwaves in these exchanges[113].

Despite the practicality of Raney nickel for deuterium exchange labeling of certain compounds, instances of its use for tritiations are few. The reason may lie in the facts that this catalyst is prepared by digestion, and normally stored in aqueous media, which is of course impractical for high specific activity tritiated water. Raney nickel's uniquely porous structure may unavoidably contain large quantities of hydrogen ions as a result of this processing. Moreover, it is much more pyrophoric than finely divided platinum or palladium when dried. Notwithstanding such issues, the pharmaceutical compounds **53** (using 20 Ci of nearly carrier-free $^3H_2O$), and **54** (with $^3HHO$ of 50 Ci/mL) have been labeled by Raney nickel-catalyzed tritium isotope exchange[91]; the sites of labeling, indicated in the structures, were determined by $^3H$ NMR.

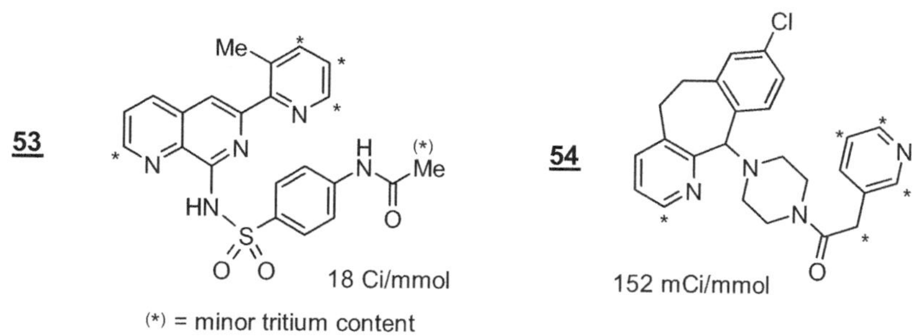


***Figure 3.21*** *Compounds labeled from tritiated water by Raney-Ni catalysis*

### 3.2.2 Other Catalysts

Only a few reports of the use of other metal catalytic systems for hydrogen isotope exchange labeling have appeared[114]; studies have generally been done with deuterium oxide and their potential for use with tritiated water of significant specific activity appears very limited.

Nafion®, a perfluoroalkyl polymer resin containing sulfonic acid groups, has been compared with trifluoromethanesulfonic acid for its regioselectivity in the labeling of model quinolines, pyridines, anilines, toluene and chlorobenzene[115]; the reaction mixtures contained substrate, resin and a small volume of tritiated water. Reactions were conducted at 90–180 °C for up to several days and gave product specific activities only a fraction of those of the tritiated water batches. Generally, both catalysts label positions subject to electrophilic attack, but Nafion behaved as a weaker acid than its homogeneous counterpart.

For example, quinoline was labeled at C5 and C8 with triflic acid but rather at C3 by Nafion, and chlorobenzene was destroyed by triflic acid but was recovered in generally labeled form from the reaction with Nafion. A more complex, alkylamino-substituted quinoline was labeled exclusively at C3, but a complex quinoline derivative possessing an alkanoyl group at C3 became labeled only in the methylene group adjacent to the carbonyl group[116]. Dowex 50W-X8, also a sulfonic acid-containing polymer, was found to be an active catalyst of tritium labeling in some simple model substrates. Some phenolic compounds have been deuterated using Nafion and $^2H_2O$[117]. Three base-functionalized polymeric resins were tested[118] for their ability to catalyze tritiation of some model compounds having acidic C–H bonds; the results were essentially the same as would be achieved under conventional homogeneous conditions (Section 3.1.3).

Based on studies to date, there are few advantages in using acid- or base-functionalized resins compared with conventional homogeneous acid or base catalysts.

One report[119] describes a study of HNaY zeolite for the catalysis of tritium exchange into simple aromatic compounds. In analogy with the use of the water-sensitive $EtAlCl_2$ (Section 3.1.2), the anhydrous active centers of the thermally activated zeolite were exposed to small quantities of tritiated water, and subsequent heating to 175 °C with substrate induced the exchange-labeling of the latter. Success was limited to aromatic compounds without bulky substituents or electron-withdrawing substituents, and the pattern of label incorporation was that of electrophilic substitution.

Neutral alumina was used[120] to mediate the exchange-labeling of the sensitive compound **55** by treatment with carrier-free $^3H_2O$ followed by mixing with a benzene solution of the substrate for 35 min at room temperature. The recovered [$^3$H]**55** was labeled (by analogy with a deuterated sample prepared the same way) exclusively at the indicated site at a specific activity of 7 Ci/mmol.

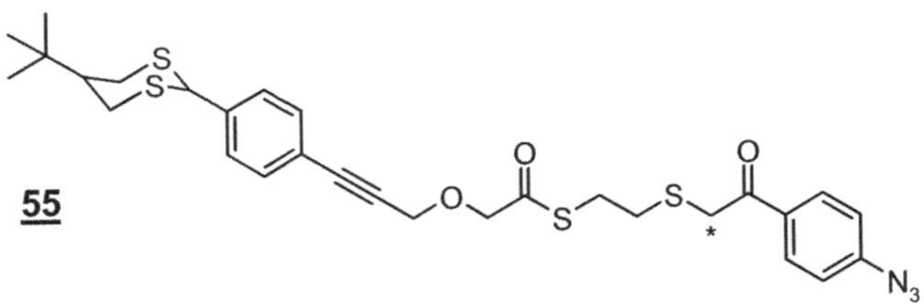


**55**

***Figure 3.22*** *Substrate labeled from tritiated water by alumina catalysis*

## 3.3 Heterogeneous Catalysis in Solution with Tritium Gas

Tritium gas is routinely available at near 100% isotopic abundance, so that the limitation on specific activity associated with the use of tritiated water does not apply. The use of heterogeneous catalysts with tritium gas as the isotope source has been shown to be effective with a range of compound types, with important exceptions. All of the catalysts capable of facilitating C–H to C–$^3$H exchange in the presence of tritium gas also have a significant propensity to hydrogenate unsaturated functions and to hydrogenolyze functions such as aryl halide. Even so, this general method has been shown to be compatible with the presence of reducible functions in some cases.

### 3.3.1 Metal Catalysts with Nonreducible Substrates in Aqueous Solution

The first extensive study of the metal-catalyzed exchange of substrates in solution with tritium gas was reported in 1974[121], though such phenomena had been the subject of occasional reports in previous years. An early review has been published[122]. The catalysts found to be effective were generally Group 8 metals or their oxides (Pd being most effective) supported on $BaSO_4$ or $CaCO_3$, and the results varied considerably with different batches of the same catalyst. Exchange reactions were generally conducted at room temperature in aqueous or mixed aqueous–organic solvents maintained at neutral or basic pH to slow the exchange between gaseous tritium and water; even so, as much as 45% of the tritium would end up in the aqueous medium in as short a period as 30 min[121]. Labeling was found to occur selectively or exclusively at benzylic positions of various compounds (e.g., amino acids, aromatic steroids), the C8 position of purine nucleosides and nucleotides, the aldehyde carbon of carbohydrates, and sometimes at $sp^2$ carbons activated by neighboring atoms. Examples include <u>56</u>–<u>59</u> in Figure 3.23. Imidazoles are often labeled, C2 being the most prone to exchange and palladium being the most active metal. The ring-C2 position of histidine is such a site, and even histidine-containing peptides of significant length have been tritiated to reasonable specific activities[123]. In contrast to purines and imidazoles, pyrimidines such as cytosine, uridine and thymine and their derivatives generally undergo reduction at the 5,6-double bond.

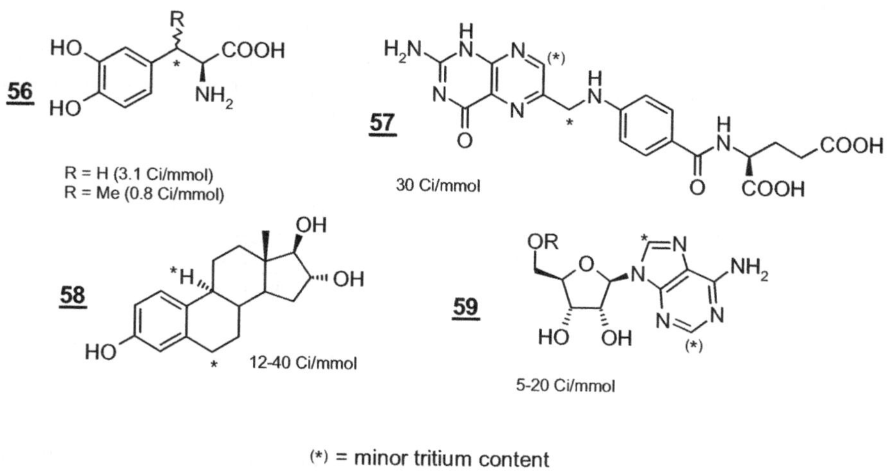


***Figure** 3.23 Regioselectivities of exchange labeling in aqueous solution from tritium gas catalyzed by supported palladium*

Product recoveries can be variable because of the large amounts of catalyst sometimes required; and catalyst, solvent composition and pH generally must be optimized for each substrate in order to obtain the best results. Tritium incorporation was noted[121] to increase over time to a maximum after which it declined, probably because of continuing exchange between tritium gas and the aqueous solvent.

Optimization of the exchange labeling of glucose has been reported[124] by testing of a number of catalysts and reaction variables; a procedure utilizing Pd/$BaSO_4$ and a pH 7 buffer gave [1-$^3$H]glucose of specific activity as high as 24 Ci/mmol. Exposure of the catalyst to tritium gas prior to using it in a reaction (gas washing) was found to have a positive effect on product specific activity.

### 3.3.2 Metal Catalysts with Nonreducible Substrates in Organic Solvents

Considering the obvious complications that can result from the use of protic solvents for such exchange reactions, and the fact that organic substrates are more easily dissolved in organic solvents, it is no surprise that effective exchange procedures using aprotic media have been sought. A study of Pd/C-catalyzed deuterium gas exchange of 1,2-diphenylethane in various solvents[125] showed dioxane, ethyl acetate and cyclohexane to be good media for exchange, whereas in methanol and benzene deuterium incorporation into this substrate was much lower. The effect of functional group substitution in substrates was shown using dioxane as solvent. Benzyl alcohol, benzylamine, dibenzylamine and dibenzyl ether were all labeled in their benzyl positions to high levels, whereas dibenzyl sulfide and phenylacetonitrile were not. The authors also provided evidence that hydrogen present on commercial catalysts represents a pool which can enter the exchange process and reduce the isotopic purity of the substrate; in order to avoid this effect they suggested a gas-washing procedure.

Some 2-substituted benzimidazoles (**60**) have been reported[126] to be labeled with tritium in dioxane solution by the action of Pd/C (60 °C, 45 min). The location of label, exclusively at C4 according to $^3$H NMR analysis, was confirmed only for (R = *S*-*n*-propyl). This regioselectivity suggests neighboring group assistance by the thioether substituent; it is possible that the analogs with C5 substituents lacking a suitable coordinating function might be labeled partly or entirely at C7 instead. Partial degradation of starting substrates possessing sulfur-containing substituents was observed.

**60**

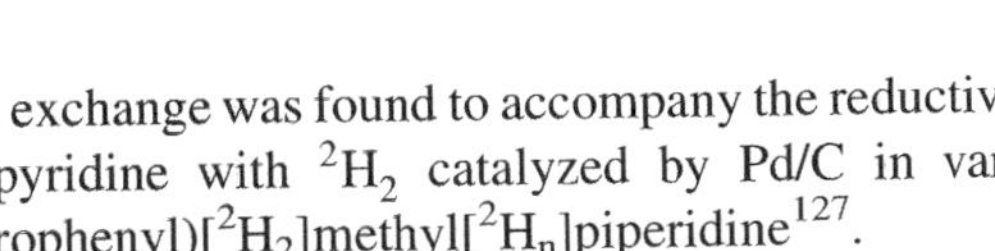


*Figure 3.24 Exchange labeling in organic solution from tritium gas catalyzed by supported palladium*

A high degree of benzyl exchange was found to accompany the reductive deuteriation of 4-(4-fluorophenyl)methylpyridine with $^2H_2$ catalyzed by Pd/C in various deuterated solvents, giving 4-(4-fluorophenyl)[$^2H_2$]methyl[$^2H_n$]piperidine[127].

Diastereoselective exchange has been observed in the labeling of acyclic benzylic sites of homochiral substrates. Upon tritiation with Pd/$BaSO_4$ in ethyl acetate solution, *N*-Boc-*O*-ethyl-D-tyrosine (**61**) was found[128] by $^3$H NMR analysis to have a 1.6/1 ratio of tritium in the two nonequivalent benzylic positions (Figure 3.25). In a separate report[129] juvenoid analog **62** was found to be labeled under similar reaction conditions to give a 11/1 ratio of tritium in the two nonequivalent benzylic positions, as shown. The preferred regioselectivity in both cases can be rationalized by a model in which the more rapid exchange occurs at the C–H bond pointed toward the metal surface in the more stable side-chain rotamer when the more exposed face of the aromatic ring is bound to the metal surface. The fact that the C9-H of estrogens is exchanged, retaining its stereochemistry, under similar conditions[130,131] was cited as support for this model.

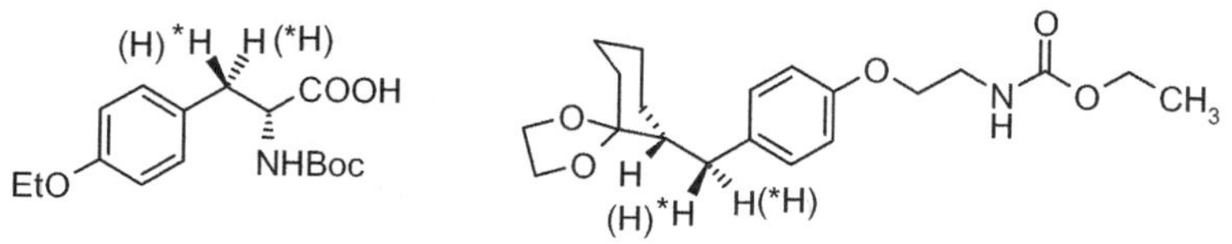


61 62

*Figure 3.25* *Diastereoselective exchange from tritium gas catalyzed by supported palladium*

Pyridines are known to be labeled C2 and C6 by a number of catalysts with heating under solvent-free conditions[132] (see also Section 3.6). In the presence of small amounts of $C^2H_3O^2H$ at room temperature and under 22 psi of $^2H_2$ for 24 h, Ru/C was shown[133] to completely exchange $^2H$ for H in the open C2 and C6 positions of several substituted pyridines and of isoquinoline; however, it is not clear whether the proximate source of isotope for these exchanges was deuterium gas or $C^2H_3O^2H$. Therefore, this method is unlikely to be practical for use with tritium.

A more recent report[134] described a much improved method for labeling pyridines, using either rhodium black, ruthenium black or 5% $Ru/Al_2O_3$ as catalyst and THF as solvent. Substantial deuterium exchange at ring positions adjacent to heterocyclic nitrogen occurs within 2 h at room temperature under 1 atm of $^2H_2$. The most generally effective of the catalysts is rhodium black. Substances resistant to labeling under Rubottom's $Ru/C-C^2H_3O^2H$ conditions, such as phthalazine and 2-bromopyridine, were labeled moderately well under the conditions published[134]. 3- and 4-acetylpyridine were labeled only at C2 and C6, demonstrating the absence of acid–base or other heteroatom-directed exchange processes. Structures 63–66 illustrate some of the results with rhodium black. This method has only been used with deuterium or a low specific activity $^3H_2/^2H_2$ mixture[134], but it should be usable without modification for tritium labeling with undiluted tritium gas.

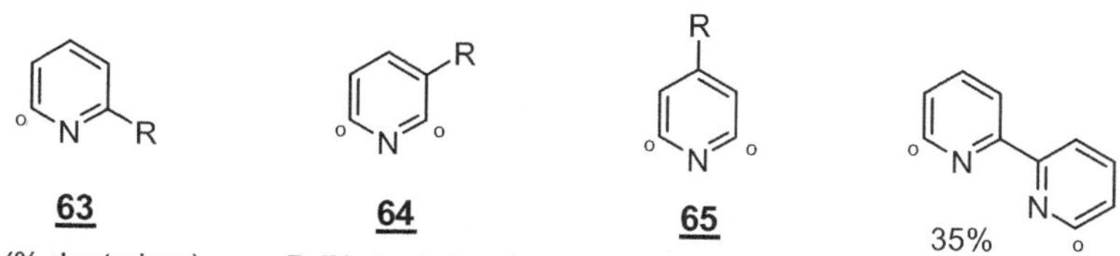


63

R (% deuterium)
benzyl (99)
Br (28)
OMe (16)
Ph (20)

64

R (% deuterium)
acetyl (reduced)
$NH_2$ (83)

65

R(% deuterium)
acetyl (99)
benzyl (99)
$NMe_2$ (100)
Me (98)

66

*Figure 3.26* *Exchange labeling of pyridines from deuterium gas catalyzed by rhodium black*

### 3.3.3 Other Catalysts

A polymer-bound form of the homogeneous iridium phosphine complex $[(1,5\text{-cyclooctadiene})Ir(PPh_3)_2]PF_6$ (67) has been prepared by stirring together [(1,5-cyclooctadiene)Ir (tricyclohexylphosphine)(pyridine)]$PF_6$ and a commercial triphenylphosphine-functionalized polystyrene in bead form. It is reported[135] to have similar activity in the presence of deuterium gas to **67** in a number of side-by-side trials with simple test substrates (e.g., *N,N*-dimethylbenzamide, benzophenone, benzanilide). Of the trials using more complex, drug-like molecules, three of the compounds were labeled to an extent comparable with that obtained with the soluble catalyst, but two were much more poorly labeled. These are only a small number of examples, so it remains to be seen whether there is any systematic difference in the activity of the polymer-bound catalyst compared with its soluble analogs. The successful labeling of three substrates (**68**–**70**) using tritium gas was reported; the high levels and the regioselectivities of labeling are not different from those expected for the analogous soluble iridium phosphine complexes, consistent with a similar exchange mechanism (see Section 3.5). The hoped-for advantage of the polymer-bound complex, that it would give cleaner reaction products requiring less intensive purification, appears well borne-out by these three examples, which, after removal of the polymer by filtration and removal of labile tritium, gave crude products of much higher radiochemical purity than is often the case with the soluble forms of this catalyst class.

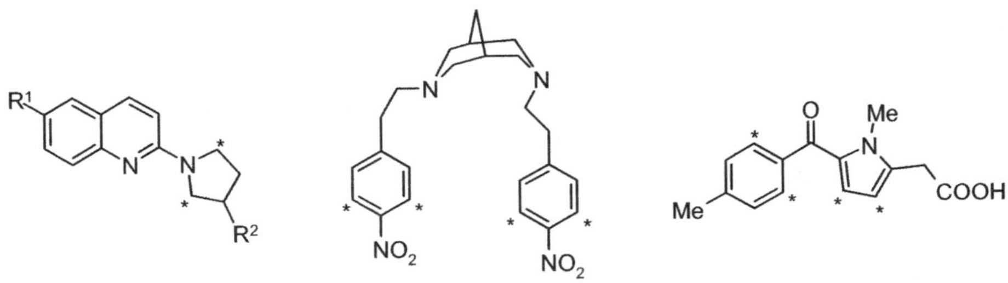


**68** 33 Ci/mmole 84% pure

**69** 33 Ci/mmole 92% pure

**70** 41 Ci/mmole 98% pure

*Figure 3.27* *Exchange labeling from tritium gas catalyzed by polystyrene-supported organoiridium complexes*

### 3.3.4 Metal Catalysts with Reducible Substrates

It is a common occurrence that the product resulting from reduction of an olefinic substrate is found to possess tritium at $sp^3$ sites adjacent to, or even farther from, the original location of the double bond, or that more tritium has been incorporated into one of the original olefinic centers than can be accounted for by simple reduction. Such isotope scrambling (see Section 4.1) reflects the well-studied phenomenon of double bond migration during catalytic hydrogenation[136,137]. More recent studies have clarified the mechanistic basis for this phenomenon[138]. The extent of any excess label incorporation is largely determined by the relative rates of isomerization and reduction, and both of these vary greatly according to reaction conditions, substrate structure and catalyst. $^3$H NMR studies[139] of the reduction of cinnamic acid, styrene and dehydroleucine demonstrated significant excess tritium incorporation into methyl and benzyl sites of the products, and the varying extent to which different catalysts produced such effects. Additional examples of excess tritium incorporation are provided in Chapter 4, Section 4.1.

In principle any except tetrasubstituted olefins could be labeled by catalytic exchange, but in practice it is difficult to find experimental conditions under which significant tritium incorporation can be catalyzed without reducing the double bond. However, if the olefinic starting material is separable from the reduced product, it is sometimes possible to find reaction conditions that disfavour complete reduction, then isolate the unreduced olefin which has become labeled. Two examples with which this has been accomplished are pleuromutilin (71) and cyclosporin A (72)[140]. Treatment with a deficiency of tritium gas for 60–90 min in the presence of 10% Pd/C in ethyl acetate (71) or DMF (72), followed by removal of reduced product by HPLC, provided the compounds, labeled as indicated, at specific activities of 10 and 19 Ci/mmol, respectively. Unfortunately, it proved to be difficult to separate the *cis/trans* mixture formed with 72, which stimulated the development of a tritide labeling approach (see Chapter 4, Section 4.3.1).

71

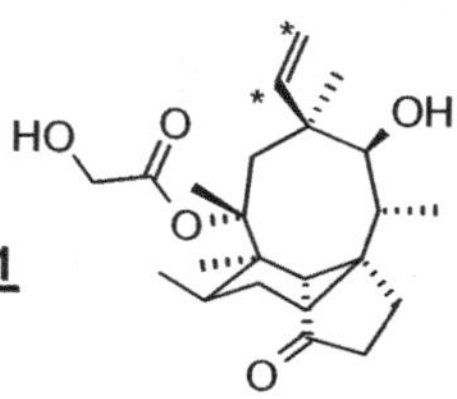


72

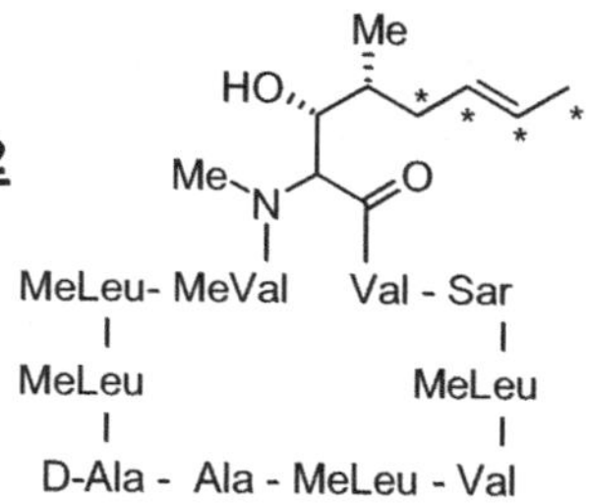


***Figure 3.28*** *Exchange labeling of olefins under conditions of incomplete reduction*

Deuterium exchange into starting material recovered from asymmetric reductions of *trans*-2-methyl-2-pentenoic acid with cinchonidine-modified $Pd/Al_2O_3$ catalysts has been observed[141].

## 3.4 Homogeneous Catalysis in Solution with Tritiated Water

Catalysts for tritium exchange labeling of compounds in solution utilizing liquid-phase tritium sources (usually tritiated water), aside from acids and bases or organic Lewis acids, are all metal based. Soluble metal catalysts are either metal salts or organometallic complexes. A recent review covers many aspects of this chemistry[142].

### 3.4.1 Exchange Catalyzed by Metal Salts

Extensive investigations were undertaken in the 1960s into catalytic hydrogen–deuterium exchange of aryl hydrogens in deuterium oxide solution by soluble metal salts such as $Na_2PtCl_4$, $Na_2PdCl_4$ and $Na_2IrCl_6$[143,144], and of alkyl hydrogens in $^2H_2O$ with platinum(II) salts[145]. The regioselectivity of exchange was found to be most strongly influenced in both cases by steric factors. The substrates studied were simple ones, such as substituted benzenes and alkanes. Unfortunately, the harsh conditions required for effective exchange (typically 130–180 °C for 1–3 days, with added mineral acid to stabilize the metal ions) are incompatible with many complex, multifunctional compounds. These catalytic exchange conditions have not been applied for preparative purposes with tritiated water in place of $^2H_2O$ because the achievement of a significant specific activity level would require large quantities of high specific activity tritiated water, posing problems in handling and disposal and also the increased risk of substrate decomposition through radiolysis.

In 1975 it was reported[146] that $RhCl_3{\cdot}3H_2O$ could catalyze similar degrees of labeling of alkylbenzenes and halobenzenes under less acidic conditions. The particular activity of the trihydrate form in this reaction is consistent with the facility with which Rh(III) undergoes ligand exchange via aquo species. Anhydrous $RhCl_3$ is polymeric and largely devoid of catalytic activity. The use of $RhCl_3{\cdot}3H_2O$ was later extended to arylcarboxylic acids[147]. Such functionalized substrates also underwent labeling at lower temperatures (e.g., 110 °C for 18 h in 2:1 DMF:$^2H_2O$); moreover, the labeling occurred almost exclusively *ortho* to the carboxylate functions. This regioselectivity was attributed[148] to the ability of rhodium(III) to form chelates of the type **73**, for which there was some precedent[149]; later studies established the ability of rhodium cyclopentadienyl complexes to form metallacycles with benzoic acid[150]. A recent review summarizes this and additional catalytic exchange systems[151].

**73**

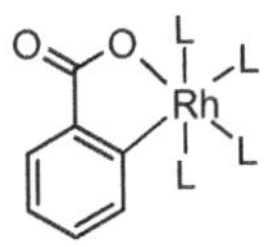


***Figure 3.29*** *Putative intermediate in* ortho-*specific exchange labeling by* $RhCl_3{\cdot}^3H_2O$ *from aqueous sources of deuterium and tritium*

Several other metal compounds have been reported[148] to catalyze *ortho*-exchange of benzoic acid, including salts [$Rh(NO_3)_3$, $RuCl_3{\cdot}3H_2O$ and $IrCl_3{\cdot}3H_2O$], which were more effective with carboxylic acids in their salt forms, and organometallics [Ru(acetylacetonate)$_3$ and $Ru(PPh_3)_3Cl_2$] which worked better on the free acids. Although arylcarboxylic esters were not labeled by $RhCl_3$, benzamides and benzylamines[148] and some anilides[152] were found to be good substrates, all being labeled in their *ortho* positions. *Meta*- and *para*-oriented chloro, nitro and cyano functions are compatible with this system. The rate of exchange is modestly enhanced by electron-withdrawing ring substituents[153,154], but in practice the amount of deuterium incorporation was not significantly affected by the electronic influence of ring substituents in any of these three substrate classes. There is relatively strong steric inhibition of labeling by *meta*-substituents in all classes, and by *N*-substituents on amides and amines.

This method has been successfully applied[155] to tritiations of a similar range of model compounds, in which typical reactions consisted of substrate, 50 mol% rhodium salt, 1 mL of DMF and 3 μL of tritiated water. Some *N*-heterocycles (such as 2-phenylpyridine and 1-phenylpyrazole) with aryl C–H bonds positioned for N,C-chelation by rhodium in a five-membered metallacycle were also found to be labeled in the expected positions, and also in some cases at C6(2) of pyridine moieties, which must proceed by a different mechanism. Labeling studies of benzanilides bearing *para*-halo substituents on the aniline ring showed label incorporation into the *ortho* positions of both rings, that into the halo-substitued ring increasing as H < F < Cl < Br[156]. The labeling under similar conditions of a variety of drugs and drug-like compounds was reported[157] (e.g., **74–76**); with $^3$HHO of specific activity 450 mCi/milliatom H (50 Ci/mL, the maximum normally commercially available), specific activities of successfully tritiated products ranged from 14 to 650 mCi/mmol. Given the possibility of using *in situ*-prepared $^3H_2O$ of higher specific activity, correspondingly higher product specific activities should be feasible. Such outcomes have been reported in two cases, that of [$^3$H]nedocromil[158] sodium (**77**) and sodium [$^3$H]cromoglycate[159] (**78**), where specific activities of 21.5 and 16.5 Ci/mmol, respectively, were achieved by heating substrate, $RhCl_3{\cdot}3H_2O$ and carrier-free $^3H_2O$ in DMF for several hours.

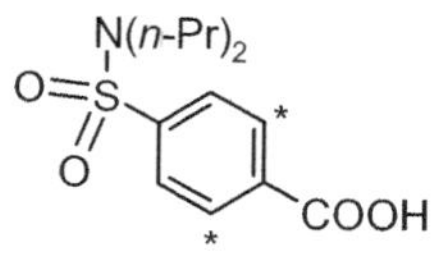


**74**

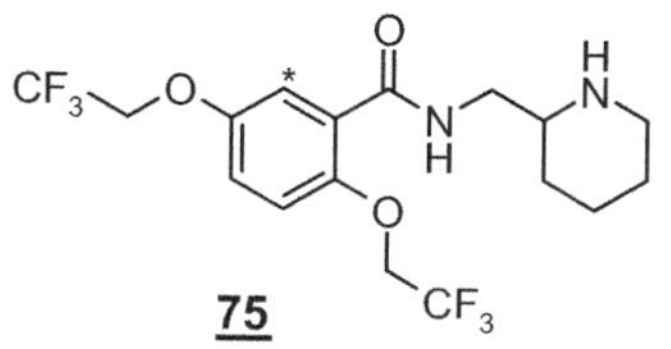


**75**

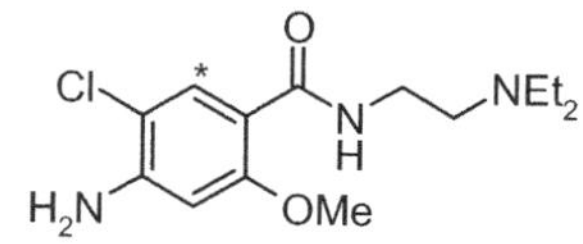


**76**

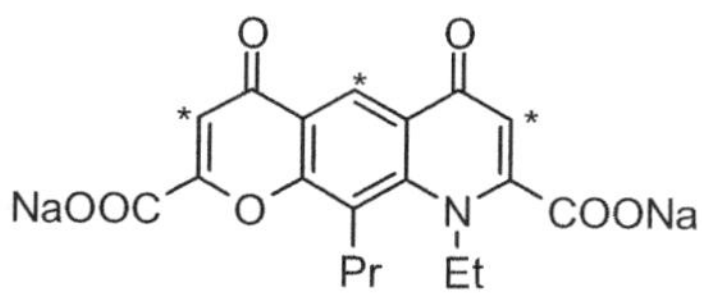


**77**

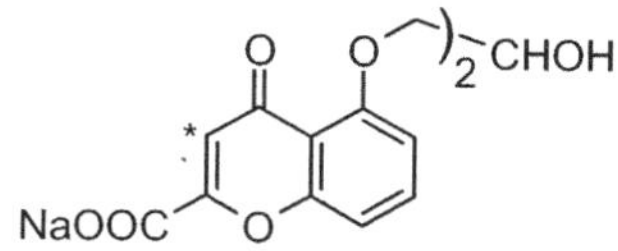


**78**

***Figure 3.30*** *Compounds labeled by exchange from aqueous tritium catalyzed by $RhCl_3{\cdot}3H_2O$*

Decomposition of complex substrates remains an occasional problem with this system, even though reaction conditions are milder than the earlier applications with substrates lacking coordinating functional groups. Nevertheless, this is a powerful labeling method which has since been applied by a number of other investigators to prepare tritiated compounds of low to moderate specific activity[160].

### 3.4.2 Exchange Catalyzed by Organoruthenium Complexes

As mentioned in the previous section, $Ru(PPh_3)_3Cl_2$ and Ru(acetylacetonate)$_3$ were reported to catalyze *ortho* exchange of $^2$H for H in arylcarboxylic acids[148]. However, the efficiency of these catalysts for this process was less than that of $RhCl_3{\cdot}3H_2O$, and the range of applicable substrates more limited.

It has long been recognized that ruthenium species are capable of oxidizing primary alcohols[161,162] and primary and secondary amines[163] under mild conditions. These reactions proceed via a ruthenium alkoxide or amine adduct, whose $\alpha$-protons can be transferred, via the metal, to a hydrogen acceptor[164]. In the presence of water or absence of a hydrogen acceptor, no net oxidation takes place, and when $^2H_2O$ is present the isotope is incorporated into C$\alpha$ by a reversible oxidation–reduction process.

Two research groups[165,166] have described reaction conditions under which alcohols can be deuterated by this process. This method has not been much used for preparing

[1-$^3$H]alcohols, both because of the limits on specific activity imposed by the aqueous source of isotope, and because there are more convenient approaches (e.g., [$^3$H]hydride reduction of the corresponding aldehyde or ketone; see Chapter 4, Section 4.3).

It is in the tritiation of amines that ruthenium-catalyzed exchange has shown the most potential. For example, the protected piperazine **79**, a synthetic intermediate, was labeled by means of $Ru(PPh_3)_3Cl_2$ and 90 atom% tritiated water to an unspecified specific activity[167]; **80** and **81** were tritiated[168,169] using $^3$HHO of specific activity 450 mCi/milliatom H by heating at 130 °C for 1 h in the presence of $Ru(PPh_3)_3Cl_2$ to give specific activities of 44 mCi/mmol to 1.2 Ci/mmol. Tritium was sometimes incorporated in both positions $\alpha$ and $\beta$ to the amine function, suggesting the occurrence of imino–enamino interconversions during the labeling process.

**79**

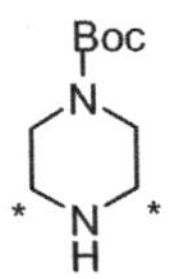


**80**

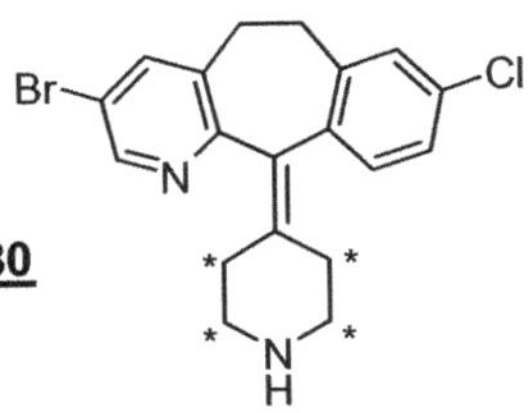


**81**

R = H or Br
X = CH or N

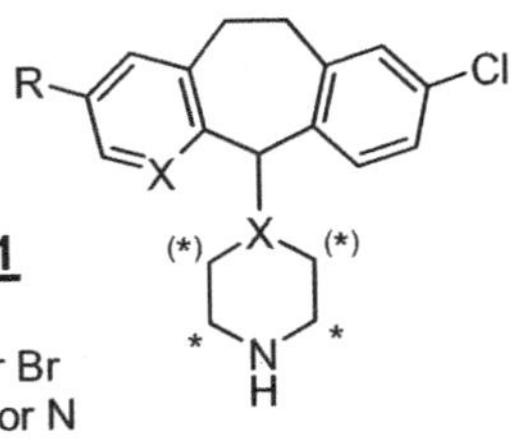


(*) = minor tritium content

*Figure 3.31* *Compounds labeled by exchange from aqueous tritium catalyzed by $Ru(PPh_3)_3Cl_2$*

A recent report[170] describes an exploration of several catalysts and substrates with $^2H_2O$ as the isotope source at 150 °C for 3 h in DMSO or DMF cosolvent. Complexes found to be most active were the already-studied $Ru(PPh_3)_3Cl_2$ and the new complexes [Ru($p$-cymene)$_3Cl_2$]$_2$, [Ru($\eta^6$-benzene)$_2Cl_2$]$_2$ and, most active, [Ru(CO)$_3Cl_2$]$_2$. DMSO was found to be superior to DMF as cosolvent, and is hypothesized to participate as a ligand for ruthenium. A variety of 1-substituted piperazines and 4-substituted piperidines were good substrates, as were dibutylamine and dioctylamine. Most substrates underwent incorporation of 2–5 mol $^2$H/mol (of an estimated maximum of 5.2 mol $^2$H/mol based on the pool of exchangeable $^2$H/H), and were labeled both $\alpha$ and $\beta$ to the secondary nitrogen. Recoveries of substrate were variable but often good.

It seems likely that $^3$HHO of very high specific activity (prepared for the event from tritium gas) could be used in such reactions to prepare amine derivatives of high specific activity. However, the elevated temperatures required may work against the general utility of this method for end-use compounds or advanced intermediates of complex structure. The development of complexes capable of operating under milder conditions would be a welcome advance.

### 3.4.3 Exchange Catalyzed by Iridium Dionates

A program of screening[171] of a wide variety of metal salts and complexes for their ability to catalyze the exchange of deuterium from $^2H_2O$ into three model substrates **82–84** identified the neutral complex (1,5-cyclooctadiene)Ir(pentane-2,4-dionate) as a competent catalyst (Figure 3.32). Modifications of the dionate ligand led to a number of complexes able to catalyze (at 20 wt% vs substrate, 2:1 DMF:$^2H_2O$, 75 °C, 1 h) the incorporation of high levels of deuterium into all these model substrates, and selectively into the *ortho* positions. The most active complexes were those having 3-phenyl- or 3-alkylpentane-2,4-dionate ligands (3-methyl analog **85** illustrated) or 1,1,1,3,3,3-hexafluoropentane-2,4-dionate (**86**).

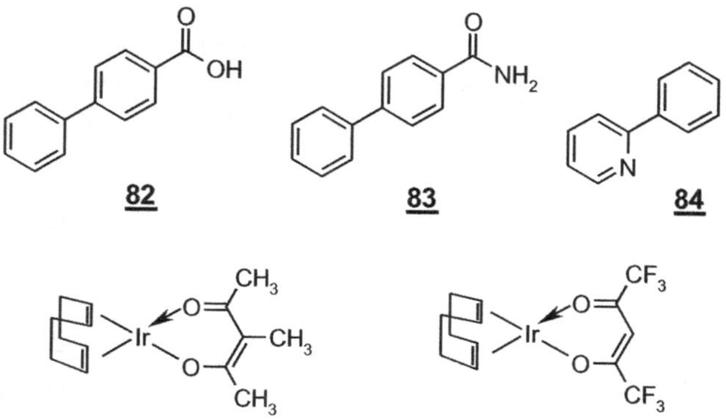


85 86

***Figure 3.32*** *Model substrates labeled from deuterium oxide and active iridium dionate complexes*

Testing[171] of catalytic complexes **85** and **86** on a number of model substrates in dimethylacetamide at 90 °C revealed that they were most active in catalyzing exchange of hydrogens *ortho* to aryl carboxyl, unsubstituted amide, sulfonamide and primary aminomethyl functions. Performing the reactions at higher temperatures and/or using microwaves enhanced the level of deuterium incorporation in otherwise poor substrates, such as *N*-substituted amides. The modest labeling of acetanilide and of the anilide ring of benzanilide signal the complex's strong preference for exchanging aryl hydrogens four bonds away from the coordinating atom, i.e., via five-membered and not six-membered metallacyclic intermediates[172]. Esters do not mediate exchange.

The method has been applied to aristolochic acids **87**[173], using tritiated water of ca 450 mCi/milliatom H, and the specific activities of the products were nearly equal to that of the $^3$HHO, indicating complete equilibration between the isotope source and the exchangeable site.

87

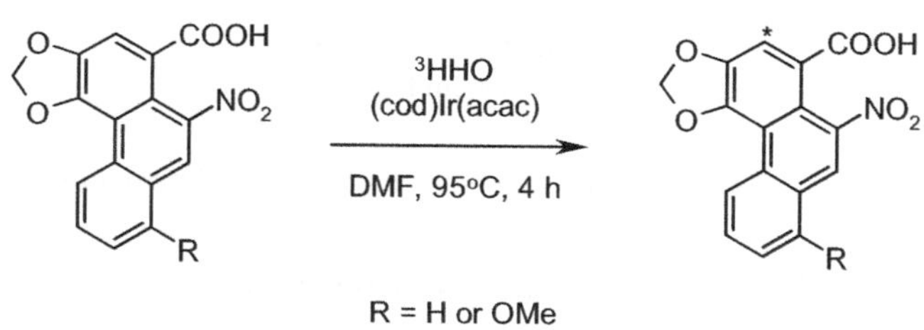


R = H or OMe

***Figure 3.33*** *Exchange labeling of aristolochic acids from tritiated water catalyzed by an iridium dionate complex*

The use of (1,5-cyclooctadiene)Ir(pentane-2,4-dionate) under similar reaction conditions has been reported[174] to lead to deuterium exchange into the $\beta$-positions of some $\alpha,\beta$-unsaturated carbonyl systems (e.g., **88**–**90**). It is likely that the regioselectivity is guided by metal coordination to the carboxyl oxygen, followed by C—H insertion to form a five-membered metallacycle. Under the same conditions, *ortho* positions of benzophenone and acetophenone oxime are exchanged to a modest extent, via what might be weaker coordination of metal to ketone carbonyl O and oxime N, respectively, compared with carboxyl. 2-Hydroxy-4-methoxyacetophenone was deuterated only in the acetyl methyl group, suggestive of base-catalyzed exchange rather than metal-catalyzed C—H activation.

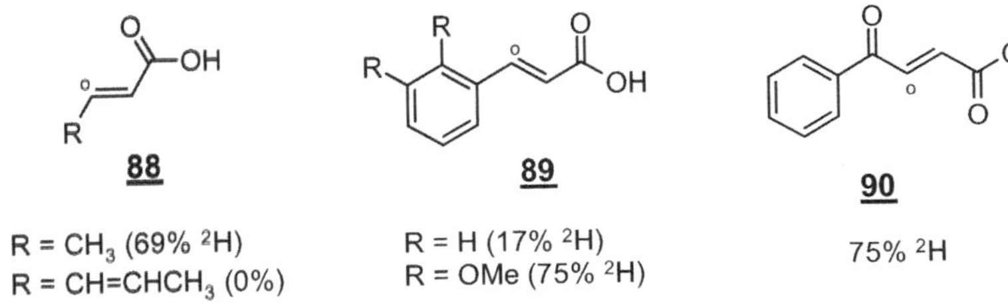


*Figure 3.34* Exchange labeling of $\alpha,\beta$-unsaturated acids from deuterium oxide catalyzed by (1,5-cyclooctadiene)Ir(pentane-2,4-dionate)

A striking difference in the regioselectivity of deuteration of certain bifunctional compounds by iridium dionate complexes was observed[175] depending on whether the isotope source was $^2H_2O$ or $^2H_2$. As discussed above, aryl carboxyl functions are excellent directors of deuterium exchange from $^2H_2O$ into their *ortho* positions. In contrast, as discussed in Section 3.6, aryl amine groups are good directors when deuterium gas is the isotope source for iridium dionate-catalyzed exchange (Figure 3.35). Therefore, it is not too surprising that in substrates such as aminobenzoic acids, the alternative labeling patterns were observed, as illustrated below. Therefore, there is the possibility for deliberate steering of tritium incorporation to different sites of more complex compounds by adjusting the reaction conditions.

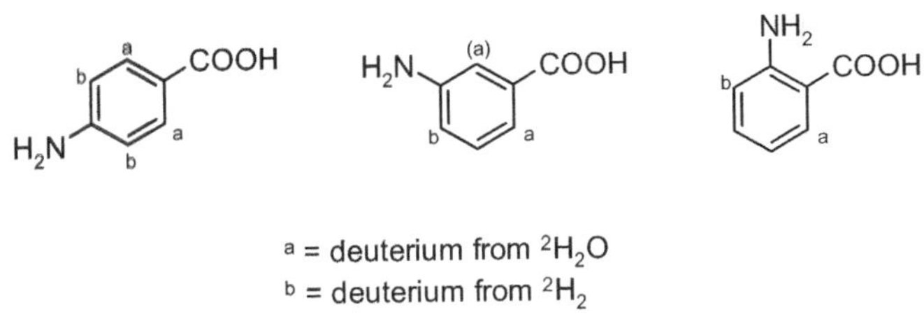


*Figure 3.35* Different labeling regioselectivities from different deuterium sources

### 3.4.4 Exchange Catalyzed by Iridium Cyclopentadienides

The first report of the use of the complex Cp*(PMe$_3$)IrCl$_2$ (*Cp = pentamethylcyclopentadienyl) for the deuteration (135 °C, 40 h) of test substrates, including **91**–**93**, in $^2H_2O$ as both isotope source and solvent were published[176] in 2002. The ability of this complex to form alkyl and aryliridium complexes by C–H activation in the absence of water had previously been studied in detail[177].

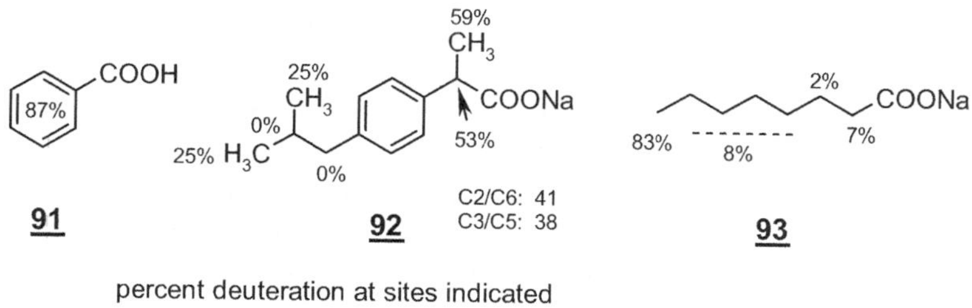


***Figure 3.36*** *Regioselectivities of deuterium labeling from deuterium oxide mediated by Cp* (PMe₃)IrCl₂*

The regioselectivity seems to be largely determined by steric factors, although labeling *alpha* to carboxyl groups may be the result of electrophilic exchange. Labeling in benzoic acid is distributed around the ring, rather than being only in the *ortho*-positions as with $RhCl_3{\cdot}3H_2O$ systems and the iridium dionate complexes. These results are similar to those obtained with $[Cp^*Ir(PMe_3)(Me)]B(Ar_f)_4$ and deuterium gas as the isotope source (Section 3.5) in being primarily sterically controlled, but the latter was more selective for aryl positions. In the $^2H_2O$ experiments the active iridium complex was unstable, losing activity apparently by ligand exchange. Testing of analogous complexes possessing cyclopentadienyl-phosphine chelating ligands[178] showed them to be similarly unstable, although two of these complexes were about as active as $Cp^*(PMe_3)IrCl_2$. To date, no labeling experiments have been reported on substrates possessing functional groups more highly coordinating than carboxyl, but the reactivity of these iridium cyclopentadienide catalysts suggests that substrate-induced ligand exchange may compromise their catalytic activity.

There are no reports of the use of iridium cyclopentadienide complexes with tritiated water. Such applications would be expected to have disadvantages similar to $RhCl_3{\cdot}3H_2O$ and related systems (Section 3.4.1) in terms of specific activity limits and the possibility of thermal decomposition of substrates, but with the additional uncertainty of its applicability to a range of substrate structures and functions.

## 3.5 Homogeneous Catalysis with Tritium Gas

To date, the only soluble catalysts reported to have significant activity in mediating exchange between atmospheric (isotopic) hydrogen and substrates in solution are organometallic complexes of iridium. They can be considered as consisting of three classes: iridium phosphine complexes, iridium dionate complexes, and iridium cyclopentadiene complexes.

### 3.5.1 Iridium Phosphines

The class first to be investigated, and to date the most extensively studied, are iridium phosphines. The first study of labeling by this method[179] utilized the complex $[Ir(H)_2(acetone)_2(PPh_3)_2]BF_4$. Test substrates became labeled to high isotopic abundance at room temperature by treatment with 1.5–2.5 mol% catalyst loading and limited amounts of deuterium gas. The labeling was regioselective, deuterium being incorporated exclusively *ortho* to particular functional groups including esters, ketones, amides and heterocycles attached directly to the ring, as exemplified by **94** to **99**. Substrates (such as **100**) with functional groups more distant from the ring were not labeled. *N,N*-Dimethylamides (e.g., **99**) were labeled in the methyl groups; however, the analogous positions of the esters or ketones were not. Benzoquinoline (**98**) was labeled both at C2 and C10 (it was also partially reduced to 1,2-dihydrobenzoquinoline). Labeling was very efficient as indicated by the incorporation levels indicated. $[Ir(H)_2(acetone)_2(PPh_3)_2]BF_4$ is an air- and temperature-stable crystalline solid prepared in two steps from commercially available $[(cod)IrCl]_2$ (cod = 1,5-cyclooctadiene)[180].

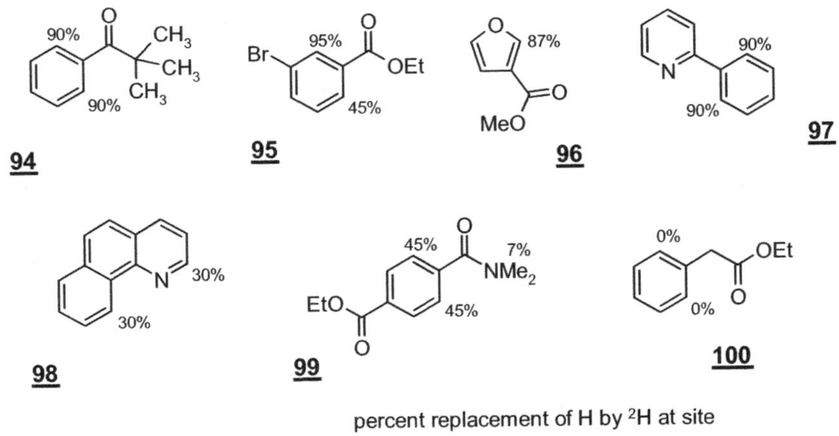


***Figure 3.37*** *Regioselectivities in exchange labeling of model compounds from deuterium gas catalyzed by* $[Ir(H)_2(acetone)_2(PPh_3)_2]BF_4$

This study was based on the finding[181] that the indicated methyl groups of 8-methylquinoline (**101**) and caffeine (**102**) were labeled from atmospheric deuterium in the presence of the iridium complex (Figure 3.38). Crabtree's group determined the structure of an 8-methylquinoline iridium complex by X-ray crystallography, and it showed the metal center to be coordinated to the quinoline nitrogen and bound in agostic fashion to one of the methyl group C–H bonds. Later studies of the reaction of $[Ir(H)_2(acetone)_2(PPh_3)_2]BF_4$ with benzoquinoline[182] and the structural determination of the adduct[183] showed that the benzoquinoline C10–H bond (positioned analogously as the methyl C–H in 8-methylquinoline) had been cleaved to give a five-membered metalacycle (**103**). Therefore, the regioselectivity of labeling could be explained by reversible coordination of the metal center to an $sp^2$ O or N, followed by reversible cleavage of a C–H bond the appropriate distance away to give a five-membered metallacycle. Scrambling of metal hydrides with atmospheric deuterium before reversal of the process would result in replacement of H by $^2H$ at the site four bonds away from the coordinating heteroatom. In due course, low-temperature proton NMR studies of reaction mixtures and additional X-ray crystal structures of iridium–substrate adducts using later generation complexes and new model

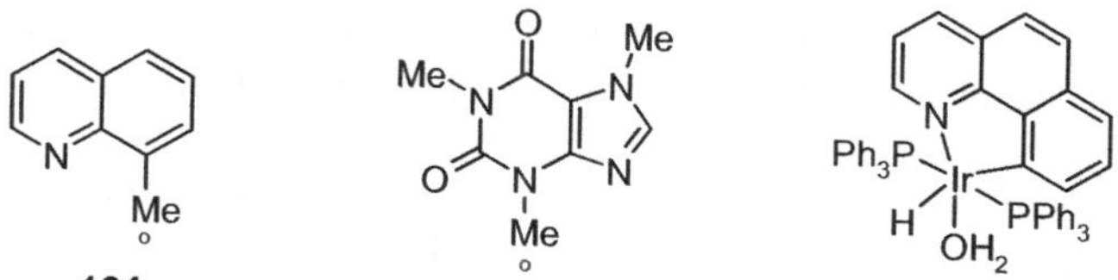

**101** **102** **103**

***Figure 3.38*** *First observations of deuterium exchange catalyzed by [Ir(H)$_2$(acetone)$_2$(PPh$_3$)$_2$]BF$_4$; metalacycle formed by interaction of [Ir(H)$_2$(acetone)$_2$(PPh$_3$)$_2$]BF$_4$ and benzoquinoline*

substrates[184,185] confirmed that the same process was occurring to label these other compounds.

With this simple mechanistic model, it became possible to anticipate whether and where any new compound might be labeled by this method. The isotopes introduced by this method are generally introduced into positions where they are chemically stable and in most cases metabolically stable, making the method valuable for biological and pharmaceutical research applications. Few other isotope exchange methods possess this degree of predictability.

Substrate ancillary functional groups such nitro, nitrile, and aryl and alkyl halogens were unaffected, a pattern of reactivity which makes this method complementary to that of heterogeneous catalytic methods, which usually reduce these functions. But as with the latter, [Ir(H)$_2$(acetone)$_2$(PPh$_3$)$_2$]BF$_4$ tends to reduce simple olefins, although as described below conjugated double bonds are often tolerated. Exchange rate studies with substituted benzoate esters and benzophenones[186,187] showed a modest sensitivity to electronic substituent effects. Methylene chloride is the optimum solvent, but acetone or even THF can be used with substrates having stronger ligating abilities, such as amides and nitrogen heterocycles[179]. Insignificant amounts of deuterium were incorporated into methylene chloride and acetone solvents. Later work has shown some success with the use of chlorobenzene, ethyl acetate[188] and methanol[189] as solvent. Also, one report[190] has appeared describing the use of ionic liquids as solvents for substrates not sufficiently soluble in the other solvents; ionic liquids are also nonvolatile.

The first reported[191] application of the new method with tritium was on the steroid **104**.

An important feature of the iridium phosphine class of complexes is that the triphenylphosphine ligands can be replaced with a wide variety of phosphine and related donor

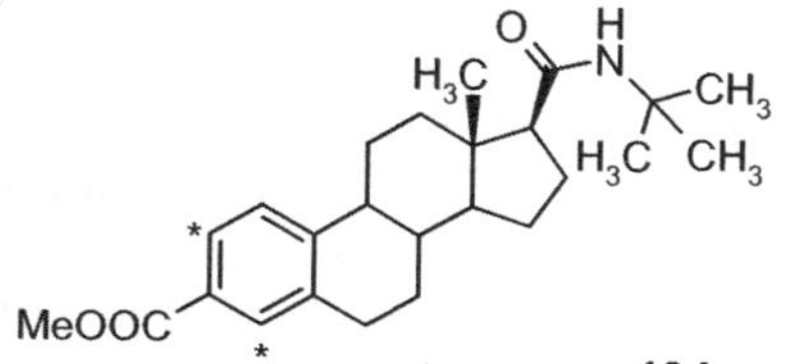

**104**

***Figure 3.39*** *Steroid labeled from tritium gas catalyzed by [Ir(H)$_2$(acetone)$_2$(PPh$_3$)$_2$]BF$_4$*

ligands, thereby providing new complexes which may have different profiles of activity in hydrogen isotope exchange. A number of developments in this direction have been reported.

The commercially available complex [Ir(cod)($PCy_3$)(py)]$PF_6$ ($PCy_3$ = tricyclohexylphosphine, py = pyridine) was shown to catalyze the deuteration of ring-*ortho* positions of various acetanilides[192] (e.g., **105**) under similar conditions as [Ir$(H)_2$(acetone)$_2$($PPh_3$)$_2$]-$BF_4$, and tritiation of the pharmaceutical compound **106** to high specific activity was reported[193]. Thus, [Ir(cod)($PCy_3$)(py)]$PF_6$ is capable of labeling compounds one bond farther away from the coordinating heteroatom (carbonyl in these cases) than is [Ir$(H)_2$(acetone)$_2$($PPh_3$)$_2$]$BF_4$. [Ir(cod)($PCy_3$)(py)]$PF_6$ can also label positions four bonds away from the coordinating heteroatom, so this complex is either more versatile or less selective, depending on the investigator's point of view. These results were obtained with catalytic loads of iridium complexes and simple substrates; results can differ under other conditions as described later.

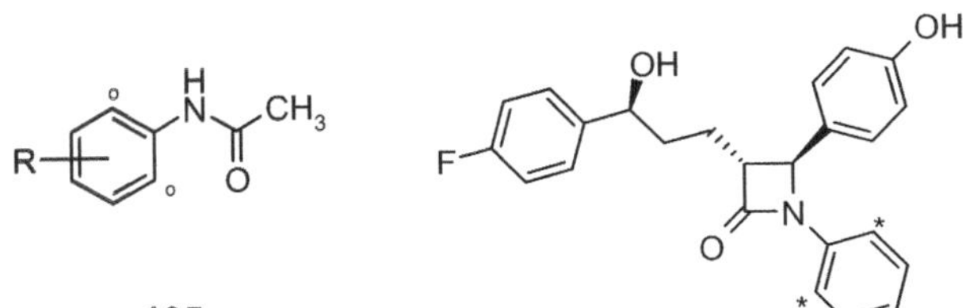


**105** **106**

***Figure 3.40*** *Acetanilides labeled from tritium gas catalyzed by [Ir(cod)($PCy_3$)(py)]$PF_6$*

Subsequent investigations of modified catalysts have for convenience usually utilized catalysts of the type [Ir(cod)$L_2$]X ($L_2$ = two monodentate or one bidentate ligand, X = noncoordinating counterion). Upon exposure to (isotopic) hydrogen gas *in situ*, they are converted to active species analogous to the dihydrido–diacetone complex first investigated. A few such complexes are commercially available and most others can be prepared in one step by treatment of a solution of [(cod)IrCl]$_2$ with L then replacing chloride ion by treatment with AgX[194]. The [Ir(cod)$L_2$]X can be isolated or, if prepared in methylene chloride, used without isolation. This is an advantage for complexes that are unstable in isolated form.

An early study[195] of the deuteration of model compounds using catalytic amounts of [Ir(cod)$L_2$]$BF_4$ (where L = a number of different phosphines and $L_2$ = $Ph_2P(CH_2)_nPPh_2$, $n = 2$ and 4) established that the complexes with bidentate ligands label sites both four and five bonds away from substrates' coordinating heteroatoms (as does [(cod)Ir($PCy_3$)(py)]$PF_6$), whereas all the monodentate phosphine complexes were selective for sites four bonds distant.

In accordance with these results it could be predicted, for example, that **107** (Figure 3.41) would be tritiated only in the right-hand ring when [(cod)Ir($PPh_3$)$_2$]$BF_4$ was used as the catalyst, and indeed this was the case (20 Ci/mmol specific activity), and that both indicated aromatic sites in **108** and **109** would be tritiated with [(cod)Ir($Ph_2PCH_2CH_2PPh_2$)]$BF_4$ (48.7 Ci/mmol and 70 Ci/mmol, respectively); distributions of isotope as indicated.

In **109**, the C2 position of the anilide might be expected to be labeled as is C6, but is not. This suggests that the catalyst is sensitive to steric blockage. An illustration[196] that this

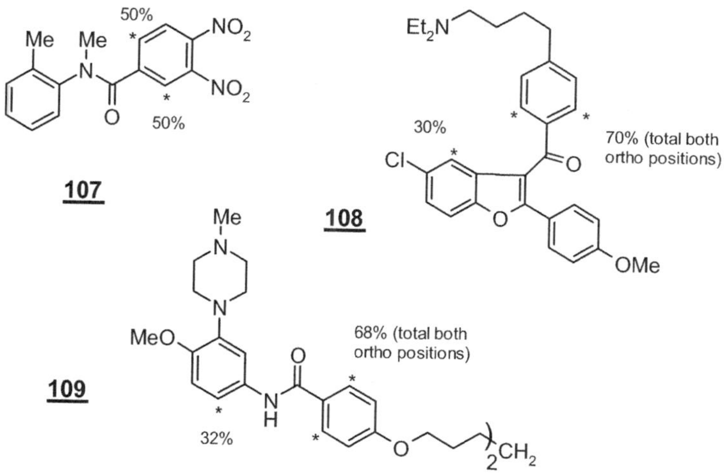


***Figure 3.41*** *Differing regioselectivities of [(cod)Ir(PPh$_3$)$_2$]BF$_4$ and [(cod)Ir(Ph$_2$PCH$_2$CH$_2$PPh$_3$)] BF$_4$*

effect is indeed operative is the series **110–112** (partial structures; remainder of label is in R). Labeled under identical conditions with the complex [(cod)Ir(PCy$_3$)(py)]PF$_6$, the distribution of tritium in each as assessed by $^3$H NMR suggests that in **111** it must be steric blockage by the C3-methyl that is preventing the labeling of the pyridine C4 position, and in **110** and **111** the C6 trifluoromethyl must be blocking tritiation at the C7 positions of the hydrindole moieties.

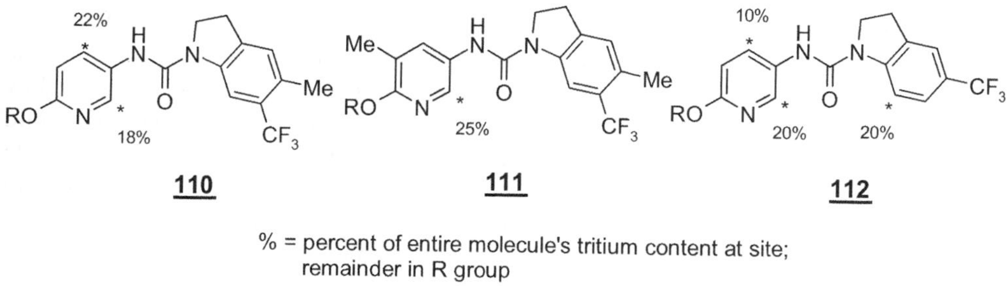


***Figure 3.42*** *Influence of steric blockage on regioselectivity of exchange labeling catalyzed by organoiridium complexes*

On the other hand, nearby groups with electron lone pairs can have positive neighboring group effects on labeling. Such a case is in **95** above and its methoxy and nitro analogs[179] (with deuterium). Other examples include **113**[197] and several other benzophenone derivatives[187] (Figure 3.43).

A systematic study[198] of neighboring group effects in a series of substrates **114** labeled by the catalysts [(cod)Ir(PPh$_3$)$_2$]BF$_4$ and [(cod)Ir(PCy$_3$)(py)]PF$_6$ has shown that, while the C3 substituents CH$_3$ and CF$_3$ invariably reduced the rate of labeling at C2 compared with C6, as expected on grounds of steric inhibition, the substituents F, OH, Cl, OCH$_3$, OCF$_3$, Br, and I

**113**

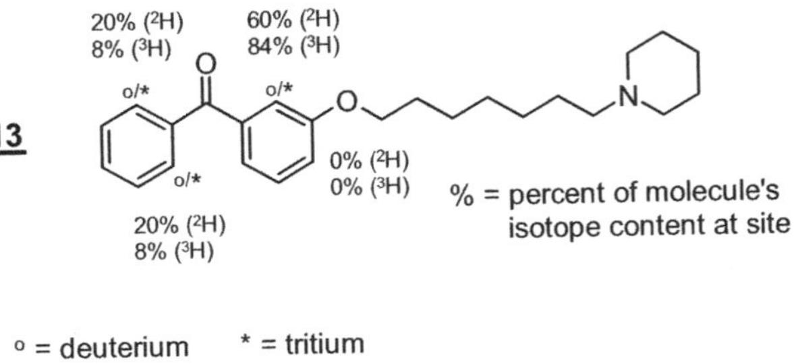


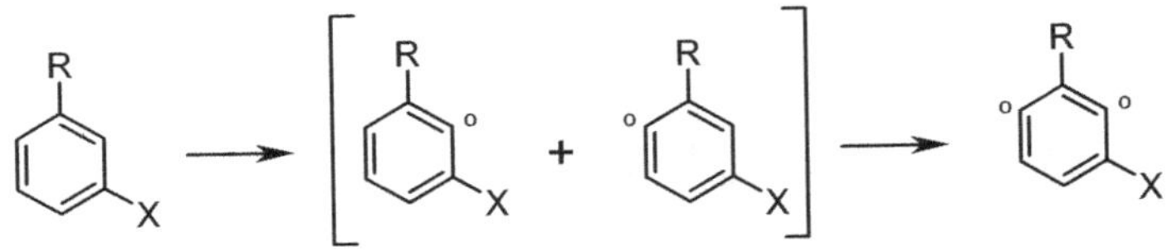

**114**

R = $COCH_3$, $CONMe_2$, $NHCOCH_3$
X = F, Cl, Br, I, OH, OMe, $OCF_3$, $CH_3$, $CF_3$

***Figure 3.43*** *Influence of* meta-*substituents on labeling regioselectivity*

all *accelerated* the rate of deuterium exchange at C2 compared with C6 (or slowed C2 labeling less than expected on steric grounds). The potency of this positive *meta*-effect decreased in the order listed, fluoro having the greatest effect and iodo the least. The experimental results indicate that the basis of this positive *meta*-effect of substituents possessing lone pairs is not so simple as their ability to coordinate with the metal center.

Although iridium phosphine complexes generally tend to reduce carbon–carbon double bonds[199], there are a number of examples in which they survive. These are usually highly substituted, conjugated or otherwise hindered. A clear example[200] is in the tritiations of paclitaxel and three photophore-containing derivatives (**115**) catalyzed by $[(cod)Ir(PPh_3)_2]BF_4$ and $[(cod)Ir(PCy_3)(py)]PF_6$ (Figure 3.44). Tritiation under mild conditions (30 mol% catalyst) gave products labeled exclusively in the benzamide side chain to specific activities of 53–71 Ci/mmol and under stronger conditions (up to 300 mol% catalyst) to specific activities of 82–195 Ci/mmol, labeled additionally in the benzoyloxy group and at C3 of the aminoester side chain. In none of these cases was significant double bond reduction observed. Little or no double bond reduction was observed with substrate **116**[201], and a small amount of double bond reduction was observed in two analogs of **117**[197]; in each case reduction products were easily separated by HPLC (Section 3.5.1.3). Recently, the protection of an acetylenic bond by formation of the bis-tricarbonyl cobalt complex prior to exchange labeling of the C2 position of an adjacent pyrid-4-yl ring by $^3H_2$ and $[(cod)Ir(PCy_3)(py)]PF_6$ has been reported[202].

The levels of catalyst loading required for labeling appear to depend upon the coordinative affinity of ancillary functional groups in substrate molecules, although no systematic study of this subject has been published. Most early studies using simple model compounds[179,186,187,192,195] gave high levels of labeling with catalyst loadings below 5 mol% (see structures **94**–**100** and **105** above). In contrast, early examples of tritium labeling of multifunctional compounds[197] often found that higher loadings (30–200 mol%) were more

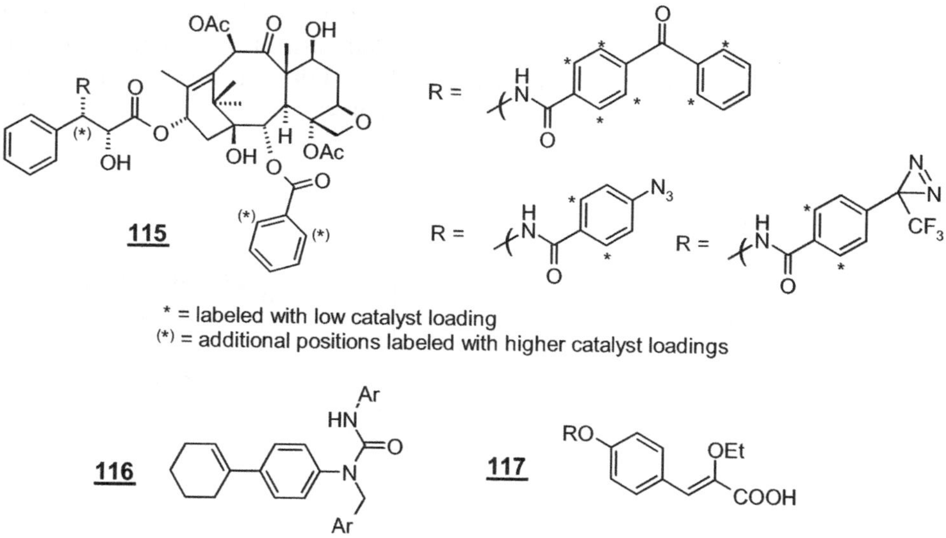


***Figure 3.44*** *Direct exchange labeling of paclitaxel and photoaffinity-labeling derivatives and toleration of olefin functions in catalysis of exchange*

effective. This is not unexpected, as a certain degree of coordination between the iridium center and the guiding functional group of the substrate is required for labeling, there is nothing to prevent unproductive coordination of the iridium center to other functional groups. Empirical observations suggest that the trend of increasing effectiveness in mediating exchange labeling in *ortho*-positions of aryl groups for the more common substituents is

$$\text{pyrid-2-yl} > \text{C(O)NR}^1\text{R}^2 > \text{C(O)R} > \text{NHC(O)R} \sim \text{COOR} > \text{NO}_2$$

but the trend may vary depending upon the structure of the catalyst. Based on general principles of coordination chemistry, the affinity of non-mediating polar functions such as nitrile, amino, etc. would be expected to interfere with catalytic activity by coordinating with the iridium center. However, as the success of a particular exchange process depends on a delicate balance of reversible coordination, bond-breaking and bond-making events, it is not possible to predict the potential for interference by ancillary functional groups. Therefore, most investigators have routinely used large catalyst loads for practical expediency, rather than optimizing conditions for a particular catalyst-substrate combination.

Several new classes of iridium phosphine related complexes have recently been reported to be effective exchange catalysts. These include complexes containing an *N*-heterocyclic carbene in place of one phosphine (**118**[203] and **119**[204]), complexes with pincer ligands (**120**[205] and **121**[206]) and P,N ligands[207] (**122**, **123**) (Figure 3.45). Only the most preliminary results are available, and only for model studies using deuterium gas. For example, complex **118** (L = P*n*-$Bu_3$), when used at loadings of 220 mol% and deuterium gas pressures of 100 mbar for 2.5 h, labeled six model substrates to between 40% and 95%

of theoretical[203], but <5% deuterium incorporation was reported[208] for the same catalyst on two similar substrates when used at 5 mol% with 1000 mbar of deuterium for 16 h. In contrast, under the latter reaction conditions complexes **119** gave >95% of theoretical deuterium incorporation in several other model substrates[204].

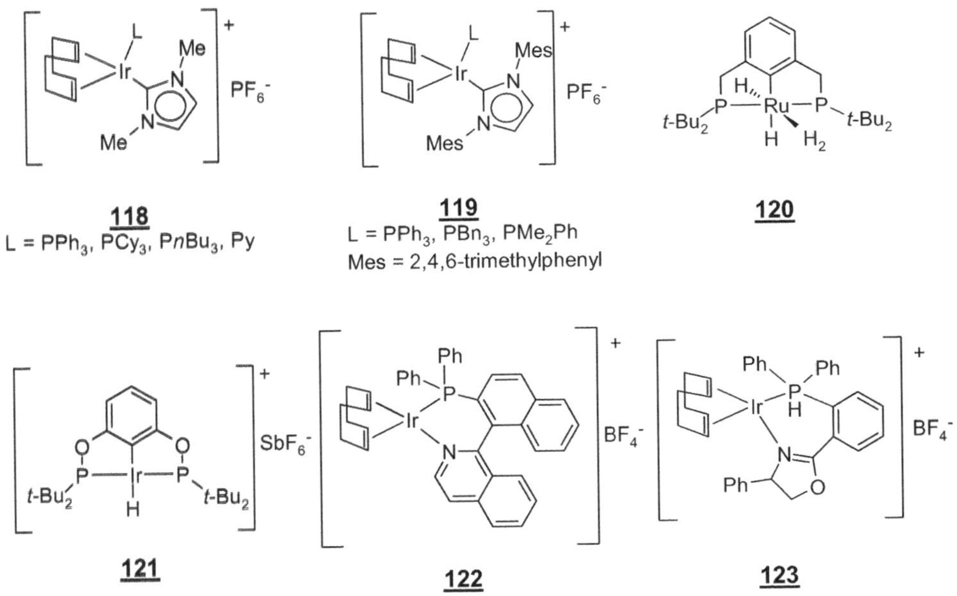


*Figure 3.45* *Newer classes of phosphine and related complexes of iridium investigated for utility in catalytic exchange labelling*

To date only limited studies using deuterium gas and simple model compounds have been reported for these complexes, so the extent of their utility for tritiations of complex organic molecules is not yet known.

The subsequent literature on iridium phosphine complexes is quite broad, and can perhaps be most clearly summarized according to the regio- and chemoselectivity that they display. Most of the labeling discussed so far is at aryl carbons *ortho* to directing functions via either five- or six-membered metallacyclic intermediates; we will call this *ortho* labeling. Other kinds of labeling have been observed with this class of catalysts, some mentioned already. They are: labeling at $sp^2$ carbon sites other than *ortho* positions, labeling at $sp^3$ carbons (such as the *N*-methyl groups of benzamides), and labeling mediated by $sp^3$ nitrogen rather than $sp^2$ N or O.

### *Ortho Labeling*

Already mentioned have been labeling of aromatic (and some heteroaromatic) rings *ortho* to carbonyl-based functions such as ketones, esters, amides and anilides, and heteroaromatic functions such as 2-pyridyl and the fused analog benzoquinoline. Modest extensions of the range of carbonyl-based functions to tritiation of ureas have been reported[196], including **124**

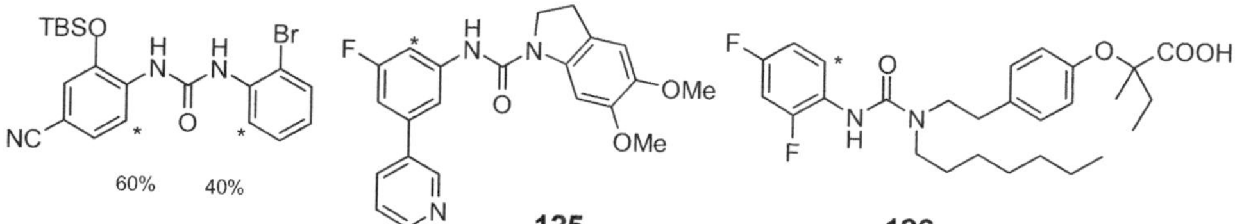


124 125 126

*Figure 3.46* Representative acylanilide substrates subject to ortho-tritiation

(to 29 Ci/mmol) and **125** (to 14 Ci/mmol) with [(cod)Ir(dppe)]$BF_4$ (dppe = 1,2-bis(diphenylphosphino)ethane), and **126** (to 9.6 Ci/mmol) with [(cod)Ir$(PCy_3)_2$]$PF_6$.

Nitro groups have been shown to mediate labeling in experiments with deuterium[179] and tritium[196] (e.g., **127**, 32 Ci/mmol) using the above catalysts and, in deuterium model studies, especially well with complexes **119**[204]. Other highly effective directing groups for phenyl *ortho* labeling by 50–110 mol% of [(cod)Ir$L_2$]X ($L_2$ = $(PCy_3)$(py), $(PPh_3)_2$ or $(PMePh_2)_2$) have been reported, but only with deuterium[189,209]. They include pyrazol-1-yl, pyrazol-3-yl, isoxazol-3-yl, thiazol-2-yl, thiazol-4-yl, imidazol-4-yl, tetrazol-5-yl, pyrazinyl and pyrimidin-4-yl, oximyl, amidinyl and hydrazonyl, some in derivatized form. These groups all have in common the features sketched in **128** and **129**, in which an $sp^2$ nitrogen is four bonds away from the *ortho* Hs. Several other related functional groups were active in mediating exchange to a lesser degree.

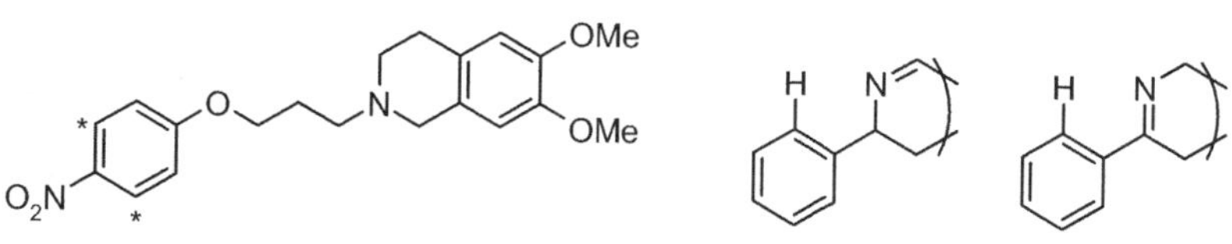


127 128 129

*Figure 3.47* Representative nitro substrate subject to ortho-tritiation and schematic drawings of other nitrogen-bases exchange-mediating functions

The *ortho* deuteration of benzenesulfonamides and methyl phenyl sulfone has been reported[210] using an *in situ* catalyst prepared by mixing $[(cod)IrCl]_2$ and two equivalents of pyridine; metallic iridium was deposited during these reactions. This is the only system that appears to be effective in labeling via mediation of sulfur(VI) functions; further investigations needed to determine its possible utility for general use with tritium have not appeared.

Examples of *ortho* labeling by coordinating functions once removed from the ring are rarer, with the exception of anilides, as mentioned above (Section 3.5.1). Deuterium labeling[191] of the phenylmethyl and phenoxy rings of **130** and **131** (Figure 3.48) (as well as the anilide-like rings) using Ir(I) 1,5-cyclooctadiene precatalysts containing the ligands dppe, $(PMePh_2)_2$ and/or $(PCy_3)$(py) has been reported[195], as has that of **132** and **133** with analogous catalysts containing the ligands $(Cy_2PCH_2CH_2PCy_2)$, $(AsPh_3)_2$ and $(Ph_2AsCH_2CH_2AsPh_2)$[211].

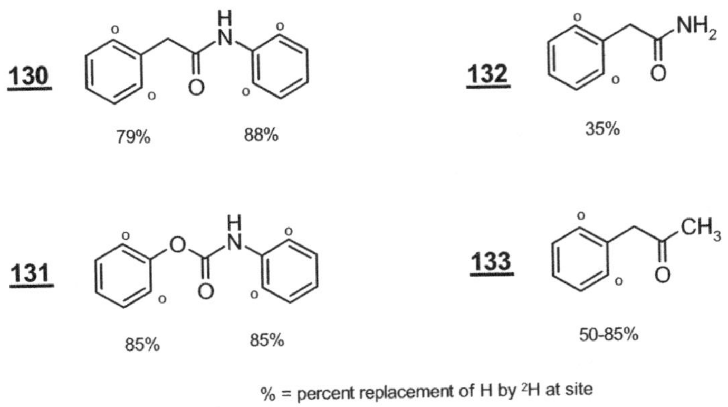


*Figure 3.48* *Examples of labeling mediated by carbonyl groups separated from the ring by a C, N or O atom*

Indoles **134** have been labeled[189] at C3 with deuterium using $[(cod)Ir(PCy_3)(py)]PF_6$ and indoles **135** at C7 with tritium[192] ($[(cod)Ir(dppe)]BF_4$) and deuterium[188]. Derivatives of camptothecin (**136**) were tritiated[212] at C5, as one of two sites (see Section 3.5.1.3), by the action of $[(cod)Ir(PCy_3)(py)]PF_6$.

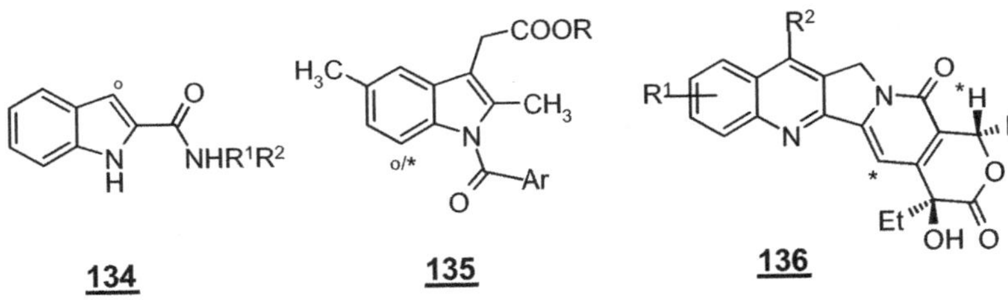


*Figure 3.49* *Labeling of differently substituted indoles and camptothecin*

A number of organoiridium complexes, possessing more complex bidentate phosphine ligands, have been tested[188] for their activity with deuterium gas toward a substrate (**137**) designed to assess preferences for four- vs five-bond labeling of *ortho* positions, and alkyl vs aryl labeling (Figure 3.50). The most active complexes were **138** and **139**, which labeled all three sites more efficiently than any other phosphine complexes to date. Among the other new complexes were some that showed selectivity for four-bond labeling over the other sites, but none showed a selectivity pattern different from previously studied complexes.

*Labeling of sp² Sites Other Than Ortho Positions*

Complexes that catalyze exchange labeling of aryl *ortho* positions through a five-membered metallacycle might be expected to do the same at the $\beta$-positions of $\alpha,\beta$-unsaturated carbonyls and equivalent sites. However, complexes of the iridium phosphine class tend to reduce olefins, so enone $\beta$-exchange can only be detected if exchange is faster than hydrogenation, and only useful as an exchange process if labeled unreduced compound

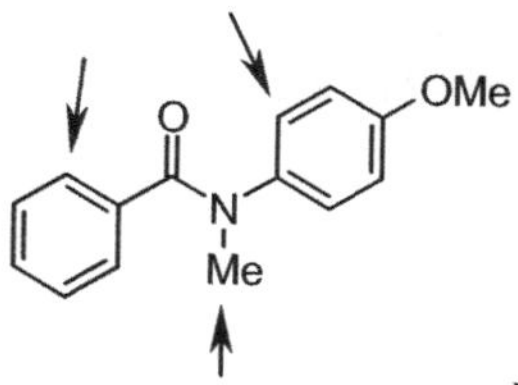


**137**

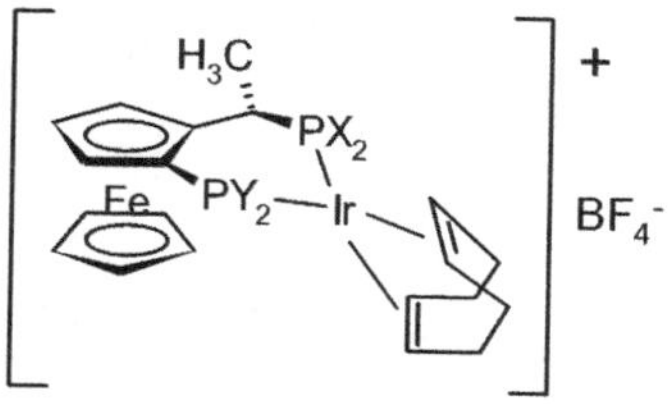


X = cyclohexyl
Y = 4-methoxy-3,5-dimethylphenyl

**138**

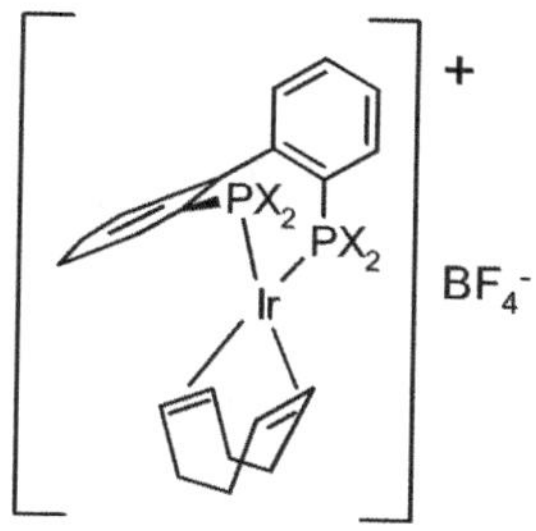


X = 3,5-di(*t*-butyl)-4-methoxyphenyl

**139**

***Figure 3.50*** *Complex active organoiridium complexes and test substrate*

can be recovered. Two such examples have been reported[213], namely aminoenone **140** and *syn*-oxime **141**, utilizing [(cod)Ir(PCy$_3$)(py)]PF$_6$ and deuterium gas. Other enones tested were reduced. That exchange and reduction processes compete is suggested by the finding of an excess of deuterium at the $\beta$-position of saturated ketones resulting from the deuterium gas reduction of enones using this catalyst. The *anti*-isomer of **141** was labeled in the adjacent free *ortho* position instead of at the $\beta$-position, reflecting the directional constraint imposed by structure on the nitrogen lone pair available for coordination to iridium.

**140**

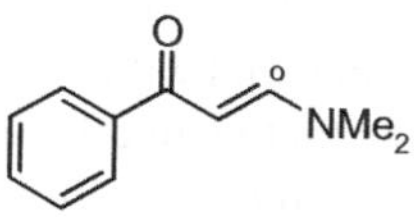


**141**

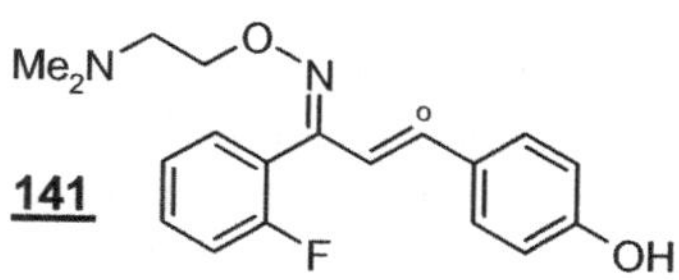


***Figure 3.51*** *β-labeling of nonreducible α,β-unsaturated ketone and oxime*

In studies of the *ortho*-labeling of phenyl-substituted heterocycles using [(cod)Ir(PCy$_3$)-(py)]PF$_6$, labeling, apparently catalyzed by iridium, was also observed[189] in some of the heterocycles, including the C2(5) positions of furan and thiophene, C2 of imidazole and C2 or C4 of thiazole, depending on other substitution. The exchange labeling of imidazoles at C2 by a variety of heterogeneous catalysts is well established (Section 3.3).

Several cases of labeling of pyridine derivatives adjacent to nitrogen have been reported. These include model studies on 2-phenylpyridine using deuterium gas and [(cod)Ir(PCy$_3$)-(py)]PF$_6$[189], slight labeling of C6 of certain pyridines by bis-phosphine complexes[214], and in three more complex compounds[196] **142** (Figure 3.52) where 18–25% of the tritium label was incorporated into the indicated site. The possibility that this type of labeling occurs via a reversible 1,2-reduction is suggested by the earlier observation[179] of both C2-labeling and partial 1,2-reduction of benzoquinoline with [IrH$_2$(acetone)$_2$(PPh$_3$)$_2$]BF$_4$. Other examples of pyridine C2(6)-labeling already mentioned (Sections 3.1.2 and 3.4.1) are perhaps more likely than these examples to occur through acid–base catalysis; however, no system has been examined in sufficient detail to support mechanistic conclusions, and there is as yet no clear correlation of its occurrence with any experimental variables.

**142**

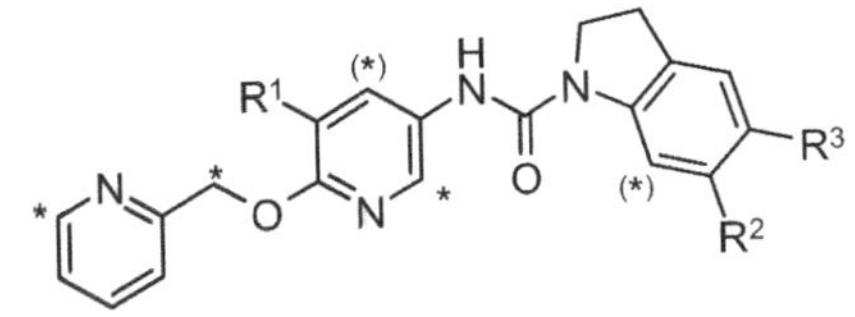


$R^1$ = H or Me
$R^2$ = H or $CF_3$
$R^3$ = Me or $CF_3$

**Figure 3.52** *Pyridine-C2(6) labeling by organoiridium complexes*

Exchange of formyl hydrogens for tritium is observed to occur in both aryl aldehydes[215] (with concurrent *ortho* labeling in appropriate cases) and aliphatic aldehydes[216] using $[(cod)Ir(PCy_3)(py)]PF_6$. Partial reduction of aldehydes to alcohols may occur in some cases. Labeling by other iridium phosphine catalysts has not been reported but is likely to occur. This type of catalytic activity, which likely involves reversible oxidative addition of the iridium center into the formyl C–H bond, is different in outcome from that of organorhodium complexes, whose insertion into formyl C–H bonds proceeds instead to decarbonylation.

*Labeling at $sp^3$ Carbons*

Deuteration of the methyl groups of *N,N*-dimethylbenzamide (**99**, Figure 3.37), 8-methylquinoline (**101**) and caffeine (**102**) (Figure 3.38) has already been noted, catalyzed by $[Ir(H)_2(acetone)_2(PPh_3)_2]BF_4$. Labeling of amide *N*-methyl groups has been observed with a number of other iridium phosphine catalysts. A model substrate, *N*-methyl-*N*-phenylbenzamide was used in catalyst screening studies[186] in which some highly active bidentate phosphine complexes were identified at 5 mol% loading, and with monodentate phosphine ligands at 50 mol% loading[214]. In summary, catalysts that were effective at *ortho* labeling specifically via five-membered metallacycles also tended to label amide *N*-methyl groups, better than catalysts that also labeled *ortho* positions via six-membered metallacycles. A study[196] comparing catalysts $[(cod)IrL_2]BF_4$ with mono- and bidentate ligands using *N,N*-dimethyl-4-nitrobenzamide and catalyst loadings of 4–5 mol% showed much lower alkyl/aryl labeling ratios for the catalysts $L_2$ = dppe and $L_2 = (PCy_3)(py)$ ligands compared with $L_2 = (PPh_3)_2$ and $L_2 = (PMePh_2)_2$ ligands. Other examples of tritium labeling of N–C–H labeling of amides include structure **136** above (using $[(cod)Ir(PPh_3)_2]PF_6$, 2–34 Ci/mmol, a minority at the methylene C1$\beta$), compound **143**[201] ($[(cod)Ir(PCy_3)(py)]PF_6$, 95 Ci/mmol), and **144**[197] ($[(cod)Ir(PPh_3)_2]BF_4$, 22–42 Ci/mmol) (Figure 3.53).

Heteroaryl nitrogen-mediating functions are more effective than amide carbonyls in mediating isotope exchange at $sp^3$ carbons. Tritiations of **145** and **146** with 150 mol% $[(cod)Ir(PPh_3)_2]BF_4$ and 17 equivalents of tritium gas[197] gave a statistically equal degree of labeling at both methyl and methylene positions in each compound, and tritiations of three related compounds **147** (313 mol% $[(cod)Ir(PCy_3)(py)]PF_6$, 11 equivalents of tritium gas) gave[196] significant labeling of the indicated methylene group (Figure 3.54).

Model studies[214] using deuterium gas and 50 mol% of iridium complexes having a variety of monodentate phosphine ligands revealed several that gave high deuterium incorporation into the methyl groups of 2-dimethylaminopyridine, but the most effective ligands were of types already generally characterized. Recent model studies[217] (Figure 3.55) with a 2-piperazinylpyridine (**148**) and several alkylaminopyrazines

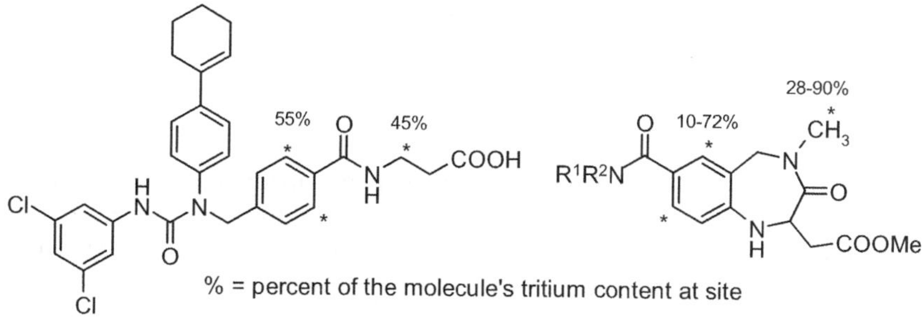


**143** **144**

***Figure 3.53*** *Examples of exchange labeling at sp*$^3$ *sites catalyzed by organoiridium complexes*

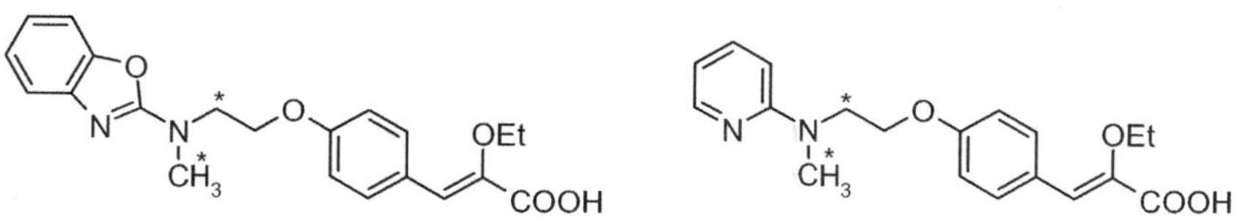


**145** (126 Ci/mmole) **146** (99 Ci/mmole)

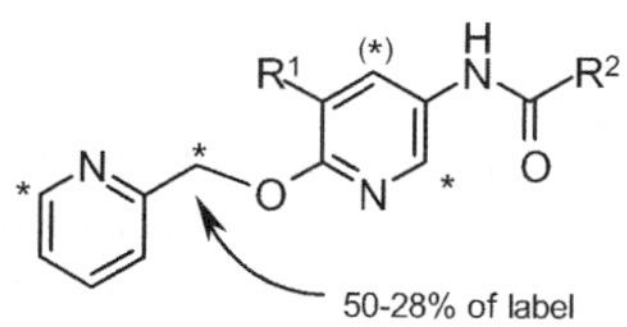


**147** (49-69 Ci/mmole)

***Figure 3.54*** *Sp*$^3$ *site tritium labeling facilitated by nitrogen heterocycle mediators*

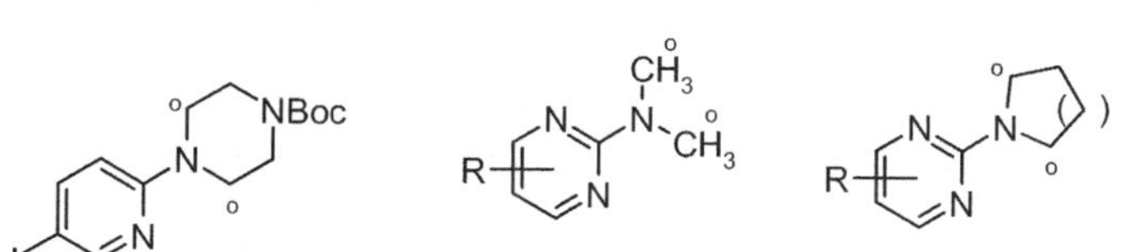


**148** **149** **150**

R = I, Cl, OMe, alkyl
( ) = -N(Boc)$CH_2$- , -$CH_2$-

***Figure 3.55*** *Model studies of sp*$^3$ *site deuterium labeling facilitated by nitrogen heterocycle mediators*

(**149**, **150**) and 50 mol% loadings of $[(cod)Ir(PCy_3)(py)]PF_6$ gave high levels of deuterium labeling in the methyl or methylene groups attached to nitrogen.

*Labeling Mediated by sp$^3$ Nitrogen*

There are a few examples of iridium phosphine complexes mediating the *ortho*-labeling of benzylamines, but in general amines have not been shown to be very effective directing groups for this class of catalyst. One example[218] of a tritiation is that of **151**, which was labeled to 17.5 Ci/mmol with 13 mol% $[(cod)Ir(PCy_3)(py)]PF_6$.

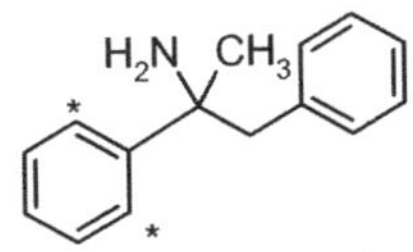


**151**

***Figure 3.56*** *Example of amino group-mediated tritium exchange labeling catalyzed by [(cod)Ir(PCy$_3$)(py)]PF$_6$*

Modest success has been achieved in some model studies, mostly using the same catalyst[219,220], but positive results have been sporadic and no particular system has been found to be widely useful. The iridium dionate class of catalysts (next section) are much more effective for labeling such substrates.

### 3.5.2 Iridium Dionate Complexes

In what may be a mechanistically different process, mono-, di-, and even trisubstituted benzylamines, as well as anilines, were found to be deuterated[219,220] in *ortho* positions from deuterium gas by use of the catalyst precursor cyclooctadienyliridium 1,1,1,5,5,5-hexafluoropentane-2,4-dionate ((cod)Ir(hfpd)). This neutral complex, with an anionic *O,O'*-bidentate ligand, was the most active of three related ones tested. This class of complex had been discovered in screening for catalysis of substrate labeling from a $^2H_2O$ isotope source[221]. The hfpd complex is active in DMF and DMA solvents, which makes the reaction conditions usable with substrates insoluble in the less polar solvents required for the iridium phosphine class of catalysts.

Model studies included a wide range of benzylamines (**152**) and several anilines (**153**) (Figure 3.57). The effectiveness of labeling in both the benzylamines and anilines was rather insensitive to the electronic effects of phenyl ring substituents and mildly sensitive to *N*-substitution. These patterns are similar to those of the iridium phosphine class of catalysts in labeling mediated by their respective coordinating groups.

So far, the only published use of (cod)Ir(hfpd) in tritium labeling has been a tracer experiment with a model compound, but the method should work generally with tritium gas. It remains to be seen how broad the applicability will be to complex substrates. The utility of this catalyst class appears to be largely complementary to that of the iridium phosphine class, mainly because of its apparently general ability to label benzylamines and anilines and its activity in DMF and DMA solvents. Nitro groups tend to be reduced. But as with iridium phosphine complexes, (cod)Ir(hfpd) tends to reduce double bonds but

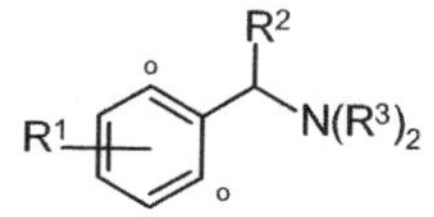


**152**

R¹ = halo, methoxy, methyl, $CF_3$, phenyl, 3,4-benzo
R² = H, Me, Bn
R³ = H or Me independently

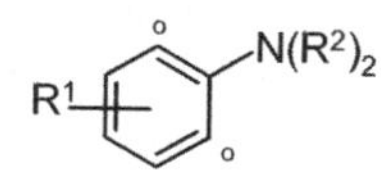


**153**

R¹ = Me, COOH, COMe, 3,4-benzo
R² = H or Me

***Figure 3.57*** *Model studies of amino group-mediated tritium exchange labeling catalyzed by (cod)Ir(hfpd)*

not hydrogenolyze halo substituents, and is capable of catalyzing the labeling of compounds possessing strong directors, such as 2-phenylpyridine and 2-phenylimidazole. It is reported not to be active, or as active, with acids, amides or esters, which may be attributed to diminished electrophilicity at the metal center compared with the phosphine class of catalysts, which are cationic.

The regioselectivity of the labeling of aminobenzoic acids and aminoacetophenones by (cod)Ir(hfpd) from deuterium gas – adjacent to the amino functions – is complementary to the locations labeled – adjacent to the carbonyl-containing function – when $^2H_2O$ is the isotope source, reinforcing the probability that different reaction mechanisms are at work under different conditions (Section 3.4.3).

### 3.5.3 Iridium Cyclopentadienide Complexes

The C–H bond activation by iridium(III) complexes such as **154** have been well studied[222]. Complexes of this type require relatively sophisticated synthesis procedures and are air-sensitive. Complex **154** can undergo rapid oxidative addition to $sp^2$ and $sp^3$ C–H bonds of substrate compounds, with concurrent loss of methane, to give new aryl- or alkyliridium (III) complexes. The iridium–carbon bonds of these can, in turn, be hydrogenolyzed *in situ* to release the original substrate. The first use of this process with deuterium in place of hydrogen, to achieve deuterium exchange labeling of substrates, was reported[223] on a few

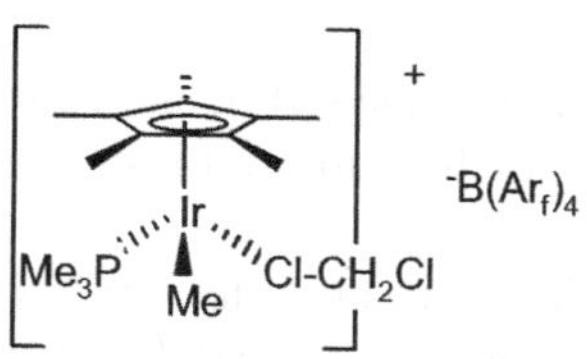


$Ar_f$ = 3,5-bis(trifluoromethyl)phenyl

**154**

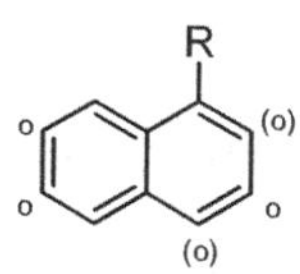


R = OH, OMe, COOH

**155**

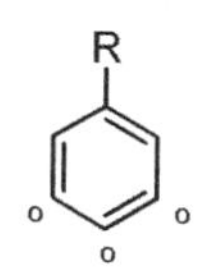


R = NHAc, $SO_2Me$

**156**

***Figure 3.58*** *Structure of active iridium cyclopentadienide complex; regioselectivity of deuterium labeling in model substrates*

test compounds, and expanded in a subsequent study[224]. Complex **154** can be prepared[225] *in situ* by treatment of the air-sensitive precursor Cp*Ir($PMe_3$)(Me)OTf (Cp* = pentamethylcyclopentadiene) or the stable precursor Cp*Ir($PMe_3$)(Me)Cl with NaB($Ar_f$)$_4$, and is most active in $CH_2Cl_2$ solvent. Labeling of model substrates **155** and **156** in positions distant from substituents suggests that steric effects, and not electronic effects or coordinating events, predominate in determining regioselectivity. The exception to this trend is that fluoro substituents have been found to be neutral to, or slightly facilitative of, labeling in adjacent positions[226].

Strongly coordinating substituents such as amino and thio poison the complex, and a quinoline failed to undergo labeling. More complex compounds such as **157**–**159** were found to be good substrates; tritiation of **158** under the same conditions gave similar results.

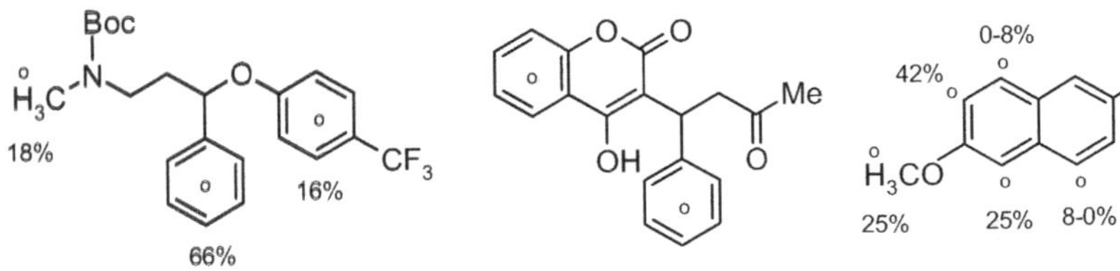


**157** **158** **159**

***Figure 3.59*** *Application of iridium cyclopentadienide complex-mediated deuterium labeling of more complex substrates*

This labeling process is not catalytic; rather, an initial iridium–substrate adduct is formed and the Ir–C bond subsequently deuteriolyzed. However, the products labeled by this procedure usually include a significant proportion of dideuterated and sometimes trideuterated species (in addition to $^2H_0$ and $^2H_1$ species). As Ir–C deuteriolysis is expected to install one deuterium atom in the substrate, the presence of additional deuterium suggests the occurrence of exchange as well. This was studied in some detail[227]. When *ortho-, meta-* and *para-* isomers of Cp*Ir($PMe_3$)(tolyl)OTf were prepared then separately deuteriolyzed, significant additional amounts of deuterium (up to 1.9 mol $^2$H/mol in excess of that resulting from direct deuteriolysis of the Ir–C bond) was incorporated in tolyl positions 3, 4 and/or 5. This was hypothesized to be the result of addition–elimination processes prior to the departure of toluene from the coordination sphere. The excess labeling did not occur with three disubstituted phenyl analogs Cp*Ir($PMe_3$)(3,5-dialkylphenyl)OTf, indicating that additional isotope incorporation by exchange appears to be subject to strong steric hindrance.

The latter procedure (involving isolation of the organoiridium–substrate complex) is not practical for use in the tritiation of most substrates, but the former may be if deactivation of the complex by polar functional groups is avoided. The regioselectivity that these iridium cyclopentadienide complexes display is largely complementary to those of iridium phosphines and iridium dionates; therefore cyclopentadienide complexes could be valuable additions to the repertory. Unfortunately the air-sensitivity of these complexes and their immediate precursors is a hindrance to their use. Further development of this class would be highly desirable.

Meanwhile, iridium cyclopentadienide complexes have been found to be catalytic in their mediation of exchange of deuterium from $^2H_2O$ (see Section 3.4).

## 3.6 Solvent-Free Catalytic Exchange

Solvent-free methods of hydrogen isotope exchange labeling usually utilize tritium gas as the isotope source, and the substrate is used either neat, as a solid, or as a component of a solid phase. Therefore in contrast to the methods of Sections 3.1–3.5 this class of methods are, at least at a gross level, gas–solid two-phase systems.

### 3.6.1 High-Temperature Solid-State Catalytic Isotope Exchange

The most commonly used method of solvent-free exchange labeling, called high-temperature solid-state catalytic isotope exchange (HSCIE), came into use in the late 1980s[228,229]. It involves exposing a mixture of substrate, support and a metal catalyst to tritium gas and heat for a short period of time. Intimate mixing by evaporation of solvent from a suspension of support and/or supported catalyst in a solution of substrate results in a structured chemisorbed layer of substrate on the support. Tritium activated by the metal catalyst migrates ('spillover hydrogen' phenomenon[230]) to the support material and thence to the substrate with which it then interacts. Studies of spillover hydrogen in this context and in other areas of chemical catalysis indicate that several forms of activated hydrogen (deuterium, tritium) are produced but the most important are usually atomic or positively charged species; this is consistent with the observed distribution of isotope in HSCIE labeling, which is similar to that expected from attack of an electrophilic species. Theoretical calculations indicate that the carbon sites most highly labeled are usually those with higher proton affinity. *Ab initio* analyses and reaction modelling[231] also suggest that exchange progresses via a three-center complex of the C–H with the activated tritium atom/ion, leading to replacement of H by T at the same face of C; i.e., without inversion of tetrahedral carbon. This is consistent with observations that C$\alpha$ labeling of amino acids usually proceeds with retention of configuration[232,233]. Extensive overviews have been published[234,235].

A wide variety of compounds have been tritiated using the HSCIE method. Nevertheless, in spite of the accumulation of a large body of empirical data and mechanistic and theoretical study, there exists no standard set of reaction conditions, nor is it yet possible to make informed selection of optimum conditions based on compound structure. Rather, each new labeling substrate requires extensive experimentation to optimize its labeling. Several variables can strongly influence the outcome of labeling experiments and it remains difficult to predict what will work. Important variables include the metal (most frequently Pd, Rh or Pt), metal support (C, $Al_2O_3$, $BaSO_4$, $CaCO_3$), percent loading of metal on support, ratio of catalyst to substrate, reaction temperature, tritium pressure and exposure time. Additional factors that may in some cases improve outcomes are the use of a secondary support (Norit, fibrous carbon, $Al_2O_3$, $BaSO_4$ or $CaCO_3$) mechanically mixed with the catalyst–substrate blend[236,237] and other additives[238,239]. The rate and extent of tritium incorporation increases with temperature, but so does the extent of substrate decomposition, so the optimum result is a tradeoff between specific activity on the one hand, and yield and difficulty of purification on the other[235]. Indeed, purification is sometimes challenging. The labeling process is also importantly affected by the physical state of the substrate at the reaction temperature. If the substrate behaves as a liquid it can migrate through the support and interact with catalyst particles, but if it remains a solid, individual substrate and catalyst particles remain at fixed distances from each other across the gradient of spillover tritium[240,241]. The effects of the degree of substrate migratory behavior on labeling outcome are apparent for example with stearic acid, whose specific activity rises to a maximum and then declines as the temperature or reaction time increases[242]. Lastly, as the HSCIE conditions effectively promote hydrogenation, it is not surprising that reducible functions such as double bonds, aryl halogen, etc., of substrates often undergo chemical reaction. Indeed, solid phase hydrogenation can be a good alternative to the solution-phase method (see Section 4.1).

Notwithstanding all these complexities, use of the method has given some excellent results with otherwise difficult-to-label compounds. Success has been demonstrated with compounds from a variety of classes, including fatty acids and esters[242], purines and pyrimidines[243,244], nucleosides and nucleotides[245,246], fungicides[247], plant hormones[248], cyclodextrins[249], dextran and inulin[250], and other biologically active componds[251–254]. Some examples will illustrate the range of the method. Stearic acid ($CH_3(CH_2)_{16}COOH$) was tritiated[240] to a specific activity of up to 350 Ci/mmol upon exposure to ca 10 Ci of tritium gas (1000 wt% of 5% Pt/C, 210 °C, 15 min) in 32% yield. 9-Nitrocamptothecin (**160**) was labeled[254] to a specific activity of 14–16 Ci/mmol (1100 wt% of 5% Pd/$BaSO_4$, 190 °C, 15 min) in 3–5% yield. Tiazofurin (**161**) was labeled[245] to a specific activity of 17 Ci/mmol (1000 wt% of 5% Pd/$BaSO_4$, 155 °C, 15 min) in ca 50% yield. 7-Thia-8-oxoguanosine (**162**), possibly because of its 'nonaromatic' sulfur only reached[246] the level of 240–320 mCi/mmol (1000 wt% of 5% Pd/$CaCO_3$, 170 °C, 15 min) in 10–15% yield. The locations of labels in these products were not reported.

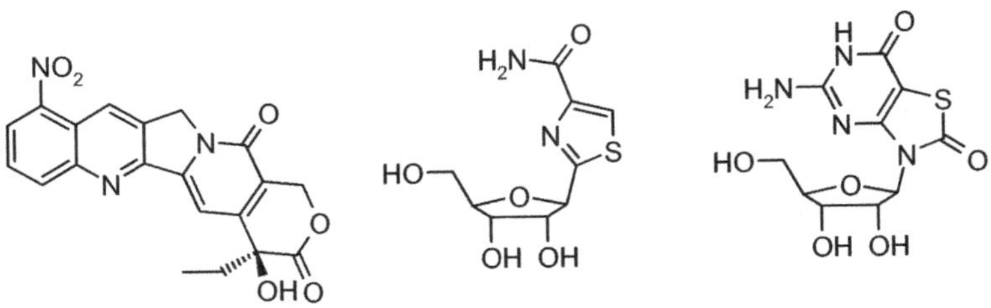


**160** **161** **162**

*Figure 3.60 Some substrates tritiated by means of solid-state catalyzed hydrogen isotope exchange (1)*

The effectiveness of this method with more complex molecules was demonstrated[255] by the labeling of coenzyme A (**163**) to 3.9 Ci/mmol (5% Pd/$BaSO_4$, 180 °C, 30 min) in 28% yield (90% of tritium was in the adenine moiety) and nicotinamide adenine dinucleotide (**164**) to 111 Ci/mmol (5% Pd/$CaCO_3$, 110 °C, 30 min) in 13% yield (most label in the nicotinamide moiety) (Figure 3.61).

Sugars have also been tritiated[256]; for example, *N*-acetylneuraminic acid $\alpha$-methylglycoside (**165**) to 12.5 Ci/mmol (2500 wt% of 5% Pd/C, 200 °C, 5 min) in 20–25% yield, and

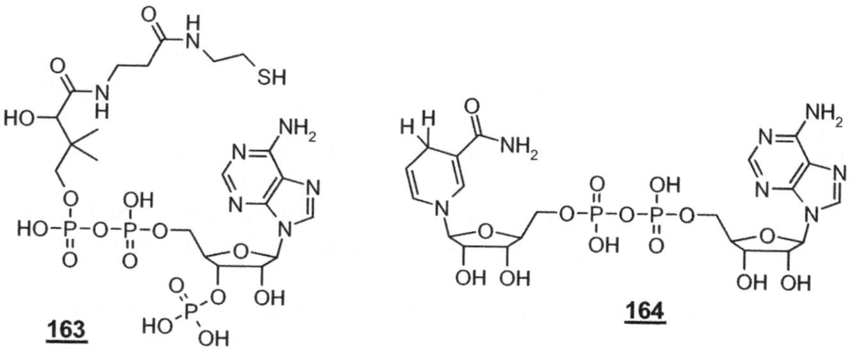

163 164

*Figure 3.61* Some substrates tritiated by means of solid-state catalyzed hydrogen isotope exchange (2)

3′-sialyllactosamine (**166**) to 8–9 Ci/mmol (2500 wt% of 5% Pd/$CaCO_3$, 160 °C, 20 min) in 10–15% yield. The locations of labels in these compounds were not reported.

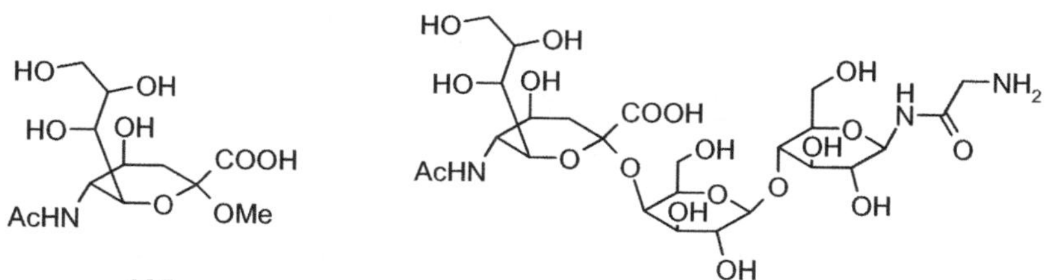

165 166

*Figure 3.62* Some substrates tritiated by means of solid-state catalyzed hydrogen isotope exchange (3)

Single amino acids could be labeled[257] (2:1 w/w mix of amino acid and $RhCl_3$ on 20 parts of 5% Pd/$CaCO_3$, 35 kPa $^3H_2$, 150 °C, 20–90 min) at rates within about one order of magnitude of each other (5–63 Ci/mmol), in spite of differences in the number of hydrogens they contain, arginine becoming labeled to the highest degree and methionine the lowest. The sites reported to be most highly labeled in aliphatic amino acids are C$\alpha$, methyl groups and carbons adjacent to O or N; benzyl methylenes of aromatic amino acids, and aryl sites activated for electrophilic substitution. Optical purities were largely retained even at high C$\alpha$ isotope incorporation[232,258]. In five 3–7mer peptides the inter-residue tritium distribution (totalling 50–110 Ci/mmol) was not too dissimilar from that expected for individual amino acids[259] (1:10 coating of peptide on $Al_2O_3$ mechanically mixed with 10 parts 5% Rh/$Al_2O_3$, 140–190 °C, 10–60 min). However, with larger peptides such as the acetylated 17mer zervamicin IIB (70 Ci/mmol) and $\alpha$-conotoxin G1 (35 Ci/mmol), a 13mer with two disulfide linkages, the labeling of certain groups of amino acids was nil or much lower than the others[260]. This was attributed to shielding of these residues by virtue of their residence within relatively immobile parts of the (known) secondary structures. Based on such findings it was hypothesized that inferences could be drawn about the secondary structures of complex molecules of unknown conformation, based on the patterns of labeling along the amino acid chain. Extended further[261] the degree of labeling of endopeptidase-generated fragments of the $\beta$-galactosidase protein (83 kDa) of known structure from *Thermoanaerobacter ethanolicus* (1400 Ci/mmol; 1:50 coating on $CaCO_3$ mechanically mixed with 20 parts of 5% Pd/$BaSO_4$, 300 Torr $^3H_2$, 120 °C, 10–60 min) were found to correlate inversely with their depth within the globular protein, thereby lending credence to the hypothesis that surface residues of proteins could be distinguished from internal residues by assessment of their differential labeling under HSCIE conditions. This is similar to the data obtained by use of bombardment with thermally produced tritium atoms (Section 3.6.2).

Finally, although HSCIE is a strongly reductive system, it has on occasion been used to label compounds sensitive to hydrogenation. One example is the adenine derivative *trans*-zeatin (**167**). Although it had previously been labeled[262] from high specific activity tritiated water (5% Pd/$CaCO_3$, dioxane, 22 h, rt) to a specific activity of 18 Ci/mmol (67% yield), careful selection of HSCIE conditions[263] provided *trans*-[$^3$H]zeatin with a specific activity up to 40 Ci/mmol (6% yield); the balance of the material was reduced to [$^3$H]dihydrozeatin which could be separated chromatographically.

**167**

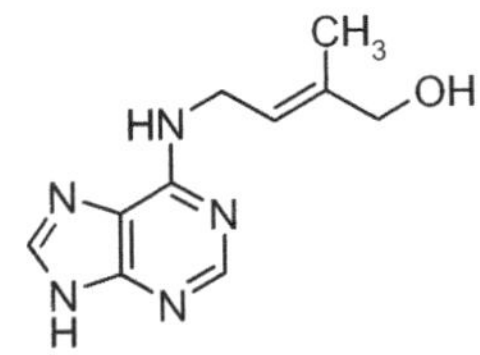


***Figure 3.63*** *HSCIE-labeling of the reducible substrate* trans-*zeatin*

Much less studied is the use of zeolites as catalytic entities at elevated temperatures for solvent-free exchange tritiation with tritium gas. Of the many types of zeolites, only metal-containing ones have been found[264] to catalyze exchange from tritium gas. Published information in this area consists of pilot studies and the method has not been developed for routine practical use. The mechanism of labeling is evidently similar to that of HSCIE, in that tritium activated by the metal spills over via the zeolite to interact with substrate as an electrophilic entity. The most thorough study yet published[265] includes results for labeling of alkyl-substituted aromatics, halobenzenes, saturated hydrocarbons and several simple heterocycles.

### 3.6.2 Thermal Tritium Atom Bombardment

The tritium atom bombardment method (TA) involves the exposure of a thin film of substrate to tritium atoms formed by dissociation of tritium gas on a heated filament. The procedure involves freezing then lyophilizing a solution of the substrate onto the walls of a rotating vessel, inserting a tungsten filament, fixing the atmosphere within to a low pressure of tritium gas, then with external liquid $N_2$ cooling of the vessel walls, heating the filament to 1000–2500 K for seconds to minutes. The description of a purpose-built manifold is available[266]. Use of this method for labeling organic compounds of various types has been reported, but the achievable specific activity levels are significantly lower than for the HSCIE procedure. For example[267], optimization of sample layer thickness, filament temperature, and exposure time for labeling of glycine, serine, proline and valine on 20 $\mu$mol scales gave maximum specific activities of approximately 54, 68, 270 and 1080 mCi/mmol, respectively; yield and optical purity results were not reported. A comparison was made[268] of TA and HSCIE for labeling of three compounds, individually optimized in each method; the results are summarized in Table 3.1.

***Table 3.1*** *Comparison of the effectiveness of thermal atom bombardment and high-temperature solid-state catalytic isotope exchange on selected substrates*

| | TA | | HSCIE | |
|---|---|---|---|---|
| | Spec. Ac. (Ci/mmol) | Yield (%) | Spec. Ac. (Ci/mmol) | Yield (%) |
| D-Ribose | 1.14 | 35 | 8.4 | 74 |
| Sucrose | 0.32 | 55 | 59 | 92 |
| Adenine | 0.07 | NR | up to 146 | NR |

NR = not reported

The major factor limiting the attainment of high specific activities in the TA method is the low energies of the tritium atoms generated by the hot filament, about 0.3 eV[267]. Particle energies in this range result in efficient $^3$H-for-H replacement without causing an unreasonable degree of substrate decomposition through side reactions. However, tritium gas pressures must be low (less than 5 Pa, or 25–100 mCi for a 500 mL vessel) to reduce atmospheric quenching of activated tritium, and the average penetration of tritium atoms into the condensed phase (substrate layer) is only about 0.4 nm, so that only a small quantity of substrate (approximately 60 ng/cm$^2$) is accessible to labeling.

This limitation in depth of labeling under TA conditions has been exploited in the study of proteins and protein assemblages, where the method is sometimes called tritium planigraphy. Such macromolecular systems, when treated under TA conditions are tritiated only in parts of molecules located at or very near the surface, leaving interior parts unlabeled. Subsequent dissection of the substrate, for example by separation of protein subunits or tryptic hydrolysis of protein followed by radioanalysis, can form the basis for inferences about the structure of the substrate. Applications of the TA method in this area include studies of the organization of the coat proteins of potato virus X[269], conformational dynamics of ribonuclease A[270], spatial localization of coenzyme FAD in the quaternary structure of yeast alcohol dehydrogenase[271] and subunit organization of the bacterial ribosome[272–274].

### 3.6.3 Other Radiation-Induced Labeling Methods

There are several other labeling methods that utilize some form of energy to activate tritium so that it readily exchanges with hydrogens of substrates, each of which has at some time had a degree of importance for practical labeling. However, with the subsequent development of improved labeling methods, none of these is any longer used to a significant extent. The original exchange labeling method, involving simple exposure of solid substrate to an atmosphere of tritium gas[275] relied upon tritium self-activation by radioactive decay. Some photochemical procedures have been reported, but only with deuterium gas[276]. Other methods involved tritium activated by electric discharges, ion beams, or microwave discharge to label solid substrates using tritium gas. An extensive review of these methods has been published[277].

## References


1. Atzrodt, J.; Derdau, V.; Fey, T. and Zimmermann, J., *Angew. Chem. Int. Edn Engl.*, ***2007***, *46*, 7744–7765.
2. a. Melander, L. and Saunders, Jr., W.H., *Reaction Rates of Isotopic Molecules;* Robert E. Krieger: Malabar, FL, ***1987***; b. Happel, J., *Isotopic Assessment of Heterogeneous Catalysis;* Academic Press: Orlando, FL, ***1986***.
3. Zoltewicz, J.A. and Smith, G.L.S., *J. Am. Chem. Soc.*, ***1967***, *89*, 3358–3359.
4. a. Werstiuk, N.H. and Ju, C., *Can. J. Chem.*, ***1989***, *67*, 5–10; b. Werstiuk, N.H. in *Isotopes in the Physical and Biomedical Sciences*, Vol. 1, Part A, Buncel, E. and Jones, J.R., Eds.: Elsevier: Amsterdam, ***1987***, pp. 122–155.
5. a. Werstiuk, N.H. and Ju, C., *Can. J. Chem.* ***1989***, *67*, 812–815; b. Werstiuk, N.H. and Timmins, G., *Can. J. Chem.*, ***1981***, *59*, 1022–1024.
6. Funke, C.W.; Kaspersen, F.M.; Wallaart, J. and Wagenaars, G.N., *J. Label. Compd. Radiopharm.*, ***1983***, *20*, 843–853.
7. Hesselbo, T.; Slade, S.W.; Villani, A.J. and Petka, L., *J. Label. Compd. Radiopharm.*, ***1978***, *16*, 463–472.
8. Krieger, M.; Koeppe, R.E. and Stroud, R.M., *Biochemistry*, ***1976***, *15*, 3458–3464.
9. Eidinoff, M.L. and Knoll, J.E., *J. Am. Chem. Soc.* ***1953***, *75*, 1992–1993.
10. Schweizer, M.P.; Chan, S.I.; Helmkamp, G.K. and Ts'o, P.O.P., *J. Am. Chem. Soc.*, ***1964***, *86*, 696–700.
11. Wong, J.L. and Keck Jr., J.H., *Chem. Commun.*, ***1975***, 125–126.
12. Jones, J.R. and Taylor, S.E., *Chem. Soc. Rev.*, ***1981***, *10*, 329–344.
13. Sidorov, G.V. and Myasoedov, N.F., *Russ. Chem. Rev.*, ***1999***, *68*, 229–240; Lauschke, U.; Lodemann, E. and Wacker, A., *J. Label. Compd*, ***1971***, *7*, 441–444.
14. Shelton, K.R. and Clark Jr., J.M., *Biochemistry*, ***1967***, *6*, 2735–2739.
15. Shelton, K.R. and Clark Jr., J.M., *Biochem. Biophys. Res. Commun.*, ***1968***, *33*, 850–854.
16. Ido, T.; Tatora, M. and Kasida, V., *Int. J. Appl. Radiat. Isotop.*, ***1973***, *24*, 81–86.
17. Graham, M.J.; Freier, S.M.; Crooke, R.M.; Ecker, D.J.; Maslova, R.N. and Lesnik, E.A., *Nucleic Acids Res.*, ***1993***, *21*, 3737–3743.
18. a. Gamble, R.C.; Schoemaker, H.J.P.; Jekowsky, E. and Schimmel, P.R., *Biochemistry*, ***1976***, *15*, 2791–2799; b. Schoemaker, H.J.P.; Gamble, R.C.; Budzik, G.P. and Schimmel, P.R., *Biochemistry*, ***1976***, *15*, 2800–2803.
19. de Keczer, S.A.; Lane, T.S. and Masjedizadeh, M.R., *J. Label. Compd. Radiopharm.*, ***2004***, *47*, 733–740.
20. Evans, E.A., *Tritium and Its Compounds*, 2nd Edn, John Wiley & Sons, Inc.: New York, ***1974***, pp. 282–293.
21. Hesk, D.; Koharski, D.; Saluja, S.; McNamara, P.; Magatti, C.; Cesarz, D.; Hendershot, S. and Jones, J.R., in *Synthesis and Applications of Isotopically Labeled Compounds 1997*, Heys, J.R. and Melillo, D.G., Eds.; John Wiley & Sons, Ltd: Chichester, ***1998***, pp. 439–443.
22. Garnett, J.L. and Long, M.A., in *Isotopes in the Physical and Biomedical Sciences*, Vol. 1, Part A, Buncel, E. and Jones, J.R., Eds., Elsevier: Amsterdam, ***1987***, pp. 86–121.
23. Bissel, U.; Katritzky, A.R. and Lea, J.R., *J. Chem. Soc. B*, ***1971***, 4–10.
24. Best, A.P. and Wilson, C.L., *J. Chem. Soc.* ***1938***, 28–29.
25. Baba, S. and Sasaki, Y., *Radioisotopes*, ***1987***, *36*, 628–632.
26. Sangster, N.C.; Lacey, E.; Than, C. and Long, M.A., *J. Label. Compd. Radiopharm.*, ***1989***, *27*, 1069–1078.
27. Zimmerman, W.T., *J. Label. Compd. Radiopharm.*, ***2000***, *43*, 767–772.

28. Yamada, S.; Hongo, C.; Yoshioka, R. and Chibata, I., *J. Org. Chem.*, ***1983***, *48*, 843–846.
29. a. Brundish, D.E.; Elliott, D.F. and Wade, R., *J. Label. Compd*, ***1971***, *7*, 473–493; b. Hochreiter, M.C. and Schellenberg, K. A., *J. Label. Compd*, ***1969***, *5*, 270–274; c. Upson, D.A. and Hruby, V. J., *J. Org. Chem.*, ***1977***, *42*, 2329–2330; d. Fujihara, H. and Schowen, R.L., *J. Org. Chem.*, ***1984***, *49*, 2819–2820.
30. Matsuo, H.; Fujimoto, Y. and Tatsuno, T., *Tetrahedron Lett.*, ***1965***, 3465–3472.
31. Mitulovi, G.; Lämmerhofer, M.; Maier, N.M. and Lindner, W., *J. Label. Compd. Radiopharm.*, ***2000***, *43*, 449–461.
32. Abbott, E.H. and Martell, A.E., *Chem. Commun.*, ***1968***, 1501–1503.
33. Stipanovic, R.D.; Williams, H.J.; Muehleisen, D.P. and Plapp, Jr., F.W., *J. Label. Compd. Radiopharm.*, ***1986***, *24*, 741–743.
34. Evchenko, S.V.; Kamounah, F.S. and Schaumburg, K., *J. Label. Compd. Radiopharm.*, ***2005***, *48*, 209–218.
35. Hanzlik, R.P.; Wiley, R.A. and Gillesse, T.J., *J. Label. Compd. Radiopharm.*, ***1979***, *16*, 523–529.
36. Magatti, C.V.; Hesk, D.; Lauzon, M.J.; Saluja, S.S. and Wang, X., *J. Label. Compd. Radiopharm.*, ***1998***, *41*, 731–739.
37. Kingston, L.P.; Lockley, W.J.S.; Mather, A.N. and Wilkinson, D.J., in *Synthesis and Applications of Isotopically Labeled Compounds*, Vol. 7, Pleiss, U. and Voges, R., Eds., John Wiley & Sons, Ltd: Chichester, ***2001***, pp. 105–108.
38. McGeady, P. and Croteau, R., *Chem. Commun.*, ***1993***, 774–776.
39. Yavorsky, P.M. and Gorin, E., *J. Am. Chem. Soc.*, ***1962***, *84*, 1072–1073.
40. Werner, T.F. and Olsson, L.-I., *J. Label. Compd. Radiopharm.*, ***1987***, *24*, 29–39.
41. Tanga, M.J.; Bupp, J.E. and Bradford, W.W., *J. Label. Compd. Radiopharm.*, ***2001***, *44*, 405–411.
42. Leppälä, E. and Wähälä, K., *J. Label. Compd. Radiopharm.*, ***2004***, *47*, 25–30.
43. Wähälä, K.; Rasku, S.; Kiuru, P; Parikka, K; Leppälä, E; Jokela, T; Salakka, A. and Lewis, P., in *Synthesis and Applications of Isotopically Labeled Compounds*, Vol. 8, Eds. Dean, D.C.; Filer, C.N. and McCarthy, K.E.; John Wiley & Sons, Ltd: Chichester, ***2004***, pp. 43–46.
44. Wähälä, K. and Rasku, S., *Tetrahedron Lett.* ***1997***, *38*, 7287–7290.
45. Wähälä, K.; Rasku, S. and Parikka, K., *J. Chromatogr. B*, ***2002***, *777*, 111–122.
46. Rasku, S.; Wähälä, K.; Koskimies, J. and Hase, T., *Tetrahedron*, ***1999***, *55*, 3445–3454.
47. Garnett, J.L.; Long, M.A. and Lukey, C.A., *J. Label. Compd. Radiopharm.*, ***1985***, *22*, 641–647.
48. a. Garnett, J.L.; Long, M.A.; Vining, R.F.W. and Mole, T., *J. Am. Chem. Soc.*, ***1972***, *94*, 5913–5914; b. Garnett, J.L.; Long, M.A.; Vining, R.F.W. and Mole, T., *J. Am. Chem. Soc.*, ***1972***, *94*, 8632–8633; c. Garnett, J.L.; Long, M.A.; Vining, R.F.W. and Mole, T., *Chem. Commun.*, ***1972***, 1172–1173.
49. a. Ruehle, P.H.; Susàn, A.B. and Wiley, Jr., J.C., *J. Label. Compd. Radiopharm.*, ***1981***, *18*, 607–610; b. Duncan, W.P.; Ogilvie, E.J. II and Engel, J.F., *J. Label. Compd*, ***1975***, *11*, 461–463.
50. Long, M.A.; Garnett, J.L. and Vining, R.F.W., *J. Chem. Soc. Perkin Trans. II*, ***1975***, 1298–1303.
51. Garnett, J.L.; Long, M.A.; Vining, R.F.W. and Mole, T., *Tetrahedron Lett.*, ***1973***, *41*, 4075–4078.
52. Brooks, R.A.; Long, M.A. and Garnett, J.L., *J. Label. Compd. Radiopharm.*, ***1982***, *19*, 659–667.
53. Thomas, A.F., *Deuterium Labeling in Organic Chemistry;* Appleton: New York, ***1971***, pp. 169–229.
54. Elvidge, J.A.; Jones, J.R.; O' Brien, C.; Evans, E.A. and Sheppard, H.C., *Adv. Heterocycl. Chem.*, ***1974***, *16*, 1–31.
55. See for example for steroids a. Stoilov, I.L.; Thompson, J.E.; Cho, J.-H. and Djerassi, C., *J. Am. Chem. Soc.*, ***1986***, *108*, 8235–8241; b. Parish, E.J. and Schroepfer, Jr., G.J., *J. Label. Compd. Radiopharm.*, ***1980***, *18*, 1429–1436; and for terpenes a. Lischewski, M.; Adam, G.; Liebisch, H.-W. and Pleiβ, U., *J. Label. Compd. Radiopharm.*, ***1982***, *19*, 725–733; b. Pattenden, G. and Storer, R., *J. Label. Compd. Radiopharm.*, ***1976***, *12*, 551–556; c. Hedden, P.; MacMillan, J. and Phinney, B.O., *J. Chem. Soc. Perkin Trans. I*, ***1974***, 587–592; d. Bearder, J.R.; MacMillan, J. and Phinney, B.O., *Phytochemistry*, ***1973***, *12*, 2173–2179.
56. De Keczer, S.A.; Land, T.S.; Voronin, T. and Masjedizadeh, M.R., *J. Label. Compd. Radiopharm.*, ***2005***, *48*, 1013–1023.

57. Ciceri, D.; Pierik, A.J.; Hartrampf, G.; Bröker, G.; Speranza, G.; Buckel, W.; Cornforth, J. and Golding, B.T., *Helv. Chim. Acta*, ***2000***, *83*, 2550–2561.
58. Scheigetz, J.; Berthelette, C.; Li, C. and Zamboni, R.J., *J. Label. Compd. Radiopharm.*, ***2004***, *47*, 881–889.
59. Kolbe, A.; Schneider, B.; Voigt, B. and Adam, G., *J. Label. Compd. Radiopharm.*, ***1997***, *41*, 131–137.
60. Kurosaki, F.; Itoh, M. and Nishi, A., *Anal. Biochem.*, ***1993***, *210*, 418–419.
61. Holland, H.L.; Brown, F.M. and Tan, L., *J. Label. Compd. Radiopharm.*, ***1989***, *27*, 629–633.
62. Feldman, D. and Rabinovitz, M., *J. Org. Chem.*, ***1988***, *53*, 3779–3784.
63. Gutman, A.L. and Boltanski, A., *J. Chem. Soc. Perkin Trans. I*, ***1989***, 47–49.
64. Iwata-Reuyl, D.; Basak, A. and Townsend, C.A., *J. Am. Chem. Soc.*, ***1999***, *121*, 11356–11368.
65. Ponsart, S.; Coudane, J.; Morgat, J.-L. and Vert, M., *J. Label. Compd. Radiopharm.*, ***2000***, *43*, 271–282.
66. Ponsart, S.; Coudane, J.; Morgat, J.-L. and Vert, M., *J. Label. Compd. Radiopharm.*, ***2001***, *44*, 677–687.
67. Laube, T.; Dunitz, J.D. and Seebach, D., *Helv. Chem. Acta*, ***1985***, *68*, 1373–1393.
68. Stork, G. and Hudrlik, P., *J. Am. Chem. Soc.*, ***1968***, *90*, 4462–4464.
69. a. Coumbarides, G.S.; Eames, J. and Weerasooriya, N., *Tetrahedron Lett.*, ***2000***, *41*, 5753–5756; b. Coumbarides, G.S.; Eames, J. and Weerasooriya, N., *J. Label. Compd. Radiopharm.*, ***2003***, *46*, 515–530.
70. Archer, W.J.; Cook, R. and Taylor, R., *J. Chem. Soc. Perkin Trans. II*, ***1983***, 813–820.
71. Smith, M.B. and March, J., *March's Advanced Organic Chemistry*, 5th Edn; John Wiley & Sons, Inc.: New York, ***2001***.
72. a. Snieckus, V., *Chem Rev.*, ***1990***, *90*, 879–933; b. Kremer, T.; Junge, M. and Schleyer, P. von R., *Organometallics*, ***1996***, *15*, 3345–3359.
73. Ciszewska, G.; Pfefferkorn, H.; Tang, Y.S.; Jones, L.; Tarapata, R. and Sunay, U.B., *J. Label. Compd. Radiopharm.*, ***1997***, *39*, 651–668.
74. Biot, C.; Caron, S.; Maciejewski, L.A. and Brocard, J.S., *J. Label. Compd. Radiopharm.*, ***1998***, *41*, 911–918.
75. Henot, F.; Thellend, A.; Pillon, F.; Rousseau, B. and Gaudry, M., *J. Label. Compd. Radiopharm.*, ***1996***, *38*, 579–583.
76. a. Seltzman, H.H.; Odear, D.F.; Carroll, F.I. and Wyrick, C.D., *Chem. Commun.*, ***1992***, *24*, 1757–1758; b. Seltzman, H.H.; Odear, D.F.; Laudeman, C.P.; Carroll, F.I. and Wyrick, C.D., *J. Label. Compd. Radiopharm.*, ***1994***, *34*, 367–375; c. Seltzman, H.H.; Fleming, D.N.; Laudeman, C.P.; Carroll, F.I. and Wyrick, C.D., in *Synthesis and Applications of Isotopically Labeled Compounds*, ***1994***, Allen, J. and Voges, R., Eds.; John Wiley & Sons, Ltd: Chichester, ***1995***, pp. 61–66.
77. Buncel, E. and Menon, B.C., *Can. J. Chem.*, ***1976***, *54*, 3949–3954.
78. a. Seltzman, H.H.; Carroll, F.I.; Burgess, J.P.; Wyrick, C.D. and Burch, D.F., *Chem. Commun.*, ***1995***, *15*, 1549–1551; b. Seltzman, H.H.; Carroll, F.I.; Burgess, J.P.; Wyrick, C.D. and Burch, D. F., *J. Label. Compd. Radiopharm.*, ***2002***, *45*, 59–70.
79. a. Cintrat, J.-C. and Rousseau, B., *J. Label. Compd. Radiopharm.*, ***1998***, *41*, 1017; b. Cintrat, J.-C.; Pillon, F. and Rousseau, B., *Tetrahedron Lett.*, ***2001***, *42*, 5001–5003.
80. Evans, E.A., *Tritium and its Compounds*, 2nd Edn; John Wiley & Sons, Inc.: New York, ***1974***, pp. 293–309.
81. Buchman, O.; Pri-Bar, I.; Shimoni, M. and Smoskowitz, L., *J. Label. Compd*, ***1974***, *10*, 345–354.
82. Garnett, J.L. and Long, M.A., in *Isotopes in the Physical and Biomedical Sciences*, Vol. 1, Part A, Buncel, E. and Jones, J.R., Eds.; Elsevier: Amsterdam, ***1987***, pp. 86–121.
83. Pri-Bar, I. and Buchman, O., *Chem. Commun.*, ***1970***, 1631.
84. Evans, E.A.; Sheppard, H.C.; Turner, J.C. and Warrell, D.C., *J. Label. Compd*, ***1974***, *10*, 569–587.
85. Garnett, J.L., *Catal. Rev.*, ***1971***, *5*, 229.
86. Williams, P.G.; Lukey, C.A.; Long, M.A. and Garnett, J.L., *J. Label. Compd. Radiopharm.*, ***1991***, *29*, 175–192.

87. Shevchenko, V.P.; Nagaev, I.Yu. and Myasoedov, N.F., *Radiochemistry*, ***2005***, *47*, 403–409.
88. Conner Jr., W.C. and Falconer, J.L., *Chem. Rev.*, ***1995***, *95*, 759–788.
89. Maegawa, T.; Hirota, K.; Tatematsu, K.; Mori, Y. and Sajiki, H., *J. Org. Chem.*, ***2005***, *70*, 10581–10583.
90. Ito, N.; Watahiki, T.; Maesawa, T.; Maegawa, T. and Sajiki, H., *Adv. Synth. Catal.*, ***2006***, *348*, 1025–1028.
91. Hesk, D.; Koharski, D.; Saluja, S.; McNamara, P.; Magatti, C.; Cesarz, D.; Hendershot, S. and Jones, J.R., in *Synthesis and Applications of Isotopically Labeled Compounds 1997*, Heys, J.R. and Melillo, D.G., Eds.; John Wiley & Sons, Ltd: Chichester, ***1998***, pp. 439–443.
92. Hesk, D.; Calvert, M. and McNamara, P., in *Synthesis and Applications of Isotopically Labeled Compounds*, Vol. 8, Dean, D.C.; Filer, C.N. and McCarthy, K.E., Eds.; John Wiley & Sons, Ltd: Chichester, ***2004***, pp. 51–54.
93. Hesk, D.; Duelfer, T.; Hickey, S.; Hochman, D.; Koharski, D.; McNamara, P. and Saluja, S.; *J. Label. Compd. Radiopharm.*, ***1994***, *34*, 681–689.
94. Li, B.; Robinson, D.H. and Birt, D.F., *J. Pharm. Sci.*, ***1997***, *86*, 721–725.
95. Shevchanko, V.P.; Nagaev, I.Yu.; Myasoedov, N.F. and Susàn, A.B., in *Synthesis and Applications of Isotopically Labeled Compounds*, Vol. 7., Plei*β*, U. and Voges, R., Eds.; John Wiley & Sons, Ltd: Chichester, ***2001***, pp. 84–88.
96. Shevchenko, V.P.; Nagaev, I.Yu., Potapova, A.V. and Myasoedov, N.F., *Radiochemistry (Moscow)*, ***1998***, *40*, 74–78.
97. Shevchenko, V.P.; Nagaev, I.Yu.; Myasoedov, N.F. and Susàn, A.B., *J. Label. Compd. Radiopharm.*, ***2002***, *45*, 911–915.
98. Shevchenko, V.P.; Nagaev, I.Yu., Myasoedov, N.F.; Andres, H.; Moenius, T. and Susàn, A.B., *J. Label. Compd. Radiopharm.*, ***2004***, *47*, 407–414.
99. Shevchenko, V.P.; Nagaev, I.Yu. and Myasoedov, N.F., *Radiochemistry (Moscow)*, ***1999***, *41*, 531–532.
100. Shevchenko, V.P.; Nagaev, I.Yu. and Myasoedov, N.F., *Russ. Chem. Rev.*, ***1999***, *68*, 859–879.
101. Esaki, H.; Ito, N.; Sakai, S.; Maegawa, T.; Monguchi, Y. and Sajiki, H., *Tetrahedron*, ***2006***, *62*, 10954–10961.
102. a. Derdau, V.; Atzrodt, J. and Holla, W., *J. Label. Compd. Radiopharm.*, ***2007***, *50*, 295–299; b. Derdau, V. and Atzrodt, J., *Synlett.*, ***2006***, 1918–1922.
103. a. Sajiki, H.; Ito, N.; Esaki, H.; Maesawa, T.; Maegawa, T. and Hirota, K., *Tetrahedron Lett.*, ***2005***, *46*, 6995–6998; b. Sajiki, H.; Esaki, H.; Aoki, F.; Maegawa, T. and Hirota, K., *Synlett.*, ***2005***, 1385–1390; c. Maegawa, Tl; Akashi, A.; Esaki, H.; Aoki, F.; Sajiki, H. and Hirota, K., *Synlett.*, ***2005***, 845–847; d. Sajiki, H.; Aoki, F.; Esaki, H.; Maegawa, T. and Hirota, K., *Org. Lett.*, ***2004***, *6*, 1485–1487.
104. Sajiki, H.; Kurita, T.; Esaki, H.; Aoki, F.; Maegawa, T. and Hirota, K., *Org. Lett.*, ***2004***, *6*, 3521–3523.
105. Maegawa, T.; Hirota, K.; Tatematsu, K.; Mori, Y. and Sajiki, H., *J. Org. Chem.*, ***2005***, *70*, 10581–10583.
106. Lockley, W.J.S. and Kershaw, G., in *Synthesis and Applications of Isotopically Labeled Compounds 1991*, Buncel, E. and Kabalka, G.W., Eds.; Elsevier: Amsterdam, ***1992***, pp. 604–607.
107. Macdonald, C.G. and Shannon, J.S., *Tetrahedron Lett.*, ***1964***, 3351–3354.
108. Garman, R.N.; Getvoldsen, G.S.; Jones, J.R.; Lawrie, K.W.M.; Lu, S.Y.; Marsden, P.; Russell, J. C. and Seddon, K.R., in *Synthesis and Applications of Isotopically Labeled Compounds*, Vol. 7, U. Pleiss and Voges, R.; John Wiley & Sons, Ltd: Chichester, ***2001***, pp. 97–100.
109. a. Koch, H.J. and Stuart, R.S., *Carbohydrate Res.*, ***1978***, *67*, 341–348; b. Koch, H.J. and Stuart, R.S., *Carbohydrate Res.*, ***1978***, *64*, 127–134.
110. Cioffi, E.A.; Alston, K.E. and Patel, A.M., *Tetrahedron Lett.*, ***2002***, *43*, 8985–8987.
111. Cioffi, E.A.; Bell, R.H. and Le, B., *Tetrahedron: Asymmetry*, ***2005***, *16*, 471–475.
112. Yau, W.-M. and Gawrisch, K., *J. Label. Compd. Radiopharm.*, ***1999***, *42*, 709–713.
113. Anto, S.; Getvoldsen, G.S.; Harding, J.R.; Jones, J.R.; Lu, S.Y. and Russell, J.C., in *Synthesis and Applications of Isotopically Labeled Compounds*, Vol. 8, Dean, D.C.; Filer, C.N. and McCarthy, K.E., Eds.; John Wiley & Sons, Ltd: Chichester, ***2001***, pp. 93–96.

114. a. Brown, W.G.; Calf, G.E. and Garnett, J.L., *Chem. Commun.*, ***1967***, 1066–1067; b. Logan, S.R., *J. Mol. Catal.*, ***1994***, *89*, 365–370; c. Calf, G.E. and Garnett, J.L., *Chem. Commun.*, ***1967***, 306–307.
115. Brewer, J.R.; Jones, J.R.; Lawrie, K.W.M.; Saunders, D. and Simmonds, A., *Chem. Commun.*, ***1990***, 1566–1567.
116. Brewer, J.R.; Jones, J.R.; Lawrie, K.W.M.; Saunders, D. and Simmonds, A., *J. Label. Compd. Radiopharm.*, ***1994***, *34*, 391–400.
117. Tuck, K.L. and Hayball, P.J., *J. Label. Compd. Radiopharm.*, ***2001***, *44*, 1005–1011.
118. Brewer, J.R.; Jones, J.R.; Lawrie, K.W.M.; Saunders, D. and Simmonds, A., *J. Label. Compd. Radiopharm.*, ***1994***, *34*, 787–794.
119. Long, M.A.; Garnett, J.L.; Williams, P.G. and Mole, T., *J. Am. Chem. Soc.*, ***1981***, *103*, 1571–1572.
120. Li, Q.X. and Casida, J.E., *Bioorg. Med. Chem.*, ***1995***, *3*, 1675–1684.
121. Evans, E.A.; Sheppard, H.C.; Turner, J.C. and Warrell, D.C., *J. Label. Compd*, ***1974***, *10*, 569–587.
122. Evans, E.A., *Tritium and its Compounds*, 2nd Edn; John Wiley & Sons, Inc.: New York, ***1974***, pp. 310–317.
123. Brundish, D.E.; Combe, M.G. and Wade, R., *J. Label. Compd. Radiopharm.*, ***1983***, *20*, 869–886.
124. Long, M.A.; Morimoto, H. and Williams, P.G., *J. Label. Compd. Radiopharm.*, ***1995***, *36*, 1037–1049.
125. Azran, J.; Shimoni, M. and Buchman, O., *J. Catal.*, ***1994***, *148*, 648–653.
126. Lacey, E.; Dawson, M.; Long, M.A. and Than, C., *J. Label. Compd. Radiopharm.*, ***1989***, *27*, 1416–1427.
127. Proszenyák, A.; Agai, B.; Tárkányi, G.; Vida, L. and Faigl, F., *J. Label. Compd. Radiopharm.*, ***2005***, *48*, 421–427.
128. Landvatter, S.W.; Heys, J.R. and Senderoff, S.G., *J. Label. Compd. Radiopharm.*, ***1987***, *24*, 389–396.
129. Elbert, T.; Černý, B.; Wimmer, Z. and Sergent, L., *Coll. Czech. Chem. Commun.*, ***1993***, *58*, 1164–1168.
130. Al Rawi, J.M.A.; Bloxsidge, J.P.; Elvidge, J.A.; Jones, J.R.; Chambers, V.E.M. and Evans, E.A., *Steroids*, ***1976***, *28*, 359–375.
131. Funke, C.W.; Kaspersen, F.M.; Wallaart, J. and Wagenaars, S.N., *J. Label. Compd. Radiopharm.*, ***1983***, *20*, 843–853.
132. Garnett, J.L. and Long, M.A., in *Isotopes in the Physical and Biomedical Sciences*, Vol. 1, Part A, Buncel, E. and Jones, J.R., Eds.; Elsevier: Amsterdam, ***1987***, pp. 86–121.
133. Rubottom, G.M. and Evain, E.J., *Tetrahedron*, ***1990***, *46*, 5055–5064.
134. Alexakis, E.; Jones, J.R. and Lockley, W.J.S., *Tetrahedron Lett.*, ***2006***, *47*, 5025–5028.
135. Hickey, M.J.; Kingston, L.P.; Lockley, W.J.S.; Allen, P.; Mather, A. and Wilkinson, D.J., *J. Label. Compd. Radiopharm.*, ***2007***, *50*, 286–289.
136. Rylander, P. *Catalytic Hydrogenation in Organic Synthesis;* Academic Press: New York, ***1979***, pp. 31–63.
137. Bartók, M. *Stereochemistry of Heterogeneous Metal Catalysis;* John Wiley & Sons, Inc.: New York, ***1985***.
138. Yu, J.; Whitney, P.-S. and Spencer, J.B., *J. Mol. Catal. A: Chem.*, ***1999***, *146*, 199–210.
139. Guilford, G.L.; Evans, E.A.; Warrell, D.C.; Jones, J.R.; Elvidge, J.A.; Lenk, R.M. and Tang, Y.S., in *Synthesis and Applications of Isotopically Labeled Compounds*, Duncan, W.P. and Susàn, A.B., Eds.; Elsevier: Amsterdam, ***1983***, pp. 327–330.
140. Voges, R.; von Wartburg, B.R. and Loosli, H.R., in *Synthesis and Applications of Isotopically Labeled Compounds 1985*, Muccino, R.R., Ed.; Elsevier: Amsterdam, ***1986***, pp. 371–376.
141. Solladié-Cavallo, A.; Hoernel, F.; Schmitt, M. and Garin, F., *J. Mol. Catal. A*, ***2003***, *195*, 181–188.
142. Lockley, W.J.S., *J. Label. Compd. Radiopharm.*, ***2007***, *50*, 779–788.
143. Garnett, J.L. and Long, M.A., in *Isotopes in the Physical and Biomedical Sciences*, Vol. 1, Part A, Buncel, E. and Jones, J.R., Eds.; Elsevier: Amsterdam, ***1987***, pp. 88–109 and references cited therein.

144. Jones, J.R., in *Isotopes: Essential Chemistry and Applications*, Elvidge, J.A. and Jones, J.R., Eds.; The Chemical Society: London (Special publications No. 35, ISSN 0577-618X), ***1981***, pp. 370–376 and references cited therein.
145. Shilov, A.E. and Shul'pin, G.B., *Chem. Rev.*, ***1997***, *97*, 2879–2932 and references cited therein.
146. Blake, M.R.; Garnett, J.L.; Gregor, I.K.; Hannan, W.; Hoa, K. and Long, M.S., *Chem. Commun.*, ***1975***, 930–931.
147. Lockley, W.J.S., *Tetrahedron Lett.*, ***1982***, *23*, 3819–3822.
148. Lockley, W.J.S., *J. Label. Compd. Radiopharm.*, ***1984***, *21*, 45–57.
149. Bruce, M.I., *Angew. Chem. Int. Edn Engl.*, ***1977***, *16*, 73–86.
150. Kiseny, J.M.; Sunley, G.J.; Cabeza, J.A.; Smith, A.J.; Adams, H.; Salt, N.J. and Maitlis, P.M., *J. Chem. Soc. Dalton Trans.*, ***1987***, 2459–2460.
151. Lockley, W.J.S., *J. Label. Compd. Radiopharm.*, ***2007***, *50*, 779–788.
152. Lockley, W.J.S., *J. Label. Compd. Radiopharm.*, ***1985***, *22*, 623–630.
153. Hesk, D.; Jones, J.R.; Nolan, K.B. and Lockley, W.J.S., in *Synthesis and Applications of Isotopically Labeled Compounds 1991*, Buncel, E. and Kabalka, G.W., Eds.; Elsevier, Amsterdam, ***1992***, pp. 600–603.
154. Oohashi, K. and Udo, M., *Radioisotopes (Tokyo)*, ***1992***, *41*, 357–361.
155. Hesk, D.; Jones, J.R. and Lockley, W.J.S., *J. Label. Compd. Radiopharm.*, ***1990***, *28*, 1427–1436.
156. Oohashi, K.; Kitayama, N. and Shinoda, Y., *Radioisotopes (Tokyo)*, ***1992***, *41*, 547–552.
157. Hesk, D.; Jones, J.R. and Lockley, W.J.S., *J. Pharm. Sci.*, ***1991***, *80*, 887–890.
158. Wilkinson, D.J. and Lockley, W.J.S., *J. Label. Compd. Radiopharm.*, ***1985***, *22*, 883–892.
159. Lockley, W.J.S. and Wilkinson, D.J., *J. Label. Compd. Radiopharm.*, ***1984***, *21*, 363–373.
160. a. Lambert, D.M. and Gallez, B., *J. Label. Compd. Radiopharm.*, ***1996***, *38*, 897–905; b. Oohashi, K.; Nakamura, T. and Aoyama, Y., *J. Radioanal. Nucl. Chem. Lett.*, ***1990***, *145*, 29–37; c. Oohashi, K. and Seki, T., *J. Radioanal. Nucl. Chem. Lett.*, ***1994***, *187*, 303–311; d. Nakagawa, K.; Hasegawa, H.; Seki, H. and Oohashi, K., *Radioisotopes (Tokyo)*, ***1997***, *46*, 513–519.
161. Koeckritz, A.; Sebek, M.; Dittmar, A.; Radnik, J.; Brückner, A.; Bentrup, U.; Pohl, M.-M.; Hugl, H. and Mägerlein, W., *J. Mol. Catal. A*, ***2006***, *246*, 85–99.
162. Murahashi, S.-I. and Komiya, N., *Catal. Today*, ***1998***, *41*, 339–350.
163. Diamond, S.E.; Tom, G.M. and Taube, H., *J. Am. Chem. Soc.*, ***1975***, *97*, 2661–2664.
164. Bäckvall, J.E., *J. Organometal. Chem.*, ***2002***, *652*, 105–111.
165. Regen, S.L., *J. Org. Chem.*, ***1974***, *39*, 260–261.
166. Al Rawi, J.M.A.; Elvidge, J.A.; Jones, J.R.; Mane, R.B. and Saieed, M., *J. Chem. Res. (S)*, ***1980***, 298.
167. Wan, Y.; Jakway, J.P.; Qiu, H.; Shah, H.; Garlisi, C.G.; Tian, F.; Ting, P.; Hesk, D.; Egan, R.W.; Billah, M.M. and Umland, S.P., *Eur. J. Pharmacol.*, ***2002***, *456*, 1–10.
168. Hesk, D.; Koharski, D.; Saluja, S.; McNamara, P.; Magatti, C.; Cesarz, D.; Hendershot, S. and Jones, J.R., in *Synthesis and Applications of Isotopically Labeled Compounds 1997*, Heys, J.R. and Melillo, D.G., Eds.; John Wiley & Sons, Ltd: Chichester, ***1998***, pp. 439–443.
169. Hesk, D; Cesarz, D.; Magatti, C.; Voronin, K.; Lavey, C.; McNamara, P.; Koharski, D.; Saluja, S.; Hendershot, S.; Pham, H. and Truong, V., *J. Label. Compd. Radiopharm.*, ***2005***, *48*, 11–23.
170. Alexakis, E.; Hickey, M.J.; Jones, J.R.; Kingston, L.P.; Lockley, W.J.S.; Mather, A.N.; Smith, T. and Wilkinson, D.J., *Tetrahedron Lett.*, ***2005***, *46*, 4291–4293.
171. Kingston, L.P.; Lockley, W.J.S.; Mather, A.N.; Spink, E.; Thompson, S.P. and Wilkinson, D.J., *Tetrahedron Lett.*, ***2000***, *41*, 2705–2708.
172. McAuley, B.; Hickey, M.J.; Kingston, L.P.; Jones, J.R.; Lockley, W.J.S.; Mather, A.N.; Spink, E.; Thompson, S.P. and Wilkinson, D.J., *J. Label. Compd. Radiopharm.*, ***2003***, *46*, 1191–1204.
173. Ray, T.; Wu, A. and Allentoff, A.J., *J. Label. Compd. Radiopharm.*, ***2007***, *50*, 433–434.
174. Krüger, J.; Manmontri, B. and Fels, G., *Eur. J. Org. Chem.*, ***2005***, 1402–1408.
175. Garman, R.N.; Hickey, M.J.; Kingston, L.P.; McAuley, B.; Jones, J.R.; Lockley, W.J.S.; Mather, A.N. and Wilkinson, D.J., *J. Label. Compd. Radiopharm.*, ***2005***, *48*, 75–84.
176. Klei, S.R.; Golden, J.T.; Tilley, T.D. and Bergman, R.G., *J. Am. Chem. Soc.*, ***2002***, *124*, 2092–2093.
177. Klei, S.R., Golden, J.T.; Burger, P. and Bergman, R.G., *J. Mol. Catal. A*, ***2002***, *189*, 79–94.

178. Klei, S.R.; Tilley, T.D. and Bergman, R.G., *Organometal.*, ***2002***, *21*, 4905–4911.
179. Heys, J.R., *Chem. Commun.*, ***1992***, 680–681.
180. Shapley, J.R.; Schrock, R.R. and Osborn, J.A., *J. Am. Chem. Soc.*, ***1969***, *91*, 2816–2817.
181. Crabtree, R.H.; Holt, E.M.; Lavin, M. and Morehouse, S.M., *Inorg. Chem.*, ***1985***, *24*, 1986–1992.
182. Crabtree, R.H. and Lavin, M., *J. Chem. Soc. Chem. Commun.*, ***1985***, 794–795.
183. Lavin, M.; Holt, E.M. and Crabtree, R.H., *Organometal.*, ***1989***, *8*, 99–104.
184. Shu, A.Y.L. and Heys, J.R., in *Synthesis and Applications of Isotopically Labeled Compounds 1997*, Heys, J.R. and Melillo, D.G., Eds.; John Wiley & Sons, Ltd: Chichester, ***1998***, pp. 223–226.
185. Heys, J.R., *J. Label. Compd. Radiopharm.*, ***2007***, *50*, 770–778.
186. Heys, J.R.; Shu, A.Y.L.; Senderoff, S.G. and Phillips, N.M., *J. Label. Compd. Radiopharm.*, ***1993***, *33*, 431–438.
187. Heys, J.R.; Shu, A.Y.L. and Nice, L.E., in *Synthesis and Applications of Isotopically Labeled Compounds 1994*, Allen, J. and Voges, R., Eds.; John Wiley & Sons, Ltd: Chichester, ***1995***, pp. 175–180.
188. Salter, R.; Chappelle, M.; Morgan, A.; Moenius, Th.; Ackermann, P.; Studer, M. and Spindler, F., in *Synthesis and Applications of Isotopically Labeled Compounds*, Vol. 7, Pleiss, U. and Voges, R., Eds.; John Wiley & Sons, Ltd: Chichester, ***2001***, pp. 63–67.
189. Ellames, G.J.; Gibson, J.S.; Herbert, J.M. and McNeill, A.H., *Tetrahedron*, ***2001***, *57*, 9487–9497.
190. Salter, R. and Bosser, I., *J. Label. Compd. Radiopharm.*, ***2003***, *46*, 489–498.
191. Shu, A.Y.L. and Heys, J.R., *J. Label. Compd. Radiopharm.*, ***1994***, *34*, 587–596.
192. Hesk, D.; Das, P.R. and Evans, B., *J. Label. Compd. Radiopharm.*, ***1995***, *36*, 497–502.
193. Hesk, D.; Bignan, G.; Lee, J.; Yang, J.; Voronin, K.; Magatti, C.; McNamara, P.; Koharski, D.; Hendershot, S.; Saluja, S. and Wang, S., *J. Label. Compd. Radiopharm.*, ***2002***, *45*, 145–155.
194. Haines, L.M. and Singleton, E., *J. Chem. Soc. Dalton Trans.*, ***1972***, 1891–1896.
195. Shu, A.Y.L.; Chen, W. and Heys, J.R., *J. Organometal. Chem.*, ***1996***, *524*, 87–93.
196. Shu, A.Y.L.; Saunders, D.; Levinson, S.H.; Landvatter, S.W.; Mahoney, A.; Senderoff, S.G.; Mack, J.F. and Heys, J.R., *J. Label. Compd. Radiopharm.*, ***1999***, *42*, 797–809.
197. Chen, W.; Garnes, K.T.; Levinson, S.H.; Saunders, D.; Senderoff, S.G.; Shu, A.Y.L.; Villani, A.J. and Heys, J.R., *J. Label. Compd. Radiopharm.*, ***1997***, *39*, 291–298.
198. Heys, J.R. and Elmore, C.S., *J. Label. Compd. Radiopharm.*, ***2009***, in press.
199. Crabtree, R.H.; Demou, P.D.; Eden, D.; Mihelcic, J.M.; Parnell, C.A.; Quirk, J.M. and Morris, G. E., *J. Am. Chem. Soc.*, ***1982***, *104*, 6994–7001.
200. Shu, A.Y.L. and Heys, J.R., *Tetrahedron Lett.*, ***2000***, *41*, 9015–9019; Shu, A.Y.L. and Heys, J.R., in *Synthesis and Applications of Isotopically Labeled Compounds*, Vol. 7, Pleiss, U. and Voges, R., Eds.; John Wiley & Sons, Ltd: Chichester, ***2001***, pp. 68–71.
201. Johansen, S.L.; Wagner, S.H.; Jessen, C.U.; Valsborg, J.S. and Martiny, L., *J. Label. Compd. Radiopharm.*, ***2007***, *50*, 466–467.
202. Huguenin, P., New experience in the Ir-catalyzed H/T-exchange reaction using Crabtree's catalyst, Presentation at the RC Tritec Workshop, Teufen, Switzerland, June 18–19, ***2007***.
203. Powell, M.E.; Elmore, C.S.; Dorff, P.N. and Heys, J.R., *J. Label. Compd. Radiopharm.*, ***2007***, *50*, 523–525.
204. a. Brown, J.A.; Irvine, S.; Kennedy, A.R.; Kerr, W.J.; Andersson, S. and Nilsson, G.N., *Chem. Commun.*, ***2008***, 1115–1117; b. Brown, J.A.; Irvine, S.; Kennedy, A.R.; Kerr, W.J.; Andersson, S. and Nilsson, G.N., *J. Label. Compd. Radiopharm.*, ***2008***, *51*, 257.
205. Prechtl, M.H.G.; Hölscher, M.; Ben-David, Y.; Theyssen, N.; Loschen, R.; Milstein, D. and Leitner, W., *Angew. Chem. Int. Edn Engl.*, ***2007***, *46*, 2269–2272.
206. Träff, A.; Nilsson, G.N.; Szabó, K. and Eriksson, L., *J. Organometal. Chem.*, ***2007***, *692*, 5529–5531.
207. Herbert, J.M.; Kohler, D. and McNeill, A.H., *J. Label. Compd. Radiopharm.*, ***2005***, *48*, 285–294.
208. Irvine, S. and Kerr, W.J., personal communication to JRH, ***2008***.
209. Ellames, G.J.; Gibson, J.S.; Herbert, J.M.; Kerr, W.J. and McNeill, A.H., *Tetrahedron Lett.*, ***2001***, *42*, 6413–6416.

210. Cross, P.W.C.; Ellames, G.J.; Gibson, J.S.; Herbert, J.M.; Kerr, W.J.; McNeill, A.H. and Mathers, T.W., *Tetrahedron*, ***2003***, *59*, 3349–3358.
211. Herbert, J.M.; Kohler, A.D. and McNeill, A.H., *J. Label. Compd. Radiopharm.*, ***2005***, *48*, 285–294.
212. Ray, T.; Wu, A. and Allentoff, A.J., *J. Label. Compd. Radiopharm.*, ***2007***, *50*, 459–462.
213. Gibson, J.S. and Nerbert, J.M., *J. Label. Compd. Radiopharm.*, ***2003***, *46*, 531–537.
214. Ellames, G.J.; Gibson, J.S.; Herbert, J.M.; Kerr, W.J. and McNeill, A.H., *J. Label. Compd. Radiopharm.*, ***2004***, *47*, 1–10.
215. Chappelle, M.R.; Morgan, A.D.; Tomkinson, N.C.O. and Willson, T.M., in *Synthesis and Applications of Isotopically Labeled Compounds*, Vol. 7, Pleiss, U. and Voges, R., Eds.; John Wiley & Sons, Ltd: Chichester, ***2001***, pp. 80–83.
216. Chappelle, M.R. and Morgan, A.D., in *Synthesis and Applications of Isotopically Labeled Compounds*, Vol. 8, Dean, D.C.; Filer, C.N. and McCarthy, K.E., Eds.; John Wiley & Sons, Ltd: Chichester, ***2004***, pp. 55–58.
217. Bushby, N. and Killick, D.A., *J. Label. Compd. Radiopharm.*, ***2007***, *50*, 519–520.
218. Coombs, M.E.; Dawson, G.E.; Fedorchuk, M.; Kingston, L.P.; Lockley, W.J.S.; Mather, A.N.; McLachlan, T.R.B.; Morlin, A.J.G.; Spink, E. and Wilkinson, D.J., *J. Label. Compd. Radiopharm.*, ***2000***, *43*, 533–543.
219. Hickey, M.J.; Jones, J.R.; Kingston, L.P.; Lockley, W.J.S.; Mather, A.N.; McAuley, B.M. and Wilkinson, D.J., *Tetrahedron Lett.*, ***2003***, *44*, 3959–3961.
220. Garman, R.N.; Hickey, M.J.; Kingston, L.P.; McAuley, B.; Jones, J.R.; Lockley, W.J.S.; Mather, A.N. and Wilkinson, D.J., *J. Label. Compd. Radiopharm.*, ***2005***, *48*, 75–84.
221. Kingston, L.P.; Lockley, W.J.S.; Mather, A.N.; Spink, E.; Thompson, S.P. and Wilkinson, D.J., *Tetrahedron Lett.*, ***2000***, *41*, 2705–2708.
222. Klei, S.R.; Golden, G.T. and Bergman, R.G., *J. Mol. Catal. A*, ***2002***, *189*, 79–94.
223. Salter, R.; Moenius, T.; Ackermann, P.; Studer, M.; Morgan, A. and Chappelle, M., *J. Label. Compd. Radiopharm.*, ***2001***, *44*, 956–957.
224. Skaddan, M.B.; Yung, C.M. and Bergman, R.G., *Org. Lett.*, ***2004***, *6*, 11–13.
225. Buchanan, J.M.; Stryker, J.M. and Bergman, R.G., *J. Am. Chem. Soc.*, ***1986***, *108*, 1537–1550.
226. Skaddan, M.B. and Bergman, R.G., *J. Label. Compd. Radiopharm.*, ***2006***, *49*, 623–634.
227. Yung, C.M.; Skaddan, M.B. and Bergman, R.G., *J. Am. Chem. Soc.*, ***2004***, *126*, 13033–13043.
228. Zolotarev, Yu.A.; Kozik, V.S.; Doroklova, E.M.; Ziatsev, D.A.; Rosenberg, G.S. and Myasoedov, N.F., in *Synthesis and Applications of Isotopically Labeled Compounds 1991*, Buncel, E. and Kabalka, G.W., Eds.; Elsevier: Amsterdam, ***1992***, pp. 687–692.
229. Zolotarev, Yu.A.; Kozik, V.S.; Zaitsev, D.A.; Dorodhova, E.M. and Myasoedov, N.F., *Dokl. Akad. Nauk. SSSR*, ***1989***, *308*, 1145–1151.
230. Conner, W.C. and Falconer, J.L., *Chem. Rev.*, ***1995***, *95*, 759–788.
231. Zolotarev, Yu.A.; Borisov, Y.A. and Myasoedov, N.F., *J. Phys. Chem. A*, ***1999***, *103*, 4861–4864.
232. Zolotarev, Yu.A.; Dadayan, A.K. and Borisov, Yu.A., *Russ. J. Bioorg. Chem.*, ***2005***, *31*, 1–17.
233. Zolotarev, Yu.A.; Kozic, V.S.; Zaitsev, D.A.; Dorokhova, E.M. and Myasoedov, N.F., *J. Radioanal. Nucl. Chem.*, ***1992***, *162*, 3–14.
234. Myadoedov, N.F., *J. Label. Compd. Radiopharm.*, ***2007***, *50*, 831–847.
235. Shevchenko, V.P.; Nagaev, I.Yu. and Myasoedov, N.F., *Russ. Chem. Rev.*, ***2003***, *72*, 423–446.
236. Shevchenko, V.P.; Nagaev, I.Yu.; Myadoedov, N.F.; Andres, H.; Moenius, T. and Susàn, A., *J. Label. Compd. Radiopharm.*, ***2004***, *47*, 407–414.
237. Shevchenko, V.P.; Nagaev, I.Yu.; Shevchenko, K.V. and Myasoedov, N.F., *Radiochemistry*, ***2002***, *44*, 593–597.
238. Shevchenko, V.P.; Nagaev, I.Yu. and Myasoedov, N.F., *Radiochemistry*, ***2003***, *45*, 81–86.
239. Shevchenko, V.P.; Nagaev, I.Yu.; Myasoedov, N.F.; Popova, N.N. and Pirogova, G.N., *Radiochemistry*, ***2002***, *44*, 588–592.
240. Shevchenko, V.P.; Faradzheva, S.V.; Nagaev, I.Yu. and Myasoedov, N.F., *Radiochemistry*, ***1998***, *40*, 88–92.
241. Nagaev, I.Yu.; Shevchenko, V.P. and Myasoedov, N.F., *Radiochemistry*, ***1999***, *41*, 305–317.
242. Shevchenko, V.P.; Nagaev, I.Yu.; Potapova, A.V. and Myasoedov, N.F., *Radiochemistry*, ***1998***, *40*, 70–74.

243. Myadoedov, N.F. and Sidorov, G.V., *J. Label. Compd. Radiopharm.*, ***1998***, *41*, 993–1003.
244. Sidorov, G.V. and Myasoedov, N.F., *J. Label. Compd. Radiopharm.*, ***1994***, *34*, 353–358.
245. Shevchenko, V.P.; Nagaev, I.Yu. and Myasoedov, N.F., *Radiochemistry*, ***1998***, *40*, 83–87.
246. Shevchenko, V.P.; Nagaev, I.Yu. and Myasoedov, N.F. *Radiochemistry*, ***2002***, *44*, 75–80.
247. Shevchenko, V.P.; Nagaev, I.lYu.; Myasoedov, N.F. and Susàn, A.B., in *Synthesis and Applications of Isotopically Labeled Compounds*, Vol. 7, Pleiss, U. and Voges, R., Eds.; John Wiley & Sons, Ltd: Chichester, ***2001***, pp. 84–88.
248. Sidorov, G.V. and Myasoedov, N.F., *Radiochemistry*, ***1999***, *41*, 574–577.
249. Shevchenko, V.P.; Nagaev, I.Yu. and Myasoedov, N.F., *Radiochemistry*, ***1999***, *41*, 80–85.
250. Akulov, G.P.; Kaminskii, Tu.L.; Korsakova, N.A. and Kudelin, B.K., *J. Label. Compd. Radiopharm.*, ***1992***, *31*, 227–230.
251. Dos Santos, I.; Morgat, J.-L. and Vert, M., *J. Label. Compd. Radiopharm.*, ***1998***, *41*, 1005–1015.
252. Shevchenko, V.P.; Nagaev, I.Yu. and Myasoedov, N.F., *Radiochemistry*, ***2001***, *43*, 417–423.
253. Shevchenko, V.P.; Nagaev, I.Yu. and Myasoedov, N.F., *Radiochemistry*, ***1998***, *40*, 79–82.
254. Dos Santos, I.; Morgat, J.-L. and Vert, M., *J. Label. Compd. Radiopharm.*, ***1999***, *42*, 1093–1101.
255. Sidorov, G.V.; Zverkov, Tu.B.; Shram, S.I.; Lazurkina, T.Yu. and Myasoedov, N.F., *J. Label. Compd. Radiopharm.*, ***2003***, *46*, 465–473.
256. Shevchenko, V.P.; Nagaev, I.Yu.; Myasoedov, N.F. and Bovin, N.V., *Radiochemistry*, ***2003***, *45*, 172–175.
257. Zolotarev, Yu.A.; Laskatelev, E.V.; Rozenberg, S.G.; Borisov, Yu.A. and Myasoedov, N.F., *Russ. Chem. Bull.*, ***1997***, *46*, 1536–1542.
258. Zolotarev, Yu.A.; Kozic, V.S.; Zaitsev, D.A.; Dorokhova, E.M. and Myasoedov, N.F., *J. Radioanal. Nucl. Chem.*, ***1992***, *162*, 3–14.
259. Zolotarev, Yu.A.; Dadayan, A.K.; Dolotov, O.V.; Kozik, V.S.; Kost, N.V.; Sokolov, O.Yu.; Dorokhova, E.M.; Meshavkin, V.K.; Inozemtseva, L.S.; Gagaeva, M.V.; Andreeva, L.A.; Alfeeva, L.Yu.; Pavlov, T.S.; Badmaeva, K.E.; Badmaeva, S.E.; Bakaeva, Z.V.; Kopylova, G.N.; Samonina, G.E.; Vaskovsky, B.V.; Grivennikov, I.A.; Zozulya, A.A. and Myasoedov, N.F., *Russ. J. Bioorg. Chem.*, ***2006***, *32*, 166–173.
260. Zolotarev, Yu.A.; Dadayan, A.K.; Bocharov, E.V.; Borisov, Yu.A.; Vaskovsky, B.V.; Dorokhova, E.M. and Myasoedov, N.F., *Amino Acids*, ***2003***, *24*, 325–333.
261. Zolotarev, Yu.A.; Dadayan, A.K.; Borisov, Yu.A.; Dorokhova, E.M.; Kozik, V.S.; Vtyurin, N.N.; Bocharov, E.V.; Ziganshin, R.N.; Lunina, N.A.; Kostrov, S.V.; Ovchinnikova, T.V. and Myasoedov, N.F., *Bioorg. Chem.*, ***2003***, *31*, 453–463.
262. Myasoedov, N.F.; Sidorov, G.V.; Kramerov, V.N. and Mishin, V.I., *J. Label. Compd. Radiopharm.*, ***1999***, *42*, 859–866.
263. Sidorov, G.V. and Myasoedov, N.F., *Radiochemistry*, ***2006***, *48*, 517–519.
264. Garnett, J.L. and Long, M.A., in *Isotopes in the Physical and Biomedical Sciences*, Vol. 1, Part A, Buncel, E. and Jones, J.R., Eds.; Elsevier: Amsterdam, ***1987***, pp. 115–118.
265. Long, M.A. and Than, C., *J. Label. Compd. Radiopharm.*, ***1994***, *34*, 549–556.
266. Yusupov, M.M. and Spirin, A.S., *Methods Enzymol.*, ***1988***, *164*, 426–439.
267. Badun, G.A.; Lukashina, E.V. and Fedoseev, V.M., *Radiochemistry*, ***2001***, *43*, 306–310.
268. Sidorov, G.V.; Badun, G.A.; Baitova, E.A.; Baitov, A.A.; Platoshina, A.M.; Myasoedov, N.F. and Fedoseev, V.M., *Radiochemistry*, ***2005***, *47*, 311–315.
269. Baratova, L.A.; Efimov, A.V.; Dobrov, E.N.; Federova, N.V.; Hunt, R.; Badun, G.A.; Ksenofontov, A.L.; Torrance, L. and Järvekülg, J., *J. Virol.*, ***2001***, *75*, 9696–9702.
270. Volynskaya, A.V.; Kasumov, E.A. and Gol'danskii, V.I., *Int. J. Biol. Macromol.*, ***2006***, *39*, 256–264.
271. Averbakh, A.Z.; Pekel, N.D.; Seredenko, V.I.; Kulikov, A.V.; Gvozdev, R.I. and Rudakova, I.D., *Biochem. J.*, ***1995***, *310*, 601–604.
272. Agalarov, S.Ch.; Kalinichenko, A.A.; Komer, A.A. and Spirin, A.S., *FEBS Lett.*, ***2006***, *580*, 6797–6799.
273. Agafonov, D.E.; Vyacheslav, A.K.; Nazimov, I.V. and Spiring, A.S., *Proc. Natl Acad. Sci. USA*, ***1999***, *96*, 12345–12349.
274. Agafonov, D.E.; Kolb, V.A. and Spirin, A.S., *Proc. Natl Acad. Sci. USA*, ***1997***, *94*, 12892–12897.

275. Wilzbach, K.E., *J. Am. Chem. Soc.*, ***1957***, *79*, 1013–1014.
276. a. Saito, I.; Muramatsu, S.; Sugiyama, H.; Yamamoto, A. and Matsuura, T., *Tetrahedron Lett.*, ***1985***, *26*, 5891–5894; b. Saito, I.; Muramatsu, S.; Sugiyama, H.; Yamamoto, A. and Matsuura, T., *J. Am. Chem. Soc.*, ***1984***, *106*, 4286–4287; c. Defoin, A.; Defoin-Straatmann, R. and Kuhn, H.J., *J. Label. Compd. Radiopharm.*, ***1982***, *19*, 891–898; d. Defoin, A.; Defoin-Straatmann, R. and Kuhn, H.J., *Tetrahedron*, ***1984***, *40*, 2651–2662.
277. Peng, C.T., in *Isotopes in the Physical and Biomedical Sciences*, Vol. 1, Part A, Buncel, E. and Jones, J.R., Eds.; Elsevier: Amsterdam, ***1987***, pp. 6–51.

# 4

# Preparation of Tritium-Labeled Compounds by Chemical Synthesis

In contrast to isotope exchange methods, which utilize the target compound or a late intermediate as substrate for labeling, chemical synthesis involves transformation of some precursor chemical with concurrent introduction of tritium. Four basic types of transformations are distinguished for discussion each in turn. They are *catalytic tritiations* (Section 4.1), which usually involve the reduction of a precursor compound using a catalyst and tritium gas; *catalytic tritiolyses* (Section 4.2), involving the replacement of functional groups by tritium, again using a catalyst and tritium gas; *tritide reductions* (Section 4.3), utilizing tritiated versions of hydridic reducing reagents; and the use of *small tritiated building blocks* (Section 4.4) in synthetic operations to construct target molecules.

A phenomenon strongly relevant to the catalytic methods discussed in Sections 4.1 and 4.2, and of some relevance also for tritide chemistry (Section 4.3) is *solvent exchange*. Most catalytic tritiations and tritiolyses are conducted in solution, and a wide variety of solvents have been used, ranging from nonpolar ones to highly polar, hydroxylic media. The properties of the solvent, however, influence not only the course and rate of a heterogeneously catalyzed reaction, but also the extent to which solvent hydrogen may exchange with tritium gas as a side reaction and thereby dilute the isotope introduced into the substrate: solvent exchange. It is commonly recognized that the use of protic solvents for tritiations and tritiolyses increases the risk of obtaining products of less than maximal specific activity, although it is equally clear that such reactions often succeed without dilution of label. The crucial factor in each experiment is the *relationship* between the rate of the tritium-incorporation reaction and the rate(s) of any tritium-diluting reaction(s). It is documented that the rate of metal-catalyzed exchange between elemental hydrogen and water is significant, especially at low pH[1], but little quantitative data are available to guide practice. Recently a turnover number of 17.3 $h^{-1}$ was reported for the exchange between $H_2$ and $^2H_2O$ catalyzed by 10% Pd/C at room temperature and a pressure of 1 atm[2]; this is rapid enough to provide the basis for an economical method of catalytic deuterium exchange

*Preparation of Compounds Labeled with Tritium and Carbon-14* Rolf Voges, J. Richard Heys and Thomas Moenius

labeling in which $^2H_2$ is generated *in situ* from $^2H_2O$[3]. Earlier, studies of $^2HHO$-$H_2$ exchange catalyzed by supported[4] and unsupported[5] metals showed that their catalytic potency increased in the order Pd < (Ir, Rh, Ru, Os) < Pt; the potencies of the metals in parentheses varied relative to each other depending on the support. Although palladium was less efficient than platinum by two to three orders of magnitude, even palladium's rate, 0.02–0.03 mmol $^2HH$/mg of metal per hour at 160 °C, was high enough to be significant when extrapolated to the context of some catalytic labeling reactions using tritium gas. Studies on the exchange from tritium gas into the solvents dioxane, THF, ethyl acetate, DMF, DMA, DMSO and acetic anhydride by the action of Pd/$BaSO_4$ and freshly prepared unsupported Pd(0) and Pt(0) showed that in most cases the tritium entering the liquid phase was in the form of $^3HHO$, generated by reduction of oxygen chemisorbed on the catalyst or by catalytic exchange of $^3H_2$ with trace moisture contained in the solvents. The latter could be largely overcome by drying with molecular sieves. There was also slow incorporation of tritium into DMF and $Ac_2O$ (with Pd) and into THF and EtOAc (with Pt)[6] but these were manifested mostly in consumption of tritium gas rather than dilution of it with hydrogen. For example, an investigation has shown significant isotope exchange with methanol solvent in the Pd/C-catalyzed deuteration of a 1,1-diphenyl-2-alkyl olefin[7]. While 81% of the theoretical level (2 mol $^2H$/mol) was present in the reduced product when methanol was used as the solvent, the level was 95% with dioxane and 100% with $CH_3O^2H$.

In the end, however, it is the *relative* rates of tritium gas exchange with water or (protic) solvent and the desired reaction that are important. Fast tritiations are less likely to be adversely affected by water, protic solvents or even low pH. But not all slow tritiations are necessarily compromised: To undergo exchange with the tritium atoms chemisorbed on the catalyst, water or protic solvent must compete with the substrate for physical adsorption to catalyst binding sites, so that even slowly reacting substrates may give high specific activity products in the presence of protic solvents or water, if their affinity for catalyst surface is high enough to saturate the binding sites and exclude the protic species.

## 4.1 Catalytic Tritiations[8,9]

Of the many methods available for the selective introduction of tritium into compounds for investigations in life sciences, catalytic tritiation of multiple bonds is one of the most convenient and frequently used. The reaction is easily effected in glassware by stirring the substrate together with the catalyst in a suitable solvent in the presence of tritium gas, normally at subatmospheric pressures but sometimes up to slightly more than 1 atm. Significantly higher tritium pressures can be achieved on commercially available tritium manifolds (Chapter 1, Section 1.5), but specially designed reaction vessels are required. After the reaction is complete, the solvent is lyophilized off into a separate glass ampoule, and labile tritium is removed by two or more cycles of methanol addition, redissolution and relyophilization.

Carbon–carbon multiple bonds are most commonly reduced by means of catalytic tritiation methods. Polarized multiple bonds such as those of carbonyl, carboxyl and nitrile groups are less often reduced by catalytic tritiation, since the use of tritide reagents is often more convenient and can provide greater selectivity (see Section 4.3).

### 4.1.1 Tritiation of Carbon–Carbon Multiple Bonds

The activity of heterogeneous metal catalysts in catalytic hydrogenation (and equally for tritiation) is highly dependent upon substrate structure, but generally decreases in the order Pd > Rh > Pt > Ni > Ir, Ru[8,9]. The activity of metals dispersed on supports normally diminishes in the order alumina > carbon > barium sulfate > calcium and strontium carbonate[9]. The activity of a given heterogeneous catalyst is usually lowest in aprotic, apolar solvents (THF, dioxane), higher in polar aprotic solvents (dimethylformamide, dimethylacetamide, ethyl acetate), and highest in protic polar solvents (ethanol, methanol, water, acetic acid).

It is frequently observed that catalytic tritiation (or deuteration) of unsaturated compounds results in incorporation of the isotope into sites besides those of the original carbon–carbon multiple bonds, or that the incorporation of isotope into one or more of the originally unsaturated carbon positions is nonstoichiometric. We use the term *isotopic scrambling* to refer to such labeling, when it is associated with the site of hydrogenation. We distinguish isotopic scrambling from catalytic hydrogen isotope exchange unrelated to the functional group being reduced (see Chapter 3, Section 3.3), and from dilution of tritium by exchange with solvent protons (see solvent exchange above). Experiments using deuterium or tritium in place of hydrogen in studies on the mechanism of heterogeneous hydrogenation have shown that the process is often not simply the addition of the isotopic hydrogen to the double bond[10]. It is generally agreed[11] that during catalytic reductions of carbon–carbon multiple bonds, isomerizations and migrations occur that involve reversible hydrometallations, as depicted in Figure 4.1 (M represents bulk metal surface or in certain cases the metal center of an organometallic complex). It is through these processes that isotope scrambling occurs. The orientation of hydrometallation, the opportunity for internal rotation of the metal–alkyl complex, the tendency for reverse hydrometallation to return labeled isomeric

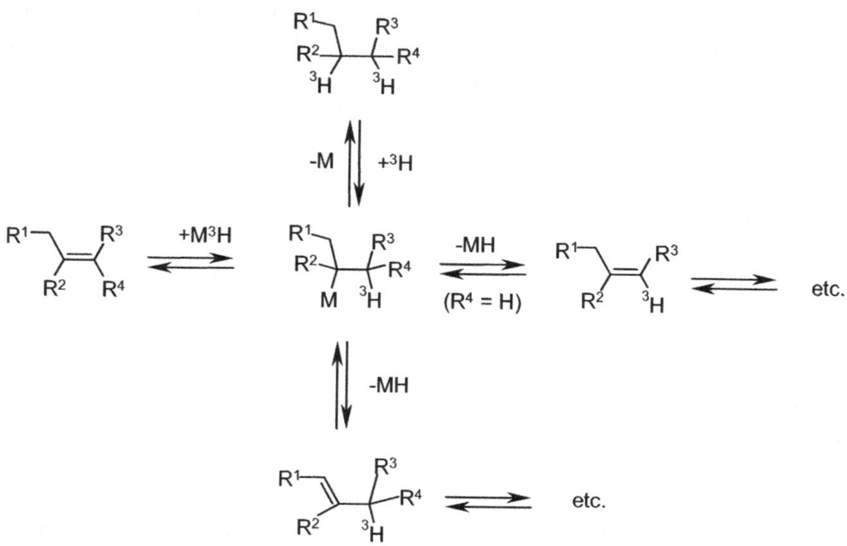


**Figure 4.1** *Mechanism for isotope scrambling*

olefin, and the relative rates of these processes vs reductive elimination of metal to give labeled product, all combine to determine the extent and location of isotopic scrambling[12]. The extent of isotopic scrambling varies widely by substrate structure, catalyst and reaction conditions. The tendency to promote double bond isomerization (which correlates with isotope scrambling) generally decreases in the order Pd > Ni ≫ Rh ≈ Ru > Os ≈ Ir ≈ Pt[9b].

A vivid illustration of this phenomenon is in the reduction of the $\beta,\gamma$-unsaturated amide **1** with tritium gas in the presence of 10% Pd/C to provide **2**, a tritium-labeled phospholipase $A_2$ inhibitor[13]. Product **2** had a specific activity of 85 Ci/mmol, in contrast to the theoretical maximum of 58 Ci/mmol from direct addition of two tritium atoms to the double bond. $^3H$ NMR analysis showed that the tritium was distributed along the entire carbon chain as indicated in the structure. This result is likely due to a combination of double bond isomerization, migration and saturation processes. Such greater-than-normal incorporation of label can be beneficial in enabling improved sensitivities of detection in the use of the labeled compounds, but only if the regioselectivity of labeling is not detrimental.

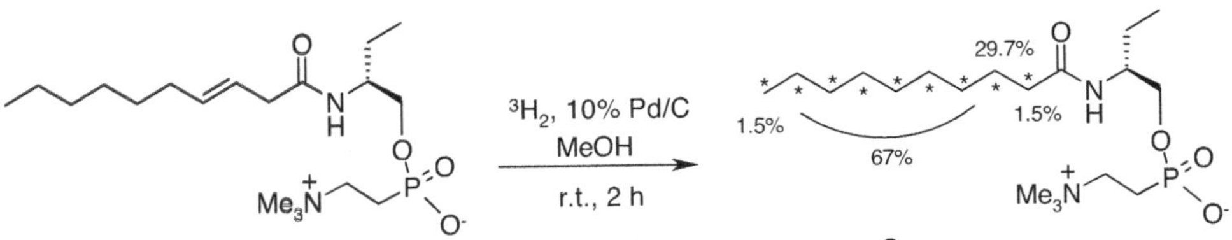


**Figure 4.2** *Isotope scrambling in the palladium-catalyzed tritiation of a nonconjugated olefin*

This tendency for olefins with vinylic hydrogens to incorporate more than stoichiometric amounts of label during reduction under certain conditions is especially common for terminal olefins. Thus, tritiation of (−)-allyl-norapomorphine hydrochloride (**3**) with 10% Pd/C afforded (−)-*N*-[$^3$H]propylnorapomorphine **4** at a specific activity of 60–80 Ci/mmol a majority of tritium located at the terminal methyl group[14].

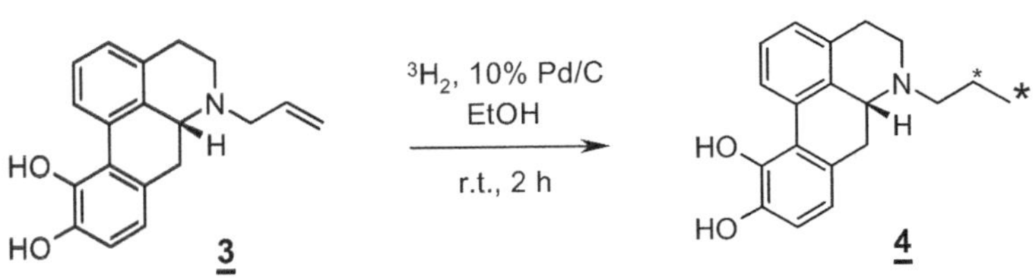


**Figure 4.3** *Additional labeling at the terminal methyl group in the tritiation of a terminal olefinic bond*

Of even more practical relevance is the incorporation of excess tritium during the catalytic tritiation of 4,5-dehydroleucine[15] ($\Delta^{4,5}$-Leu) and of peptides containing this precursor of [$^3$H]leucine, leading to specific activities of up to 200 Ci/mmol. *N*-Boc-$\Delta^{4,5}$-Leu (**5**) has been subjected to catalytic tritiation using palladium catalysts then the amino acid (6) incorporated into peptides by synthesis, or peptides containing $\Delta^{4,5}$-Leu in place of Leu are prepared, then catalytically reduced with a palladium catalyst and tritium to give the [$^3$H]Leu peptide, as with [$^3$H]urocortin (110 Ci/mmol) and [$^3$H]noceceptin (160 Ci/mmol)[16a], the enkephalin pentapeptides [$^3$H]DALE (198 Ci/mmol) and [$^3$H]DADLE (201 Ci/mmol)[16b] and even methionine-containing peptides such as the $\beta$-amyloid fragments 5-mer (122 Ci/mmol) and 10-mer (119 Ci/mmol)[16c]. Labeling of complex peptides or those containing S–S bridges may start with the preparation of an appropriately protected [$^3$H]Leu followed by its coupling to an appropriate pre-formed fragment, as reported for [$^3$H-Leu$^4$]salmon calcitonin having a specific activity of 100 Ci/mmol[17]. The high specific activity of [$^3$H]Leu prepared from $\Delta^{4,5}$-Leu derivatives results from both excess labeling of its terminal olefinic carbon and its neighboring methyl group as well. For example, [$^3$H]-DALE and [$^3$H]DADLE mentioned above had isotopomer distributions containing between one and seven tritium atoms per molecule[16a]. Similar results are frequently obtained with palladium-catalyzed tritiations of 2-methyl-2-alkenes, common in isoprenoids. In contrast, catalytic tritiation of 7 using platinum (which has a lower tendency to mediate isomerizations) gave 8 at a specific activity of 53.3 Ci/mmol[18], consistent with a limited amount of isotopic scrambling.

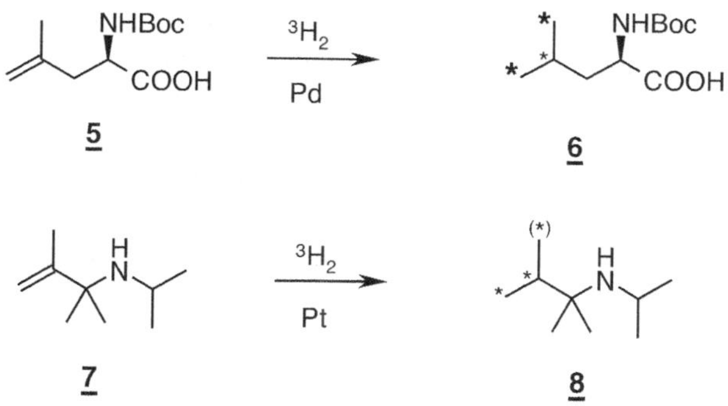


***Figure 4.4*** *Contrast in degree of isotope scrambling induced by palladium and platinum catalysts*

Other dehydro amino acids conveniently used in the same way include $\Delta^{3,4}$-proline[19] and 2,6-diaminohex-3-ynoic acid (for [$^3$H]lysine)[20]. These residues usually give more nearly stoichiometric tritium incorporation upon reduction than does $\Delta^{4,5}$-leucine. The tritiation of $\Delta^{2,3}$-amino acids (or peptides containing them) would require stereoselective reductions in order to obtain the desired (*R*)- or (*S*)-forms (as discussed later in this section).

Aside from isotopic scrambling, extraneous labeling by catalytic exchange processes (see Chapter 3, Section 3.3) is common for substrates possessing benzylic groups and certain sp$^2$ C–H bonds near heteroatoms. For example, when $\Delta^{16,17}$-estrone enol acetate was catalytically reduced with tritium gas over Pd/C, a large proportion of the label was incorporated at the benzylic positions C6 and C9[21]. Nor will the use of metals other than palladium guarantee the absence of extraneous exchange labeling: a highly unfavorable distribution of tritium, for example, was observed when methyl $\Delta^{7,8}$-lysergate (9) was catalytically reduced over Pt in the presence of catalytic amounts of perchloric acid. Only 33% of the tritium incorporated into methyl 8$\alpha$-9,10-dihydro[$^3$H]lysergate (**10**) was located at the position of the original double bond (7% at C7 and 26% at C8); 42% of the tritium was at C2 and 25% at C14, both as a result of Pt-catalyzed exchange. Since the C2 and C14 sites were known to be susceptible to metabolic processes, **10** was inappropriate as a precursor for the preparation of labeled 8$\alpha$-aminoergolines such as mesulergine (**11**) for *in vivo* investigations in animal and man[22].

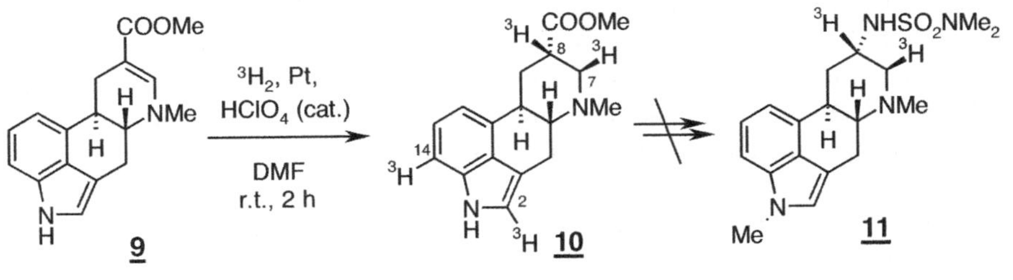


***Figure 4.5*** *Ancillary exchange labeling during tritiation of an olefin*

In rare cases advantage has been taken of isotopic scrambling to introduce tritium at or adjacent to double bonds without (entirely) reducing them. This was exploited for the preparation of labeled pleuromutilin ([$^3$H]**12**) and cyclosporin A ([$^3$H]**13**). Treatment of the unlabeled drug substances **12** and **13** with a deficiency of tritium gas over 10% Pd/C, followed by HPLC separation of the unsaturated and the saturated reaction products, provided the two isotopomers with specific activities of 10 Ci/mmol and 19 Ci/mmol, respectively[23]. The label proved to be biologically stable for both compounds, so that they could be used for a variety of *in vivo* investigations.

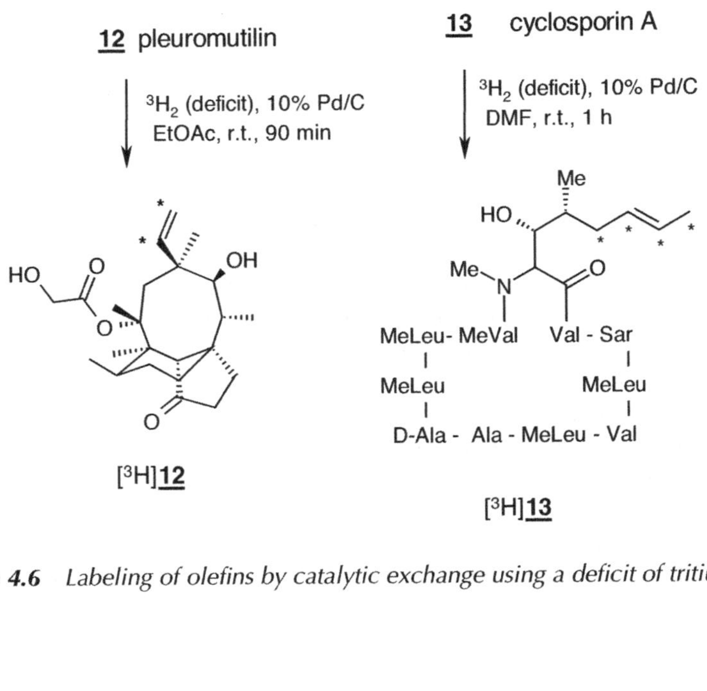


***Figure 4.6*** *Labeling of olefins by catalytic exchange using a deficit of tritium gas*

The ease of olefin tritiation is dependent on the number and size of the substituents. Hydrogenation rates generally decrease in the order $CH_2{=}CHR^1 > CH_2{=}CR^1R^2 > cis\text{-}R^1CH{=}CHR^2 > trans\text{-}R^1CH{=}CHR^2 > R^1CH{=}CR^2R^3 > R^1R^2C{=}CR^3R^4$. Tetrasubstituted double bonds normally resist reduction using heterogeneously catalyzed tritiation methods. This range of reactivity sometimes allows selective reduction of one out of several double bonds in the same substrate, especially when only a slight excess of tritium gas is employed and reaction times are kept short. Illustrative examples are the preparation of dihydro[35,36-$^3H_2$]FK506 (**14**)[24], in which only the monosubstituted double bond was reduced, and of (17*R*)-17,20-dimethyl-7-thia[19,20-$^3H_2$]PGE methyl ester (**15**). In the latter case the tritiation rates for the *cis* and *trans* double bonds differed sufficiently to allow a selective reduction of the *cis*-double bond at C19,C20[25]. The preparation of **16** illustrates not only discrimination between tri- and tetrasubstituted olefinic bonds, but also shows that in most cases the addition of tritium is *syn* and takes place from the less-hindered side of the double bond[26].

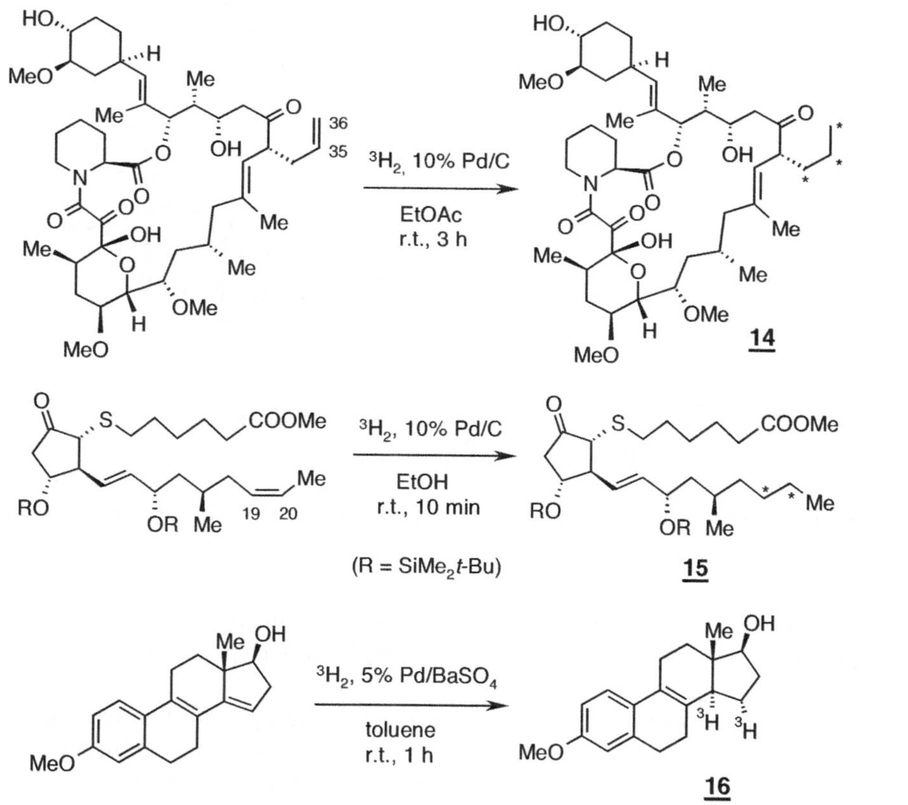


***Figure 4.7*** *Selective tritiation of olefinic bonds*

Fine-tuning of reactions to achieve desired regioselectivities can be accomplished by varying the tritium pressure or the catalyst activity. For example, PdO-catalyzed tritioreduction of the spirooxindolamine **17** at 450 mbar provided **18** at a specific activity of

30.8 Ci/mmol, in which only the terminal double bond was reduced. The application of a pressure of 770 mbar, however, led to the reduction of both double bonds[27]. And in the tritiation of the digoxigenin derivative **19**, it proved necessary to deactivate the catalyst slightly by addition of a small amount of diisopropylethylamine, in order to avoid over-reduction of the furyl system[28].

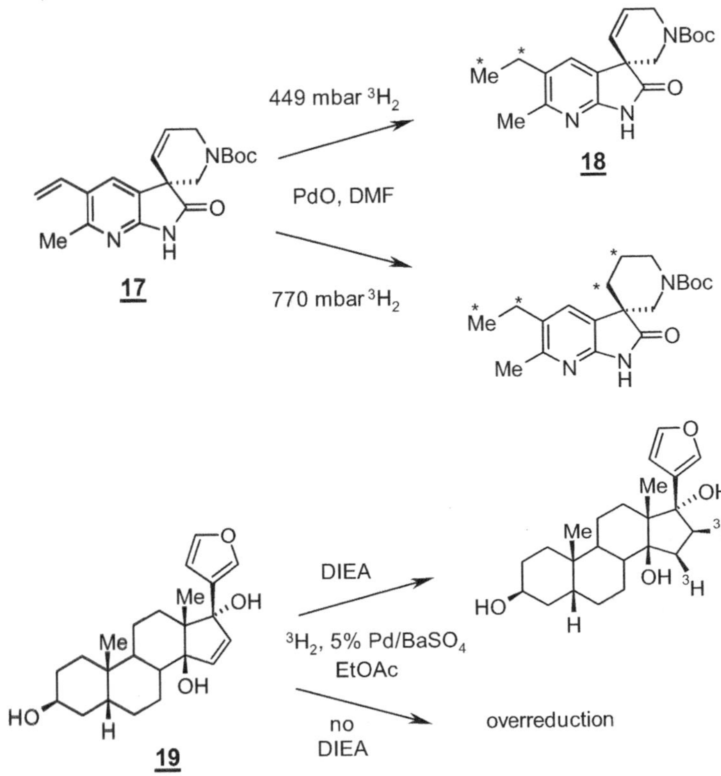


***Figure 4.8*** *Influence of pressure and catalyst poisoning on the selectivity of tritium addition*

Many examples exist, however, in which competitive tritiation rates prevent easy discrimination between different double bonds in a substrate. In such cases, it is necessary to optimize conditions specifically for the substrate by varying parameters such as catalyst, tritium pressure, solvent and reaction time.

For study of the molecular mechanisms of action of insect juvenile hormones and related analogs such as (7*S*)-methoprene (**22**) (Figure 4.9), selectively tritiated material of high specific activity was needed. The synthetic approach required discrimination between the isolated double bond at C9,C10 and the one at C4,C5 belonging to the dienoate moiety of the precursor molecule **20**. This was accomplished through blocking of the dienoate system as an iron tricarbonyl diene complex (**21**), which was easily accessed through reaction of the substrate with triiron dodecacarbonyl. Subsequent tritiation of the C9,C10-double bond with 5% Rh/C and oxidative deprotection of the dienoate system provided material with a specific activity of 84 Ci/mmol[29]. Trials with Pd/C or Pt/C in the hydrogenation step were unsuccessful; experiments using Wilkinson's catalysts failed to effect the C9,C10-hydrogenation of the $\alpha$-alkoxy substituted alkene (see Section 4.1.3).

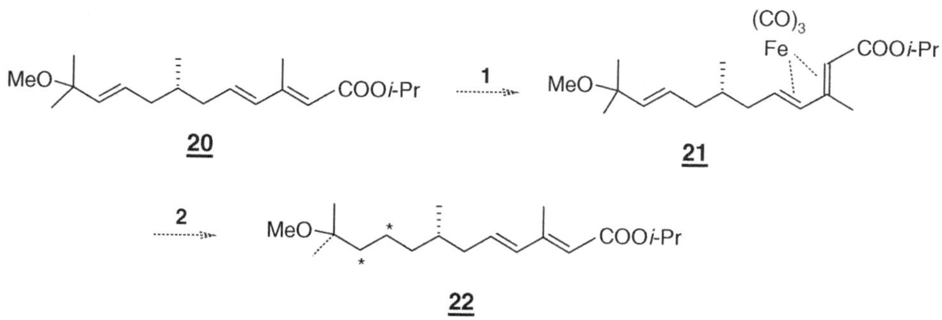


**Reaction conditions**: **1**. $Fe_3(CO)_{12}$, benzene; reflux, 7 h; **2a**. $^3H_2$,5% Rh/C, EtOAc; r.t., 3 h; **2b**. $(NH_4)_2Ce(NO_3)_6$, $CH_3CN$; 0 °C, 4 h

***Figure 4.9*** *Selective tritiation of an isolated double bond in the presence of a dienoate moiety*

Owing to the greater reactivity of carbon–carbon triple bonds, it is generally very difficult to selectively reduce double bonds present in the same molecule. For terminal alkyne groups, however, this selectivity pattern may be inverted by blocking the approach of the triple bond to the catalyst's surface through a temporary bulky protecting group. Thus, catalytic tritiation of 4-allyl-1-(4-trimethylsilylethinyl-4-phenyl)-2,6,7-trioxa-bicyclo[2.2.2]octane (23) over 5% Pt/C in ethyl acetate in the presence of traces of triethylamine provided 24 in nearly quantitative yield and with a specific activity of 58.4 Ci/mmol, in which only the olefinic bond was attacked. Subsequent treatment of 24 with $NBu_4^+F^-$ removed the trimethylsilyl protecting group to give the desired 25[30].

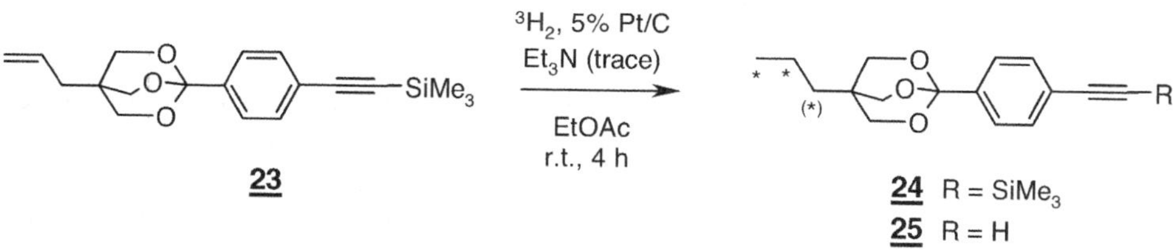


***Figure 4.10*** *Strategy for selective tritiation of an olefinic bond in the presence of a terminal alkyne moiety*

Internal triple bonds can be selectively blocked as an acetylene–dicobalt hexacarbonyl complex through reaction of dicobalt octacarbonyl and the alkyne at room temperature in a hydrocarbon solvent following a protocol similar to the protection of the dienoate moiety mentioned above[31]. Unfortunately the unprotected olefinic bond cannot be reduced by heterogeneous catalysis, since the cobalt complexes have proved to be catalyst poisons. It can, however, be reduced using [$^3$H]diborane or [$^3$H]hydrazine. Furthermore, such a complex was used successfully in organoiridium-catalyzed $^3H/^1H$-exchange procedures, so that use of homogeneous catalysts in selective olefin tritiations may be promising[32].

Selective tritiation of carbon–carbon double bonds in compounds containing *other* reducible groups can often be accomplished. $\alpha,\beta$-Unsaturated ketones and esters, for example, can readily be reduced to give the corresponding saturated derivatives, as illustrated by the tritiations of **26**, catalyzed by PdO, and of **28**, catalyzed by Raney nickel. Subsequent removal of the $\alpha$-tritons from the products by base-catalyzed exchange, and in the second case simultaneous epimerization at the $\alpha$-position to a 60:40 mixture, afforded [8-$^3$H]oxymorphone (**27**)[33] and the saturated tricyclic ester **29**[23].

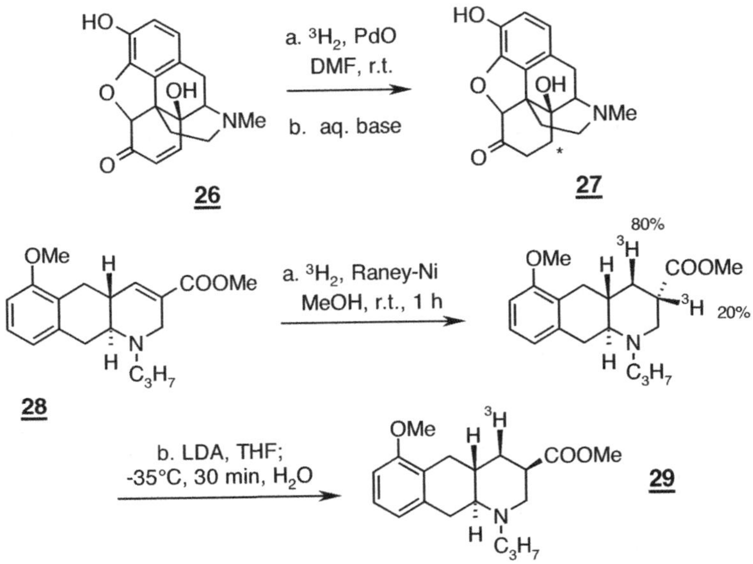


***Figure 4.11*** *Selective tritiation of $\alpha,\beta$-unsaturated carbonyl systems*

Palladium- and platinum-catalyzed tritiations of olefinic substrates containing halogen substituents are frequently accompanied by tritiolytic cleavage of the halogen (I > Br > Cl > F; see Section 4.2). This competing reaction may sometimes be suppressed sufficiently by omitting base in the reaction[34]; the most promising results can be expected for fluorine and chlorine substituents, as illustrated by the preparation of [$^3$H]CB34 (**30**) at a specific activity of 111 Ci/mmol[34] (Figure 4.12).

The nitro group is one of the most easily reducible groups, so that it is rarely possible to selectively reduce a double bond in its presence[35]. Nevertheless, the synthesis of [propyl-2,3-$^3H_2$]MN-9201 (**32**) of a specific activity of 56.8 Ci/mmol tritio-reduction of the unsaturated nitroaryl precursor **31** demonstrates that there are exceptions.

Enamines behave much as do other olefins in heterogeneous catalytic tritiations[8b]. They have value as precursors for tritiated ligands containing a tertiary amine function, especially because they can be readily prepared from the saturated parent compound through $Hg(OAc)_2$ oxidation. Accordingly, $PdO/BaSO_4$-promoted tritiation of $\Delta^{15,16}$-buprenorphine provided the corresponding opioid ligand **33** with a specific activity of

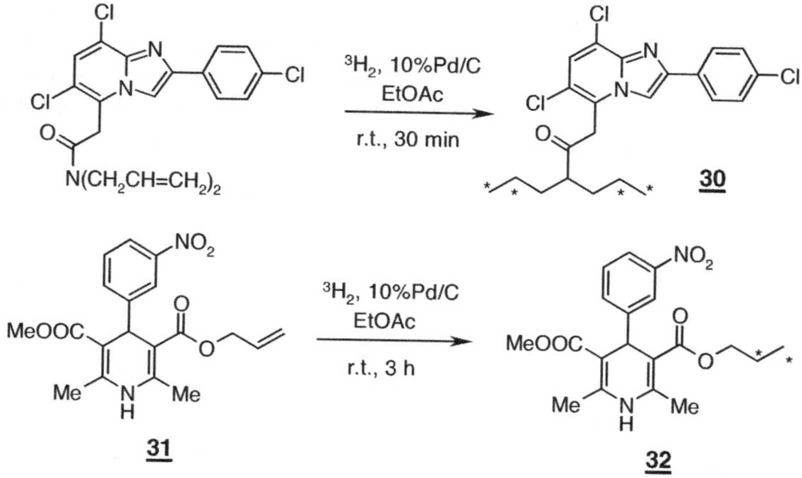


*Figure 4.12* *Selective tritiation of olefinic bonds in the presence of aryl chloro and nitro groups*

64 Ci/mmol[36]. Interestingly, application of the same reaction under slightly modified reaction conditions to $\Delta^{15,16}$-etorphine (34) resulted in the formation of [15,16-$^3H_2$]etorphine (35), in which again the enamine double bond was reduced, but the etheno bridge was not affected[37].

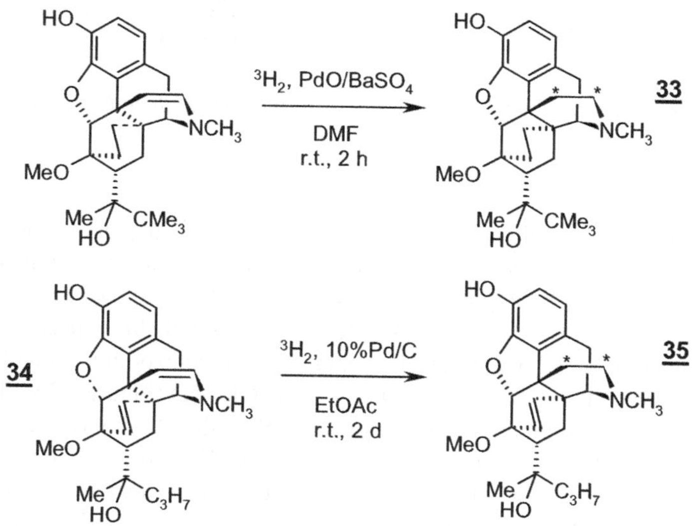


*Figure 4.13* *Reduction of an enamine in the presence of an isolated olefinic double bond*

Oxygen and nitrogen functions are rarely of concern in catalytic tritiations of olefinic substrates, as long as they are not located in benzylic or allylic positions. If they are, tritiolytic cleavage may occur, but this can often be avoided by utilizing rhodium or ruthenium catalysts in lieu of palladium. Although these cleavages may be undesired side reactions, as in the case of **36** when the fragment split off is part of the substrate's structure[38], they may facilitate the use of a protecting group strategy, improve the solubility of a substrate or lessen the potential for isotope dilution, as in **37** where the benzyl ester is cleaved concurrently with double bond tritiation to give product **38**[39].

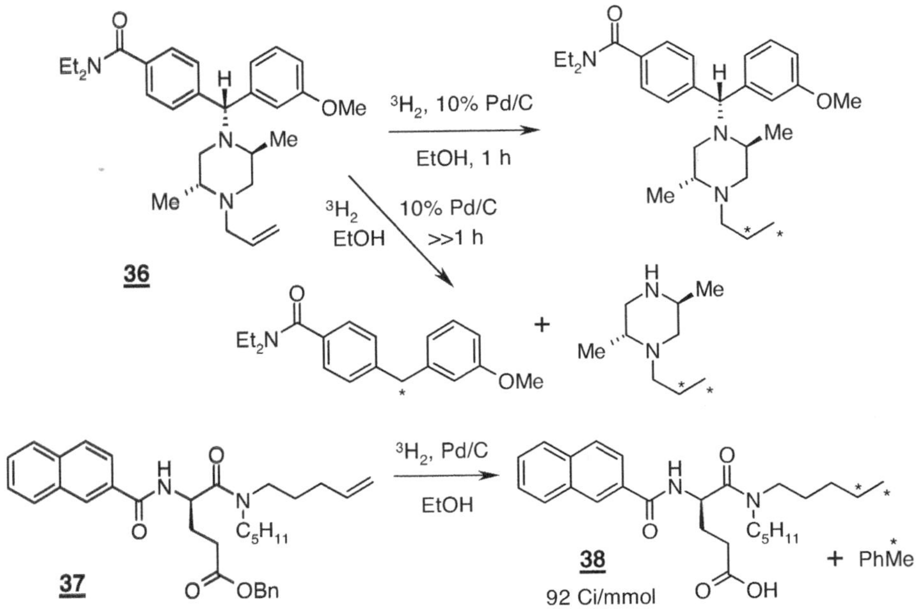


***Figure 4.14*** *Competetive tritiolysis reactions in the Pd-catalyzed tritiations of olefinic substrates*

Catalytic tritiations of alkynes is a two-step process and, depending on the reaction conditions, either the labeled alkene or alkane derivative can be isolated. Complete reduction to the saturated compound is easily accomplished over palladium or platinum catalysts. This is of interest whenever substrates of exceptionally high specific activities are needed or when only small quantities of an appropriate precursor are available. The latter may be illustrated by the synthesis of [4,4,5,5-$^3$H]TLCK (**41**) (Figure 4.15). Treatment of doubly protected (*S*)-2,6-diamino-4-hexynoic acid **39** with tritium gas in the presence of 10% Pd/C provided $N_\varepsilon$-Boc,$N_\alpha$-tosyl-[4,4,5,5-$^3$H]lysine (**40**) with a specific activity of 123 Ci/mmol. Subsequent dilution of the specific activity with cold carrier, chain elongation with $CH_2N_2$ and selective acid-catalyzed $N_\varepsilon$-deprotection converted **40** to **41**[40]. Lysine analogs such as this and $N_\alpha$,$N_\varepsilon$-diprotected (*S*)-diamino-4-hexynoic acid (e.g. **42**) are common labeling precursors in the synthesis of tritiated peptides, as demonstrated with the synthesis of the [$^3$H]antamanide analog **43**[20c,d]. Furthermore, saturation of *N*-propargyl derivatives can be regarded as an attractive route whenever a tritiated ligand containing an *N*-propyl group is required at a specific activity above 58 Ci/mmol. This approach was applied to the tritiation of **44** to give a 1H-benz[c]indole-type $5HT_{1A}$-receptor agonist with a

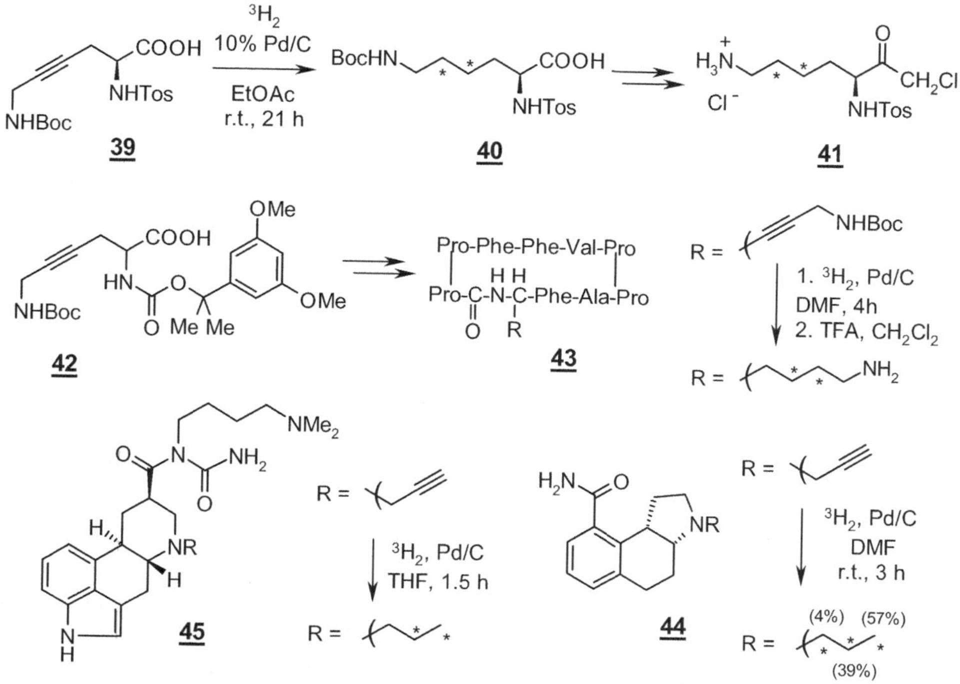


*Figure 4.15* Tritiations of acetylenic substrates

specific activity of 86.7 Ci/mmol[41], and of **45** to furnish [$^3$H]dihydrocarbergoline, a dihydrolysergic acid derivative, at 109 Ci/mmol[42].

Very useful from a synthetic point of view is the partial tritiation of alkynes to [$^3$H] alkenes. This reaction can be effected in high yield with a palladium–calcium carbonate catalyst partially deactivated by addition of lead acetate and/or quinoline (Lindlar catalyst)[43]. Lindlar reductions are clean 1,2-additions and are not accompanied by isotope scrambling. The Lindlar method was utilized in the preparation of tritiated 1,2-anhydroglucitol-6-phosphate (**48**), an inhibitor of glucose-6-phosphate isomerase (Figure 4.16). Reduction of the acetylenic precursor **46** afforded the olefinic key intermediate **47** with a specific activity of 59 Ci/mmol. Subsequent epoxidation and hydrogenolytic cleavage of the protecting group converted this intermediate into the target molecule[44]. The same approach has been used for ligands containing *N*-allyl groups, for example [*N*-allyl-2,3-$^3H_2$]naloxone (**49**)[45], [*N*-allyl-2,3-$^3H_2$]SKF 10,047 (**50**)[46] and [*N*-allyl-2,3-$^3H_2$]carbergoline (**51**)[42]. The specific activities typically achieved are 30–60 Ci/mmol.

Semireduction of internal alkynes with Lindlar catalysts invariably gives *cis*-olefins. Analysis by $^3$H NMR of the Japanese beetle pheromone **52** prepared from the corresponding alkyne showed clean signals consistent with two *cis*-disposed vinyl tritons[47] (Figure 4.17). The lack of signals from isomeric olefins or tritium on sp$^3$ carbon was also consistent with the general finding that no isotope scrambling occurs during Lindlar-catalyzed alkyne semitritiation. The use of Lindlar catalysts permits discrimination between a carbon–carbon triple

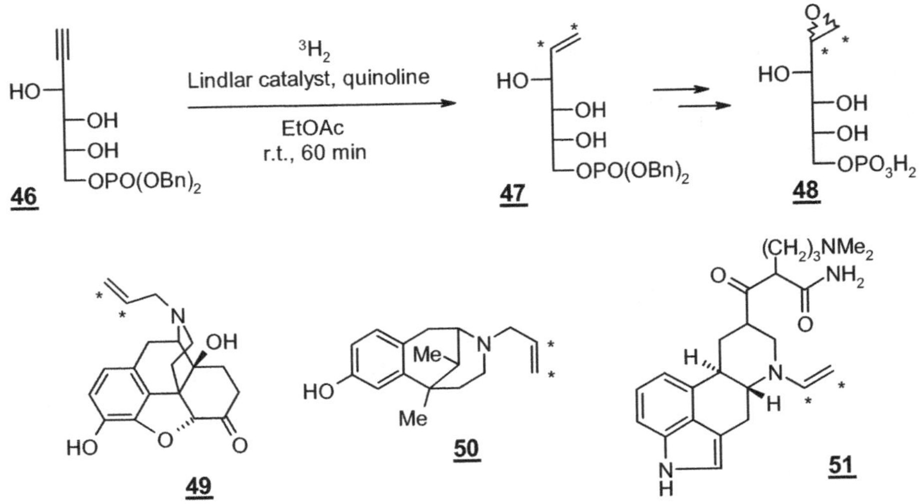


*Figure 4.16* *Semitritiations of terminal alkynes to afford tritium-labeled alkenes*

bond and carbon–carbon double bonds within the same molecule. This has been exploited for the preparation of several tritiated leukotriene derivatives[48], retinoids[49] and prostaglandin analogs[50]. Treatment of 5,6-epoxy-14,15-dehydro-LTA methyl ester (**53**), for example, in hexane solution with carrier-free tritium gas over Lindlar catalyst provided 5,6-epoxy[14,15-$^3H_2$]LTA4 (**54**). Subsequent ring opening with (*S*)-cysteine methyl ester yielded (5*S*,6*R*)-[14,15-$^3H_2$]LTE4 dimethyl ester (**55**) at a specific activity of 40 Ci/mmol[48a]. However, terminal carbon–carbon triple bonds are more reactive towards catalytic reductions than internal ones, so that it may prove difficult to isolate the corresponding terminal alkene.

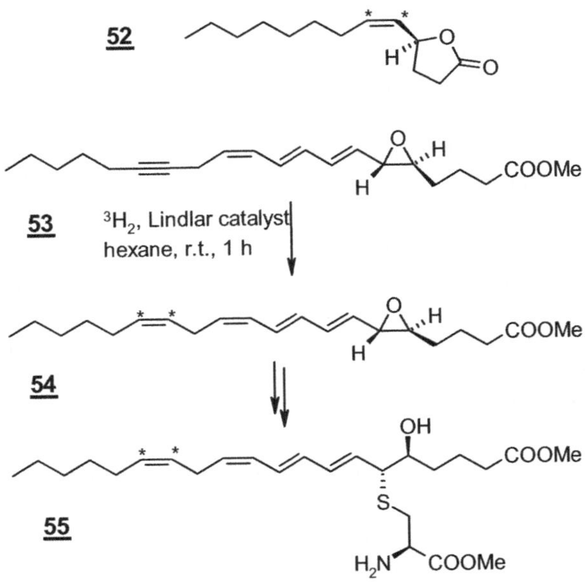


*Figure 4.17* *Semitritiations of internal alkynes and selectivity versus olefinic functions*

The different reactivities of internal and terminal carbon–carbon triple bonds were exploited for the synthesis of [$^3$H]LTC4 (**58**), an investigational drug of the leucotriene type. In order to shorten the synthetic pathway, which would have consumed large quantities of radioactivity, diyne **56** was submitted to Lindlar catalyst-promoted hydrogenation in ethyl acetate, which gave exclusively the tetradehydro derivative **57**. However, running the reaction in methanol in the presence of 4-hydroxy-TMPO as a radical scavenger provided a mixture of partially reduced compounds **58–60**, of which in the radioactive case the desired **58** could be isolated in 17% yield after HPLC separation with a specific activity of 180 Ci/mmol[51].

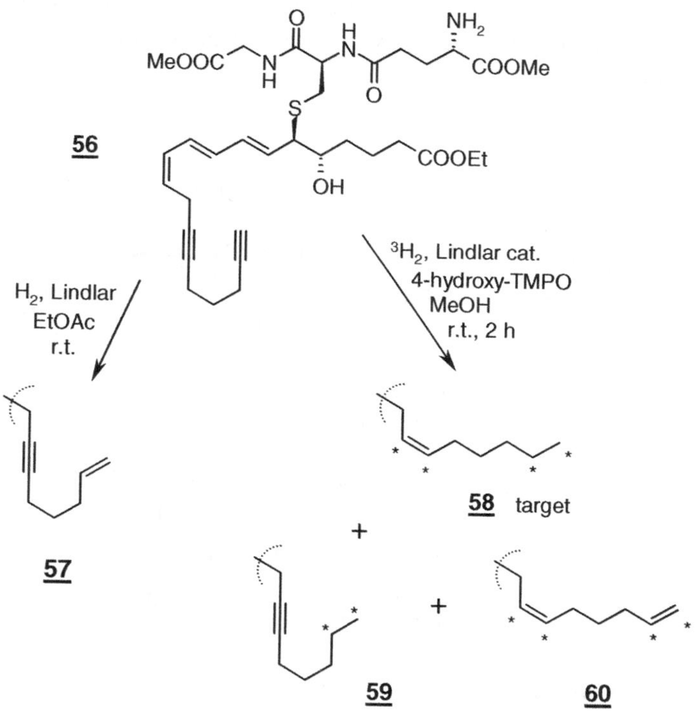


*Figure 4.18 Chemoselectivity in the Lindlar's catalyst promoted tritiations of terminal and internal alkynyl units*

Aromatic and heteroaromatic systems are less easily reduced by heterogeneous catalysts than ordinary carbon–carbon multiple bonds. In many cases they require higher temperatures. Platinum, palladium and rhodium catalysts are most frequently used, of which rhodium is the most active. Rhodium and ruthenium are employed when hydrogenolysis of carbon–oxygen and carbon–nitrogen bonds are to be avoided. Although this approach might be an effective alternative to methods used in some cases for compounds of high specific activity, only a few examples are mentioned in the literature. One is the 5% Pd/C-catalyzed tritio-reduction of phenol which reacted under atmospheric pressure to give tritiated cyclohexanone (**61**) in 27% yield with a specific activity of 96.9 Ci/mol[52] (Figure 4.19). It was directly treated with triethyl phosphonoacetate anion to give ethyl

[$^3$H]cyclohexylidene acetate, which served as precursor to 2-(S-(*p*-methyl)benzylmercapto)-2,2-[$^3$H]cyclopentamethylene propionic acid (**62**).

As is the case with many arene hydrogenations, the reduction of the hydrochloride salt of benzylamine (**63**) with tritium gas over rhodium catalyst to give [$^3$H]cyclohexylmethylamine (**64**), 25% yield, specific activity of 115–120 Ci/mmol) required an acidic medium to progress the reaction. The nonoptimal specific activity may have resulted from competing tritium gas exchange with solvent protons under these conditions (see solvent exchange at the beginning of this chapter). In an investigation aimed at achieving higher specific activity, an optimized solid-phase method was identified, in which benzylamine was absorbed on $Rh/Al_2O_3$ and heated at 60 °C under tritium gas for 3 h. This gave up to 45% yield of **64** at a specific activity of 200–210 Ci/mmol. Here, the advantage of obtaining a very high specific activity was offset by the need for more extensive optimization of reaction conditions[53].

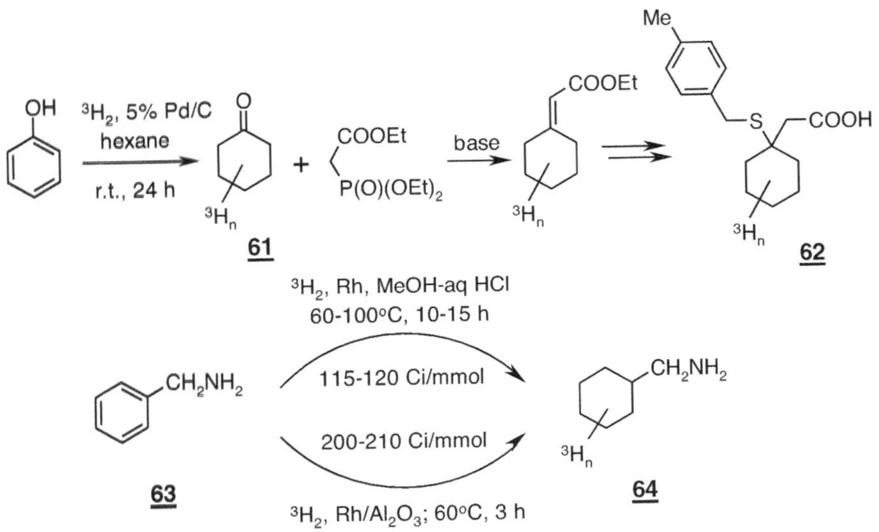


***Figure 4.19*** *Catalytic tritiations of aromatic systems*

Heteroaromatic systems of the pyridine type are also readily reduced. For example, reduction of 2-propylpyridine took place under mild conditions with platinum(0) (prepared *in situ* by reduction of $PtO_2$ with hydrogen gas) and a tritium–hydrogen mixture, affording racemic tritiated coniin (**65**) in 54% chemical yield and 33% radiochemical yield at a specific activity of 385 mCi/mmol (theor. 610 mCi/mmol)[54] (Figure 4.20). The presence of acetic acid was necessary for the hydrogenation to progress; the reaction procedure was carefully designed to minimize the exchange of tritium gas into solvent. Treatment of isonicotonic acid hydrochloride (**66**) with tritium gas in DMF in the presence of $PtO_2$ gave [$^3$H]isonipecotic acid hydrochloride (**67**), which was converted via several steps to enantiomerically pure amino[$^3$H]quinuclidine (**68**) of specific activity 66 Ci/mmol, an essential component of various highly tritiated seretonin receptor ligands[55]. In the last example, the solid-state tritiation of 3-(4-pyridyl)-1,4-dihydro-2-oxo-3(2H)quinazoline

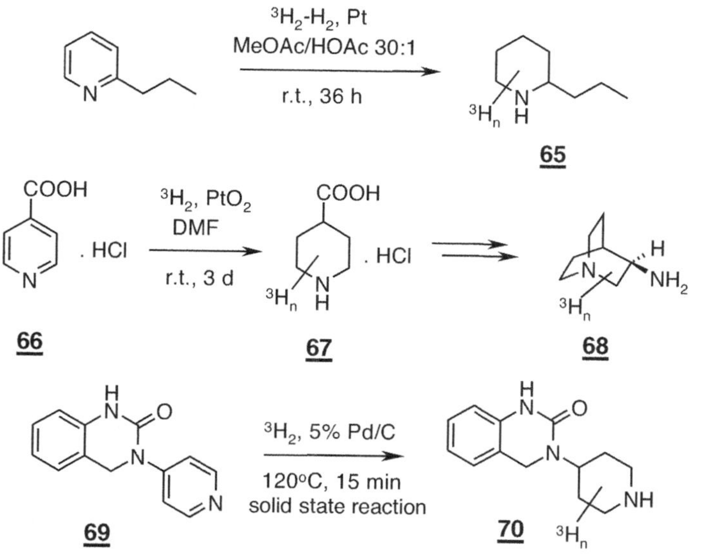


***Figure 4.20*** *Catalytic tritiation of heteroaromatic systems*

(69) provided **70** at 180 Ci/mmol, the phenyl ring remaining intact; various attempts at the same reaction in solution gave much lower specific activities[56].

### 4.1.2 Tritiation of Carbon–Heteroatom Multiple Bonds

As already mentioned, reductions of functional groups such as aldehyde, ketone, imine and nitrile are usually carried out using tritide reagents. The polar nature of the carbon–heteroatom multiple bonds makes them more favorable substrates for reduction with complex tritides than for catalytic reduction. However, there are sometimes advantages to the catalytic approach either because of a particular stereoselectivity or for the sake of convenience. Catalytic reduction of aldehydes and ketones is generally easier than that of aromatic systems but less easy than that of carbon–carbon multiple bonds, so that selective tritio-reduction of carbonyl groups in the presence of olefinic or acetylenic bonds is difficult to achieve. Platinum, palladium and active forms of Raney nickel are the most commonly used catalysts[9]. The synthesis of D-[1,3,5-$^3H_3$]ribose (74), a key intermediate for the enzymatic preparation of several highly tritiated nucleotides may serve as an example (Figure 4.21). Palladium-catalyzed reduction of 1,2:5,6-di-*O*-isopropylidene-$\alpha$-D-ribo-3-hexo-furanosulose (71) provided 1,2:5,6-di-*O*-isopropylidene-$\alpha$-D-[3-$^3H$]allofuranose (72). Subsequent selective cleavage of the 5,6-isopropylidene protecting group, glycol cleavage of the liberated diol function with periodic acid, another catalytic tritio-reduction of the resulting aldehyde derivative **73** and final acid-catalyzed deprotection furnished **74** with a specific activity of 80 Ci/mmol, accompanied by a simultaneous H/$^3H$-exchange at C1 resulting in the incorporation of additional tritium which was detected in the analysis of **74**[57].

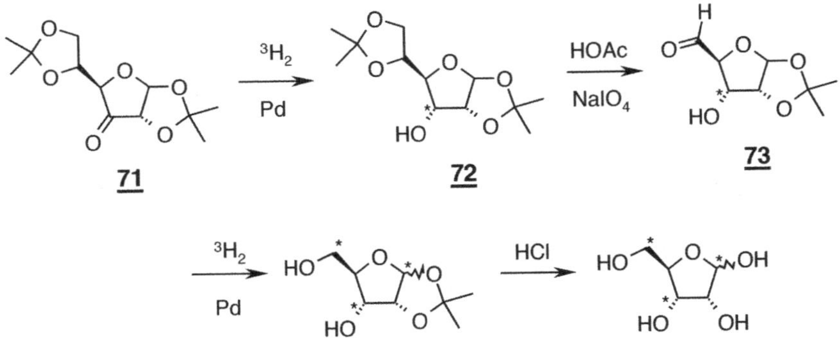


*Figure 4.21* *Catalytic tritiation of a carbonyl function in the synthesis of tritiated D-ribose*

In catalytic reductions of aromatic aldehydes and ketones, excess tritium and prolonged reaction times should be avoided since they may result in tritiolysis of the carbon–oxygen bonds, producing labeled methyl or methylene groups[58]. Examples of tritio-deoxygenations can be found in Section 4.2.2.2.

Compounds containing carbon–nitrogen multiple bonds, e.g. imines, oximes and nitriles are also readily reduced by catalytic tritiation. Platinum and palladium under neutral or slightly acidic conditions have proven to be highly efficacious, as has Raney nickel under basic conditions. Two examples illustrate the Pd-catalyzed tritio-reduction of imines or the corresponding iminium salts: the preparation [$^3$H]pinoline (**75**)[59] and *N,N*-di[$^3$H]methyl-*N'*-Boc-*p*-phenylendiamine **76**[60] at specific activities of 35.9 Ci/mmol and 47 Ci/mmol, respectively (Figure 4.22). The enhanced specific activity of **75** indicates the occurrence of isotopic exchange at the position $\alpha$ to nitrogen prior to the reduction. The synthesis of the (1*R*,2*S*)-(−)-2-amino-1,2-diphenyl[2-$^3$H]ethanol **77** at 92% e.e. through 5% Pd/C-catalyzed reduction of (*S*)-benzoin oxime hydrochloride is of special interest because of the high degree of stereoselectivity induced at the amino carbon by the adjacent stereogenic center[61]. Compound **78**, accessible through Raney nickel tritio-reduction of the corresponding oxime at 13 Ci/mmol, was a key intermediate for the preparation of ABT-[2-$^3$H]-BMPA (**79**), a photoaffinity labeling reagent for exofacial glucose transporters[62].

Rh/$Al_2O_3$, in addition to platinum and palladium, is highly recommended for reductions of nitriles, which need to be performed in the presence of an acylating agent such as $Ac_2O$ or $(CF_3CO)_2O$, or in acidic solution, in order to suppress the formation of secondary amine coupling products. This method has occasionally been used for the synthesis of highly tritiated primary amines such as [$^3$H]baclofen (**80**)[63] and 5-trifluoroacetylamino[$^3H_2$]-methyl-(3′,5′-di-*O*-acetyl)-2′-deoxyuridine (**81**)[64] (Figure 4.23).

### 4.1.3 Homogeneously Catalyzed Reactions

Soluble organic complexes of rhodium, ruthenium and iridium (e.g. $(Ph_3P)RhCl$, RuH(Cl)$(Ph_3P)_3$, [Ir(cod)($PCy_3$)(py)]$PF_6$) have advantages over heterogeneous catalysts in certain cases where the latter induce tritiolytic cleavage of a heteroatom, lack the required selectivity, or result in nonspecific labeling through double bond migration or tritium/hydrogen exchange. RuH(Cl)$(Ph_3P)_3$, prepared *in situ* from $(Ph_3P)_3RuCl_2$, has been found

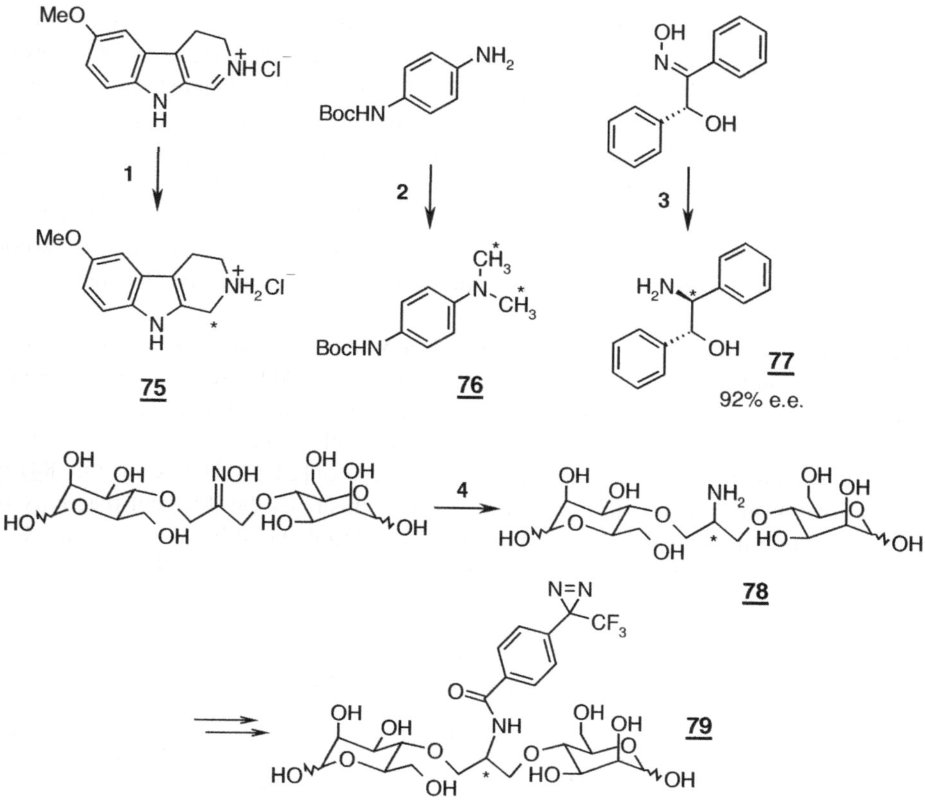


**Reaction conditions:** **1**. $^3H_2$, 10 % Pd/C, EtOH; r.t., 1.5 h; **2**. $^3H_2$, HCHO, 10 % Pd/C, EtOAc; r.t., 20-36 h; **3**. $^3H_2$, 5 % Pd/C, 1 equiv. HCl, EtOH; r.t., 2 h; **4**. $^3H_2$, Raney Ni, MeOH, conc. aq. $NH_3$; r.t., 1 h

***Figure 4.22*** *Catalytic tritiations of imines and oximes*

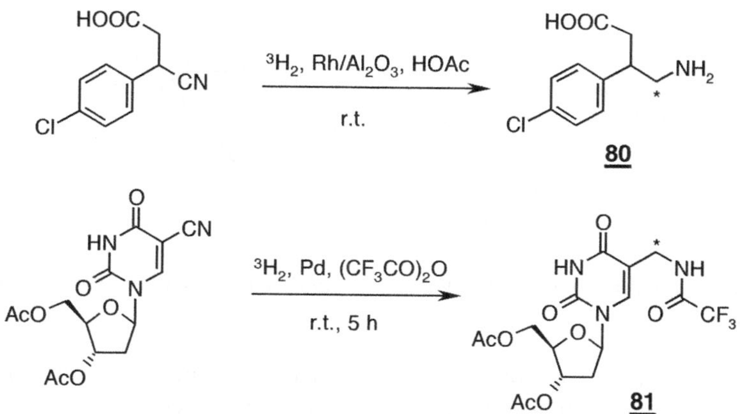


***Figure 4.23*** *Catalytic tritiations of nitriles*

especially suitable for reducing monosubstituted alkenes; $(Ph_3P)_3RhCl$ (Wilkinson's catalyst) can, in addition, reduce disubstituted olefins (*cis* > *trans*). [(cod)Ir $(PCy_3)(py)$]-$PF_6$ (Crabtree's catalyst) is able to reduce tri- and tetrasubstituted double bonds as well[65]. Surprisingly, only Wilkinson's catalyst has been used so far for tritiations of carbon–carbon multiple bonds, whereas Crabtree's catalyst has recently found broad application in tritium exchange labeling of benzamides, acylanilides, aryl ketones, and a variety of aryl-substituted heterocycles (see Chapter 3, Section 3.5). The major advantage of Wilkinson's catalyst is that it usually gives specific *syn*-addition of tritium across carbon–carbon multiple bonds and is compatible with aromatic halogen, nitrile, nitro, and *O*- and *N*-benzyl groups, all of which are inert.

The tritiation of the $\Delta^{3,4}$-PLA2 inhibitor **1** over Wilkinson's catalyst may serve as an illustrative example. In contrast to the reduction over 10% Pd/C (see Figure 4.2 above), Wilkinson's catalyst incorporated tritium exclusively at C3 and C4, although the specific activity of the resulting **82** (57 Ci/mmol) was significantly lower than that of the analogous **2**, reduced with Pd/C[13] (85 Ci/mmol). The synthesis of [22,23-$^3H_2$]ivermectin (**84**) by tritiation of avermectin $B_1a$ (**83**) over $(Ph_3P)_3RhCl$ illustrates the different reactivity of various olefinic functions. Only the disubstituted *cis*-double bond between C22 and C23 was reduced, and neither the diene system nor the trisubstituted double bond reacted[66].

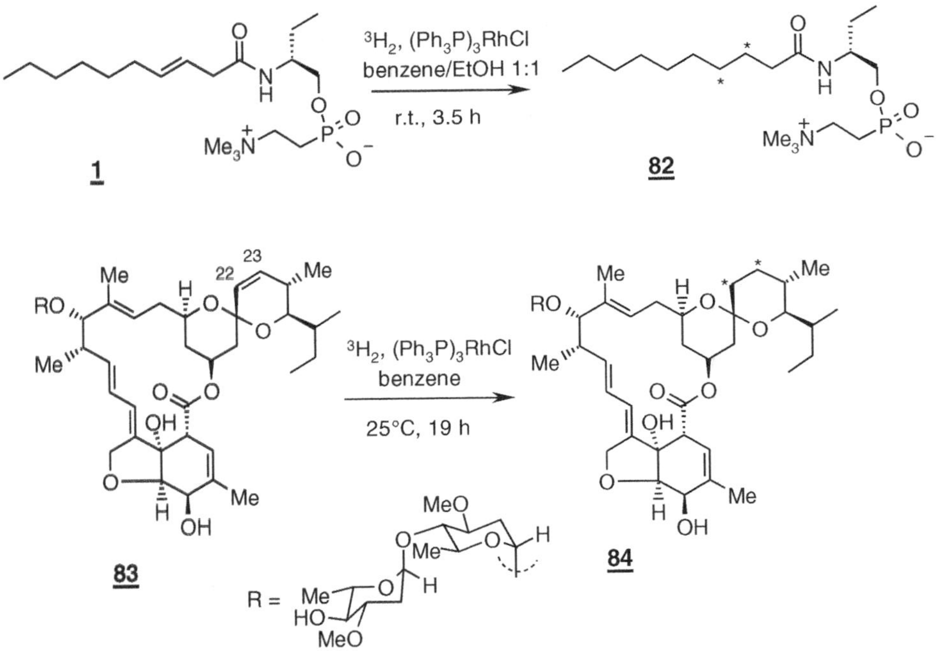


*Figure 4.24 Tritiation of olefins with Wilkinson's catalyst and an example of selectivity among double bonds*

$(Ph_3P)_3RhCl$ is also capable of catalyzing the reduction of carbon–carbon triple bonds, as illustrated by the preparation of tritiated phosphoryl choline-containing linear saturated polymers **86** from the respective acetylenic precursors **85**. $^3H$-NMR spectroscopic

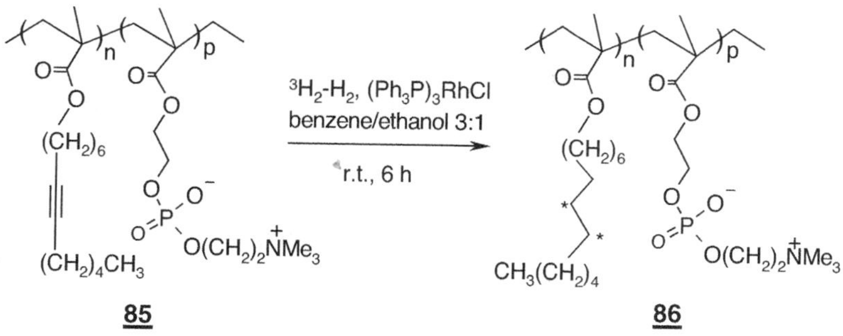


*Figure 4.25* Tritiations of alkynes with Wilkinson's catalyst

investigations using lanthanide shift reagent confirmed that 98% of the label was located at the original acetylenic positions[67].

The five-to-ten-fold difference in reactivity of *cis-* and *trans*-olefinic bonds towards Wilkinson's catalyst-mediated reductions[68] allows discrimination between them. This was exploited for the synthesis of [5,6-$^3H_2$]prostaglandin $E_1$ (**87**)[69] and [17,18-$^3H_2$]isocarbacyclin methyl ester (**88**)[70] from their respective unsaturated precursors. The specific activity of **88** amounted to 36.7 Ci/mmol, when carrier-free tritium gas was used.

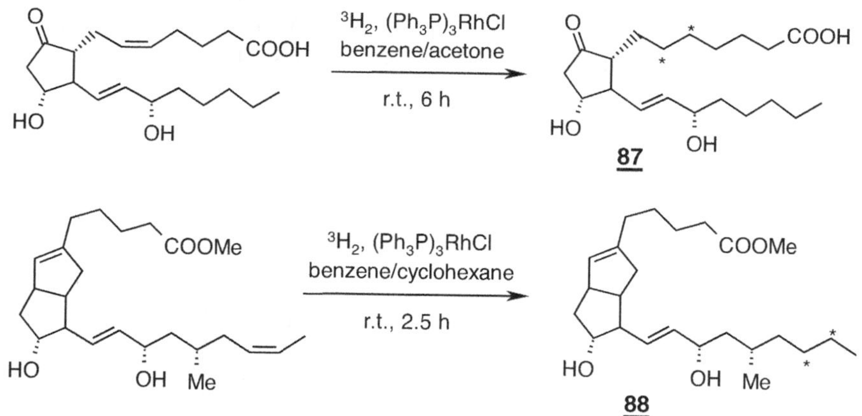


*Figure 4.26* Discrimination between disubstituted olefinic bonds in tritiations catalyzed by Wilkinson's catalyst

In contrast to its failure to catalyze the tritiation of the iron carbonyl-protected methoprene (**21**, Section 4.1.1 above), Wilkinson's catalyst was effective in the selective tritiation of the olefinic double bond of the analogously protected hydroprene (**89**) to give [$^3$H]hydroprene (**90**) at a specific activity of 65.5 Ci/mmol after deprotection, as shown in Figure 4.27[71].

Although selectivity is usually very high with Wilkinson's catalyst, this is not always the case. $(Ph_3P)_3RhCl$-catalyzed tritiation of the 1,2-double bond of androsta-1,4-diene-3-one-

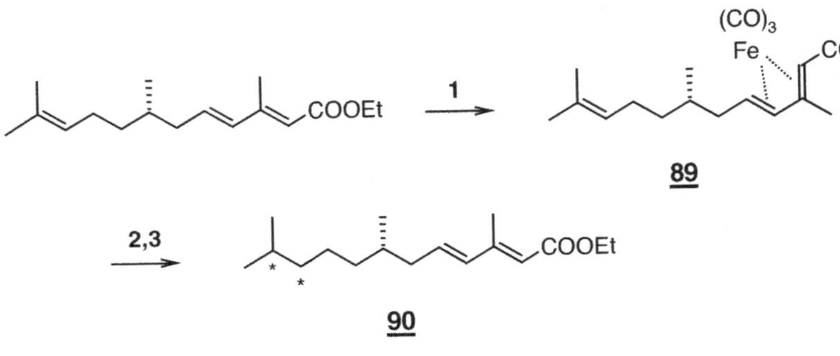


**Reaction conditions**: **1**. $Fe_3(CO)_{12}$, benzene; reflux, 7 h; **2**. $^3H_2$,Wilkinson´s catalyst, benzene; r.t., 15 h; **3**. $(NH_4)_2Ce(NO_3)_6$, $CH_3CN$; 0 °C, 4 h

***Figure 4.27*** *Selective tritiation of an isolated double bond in the presence of a dienoate moiety*

17$\beta$-carboxylic acid (**91**), an initial step in the labeling of numerous cortocoid derivatives[72], provided [1,2-$^3H_2$]androst-4-ene-3-one-17$\beta$-carboxylic acid with a specific activity of 18.7 Ci/mmol. Careful $^3$H NMR analysis showed that a 53% proportion of the label was incorporated as expected in the 1$\alpha$ and 2$\alpha$ positions (**92a**) and 8% 1$\beta$,2$\beta$ (**92b**), corresponding to a *cis*-addition on the $\alpha$-face and $\beta$-face, respectively; however, 20% resided in the 1$\alpha$,2$\beta$-positions (**92c**), a *trans*-addition product. In contrast, 10% Pd/C reportedly gives preferentially the 1$\beta$,2$\beta$-isotopomer[73]. Incomplete *cis*-selectivity was also observed in the tritiation of methyl 4-*N*-phthalimidobut-2-ynoate (**93**). Workup of the reaction product provided a 3:1-mixture of the labeled *cis* and *trans* isomers **94** and **95**[74].

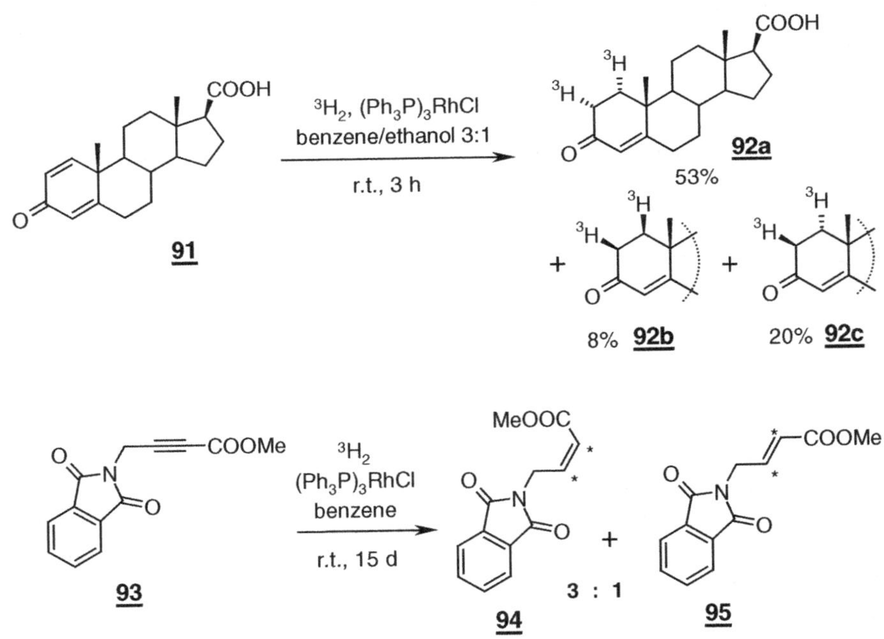


***Figure 4.28*** *Incomplete stereoselectivity in tritiations using Wilkinson's catalyst*

Unfortunately, isotopic scrambling with Wilkinson's catalyst cannot be excluded, although it is usually much less pronounced than in heterogeneously catalyzed reactions. Changing the catalyst system for the tritiation of cyclosporin A from 10% Pd/C to $(Ph_3P)_3RhCl$ resulted in a lower but still significant incorporation of tritium into the terminal methyl group of [$^3H_2$]dihydrocyclosporin A (dihydro- **13**, Figure 4.6; 30% vs 60%)[75]. Finally, the tritiation of (−)-$\Delta^{1,2}$-$\alpha$-acetylmethadol (**96**) was found to be accompanied by a 20 % exchange into both $\beta$-positions of the amino function of $\alpha$-[$^3$H]acetylmethadol (**97**)[76].

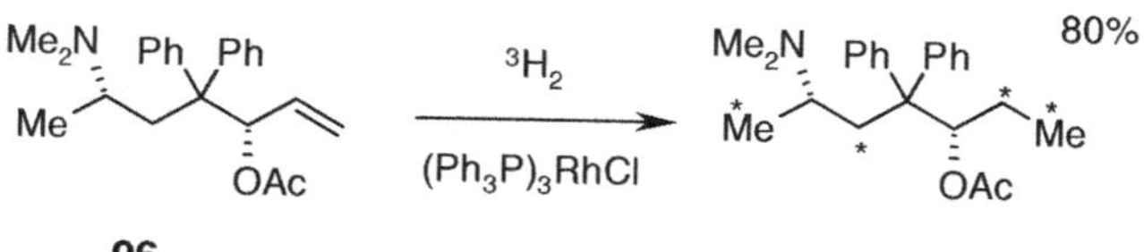


*Figure 4.29* Exchange side reaction in a tritiation using Wilkinson's catalyst

Surprisingly, the rapidly increasing variety of enantioselective procedures in organic synthesis[77] have found only very limited use in tritium chemistry. The only method that has gained some importance so far, including in commercial applications[78], is the enantioselective tritiation of prochiral *N*-acylated *Z*-$\Delta^{2,3}$-amino acids, esters or amides over rhodium catalysts containing chiral phosphine ligands. Ligands include (*R*)-PROPHOS (**98**), (*S,S*)-BPPM (**99**), (*R,R*)- or (*S,S*)-DIOP (**100**), (*R,R*)- or (*S,S*)-DIPAMP (**101**), (*R,R*) and (*S,S*)-CHIRAPHOS (**102**).

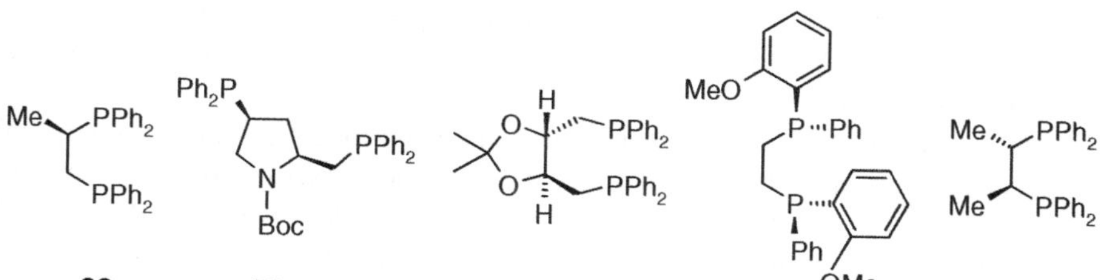


*Figure 4.30* Chiral ligands used for enantioselective tritiations of $\Delta^{2,3}$-amino acids

The commercial products (*S*)-[2,3-$^3H_2$]alanine, (*S*)-[2,3-$^3H_2$]phenylalanine, (*S*)-[2,3-$^3H_2$]tryptophan, (*S*)-[2,3-$^3H_2$]tyrosine, and (*S*)-[2,3-$^3H_2$]DOPA are reported to be produced by tritiation of the corresponding *N*-acetylamino acrylic acid precursors over [Rh(NBD)((*R*)-PROPHOS)]$ClO_4$[79].

*N*-acetyl-(*S*)-[2,3-$^3H_2$]phenylalanine (**104a**)[80], *N*-acetyl-(*S*)-[2,3-$^3H_2$]naphthylalanine (**104b**)[80] and *N*-acetyl-(*S*)-tryptophanamide (**104c**)[81], the first (*S*)-amino acid derivatives for which experimental details have been published, were obtained by tritiation (0.8–1.2 bar tritium gas) of the dehydro precursors in methanol over [Rh(cod)((*S,S*)-BPPM)]Cl at 25 °C (**103a,b**) and [Rh(cod)((*R,R*)-DIPAMP)]$BF_4$ at 40 °C (**103c**). Products **104a** and **104b** were obtained in 96% e.e. and about 44 Ci/mmol and **104c** in 94.4% e.e. and 58 Ci/mmol. Precursors **103a** and **103b** were readily prepared following routine Erlenmeyer procedures by reaction of the respective aldehyde with *N*-acetylglycine in the presence of NaOAc-HOAc or DCC and subsequent hydrolytic cleavage of the intermediately formed azlactones. Substrate **103c** was prepared by enzymatic dehydrogenation of *N*-acetyl-(*S*)tryptophanamide with tryptophan-side chain oxidase from *pseudomonas* or *chromobacterium violaceum*[81].

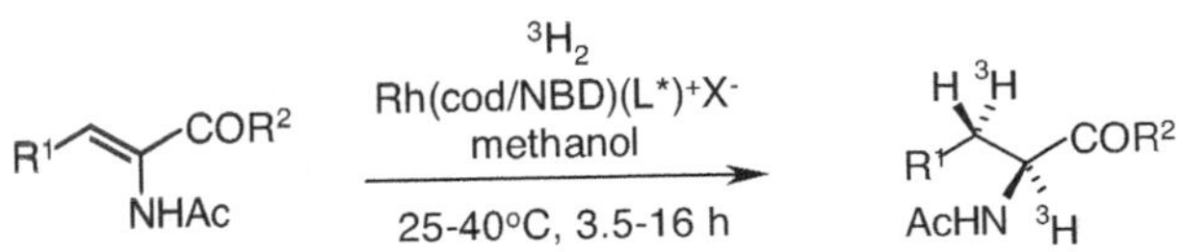


**103** **104**

**a**: R¹ = phenyl, R² = OH, cat. = [Rh(cod)((*S,S*)-(BPPM)]Cl
**b**: R¹ = naphthyl, R² = OH, cat. = [Rh(cod)((*S,S*)-(BPPM)]Cl
**c**: R¹ = indolyl, R² = $NH_2$, cat. = [Rh(cod)((*R,R*)-DIPAMP]$BF_4$

***Figure 4.31*** *Selected tritiated amino acid derivatives prepared by enantioselective olefin tritiations*

Satisfactory results were also obtained when dehydrodipeptide derivatives *N*-acetyl-ΔPhe-(*S*)-Phe-OMe and *N*-acetyl-ΔTrp-(*S*)-Phe-OMe were employed as substrates[82]. Tritiation over [Rh(cod)((*R,R*)-DIOP)]Cl and [Rh(cod)((*R,R*)-DIPAMP)]$BF_4$, respectively, furnished *N*-acetyl-(*S*)-[2,3-$^3H_2$]Phe-(*S*)-Phe-OMe and *N*-acetyl-(*S*)-[2,3-$^3H_2$]Trp-(*S*)-Phe-OMe with diastereomeric excesses >82%. Changing the chirality of the ligand of the first catalyst to (*S,S*)-DIOP resulted in the formation of the corresponding (*R,S*)-diastereomer in 80% d.e.

Exploratory studies with higher or homologous dehydropeptides and different chiral rhodium catalysts revealed that in some cases diastereomeric excesses did not even exceed 20%[83]. Furthermore, tritiation of two dehydropentapeptides over Rh-(*R,R*)-DIPAMP showed contradictory results. (ΔTrp)-LH-RH$^{1-5}$ gave an excellent selectivity of 99.5:0.5, whereas the diastereomeric excesses dropped to 0–20% when the same reaction conditions were applied to (ΔTrp)-pentagastrin. Accordingly, the diastereoselectivity of the tritiation of *N*-acylaminoacrylic acid precursors seems to be largely determined by the conformation of the substrate–catalyst complex. Unfortunately, at the present time there is no clear and practical predictive model available for matching the catalyst ligand with the intrinsic substrate stereofacial selectivity.

## 4.2 Catalytic Tritiolyses

In contrast to tritiations (reductions in which tritium is added to unsaturated substrates) tritiolyses are reductive reactions in which tritium replaces another atom or group. The most common class of tritiolyses is replacement of halogen (tritiodehalogenations), but some use has also been made of the replacements of oxygen and sulfur functions (tritiodeoxygenations and tritiodesulfurizations, respectively).

### 4.2.1 Tritiodehalogenations[84,85]

Catalytic dehalogenation of organic halides with tritium gas is one of the most frequently used methods for introduction of tritium into target molecules. A vast number of synthetic, bioactive compounds and research tools have been prepared this way. Advantages to this method include the ready accessibility of halogenated derivatives[86], the versatility of the method, achievement of high specific activities and the relative certainty as to the location of tritium so introduced. The most serious drawback of the method is that other functionalities sensitive to catalytic reduction (such as other halogens, multiple bonds, nitro and benzyl groups, etc.) may be competitively reduced. Issues of selectivity are addressed later in this Section.

Aryl halides are the most common type of substrate, although on occasion compounds containing $sp^3$ carbon–halogen bonds have been found to undergo dehalogenation, especially if they are activated by adjacent functionality such as carbonyl or aryl. Specific activities typically achieved are in the range of 20–28 Ci/mmol for tritiodeiodinations, 14–25 Ci/mmol for tritiodebrominations, and 8–16 Ci/mmol for tritiodechlorinations[84c]. Accordingly, although chlorine, bromine and iodine can be readily replaced by tritium, as demonstrated for $[^3H]$NGD 94-1 (**105**)[87], organic bromides and iodides undergo the reaction significantly more easily. Since the carbon–fluorine bond is highly resistant to tritiolysis, fluorinated compounds are poor substrates[85]. In general, alkyl halides are cleaved more slowly than aryl, allyl and vinyl halides [85b,c].

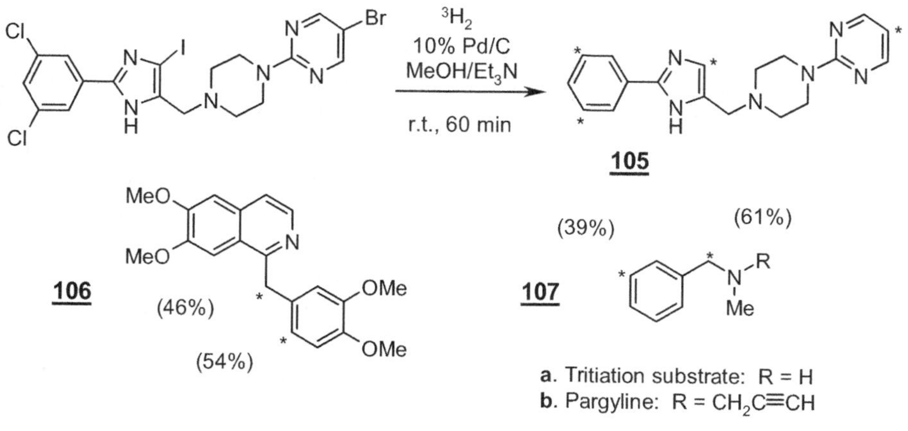


***Figure 4.32*** *Multihalogen tritiolysis; catalytic exchange as a side reaction*

The most active catalysts for many tritiolyses are palladium metal (freshly prepared from PdO) or palladium supported on charcoal, barium sulfate or aluminum oxide, but palladium catalysts are sometimes associated with side reactions such as *O*- and *N*-debenzylations, isomerizations and labeling of certain sites by catalytic exchange (see Chapter 3, Section 3.3). For example, in the case of $[^3H]$papaverine (**106**)[88] and the $[^3H]$pargyline precursor

(**107a**)[89], both prepared by tritiolysis of bromoaryl precursors, much of the tritium incorporated proved to be located at the benzylic positions. If selectivity in labeling is important, alternative catalysts must be considered. In general, however, extraneous labeling is less common in tritiodehalogenations than in olefin reductions, where isotopic scrambling is an additional risk (Section 4.1.1).

The supported catalysts are especially recommended, since they are more resistant to poisoning and offer a larger effective active metal surface area. Raney nickel has also found some applications, although it is more widely used in tritiodesulfurization reactions. Platinum and rhodium catalysts are much less effective in tritiodehalogenations[84a,85,90]. Best results are obtained if catalysts are pre-reduced with hydrogen or tritium then dried, and solvents rigorously dried; otherwise $^{3}HHO$ may be formed during the reaction, which together with the catalyst may induce $^{3}H/H$ exchange reactions (Chapter 3, Section 3.2)[84a,90].

Tritiodehalogenations are normally performed in the presence of a base such as triethylamine, diisopropylethylamine (Hunig's base), quinoline, alkali metal hydroxide, sodium or potassium carbonate, phosphate buffer pH 8.6 etc., in order to neutralize the tritium halide formed, thereby reducing its poisoning effect on the catalyst and enhancing the rate of carbon–halogen bond cleavage[84a,90]. Simultaneously, base suppresses potential acid-catalyzed $^{3}H/H$ exchange reactions.

As with tritiations of carbon–carbon multiple bonds, catalytic tritiodehalogenations are favored by polar and protic solvents, such as methanol, ethanol, and water. On the other hand, catalyst-mediated $^{3}H/H$ exchange between solvent and tritium gas may give rise to a significant reduction of the specific activity of the substrate (see solvent exchange at the beginning of this chapter). Even though this effect is less pronounced than for tritiations, aprotic solvents (e.g., DMF, THF, dioxane, ethyl acetate) should be used whenever high specific activities are required.

Palladium-catalyzed tritiodehalogenation of peptides containing 3,5-diiodo-(*S*)-tyrosine, 2,5-diiodo-(*S*)-histidine or 4-iodo/bromo-(*S*)-phenylalanine residues is one of the standard procedures for the preparation of labeled peptides of high specific activity[16a,91–96]. Depending on their structure and the presence or absence of sulfur-containing amino acids, tyrosine- and/or histidine-containing peptides or fragments can be directly iodinated with ICl or NaI and Chloramine-T®, $KI_3$ or $I_2$–$CF_3COOAg$ in TFA–dichloromethane[92,86e]. In an alternative approach, which has been applied to e.p. 4-iodo/bromo-Phe, the label is introduced either via a *de novo* synthesis starting with the labeled amino acid ([2,5-$^{3}H_2$]-Tyr[94]; [2,5-$^{3}H_2$]His[95], [4-$^{3}$H]Phe[95,96]) or by coupling the halogenated analog to an appropriate small fragment followed by tritiodehalogenation and construction of the respective labeled peptide through fragment coupling[93–96]. Sulfur-containing peptides or fragments tend to poison the catalyst, thereby slowing the tritiodehalogenation process and favoring dilution of tritium gas by catalyst-mediated $^{3}H/H$ exchange processes, thereby giving rise to products of low specific activity. In addition, tritiodesulfurization of cysteine may occur, producing labeled alanine. Occasionally, extraneous labeling may occur by catalytic exchange, such as in the susceptible imidazole-C2 position of histidine during the tritiodebromination of a dibromotryptophan-containing hexapeptide[97].

Tritiodehalogenation of appropriately chlorinated, brominated or iodinated precursors over PdO or Pd/$CaCO_3$ has emerged as a main source for commercially available tritiated nucleosides and nucleotides of high specific activities such as [2-$^{3}$H]adenosine (20–25

Ci/mmol), [8-$^3$H]adenosine (20–30 Ci/mmol), [5-$^3$H]cytidine (10–30 Ci/mmol), 2′-deoxy [5-$^3$H]cytidine (10–30 Ci/mmol), 2′-deoxy[5-$^3$H]uridine, (5–15 Ci/mol), [8-$^3$H]guanosine (10–30 Ci/mmol), [methyl-$^3$H]thymidine (25–30 Ci/mmol), and [5-$^3$H]uridine[84a,98]. In some cases simultaneous incorporation of tritium into remote positions by $^3$H/H exchange processes has been reported. For example, incorporation exclusively into adenine-C2 has recently been accomplished during the tritiodehalogenation of a bromoadenosine-5′-triphosphate in phosphate buffer pH 8.6 in the presence of PdO as a catalyst[99].

Since most ligands employed in receptor binding studies contain aromatic or heteroaromatic systems, it is not surprising that the preparation of substrates of high specific activity starting from polyhalides is common. Depending on the number of halogen atoms incorporated, specific activities up to 200 Ci/mmol have been accomplished. Tritiodebromination of **108** (X = Br; $^3H_2$, 10% Pd/C, $Et_3N$, DMF–THF), for example, furnished **108** (X = $^3$H), an adenosine receptor antagonist[100], and of **109** (X = Br → $^3$H), a phencyclidine receptor ligand[101], with specific activities of 69 Ci/mmol and 97 Ci/mmol, respectively. Via the introduction of more than eight bromine atoms into **110**, an angiotension II inhibitor, a specific activity of 209 Ci/mmol was achieved upon exhaustive tritiodebromination ($^3H_2$, 10% Pd/C, $Et_3N$, EtOAc)[102].

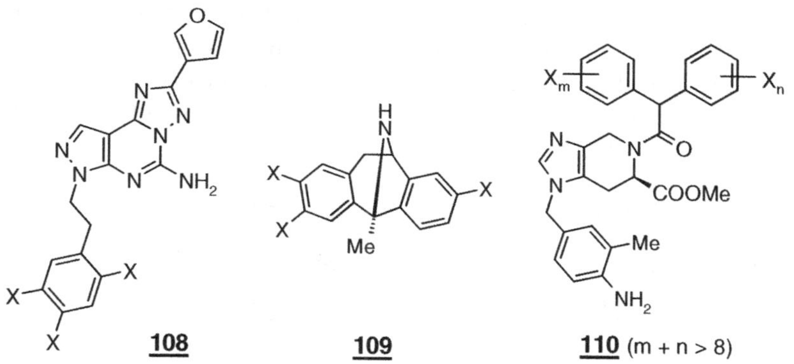


108    109    110 (m + n > 8)

X = Br→ $^3$H in all compounds

***Figure 4.33*** *Tritiodehalogenation of polyhalides*

Since cytochrome P450-mediated hydroxylation is one of the main metabolic pathways of aromatic systems[103], many compounds labeled by tritiodehalogenation in these subunits suffer from insufficient biological stability of the label and can therefore not be used for registration studies on pharmacokinetics and metabolism in animal and man. Their area of application is mostly restricted to *in vitro* investigations and, to some extent, to early pharmacokinetic studies on investigational drugs, in order to gain some preliminary information on their *in vivo* behavior.

While catalytic tritiodehalogenation of aromatic halides plays a dominant role in the preparation of substrates of high specific activity, aliphatic halides are more often treated with strong tritide reagents (see Section 4.3). However, sp$^3$ halides activated by a neighboring functional group such as aryl-, heteroaryl, or carbonyl are often viable substrates for catalytic treatment. Among those are benzylic[104] and allylic halides[105] Others activated by enolizing functions can be converted to the corresponding tritiated compounds, but care has to be taken lest the label is lost via subsequent acid/base-catalyzed enolization. For example, [3-$^3$H]-camphor (**111**) was prepared by catalytic tritiolysis of 3-bromocamphor then immediately treated with phenylmagnesium bromide/ether to give [3-$^3$H]phenylborneol (**112**); the material had a specific activity of approximately 13 Ci/mmol[106].

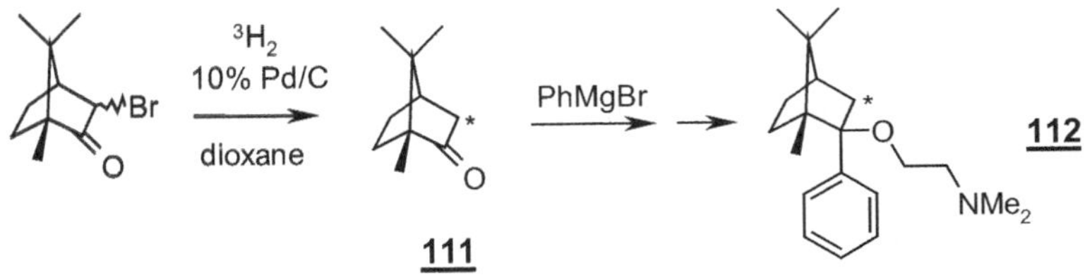


*Figure 4.34* *Tritiodehalogenation of an activated alkyl halide*

Methods for the preparation of tritiated methyl iodide in the laboratory use variations of the sp$^3$ tritiodehalogenation approach. Palladium-catalyzed tritiodechlorinations of chloromethyl ester **113** and of chloromethyl sulfide **115**[107] have been developed as less hazardous alternatives to the older laboratory procedure for $C^3HH_2I$ at 15–22 Ci/mmol starting from bis(chloromethyl)ether[108]. The initially formed nonvolatile monotritiated **114** and **116** can be conveniently isolated, purified and finally subjected to ester cleavage with either lithium iodide or hydriodic acid, or, in the second case, by thermolysis of the intermediately generated sulfonium salt **117**.

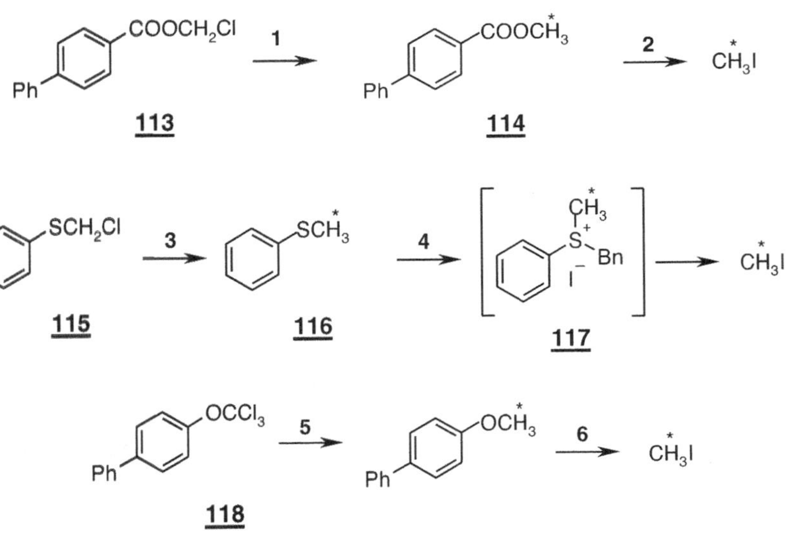


**Reaction conditions:** **1.**$^3H_2$, 10 % Pd/C, Hunig's base, DMF; r.t., 1–12 h; **2.**LiI, DMF, 140 °C, 2-3 h **or** HI, $H_3PO_2$, $Ac_2O$; 0 °C, 30 min, 125 °C, 3 h; **3.**$^3H_2$, 30 % Pd/C, $Et_3N$, EtOAc; r.t., 1 h; **4.** BnI, EtOAc, 140 °C, 24 - 48 h; **5.**$^3H_2$, 30 % Pd/C, proton sponge, EtOAc; r.t., 2 h; **6.** HI, HOAc; 130 °C, 2 h

*Figure 4.35* *Preparations of tritiated methyl iodide*

Tritiated methyl iodide of much higher specific activity can be prepared by the related procedure involving catalytic tritiodehalogenation of 4-trichloromethoxybiphenyl (**118**). Conducted as exemplified in Figure 4.35, it provided [$^3$H]methyl iodide at 56 Ci/mmol (theoretical maximum 86.3 Ci/mmol)[109]. This method is a laboratory-friendly alternative to the commercial process, which involves the high-temperature, high-pressure catalytic reduction of $CO_2$ to $C^3H_3OH$[110].

[3-$^3$H]Propionic acid (**119**) is an additional example of a highly useful low molecular tritiated building block prepared by sp$^3$ tritiodehalogenation. It is conveniently accessible at a specific activity of 18.9 Ci/mmol by 10% Pd/C catalyzed tritiolysis of 3-bromopropionic acid in the presence of triethylamine. Its applications include the synthesis of 2-(2,4-dichlorophenoxy)-[3-$^3$H]propionic acid (**120**), a fine herbicide[111], and the biosynthetic preparation of [$^3$H]geldanamycin (**121**) using biochemical techniques[112] (see Chapter 12).

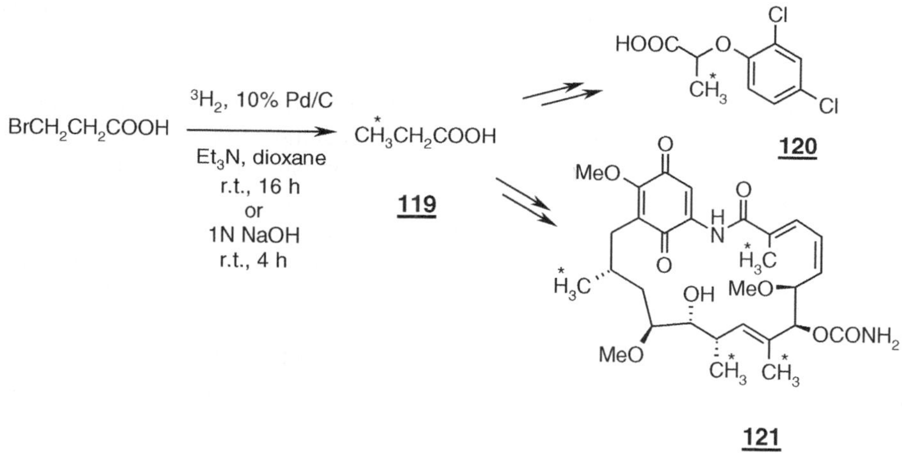


***Figure 4.36*** *Tritiodehalogenation at an alkyl carbon*

A strategy to avoid a potentially hazardous release of highly tritiated volatile building blocks is to operate on nonvolatile derivatives from which they can be released as required. This is illustrated for high specific activity acetic acid and acetaldehyde by the catalytic tritiodehalogenation of derivatives **122** (X = Br → $^3$H)[113] and **123** (X = Cl → $^3$H)[114], respectively. By following this approach, the handling of volatile, highly radioactive [$^3H_2$]-acetic acid and [$^3H_3$]acetaldehyde could be avoided.

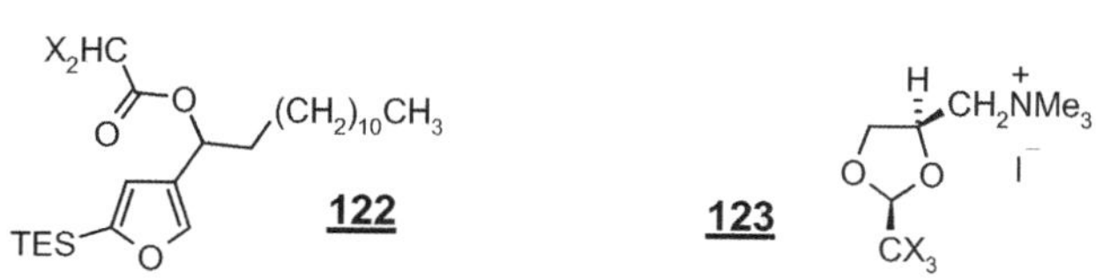


***Figure 4.37*** *Devolatilized building blocks*

*Interhalogen Selectivity*

In tritiodehalogenations, various degrees of selectivity are achievable, both between different halogens and between halogens and other reducible functions in the same molecule.

Aryl carbon–halogen bond dissociation energies increase in the order I (223 kJ/mol) < Br (281 kJ/mol) < Cl (340 kJ/mol) < F (453 kJ/mol), and halogens' electronegativities increase as I < Br < Cl < F; consequently the ease with which halogen atoms are replaced by tritium should be in the order I > Br > Cl > F, assuming identical environments[84]. However, measured dehalogenation rates of different halo groups differ substantially both in magnitude and in relation to each other depending upon conditions. In mixtures of substrates reduced with hydrogen (or isotopic hydrogen) gas, the order has been found to be I > Cl > Br in one study[84c], and I > Br > Cl in another[115]. Reductions of individual halobenzenes under identical conditions (1 atm $H_2$, 10% Pd/C, MeOH, $Et_3N$) went to completion in 6 min (PhCl), 14 min (PhBr) and 40 min (PhI)[115]. Different orders have been found under other hydrogenolysis conditions[116]. The explanation for these discrepancies is that factors such as substrate adsorption and desorption rates and hydrogen availability on the catalyst surface are more important than bond strengths in determining the dehalogenation rates of different halogens, as well as the rates of any competitive hydrogenations (such as the double bonds of halo-alkenes). The adsorption constant for simple halobenzenes to palladium increases as PhCl < PhBr < PhI[117], and the catalyst surface in chloroarene reduction is hydrogen rich whereas that in iodoarene reduction is hydrogen poor[84]. These factors are of minor importance in the tritiolyses of monohalogenated substrates, as the rates for all three halogens are relatively fast. It does become important, however, when there are two or more different halogens, whether in the same molecule or different molecules in the same solution. In competitive reductions of mixtures of PhI, PhBr and PhCl (1 : 1 : 1), reactions of PhBr and PhCl do not begin until PhI is reduced, and PhCl not until PhBr is nearly consumed[116,118]. In tritiolyses of multihalogenated substrates, this pattern of inhibition can account for the selectivity frequently observed. There are numerous examples of selective tritiodeiodination in the presence of aryl chloro functions, as illustrated by **124**[119] and **125**[120], and there are also examples, such as **126**[121], of selective tritiodeiodination in the presence of an activated $sp^3$ chloro function.

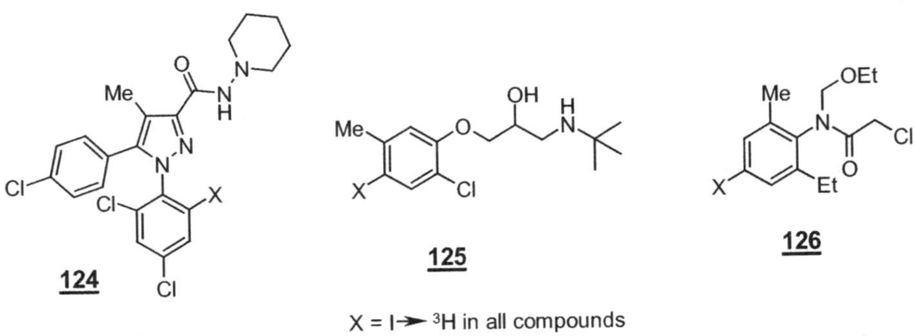


**Figure 4.38** *Selective tritiodeiodinations*

Although the presence or absence of base is often unimportant for iodo-vs-chloro selectivity, it is often crucial for bromo-vs-chloro selectivity. Selective tritiodebromination in the presence of chloro substituents was observed in the tritiolyses of **127**[122] ($^3H_2$, EtOH/THF, 10% Pd/C) and of **128**[123] ($^3H_2$, DMF, 10% Pd/C) in the absence of added base, giving monotritiated dichloro products. However, when the reactions were performed with added triethylamine, the carbon–chlorine bonds also underwent tritiolysis. Similarly, with ethanol as solvent and no added base, treatment of **129** ($^3H_2$, EtOH, 10% Pd/C) gave only tritiodebromination and the chloro substituent was preserved[124]; the effect (if any) of added base was not reported in this case. An alternative approach recommended the use of $Pd/BaSO_4$ in the absence of a base to achieve high bromo-vs-chloro selectivity for a series of compounds including **130** and **131** ($^3H_2$, 5% $Pd/BaSO_4$, EtOH or DMF)[125].

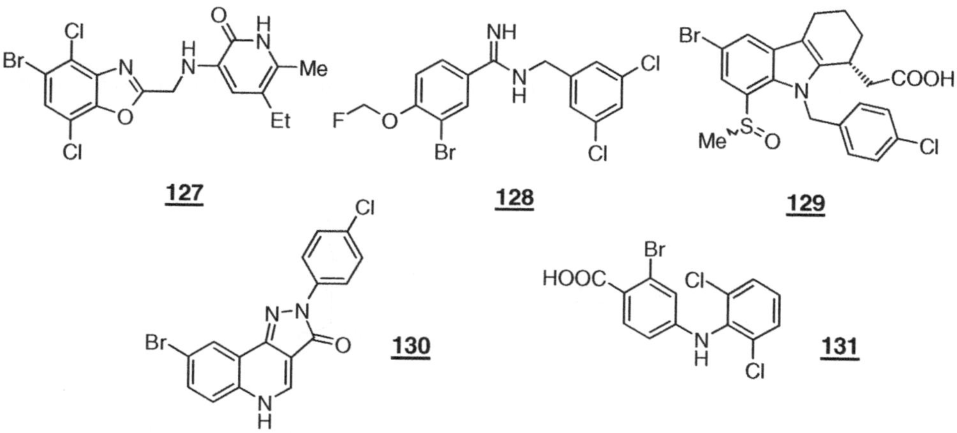

*Figure 4.39 Bromine-vs-chlorine selectivity in polar/protic solvents is often better in absence of base*

On the other hand, in the presence of base, Br-vs-Cl selectivity was observed for compounds **132**[126], **133**[127] and **134**[128] ($^3H_2$, Pd/C, $Et_3N$, THF) and **135**[129], ($^3H_2$, Pd/C, *i*-$Pr_2NH$, THF), all giving products that retained their chloro groups (Figure 4.40). The combination of base and highly polar solvent (protic or DMF) in the former cases (Figure 4.39) seems to accelerate tritiodechlorination too much to allow chemoselectivity, whereas in less polar (ethereal) solvents (Figure 4.40) tritiodehalogenations are slower, so that selective cleavage of C–Br bonds is possible even in the presence of base. In any case, prior testing of reaction conditions using hydrogen or deuterium is recommended before embarking on tritiolyses of new substrates.

While there are numerous reports of successful bromo-vs-chloro selectivity in tritiolyses, selective tritiodeiodination in the presence of bromine seem to be more difficult and less promising. So far only the Raney nickel- promoted tritiodeiodination of 4-bromo-2-iodo-6-trifluoromethylaniline (**136**) has been published ($^3H_2$, Raney nickel, NaOH)[130] (Figure 4.41).

Selectivity between chloro and fluoro in tritiolyses is not a problem because of the inertness of the latter halogen to catalytic reduction under mild conditions. A simple example is (4*S*)-8-chloro-6-fluorospiro[chroman-4,4′-imidazolidine]-2′,5′-dione (**137**), reported to have been tritiodechlorinated ($^3H_2$, 10% Pd/C, 15 equiv. $Et_3N$, EtOH; r.t., 18 h) to give the corresponding 6-fluoro[8-$^3H$]derivative at a specific activity of about 10 Ci/mmol[131].

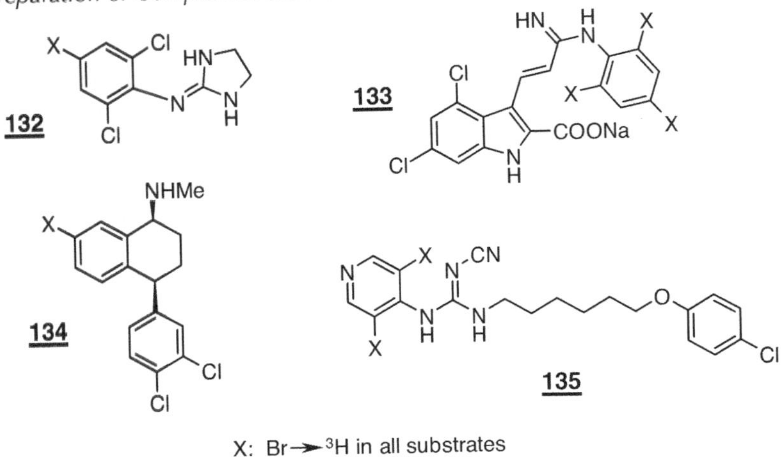


*Figure 4.40* *Bromine-vs-chlorine selectivity in ethereal solvents is often insensitive to the presence of base*

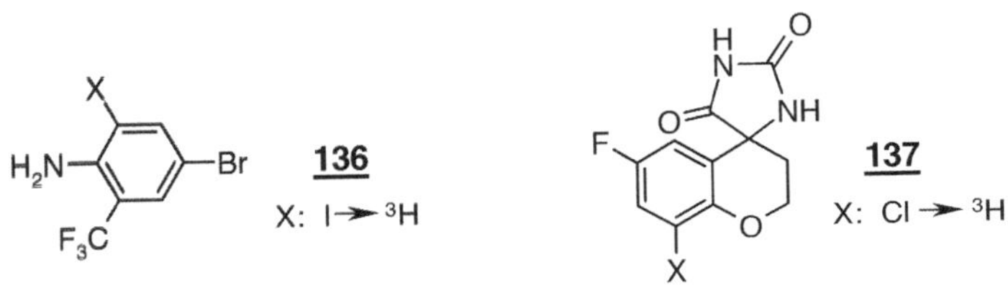


*Figure 4.41* *Iodo-bromo and chloro-fluoro selectivity in catalytic tritiolyses*

*Selectivity vs Other Reducible Functions*

The easiest halogen to remove selectively in the presence of other reducible functionality is iodine, which follows from the higher propensity for this halogen to bind catalyst active sites, as discussed in the previous section. One survey has determined that 10% Pd/C and methanol solvent with ten equivalents of amine base gives excellent chemical yields and isotope incorporations in the selective fission of aryl C–I bonds in the presence of other reducible functions such as carbonyl, diazirenyl, azido and aryl nitro; among those reported for isotopic hydrogen were **138**[115], **139** and **140**[132a], and **141**[132b] (Figure 4.42). All of these compounds are photophores, useful as reagents for photoaffinity labeling of proteins. Related reaction systems in which selective tritiodeiodination was achieved in the presence of carbon– carbon double bonds include the aryl iodo compounds **142**[133] and **143**[134], the primary alkyl iodo compound **144**[135], and the iodouracil derivative **145**[136].

The existence of an inhibitory effect of triethylamine on catalyst activity, upon which the feasibility of this discrimination of reductive reactivity is postulated to rest, is evident in the tritiodeiodination of the olefinic double bond-containing [β-iodo-Abu]cyclosporin A (**146**, R = I)[137] (Figure 4.43). Only traces of [β-$^3$H-Abu]cyclosporin A (**146**, R = $^3$H) could be

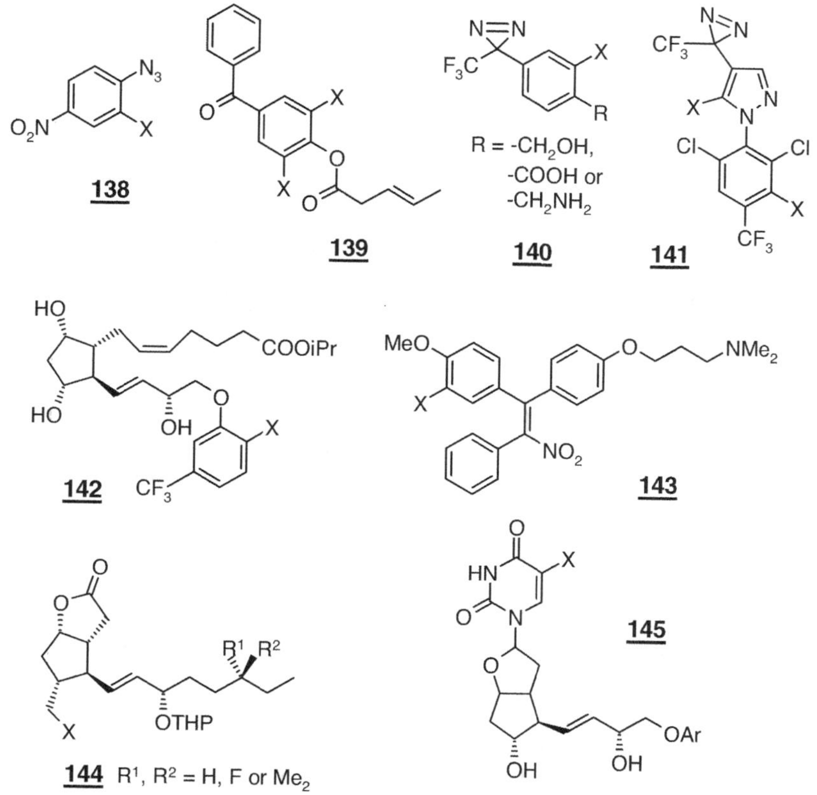


all substrates: X = I; products **138**, **139**: X = ²H; products **140**-**144**: X = ³H

**Reaction conditions:** **138** and **139**: $^2H_2$, 10 % Pd/C, $Et_3N$ (10 equiv.), MeOH; r.t.; **140**: $Et_3N$, $^3H_2$ (64–68 % chemical yields, 25–26 Ci/mmole); **141**: $^3H_2$, 10 % Pd/C, $Et_3N$ (30 equiv.), EtOAc; r.t., 4h (15 Ci/mmol); **142**: $^3H_2$, 10 % Pd/C, EtOAc,. $Et_3N$ (2 equiv.), $K_2CO_3$ ; 20 min (23 Ci/mmol); **143**: $^3H_2$, 5 % Pd/$Al_2O_3$, DMF with no added base; r.t., 25 min (27.7 Ci/mmol); **144**: $^3H_2$, 10 % Pd/C, MeOH, $Et_3N$ (3 equiv.); 35 min (10–15 Ci/mmol); **145**: $^3H_2$, Pd/C, dioxane, NaOH (24 Ci/mmol)

***Figure 4.42*** *Deuteriolyses and tritiolyses in the presence of other reducible functions*

isolated when the iodo substrate was treated with tritium gas over 10% Pd/$CaCO_3$ in the presence of equimolar amounts of triethylamine. Increasing the amount of base 10-fold, however, provided **146** (R = $^3$H) in 37% yield at a specific activity of 7.3 Ci/mmol; greater amounts of base completely halted the tritiodeiodination. On the other hand, tritiodebromination of the more reactive allylic bromo compound [ω-bromo-Mebmt]Cyclosporin A (**147**, R = Br)[138] proceeded satisfactorily in the presence of a much larger amount of base (10% Pd/C, MeOH, 67 equiv. of $Et_3N$, 7 min) to give a good yield of [ω-$^3$H-Mebmt] Cyclosporin A (**147**, R = $^3$H) having a specific activity of 4.3 Ci/mmol (using a 15 : 85 mixture of $^3H_2$ : $H_2$).

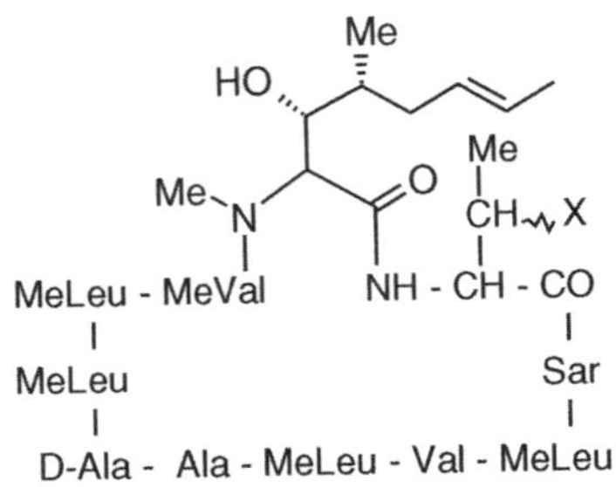


**146** X = I → $^3$H

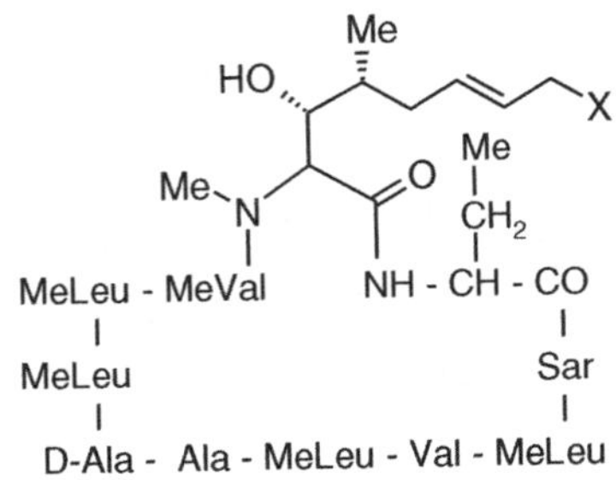


**147** X = Br → $^3$H

**148**
X = Br → $^3$H

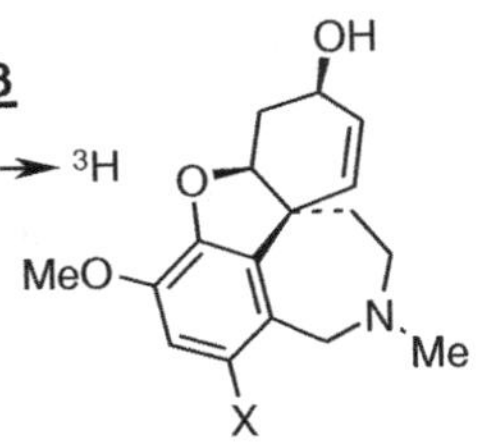


**149**
X = Br → $^3$H

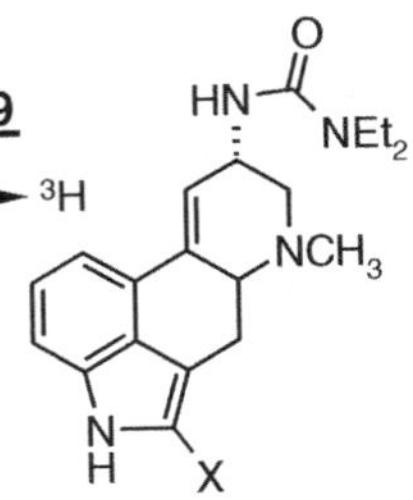


*Figure 4.43* Selective tritiolyses in olefinic substrates

Other partially poisoned catalyst systems that allow selective tritiodebrominations in the presence of olefinic double bonds include 10% Pd/C doped with triethylamine and butanethiol and the Lindlar catalyst (Pd–$BaSO_4$–PbO), which were exploited for the synthesis of [1-$^3$H]galantamine (**148**, $^3H_2$, 10% Pd/C, 4 equiv. $Et_3N$, 0.2 equiv. *n*-BuSH, THF; 48°C; 2.5 Ci/mmol)[139] and [2-$^3$H]lisuride (**149**, $^3H_2$, Lindlar catalyst; $Et_3N$, THF)[140].

It should be expected that $\alpha,\beta$-unsaturated carbonyl systems, because of their lower reactivity, could be preserved even under more vigorous tritiodehalogenation conditions. This indeed was the case for the multi-tritiodeiodinations of substrates **150** (5% Pd/$Al_2O_3$, DMF, 1 atm. $^3H_2$, 3–4 h)[105], and **151** ($^3H_2$, Pd/$CaCO_3$, benzene; r.t., 1 h)[141] (Figure 4.44). Alternatively, for allylic alcohol derivatives steric blocking of the hydroxy group can be used to slow the reduction of the double bond. Thus with **152**, the olefinic bond was reduced in competition with the aryl iodo substituent ($^3H_2$, 10% Pd/C, $Et_2O$ or EtOH, 11–13 equiv. $Et_3N$) even at short reaction times, but in **153** it was reduced only to the extent of 3% or less in reactions of 3 h duration, regardless of the substitution at the phenolic hydroxyl[142].

Selective tritiodehalogenations in the presence of carbonyl functions are usually not problematic when palladium catalysts and basic or neutral media are used. Examples of selective tritiodehalogenations include aryl bromides **154** ($^3H_2$, 10% Pd/C, xs $Et_3N$; r.t., 2h; 50.7 Ci/mmol)[143] and **155** ($^3H_2$, 10% Pd/C, 3 equiv. $Et_3N$, THF; r.t., 1 h; 38 Ci/mmol)[144], benzylic bromide **156** ($^3H_2$, 10% Pd/C, MeOH; r.t., 4 h; 12.5 Ci/mmol)[145] and substrate **139** above. The specific activity of **155** indicates that tritiation had also occurred elsewhere in the molecule, probably in the benzylic position.

Although it is generally difficult to discriminate between catalytic tritiodehalogenations of aryl bromides and the reduction of aromatic and allylic nitro groups (in contrast to iodo-nitro selectivity as described earlier in this section), there are some examples demonstrating

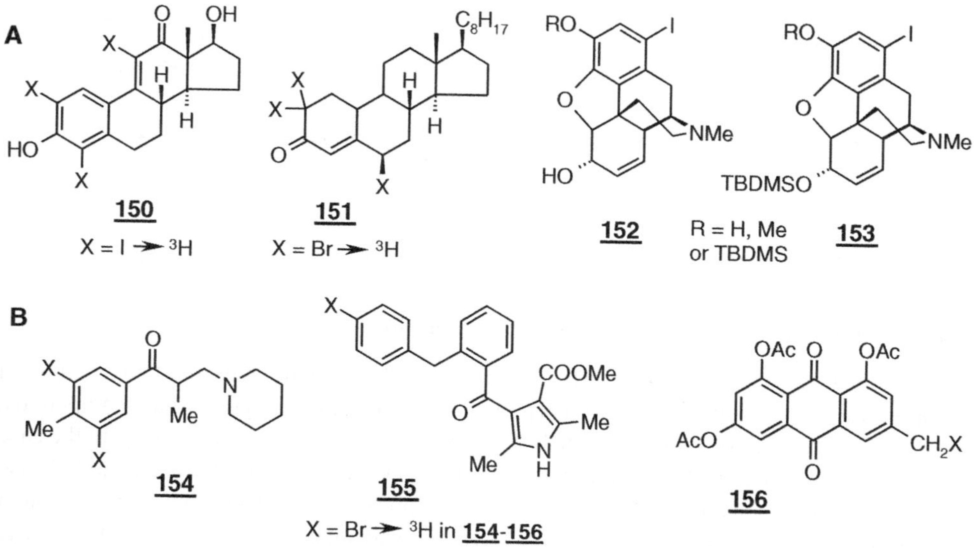


***Figure 4.44*** *A. Facile discrimination between tritiodehalogenations and conjugated or sterically hindered olefins olefinic bonds; B. compatibility of carbonyl groups to catalytic tritiodebrominations*

that, at least in special cases, selective debrominations can be accomplished. Treatment of 4′-bromonimodipine (157)[146] and of 2′,4′-dibromo-BAYx9228 (158)[147] with tritium gas over 20% $Pd(OH)_2/C$ (Pearlman's catalyst) in the presence of a large excess of triethylamine furnished the corresponding tritiated compounds in 30–40% yield with specific activities of 16.4 Ci/mmol (using 1.8 equivalents of tritium gas at 44 kPa) and 47.9 Ci/mmol (using 14 equivalents of tritium at 65 kPa), respectively.

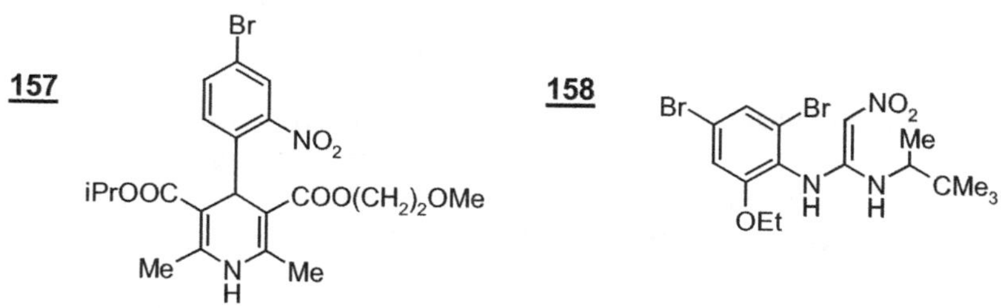


***Figure 4.45*** *Selective tritiodebrominations in the presence of nitro functions*

$Pd(PPh_3)_4$ with sodium formate, already known for its ability to dehalogenate aryl halides in the presence of other reducible functions[148], has been shown[149] to selectively deuteriodehalogenate several test substrates (Figure 4.46) using deuterium gas. Yields were 67–96% and deuterium incorporation levels were around 90%. Use of this method for tritiodehalogenation apparently has not been reported, but as it utilizes a stream of deuterium gas through a reaction maintained at 100 °C, significant procedural modifications would probably be required to make it practical for tritiations.

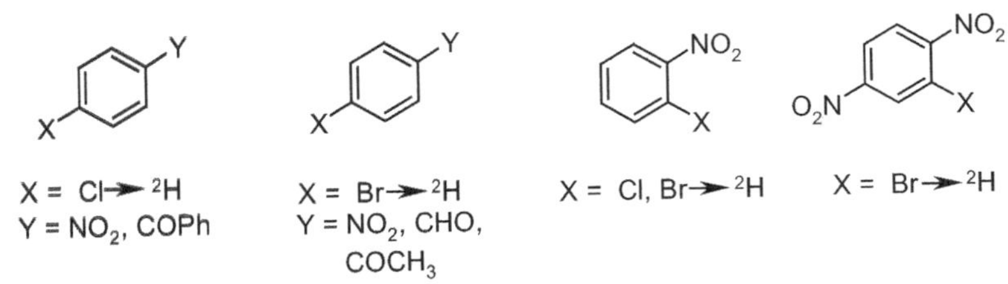


*Figure 4.46* Selective deuteriodehalogenations catalyzed by $Pd(PPh_3)_4$

### 4.2.2 Tritiolyses of Benzylic *N*- and *O*-Functions

Palladium catalysts are well known to cleave *N*- and *O*-benzyl moieties; this can either be an opportunity or a serious side reaction in palladium-catalyzed tritiolyses of halo compounds. Again, the presence of an appropriate base may enhance the desired selectivity. In the case of **159** (X = Br), tritiodebromination over 5% Pd/C in ethanol in the absence of added base produced significant amounts of debenzylated bromo compound **160**, whereas reduction in pyridine provided exclusively **159** (X = $^3$H) in excellent yield and at high specific activity[150]. Analogous results were obtained with the compound **134** above. In contrast, conditions giving satisfactory selectivity between tritiodebromination of **161** could not be found and the more reactive iodo analog **162** had to be resorted to (6 kPa $^3H_2$, $Pd(OH)_2$, EtOH, 4h)[151].

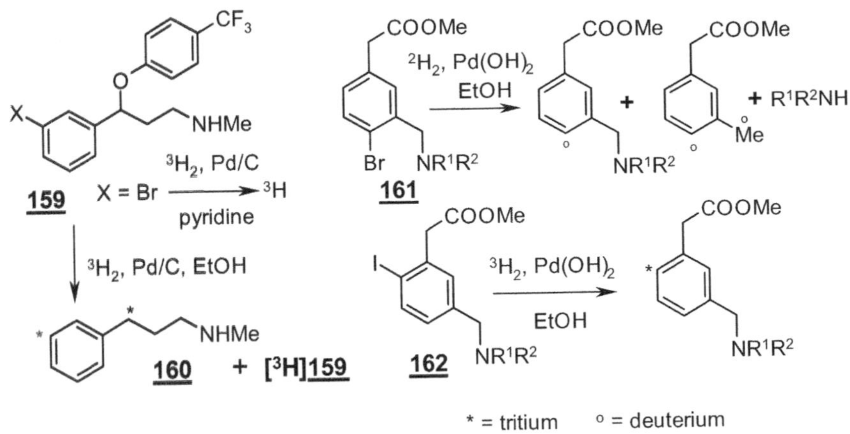


*Figure 4.47* Selective tritiodehalogenations in the presence of benzyl functions

Palladium catalysts are generally recommended for the hydrogenation of aldehyde and ketone functions adjacent to aromatic and heteroaromatic rings. Continuation of reactions beyond uptake of one molar equivalent of hydrogen can lead to reductive cleavage of the carbon–oxygen bond, especially at elevated temperature and/or in the presence of acid. With tritium gas, the introduction of up to two tritium atoms in the resulting methyl or methylene group is expected, but the protic solvents and vigorous reaction conditions needed for the reaction may result in reduced specific activities because of tritium–protium exchange with solvent. Unfortunately, examples to guide expectations are few. In one such, reduction of **163** furnished **164**, a selective intracellular histamine receptor ligand, with a specific activity of 30.5 Ci/mmol, about half of the theoretical level[152].

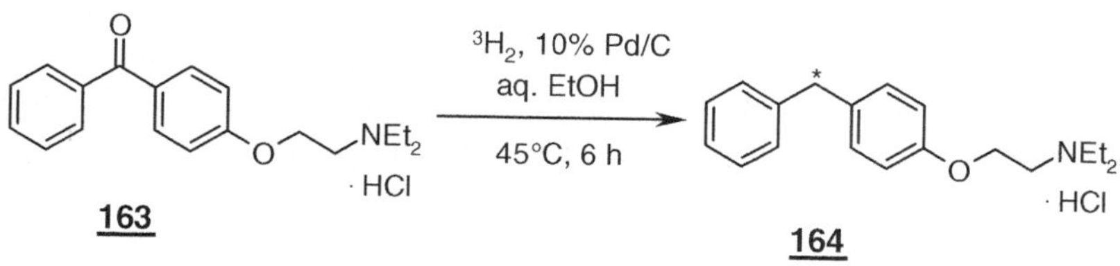


*Figure 4.48* *Tritiodeoxygenation of a benzophenone derivative*

Of commercial interest is the palladium-catalyzed tritiodeoxygenation of 5-formyluracil (**165**) to [methyl-$^3H_n$]thymine (**167**)[153]. Aldehyde **165** could be reduced directly to **167** in the presence of acid, but the specific activity of the produce was only 29.7 Ci/mmol, about half of the theoretical level. The low pH of the protic medium probably contributed to this through relatively rapid tritium gas exchange with solvent. When the reaction sequence was carried out in two steps, by reduction of **165** to **166** in pH 7 buffer followed by tritideoxygenation in dioxane in the presence of $PdCl_2$ to form the requisite acid *in situ*, [methyl-$^3H_n$]thymine was obtained with a specific activity of 68.9 Ci/mmol, even higher than the 58 Ci/mmol expected for incorporation of two tritium atoms. In this case, $^3H/H$ exchange of the formyl proton during the first step was proffered as the explanation for the increased tritium content of the product.

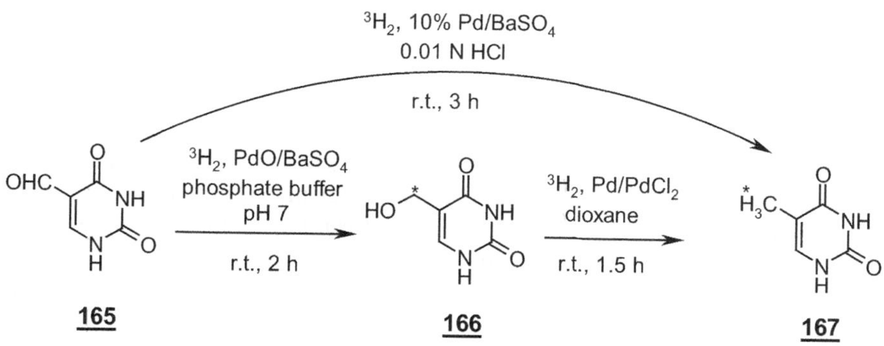


*Figure 4.49* *Preparations of tritiated thymine by tritiodeoxygenation of 5-formyluracil*

### 4.2.3 Tritiodesulfurizations

Although nearly all unsaturated groups can be reduced with Raney nickel, this catalyst is most frequently used for the hydrogenolysis of sulfur compounds such as disulfides, thioacetals and thioamides[154]. When freshly prepared, it contains 20–100 mL absorbed hydrogen per gram of nickel. The bound hydrogen can be partially replaced with tritium by treating the catalyst with tritium gas, but the specific activities of reduced products are significantly lower than for Pd-catalyzed reactions. For example, Raney nickel treated with tritium gas at a pressure of 72 kPa for 30 min gave products with specific activities of 2.5–7.5 Ci/mmol per incorporated tritium atom (vs the theoretical 28.8 Ci/mmol)[155]. In spite of the expediency of Raney nickel-catalyzed tritiodesulfurization for rapid labeling of certain compounds, very few examples of its use have been published so far.

Applied to 5-(1,3-dithiacyclopent-2-yl)uracil (**168**) and 3β-acetoxy-7-spiro(1′,3′-dithiacyclopent-2′-yl)cholest-5-ene (**170**), Raney nickel tritiolyses furnished [5-methyl-$^3$H]-thymine (**169**) and [7-$^3$H]cholesteryl acetate (**171**) in moderate yields with specific activities of 11.6 Ci/mmol and 8.1 Ci/mmol, respectively[155].

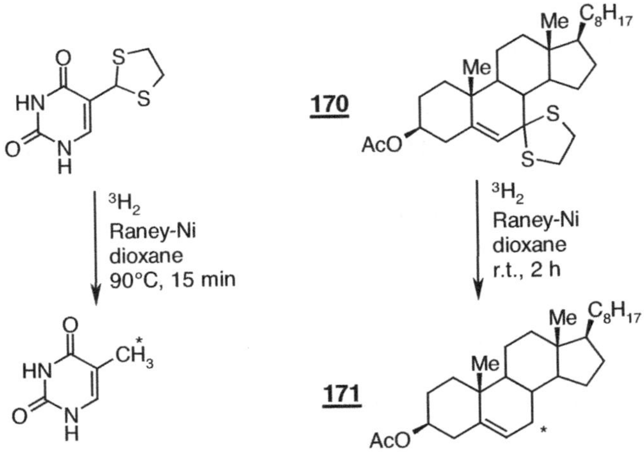


*Figure 4.50 Tritiodesulfurizations of a dithioacetal and a dithioketal*

## 4.3 Tritide Reductions

Owing to the availability of a variety of tritide reagents each having unique reactive properties, tritide reductions of functional groups offer much more control of chemoselectivity and regioselectivity in the introduction of tritium, compared with the catalytic processes discussed in the preceding sections. Products of tritide reductions very rarely are found to have suffered from isotope scrambling or exchange side reactions. However, isotope dilution by solvent or adventitious water and hydrogen–tritium kinetic isotope effects can be significant. The impacts of both depend a great deal on experimental variables; a brief discussion of these phenomena is provided below. The simple tritide reagents reduce carbonyls and imines to give racemic products, or may produce diastereomeric products in substrates possessing ancillary chiral centers. However, higher degrees of stereocontrol can often be achieved by using bulky reagents (e.g., Li T-Selectride), chiral tritides, or common achiral tritides in the presence of appropriate chiral catalysts (e.g., Corey–Itsuno).

As for the utility of compounds labeled by tritide reduction, the label, being introduced adjacent to a heteroatom, can be vulnerable to metabolic or other biological oxidative reactions (such as enzymatic oxidative deamination), and therefore may be unsuitable for certain studies. Pilot studies *in vitro* or *in vivo* may be indicated to determine whether the site in question is subject to metabolic activity (Chapter 2). Moreover, should such compounds be used in biological systems where metabolism at the labeled site takes place, it should be recognized that significant hydrogen–tritium isotope effects are to be expected, possibly leading to metabolic switching. In any case, tritide reduction can be an efficient way to prepare synthetic intermediates for subsequent transformations that abrogate the metabolic liability.

Beginning around 1952, standard methods have been developed for the convenient preparation of those tritides most commonly used; these include $LiB^3H_4$, $NaB^3H_4$, $LiAl^3H_4$, $NaB^3H_3CN$ and $B_2{}^3H_6$†. For reasons of stability, $NaB^3H_4$ and $NaB^3H_3CN$ are the only tritide reagents to have become commercially available (although some vendors have offered very low specific activity lithium aluminum tritide). Exchange labeling of $LiBH_4$ and $NaBH_4$, respectively, with tritium gas at elevated temperatures was shown to furnish $LiB^3H_4$ and $NaB^3H_4$ at specific activities of 60–80 Ci/mmol[156]; this thermal $^3H/H$ exchange process was recently re-investigated and optimized[157]. Treatment of the respective borohydrides with 15 equivalents of tritium gas at 270 °C (for $LiB^3H_4$), 450 °C (for $NaB^3H_4$), or 500 °C (for $KB^3H_4$) for 4–6 h provides the desired tritides with specific activities in the range of 80–95 Ci/mmol as determined by mass spectrometry and $^1H$, $^3H$ and $^{11}B$-NMR spectroscopy.

*Tritide exchange with protic media.* In a series of experiments with deuterated reagents or solvents[158], it was demonstrated that isotope exchange can occur between the hydrogens of certain hydroxylic solvents and the deuteride equivalents of borodeuteride anions during their hydrolysis. Extensive deuterium–hydrogen exchange occurs between sodium borodeuteride and ethanol, but much less occurs with isopropanol[159a]. Exchange is also less prone to occur with water, in which (isotopic) sodium borohydride is either stable (high pH) or hydrolyzes to release (isotopic) hydrogen gas (low pH). Lithium borohydride is more prone to exchange than the sodium salt, and does so with water as well as with other hydroxylic solvents. Cyanoborohydride anion readily exchanges under a variety of conditions[160]. This phenomenon may account for the frequent obtention of product specific activities lower than expected by consideration of stoichiometry alone, and would be expected to be more of a problem with more slowly reduced substrates.

*Kinetic isotope effects in tritide/hydride transfer.* Hydrogen–tritium kinetic isotope effects (KIEs) may also be a factor in lower-than-expected product specific activities. In a study[159a] of the reductions of ketones by sodium borotritide in isopropanol solution at room temperature, the hydrogen–tritium KIEs varied from about 3.2 to 5.8 for a set of model substrates. Analogous reductions using borane in THF gave hydrogen–tritium KIEs that varied from about 6 to 10. For both hydrides, the magnitude of the isotope effect increases with steric congestion around the carbonyl function but is relatively independent of the rate of reaction. The KIEs for hydroboration of alkenes vary from about 3–10, but trend the

---

†For the sake of convenience, tritide reagents designated by their formulas will always be written, as here, as if they contained only tritium. In reality, all reagents contain a mixture of tritium and hydrogen that varies with the specific activity of the reagent. For example, lithium borotritide of specific activity 80 Ci/mmol has the average formula approximately $LiB^3H_{2.7}H_{1.3}$, and probably consists of a mixture of all possible isotopomers $LiBH_4$, $LiB^3HH_3$, $LiB^3H_2H_2$, $LiB^3H_3H$ and $LiB^3H_4$.

opposite way, KIEs decreasing with increasing substitution around the olefin[159b]. In contrast to these data, an examination of literature reports for a number of reductions of aldehydes and ketones with sodium borotritide, tritiated borane and lithium borotritide reveals substantial inconsistencies, although in general the KIEs actually encountered in everyday practice seem to be lower than those observed in controlled studies. Nevertheless, their effect is almost always to give lower than statistical isotope incorporation into substrate, and lower specific activities than expected on purely statistical considerations. Experimental variables clearly play a role in the extent to which a KIE is operative.

Reaction conditions expected to exclude an isotope effect can be defined *a priori*, and should be taken into account when designing reactions. First, the use of a tritide reagent in carrier-free form should (barring solvent exchange) give a product having a specific activity of about 29 Ci/mmol times the number of tritide atoms incorporated. No KIE is expected, as there is no hydride to compete with tritide. Second, for tritide reagents of any specific activity (i.e., tritium-to-hydrogen ratio) the use of one equivalent or less of tritide reagent (based on the number of tritides per molecule of reagent) compared with substrate will exclude an isotope effect. Providing the reagent reacts fully, all tritide/hydride equivalents will be transferred to the substrate, precluding a hydrogen–tritium KIE.

It is only when an excess of tritide reagent is used that a KIE may be expected. The availability in the reaction of more tritide/hydride equivalents than are needed to reduce all substrate molecules provides the opportunity for hydrides to outcompete tritides, and tritium incorporation into substrate will be less than statistical.

### 4.3.1 Sodium Borotritide ($NaB^3H_4$)

Sodium borotritide, because of its versatility, stability and commercial availability at high specific activity (up to 80 Ci/mmol), is by far the most commonly used tritide reagent. It is a mild reducing agent, insoluble in diethyl ether, slightly soluble in THF and isopropanol, but readily soluble in diglyme, DMF, ethanol, and water. It reacts with water and methanol at an appreciable rate, but only slowly with ethanol and isopropanol, which are the preferred protic reaction solvents. In hydroxylic solvents and under mild conditions it reduces aldehydes, ketones and acyl chlorides to $\alpha$-tritiated alcohols, and azides, imines and iminium salts to the respective labeled amines. Under more vigorous reaction conditions (elevated temperatures, prolonged reaction times), methyl esters may be reduced, although esters as a class are generally resistant. In aprotic solvents $NaB^3H_4$ has been used to reduce polarized carbon–carbon double bonds. With $\alpha,\beta$-unsaturated carbonyl compounds substantial amounts of saturated alcohols may be formed. Under solvolytic conditions secondary and tertiary halides, sulfonates or triflates are reduced to the corresponding labeled hydrocarbons[161,162]. $NaB^3H_4$ reductions frequently can be carried out selectively in the presence of other functional groups such as ester, lactone, epoxy, acid, amide, nitrile, nitro functions as well as nonpolar carbon–carbon double and triple bonds. Modified by the addition of a variety of metal salts, sodium borotritide's reactivity can be extended, and it can even be used as an *in situ* source of tritium gas. The broad utility of sodium borotritide is illustrated by a discussion of a number of examples.

The $NaB^3H_4$ reduction of carbonyl compounds is a standard procedure for the synthesis of [1-$^3$H]alcohols, most usefully in the presence of various other functional groups, as for the precursors to [11-$^3$H]ABT-229 (**<u>172</u>**)[163], (8*S*)-20-hydroxy[20-$^3$H]hepoxilin $A_3$ methyl ester (**173**)[164a] and the hypoxia-activated antitumor prodrug intermediate for [$^3$H]PR-104 (**174**)[165]. For further examples see [163b,c]. Conversely it has already been applied to the synthesis of [1-$^3$H]aldehydes by reoxidation of the [1-$^3$H]alcohols, as exemplified with the preparation of all-*trans*-8′-apo-*β*-[10-$^3$H]carotenoic acid **175**[166]. For information on isotope effects in such oxidations, see p. 483.

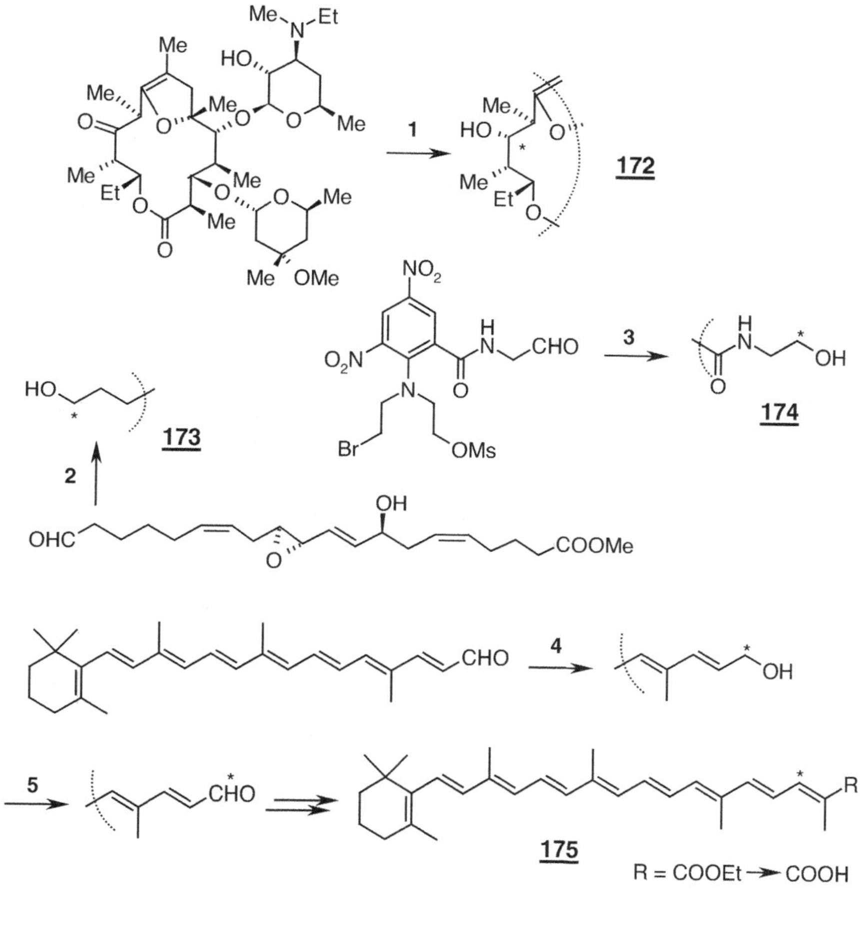


**Reaction conditions:** **1**. NaB$^3$H$_4$, EtOH; 0 °C, 3 h; r.t., 16 h; **2**. NaB$^3$H$_4$, MeOH-phosphate buffer pH 4; 20 °C, 2 h; **3**. NaB$^3$H$_4$, EtOH–THF; 20 °C, 20 min; **4**. NaB$^3$H$_4$, THF–MeOH 1:1; r.t ., 35 min; **5**. MnO$_2$, CH$_2$Cl$_2$; 0 °C, 5 h

***Figure 4.51*** *Reductions of aldehydes and ketones with sodium borotritide*

Acyl chlorides **176** and **177** have been reduced to primary alcohols containing two tritium equivalents, as with intermediates in the preparation of the glutamate-gated chloride channel modulator nodulisporic [$^3$H]ethanolamide (22 Ci/mmol, **177**)[167] and the NADH:ubiquinone oxidoreductase photoaffinity probe **179** (4.7 Ci/mmol)[168], respectively.

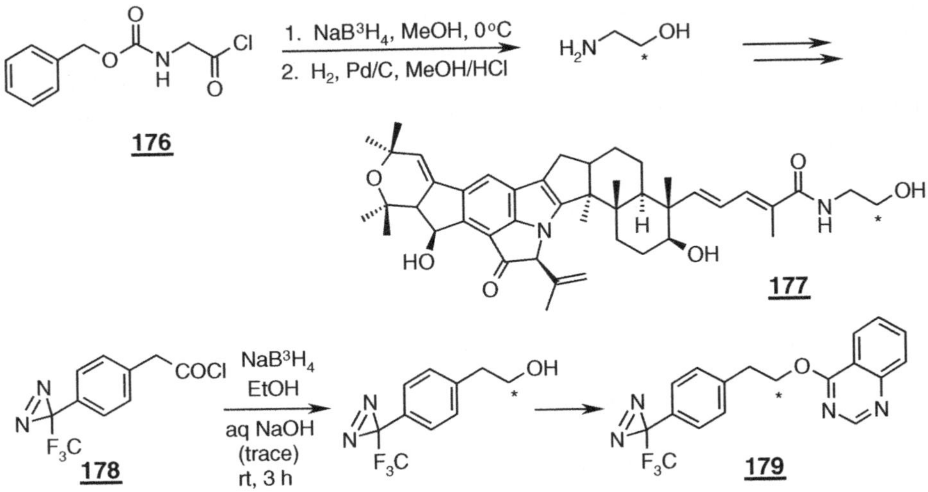


***Figure 4.52*** *Reductions of acyl halides with sodium borotritide*

Methyl esters have been reduced with sodium borotritide at room temperature or at elevated temperature, providing an alternative to the less readily available and less stable $LiAl^3H_4$ for the preparation of doubly labeled primary alcohols. High yields of products were reported for the reduction of ester **180** to the antitumor compound [$^3$H]CEP-2563 (**181**)[169], and of methyl benzyloxyacetate to give [$^3H_2$]alcohol **182**, an intermediate in the preparation of the tritiated 5-$HT_2$ antagonist deramciclane (**183**)[106].

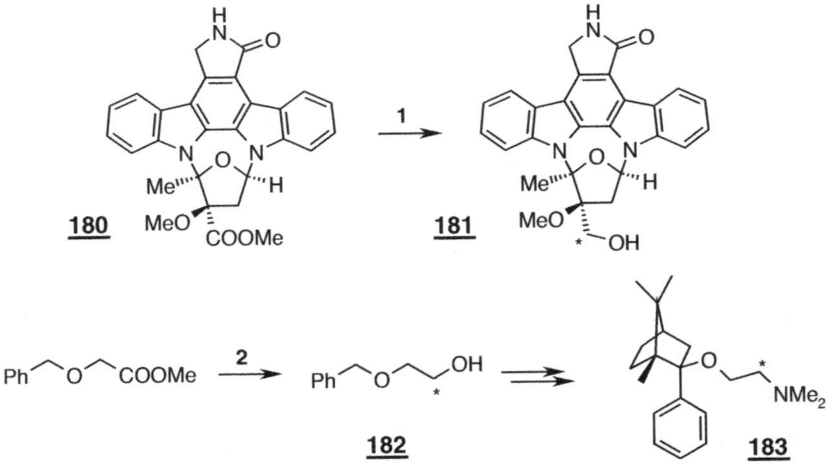


**Reaction conditions: 1** . $NaB^3H_4$ (10 equiv., 80 Ci/mmol), MeOH–THF; 24 °C, 18 h (product 35.3 Ci/mmol); **2**. $NaB^3H_4$ (1.4 equiv, 633 mCi/mmol), MeOH–*tert*-BuOH; reflux, 1 h (product 221 mCi/mmol)

***Figure 4.53*** *Reductions of methyl esters with sodium borohydride*

The examples in Figure 4.54 demonstrate that $NaB^3H_4$ can be a valuable reagent for the tritiation of polarized carbon–carbon double bounds through 1,4-tritide addition, giving saturated $\beta$-tritio products and leaving the adjacent functional group intact. Applied to the $\alpha,\beta$-unsaturated sulfonate **184** and the nitroolefin **185** such reductions gave high specific activity products that were used as precursors to an aminopeptidase A inhibitor[170] and a proposed biosynthetic intermediate to the cyanogenic glucoside dhurrin[171], respectively. Although enones often undergo smooth 1,2-reduction to allylic alcohols, as in the case of the avermectin affinity label **188**[172], the exocyclic enone of the mitomycin derivative **186** underwent 1,4-reduction to give **187**, albeit at a lower specific activity than might have been expected (16.2 mCi/mmol from sodium borotritide of specific activity 500 mCi/mmol)[173]. In 1,4-reductions, the reaction does often not stop at the carbonyl stage but produces saturated [1,3-$^3H_2$]alcohols, as illustrated by the conjugate $NaB^3H_4$ reduction of enone **189** giving rise to a 2:1 mixture of the epimeric [1,3-$^3$H]gibberellin $A_1$-13-acetate methyl esters **190** and minor amounts of the 1,2 reaction product[174].

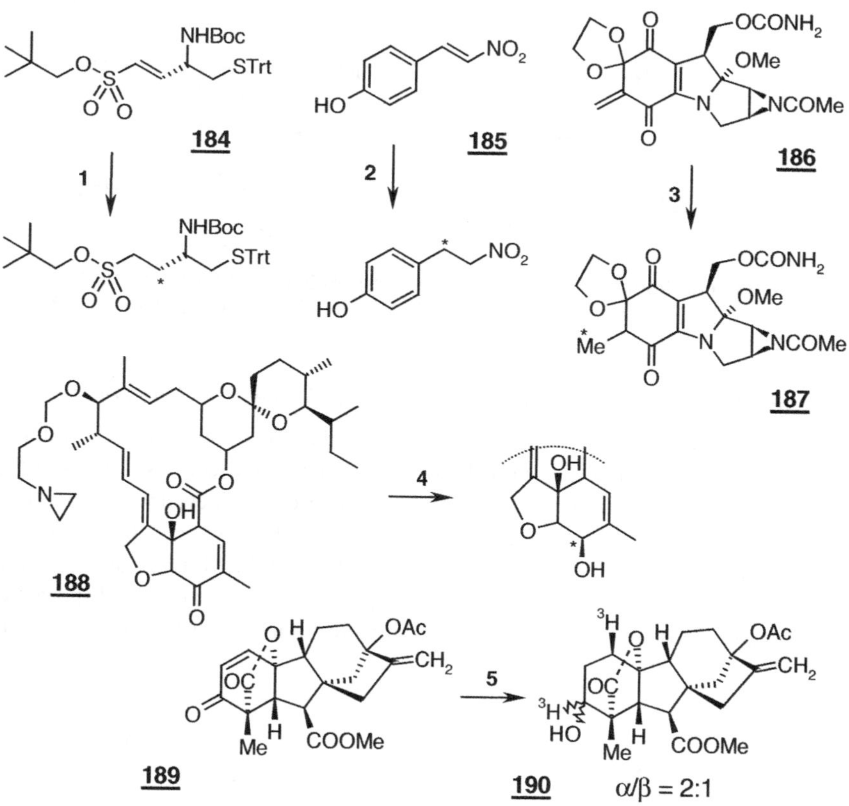


**Reaction conditions:** **1.** $NaB^3H_4$, EtOH; r.t., overnight; **2.** $NaB^3H_4$, EtOH; 40 °C, 3.5 h; **3.** $NaB^3H_4$, $CHCl_3$/py/EtOH; 20 °C, 15 min; **4.** $NaB^3H_4$, MeOH; 0 °C → r.t., 10 min; **5.** $NaB^3H_4$, LiBr, diglyme; 0 °C, 1 h

*Figure 4.54* *Reductions of polarized olefins with sodium borotritide*

The sodium borotritide-promoted reduction of imines has been widely exploited as a standard procedure for labeling complex peptides and proteins by reductive methylation[175]. The requisite imines (Schiff bases) **191** generated *in situ* from HCHO and ε-amino groups of lysine residues or free amino termini are readily reduced to give the corresponding *N*-[$^3$H]methyl derivatives **192**. The tritium labeling of SK&F 105685 (**193**) illustrates the $NaB^3H_4$ reduction of an iminium ion, which in this case was generated from unlabeled **189** via $Hg(OAc)_2$-mediated oxidation of the tertiary amine and precipitated upon treatment with 10% perchloric acid. Though an equilibrium between different isomeric iminium species was possible and, therefore, a nonspecific distribution of label, $^3$H NMR analysis of the product proved the site of tritium labeling to be as depicted in [$^3$H]**193**[176].

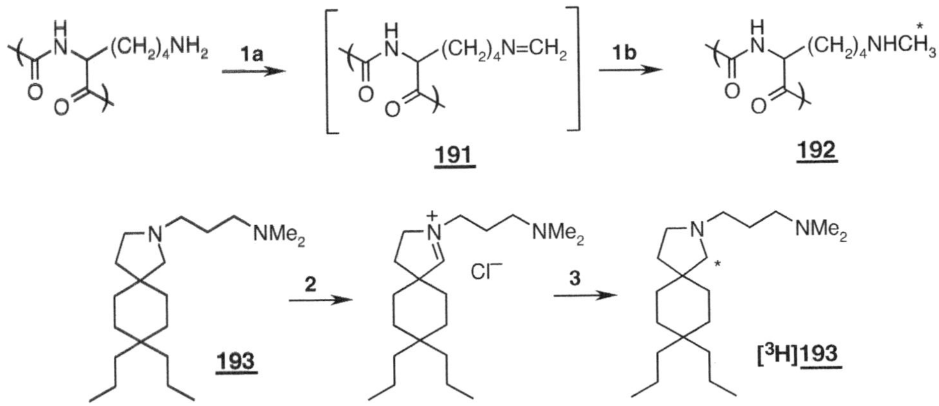


**Reaction conditions: 1a**. HCHO, borate buffer pH 9; 0 °C, 10 min, **b**. $NaB^3H_4$, 0 °C, 5 min; **2a**. $Hg(OAc)_2$, HOAc; 100°C, 2 h **b**. 70% aq.$HClO_4$-EtOH 1:1; **3**. $NaB^3H_4$, EtOH; r.t., 1.25 h

*Figure 4.55 Reductions of imines and iminium ions with sodium borotritide*

Although the cleavage of C-halogen, tosylate or triflate functions is mostly a domain of catalytic procedures (Section 4.2) or $LiEt_3B^3H$ or $Bu_3Sn^3H$ (Sections 4.3.6, 4.3.11), the example in Figure 4.56 demonstrates that tritiolyses can also be achieved with $NaB^3H_4$. Thus, the secondary alkyl triflate of ginkgolide C (**194**) could be displaced by use of sodium borotritide of specific activity 100 Ci/mmol to give [$^3$H]ginkgolide B (**195**) at 18 Ci/mmol[177]. At least one equivalent of the reagent is normally required for this kind of $S_N2$-displacement; in the present example five molar equivalents of tritide were required in order to give an acceptable yield.

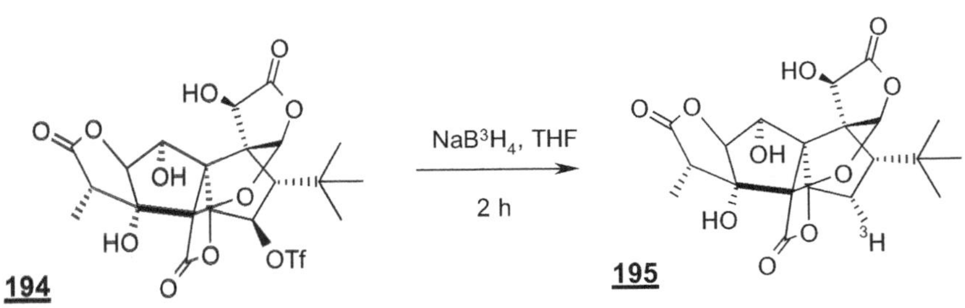


*Figure 4.56 Displacement of a triflate with sodium borotritide*

The addition of appropriate metal salts can help to modify the reactivity of $NaB^3H_4$ and the chemo- and regioselectivity of its reductions. For example, treatment of methyl 2-trifluoroacetamido-2-vinylnonadecanoate-3-one (**196**) with $Ca(B^3H_4)_2$, prepared *in situ* from $NaB^3H_4$ and $CaCl_2$, caused reduction of both the carbonyl and ester groups to provide 2-vinyl[1,1,3-$^3$H]dihydrosphingosine in a radiochemical yield of 5% and a specific activity of 14% of the theoretical value[178]. Use of $CeCl_3$ enables discrimination between ketonic and aldehydic carbonyl groups, and ensures the regioselective 1,2-reduction of $\alpha,\beta$-unsaturated ketones to allylic alcohols, whereas the unmodified hydride or tritide might first reduce the double bond via a 1,4-attack. This potential was exploited for the regioselective tritiation of the $\Delta^{20,21}$ double bond of the gestagenic steroid **197**. $CeCl_3$-modified $NaBH_4$ reduction of the $\alpha,\beta$-unsaturated carbonyl group gave the allylic alcohol, whose C20,21-double bond could be tritiated by Wilkinson's catalyst (see Section 4.1) to give saturated alcohol **198**[179]. In the second example the $NaB^3H_4$–$CeCl_3$ combination was used to ensure that 1,4-addition to the ancillary enone system of **199** would not compete with reduction at the intended keto function. Moreover the stereoselectivity of the carbonyl reduction was improved[180]. Finally, 3:1 mixtures of freshly prepared solutions of sodium borohydride and aluminium chloride have been shown to give a clear solution of complex sodium aluminium hydrides that has powerful reducing properties. It reduces aldehydes, ketones, esters, tertiary amides, nitrile etc. to the corresponding alcohols and amines[161]. This reactivity was exploited for the preparation of 2-amino[1-$^3$H]ethanol through treatment of glycine methyl ester hydrochloride with $NaB^3H_4$–$AlCl_3$ 3:1 in bis-(2-methoxyethyl) ether[181].

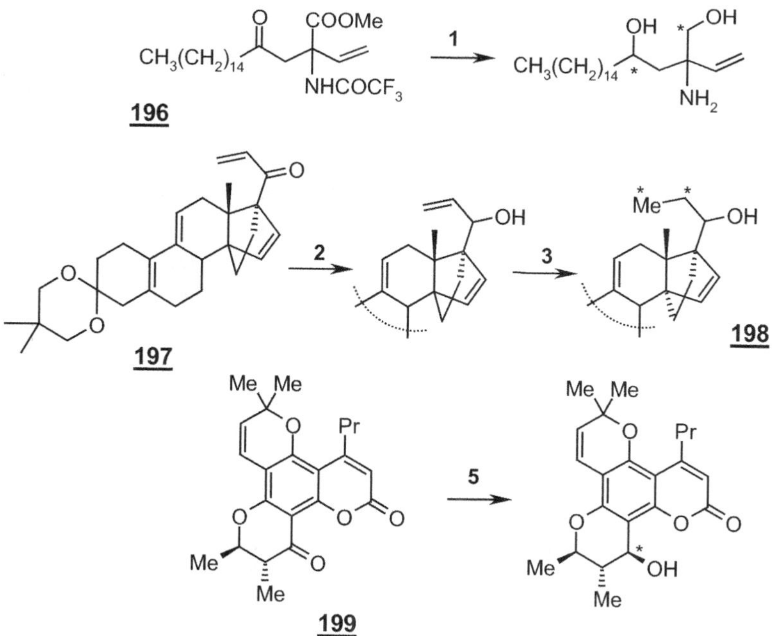


**Reaction conditions: 1**. $NaB^3H_4$, $CaCl_2$, THF–EtOH 1:1; –20 °C → r.t.; r.t., 20 h; **2**. $NaBH_4$, $CeCl_3$, THF; **3**. $^3H_2$, $(PPh_3)_3RhCl$, toluene, r.t.; **4**. PCC, $CH_2Cl_2$; **5**. $NaB_3H_4$, $CeCl_3$, MeOH; 0 °C, 5 min

***Figure 4.57*** *Use of sodium borotritide modified by $CaCl_2$ and $CeCl_3$*

Nickel(II) and Co(II) salts are well known to enhance the ability of sodium borotritide to reduce carbon–carbon double bonds and nitriles[162], probably by *in situ* generation of tritium gas and either nickel or cobalt boride. This approach has been used with $NaB^3H_4$ to reduce, for example, compounds **200**[183] and **201**[182]. The $^3H$ NMR spectroscopic analysis of **201** confirmed that in this case the $NaB^3H_4$–$NiCl_2$ system provided excellent chemoselectivity and high specific activity in both of the original olefinic positions[182]. This system was also used to doubly label an $\alpha$-bromoketone (**202**) by both C–Br bond tritiolysis and carbonyl reduction[184]. Under aprotic conditions (e.g. THF), only the corresponding bromohydrin is formed. The $NaBH_4$–$CoCl_2$ system has been especially recommended for the reduction of nitriles to primary amines[162]. In the example **203**, it is not only the nitrile function but also the conjugated double bond that was reduced. Radiochemical yield was found to be 20%, the specific activity was 40% of the theoretical[185].

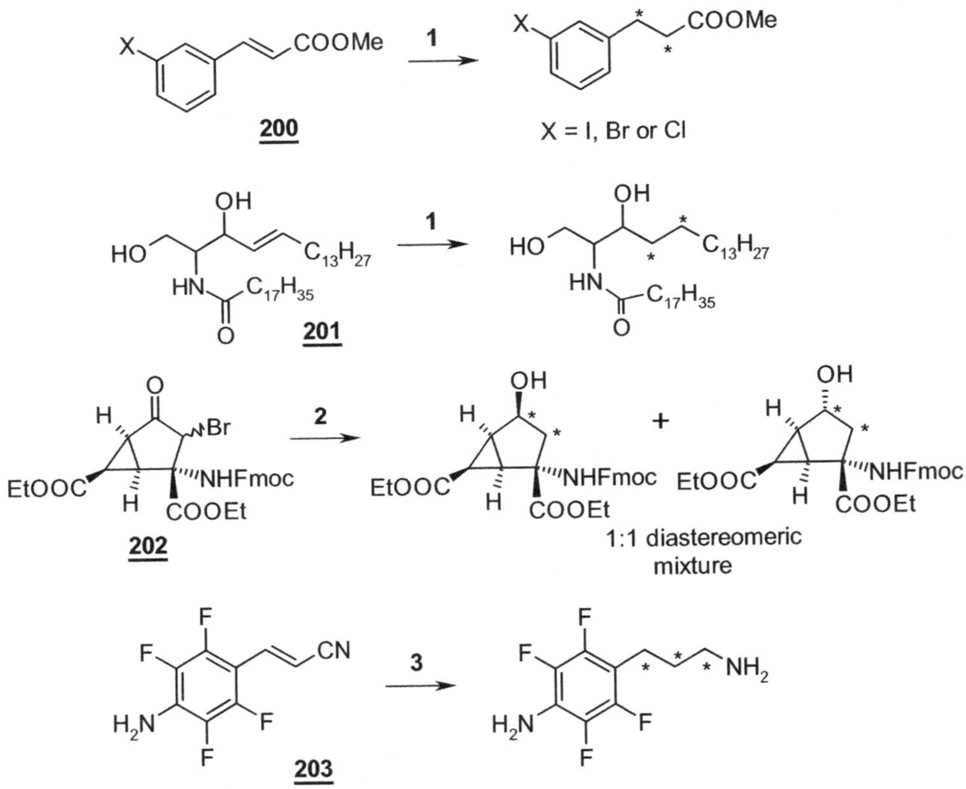


**Reaction conditions:** **1.** $NaB^3H_4$, $NiCl_2$, MeOH; 2. $NaB^3H_4$, cat. $Ni(OAc)_2{\cdot}4H_2O$, EtOH–THF 8:1; r.t., 20 min **3.** $NaB^3H_4$, $CoCl_2{\cdot}6H_2O$, MeOH; 0 °C, 30 min; r.t., 20 h

***Figure 4.58*** *Nickel(II) and cobalt(II)-modified reductions with sodium borotritide*

Modification of sodium borotritide reactivity in THF solvent with 0.2–0.5 equivalents of palladium(II) acetate has been exploited for tritioreduction of $\mu$mol quantities of olefinic and acetylenic compounds, as illustrated for **204, 205**[186a] and **206**[186b]. The specific activities of the resulting products at each position were 40–90% of those of the tritide equivalents of $NaB^3H_4$ employed. The radiochemical yields, however, varied from 1% (**204, 205**) to 38% (**206**). By contrast, in methanol solvent, tritiolyses of aryl halo substituents are possible[183], with selectivity among halogens (I > Br > Cl) *without* reducing double bonds (e.g., **207–209**). In these cases the specific activities that can be expected for products of C–I, C–Br and C–Cl bond cleavages are about ≥90%, 75% and 50%, respectively, of the $NaB^3H_4$ (per tritide equivalent)[183]. This methodology gave a 10% radiochemical yield for **210**[187], and was superior to catalytic triolysis using $^3H_2$ and Pd catalysts, which resulted in additional tritium incorporation into the methylene position and the pyridrine ring. Aryl tritio-debrominations have also been carried out using $PdCl_2$ as the tritide modifier[188].

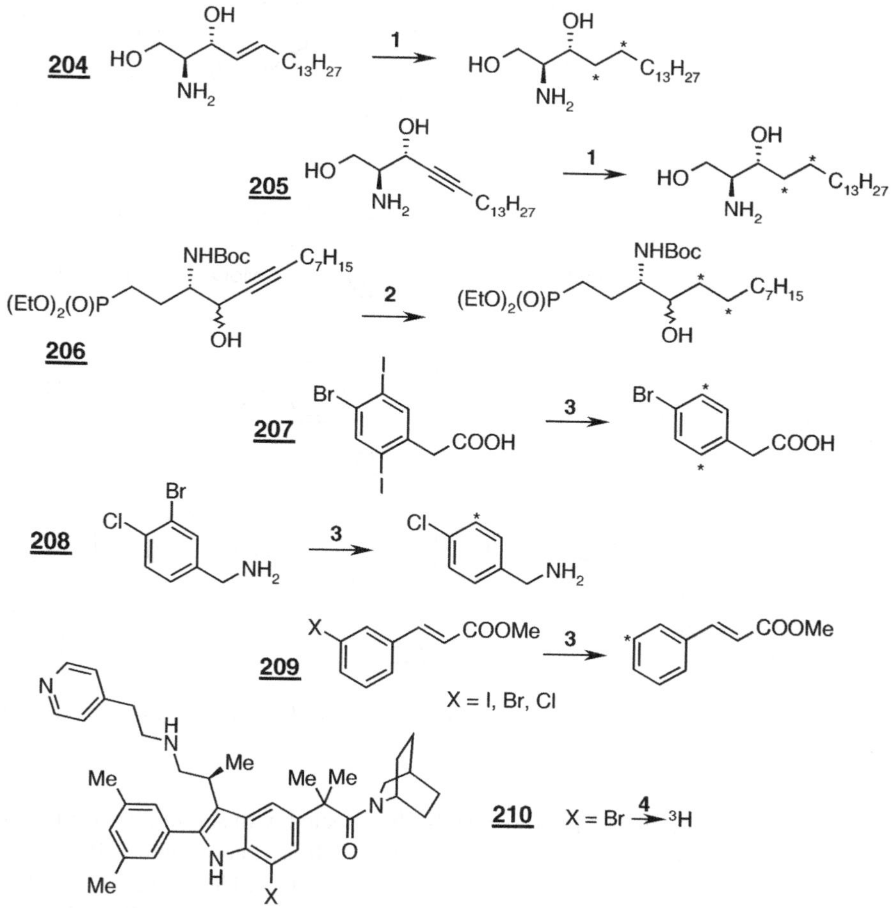


**Reaction conditions:** **1**. $NaB^3H_4$ (4.5 equiv.), $Pd(OAc)_2$ (1 equiv.), THF-1 N NaOH 5:1; r.t., 2 h; **2**. 2 equiv. $NaB^3H_4$, 1 equiv. $Pd(OAc)_2$, THF-1 N NaOH 2:1; r.t., 48 h; **3**. 1 equiv. $NaB^3H_4$, cat. $Pd(OAc)_2$, MeOH; r.t., 30 min; **4**. $NaB^3H_4$, $Pd(OAc)_2$, *i*-PrOH; r.t.

***Figure 4.59*** *Uses of palladium(II)-modified sodium borotritide*

A detailed study has reported[189] on the utility of $NaB^3H_4$ and $Pd(PPh_3)_4$ heated in DMF solvent for the tritiodehalogenation of aryl iodides, bromides and triflates, some in the presence of olefinic functions (see for example **211** and **212**). Both this system and the Pd(II)–MeOH system above can be good alternatives to catalytic tritiolysis (Section 4.2), not only in their selectivity for halogen over olefin, but sometimes also for the achievability of higher specific activity, and the obviation of extraneous labeling sometimes seen with heterogeneous palladium-catalyzed tritiations. Moreover, treatment of model compounds under the same conditions was shown to result in replacement of *alkyl* bromides, iodides, mesylates and tosylates with tritium, as in **213**, extending the utility of this reaction compared with catalytic tritiolysis.

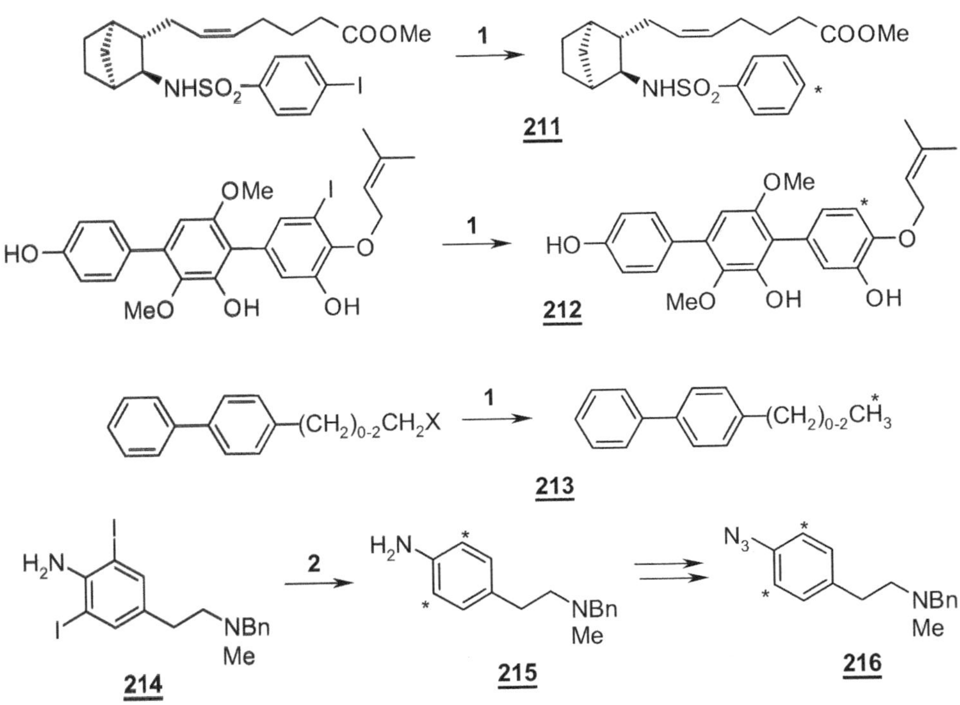


***Figure 4.60*** *Palladium(0) modification of sodium borotritide for tritiolyses of aryl and alkyl halides*

Tritiodehalogenation of aryl iodides can also be effected with sodium borotritide in the presence of Pd/C as an *in situ* generator of tritium gas. Reaction of diiodo-$BPNH_2$ (**214**), with a 1.5–2.5-fold excess of $NaB^3H_4$ provided [$^3$H]BP–$NH_2$ (**215**) which could be readily converted into [$^3$H]BP–$N_3$ (**216**), a photoaffinity probe for cytochrome P450 B1, with a specific activity of more than 60% of the potential value[190]. Separately, a similar procedure utilizing acetic acid in place of Pd/C, and an apparatus for conducting such work, have been described[191], and yet another, using $CoCl_2$[192].

In addition to these findings, it should be noted that recently updated routine procedures exist for the *in situ* generation of further tritides such as, $B_2{}^3H_6$ and *n*-$Bu_3Sn^3H$ by reaction of commercially available $NaB^3H_4$ with $BF_3{\cdot}OEt_2$ and *n*-$Bu_3SnCl$, respectively (see

Sections 4.3.8 and 4.3.11). These results not only document the broad scope of $NaB^3H_4$ applications, especially when only small amounts of radioactivity are required, but they also provide an economical and convenient alternative to procedures normally requiring specialized laboratories equipped for handling tritium gas and larger quantities of high specific activity reactive tritides.

### 4.3.2 Sodium Cyanoborotritide ($NaB^3H_3CN$)

$NaBH_3CN$, because of its strong electron-withdrawing cyano group, is a milder and more selective reagent than $NaBH_4$. Aldehydes, ketones and oximes are not reduced in hydroxylic solvents (methanol, water) under neutral conditions, but their reduction is rapid at pH 3–4. Amide, ester, lactone, nitrile, nitro and epoxy functions are inert towards this reagent. Imines and iminium ions, however, even if generated *in situ* from aldehydes or ketones and ammonia or primary and secondary amines, are rapidly reduced under acid catalysis (pH 6–8) to provide primary, secondary and tertiary amines, respectively. This reactivity pattern makes (isotopic) $NaBH_3CN$ ideal for the reductive alkylation of amines by combining them with carbonyl-containing reactants in the presence of the hydride[193a,b]. In HMPA solution the reagent rapidly and selectively removes iodo, bromo, tosyloxy functions (and thereby, indirectly, hydroxy groups) at 50–70 °C in chemical yields of 60–90%[193c]. Sodium cyanoborotritide is commercially available at high specific activity levels, but it can be prepared in the laboratory owing to the fact that the hydrogens in $NaBH_3CN$ are readily exchanged for tritium upon repeated treatment with tritiated water at pH 3 and ambient temperature; 10–20 Ci/mmol is the practical maximum specific activity achievable with this procedure. However, this ready exchangeability can also lead to loss of tritium from $NaB^3H_3CN$ if it is used in acidic protic media[194]. According to the reactivity of this tritide reagent, the synthesis of $\alpha$-tritiated amines, the labeling of complex peptides and proteins by reductive tritio-methylation[194,195], and the reduction of carbonyl groups in the presence of additional functional groups sensitive to $NaB^3H_4$, are the domain of $NaB^3H_3CN$ in tritium chemistry[193b,194].

Examples of reductive alkylations using $NaB^3H_3CN$ include preparations of **217**, an intermediate in the preparation of the $5HT_{1a}$ agonist 8-hydroxy[2-$^3$H]DPAT[196] and of **218**, a series of candidate inhibitors of ceramide glucosyltransferase[197].

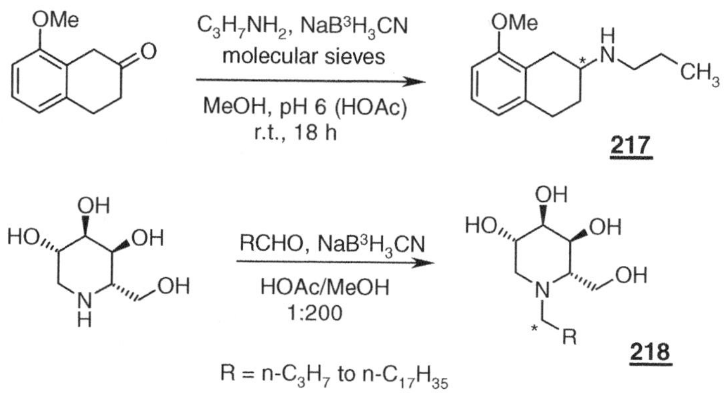


**Figure 4.61** *Reductive alkylations using sodium cyanoborotritide*

The use of $NaB^3H_3CN$ for reductive methylations of peptides and proteins using formaldehyde has significant advantages over $NaB^3H_4$ in that the former reagent does not reductively cleave disulfide bridges or react with formaldehyde under neutral reaction conditions, gives higher yields of lysine-ε-*N*-[$^3$H]methyl and -[$^3$H]dimethyl derivatives, and produces fewer byproducts[195]. The disadvantage is that $NaB^3H_3CN$ is not routinely available at specific activities as high as those of $NaB^3H_4$.

Enamine reductions are initiated by a rapid and reversible protonation at the β-carbon, generating a readily reducible iminium salt which is then attacked by $NaB^3H_3CN$ at the α-position, as demonstrated by methyl 1-benzoyl-2,3-dihydro-$\Delta^{5,10}$-lysergate (**219**). The mixture of epimeric iminium chlorides (**220**) derived from **219** was reduced to give a tritiated product mixture consisting of the desired 5,10-*trans-syn* isomer **221** together with the 5,10-*cis-syn* and 5,10-*cis-anti* isomers in a ratio of 83:3:14 in 80% yield. Intermediate **221** was separated by HPLC and used as key intermediate for various 8α-amino[5-$^3$H] ergolines. Reduction of **220** with $NaB^3H_4$ gave a lower yield and the predominant isomer was 5,10-*cis-syn* isomer[198].

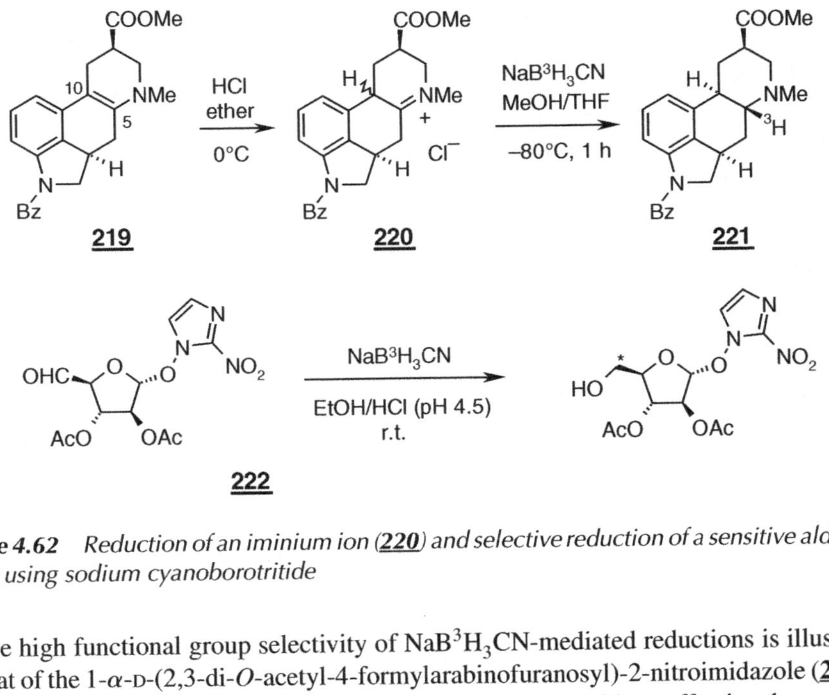


**219** **220** **221**

**222**

***Figure 4.62*** *Reduction of an iminium ion (**220**) and selective reduction of a sensitive aldehyde (**222**) using sodium cyanoborotritide*

The high functional group selectivity of $NaB^3H_3CN$-mediated reductions is illustrated by that of the 1-α-D-(2,3-di-*O*-acetyl-4-formylarabinofuranosyl)-2-nitroimidazole (**222**) to the corresponding [1-$^3$H]alcohol in 50% radiochemical yield without affecting the ancillary nitro and ester functions. In contrast, $NaB^3H_4$ was found to cleave the acetate protecting groups[199].

In the presence of $ZnI_2$, dichloromethane or dimethoxyethane solutions of $NaBH_3CN$ have been found to deoxygenate aryl aldehydes and ketones as well as benzylic, allylic and tertiary alcohols in moderate to good yields. In spite of the unfavorably large excess of the reagent, the procedure is simple and the substrates require no previous transformations to halides or tosylates, so that the reagent can also be recommended for the synthesis of tritiated compounds. The use of $NaB^3H_3CN$ in the synthesis of [4-$^3$H]SDZ CQP201-403 (**224**) at 3.2 Ci/mmol (80% of the potential value) by tritio-deoxygenation of the corresponding 4-hydroxy precursor **223** may serve as an illustrative example[200].

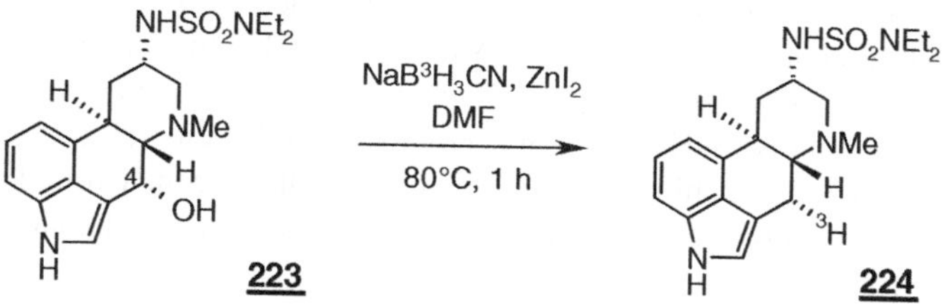


*Figure 4.63* Tritiodeoxygenation using zinc-modified sodium cyanoborotritide

### 4.3.3 Sodium/Tetramethylammonium Triacetoxyborotritide [$Na/NMe_4B^3H(OAc)_3$]

Both $NaB^3H(OAc)_3$ and $Me_4NB^3H(OAc)_3$ have been introduced as mild and selective agents for the reductive amination of aldehydes and ketones, as well as for the reduction of β-hydroxy ketones. In the case of $NaB^3H(OAc)_3$, the configuration of the newly formed

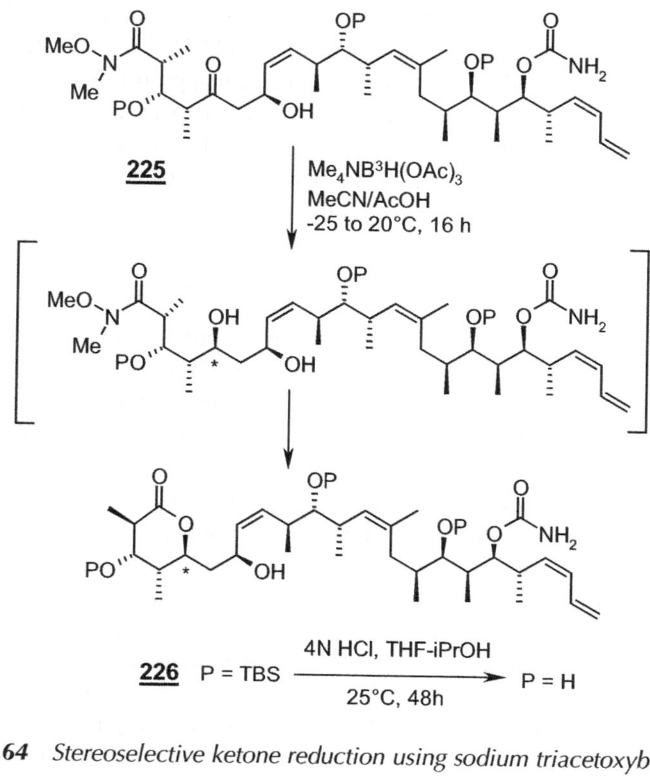


*Figure 4.64* Stereoselective ketone reduction using sodium triacetoxyborotritide

hydroxy group has often proved to be opposite to that resulting from the reduction with $NaB^3H_4$. In β-hydroxy ketones the different stereoselectivity is attributed to coordination of the substrate's β-hydroxy group with the boron center of the reagent (displacing one of the acetoxy groups) thereby guiding delivery of the tritide. This effect is greater with the counterion $Me_4N^+$, which is less prone to interact with nearby functional groups. $NaB^3H(OAc)_3$ is readily prepared by treatment of $NaB^3H_4$ with 3.3 equivalents of acetic acid in benzene solution at room temperature. Exchange of the counterion is effected by dissolving $Me_4NCl$ and $NaB^3H(OAc)_3$ in acetonitrile:acetic acid 1:1 at 0 °C and separating the precipitated sodium chloride. $NMe_4B^3H(OAc)_3$ has recently found application in the stereoselective reduction of **225** to **226** (P = TBS), a key step in the labeling sequence designed for [$^3$H]discodermolide (**226**, P = H), a cytotoxic natural polyketide derivative[201].

### 4.3.4 Lithium Tritide ($Li^3H$)

Lithium tritide has not been used so far as a reducing agent in tritium chemistry but has served as a central precursor to valuable complex tritides. Until 1990 it was prepared by reaction of lithium hydride with tritium gas at 330 °C for 16 h[202]. Following these conditions, 3.0 mmol LiH and 110 Ci (1.9 mmol) of carrier-free tritium gas provided $Li^3H$ at a specific activity of 11.5 Ci/mmol (40% of the maximum theoretical value)[202]. Values of this magnitude are generally sufficient for biological investigations, but applications in radioimmunoassay techniques and receptor research often require higher specific activities. However, although $Li^3H$ prepared by tritium–hydrogen exchange smoothly reacts with $AlCl_3$ and $AlBr_3$ in diethyl ether or THF to produce $LiAl^3H_4$[203], its reaction with weaker Lewis bases such as $Et_3B$ and *sec*-$Bu_3B$ is sluggish and inappropriate for the preparation of the corresponding borotritide reagents especially at high specific activities[204].

This problem was overcome when highly reactive LiH became accessible upon hydrogenation of a commercially available solution of *n*-BuLi in hexane in the presence of TMEDA (1–1.2 equivalents) under very mild conditions (r.t., 30–60 min)[205]. Applied to tritium[206], the reaction proceeds according to Equation (4.1).

$$n\text{-BuLi} + {}^3H_2 \rightarrow Li^3H + [1\text{-}^3H]\text{butane} \quad (4.1)$$

The analogous preparation of NaH and KH was found to require the prior *in situ* generation of highly active BuNa and BuK, respectively, by reaction of sodium or potassium *tert*-butoxide with *n*-butyllithium; the commercial hydrides are less active and insoluble in hexane. The utility of the superactive forms of these hydrides in reductions and metallations has been described[207].

This newer procedure was adapted to the preparation of $Li^3H$ in the radiochemistry laboratory and optimized for the preparation of tritides of maximum specific activities and 0.05–2 mmol scales (Scheme 4.1). It is conducted[208] on a tritium handling manifold by charging an attached flask, under inert gas, with *n*-butyllithium in hexane and TMEDA. The resulting solution is carefully degassed, tritium gas is introduced to an initial pressure of 50–90 kPa, and the reaction mixture is stirred at ambient temperature for 0.5–2 h to give a white creamy precipitate of $Li^3H$. Excess tritium gas is removed, and all volatile components including [$^3$H]butane and TMEDA are carefully lyophilized off and separated for disposal. The residual highly reactive $Li^3H$ is suspended in THF and treated with the

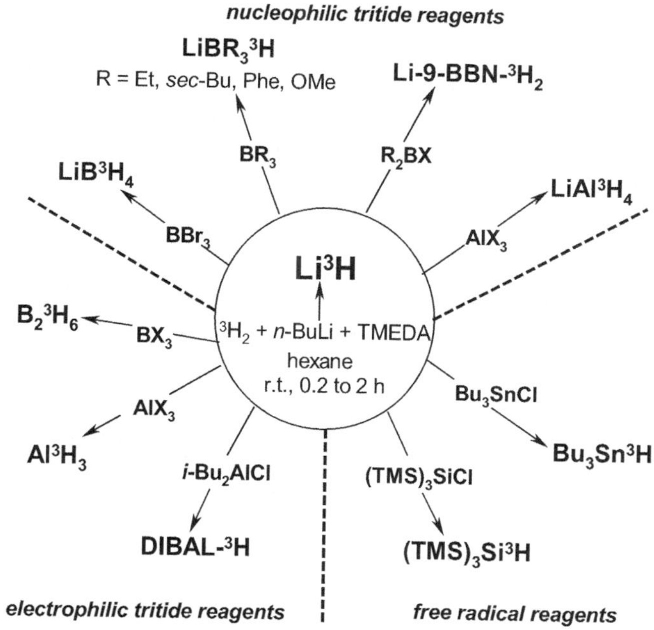


**Scheme 4.1** *The range of tritide reagents available from highly reactive lithium tritide*

appropriate Lewis acid (e.g., $BR_3$, $BF_3{\cdot}OEt_2$, $AlBr_3$, $R_3SnX$, $Cp_2ZrCl_2$) to furnish the requisite tritide reagent at temperatures ranging from −15 °C to 70 °C for 5–60 min. The reaction conditions required are strongly dependent on the nature of the Lewis acid employed, and are described in the following Sections. In most cases the resulting solution or suspension of the tritide reagent can be used immediately.

### 4.3.5 Lithium Borotritide ($LiB^3H_4$)

Since $Li^+$ is a stronger Lewis acid than $Na^+$, $LiB^3H_4$ is a more powerful reducing agent than $NaB^3H_4$, but it is still less reactive than $LiAl^3H_4$. Thus it reduces aldehydes, ketones, acid chlorides, esters and lactones without affecting acids, primary or secondary amides, nitriles or nitro groups. Tertiary amides may be reduced to give either labeled primary alcohols or amines depending on the solvent used[209]. In contrast to $NaB^3H_4$ it is soluble in diethyl ether and THF, so that homogenous reductions with $LiB^3H_4$ in these solvents are possible. However, while $LiB^3H_4$ exhibits a higher diastereoselectivity than $NaB^3H_4$, it has been less frequently used, probably because of its greater propensity to hydrolyze. As a result it has never been made available commercially. Its laboratory preparation from lithium tritide, however, is straightforward. After lyophilization of the volatiles away from the $^3H_2$/*n*-BuLi reaction by which $Li^3H$ is formed (previous section), the light powder is suspended in THF and 0.25 molar equivalent of $BBr_3$ is added and the solution heated for 30 min at 70 °C[210]. The yield of $LiB^3H_4$ is about 75%; an earlier method using $BF_3$ etherate instead of $BBr_3$ gave lower yields[203,211]. In either case the tritide reagent is usually used immediately *in situ*. Earlier methods for the preparation of $LiB^3H_4$ gave material of lower specific activity[212].

Early applications of $LiB^3H_4$ were commonly in the carbohydrate field, where it was used to selectively reduce aldonic lactones to give [1-$^3$H]aldoses. An example of such a reaction is the conversion of **227** to **228**[213]; the reaction requires pyridine as solvent, in order to minimize further reaction to the corresponding [1-$^3H_2$]alditol.

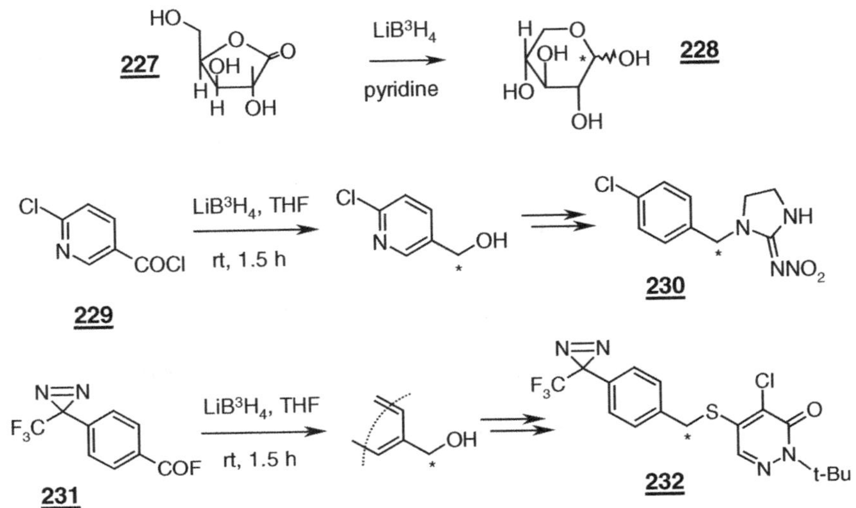


***Figure 4.65*** *Use of lithium borotritide for reductions of a sugar lactone and acyl halides*

Since $LiBT_4$ at maximum specific activity has become accessible, it has gained more importance especially in the synthesis of highly tritiated ligands. Reduction of the respective carboxylic acid halides **229** and **231** gave the respective [$\alpha$,$\alpha$-$^3H_2$]carbinols; subsequent conversion of them to the corresponding benzyl halides and halide displacements with nitrogen- and sulfur nucleophiles afforded, respectively, the insecticidal *N*-acetylcholinesterase inhibitor [$^3H_2$]CH-IMI (**230**)[214] and the photoaffinity probe (trifluoromethyl)diazirinyl[*S*-methylene-$^3H_2$]pyridaben (**232**)[215] in 30–50% radiochemical yields and at specific activities in the range of 55–56 Ci/mmol.

As $LiBH_4$ has been reported[216] to display greater diastereoselectivity than $NaBH_4$, $LiB^3H_4$ was selected for the reduction of 2-hydroxy-5-methyl-*N*-(1,3,3-trimethyl-5-oxocyclohexyl)methyl benzamide (**233**) (Figure 4.66). It provided the desired *cis*-alcohol **234** in better than 20 : 1 ratio vs its diastereomer[217], and with a specific activity of 28 Ci/mmol. This result buttressed the case for the utility of $LiB^3H_4$ in diastereoselective reductions. Differences in the reactivity of various functional groups towards $LiB^3H_4$ facilitated the selective reduction of the mixed anhydride moiety of the aspartic acid derivative **235** to doubly protected $\gamma$-hydroxy-$\alpha$-amino[$\gamma$-$^3H_2$]butyric acid **236**, a key intermediate in the synthesis of [$\gamma$-$^3H_2$]azoxybacilin **237** at 57 Ci/mmol[218].

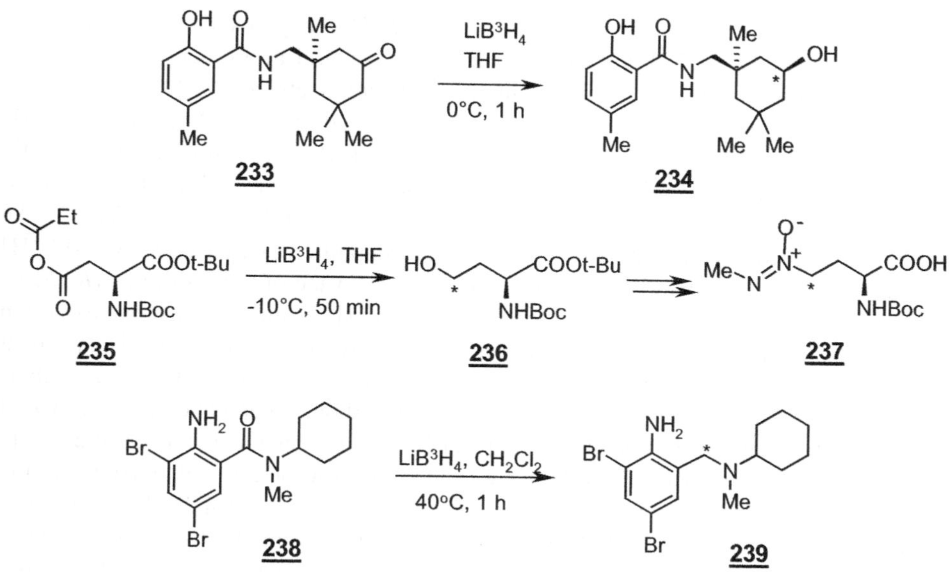


*Figure 4.66 Additional applications of lithium borotritide*

As mentioned above, $LiB^3H_4$ has been used very successfully for reduction of a tertiary amide to an amine. Whereas treatment in trial reactions of **238** with $LiBH_4$ in THF solution at 65 °C gave the benzyl alcohol, the outcome in methylene chloride as solvent was different. Namely, reduction of **238** with $LiB^3H_4$ in $CH_2Cl_2$ at 40 °C gave **239**, the mucolytic agent bromhexine, in 43% yield with a specific activity of 54 Ci/mmol[219].

### 4.3.6 Lithium Triethylborotritide ($LiEt_3B^3H$, Li-Super-Tritide)

Lithium triethylborotritide is a highly nucleophilic and even more powerful reducing agent than $LiB^3H_4$ and $LiAl^3H_4$. With the exception of free acids and the corresponding salts nearly all other reducible functional groups are attacked. Besides the tritio-reduction of esters, aldehydes and ketones, it is very effective in tritio-dehalogenations of alkyl halides and tritio-deoxygenations of alcohols through reduction of their tosylates, both with clean inversion of configuration at the reduced site, and reductive cleavage of epoxides to furnish the corresponding labeled Markovnikov alcohols[220]. In contrast to other tritides such as $LiAl^3H_4$ and $NaB^3H_4$, where only a fraction of the available tritide is often exploited, most of the tritium in Super-Tritide can be effectively incorporated into product[221].

$LiEt_3B^3H$ at nearly 100% abundance tritium is readily produced from $Li^3H$ (as prepared in Section 4.3.4) and $Et_3B$ (THF; r.t., 1 min)[206]. Earlier methods using less reactive crystalline $Li^3H$ at lower specific activities[221] had to be more drastic (r.t., 2 days or reflux, 2 h) and are therefore less appropriate for the preparation of a highly tritiated reagent. Despite its excellent synthetic potential only a few applications have been published so far, aside from test reactions on model substrates[221]. Among the simpler but elegant were tritio-deoxygenation of (*R*)- and (*S*)-[1-$^2$H]ethyl and [1-$^2$H]butyl tosylates to give (*R*)- and (*S*)-[1-$^2$H,1-$^3$H]ethane[222] and (*R*)- and (*S*)-[1-$^2$H,1-$^3$H]butane[223], respectively, for use in enzyme mechanistic studies; and of 3,5-dimethoxy[7-$^2$H]benzyl tosylate to give (*R*)- and (*S*)-3,5-dimethoxy[7-$^2$H,7-$^3$H]toluene, a precursor of 'chiral acetic acid' (strictly, only those molecules containing tritium are chiral)[224]. 'Chiral acetic acid' (**241**) at a specific activity of 3 Ci/mmol was also prepared through reduction of racemic *O*-benzyl-*cis*-[3-$^2$H] glycidol (**240**) by LiBEt$_3$$^3$H (diluted eightfold with LiBEt$_3$H) followed by synthetic elaboration to *N*-acetyl[3-$^2$H,3-$^3$H]alanine, enzymatic resolution with acylase and concluding chromic acid oxidation of the resulting L-(2*S*,3*R*)-[3-$^2$H,3-$^3$H]alanine (**241**)[203]. An alternative procedure for **242** started with the nucleophilic opening of *E*-(*S*,*S*,)-[2-$^2$H] styrene oxide with LiBEt$_3$$^3$H followed by oxidative cleavage of the resulting (1*R*,2*S*)-α-phenethanol. It was further converted to (*S*)-[3-methyl-$^2$H,$^3$H]mevalonolactone for use in biosynthetic studies[225]. Recently the reagent has been used for nucleophilic displacement of the carbamyl group of **243** to give e.p. (1*S*)-(1-dimethylphenylsilyl)[$^2$H,$^3$H]methylboronate (**244**) with inversion of the configuration at the carbon center. Oxidative cleavage of the B–C bond followed by base-mediated [1,2]-Brook rearrangement converted **244** into chiral (*R*)-[$^2$H,$^3$H]methanol at >98% e.e. and a specific activity of 306 mCi/mmol, subsequently trapped with 3,5-dinitrobenzoyl chloride to give (*R*)-[$^2$H,$^3$H]methyl 3,5-dinitrobenzoate.[226]. A further application includes the reduction of 4,6-*O*-benzylidene-D-gluconolactone (**245**) followed by acid-catalyzed deacetalization to afford D-[1-$^3$H]glucose (**246**) with a specific activity of 24.5 Ci/mmol[227].

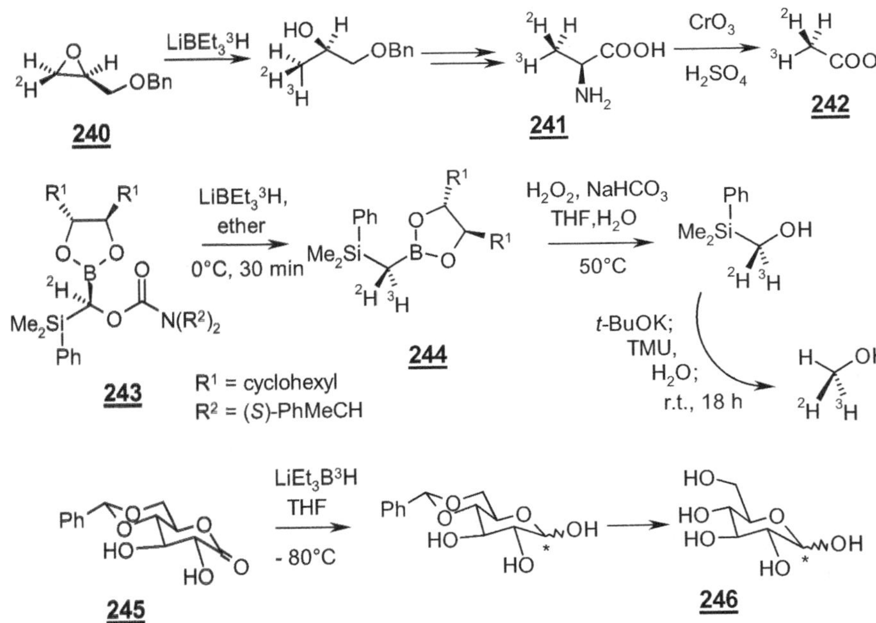


*Figure 4.67* Applications of Lithium Super-Tritide (1)

In a more complex situation, LiEt$_3$B$^3$H was employed in the synthesis of (2-hydroxy-[1,1,2,2-$^3$H]ethyl)everolimus **249**, an immunosuppressant rapamycin derivative[228]. Reduction of methyl *O*-*tert*-butyldiphenylsilyl glycolate **247** followed by purification on silica gel gave monoprotected ethylene glycol **248**. According to its $^3$H-NMR spectrum **248** proved to be labeled in both methylene positions due to a silica gel-mediated migration of the TBDPS group during the chromatographic purification. Coupling of **248** via its triflate to rapamycin followed by acid-catalyzed cleavage of the TBDPS group furnished **249** with a specific activity of 55.6 Ci/mmol.

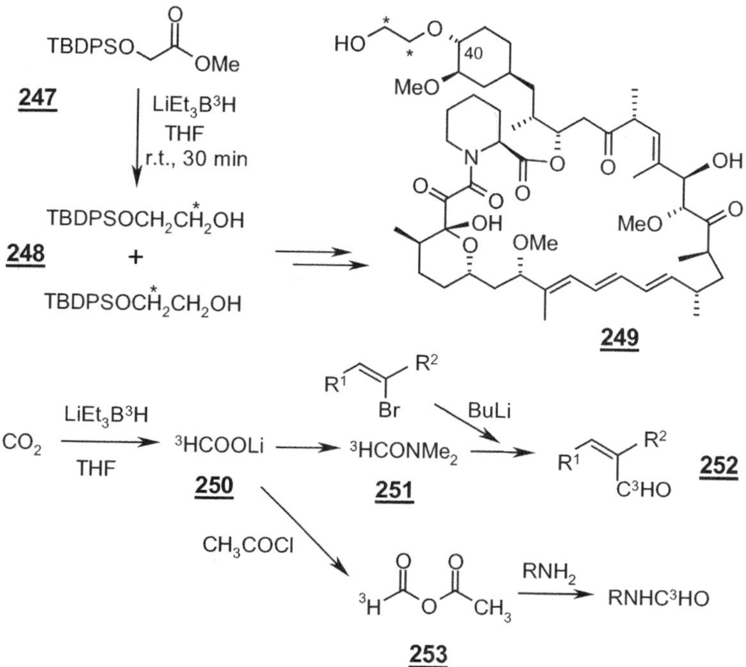


*Figure 4.68 Applications of Lithium Supertritide (2)*

$LiEt_3B^3H$ has been shown to reduce excess $CO_2$ to afford lithium [$^3$H]formate, thereby offering access to some valuable tritium-labeled $C_1$-building blocks. For example, esterification of (**250**) and amide formation converted it into highly tritiated dimethyl[$^3$H]-formamide (**251**)[229]. The latter can be used as a low molecular weight building block for the [$^3$H]formylation of alkenyl- and aryllithium compounds to give the respective [1-$^3$H]-aldehydes e.g., **252**. Alternatively, [$^3$H]formate can react with acetyl chloride to give acetic [$^3$H]formic anhydride (**253**) of high specific activity, a powerful reagent for one-carbon extension reactions and *N*-[$^3$H]formylations of peptides[230]. Its reaction with the synthetic tripeptide MetLeuPhe, for example, afforded the chemotactic peptide *N*-[$^3$H]formyl-MetLeuPhe at a specific activity of 28.6 Ci/mmol.

### 4.3.7 Lithium Tri-*sec*-Butylborotritide [Li(*sec*-$Bu_3$)$B^3$H, Li T-Selectride]

Li(*sec*-Bu)$_3$$B^3$H is accessible in an manner analogous to $LiEt_3B^3H$, from $Li^3H$ and *sec*-$Bu_3B$[231,232]. It has approximately the same reducing potential as $LiEt_3B^3H$. However, since the stereoselectivity of hydride transfer from alkylboron hydrides increases with the size of the alkyl groups, Li-*sec*-butylborotritide is highly recommended for the reduction of cyclic ketones, where it gives excellent stereoselectivity. This property was exploited for the labeling of sanglifehrin (**255**), a macrolide compound containing several reducible functional groups (ketone, ester, amide, conjugated carbon–carbon double bonds)[231b]. Chemoselective and stereoselective reduction of the corresponding ketone precursor **254** furnished tritiated **255** at theoretical specific activity with virtually no formation of the epimeric alcohol. The same excellent steric control was observed when (−)-narwidin (**256**) was reduced with Li-*sec*-$Bu_3B^3H$ to (−)-[8-$^3$H]galanthamine (**257**), an acetylcholinesterase inhibitor, at a specific activity of 23.8 Ci/mmol without any detectable amount of its stereoisomer[232].

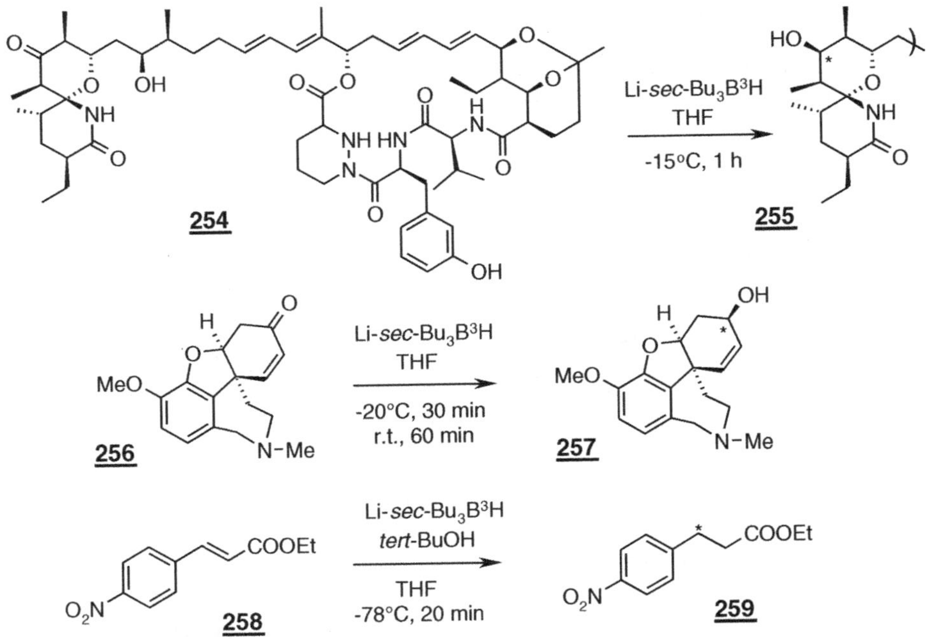


***Figure 4.69*** *Applications of lithium tri-sec-butylborotritide*

Li-*sec*-$Bu_3B^3H$ is also effective in the 1,4-reduction of $\alpha,\beta$-unsaturated esters. The label is incorporated cleanly into the chemically and metabolically stable position $\beta$ to the carboxyl function. The reaction has to be performed at low temperature (−78 °C) in the presence of *tert*-butanol as a proton donor in order to trap the enolate intermediate and prevent its undergoing Claisen condensation. Under these conditions even aromatic nitro groups are tolerated, as documented by the reduction of **258** to **259**[203].

### 4.3.8 Lithium [$^3H_2$]Boratabicyclo[3.3.1]nonane

Lithium [$^3H_2$]boratabicyclo[3.3.1]nonane (Li-BBNT, **262**) is the tritiated isotopomer of a borohydride that has reductive characteristics intermediate between those of $LiBH_4$ and the more powerful $LiEt_3BH$[233]; therefore it is particularly useful for selective reductions. For example, it reduces esters or lactones selectively in the presence of other functional groups such as carboxylic acid, amide, nitrile and halide. The tritiated reagent is prepared[234] by treatment of $Li^3H$ with 9-methoxy-BBN, readily accessible from 9-BBN and methanol. The initially formed Li-methoxy-BBNT (**261**) disproportionates to give **262** and lithium dimethoxydialkylboronate (**263**). The latter reaction is catalyzed by **260**, which is employed in 10% excess in the initial sequence depicted below. The lithium dimethoxyboronate (**263**) is apparently inert because it did not interfere with test reactions using the solution as-is for tritioreductions. Li-BBNT was tested[234] with selected ketones and esters such as estrone and ethyl 1-naphthoate, giving [7-$^3$H]estradiol at a specific activity of 26 Ci/mmol, and 1-(hydroxy[$^3H_2$]methyl)naphthaline at 55 Ci/mmol in moderate radiochemical yields.

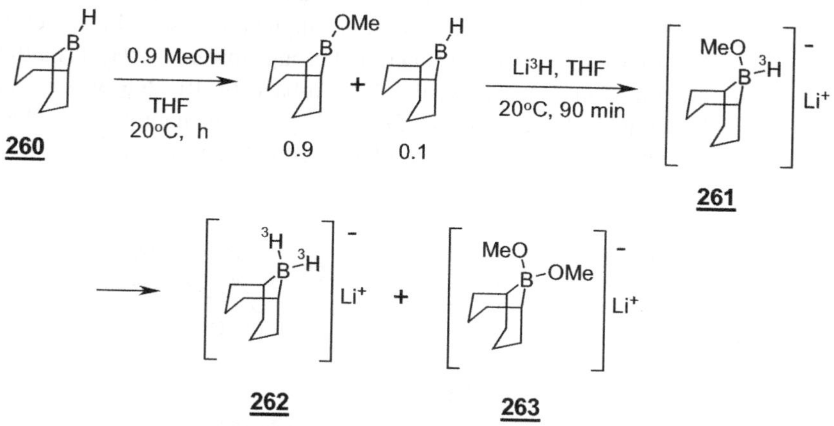


*Figure 4.70* *Preparation of lithium [$^3H_2$]boratabicyclo[3.3.1]nonane (Li-9BBNT)*

### 4.3.9 Tritiated Borane (THF-Complex) ($B_2{}^3H_6$; $B^3H_3 \cdot THF$)

In contrast to the aforementioned nucleophilic complex tritides, both, $B_2{}^3H_6$ and $B^3H_3 \cdot THF$, are Lewis acids that attack functional groups at the electron rich centers. Acids, aldehydes and ketones are readily reduced to $\alpha$-tritiated alcohols; nitriles and amides to $\alpha$-tritiated amines. Reduction of unsymmetrical epoxides preferentially forms the less substituted carbinols, in contrast to the nucleophilic tritides. Alkyl halides, aryl halides, and nitro groups do not react. Esters are much less readily reduced than carboxylic acids, allowing differentiation between these functional groups. *Syn*-addition to carbon–carbon double or triple bonds, followed by protonolytic or oxidative cleavage of the resulting carbon–boron bond, allows the synthesis of labeled alkanes, *Z*-alkenes, alcohols or aldehydes[235].

Tritioborane of maximal isotopic purity is prepared by the treatment of a THF suspension of $Li^3H$ (Section 4.3.3) with boron trifluoride etherate[236], according to the stoichiometry

$$3Li^3H + 4BF_3 \rightarrow B^3H_3 + 3LiBF_4$$

This reaction gives a yield of about 60%. If cross-reactions between the $LiBF_4$ byproduct and the reaction components are a concern, the $B^3H_3$·THF solution can be vacuum transferred into a separate reaction flask. Tritio-borane can also be prepared by the reaction between $NaB^3H_4$ and $BF_3$·$Et_2O$ according to the stoichiometry

$$3NaB^3H_4 + 4BF_3 \rightarrow 4B^3H_3 + 3NaBF_4$$

but the specific activity is limited by that of the sodium borotritide used. This reaction can either be conducted in THF solvent and the reagent used *in situ*, or conducted in diglyme solvent and the $B^3H_3$ extracted on a stream of inert gas, then trapped in THF if desired. The latter procedure and an apparatus for this purpose have been described[237].

In alkene and alkyne tritioborations, the boron adds preferentially to the less hindered side of the multiple bond, so that the label should be located mostly at the adjacent atom; this is evident in the isotope distribution in [$^3$H]safrole (**264**) and [$^3$H]isosafrole (**265**)[238], resulting from reactions with the respective isomeric alkyne precursors. In a reaction that nicely exploited the differing reactivity of functional groups towards $B^3H_3$·THF, the $\alpha,\beta$-unsaturated carbonyl function of 7-triethylsilyl-13-oxo-baccatin III (**266**) was selectively reduced to 7-triethylsilyl[13-$^3$H]baccatin III (**267**), a key intermediate in the synthesis of tritiated paclitaxel (**268**), an efficacious and widely used anticancer drug[239].

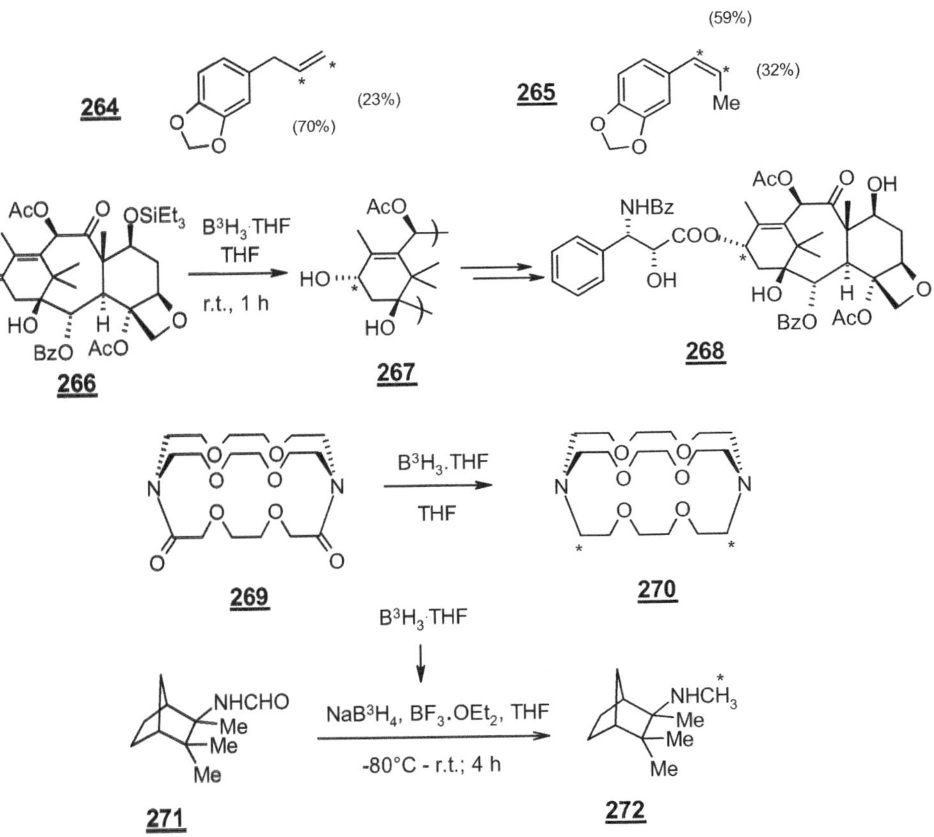


**Figure 4.71** *Applications of tritiated borane*

The ability of $B^3H_3{\cdot}THF$ to effectively reduce amides is illustrated with the reduction of diamide **269** to give the tritiated cryptand **270** in chemical yield of about 50%[237]. The reagent employed in this case was prepared from $NaB^3H_4$ and $BF_3{\cdot}Et_2O$ (see also Chapter 11). The $B^3H_3{\cdot}THF$-mediated reduction of *N*-formyl compounds to high specific activity *N*-[$^3$H]methylamines is of particular interest as an alternative to *N*-alkylation with [$^3$H]-methyl iodide. Whereas the preparation of [$^3$H]mecamylamine (**271**) by the latter method was not feasible owing to steric hindrance, application of the $B^3H_3{\cdot}THF$ method to **271** furnished **272** at a specific activity of 35 Ci/mmol[240].

Surprisingly, no publication exists so far reporting on the use of $B^3H_3{\cdot}THF$ in 1,3,2-oxazaborolidine-catalyzed stereoselective reductions[241] of prochiral ketones with e.e.'s ranging from 85–99% as has been employed in carbon-14 synthesis[242].

### 4.3.10 Tritiated Alkylboranes

Two tritiated alkylboranes have been reported to date. They have been synthesized at very high specific activities via $Li^3H$ in one-pot procedures and can be used *in situ*.

*1,3-Dithian-2-ylmyrtanyl[$^3$H$_2$]borane (MDBT$_2$)* (**273**) is the only chiral borane so far reported in radiolabeled form. It is prepared by treatment of $Li^3H$ with $AlCl_3$ followed by reaction of the resulting $Al^3H_3$ *in situ* with [2-(1,3-dithianyl)]myrtanylcatecholborane[243].

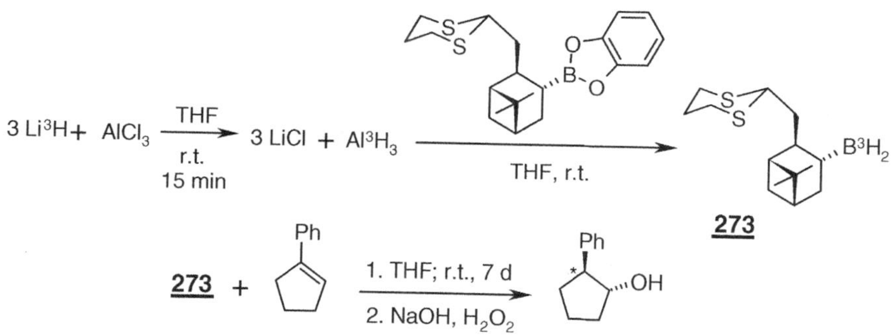


*Figure 4.72* Preparation of 1,3-dithian-2-ylmyrtanyl[$^3$H$_2$]borane

MDBT$_2$ is of value for preparation of chiral $\beta$-tritiated alcohols in moderate to good enantiomeric excesses by tritioboration of prochiral olefins followed by oxidative cleavage of the B–C bond with $H_2O_2$. Its efficacy was tested in a model reaction with 1-phenylcyclopentene. Oxidative cleavage of the intermediately generated borane with alkaline $H_2O_2$ after seven days provided *trans*-2-phenyl[2-$^3$H]cyclopentanol with 75% enantiomeric excess and a specific activity of 25 Ci/mmol. The enantiomeric excess obtained with MDBT$_2$, when compared with the corresponding values for the unlabeled and the deuterated isotopomers in the same reaction (85% and 80%, respectively), seems to hint at an isotope effect disfavoring the heavier element.

*Methyl[$^3$H$_2$]borane*, prepared in the same way as MDBT$_2$ except using methylcatecholborane, was tested in a hydroamination reaction with 1-phenylcyclopentene. Cleavage of the B–C bond of the resultant alkyl borane with $H_2NOSO_3H$ afforded *trans*-2-phenyl[2-$^3$H]cyclopentylamine in 20–40% yield[244].

### 4.3.11 Lithium Aluminum Tritide ($LiAl^3H_4$)

Lithium aluminum tritide is a more powerful reagent than sodium or lithium borotritide and reduces most of the commonly encountered functional groups. Aldehydes, ketones, esters, lactones, carboxylic acids are smoothly converted to $\alpha$-tritiated alcohols under mild conditions. Substituted amides are reduced to [$\alpha$-$^3$H]amines or [carbonyl-$^3$H]aldehydes depending on the nature of the respective substrate and the reaction conditions. Alkyl halides and tosylates give the corresponding labeled alkanes[245]. Unlike alkali borotritides, solid $LiAl^3H_4$ is already unstable at a specific activity of 100 mCi/mmol[202]. Although high specific activity $LiAl^3H_4$ has been commercially available as a THF solution shipped in dry ice, it should be used with caution as its purity or potency cannot be easily assessed prior to use. Accordingly, even material of moderate specific activity is usually prepared *in situ* by reaction of $Li^3H$ with $AlCl_3$ or preferably with $AlBr_3$ in diethyl ether or THF[206] (Figure 4.73).

Because of its lack of selectivity and undependable commercial availability $LiAl^3H_4$ has usually been used only in cases where reductions with $NaB^3H_4$ alone or in the presence of

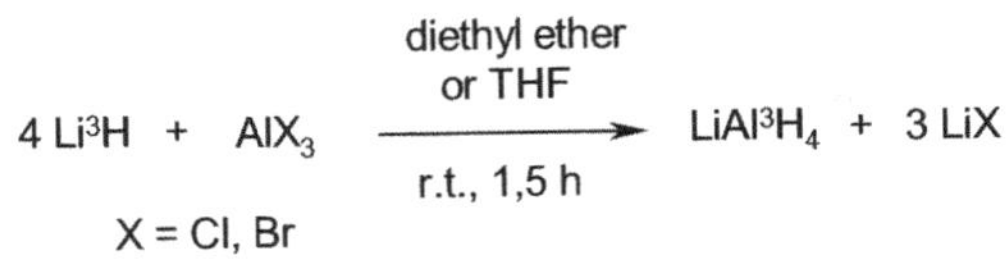


*Figure 4.73* Preparation of lithium aluminum tritide

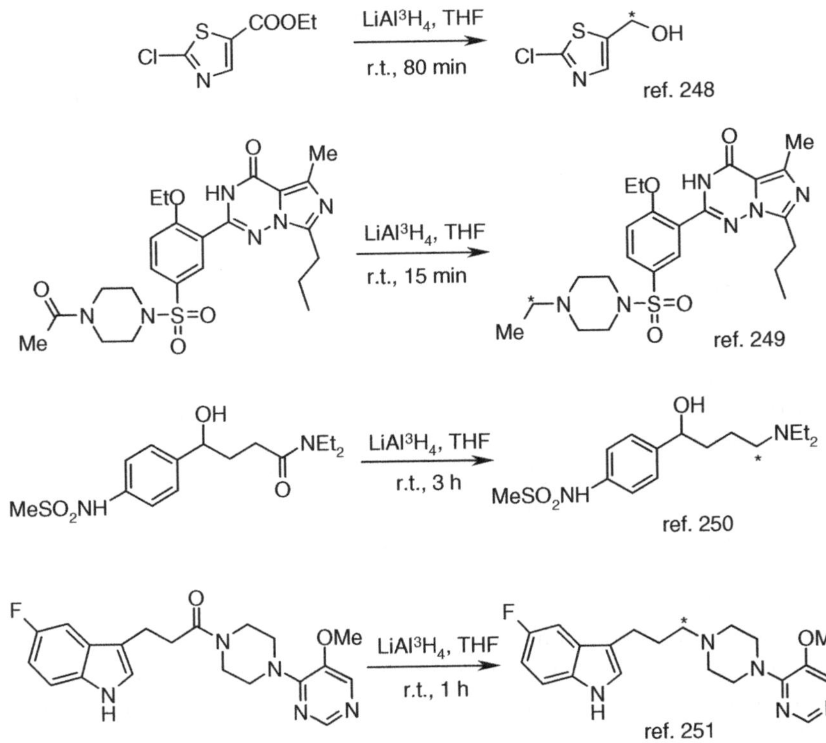


*Figure 4.74* Applications of high specific activity lithium aluminum tritide (1)

activating metal salts are not satisfactory. It has most often been used for reduction of amides to [$^3$H]amines, and esters to [$^3$H]alcohols, sometimes with good selectivity in the presence of other functional groups (see Figure 4.74). Product specific activities are generally near those of the tritide reagent, on a per-tritide basis; in the examples below they were 52–56 Ci/mmol using carrier-free $LiAl^3H_4$ prepared *in situ*.

The ability of $LiAl^3H_4$ to reduce carbamates to labeled methyl groups, such as in the synthesis of [*N*-methyl-$^3$H]nortriptylin (**274**)[230b], provides an alternative to *N*-[$^3$H]methylation with [$^3$H]methyl iodide for the preparation of high specific activity tracers for *in vitro* binding studies (see also Chapter 4.4.3). An example of a uniquely useful carbamate reduction is that of *O*-benzyl carbamate to give [$^3$H]methylamine at a specific activity of >70 Ci/mmol. The latter was further reacted with 3-azido-4-chlorophenylisothiocyanate to furnish the thiourea **275**, an efficient photoaffinity probe for the urea carrier[230b].

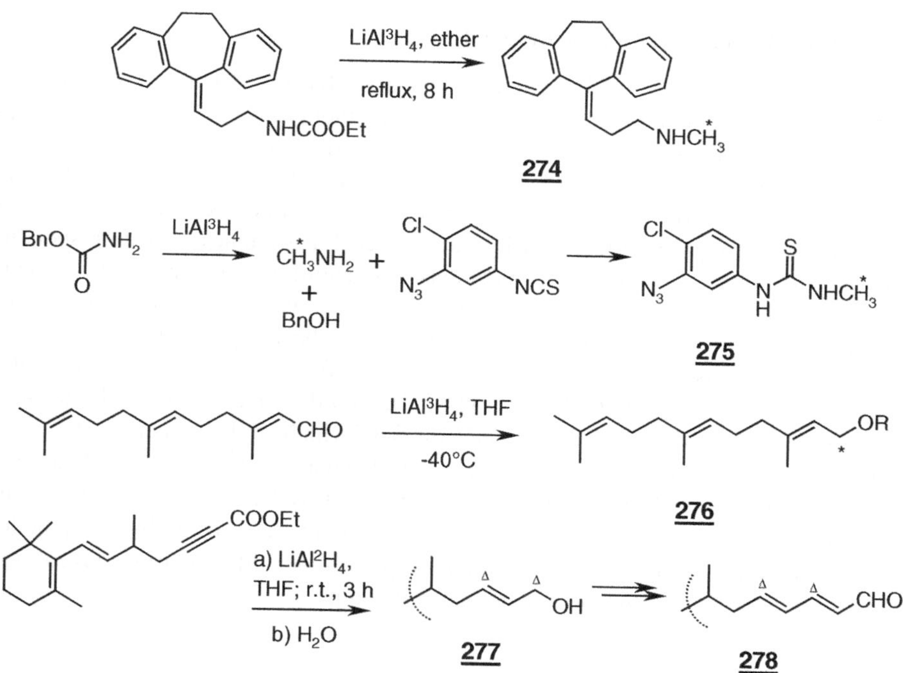


**Figure 4.75** *Applications of high specific activity lithium aluminum tritide (2)*

$LiAl^3H_4$ reduces $\alpha,\beta$-unsaturated aldehydes to the corresponding allylic alcohols in a 1,2-addition mode, as demonstrated with the synthesis of *E,E*-[1-$^3$H]farnesol (**276**, R = H), a precursor of *E,E*-[1-$^3$H]farnesyl pyrophosphate (**276**, R = $P_2O_6^{3-}$)[230b,250]. On the other hand propargylic alcohols and carboxylic acid esters provide the corresponding *E*-allylic alcohols with deuterium at the $\beta$-position (with $LiAl^2H_4$), and hydrogen or deuterium at the $\alpha$-position, depending on whether the resulting Al-complex is hydrolyzed with $H_2O$ or $^2H_2O$. The latter reaction was exploited for the preparation of the *E*-allylic [1,3-$^2H_2$]alcohol **277**, which was converted to 13-desmethyl[11,13-$^2H_2$]retinal (**278**)[251].

The availability of high specific activity $LiAl^3H_4$ stimulated the development of a new laboratory method[252] for preparing high specific activity tritiated methyl iodide, previously produced by catalytic methods (Section 4.4.3). The new method is analogous to that for [$^{14}$C]methyl iodide using $^{14}CO_2$ and $LiAlH_4$ and is illustrated in Figure 4.76. Since its publication it has already been exploited for various *N*, *O*, and *S*-tritio-methylations (Section 4.4.3).

$$^3H_2 \xrightarrow{n\text{-BuLi}} Li^3H \xrightarrow[\text{r.t., 1.5 h; X = Cl or Br}]{AlX_3,\ Et_2O} LiAl^3H_4 \xrightarrow[\text{0°C, 60 min, r.t., 1 h}]{CO_2,\ \text{ether}} M(OC^3H_3) \xrightarrow[\text{80°C, 30 min}]{HI} C^3H_3I \qquad M = Li, Al$$

*Figure 4.76* *Preparation of high specific activity tritiated methyl iodide via lithium aluminum tritide*

### 4.3.12 Tri-*n*-Butyltin Tritide (*n*-$Bu_3Sn^3H$)

*n*-$Bu_3SnH$ is a moderately stable, commercially available reagent that has become an important tool in synthetic organic chemistry for a plethora of chemical transformations, such as radical dehalogenations, deoxygenations of alcohols (through reduction of their xanthates, aryloxythiocarbonates or imidazolylthiocarbonyl derivatives), denitrations, decarboxylations of carboxylic acids via their *N*-hydroxy-2-thione esters, $Pd(Ph_3)_4$-catalyzed reductions of acyl chlorides to aldehydes and of enals and enones to the corresponding saturated derivatives, free radical cyclizations and additions to vinylogous carbonyl systems[253]. Although all these reactions show a high potential also for synthesis of tritiated compounds, only a few applications of *n*-$Bu_3Sn^3H$ have been reported so far.

Tributyltin tritide is prepared[254–257] at maximal specific activity by the reduction of *n*-$Bu_3SnCl$ with freshly prepared $Li^3H$ (Section 4.3.4), according to equation 1 in Figure 4.77. It has also been prepared by the reduction of *n*-$Bu_3SnCl$ with $NaB^3H_4$ (equation 2)[258–260] or the tritiolysis of *n*-$Bu_3SnLi$ or *n*-$Bu_3SnMgX$ by $^3H_2O$ (equation 3)[258,261,262]. Both the latter methods produce materials whose specific activities are less than carrier-free – in the first case limited by the specific activity of $NaB^3H_4$ and in the second by the practical impossibility of preparing and manipulating $^3H_2O$ without dilution from adventitious protons – and problems with side reactions and workups.

$$Li^3H + n\text{-}Bu_3SnCl \xrightarrow[\text{r.t., 15 min}]{\text{THF (cat. }Et_3B)} n\text{-}Bu_3Sn^3H + LiCl \qquad \text{eq. 1}$$

$$NaB^3H_4 + n\text{-}Bu_3SnCl \xrightarrow[\text{2. EtOH, r.t., 30 min}]{\text{1. DME, r.t., 45 min}} n\text{-}Bu_3Sn^3H + NaCl + B(OEt)_3 + 3\,^3HH \qquad \text{eq. 2}$$

$$n\text{-}Bu_3SnCl \xrightarrow[\text{2. }^3H_2O]{\text{1. 2 Li, THF}} n\text{-}Bu_3Sn^3H + LiCl + LiO^3H \qquad \text{eq. 3}$$

*Figure 4.77* *Preparative methods for tri-n-butyltin tritide*

Tributyltin tritide was used for the $Pd(Ph_3)_4$-catalyzed reduction of the $\alpha,\beta$-unsaturated carbonyl system of $\Delta^{6,7}$-thiaprostaglandin $E_1$ methyl ester (**279**) to give the [7-$^3$H] derivative **280** in 58% radiochemical yield based on $Bu_3Sn^3H$ generated *in situ* (in 14.6% yield based on the $NaB^3H_4$ employed)[259]. The reagent similarly prepared was also applied to the selective tritiodehalogenation of the $\alpha$-bromosteroids **281** and **282** possessing other potentially reducible substituents (F, Cl, saturated and $\alpha,\beta$-unsaturated carbonyl systems)[258], a series of 23-iodo-15-ketosterols[263a] (**283**), alkyl and benzyl bromides[261,263], *N*-iodoacetoxyphthalimide (**284**, see Section 4.4.8)[263b] and, more recently, of $\alpha$-bromo aldehyde **285**, key intermediate in the synthesis [$^3$H]DCG-IV (**286**), a metabotropic glutamate receptor agonist labeled at a specific activity of 17.5 Ci/mmol[257].

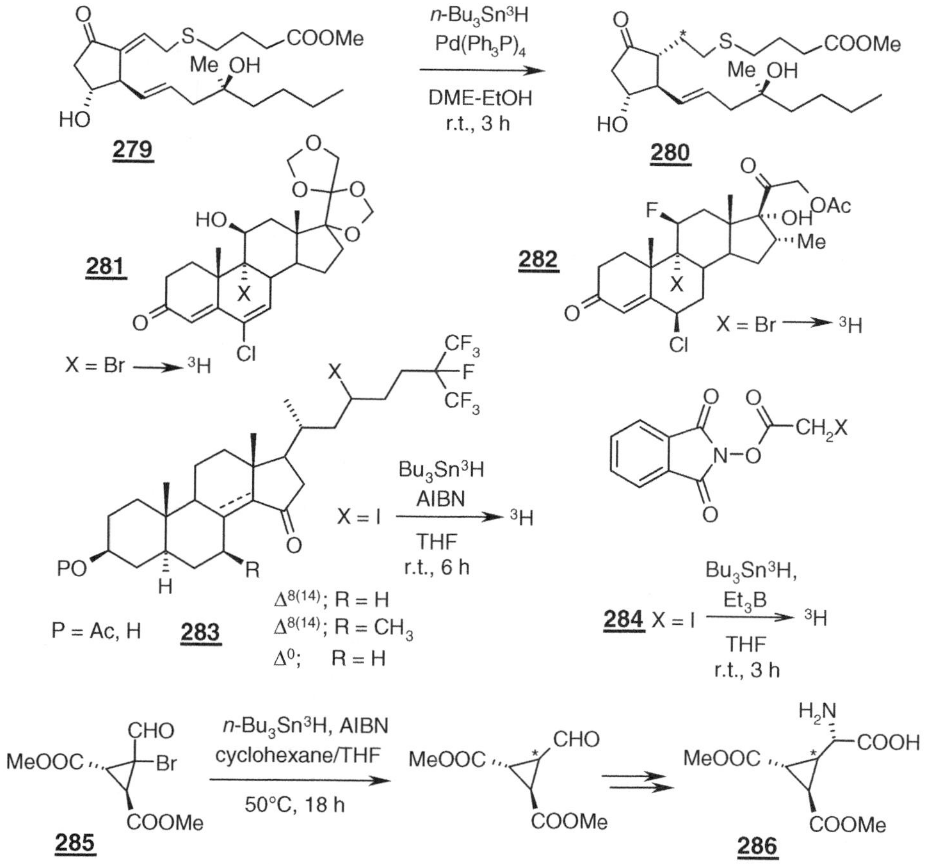


***Figure 4.78*** *Applications of tri-n-butyltin tritide in conjugate reductions and tritiodehalogenations*

The $Bu_3Sn^3H$-mediated tritiodeoxygenation of aryloxythiocarbonate and alkyldithiocarbonate esters of hydroxyl-substituted substrates (Barton deoxygenation) has been reported[260,263] to be a good general method for the preparation of methylene-labeled compounds from the corresponding secondary alcohols. For example 2′-*O*-phenoxythiocarbonate esters of C-ribonucleosides have been converted to their corresponding 2′-desoxy [2′-$^3$H]-derivatives. This type of reductive cleavage reaction is, as expected of a radical reaction, not stereoselective. Applied to the 4-*p*-fluorophenoxythiocarbonate ester of cholesterol (**287**), the reduction led to the formation of the same ratio of [3-$^3$H]cholest-5-en isomers **288** (22 Ci/mmol, 70% 3$\alpha$, 30% 3$\beta$) independent of the stereochemistry of the starting ester[263]. The method has recently been applied to the preparation of tritiated taxanes for use in investigations of paclitaxel biosynthesis[264]. Reaction of bis(phenoxythiocarbonate) **289** and the methyldithiocarbonate ester **291** with mixtures of $Bu_3SnH$ and $Bu_3Sn^3H$ provided **290** and **292**, respectively, in fair to good chemical yields. The specific activities, however, were much lower than expected, which may have been caused at least in part by a kinetic isotope effect in the hydrogen/tritium transfer from reagent to substrate.

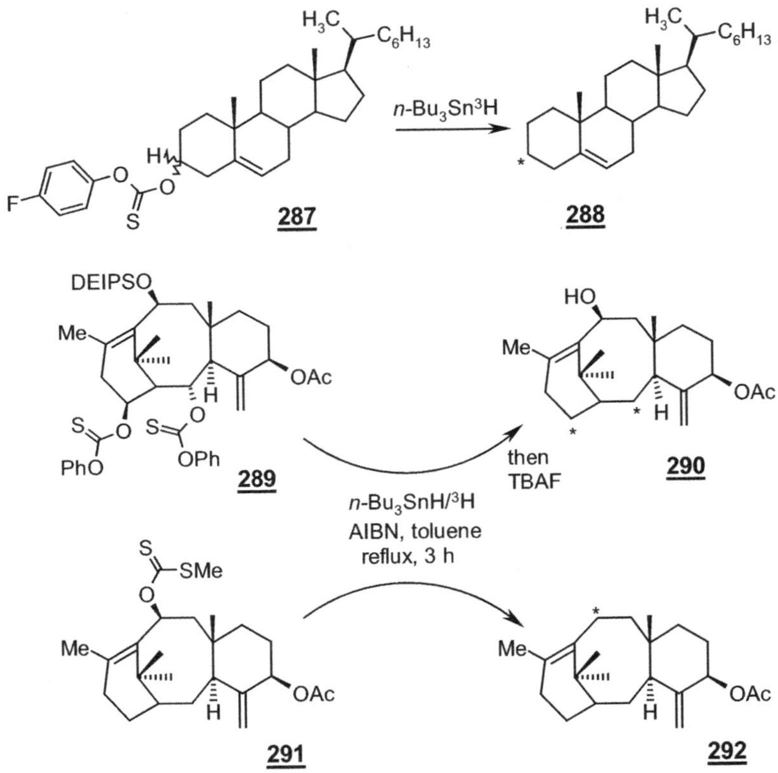


*Figure 4.79* *Use of tri-n-butyltin tritide in tritiodeoxygenations*

*Tritio-denitration*: In the presence of azoisobutyronitrile as radical initiator, tri-*n*-butyltin tritide can tritio-denitrate secondary and tertiary nitro compounds to give the respective labeled derivatives. Primary nitro groups can only be denitrated if they are located adjacent to keto groups[265]. The carbanion-stabilizing properties of nitro groups facilitate the introduction of electrophiles such as alkyl halides, aldehydes and ketones at the $\alpha$-position. Furthermore, nitroolefins are excellent Michael acceptors as well as dienophiles, which enables the regioselective and chemoselective synthesis of a broad range of acyclic and cyclic nitro compounds. Subsequent tritio-denitration of such compounds can give access to the selectively labeled compounds. So far, only the two examples in Figure 4.80 below have so far been published; they were investigated only in an exploration of the range of this methodology. In the first case radical tritio-denitration of $\alpha$-nitromethyl adamantyl ketone **293** with tri-*n*-butyl trn tritide at a specific activity of about 10 Ci/mmol afforded [$^3$H]methyl adamantyl ketone **294** with a specific activity of 6.7 Ci/mmol in 67% yield. Application of this methodology to **295**, accessible through Michael-addition of 2-nitrocyclododecan-1-one to acrolein, followed by deuterio-denitration with tri-*n*-butyltin deuteride provided the 1,5-dicarbonyl compound **296** specifically labeled at C5[266,250].

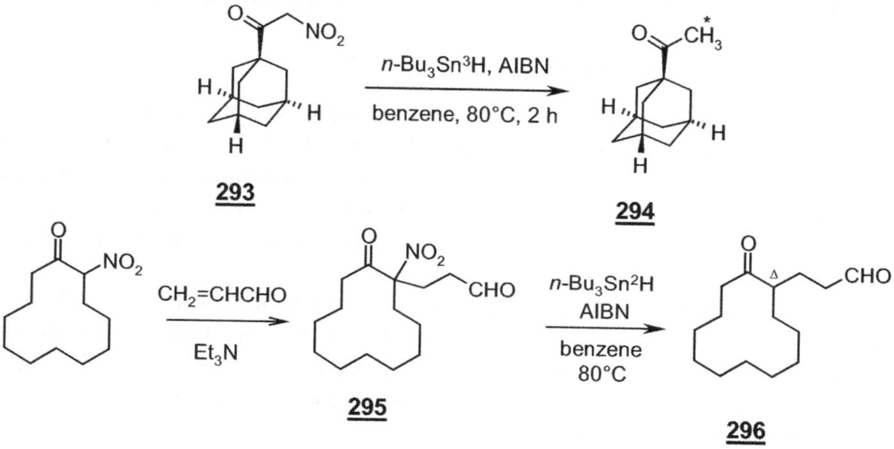


**Figure 4.80** *Use of tri-n-butyltin tritide in tritiodenitrations*

*Protecting radical translocation:* A protecting group, suitably designed also to be a radical precursor, can function as a hydrogen isotope labeling agent if, upon generation of a

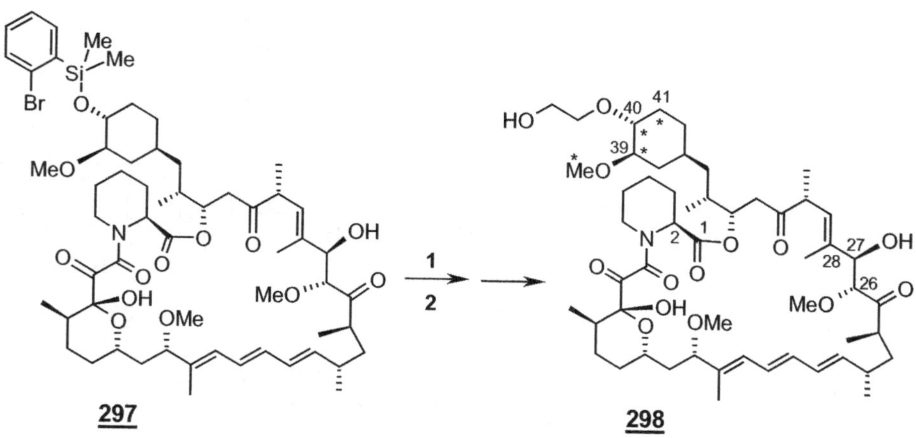


**Reaction conditions: 1.** *n*-Bu$_3$Sn$^3$H, THF, Et$_3$B (0.2 equiv); 23 °C, 2 h; **2.** aq. AcOH, THF, 0 °C, 2.5 h

**Figure 4.81** *Application of tri-n-butyltin tritide in labeling by protecting radical translation*

radical in the group, intramolecular hydrogen transfer occurs, translocating the radical to the substrate where it can be quenched by $n$-$Bu_3Sn^2H$ or $n$-$Bu_3Sn^3H$. This protecting radical translocation strategy[267], which appeared to be useful for the labeling of various macrolides, was successfully applied to the synthesis of tritiated everolimus (**298**), a rapamycin derivative with improved immunosuppressive properties[255]. Thus, rapamycin (**297**), selectively protected with (2-bromophenyl)dimethylchlorosilane at the C40-hydroxy group, was treated with $n$-$Bu_3Sn^3H$. Subsequent cleavage of the protecting group with AcOH and hydroxyethylation of the released hydroxy group converted the monosilyl ether into the tritiated drug substance at a specific activity of 10 Ci/mmol, the label being located at the C39,40 (80%), C41 (18%) and the C24-*O*-methyl group (9%).

### 4.3.13 Tritiated Schwartz's Reagent ($ZrCp_2Cl^3H$)

Schwartz's reagent ($ZrCp_2ClH$) has emerged as a highly versatile tool for the stereoselective hydrozirconation of carbon–carbon multiple bonds[268] and the reduction of $\beta$-keto esters to $\alpha,\beta$-unsaturated esters[269]. It shows good functional group tolerance. Since these applications either start with *cis*-addition of $ZrCp_2ClH$ to the respective multiple bonds or involve an intramolecular hydride transfer, product stereochemistries are complementary to other, more commonly used, reductive procedures.

$ZrCp_2Cl^3H$ is prepared[270] (Figure 4.82) by *in situ* reduction of $ZrCp_2Cl_2$ with either $Li^3H$ or $LiEt_3B^3H$, prepared as described earlier in this chapter. Care has to be taken against two eventualities: first that the insoluble $Li^3H$ has indeed been converted to the equally insoluble zirconium complex, and second that the byproduct borane is removed in cases where it might react with substrate to be added.

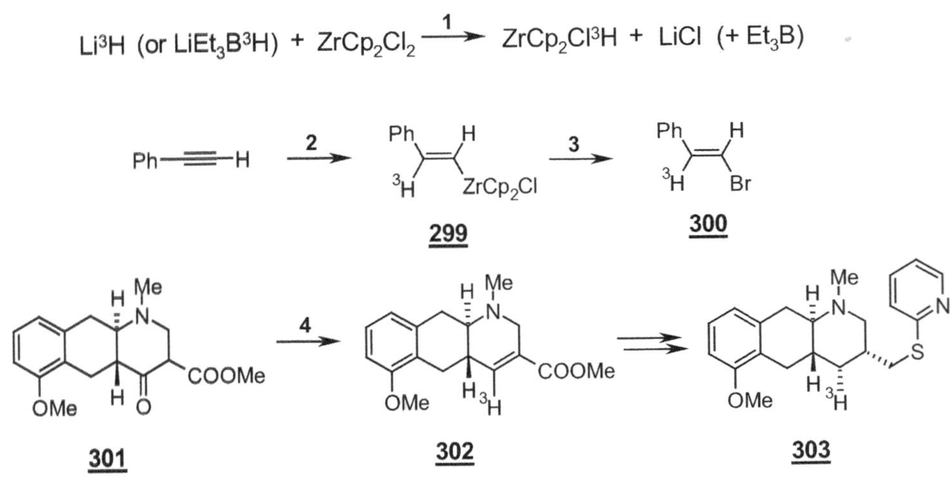


**Reaction conditions:** **1**. THF; 0 °C, 5 min; **2**. $ZrCp_2Cl^3H$, THF; r.t., 20 min; **3**. NBS; r.t., 35 min; **4a**. LiHMDS, THF; –20 °C, 2 h, **b**. $ZrCp_2Cl^3H$, THF; r.t., 1–2 h

***Figure 4.82*** *Preparation and applications of tritiated Schwartz's reagent*

The utility of tritiozirconations of acetylenes to give tritiated *trans*-olefins has been demonstrated[270] in a model reaction with phenylacetylene using $Li^3H$ prepared from a

1 : 10 mixture of tritium gas and hydrogen gas. Phenylacetylene was added to the solution in which $ZrCp_2Cl^3H$ had been prepared, producing **299**, followed by *N*-bromosuccinimide as electrophile to quench the vinylzirconium intermediate. The reaction afforded *trans*-$\beta$-bromo[$\alpha$-$^3$H]styrene (**300**) with a specific activity of 2.8 Ci/mmol and with very high stereo- and regioselectivity, as determined by $^3$H NMR analysis. This result indicated that the tritiozirconation occurred as expected, in a *cis*-mode with the metal attacking the less hindered carbon. Besides NBS, intermediates such as **299** can be quenched with other electrophiles such as $I_2$, $H_3O^+$, *t*-BuOOH, or, after an appropriate transmetallation, with carbon electrophiles to give internal *trans*-olefins labeled with tritium at the vinyl position. This procedure, achievable in one pot from tritium gas, provides a route to tritiated *trans*-olefins that are otherwise difficult to prepare.

The reduction of lithium enolates of $\beta$-keto esters with $ZrCp_2Cl^3H$ to give $\beta$-tritiated $\alpha$,$\beta$-unsaturated esters has been demonstrated[271] by the transmetallation of the lithium enolate of **301**, followed by an intramolecular tritide transfer. Spontaneous elimination of $O{=}ZrCp_2$ converted the intermediate zirconium enolate into **302**, a key intermediate in the radiosynthesis of NVP GLC756 (**303**). The specific activity of **302** was 24 Ci/mmol[272].

### 4.3.14 Tritiated Triethylsilane and Trihexylsilane

$Et_3Si^3H$ and $Hex_3Si^3H$ are much milder tritide reagents than the others in this chapter. Because of the relatively small difference in the electronegativities of tritium and silicon, the tritium bond in each of these reagents is only slightly 'hydridic' in nature, which means that they can only function as tritide donors towards highly electrophilic species[273]. They are stable towards air and moisture and tolerate even strong acids ($CF_3COOH$, $CF_3SO_3H$); therefore they can be used for the reductive labeling of substrates that are able to form carbonium ions under such conditions ('ionic hydrogenation/tritiation'). Accordingly, the reduction of tri- and tetra-substituted olefins, aryl alkyl ketones, diaryl ketones and tertiary alcohols should furnish labeled alkane derivatives. $\alpha$,$\beta$-Unsaturated ketones should be converted into $\beta$-tritiated saturated ketones. Nitrilium ions, routinely prepared by N-alkylation of nitriles, should be converted into the corresponding [1-$^3$H]imines, which can be readily hydrolyzed to [1-$^3$H]-aldehydes. Other reducible groups such as $NO_2$, CN, COOR, and halogen are unaffected.

In the presence of rhodium catalysts, both of these reagents add to terminal alkynes to give labeled *E*-vinylsilanes. [1,1′]-Bis(diphenylphosphine)ferrocene]palladium(II) chloride catalyzes the reduction of vinyl triflates to monotritiated olefins.

Both reagents are prepared by reaction of $Li^3H$ with the respective chlorosilane (Figure 4.83). Since some of the reactions mentioned above do not work well in THF, the volatile $Et_3Si^3H$ must be generated in triglyme, from which it is isolated by vacuum transfer. Unfortunately, traces of TMEDA (carried over from the preparation of $Li^3H$) can co-distil and poison the transition metal catalysts. Alternatively, $Hex_3Si^3H$ can be employed in this type of reaction, since it can be easily purified by filtration over silica gel[274].

Despite its interesting synthetic potential, $Hex_3Si^3H$ has only been investigated in a few model reactions[274], such as the ionic tritio-deoxygenation of 1,1-diphenylethanol (Figure 4.83, Equation 1), the $Pd(OAc)_2$/dppf-catalyzed deoxygenation of $\alpha$-tetralone (via its enol triflate) (Equation 3) and the stereoselective addition to triple bonds to provide either the labelled *E*-alkene, when catalyzed by $[Rh(COD)_2]BF_4$ or the corresponding Z-alkene, when $AlCl_3$ is used as a catalyst (Equations 2)[274]. Only one application of $Et_3Si^3H$ has been

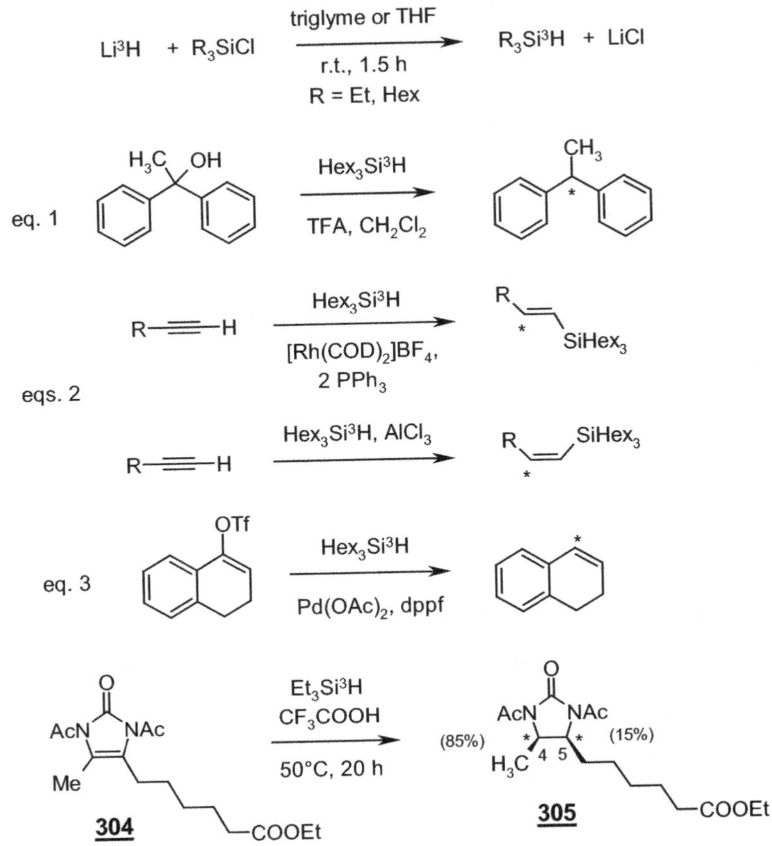


*Figure 4.83* Preparation and applications of tritiated trialkylsilanes

published so far[275]. Treatment of 1,3-diacetyl-4-methyl-5-(ω-carboethoxyamyl)-2-imidazolone (**304**) with $Et_3Si^3H$ in TFA gave di-protected *cis*-[4,5-$^3$H]dethiobiotin (**305**) in 95% stereoselectivity and 70% yield; 85% of the label was located at C4, the residual 15% at C5.

An indication of further possibilities with tritium is provided by the use of $Et_3Si^2H$ in the Lewis acid-catalyzed rearrangement of phenyl-substituted cyclic acetals to isochromans[276]. The use of $Et_3Si^2H$ instead of $Et_3SiH$ in the reaction with acetal **306** produced [4-$^2$H]isochroman **307** in high chemical and isotopic yields and high stereoselectivity (Figure 4.84). Interestingly, [2-$^2$H]**306** prepared from the methoxy analog by reduction with $n$-$Bu_3Sn^2H$, and reacted with nonisotopic silane, produced the isotopomeric [1-$^2$H]**307**[277].

## 4.4 Small Tritiated Building Blocks

The tritiated compounds discussed in this section are selected because of their broad or unique utility or potential in the preparation of tritium-labeled compounds. There are a number of reviews[278] that discuss various aspects of this topic. We distinguish them from

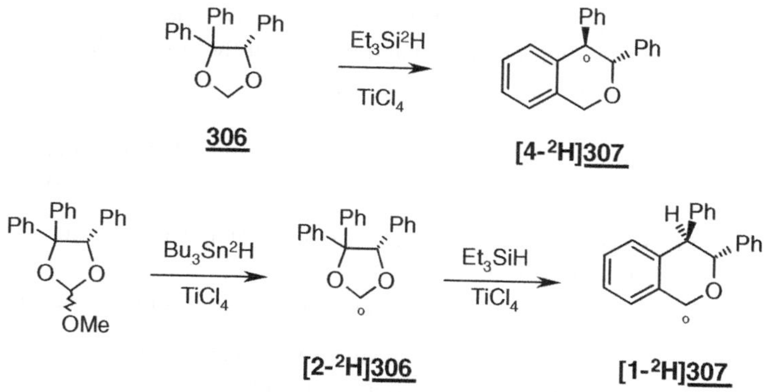


***Figure 4.84*** *Isomeric labeling by means of isotopic hydrides*

the many other, commercially available, tritiated compounds that serve as starting materials, building blocks or intermediates for synthetic elaboration.

### 4.4.1 Tritiated Water ($^3H_2O$; $^3HHO$)

Despite its high radiotoxicity[279] and its instability – it decomposes by autoradiolysis into its elements even at low isotopic abundance and liquid nitrogen temperatures[280] – tritiated water is the most common low molecular weight reagent in the synthesis of tritium-labeled compounds. Two procedures have been published for the preparation of tritiated water free of solvent at specific activities ranging from 40 to 58 Ci/mmol (about 2–3 Ci/$\mu$L). In the first case tritium gas is oxidized by copper oxide at elevated temperature[281], whereas the second process uses $PtO_2$ (Adams' catalyst) as oxidant. The reaction of $PtO_2$ with $^3H_2$ proceeds rapidly at room temperature and is exothermic; the progress of the reaction can be monitored by manometric observation, providing the $^3H_2O$ vapor is first condensed by cooling the reaction vessel[282]. A slight modification of this procedure has been recently published[283]. Vacuum transfer of neat tritiated water is difficult, owing to the combination of its high specific heat of vaporization, high surface tension and relatively high boiling point.

Preparation of tritiated water as a solution in a dry organic solvent makes it easier to manipulate and is preferable from the radiation safety perspective. Stirring a suspension of PdO in a dry organic solvent such as THF or dioxane in the presence of tritium gas produces tritiated water at specific activities of 10–45 Ci/mmol. To minimize decomposition, it should be prepared at a concentration of less than 0.05% wt/wt and used shortly thereafter. It can be introduced into the reaction mixture by vacuum transfer[284,285]. Higher specific activities are difficult to achieve due to the omnipresence of water in solvents, catalyst and on glassware surfaces. The tritiated water produced under these conditions is apparently safe enough that it has been commercially available (0.01 M solution in THF at a specific activity of 47 Ci/mmol)[286].

There are numerous reports and review articles on the use of tritiated water in acid, base and heterogeneous or homogeneous metal catalyzed tritium–hydrogen exchange reactions and their importance for the preparation of labeled physiologically active compounds in support of drug research and early development[285,287]. Modern synthetic methods such as directed *ortho* metallation[288] developed during the last two decades have significantly expanded the scope of applications of tritiated water. All these procedures, which do not change the substrate's structure except the replacement of hydrogen by tritium are discussed in Chapter 3.

Besides these procedures, there are others that work in cases where classical methods fail. They have been little used so far, but have unique advantages in certain situations.

- Deoxygenation of ketones to tritiated alkenes via the Shapiro-reaction;
- Fluoride-induced tritiodesilylation;
- Zn–Cu/$^3$HHO-mediated tritiodehalogenation of alkyl, vinyl and aryl halides.

The *Shapiro reaction* enables the conversion of ketones containing at least one $\alpha$-hydrogen into tritiated alkenes. The dianion **309**, generated by deprotonation of the corresponding tosylhydrazone **308** with *tert*-BuLi at −78 °C, decomposes upon warming from 0 to 25 °C to the alkenyllithium species **310**, which can then be trapped with tritiated water to give the tritioalkene **311** in excellent yields and with specific activities of 14–24 Ci/mmol[289]. Thus, cholest-4-en-3-one (**312**) and estrone (**314**), for example, could be converted to [3-$^3$H]cholesta-2,4-diene (**313**) and [17-$^3$H]estra-1,3,5,16-tetraen-3-ol (**315**), respectively, at specific activities of 14.3–25.3 Ci/mmol and in about 65% chemical yield. With unsymmetrical ketones, the regioselectivity of the reaction depends on the geometry of the initial tosylhydrazone and on the solvent used. In ether solvents, the proton nearest to the nitrogen-centered anion is removed.

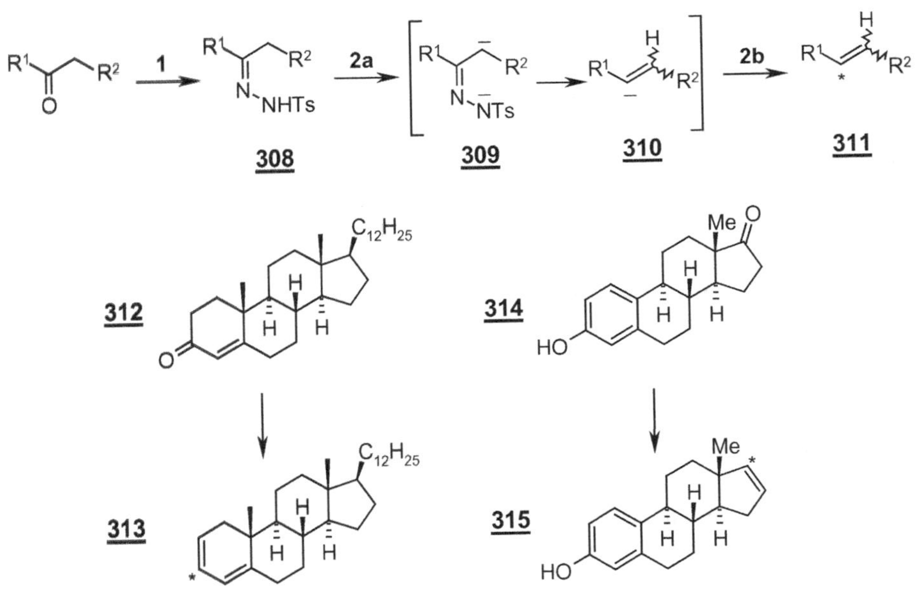


**Reaction conditions: 1.** tosylhydrazide, HOAc; 60 °C; **2a.** *tert*-BuLi, THF; −78 °C, 2 h; 0 °C, 5 min, **b.** $^3H_2O$, THF; 0 ° → r.t., 5 min

**Figure 4.85** Applications of the Shapiro reaction with tritiated water

*Fluorodesilylation* of organosilanes with CsF or $n$-$Bu_4NF$ is an effective method for generating carbanions in the absence of strongly basic conditions[290]. Anions so produced can be trapped *in situ* with tritiated water, making this an attractive method for introduction of tritium into selected positions at high specific activity. Only two reports of the use of this method have been published so far: treatment of (trimethylsilyl)diazomethane (**316**) with $n$-$Bu_4NF$ and $^3HHO$ to give [$^3H$]diazomethane (**217**) at a specific activity of 20 Ci/mmol (see Section 4.4.3) and a slightly modified procedure applied to *N*-(trimethylsilyl)acetoxy phthalimide (**318**) which furnished **319** at a specific activity of 12.5 Ci/mmol. The latter is an acylation reagent for amino groups (see Section 4.4.5).

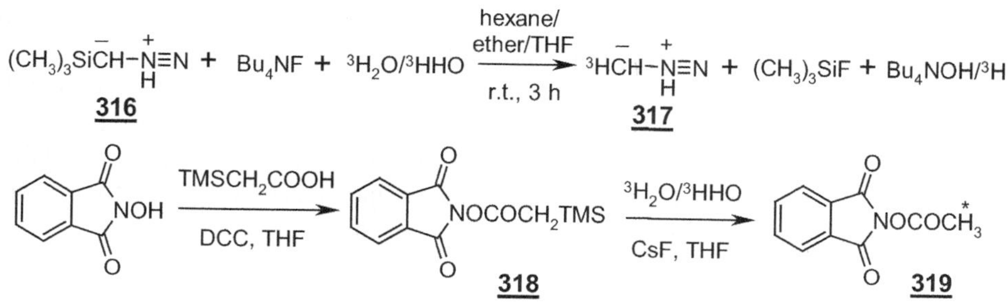


***Figure 4.86*** *Tritiation via fluorodesilylation*

Finally, *Zn–Cu couple/tritiated water* in dioxane tritiodehalogenates alkyl, vinyl and aryl halides[291]; the reaction is compatible with a wide range of sensitive functional groups. For example, treatment of 19-bromoherbimycin A (**320**) with excess reagent, followed by reoxidation of the hydroquinone intermediate, afforded the tritiated antibiotic **321** in low, but adequate radiochemical yield with a specific activity of 4 Ci/mmol (from $^3HHO$ of 20 Ci/mmol). Attempts to reduce **320** with $Bu_3Sn^3H$ failed. The Zn–Cu procedure is a modification of one using Zn, NaI and $^3HHO$ in dimethoxyethane, a reaction that was designed to replace primary and secondary hydroxy groups while tolerating a broad spectrum of reducible functionalities such as $\alpha,\beta$-enones, ketones and esters[292].

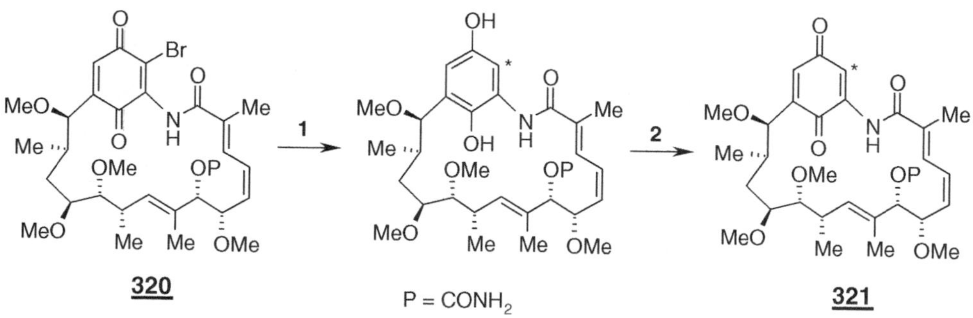


**Reaction conditions: 1**. Zn–Cu, $^3HHO$, dioxane; 80 °C, 3 h; **2**. $MnO_2$, EtOAc; r.t., 30 min

***Figure 4.87*** *Application of Zn–Cu couple and tritiated water*

### 4.4.2 Tritiated Diimide ($^3HN = N^3H$)

Diimide is a highly selective reagent that reduces symmetrical multiple bonds (C≡C, C=C, N=N) by transfer of its two hydrogens in a *cis* manner; to olefinic double bonds the addition takes place from the less hindered side. *Trans*-olefins are usually reduced slightly more rapidly than *cis*-olefins and the ease of reduction diminishes with increasing alkyl substitution[293]. Unsymmetrical multiple bonds such as C=N, C≡N, C=O and S=O and functional groups such as nitro, bromo, azide and alkoxycarbonyl remain unaffected. A synchronous transfer through a six-membered transition state is regarded as the most likely mechanism; this would predict regiospecific labeling and a lack of isotopic scrambling when carried out with tritiated diimide. Accordingly, in many cases $^3HN{=}N^3H$ reduction offers a useful alternative to catalytic tritiation and tritide reduction, in particular in late stages of 'hot' syntheses of highly functionalized substrates.

The reagent is prepared *in situ* at high specific activity by decomposition of azodi[$O$-$^3H$]-carboxylic acid in the presence of the unsaturated substrate[294] (Figure 4.88). The acid is generated by treatment of its potassium salt with $CH_3COO^3H$, which in turn is available from carrier-free tritiated water and acetic anhydride upon reaction at 60 °C for 30 min. Acetonitrile has been found to be a favorable medium for this reaction[294].

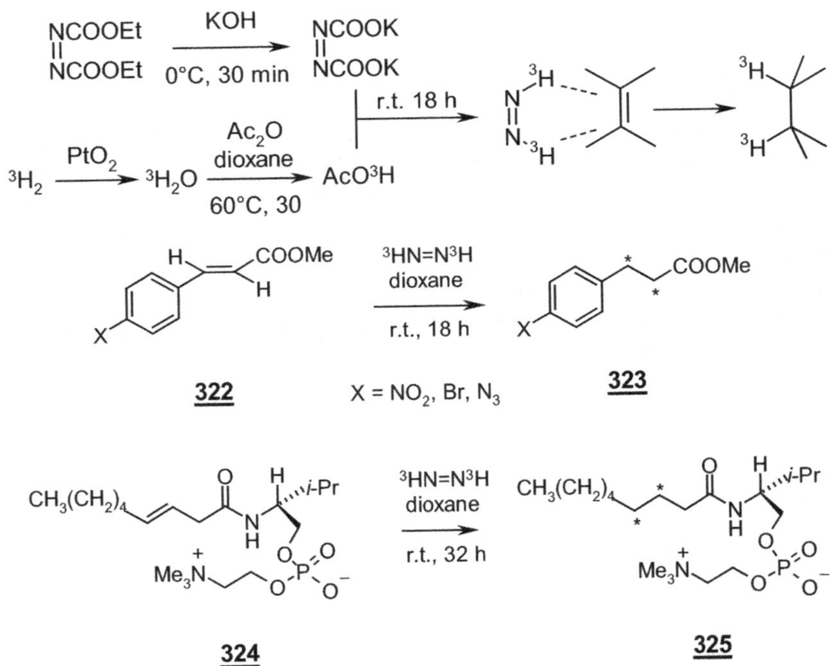


**Figure 4.88** *Preparation and applications of tritiated diimide*

Model reactions with *p*-substituted methyl cinnamates **322** furnished the corresponding 3-aryl[2,3-$^3H$]propionic acid methyl esters **323** in chemical yields of 50–65% but in specific activities only 17–25% of theoretical. This outcome may be the result of the infeasibility of excluding traces of moisture (especially from the dipotassium decarboxylate) and/or primary isotope effects. As the [$^3$H]diimide is a mixture of $^3H_2$, $^3$HH and $H_2$ isotopomers, the product specific activities may be the result of a primary isotope effect at one or more of the stages in the preparation of tritiated diimide or the transfer of tritium to the olefinic bond. Overall isotopic dilution may be reduced significantly by using di-*t*-butyl azodicarboxylate in place of dipotassium azidodicarboxylate and trifluoro[*O*-$^3$H]acetic acid of higher specific activity, which can be prepared by reduction of dry $(CF_3COO)_2Pd$ with tritium gas[278a]. The distribution of tritium in **323** was found to be slightly uneven with 17–20% more tritium at C3, even though the mechanism of the reaction is postulated to involve synchronous tritium transfer.

Unfortunately, reaction with the olefinic precursor of the $PLA_2$-inhibitor **324**[13] (see also Section 4.1.1) was found to be incomplete even after 36 h. The radiochemical yield was very low, with only 74 mCi of **325** obtained from the use of 10.9 Ci of carrier-free tritiated water. This may result from partial loss of [$^3$H]diimide through disproportionation to [$^3$H]-hydrazine and nitrogen, so that an excess of diimide may be required to achieve good product yields[293].

### 4.4.3 Tritiated Methyl Iodide ($C^3H_3I$; $C^3HH_2I$)

[$^3H_n$]Methyl iodide ($n = 1$ or 3), the most common tritiated building block, has been extensively employed in various *C*-, *N*-, *O*- and *S*-tritiomethylation reactions[278e,295]. Because of the high metabolic instability of the label in *N*- and *O*-tritiomethylated compounds, they are usually prepared for *in vitro* investigations such as receptor binding studies, photoaffinity labeling and radioimmunoassays, where metabolism is unlikely to be an issue[103]. On the other hand, [$^3H_n$]methyl groups attached to aliphatic carbon show a much higher biological stability and usually can be safely used for *in vivo* studies as an alternative to carbon-14 labeling when specific activities above 60 mCi/mmol are required. Consequently, *N*- and *O*-tritiomethylations usually utilize $C^3H_3I$ for maximum specific activities, whereas for *C*-tritiomethylations incorporation of one tritium atom may be sufficient. Aromatic $CH_3$ groups, however, are more frequently oxidized metabolically to give $CH_2OH$ or COOH, so that the respective substrates are also more suited for *in vitro* investigations or early exploratory *in vivo* studies.

The preparation of monotritiomethyl iodide by catalytic tritiodehalogenation of bis-chloromethyl ether, chloromethyl 4-phenylbenzoate or chloromethyl phenyl sulfide with tritium gas over 10% Pd/C in ethyl acetate or DMF in the presence of a tertiary amine, and subsequent cleavage of the resulting [$^3$H]methyl ether, ester or sulfide is addressed in Section 4.2.1 [107,108]. Yields are usually in the range of 30–60%.

$C^3H_3I$ is commercially available at a specific activity of 80–85 Ci/mmol; it is generally produced by catalytic reduction of CO or $CO_2$ with tritium gas over Cu–Zn–Cr at 220–230 °C for 24 h, followed by reaction of the resultant $C^3H_3O(^3H/H)$ with HI[110]. Two alternative procedures for preparing high specific activity $C^3H_3I$ on a laboratory scale were introduced in Sections 4.2.1 and 4.3.10. They are tritiodehalogenation of 4-trichloromethoxybiphenyl over 30% Pd/C with carrier-free tritium gas in the presence of proton sponge,[109] and reduction of $CO_2$ with *in situ* prepared $LiAl^3H_4$[296,297]. Subsequent release upon treatment with HI furnishes the reagent in 20–35% yield and with specific activities in the range of 60 and 80 Ci/mmol, respectively.

As many ligands contain *N*-methyl groups, numerous examples of *N*-tritiomethylations have been published, as illustrated by the compounds **326, 327** and **328**,[298] which belong to very different structural classes. *N*-[$^3$H]methylations are usually performed in a polar solvent in the presence of a base such as Hunig's base, triethylamine, proton sponge, $Ag_2O$, $K_2CO_3$, $NaHCO_3$ or $K_2CO_3$. The desmethyl precursor may be prepared by multistep synthesis, as for **326**, **327** and **328**, or, more conveniently, by *N*-demethylation of the unlabeled target compound using cyanogen, vinyl chloroformate or 1-chloroethyl chloroformate[299]. The precursors for [methyl-$^3H_3$]mianserin (**329**)[278e] and 2-bromo-9,10-dihydro[methyl-$^3H_3$]bromocryptine (**330**)[300], a radioligand for the RIA-determination of bromocryptine, were prepared in this fashion.

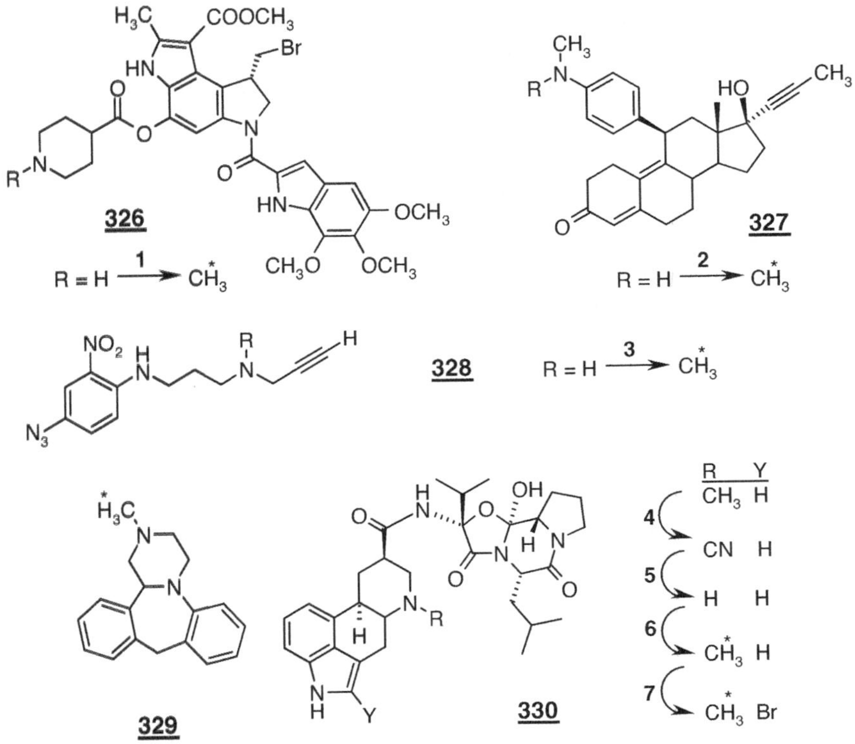


**Reaction conditions:** **1**. $C^3H_3I$, $NaHCO_3$, MeOH, acetone; 35 °C, 10 h; **2**. $C^3H_3I$, DMF, toluene; **3**. $C^3H_3I$, DMF, DMSO; 60 °C, 5 min; **4**. BrCN, $CH_2Cl_2$; r.t., 6 h; **5**. $H_2$, Raney Ni, DMF; r.t., 24 h; **6**. $C^3H_3I$, i-$Pr_2$NEt, $CH_3CN$; r.t., 24 h; **7**. NBS, $CHCl_3$; 50 °C

***Figure 4.89*** *Preparations of N-tritiomethyl compounds at 80 Ci/mmol*

A different approach, investigated for the labeling of **331**[301] with tritium, and of some structurally related compounds with [$^{13}$C,$^2$H]methyl iodide, involves (see generic reaction sequence in Figure 4.90) *N*-quaternization of the target compound using tritiated methyl iodide, followed by demethylation of the quaternary salt by heating with diazabicyclooctane (DABCO) in DMF, or by displacement by $LiEt_3BH$. This approach has the advantages of the ability to use the unlabeled target as the starting material and of avoiding the formation of reaction mixtures resulting from incomplete or over-methylation. A significant disadvantage is the partial loss of label in the demethylation, which shows no preference between labeled and nonlabeled methyl groups. Thus in trial reactions using $C^2H_3I$, one-third of the label was lost in demethylation of quaternary salts from reaction of $C^2H_3I$ with dimethylalkylamines, and one-half with methyldialkylamines. In the case of compound <u>**331**</u>, the specific activity obtained was less than expected (39 Ci/mmol from $C^3H_3I$ of 80 Ci/mmol; $2/3 \times 80 = 53$), possibly owing to incomplete quaternization of the starting material, but the product specific activity was improved by exploiting the phenomenon of chromatographic fractionation between the labeled and unlabeled isotopomers (see references within[301a]). The problem resulting from the nonselective demethylation could be avoided by employing benzyl iodide for the initial quaternization of the unlabeled target. Subsequent selective demethylation upon treatment of the quaternary salt with lithium trisiamylborohydride, chromatographic separation of the resultant mixture of tertiary benzyl- and methylamines, re-quaternization of the benzylamine derivative with $C^3H_3I$ and hydrogenolytic cleavage of the benzyl group over 10% Pd/C provided the desired *N*-$C^3H_3$ derivative in low, but for many purposes sufficient, yields[301].

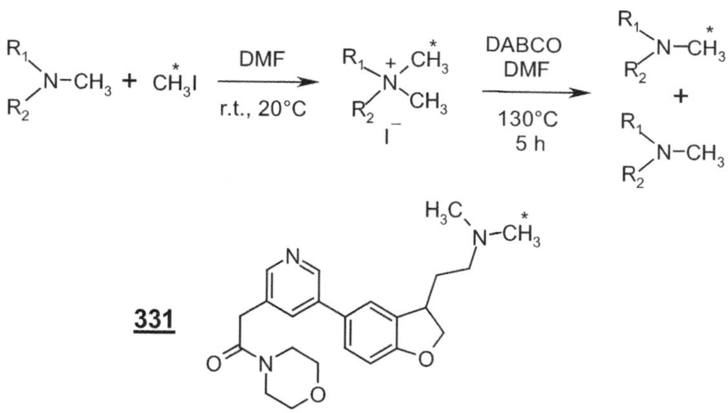


***Figure 4.90*** *Preparation of high specific activity N-tritiomethyl compounds via N-quaternization*

Only a few *N*-tritiomethylation reactions of amides have been published so far. As the amide nitrogen is less nucleophilic, the amide function must be activated by deprotonation (NaH, thallium(I) ethoxide etc.) as illustrated by the labeling of <u>**332**</u>[108] and <u>**333**</u>[302], obtained in 40–75% yields (Figure 4.91).

Acetosyringone (<u>**334**</u>) and compound <u>**335**</u>, a potent GAT-3 transporter ligand[303], are examples in which a phenolic hydroxy group of an appropriate precursor reacted with tritiomethyl iodide in the presence of either $K_2CO_3$, $Na_2CO_3$ or NaOH to provide a successful synthetic route for *O*-[$^3H_n$]methyl aryl ethers of high specific activity. An

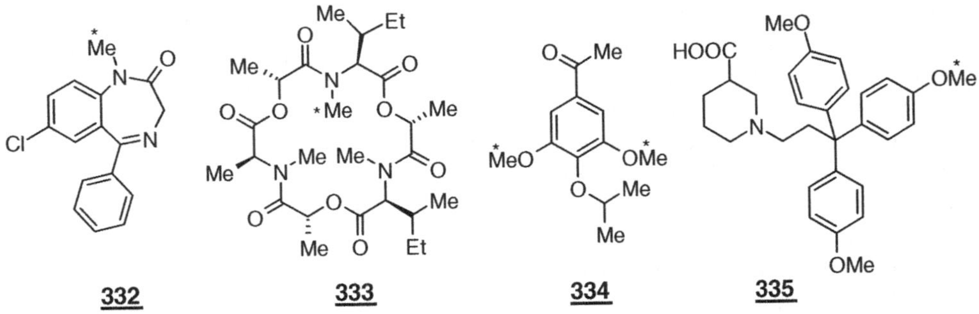

**Figure 4.91** *Tritiomethylations of amides and phenols*

indirect approach, developed[304] to achieve an *O*-[$^3$H]methylation in the presence of a reactive ancillary amino function, is described in Chapter 10, Section 10.2.1.2.

So far only one example (**336**) of the synthesis of an arylcarboxylic acid [$^3$H]methyl ester through [$^3$H]methylation of the alkali metal salt has been reported[278e] (Figure 4.92). Much more important are the [$^3$H]methyl esters of arylsulfonic acids, which have been described[308] as more stable and less volatile substitutes for, or storage forms of, [$^3$H]methyl iodide, retaining much of the latter's range of reactivity. Examples of these reagents and their applications are illustrated in Figure 4.92. Reagents **337a** and the more reactive **337b** can be stored in hexane–ethyl acetate at 4 °C for months without significant decomposition. The relative reactivities of **337b** and unlabeled methyl iodide were tested by stirring an equimolar quantity of each with potassium naphth-2-ylacetate in DMF solution. The specific activity of the resulting [$^3$H]methyl ester was approximately the same as that of pure **337b**, demonstrating the higher reactivity of the latter[308a]. The utility of **337b** is illustrated by its use for the esterification of sodium carfentanil carboxylate, the etherification of *O*-desmethyl raclopride and alkylation of *N*-desmethyl nicotine, affording the respective labeled targets **338, 339, 340** at specific activities of at least 80 Ci/mmol in good to excellent yields[309].

Reaction of [$^3H_n$]methyl iodide with homocysteine (**341**) in liquid ammonia to give [methyl-$^3H_n$]methionine (**342**) is the most prominent and commercially exploited *S*-tritiomethylation process[296] (Figure 4.93). [methyl-$^3H_n$]Methionine ($n = 1,3$) has frequently been used as a starting material for the synthesis of tritiated peptides and as a highly valuable precursor for the biological labeling of highly complex molecules such as [$^3$H]cyclosporin A[305] and [$^3$H]SCH 27899[306] (see Chapter 12). An *S*-[$^3$H]methyl exchange process has been implemented as a useful alternative to *de novo* or fragment synthesis, and it permits labeling of methionine residues in intact peptides and proteins. Reaction of appropriate peptides or proteins (**343**) with $C^3H_3I$ at pH 4 and 37 °C for 24 h converts the methionine residues into the corresponding [$^3H_3$]methylsulfonium iodide intermediates (**344**). Gel chromatographic separation of the non-methylated substrate followed by demethylation by treatment with mercaptoethanol at pH 8 regenerates the original peptide or protein in labeled form with half of the specific activity of $C^3H_3I$ initially employed (providing all *S*-methyl groups had reacted)[307]. Applied to the tritium labelling of apolipoprotein A-I and C-I[307b] and calcitonin[307a] products having specific activities of 1.5–15 Ci/mmol could be isolated with full retention of biological properties. The synthesis

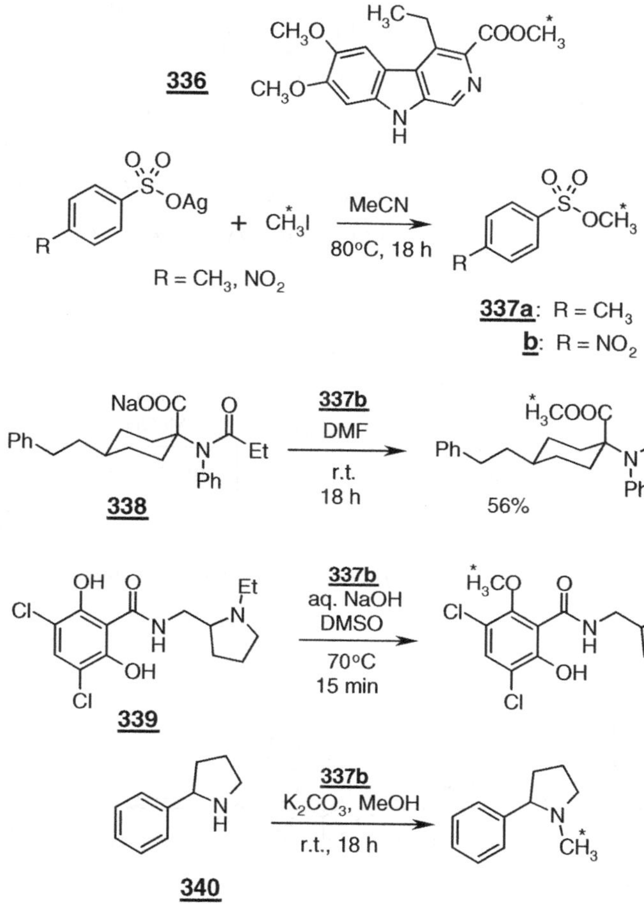


*Figure 4.92 Preparation and applications of [$^3$H]methyl esters and [$^3$H]methyl arylsulfonates*

of [$^3$H]LY274601 (**345**), a serotonin 5HT$_{1A}$ receptor agonist, through *S*-[$^3$H]methylation of the respective thiolate, *in situ* generated by hydrolysis of its thioester, with C$^3$H$_3$I at 80 Ci/mmol may serve as an additional example [310].

Apart from the alkylation of deprotonated C–H acidic substrates, such as the α-deprotonated Δ$^{8,9}$-15-oxo steroid **346**[311] and the hexahydropyrrolo[2,3-b]indole ester **347**, obtained by cyclization of *N*-protected tryptophan methyl ester[312], C-tritiomethylations are experimentally more demanding than the analogous one-step *N*-, *O*- or *S*-functionalizations (Figure 4.94).

For most C-tritiomethylations the electrophilic [$^3$H$_n$]methyl iodide must first be converted into a nucleophilic species such as [$^3$H$_n$]MeLi, [$^3$H$_n$]MeMgI, or [$^3$H$_n$]methyltriphenylphosphonium iodide (labeled Wittig reagent). [$^3$H$_3$]MeMgI, for example, was freshly prepared from magnesium and C$^3$H$_3$I (85 Ci/mmol) in ether, then subjected to reaction with aldehyde **348** to give the corresponding secondary alcohol **349** (Figure 4.95). Oxidation with MnO$_2$ followed by a Horner–Wadsworth–Emmons-type two-carbon

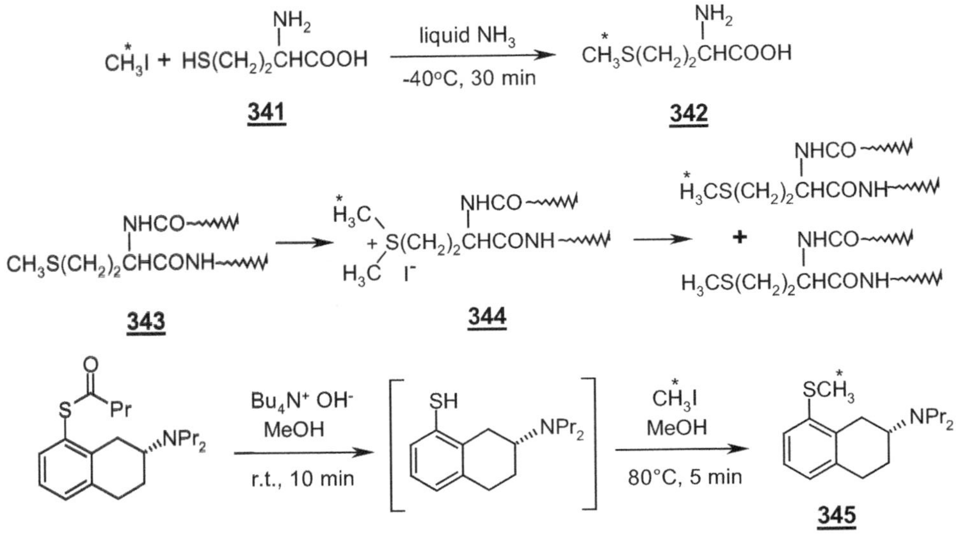


*Figure 4.93* Preparation of S-tritiomethyl derivatives

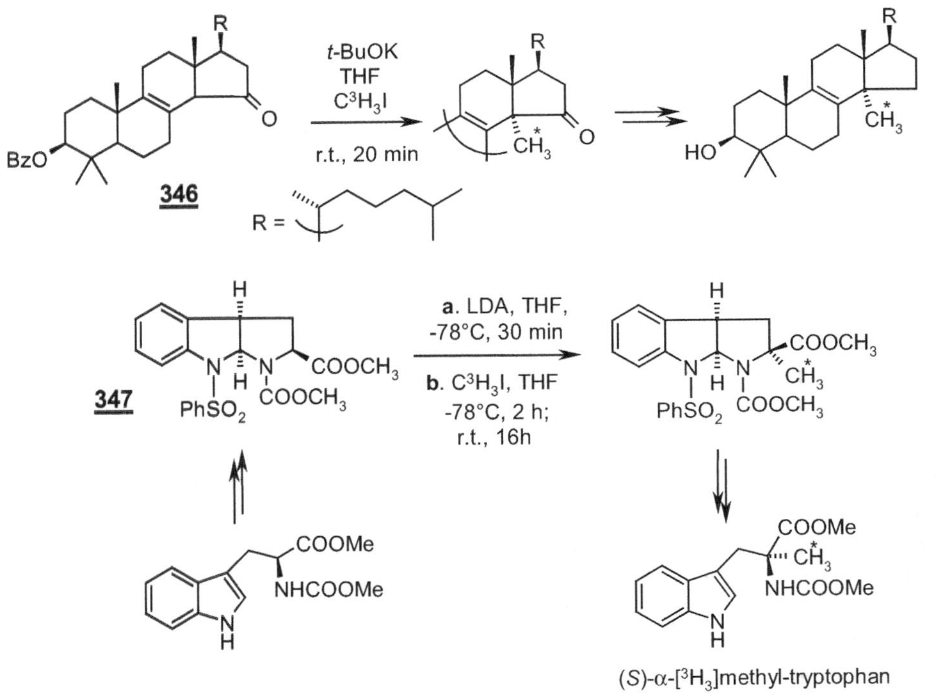


*Figure 4.94* C-Tritiomethylation of a-deprotonated ketones and esters at high specific activities

homologation and ester hydrolysis converted **349** into 9-*cis*-[20-$^3H_3$]retinoic acid (**350**) having a specific activity of 73 Ci/mmol[313]. Analogously, [$^3H_3$]MeMgI reacted with the methyl ester function of the *O*-protected pregna-5,7-diene derivative **351** to give the respective labeled tertiary alcohol **352** which, after deprotection and UV irradiation, provided tritiated 1$\alpha$,25-dihydroxy-2$\beta$-(3-hydroxypropoxy]vitamin D3 (**353**) at 136 Ci/mmol in a radiochemical yield of 0.5%[314].

In a related reaction but using deuterium instead of tritium, epoxide **354** was treated with the cuprate reagent prepared from *in situ* generated [$^2H_3$]methyllithium and copper(I) cyanide to provide **355** (major isomer, 65% yield), an intermediate in the preparation of L-(2*S*,3*S*)-[4,4,4-$^2H_3$]valine[315].

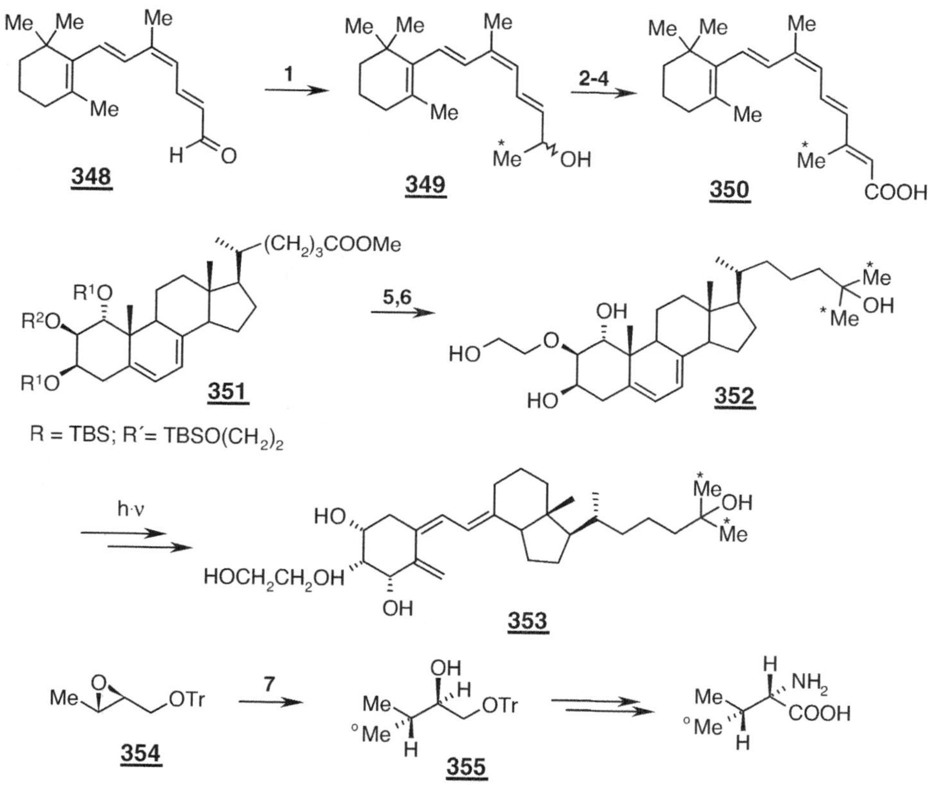


**Reaction conditions: 1**. $C^3H_3MgI$, ether–THF 4:1; r.t, 2 h; **2**. $MnO_2$, $CH_2Cl_2$; r.t., 1 h; **3a**. $(EtO)_2P(O)CH_2COOMe$, NaH, THF; r.t, 15 min, then ketone, r.t., 18 h; **4**. aq. KOH, MeOH; refux, 2h; **5a**., $C^3H_3MgI$, ether; 0 °C, 1 h; **6**. $Bu_4NF$, THF; 65 °C, 15h; **7**. $C^2H_3Li$, LiI, CuCN, THF

***Figure 4.95*** *Applications of nucleophilic tritiomethyl reagents*

The ylid [$^3H_2$]methylenetriphenylphosphine (**356**), readily accessible by reaction of triphenylphosphine and $C^3H_3I$ (85 Ci/mmol) followed by deprotonation with *n*-BuLi, was employed as a reagent in a synthesis of (−)-$\alpha$-[vinylidene-$^3H_2$]kainic acid (**358**), through its reaction with ketone **357**. The resulting protected pyrrolidine intermediate was converted, by successive *N*-deprotection and saponification of the esters, into the free target molecule **358** with a specific activity of 58 Ci/mmol[316].

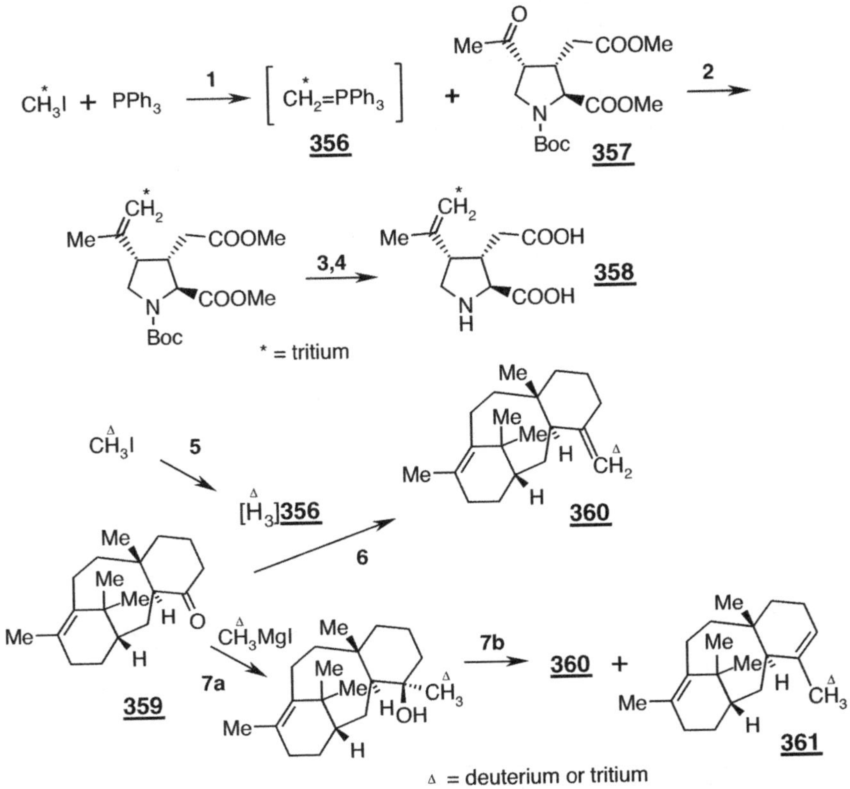


**Reaction conditions: 1a**. $Ph_3P$; THF; r.t., 16h, **b**. *n*-BuLi, THF; r.t.; **2**. THF; r.t., 2 h; **3**. TFA; r.t., 16 h; **4**. 1 *N* NaOH; r.t., 16h; **5a**. $Ph_3P$, toluene; r.t., 23 h **b**. NaHMDS, THF; reflux, 1 h; **6**. THF; reflux 18h; **7a**. $CeCl_3$, THF; 0 °C, 1 h, **b**. $MeO_2CNSO_2NEt_3$ (Burgess reagent), toluene; reflux, 10 min

***Figure 4.96*** *Applications of tritiomethyl phosphorus ylids*

Both abovementioned building blocks were exploited for the synthesis of isotopically labeled taxadiene **360**, for use in biosynthetic studies on paclitaxel. In one approach, the tricyclic ketone **359** was subjected to a Wittig reaction starting with $C^2H_3I$ or low specific activity tritiated methyl iodide to give taxadiene **360** in overall yields of about 50% from the iodide reagent (~100% based on ketone). In an interesting contrast, Grignard reaction of the ketone with isotopic methylmagnesium iodide followed by dehydration of the tertiary alcohol with Burgess reagent gave a mixture of **360** (39%) and its endocyclic isomer **361** (55%)[317].

### 4.4.4 Tritiated Diiodomethane ($C^3HHI_2$)

Diiodomethane/zinc–copper couple and diiodomethane–triisobutylaluminum complex are particularly valuable reagents for the insertion of a methylene group into olefinic double bonds to give cyclopropane derivatives (Simmons–Smith reaction[318]). In order to make use of this reaction for the construction of tritiated cyclopropanes, a previous crude preparation procedure for $C^3HHI_2$ was recently re-investigated and its suitability confirmed by the regiospecific addition of the isolated reagent to the unhindered double bond of perillyl alcohol (**362**), giving a mixture of stereoisomeric labeled cyclopropyl derivatives. In the earlier procedure the reagent was obtained by base-catalyzed tritium–hydrogen exchange of diiodomethane in the presence of tritiated water at 100 °C[319]. According to the more recent procedure[320], iodoform is reduced with sodium arsenite in the presence of tritiated water under much milder conditions. The reagent is extracted from the reaction mixture with chloroform in 74% radiochemical yield. After drying over $MgSO_4$ the solution was directly employed in the Simmons–Smith reaction. Unfortunately, in both cases low-level tritiated water was used and therefore no information is available regarding the suitability of the reaction conditions for the preparation of the reagent at high specific activity.

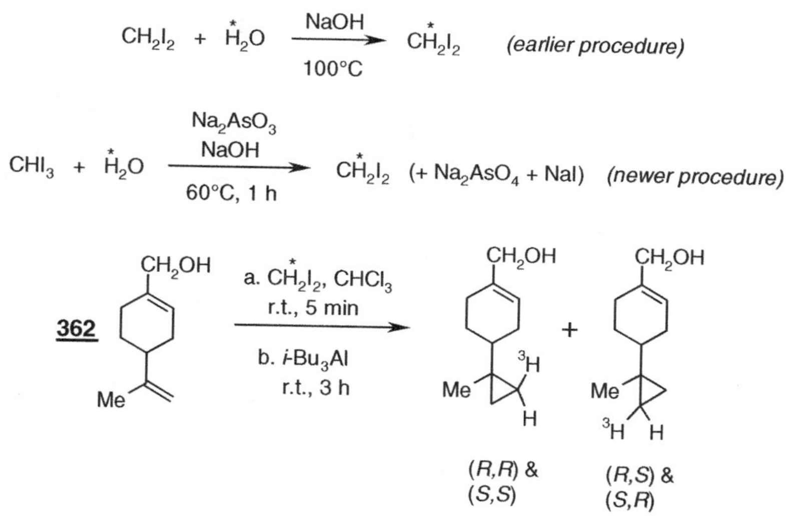


*Figure 4.97* *Preparation and applications of tritiated diiodomethane*

### 4.4.5 Tritiated Formaldehyde ($^3$HCHO, $^3$HC$^3$HO)

Tritiated formaldehyde can be prepared at high specific activity *in situ* either by catalytic reduction of hydrogen cyanide with tritium gas over 5% Pd/$BaSO_4$ in aqueous methanol[311], or by reduction of $CO_2$ in THF with $ZrCp_2Cl^3H$ (tritiated Schwartz Reagent)[322] (Figure 4.98) (see Section 4.3.12 and Reference 272a). The latter procedure can also be used to isolate the reagent in free form by vacuum transfer if THF is replaced by a high-boiling ether[278a]. Achievable specific activities range from 13 Ci/mmol (procedure 1) to 58 Ci/mmol (procedure 2). Tritiated formaldehyde together with an unlabeled hydride reagent can be used in reductive methylations (Equation 1) as an alternative to $NaB^3H_3CN$ or $NaB^3H_4$ for tritiomethylation of peptides and proteins in the $\varepsilon$-amino groups of lysine residues or free amino terminals (see Section 4.3.2).

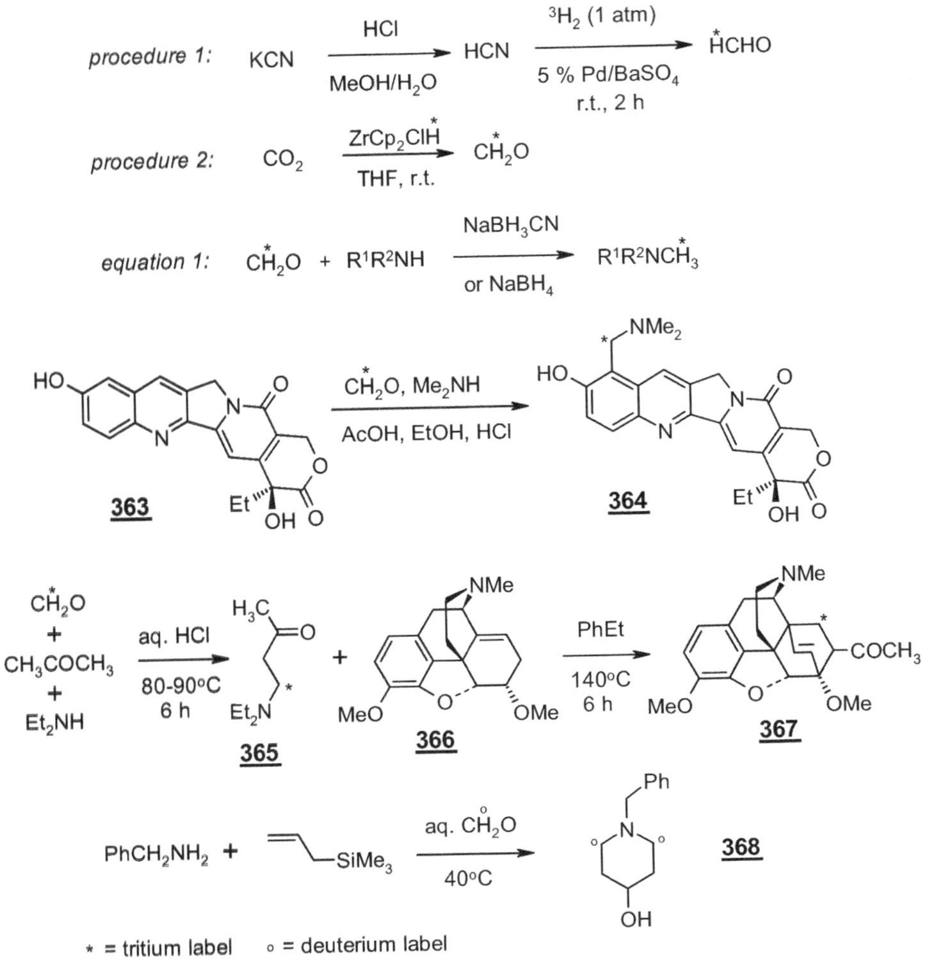


***Figure 4.98*** *Preparations and applications of tritiated formaldehyde*

Tritiated formaldehyde is also useful in Mannich reactions[323]. Examples of such applications include the preparation of the topoisomerase-1 inhibitor **364**[324] by condensation of tritiated formaldehyde and dimethylamine with 10-hydroxycamptothecin (**363**), the reaction with acetone and diethylamine to give 4-diethylamino-4-[$^3H_2$]butan-2-one **365** which was further converted upon heating with thebaine (**366**) to the morphine derivative **367** (specific activity of 46 Ci/mol)[325]. Furthermore, condensation of aqueous deuterated formaldehyde with allyltrimethylsilane to prepare the 2,6-doubly labeled piperidine **368**, as indicated[326], would be expected to work equally well with tritiated formaldehyde.

### 4.4.6 Dimethyl[$^3$H]formamide ($^3HCONMe_2$), Acetic [$^3$H]Formic Anhydride ($^3HCOOCOMe$)

The preparation of high specific activity dimethyl[$^3$H]formamide[229] and acetic [$^3$H]formic anhydride [278b,c] by reduction of excess $CO_2$ with freshly prepared $LiEt_3B^3H$, followed either by esterification and subsequent reaction with dimethylamine or by acylation of the resulting lithium [$^3$H]formate with acetyl chloride was discussed in Section 4.3.6. Dimethyl [$^3$H]-formamide is a valuable synthon for the *C*-[$^3$H]formylation of heteroaromatic and vinylic organolithium compounds, as well as of activated aromatic substrates, if promoted by phosphorous oxychloride (Vilsmeyer–Haack reaction), to give the respective [formyl-$^3$H]aldehydes[327].

For the production of pure acetic [$^3$H]formic anhydride, strictly equimolar amounts of lithium [$^3$H]formate and acetyl chloride are necessary; otherwise it is contaminated with [$^3$H]formic anhydride or acetic anhydride. The reagent, which is normally obtained in yields of 60–90%, should only be used *in situ*, since it is somewhat unstable and decomposes slowly at room temperature. Decomposition is accelerated by strong acids, tertiary amines and sodium acetate[328]. Acetic [$^3$H]formic anhydride can be a versatile reagent for *O*-, *N*- and *C*-tritio-formylations. Alcohols and phenols in the presence of tertiary amines generally give [$^3$H]formates; aliphatic and aromatic amines give mono-*N*-[$^3$H]formylated amines. The latter can be exploited for the preparation of mono-*N*-[$^3$H]-methylamines through reduction with borane–methyl sulfide. *N*-Formylation of amino acids and peptides should occur in good yields with retention of configuration. Reaction with the synthetic tripeptide Met-Leu-Phe has been reported to give *N*-[$^3$H]formylMet-Leu-Phe[326]. With organolithium and Grignard compounds, the respective [carbonyl-$^3$H]-aldehydes can be expected. Finally, several selectively labeled heterocyclic systems such as [2-$^3$H]chromenes, [2-$^3$H]adenines and [2-$^3$H]benzothiazoles should become accessible by subsequent cyclodehydration of the initially formed *O*-, *N*-, or *S*-[$^3$H]formylated intermediates, as demonstrated by the synthesis of isotopomeric 8-aza[2-$^{14}$C]adenine (**369**)[329].

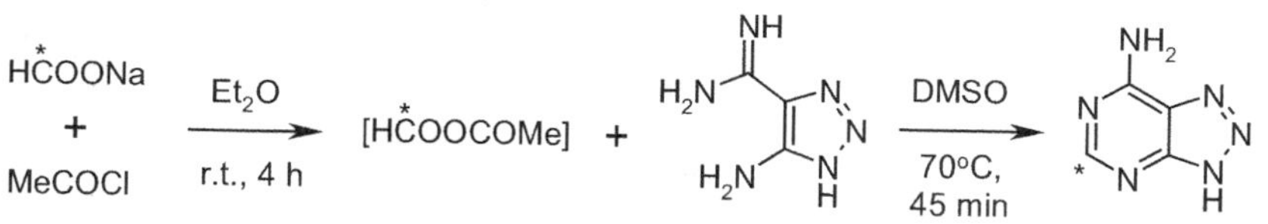


**369**

***Figure 4.99*** *Application of labeled acetic formic anhydride*

### 4.4.7 Tritiated Diazomethane ($C^3HHN_2$)

Diazomethane is among the most common and versatile reagents in organic synthesis[330]. It converts carboxylic acids into methyl esters, and alcohols and phenols into methyl ethers. It reacts with aldehydes to give methyl ketones, and converts ketones and acyl halides into homologous ketones and carboxylic acid derivatives, respectively; cyclic ketones expand by one carbon unit. Its photolysis generates a singlet carbene, which adds to olefinic double bonds to give cyclopropane derivatives.

In spite of its broad synthetic potential and in contrast to its carbon-14 isotopomer, diazo [$^3$H]methane has so far found only very few applications. It was originally prepared by basic hydrolysis of *N*-nitroso-*N*-[$^3$H]methyl urea in specific activities of 0.5–2 Ci/mmol, following classical synthesis procedures. A newer procedure that makes use of the tritiodesilylation of (trimethylsilyl)diazomethane with tetrabutylammonium fluoride in the presence of tritiated water is discussed in Section 4.4.1. Diazo[$^3$H]methane thus generated *in situ* can be immediately employed in any number of reactions. For example, in a pilot study *p*-nitrobenzoic acid was converted into its [$^3$H]methyl ester; subsequent analyses demonstrated that specific activities up to 20 Ci/mmol could be achieved by such methods[278b,c]. Because of the ease and safety of this procedure, broader application of the reagent might be expected in the future.

### 4.4.8 *N*-Tritioacetoxyphthalimide

*N*-Tritioacetoxyphthalimide (**371**) at high specific activity has been introduced as an alternative to commercially available [$^3$H]acetic anhydride as a labeling reagent for peptides and proteins[269b,c]. It eliminates some disadvantages of using the anhydride, such as its volatility and the unavoidable loss of half of the radioactivity employed. Moreover, little more than stoichiometric amounts of **371** are sufficient to achieve rapid and complete acylation under mildly basic conditions, whereas the anhydride undergoes rapid hydrolysis even at neutral pH and therefore must be used in large excess to give high yields. In acylations of proteins, this may account for the lack of selectivity with the use of [$^3$H]acetic anhydride[331].

The reagent has been prepared at a specific activity of 13–17 Ci/mmol by radical tritiodehalogenation of *N*-iodoacetoxyphathalimide (**370**) with tri-*n*-butyltin tritide in the presence of 0.1 equivalent of triethylborane (Section 4.3.11). Alternative methods, e.g. desilylation of (trimethylsilyl)acetoxyphthalimide with tetrabutylammonium fluoride in the presence of tritiated water, afforded somewhat lower specific activities (see Section 4.4.1). Pd-catalyzed tritiodehalogenation was not investigated, since analogous succinimide and naphthalimide derivatives had shown very low tritium incorporation. The requisite intermediate **370** is conveniently accessible by DCC-mediated coupling of iodoacetic acid and *N*-hydroxyphthalimide. The phthalimide derivative has been given preference to the succinimide analogue in order to improve its monitoring by HPLC.

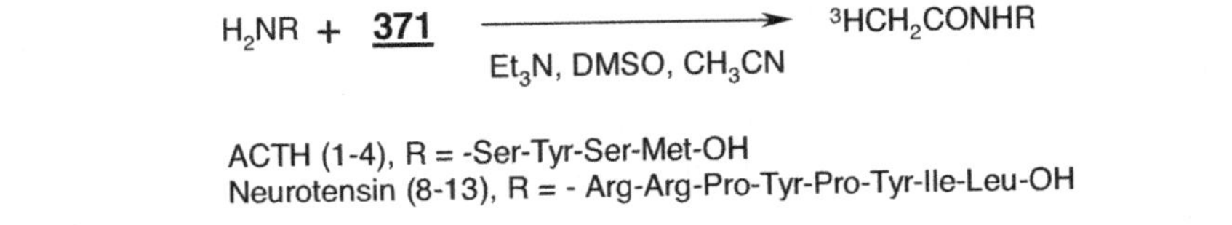


**370** **371**

$H_2NR$ + **371** $\xrightarrow{Et_3N,\ DMSO,\ CH_3CN}$ $^3HCH_2CONHR$

ACTH (1-4), R = -Ser-Tyr-Ser-Met-OH
Neurotensin (8-13), R = - Arg-Arg-Pro-Tyr-Pro-Tyr-Ile-Leu-OH

***Figure 4.100*** *Preparation of N-tritioacetoxyphthalimide*

[$^3$H]Acetylation of peptides with **371** can be performed under mildly basic conditions (18 equivalents of triethylamine) at ambient temperatures in a wide range of solvents (acetonitrile, dimethylsulfoxide, methanol, dioxane, water). In comparison to *N*-succinimidyl-[2,3-$^3$H]propionate ([$^3$H]NSP, see next topic), the commercially available standard labeling reagent, conjugation labeling with *N*-tritioacetoxyphthalimide is expected to have lesser effects on the biological activity of peptides, and to show improved selectivity between amino, hydroxyl and thiol groups. This is evidenced by the labeling of terminal amino acids of peptides such as ACTH(1–4) and neurotensin(8–13)[278b,c,332].

### 4.4.9 *N*-Succinimidyl [2,3-$^3$H]Propionate ([$^3$H]NSP)

*N*-Succinimidyl [2,3-$^3$H]propionate ([$^3$H]NSP, **373**) is presently the reagent of choice for conjugation labeling of free amino groups in peptides and proteins[16a]. Free thiol groups will also react, whereas hydroxy groups do not. Since the propionyl group is small, there is usually minimal alteration of the peptide and protein structure. Accordingly, a large number of peptides and proteins have been reported to retain their biological activity after having been labeled with [$^3$H]NSP. Labeling can be performed at a pH in the range of 7.5–8.0 at room temperature for 1–2 h or even for longer periods at 0 °C. This also allows the labeling of lysine residues, which to some extent may be sterically hindered in complex peptides and proteins[333].

Two procedures have been published for the preparation of [$^3$H]NSP at high specific activity. In the first, either acrylic acid[334] or propiolic acid[335] is coupled to *N*-hydroxysuccinimide in the presence of DCC and the resulting *N*-acryloyloxy or *N*-propioyloxy derivative **372** reduced with tritium gas over 10% Pd/C to give the reagent at a specific activity of 30–40 Ci/mmol and 50–100 Ci/mmol, respectively. Alternatively, propiolic acid is initially reduced and the resulting [2,3-$^3$H]propionic acid coupled in a second step to *N*-hydroxysuccinimide[336].

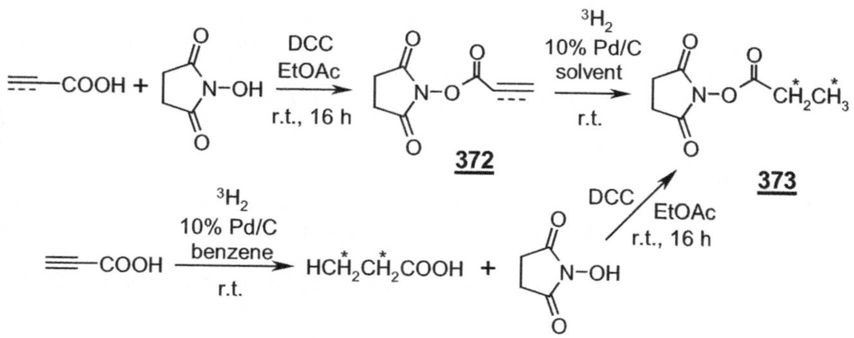


***Figure 4.101*** *Preparations of N-succinimidyl [2,3-$^3$H]propionate*

## References


1. Evans, E.A., *Tritium and its Compounds,* 2nd Edn; John Wiley & Sons, Inc.: New York, ***1974***, p. 310.
2. Sajiki, H.; Kurita, T.; Esaki, H.; Aoki, F.; Maegawa, T. and Hirota, K., *Org. Lett.*, ***2004***, *6*, 3521–3523.
3. Esaki, H.; Aoki, F.; Umemura, M.; Kato, M.; Maegawa, T.; Monguchi, Y. and Sajiki, H., *Chem. Eur. J.*, ***2007***, *13*, 4052–4063 and references cited therein.
4. Sagert, N.H. and Pouteau, R.M.L., *Can. J. Chem.*, ***1973***, *51*, 4031–4037.
5. Sagert, N.H. and Pouteau, R.M.L., *Can. J. Chem.*, ***1974***, *52*, 2960–2967.
6. a. Hanus, J.; Cerny, B. and Benes, J., *J. Label. Compd*, ***1974***, *10*, 523–531; b. Janssen, C.G.; Verluyten, W.L.M. and Vliegen, M., *J. Label. Compd. Radiopharm.*, ***2006***, *49*, 219–225.

7. Moe, S.T.; Smith, D.L. and Filer, C.N., *J. Label. Compd. Radiopharm.*, ***1998***, *41*, 535–543.
8. a. Evans, E.A. *Tritium and its Compounds*, 2nd Edn; John Wiley & Sons, Inc.: New York, ***1974***, pp. 320–326; b. Filer, C.N., *Isotopes in the Physical and Biomedical Sciences*, Vol. 1, Part A, Buncel, E. and Jones, J.R., Eds.; Elsevier: Amsterdam, ***1987***, pp. 169–178; c. Nagaev, I.Y.; Shevchenko, V.P. and Myasoedov, N.F., *Radiochemistry*, ***1999***, *41*, 305–317.
9. a. Siegel, S., Heterogeneous catalytic hydrogenation of C=C and C≡C, in *Comprehensive Organic Synthesis*, Vol. 8, Trost, B.M. and Fleming, I., Eds.; Pergamon: Oxford, ***1992***, pp. 418–442; b. Rylander, P., *Catalytic Hydrogenation in Organic Synthesis;* Academic Press: New York, ***1979***, pp. 31–63; c. Carruthers, W., Catalytic Hydrogenation, in *Some Modern Methods of Organic Synthesis*, 3rd Edn; Cambridge University Press: Cambridge, ***1986***, pp. 411–425; d. Smith, M.B. and March, J., *March's Advanced Organic Chemistry*, 5th Edn; John Wiley & Sons, Inc.: New York, ***2001***, pp. 1002–1009; e. House, H.O., Catalytic hydrogenation and dehydrogenation, in *Modern Synthetic Reactions*, 2nd Edn; W.A. Benjamin: Menlo Park, CA, ***1972***, pp. 1–44.
10. Williams, P.G.; Morimoto, H. and Wemmer, D.E., *J. Am. Chem. Soc.*, ***1988***, *110*, 8038–8044.
11. Bartok, M., *Stereochemistry of Heterogeneous Metal Catalysis;* John Wiley & Sons, Inc.: New York, ***1985***.
12. Yu, J.; Whitney, P.-S. and Spencer, J.B., *J. Mol. Catal. A: Chemical*, ***1999***, *146*, 199–210.
13. Culf, A.S.; Morimoto, H.; Williams, P.G.; Lockley, W.S.J.; Primrose, W.U. and Jones, J.R., *J. Label. Compd. Radiopharm.*, ***1996***, *34*, 373–384.
14. Filer, C.N.; Ahern, D.G.; Granchelli, F.E.; Neumeyer, J.L. and Law, S.J., *J. Org. Chem.*, ***1980***, *45*, 3465–3467.
15. a. Hardy, P.M.; Sheppard, P.W.; Brundish, D.E. and Wade, R., *Peptides, Proceedings Eur. Peptide Symp. 1982*, Blaha, E. and Malone, J.A., Eds., Walter de Guyter & Co.: Berlin, ***1983***, pp. 297–300; b. Evans, E.A.; Warrell, D.C.; Elvidge, J.A. and Jones, J.R., *Handbook of Tritium NMR Spectroscopy and Applications;* John Wiley & Sons, Ltd: Chichester, ***1985***, pp. 34–42.
16. a. Kent, B. and James, M., in *Synthesis and Applications of Isotopically Labeled Compounds 1997*, Heys, J.R. and Melillo, D., Eds.; John Wiley & Sons, Inc.: New York, ***1998***, pp. 227–230; b. Hasegawa, H.; Shinohara, Y. and Baba, S., *J. Chem. Soc. Perkin Trans. I*, ***1990***, 2641–2644; c. Gulyás, E.C.; Soos, K.; Varga, J.; Toth, G. and Penke, B., *J. Label. Compd. Radiopharm.*, ***1998***, *41*, 763–771; d. Baba, S.; Hasegawa, H. and Shinohara, Y., *J. Label. Compd. Radiopharm.*, ***1989***, *27*, 1359–1369.
17. Birtle, T.; White, A.J. and Woodhouse, D.P., *J. Label. Compd. Radiopharm.*, ***1994***, *34*, 401–410.
18. Zhang, C.; Yang, R.; Chen, L.; Zhong, B.; Yun, L. and Wang, H., *J. Label. Compd. Radiopharm.*, ***2004***, *47*, 583–590.
19. a. Fontana, E.; Giribone, D.; Pignatti, A.; Galantino, M.; Forina, R. and Dostert, P., *J. Label. Compd. Radiopharm.*, ***1996***, *38*, 483–488; b. Kozlowski, E.S.; Dischino, D.D. and Johnson, G., *J. Label. Compd. Radiopharm.*, ***1995***, *31*, 871–876; c. Oehlke, J.M; Brankoff, T.; Schmidt, M.; Brudel, M.; Kertscher, U. and Berger, H., *J. Label. Compd. Radiopharm.*, ***1993***, *33*, 161–169.
20. a. Sasaki, A.N. and Bricas, E., *Tetrahedron Lett.*, ***1980***, *21*, 4263–4264; b. Sasaki, A.N.; Morgat, J.L. and Bricas, E., *Peptides, Proceedings Eur. Peptide Symp. 1980*, Brunfeldt, K., Ed.; Walter de Gruyter & Co: Berlin, ***1981***, 224–228; c. Nasal, M.; Buc, P. and Wieland, T., *Liebigs Ann. Chem.*, ***1983***, 1524–1532; d. Nasal, M.; Buc, P. and Wieland, T., Chemistry of peptides and proteins, in *Proceedings of the USSR–FRG Symposium, 1982*, Voelter, W., Ed., Walter de Gruyter & Co.: Berlin, ***1984***, pp. 71–78; e. Labbé-Jullié, C.; Blanot, D.; Morgat, J.L.; Kitabgi, P. and Van Rietschoten, J., *Biochimie*, ***1983***, *65*, 553–562.
21. a. Vater, J.; ter Meer, A.; Hilberink, P.; Sperling, E.; Kaspersen, F.; Martens, F. and Wagenaars, G., in *Synthesis and Applications of Isotopically Labeled Compounds 1994*, Allen, J. and Voges, R., Eds.; John Wiley & Sons, Ltd: Chichester, ***1995***, pp. 117–120; b. Vader, J.; Sperling, E.; Kaspersen, F.; Wagenaars, G. and Funke, C., *Synthesis and Applications of Isotopically Labeled Compounds 1994*, Buncel, E. and Kabalka G.W., Eds.; Elsevier Science Publishers B.V.: Amsterdam, ***1992***, 72–73.
22. Voges, R.; Andres, H.; Loosli, H.R. and Schreier, E. *Synthesis and Applications of Isotopically Labeled Compounds*, Duncan, W.P. and Susàn, A.B., Eds.; Elsevier: Amsterdam, ***1983***, pp. 331–336.

23. Voges, R.; von Wartburg, B.R. and Loosli, H.R.,in *Synthesis and Application of Isotopically Labeled Compounds 1985*, Muccino, R.R. Ed.; Elsevier: Amsterdam, ***1986***, pp. 371–376.
24. Kingston, L.P.; Lockley, W.J.S.; Mather, A.N. and Wilkinson, D.J., in *Synthesis and Applications of Isotopically Labeled Compounds 1994*, Allen, J. and Voges, R., Eds.; John Wiley & Sons, Ltd: Chichester, ***1995***, pp. 57–60.
25. Tanaka, T.; Bannai, K.; Manabe, K. and Kurozumi, S., *J. Label. Compd. Radiopharm.*, ***1991***, *29*, 933–944.
26. Droescher, P. and Römer, J., *J. Label. Compd. Radiopharm.*, ***1995***, *36*, 111–119.
27. a. Ström, P.; Malmquist, J.; Hansson, J.; Yngve, U. and Claesson, A., in *Synthesis and Applications of Isotopically Labeled Compounds*, Vol. 8, Dean, D.C.; Filer, C.N. and McCarthy, K.E., Eds.; John Wiley & Sons, Ltd: Chichester, ***2004***, pp. 285–288; b. Dorokhova, E.M.; Zolotarev, Yu.A.; Kozik, V.S.; Penkina, V.I. and Myasoedov, N.F., *Radioisotopy*, ***1990***, 40–47.
28. Angelini, G.; Cerri, A.; Filippi, A.; Lilla, G.; Margonelli, A. and Melloni, P., in *Synthesis and Applications of Isotopically Labeled Compounds 1997*, Heys, J.R. and Melillo, D.G.,Eds.; John Wiley & Sons, Ltd: Chichester, ***1998***, pp. 499–502.
29. Boehm, M.F. and Prestwich, G.D., *J. Label. Compd. Radiopharm.*, ***1988***, *25*, 653–659.
30. Palmer, C.J. and Casida, J.E., *J. Label. Compd. Radiopharm.*, ***1991***, *29*, 829–839.
31. a. Nicholas, K.M. and Pettit, R., *Tetrahedron Lett.*, ***1971***, 3475–3478; b. Seyferth, D.; Nestle, M. O. and Wehman, A.T., *J. Am. Chem. Soc.*, ***1975***, *97*, 7417–7426
32. Huguenin, P.N., Hoffmann-La Roche, CH-4002 Basel, presentation at the R.C.TRITEC Workshop 2005; personal communication.
33. Varga, E.; Toth, G.; Benyhne, S.; Hosztafi, S. and Borsodi, A., *Life Sci.*, ***1987***, *40*, 1579–1588.
34. a. Latrofa, A.; Trapani, G.; Franco, M.; Harris, M.J.; Serra, M.; Bigio, G. and Liso, G., *J. Label. Compd. Radiopharm.*, ***2001***, *44*, 521–528; b. Thurkauf, A.; de Costa, B. and Rice, K.G., *J. Label. Compd. Radiopharm.*, ***1989***, *27*, 577–582.
35. Wu, D.; Huang, W. and Wang, W., *J. Label. Compd. Radiopharm.*, ***1998***, *41*, 785–789.
36. Otvös, F.; Hosztali, S.; Simon, C.; Toth, G., *J. Label. Compd. Radiopharm.*, ***1995***, *36*, 79–83
37. Lewis, J.W.; Rance, M.J. and Young, G.R., *J. Med. Chem.*, ***1974***, *17*, 465–466.
38. Calderon, S.N.; Bertha, C.M.; Gutkind, J.S.; Xu, H.; Partilla, J.S. and Rothman, R.B., *J. Label. Compd. Radiopharm.*, ***1996***, *38*, 847–850.
39. Malone, J.A.; Reidelberger, R.D. and Hulce, M., *J. Label. Compd. Radiopharm.*, ***2000***, *43*, 77–90.
40. Villani, A.J. and Heys, J.R., *J. Label. Compd. Radiopharm.*, ***1997***, *39*, 379–386.
41. Stolle, W.T.; Easter, J.A.; Ackland, M.J.; Haadsma-Svensson, S.R.; Lin, C.H. and Hsi, R.S.P., *J. Label. Compd. Radiopharm.*, ***1997***, *39*, 891–900.
42. Mantegani, S.; Brambilla, E. and Ermoli, A., *J. Label. Compd. Radiopharm.*, ***1991***, *29*, 519–533
43. Adlof, R.O.; Emken, E.A., *J. Label. Compd. Radiopharm.*, ***1984***, *21*, 75–79.
44. Leriche, C.; Rene, L.; Derouet, F.; Rousseau, B. and Badet, B., *J. Label. Compd. Radiopharm.*, ***1995***, *36*, 1115–1121.
45. Toth, G.; Kramer, M.; Sirokman, F.; Borsodi, A. and Ronai, A., *J. Label. Compd. Radiopharm.*, ***1982***, *19*, 1021–1030.
46. Filer, C.N.; Ahern, D.G.; Faszio, R. and Seguin, R.J., *J. Org. Chem.*, ***1981***, *46*, 4968–4978.
47. Prestwich, G.D., in *Isotopes in the Physical and Biomedical Sciences*, Vol 1, Part B, Buncel, E. and Jones, J.R., Eds.; Elsevier: Amsterdam, ***1991***, pp. 137–178.
48. a. Cohen, N.; Banner, B.L.; Lopresti, R.J.; Wong, F.; Rosenberger, M.; Liu, Y.-Y.; Thom, E. and Liebman, A.A., *J. Am. Chem. Soc.*, ***1983***, *105*, 3661–3672; b. Shevchenko, V.P.; Nagaev, I.Yu.; Myasoedov, N.F.; Guy, A.; Durand, T.; Vidal, A.; Vidal, J.-P.; Rossi, J.-C. and Bezuglov, V.V., *J. Label. Comp. Radiopharm.*, ***1999***, *42*, 663–672; c. Liu, Y.-Y.; Thom. E. and Liebman, A.A., in *Synthesis and Applications of Isotopically Labeled Compounds*, Duncan, W.P. and Susan, A.B., Eds.; Elsevier: Amsterdam, ***1983***, pp. 433–434; d. Demin, P.M.; Pivnitzky, K.K.; Vasiljeva, L.L. and Pace-Asciak, C.R., *J. Label. Compd. Radiopharm.*, ***1994***, *34*, 221–230; e. Leblanc, Y.; Fitzsimmons, B.J.; Zamboni, R. and Rokach, J., *J. Org. Chem.*, ***1988***, *53*, 265–267; f. Delorme, D.; Girard. Y and Rokach, J., *J. Org. Chem.*, ***1989***, *54*, 3635–3640; g. Shevchenko, V.P.; Demin, P.M.; Nagaev, I.Y.; Pivnitzky, K.K.; Myagkova, G.I. and Myasoedov, N.F., in *Synthesis and Applications of Isotopically Labeled Compounds 1991*, Buncel, E. and Kabalka, G.W., Eds.; Elsevier: Amsterdam, ***1992***, pp. 482–485.

49. a. Liebman, A.A.; Burger, W.; Malarek, D.H.; Serico, L.; Muccino, R.R.; Perry, C.W. and Choudry, S.C., *J. Label. Compd. Radiopharm.*, ***1990***, *28*, 525–541; b. Coe, J.W., Hawes, C.R. and Towers, P., *J. Label. Compd. Radiopharm.*, ***1995***, *36*, 587–594.
50. Parnes, H., *J. Label. Compd. Radiopharm.*, ***1990***, *28*, 29–40.
51. Belley, M.L.; Gareau, Y. and Zamboni; R.J., *J. Label. Compd. Radiopharm.*, ***1995***, *36*, 599–608
52. Landvatter, S.W. and Heys, J.R., *J. Label. Compd. Radiopharm.*, ***1987***, *24*, 689–694.
53. Shevchenko, V.P.; Nagaev, I.Yu. and Myasoedov, N.F., in *Synthesis and Applications of Isotopically Labeled Compounds 1994*, Allen, J. and Voges, R. Eds.; John Wiley & Sons, Ltd: Chichester, ***1995***, pp. 79–82.
54. Just, W.W.; Böhnisch, V. and Werner, G., *J. Label. Compd. Radiopharm.*, ***1976***, *12*, 287–291.
55. a. Masjedizadeh, M.R. and Parnes, H., *J. Label. Compd. Radiopharm.*, ***1995***, *38*, 53–60; b. Masjedizadeh, M.R. and Parnes, H. in *Synthesis and Applications of Isotopically Labeled Compounds 1994*, Allen, J. and Voges, R., Eds.; John Wiley & Sons, Ltd: Chichester, ***1995***, pp. 97–100.
56. Shevchenko, V.P.; Nagaev, I.Yu.; Myasoedov, N.F.; Susàn, A.B.; Switek, J.-H. and Braunger, H., *J. Label. Compd. Radiopharm.*, ***2006***, *49*, 421–427.
57. Myasoedov, N.F., *Tritium Labeled Physiologically Labeled Compounds*, presentation at the IIS–CED Symposium, Bad Soden, Germany, ***1996***.
58. Rylander, P.N., *Catalytic Hydrogenation in Organic Synthesis;* Academic Press: New York, ***1979***, pp. 77–100.
59. Callaway, J.C.; Morimoto, H.; Gynther, J.; Airaksinen, M.M. and Williams, P.G., *J. Label. Compd. Radiopharm.*, ***1992***, *31*, 355–364.
60. Autelitano, F.; Labat-Alcaraz, M.L.; Sobrio, F.; Goeldner, M. and Ilien, B., *J. Label. Compd. Radiopharm.*, ***1996***, *38*, 567–578.
61. Egan, J.A. and Filer, C.N., *J. Label. Compd. Radiopharm.*, ***2001***, *44*, 43–45.
62. Sher, P.M. and Kronenthal, D.R., *J. Label. Compd. Radiopharm.*, ***1997***, *39*, 21–27.
63. Filer, C.N., in *Isotopes in the Physical and Biomedical Sciences* Vol. 1, Part A, Buncel, E. and Jones, J.R., Eds.; Elsevier: Amsterdam, ***1987***, p. 178.
64. Bärwolff, D. and Murawski, D.; Akad. der Wissenschaften DDR; Patentschrift DD-136-957, ***1979***.
65. Carruthers, W., *Some Modern Methods of Organic Synthesis*, 3rd Edn; Cambridge University Press: Cambridge, ***1986***, pp. 425–431.
66. Toth, G.; Kardos, J.; Fodor, A. and Sirokman, J., *J. Label. Compd. Radiopharm.*, ***1987***, *24*, 683–688.
67. a. Russel, J.C.; Jones, J.R.; Vick, T.A. and Stratford, P.W., *J. Label. Compd. Radiopharm.*, ***1993***, *33*, 957–964; b. Russel, J.C.; Jones, J.R.; Vick, T.A. and Stratford, P.W., in *Synthesis and Applications of Isotopically Labeled Compounds 1994*, Allen, J. and Voges, R., Eds.; John Wiley & Sons, Ltd: Chichester, ***1995***, pp. 149–155.
68. Takaya, H; Noyori, R., Homogeneous catalytic hydrogenation of C=C and C≡C bonds in *Comprehensive Organic Synthesis*, Vol. 8, Trost, B.M. and Fleming, I., Eds.; Pergamon: Oxford, ***1992***, pp. 443–469.
69. Koch, G. and Dalenberg, J.W., *J. Label. Compd*, ***1970***, *6*, 395–398.
70. Manabe, K.; Tanaka, T.; Kurozumi, S. and Kato, Y., *J. Label. Compd. Radiopharm.*, ***1991***, *29*, 1107–1119.
71. Boehm, M.F. and Prestwich, G.D., *J. Org. Chem.*, ***1987***, *52*, 1349–1351.
72. a. Hesk, D.; Delduca, P.; Koharski, D.; McNamara, P.; Magatti, C.; Saluja, S. and Thomas, L., *J. Label. Compd. Radiopharm.*, ***1993***, *33*, 439–442; b. Fagart, J.; Sobrio, F. and Marquet, A., *J. Label. Compd. Radiopharm.*, ***1997***, *39*, 791–795.
73. Evans, E.A., in *Isotopic Labeling with Carbon-14 and Tritium: Essential Chemistry and Applications*, Elvidge, J.A. and Jones, J.R., Eds.; The Chemical Society: Burlington House, London, ***1980***, pp. 47–48.
74. Duke, R.K.; Allan, R.D.; Drew, C.A.; Johnston, G.A.R. and Mewett, K.N., *J. Label. Compd. Radiopharm.*, ***1993***, *33*, 527–540.
75. Voges, R.; von Wartburg, B. and Loosli, H.R., *$^{3}$H-NMR Spectroscopy/CAMP Coupling, Application to Tritiated Compounds for in vivo Studies*, American Nuclear Society, Winter Meeting, San Francisco, USA, ***1983***.

76. Morehead, A.T.; Lamb, P.; Lewin, A.H. and Carroll, F.I., in *Synthesis and Application of Isotopically Labeled Compounds 1991*, Buncel, E. and Kabalka, G.W., Eds.; Elsevier: Amsterdam, ***1992***, pp. 82–84.
77. a. Pfaltz, A.; Blankenstein, J.; Hilgraf, R.; Hörmann, E.; McIntyre, S.; Menges, F.; Schönleber, M.; Smidt, P.; Wüstenberg, B. and Zimmermann, N., Iridium-Catalyzed Enantioselective Hydrogenation of Olefins in *Adv. Synth. Catal.* ***2003***, *345*, 33–43; b. Nagel, U. and Albrecht, J.; Enantioselective hydrogenation of *N*-acyl-dehydroamino acids, in *Top. Catal.*, ***1998***, *5*, 3–23; c. Noyori, R., Asymmetrische Katalyse: Kenntnisstand und Perspekiven in *Angew. Chem.* ***2002***, *114*, 2108–2123.
78. a. Andres, H., in *Synthesis and Application of Isotopically Labeled Compounds*, Vol. 7, Pleiss, U. and Voges, R., Eds.; John Wiley & Sons, Ltd: Chichester, UK, ***2001***, pp. 49–51; b. Voges, R., in *Synthesis and Application of Isotopically Labeled Compounds 1994*, Allen, J. and Voges, R., Eds.; John Wiley & Sons, Ltd: Chichester, UK, ***1995***, pp. 1–28.
79. Kent, B., Amersham Pharmacia Biotech, personal communications to R.V.
80. a. Parnes, H. and Shelton, E.J., *Int. J. Peptide Protein Res.*, ***1986***, *27*, 239–244; b. Parnes, H. and Shelton, E.J., in *Synthesis and Application of Isotopically Labeled Compounds 1985*, Muccino, R.R., Ed.; Elsevier: Amsterdam, ***1986***, pp. 159–160.
81. a. Hammadi, A.; Meunier, G. and Genet, R., in *Synthesis and Application of Isotopically Labeled Compounds 1994*, Allen, J. and Voges, R., Eds; John Wiley & Sons, Ltd: Chichester, UK, ***1995***, pp. 131–136; b. Pinto-Alphandary, H.; van Craenest-Jimonet, C.; Morgat, J.L. and Fromageot, P., *J. Label. Compd. Radiopharm.*, ***1988***, *25*, 1273–1279.
82. a. Levine-Pinto, H.; Fromageot, P.; Meyer, D.; Poulin, J.-C. and Kagan, H.B., *Tetrahedron*, ***1982***, *38*, 119–123; b. Meyer, D.; Poulin, J.-C.; Kagan, H.B.; Levine-Pinto, H.; Morgat, J.L. and Fromageot, P., *J. Org. Chem.*, ***1980***, *45*, 4680–4682.
83. Hammadi, A.; Meunier, G.; Tarride, J.L.; Mênez, A. and Genet, R, in *Synthesis and Application of Isotopically Labeled Compounds 1997*, Heys, J.R. and Melillo, D.G., Eds.; John Wiley & Sons, Ltd: Chichester, UK, ***1998***, pp. 197–201.
84. a. Evans, E.A., Catalytic Halogen–Tritium Replacement in *Tritium and its Compounds*, 2nd Edn; John Wiley & Sons, Inc.: New York, ***1974***, pp. 326–330; b. Filer, C.N., Catalytic Dehalogenations with Tritium in *Isotopes in the Physical and Biomedical Sciences*, Vol. 1, Part A, Buncel, E. and Jones, J.R., Eds.; Elsevier: Amsterdam, ***1987***, pp. 163–168; c. Wilkinson, D.J.; Hickey, M.J.; Kingston, L.P. and Mather, A.N., Tritio-dehalogenation: New variants of an old theme, in *Synthesis and Applications of Isotopically Labeled Compounds*, Vol. 8, Dean, D. C.; Filer, C.N. and McCarthy, K.E., Eds.; John Wiley & Sons, Ltd: Chichester, ***2004***, pp. 47–50.
85. a. Pinder, A.R., *Synthesis*, ***1980***, 425–452; b. Rylander, P.N., Catalytic dehydrohalogenation, in *Catalytic Hydrogenation in Organic Syntheses*; Chapter 3, Academic Press: New York, ***1979***, pp. 235–248; c. House, H.O., *Modern Synthetic Reactions*, 2nd Edn; W.A. Benjamin: Menlo Park, CA, ***1972***, pp. 14–15.
86. a. Hickey, M.J.; Kingston, L.P.; Jones, J.R.; Lockley, W.J.S; Mather, A.N., Wilkinson, D.J., *Synthesis and Applications of Isotopically Labeled Compounds*, Vol. 8, Dean, D.C.; Filer, C.N. and McCarthy, K.E., Eds.; John Wiley & Sons, Ltd: Chichester, ***2004***, pp. 243–244; b. Barluenga, J.; Gonzalez, J.M.; Garcia-Martin, M.A.; Campos, P.J.; Asensio, G., *J. Org. Chem.* ***1993***, *58*, 2058–2060; c. DeMarinis, R.M.; Villani.; Brungard, D.; Meier, J., *J. Label. Compd. Radiopharm.*, ***1982***, *19*, 267–270; d. Janssen, D.E.; Wilson, C.V., *Organic Syntheses*, Vol. 4, John Wiley & Sons, Inc.: New York, ***1963***, pp. 547–549; e. Weaner, L.E.; Yim, N.C.F.; Hoerr, D. C., *Synthesis and Applications of Isotopically Labeled Compounds*, Vol. 7, Allen, J. and Voges, R., Eds.; John Wiley & Sons, Ltd: Chichester, ***2001***, pp. 137–140; f. Walker, D.G.; Hoerr, D.C.; Leister, W.H. and Weaner, L.E., *J. Label. Compd. Radiopharm.*, ***1997***, *39*, 147–157; g. Johnsson, R.; Meijer, A. and Ellervik, U., *Tetrahedron*, ***2005***, *61*, 11657–11663.
87. Thurkauf, A., *J. Label. Compd. Radiopharm.*, ***1997***, *39*, 123–128.
88. Devillers, J.; Winand, M. and Bettens, B., *J. Label. Compd. Radiopharm.*, ***1976***, *12*, 219–229
89. a. Jahngen, E.G.E.; Shelton, E.J. and Young, R.L., Abstracts 8th Northeast Regional Meeting of the American Chemical Society, Boston, MA, MED 18, ***1978***; b. Filer, C.N., in *Isotopes in the Physical and Biomedical Sciences*, Vol. 1, Part A, Buncel, E. and Jones, J.R., Eds.; Elsevier: Amsterdam, ***1987***, pp. 163–168.

90. Oehlke, J.; Niedrich, H.; Zöpfl, H.J. and Franke, P., *J. Label. Compd. Radiopharm.*, ***1990***, *28*, 387–393.
91. a. Morgat, J.L., in *Synthesis and Application of Isotopically Labelled Compounds 1988*, Baillie, T.A. and Jones, J.R., Eds.; Elsevier: Amsterdam, ***1989***, pp. 67–72; b. Morgat, J.L.; Fromageot, P., *Pept. Chem.* ***1988***, 195–198; c. Morgat, J.L.; Fromageot, P., Preparation of tritium labelled peptide hormones of high specific radioactivity, in *Proceedings of the Symposium on the Development in Radiopharmaceuticals and Labelled Compounds*, Vol. II, 1973; International Atomic Energy Agency, Vienna, ***1973***, 109–119.
92. a. Darula, Z,; Peter, A. and Toth, G., *J. Label. Compd. Radiopharm.*, ***1997***, *39*, 817–826; b. Morgat, J.L.; Hung, L.T.; Cardinaud, R.; Fromageot, P.; Bockardt, J.; Imbert, M. and Morel, F., *J. Label. Compd. Radiopharm.*, ***1970***, *6*, 276–280.
93. a. Hasegawa, H.; Shinohara, Y.; Baba, S., *J. Label. Compd. Radiopharm.*, ***1996***, *38*, 825–833; b. Toth, G.; Nevin, S.; Borsodi, A.; Nguyen, T.M.D. and Schiller, P., in *Synthesis and Applications of Isotopically Labelled Compounds* 1994, Allen, J. and Voges, R., Eds.; John Wiley & Sons, Ltd: Chichester, ***2001***, pp. 141–144; c. Brundish, D.E. and Wade, R., *J. Chem. Soc. Perkin Trans. I*, ***1981***, 318–321; d. Brundish, D.E. and Wade, R., *Biochem. J.*, ***1977***, *165*, 169–171; e. Eberle, A.N. and Zeller, A., *Helv. Chim. Acta*, ***1985***, *68*, 1880–1892; f. Houghten, R.A.; Chang, W.C. and Li, C.H., *Int. J. Peptide Protein Res.*, ***1980***, *16*, 311–320.
94. a. Brundish, D.E. and Wade, R., Peptides, structure and biological functions, in *Proceedings Am. Pept. Symp.* 1979, Gross, E. and Meienhöfer J., Eds., ***1979***, 483–486; b. Allen, M.C.; Brundish, D.E. and Wade, R., *J. Chem. Soc. Perkin Trans. I*, ***1979***, 2057–2063; c. Menez, A.; Morgat, J.L.; Fromageot, P.; Ronseray, A.M. Bouquet, P. and Changeux, J.P., *FEBS Lett.*, ***1971***, *17*, 333–335.
95. a. Tömböly, C.; Dixit, R.; Lengyel, I.; Borsodi, A. and Toth, G., *J. Label. Compd. Radiopharm.*, ***2001***, *44*, 355–363; b. Toth, G.; Farkas, J.; Kertesz, I.; Tomboly, C.; Darula, Z. and Peter, A., *Peptides*, ***1998***, 636–637; c. Allen, M.C.; Brundish, D.E.; Fullerton, J.D. and Wade, R., *J. Chem. Soc. Perkin Trans. I*, ***1986***, 989–1003; d. Allen, M.C.; Brundish, D.E.; Martin, J.R. and Wade, R., *J. Chem. Soc. Perkin Trans. I*, ***1981***, 2040–2048.
96. a. Sandberg, B.E.B.; Hanley, M.R.; Watson, S.P.; Brundish, D.E. Wade, R. and Eison, A.S., *FEBS Lett.*, ***1982***, *137*, 236–240; d. Leduc, R.; Bernier, M. and Escher, E., *Helv. Chim. Acta*, ***1983***, *66*, 960–970; e. Escher, E.; Mayhow, H.; Guillemette, G.; Bernier, M. and Resoli, D., *Proc. Am. Pept. Symp.*, ***1979***, 153–156.
97. Shu, A.Y.L. and Heys, J.R., *Intl. J. Pept. Prot. Res.*, ***1993***, *42*, 432–439.
98. a. Myasoedov, N.F. and Sidorov, G.V., *Radiokhimiya*, ***1980***, *4*, 574–578; b. Asano, T. and Kiritani, R., *J. Label. Compd. Radiopharm.*, ***1994***, *34*, 603–616.
99. Jaiswal, D.K.; Morimoto, H.; Trump, E.L.; Williams, P.G. and Wemmer, D.E., *J. Label. Compd. Radiopharm.*, ***1996***, *38*, 743–752.
100. Baraldi, P.G.; Cacciari, B.; Dionisotti, S; Egan, J.; Spalluto, G. and Zocchi, S., *J. Label. Compd. Radiopharm.*, ***1996***, *38*, 725–732.
101. Keana, J.F.W.; Scherz, M.W.; Quarum, M.; Sonders, M.S. and Weber, E., *Life Sciences*, ***1988***, *43*, 965–973.
102. Gong, L. and Parnes, H., *J. Label. Compd. Radiopharm.*, ***1995***, *38*, 31–40.
103. Parkinson, A., Biotransformations of Xenobiotics, in *Casarett & Doull's Toxicology, the Basic Science of Poisons*, 5th Edn, Klaassen, C.D., Ed.; McGraw-Hill: New York, ***1996***, pp. 113–186.
104. Hamdri, A.H.; Vidal-Cros, A.; Gaudry, M.; Sobrio, F. and Rosseau, B., *J. Label. Compd. Radiopharm.*, ***1995***, *36*, 795–799.
105. Katzenellenbogen, J.A.; Carlson, K.E.; Bindal, R.D.; Neeley, R.L.; Martin, P.M. and Magdelena, H.P., *Anal. Biochem.*, ***1986***, *159*, 336–348.
106. Szammer, J.; Simon-Trompler, E.; Banka, Z.; Szunyog, J. and Klebovich, Y., *J. Label. Compd. Radiopharm.*, ***2005***, *48*, 693–700.
107. a. Saljoughian, M.; Morimoto, H.; Dorsky, A.M. and Rapoport, H., *J. Label. Compd. Radiopharm.*, ***1989***, *27*, 767–776; b. Saljoughian, M.; Morimoto, H. and Rapoport, H., *Synthesis and Application of Isotopically Labelled Compounds 1988*, Baillie, T.A. and Jones, J.R., Eds.; Elsevier: Amsterdam, ***1989***, pp. 517–520.
108. Schwob, R. and Wuersch, J., *J. Label. Compd. Radiopharm.*, ***1978***, *14*, 355–360.

109. Saljoughian, M.; Morimoto, H. and Williams, P.G., *J. Chem. Soc. Perkin Trans. I*, ***1990***, 1803–1808.
110. Ott, D.G.; Kerr, V.N.; Whaley, T.W.; Benziger, T. and Rohwer, R.K., *J. Label. Compd. Radiopharm.*, ***1974***, *10*, 315–324.
111. Buchman, O.; Pri-Bar, I.; Shimoni, M. and Azrau, J., *J. Label. Compd. Radiopharm.*, ***1992***, *31*, 539–544.
112. Taylor, G.F.; Foarde, K.K.; Webber, T.D. and Kepler, J.A., *J. Label. Compd. Radiopharm.*, ***2001***, *44*, 165–168.
113. Senderoff, S.G; Shu, A.Y.L.; Lawrie, K. and Heys, J.R., *J. Label. Compd. Radiopharm.*, ***1992***, *31*, 519–538.
114. Triggle, D.J. and Belleau, B., *Can. J. Chem.*, ***1962***, *40*, 1201–1215.
115. a. Faucher, N; Ambroise, Y.; Cintrat, J.C.; Doris, E.; Pillon, F. and Rousseau, B., *J. Org. Chem.*, ***2002***, *67*, 932–934; b. Ambroise, Y.; Miokowski, C.; Djega-Mariadassou, G. and Rousseau, B., *J. Org. Chem.* ***2000***, *65*, 7183–7186.
116. Zinovyev, S.S.; Perosa, A. and Tundo, P., *J. Catal.*, ***2004***, *226*, 9–15 and references therein
117. Wiener, H.; Blum, J. and Sasson, Y., *J. Org. Chem.*, ***1991***, *56*, 6145–6148.
118. Marques, C.A.; Selva, M. and Tundo, P., *J. Org. Chem.*, ***1993***, *58*, 5256–5260.
119. Seltzman, H.H.; Carroll, F.I.; Burgess, J.P.; Wyrick, C.D. and Burch, D.F., *J. Label. Compd. Radiopharm.*, ***2002***, *45*, 59–70.
120. Morris; T.H.; Sandrock, K. and Kaumann, A.J., *Naunyn Schmiedeberg's Arch. Pharmacol.*, ***1981***, *317*, 19–25.
121. Latli, B. and Casida, J.E., *J. Label. Compd. Radiopharm.*, ***1995***, *36*, 147–156.
122. Hamill, T.G.; Brenner, N.J.; Eng, W.S. and Burns, H.D., *J. Label. Compd. Radiopharm.*, ***1998***, *41*, 319–327.
123. Hamill, T.G.; McCauley, J.A. and Burns, H.D., *J. Label. Compd. Radiopharm.*, ***2005***, *48*, 1–10.
124. Berthelette, C.; Boyd, M.J.; Lachance, N.; Roy, B.; Sturino, C.F.; Scheigetz, J. and Zamboni, R. J., *J. Label. Compd. Radiopharm.*, ***2004***, *47*, 107–113.
125. Brundish, D.E. and Kane, P.D., *J. Label. Compd. Radiopharm.*, ***1988***, *25*, 1361–1369.
126. Stiasni, M. and Stähle, H., *J. Label. Compd. Radiopharm.*, ***1978***, *14*, 51–58.
127. Almi, M., in *Synthesis and Applications of Isotopically Labeled Compounds*, Vol. 8, Dean, D.C.; Filer, C.N. and McCarthy, K.E., Eds.; John Wiley & Sons, Ltd: Chichester, ***2004***, pp. 215–218
128. Welch, W.M. and Williams, M.T., *J. Label. Compd. Radiopharm.*, ***1993***, *33*, 119–125.
129. Schou, S.C.; Hunneche, C.S.; Binderup, E. and Grue-Sørensen, G., *J. Label. Compd. Radiopharm.*, ***2005***, *48*, 789–795
130. Blackburn, D.W. and Levinson, S.H., *Catalysis in Organic Syntheses, Conf. 5$^{th}$*, Boston, 1975, Academic Press: New York, ***1976***, pp. 75–99.
131. Sarges, R.;US Patent No. 4474967, ***1984***.
132. a. Ambroise, Y.; Pillon, F.; Mioskowski, C.; Valleix, A. and Rousseau, B., *Eur. J. Org. Chem.*, ***2001***, 3961–3964; b. Sammelson, R.E. and Casida, J.E., *J. Org. Chem.*, ***2003***, *68*, 8075–8079.
133. Selliah, R.; Dantanarayana, A.; Haggard, K.; Egan, J.; Do, E.U. and May, J.A., *J. Label. Compd. Radiopharm.*, ***2001***, *44*, 173–183.
134. Blackburn, C.E., *J. Label. Compd*, ***1972***, *8*, 279–292.
135. Muccino, R.R.; Cupano, J. and Liebman, A.A., *Prostaglandins*, ***1981***, *21*, 615–622.
136. Postolache, C.; Tanase, C.; Matei, L. and Serban, V., *J. Label. Compd. Radiopharm.*, ***2007***, *50*, 609–610.
137. Voges, R.,Cyclosporin A, labeling with tritium, specifically labeled at the $\beta$-position of the $\alpha$-aminobutyric acid moiety, ***1984*** (June); Report: Sandoz Ltd.
138. Cerny, B.; Jegorov, A.; Polivkova, J.; Sedmera, P. and Havlicek, V., *J. Label. Compd. Radiopharm.*, ***1998***, *41*, 267–272.
139. Janssen, C.G.M.; Thijsen, J.B.A. and Verluyten, W.L.M., *J. Label. Compd. Radiopharm.*, ***2002***, *45*, 841–855.
140. Filer, C.N.; Hainley, C. and Nugent, R.P.; *J. Radioanal. Nucl. Chem.*, ***2006***, *267*, 345–348.
141. Roemer, J.; Scheller, D., Germany (East) Patent: DD 237067 A3, ***1986***.
142. Seltzman, H.H.; Roche, M.J.; Laudeman, C.P.; Wyrick, C.D. and Carroll, F.I., *J. Label. Compd. Radiopharm.*, ***1998***, *41*, 811–821.

143. Dietrich, A. and Fels, G., *J. Label. Compd. Radiopharm.*, ***1999***, *42*, 1125–1134.
144. Foged, C. and Madsen, P., *J. Label. Compd. Radiopharm.*, ***1994***, *34*, 659–662.
145. Hadj-Hamdri, A.; Vidal-Cros, A.; Gaudry, M.; Sobrio, F. and Rousseau, B., *J. Label. Compd. Radiopharm.*, ***1995***, *36*, 795–799.
146. Pleiss U. and Schmitt P., *J. Label. Compd. Radiopharm.*, ***1991***, *29*, 1091–1094.
147. Pleiss, U.; Lenfers, J.-B. and Schmitt, P., *J. Label. Compd. Radiopharm.*, ***1993***, *33*, 681–686.
148. a. Helquist, P., *Tetrahedron Lett.*, ***1978***, 1913–1914; b. Zask, A. and Helquist, P., *J. Org. Chem.*, ***1978***, *43*, 1619–1620.
149. Akita, Y.; Inoue, A.; Ishida, K.; Terui, K. and Ohta, A., *Synth. Commun.*, ***1986***, *16*, 1067–1072.
150. Hsi, R.S.P. and Stolle, W.T., *J. Label. Compd. Radiopharm.*, ***1996***, *38*, 1147–1149.
151. Allen, P.H.; Hickey, M.J.; Kingston, L.P. and Wilkinson, D.J., *J. Label. Compd. Radiopharm.*, ***2007***, *50*, 558–560.
152. Kovalainen, J.T.; Morimoto, H.; Williams, P.G.; Yepsäläinen, J.; Reijonen, A. and Gynther, J., *J. Label. Compd. Radiopharm.*, ***1995***, *36*, 1147–1156.
153. Pleiss, U. and Römer, J., *J. Label. Compd. Radiopharm.*, ***1983***, *20*, 1355–1362.
154. a. Block, E., in *The Chemistry of Functional Groups, Supplement E, Part 1*, Patai, S., Ed.; Wiley & Sons, Inc.: New York, ***1980***, p. 585; b. Belen'kii, L. I., *Chemistry of Organosulfur Chemistry;* Ellis Hoorwood: Chichester, ***1990***, p. 193; c. Protiva, J.; Ocadlik, R.; Klinotova, E.; Klinot, J. and Vystrcil, A., *Coll. Czech. Chem. Commun.*, ***1987***, *52*, 501–507; d. Raj, R.K. and Hutzinger, O., *J. Label. Compd*, ***1970***, *6*, 399–400.
155. a. Pleiss, U., *Radiochem. Radioanal. Lett.*, ***1982***, *51*, 363–372; b. Pleiss, U. and Römer, J., Wirtschaftspatent, Deutsche Demokratische Republik, ***1981***, C 07 B/2313356.
156. Brown, W.G.; Kaplan, L. and Wilzbach, K.E., *J. Am. Chem. Soc.*, ***1952***, *74*, 1343–1344.
157. Than, C.; Morimoto, H.; Andres, H. and Williams, P.G., *J. Label Compd Radiopharm.*, ***1996***, *38*, 693–711.
158. Cornforth, R.H., *Tetrahedron*, ***1970***, *26*, 4635–4640.
159. a. Pasto, D.J. and Lepeska, B., *J. Am. Chem. Soc.*, ***1976***, *98*, 1091–1095; b. Pasto, D.J.; Lepeska, B. and Cheng, T.C., *J. Am. Chem. Soc.*, ***1972***, *94*, 6083–6090.
160. a. Kreevoy, M.M. and Hutchins, J.E.C., *J. Am. Chem. Soc.*, ***1963***, *85*, 487–492; b. Levine, A.L. and Kreevoy, M.M., *J. Am. Chem. Soc.*, ***1972***, *94*, 3346–3349.
161. Hajos, A., *Complex Hydrides and Related Reducing Agents in Organic Synthesis;* Elsevier: Amsterdam, ***1979***, pp. 46–62.
162. Verstraeten, W., The use of sodium borohydride in synthesis, in *Proc. 2nd Int. Symp. on $NaBH_4$–Applications in Synthesis*, Milano; Morton Thiokol, Ventron Division, June ***1988***, pp. 1–32.
163. Faghih, R.; Burnell-Curty, C.; Surber, B.; Shoghi, S.; Borre, A.; Yao, Y.; Lartey, P.A. and Nellans, H.N., *J. Label. Compd. Radiopharm.*, ***1996***, *38*, 687–692.
164. a. Demin, P.M.; Vasiljeva, L.L.; Kochev, D.M.; Pivnitsky, K.K. and Pace-Asciak, C.R., *J. Label. Compd. Radiopharm.*, ***1999***, *42*, 773–779; b. Brown, A.J.; Perlman, K.; Schnoes, H.K. and DeLuca, H.F., *Anal. Biochem.* ***1987***, *164*, 424–429; c. Davisson, V.J.; Zabriskie, T.M. and Poulter, C.D., *Bioorg. Chem.* ***1986***, *14*, 46–54.
165. Atwell, G.J. and Denny, W.A., *J. Label. Compd. Radiopharm.*, ***2007***, *50*, 7–12.
166. a. Reddy, P.V.; Rabago-Smith, M. and Borhan, B., *J. Label. Compd. Radiopharm.*, ***2002***, *45*, 79–89; b. Ceruti, M.; Grosa, G.; Rocco, F.; Dosio, F. and Cattel, L., *J. Label. Compd. Radiopharm.*, ***1994***, *34*, 577–585; c. Furth, P.S. and Robinson, C.H., *Biochemistry*, ***1989***, *28*, 1254–1259; d. Yang, D.; Shih, Y. and Liu, H., *J. Org. Chem.*, ***1991***, *56*, 2940–2946.
167. Smith, M.M.; Warren, V.A.; Thomas, B.S.; Brochu, R.M.; Ertel, E.A.; Rohrer, S.; Schaeffer, J.; Schmatz, D.; Petuch, B.R.; Tang, Y.S.; Meinke, P.T.; Kaszorowski, G.J. and Cohen, C.J., *Biochemistry*, ***2000***, *39*, 5543–5554.
168. Latli, B.; Wood, E. and Casida, J.E., *Chem. Res. Toxicol.*, ***1996***, *9*, 445–450.
169. Egan, J.A.; Filer, C.N. and Hudkins, R.L., *J. Label. Compd. Radiopharm.* ***2005***, *48*, 997–1001.
170. Inguimbert, N.; Coric, P.; Dhotel, H.; Llorens-Cortes, C.; Fournié-Zaluski, M.-C. and Roques, B.P., *J. Label. Compd. Radiopharm.*, ***2004***, *47*, 997–1005.
171. a. Halkier, B.A.; Olsen, C.E. and Marcussen, J., *J. Label. Compd. Radiopharm.*, ***1991***, *29*, 1–7; b. Rotman, A.; Daly, J.W. and Creveling, C.R., *J. Label. Compd. Radiopharm.*, ***1975***, *11*, 445–452.
172. Meinke, P.T.; Rohrer, S.P.; Hayes, E.C.; Schaeffer, J.M.; Fisher, M.H. and Mrozik, H., *J. Med. Chem.*, ***1992***, *35*, 3879–3884.

173. Kanda, Y.; Akinaga, S. and Kasai, M., *J. Label. Compd. Radiopharm.*, ***1990***, *28*, 1033–1036.
174. Beale, M.H.; Gaskin, P.; Kirkwood, P.S. and MacMillan, J., *J. Chem. Soc. Perkin I*, ***1980***, 885–891.
175. a. Tack, B.F. and Wilder, R.L., *Methods Enzymol.*, ***1981***, *73*, 138–147; b. Tack, B.F.; Dean, J.; Eilat, D.; Lorenz, P.E. and Schecher, A.N., *J. Biol. Chem.*, ***1980***, *255*, 8842–8847; c. Nelles, L.P. and Bamburg, J.R., *Anal. Biochem.*, ***1979***, *94*, 150–159.
176. Landvatter, S.W.; Senderoff, S.G. and Heys, J.R., *J. Label. Comp. Radiopharm.*, ***1993***, *33*, 1113–1117.
177. Strøgaard, K.; Suehiro, M. and Nakanishi, K., *Bioorg. Med. Chem. Lett.*, ***2004***, *14*, 5673–5675.
178. Boumendjel, A. and Miller, S.P.F., *J. Label. Comp. Radiopharm.*, ***1995***, *36*, 337–383.
179. Gay, J. and Rohlfs, G., *J. Label. Comp. Radiopharm.*, ***1999***, *42*, 994–995.
180. Rehder, K.S.; Hristova-Kazmierski, M.K. and Kepler, J.A., *J. Label. Comp. Radiopharm.*, ***1996***, *38*, 1077–1081.
181. Elvidge, J.A.; Jones, J.R.; Saieed, M.S.; Evans, E.A. and Warrell, D.C., *J. Chem. Res. (S)*, ***1981***, 288–289.
182. Tang, Y.S.,Uses of sodium borotritides as a tritiation reagent, presented at the IIS-Symposium, Boston 2003, personal communication.
183. Tang, Y.S.; Liu, W.; Chaudhary, A.; Melillo, D.G. and Dean, D.C., in *Synthesis and Applications of Isotopically Labeled Compounds*, Vol. 8, Dean, D.C.; Filer, C.N. and McCarthy, K.E.; John Wiley & Sons, Ltd: Chichester, ***2004***, pp. 71–74.
184. Kuo, F.; Kulanthaivel, P.; Rener, G.A.; Yi, P. and Wheeler, W.J., *J. Label. Compd. Radiopharm.*, ***2004***, *47*, 571–581.
185. Branchini, B.R.; Adams, E. W.; Egan, L.A. and Murtiashaw, M.H., *J. Label. Compd. Radiopharm.*, ***1992***, *31*, 387–395.
186. a. Bielawska, A.; Hannum, Y.A. and Szulc, Z., *Methods Enzymol.*, ***1999***, *311*, 480–498; b. Schick, A.; Schwarzmann, G.; Kolter, T. and Sandhoff, K., *J. Label. Comp. Radiopharm.*, ***1997***, *39*, 441–451.
187. Chaudhary, A.; Tang, Y.S; Dean, D.C; Ashton, W. and Melillo, D.G., in *Synthesis and Applications of Isotopically Labeled Compounds*, Vol. 8, Dean, D.C.; Filer, C.N. and McCarthy, K.E., Eds.; John Wiley & Sons, Ltd: Chichester, ***2004***, pp. 425–428.
188. Lacey, E., *J. Label. Compd. Radiopharm.*, ***1985***, *22*, 893–902.
189. Nagasaki, T.; Sakai, K.; Segawa, M.; Katsuyama, Y.; Haga, N.; Koike, M.; Kawada, K. and Takechi, S., *J. Label. Compd. Radiopharm.* ***2001***, *44*, 993–1004.
190. Hodek, P. and Strobel, H.W., *J. Label. Compd. Radiopharm.*, ***1995***, *36*, 321–329.
191. Lockley, W.J.S., *J. Label. Compd. Radiopharm.*, ***1987***, *24*, 1509–1515.
192. Malik, S.; Kenny, M.; Doss, G. and John, V., *J. Label. Compd. Radiopharm.*, ***1994***, *34*, 471–473.
193. a. Hutchins, R.O.; Natale, N.R., Cyanoborohydride. Utility and applications in organic synthesis, in *Org. Prep. Proc. Int.*, ***1979***, 11, 201–246; b. Lane, C.F., *Synthesis*, ***1974***, 135–146; c. Hutchins, R.O.; Maryanoff, B.E.; Milewski, C.A., *Chem. Commun.*, ***1971***, 1097–1098.
194. a. Borch, R.F.; Bernstein, M.D.; Durst, H.D., *J. Am. Chem. Soc.*, ***1971***, *93*, 2897–2904; b. Kreevoy, M.M.; Hutchins, J.E.C., *J. Am. Chem. Soc.*, ***1969***, 4329–4330.
195. a. Jentoft, N. and Dearborn, D.G., *J. Biol. Chem.*, ***1979***, *254*, 4359–4365; b. March, J.W., Nahum, A. and Steiner, D.F., *Int. J. Peptide Protein Res.*, ***1983***, *22*, 39–49.
196. Ackland, M.J.; Dring, L.G.; Jones, J.R. and Gilon, N., *J. Label. Compd. Radiopharm.*, ***1991***, *29*, 909–918.
197. Mellor, H.R.; Nolan, J.; Pickering, L.; Wormald, M.R.; Platt, F.M.; Dwek, R.A.; Fleet, G.W.J. and Butters, T.D., *Biochem. J.*, ***2002***, *366*, 225–233.
198. Voges, R. and Kohler, B., in *Synthesis and Applications of Isotopically Labeled Compounds 1991*, Buncel, E. and Kabalka, G.W., Eds.; Elsevier: Amsterdam, ***1992***, pp. 74–81.
199. Kumar, P.; Stypinski, D.; Xia, H.; McEwan, A.J.B.; Machula, H.-J. and Wiebe, L.I., *J. Label. Compd. Radiopharm.*, ***1999***, *42*, 3–16.
200. a. Voges, R., in *Synthesis and Applications of Isotopically Labeled Compounds 1988*, Baillie, T.A. and Jones, J.R., Eds.; Elsevier: Amsterdam, ***1989***, pp. 33–40; b. Kim, S.; Oh, C.H.; Ko, J.S.;

Ahn, K.H. and Kim, Y.J., *J. Org. Chem*, ***1985***, *50*, 1927–1932; c. Kim, S.; Kim, Y.J. and Ahn, K. H., *Tetrahedron Lett.*, ***1983***, *24*, 3370–3373.

201. a. Andres, H., *J. Label. Compd. Radiopharm.*, ***2001***, *44*, 950–951; b. Evans, D.A.; Chapman, K. T. and Carreira, E.M., *J. Am. Chem. Soc.*, ***1988***, *10*, 3560–3562.
202. Wilzbach, K.E. and Kaplan, L., *J. Am. Chem. Soc.*, ***1950***, *72*, 5795–5796.
203. Andres, H.; Morimoto, H. and Williams, P.G., in *Synthesis and Applications of Isotopically Labeled Compounds 1991;* Buncel, E. and Kabalka, G.W., Eds.; Elsevier: Amsterdam, ***1992***, pp. 40–45.
204. Brown, H.C.; Krishnamurthy, S. and Hubbard, J.L., *J. Am. Chem. Soc.*, ***1978***, *100*, 3343–3349.
205. Klusener, P.A.A.; Brandsma, L.; Verkruijsse, H.D.; Schleyer, P.v.R.; Friedl, T. and Pi, R., *Angew. Chem. Int. Edn Engl.*, ***1986***, *25*, 465–466.
206. a. Andres, H., Morimoto, H. and Williams, P.G., *J. Chem. Soc. Chem. Commun.*, ***1990***, 627–628; b. Andres, H.; Morimoto, H. and Williams, P.G., U.S. Patent 5,186,868, Feb. 16, ***1993***.
207. Pi, R.; Friedl, T.; Schleyer, P.v.R.; Klusener, P. and Brandsma, L., *J. Org. Chem.*, ***1987***, *52*, 4299–4303.
208. Zippi, E.M.; Andres, H.; Morimoto, H. and Williams, P.G., *Synth. Commun.*, ***1995***, *25*, 2685–2693.
209. Hajos, A., *Complex Hydrides and Related Reducing Agents in Organic Synthesis*, Elsevier: Amsterdam, ***1979***, pp. 43–44.
210. Than, C.; Morimoto, H.; Andres, H. and Williams, P.G., *J. Org. Chem.*, ***1996***, *61*, 8771–8774.
211. Andres, H.; Morimoto, H.; Williams, P.G. and Zippi, E.M., in *Synthesis and Applications of Isotopically Labeled Compounds 1994*, Allen, J. and Voges, R., Eds.; John Wiley & Sons, Ltd: Chichester, ***1995***, pp. 83–90.
212. a. Than, C.; Morimoto, H.; Andres, H. and Williams, P.G., *J. Label. Compd. Radiopharm.*, ***1996***, *38*, 693–711; b. Atkinson, J.G.; MacDonald, D.W.; Stuart, R.S. and Tremaine, P., *Can. J. Chem.*, ***1967***, *45*, 2583–2588; c. Schlesinger, H.I.; Brown, H.C.; Hoekstra, H.R. and Rapp, L.R., *J. Am. Chem. Soc.*, ***1953***, *75*, 199–204; d. Brown, W.G.; Kaplan, L. and Wilzbach, K.E., *J. Am. Chem. Soc.*, ***1952***, *74*, 1343–1344; e. Schlesinger, H.I. and Brown, H.C., *J. Am. Chem. Soc.*, ***1940***, *62*, 3429–3435.
213. Evans, E.A., *Tritium and its Compounds*, 2nd Edn; John Wiley & Sons, Inc.: New York, ***1974***, pp. 369–378.
214. Latli, B.; Than, C.; Morimoto, H.; Williams, P.G. and Casida, J.E., *J. Label. Compd. Radiopharm.*, ***1996***, *38*, 971–978.
215. Latli, B.; Morimoto, H.; Williams, P.G. and Casida, J.E., *J. Label. Compd. Radiopharm.*, ***1998***, *41*, 191–199.
216. Maitra, U. and Mathivanan, P., *Tetrahedron: Asymmetry*, ***1994***, *5*, 1171–1174.
217. Dischino, D.D.; Meanwell, N.A.; Morimoto, H.; Pearce, B.C.; Than, C.; Torri, A.F.; Williams, P.G. and Yu, K.-L., *J. Label. Compd. Radiopharm.*, ***1999***, *42*, 1061–1068.
218. Huguenin, P.N. and Shimma, N., *J. Label. Compd. Radiopharm.*, ***1996***, *38*, 999–1005.
219. Larsen, U.S.; Begtrup, M. and Martiny, L., *J. Label. Compd. Radiopharm.*, ***2005***, *48*, 429–434.
220. Brown, H.C.; Kim, S.C. and Krishnamurthy, S., *J. Org. Chem.*, ***1980***, *45*, 1–12.
221. a. Hegde, S.; Coates, R.M. and Pearce, C.J., *J. Chem. Soc. Chem. Commun.*, ***1983***, 1484–1485; b. Coates, R.M.; Hegde, S. and Pearce, C.L., in *Synthesis and Applications of Isotopically Labeled Compounds*, Duncan, W.P. and Susàn, A.B., Eds.; Elsevier: Amsterdam, ***1983***, pp. 429–430.
222. a. Wilkinson, B.; Zhu, M.; Priestley, N.D.; Nguyen, H.-H.T.; Morimoto, H.; Williams, P.G.; Chan, S.I. and Floss, H.G., *J. Am. Chem. Soc.*, ***1996***, *118*, 921–922; b. Priestley, N.D.; Floss, H. G.; Froland, W.A.; Lipscomb, J.D.; Williams, P.G. and Morimoto, H., *J. Am. Chem. Soc.*, ***1992***, *114*, 7561–7562.
223. Valentine, A.M.; Wilkinson, B.; Liu, K.E.; Komar-Panicucci, S.; Priestley, N.D.; Williams, P.G.; Morimoto, H.; Floss, H.G. and Lippard, S.J., *J. Am. Chem. Soc.*, ***1997***, *119*, 1818–1827.
224. Kobayashi, K.; Jadhav, P.K.; Zydowski, T.M. and Floss, H.G., *J. Org. Chem.*, ***1983***, *48*, 3510–3512.
225. Coates, R.M.; Hegde, S.; Koch, S.C. and Chakraborty, P.K.,in *Synthesis and Applications of Isotopically Labeled Compounds 1985*, Muccino, R.R.,Ed.; Elsevier: Amsterdam, ***1986***, pp. 207–212.

226. Simov, B.P.; Wuggenig, F.; Mereiter, K.; Andres, H.; France, J.; Schnelli, P. and Hammerschmidt, F., *J. Am. Chem. Soc.*, ***2005***, *127*, 13934–13940.
227. Saljoughian, M.; Morimoto, H.; Williams, P.B. and Than, C., *J. Label. Compd. Radiopharm.*, ***2001***, *44*, 954–955.
228. Moenius, Th.; Andres, H.; Acemoglu, M.; Kohler, B.; Schnelli, P. and Zueger, C., *J. Label. Compd. Radiopharm.*, ***2000***, *43*, 113–120.
229. Liu, Y.-Y.; Thom, E. and Burger, W., in *Synthesis and Applications of Isotopically Labeled Compounds 1994*, Allen, J. and Voges, R., Eds.; John Wiley & Sons, Ltd: Chichester, ***1995***, pp. 93–95.
230. a. Saljoughian, M.; Morimoto, H. and Williams, P.G., in *Synthesis and Applications of Isotopically Labeled Compounds*, Vol. 7, Pleiss, U and Voges, R, Eds.; John Wiley & Sons, Ltd: Chichester, ***2001***, pp. 167–169; b. Saljoughian, M., *Synthesis*, ***2002***, *13*, 1783–1784.
231. a. Andres, H.; Than, C.; Morimoto, H. and Williams, P.G., in *Synthesis and Applications of Isotopically Labeled Compounds 1997*, Heys, J.R. and Melillo, D.G., Eds.; John Wiley & Sons, Ltd: Chichester, ***1998***, pp. 213–217; b. Wagner, J.; Andres, H.; Rohrbach, S.; Wagner, D.; Oberer, L. and France, J., *J. Org. Chem.*, ***2005***, *70*, 9588–9590.
232. Linnemann, E. and Fels, G., *J. Label. Compd. Radiopharm.*, ***2001***, *44*, 661–669.
233. Brown, H.C. and Mathew, C.P., *J. Org. Chem.*, ***1984***, *49*, 3091–3097.
234. a. Rasset, C. and Rousseau, B., *J. Label. Compd. Radiopharm.*, ***1994***, *34*, 523–528; b. Rasset, C. and Rousseau, B. in *Synthesis and Applications of Isotopically Labeled Compounds 1994*, Allen, J. and Voges, R., Eds.; Elsevier: Amsterdam, ***1995***, pp. 91–92.
235. a. Carruthers, W., *Some Modern Methods of Organic Chemistry*, 3rd Edn; Cambridge University Press: Cambridge, ***1986***, pp. 478–482; b. Hajos, A., *Borane and its Derivatives in Complex Hydrides and Related Reducing Agents in Organic Synthesis*, Chapter 3; Elsevier: Amsterdam, ***1979***, pp. 24–35.
236. Than, C.; Morimoto, H.; Andres, H. and Williams, P.G., *J. Org. Chem.*, ***1995***, *60*, 7503–7507.
237. Garmestani, K.; Link, J.M. and Krohn, K.A., *J. Label. Compd. Radiopharm.*, ***1990***, *28*, 1171–1178.
238. Evans, E.A., *Tritium and its Compounds*, 2nd Edn; John Wiley & Sons, Inc.: New York, ***1974***, pp. 402–406.
239. Taylor, G.F.; Thornton, S.S.; Tallent, C.R. and Kepler, J.A., *J. Label. Compd. Radiopharm.*, ***1993***, *33*, 501–514.
240. Filer, C.N.; Nugent, R.P. and Shelton, E.J., *Trans. Am. Nucl. Soc.*, ***1989***, 36–38.
241. a. Corey, E.J. and Helal, C.J., *Angew. Chem. Int. Ed.*, ***1998***. *37*, 1986–2012; b. Corey, E.J., *Pure Appl. Chem.*, ***1990***, *62*, 1209–1216.
242. Hesk, D.; Bignan, G.; Lee, J.; Yang, J.; Voronin, K.; Magatti, C.; McNamara, P.; Koharski, D.; Hendershot, S.; Saluja, S. and Wang, S., *J. Label. Compd. Radiopharm.*, ***2002***, *45*, 145–155.
243. Burgos, A.; Kamenka, J.-M.; Moustier, A.-M. and Rousseau, B., *J. Label. Compd. Radiopharm.*, ***1991***, *29*, 1061–1064.
244. Burgos, A.; Kamenka, J.-M. and Rousseau, B., *J. Label. Compd. Radiopharm.*, ***1991***, *29*, 1347–1350.
245. Hajos, A.; Metal aluminum hydride, in *Complex Hydrides and Related Reducing Agents in Organic Synthesis*, Chapter 6, Elsevier: Amsterdam, ***1979***, pp. 83–175.
246. Liu, M.-Y.; Latli, B. and Casida, J.E., *Pest. Biochem. Physiol.*, ***1994***, *50*, 171–182.
247. Pleiss, U., *J. Label. Compd. Radiopharm.*, ***2003***, *46*, 1241–1247.
248. Stolle, W.T.; Stelzer, L.S.; Hester, J.B.; Perricone, S.C. and Hsi, R.S.P., *J. Label. Compd. Radiopharm.*, ***1994***, *34*, 929–942.
249. Dischino, D.D.; Combrink, K.D.; Doweyko, D.; Morimoto, H.; Pearce, B.C.; Williams, P.G. and Yevich, J.P., *J. Label. Compd. Radiopharm.*, ***1995***, *36*, 789–794.
250. Saljoughian, M.; Morimoto, H. and Williams, P.G., *Fusion Technol.*, ***1992***, *21*, 318.
251. a. van den Berg, E.M.M.; van der Bert, A. and Lugtenburg, J., *Recl. Trav. Chim. Pays-Bas*, ***1990***, *109*, 160–167; b. Goerger, M.M. and Hudson, B.S., *J. Org. Chem.*, ***1988***, *53*, 3148–3153.
252. a. Lee, S.; Morimoto, H. and Williams, P.G., *J. Label. Compd. Radiopharm.*, ***1997***, *39*, 461–470; b. Liu, Y.-Y. and Chen, L., *J. Label. Compd. Radiopharm.*, ***1995***, *38*, 71–76.

253. a. Neumann, W.P., *Synthesis*, ***1987***, 665–683; b. Kuivila, H.G., *Synthesis*, ***1970***, 499–509; c. Hartwig, W., *Tetrahedron*, ***1983***, *39*, 2609–2645; d. Miura, K.; Ichinose, Y.; Nozaki, K.; Fugami, K.; Oshima, K. and Utimoto, K, *Bull. Chem. Soc. Jpn.*, ***1989***, *62*, 143–147.
254. Moenius, Th.; Andres, H. and Neu, H., *J. Label. Compd. Radiopharm.*, ***2000***, *43*, 1253–1253.
255. Moenius, Th.; Andres, H.; Acemoglu, M.; Kohler, B.; Schnelli, P. and Zueger, C., *J. Label. Compd. Radiopharm.*, ***2000***, *43*, 113–120.
256. Jaiswal, D.K.; Andres, H.; Morimoto, H. and Williams, P.G., *J. Chem. Soc. Chem. Commun.*, ***1993***, 907–909.
257. Wichmann, J.; Huguenin, P. and Adam, G., *J. Label. Compd. Radiopharm.*, ***2000***, *43*, 1–10.
258. a. Parnes, H. and Pease, J., *J. Org. Chem.*, ***1979***, *44*, 151–152; b. Birnbaum, E.R. and Javora, P.H., *J. Organomet. Chem.*, ***1967***, *9*, 379–382.
259. Sugiura, S.; Tanaka, T.; Bannai, K. and Kurozumi, S., *J. Label. Compd. Radiopharm.*, ***1991***, *29*, 1041–1050.
260. Saljoughian, M.; Morimoto, H.; Than, C. and Williams, P.G., *J. Label. Compd. Radiopharm.*, ***1998***, *41*, 1027–1028.
261. Lee, S.Y., *Bull. Sing. N. I. Chem.*, ***1990***, *18*, 115–120.
262. Tamborski, C.; Ford, F.E. and Soloski, E.J., *J. Org. Chem.*, ***1963***, *28*, 237–239.
263. a. Su, X.; Siddiqui, A.; Swaminathan, S.; Wilson, W.K. and Schroepfer, G.J., *J. Label. Compd. Radiopharm.*, ***1998***, *41*, 63–74; b. Saljoughian, M.; Morimoto, H.; Williams, P.G.; Than, C. and Seligman, S.J., *J. Org. Chem.*, ***1996***, *61*, 9625–9628.
264. Horiguchi, T.; Rithner, C.D.; Croteau, R. and Williams, R.M., *J. Label. Compd. Radiopharm.*, ***2008***, *51*, 325–328.
265. Ono, N. and Kaji, A., *Synthesis*, ***1986***, *693*–704.
266. Saljoughian, M.; Morimoto, H.; Than, C. and Williams, P.G., in *Synthesis and Applications of Isotopically Labeled Compounds 1997*, Heys, J.R. and Melillo, D.G., Eds.; John Wiley & Sons, Ltd.: Chichester, ***1998***, pp. 193–196.
267. Curran, D.P.; Somayajula, K.V. and Yu, H., *Tetrahedron Lett.*, ***1992***, *33*, 2295–2298; Curran, D. P.; Kim, D.; Liu, H.T. and Shen, W., *J. Am. Chem. Soc.*, ***1988***, *110*, 5900–5902.
268. a. Kalesse, M., *Acros Organics Acta*, ***1995***, *1*, 29–31; b. Schwartz, J. and Labinger, J.A., *Angew. Chem. Int. Ed.*, ***1976***, *15*, 333–340; c. Schwartz, J.; Loots, M. J.; Kosugi, H.J., *J. Am. Chem. Soc.*, ***1980***, *102*, 1333–1340; d. Hart, D.W.; Blackburn, T.F. and Schwartz, J., *J. Am. Chem. Soc.*, ***1975***, *97*, 679–680.
269. Godfrey, A.G. and Ganem, B., *Tetrahedron Lett.*, ***1992***, *33*, 7461–7464.
270. Zippi, E.M.; Andres, H.; Morimoto, H. and Williams, P.G., *Syn. Commun.*, ***1994***, *24*, 1037–1044.
271. Andres, H.; Morimoto, H.; Williams, P.G. and Zippi, E.M., in *Synthesis and Applications of Isotopically Labeled Compounds 1994*, Allen, J. and Voges, R., Eds.; John Wiley & Sons, Ltd: Chichester, ***1995***, pp. 83–90.
272. Moenius, Th.; Andres, H.; Nozulak, J.; Salter, R.; Ray, T.; Burtscher, P.; Schnelli, P.; Zueger, C. and Voges, R., *J. Label. Compd. Radiopharm.*, ***2007***, *50*, 616–619.
273. Hiyama, T. and Kusumoto, T., in *Comprehensive Organic Synthesis*, Vol. 8, Trost, B.M. and Fleming, I., Eds; Pergamon: Oxford, ***1992***, pp. 546–557.
274. Neu, H. and Andres, H., *J. Label. Compd. Radiopharm.*, ***1999***, *42*, 992–993.
275. Guillerm, G.; Frappier, F.; Tabet, J.-C. and Marquet, A., *J. Org. Chem.*, ***1977***, *42*, 3776–3778.
276. Giles, R.G.F.; Rickards, R.W. and Senanayake, B.S., *J. Chem. Soc. Perkin Trans. I*, ***1997***, 3361–3370.
277. Guyard, A.-L.; Valleix, A.; Rousseau, B. and Cintrat, J.-C., *J. Label. Compd. Radiopharm.*, ***2001***, *44*, 529–539.
278. a. Andres, H., in *Synthesis and Applications of Isotopically Labeled Compounds*, Vol. 7, Pleiss, U. and Voges, R., Eds.; John Wiley & Sons, Ltd.: Chichester, ***2001***, pp. 49–62; b. Saljoughian, M. and Williams, P.G., *Curr. Pharm. Design*, ***2000***, *6*, 1029–1056; c. Saljoughian, M.; Morimoto, H. and Williams, P.G., *Synthesis and Applications of Isotopically Labeled Compounds*, Vol. 7, Pleiss, U. and Voges, R., Eds.; John Wiley & Sons, Ltd.: Chichester, ***2001***, pp. 167–169; d. Saljoughian, M.; Morimoto, H.; Than, C. and Williams, P.G., in *Synthesis and Applications of Isotopically Labeled Compounds 1997*, Heys, J.R. and Melillo, D.G., Eds.; John Wiley & Sons,

Ltd.: Chichester, ***1998***, pp. 193–198; e. Filer, C.N., in *Isotopes in the Physical and Biomedical Sciences*, Vol. 1, Part A, Buncel, E. and Jones, J.R., Eds.; Elsevier: Amsterdam, ***1987***, pp. 178–181.

279. Tritium and other radionuclide labeled compounds incorporated in genetic material, NCRP Report No. 63, National Council on Radiation Protection and Measurements: Washington, DC, ***1979***.
280. Evans, E.A., *Tritium and Its Compounds*, 2nd Edn; John Wiley & Sons, Inc.: New York, ***1974***, pp. 701–702.
281. a. Wilson, E.J., *Vacuum*, ***1954***, *4*, 303; b. Wilson, E.J., A.E.R.E. Harwell Report 28, ***1953***, AERE I/M; c. Evans, E.A.,Tritium and Its Compounds, 2nd Edn; John Wiley & Sons, Inc.: New York, ***1974***, pp. 318–319.
282. Pleiss, U. and Römer, J., GDR-Patent 248680/8, ***1983***.
283. a. Jaiswal, D.K.; Morimoto, H.; Saljoughian, M. and Williams, P.G., in *Synthesis and Applications of Isotopically Labeled Compounds 1991*, Buncel, E. and Kabalka, G.W. Eds.; Elsevier: Amsterdam, ***1992***, pp. 46–51; b. Morimoto, H. and Williams, P.G., *Fusion Technol.*, ***1992***, *21*, 256–261.
284. Henot, F.; Thellend, A.; Pillon, F.; Rousseau, B. and Gaudry, M., *J. Label. Compd. Radiopharm.*, ***1996***, *38*, 579–583.
285. Myasoedov, N.F.; Sidorov, G.V.; Kramerov, V.N. and Mishin, V.I., *J. Label. Compd. Radiopharm.*, ***1999***, *42*, 859–866.
286. Blizzard, T.A.; Rosegay, A.; Mrozik, H. and Fisher, M.H., *J. Label. Compd. Radiopharm.*, ***1990***, *28*, 461–464.
287. Sidorov, G.V.; Lushkina, O.V.; Myasoedov, N.F.; Kramerov, V.N. and Mishin, V.I., in *Synthesis and Applications of Isotopically Labeled Compounds 1991*, Buncel, E. and Kabalka, G.W. Eds.; Elsevier: Amsterdam, ***1992***, pp. 452–455.
288. Snieckus, V., *Chem. Rev.*, ***1990***, *90*, 879–933.
289. a. Saljoughian, M.; Morimoto, H.; Than, C. and Williams, P.G., *Tetrahedron Lett.*, ***1996***, *37*, 2923–2926; b. Adlington, R.M. and Barrett, A.G.M., *Acc. Chem. Res.*, ***1983***, *16*, 55–59.
290. a. Kessar, S.V., *Pure Appl. Chem.*, ***1990***, *62*, 1397–1400; b. Brooks, A.G., *Acc. Chem. Res.*, ***1974***, *7*, 77–84.
291. Dow, R.L.; Bechle, B.M.; Chalabi, P.M.; Windels, J.H. and Roth, R.W., *J. Label. Comp. Radiopharm.*, ***1992***, *31*, 333–339.
292. Turecek, F.; Veres, K.; Kocovsky, P.; Pouzar, V. and Fajkos, J., *J. Org. Chem.*, ***1983***, *48*, 2233–2237.
293. a. Smith, M.B. and March, J., *Advanced Organic Chemistry*, 5th Edn; John Wiley & Sons, Inc.: New York, ***2001***, pp. 1007–1008; b. House, H.O., *Modern Synthetic Chemistry*, 2nd Edn; W.A. Benjamin: Menlo Park, California, ***1972***, pp. 248–256.
294. a. Saljoughian, M.; Williams, P.G.; Morimoto, H.; Goodlett, D.R. and van Bremen, R.B., *J. Chem. Soc. Chem. Commun.*, ***1993***, 414–416; b. Koch, G.K., Specific $^3$H-Labelling by diimide reduction of unsaturated bonds, Parts 1 and 2, *J. Label. Compd*, ***1969***, *5*, 99–121; c. Paliokas, A. M. and Schroepfer, G.J., *Biochem. Biophys. Res. Commun.*, ***1967***, *26*, 736–737; d. Evans, E.A., *Tritium and its Compounds*, 2nd Edn; John Wiley & Sons, Inc.: New York, ***1974***, pp. 337–338.
295. a. Bolton, R., *J. Label. Compd. Radiopharm.*, ***2001***, *44*, 701–736.
296. Liu, Y.Y. and Chen, L., *J. Label. Compd. Radiopharm.*, ***1995***, *38*, 71–76.
297. Lee, S.; Morimoto, H. and Williams, P.G., *J. Label. Compd. Radiopharm.*, ***1997***, *39*, 461–470.
298. a. Nagamura, S.; Kinugawa, M.; Ogasa, T. and Saito, H., *J. Label. Compd. Radiopharm.*, ***1997***, *39*, 471–477; b. Mais, D.E.; Chen, L.Z.; Wagoner, M.A.; Hayes, J.S. and Wang, M.W., *J. Label. Compd. Radiopharm.*, ***1995***, *36*, 1199–1203; c. Backwell, F.R.C. and Williams, C.M., *J. Label. Compd. Radiopharm.*, ***1990***, *29*, 421–426.
299. Cooley, J.H. and Evain, E.J., *Synthesis*, ***1989***, 1–7.
300. Rosenthaler, J.; Munzer, H. and Voges, R., *Methodological Surveys in Biochemistry 12, Drug Metabolite Isolation and Determination*, Reid, E. and Leppard, J.P., Eds.; Plenum: New York, ***1983***, pp. 1–9.

301. a. Fellows, I.; Carr, R.M.; deBoeck, N.; Montgomery, S. and Waterhouse, I., *J. Label. Compd. Radiopharm.*, ***1998***, *41*, 1127–1143; b. Fellows, I.; Carr, R.M.; deBoeck, N.; Montgomery, S. and Waterhouse, I., *Synthesis and Applications of Isotopically Labeled Compounds, 1997*, Heys, J.R. and Melillo, D.G.,Eds.; John Wiley & Sons, Ltd: Chichester, ***1998***, pp. 507–511.
302. Pleiss, U.; Turberg, A.; Harder, A.; Londershausen, M.; Jeschke, P. and Boheim, G., *J. Label. Compd. Radiopharm.*, ***1996***, *38*, 651–659.
303. Schirrmacher, R.; Hamkes, W.; Lüddens, H. and Rösch, F., *J. Label. Compd. Radiopharm.*, ***2000***, *43*, 1127–1134.
304. Senderoff, S.G.; Landvatter, S.W. and Heys, J.R., *J. Label. Compd. Radiopharm.*, ***2000***, *43*, 1283–1288.
305. a. Kobel, H.; Loosli, H.R. and Voges, R., *Experientia*, ***1983***, *39*, 873–883; b. Moenius, T.; Baumann, K.; Bulusu, M.; Schweitzer, A. and Voges, R., in *Synthesis and Applications of Isotopically Labeled Compounds*, Vol. 7, Pleiss, U. and Voges, R., Eds.; John Wiley & Sons, Ltd: Chichester, ***2001***, pp. 424–429.
306. Hesk, D.; Gunnarson, I.; Hendershot, S.; Koharski, D.; McNamara, P.; Schwartz, J.L.; Thonoor, M. and Wirth, M., *J. Label. Compd. Radiopharm.*, ***1999***, *42*, 159–167.
307. a. Barling, P.M.; Preston, J.R.; Bibby, N.J. and Wilson, T., *Anal. Biochem.*, ***1985***, *144*, 542–552; b. Handcock, W.S.; Harding, D.R.K.; Barling, P.M. and Sparrow, J.T., *Int. J. Peptide Protein Res.*, ***1985***, *26*, 105–108.
308. a. Pounds, S.,in *Synthesis and Applications of Isotopically Labeled Compounds*, Vol. 8, Dean, D.C.; Filer, C.N. and McCarthy, K.E., Eds.; John Wiley & Sons, Ltd: Chichester, ***2004***, pp. 63–66; b. Pounds, J.S., *PCT Int. Appl.*, ***2004***, WO2004108636.
309. Pounds, S., Preparation of Carrier-free [methyl-$^{3}$H]methyl nosylate, presentation at the IIS–CED Symposium, Bad Soden, Germany, September 25–26, ***2003***; personal communication.
310. Suehiro, M.; Wang, T.S.; Yatabe, T.; Van Heertum, R.L. and Mann, J.J., *J. Label. Compd. Radiopharm.*, ***1998***, *41*, 725–730.
311. DeKeczer, S.; Kertesz, D. and Parnes, H., *J. Label. Compd. Radiopharm.*, ***1993***, *33*, 219–231.
312. Venkatachalam, T.K.; Mzengeza, S. and Diksic, M., *J. Label. Compd. Radiopharm.*, ***1993***, *33*, 1029–1038.
313. Tadikonda, P.K.; Lacy, J.M.; Rigdon, M.G. and DeLuca, H.F., *J. Label. Compd. Radiopharm.*, ***1997***, *39*, 1–10.
314. Watanabe, H.; Kawase, A.; Okano, K.; Mitsui, T; Ishitani, Y.; Morikawa, K. and Kubodera, N., *J. Label. Compd. Radiopharm.*, ***1999***, *42*, 519–525.
315. Shattuck, J.C. and Meinwald, J., *Tetrahedron Lett.*, ***1997***, *49*, 8461–8464.
316. Ahern, D.G.; Seguin, R.J. and Filer, C.N., *J. Label. Compd. Radiopharm.*, ***2002***, *45*, 401–405.
317. Rubenstein, S.M.; Vazquez, A.; Sanz-Cevera, J.F. and Williams, R.M., *J. Label. Compd. Radiopharm.*, ***2000***, *43*, 481–491.
318. a. Simmons, H.E.; Cairns, T.L.; Vladuchick, S.A. and Hoiness, C.M., *Org. React.*, ***1973***, *20*, 1–49; b. Liu, H.W. and Walsh, C.T., in *The Chemistry of Cyclopropyl Group*, Rappoport, Z., Ed.; John Wiley & Sons, Inc.: New York, ***1987***, pp. 959–1025; c. Carpenter, B.K., in *The Chemistry of Cyclopropyl Group*, Rappoport, Z., Ed.; John Wiley & Sons, Inc.: New York, ***1987***, pp. 1027–1082.
319. a. Atkinson, J.G.; Cillis, D.W.; Gsakvary, J.J.; Morse, A.T. and Stuart, R.S., *Proc. 2$^{nd}$ Int. Conf. on Methods of Preparing and Storing Labeled Compounds, Brussels, 1966*, European Atomic Energy Community Euratom Pub EUR 3746d-f-e, ***1968***, pp. 657–683; b. Evans, E.A., *Tritium and its Compounds*, 2nd Edn; John Wiley & Sons, Inc.: New York, ***1974***, p. 277.
320. Saljoughian, M.; Morimoto, H.; Williams, P.G. and DeMello, N., *J. Chem. Soc. Chem. Commun.*, ***1990***, 1652–1653.
321. Coppo, M.; Rosseau, B. and Beaucourt, J.P., *J. Label. Compd. Radiopharm.*, ***1988***, *25*, 921–925.
322. Gambarotta, S.; Strologo, S.; Floriani, C.; Chiesa-Villa, A. and Guastini, C., *J. Am. Chem. Soc.*, ***1985***, *107*, 6278–6282.
323. For a review, see Tramontini, M.And Angiolini, L., *Tetrahedron*, ***1990***, *46*, 1791–1837.
324. Lawrie, K.W.M.; Heys, J.R.; Garnes, K.; Villani, A.J. and Landvatter, S., *J. Label. Compd. Radiopharm.*, ***2007***, *50*, 530–531.

325. Lane, A.C.; McCoubrey, A. and Peaker, R., *J. Label. Compd*, ***1966***, *2*, 284–288.
326. Hesk, D.; Voronin, K.; McNamara, P.; Royster, P.; Koharski, D.; Hendershot, S.; Saluja, S.; Truong, V. and Chan, T.M., *J. Label. Compd. Radiopharm.*, ***2007***, *50*, 131–137.
327. For a review, see Jutz, C., *Adv. Org. Chem.*, ***1976***, *9*, pt. 1, pp. 225–342.
328. Strazzolini, P.; Giumanini, A.G. and Cauci, S., *Tetrahedron*, ***1990***, *46*, 1081–1118.
329. Montgomery, J.A. and Thomas, H.J., *J. Label. Compd. Radiopharm.*, ***1978***, *15*, 727–730.
330. a. Black, T.H., *Aldrichimica Acta*, ***1983***, *16*, 3–12; b. Pizey, J.S., in *Synthetic Reagents*, Vol. VII, Ellis Horwood Limited: Chichester, ***1974***, pp. 65–142; c. Gutsche, C.D., in *Organic Reactions*, Vol. 8, John Wiley and Sons: New York, ***1954***, pp. 364–429.
331. a. Lindsay, D.G. and Shall, S., *Biochem. J.*, ***1971***, *121*, 737–745; b. de Groot, N.; Lapidot, Y.; Panet, A. and Wolman, Y., *Biochem. Biophys. Res. Commun.*, ***1965***, *25*, 17–22.
332. Saljoughian, M.; Morimoto, H.; Williams, P.G.; Than, C. and Seligman, S.J., *J. Org. Chem.*, ***1996***, *61*, 9625–9628.
333. a. Kas, J. and Rauch, P., Labeled proteins, their preparation and applications, in *Top. Curr. Chem.*, ***1983***, *112*, 163–230; b. Protein Labeling, Technical Bulletin *S92/85*, Amersham International plc, ***1985***, pp. 8–9; c. Tang, Y.S.; Davis, A.M. and Kitcher, J.P., *J. Label. Compd. Radiopharm.*, ***1983***, *20*, 277–284; d. Corringer, P.J.; Durieux, C.; Ruiz-Gayo, M. and Roques, P. B., *J. Label. Compd. Radiopharm.*, ***1992***, *31*, 459–468; e. Begley, D.J. and Chain, D.G., *J. Physiol.*, ***1984***, 8–9; f. Dolly, J.O.; Nockles, E.A.V.; Lo, M.M.S. and Barnard, E.A., *Biochem. J.*, ***1981***, *193*, 919–923; g. Caras, I.W.; Friedlander, E.J. and Bloch, K., *J. Biol. Chem.*, ***1980***, *255*, 3575–3580; h. Müller, G.H., *J. Cell. Sci.*, ***1980***, *43*, 319–320; i. Kummer, U., *Methods Enzymol.*, ***1986***, *121*, 670–679.
334. Zhang, Z.; Li, Z. and Gong, X., *Hejishu*, ***1989***, *12*, 747–749.
335. Shestakov, A.D.; Rumyantsev, S.E.; Fateer, A.N. and Kaminskii, Y.L., *Radiokhimiya*, ***1991***, *33*, 113–117.
336. Tang, Y.S., *J. Label. Compd. Radiopharm.*, ***1983***, *20*, 277–284.

# 5

# Barium [$^{14}$C]Carbonate and the Preparation of Carbon-14-Labeled Compounds via One-Carbon Building Blocks of the [$^{14}$C]Carbon Dioxide Tree

Barium [$^{14}$C]carbonate (Ba$^{14}$CO$_3$) is the universal starting material from which all other carbon-14-labeled compounds are prepared. The carbon-14 nuclide is produced in a $^{14}$N(n, p)$^{14}$C reaction by thermal neutron irradiation of solid beryllium or aluminum nitride, or a saturated solution of ammonium nitrate, in a nuclear reactor for periods ranging from one to three years. Afterwards the target is dissolved in half-concentrated sulfuric acid and the effluent gases are oxidized over an appropriate catalyst. The resulting [$^{14}$C]carbon dioxide is absorbed by an aqueous solution of sodium hydroxide and Ba$^{14}$CO$_3$ is precipitated by addition of aqueous barium hydroxide. Because of the omnipresence of environmental carbon the isotopic purity is normally in the range of 80–90%, corresponding to a specific activity of 50–56 mCi/mmol. Material of higher specific activity approaching the theoretical limit of 62 mCi/mmol is commercially available, but it is considerably more expensive and is only needed in exceptional cases.

Ba$^{14}$CO$_3$ can be regarded as a storage form for [$^{14}$C]carbon dioxide, which can be released on demand by reaction with concentrated sulfuric acid[1] or PbCl$_2$[2]. A family tree of building blocks is derived from $^{14}$CO$_2$ and is the topic of this Chapter and the next. This Chapter covers the one-carbon building blocks of the [$^{14}$C]carbon dioxide tree, and Chapter 6 covers the corresponding building blocks consisting of two to four carbons.

In addition, $^{14}$CO$_2$ is the precursor for additional fundamental one carbon building blocks, each of which is a starting material for its own family of low molecular weight carbon-14-labeled intermediates (Figure 5.1). Its conversion to alkali metal [$^{14}$C]cyanides

*Preparation of Compounds Labeled with Tritium and Carbon-14* Rolf Voges, J. Richard Heys and Thomas Moenius

by direct reduction with potassium azide[3] or via reduction to [$^{14}$C]methane[4] and the cyanide tree are the topics of Chapter 7. Heating of Ba$^{14}$CO$_3$ with barium metal shavings produces Ba$^{14}$C$_2$, which on treatment with water releases [$^{14}$C$_2$]acetylene[5]; it and the intermediates derived from it, including the vast majority of [ring-$^{14}$C]aromatic compounds, are covered in Chapter 8. Finally, reaction of Ba$^{14}$CO$_3$ with gaseous ammonia at 850 °C followed by hydrolysis of the resulting BaN$^{14}$CN provides [$^{14}$C]cyanamide[6], whose chemistry, including routes to many [$^{14}$C]heterocycles, is the subject of Chapter 9.

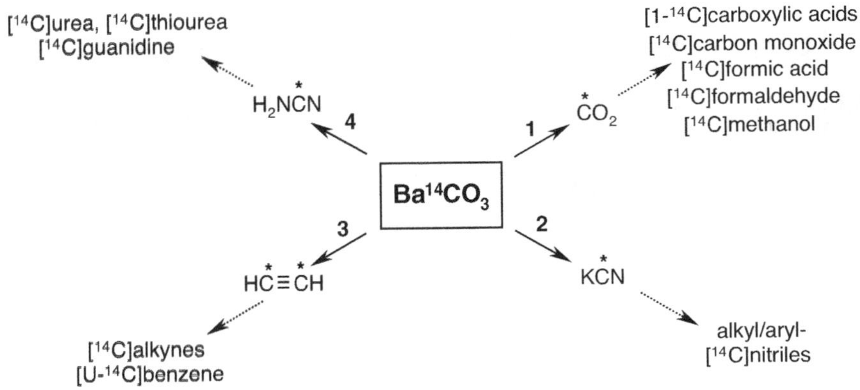


**Reaction conditions: 1.** $H_2SO_4$; r.t.[1] or $PbCl_2$; heat (exothermic fusion)[2]; **2.** $KN_3$, sea sand, 350–700 °C (90–98 %)[3] or **a.** $H_2$, Ni, 350 °C, **b.** $NH_3$, Pt, 1150 °C, **c.** aq. KOH or MeOH (~100 %)[4]; **3a.** Ba, heat to fuse, **b.** $H_2O$[5]; **4a.** $NH_3$ (stream), 850°C, **b.** $H_2SO_4$(70–80%)[6]

***Figure 5.1*** *Barium [$^{14}$C]carbonate and the first generation of one- and two-carbon [$^{14}$C] building blocks*

## 5.1 [$^{14}$C]Carbon Dioxide ($^{14}$CO$_2$)

By far the most common use of $^{14}$CO$_2$ is the carboxylation of organomagnesium or organolithium compounds to give [1-$^{14}$C]carboxylic acids. Another application is in the biological labeling of amino acids, nucleotides, saccharides, lipids or proteins using green algae or growing plants ('isotope farming'). This application is covered in Chapter 12. [$^{14}$C]Carbon dioxide is also the precursor for preparation of low molecular weight labeled building blocks, such as [1-$^{14}$C]acetic acid, [$^{14}$C]formic acid, [$^{14}$C]carbon monoxide, [$^{14}$C] formaldehyde, and [$^{14}$C]methanol, the precursor of [$^{14}$C]methyl iodide.

### 5.1.1 [$^{14}$C]Carboxylations of Organometallic Compounds

The organometallic precursors required for the introduction of $^{14}$CO$_2$ are preparatively accessible through reaction of organic halides with metallic lithium or magnesium, metal–halogen exchange, *ortho*-lithiation, transmetallation or by deprotonation of CH-acidic sites, as described in the sections following.

*Via Reaction of Organic Halides with Metals*

Reaction of $^{14}CO_2$ with Grignard reagents is one of the two most common methods for the preparation of [1-$^{14}$C]carboxylic acids, the other being via metal–halogen exchange (Section 5.1.1). Alkyl and alkenyl Grignard reagents are easily prepared from alkyl/alkenyl halides and magnesium turnings, which is by far the most common method for the preparation of aliphatic [1-$^{14}$C]carboxylic acids [7]. Primary, secondary and even tertiary halides[7f] can be used as starting materials. The order of halide activity towards magnesium is I > Br > Cl; fluorides normally do not react. Bromides are preferred, since with iodides Wurtz-coupling is a more severe side reaction. Grignard formation is strongly affected by the presence of other functional groups that can interact, such as C=O, C≡N, $NO_2$ and COOR. In these cases cyanodehalogenation with $K^{14}CN$ is usually the alternative of choice (see Chapter 7, Section 7.1.2). Sometimes protecting group chemistry has been exploited, as illustrated by the synthesis of 5-(triphenylphosphoranylidene)[5-$^{14}$C]pentanoic acid (2), which served as key intermediate for the carbon-14 labeling of prostaglandins [8]. The route proceeded via the 2,8,9-trioxaadamantyl derivative of 4-bromobutyric acid (1). This protection of the carboxyl group allowed the selective introduction of a second, labeled, carboxyl group and its subsequent transformation through reduction and bromo-dehydroxylation.

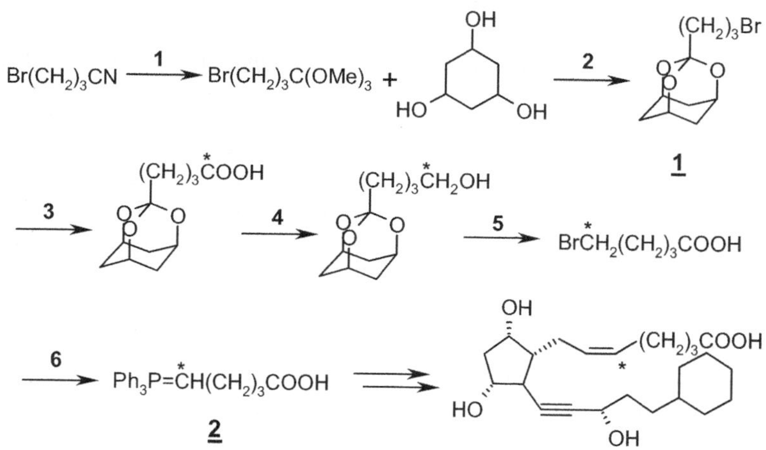


**Reaction conditions**: **1**. $HCl_{gas}$, MeOH; r.t.; **2**. $BF_3 \cdot OEt_2$, MeOH; **3a**. Mg, THF; 45°C, 1 h, **b**.$^{14}CO_2$; −78 °C, 1.5 h; r.t., 1 h; **4**. $LiAlH_4$, ether; r.t., 2 h; **5**. 48 % HBr/conc. $H_2SO_4$ 98:2; 130 °C, 2.5 h; **6a**. $Ph_3P$, no solvent; 135–140 °C, 3 h, **b**. KH,15 min THF; r.t.,

***Figure 5.2*** *Use of a protecting group in the [$^{14}$C]carboxylation of a substituted Grignard reagent*

Double bonds, internal triple bonds, ether and tertiary amino groups are tolerated, with the exception of ether groups $\beta$ to the halide, which undergo elimination upon reaction with magnesium. Grignard compounds from $\alpha$-haloethers (e.g., 3) are stable, at least in solution. They react with $^{14}CO_2$ to produce (after deprotection) [1-$^{14}$C]glycolates **4**; if there is excess Grignard reagent relative to $^{14}CO_2$, at room temperature it can react with the initial carboxylation product to produce bis-adducts, such as **5**, a dihydroxy[2-$^{14}$C]-acetone derivative. This was exploited for the preparation of [2-$^{14}$C]glycerol (**6**) through reduction of the carbonyl group of **5** and removal of the protecting groups[9]. The formation of symmetrical [$^{14}$C]ketones tends to occur more readily with organolithium compounds but, as with Grignard reagents, this reaction can be suppressed by maintaining low reaction temperatures and avoiding large excesses of organometallic reagent.

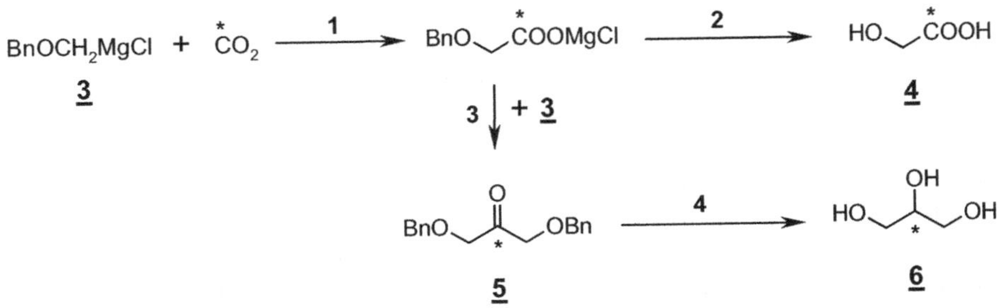


**Reaction conditions: 1**. THF/$Et_2O$; −20 °C, 1 h; **2a**. $H_3O^+$, **b**. $H_2$, Pd/C, EtOH/r.t.; **3**. **3**; r.t., 20h; EtOH; **4a**. $LiAlH_4$, $Et_2O$, **b**. $H_2$, Pd/C, r.t., 2h

***Figure 5.3*** *Mono- and bis-adducts of α-alkoxy Grignard reagents with* $^{14}CO_2$

Di-Grignard reagents are accessible, if the halogens are at least three carbon atoms apart. With 1,3, 1,4 and 1,5-dimagnesio compounds cyclic ketones can be obtained, unless conformational or entropic factors disfavor the intramolecular cycloaddition[10a,b]. $BrMg(CH_2)_5MgBr$, for example, reacts with $^{14}CO_2$ to give [1-$^{14}$C]cyclohexanone, which serves as a valuable intermediate for [$^{14}C_1$]benzene and alkyl[1-$^{14}$C]benzenes such as [1-$^{14}$C]toluene[10a–c]. Higher homologues give the corresponding di[1, n-$^{14}C_2$]carboxylic acids as exploited for the synthesis of [1, 8-$^{14}C_2$]suberic acid by [$^{14}$C]carboxylation of the di-Grignard reagent prepared from $Br(CH_2)_6Br$[11].

Aryl and heteroaryl bromides and iodides react analogously[7a,12]. The formation of arylmagnesium chlorides, however, requires more drastic reaction conditions such as higher reaction temperatures and reaction times for completion. This allows discrimination between Cl- and Br- or I-substituents, as demonstrated by the syntheses of 2-chloro- and 2,6-dichlorobenzoic [$^{14}$C]acids from 2-bromochlorobenzene[12d,e] and 2,6-dichloroiodo-benzene, respectively[12f].

*α,β*-Unsaturated acids such [1-$^{14}$C]acrylic acid and [1-$^{14}$C]methacrylic acid, which result from carboxylation of the respective alkenylmagnesium bromide[7c–e], have a great tendency to undergo radiation-induced polymerization. Although it has been reported that [1-$^{14}$C]acrylic acid at maximum specific activity can be successfully prepared, the raw product had to be used immediately in a further reaction[13]. The feasibility of isolating and purifying [1-$^{14}$C]acrylic acid (through fractional distillation in the presence of hydroquinone as a radical scavenger) at specific activities of 5–13 mCi/mmol has been reported[7c,d], as has that for [1-$^{14}$C]methacrylic acid at a specific activity of 3 mCi/mmol[7e,14].

Water soluble and volatile [1-$^{14}$C]carboxylic acids of up to seven carbon atoms are routinely isolated as their sodium or potassium salts, from which they can be liberated through treatment with dry HCl gas and subsequent static vacuum transfer or fractional distillation. Acids with a chain length of more than seven carbon atoms can be extracted from the acidified reaction mixture.

### *Via Metal–Halogen Exchange*

Metal–halogen exchange on aryl and heteroaryl halides using *n*-butyllithium in ether or THF is the preferred method for preparation of sp$^2$-organometallic species suitable for [$^{14}$C]carboxylations[12e,15]. If Wurtz-type couplings of butyl bromide with aryllithiums–the initial products of these metal–halogen exchanges–become problematic side reactions, they can be suppressed by using two equivalents of *tert*-butyllithium instead of one equivalent of *n*-butyllithium. The second equivalent of *tert*-butyllithium rapidly reacts with the *tert*-butyl halide formed in the initial metal–halogen exchange to give lithium halide and isobutene[16]. In general, metal–halogen exchange works well with bromo and iodo substrates but is less satisfactory with chloro derivatives. This allows discrimination between bromo and chloro substituents in polyhalogenated aromatic compounds, as illustrated by the examples **7**,[17] **8**[18] and **9**[18]. Fluoro substituents are unreactive, so their presence is immaterial. Iodo substituents, however, are highly reactive, so that it is not feasible to prepare iodoaryl

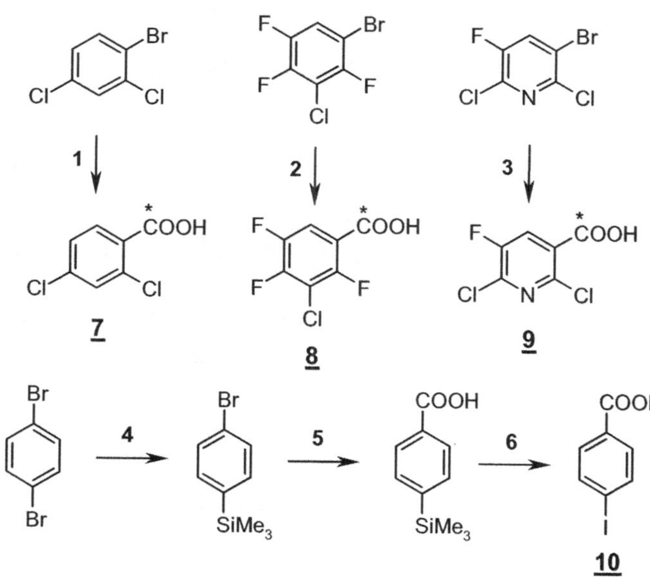

**Reaction conditions: 1a**. *n*-BuLi, ether; −78 °C, 20 min, **b**. $^{14}CO_2$;−78 °C, min; **2a**. *n*-BuLi, THF; −92 °C, 45 min, **b**. $^{14}CO_2$;−78 °C, 16 h; **3a**. *n*-BuLi,ether;−78 °C, 10 min, **b**. $^{14}CO_2$;−78 °C, 45 min; **4a**. *n*-BuLi, THF; −78 °C,105 min, **b**. $Me_3SiCl$; −78°C to r.t.; **5a**. *n*-BuLi, THF; −78 °C, 40 min, **b**. $CO_2$; −78 °C, 15 min then to r.t.; **6**. ICl, $CCl_4$; r.t., 30 min then reflux, 1 h

**Figure 5.4** *Selective lithium–halogen exchange on multisubstituted aryl halides*

carboxylic [$^{14}$C]acids by this process. A remedy for such cases may be an alternative method which makes use of the ICl-mediated replacement of silyl groups by iodine. By this method, the metal–halogen exchange is conducted on a substrate possessing a silyl substituent at the position where iodo is desired in the final product. After the [$^{14}$C]carboxylation step, the iodo-desilylation is conducted as illustrated in a cold model synthesis for *p*-iodobenzoic [$^{14}$C]acid (**10**)[19].

Metal–halogen exchange reactions often proceed very rapidly even at temperatures as low as −100 °C, so that various critical functional groups such as $COO^-$, CN, NHCHO and provisionally COOR may be tolerated. According to recent reports[20], the range of compatible functional groups may be extended to amide, *ortho*-nitro and, more safely, ester substituents by employing aryl iodides as precursors and PhMgCl or *i*-PrMgBr instead of *n*-BuLi as metallation reagents. *Meta-* and *para*-substituted iodonitroarenes lead to nonselective reactions with addition of *i*-PrMgBr to the nitro group. This new method has also been reported to allow discrimination between $C_{Ar}$—I and $C_{Ar}$—Br bonds.

*Via Directed ortho-Metallation (DoM)*

*Ortho*-lithiation by a strong lithium base (*sec*-BuLi, *tert*-BuLi, *sec*-BuLi-TMEDA or *tert*-BuLi-TMEDA), guided by an substituent with a sufficient coordinative affinity for

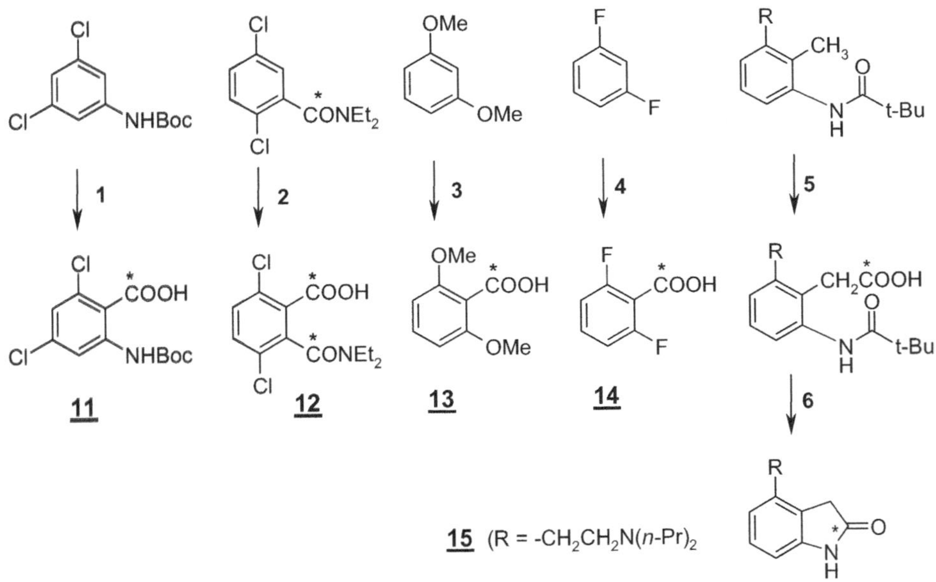


**Reaction conditions:** **1a**. *tert*-BuLi, TMEDA, THF; −95 °C, 2 h, **b**.$^{14}CO_2$; −75 °C to r.t., 5 h (25 %); **2a**. *s*-BuLi, TMEDA, ether; −70 °C, 30 min, **b**.$^{14}CO_2$; −78 °C to r.t., 16 h (73 %); **3a**.*n*-BuLi, THF; −30 °C, 4.5 h, **b**.$^{14}CO_2$; −70 °C, 90 min (56 %); **4**.*n*-BuLi, THF; −78 °C, 3.5 h, **b**.$^{14}CO_2$;−78 °C, 1 h; **5a**. 2 equiv.*n*-BuLi, THF; −78 °C, **b**.$^{14}CO_2$; 0 °C, 1 h; **6**. aq. HCl; reflux 41 h (40 % from $Ba^{14}CO_3$)

**Figure 5.5** *Regioselective metallation of benzenes by ortho-directing substituents*

lithium, has become the established method for the preparation of *o*-substituted aromatic and heteroaromatic [$^{14}$C]carboxylic acids. The *ortho*-directing power of the guiding substituents is in the approximate order $OCONEt_2$ > $CONR_2$ > 3,3-dimethyloxazolinyl > $SO_2NEt_2$ > OMEM > OMe > NHBoc > $NR_2$[21]. The order varies significantly with the nature of any other *ortho*-directing substituents present, as well as with the metallation conditions. Some of the potential of [$^{14}$C]carboxylations via directed *ortho*-metallation is exemplified for compounds **11**[22], **12**[23], **13**[24], **14**[25] and **15**[26]. The transformation to give **14** shows that *ortho*-lithiation of arenes can also be achieved if the resulting aryllithium intermediate is stabilized by adjacent fluoro (or chloro) substituents[22]. Bromo and iodo substituents, although they can coordinate more strongly with lithium ion, are of less practical use because they undergo metal–halogen exchange. In addition, alkyl groups *o*-located to D*o*M-groups, as in the case of **15**[26], can also be metallated and undergo [$^{14}$C]carboxylation.

### *Via Transmetallation*

Transmetallation procedures are generally applied only when the organolithium compounds needed for the [$^{14}$C]carboxylation reaction are not easily accessible. Accordingly, only two examples have been reported so far in carbon-14 synthesis, which, however, show broader synthetic potential. They are the preparation of *O*-protected [1-$^{14}$C]glycolic acid **16** and of *N*-protected *N*-methyl[1-$^{14}$C]glycine **17**. In these cases, organotin intermediates were prepared by reaction of tributyltin hydride with benzyl chloromethyl ether[27] and of tributyl(iodomethyl)tin with deprotonated *N*-Boc-*N*-methylamine[28], respectively. Sn/Li-replacement with *n*-BuLi or MeLi and trapping of the organolithium intermediate with $^{14}CO_2$ provided **16** and **17** in radiochemical yields of about 80%. Compound **16** was used as a starting material for the stereoselective synthesis of *O*-protected (*R*)-[2-$^{14}$C]glycidol (see also Chapter 11, Section 11.2.1). The broader synthetic potential of the Sn/Li transmetallation derives from the fact that a variety of organotin intermediates are conveniently accessible[29] by treatment of aldehydes with $Bu_3SnLi$ and subsequent protection of the hydroxyl group of the resulting $\alpha$-hydroxystannane.

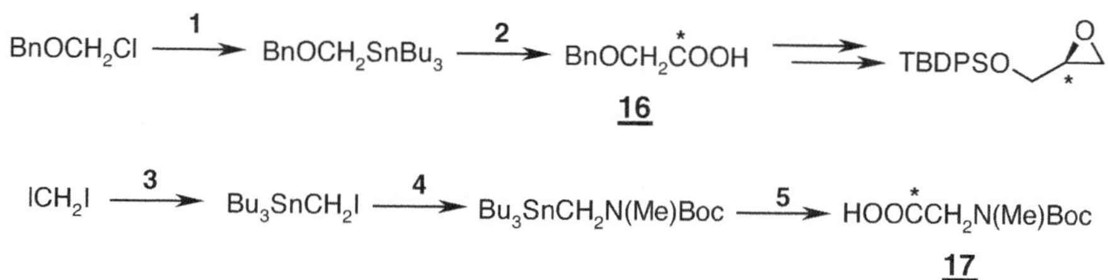


**Reaction conditions: 1a.** $Bu_3SnH$, LDA, THF; 0 °C, 15 min, **b.** $BnOCH_2Cl$, −78 °C, 45 min; **2a.** *n*-BuLi, THF; −78 °C; 10 min, **b.** $^{14}CO_2$,−78 °C,15 min; −75 °C to r.t., 1 h; **3a.** Zn, $Cu(OAc)_2$(cat.); THF; 40°C, **b.** $Bu_3SnCl$; 40 °C, 10 h **4a.** *N*-Boc-*N*-methamine, NaH, DMF; 65 °C, 2 h, **b.** $Bu_3SnCH_2I$, 60 °C, 1 h; **5a.** MeLi, THF, − 95 °C, 15 min, **b.** $^{14}CO_2$,−78 °C, 20 min

***Figure 5.6*** *Application of transmetallation reactions in [14C]carboxylations*

*Via Deprotonation of C–H Acidic Substrates*

Various [1-$^{14}$C]carboxylic acids have been prepared by carboxylation of carbanions stabilized by –I or –M substituents. The examples in Figure 5.7 have been selected here as prototypes, since they are of strategic interest. Deprotonation of diethyl methylphosphonate (**18**) with *n*-BuLi followed by [$^{14}$C]carboxylation and esterification with diazoethane provided triethyl phosphono[1-$^{14}$C]acetate (**19**) in an overall radiochemical yield of 62%[30]. Compound **19** has been widely exploited for chain extensions by a labeled two-carbon unit via its alkenylation reaction with carbonyl compounds (Horner–Wadsworth–Emmons reaction). Similarly, reaction of alkyl halides, tosylates or carbonyl compounds with LiC≡CH or LiC≡CH·$H_2NCH_2CH_2NH_2$ followed by deprotonation and [$^{14}$C]carboxylation of the resulting terminal alkynes has been used as a strategic tool for the incorporation of a labeled three-carbon unit, as exemplified in a steroid platform (**20** to **21**)[31]. This chemistry provides outcomes complementary to those using [$^{14}C_2$]acetylene (Chapter 8, Section 5.1). Finally, the [$^{14}$C]carboxylation of lithiated dimethylsulfide provided an alkylthio[1-$^{14}$C]acetic acid **22** and thence a functionalized 2-alkylthio[1-$^{14}$C]ethyl derivative **23** useful, in this case, for elaboration into e.p. [3-$^{14}$C]methionine[32].

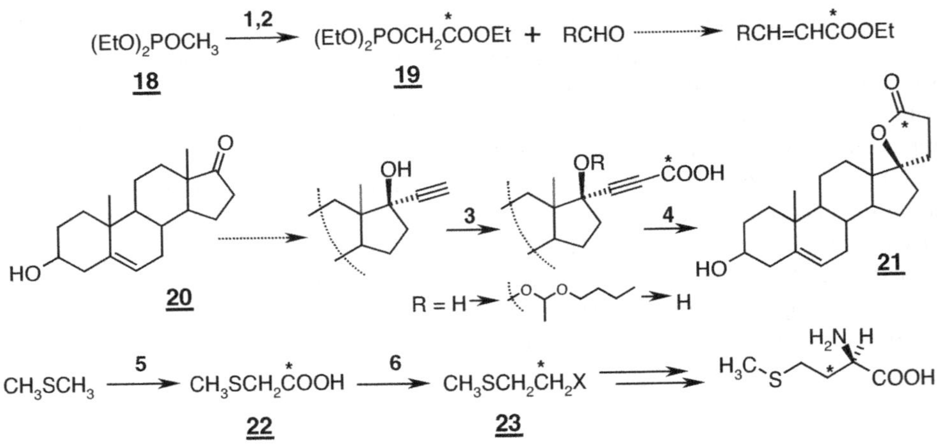


**Reaction conditions:** **1a**. *n*-BuLi, THF; −78 °C, 30 min, **b**. $^{14}CO_2$;−70 °C, 2.5 h, -70°C to r.t.,1 h; **2**. $CH_3CHN_2$, $CH_2Cl_2$/ether, r.t., 16 h; **3a**. protect, **b**. MeMgCl, THF; reflux 1.5 h, **c**. $^{14}CO_2$,−78 °C, 20 min; 0 °C, 90 min, **d**. deprotect; **4a**. $H_2$, 5 % Pd/$CaCO_3$, MeOH; r.t., 15 min, **b**. 8N aq. HCl, r.t. to reflux; **5a**. *n*-BuLi, TMEDA, hexane; r.t., **b**. $^{14}CO_2$, THF; −80 °C to r.t.; **6a**. $LiAlH_4$, ether; reflux 3 h, **b**. $(C_8H_{17})_3P$, $CCl_4$; r.t.

**Figure 5.7** *Syntheses beginning with [$^{14}$C]carboxylation of C–H acidic compounds*

## 5.1.2 Manipulations of [$^{14}$C]Carboxylation Products

Aside from standard radiation safety procedures, the methods used for the transformation of [1-$^{14}$C]carboxylic acids containing more than seven carbon atoms correspond to those applied in normal organic synthetic chemistry on the same scale. But for those with fewer than four carbon atoms, the use of special techniques can facilitate safe transfers and high-yield transformations. For example, [1-$^{14}$C]carboxylic acids from acetic to pentanoic, or their [1-$^{14}$C]acyl chlorides, are preferably reduced at 60–80 °C with $LiAlH_4$ in an inert high boiling solvent (such as diethylene glycol diethyl ether or 2(3)-(tetrahydrofurfuryloxy)pyran). The reaction is then quenched by addition of 2-phenoxyethanol, and the liberated $C_2$–$C_5$ [1-$^{14}$C]alcohol is expelled into a cooled receiver using a stream of helium[33]. Reduction of $C_5$–$C_7$ acids is performed either with $LiAlH_4$ in refluxing ether or with $BH_3$ in THF, the reaction quenched with dilute mineral acid and the [1-$^{14}$C]alcohol product isolated by (continuous) extraction with ether followed by fractional distillation. [1-$^{14}$C]Acyl chlorides of up to four carbon atoms can be prepared by reaction of the free acid with phthaloyl dichloride or benzoyl chloride at temperatures ranging from 80–100 °C and isolated by vacuum distillation[34]. In an alternative procedure, which is recommended for small quantities, the conversion can be accomplished by treatment of the initially formed crude magnesium [1-$^{14}$C]carboxylate with excess oxalyl chloride. After completion of the reaction, excess reagent is destroyed with potassium stearate and the volatile [1-$^{14}$C]acyl chloride distilled off[35]. Acids of five to seven carbons are converted with oxalyl chloride or phosphorous pentachloride and isolated by fractional distillation. Ethyl esters of $C_1$–$C_4$ acids are obtained through heating of their sodium salts with triethyl phosphate. The corresponding methyl esters are conveniently accessible upon reaction with diazomethane. [1-$^{14}$C]Acetic anhydride, which is used as a labeling reagent in peptide and protein chemistry, is formed by the action of *p*-toluenesulfonyl chloride on sodium [1-$^{14}$C]acetate at 200 °C.

### 5.1.3 *N*-[$^{14}$C]Acyl Building Blocks

Curtius degradation of [1-$^{14}$C]acyl azides and [$^{14}$C]carbonylation of aliphatic and aromatic primary amines have become established methods for the synthesis of [$^{14}$C]isocyanates, key precursors for the preparation of labeled ureas, carbamates and thiocarbamates (see also Chapter 7, Section 7.4). Phosgene, classically the most common reagent for preparing such compounds in normal organic synthesis, is less attractive for use in carbon-14-labeled form because it is difficult to prepare in the laboratory (see Section 5.2), it is costly to purchase,

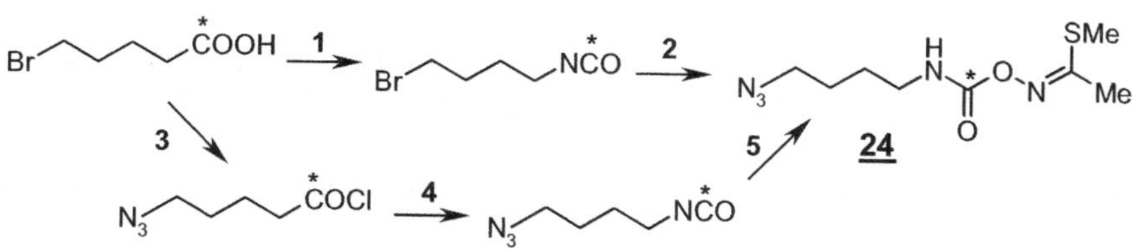


**Reaction conditions:** **1**. $(PhO)_2P(O)N_3$, benzene, r.t., 6 h; reflux 1 h; **2a**. $CH_3(SCH_3)$=NOH, $Et_3N$, $CH_2Cl_2$, r.t., 24 h, **b**. $NaN_3$, EtOH–$H_2O$ 9:1; reflux 6 h; **3a**. TBDMSCl, imidazole, DMF; r.t., 18 h, **b**. $NaN_3$, DMF; 70 °C, 3 h, **c**. oxalyl chloride, $CH_2Cl_2$; 0 °C, 18 h; **4**. $Me_3SiN_3$, benzene; r.t., 2 h; reflux, 30 min; **5**. $CH_3(SCH_3)$=NOH, $Et_3N$, $CH_2Cl_2$, r.t., 24 h

***Figure 5.8*** *Synthesis of [carbonyl-$^{14}$C]isocyanates from [1-$^{14}$C]carboxylic acids*

it has limited radiolytic stability, and finally, because it must often be employed in large excess in order to minimize the formation of symmetrical urea byproducts through reaction with two amine equivalents. The Curtius degradation procedure follows the routine protocol for the thermolysis of [1-$^{14}$C]acyl azides, which are routinely available from the reaction of [1-$^{14}$C]carboxylic acids and diphenylphosphoryl azide or, in a two step sequence, from [1-$^{14}$C]acyl halides and sodium or trimethylsilyl azide. Each of these procedures is exemplified by the labeling of the insecticide methomyl **24**[36]. This very general method can be applied to almost all carboxylic acids including $\alpha,\beta$-unsaturated analogs.

The second approach makes use of the [$^{14}$C]carbonylation of primary and secondary amines in the presence of a second equivalent of the amine component[37] or, preferably, of an appropriate ternary base such as triethylamine or a combination of pyridine and CyTMG (*N*-cyclohexyl-*N'*, *N'*, *N''*, *N''*-tetramethyl guanidine)[38]. The resulting ammonium or guanidinium [$^{14}$C]carbamates **25** can either be directly converted into [$^{14}$C]isocyanates **26** (in the case of primary amines) or [$^{14}$C]carbamoyl chlorides **27** (secondary amines) upon treatment with $POCl_3$, $PCl_5$ or $SOCl_2$. Alternatively, they can be trapped with trimethylsilyl chloride to give trimethylsilyl [$^{14}$C]carbamates **28**, which then can be converted to **26** or **27** under similar reaction conditions. For both alternatives, yields are strongly dependent on the ability of the amine to trap $^{14}CO_2$, as well as on the rate of the side reaction of **26** with as-yet unreacted amine to produce symmetrical [$^{14}$C]ureas. Satisfactory yields were achieved with aliphatic amines and unsubstituted aniline, as well as with anilines substituted by +M substitutents. Anilines containing −M substituents in the *para*-position (e.g., −COOMe, $-NO_2$) were unreactive or gave low yields of the respective [$^{14}$C]isocyanates. As far as the base is concerned, optimal results were obtained when CyTMG was employed.

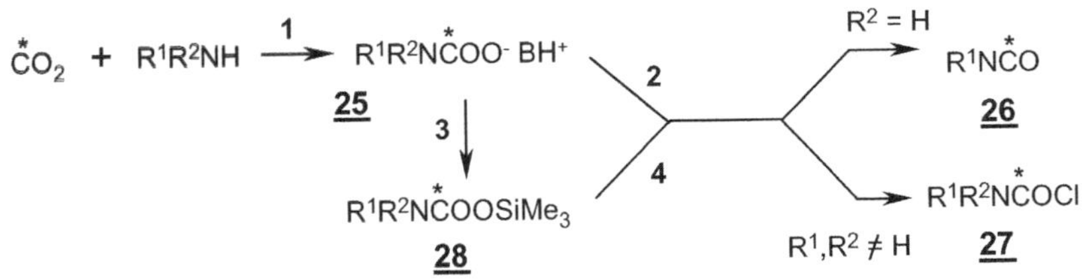


**Reaction conditions: 1**. $ET_3N$, $CH_2Cl_2$ or py, CyTMG, toluene; −20 °C or −10°C,20–30 min; **2**. $POCl_3$, $CH_2Cl_2$ or $SOCl_2$, toluene; −20 °C or −10 °C, 20 min

***Figure 5.9*** *[$^{14}$C]Carbonylation of primary and secondary amines: an alternative route to carbon-14-labeled isocyanates, ureas, and thiocarbamates*

The utility of this approach is exemplified by the synthesis of [carbonyl-$^{14}$C]L-740,093 (**29**)[38], a urea type CCKB antagonist, and of [carbonyl-$^{14}$C]EPTC **30**[37], a thiocarbamate type herbicide. Furthermore the synthesis of **31** demonstrates that it can also be exploited for the preparation of cyclic [$^{14}$C]ureas[38b].

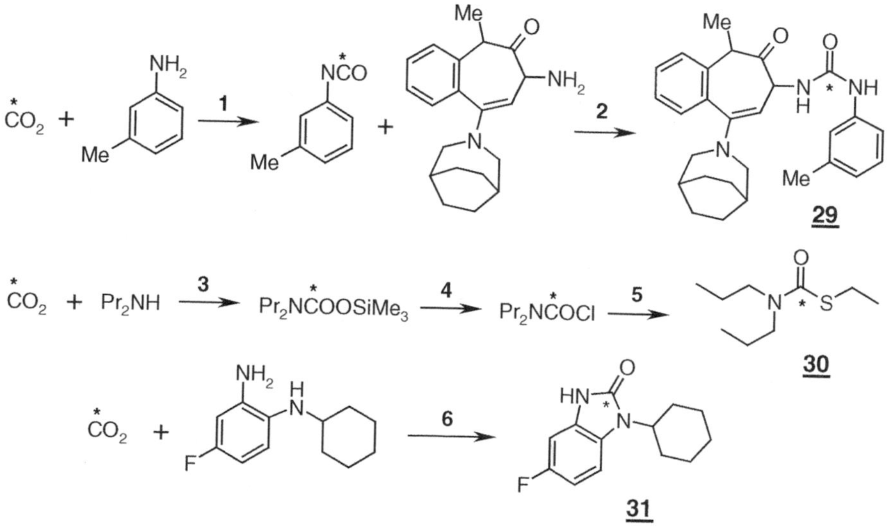


**Reaction conditions:** **1a**. $Et_3N$, $CH_2Cl_2$;−78 °C, 15 min, −20 °C, 30 min, **b**. $POCl_3$, $CH_2Cl_2$;−78 °C, 10 min; −20 °C, 1 h; **2**. $CH_2Cl_2$; r.t.; **3a**. $CH_2Cl_2$;−78 °C to 40 °C, **b**. $Me_3SiCl$, $CHCl_3$;−40 °C, 30 min; **4a**. $PCl_5$, $CHCl_3$; 0 °C, 30 min; r.t., 1 h, **b**. $Pb(SEt)_2$, chlorobenzene; 140–150 °C, 3 h; **6a**. $Et_3N$ $CH_2Cl_2$; −78 °C, 15 min; −20 °C, 30 min, **b**. $POCl_3$, $CH_2Cl_2$;−78 °C, 10 min; −20 °C, 1 h

***Figure 5.10*** *Applications of $^{14}CO_2$ in the synthesis of [$^{14}$C]urea and [$^{14}$C]thiocarbamate derivatives*

### 5.1.4 Preparation of Other Building Blocks from [$^{14}$C]Carbon Dioxide

The [$^{14}$C]carboxylation of MeMgI to [1-$^{14}$C]acetate and the reduction of $^{14}CO_2$ to [$^{14}$C]-methanol, [$^{14}$C]formic acid and [$^{14}$C]formaldehyde play crucial roles in the synthesis of a broad variety of building blocks essential for the construction of labeled complex target molecules (Figure 5.11). The preparation of [1-$^{14}$C]acetic acid corresponds to the general procedure for aliphatic [1-$^{14}$C]carboxylic acids[39] discussed earlier in this chapter. Carbon-14-labeled acetic acid serves as a central intermediate for further building blocks such as labeled haloacetic acid (ester), cyanoacetic acid (esters), $\beta$-keto acid (esters), malonic acid (esters), and acetone. These multicarbon building blocks are discussed in Chapter 6. Two procedures are available for the preparation of alkali metal [$^{14}$C]formates, namely reduction of $^{14}CO_2$ with lithium triethylborohydride[40] or with sodium trimethoxyborohydride[41]. A third method, which makes use of the hydrolysis of potassium [$^{14}$C]cyanide, is discussed separately (see Chapter 7). [$^{14}$C]Carbon monoxide, used classically for preparing [$^{14}$C]phosgene, has more recently found good use in palladium-catalyzed carbonylation reactions. It is available by high-temperature reduction of $^{14}CO_2$ or $Ca^{14}CO_3$ with zinc, which both require specialized equipment[42], or by a procedure more suitable for radiosynthesis laboratories, involving dehydration of [$^{14}$C]formic acid prepared *in situ* from $^{14}CO_2$[40,43]. Three independent methods have been reported for the production of [$^{14}$C]formaldehyde: reduction of $^{14}CO_2$ with equimolar amounts of $LiAlH_4$ at low

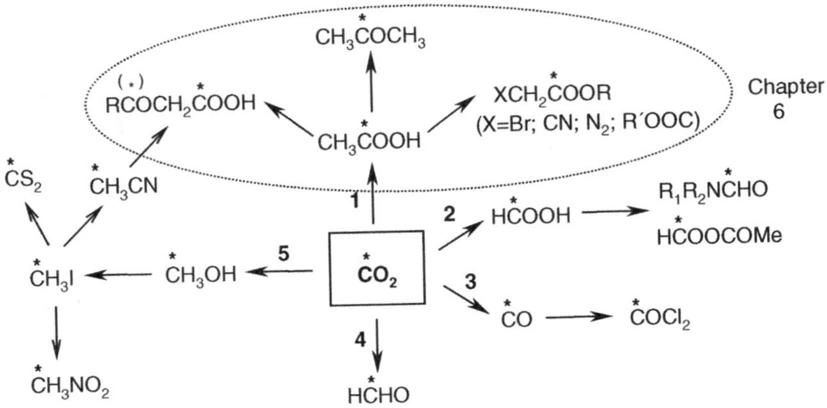


**Reaction conditions: 1a**. MeMgI, ether; −170 °C to −20 °C; −20 °C, 30 min, **b.** 20 % NaOH, **c.** HCl$_{gas}$; **2.** LiEt$_3$BH, THF; −170 °C to 0 °C; 0 °C, 2 h; 20 °C, 2 h, **or** NaBH(OMe)$_3$, THF; r.t., 7 d, **or** K$_{14}$CN, H$_2$O; 150 °C, 4 h; **3.** Zn (vacuum manifold); 385 °C; **4.** LiAlH$_4$, THF; 0 °C; 15 min; **5a.** LiAlH$_4$, diethylene glycol diethyl ether, 0 °C; 60 min, **b.** BnOCH$_2$CH$_2$OH; 0 °C, **c.** He stream, 110 °C; 2–4 h

**Figure 5.11** *Conversion of $^{14}CO_2$ into additional [$^{14}$C]building blocks*

temperature[44], $Fe_2O_3$·$MoO_3$-catalyzed oxidation of [$^{14}$C]methanol with dry air[45] and $LiAlH_4$ reduction of *N*-methyl[carbonyl-$^{14}$C]formanilide[44c]. [$^{14}$C]Methanol is routinely prepared by reduction of $^{14}CO_2$ with $LiAlH_4$ in diethylene glycol dimethyl ether[33b,d,46]. Despite being only rarely used for the labeling of complex molecules, it is very important as a precursor of [$^{14}$C]methyl iodide, one of the key building blocks in carbon-14 synthesis.

## 5.2 [$^{14}$C]Carbon Monoxide ($^{14}$CO)

Three routes for the preparation of $^{14}$CO have been described. Heating a mixture of Ca$^{14}$CO$_3$ and zinc dust to 700–750 °C is reported to give CaO and ZnO, and $^{14}$CO as the only volatile product in essentially quantitative yield[42d,47]. In a chemically related method $^{14}CO_2$ was circulated over zinc dust maintained at 385 °C[42a,48]. This method is less convenient because the progress of the reaction must be assessed, and any residual unreacted $^{14}CO_2$ removed before the $^{14}$CO is used; it therefore requires a more complicated apparatus than the first method. The third and most laboratory-friendly method involves dehydration of [$^{14}$C]formic acid by treatment with concentrated sulfuric acid at 70 °C[40,49].

[$^{14}$C]Carbon monoxide has not been used as widely in carbon-14 synthesis as its carbon-11 and unlabeled counterparts in short-lived isotope and conventional synthetic chemistry. Nonlabeled CO has been extensively exploited for transition metal-catalyzed carbonylations of aryl, allyl, alkenyl halides or enol triflates and more powerful multicomponent amino- and alkoxycarbonylations, as well as for the hydroformylation of alkenes. The infrequent use of $^{14}$CO is probably due to two common perceptions, that it is difficult to prepare and handle, that vast excesses and/or enhanced pressures are needed for efficient reactions. The first was true in the past, but convenient experimental procedures have recently been developed that utilize simple glassware setups[40,49]. The second is true only for some reactions. For example, the [$^{14}$C]- and [$^{13}$C]carbonylations of organoboranes in the presence of $K(O\textit{i}\text{-}Pr)_3BH$, giving [$^{13/14}$C]aldehydes and [$^{13/14}$C]acids (upon oxidation with $Ag_2O$), works well in excellent yields at 1 atm (see Figure 5.12)[50]. The starting materials were readily accessible through hydroboration of the respective olefinic substrates. Unfortunately, reported yields were based on the organoboranes employed and no data on the amount of $^{13/14}CO$ used were published.

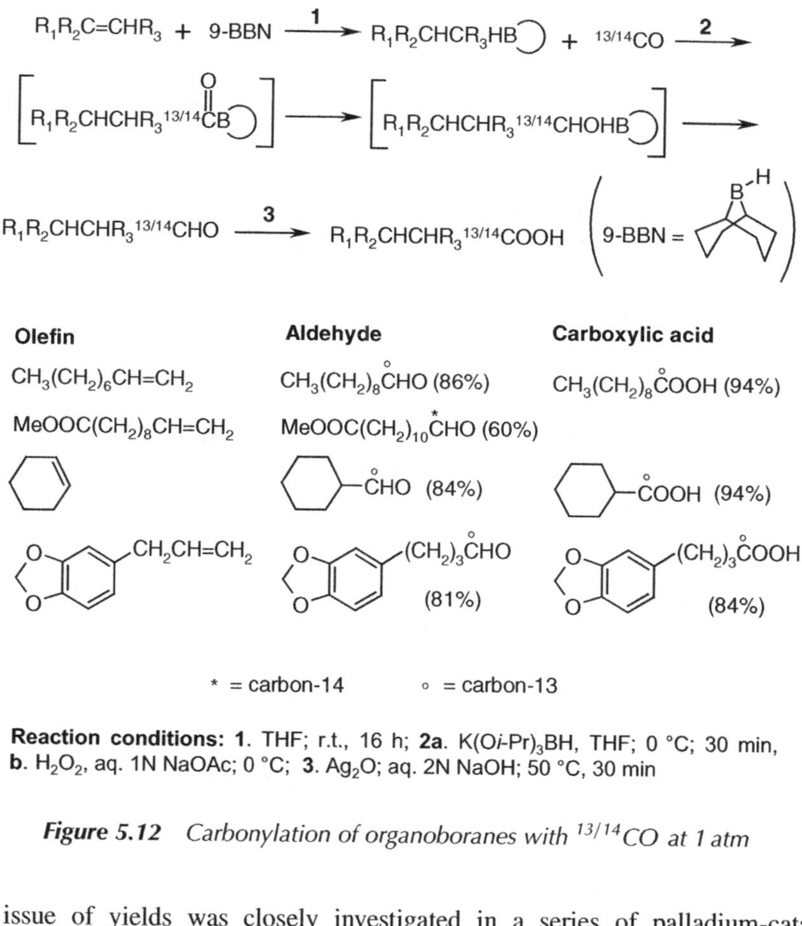


**Reaction conditions: 1**. THF; r.t., 16 h; **2a**. $K(O\textit{i}\text{-}Pr)_3BH$, THF; 0 °C; 30 min, **b**. $H_2O_2$, aq. 1N NaOAc; 0 °C; **3**. $Ag_2O$; aq. 2N NaOH; 50 °C, 30 min

***Figure 5.12*** *Carbonylation of organoboranes with $^{13/14}CO$ at 1 atm*

The issue of yields was closely investigated in a series of palladium-catalyzed carbonylations of aryl/heteroaryl halides and triflates with [$^{14}$C]carbon monoxide, prepared through treatment of well-defined quantities of lithium [$^{14}$C]formate with concentrated sulfuric acid. The experiments were carried out in a dual-chamber reaction vessel, one section for the generation of $^{14}CO$, the other for the carbonylation reaction. This arrangement eliminated the need to manipulate the $^{14}CO$ released but instead relied on its being drawn from the first chamber into the second as it was consumed. The reaction sequence was typically performed with two equivalents of $Ba^{14}CO_3$, which corresponded to 1.6 equivalents of $^{14}CO$, since the $LiEt_3BH$ reduction was shown to proceed with about 80% yield[40]. As illustrated in Figure 5.13, yields of isolated aryl/heteroaryl [$^{14}C$]carboxylic acids ranged from 41% for 1-methyl-3-indole[$^{14}C$]carboxylic acid to 72–78% for benzoic [$^{14}C$]acid (based on the aryl/heteroaryl halide), which corresponds to about 25–50% to $^{14}CO$ consumed. It was also found that there was little difference in yields whether $^{14}CO$ was generated from dry lithium [$^{14}C$]formate or from the wet crude material obtained directly from the reduction of $^{14}CO_2$ with $LiEt_3BH$ followed by aqueous quench.

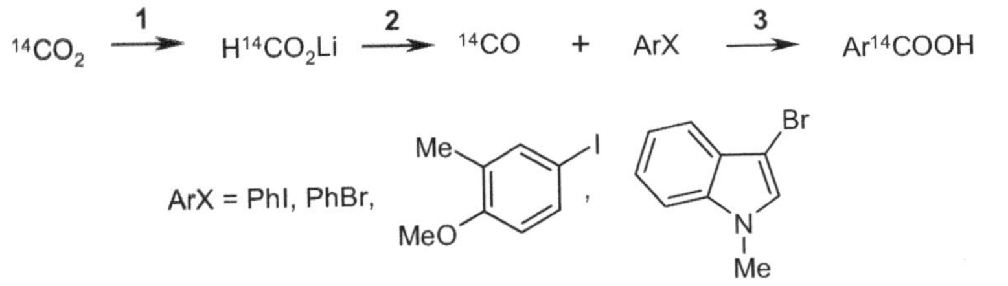


**Reaction conditions:1**. $LiEt_3BH$, THF; 0 °C, 2 h; r.t., 2 h; then $H_2O$; **2**.conc. $H_2SO_4$; 70 °C, 1 h; **3**. $Pd(dppf)_2Cl_2$ or $Pd(Ph_3)_4$, KOAc, DMSO or DMF; 70 °C, 16 h

***Figure 5.13*** *Synthesis and use of $^{14}CO$ in the transition metal-catalyzed carbonylation of aryl/heteroaryl halides*

Trapping of the [$^{14}C$]acylpalladium complexes intermediately generated by addition of MeOH, $Bu_3SnH$ or $PhSnMe_3$ allowed for the formation of aryl[$^{14}C$]carboxylic acid methyl esters, [carbonyl-$^{14}C$]aldehydes and [carbonyl-$^{14}C$]benzophenone derivatives as depicted in Figure 5.14.

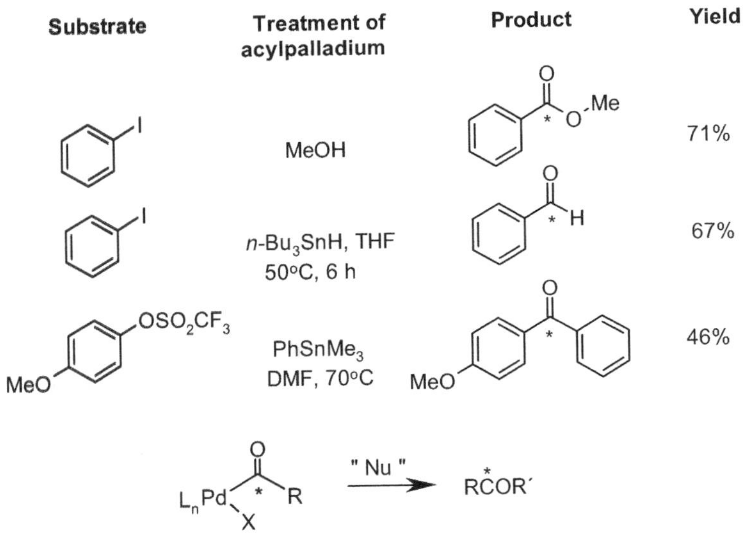


***Figure 5.14*** *Trapping of in situ generated [$^{14}C$]acylpalladium complexes with appropriate nucleophiles*

The same method was applied to the synthesis of [$^{14}$C]carboxylic acid **33** from the more complex substrate *o*-chloroaryl iodide **32** to provide the product in 87% chemical yield and 32% overall radiochemical yield from barium [$^{14}$C]carbonate[49]. The acid **33** was a key intermediate in the preparation of a series of labeled GnRH receptor agonists of the amide type. The tolerability of this methodology towards COOR, CHO, CN, $NO_2$, OR, $NH_2$, $NR_2$, Cl, $CF_3$, COR demonstrates that in cases like the above procedure it can be superior to classical [$^{14}$C]cyanations or [$^{14}$C]carboxylations. For multifunctional substrates such as **32**, it enables the introduction of carbon-14 at a later synthetic stage than would otherwise be possible.

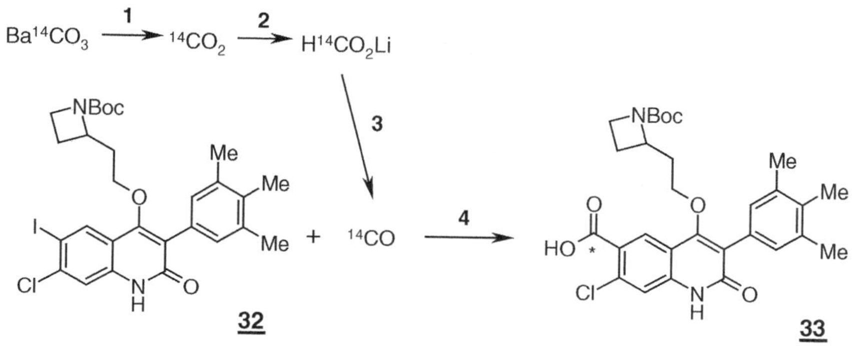


**Reaction conditions: 1.** $PbCl_2$, heat; **2.** $LiEt_3BH$; **3.** conc. $H_2SO_4$, 70 °C; **4.** $Pd(dppf)_2Cl_2$, KOAc, DMSO; 70 °C, 16 h

***Figure 5.15*** *Palladium(0)-catalyzed [$^{14}$C]carbonylation of a multi-functionalized aryl iodide*

The entire sequence depicted in Figure 5.15 can be performed in a nearly continuous procedure using a glassware setup of relative simplicity. Moreover, as is clear from the studies described, it is *not* necessary to use a large excess of $^{14}$CO to obtain good yields of carbonylation products, so that economy in the use of radioactivity is possible. As such, and because of the variety of [$^{14}$C]acyl derivatives that can be prepared in one pot, this methodology has significantly more potential for synthesis with carbon-14 than has been exploited to date.

Perhaps surprisingly, it is with $^{11}$CO that more of this potential has been reduced to practice. In spite of the very much smaller scales and the severe time constraints imposed by the 20 min half-life of carbon-11, a large number of effective applications have been devised. The carbon-11 procedures generally use a microautoclave to confine the $^{11}$CO with the substrate and reagents in solution, and utilize large substrate excesses in order to complete the reactions within a few minutes. In contrast, $^{14}$CO procedures can be applied with reaction times as long as desired, so that the reactions shown to be feasible with $^{11}$CO are also likely to work, if carried out with $^{14}$CO in a simple glass dual-chamber vessel as the above carbon-14 examples suggest. Moreover, the recent investigations into the utility of $^{11}$CO-complexing agents, such as borane[51] and copper(I)tris(pyrazolyl)borate[52], which trap and solubilize the gas, but make it available for reactions, may prove as a valuable alternative to autoclaving methods thereby providing additional means to facilitate the use of $^{14}$CO.

A recent review on the chemistry of $^{11}CO$ discusses the range of compound types prepared from this building block[53]. A variety of [$^{11}$C]carboxylic acids have been prepared from aryl halides, aryl triflates and benzyl halides in combination with either tetrabutylammonium hydroxide or trimethylphenylammonium hydroxides as powerful nucleophilic substrates in radiochemical yields of 20–85%[54]. Examples include radiolabeled carboxylic acids **34–37**.

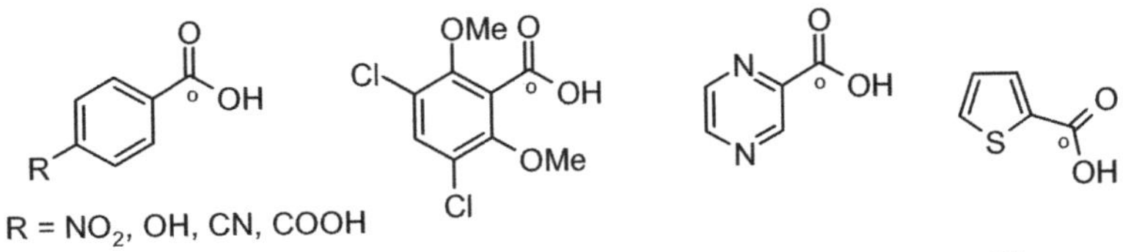


**34** **35** **36** **37**

**Reaction conditions:** Aryl halide or triflate, $Pd(PPh_3)_4$, $Me_4NOH$ or $Me_3PhNOH$, $THF/H_2O$, $^{11}CO$, 180 °C, 5 min

***Figure 5.16*** *Aryl and heteroaryl [$^{11}$C]carboxylic acids via palladium(ϕ)-catalyzed [$^{11}$C]-carbonylations*

Replacing the hydroxides for reactive amines provided the corresponding [$^{11}$C]carboxamides in satisfying yields[55]. Amines, which had shown so far low reactivity in the palladium-catalyzed carboxyamination, could be activated through *in situ* deprotonation with LiHMDS and employment as lithium amides in the reaction, thereby raising the yields from 27% to 82% for **38**, from <0.1% to 41% for **39** and from <0.1% to 33% for **40**. Satisfactory yields of [$^{11}$C]carboxamides were also obtained when aryl triflates[56] were used in the above reaction.

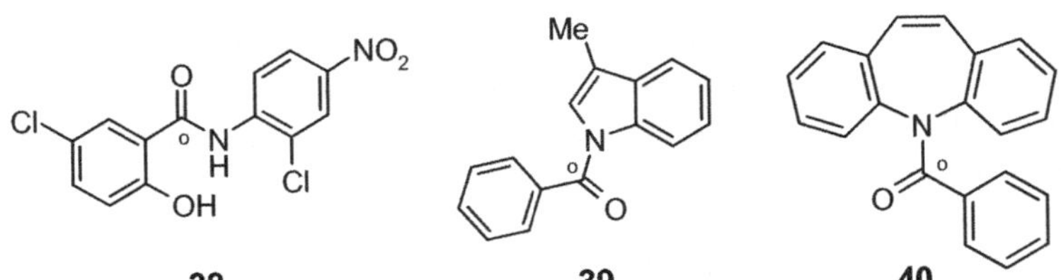


**38** **39** **40**

**Reaction conditions:** Aryl halide, $Pd(PPh_3)_4$, RR′NLi $^{11}CO$, THF, 150 °C, 5 min

***Figure 5.17*** *[$^{11}$C]Amides via palladium(ϕ)-catalyzed [$^{11}$C]carbonylations*

Use of amides as reaction partner in the Pd(0)-catalyzed carboxamination reaction allowed for the synthesis of [carbonyl-$^{11}$C]imides as exemplified by **41–43**[57].

Palladium(0)-catalyzed cross-coupling reactions were also extensively exploited for the preparation of carbon-11-labeled target compounds. Thus a range of aryl/heteroaryl

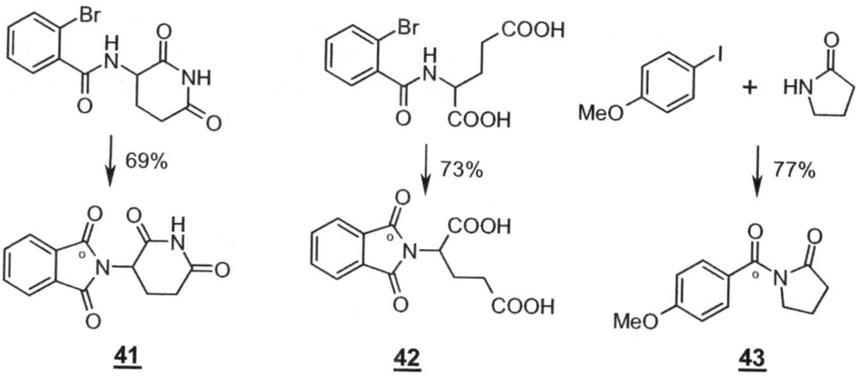


**Reaction conditions:** $Pd(PPh_3)_4$, $^{11}CO$, THF; 140–200 °C, 5 min

***Figure 5.18*** *[carbonyl-$^{11}$C]Imides via palladium(ϕ)-catalyzed [$^{11}$C]carbonylations*

methyl [$^{11}$C]ketones and aryl heteroaryl [$^{14}$C]ketones were obtained by carbonylative Stille cross-coupling between aryl/heteroaryl halides and tetramethyl stannanes or trialkyl arylstannanes and methyl iodide (e.g., **44** and **45**)[58], as well as by Suzuki reaction utilizing organoboranes[59], as illustrated by the examples **46** and **47** in Figure 5.19.

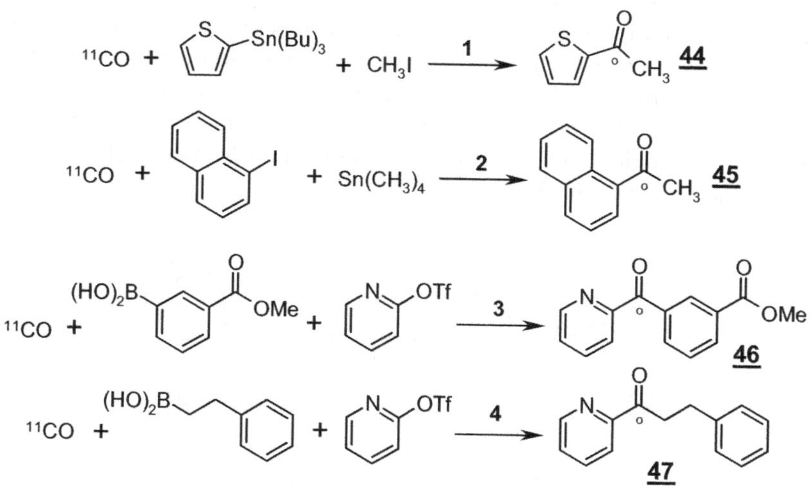


**Reaction conditions: 1a**. MeI, $Pd_2(dba)_3$, P(*o*-tol)$_3$, DMSO; r.t., 5 min, **b**. 2-tributylstannylthiophene, $^{11}CO$, 100 °C, 5 min (98 %); **2a**. 1-iodonaphthalene, $Pd_2(dba)_3$, P(*o*-tol)$_3$, DMSO; r.t., 5 min, **b**. $Me_4Sn$, $^{11}CO$, 150 °C, 5 min (63 %); **3/4a**. heteroaryl triflate, $Pd(PPh_3)_4$, LiBr, THF; r.t., **b**. aryl/alkyl boronic acid, $^{11}CO$, KO*tert*-Bu, THF; 170 °C, 5 min in microautoclave (**46**, 49 %; **47**, 43 %)

***Figure 5.19*** *Preparation of aryl/heteroaryl [$^{11}$C]ketones by Pd(0)-catalyzed carbonylative cross couplings*

Subsequent sequential treatment of the resulting crude reaction mixtures with $TiCl_4$, primary or secondary amines and $NaBH_3CN$ rapidly converts the [$^{11}$C]ketones into the corresponding secondary and tertiary amines[60]. Examples of compounds prepared this way include **48–49**.

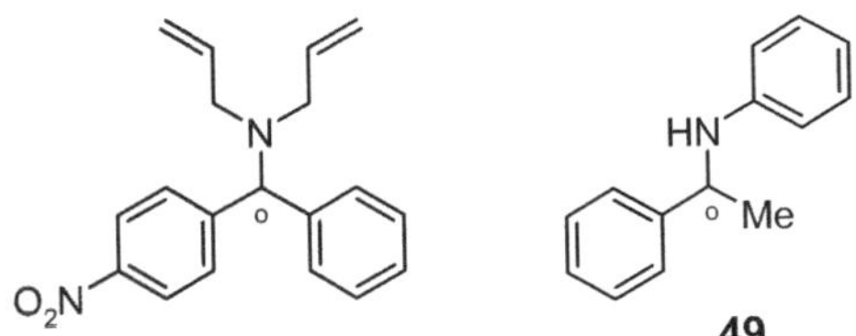

**48** **49**

**Reaction conditions:** [$^{11}$C]ketone reaction mixtures as obtained from Suzuki couplings as for **46** and **47** above, $TiCl_4$, amine; 60 °C, 3 min then $NaBH_3CN$; 60 °C, 3 min (**48**, 36 %; **49**, 39 %)

***Figure 5.20*** *[α-$^{11}$C]Amines prepared insitu by reductive amination of [$^{11}$C]ketones*

In all the reactions with $^{11}$CO, the high reaction temperatures and consequent need for autoclave confinement is due to the extremely small reaction scales and the desire for reaction times of just a few minutes. It is likely that the use of reaction conditions analogous to those for the $^{14}$CO examples early in this section will succeed in these cases.

Unfortunately, the use of palladium(0)-catalyzed carbonylation reactions in carbon-11 chemistry is restricted by problems related to the competing β-hydride elimination for electrophiles bearing a hydrogen in the β-position. In order to overcome this restriction, uncatalyzed photoinitiated carbonylation reactions have been more closely investigated using alkyl iodides, amines and alcohols as precursors. The first results were promising, providing esters and amides in the majority of cases in radiochemical yields of 40–70% as indicated for the examples **50–55**[61].

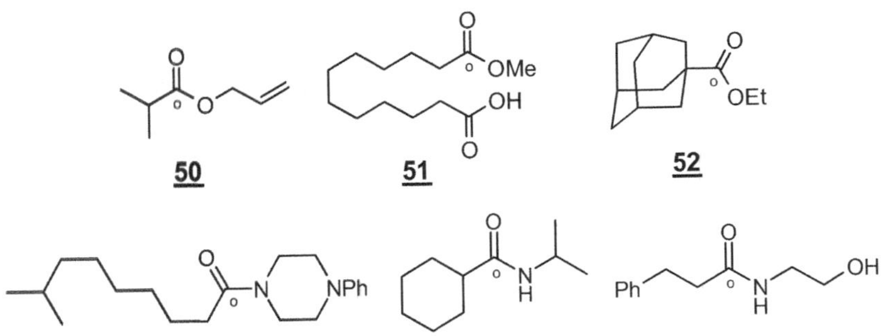

**50** **51** **52**

**53** (37%) **54** (61%) **55** (46%)

**Reaction conditions:** Alkyl iodide, $Pd(PPh_3)_4$, alcohol or amine, base, $^{11}$CO, hv, THF, 35 °C, 6 min

***Figure 5.21*** *Use of photoinitiated carbonylations in the synthesis of carbon-11-labeled aliphatic esters and amide*

### 5.2.1 [$^{14}$C]Phosgene

[$^{14}$C]Carbon monoxide has classically been the source of [$^{14}$C]phosgene, which is prepared in nearly quantitative yield by treatment of [$^{14}$C]carbon monoxide with chlorine gas in the presence of strong visible light[62,63]. It is commercially available in gaseous form as well as a solution in toluene. Its high cost, its instability and the inconvenience that transfer and subsequent reactions have to be carried out in closed systems using vacuum manifold techniques has limited its use and stimulated the development of surrogates such as 1,1′-[$^{14}$C]carbonyldiimidazole ([$^{14}$C]CDI) and 1,1′-[$^{14}$C]carbonyldi-1,2,4-triazole ([$^{14}$C]CDT) as well as ethyl and benzyl chloro[$^{14}$C]formate, which are all accessible from [$^{14}$C]phosgene.

Despite of its disadvantages it has found broad application in the synthesis of unsymmetrical linear and cyclic [carbonyl-$^{14}$C]ureas, linear and sulfonyl[carbonyl-$^{14}$C]ureas and

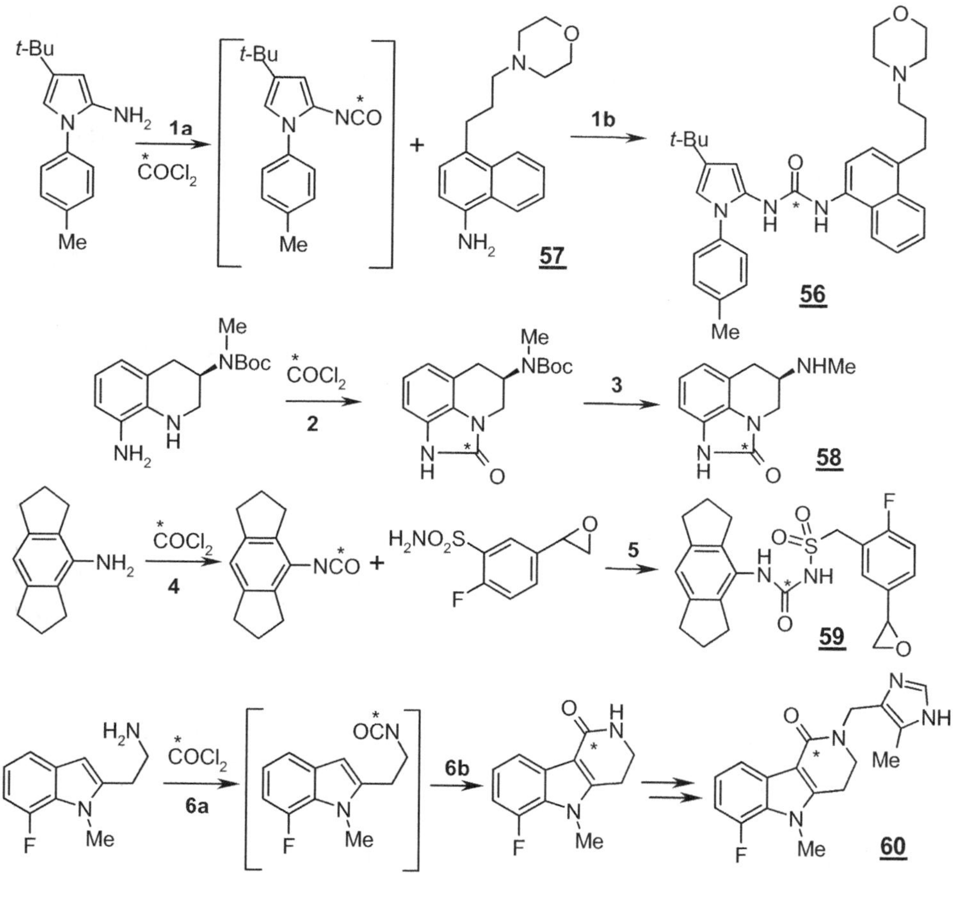


**Reaction conditions**: **1a**. sat. aq. $NaHCO_3$, $CH_2Cl_2$/toluene 10:1; 0 °C, 4 h, **b**.r.t., 16 h; **2**. $Et_3N$, THF; r.t., 3.5 h; **3**. HCl/MeOH; 70 °C, 3.5 h; **4**. $Et_3N$, toluene; r.t., 30 min; **5**. NaH, THF; r.t., 15 min; **6a**. $Et_3N$, $CH_2Cl_2$; 0 °C, 1 h, **b**. $AlCl_3$, $CH_2Cl_2$ ; reflux, 4 h

***Figure 5.22*** *Use of [$^{14}$C]phosgene in the synthesis of linear unsymmetrical and cyclic [$^{14}$C] ureas, sulfonyl[$^{14}$C]ureas and cyclic [$^{14}$C]amides*

cyclic [carbonyl-$^{14}$C]amides as illustrated by some recent examples depicted in Figure 5.22. These include [$^{14}$C]BIRB 796 (**56**), a MAP-kinase inhibitor[64]; **58**, a dopamine agonist[65]; [$^{14}$C]CP-470947 (**59**), an inflammatory drug candidate[66]; and [$^{14}$C]alosetron (**60**), a selective 5-HT$_3$-receptor agonist developed for the treatment of psychiatric diseases[67]. The procedure includes two separate steps with the initial formation of an [$^{14}$C]isocyanate with the first amine component, which is then trapped by the second amine. A problem which may arise in some cases is formation of the undesired symmetrical urea, resulting from the reaction of the first amine with the isocyanate generated *in statu nascendi*. This may depend on the nucleophilicity of the first amine as well as the base employed for the neutralization of HCl formed. In the case of **56** it was reported that reverse addition of the two amines (i.e., starting with amine **57**) gave exclusively the symmetrical naphthyl urea[64]. Accordingly, in each case the procedure has to be optimized with regards to the experimental conditions, the amount of [$^{14}$C]phosgene and the order of the amines to be added. In the last example the *in situ* generated isocyanate was exploited for an $AlCl_3$-catalyzed Friedel–Crafts acylation to provide [$^{14}$C]alosetron (**60**)[67].

Replacing the first amine with aliphatic or aromatic alcohols opens access to labeled carbamates, as illustrated with the synthesis of [$^{14}$C]mitomycin C (**61**)[68] and [$^{14}$C]CY 223 (**62**)[69].

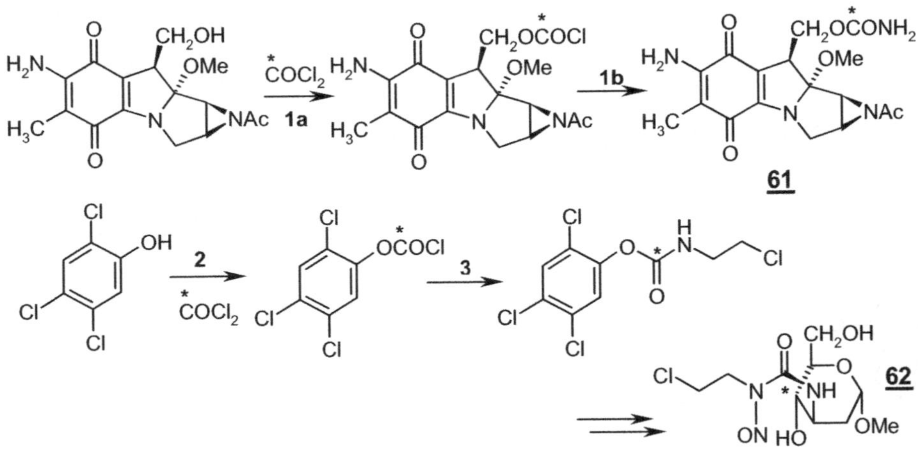

**Reaction conditions:** **1a**.*N*,*N*-dimethylaniline, $CH_2Cl_2$; r.t., 1 h, **b**. $NH_3$/MeOH; r.t., 15 h; **2**.*N*,*N*-dimethylaniline, toluene; 20 °C, 75 h; **3**. $H_2NCH_2CH_2Cl$, $Et_3N$, $CH_2Cl_2$; 20 °C, 18 h

*Figure 5.23* Use of [$^{14}$C]phosgene in the synthesis of [$^{14}$C]carbamates

[$^{14}$C]CDI and [$^{14}$C]CDT, both of which may serve as [$^{14}$C]phosgene surrogates, can be more easily handled as solids and stored for longer periods of time when kept away from moisture. They are conveniently accessible in nearly quantitative yields through treatment of *N*-(trimethylsilyl)imidazole **63**[70] or 1,2,4-triazole **65**[71] with a solution of [$^{14}$C]phosgene in toluene. In the first case the trimethylsilyl chloride formed is removed by careful evaporation in the latter case the HCl released is trapped by addition of a fourfold excess of 1,2,4-triazole yielding the respective hydrochloride, which precipitates and can be separated by filtration. Different from the [$^{14}$C]phosgene procedure [$^{14}$C]imidazolides and [$^{14}$C]triazolides, respectively, are generated and readily react *in situ* with the second *N*-nucleophile as illustrated with the formation of the semicarbazide **64**[70] and the linear [$^{14}$C] urea [$^{14}$C]BIBN 4096 (**66**)[71] (Figure 5.24). [$^{14}$C]CTI can also be exploited for the ring closure of diamines to provide cyclic [$^{14}$C]ureas such as **67**[71].

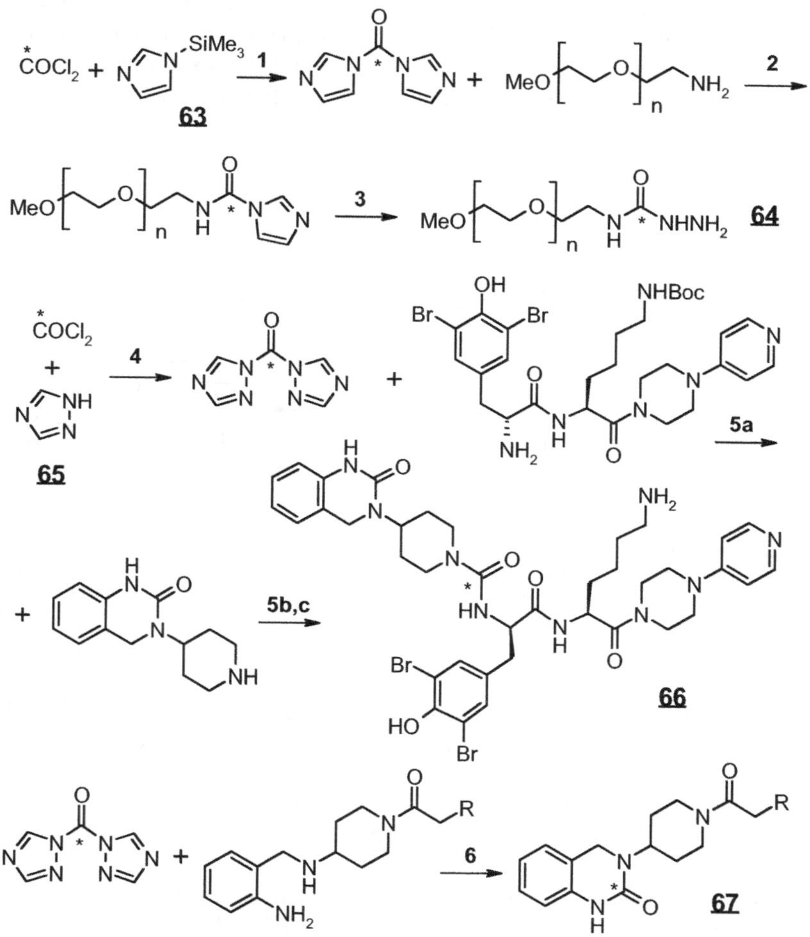


**Reaction conditions:** **1**. toluene; −90 °C to r.t.; r.t., 3.5 h; **2**. MPEG–amine, $CH_2Cl_2$; r.t., 2.5 h; **3**. $H_2NNH_2$, toluene; 50 °C, 3 h; **4**. toluene/THF; r.t., 20 min; **5a**. DMF; 0 °C, **b**. DMF; 65 °C, **c**. TFA, r.t. **6**. DMF; 65°C

**Figure 5.24** *Synthesis and use of [$^{14}$C]phosphene surrogates, [$^{14}$C]CDI and [$^{14}$C]CDT, in the synthesis of [$^{14}$C]semicarbazides and [$^{14}$C]ureas*

Aryl or alkyl chloro[$^{14}$C]formates may act as further alternatives to [$^{14}$C]phosgene in the preparation of [$^{14}$C]ureas and [$^{14}$C]carbamates. Some of them are commercially available or can be prepared in substance or *in situ* upon reaction of [$^{14}$C]phosgene with the respective alcohol in the absence[72] or presence of a base[73–75]. Further reaction with amines provide [$^{14}$C]carbamates, which may serve as intermediates for the preparation of [$^{14}$C]ureas, as demonstrated with the synthesis of **68**[73]. The [$^{14}$C]carbamate intermediates can also be reduced with complex hydrides, following one of the standard preparative procedures for secondary *N*-[$^{14}$C]methylamines as exemplified with the synthesis of nor[*N*-methyl-$^{14}$C]triptyline (**69**)[74]. The reaction of appropriate chloro[$^{14}$C]formates with *C*-nucleophiles has been a valuable method for the introduction of a [$^{14}$C]carboxyl function. In the example **71** depicted in Figure 5.25 this mode of action was exploited for the replacement of an unlabeled carboxyl group for a labeled one via formation of a malonate derivative **70** followed by a selective saponification and decarboxylation of the unlabeled functional group[75]. The use of this strategy to prepare a carbon-14-labeled target compound from the unlabeled compound is discussed in Chapter 10.

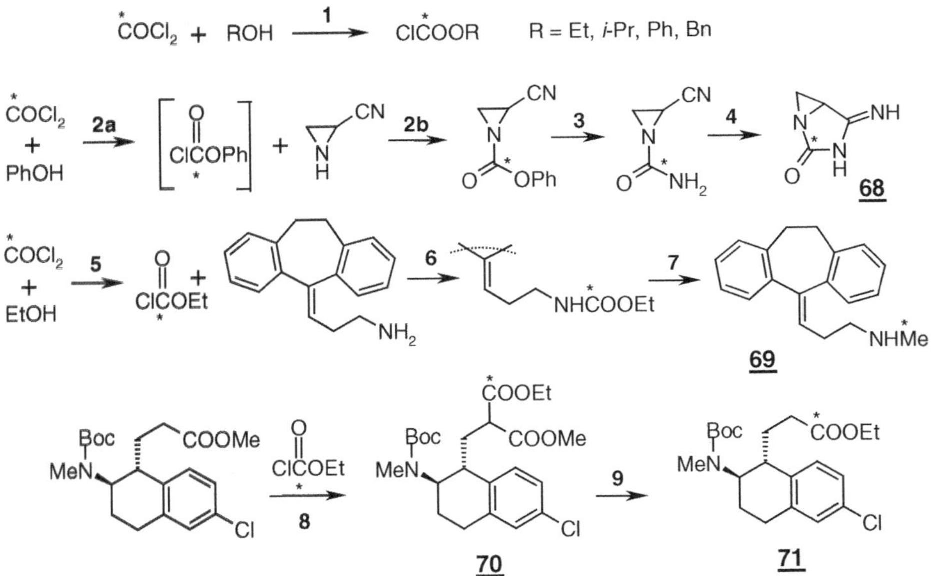


**Reaction conditions:** **1**. $Et_3N$, toluene; 0°C,4h, r.t., 2 h **2a**. 1N NaOH/toluene 1:1; 20 °C to r.t., 2 h, **b**. aq. 2N $Na_2CO_3$, ether; r.t., 2 h; **3**. liq. $NH_3$/ether; −35 °C, 2 h; 35 °C to r.t.; **4**. Triton B, EtOH; r.t., 19 h; **5**. collidine, toluene; 0 °C, 2 h; r.t., 18 h; **6**. collidine, toluene; r.t., 2 h; **7**. $LiAlH_4$, ether; reflux, 16 h; **8a**. LiN(*i*-Pr)(cyclohexyl), THF; −78 °C, **b**. $Cl^{14}COOEt$, −78 °C,1.5 h; **9**. NaCN, DMPU; 85 °C, 8 h

***Figure 5.25*** *Preparation and use of alkyl/aryl [$^{14}$C]formates in the aryl/alkyloxycarboxylation of N- and C-nucleophiles*

## 5.3 [$^{14}$C]Formic Acid ($H^{14}COOH$)

The classical syntheses of [$^{14}$C]formic acid or its salts involve catalytic hydrogenation of aqueous sodium carbonate[76,77] or hydrolysis of $K^{14}CN$ in water at 150 °C[76a,c,77]. In the former, careful pH control and high pressures are required to obtain good yields and in the latter care should be taken during the acidic workup to avoid exposure to $H^{14}CN$ in case the hydrolysis is not quantitative. A more recent method involves reduction of $^{14}CO_2$ with a solution of $NaBH(OMe)_3$ in THF at ambient temperature to give sodium [$^{14}$C]formate in 95% radiochemical yield after seven days reaction time[41]. This procedure was significantly improved[78] by replacing $NaBH(OMe)_3$ with $LiEt_3BH$ and running the reaction at 0 °C for 3 h (Figure 5.26). Subsequent acidification, careful removal of THF and extraction of the aqueous phase with ether provided the free acid in 82% radiochemical yield. This method is both economical and convenient for use in standard radiochemistry facilities.

[$^{14}$C]Formic acid and its chemically activated forms have become valuable one-carbon [$^{14}$C]formyl synthons for a variety of carbon-14-labeled heterocyclic compounds, mainly via *N*- or *S*-[$^{14}$C]formyl species, as indicated in Figure 5.26. In addition they have become reagents of choice for the preparation of aromatic, heteroaromatic and α-keto[carbonyl-$^{14}$C]aldehydes.

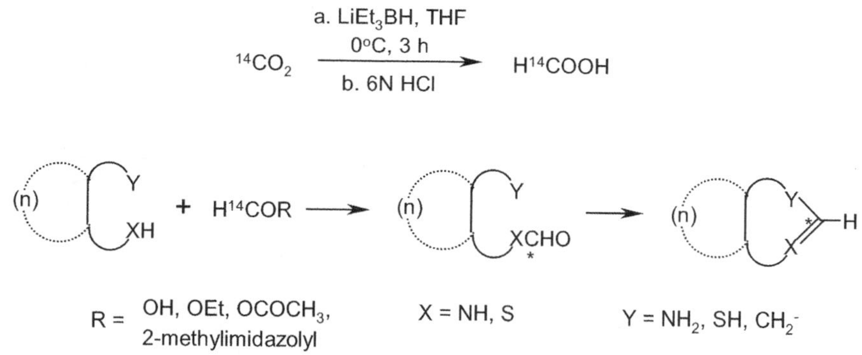


***Figure 5.26*** *Newer preparation of [$^{14}$C]formic acid and it application to the preparation of heterocyclic systems*

Classically, [$^{14}$C]formic acid has been the most extensively utilized starting material for the carbon-14 labeling of purines (e.g., **72, 73**)[79] and 3-deazapurines (e.g., **74**)[80] by cyclocondensations with appropriate diamino reaction partners. Purines labeled at C8, such as [8-$^{14}$C]adenine[81], [8-$^{14}$C]-guanine[82], [8-$^{14}$C]hypoxanthine[83], [8-$^{14}$C]theophylline[79b,84], [8-$^{14}$C]caffeine[79b] and 3-deaza[8-$^{14}$C]adenosine[80], have been prepared as illustrated in Figure 5.27. The two-step procedure starts with the reaction of [$^{14}$C]formic acid and appropriately substituted 3,4-diaminopyrimidines at temperatures in the range of 20–80 °C, followed by cyclization of the initially formed 5-[$^{14}$C]formamides upon heating to 210–220 °C. For the 3-deaza[8-$^{14}$C]adenosine derivative **74**, a key intermediate in the synthesis of 3-deaza[8-$^{14}$C]adenosine, and 2-amino-[5-$^{14}$C][1,3,4]thiadiazole (**75**)[85] the cyclization step requires much less harsh conditions, so that the two step sequence could be accomplished directly at 80–85 °C.

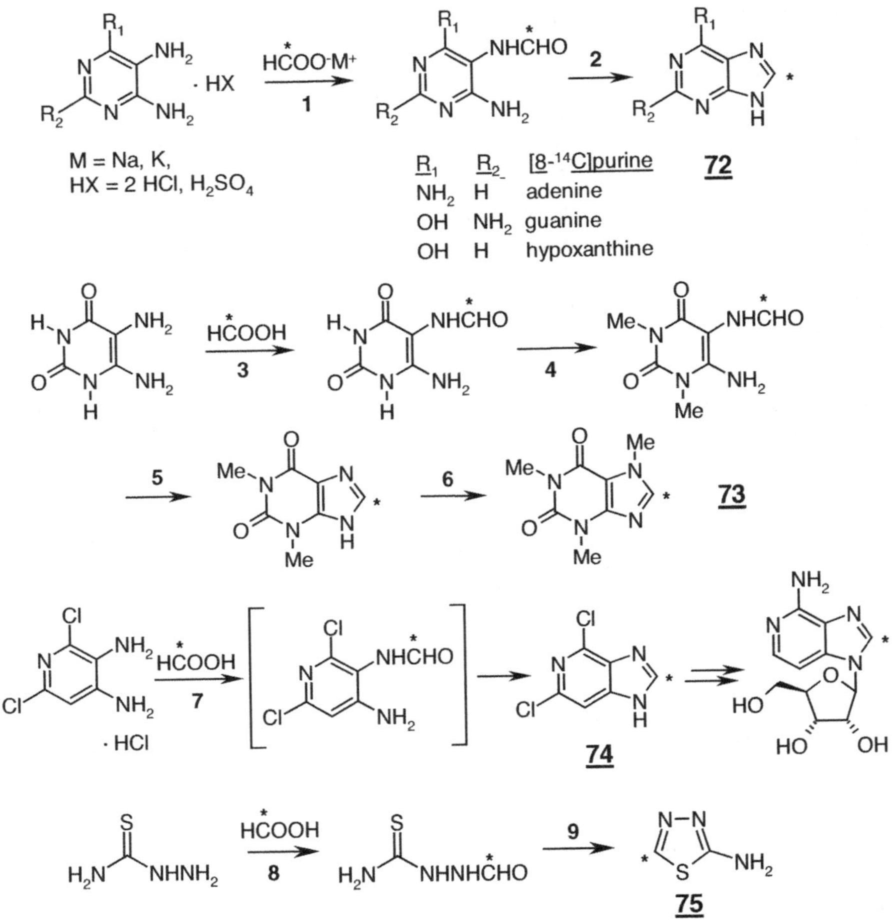


**Reaction conditions:** **1**.aq. HCl or aq. HCl/ethylene glycol; 50–80 °C, 1–2 h; **2**. diethanolamine or ethylene glycol; 200–220 °C, 1–2 h; **3**. water; reflux, 1 h; **4**. $Me_2SO_4$, NaOH (pH 8.5–9); 25–35 °C, 1 h; **5**. NaOH, 100 °C, 15 min; **6**. $Me_2SO_4$, NaOH; r.t., 75 min; 80 °C, 10 min; **7**. conc. HCl, ethylene glycol; 80–85°C, 4 d; **8**. neat mixture; 80–90 °C; **9**. conc. HCl; 85 °C

***Figure 5.27*** *Use of [$^{14}$C]formic acid in the formation of [2-$^{14}$C]imidazole and [5-$^{14}$C] thiadiazole ring systems*

The initial formylation can sometimes be facilitated by prior conversion of [$^{14}$C]-formic acid into a more powerful derivative having better formylating properties, as exploited in the preparation of *N*-[$^{13}$C]formylmorpholine (**76**)[86] which reacted *in situ* with 4,5,6-triaminopyrimidine to give [8-$^{13}$C]adenine and with 2,4,5-triamino-6-hydroxypyrimidine to give [8-$^{13}$C]guanine (77) in 85% yield (Figure 5.28). Nor is the construction of carbon-14-labeled heterocyclic systems by an intramolecular attack of an appropriate nucleophilic center on a [$^{14}$C]formyl unit restricted to the preparation of five-membered [2-$^{14}$C]imidazole and [5-$^{14}$C][1,3,4]thiadiazole systems. The synthesis of 8-chloro-6-phenyl-4*H*-*s*-[1-$^{14}$C]triazolo[4,3a][1,4]benzodiazepine (81) through treatment of hydrazine derivative 79 with [1-$^{14}$C]formyl-5-methylimidazole (78), prepared *in situ* from [$^{14}$C]formic acid and 1,1′-carbonyldi(5-methylimidazole), followed by ring closure of the resulting *N*-[$^{14}$C]formyl derivative 80 upon heating in refluxing pyridine[87] embodies this strategy, as does the Madelung synthesis of [2-$^{14}$C]indole (82) from ethyl [$^{14}$C]formate via reaction of 2-methyl[carbonyl-$^{14}$C]formanilide and potassium *tert*-butylate at elevated temperature[88]. The latter reaction demonstrates that also carbanions, generated *in situ*, are effective.

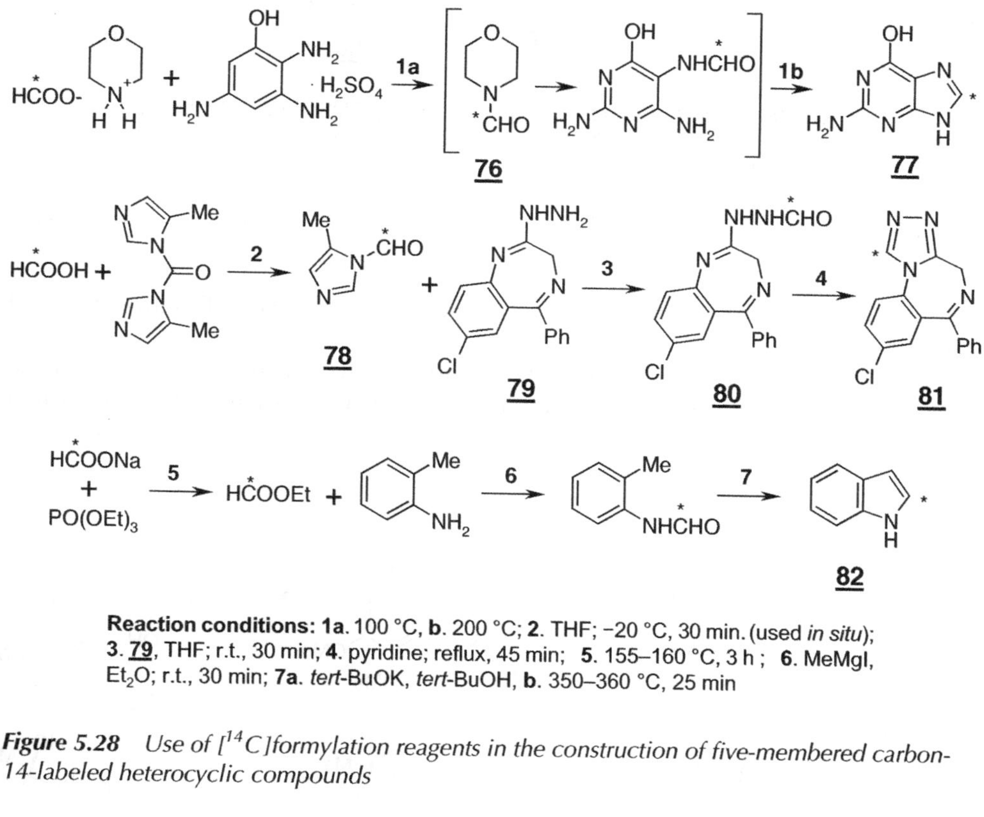


**Reaction conditions:** **1a.** 100 °C, **b.** 200 °C; **2.** THF; −20 °C, 30 min. (used *in situ*); **3.** 79, THF; r.t., 30 min; **4.** pyridine; reflux, 45 min; **5.** 155–160 °C, 3 h; **6.** MeMgI, $Et_2O$; r.t., 30 min; **7a.** *tert*-BuOK, *tert*-BuOH, **b.** 350–360 °C, 25 min

***Figure 5.28*** *Use of [$^{14}$C]formylation reagents in the construction of five-membered carbon-14-labeled heterocyclic compounds*

The methodology presented can also be exploited for the preparation of six-membered heterocyclic systems, as illustrated in Figure 5.29. For example, [$^{14}$C]formylation of the precursors 84 and 88 with acetic [$^{14}$C]formic anhydride (A[$^{14}$C]FA, 83) and dimethyl[$^{14}$C]-formamide dimethyl acetal (87), respectively, provided 8-[2-$^{14}$C]azaadenine 85[89] and 5-[6-$^{14}$C]azacytidine 89[90] in excellent yields. Compound 85 was used as a key intermediate for the synthesis of 9-α-D-arabinofuranosyl-8-deaza[2-$^{14}$C]adenine (86)[89]. A[$^{14}$C]FA was prepared *in situ* from sodium [$^{14}$C]formate and acetyl chloride, whereas 87 had to

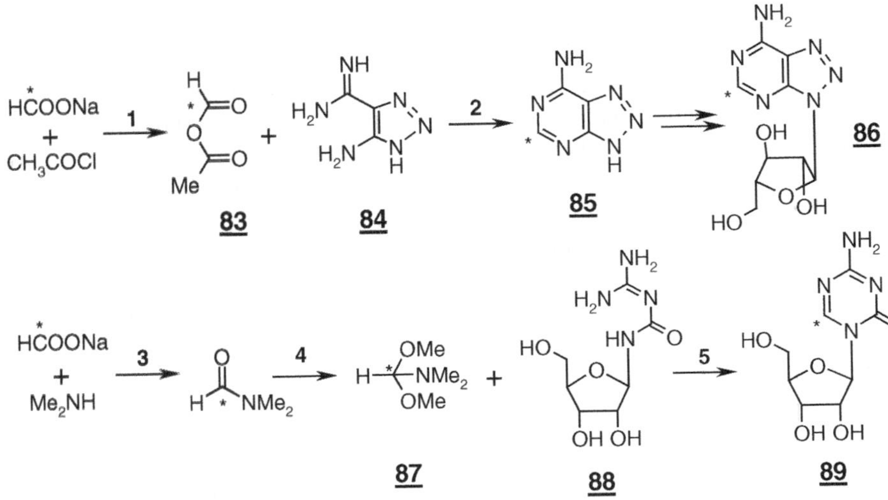

**Reaction conditions:** **1a**. ether; r.t., 4 h, **b**. **84**, DMSO; 70°C, 45 min **2**. DMSO; 70 °C, 45 min; **3**. $Me_2SO_4$; 70 °C, 3 h; **4**. NaOMe, MeOH; distil; **5**. MeOH; r.t., 6 h

***Figure 5.29*** *Use of [$^{14}$C]formylation reagents in the construction of six-membered heterocyclic systems*

be prepared separately in two steps from sodium [$^{14}$C]formate, dimethylamine and dimethylsulfate.

Ethyl [$^{14}$C]formate, which had served as a starting material in the Madelung synthesis, has also been used for the $\alpha$-[$^{14}$C]formylation of ketones and nitriles through Claisen

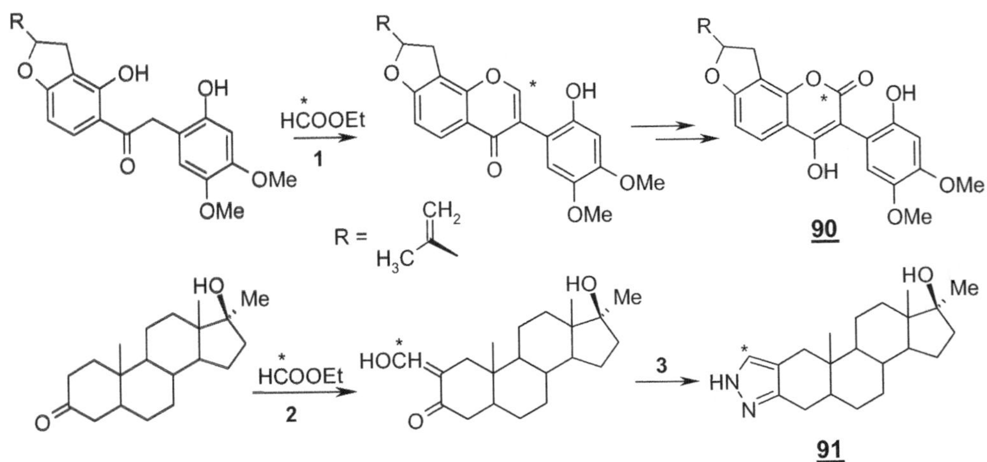

**Reaction conditions:** **1**. Na, ether; r.t., 16 h; **2**. NaH, pyridine; 0 °C, 75 min; r.t., 24 h; **3**. $N_2H_4.H_2O$, EtOH; reflux, 6 h

***Figure 5.30*** *Use of ethyl [$^{14}$C]formate in the synthesis of $\alpha$-keto[carbonyl-$^{14}$C]aldehydes*

condensation promoted by Na, NaH or $NaNH_2$, as illustrated with the carbon-14 labeling of 4-hydroxycoumarin derivative **90**, an insecticide[91], and stanozolol (**91**), an anabolic steroid[92].

Aside from the [$^{14}$C]formylating reagents mentioned above, others have been prepared in order to modify their physical properties and tune the reactivity of the [$^{14}$C]formyl synthons. Figure 5.31 illustrates the various methods used, including the dehydration of the corresponding ammonium formates upon heating (dimethyl [$^{14}$C]formamide, **92** and *N*-methyl-*N*-phenyl[$^{14}$C]formamide, **93**)[93,94], removal of the water formed by azeotropic distillation (*N*-methyl-*N*-*p*-tolyl[$^{14}$C]formamide, **94**)[95] or through treatment with $TiCl_4$ (**93**)[96] as well as by activation of [$^{14}$C]formic acid as an acylurea or a mixed anhydride (EDC[97], DCC[98] or $CH_3COCl$[99]) and subsequent coupling to the respective amine (dibutyl amine for **95** and *N*-methyl-*N*-phenyl[$^{14}$C]formamide, **93** or 2-(*N*-methyl-*N*-[$^{14}$C]formyl)-pyridine for **96**) and, finally, by aminolysis of *O*-[$^{14}$C]formylglycolate (**97**)[100] for the preparation of *N*-[$^{14}$C]formyl-8-aza-[4,5]spirodecane (**98**). These formyl derivatives have

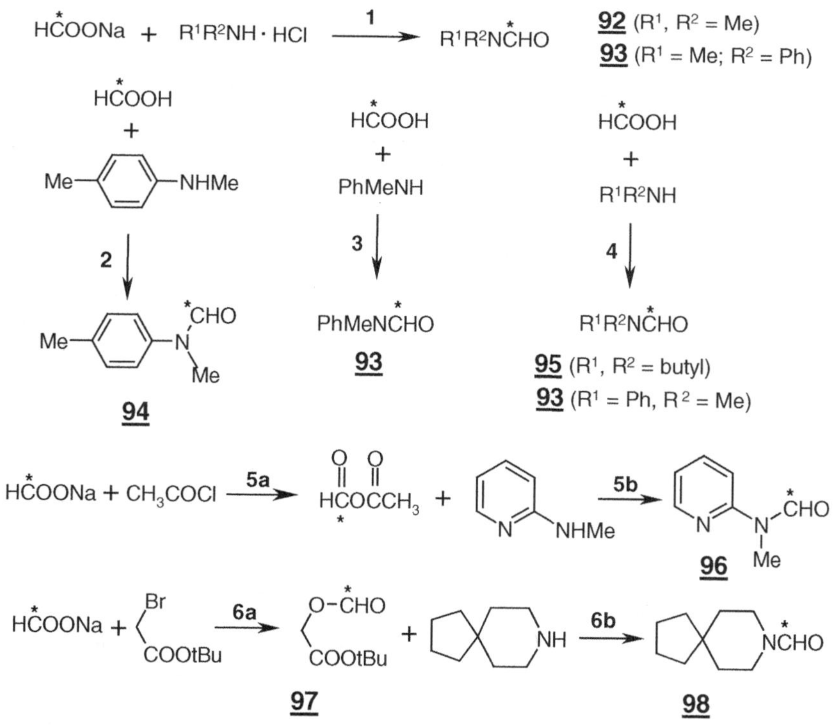


**Reaction conditions:** **1**. removal of water by heating; **2**. toluene; reflux; **3**. $TiCl_4$, THF; r.t.; **4**. for $R^1$,$R^2$= dibutyl: EDC, $Et_3N$, DMAP, $CH_2Cl_2$; r.t., 12 h; for $R^1$=Ph, $R^2$= Me: DCC, MED; 0 °C, 2 h; **5a**. ether; r.t., 6 h; **5b**. ether; r.t.; 16 h; **6a**. DMF; 55 °C; **b**. DMF; 55 °C

***Figure 5.31*** *Preparation of additional formamide derived [$^{14}$C]formylating reagents*

enhanced lipophilicity, lower water solubility and lower volatility than dimethyl[$^{14}$C]formamide, and so are more convenient to purify and handle and allow more straightforward reaction workups.

Several of these *N,N*-disubstituted [$^{14}$C]formamides have demonstrated utility in C-[$^{14}$C]formylations of electron-rich aromatic and heteroaromatic substrates via the Vilsmeyer–Haack reaction[101]. In the presence of $POCl_3$, $Cl_2(O)POP(O)Cl_2$, $(COCl)_2$ or $(CF_3SO_2)_2O$ they can C-formylate benzene and naphthalene derivatives that possess an electron-releasing substituent (*O*-alkyl, *S*-alkyl, *N*-dialkyl) as well as anthracene, pyrrole,

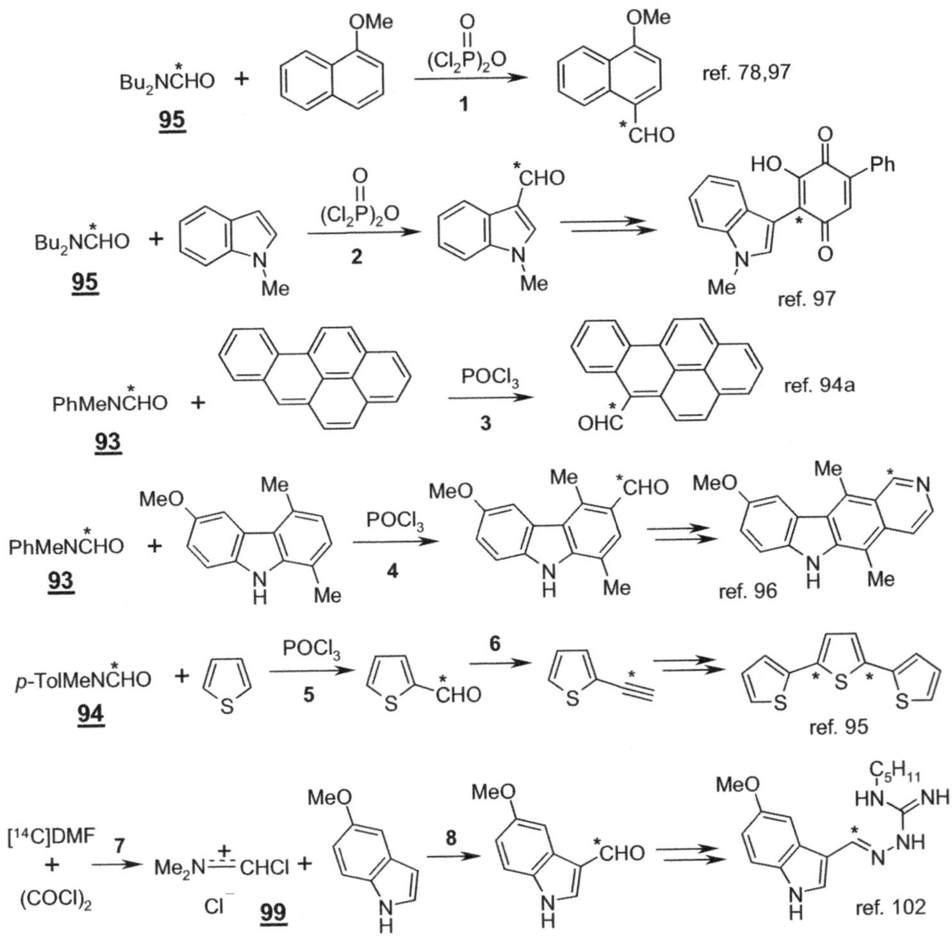


**Reaction conditions:** **1**. toluene; 80 °C (82 %); **2**. toluene; 80 °C (78 %); **3**. neat; 100 °C, 2.5 h (88 %); **4**. *o*-$Cl_2C_6H_4$; 100 °C, 2.5 h (73 %); **5**. neat; 35–40 °C, 2 h then r.t. 16 h (94 %); **6a**. $PPh_3$, $CBr_4$, $CH_2Cl_2$; r.t., 1 h, **b**. *n*-BuLi, $Et_2O$; −78 °C to r.t. (53 % in two steps); **7**. MeCN; −40 °C to r.t. then remove all volatiles under high vacuum; **8**. MeCN; r.t. (>90 %)

***Figure 5.32*** *Vilsmeyer–Haack [$^{14}$C]formylations of electron-rich aromatic and heteroaromatic systems*

furan, thiophene, indoles etc., as illustrated with the examples in Figure 5.32. Benzene and naphthalene are not sufficiently electron rich and therefore do not react.

Although the actual electrophilic reagent in Vilsmeyer reactions, a chloromethylene-iminium cation such as **99**, is normally generated *in situ*, it can also be isolated as a salt,

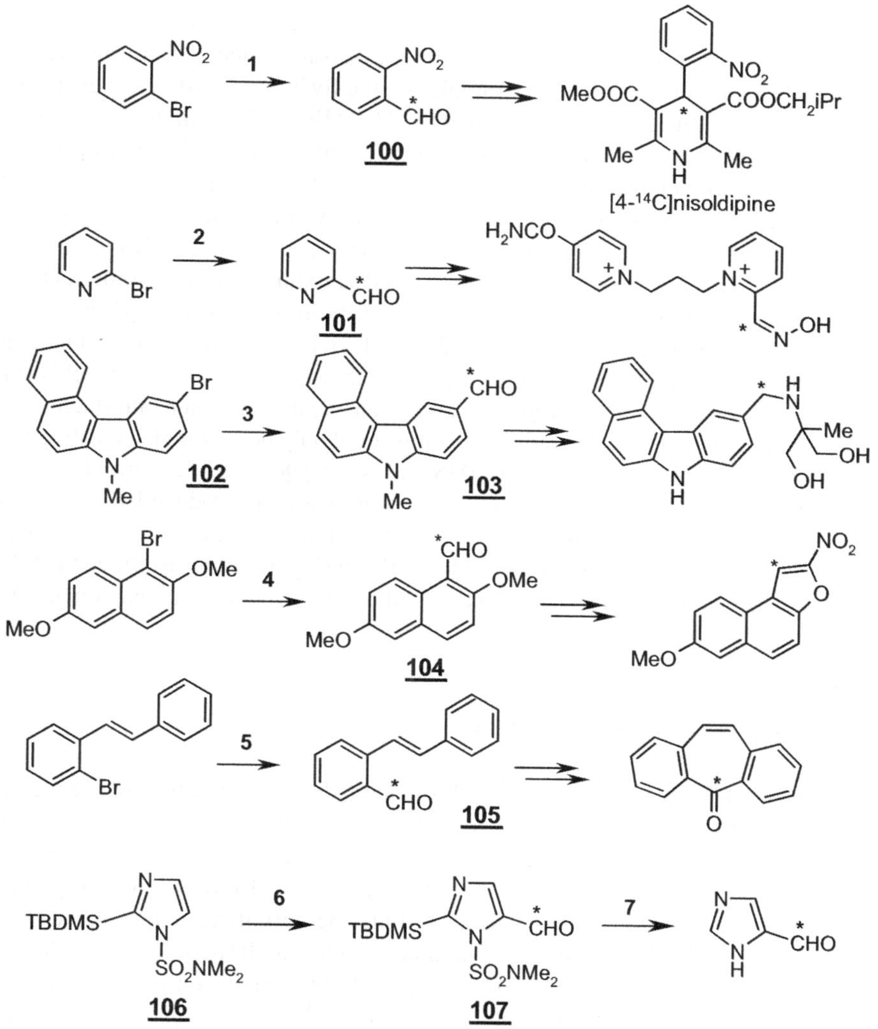

**Reaction conditions**: **1a**. PhLi, THF; −100 °C, 1.5 h, **b**. [$^{14}$C]DMF; −100 °C, 45 min,−78 °C, 1.5 h (22–78 %); **2a**. *n*–BuLi, ether; −50 °C, 30 min, **b**. *N*-methyl [$^{14}$C]formanilide;−60 °C, 3 h (80 %); **3a**. *n*-BuLi, DME; −60 °C, **b**. [$^{14}$C]DMF;−60 °C to 10 °C; **4a**. *n*-BuLi, THF; −60 °C, 3 min, **b**. *N*-methyl[$^{14}$C]formanilide;−60 °C to 0 °C, 1 h (70 %); **5a**. *n*-BuLi, $Et_2O$;−12 °C, 80 min, **b**. [$^{14}$C]DMF; r.t., 16 h (83 %); **6a**. *n*-BuLi; THF; −78 °C, 30 min, **b**. TMSCl; −78 °C to r.t., 19 h; **c**. *sec*-Buli, hexane/cyclohexane; −78 °C, **d**. 2 (*N*-methyl-*N* [$^{14}$C] formylamino)pyridine (**96**), THF; −78 °C to 0 °C; 0 °C; **7**. 2N HCl, r.t.; 1 h

**Figure 5.33** *Formylation of organolithium compounds using various [$^{14}$C]formamides ([$^{14}$C] DMF, N-methyl-N-phenyl[$^{14}$C]formamide; 2-(N-methyl-N-[$^{14}$C]formylamino)pyridine)*

when either the usual solvents for Vilsmeyer reactions (toluene, $CH_2Cl_2$, $CHCl_3$) are not suitable, or excess chlorination reagent ($POCl_3$, $(COCl)_2$) has to be removed by evaporation. Such was the case in the [$^{14}$C]formylation of 5-methoxyindole in the last example of Figure 5.33, which gave good results only in acetonitrile[102].

C-Formylation of electron-poor aromatic and heteroaromatic substrates that are not appropriate for the Vilsmeyer–Haack reaction can be accomplished by treating their organolithium or -magnesium derivatives with *N,N*-disubstituted [$^{14}$C]formamides. Although nitrobenzene and pyridine, for example, are totally inert towards Vilsmeyer–Haack conditions, their 2-lithio derivatives react with [$^{14}$C]DMF and N-methyl[$^{14}$C]formanilide, respectively, even at temperatures as low as $-100\,°C$ to give the [$^{14}$C]aldehydes **100**[103a] and **101**[94b]. The use of organometallic reaction partners with [$^{14}$C]formylating reagents of this type can also permit control of regioselectivity in cases where an electrophilic substitution would give an undesired isomer or a mixture of isomers. Electrophilic formylation of *N*-methylbenzo[c]carbazole (*des*-bromo **102**) produced a mixture of formyl isomers, but [$^{14}$C] formylation of the lithio species derived from the 10-bromo compound **102** provided 10-[$^{14}$C]formyl-*N*-methylbenzo[c]carbazole **103** in high yield[105]. Similar circumstances hold for the preparation of aldehydes **104** and **105**[104], which were obtained in high yields via the organolithium species. Imidazole **106** may serve as an example in which introduction of the [$^{14}$C]formyl function into the desired C5 position could only be accomplished after having previously blocked C2 with *n*-BuLi/TBDMSCl. Use of **106** as a [$^{14}$C]formylating agent proved to be superior, since only one equivalent was necessary to produce **107** in 51% yield, whereas the other reagents discussed so far had to be employed in excess to get comparable yields. The better reactivity of **106** may be due to the pyridyl nitrogen, to which the lithium could coordinate in a six membered complex and thereby facilitate the carbanion transfer to the formamide carbonyl.

## 5.4 [$^{14}$C]Formaldehyde ($H^{14}CHO$)

The variety of published syntheses of [$^{14}$C]formaldehyde is evidence that no one method is sufficiently robust and convenient to have become routinely used in radiochemistry laboratories. As $H^{14}CHO$ is commercially available in various forms, usually it is only when large quantities are needed that it may be economically reasonable to prepare it in house. Reported methods (Figure 5.34) include the partial reduction of $^{14}CO_2$ with $LiAlH_4$ in THF[44a,b,c,106] or ether[107] (Method 1), the $LiAlH_4$-reduction of *N*-methyl [carbonyl-$^{14}$C]formanilide[44a,c] (Method 2), the Pd/$BaSO_4$-catalyzed partial reduction of $H^{14}CN$[108] (Method 3), the ferric-molybdenum oxide-[44a–c] or silver wool-catalyzed[109] oxidation of [$^{14}$C]methanol with air (Method 4), the hydrolysis of chloro[$^{14}$C]methyl acetate[110] (Method 5), and finally, the $HIO_4$- or $Pb(OAc)_4$-mediated oxidative cleavage of [1,2-$^{14}$C]ethylene glycol[111], which has to be prepared beforehand in three steps from [1,2-$^{14}C_2$]acetylene (Method 6). Despite moderate yields, Method 1 seems to be the one most frequently used; the ether variant has been shown to work optimally at temperatures around $-50\,°C$ in the presence of two equivalents of $LiAlH_4$. At higher temperatures [$^{14}$C]methanol is preferentially formed. Method 4 promises even better yields, but it suffers from two disadvantages: the need to prepare [$^{14}$C]methanol in the first step, and the requirement for optimization of the reaction conditions for the catalytic oxidation.

Method 6 is the only method that gives $H^{14}CHO$ directly in anhydrous form, and although it requires several steps, the [$^{14}C_2$]ethylene glycol precursor is easily stored and portions converted as needed. The aqueous solutions of $H^{14}CHO$ produced by the other methods can, if desired, be processed into anhydrous form through acid-catalyzed precipitation as [$^{14}$C]paraformaldehyde, subsequent drying and depolymerisation upon heating to 100–120 °C.

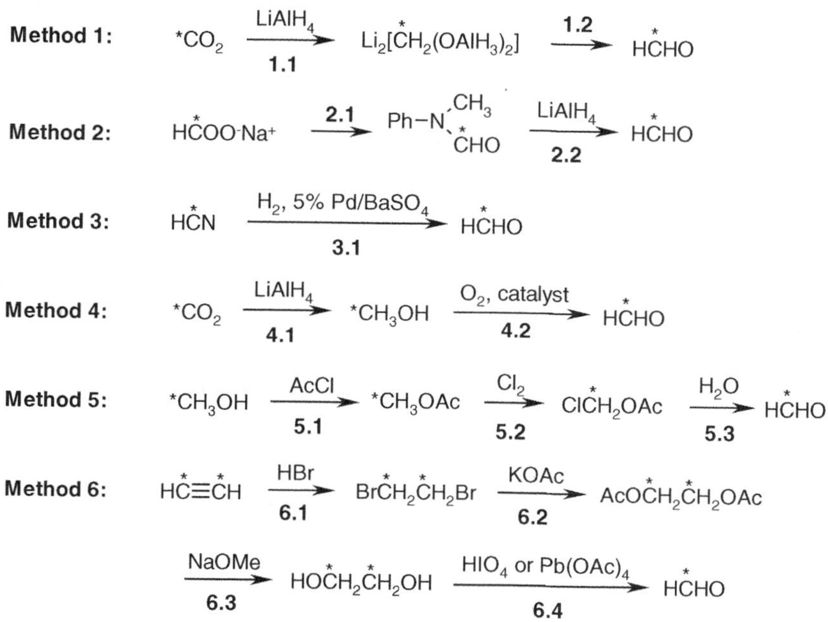

**Reaction conditions: 1.1**. $LiAlH_4$ (0.5 equiv), ether; −55 to −45 °C, 5–10 min; **1.2**. $H_2SO_4$, $H_2O$;−5 °C (40–50 %) **or** $LiAlH_4$ (0.5 equiv), THF; 0 °C, 15 min; **1.2**. 6N $H_3PO_4$;−80 °C (60–80 %); **2.1**.*N*-methylaniline.HCl, 150 °C; **2.2a**. $LiAlH_4$ (0.5 equiv), THF; −20 °C, 1 h; **b**. aq. $H_2SO_4$;−20 °C (60–70 %); **3.1**. $H_2$ (1atm), 5 % Pd–$BaSO_4$, $H_2O$; 5 °C to r.t., 8 h (80 %); **4.1a**. $LiAlH_4$ (4 equiv), diethylene glycol; r.t., 1 h; 75 °C, 2 h, **b**. tetrahydrofurfuryl alcohol; r.t. to110 °C, $N_2$; **4.2**. Ag; 687 °C, 1–7 h (75–80%) **or** $O_2$, $Fe_2O_3Mo_2O_3$; 372 °C, 30 min (75–80 %); **5.1**. AcCl; 40–50 °C, 45 min; **5.2**. $Cl_2$; 0 °C, 1–2 h; **5.3**. $H_2O$; 100 °C,30 min (60 %); **6.1**. $HBr_{gas}$, $O_2$(trace), UV; **6.2**. KOAc, AcOH; 190 °C, 16 h; **6.3**. NaOMe, MeOH; reflux, 2 h; **6.4**. $HIO_4$, $H_2O$; 0–5 °C, 10 min **or** $Pb(OAc)_4$, benzene; r.t.

***Figure 5.34*** *Preparations of [14C]Formaldehyde*

The utility of [$^{14}$C]formaldehyde is extensive. It has been widely employed in carbanion-mediated hydroxy[$^{14}$C]methylations and [$^{14}$C]methylenations (aldol reaction, Knoevenagel reaction) (Section 5.4.1 below), in acid-catalyzed hydroxy[$^{14}$C]methylenations and halo[$^{14}$C]methylation (5.4.2), in amino[$^{14}$C]methylations (Mannich reaction and variants) (5.4.3), in reductive [$^{14}$C]methylations (5.4.4), in polycondensation reactions with primary amines (5.4.5), and in thio[$^{14}$C]methylations (5.4.6).

### 5.4.1 Carbanion-Mediated Hydroxy[$^{14}$C]methylation and [$^{14}$C]Methylenenation

In the first example of Figure 5.35, base-catalyzed reaction of two equivalents of aqueous $H^{14}CHO$ with diethyl malonate furnished diethyl bis(hydroxy[$^{14}$C]methyl)malonate **108**. Concurrent bis-bromodehydroxylation, ester hydrolysis and decarboxylation of **108**, followed by re-esterification and dehydrobromination provided ethyl 2-bromo[$^{14}$C]-meth[3-$^{14}$C]acrylate **109**, a key intermediate in the Reformatsky reaction with ketones, which give $\alpha$-[$^{14}$C]methylene $\gamma$-[$\beta$-$^{14}$C]butyrolactones **110**[112].

The corresponding monohydroxy[$^{14}$C]methyl derivatives, accessible through reaction of equimolar amounts of $H^{14}CHO$ and alkyl cyanoacetates or dialkyl malonates readily eliminate water to give alkyl 2-cyano[3-$^{14}$C]acrylates (**111**, $R^1 = CN$) and dialkyl [$^{14}$C]-methylenemalonates (**111**, $R^1 = COOR^2$), respectively, which are highly valuable intermediates in Michael additions, Diels–Alder reactions and epoxidations. Unfortunately, carbon-14-labeled acrylates are unstable towards radiation induced polymerization. In order to circumvent this undesirable side effect, it has been recommended to run the formation in the presence of anthracene in order to trap the [$^{14}$C]acrylates *in situ* as stable Diels–Alder adducts **112**. These can be purified and stored briefly. Their subsequent thermolysis in the presence of maleic acid anhydride re-generates the acrylates **111** for immediate use in further reactions[113].

[$^{14}$C]Formaldehyde and nitromethane react analogously. With two equivalents of $H^{14}CHO$, for example, 1,3-dihydroxy-2-nitro[1,3-$^{14}C_2$]propane (**113**) was formed.

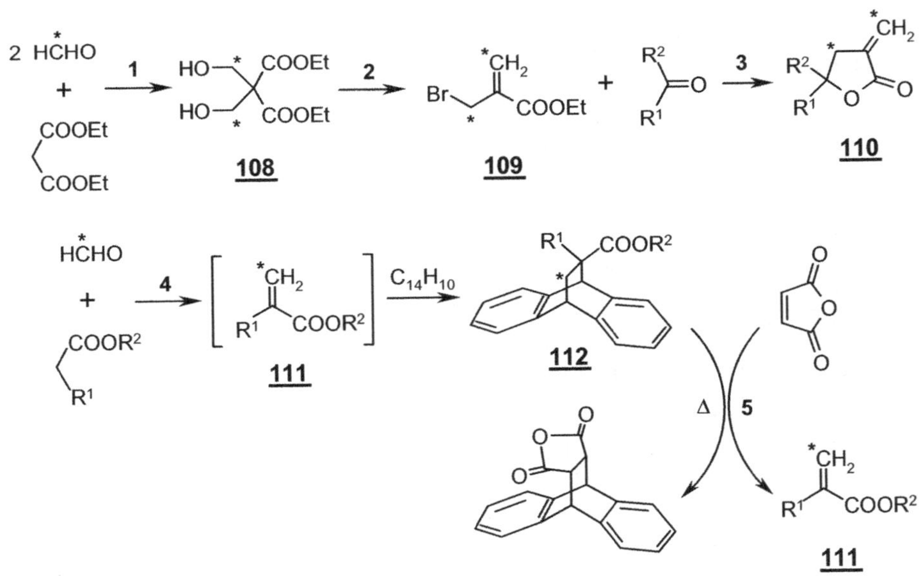


**Reaction conditions:** **1**. $K_2CO_3$, $H_2O$; r.t., 2 h; **2a**. HBr (48 %); reflux, 2 h, **b**. EtOH,*p*-TosOH, 12 h, **c**. NaH, ethylene glycol, THF; 40–45 °C, 1.5 h; **3**. Zn, THF, reflux 4 h; **4**. $NCCH_2COO$*i*-Bu, anthracene, piperidine, benzene; 115 °C, 20 h (38 %) **or** $CH_2(COOEt)_2$, anthracene, $Cu(OAc)_2$, HOAc, butanone, 95 °C, 2 h; 130 °C, 3 h (45 %); **5**. maleic anhydride, $SO_2$, high boiling mineral oil; 220 °C, 45 min (40 %)

***Figure 5.35*** *Hydroxy[$^{14}$C]methylation of malonic and cyanoacetic acid esters*

Hydrolysis of **113** produced 1,3-dihydroxy[1,3-$^{14}C_2$]acetone, which upon reduction gave [1,3-$^{14}C_2$]glycerol (**114**) with a specific activity of 120 mCi/mmol. This was converted to [$^{14}$C]glycerol trinitrate[114]. On the other hand, equimolar amounts of H$^{14}$CHO reacted with nitromethane to give 2-nitro[1-$^{14}$C]ethanol (**115**). Dehydration of **115** furnished 1-nitro-[2-$^{14}$C]ethylene (**116**), a valuable Michael acceptor and Diels–Alder dienophile. Its reaction with cyclopentadiene, for example, gave predominantly *endo*-2-nitro[3-$^{14}$C]-norcamphane (**117**). Subsequent catalytic reduction of the double bond, Fe-reduction of the nitro group and concluding acyloxy-deamination converted **117** into *exo*-[$^{14}$C]norbornyl acetate (**118**)[115].

With $\beta$-keto esters HCHO is known to undergo Knoevenagel reaction to give 2-acylacrylates, key intermediates in the Hantzsch dihydropyridine synthesis. Application

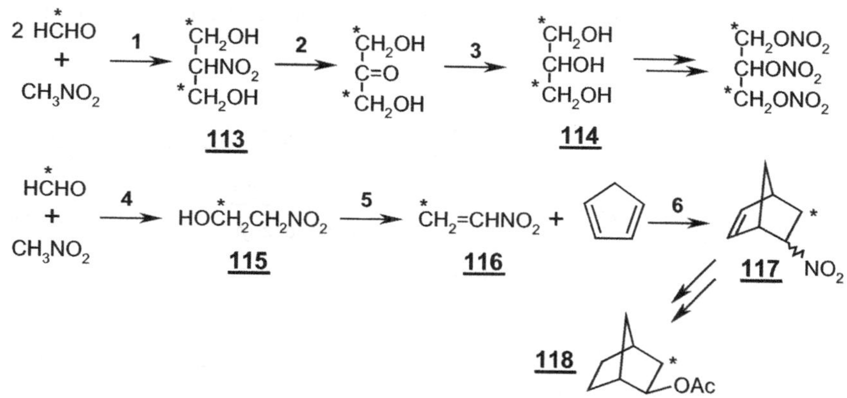


**Reaction conditions:** **1**. $H_2O$; r.t., 4 h; **2a**. NaOMe, MeOH, r.t., **b**. aq. $H_2SO_4$; r.t., 10 min; **3**. $NaBH_4$, MeOH; r.t.; **4**. aq. citric acid, NaOH, dioxane; reflux, 2.5 h;**5**. phthalic anhydride; 140–180 °C;**6**. ether; r.t., 18 h

*Figure 5.36* *Hydroxy[$^{14}$C]methylations of nitromethane*

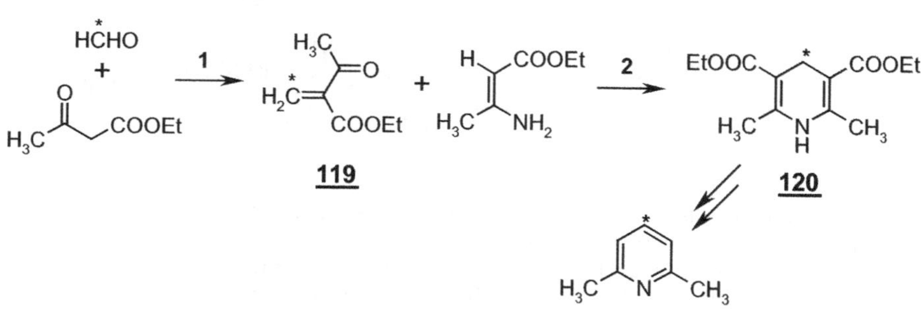


**Reaction conditions:** **1**. $Et_2NH$ (trace), $H_2O$; 0 °C, 6 h; r.t., 1–2 d; **2**. ethyl acetoacetate, $NH_{3,gas}$; 0 °C, 1 h; r.t., 1–2 d

*Figure 5.37* *[$^{14}$C]Methylenation of alkyl $\beta$-keto acetates: Hantzsch synthesis of 2,6-dialkyl [4-$^{14}$C]pyridines*

of this sequence to ethyl acetoacetate and $H^{14}CHO$ produced 2-acetyl[3-$^{14}$C]acrylate (**119**). This intermediate reacted with 3-aminocrotonate (formed *in situ* from the reaction of ammonia with an extra equivalent of ethyl acetoacetate) to afford diethyl 1,4-dihydro-2,6-dimethyl[4-$^{14}$C]pyridine-3,5-dicarboxylate (**120**). Oxidation, saponification and decarboxylation converted **120** into 2,6-dimethyl[4-$^{14}$C]pyridine (Chapter 6, Section 6.4.4)[116].

The reaction of $H^{14}CHO$ with *N*-acyl protected dialkyl aminomalonates and $\alpha$-aminoketones stops at the hydroxymethylation step. This behavior could be exploited for the synthesis of (*R,S*)-, (*R*)- and (*S*)-[3-$^{14}$C]serine, (*R,S*)-[3-$^{14}$C]cysteine and e.p. [1-$^{14}$C]-chloramphenicol, as illustrated in Figure 5.38. In the [3-$^{14}$C]serine preparations, saponification and decarboxylation converted the hydroxy[$^{14}$C]methyl derivatives **121** (R = acetyl or *p*-nitrobenzoyl), generated from $H^{14}CHO$ and diethyl *N*-acylaminomalonate, to racemic *N*-acyl[3-$^{14}$C]serine (**122**). Optical resolution of **122** with quinine and brucine followed by acid-catalyzed cleavage of the *N*-acyl groups afforded the enantiomerically pure

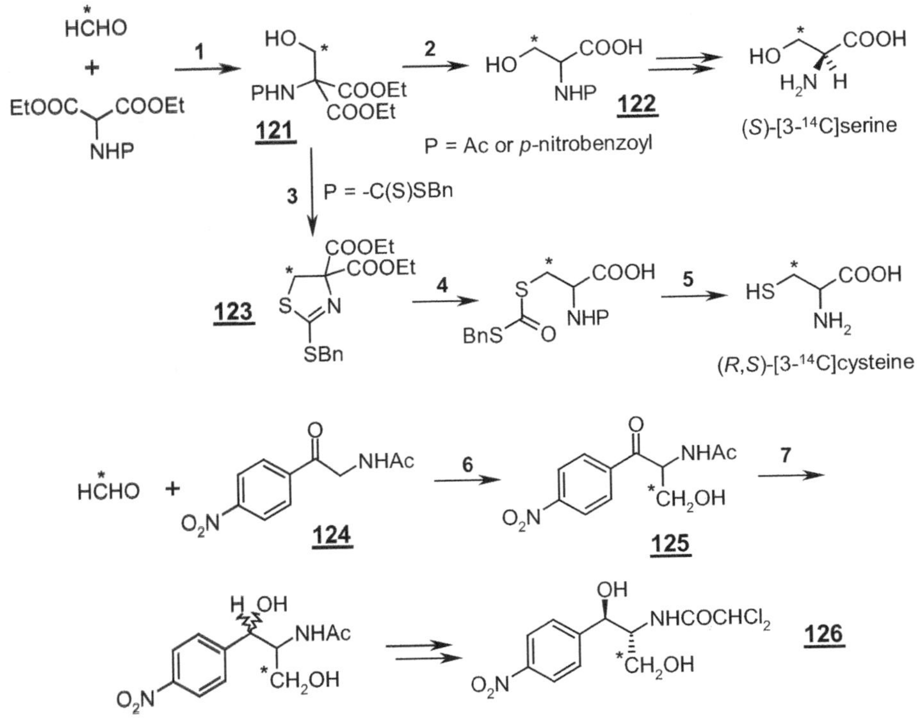


**Reaction conditions: 1a**. aq. $Ba(OH)_2$; r.t., 24 h; **b**. aq. AcOH; reflux, 1 h (P=Ac) **or 1**. pyridine/$H_2O$ 1:1; r.t., 20 h (P = *p*-nitrobenzoyl) **or 1**. pyridine/$H_2O$ 1:2; r.t., 16 h (P = –C(S)SBn); **2a**. $Ba(OH)_2$ $H_2O$; r.t., 12 h, **b**. HoAc,$H_2O$; reflux, 1 h (P =Ac) **or 2a**. 2N NaOH, **b**. aq. HCl; reflux, 1.5 h (P = *p*-nitrobenzoyl); **3**. $SOCl_2$, $Et_2O$; 0°C, 1 h; **4a**. 5N ethanolic NaOH, **b**. 3N HCl; reflux, 2 h; **5a**. 0.5 NaOH; 30 min, **b**. conc. HCl; reflux 24 h; **6**. $NaHCO_3$, EtOH; 35 °C, 2 h; **7**. Al(*i*-OPr)$_3$,*i*-PrOH; reflux, 7 h

***Figure 5.38*** *Syntheses initiated with hydroxy[$^{14}$C]methylation of dialkyl N-acylaminomalonates and N-acyl aminoketones*

amino acids in overall radiochemical yields of 20% and 15%, respectively[117]. With diethyl *N*-benzyldithiocarbamate-protected aminomalonate, intermediate **121** (P=–C(S)SBn) was obtained, which gave access to (*R*,*S*)-[3-$^{14}$C]cysteine and (*R*,*S*)-[3-$^{14}$C]cysteine in three steps upon treatment with thionyl chloride and subsequent decarboxylation and ring opening of the intermediately generated thiazoline system **123**[118].

Hydroxy[$^{14}$C]methylation of *N*-(*p*-nitrophenacyl)acetamide (**124**), a *N*-acetyl protected $\alpha$-aminoketone, provided [$^{14}$C]hydroxymethyl derivative **125**, which was converted to [1-$^{14}$C]chloramphenicol (**126**) by means of straightforward transformations, e.g. Meerwein–Ponndorf–Verley reduction, hydrolytic cleavage of the *N*-acetyl group, optical resolution with (−)-dibenzoyltartaric acid and re-acylation of the amino group with methyl dichloroacetate[119].

Fluorodesilylation is a valuable alternate method for the *in situ* generation of carbanions, which can be trapped as hydroxy[$^{14}$C]methyl derivatives through treatment with H$^{14}$CHO. This sequence, applied to diethyl difluoro(trimethylsilyl)methylphosphonate (**127**), afforded the *O*-silylated intermediate **128**, which served as a precursor for the preparation of 2,2-difluoro[1-$^{14}$C]ethylene. This was conveniently accomplished through thermolysis of **128** in the presence of catalytic amounts of CsF[120].

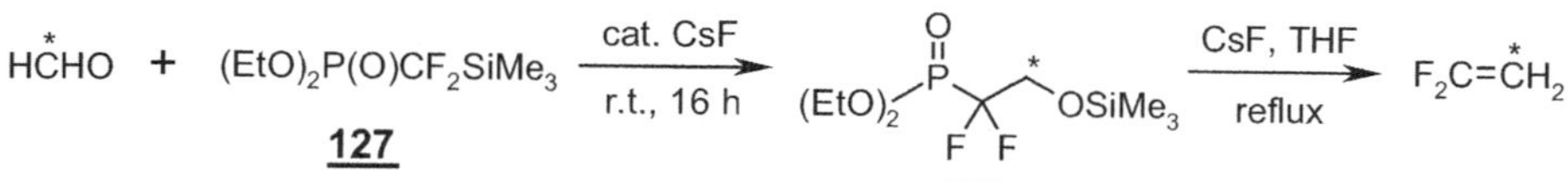


**127** **128**

***Figure 5.39*** *Trapping of [$^{14}$C]formaldehyde-derived carbanions through fluorodesilylation*

The reactivity of H$^{14}$CHO allows also for the *ortho*-[$^{14}$C]formylation of weak nucleophiles such as aryloxymagnesium halides in the presence of HMPA[121]; unfortunately two equivalents of formaldehyde are required, so radiochemical yields are not high. This method is complementary to some *C*-[$^{14}$C]formylation reactions with [$^{14}$C]formylating agents prepared from [$^{14}$C]formic acid (see Section 5.3). The resulting *ortho*-hydroxybenz [$^{14}$C]aldehydes can be readily converted into the corresponding [$^{14}$C]methylarenes through hydrogenolytic deoxygenation, as illustrated by the preparation of **129**[122].

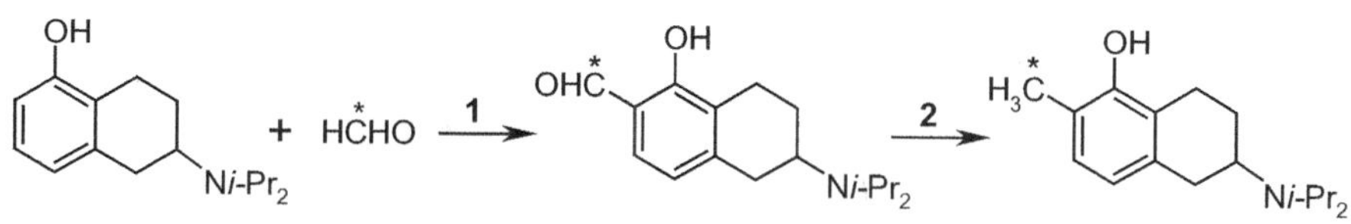


**129**

**Reaction conditions: 1**. EtMgBr, HMPA, benzene; 20–80 °C, 4 h; **2**. H$_2$, Pd/C, HCl, EtOH

***Figure 5.40*** *ortho-[$^{14}$C]Formylation of aryloxymagnesium halides*

### 5.4.2 Acid-Catalyzed Hydroxy[$^{14}$C]methylations

Under acid catalysis [$^{14}$C]formaldehyde reacts with benzene, alkyl, alkoxy, and halobenzenes to give aryl[$^{14}$C]methanols. Depending on the reactivity of the arene and the nature of the acid catalyst, the initially formed hydroxy[$^{14}$C]methyl intermediates may further react to give either the corresponding diaryl[$^{14}$C]methylene species or aryl[$^{14}$C]methyl halides (Blanc chloromethylation). These outcomes are illustrated in Figure 5.41.

The reaction of desmethyl-$\alpha$-tocopherol **130** and para[$^{14}$C]formaldehyde, for example, gave the hydroxy[$^{14}$C]methyl derivative **131**. Hydrogenolytic cleavage of the alcoholic hydroxyl group followed by acylation of the phenolic hydroxyl provided $\alpha$-[5-methyl-$^{14}$C]-tocopherol (**132**) in an overall radiochemical yield of 48%[123]. Chloro[$^{14}$C]methylations are illustrated by the conversion of substrates **133** to **134** with H$^{14}$CHO under the influence of concentrated HCl. Subsequent transformation to [$^{14}$C]methyl **135**[124] and to aminoxy[$^{14}$C]-methyl **136**, the latter obtained through reaction with *N*-hydroxyphthalimide and

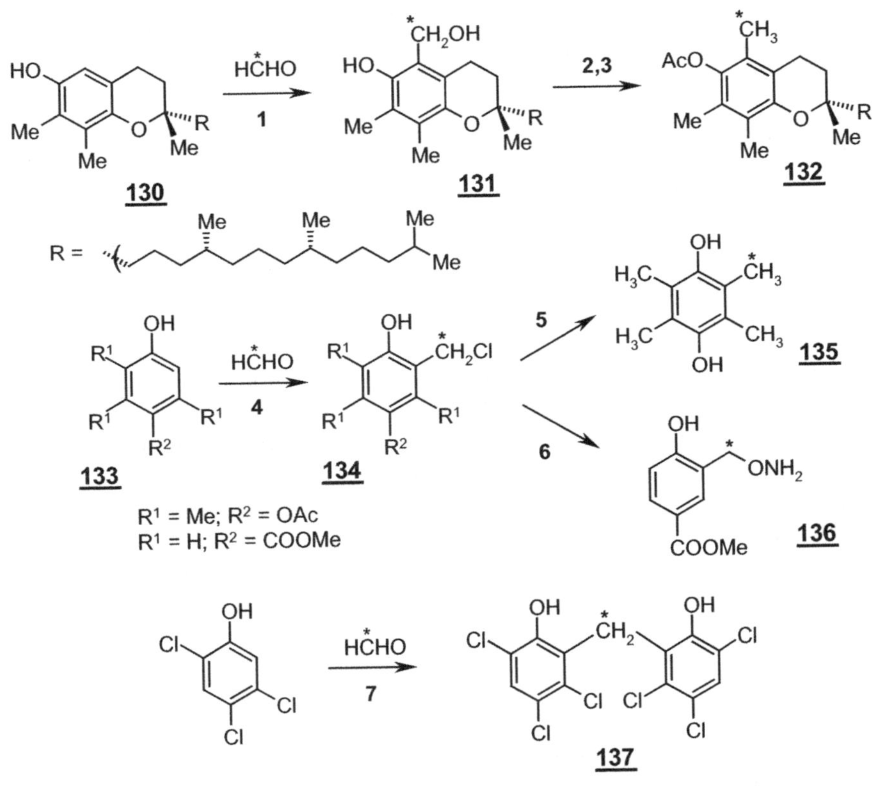

**Reaction conditions: 1**. $H_3BO_3$, toluene; 100 °C, 3 h; **2**. Zn, HCl–HOAc; r.t., 2 h; **3**. $Ac_2O$, pyridine; 95 °C, 2 h; **4**. conc. HCl; –20 °C, 1 h, 55°C, 2.5 h; **5**. $LiAlH_4$, $Et_2O$; r.t., 30 min; **6a**. *N*-hydroxyphthalimide, $Et_3N$, DMF; r.t., 5 d, **b**. *n*-butylamine, MeOH; reflux, 3 h; **7**. $H_2SO_4$, toluene

***Figure 5.41*** *Acid-catalyzed additions of H$^{14}$CHO to phenols to produce hydroxy[$^{14}$C]-methyl, chloro[$^{14}$C]methyl or diaryl[$^{14}$C]methyl products*

subsequent aminolytic cleavage of the phthaloyl residue[125], exemplify options for further transformations of the new function. Byproducts sometimes encountered with acid-catalyzed hydroxymethylations are diarylmethanes derived from addition of two equivalents of the respective phenol to formaldehyde. They can be selected for, so as to achieve the syntheses of diarylmethanes in good yield. An example of this is the synthesis of [$^{14}$C]hexachlorophene (**137**) in a yield of 60%[126].

$H^{14}CHO$ also undergoes acid-catalyzed addition to olefinic substrates (Prins reaction). Treatment of the tetrahydropyridine derivative **138** with $H^{14}CHO$ in half-concentrated HCl provided racemic 3-hydroxy[$^{14}$C]methyl product **139**, a key intermediate in the synthesis of e.p. [7-$^{14}$C]paroxetine[127]. Similarly, Prins-hydroxy[$^{14}$C]methylation of acylfulvene **140** in the presence of 1 M $H_2SO_4$ afforded [$^{14}$C]HMAF (**141**), an anticancer agent derived from illudin S, a highly toxic sesquiterpene[128].

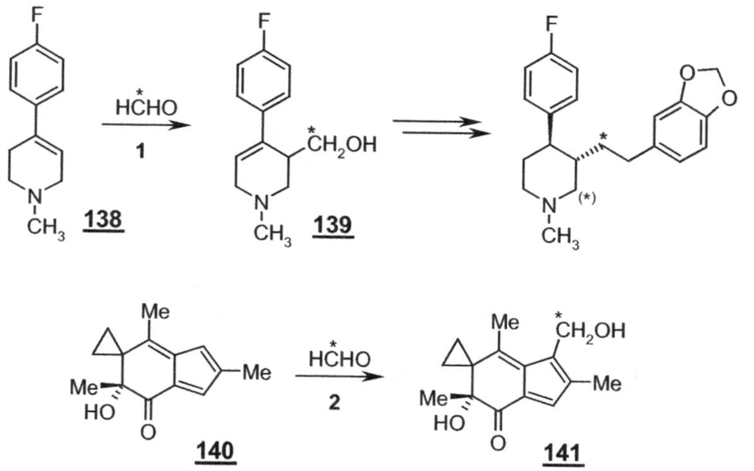


**Reaction conditions: 1.** 6N HCl; reflux; **2.** 1M $H_2SO_4$/THF 1:2 ; r.t., 48 h

***Figure 5.42*** *Acid-catalyzed hydroxy[$^{14}$C]methylation of olefinic substrates*

Under anhydrous conditions $H^{13}CHO$ reacts with benzyl alcohol and HCl gas to furnish benzyloxy[$^{13}$C]methanol. Treatment of this compound with thionyl chloride furnished, in 67% yield, benzyloxy[$^{13}$C]methyl chloride (**142**), a highly valuable electrophilic hydroxy[$^{13}$C]methylation reagent (Figure 5.43). **142** was successfully employed[129] in the diastereoselctive synthesis of e.p. (S)-[3-$^{13}$C]serine, starting with its reaction with the lithio derivative of Schöllkopf's *cyclo*-(D-Val-Gly) lactim ether. This provided d.p. intermediate **143** in moderate yield. Hydrolytic cleavage of the cyclic dipeptide followed by hydrogenolytic cleavage of the *O*-protecting group converted **143** into the e.p. amino acid in 35% overall yield. There is little doubt that this procedure should be also applicable to carbon-14 synthesis (see Chapter 11, Section 11.3.6). Analogously, reaction with methanol and HCl gas converts [$^{14}$C]-paraformaldehyde into chloro[$^{14}$C]methyl methyl ether. This was trapped with triphenylphospine to give methoxy[$^{14}$C]methyltriphenylphosphonium chloride (**144**), a highly valuable modified Wittig reagent. The use of **144** allows the chain extension of carbonyl compounds by one carbon-14 unit to furnish [carbonyl-$^{14}$C]aldehydes. This is exemplified by the conversion of ketone **145a** to [$^{14}$C]aldehyde **145b** via the intermediately formed [$^{14}$C]enol ether[130a]. The analogous chloro[$^{14}$C]methylphosphite **146b** has been successfully prepared via the hydroxy[$^{14}$C]methylphosphite **146a** and used to extend an aldehyde to a [$^{14}$C]vinylchloride[130b] as depicted in Figure 5.43.

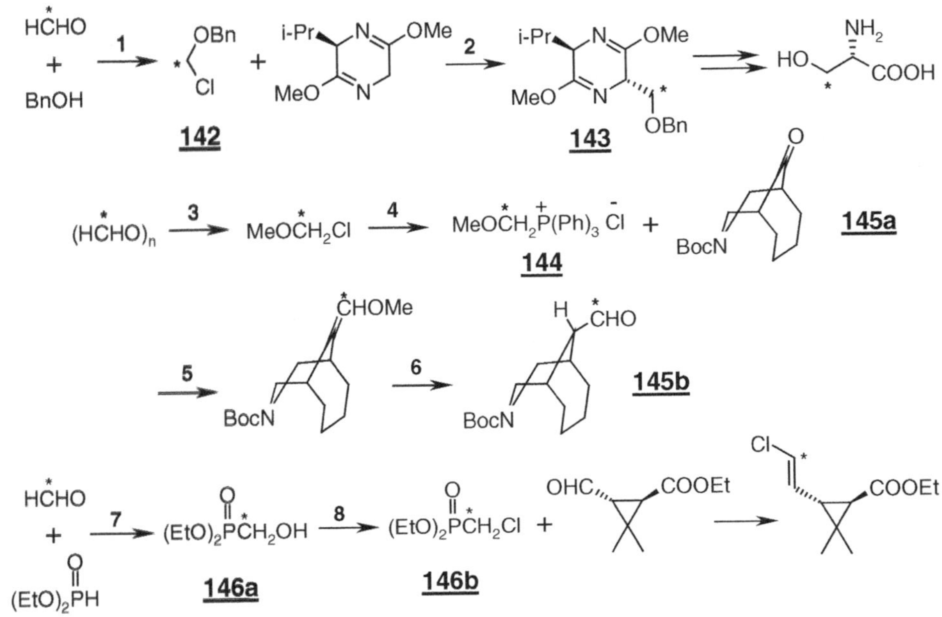


**Reaction conditions:** **1a**. $HCl_{gas}$, THF; r.t., 1 h, **b**. $SOCl_2$, THF; 5 °C; **2**.*n*-BuLi, THF; −70 °C; **3**. $HCl_{gas}$;−10 °C, 10 min; **4**. $PPh_3$, benzene; 80 °C, 16 h; **5**. NaHMDS, THF; −10 °C, 16 h; **6**. HOAc, r.t.; **7**. aq. $K_2CO_3$; r.t., 2 h; **8**. $PPh_3$, $CCl_4$; reflux, 16 h

***Figure 5.43*** *Synthesis of chloro[$^{14}$C]methyl ethers and diethyl hydroxy[$^{14}$C]methylphosphonate and their application in the one-carbon chain extension of carbonyl compounds and cyclic glycinates*

### 5.4.3 Amino[$^{14}$C]methylations

An important use of $H^{14}CHO$ has been found in amino[$^{14}$C]methylations of nucleophiles in the presence of primary and secondary amines (Mannich reaction). Nucleophilic substrates include C–H acidic aliphatic compounds (e.g. aldehydes, ketones, esters, nitroalkanes and nitriles), and reactive aromatic and heteroaromatic substrates such as indoles, furans and phenols. The amino[$^{14}$C]methylations of some C–H acidic methyl, methylene and methine substrates are exemplified in Figure 5.44. The immediate products, alkyl- or dialkylamino[$^{14}$C]methylene derivatives ('Mannich bases'), can undergo a number of useful synthetic transformations. Condensation of $H^{14}CHO$ with acetophenone and dimethylamine provided the carbon-14-labeled $\beta$-aminoketone **147** in 51% yield[131]. $\beta$-Aminocarbonyl compounds such as **147**, while stable at room temperature, can eliminate

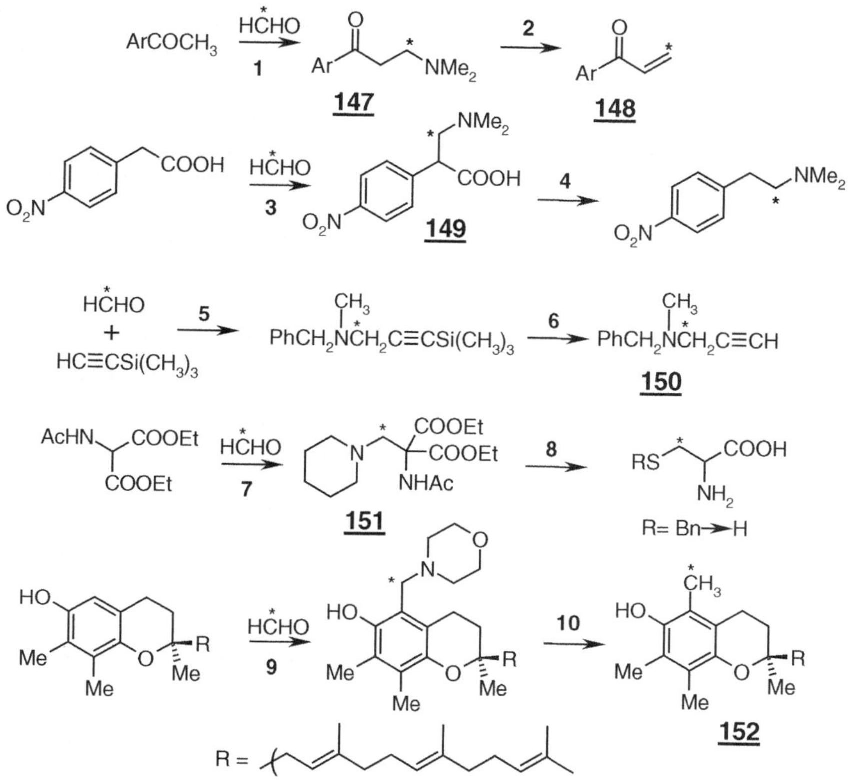

**Reaction conditions:** **1**. $Me_2NH.HCl$; 95 % EtOH, reflux, 2 h; **2**. $H_2O$; steam distillation; **3**. $Me_2NH$, $H_2O$; r.t., 4 d; **4**. $CuCO_3$, pyridine; reflux; **5**. *N*-methyl benzylamine, $Cu(OAc)_2.2H_2O$; dioxane; 85 °C, 24 h; **6**. KOH, EtOH; r.t., 5.5 h; **7**. piperidine, $H_2O$; r.t., 30 min (70 %); **8a**. MeI; r.t., 72 h, **b**. $PhCH_2SH$, EtOH; reflux, 5 h, **c**. conc. HCl; 90 °C, 5 h **d**. $Na(NH_3Cl)$; **9a**. morpholine; 75 °C, 1.5 h, **b**. 130 °C, 4.5 h; **10**. $NaCNBH_3$,*i*-BuOH; 110 °C, 5.5 h

***Figure 5.44*** *Mannich reactions of [14C]formaldehyde with C–H acidic aliphatic compounds*

dialkylamine upon heating to give $\alpha,\beta$-unsaturated carbonyl compounds such as aryl [14C]vinyl ketone **148**. In the second example, thermal decarboxylation of 2-dialkylamino [14C]methyl-2-arylacetic acids **149**, obtainable through Mannich reaction of 3-arylacetic acids with $H^{14}CHO$ and dialkylamines, offers convenient access to *N,N*-dialkyl-2-aryl-[1-14C]ethylamines, a pharmacologically interesting structural class[132]. The synthesis of [14C]pargyline (*N*-methyl-*N*-[1-14C]propargyl)benzylamine, **150**), a monoamine oxidase inhibitor, demonstrates that terminal alkynes can also undergo Mannich reactions when catalyzed by cupric acetate[133]. Furthermore, the amino[14C]methylation of acetylamidomalonate provided an intermediate **151** suitable for conversion by standard methods into 3-functionalized $\alpha$-amino[3-14C]acids, in this case racemic [3-14C]cysteine[134]. A key feature of this sequence is the facile replacement of the dialkylamino function by alternate functionalities through nucleophilic displacement of the quaternized amino group. Finally, arylmethyl amines are susceptible to reductive cleavage of the amino function either hydrogenolytically or by appropriate complex hydrides. This fact has been exploited for the selective introduction of methyl groups, as illustrated with the synthesis of $\alpha$-5-[$^{14}$C]methyl-tocotrienol (**152**)[135].

Reactive aromatic systems used as substrates for Mannich reactions with H$^{14}$CHO include substituted or unsubstituted indoles; these provided the corresponding 3-dimethylamino[$^{14}$C]methylindoles **153**[111,136,137](Figure 5.45). When the dialkylamino group is of the benzylic type, as in this case, it can undergo displacement without prior quaternization. Thus, treatment with cyanide ion produced 3-indolyl[2-$^{14}$C]acetonitrile, which could easily be hydrolyzed to 3-indolyl[2-$^{14}$C]acetic acid[136,137] (**154**). Displacement of the dimethylamino group with sodium diethyl formamidomalonate provided, in 82% yield, diethyl 2-(3-indolyl[$^{14}$C]methyl)-2-acetylaminomalonate (**155**), which was taken on in high yield to (*R*,*S*)-[alanine-3-$^{14}$C]tryptophan (**156**)[111,137,138]. Compound **156** served *inter*

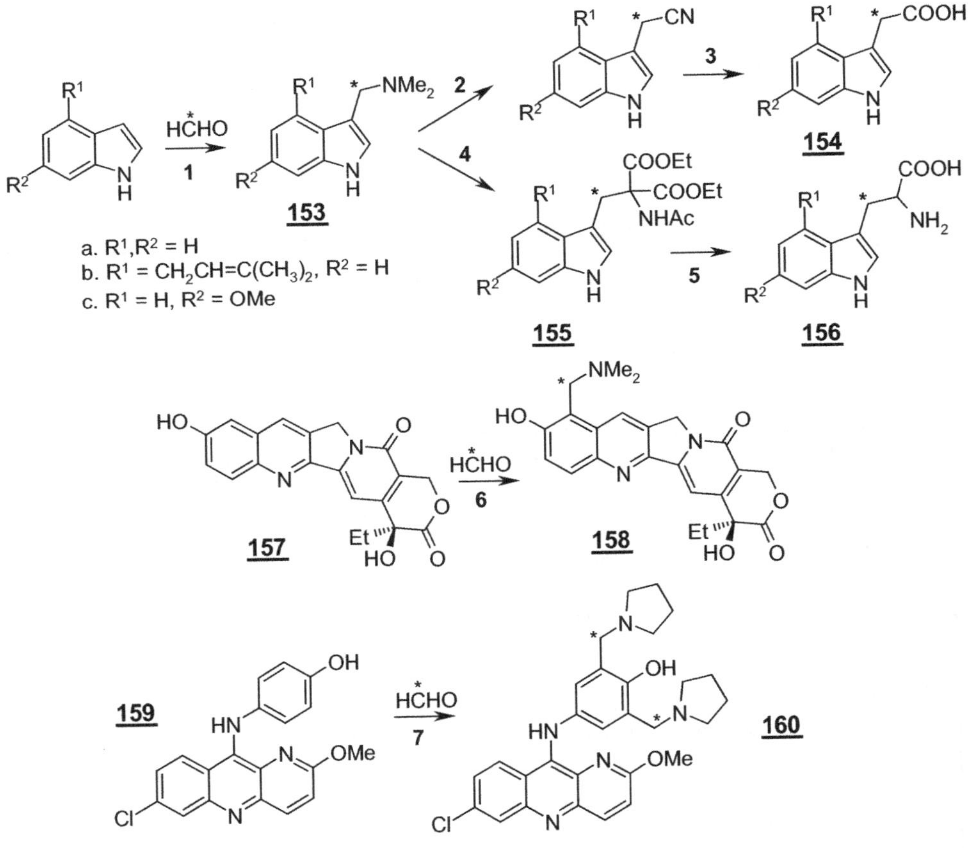


**Reaction conditions:** **1**. (H$^{14}$CHO)$_n$, Me$_2$NH, aq. AcOH; r.t., 4 to 20 h; **2**. NaCN, H$_2$O; reflux, 6 h; **3**. NaOH; reflux, 6 h (50 % over two steps); **4**. (EtOOC)$_2$CHNHAc, NaOH; toluene (xylene); reflux, 3–6 h; **5**. 2N NaOH, reflux, 2–6h followed by HOAc, reflux 3–4 h; **6**. H$^{14}$CHO, Me$_2$NH, HOAc; 75 °C (68 %); **7**. (H$^{14}$CHO)$_n$, pyrrolidine, EtOH; 45–55°C, 12–14h (36%) or 50–60°C, 1 h. Then 140 °C, 25 min using microwave irradiation (45 %)

***Figure 5.45*** *Mannich reactions of H$^{14}$CHO with reactive aromatics*

*alia* as a precursor for the carbon-14 labeling of somatostatin derivatives as well as for the biological preparation of [5-$^{14}$C]lysergic acid[111]. Subsequent esterification with methanolic HCl, enzymatic hydrolysis with chymotrypsin, re-racemization of the (*R*)-amino acid ester upon heating to 200 °C and repeated enzymatic hydrolysis has been developed as a standard procedure for the preparation of (S)-[alanine-3-$^{14}$C]tryptophan (Chapter 11, Section 11.1). A slightly modified procedure was applied to the synthesis of (*R*,*S*)-[3-$^{14}$C]tryptophans, differently substituted at the C4[138] and C6[139] of the indole system. Other substrates for Mannich reactions with H$^{14}$CHO include camptothecin **157**, giving the anticancer topoisomerase-I inhibitor [$^{14}$C]topotecan (**158**)[140], and the phenol **159**, furnishing the bis-adduct [$^{14}$C]pyronaridine (**160**), an antimalarial drug[141].

Primary amines can undergo double Mannich reactions, and this was exploited for the synthesis of several labeled bicyclic products. Thus, H$^{14}$CHO was treated with ethylamine and *N*-methylpiperidone to provide 3-methyl-7-ethyl-9-oxo-3,7-diaza[6, 8-$^{14}C_2$]bicyclo [3.3.1]nonane (**161**, X = NMe, R = Et), a key intermediate in the synthesis of [$^{14}$C]-bisaramil (**162**), an antiarrhythmic drug[142]. 4-Thiocyclohexanone reacted analogously with $(H^{14}CHO)_n$ and benzylamine to give 7-benzyl-7-aza-3-thia[6, 8-$^{14}C_2$]bicyclo[3.3.1]-nonane (**161**, X = S, R = Bn), another investigational antiarrhythmic drug substance[143].

With $\gamma,\delta$-unsaturated $\beta$-ketocarboxylic acid esters as nucleophiles, 1-alkyl-3-alkoxy-carbonyl-4-[2-$^{14}$C]piperidones can be formed via a tandem Michael addition/Mannich

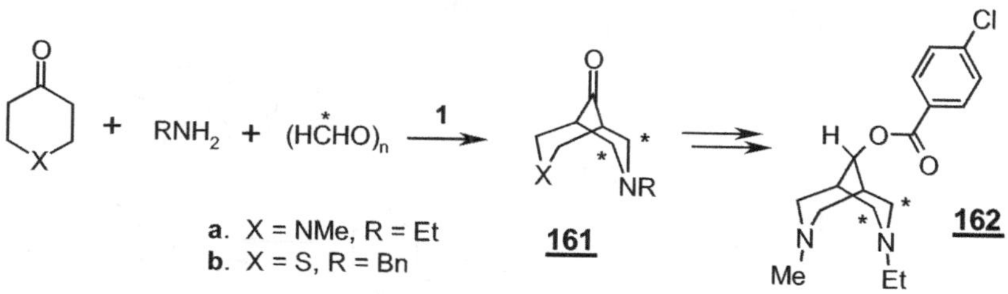


**Reaction conditions: a1**. HOAc, MeOH; reflux, 2.5 h; **b1**. HCl–HOAc, MeOH; reflux, 6 h

***Figure 5.46*** *[$^{14}$C]Formaldehyde in double Mannich reactions on primary amines*

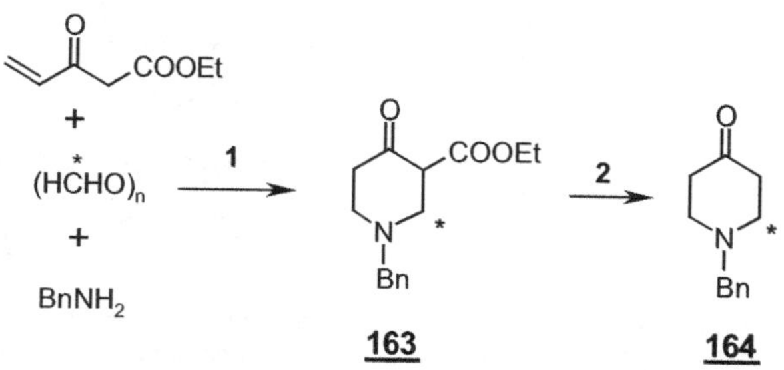


**Reaction conditions: 1**. Et3N3 molecular sieves, DMSO; 25 °C, 16 h; **2**. 20 % HCl; reflux, 1 h

***Figure 5.47*** *Preparation of [$^{14}$C]piperidines via tandem Mannich/m.chael reaction (procedure A)*

reaction; these products can be decarboxylated to the respective 1-alkyl-4-[2-$^{14}$C]piperidones, as exemplified by the preparation of **163** and **164**[144]. *N*-substituted, carbon-14-labeled, piperidones and piperidines are valuable building blocks in the radiolabeling of various pharmacologically interesting drug substance classes.

The Mannich reaction-based halide-promoted cyclization of [$^{14}$C]methylene iminium ions, generated *in situ* from $H^{14}CHO$ and appropriate 3,4-doubly or triply unsaturated butylamines (Overman reaction) offers an alternate access to C4-functionalized [$^{14}$C]-piperidines and tetrahydro[$^{14}$C]pyridines, as illustrated in Figure 5.48. The observed

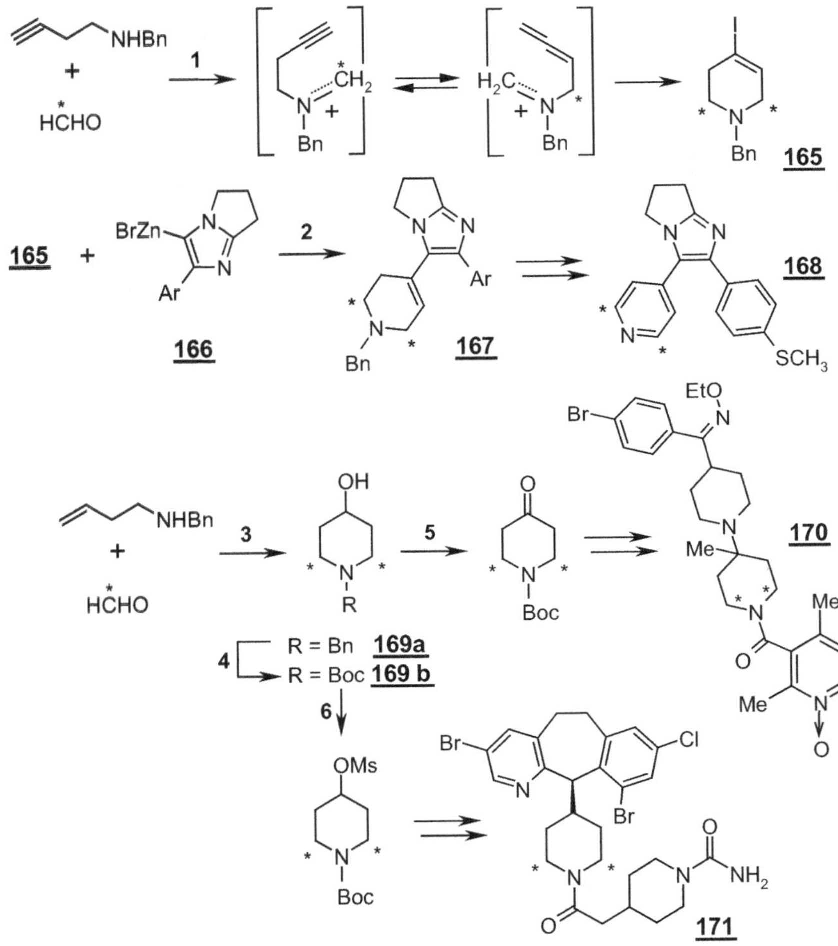


**Reaction conditions:** 1. NaI, (+)-10-camphorsulfonic acid, $H_2O$; 95 °C, 2.5 h (60 %); **2**. $Pd(Ph_3)_4$, THF; r.t., 1 h (65 %); **3**. (+)-10-camphorsulfonic acid, $H_2O$; reflux, 4 h (93 %); **4**. $H_2$, $Pd(OH)_2$, $(Boc)_2O$, EtOAc; r.t., 25 h (95 %); **5**. TPAP, NMO, mol. sieves, $CH_2Cl_2$; r.t. (95 %);**6**. MsCl, $Et_3N$, MED; 0 °C, 2.5 h

**Figure 5.48** *Preparation and application of [$^{14}$C]piperidines via Mannich–Overman reaction (procedure B)*

random distribution of the label at C2 and C6 of the 4-iodotetrahydro[$^{14}$C]piperidine (**165**) derived from the 3-butynylamine was interpreted as evidence of rapid and reversible [3,3]-sigmatropic rearrangement (aza-Cope rearrangement) of the [$^{14}$C]iminium ion, prior to its quenching by iodide in the ring-closing step[145]. Subsequent Pd(Ph$_3$)$_4$-catalyzed cross-coupling of **165** with the organozinc reagent **166** furnished **167**, key intermediate in the preparation of [$^{14}$C]SK&F 105809 (**168**). In similar fashion but in the absence of iodide, the corresponding alkenylamine provided *N*-benzyl[2, 6-$^{14}$C]piperidin-4-ol (**169a**). Switching of *N*-protecting groups converted **169a** into **169b**, which served as key intermediate to the CCR5 receptor antagonist [$^{14}$C]SCH351125 (**170**)[146a] and [$^{14}$C]-lonafarnib (**171**), an investigational anticancer drug[146b].

The classical Pictet–Spengler procedure, in which *N*-alkyl arylamines react with HCHO to give annulated tetrahydropiperidines, is an additional example of iminium ion initiated cyclizations. Being comparably weak, iminium ions require rather strong nucleophilic systems such as phenols, phenol ethers or indoles. $\beta$-[1-$^{14}$C]Carbolines emerge from cyclization onto C2 of an indole, as exemplified by conversion of substrates **172** to products **173** in moderate yields in pH = 4 aqueous solution (R = H or OMe)[147] or in acetate buffer (R = H or Cl); compound **173** (R = Cl) was an intermediate in the synthesis of the IKK inhibitor **174**[148]. Mannich condensation of 7-benzyloxy1-(4-benzyloxy-3-hydroxybenzyl)-6-methoxy-1,2.3,4-tetrahydroisoquinoline (**175**) with H$^{14}$CHO cyclized onto a phenol moiety to furnish a mixture of isomeric [8-$^{14}$C]berberines **176**. Subsequent selective *O*-methylation, separation of the isomers by crystallization and acid-mediated debenzylation converted **176** into [8-$^{14}$C]stepholidine (**177**), a dopamine receptor antagonist[149].

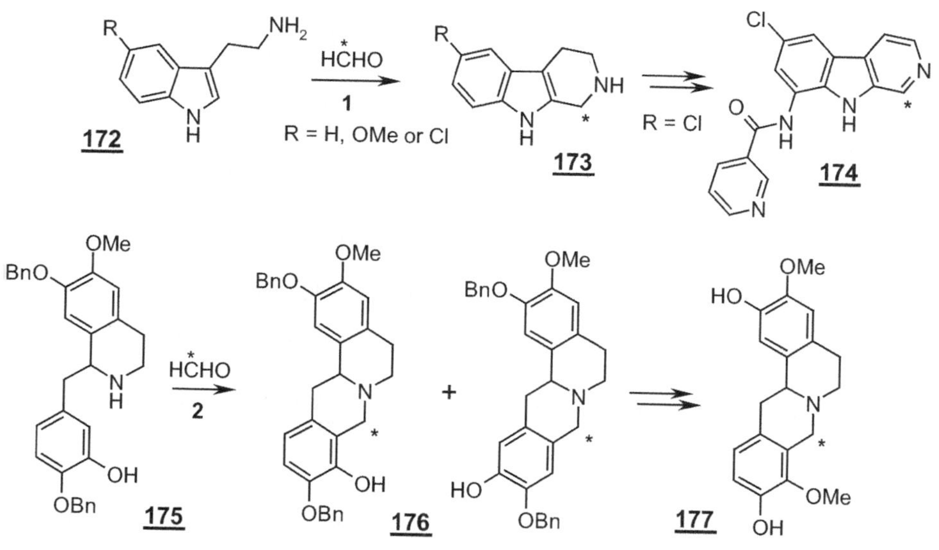


**Reaction conditons: 1.** H$_2$O at pH 4; r.t., 12 h (R = H, OMe) or acetate buffer; 105 °C, 4 h (R = H, Cl);**2.** THF/H$_2$O 1:2 at pH 6.4; r.t., 16 h

***Figure 5.49*** *Uses of [$^{14}$C]formaldehyde in Pictet–Spengler cyclizations*

### 5.4.4 Reductive Methylations

In the absence of appropriate nucleophiles, [$^{14}$C]methyleneiminium moieties resulting from condensation of H$^{14}$CHO with amine functions can be trapped either by hydride donors, formic acid or catalytic hydrogenation, giving rise to *N*-[$^{14}$C]methylated derivatives. Particularly, the reductive methylation of primary and secondary amines, peptides and proteins with H$^{14}$CHO and $NaBH_4$, or preferably $NaBH_3CN$, has become a standard method because of its mildness, rapidity and simplicity. In peptides and proteins, [$^{14}$C]methylations usually occur selectively at internal lysyl residues and at terminal amine functions. Methylations at the latter positions usually do not alter the substrates' biological activities to a significant degree. Examples of such labeling are shown for erythromycin (**178**)[150], thionin (**179**)[151], chlorpromazine (**180**)[152] and several proteins[153] are summarized in Figure 5.50.

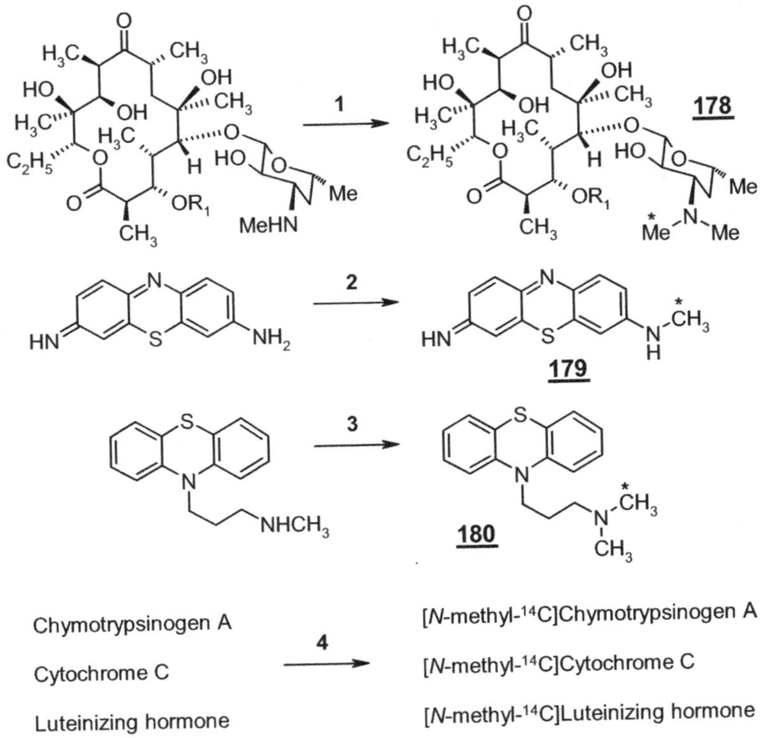


**Reaction conditions: 1.** H$_{14}$CHO, $H_2$(1 atm), Pd black, EtOH; 28 °C, 48 h; **2.** H$_{14}$CHO, $H_2$(1.6 atm), $PtO_2$, EtOH; r.t., 3 h; **3.** H$^{13/14}$CHO, $NaBH_4$, MeOH, $H_2O$; r.t., 1 h; **4.** H$_{14}$CHO, $NaBH_3CN$, phosphate buffer pH 7; 25 °C, 1 h

***Figure 5.50*** *Use of H$^{14}$CHO in the reductive [$^{14}$C]methylation of primary and secondary amine derivatives*

In contrast to alkylations using [$^{14}$C]MeI, quarternization cannot occur in reductive [$^{14}$C]methylations. In reductive [$^{14}$C]methylations of primary amines the use of a sufficiently large excess of amine can suppress formation of tertiary amine products.

Reductive [$^{14}$C]methylation procedures have found broad application in the rapid labeling of investigational drug substances for *in vitro* studies where metabolism does not occur, in early exploratory *in vivo* investigations in drug research and development, and in peptide and protein mapping. However, because *N*-dealkylation reactions are common metabolic events, compounds labeled by *N*-[$^{14}$C]methylation are usually not suitable for registration studies.

### 5.4.5 Polycondensations

In the absence of (reactive) nucleophiles, H$^{14}$CHO can undergo polycondensations with amines. Condensation with ammonia forms hexa[$^{14}$C]methylenetetramine (**181**), which has served as an intermediate in the preparation of [$^{14}$C]RDX (**182**)[154] and of [$^{14}$C]HMX (**183**)[155], potent explosives. Condensation with methyl glycinate produced the 1,3,5-trisubstituted [2, 4, 6-$^{14}C_3$]triazine **184**, from which [$^{14}$C]glyphosate (**185**), a broad-spectrum herbicide, was prepared[156]. With 1,3-propandiamine, hexahydro[2-$^{14}$C]-pyrimidine (**186**) is formed; it was further converted into the corresponding 1,3-dinitroso derivative **187**, a cyclic dialkylnitrosamine found in food and drugs[157].

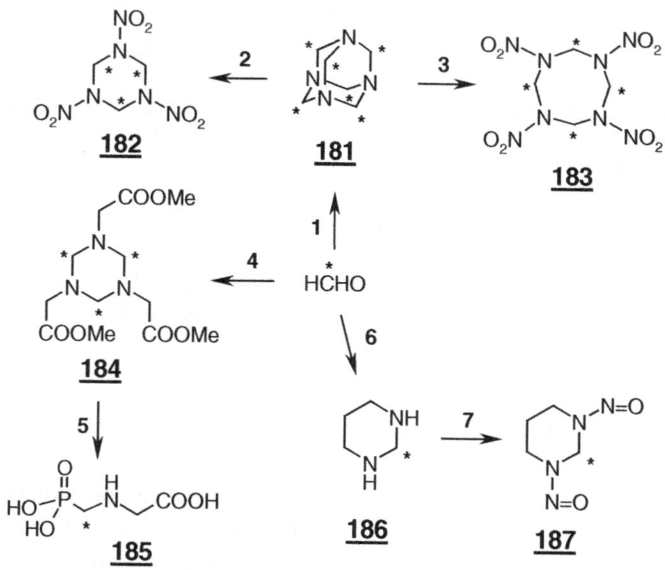

**Reaction conditions:** **1**. conc. aq. $NH_3$; 28–35 °C, 24 h; **2**.$HNO_3$; 5–10 °C, 1 h; 15–20 °C, 3 h; **3**. $NH_4NO_3$, $HNO_3$, HOAc, $Ac_2O$, $BF_3{\cdot}OEt_2$; 44 °C, 1 h; **4**. $MeOOCCH_2NH_3{+}Cl{-}$, $K_2CO_3$, $H_2O$; r.t., 24 h; **5a**. $(PhO)_3PO$, benzene; reflux, 4 h, **b**. conc. HCl–HCOOH 9:1; reflux, 6 h; **6**. $H_2N(CH_2)_3NH_2$, benzene, $H_2O$; r.t., 70 h; **7**. $NaNO_2$, HCl, $CH_2Cl_2$, $H_2O$; r.t., 1.5 h

***Figure 5.51*** *Polycondensation of H$^{14}$CHO in the presence of primary amines*

### 5.4.6 Thio[$^{14}$C]methylations

In a Mannich type reaction $H^{14}CHO$ reacts with thiophenols and C–H acidic compounds to furnish the respective thio[$^{14}$C]methylated derivatives, as demonstrated with the two examples in Figure 5.52. In the first case, treatment of $H^{14}CHO$ with 4-chlorothiophenol and nitroethane yielded 1-(4-chlorophenylthio)-2-nitro[1-$^{14}$C]propane (**188**). Subsequent reduction converted **188** into [$^{14}$C]2998-60 (**189**), an investigational antidepressive drug substance[158]. Of perhaps broader synthetic utility is the $\alpha$-[$^{14}$C]methylation of $\alpha,\beta$-unsaturated carbonyl compounds through reaction with $H^{14}CHO$ in the presence of

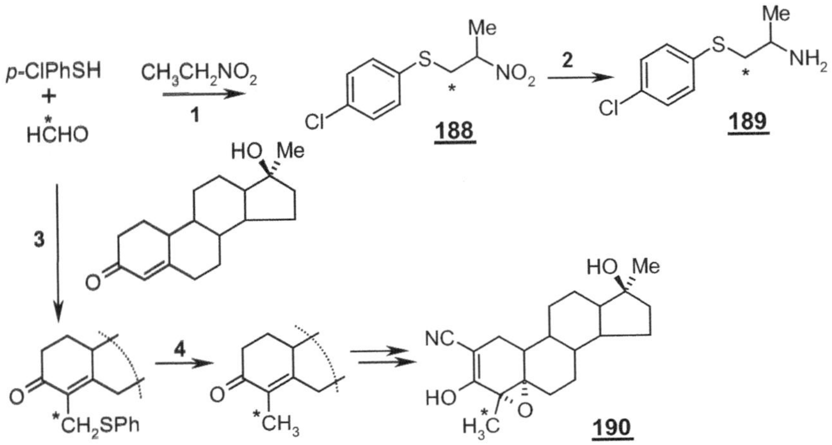


**Reaction conditions:** **1**. piperidine (cat.), dioxane; 100 °C, 12 h; **2**. $SnCl_2$, conc. HCl; 100 °C, 4 h; **3**. $H_{14}CHO$, p-ClPhSH, Et3N, aq. EtOH; reflux, 68 h; **4**. Raney-Ni, acetone; r.t., 3 h

***Figure 5.52*** *Thio[$^{14}$C]methylation of olefinic substrates*

4-chlorothiophenol and a tertiary amine, followed by Raney nickel-mediated desulfurization. This Baylis–Hillman type protocol was followed in the synthesis of [4-methyl-$^{14}$C]-epostane (**190**)[159].

## 5.5 [$^{14}$C]Methyl Iodide ($^{14}CH_3I$)

[$^{14}$C]Methyl iodide is one of the most frequently used carbon-14-labeled building blocks, being the source of both electrophilic and nucleophilic one-[$^{14}$C]carbon synthons. It is commercially available, packaged in breakseal ampoules from which it can be transferred into reaction vessels by static vacuum transfer. It is prepared by reaction of [$^{14}$C]methanol with hydriodic acid at temperatures of 40–90 °C[160]. The former is obtained by reduction of $^{14}CO_2$ with $LiAlH_4$ in diethylene glycol dimethyl ether and subsequent release from the reaction mixture through addition of tetrahydrofurfuryl alcohol and expulsion with nitrogen[160a,161].

In general *O*-, *N*-, *S*- and *C*-[$^{14}$C]methylation chemistry does not differ fundamentally from normal organic synthesis, except that for radioprotection reasons reactions are performed in sealed flasks, unlabeled reaction partners should be employed in excess when feasible, and unreacted [$^{14}$C]methyl iodide or other volatile [$^{14}$C]methyl byproducts should be vacuum transferred into a second vessel so as to avoid the release of volatile radioactivity.

### 5.5.1 [$^{14}$C]Methyl Iodide as an Electrophilic One-[$^{14}$C]Carbon Building Block

Aliphatic [$^{14}$C]methyl ethers are routinely prepared through [$^{14}$C]methylation of alkoxides formed by deprotonation of alcohols with sodium in a nonprotic solvent or under phase transfer catalysis conditions with $CsCO_3$ in the presence of tetrabutylammonium hydrogensulfate. Aromatic [$^{14}$C]methyl ethers are generally prepared using the bases sodium ethoxide or aqueous sodium hydroxide, or $K_2CO_3$ in acetone or DMF. *N*-[$^{14}$C]Methylation of amines in contrast is generally performed in the presence of a weak base such as $K_2CO_3$ or a non-nucleophilic amine. Because of their strong tendency for over-alkylation, primary amines are rarely used as substrates for reaction with [$^{14}$C]methyl iodide, and even secondary amines often cannot be cleanly [$^{14}$C]methylated. Competing quaternization can be minimized by use of a sufficient excess of the precursor amine; however, it is not always easy to separate the desired product from the excess starting material. A solution to this problem[162], at least for the preparation of tertiary amines containing an *N*-[$^{14}$C]methyl group, is to start with the analogous (tertiary) amine possessing a benzyl group. Alkylation of this substrate with $^{14}CH_3I$ will give a quaternary salt, which can be debenzylated by catalytic hydrogenation in high yield to furnish the desired product. The requisite starting

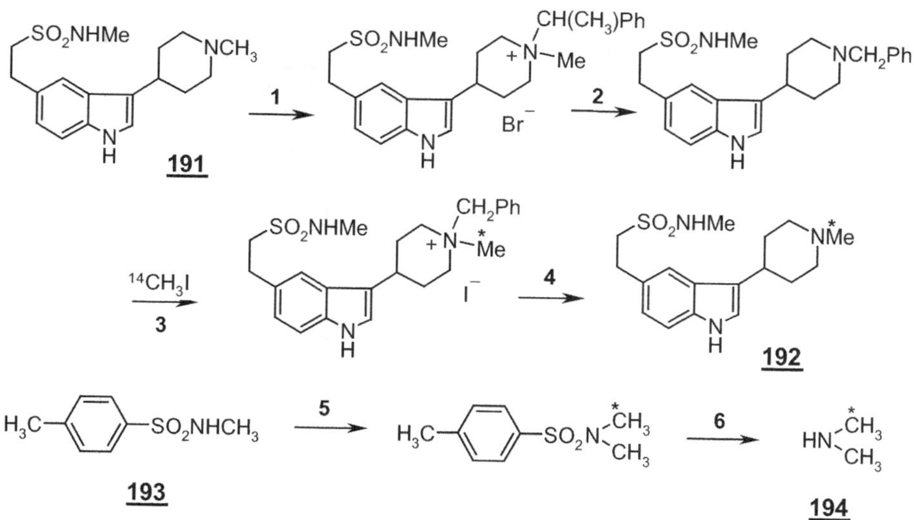


**Reaction conditions: 1.** $PhCH(CH_3)Br$, acetone; reflux, 4 d (97%) **2.** $LiBH[CH(Me)\text{-}CHMe_2]_3$, THF, reflux, 70 h; **3.** $^{14}CH_3I$, acetone; r.t.; 2–3 d (87%); **4.** $H_2$, 10 % Pd/C, amberlite resin IR400, aq. EtOH; r.t., 3 h (83%); **5.** $^{14}CH_3I$, KOH, aq. EtOH; 76 °C, 72 h; **6.** HI; 100 °C, 2 h

***Figure 5.53*** *Synthesis of tertiary and secondary amines possesing an N-[$^{14}$C]methyl group*

material is accessible through quaternization of the unlabeled target compound with benzyl bromide (or better $\alpha$-methylbenzyl bromide) followed by reductive *N*-demethylation using trisiamylborohydride (LS-Selectride). Such a sequence is illustrated in Figure 5.53 (**191** to **192**)[162]. A related strategy, suitable for preparing secondary amines containing a *N*-[$^{14}$C]methyl group, is to utilize as substrate a secondary *N*-benzyl or $\alpha$-methylbenzyl amine, or better, a secondary amide which can, after [$^{14}$C]methylation, undergo hydrolytic cleavage, as demonstrated in the sequence **193** to **194**[163].

Alternatively, in cases where the unlabeled *O*- or *N*-methyl target compound is available, the requisite starting material may be prepared from it by either ether cleavage (e.g., HI, HBr, $BBr_3$, TMSI, thiolates) or *N*-demethylation (e.g., BrCN, vinyl chloroformate, $\alpha$-chloroethyl chloroformate, MCPBA–$FeSO_4$).

With substrates possessing both hydroxyl and amino groups, selective *N*-[$^{14}$C]methylation is usually not problematic, but *O*-methylation often is. In the case of galantamine labeling[164], for example, *N*-desmethyl compound **195** reacted with $^{14}CH_3I$ under standard

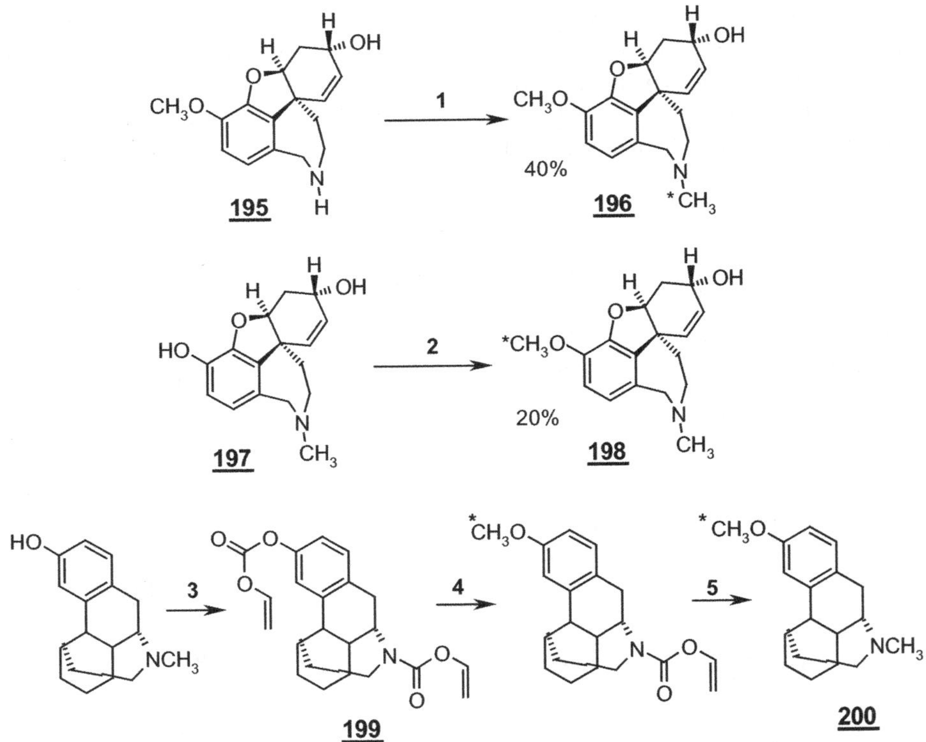


**Reaction conditions: 1.** $^{14}CH_3I$, $(i\text{-}Pr)_2NH$, MeOH; r.t., 16 h (37%); **2.** $^{14}CH_3I$, KOH, *N,N*-dimethylimidazolidine-2-one; r.t., 18 h (18%); **3.** $ClCOOCH{=}CH_2$, proton sponge,1,2-dichloroethane; 60 °C, 16 h; **4a.** NaOH, aq. dioxane; 50 °C, 4h, **b.** NaH, DMF; r.t., 1 h, **c.** $^{14}CH_3I$, r.t., 16 h; **5.** $LiAlH_4$, THF; 0 °C to r.t; 16 h

*Figure 5.54 Selective O-[$^{14}$C]methylations in presence of a competing N-methyl group*

conditions to form **196** in moderate yield, although a four-fold excess of **195** had to be used to suppress quaternary salt formation. In contrast, reaction conditions for [$^{14}$C]methylation of *O*-desmethyl compound **197** had to be carefully developed, and even then the *O*-[$^{14}$C]-methyl product **198** was obtained only in low yield. A novel strategy was used for the preparation of [*O*-methyl-$^{14}$C]dextromethorphan (**200**), which could not be effectively *O*-[$^{14}$C]methylated without extensive quaternization of the amine function. By this strategy, *O*-desmethyldextromethorphan was treated with vinyl chloroformate to give the *N,O*-bis(vinyloxycarbonyl) (VOC) derivative **199** with simultaneous *N*-demethylation. Subsequent selective removal of the *O*-VOC group by mild basic hydrolysis resulted in the generation of the protected nitrogen center inert to the subsequent *O*-[$^{14}$C]methylation, which proceeded in nearly quantitative yield. Concluding hydride reduction converted the *N*-VOC group to the required unlabeled *N*-methyl group and the desired **200**[165].

Unfortunately, *O*- and *N*-methyl groups are highly susceptible to biological degradation processes. Accordingly, in most cases *O*- and *N*-[$^{14}$C]methylated substrates are of limited use in routine pharmacokinetic, metabolic and ecological investigations. However, since they are conveniently accessible at low cost, they are frequently employed in *in vitro* experiments as well as in early stage development studies of potential drug substances.

Analogous to the preparation of [$^{14}$C]methyl ethers is the reaction of alkali thiolates with [$^{14}$C]methyl iodide to give the labeled thioethers. The synthesis of (*S*)-[methyl-$^{14}$C]-methionine (**201**), the most important representative of this substance class, may serve as an example[43,166]. It is conveniently available in 60–80% radiochemical yield by reaction of [$^{14}$C]methyl iodide with deprotonated (*S*)-homocysteine, generated *in situ* by reductive cleavage of the *S*-benzyl. It is widely used as a [$^{14}$C]methyl donor in fermentative labeling processes (see Chapter 12).

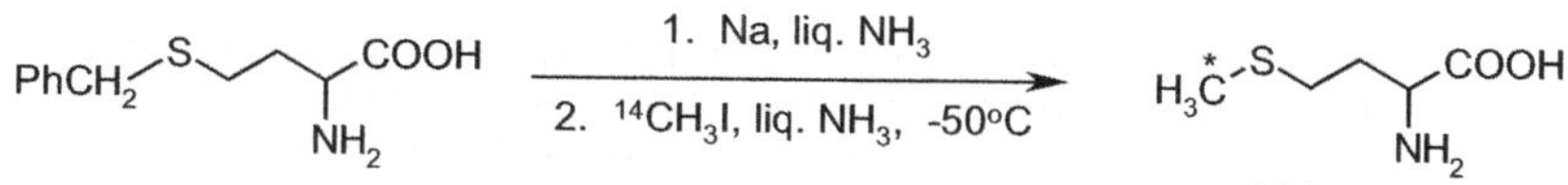


**201**

***Figure 5.55*** *Synthesis of [methyl-$^{14}$C]methionine*

Safety aspects and experimental scale apart, there are no fundamental differences in the reaction of aliphatic C–H acidic compounds with labeled or unlabeled methyl iodide. For example (Figure 5.56), the alkylation of deprotonated 2,6-dimethylcyclohexanone (**202**), key step in the synthesis of [$^{14}$C]vitamin A[167], of the *N,N*-dimethylhydrazone of dihydro-testosterone (**204**)[168], of mevinolin (**206**)[169] and tricarbethoxymethane (**208**)[170], furnishing the respective *C*-[$^{14}$C]methylated intermediates **203**, **205**, **207** and **209** in radiochemical yields ranging from 20% to 90%. The $Li_2CuCl_4$-catalyzed reaction of $^{14}CH_3I$ with alkyl Grignard reagents has been recommended as a valuable method for the one-carbon extension of the alkyl bromides and chlorides, as illustrated by the synthesis of **210**[171].

The use of $^{14}CH_3I$ in diastereoselective [$^{14}$C]methylations is relatively common. The examples depicted in Figure 5.57 represent two different approaches of asymmetric induction, either by an appropriate chiral auxiliary reversibly attached to the achiral substrate as for (*R*)-[$^{14}$C]ibuprofen (**211**, Evans–Oppolzer approach)[172], or as for $\alpha$-[$^{14}$C]methyl-(*S*)-tryptophan (**212**)[173] by at least one reversibly generated stereogenic

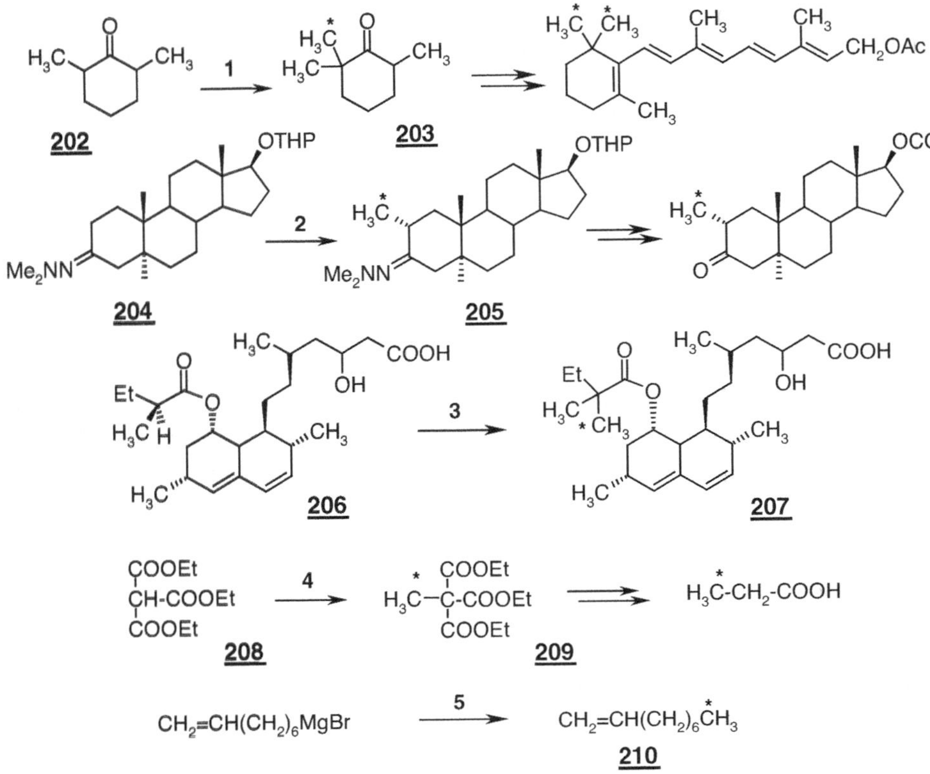


**Reaction conditions:** **1a**. KH, THF; r.t., 30 min, **b**.$^{14}CH_3I$, r.t., 40 min (90 %); **2a**. LDA, THF; r.t., 16 h, **b**. $CH_3I$, r.t. 3 h (60 %); **3a**. aq. KOH, cyclohexane, *i*–PrOH; r.t., 30 min, **b**. *n*–BuLi, pyrrolidine, THF; –50 °C to –30 °C, 2 h, **c**.$^{14}CH_3I$, –55 °C, 30 min, –45 °C, 45 min, **d**. toluene, 60 °C (20 %); **4a**. EtONa, EtOH; r.t., 30 min, **b**.$^{14}CH_3I$, r.t., 4 h (91 %); **5**. $Li_2CuCl_4$, $^{14}CH_3I$, THF; –20 °C; 1–2 h; r.t., 1–2 h

***Figure 5.56*** *[$^{14}$C]Methylation of aliphatic C–H acidic compounds*

center which is removed after the asymmetric operation. This and related methodologies will be discussed in detail in Chapter 11.

Only a few [$^{14}$C]methylations of aromatic systems have been reported to date, although this position of label has proved to be metabolically stable in numerous *in vivo* investigations. The classical approach to accomplish *C*-alkylation through reaction of aryllithium or magnesium intermediates with [$^{14}$C]methyl iodide[174] often fails because of interfering substituents or low reactivity. Organocopper reagents, which do not cross-react with COOR and CN substituents, have proved to be more efficacious. The organocuprate, for example, generated *in situ* by sequential treatment of aryl bromide **213a** (X = Br) with *tert*-BuLi and CuBr·$Me_2S$, reacted with [$^{14}$C]MeI to give **213b** (X = $^{14}CH_3$) in 30–40% radiochemical yield, whereas the corresponding aryllithium compound was unreactive[175]. Palladium-catalyzed reactions on organoboranes and organostannanes (e.g. Suzuki and

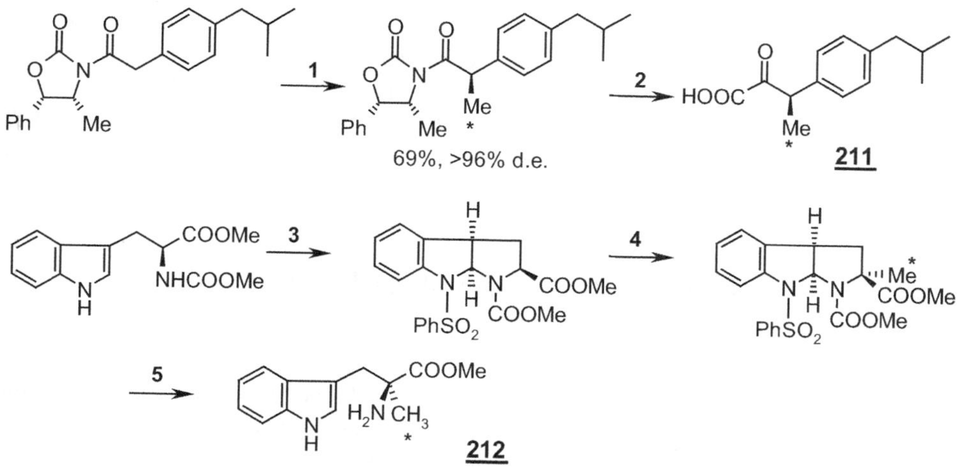


**Reaction conditions:** **1a**. NaHMDS, THF; −78 °C, 1 h, **b**.$^{14}CH_3I$, THF; −78 °C, 1 h; −40 °C, 2 h; **2**. LiOH, $H_2O_2$, THF; 0 °C, 3 h; **3a**. 85 % $H_3PO_4$,r.t., 3 h, **b**. $PhSO_2Cl$, pyridine; r.t.; 16 h; **4a**. LDA, THF; −78 °C, 1 h, **b**.$^{14}CH_3I$, THF; −78 °C, 2 h; r.t., 16 h; **5a**. $CF_3COOH$, 210 °C, 1 h, **b**. 5N KOH;210 °C, 1–2 h

***Figure 5.57*** *Use of $^{14}CH_3I$ in diastereoselective alkylations*

Stille couplings) show an even wider tolerance towards various functionalities and should therefore allow for the introduction of [$^{14}$C]methyl groups at a very late stage of synthetic pathways. Although this has been frequently exploited in carbon-11 chemistry[176], there are only a few analogous applications in carbon-14 synthesis. The palladium(II)

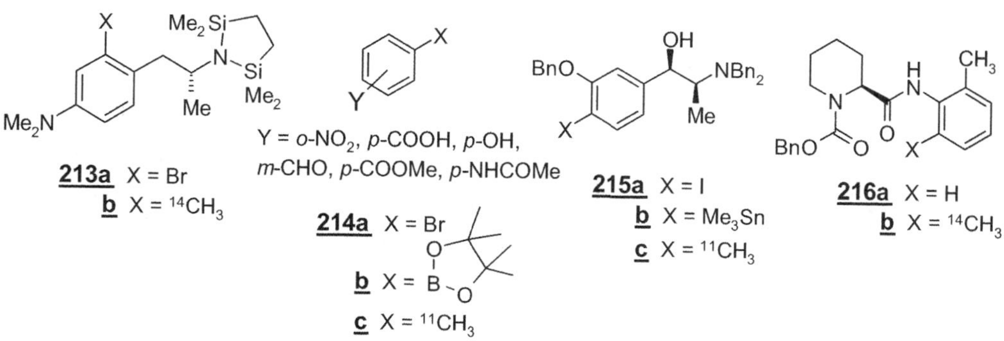


**Reaction conditions :** **213**: **a**. *tert*-BuLi;−70 °C, 30 min, **b**. CuBr. $Me_2S$;−70 °C, 30 min,**c**. $^{14}CH_3I$;−70 °C, 1 h; 3.t. 48 h (ca 50 %); **214**: **a**. Y–ArBr,[$(Me_4C_2O_2)BB(O_2C_2Me_4)$], $PdCl_2$(dppf), KOAc, DMSO ; 70 °C, 1–6 h, **b**.$^{11}CH_3I$, Pd(dppf)$Cl_2$, $K_3PO_4$, DMF, 100-120 °C, microwave, 50W, 1.5 min(35–88 %); **215**: **a**. $Sn_2Me_6$, $(PPh_3)_4Pd$, dioxane, 110 °C, 16 h, **b**. $^{11}CH_3I$,$Pd_2(dba)_3$, (*o*-Tol)$_3$P, CuCl, $K_2CO_3$, DMF; 101°C,10 min (30–40 %); **216**: $^{14}CH_3I$, $Pd(OAc)_2$, HOAc; 90 °C, 48 h (61 %)

***Figure 5.58*** *Transition metal-mediated reactions using [$^{11}$C]- and [$^{14}$C]methyl iodide*

1,1′-bis(diphenylphosphino)ferrocene dichloride-catalyzed [$^{11}$C]methylation of the organoborane **214b**[176b] as well as the analogous $Pd_2(dba)_3$-catalyzed reaction on organostanane **215b**[177] demonstrate the potential of this approach. The *ortho*-[$^{14}$C]methylation of the organopalladium species, generated *in situ* from the acetanilide derivative **216a** and 1.5 equivalents of $Pd(OAc)_2$[178] may serve as additional illustrative examples. The alternative use of reversed polarity "umgepolte" [$^{14}$C]methyl reagents are discussed in the next section.

### 5.5.2 [$^{14}$C]Methyl Iodide as a Source of Nucleophilic [$^{14}$C]Methyl and [$^{14}$C]Methylene Building Blocks

The synthetic access to a broad variety of nucleophilic [$^{14}$C]methyl and [$^{14}$C]methylene synthons has significantly widened the scope of [$^{14}$C]methyl iodide. Reaction with magnesium turnings in ether or THF produces [$^{14}$C]methylmagnesium iodide[179], which has been frequently used for the preparation of various secondary and tertiary [1-$^{14}$C]-alcohols from the corresponding aldehydes and ketones and of [$^{14}$C]methyl ketones from esters or nitriles. Its reaction with enol lactones, such as **217** (Figure 5.59), to give [2-$^{14}$C]-cyclohexen-2-ones **218** has been exploited for the synthesis of numerous A-ring-labeled

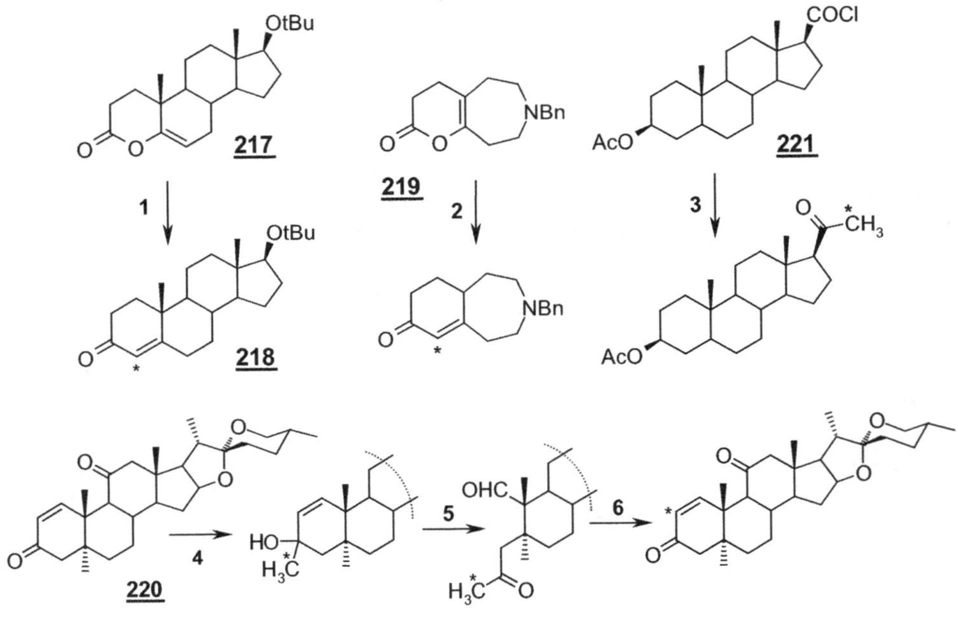


**Reaction conditions: 1a**. $^{14}CH_3MgI$, ether; 0 °C, 1 h, **b**. 30 % NaOH,EtOH 1:7, r.t., 16 h; **2a**. $^{14}CH_3MgI$, ether; −20 to −30 °C, 30 min (64 %), **b**. KOH, $H_2O$, MeOH; r.t., 3 h (94 %); **3a**. $^{14}CH_3MgI$, $CdCl_2$, ether, r.t., 2 h, **b**. benzene r.t., 18 h; 50 °C, 1 h; **4**. $^{14}CH_3MgI$, ether, THF; −78 °C to r.t.(95 %); **5**. $O_3$, $CH_2Cl_2$, MeOH; −78 °C to r.t.; **6**. NaOH, $H_2O$, MeOH,$CH_2Cl_2$; r.t., 4 h (54 % in two steps)

*Figure 5.59 Reactions of enol lactones with [$^{14}$C]methyl Grignard and acid chlorides with di[$^{14}$C]methylcadmium*

steroids[180] and related compounds such as **219**[181] using what has become known as the Fujimoto–Belleau protocol for excising C4 of fused 3(4)enones and replacing it with isotopic carbon[180b,182]. An alternative labeling procedure, involving addition of $^{14}CH_3MgI$ to the starting enone, prior to one-carbon excision and ring closing, was devised[183] for steroid **220**, which lacked the γ-hydrogen necessary for enelactone formation integral to the Fujimoto–Belleau reaction. (see also Chapter 10). Transmetallation of $^{14}CH_3MgI$ with $CdCl_2$ produces [$^{14}$C]dimethylcadmium, which has allowed the conversion of acid chlorides into the respective [$^{14}$C]methyl ketones, such as is illustrated with steroid **221**[184].

The reaction of [$^{14}$C]methyl iodide and $SmI_2$ with aldehydes and ketones in THF–HMPA to furnish secondary or tertiary [1-$^{14}$C]alcohols has been recommended as an alternative to Grignard reactions. Not only does it eliminate the separate preparation of $^{14}CH_3MgI$, but it also offers a greater tolerance to various functional groups such as COOR, $CONR_2$ and CN.

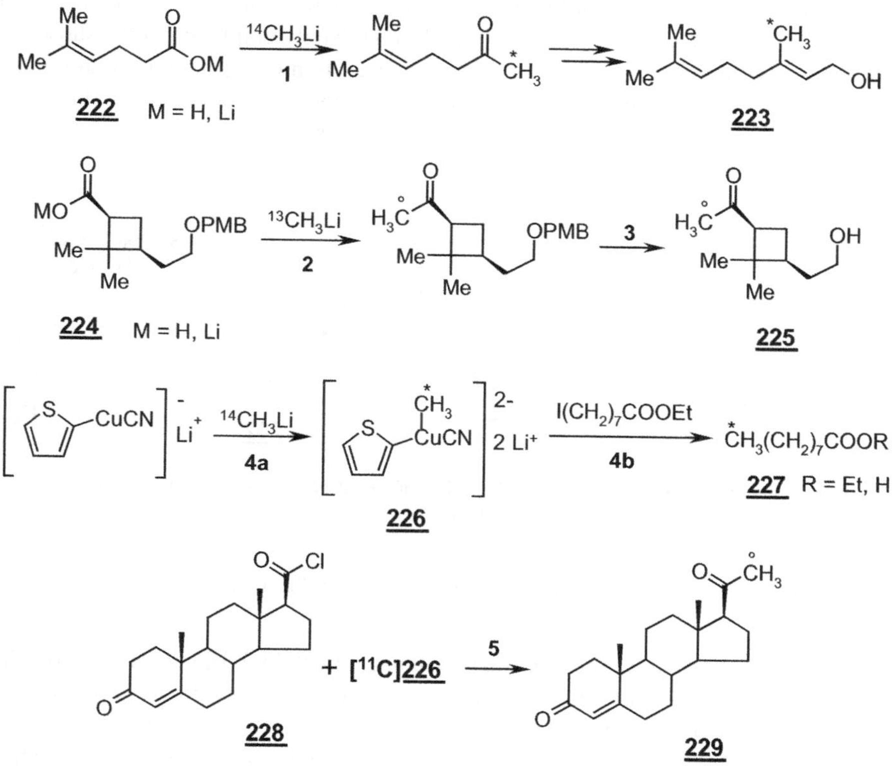


**Reaction conditions**: **1a**. MeLi (1 equiv.), ether; r.t., **b**.$^{14}CH_3Li$, ether; r.t., 3 h (53%); **2**.$^{13}CH_3Li$ (2 equiv.), THF; −78 °C, 30 min; r.t., 2 h (32 %); **3**. DDQ;$CH_2Cl_2/H_2O$ 20:1; r.t., 1 h; **4a**. THF; −78 °C, 15 min; 0 °C, 5 min, **b**. hexane; −78 °C, 1.5 h; r.t., 20–24 h, **c**. 6N KOH, 130–135 °C, 20 h (40 % overall); **5a**. as 4a but with $^{11}CHI$, **b**. **228**, THF; 50 °C, 3 min (30–35 % decay correctedyield)

***Figure 5.60*** *Use of [$^{11/13/14}$C]MeLi for the synthesis of isotopically labeled ketones and chain extension*

The reaction, which is normally complete within a few minutes, gives moderate to good yields for aliphatic aldehydes and ketones, and moderate ones for aromatic aldehydes[185].

[$^{14}$C]Methyllithium, which is conveniently available from $^{14}CH_3I$ and lithium metal[186], *n*-BuLi[187] or preferably *tert*-BuLi[16,188], has been rarely used in carbon-14 synthesis to date. The few examples published so far (Figure 5.60) include the preparation of [$^{14}$C]- and [$^{13}$C] methyl ketones such as [9-$^{14}$C]geraniol (**223**)[189] and [$^{13}$C]pinonalcohol (**225**)[188a] upon treatment of the corresponding lithium carboxylates **222** and **224** with [$^{13/14}$C]MeLi and of 2-thienyl([$^{14}$C]methyl)(cyano)cuprate (**226**), which was used in the one-carbon extension of ethyl octanoate to ethyl [9-$^{14}$C]nonanoic acid (**227**)[186a], and in carbon-11 and carbon-13-labeled form to prepare a [$^{11/13}$C]methyl ketone **229** from the corresponding carboxylic acid chloride **228**[190].

Future applications of $^{14}CH_3Li$ may increase, however, because of its use in the preparation of the lithium [$^{14}$C]methylborinate reagent **231**. This reagent is effective in the Suzuki [$^{14}$C]methylation of aryl and alkenyl halides (I > Br) and triflates, a reaction which tolerates a broad variety of functional groups. The reagent is generated *in situ* from $^{14}CH_3Li$ and *B*-methoxy-oxa-9-BBN (**230**) and immediately employed in the subsequent cross-coupling reaction, as demonstrated in the preparation of 4-[methyl-$^{14}$C]toluic acid and 4-[methyl-$^{14}$C]toluenesulfonic acid (**232**) from 4-iodobenzoic acid and 4-iodobenzene-sulfonic acid, respectively, in 50% radiochemical yield from $^{14}CH_3I$[187a].

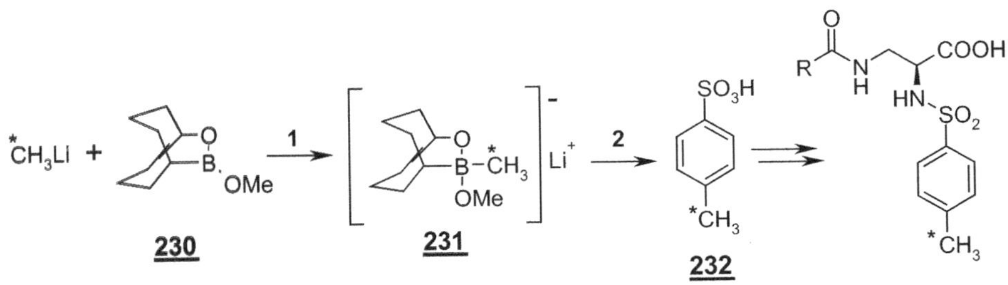


**Reaction conditions: 1.** THF; r.t., 1 h; **2.** $PdCl_2(PhCN)_2$, 4-iodobenzenesulfonic acid, $AsPh_3$, $Ba(OH)_2$8 $H_2O$, DMF; 60 °C, 15 min

***Figure 5.61*** *Suzuki coupling of lithium [$^{14}$C]methylborinate with aryl halides*

The same holds for 5-[$^{11}$C]methyl-1-aza-5-stanna bicyclo[3.3.3]undecane (**233**), which is also readily accessible for *in situ* use by treatment of the corresponding 5-chlorostannane and [$^{11}$C]MeLi[191] (Figure 5.62). Despite being employed so far only in carbon-11 syntheses, there should be no fundamental problem for its use in palladium-mediated $^{14}C-C_{aryl/vinyl}$ bond forming reactions (Stille reaction).

The [1,3-bis(diphenylphosphino)propane]nickel(II) chloride-catalyzed [$^{14}$C]methylation of **234** (Figure 5.63) using the [$^{14}$C]methylzinc species generated *in situ* from excess $^{14}CH_3I$ and Zn–Cu couple completes the spectrum of cross-coupling reactions used so far for the late stage carbon-14 labeling of complex aromatic systems[192].

[$^{14}$C]Methyltriphenylphosphonium iodide has emerged as a labeling reagent of importance comparable with $^{14}CH_3MgI$. It is conveniently prepared from [$^{14}$C]methyl iodide and triphenylphosphine in nearly quantitative yield[193] (Figure 5.64). Its ylide, which is

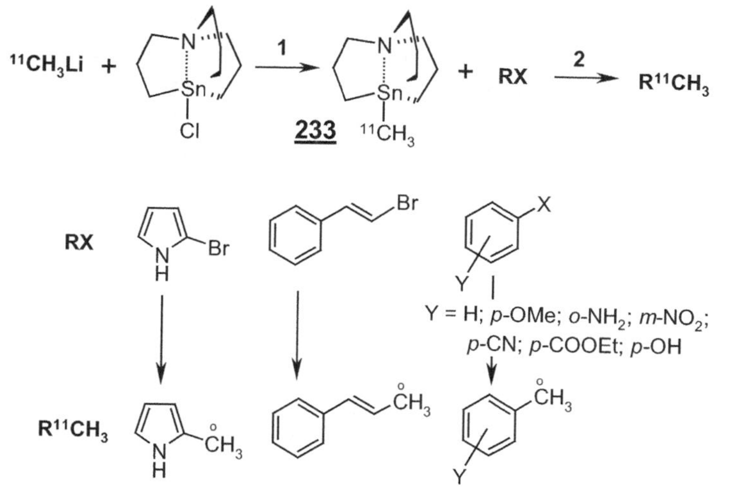


Reaction conditions: 1.THF,−78 °C, 1 min; r.t.; 5 min; 2. $Pd_2Cl_2(CH_2{=}CHCH_3)_2$ catalyst, DMF;100 °C, 2–15 min (15–60 % decay-corrected overall yields)

***Figure 5.62*** *Use of Stille reagents in the [11/14C]methylation of aromatic and vinylic substrates*

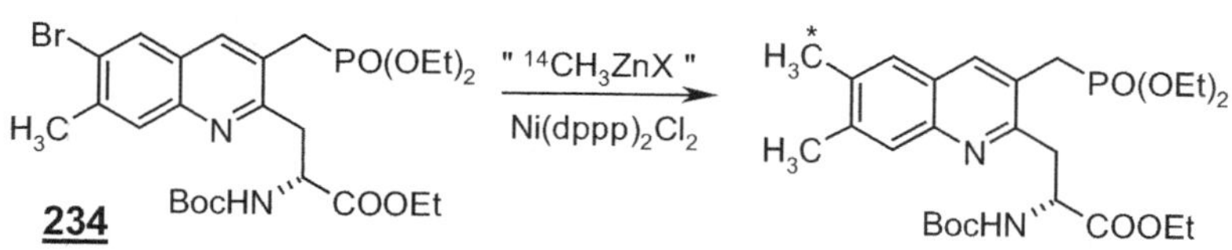


Reaction conditions: a. $^{14}CH_3I$, Zn–Cu, THF; ultrasound, 30 °C, 2 h, b.$NiCl_2(dppp)_2$; 60 °C, 16 h (37 % yield)

***Figure 5.63*** *Ni(II)(dppp)-catalyzed bromo/[14C]methyl replacement in haloarenes with in situ generated [14C]methylzinc derivatives*

generated by deprotonation with *tert*-BuOK or MeLi, has been employed in Wittig [$^{14}$C] methylenenations of aldehydes and ketones belonging to various structural classes including terpenes[194], cannabinoids[195], ergot alkaloids[196], heterocyclic compounds (e.g., **235–236**)[197] and of even highly complex molecules such as cobalt(II) mesoporphyrin[193b] (Figure 5.64). Deprotonation of $(EtO)_2P(O)^{14}CH_3$, the analogous Horner–Emmons reagent, followed by carboxylation and esterification provides triethyl phosphono[2-$^{14}$C]acetate (**237**), a powerful two carbon synthon which has been used for chain elongation of various key intermediates in the synthesis of labeled retinoids (e.g., **238**)[198] and prostaglandins. [2-$^{14}$C]-Isopropyltriphenylphosphonium iodide (**239**), a Wittig three-carbon synthon, generated *in situ* from [$^{14}$C]methyl iodide and deprotonated ethyltriphenylphosphonium iodide was employed in the stereoselective total synthesis of (3S)-2,3-epoxy[24, 30-$^{14}C_1$]squalene (**240**)[199].

The corresponding sulfonium- and sulfoxonium [$^{14}$C]methylides, generated by deprotonation of diphenyl[$^{14}$C]methylsulfonium tetrafluoroborate (**241**)[200] and

[$^{14}$C]trimethylsulfoxonium iodide ($^{14}CH_3(CH_3)_2S^+(O)I^-$)[201], were successfully employed in the synthesis of [$^{14}$C]epoxides from carbonyl precursors. Compound **241** was prepared by treatment of a suspension of diphenylsulfide and $AgBF_4$ with $^{14}CH_3I$ and used without purification. The epoxide product **242** was isolated in 50% radiochemical yield. Previous experiments using different counterions were unsuccessful, indicating that the $BF_4^-$ anion is essential. $^{14}CH_3(CH_3)_2S^+(O)\ I^-$ is conveniently available through [$^{14}$C]methylation of DMSO. This reagent suffers, however, from the limitation that a theoretical radiochemical yield of only 33% can be expected.

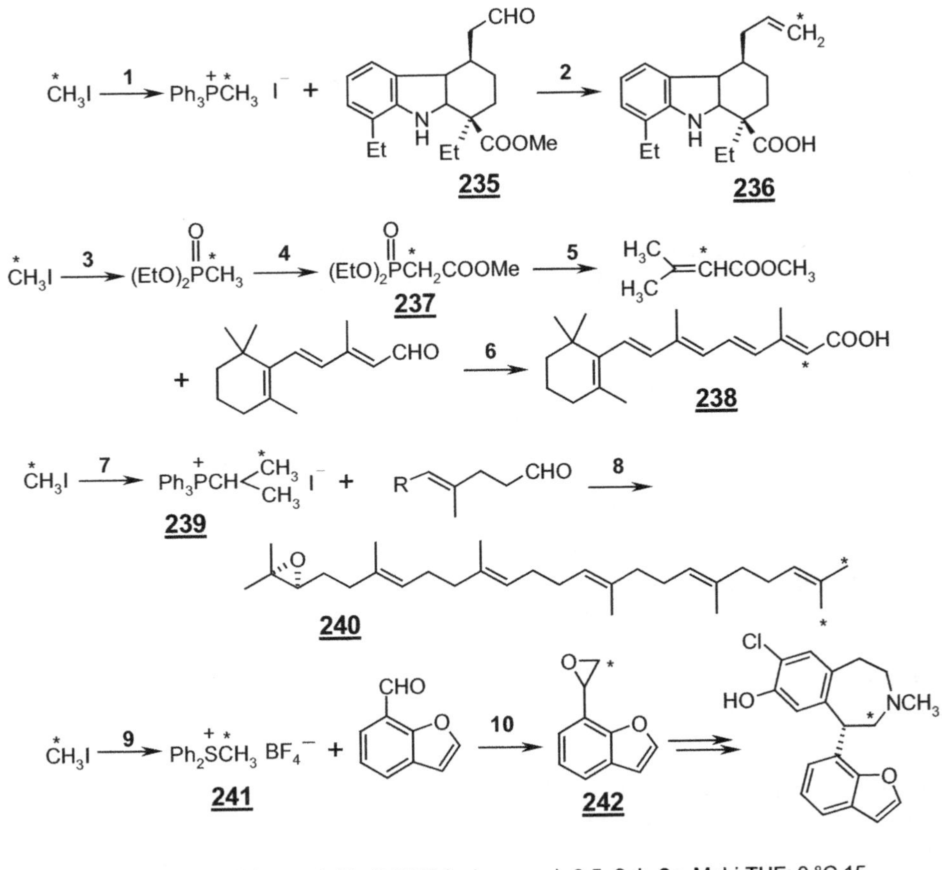


**Reaction conditions:** **1**. $Ph_3P$, THF (or benzene); 0.5–2 d; **2a**. MeLi,THF; 0 °C 15 min, r.t., 15 min, **b**. **235**, THF; 0 °C, 5 min; **3a**. $(EtO)_2P(O)H$, NaOEt, $C_6H_6$–EtOH, r.t., 1 h, **b**. $^{14}CH_3I$; r.t., 16 h; **4a**. MeLi, $CO_2$;–65 °C, 3.5 h, **b**. $CH_2N_2$, ether; r.t. **c**. IN NaOH; reflux; **5**. NaH, acetone, ether; r.t., 3,5 h; **6**. β-iononylideneacealdehyde, $KNH_2$, liq. $NH_3$;–65 °C, 1 h; r.t., 17 h; **7a**. $Ph_3P^+Et \cdot Br^-$, BuLi, THF; r.t.,1 h, **b**.$^{14}CH_3I$, THF; –78 °C, 1 h; **8**. *n*–BuLi (20*S*) -epoxytrisnorsqualenealdehyde, THF, –78 °C, 1 h, –5 °C, 20 h; **9**. $Ph_2S$,$AgBF_4$, $CH_2Cl_2$–toluene; r.t. 4 h;**10**. NaOH, acetonitrile; r.t. 4 h.

***Figure 5.64*** *Synthesis and application of carbon-14-labeled methyl- and isopropyltriphenylphosphonium iodide, diethyl phoshonoacetate and diphenylsulfonium fluoroborate from* $^{14}CH_3I$

[$^{14}$C]Methyl phenyl sulfone (**243**), another nucleophilic one-carbon synthon[202], is readily available through reaction of $^{14}CH_3I$ with sodium benzenesulfinate[203]. It is a stable, crystalline solid, so that the compound can be retained in stock and portions weighed out as needed. Treatment of **243** with appropriate amounts of potassium hydride or *n*-butyllithium generates either the monoanion or dianion. Trapping of the monoanion with alkyl halides, carboxylic esters or $\alpha,\beta$-unsaturated ketones and subsequent reductive cleavage of the phenyl sulfonyl moiety with Al–Hg or Na–Hg provides [$\omega$-methyl-$^{14}$C] alkanes, [$^{14}$C]methyl ketones or $\alpha$-[$^{14}$C]methyl substituted allylic alcohols (Figure 5.65, route A). The dianion shows an even broader spectrum of potential applications since it allows for the insertion of a [$^{14}$C]methylene unit (Route B). Sequential trapping with diethyl carbonate and ethyl chloroformate followed by reductive elimination of the phenylsulfonyl function, for example, furnishes [2-$^{14}$C]malonic diester. Use of differently esterified chloroformates should produce unsymmetrically substituted [2-$^{14}$C]malonic esters. [2-$^{14}$C]Carboxylic acids esters are obtained when alkyl halides are employed as second electrophiles.

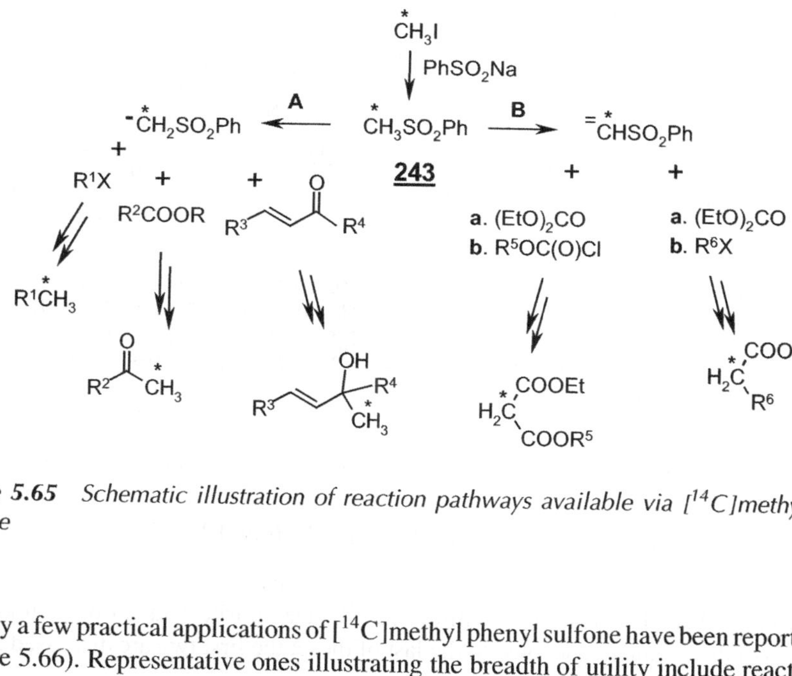


*Figure 5.65* *Schematic illustration of reaction pathways available via [$^{14}$C]methyl phenyl sulfone*

Only a few practical applications of [$^{14}$C]methyl phenyl sulfone have been reported so far (Figure 5.66). Representative ones illustrating the breadth of utility include reaction with allylic acetate in the presence of Pd(0) to give the respective $\delta,\gamma$-unsaturated derivative **244** as demonstrated with the synthesis of **245**[203a], with ester **246** to furnish [$^{14}$C]methyl ketone **247**[203b], and the sequential reaction of the dianion with a carbamate (**248**) and an alkyl halide (**249**) to afford the complex [2-$^{14}$C]amide **250**[204]. In spite of the promising potential and versatility of this synthon, the use of Na–Hg for reductive cleavage of the phenylsulfone function may be a significant disadvantage for many users, owing to regulatory requirements for the disposal of wastes containing both radioactivity and the toxic metal mercury.

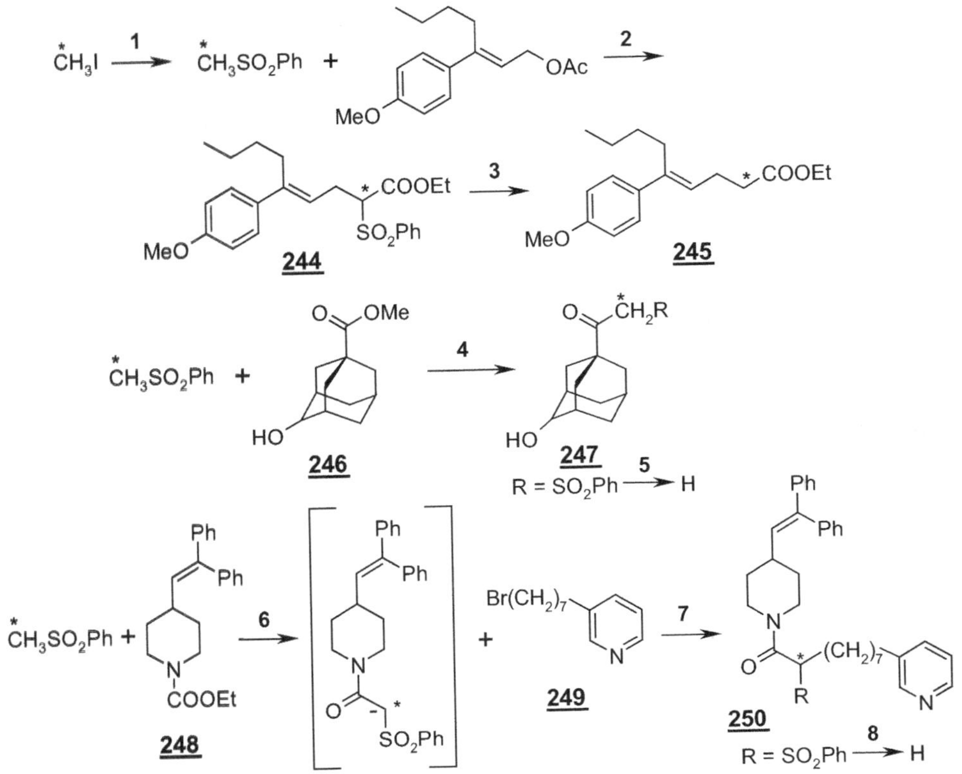

**Reaction conditions:** **1**. $PhSO_2Na$, DMF; r.t. 4d (70%); **2a**. *n*-BuLi (2.2 equiv), **b**. $OC(OEt)_2$,THF, 0 °C to r.t., 1 h, **c**. allylic acetate, Pd(0), 0 °C to r.t.,1–2h; **3**. Na–Hg,EtOH, aq. HOAc; **4**. *n*-BuLi (2 equiv), THF; 0 °C,10 min, **b**. **246**; 0°C,10 min, r.t., 30 min; **5**. Na–Hg, MeOH/aq. $Na_2HPO_4$ buffer; **6**. *n*-BuLi (2 equiv), THF; 0°C; **7**. $Bu_4NI$, THF; 65–85 °C; **8**. Na/Hg, MeOH/aq. $Na_2HPO_4$

**Figure 5.66** *Synthesis and applications of [$^{14}$C]methyl phenyl sulfone*

## 5.5.3 Further Building Blocks Derived from [$^{14}$C]Methyl Iodide

[$^{14}$C]Methyl iodide is the source of other simple molecules that can be utilized as building blocks, among them [$^{14}$C]carbon disulfide, [2-$^{14}$C]acetonitrile and nitro[$^{14}$C]methane. Section 5.6 of this chapter is devoted to the last of these; the first two are discussed briefly below.

*[$^{14}$C]Carbon Disulfide*

[$^{14}$C]Carbon disulfide is accessible in nearly quantitative yield by carrying [$^{14}$C]methyl iodide with a stream of nitrogen or helium through a matrix of phosphorus pentasulfide and sea sand at 300–350 °C[205]. The reagent has been used for the preparation of alkyl and aryl [$^{14}$C]isothiocyanates[206], 2-mercapto-[2-$^{14}$C]benzimidazoles[207] and [2-$^{14}$C]benzothiazole[208] derivatives. Its reacts with appropriate C,N,O-nucleophiles (e.g. phenolates, arylcopper compounds, metallated amines) to give dithiocarboxylates **251**, dithiocarbamates **252**, and xanthates **253**, respectively, which are easily esterified *in situ* with electrophiles such as methyl iodide. Subsequent oxidative fluoro-desulfurization of these intermediates with HF/pyridine and 1,3-diodo- or 1,3-dibromo-5,5-dimethylhydantoin (DIH, DBH) enables the trifluoro[$^{14}$C]methylation of arenes **254**, amines **255** and phenols **256** in moderate overall radiochemical yields[209]. *In situ* generated $I^+$ is speculated to be responsible for the simultaneous iodination of the xanthate **253** *ortho* to the electron-rich benzyloxy group.

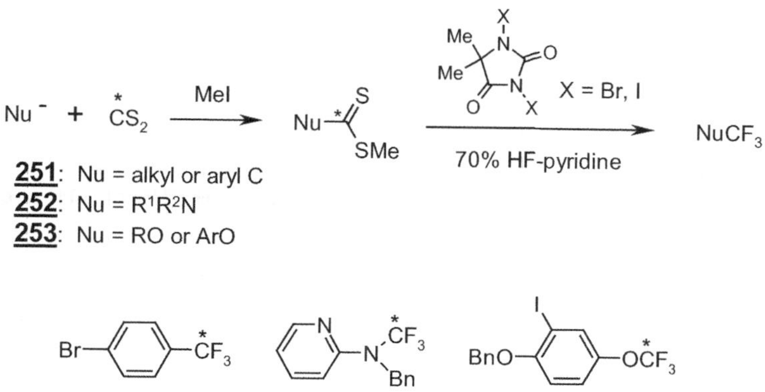


**251**: Nu = alkyl or aryl C
**252**: Nu = R¹R²N
**253**: Nu = RO or ArO

**254** **255** **256**

**Reaction conditions: 254 1a**. 1,4-dibromobenzene, *n*-BuLi, THF; −80 °C,10 min, **b**. CuBr, LiBr, THF; −65 °C, 15 min, **c**.$^{14}CS_2$,−40 °C, 30 min, **d**. MeI;−40 °C to r.t.; **2**. 70 % HF–pyridine, DBH, $CH_2Cl_2$; r.t., 1 h; **255 1a**. *N*-(2-pyridyl)benzylamine, *n*-BuLi, THF; −15 °C, 1 h, **b**.$^{14}CS_2$; r.t., 16 h, **c**.MeI; r.t., 6 h; **2**. $Bu_4N+H_2F_3$-, DHB, $CH_2Cl_2$/toluene; 0 °C, 4 h; **256 1a**. *p*-benzyloxyphenol,$^{14}CS_2$, MeI, *n*-$Bu_4NH_2PO_4$, 50 % NaOH, benzene; r.t., 8 h; **2**. DIH or DIB, 70 % HF–pyridine, $CH_2Cl_2$; 0 °C, 80 min

***Figure 5.67*** *[$^{14}$C]Trifluoromethylation of arenes, secondary amines and phenols*

### *[2-$^{14}$C]Acetonitrile*

Treatment of [$^{14}$C]methyl iodide with an aqueous solution of potassium cyanide at room temperature for 16 h furnishes [2-$^{14}$C]acetonitrile in nearly quantitative yield[210]. It can be readily hydrolyzed to give [2-$^{14}$C]acetic acid. Alternatively, it can be deprotonated with LDA in THF at −78 °C giving anion **257** (Figure 5.68) which, as demonstrated with carbon-13-labeled material, can be trapped with diethyl carbonate or ethyl chloroformate to afford ethyl cyano[2-$^{13}$C]acetate (**258**) in excellent yield, thus providing a valuable alternative to the standard Br/$^{14}$CN-replacement method that gives [$^{14}$C]cyanoacetate (see Chapter 7, Section 7.2.6). Furthermore, reaction of **257** with diethyl chlorophosphate provides diethyl phosphono[2-$^{14}$C]acetonitrile ($(EtO)_2$-$PO^{14}CH_2CN$, **259**), an additional Horner–Emmons reagent. It is useful for the chain extension of aldehydes and ketones by two-carbon units with simultaneous introduction of label, as illustrated with the [2-$^{14}$C]acrylonitrile polymer **260**[211].

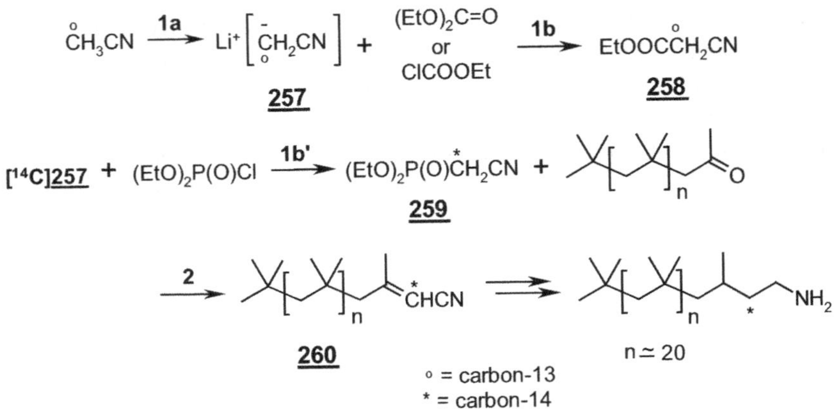


**Reaction conditions**: **1a**. LDA, THF; −78 °C to −60 °C , 30 min, **1b**. THF;−60 °C, 2 h (78 % for diethyl carbonate), **1b'**.−78 °C, 1 h; **2a**. NaH, THF; r.t.,10 min, **b**. 257; 60–65 °C, 12.5 h.

***Figure 5.68*** *Applications of [2-$^{13/14}$C]acetonitrile to the synthesis of [2-$^{13,14}$C]cyanoacetates and the two-carbon extension and labeling of carbonyl compounds*

## 5.6 [$^{14}$C]Nitromethane ($^{14}CH_3NO_2$)

[$^{14}$C]Nitromethane is a versatile building block. It is most useful as a nucleophilic labeled $C_1$-building block for major homologation processes, such as nitroaldol reactions and Michael additions, and there exist several valuable methods for the subsequent transformation of the nitro group into further interesting functionalities. In addition, nitroolefins resulting from the nitroaldol reaction and subsequent dehydration are excellent Michael acceptor molecules and Diels–Alder dienophiles.

[$^{14}$C]Nitromethane (unlabeled, b.p. 101 °C) is usually prepared either by treatment of [$^{14}$C]methyl iodide with a dry mixture of silver nitrite and sand at 60–80 °C for 2 h, followed by extracting the reaction product into ether[212], or by reaction of [$^{14}$C]methyl iodide with a vigorously stirred suspension of silver nitrite in diethyl ether (0 °C, 10 h; r.t., 14 h)[213], acetonitrile or DMF (50–100 °C)[214] and subsequently separated by fractional distillation. Yields are generally 75–85% on a 5–8 mmol scale. Concomitant *O*-alkylation results in the formation of [$^{14}$C]methyl nitrite (unlabeled, b.p. −12 °C) as major byproduct, which, however, is readily separated. Depending on the quality of the silver nitrite employed, explosive [$^{14}$C]methyl nitrate (unlabeled, b.p. 64 °C) has been detected as an impurity arising from contamination with silver nitrate[215].

The electron-withdrawing effect of the nitro group and the mesomeric stabilization of the anionic carbon strongly acidify the methyl group hydrogens (p$K$ = 10.2). Accordingly, [$^{14}$C]nitromethane can be easily monodeprotonated, and even doubly deprotonated when treated with two equivalents of *n*-BuLi in THF at −78 °C in the presence of HMPA. Whereas the monoanion (a nitronate) can react at either O or C, the *C*-nucleophilicity of the dianion is dramatically greater than that of the monoanion, making it the preferred form for *C*-alkylations to prepare higher homologues of [$^{14}$C]nitromethane[216].

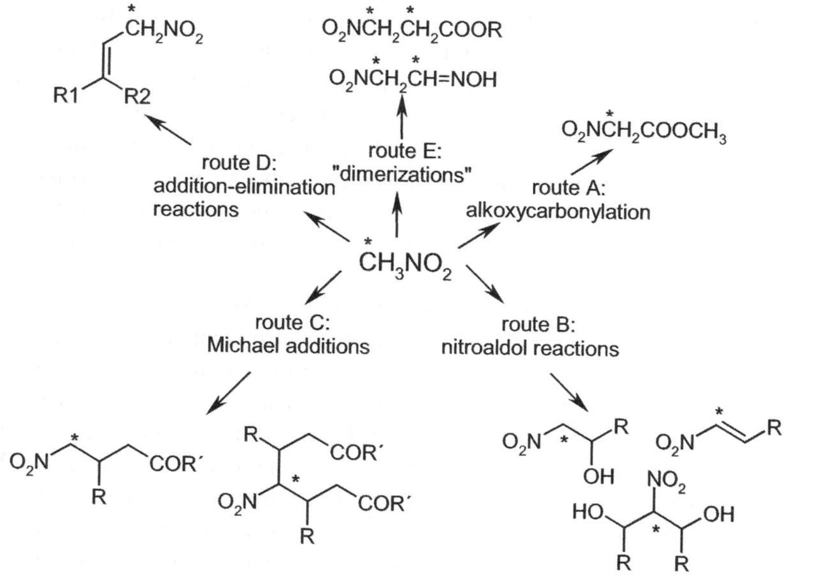


*Figure 5.69 Synthetic applications of nitromethane labeled with isotopic carbon*

Alkoxycarbonylation of the monoanion with methyl methoxy magnesium carbonate, however, has been found to occur preferentially at the C-atom, as reported for the reaction of the carbon-13 labeled nitromethane. *In situ* esterification of the resultant nitro[2-$^{13}$C]-acid provided methyl nitro[2-$^{13}$C]acetate **261** in about 30% from [$^{13}$C]nitromethane (19% overall yield from [$^{13}$C]methyl iodide)[213]. Michael addition of **261** to methyl acrylate, submission of the resultant dimethyl 2-nitro[2-$^{13}$C]glutarate to ozonolytic cleavage of the nitro group and concluding saponification converted **261** into [2-$^{13}$C]ketoglutaric acid (**262**), which was used for investigations in the biosynthesis of cephalosporins[213].

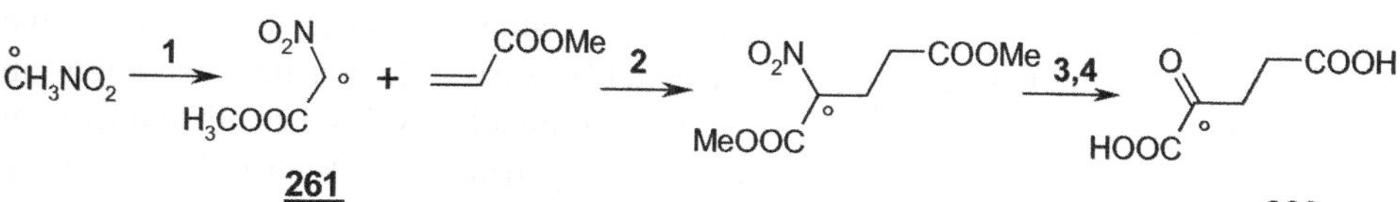


**261**

**262**

**Reaction conditions: 1a.** $CH_3OC(O)OMgOCH_3xCO_2$, DMF; 50 °C, 6 h, **b.** $CH_3COCl$, MeOH; −50 °C to r.t., 15 h; **2.** $BnMe_3NOH$, dioxane MeOH; 60–70 °C, 15h; **3a.** MeONa, MeOH; 0 °C, 10 min,**b.** $O_3$,−78 °C;**4.** 4N HCl, reflux, 2 h

*Figure 5.70 Preparation of methyl nitro[2-$^{13}$C]acetate by alkoxycarbonylation of [$^{13}$C]nitromethane and its use in a Michael addition*

Of more importance are nitroaldol reactions between mono-deprotonated [$^{14}$C]nitromethane and carbonyl compounds. Such condensations can provide $\beta$-nitro[$\beta$-$^{14}$C]-alcohols (Figure 5.69, route B), Michael additions of the nitronate anion to $\alpha$,$\beta$-unsaturated carbonyl or carboxyl compounds (Figure 5.69, route C) and the addition– elimination sequence in the reaction between the nitronate and iminium salts or amidine derivatives (Figure 5.69, route D). The 'dimerization' products formed upon treatment of labeled nitromethane with aqueous alkali hydroxides at elevated temperature are only of limited interest (Figure 5.69, route E). Despite the synthetic potential of [$^{14}$C]nitromethane, many examples of the routes discussed in the following have been carried out so far only with carbon-11 and carbon-13 derivatives. Their inclusion in this section is based on the expectation that equivalent procedures on the carbon-14 isotopomers will produce similar results.

The nitroaldol reaction of mono-deprotonated [$^{14}$C]nitromethane with carbonyl compounds is of particular interest, since the resultant $\beta$-nitro alcohols are useful for further chemical transformations such as dehydration, reduction, and removal of the nitro group under acid-catalyzed (Nef reaction) or ozonolytic cleavage conditions. The range of intermediates accessible thereby is indicated in Figure 5.71. Furthermore, [$^{14}$C]nitroalkenes resulting from dehydration are highly interesting reaction partners in Michael additions and Diels–Alder reactions. Methods recommended for nitroaldol reactions include the use of ethanolic sodium hydroxide[217], potassium fluoride–isopropanol[217] and ammonium acetate/acetic anhydride[217,218] as bases. Dehydration of the $\beta$-nitroalcohols to the respective nitroolefins can be accomplished with acetic anhydride/acetic acid[217b], trifluoroacetic anhydride/$Et_3N$ in dichloromethane[217] or DCC, CuCl in ether[217]. When aromatic aldehydes are employed, the initially generated nitroalcohols may spontaneously dehydrate to nitroolefins, owing to the formation of conjugated systems as illustrated by the first two examples in Figure 5.71.

Thus reaction of [$^{14}$C]nitromethane with veratraldehyde (**263**) (Figure 5.71) provided directly the $\beta$-nitro[$\beta$-$^{14}$C]styrene derivative **264**. Reduction and cleavage of the ether groups of **264** afforded [$\alpha$-$^{14}$C]dopamine (**265**), which has been used in numerous pharmacological investigations[219]. $\beta$-Nitro[$\beta$-$^{14}$C]styrenes **267** were formed directly when the nitroaldol reaction was applied to substituted *o*-nitroaryl aldehydes in the presence of $NH_4OAc$-HOAc[218]. In contrast, the use of methanolic NaOH allowed isolation of the initially formed $\beta$-nitroalcohols **266**, which could be dehydrated by treatment with acetic anhydride–acetic acid[217b]. Subsequent reductive cyclization was carried out to provide the corresponding substituted [2-$^{14}$C]indoles **268** under very mild reaction conditions in overall yields of 10–20%[215,217b,218]. This route to [$^{14}$C]indoles avoids the high temperature (300–350 °C) needed for the *tert*-BuOK-promoted cyclization of *N*-formyl-*o*-toluidines (Madelung's indole synthesis).

Depending on the reaction conditions, excess formaldehyde can form double adducts (**270**)[220] or triple adducts (**269**)[212b,c,221]. Both have served as key intermediates for the preparation of 1,3-dihydroxy[2-$^{14}$C]acetone (**271**) and [2-$^{14}$C]glycerol (**273**). For this purpose, **269** was first converted into **270** in a retroaldol-like reaction. Nef reaction (acid-catalyzed cleavage) of **270** gave **271** in 10% yield[212b,c]. Alternatively, [2-$^{14}$C]glycerol was prepared in 15–20% overall yield[221] beginning with catalytic reduction of **270** to 2-amino-[2-$^{14}$C]propane-1,3-diol (**272**). *N*-Protection through protonation, *O*-protection through acylation, hydroxy-deamination and removal of the protecting group furnished **273**.

In the classical Sowden–Fischer procedure for the extension of sugars by a one-carbon unit, [$^{14}$C]nitromethane is added to the aldehyde function of the next lower homolog, thereby producing a mixture of two C2-epimeric nitro alcohols **274** and **275**

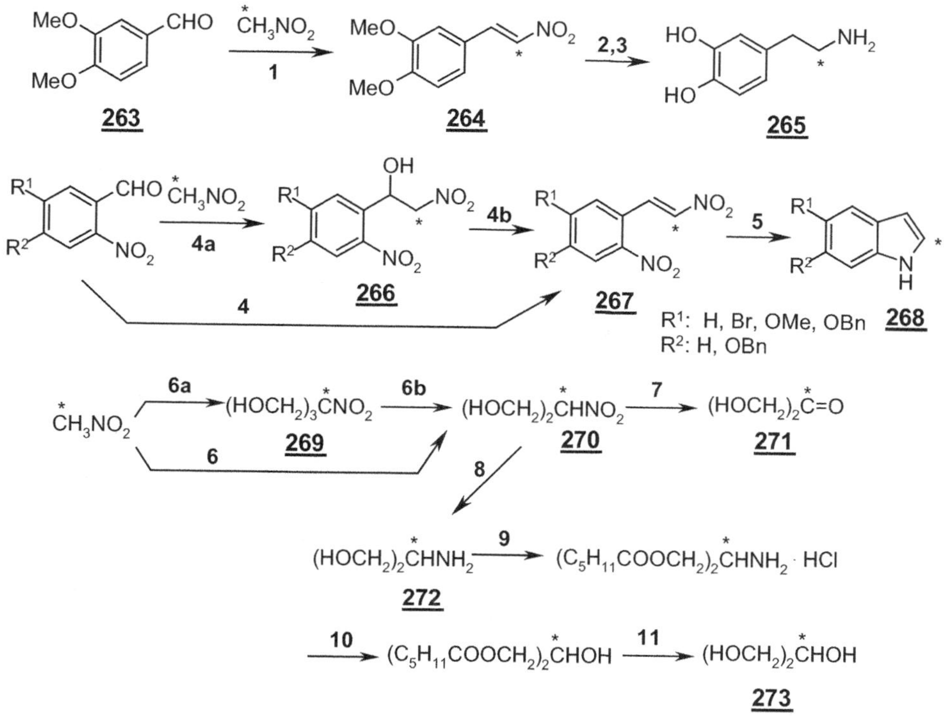


**Reaction conditions:** **1**. $MeNH_2$, MeOH; r.t., 3 d; **2**. $LiAlH_4$, ether; reflux, 20 h; **3**. conc. HCl, 140–160 °C, 2h; **4**. $NH_4OAc$, HOAc; 100 °C, 5 h; **4a**. NaOH, MeOH; **4b**. NaOAc, $Ac_2O$; **5**. Fe, $SiO_2$, toluene **or** $TiCl_3$ **or** $H_2$, 10 % Pd/C; **6**. $(HCHO)_n$ (3equiv.), KOH, MeOH; reflux, 30 min; **6a**. $(HCHO)_n$, $K_2CO_3$, $H_2O$; 50 °C, 5 min; **6b**. MeONa, MeOH, −5 °C, 6 h; **7a**. MeONa, MeOH, 0 °C, 1 h, **b**. aq. $H_2SO_4$, r.t., 10 min; **8**. $H_2$, Raney-Ni, MeOH/HOAc; r.t., 6 h; **9**. conc. HCl; **10**. $CH_3(CH_2)_4COCl$, 90 °C; **11a**. $NaNO_2$. 50 % aq. HOAc; 5 °C, 1.5 h, **b**. Amberlite IR-A-400, MeOH; r.t., 2 h

**Figure 5.71** *Use of [14C]nitromethane in nitroaldol reactions*

(Figure 5.72). Subsequent treatment of the corresponding sodium nitronates with sulfuric acid (Nef reaction) yields a mixture of the homologous C1-labeled epimeric sugars **276** and **277**, which can be separated either by partition chromatography or (on the nitro alcohol state) by fractional crystallization. In typical procedures D-arabinose was converted on a 290 mmol scale into D-[1-$^{14}$C]glucose and D-[1-$^{14}$C]mannose in crystallized yields of 6% and 13%, respectively, from [$^{14}$C]nitromethane. Correspondingly, L-[1-$^{14}$C]ribose and L-[1-$^{14}$C]arabinose were obtained from L-erythrose on a 10 mmol scale in yields up to 20%[222]. The Sowden–Fischer sequence has been largely supplanted by the higher-yielding Kiliani–Fischer procedure, which involves the use of the less expensive K$^{14}$CN (see Chapter 7, Section 7.1.3).

Nitroalkenes **278**, prepared from aldehydes and nitromethane[217], have been shown (in unlabeled pilot studies designed for later carbon-13 experiments) to add to chiral titanium enolates of *N*-acylated (2*S*)- or (2*R*)-bornane-10.2-sultams (**279**) in excellent diastereo-

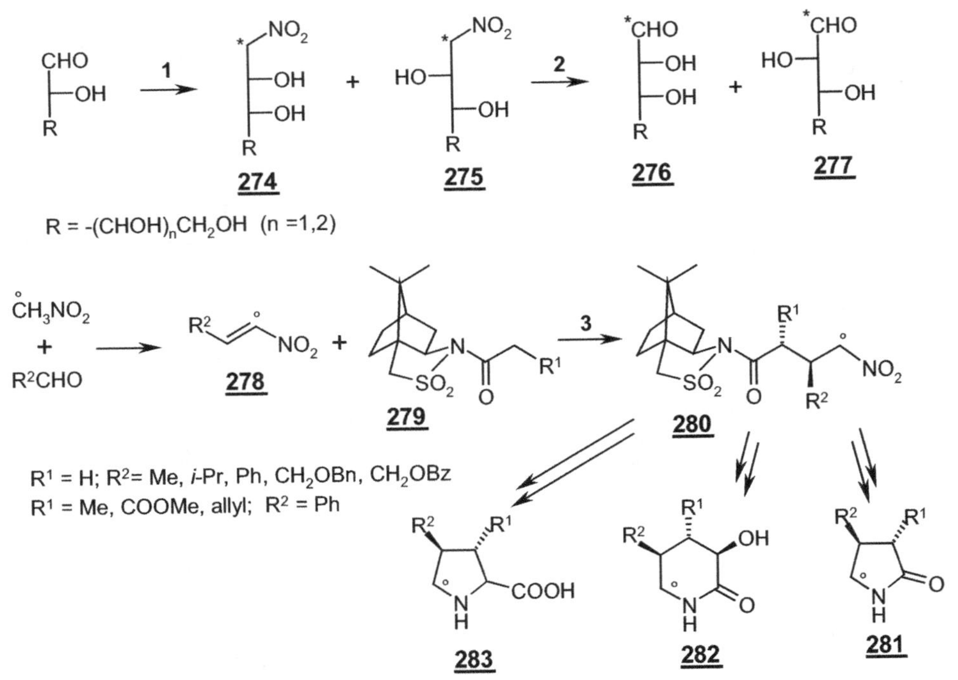

**Reaction conditions:** **1a**. MeONa, MeOH; r.t., 5 h, **b**. fractional crystallization from EtOH; **2a**. basic solution of 274, 50 % $H_2SO_4$; 10–15 °C; separation as phenylhydrazone, **b**. basic solution of 275, 50 % $H_2SO_4$; 10–15 °C; **3a**. $TiCl_4$, $Et_3N$, $CH_2Cl_2$;–78 °C,**b**. 10 % aq. $NH_4F$

*Figure 5.72* *Application of [$^{13/14}$C] nitroaldol reactions to the homologation of sugars and the synthesis of e.p. nitrogen heterocycles*

selectivities (d.e. ≥90%). The resultant Michael adducts (**280**) are valuable building blocks to e.p. isotopically labeled nitrogen heterocycles including pyrrolidinones (**281**), piperidinones (**282**) and substituted prolines (**283**)[223], as discussed in more detail in Chapter 11, Section 11.2.2.

Other major homologation processes include the Michael addition of deprotonated [$^{14}$C]nitromethane to acrylates (Figure 5.69, route C). Like the aldol reaction with

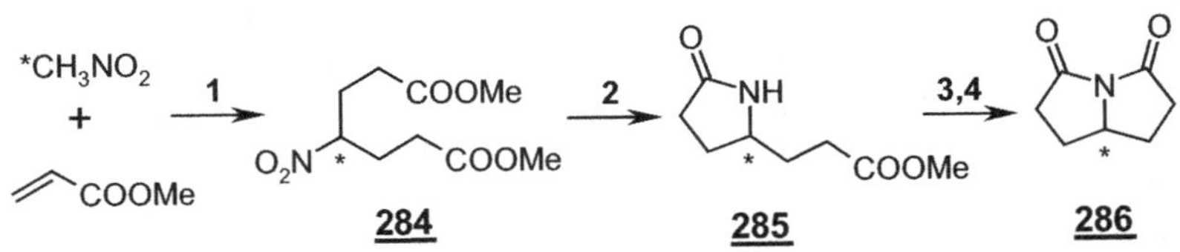

**Reaction conditions:** **1**. Triton B, MeOH, *t*-BuOH; r.t., overnight; **2**. $H_2$, Pd/C, MeOH; r.t., 5 h; **3**. NaOH, $H_2O$, MeOH; reflux, 24 h, **4**. $Ac_2O$; 95 °C, 4 h

*Figure 5.73* *Use of [$^{14}$C]nitromethane in Michael additions*

formaldehyde, reaction with excess methyl acrylate resulted in double addition to give dimethyl 4-nitro[4-$^{14}$C]heptanoate (**284**). Catalytic reduction of the nitro group triggered the intramolecular cyclization to pyrrolidinone **285**. Concluding saponification of the ester function and heating of the liberated acid with acetic anhydride resulted in a second intramolecular cyclization to give dihydro-1H-[7a-$^{14}$C]pyrrolizine-3,5-(2H,6H)dione (**286**), an alkaloid subunit and investigational cognition-activating agent[224].

Only a few examples have been published so far dealing with the nucleophilic addition of isotopically labeled deprotonated nitromethane to iminium salts or *N*-nitrosoamidines followed by elimination of the respective nucleofuge nitrogen moiety thereby creating a nitroolefinic subunit (Figure 5.69, route D).

In the first case, addition of deprotonated [$^{11}$C]nitromethane to unsubstituted and substituted 5-dimethylaminopenta-2,4-dienylidene dimethylammonium tetrafluoroborates

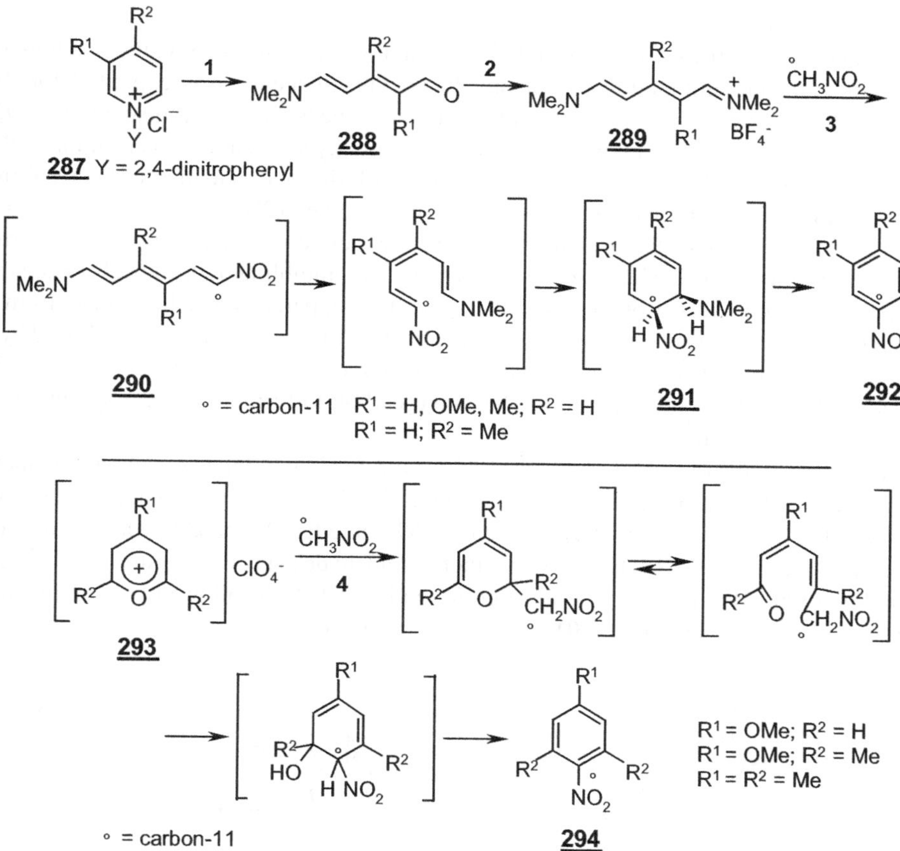


**Reaction conditions:** **1**. aq. $Me_2NH$, EtOH; 70°C, 30 min; **2**. $Me_2NH^+ BF_4^-$, EtOH; reflux, 2 h; **3**. *tert*-BuOK, HMPA; 170 °C, 7 min **or** *n*-BuLi, HMPA; 170 °C, 10 min; **4**. *n*-$Bu_4N^+F^-$, *tert*-BuOK, *tert*-BuOH, HMPA; 120 °C, 20 min **or** *tert*-BuOK, *tert*-BuOH; 120 °C, 20 min

***Figure 5.74*** *Use of [$^{11}$C]nitromethane in addition/elimination reactions; application to the synthesis of [$^{11}$C]nitrobenzenes*

**289** and elimination of dimethylamine gave the nitrotrienes **290**. Spontaneous thermal disrotatory electrocyclization afforded the cyclohexadienes **291**, which eliminated dimethylamine to give the [1-$^{11}$C]nitrobenzenes **292**. This approach has been accomplished in decay corrected radiochemical yields ranging from 60–80%. It works well for unsubstituted, *m*-substituted and for *p*-nitro[4-$^{11}$C]toluene, but it fails for *p*-nitro[4-$^{11}$C]anisole. The requisite starting materials **290** were prepared on a multigram scale by ring-opening of the respectively substituted *N*-[2,4-dinitrophenyl]pyridinium chloride **287** with dimethylamine followed by reaction of the resultant 5-dimethylaminopent-2,4-dienals **288** with dimethylammonium fluoroborate in 30–40% overall yield[225]. *p*-Methoxy- and 1,3,5-trimethyl-substituted [1-$^{11}$C]nitrobenzenes have become accessible by a formally related procedure. It involves ring opening of 3- or 1,3,5-substituted pyrylium salts **293** with deprotonated [$^{11}$C]nitromethane, followed by an intramolecular nitroaldol-like cyclization and spontaneous dehydration to afford the respective [1-$^{11}$C]nitrobenzenes **294** in 30–80% decay corrected yields[226].

A related mechanism also works for the addition of carbanions to *N*-nitrosoamidines with simultaneous elimination of diazomethane. This sequence was exploited for the *tert*-BuOK-promoted addition of [$^{13}$C]nitromethane to the bicyclic *N*-nitrosoamidine **295**, providing the nitroolefinic intermediate **296**. Catalytic hydrogenation, cyclization of the resulting diamine **297** with ethyl acetimidate hydrochloride and oxidation of the resulting tricyclic imidazoline derivative afforded [$^{13}$C]midazolam (**298**), a pharmacologically active imidazo[1,5-a][1,4]benzodiazepine derivative routinely used as a substrate for drug interaction studies. $^{13}$C-NMR analysis of **298** revealed scrambling of the label between the expected position and C3 of the diazepine ring. This apparently resulted either from a temporary hydrolytic ring opening at the imino function in the diazepine ring followed by a recyclization of the resulting equivalent methylamino groups, or by an internal nucleophilic attack of the pendant amino group on the imino carbon to form a transient symmetrical [2.2.n]bicyclic system[227].

Depending on the reaction conditions, [$^{13,14}$C]nitromethane 'dimerizes' upon treatment with alkali hydroxide to give either 2-nitro[1,2-$^{13,14}$C]acetaldehyde oxime (**299**)[228] or methyl nitro[1,2-$^{13}$C]acetate (**302**)[213] (Figure 5.69, route E). Oxime **299**, which is formed under milder conditions, reacted with *o*-aminobenzaldehyde to provide 3-nitro[2, 3-$^{13,14}$C]-quinoline (**300**). Reduction of the nitro group with $SnCl_2$/HCl followed by deamination afforded [2,3-$^{13,14}$C]quinoline (**301**) in 25% overall yield[228].

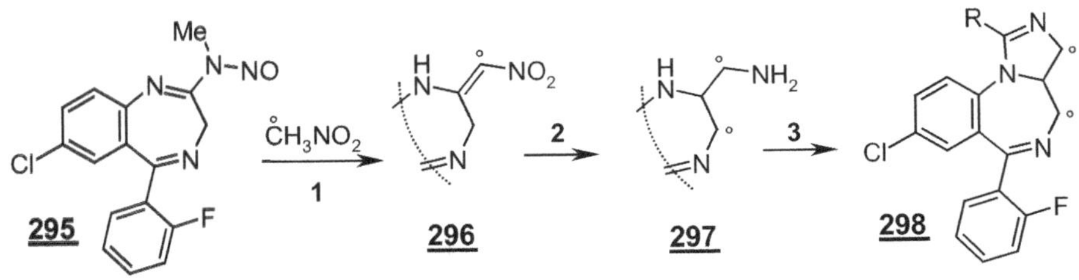


**Reaction conditions:** **1**. *tert*-BuOK, DMF; r.t., 1 h (50 %); **2**. $H_2$, RaNi, THF/MeOH; r.t., 24 h (91 %); **3**. $CH_3C(OEt)=NH.HCl$, EtOH–THF; 0 °C, 3 h (78–93 %)

**Figure 5.75** *Use of [$^{13}$C]nitromethane in the addition/elimination sequence (part 2)*

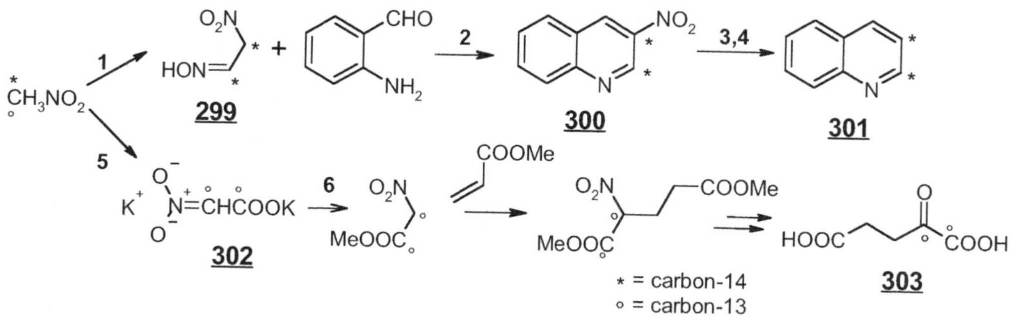


**Reaction conditions:** **1**. aq. NaOH; 50 °C, 30 min; **2**. aq. HCl, EtOH; r.t., 16 h; **3**. $SnCl_2$, conc. HCl; r.t., 3 h; **4**. $NaNO_2$, aq. $H_3PO_2$; 0 °C, 20 h; **5**. conc. KOH, r.t., 10 min, 160 °C, 70 min, **6**. HCl, MeOH

***Figure 5.76*** *'Dimerization' of [$^{13,14}$C]nitromethane and applications*

Methyl nitro[1,2-$^{13}$C]acetate (**302**) was formed in good yield by 'dimerization' at higher alkali hydroxide concentrations and higher temperature. Compare the preparation of singly labeled methyl nitroacetate from nitro[$^{14}$C]methane early in this section. Isotopomer **302** was converted into $\alpha$-keto[1,2-$^{13}$C]glutaric acid (**303**) following the reaction conditions for singly labeled isotopomer **262**[213] (Figure 5.70).

## References


1. a. Dauben, W.G.; Reid, J.C. and Yankewich, P.E., *Anal. Chem.*, ***1947***, *19*, 828–832; b. Heidelberger, C.; Brewer, P. and Dauben, W.G., *J. Am. Chem. Soc.*, ***1947***, *69*, 1389–1391; c. Murray, A. and Williams, D.L., *Organic Syntheses with Isotopes;* Interscience: New York, ***1958***, pp. 86–88.
2. a. Zweibel, N.; Turkevich, J. and Miller, W.W., *J. Am. Chem. Soc.*, ***1949***, *71*, 376–377; b. Murray, A. and Williams, D.L., *Organic Syntheses with Isotopes;* Interscience: New York, ***1958***, pp. 580–581.
3. a. Perry, C.W.; Burger, W. and Delaney, C.M., *J. Label. Compd. Radiopharm.*, ***1979***, *16*, 645–649; b. Murray, A. and Williams, D.L., *Organic Syntheses with Isotopes;* Interscience: New York, ***1958***, p. 564.
4. Bánfi D. and Mlinkó S. in *Synthesis and Applications of Isotopically Labeled Compounds 1985*, Muccino R.R., Ed.; Elsevier: Amsterdam, ***1986***, pp. 319–324.
5. a. Bubner, M. and Schmidt, J., *Die Synthese Kohlenstoff-14 markierter organischer Verbindungen;* Georg Thieme Verlag: Leipzig, ***1966***, pp. 103–104; b. Murray, A. and Williams, D., *Organic Syntheses with Isotopes;* Interscience: New York, ***1958***, pp. 45–46.
6. Murray, A. and Williams, D., *Organic Syntheses with Isotopes;* Interscience: New York, ***1958***, pp. 588–589.
7. a. Murray, A. and Williams, D., *Organic Syntheses with Isotopes;* Interscience: New York, ***1958***, pp. 94–98; b. Le, D.D.; Zhang, Y.; Nguyen, D. and Moravek, J., *J. Label. Compd Radiopharm.*, ***2000***, *43*, 1107–1111; c. Senderoff, S.G.; Shu, A.Y.L.; Lawrie, K. and Heys, J.R., *J. Label. Compd. Radiopharm.*, ***1992***, *31*, 519–538; d. Azim, M.; Madelmont, J.C.; Rapp, M.; Maurizis, J.C.; Pavia, A.; Pucci, B. and Veyre, A., *J. Label. Compd. Radiopharm.*, ***1993***, *33*, 1155–1159; e. Hasegawa, M.; Ohyabu, S.; Ueda, T. and Yamaguchi, M., *J. Label. Compd. Radiopharm.*, ***1981***, *18*, 643–650; f. Sieving, P., *J. Label. Compd. Radiopharm.*, ***1987***, *24*, 753–758.

8. a. Tschesche, R. and Wirth, W., *J. Label. Compd. Radiopharm.*, ***1981***, *18*, 433–438; b. Muccino, R.R.; Liebman, A.A.; Cupano, J. and Malarek, D.H., *J. Label. Compd. Radiopharm.*, ***1985***, *22*, 159–170.
9. Wirz, B. and Küng, W., *J. Label. Compd. Radiopharm.*, ***1983***, *20*, 635–653.
10. a. Koermeling, R., *J. Label. Compd. Radiopharm.*, ***2002***, *45*, 269–270; b. Koermeling, R., personal communication to RV, ***2005***; c. Pickering, R.E.; Wysocki, M.A.; Eisenbraun, E.J.; Pell, R.J.; Gearhart, H.L. and Hamming, M.C., *J. Label. Compd. Radiopharm.*, ***1987***, *24*, 919–924; d. Seetz, J.W.F.L.; Tol, R.; Akkerman, O.S. and Bickelhaupt, F., *Chem. Commun.*, ***1983***, 721.
11. Rhee, S.W.; Reist, E.J. and Rock, M.T, *J. Label. Compd. Radiopharm.*, ***1994***, *34*, 173–178.
12. a. Liebman, A.A.; Bader, G.J.; Kaegi, H.H. and Perry, C.W., *J. Label. Compd. Radiopharm.*, ***1986***, *23*, 693–698; b. Verny, M.; Nicolas, C. and Veyre, A., *J. Label. Compd. Radiopharm.*, ***1991***, *29*, 927–931; c. Herbert, M. and Pichat, L., *J. Label. Compd. Radiopharm.*, ***1976***, *12*, 437–441; d. Yoshitake, A.; Makari, Y. and Endo, M., *J. Label. Compd*, ***1972***, *8*, 615–621; e. Nguyen, V. B.; Herbert, M.; Pichat, L. and Nguyen, D.-X., *J. Label. Compd*, ***1973***, *9*, 545–552; f. Ellsworth R.L.; Meriwether H.T. and Mertel H.E., *J. Label. Compd. Radiopharm.* ***1989***, *27*, 1229–1241.
13. Reed G.,Amersham Biosciences, personal communication to RV, ***2005***,
14. Egli, P. and Migdalof, B.H., *J. Label. Compd. Radiopharm.*, ***1988***, *25*, 1105–1115.
15. a. Burgos, A.; Duffin, G.R.; Ellames, G.J. and Wedge, K.J., *J. Label. Compd. Radiopharm.*, ***1996***, *38*, 193–201; b. Hicks, J.L. and Heys, J.R., *J. Label. Compd. Radiopharm.*, ***1981***, *18*, 1457–1464; c. Le, D.D.; Zhang, Y.; Chien, D.H. and Moravek, J., *J. Label. Compd. Radiopharm.*, ***2000***, *43*, 1119–1125.
16. Negishi, E.; Swanson, D.R. and Rousset, C.J., *J. Org. Chem.*, ***1990***, *55*, 5406–5409.
17. Johansson, R. and Werner, T., *J. Label. Compd. Radiopharm.*, ***2000***, *43*, 1265–1273.
18. Ekhato, I.V. and Huang, C.C., *J. Label. Compd. Radiopharm.* ***1993***, *33*, 869–880.
19. Ekhato, I.V. and Huang, C.C., *J. Label. Compd. Radiopharm.*, ***1994***, *34*, 213–220.
20. a. Knochel, P.; Dohle, W.; Gommermann, N.; Kneisel, F.F.; Kopp, F.; Korn, T.; Sapountzis, I.; and Vu, V.A., *Angew. Chem. Int. Edn Engl.*, ***2003***, *42*, 4302–4320; b. Boudier, A.; Bromm, L.O.; Lotz, M.; and Knochel, P., *Angew. Chem. Int. Edn Engl.*, ***2000***, *39*, 4414–4435; c. Ricci, A.; and Fochi, M., Angew. Chem. Int. Edn Engl., ***2003***, *42*, 1444–1446.
21. Snieckus, V., *Chem. Rev.*, ***1990***, *90*, 879–933.
22. Ekhato, I.V. and Huang, C.C., *J. Label. Compd. Radiopharm.*, ***1994***, *34*, 505–512.
23. Hicks, J.L., Huang, C.C. and Showalter, H.D.H., *J. Label. Compd. Radiopharm.*, ***1987***, *24*, 1209–1220.
24. Gawell, L., *J. Label. Compd. Radiopharm.*, ***1986***, *23*, 45–50.
25. Nishioka, K.; Kamada, T. and Kanamara, H., *J. Label. Compd. Radiopharm.*, ***1992***, *31*, 553–560.
26. Heys, J.R., *J. Label. Compd. Radiopharm.*, ***1987***, *24*, 157–162.
27. a. Ekhato, I.V., *J. Label. Compd. Radiopharm.*, ***1998***, *41*, 523–533; b. Clizbe, L.A. and Reist, E. J., *J. Label. Compd. Radiopharm.*, ***1992***, *31*, 951–954; c. Ekhato, I.V. and Palazzoio, M., *Synth. Commun.*, ***1998***, *28*, 3727–3741.
28. Ekhato, I.V. and Huang, C.C., *J. Label. Compd. Radiopharm.*, ***1994***, *34*, 107–115.
29. Still, W.C., *J. Am. Chem. Soc.*, ***1978***, *100*, 1481–1487.
30. Crevoisier, M.; Guilford, W.J. and Coates, R.M., *J. Label. Compd. Radiopharm.*, ***1983***, *20*, 1213–1218.
31. Markos, C.S.; Dorn, C.R. and Zitzewitz, D.J., *J. Label. Compd. Radiopharm.*, ***1988***, *25*, 515–529.
32. Pichat, L. and Beacourt, J.P., *J. Label. Compd*, ***1974***, *10*, 103–112.
33. a. Bubner, M. and Schmidt, L., *Die Synthese Kohlenstoff-14 markierter organischer Verbindungen;* Georg Thieme Verlag: Leipzig, ***1966***, pp. 1011–1013 and 137–137; b. Cox, J.D. and Warne, R.J., *Nature*, ***1950***, *165*, 563–564; c. Murray A. and Williams D.L., *Organic Syntheses with Isotopes;* Interscience: New York, ***1958***, pp. 899–900 and 905–906; d. Cox, J.D.; Turner, H.S. and Warne, R.J., *J. Chem. Soc.*, ***1950***, 3167–3176.
34. a. Bubner, M. and Schmidt, L., *Die Synthese Kohlenstoff-14 markierter organischer Verbindungen;* Georg Thieme Verlag: Leipzig, ***1966***, pp. 122–123; b. Roberts, J.D.; McMahon, R.E. and

Hine, J.S., *J. Am. Chem. Soc.*, ***1950***, *72*, 4237–4240; c. Cox, J.D. and Turner H.S., *J. Chem. Soc.* ***1950***, 3176–3180; d. Murray, A. and Williams, D.L., *Organic Syntheses with Isotopes;* Interscience: New York, ***1958***, pp. 378–379.

35. Parnes, H. and de Keczer, S., *J. Label. Compd. Radiopharm.*, ***1985***, *21*, 287–292.
36. Aranda, G.; Herbert, M. and Pichat, L., *J. Label. Compd. Radiopharm.*, ***1986***, *23*, 257–266.
37. Volford, J.; Kanusz, D.; Meszticzky, A.; Horváth, L. and Csákvári, B., *J. Label. Compd. Radiopharm.*, ***1981***, *18*, 555–561.
38. a. Dean, D.C.; Wallace, M.A.; Marks, T.M. and Melillo, D.G., *Tetrahedron Lett.*, ***1997***, *38*, 919–922; b. Wallace, M.A.; Dean, D.C.; Marks, T.M. and Melillo, D.G., *Synthesis and Applications of Isotopically Labeled Compounds 1997,* Heys, J.R. and Melillo, D.G.,Eds.; John Wiley & Sons, Ltd: Chichester, ***1998***, pp. 179–183.
39. Murray, A. and Williams, D.L., *Organic Syntheses with Isotopes;* Interscience: New York, ***1958***, pp. 34–35.
40. Elmore, C.S.; Dean, D.C. and Melillo, D.G., *J. Label. Compd. Radiopharm.*, ***2000***, *43*, 1135–1144.
41. Parnes, H. and Shelton, E., *J. Label. Compd. Radiopharm.*, ***1986***, *23*, 677–680.
42. a. Huston, J.L. and Norris, T.H., *J. Am. Chem. Soc.* ***1948***, *70*, 1968–1969; b. Stranks, D.R., *Trans. Faraday Soc.*, ***1955***, *51*, 514–524; c. Murray, A. and Williams, D.L., *Organic Syntheses with Isotopes;* Interscience: New York, ***1958***, pp. 576–577; d. Weinhouse, S., *J. Am. Chem. Soc.*, ***1948***, *70*, 442–443.
43. Melville, D.B.; Rachele, J.R. and Keller, E.B., *J. Biol. Chem.*, ***1947***, *169*, 419–425.
44. a. Weygand, F. and Linden, H., *Angew. Chem.*, ***1954***, *66*, 174–176; b. Weygand F. and Linden H., *Z. Naturforsch.*, ***1954***, *96*, 682–684; c. Murray, A. and Williams D.L., *Organic Syntheses with Isotopes;* Interscience: New York, ***1958***, pp. 607–608; d. Bubner M. and Schmidt L., *Die Synthese Kohlenstoff-14 markierter organischer Verbindungen;* Georg Thieme Verlag: Leipzig, ***1966***, pp. 98–101.
45. a. Murray, A.; Bills, C.W. and Ronzio A.R., *J. Am. Chem. Soc.*, ***1951***, *74*, 2405–2406; b. Arnstein H.R.V., *Nature* ***1949***, *164*, 362–365; c. Arnstein H.R.V., *Biochem. J.*, ***1951***, *49*, 439–441; d. Murray, A. and Williams, D.L., *Organic Syntheses with Isotopes;* Interscience: New York, ***1958***, pp. 608–614.
46. Murray, A. and Williams, D.L., *Organic Syntheses with Isotopes;* Interscience: New York, ***1958***, pp. 899–900.
47. Murray, A. and Williams, D.L., *Organic Syntheses with Isotopes;* Interscience: New York, ***1958***, pp. 575–576.
48. Murray, A. and Williams, D.L., *Organic Syntheses with Isotopes;* Interscience: New York, ***1958***, pp. 577–578.
49. Elmore, C.S.; Dean, D.C.; DeVita, R.J. and Melillo, D.G., *J. Label. Compd. Radiopharm.*, ***2003***, *46*, 993–1000.
50. a. Kabalka, G.W.; Gooch, E.E.; Collins, C.J. and Raaen, F., *J. Chem. Soc. Chem. Commun.*, ***1979***, 607–608; b. Kabalka, G.W.; Delgado, M.C.; Sastry, U. and Sastry, K.A.R., *J. Chem. Soc. Chem. Commun.*, ***1982***, 1273–1274; c. Kabalka, G.W.; Delgado, M.C.; Kunda, U.S. and Kunda, S.A., *J. Org. Chem.*, *49*, ***1984***, 174–176; d. Kabalka, G.W., *Acc. Chem. Res.*, ***1984***, *17*, 215–221; e. Kabalka G.W., *J. Label. Compd. Radiopharm.*, ***2007***, *50*, 888–894.
51. a. Martarello, L.; Gee, A.D.; Audrain, H.; Aburel, P.S. and Bender, D., *J. Label. Compd. Radiopharm.*, ***2005***, *48*, 541; b. Audrain, H.; Gosmer, K.B.; Gee, A.D. and Bender, D., *J. Label. Compd. Radiopharm.*, ***2007***, *50(S1)*, p. S144; c. Audrain, H.; Martarello, L.; Gee, A.D. and Bender, D., *Chem. Commun.*, ***2004***, 558–559.
52. Kealey, S.; Long, N.J.; Miller, P.W.; Bennacef, I.; Audrain, H.; Bender, D. and Gee, A., *J. Label. Compd. Radiopharm.*, ***2007***, *50(S1)*, p. S131.
53. Långström, B.; Itsenko, O. and Rahman, O., *J. Label. Compd. Radiopharm.*, ***2007***, *50*, 794–810.
54. Karimi, F. and Långström, B., *J. Chem. Soc. Perkin Trans.*, ***2002***, *1*, 2256–2259.
55. Karimi, F. and Långström, B., *Org. Biomol. Chem.*, ***2003***, *1*, 541–546.
56. Rahman, O.; Kihlberg, T. and Långström, B., *J. Org. Chem.*, ***2003***, *68*, 3358–3362.
57. Karimi, F.; Kihlberg, T. and Långström, B., *J. Chem. Soc., Perkin Trans.* ***2001***, *1*, 1528–1531.

58. Karimi, F.; Barletta, J. and Långström, B., *Eur. J. Org. Chem.*, ***2005***, 2374–2378.
59. Rahman, O.; Llop, J. and Långström, B., *Eur. J. Org. Chem.*, ***2004***, 2674–2678.
60. Rahman, O.; Kihlberg, T. and Långström, B., *Org. Biomol. Chem.*, ***2004***, *2*, 1612–1616.
61. a. Itsenko, O.; Kihlberg, T. and Långström, B., *Eur. J. Org. Chem.*, ***2005***, 3830–3834; b. Itsenko, O.; Kühlberg, T. and Långström, B., *J. Org. Chem.* ***2004***, *69*, 4356–4360.
62. Calvin, M.; Heidelberger, C.; Reid, J.C.; Tolbert, B.M.; and Yankwich, P.E., *Isotopic Carbon*, New York, ***1949***, p. 155.
63. Murray, A. and Williams, D.L., *Organic Synthesis with Isotopes;* Interscience: New York, ***1958***, pp. 585–587.
64. Latli, B., *J. Lab. Comp. Radiopharm.*, ***2004***, *47*, 847–856.
65. Heier, R.F.; Moon, M.W.; Stolle, W.T.; Easter, J.A. and Hsi, R.S.P., *J. Label. Compd. Radiopharm.*, ***1996***, *38*, 1087–1098.
66. Eggler, J.F.; Gabel, C.A.; Zandi, K.; Weaver, J.D.; McKechney, H.; Dombroski, M.A.; Peurano, H. and Hawryluk, N., *J. Label. Compd. Radiopharm.*, ***2002***, *45*, 785–794.
67. Prakash, S.R.; Cable, K.M.; Correa, I.D.; Fellows, I.; Montgomery, S.; Newman, J.J.; Waterhouse, I. and Wells, G.N., *J. Label. Compd. Radiopharm.*, ***1995***, *36*, 993–1007.
68. Haynes, U.J.; Swigor, J.E. and Kaneko, J., *J. Label. Compd. Radiopharm.*, ***1990***, *28*, 241–244.
69. Sion R.; Schumer, A.; van Durme, E.; Gouyette, A.; Geslin, M.; Fournier, J.P.; Berger, Y. and Roger, P., *J. Label. Compd. Radiopharm.*, ***1990***, *28*, 653–665.
70. Walker, D.G.; Leister, W.H. and Weaner, L.E., *J. Label. Compd. Radiopharm.*, ***1995***, *36*, 661–669.
71. Switek, K.H. and Braunger, H., *J. Label. Compd. Radiopharm.*, ***2002***, *45*, 1175–1176.
72. Nassar, M.N.; Agha, B.J. and Digenis, G.A., *J. Label. Compd. Radiopharm.*, ***1986***, *23*, 667–676.
73. Jagadish. B.; Iyengar, B.S.; Solyom, A.M.; Remers, W.A.; Dorr, R.T.; Yu, J.S.; Gupta, S. and Mash, E.A., *J. Label. Compd. Radiopharm.*, ***2005***, *48*, 165–170.
74. Marshall, F.J. and McMahon, R.E., *J. Label. Compd. Radiopharm.*, ***1970***, *6*, 261–265.
75. Kuo, F. and Wheeler, W.J., *J. Label. Compd. Radiopharm.*, ***1994***, *34*, 915–920.
76. a. Murray, A. and Williams, D.L., *Organic Synthesis with Isotopes;* Interscience: New York, ***1958***, pp. 31–34; b. Spyker, J.W. and Neish A.C., *Can. J. Chem.* ***1952***, *30*, 461–465; c. Bubner, M. and Schmidt, L.H., *Die Synthese Kohlenstoff-14 marlierter organischer Verbindungen*, Georg Thieme Verlag: Leipzig, ***1966***, pp. 96–98.
77. Grant, D.G. and Turner, H.S., *Nature*, ***1950***, *165*, 153–154.
78. Ho, J.Z.; Elmore, C.S.; Wallace, M.A.; Yao, D.; Braun, M.P.; Dean, D.C.; Melillo, D.G. and Chen, C., *Helv. Chimica Acta*, ***2005***, *88*, 1040–1047.
79. a. Balaban, A.T. and Bally, I., Isotopically labeled pyrimidines and purines, in *Isotopes in the Physical and Biomedical Sciences*, Vol. 1, *Part A*., Buncel, E. and Jones J.R., Eds.; Elsevier: Amsterdam, ***1987***, pp. 290–329; b. Ayrey, G. and Yeomans, M.A., *J. Label. Compd*, ***1976***, *12*, 323–345.
80. Montgomery, J.A. and Thomas, H.J., *J. Label. Compd. Radiopharm.*, ***1978***, *15*, 539–543.
81. a. Abrams, R. and Clark, L., *J. Am. Chem. Soc.*, ***1951***, *73*, 4609–4612; b. Murray, A. and Williams, D.L., *Organic Synthesis with Isotopes;* Interscience: New York, ***1958***, pp. 752–756.
82. a. Weygand F.; and Grosskinsky, O.A., *Chem. Ber.*, ***1951***, *84*, 839–844; b. Murray, A. and Williams, D.L., *Organic Synthesis with Isotopes;* Interscience: New York, ***1958***, pp. 764–767.
83. a. Weygand, F.; Klebe, H.; Trebst, A. and Simon H., *Z. Naturforsch.*, ***1954***, *9b*, 450–455; b. Murray, A. and Williams, D.L., *Organic Synthesis with Isotopes;* Interscience: New York, ***1958***, pp. 759–760.
84. a. Lohmann, S.M. and Meich, R.P., *J. Label. Compd*, ***1975***, *11*, 515–519; b. Hignite, C.E.; Huffman, D.E. and Azamoff, D.C., *J. Label. Compd. Radiopharm.*, ***1978***, *14*, 475–478.
85. Wheeler, W.J. and Blanchard, W.B., *J. Label. Compd. Radiopharm.*, ***1992***, *31*, 495–504.
86. Sharma, M.; Alderfer, J.L. and Box, H.C., *J. Label. Compd. Radiopharm.*, ***1983***, *20*, 1219–1225.
87. Hayashi, N.; Shinozaki, K.; Kato, S.; Meguro, K. and Kuwada, Y., *J. Label. Compd*, ***1974***, *10*, 73–77.
88. a. Lautie, M.F. and Corval, M., *J. Label. Compd. Radiopharm.*, ***1982***, *19*, 1011–1020; b. Pichat L.; Audinot M. and Monnet J., *Bull. Soc. Chim. Fr.*, ***1954***, 85–88.
89. Montgomery, J.A. and Thomas, H.J., *J. Label. Compd. Radiopharm.*, ***1978***, *15*, 727–730.

90. Chan, K.K.; Staroscik, J.A. and Sadée, W., *J. Med. Chem.*, ***1977***, *20*, 598–600.
91. Chubachi, M.; Kawano, E., *J. Label. Compd. Radiopharm.*, ***1987***, *24*, 797–803.
92. a. Duncan, D.R. and Johnston, D., *J. Label. Compd. Radiopharm.*, ***1983***, *20*, 1227–1228; b. Pudles, J.; and Bloch, K., *J. Biol. Chem.*, ***1960***, *235*, 3417–3420.
93. Threadgill, M.D. and Gate, E.N., *J. Label. Compd. Radiopharm.*, ***1983***, *20*, 447–451.
94. a. Royer, R.E.; Daub, G.H. and Vander Jagt, D.L., *J. Label. Compd. Radiopharm.*, ***1976***, *12*, 377–380; b. Nicolas, C.; Mandelmont, J.C.; Maurizis, J.C.; Garrigue, H.; Meyniel, J.M.; Demerseman, P.; Sentenac-Roumanou, H. and Veyre, A., *J. Label. Compd. Radiopharm.*, ***1990***, *28*, 1375–1383; c. Madelmont, J.C.; Parry, D.; Maurizis, J.C.; Buisson, J.P.; Royer, R. and Meyniel J.M., *J. Label. Compd. Radiopharm.*, ***1984***, *21*, 639–647.
95. Beny, J.-P. and Kagan, J., *J. Label. Compd. Radiopharm.*, ***1982***, *19*, 313–320.
96. Nguyen, V.-B.; Herbert, M.; Pichat, L.; Janot, M.-M. and Nguyen, D.-X., *J. Label. Compd*, ***1975***, *11*, 241–247.
97. Elmore, C.S.; Dean, D.C.; Liu, K.; Jones, A.B. and Melillo, D.G., in *Synthesis and Applications of Isotopically labeled Compounds*, Vol. 7, Pleiss, U. and Voges R., Eds.; John Wiley & Sons, Ltd: Chichester, ***2001***, pp. 224–227.
98. Zeng, D.; Mi, Q.; Sun, H.; and Wang, H., *J. Label. Compd. Radiopharm.*, ***2004***, *47*, 167–174.
99. a. Geach, N.J.; Randall, G.P.; Winfield, C.J., *J. Label. Compd. Radiopharm.*, ***2004***, *47*, 1039–1040; b. Geach N.J., Private Communication ***2004***; c. Comins, D.L.; and Dernell, W., *Tetrahedron Lett.* ***1981***, *22*, 1085–1088; d. Amaratunga, W.; and Frechet, J.M.F., *Tetrahedron Lett.* ***1983***, *24*, 1143–1146.
100. Kendall, J.T. and May, M.K., in *Synthesis and Applications of Isotopically Labeled Compounds*, Vol. 7, Pleiss, U. and Voges, R., Eds.; John Wiley & Sons, Ltd: Chichester, ***2001***, pp. 221–223.
101. Meth-Cohn, O. and Stanforth, S.P., The Vilsmeyer Reaction, in *Comprehensive Organic Synthesis*, Trost, B.M. and Fleming I., Eds.; Pergamon: New York, ***1991***, pp. 777–794.
102. Andres, H., *J. Label. Compd. Radiopharm.*, ***1999***, *42*, 1008–1009.
103. a. Scherling, D. and Pleiss, U., *J. Label. Compd. Radiopharm.*, ***1988***, *25*, 1393–1400; b. Johnson, P.C. and Covington, R.R., *J. Label. Compd. Radiopharm.*, ***1982***, *19*, 953–958.
104. Kendall, J.T., *J. Label. Compd. Radiopharm.*, ***1999***, *42*, 477–484.
105. Hill, J.A. and Eaddy, J.F., *J. Label. Compd. Radiopharm.*, ***1994***, *34*, 697–706.
106. a. Bubner, M. and Schmidt, L., *Die Synthese Kohlenstoff-14 markierter organischer Verbindungen;* Georg Thieme Verlag: Leipzig, ***1966***, pp. 99–100; b. Nader, M.W.; Zeisler, S.K.; Theobald, A. and Oberhofer, F., *Appl. Radiat. Isotop.*, ***1998***, *49*, 1599–1603.
107. Mlinkó, S. and Szammer, J., *J. Radioanal. Nucl. Chem. Lett.*, ***1993***, *176*, 491–496.
108. a. Coppo, M.; Rousseau, B. and Beaucourt, J.P., *J. Label. Compd. Radiopharm.*, ***1988***, *25*, 921–925; b. Serianni, A.S.; Clark, E.L. and Barker, R., *Carbohydrate Res.*, ***1979***, *72*, 79–91.
109. a. Weygand, F. and Schaefer, G., *Chem. Ber.*, ***1952***, *85*, 307–314; b. Berger, G.; Maziére, M.; Sastre, J. and Comar, D., *J. Label. Compd. Radiopharm.*, ***1980***, *17*, 59–71.
110. a. Jones, A.R. and Skraba, W.J., *J. Am. Chem. Soc.*, ***1952***, *74*, 2437–2438; b. Jones, A.R. and Skraba, W.J., *Science*, ***1949***, *110*, 332.
111. Schreier, E., *Helv. Chim. Acta*, ***1976***, *59*, 585–606.
112. Schlewer, G.; Stampf, J.L. and Benezra, C., *J. Label. Compd. Radiopharm.*, ***1980***, *17*, 297–302.
113. De Keyser, J.-L.; De Cock, C.J.C.; Poupaert, J.H. and Dumont, P., *J. Label. Compd. Radiopharm.*, ***1989***, *27*, 909–916.
114. Szammer, J.; Simon-Trompler, E.; Máté, J.; Abermann, M. and Klebovich, I., *J. Label. Compd. Radiopharm.*, ***1994***, *34*, 1111–1115.
115. a. Roberts, J.D.; Lee, C.C. and Saunders, W.H. Jr., *J. Am. Chem. Soc.* ***1955***, *77*, 3034–3037; b. Roberts, J.D.; Lee, C.C. and Saunders, W.H. Jr., *J. Am. Chem. Soc.* ***1954***, *76*, 4501–4510.
116. Robson, J. and Sorby, P., Aust. A.E.C., AAEC/E440, ***1978***.
117. a. Arnstein, H.R.V., *Biochem. J.* ***1951***, *49*, 439–446; b. Levine, M. and Tarver, H., *J. Biol. Chem.*, ***1950***, *184*, 427–436; c. Kögl, F. and Municio, A.M., *Z. Physiol. Chem.*, ***1955***, *300*, 17–26; d. Murray, A. and Williams, D.L., *Organic Syntheses with Isotopes;* Interscience: New York, ***1958***, pp. 169–174.

118. Arnstein, H.R.V. and Crawhall, J.C., *Biochem. J.*, ***1953***, *55*, 280–285; b. Murray, A. and Williams, D.L., *Organic Syntheses with Isotopes;* Interscience: New York, ***1958***, pp. 206–209.
119. a. Iliceto, A., *Ann. Chim. (Rome)*, ***1953***, *43*, 638–642; b. Murray, A. and Williams, D.L., *Organic Syntheses with Isotopes;* Interscience: New York, ***1958***, pp. 382–386.
120. Ruzicka, J.A.; Qiu, W.; Baker, M.T. and Burton, D.J., *J. Label. Compd. Radiopharm.*, ***1994***, *34*, 59–65.
121. a. Casiraghi, G.; Casnati, G.; Cornia, M.; Pochini, A.; Puglia, G.; Sartori, G. and Ungaro, R., *J. Chem. Soc. Perkin I*, ***1978***, 318–321; b. Casiraghi, G.; Casnati, G.; Puglia, G.; Sartori, G. and Terenghi, G., *J. Chem. Soc. Perkin I*, ***1980***, 1862–1865.
122. Koble, D.L. and Fischer, L.J., *J. Label. Compd. Radiopharm.*, ***1984***, *21*, 519–524.
123. Nakamura, T.; Hamamura, K. and Ueda T., *J. Label. Compd. Radiopharm.*, ***2000***, *43*, 185–194.
124. Bothner-By, A.A., *J. Am. Chem. Soc.*, ***1951***, *73*, 4228–4230.
125. Engler, J. and Katztreiner, E., *J. Label. Compd*, ***1973***, *9*, 237–241.
126. Isikow H.M. and Gump, W.S., *J. Am. Chem. Soc.*, ***1952***, *74*, 2434–2435.
127. Willcocks, K.; Barnes, R.D.; Rustidge, D.C. and Tidy, D.J.D., *J. Label. Compd. Radiopharm.*, ***1993***, *33*, 783–794.
128. McMorris, T.C.; Yu, J.; Herman, D.M.; Kelner, M.J.; Dawe, R. and Minamida, A., *J. Label. Compd. Radiopharm.*, ***1998***, *41*, 279–285.
129. Karstens, W.F.J.; Berger, H.J.F.F.; van Haren, E.R.; Lugtenberg, J. and Raap, J., *J. Label. Compd. Radiopharm.*, ***1995***, *36*, 1077–1096.
130. a. Salter R., Züger C. and Schreiner E., presentation at the IIS-CED meeting Bad Soden, Germony, June 16–17, ***2004***; b. Bloom, A.J. and Winton, P.M., *J. Label. Compd. Radiopharm.*, ***1999***, *42*, 1004–1005.
131. Logan, A.V.; Huston, J.L. and Dorward, D.L., *J. Am. Chem. Soc.*, ***1950***, *73*, 2952–2953.
132. Simon, H. and Müllhofer, G., *Chem. Ber.*, ***1963***, *96*, 3167–3176.
133. Goon, D.J.W.; Elberling, J.A.; DeMaster, E.G. and Nagasawa, H.T., *J. Label. Compd. Radiopharm.*, ***1990***, *28*, 203–208.
134. Gasparini, G.M.; Pichat, L. and Baret, C., *Bull. Soc. Chim. Fr.*, ***1965***, 794–797.
135. Gu, F.; Netscher, T. and Atkinson, J., *J. Label. Compd. Radiopharm.*, ***2006***, *49*, 733–734.
136. Van de Westeringh, C. and Veldstra, H., *Rec. trav. Chim.*, ***1958***, *77*, 1114–1128.
137. Heidelberger, C., *J. Biol. Chem.*, ***1949***, *179*, 139–142.
138. a. Plieninger, H.; Fischer, R. and Liede, K., *Liebigs Ann. Chem.*, ***1964***, *672*, 223–231; b. Plieninger, H. Höbel, M. and Liede, K., *Chem. Ber.*, ***1963***, *96*, 1618–1629.
139. Leete, E., *J. Am. Chem. Soc.*, ***1960***, *82*, 6338–6342.
140. Lawrie, K.W.M.; Heys, J.R.; Garnes, K.; Villani, A.J. and Landvatter, S.W., *J. Label. Compd. Radiopharm.*, ***2007***, *50*, 530–531.
141. Park, S.H.; Pradeep, K. and Jang, S.H., *J. Label. Compd. Radiopharm.*, ***2007***, *50*, 1248–1254.
142. Szammer, J.; Simon-Trompler, E. and Mlinko, S., *J. Label. Compd. Radiopharm.*, ***1994***, *34*, 313–321.
143. Zisman, S.A.; Berlin, K.D.; Alavi, F.K.; Sangiah, S.; Clarke, C.R. and Scherlag, B.J., *J. Label. Compd. Radiopharm.*, ***1989***, *27*, 885–888.
144. Nakatsuka, I.; Kawahara, K. and Yoshitaki, A., *J. Label. Compd. Radiopharm.*, ***1981***, *18*, 495–506.
145. a. Heys, J.R.; Villani, A.J. and Mastrocola, A.R., *J. Label. Compd. Radiopharm.*, ***1996***, *38*, 761–769; b. Overman, L.E. and Sharp, M.J., *J. Am. Chem. Soc.*, ***1988***, *110*, 612–614.
146. a. Ren, S.; McNamara, P.; Royster, P.; Lee, J.; Saluja, S.S.; Koharski, D.; Hendershot, S. and Truong, V., *J. Label. Compd. Radiopharm.*, ***2007***, *50*, 643–648; b. Hesk, D.; Cesarz, D.; Magatti, C.; Voronin, K.; Lavey, C.; McNamara, P.; Koharski, D.; Saluja, S.; Hendershot, kS.; Pham, H. and Truong, V., *J. Label. Compd. Radiopharm.*, ***2005***, *48*, 11–23.
147. Gynther, J.; Huhtikangas, A.; Auriola, S. and Juvonen, H., *J. Label. Compd. Radiopharm.* ***1986***, *23*, 741–748.
148. Li, Y. and Prakash, S.R., *J. Label. Compd. Radiopharm.*, ***2005***, *48*, 323–330.
149. Liu, Y. and Xin, Z., *J. Label. Compd. Radiopharm.*, ***1988***, *25*, 569–572.
150. a. Flynn, E.H.; Murphy, H.W. and McMahon, R.E., *J. Am. Chem. Soc.*, ***1955***, *77*, 3104–3106; b. Pike, V.W.; Palmer, A.J.; Horlock, P.L.; Perun, T.J.; Freiberg, L.A.; Dunnigan, D.A. and Liss, R.H., *Int. J. Appl. Radiat. Isot.*, ***1984***, *35*, 103–109.

151. Eng, R.R. and Spitznagle L.A., *J. Label. Compd*, ***1975***, *11*, 291–294.
152. Ross, J.J.; Young, R.L. and Maass A.R., *Science*, ***1958***, *128*, 1279–1280.
153. a. Jentoft, N. and Dearborn, D.G., *J. Biol. Chem.*, ***1979***, *254*, 4359–4364; b. Dottavio-Martin, D. and Ravel, J.M., *Anal. Biochem.*, ***1978***, *87*, 562–565; c. Berger, G.; Maziere, M.; Prenant, C.; Sastre, J. and Comar, D., *Int. J. Appl. Radiat. Isot.*, ***1984***, *35*, 81–83.
154. a. Ampleman, G.; Thiboutot, S.; Lavigne, J.; Marois, A.; Hawari, J.; Jones, A.M. and Rho, D., *J. Label. Compd. Radiopharm.*, ***1995***, *36*, 559–577; b. Horváth, K. and Alworth, A.L., *J. Label. Compd. Radiopharm.*, ***1993***, *33*, 467–471.
155. Huang, C.-Y.; Ma, R.A. and Hee, S.S.Q., *J. Label. Compd. Radiopharm.*, ***1998***, *41*, 377–385.
156. Stock, M. and Schütte H.R., *Isotopenpraxis*, ***1990***, *26*, 293.
157. Braun, H.; Bertram, B. and Wiessler, M., *J. Label. Compd. Radiopharm.*, ***1977***, *13*, 375–378.
158. Maller, R.K.; Nagarajan, K. and Nair, M.D., *J. Label. Compd. Radiopharm.*, ***1985***, *22*, 577–581.
159. a. Johnston, D. and Duncan, D.R., *J. Label. Compd. Radiopharm.*, ***1989***, *27*, 45–50; b. Drewes, S.E. and Roos, G.H.P., *Tetrahedron*, ***1988***, *44*, 4653–4670.
160. a. Nagasaki, T.; Katsuyama, Y. and Yoshioka, M., *J. Label. Compd. Radiopharm.*, ***1987***, *24*, 65–71; b. Murray, A. and Williams, D.L., *Organic Syntheses with Isotopes;* Interscience: New York, ***1958***, pp. 861–864 and 900–901; c. Murray, A. and Ronzio, A.R., *J. Am. Chem. Soc.*, ***1952***, *74*, 2408–2412.
161. a. Murray, A. and Williams, D.L., *Organic Syntheses with Isotopes;* Interscience: New York, ***1958***, pp. 899–905; b. Bubner, M. and Schmidt, L.H., *Die Synthese Kohlenstoff-14 markierter Verbindungen;* Georg Thieme Verlag: Leipzig, ***1966***, pp. 102–103 and 141–142.
162. Fellows, I.; Carr, R.M.; deBoeck, N.; Montgomery, S. and Waterhouse, I., *J. Label. Compd. Radiopharm.*, ***1998***, *41*, 1127–1143.
163. Dutton A.H. and Heath D.F. *J. Chem. Soc.*, ***1956***, 1892–1893.
164. Janssen C.G.M.; Thijssen, J.B.A. and Verluyten, W.J.M., *J. Label. Compd. Radiopharm.*, ***2002***, *45*, 841–855.
165. Senderoff, S.G.; Landvatter S.W. and Heys J.R., *J. Label. Compd. Radiopharm.*, ***2000***, *43*, 1283–1288.
166. Murray, A. and Williams, D.L., *Organic Syntheses with Isotopes;* Interscience: New York, ***1958***, pp. 198–199.
167. Colwell, W.T.; SooHoo, C. and DeGraw, J.I., *J. Label. Compd. Radiopharm.*, ***1979***, *16*, 551–558.
168. Parnes H. and Pease J., *J. Label. Compd. Radiopharm.*, ***1982***, *19*, 447–453.
169. Prakash, S.R. and Ellsworth, R.L., *J. Label. Compd. Radiopharm.* ***1988***, *25*, 815–825.
170. Koltai, E. and Banfi, D., *J. Label. Compd. Radiopharm.*, ***1981***, *18*, 809–812.
171. a. Campbell, J.B.; Knagg, E.; Newby, M.; Reid, R.G.; and Wilde, P.D., *Synthesis and Applications of Isotopically Labeled Compounds 1991*, Buncel, E. and Kabalka, G.W., Eds., Elsevier: Amsterdam, ***1992***, 207–213; b. Mirviss, S.B., *J. Org. Chem.*, ***1989***, *54*, 1948–1951; c. Xie, X.-Q.; Yang, H.; Choi, G.; Lin, S. and Makriyannis, A., *J. Label. Compd. Radiopharm.*, ***2002***, *45*, 775–784.
172. Zhang, Y.; Hong, Y.; Huang, C.C. and Belmont, D., *J. Label. Compd. Radiopharm.*, ***2006***, *49*, 237–244.
173. a. Venkatachalam, T.K.; Mzengeza S. and Diksic, M., *J. Label. Compd. Radiopharm.*, ***1993***, *33*, 1029–1038; b. Chan, A.-O.; Cooksey, C.J. and Crich, D., *J. Chem. Soc. Perkin I*, ***1992***, 777–780; c. Bourne, G.T., Crich, D., Davies, J.W, and Horwell, D.C., *J. Chem. Soc. Perkin Trans. I*, ***1991***, 1693–1694.
174. a. Yoshitake, A.; Kawahara, K.; Kamada, T. and Endo, M., *J. Label. Compd. Radiopharm.*, ***1977***, *13*, 323–331; b. Thourel, P.; Vandais, A.; Noel, J.-P. and Beaucourt J.-P., *J. Label. Compd. Radiopharm.*, ***1994***, *34*, 275–279.
175. Sahlberg, C. and Gawell, L., *J. Label. Compd. Radiopharm.*, ***1987***, *24*, 1021–1027.
176. a. Antoni, G., Kihlberg, T. and Långström, B., Aspects of the synthesis of $^{11}$C-labeled compounds, in *Handbook of Radiopharmaceuticals*, Welch, M.J. and Redvanly, C.S., Eds.; John Wiley & Sons, Ltd: Chichester, ***2003***, pp. 141–194; b. Hostetler, E.D.; Terry, G.E. and Burns, G.D., *J. Label. Compd. Radiopharm.*, ***2005***, *48*, 629–634; c. Wuest, F.R. and Berndt, M., *J. Label. Compd Radiopharm.*, ***2006***, *49*, 91–100.

177. Langer, O.; Forngren, T.; Sandell, J.; Dollé, F.; Långström, B.; Någren, K. and Halldin, C., *J. Label. Compd Radiopharm.*, ***2003***, *46*, 55–65.
178. Sahlberg, C., *J. Label. Compd. Radiopharm.*, ***1987***, *24*, 529–534.
179. Bubner, M. and Schmidt, L., *Die Synthese Kohlenstoff-14 markierter organischer Verbindungen;* Georg Thieme Verlag: Leipzig, ***1966***, pp. 148–149.
180. a. Han, M.; Hayes, B.A.; Prendergast, P.T. and Gupta, S., *J. Label. Compd. Radiopharm.*, ***2000***, *43*, 1149–1157; b. Chadhuri, N.K. and Gut M., *Methods Enzymol.*, ***1969***, *15*, 305–346.
181. Heys, J.R. and Senderoff, S.G., *J. Org. Chem.*, ***1989***, *54*, 4702–4706.
182. Weill-Raynal, J., *Synthesis*, ***1969***, *1*, 49–56.
183. DeNinno, M.P. and McCarthy, K.E., *Tetrahedron*, ***1997***, *53*, 11007–11020.
184. a. Garraffo H.M.; Deluca, M.E.; Seldes, A.M. and Gros, E.G., *J. Label. Compd. Radiopharm.*, ***1985***, *22*, 673–676; b. Deluca, M.E.; Seldes A.M. and Gros E.G., *Helv. Chim. Acta*, ***1986***, *69*, 1844–1850.
185. Landvatter, S.W., *J. Label. Compd. Radiopharm.*, ***2000***, *43*, 1321–1326.
186. a. Tetrick, M.A.; Crenshaw, T.D. and Benevenga, N.J., *Anal. Biochem.*, ***1997***, *248*, 1–6; b. Gilman, H.; Zöllner, E.A. and Selby, W.M., *J. Am. Chem. Soc.*, ***1933***, *55*, 1252–1257.
187. Braun, M.P.; Dean, D.C. and Melillo, D.G., *J. Label. Compd. Radiopharm.*, ***1999***, *42*, 469–476.
188. a. Dicus, C.W.; Willenbring, D. and Nantz, M.H., *J. Label. Compd. Radiopharm.*, ***2005***, *48*, 223–229; b. Bailey, W.F. and Punzalan, E.R., *J. Org. Chem.*, ***1990***, *55*, 5404–5406.
189. Rajan, S.J.; Wemple, J. and Sahlberg, C., *J. Label. Compd. Radiopharm.*, ***1975***, *11*, 467–474.
190. Lidström, P.; Neu, H.; Langström, B. and Sahlberg, C., *J. Label. Compd. Radiopharm.*, ***1997***, *39*, 697–704.
191. Forngren, T.; Samuelson, L.; Langström, B. and Sahlberg, C., *J. Label. Compd. Radiopharm.*, ***2004***, *47*, 71–78.
192. Swahn, B.M.; Andersson, F.; Pelcman, B.; Söderberg, J.; Claesson, A. and Sahlberg, C., *J. Label. Compd. Radiopharm.*, ***1997***, *39*, 259–266.
193. a. Gill, H., *J. Label. Compd. Radiopharm.*, ***1995***, *36*, 425–430; b. Woo, P.W.K.; Hayes, R.; Beylin, V.G.; Johnson, S.; Lovdahl, M.; Huang, Y. and Huang, C.C., in *Synthesis and Applications of Isotopically Labeled Compounds 1997*, Heys, J.R. and Melillo, D.G., Eds.; John Wiley & Sons, Ltd: Chichester, ***1998***, pp. 429–432.
194. Oliver, J.E., *J. Label. Compd. Radiopharm.*, ***1977***, *13*, 349–352.
195. Pitt C.G.; Hobbs, D.T.; Schran, H.; Twine, C.E., Jr. and Williams, D.L., *J. Label. Compd*, ***1975***, *11*, 551–575.
196. Wheeler, W.J., *J. Label. Compd. Radiopharm.*, ***1988***, *25*, 667–674.
197. Hicks, D.R.; Hangeland, J.J.; Mobilio, D. and DeLange, B., *J. Label. Compd. Radiopharm.*, ***1988***, *25*, 89–95.
198. Kaegi, H.; Chew, E. and Chien, P.-L., *J. Label. Compd. Radiopharm.*, ***1980***, *17*, 745–751.
199. Xiao, X. and Prestwich, G.D., *J. Label. Compd. Radiopharm.*, ***1991***, *29*, 883–890.
200. Foged, C.; Hansen, L.B. and Halldin, C., *J. Label. Compd. Radiopharm.*, ***1993***, *33*, 747–757.
201. a. Hesk, D.; Bowlen, C.; Duelfer, T.; Koharski, D.; McNamara, P. and Saluja, S., *J. Label. Compd. Radiopharm.*, ***1992***, *31*, 445–454; b. Hesk, D.; Bowlen, C.; Duelfer, T.; Koharski, D.; McNamara, P. and Saluja, S., in *Synthesis and Applications of Isotopically Labeled Compounds, **1991***, Buncel, E. and Kabalka, G.W., Eds; Elsevier: Amsterdam, ***1992***, pp. 552–555.
202. Pine, S.H.; Shen, G.; Bautista, J.; Sutton, Jr., C.; Yamada, W. and Apodaca, L., *J. Org. Chem.*, ***1990***, *55*, 2234–2237.
203. a. Choudhry, S.C.; Serico, L.; Cupano, J. and Liebman, A.A., in *Synthesis and Applications of Isotopically Labeled Compounds 1991*, Buncel, E. and Kabalka, G.W., Eds.; Elsevier: Amsterdam, ***1992***, pp. 218–221; b. Choudhry, S.C.; Serico, L. and Cupano, J., *J. Org. Chem.*, ***1989***, *54*, 3755–3757; c. McCarthy, K.E.; Schildtknegt, K.S. and Zandi, K.S., in *Synthesis and Applications of Isotopically Labeled Compounds 1997*, Heys, J.R. and Melillo, D.G., Eds.; John Wiley & Sons, Ltd: Chichester, ***1998***, pp. 165–168.
204. White, J.D.; Blakemore, P.R.; Milicevic, S.; Choudhry, S.C.; Cupano, J. and Serico, L., *Org. Lett.*, ***2002***, *4*, 1893–1896.
205. Perry, C.W. and Burger, W., *J. Label. Compd. Radiopharm.*, ***1977***, *13*, 113–117.

206. Sakai, T.; Fujinami, T. and Aizawa, T., *Bull. Chem. Soc. Jpn.*, ***1975***, *48*, 2981–2982.
207. a. Tagami, K.; Chiku, S. and Sohda, S., *J. Label. Compd. Radiopharm.*, ***1993***, *33*, 849–852; b. Stolle, W.T.; Sih, J.C. and si, R.S.P., *J. Label. Compd. Radiopharm.*, ***1988***, *25*, 891–900; c. Doerge, D.R., *J. Label. Compd. Radiopharm.*, ***1988***, *24*, 985–990.
208. Tanaka, A.; Fukuoka, M.; Adachi, T. and Yamaha, T., *J. Label. Compd. Radiopharm.*, ***1986***, *23*, 405–413.
209. Raab, C.E.; Dean, D.C. and Melillo, D.G., *J. Label. Compd. Radiopharm.*, ***2001***, *44*, 815–829.
210. Murray, A. and Williams, D.L., *Organic Syntheses with Isotopes;* Interscience: New York, ***1958***, pp. 40–41.
211. Schlecker, R., *J. Label. Compd. Radiopharm.*, ***1997***, *39*, 299–302.
212. a. Bubner, M. and Schmidt, L.H., *Die Synthese Kohlenstoff-14 markierter Verbindungen;* Georg Thieme Verlag: Leipzig, ***1966***, pp. 152–153; b. Arnstein, H.R.V. and Bentley, R., *J. Chem. Soc.*, ***1951***, 2385–2387; c. Murray, A. and Williams, D.L.W., *Organic Syntheses with Isotopes;* Interscience: New York, ***1958***, pp. 637–639.
213. Baldwin, J.E.; Adlington, R.M.; Russell, M.A.; Schofield, J.C. and Wood, M.E., *J. Label. Compd. Radiopharm.*, ***1989***, *27*, 1091–1099.
214. Schoeps, K-O.; Halldin, C.; Stone-Elander, S.; Langström, B. and Greitz, T., *J. Label. Compd. Radiopharm.*, ***1988***, *25*, 749–758.
215. Hendry D.; Nixon N.S.; Roughley B.S.; Skagestad P. and Winton P.M., *Synthesis and Applications of Isotopically Labeled Compounds* Vol. 8, Dean, D.C.; Filer, C.N. and McCarthy K. E., Eds.; John Wiley & Sons, Ltd: Chichester, ***2004***, pp. 25–28.
216. a. Seebach, D. and Lehr, F., *Angew. Chem. Int. Edn Engl.*, ***1976***, *15*, 505–506; b. Seebach, D.; Hening, R.; Lehr, F. and Gonermann, J., *Tetrahedron Lett.*, ***1977***, 161–164.
217. a. Lucet, D.; Sabelle, S.; Kostelitz, O.; Le Gall, T. and Mioskowski, C., *Eur. J. Org. Chem.*, ***1999***, 2583–2591; b. Zessin, J. and Steinbach, J., *J. Label. Compd. Radiopharm.*, ***1998***, *41*, 669–676.
218. Reinheimmer, P.; Hirschinger, J.; Granger, P.; Breton, P., Lagrange, A.; Gilard, P.; Lefebvre, A. and Goetz, N., *Biochim. Biophys. Acta*, ***1999***, *1472*, 240–249.
219. a. Swan, G.A. and Wright, D., *J. Chem. Soc.*, ***1954***, 381; b. Murray, A. and Williams, D.L.W., *Organic Syntheses with Isotopes;* Interscience: New York, ***1958***, pp. 520–521.
220. Gensler, W.J. and Dheer, S.K., *J. Org. Chem.*, ***1981***, *46*, 4051–4057.
221. a. Schlenk, H. and DeHaas, B.W., *J. Am. Chem. Soc.*, ***1951***, *73*, 3921; b. Murray, A. and Williams, D.L.W., *Organic Syntheses with Isotopes;* Interscience: New York, ***1958***, pp. 934–938.
222. a. Sowden, J.C. and Fischer, H.O.L., *J. Am. Chem. Soc.*, ***1947***, *69*, 1963–1965; b. Sowden, J.C., *J. Biol. Chem.*, ***1949***, *180*, 55–58; c. Sowden, J.C., *Science*, ***1949***, *109*, 229; d. Rappoport, D.A. and Hassid, W.Z., *J. Am. Chem. Soc.*, ***1951***, *73*, 5524–5525; e. Murray, A. and Williams, D., *Organic Syntheses with Isotopes;* Interscience: New York, ***1958***, pp. 975–978; f. Murray, A. and Williams, D., *Organic Syntheses with Isotopes;* Interscience: New York, ***1958***, pp. 990–998.
223. Clare, J.E.; Lawrie, K.W.M.; Willis, C.L. and Yuen, J., in *Synthesis and Applications of Isotopically Labeled Compounds*, Vol. 8, Dean, D.C.; Filer, C.N. and McCarthy, K.E., Eds.; John Wiley & Sons, Ltd: Chichester, ***2004***, pp. 1–4.
224. Hartman, J.D.; Dodd, J.H.; Hicks, J.L.; Hershenon, F.M.; Huang, C.C. and Butler, D.E., *J. Label. Compd. Radiopharm.*, ***1985***, *22*, 583–594.
225. a. Mäding, P.; Steinbach, J. and Johannsen, B., *J. Label. Compd. Radiopharm.*, ***1997***, *39*, 585–599; b. Steinbach, J.; Mäding, P.; Füchtner, F. and Johansen, B., *J. Label. Compd. Radiopharm.*, ***1995***, *36*, 33–41.
226. Mäding, P.; Steinbach, J. and Johannsen, B., *J. Label. Compd. Radiopharm.*, ***2000***, *43*, 565–583.
227. a. Zhang, Y.; Woo, P.W.K.; Hartman, J.; Colbry, N.; Huang, Y. and Huang, C.C., *Tetrahedron Lett.*, ***2005***, *46*, 2087–2091; b. Walser, A.; Benjamin, L.E.; Flynn, T.; Mason, C.; Schwartz, R. and Fryer, R.I., *J. Org. Chem.*, ***1978***, *43*, 936–944.
228. Leete, E., *Biorg. Chem.*, ***1977***, *6*, 273–286.

# 6

# Preparation of Carbon-14-Labeled Compounds via Multi-Carbon Building Blocks of the [$^{14}$C]Carbon Dioxide Tree

Chapter 5 introduces carbon-14 the isotope, barium [$^{14}$C]carbonate the main form for storage and commerce, and [$^{14}$C]carbon dioxide as a building block in radiochemical synthesis. The chapter also describes the conversion of $^{14}CO_2$ into one-[$^{14}$C]carbon building blocks and their uses. The subject of the next chapter is the conversion of $^{14}CO_2$ into carbon-14-labeled building blocks of two or more carbons, and their uses in organic radiochemical synthesis.

## 6.1 [$^{14}$C]Acetic Acid and Its Derivatives

Besides their importance as acylating agents and as reaction partners in ester condensations in carbon-14 syntheses, [$^{14}$C]acetic acid and its derivatives are indispensible precursors for a broad spectrum of low molecular weight-labeled building blocks such as alkyl halo[$^{14}$C] acetates, alkyl phosphono[$^{14}$C]acetates, alkyl cyano[$^{14}$C]acetates, dialkyl [$^{14}$C]malonates, and alkyl [$^{14}$C]acetoacetates. In addition to [$^{14}$C]acetic acid, this section covers the primary derivatives [$^{14}$C]acetyl chloride, [$^{14}$C]acetic anhydride and esters of [$^{14}$C]acetic acid and their uses.

### 6.1.1 [$^{14}$C]Acetic Acid

Depending on the position of label, monolabeled acetic acid is prepared either by carboxylation of methylmagnesium iodide (Figure 6.1, Procedure A)[1], if the label is to

*Preparation of Compounds Labeled with Tritium and Carbon-14* Rolf Voges, J. Richard Heys and Thomas Moenius

be introduced into C1, or through reaction of [$^{14}$C]methyl iodide with potassium cyanide (Procedure B), when [2-$^{14}$C]acetic acid is required[2]. In Procedure A, $^{14}CO_2$ is produced by treatment of barium [$^{14}$C]carbonate with concentrated sulfuric acid[1b,3a,b], perchloric acid[2b] or lead(II) chloride[3a,c] and condensed into the ethereal solution of methylmagnesium iodide using high vacuum techniques. The resulting reaction mixture is acidified with sulfuric acid and silver(I) sulfate is added to precipitate the iodide as AgI, thereby destroying any volatile HI. Separation of the resulting [1-$^{14}$C]acetic acid by steam distillation and adjustment of the distillate to pH 8.8 with alkali hydroxide allows isolation of the reagent as alkali [1-$^{14}$C]acetate, from which the free acid can be regenerated upon treatment with a slight excess of dry HCl gas. [2-$^{14}$C]Acetic acid, generated according to Procedure B, is isolated analogously.

Both procedures can be modified so as to obtain acetic acid isotopomers labeled at the alternate positions; i.e. $^{14}CH_3MgI$ (or $^{14}CH_3Li$) and $CO_2$ in Procedure A[4] and $K^{14}CN$ and $CH_3I$ in Procedure B[2b,4a]. The organometallic component in the first case is accessible through reaction of magnesium with $^{14}CH_3I$ or, more conveniently through metal–halogen exchange by reaction of $^{14}CH_3I$ with two equivalents of *tert*-butyllithium[5]. Doubly labeled alkali acetate is readily available following Procedure B starting from $^{14}CH_3I$ and $K^{14}CN$[2a,b,4a]. Alternatively, it can be prepared by Hg(II)-mediated hydration and oxidation of [$^{14}C_2$]acetylene with ammonium persulfate according to procedure C[6] (Chapter 8, Section 6.3).

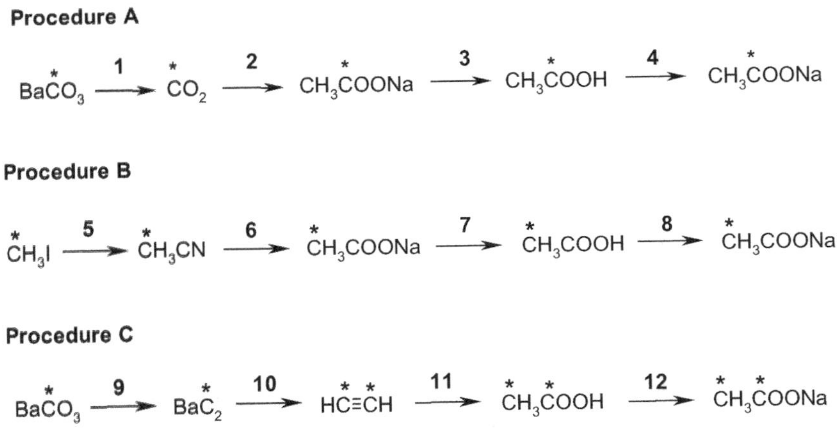


**Reaction conditions:** **1**. 100 % $H_2SO_4$; r.t. to 100 °C, 1 h **or** 60 % $HClO_4$; –78 °C to r.t. **or** $PbCl_2$, Δ (microburner); **2a**. $CH_3MgI$, ether; –78 °C, 5 min, **b**. aq. NaOH, **c**. $Ag_2SO_4$, $H_2O$; **3**. 50 % $H_2SO_4$, steam distillation; **4**. 0.1 NaOH to pH 8.8, evaporation, vacuum drying; **5**. KCN, $H_2O$; r.t., 16 h; **6**. 20 % NaOH; 80 °C, 4 h; **7**. conc. $H_2SO_4$, $Ag_2SO_4$; steam distillation; **8**. 1N NaOH to pH 8.8, evaporation, vacuum drying; **9**. Ba; 800–830 °C; **10a**. $HClO_4$–$H_2O$ 1:1; reflux, **b**. expulsion with hydrogen; **11a**. HgO, $(NH_4)_2S_2O_8$, aq. $H_2SO_4$; r.t., 16 h, **b**. $Cu_2O$, 40 °C, C.steam distillation; **12**. 1N NaOH to pH 8.8, evaporation, vacuum drying

**Figure 6.1** *Standard procedures for the preparation of sodium [1-$^{14}$C]acetate, sodium [2-$^{14}$C] acetate and sodium [1, 2-$^{14}C_2$]acetate*

Both positions of [$^{14}$C]acetic acid and its derivatives are readily accessible to chemical transformations. Activation of the carboxyl group (formation of [$^{14}$C]acetyl chloride, acetic anhydride, mixed anhydrides, imidazolides) allows for acylation of organometallic reagents, C–H acidic compounds, aromatic and heteroaromatic substrates, amines, amides, imides, alcohols, thiols, etc., to give access to labeled methyl alkyl or aryl ketones, alkyl *β*-acetoacetates, acetamides, acetimides, alkyl acetates, thioacetate, etc. Reduction of the carbonyl function, either catalytically or with appropriate (complex) metal hydrides, affords [$^{14}$C]ethanol, [$^{14}$C]acetaldehyde and [$^{14}$C]ethylamine, all frequently used low molecular weight building blocks in carbon-14 syntheses. In contrast, deprotonation of the methyl group of alkyl or aryl acetates or imides generates reactive ester and imide enolates which undergo condensation reactions with aldehydes, esters, and Michael acceptors to give labeled *β*-hydroxy and keto esters.

Apart from differences in the handling of these materials, their reactivity and applications are basically the same as that of their unlabeled counterparts. However, whereas synthetic organic chemists rarely consider synthesizing compounds that are so routinely available commercially, it is often important for isotope chemists to do so, whether for reasons of economy, timely availability, stability or simple expediency. Therefore, a selection of characteristic examples will be presented in the following sections.

### 6.1.2 [$^{14}$C]Acetyl Chloride

[$^{14}$C]Acetyl chloride is routinely prepared by reaction of [$^{14}$C]acetic acid or sodium [$^{14}$C] acetate with phthaloyl dichloride[7], benzoyl chloride[8], phosphoryl chloride[9] or oxalyl chloride[10] followed by fractional distillation or vacuum transfer (Figure 6.2). The reaction with phthaloyl dichloride and oxalyl chloride may serve as illustrative examples.

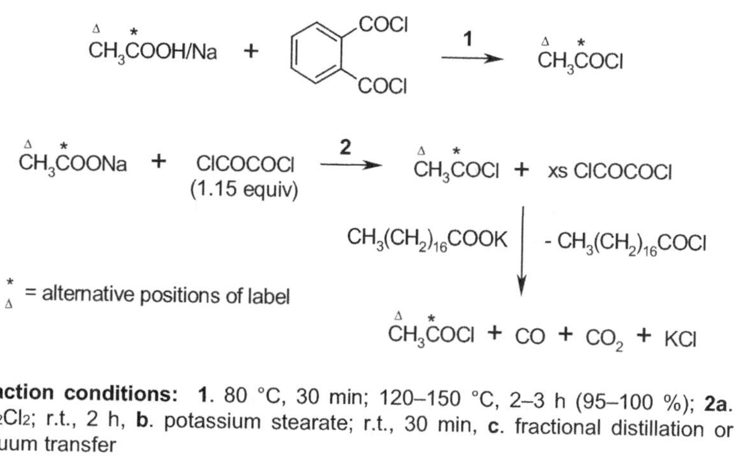


***Figure 6.2*** *Selected standard procedures for the preparation of [$^{14}$C]acetyl chloride*

With appropriate reaction partners and conditions, [$^{14}$C]acetyl chloride may be caused to react *in situ* immediately after it is generated (e.g. sodium [$^{14}$C]acetate, oxalyl chloride, dichloromethane; 0 °C, 3 h[11] or sodium [$^{14}$C]acetate, phosphorus oxychloride, dichloromethane; 0 °C, 20 min; 40 °C, 2 h[12]).

The less frequently used [$^{14}$C]acetyl bromide is accessible similarly upon treatment of [$^{14}$C]acetic acid with benzoyl bromide or phosphorus tribromide[8].

Reduction of [$^{14}$C]acetyl chloride with $LiAlH_4$ in diethylene glycol diethyl ether is the standard procedure for the preparation of [$^{14}$C]ethanol. The [$^{14}$C]ethoxide formed under these conditions is protonated by addition of ethylene glycol monophenyl ether, and [$^{14}$C] ethanol is expelled from the reaction mixture by a stream of nitrogen then further purified if desired by fractional distillation (Figure 6.3)[13]. Heating [$^{14}$C]ethanol with hydroiodic acid under reflux provides [$^{14}$C]ethyl iodide in nearly quantitative yield[14]. Both reagents are frequently used as low molecular weight building blocks.

The use of tributyltin hydride and catalytic amounts of $Pd(PPh_3)_4$ allows the reduction of [$^{14}$C]acetyl chloride to be stopped at the aldehyde level. The reaction can be performed either in toluene, so that [$^{14}$C]acetaldehyde can be isolated as a toluene solution by vacuum transfer, or in 1-methylnaphthalene, from which it can be expelled in chemically pure form (Figure 6.3)[15].

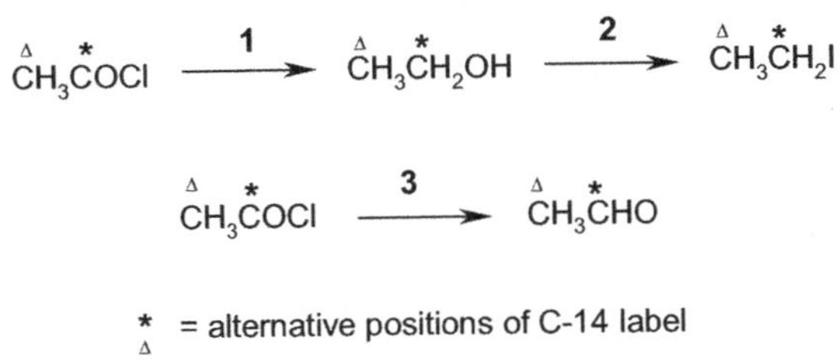


**Reaction conditions:** **1a**. $LiAlH_4$, diethylene glycol diethyl ether; 0 °C, 2 h, **b**. ethylene glycol monophenyl ether, 90 °C, 15 min; 100 °C, 2 h (95 %); **2a**. HI; 80–130 °C, 2–3 h, **b**. fractional vacuum distillation (96–99 %); **3**. $Bu_3SnH$, $Pd(PPh_3)_4$, toluene or 1-methylnaphthalene, 0–20 °C, 10–30 min (80 %)

***Figure 6.3*** *Standard syntheses of [$^{14}$C]ethanol, [$^{14}$C]ethyl iodide and [$^{14}$C]acetaldehyde from [$^{14}$C]acetyl chloride*

The reaction of [$^{14}$C]acetyl chloride with dialkylcadmium reagents or with Grignard reagents in the presence of copper(I) salts has found some utility for synthesis of labeled methyl alkyl or aryl ketones. This methodology is limited to aryl and primary alkyl derivatives, since secondary and tertiary alkylcadmium reagents are insufficiently stable. The synthesis of [2-$^{14}$C]acetone and methyl ethyl [$^{14}$C]ketone may serve as illustrative examples (Figure 6.4)[16]. The reaction has been performed in diethyleneglycol dimethyl ether (b.p. 180–190 °C) to allow separation of the resulting [$^{14}$C]ketone by static vacuum transfer or fractional distillation. It is, however, of minor importance for [$^{14}$C]acetone, which is generally prepared by pyrolysis of lithium or barium [$^{14}$C]acetate (Section 6.3). The same procedure has also been applied to the synthesis of [carbonyl-$^{14}$C]acetophenone, but due to the high boiling point of this product, benzene was used as the solvent.

Unsymmetrical methyl ketones containing secondary or tertiary alkyl groups are obtained when [$^{14}$C]acetyl chloride is treated with alkyllithium or alkylmagnesiun halides in the presence of copper(I) salts. The mixed organocopper reagent generated *in situ* prevents further addition of organolithium or organomagnesium reagent, which would give the corresponding tertiary alcohol. This methodology has also been successfully applied to the synthesis of labeled methyl aryl ketones, as illustrated by syntheses of 3,3-dimethyl[2-$^{14}$C]butan-2-one (methyl *tert*-butyl [$^{14}$C]ketone, [carbonyl-$^{14}$C]pinacolone)[17] and methyl 2-naphthyl [$^{14}$C]ketone (2-[carbonyl-$^{14}$C]acetonaphthone)[18] (Figure 6.4).

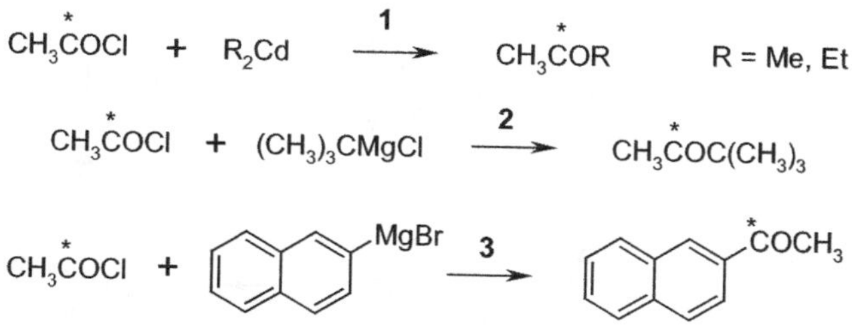


**Reaction conditions:** **1**. diethyleneglycol dimethyl ether; 50 °C, 15 min; 70 °C, 30 min (50–75 %); **2**. CuCl(0.05 equiv.), ether; 0 °C, 30 min (52 %); **3**. CuCl (0.33 equiv), Cu, THF; 2–3 °C, 2.5 h (51 %)

***Figure 6.4*** *Synthesis of carbon-14-labeled alkyl and aryl methyl ketones*

[$^{14}$C]Acetylation of C–H acidic compounds plays an important role in the preparation of alkyl [$^{14}$C]acetoacetates and is described in more detail in Section 6.4 below. By way of example, however, reaction of [1-$^{14}$C]acetyl chloride with Meldrum's acid (1), a masked malonic acid building block followed by alcoholysis with isopropanol gives isopropyl [3-$^{14}$C]acetoacetate. This was used as an intermediate in the synthesis of [$^{14}$C]nimodipine® (2), a calcium antagonist of the dihydropyridine type[19].

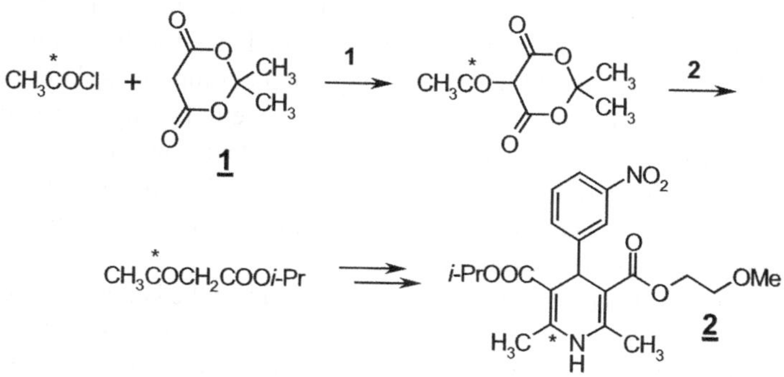


**Reaction conditions:** **1**. pyridine, $CH_2Cl_2$; 2 °C, 30 min; r.t., 40 min; **2**. isopropanol; reflux, 60 min

***Figure 6.5*** *Preparation and utility of alkyl [2-$^{14}$C]acetoacetates, prepared from [1-$^{14}$C]acetyl chloride*

While the preceding malonic ester synthesis enables elongation of the labeled acetyl carbon chain by two carbon atoms, extension by one carbon is achieved when [$^{14}$C]acetyl chloride is treated with diazomethane followed by quenching of the resulting diazoketone with gaseous HCl. The product in this example, 1-chloro[2-$^{14}$C]propan-2-one (3), may serve as a building block for the preparation of various *N*- and *O*-protected $\alpha$-amino [carbonyl-$^{14}$C]ketones **4**, **5** and **6**. These have been employed in the carbon-14 labeling of CI-918[20].

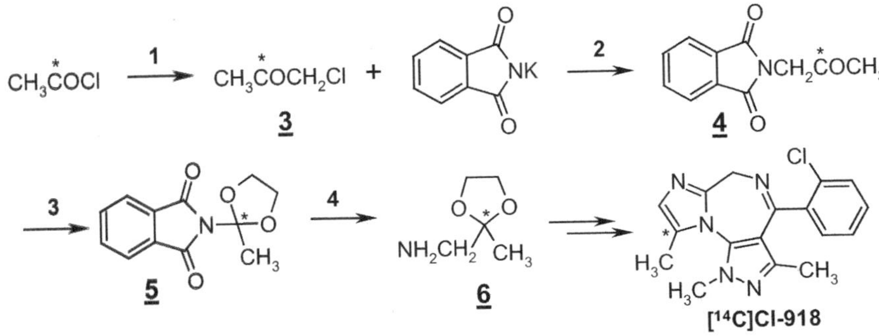

**Reaction conditions:** **1a**. $CH_2N_2$, ether; 0 °C to r.t., 4.5 h, **b**. $HCl_{gas}$, ether; 0 °C to r.t., 16 h; **2**. DMF; 60 °C, 2 h; **3a**. 1,2-ethanediol, *p*-TsOH, heptane; reflux, 5 h, **b**. hydrazine, THF; reflux, 10 h

**Figure 6.6** *Application of the $C_1$-extension of [$^{14}$C]acetyl chloride in the synthesis of 1-amino [2-$^{14}$C]propan-2-ones*

One of the most common uses of [$^{14}$C]acetyl chloride is its Lewis acid- ($AlCl_3$, $SnCl_4$) catalyzed Friedel–Crafts reaction with aromatic or heteroaromatic substrates to produce labeled aryl/heteroaryl methyl ketones. As these intermediates are subject to several types of transformations, they have been used as key intermediates for the synthesis of a wide variety of $\alpha,\beta$-functionalized aryl/heteroaryl alkyl target compounds. For example, aryl/heteroaryl methyl ketones can be (a) halogenated in the methyl group to provide substrates for reaction with carbon or nitrogen nucleophiles, (b) deprotonated so as to react with appropriate electrophilic partners, (c) subjected to stereoselective carbonyl group reduction to alcohols, or (d) reduced to aryl/heteroaryl alkyls. Such transformations can be conducted sequentially in many combinations.

Labeled $\alpha$-bromomethyl derivatives can be prepared from the ketones by treatment with pyridinium tribromide[21], elementary bromine[22], or $CuBr_2$[23]. One example of the preparation and use of such intermediates is the extension of the aliphatic side chain by two carbon atoms using the malonic ester synthesis to produce the 4-aryl-4-oxo[4-$^{14}$C]butyrate (**7**) in good yield[21] (Figure 6.7).

Reaction of labeled aryl halomethyl ketones with amines furnishes $\beta$-amino[$^{14}$C] ketones. If e.p. amines are employed, subsequent reduction of the keto group with achiral reducing agents produces diastereomeric mixtures of $\beta$-amino alcohols. The individual diastereomers can usually be separated by achiral chromatographic methods, or if these fail chiral HPLC procedures or salt formation with a chiral acid (e.g. dibenzoyl-L-tartaric acid) may prove successful. This pathway was followed for the preparation of [$^{14}$C]dilevalol (**12**), an antihypertensive agent[22] (Figure 6.8). Within this synthetic sequence the reaction of the $\alpha$-bromomethyl ketone **8** with e.p. (*R*)-*N*-benzyl-(1-methyl-3-phenyl)propylamine (**9**)

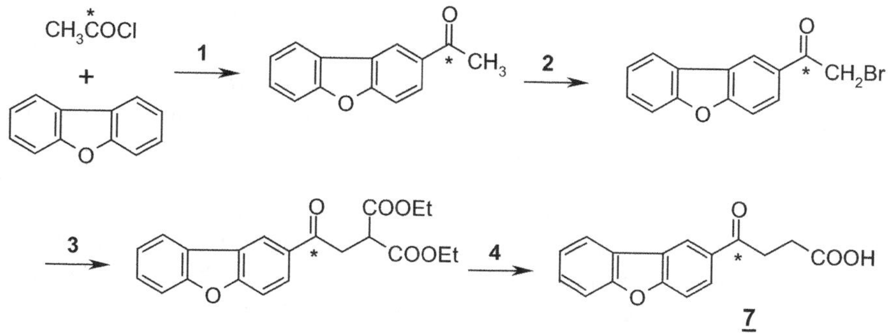


**Reaction conditions:** **1**. $AlCl_3$, $CH_2Cl_2$; 0 °C, 15 min; r.t., 4 h (37–77%); **2**. pyridinium tribromide, $CHCl_3$; 0 °C, 15 min; r.t., 15 min (92 %); **3**. $CH_2(COOEt)_2$, NaH, DMF; r.t., 2 h (51 %); **4a**. KOH, MeOH/$H_2O$ 100:1; r.t., 15 h, **b**. HOAc, conc. $H_2SO_4$ (cat.), reflux, 15 h (ca. 50 %)

***Figure 6.7*** *Chain extension of aryl methyl [$^{14}$C]ketones via malonic ester synthesis*

afforded amino ketone **10**. Reduction of the carbonyl function with $NaBH_4$ followed by removal of the *N*-protecting group yielded a 9:1 mixture of the diastereomeric amino alcohols **11**. Finally, separation through salt formation with dibenzoyl-*R*, *R*-tartaric acid gave labeled e.p. **12** in an overall radiochemical yield of 2.4%.

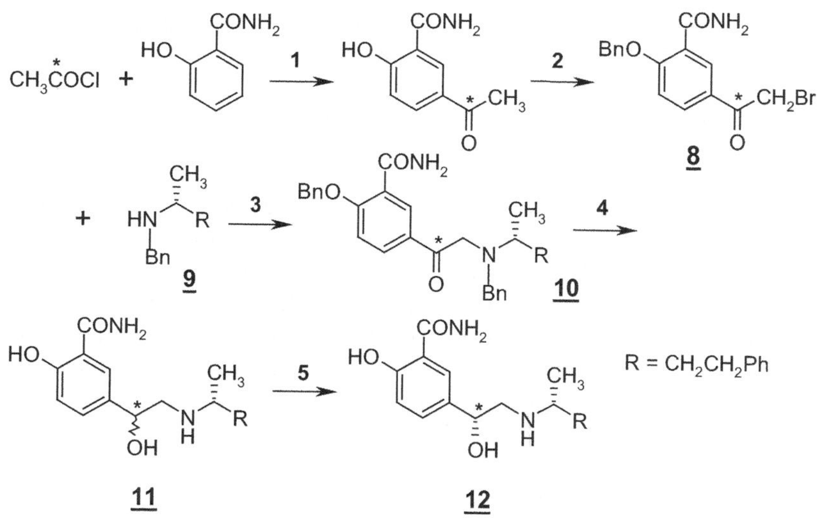


**Reaction conditions:** **1**. $AlCl_3$, nitrobenzene; r.t., 2 h (51 %); **2**. $C_6H_5CH_2Cl$, NaOMe, DMF; 90 °C, 7 h (83%); **b**. $Br_2$, 2% EtOH benzene; r.t., 2.5 h (70%); **3**. DMF, $K_2CO_3$; r.t., 16 h (74%); **4a**. $NaBH_4$, EtOH; 0 °C, 3 h; r.t., 16 h (71%); **b**. $H_2$, 5% Pd–C, EtOH; r.t., 16 h (60 %); **5a**. dibenzoyl-L-tartrate, EtOH; 0 °C, 16 h (separation of the salts) (60%), **b**. $HCl_{gas}$, *i*-PrOH-EtOH (70 %)

***Figure 6.8*** *Synthesis and separation of diastereomeric amino [$^{14}$C]alcohols from methyl aryl [$^{14}$C]ketones*

Borane reduction of the carbonyl group of $\alpha$-haloketones gives halohydrins. Chiral borane reagents (e.g. (*R*) or (*S*)-*B*-chlorodiisopinocampheylborane[24] or DIP chloride) or $BH_3$·THF in the presence of an appropriate chiral catalyst (e.g. (*R*)-1,3,2-oxazaborolidine (CBS) catalyst[25]) gives enantiomerically enriched derivatives, which can be readily converted to labeled oxiranes of excellent optical purity. For example, reduction of $\alpha$-chloro[carbonyl-$^{14}$C]acetophenone (Figure 6.9) with the CBS catalyst provided a 94:6 (*R*)/(*S*) mixture of halohydrins 13 in quantitative yield[25a]. Under basic conditions 13 was converted to the aryloxirane, which reacted with ammonia to give the (*R*)-aminoalcohol 14, isolated in an overall radiochemical yield of 41% and 98% e.e. This compound was used as a key intermediate in the synthesis of [$^{14}$C]NVP VID400 (15)[25a].

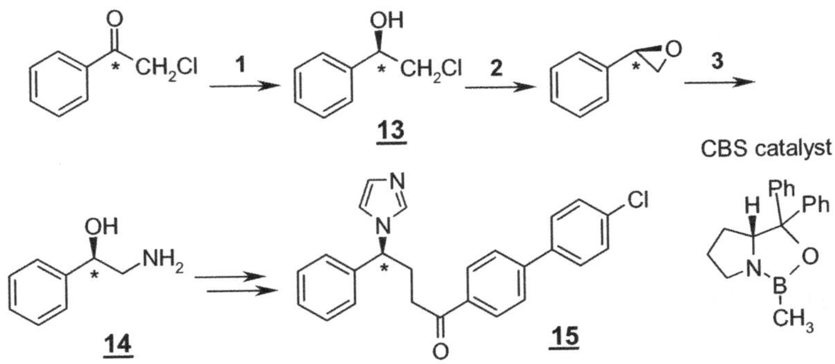


**Reaction conditions: 1.** $BH_3$.THF, CBS-catalyst, THF r.t., 1 h; **2.** 2N NaOH, ether 1:1; r.t., 1 h; **3.** aq. $NH_3$, MeOH 3:1; 40 °C, 3.5 h

***Figure 6.9*** *Synthesis of an e.p. aryl[$^{14}$C]oxirane through enantioselective reduction of a halomethyl aryl ketone with $BH_3$·THF mediated by CBS catalyst*

The carbonyl function of $\alpha$-unsubstituted aryl methyl ketones can be readily reduced to 1-aryl/heteroaryl ethanols by a number of reagents, among which complex hydrides are the most commonly employed. Of special interest are enantioselective reductions using one of the aforementioned reducing systems (i.e. DIP chloride[24] or $BH_3$·THF in the presence of a chiral oxaborolidine catalyst[25]), which usually provide the corresponding carbinols in high yields and excellent stereoselectivities. Borane reduction of various substituted [carbonyl-$^{14}$C]acetophenones (16), for example (Figure 6.10), in the presence of (*S*)-(−)-$\alpha$,$\alpha$-diphenyl- 2-pyrrolidinemethanol has been shown to give (*R*)-1-aryl[1-$^{14}$C]ethanols (17) in chemical yields of 97% and enantiomeric excesses of at least 93% (see also Chapter 11). Subsequent Mitsunobu reaction with $HN_3$ followed by reduction of resulting azide converts the carbinol into the respective e.p. 1-aryl[1-$^{14}$C]ethylamine (18) with inversion of the configuration at the chiral center, as illustrated by the synthesis of [$^{14}$C]rivastigmine (19), an acetylcholine esterase inhibitor for treatment of Alzheimer's disease[26]. The first two steps of this sequence may be reversed. As illustrated by steps 3 and 4, the amine function can be introduced in the first step through reaction of methyl aryl [$^{14}$C]ketones with methoxylamine, followed in the second step by enantioselective reduction of the resulting oxime ether. In this example the reducing agent was $NaBH_4/ZrCl_4$ in the presence of (*R*)-2-amino-3-methyl-1,1-diphenyl-1-butanol. However, enantiomeric excesses of only 90%

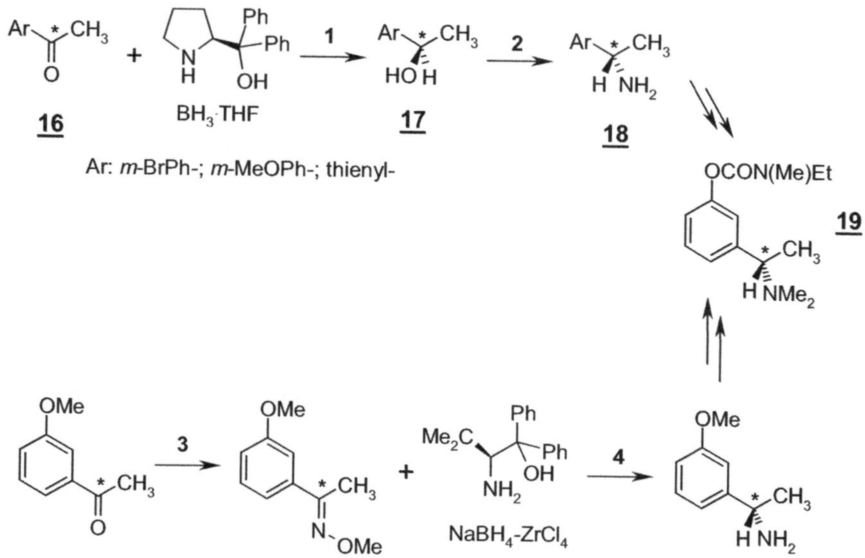

**Reaction conditions:** **1**. $BH_3$·THF; 0 °C, 15 min; r.t., 20 min; **2a**. $HN_3$, $PPh_3$, DEAD; toluene; 0 °C, 5 min; r.t., 10 min, **b**. $Ph_3P$, 2N HCl; r.t., 5 d; **3**. $H_2NOMe$. HCl, NaOAc, MeOH–$H_2O$; r.t., 16 h; **4**. THF; r.t., 70 h

***Figure 6.10*** *Synthesis and use of e.p. 1-aryl/heteroaryl[1-$^{14}$C]ethylamines*

could be achieved, so the undesired enantiomer had to be removed in one of the subsequent steps through diastereomeric salt formation with (+)-di-*p*-toluoyl-D-tartaric acid[27].

The reduction of aryl methyl ketones to arylethanes with metal hydrides in the presence of an appropriate Lewis acid (e.g. $NaBH_4$–$AlCl_3$) or with triethylsilane–trifluoroacetic acid has also found some applications in carbon-14 synthesis. This may be documented with the synthesis of 3-[1-$^{14}$C]ethylazulene derivative **20** (Figure 6.11), which was obtained when the [$^{14}$C]acetyl analog was treated with $NaBH_3CN$–$BF_3$·$OEt_2$. Compound **20** served as penultimate intermediate in the synthesis of 3-[1-$^{14}$C]ethyl-7-isopropyl-1-azulenesulfonate (**21**), an anti-inflammatory drug substance[12].

Treatment of [$^{14}$C]methyl aryl ketones with thallium(III) nitrate in acidic methanol (McKillop rearrangement)[11,28] or with ammonium polysulfide or sulfur and morpholine at elevated temperatures (Willgerodt–Kindler reaction)[29] provides labeled arylacetic acid derivatives. In the first case the aryl group migrates from the carbonyl position into the former methyl position, so that the label is shifted by one position. For example (Figure 6.12), in the synthesis of [6-$^{13/14}$C]porphobilinogen (**24**)[28a], the pyrrole **22** was acetylated with [2-$^{13/14}$C]acetyl chloride, and subsequent McKillop rearrangement gave the product (**23**), in which the labeled carbon was now attached to the ring. This synthesis demonstrates that the reaction can be applied also to heteroaromatic substrates, but amino substituents have to be blocked in advance by acylation, otherwise the reaction fails because of complexation of the thallium species. The Willgerodt–Kindler reaction requires high temperatures and the primary products are thiomorpholides, which require harsh conditions to free the carboxylic

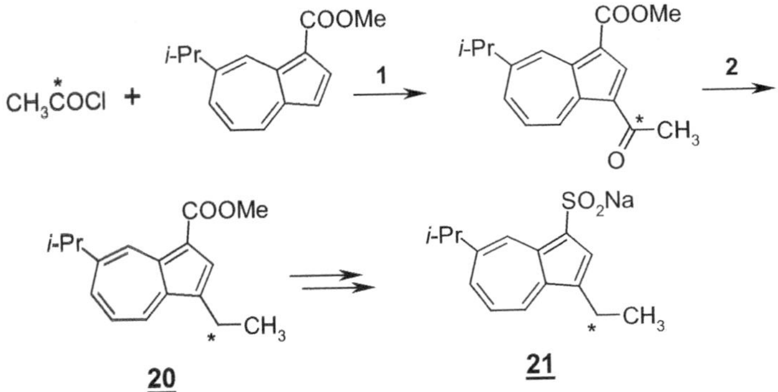

**Reaction conditions:** **1**. $SnCl_4$, $CH_2Cl_2$, 0 °C, 20 min; r.t., 30 min; **2**. $NaBH_3CN$, $BF_3{\cdot}OEt_2$,ether–diglyme; r.t., 30 min

*Figure 6.11* Lewis acid-catalyzed reduction of aryl methyl [$^{14}$C]ketones with complex hydrides

acid product; therefore the overall yields are usually lower than with the McKillop reaction. In addition, the acetyl carbons do not undergo rearrangement of the carbon skeleton in the Willgerodt–Kindler reaction as they do under McKillop conditions. This difference is illustrated by the conversion of carbonyl-$^{14}$C substrate 25 to aryl [2-$^{14}$C]acetic acid derivative 26, both intermediates in the synthesis of [α-$^{14}$C]homogentisic acid-γ-lactone (27)[29].

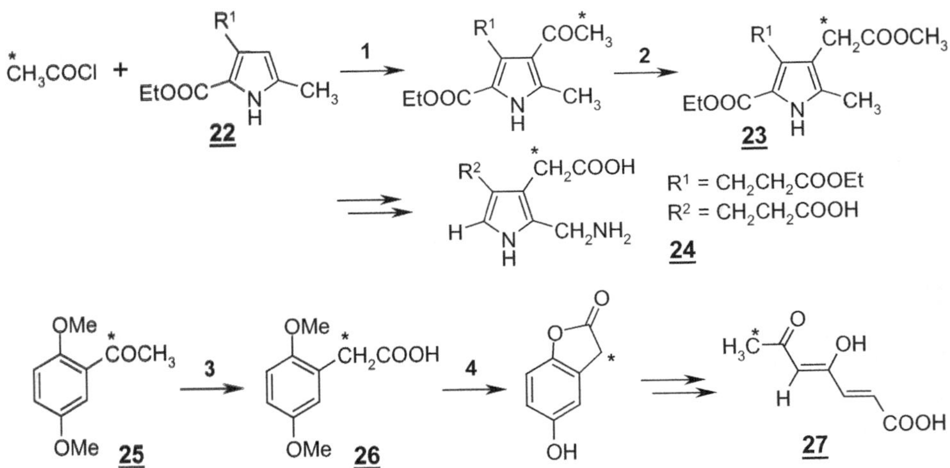

**Reaction conditions:** **1**. $AlCl_3$, $CH_2Cl_2$; 50 °C, 1 h; 25 °C, 16 h; **2**. $Tl(NO_3)_3$ $3H_2O$, $HNO_3$, MeOH; r.t., 48 h (79 % yield); **3a**. S, morpholine; reflux 7 h, **b**. KOH, EtOH; reflux (32 %); **4**. 48 % HBr; reflux, 4 h

*Figure 6.12* McKillop and Willgerodt–Kindler rearrangements of [$^{14}$C]acetyl aryls and heteroaryls

The *N*-[[14]C]acetyl group is a less common subunit in compounds selected for long-term *in vivo* investigations in life science, even though enzymatic hydrolysis is slower than for ester groups[30]. [1-[14]C]Acetylpractolol (**29**), for example, has been found in animal and human *in vivo* studies to be deacylated by less than 5%[31]. It was prepared in one step upon treatment of the etherified *p*-anisidine **28** with [1-[14]C]acetyl chloride in about 75% radiochemical yield by maintaining the reaction at pH 4–6[32,33].

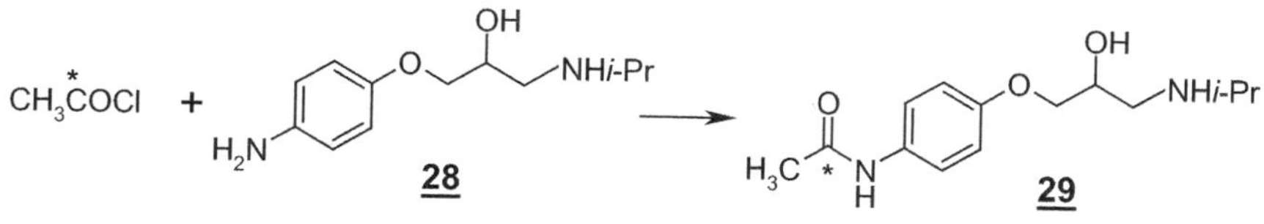


**Reaction conditions**: acetone/water (pH 4–6); 0 °C, 1 h

***Figure 6.13*** *Synthesis of [[14]C]practolol through [[14]C]acetylation of an O-etherified anisidine*

On the other hand, *N*-[[14]C]acetylation procedures have found wide application for substances needed either for *in vitro* studies (for carbon-14 labeling of proteins; see next section) or as valuable synthetic intermediates. In the latter case, 3-substituted *N*-[1-[14]C]-acetylated toluidine **30** served as a key intermediate for the synthesis of [2-[14]C]mepindolol

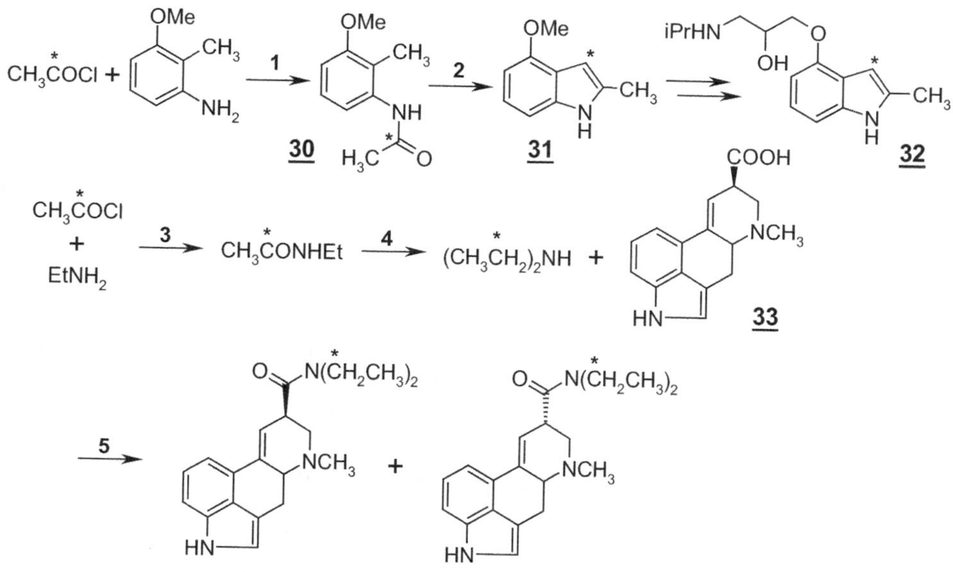


**Reaction conditions: 1.** $NEt_3$, toluene; 50 °C, 3 h, r.t., 16 h; **2.** $NaNH_2$; 250 °C, 1.5 h; **3.** excess ethylamine, ether; 0 °C; **4.** $LiAlH_4$, ether; r.t., 16 h reflux, 3 h **5.** *N,N′*-carbonyldiimidazole, DMF–ether; r.t., 2.5 h in the dark

***Figure 6.14*** *Further preparations and applications of amides from [[14]C]acetyl chloride*

(**32**), a $\beta$-blocking agent. Compound **30** cyclized upon treatment with $NaNH_2$ to give 4-methoxy-2-methyl[2-$^{14}$C]indole (**31**), which was converted by standard procedures into the target molecule **32**[34]. In the second example *N*-ethyl[1-$^{14}$C]acetamide, accessible from [1-$^{14}$C]acetyl chloride and ethylamine, was reduced with $LiAlH_4$, and the resulting [1-$^{14}$C]diethylamine was coupled to lysergic acid (**33**) with the aid of carbonyldiimidazole. The resulting epimeric mixture of the lysergic acid [1-$^{14}$C]diethylamides was separated chromatographically[35].

See next section (Section 6.1.3) for examples of the *N*-[$^{14}$C]acetylation of amides and sulfamides (preparation of *N*-[1-$^{14}$C]acetylbornane-10.2-sultam).

The reaction of [$^{14}$C]acetyl chloride and sodium azide forms [$^{14}$C]acetyl azide, which upon heating rearranges to give methyl [$^{14}$C]isocyanate (**34**). The latter was used for the carbon-14 labeling of eserine (**35**), a potent acetylcholinesterase inhibitor of the carbamate type, and of IBMU (**36**), a urea type pesticide[36].

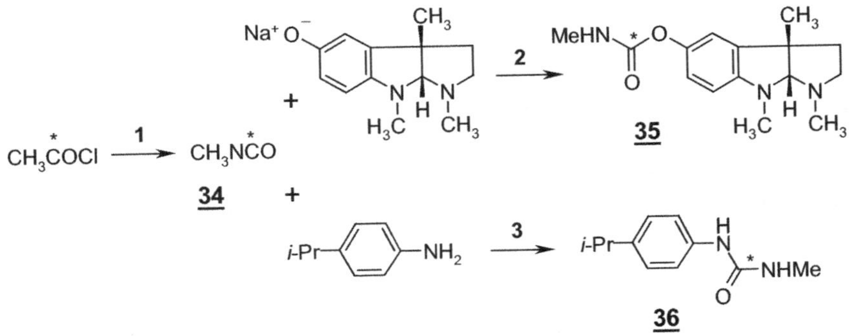


**Reaction conditions: 1.** sodium azide, benzene; r.t.,15 min; 100 °C 3.5 h; **2.** diisopropyl ether, r.t., dark, 1 h; **3.** $NEt_3$, $CH_2Cl_2$; r.t., 2 h

***Figure 6.15*** *Preparation and use of methyl [$^{14}$C]isocyanate for the carbon-14 labeling of carbamates and ureas*

### 6.1.3 [$^{14}$C]Acetic Anhydride

In mainstream organic synthesis acetylation is usually accomplished using acetic anhydride rather than the more reactive acetyl chloride. However, acetic anhydride has been much less used in carbon-14 synthesis because of the unavoidable loss of 50% of the radioactivity. Therefore, its use is limited to the labeling of substrates that have to be acylated either under very mild conditions or in very small quantities (e.g. proteins[37]), or to Perkin reactions, where its active methylene group is involved.

[$^{14}$C]Acetic anhydride is routinely obtained in a radiochemical yield of 80% upon heating sodium [$^{14}$C]acetate with excess *p*-toluenesulfonyl chloride at 200–260 °C and collecting the distillate in a cold receiver[38] (Figure 6.16). It has also been prepared *in situ* by treatment of sodium [$^{14}$C]acetate with an equimolar quantity of acetyl chloride at room temperature for 16 h[39]. In the latter case the specific activity of product [$^{14}$C]acetic anhydride is equal to that of the sodium [$^{14}$C]acetate starting material, or *diluted by half on a per-acetyl basis.*

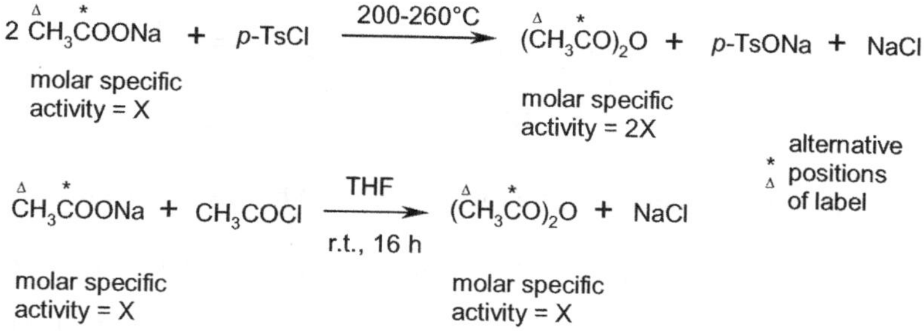


***Figure 6.16*** *Standard procedures for the preparation of [$^{14}$C]acetic anhydride*

The preparation of [$^{14}$C]paclitaxel 37, by *O*-acylation is one of the few examples in which the best results were reportedly achieved with [$^{14}$C]acetic anhydride. The reaction was carried out with three equivalents of [$^{14}$C]acetic anhydride in pyridine in the presence of catalytic amounts of 4-dimethylaminopyridine to give the target compound 38 in 89% chemical yield[39]. [1-$^{14}$C]Acetic anhydride was also preferred in the synthesis of [1-$^{14}$C]acetyl choline iodide (40) because the use of [1-$^{14}$C]acetyl chloride for the acetylation of *β*-dimethylaminoethanol produced byproducts that made purification of the intermediate acetate 39 difficult[40].

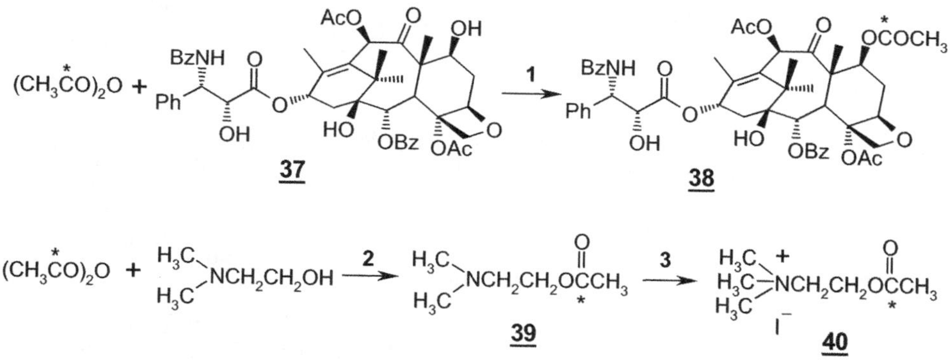


**Reaction conditions**: **1**. DMAP, pyridine; r.t., 20 h; **2**. benzene; reflux, 12–15 h; **3**. MeI, ether; refux, 6 h; r.t., 2 d

***Figure 6.17*** *Use of [$^{14}$C]acetic anhydride in O-[$^{14}$C]acetylation of the 7-hydroxy group of paclitaxel and of the hydroxy group of N,N-*dimethylaminoethanol

Useful but rarely exploited is the use of [$^{14}$C]acetic anhydride as an active methylene compound with aromatic aldehydes to produce labeled cinnamic acids (Perkin reaction, Figure 6.18). This adol-type reaction is limited to aromatic aldehydes, and in many cases malonic acid ester methods are superior, being more flexible and giving higher yields. The reaction of [1-$^{14}$C]acetic anhydride with *o*-nitrobenzaldehyde is one of the very few examples published in the literature[41].

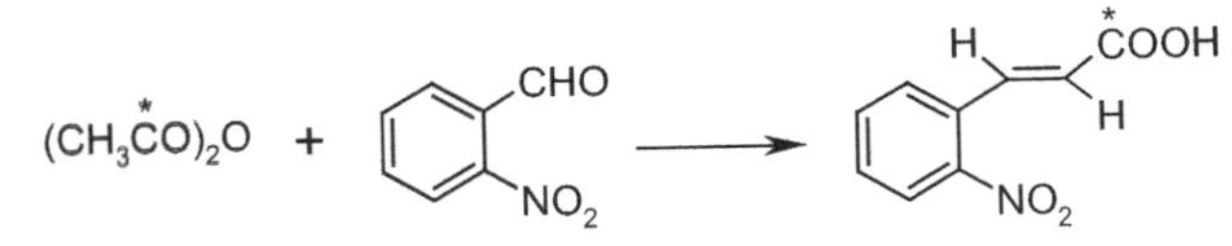


**Reaction conditions**: sodium acetate; 160 °C, 1 h; 170–180 °C, 2 h

***Figure 6.18*** *Preparation of (E)-[$^{14}$C]cinnamic acids from [1-$^{14}$C]acetic anhydride*

Mixed anhydrides such as isobutyl [$^{14}$C]acetoxyformate (e.g., **41**)[42], [$^{13/14}$C]acetyl pivaloyl anhydride (**45**)[43] and *N*-[$^{14}$C]acetyl imidazolide (e.g., **47**)[44] have each been recommended as alternatives to [$^{13/14}$C]acetic anhydride, in order to circumvent the loss of 50% of the radioactivity employed and to avoid the inherent problems of handling volatile [$^{14}$C]acetyl chloride. All three reagents can be prepared and immediately used *in situ* for further reactions, as illustrated by the three examples in Figure 6.19. In the first

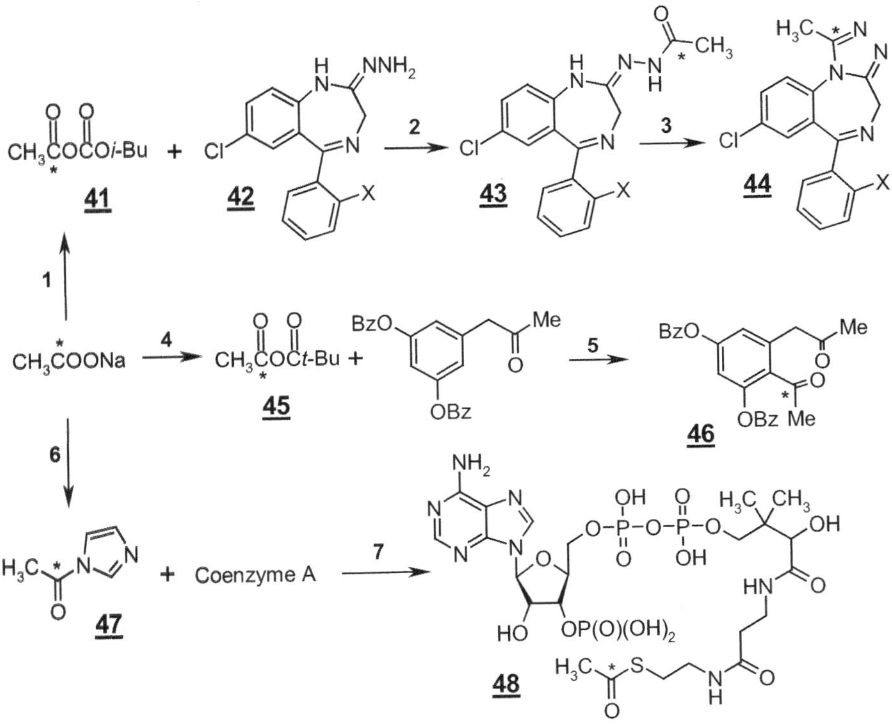


**Reaction conditions: 1**. isobutyl chloroformate, triethylamine, MeCN/THF; –20°C, 45 min; **2**. THF; –45 °C to r.t., 4 h; **3**. 205 °C (*in vacuo*), 15 min; **4**. $(CH_3)_3CCOCl$, ether, reflux, 5 h; **5a**. $HBF_4$, ether, $CH_2Cl_2$; 0 °C, **b**. basic $Al_2O_3$, $CH_2Cl_2$; r.t.; **6a**. 2,6-dichlorobenzoic acid, THF; 67–70 °C, 20 min, **b**. 1,1′-carbonyldiimidazole, THF; 67–70 °C, 1 h; **7**. buffer pH 7.0; r.t., 25–30 min

***Figure 6.19*** *Preparation and use of mixed [$^{14}$C]anhydrides and [$^{14}$C]acetyl imidazolide as alternatives to [$^{14}$C]acetyl chloride and [$^{14}$C]acetic anhydride*

example reagent **41**, obtained from sodium [1-$^{14}$C]acetate and isobutyl chloroformate, reacted with the 1,4-benzodiazepine-2-ylhydrazine **42** to give [1-$^{14}$C]acetylhydrazide **43**, which upon thermolysis cyclized to form the [$^{14}$C]triazolobenzodiazepine **44**[42]. Reagent **45** has been prepared by treatment of sodium [1-$^{13/14}$C]acetate with pivaloyl chloride in THF at room temperature[43a,c,d] or in boiling ether[43a] and used, for example, for the preparation of 1-(2′-[1-$^{14}$C]acetyl-3′,5′-dibenzoyloxyphenyl)propan-2-one **46**[43a]. The *N*-[1-$^{14}$C]acetyl imidazolide reagent (**47**) has found several applications in the labeling of ordinary low molecular weight drug substances[44c] as well as of proteins through [$^{14}$C]acylation of lysine, cysteine, serine, threonine and *N*-terminal residues[45]. The procedure displayed in Figure 6.19 has been especially developed for the preparation of [1-$^{14}$C]acetylCoenzyme A (**48**), since the commonly used acetic anhydride method gives highly variable yields. [1-$^{14}$C]acetylCoenzyme A has found broad application in investigations on biosynthetic pathways of fatty acids and steroids[44a,b].

### 6.1.4 [$^{14}$C]Acetic Acid Alkyl/Aryl Esters

Ethyl [$^{14}$C]acetate is routinely prepared through heating of sodium [$^{14}$C]acetate with triethyl phosphate at 170–200 °C and separation by fractional distillation[46] (Figure 6.20). The same procedure is applicable to the synthesis of the homologous propyl and butyl esters. Because of their higher boiling points and greater stability at these high temperatures, trialkylphosphates are recommended as more appropriate reaction media than the alternatively used dialkyl sulfates[13b,46,47]. The procedure cannot be applied to the synthesis of *tert*-butyl [$^{14}$C]acetate, which is obtained by treatment of [$^{14}$C]acetic acid with *tert*-butanol in the presence of dicyclohexylcarbodiimide and catalytic amounts of 4-(1-pyrrolidinyl)pyridine in ether[48].

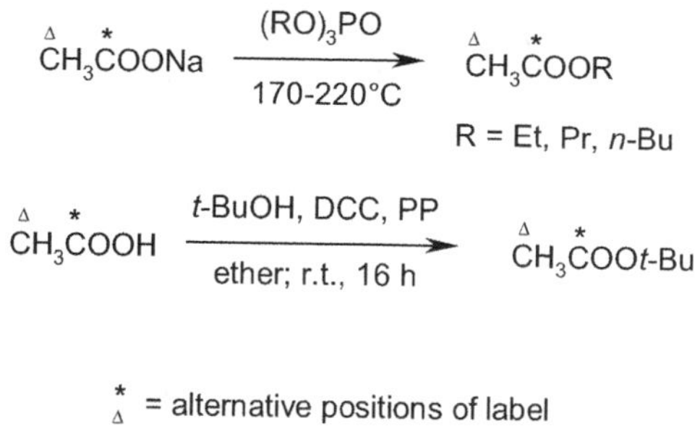


***Figure 6.20*** *Standard procedure for the preparation of $C_1$–$C_4$ alkyl [$^{14}$C]acetates and tert-butyl [$^{14}$C]acetates*

Higher alkyl or aryl esters are prepared either by reaction of [$^{14}$C]acetyl chloride and the respective alcohol in an inert solvent or by alkylation of sodium or triethylammonium salts of acetic acid with primary or secondary alkyl halides, as documented by the procedures for phenyl [$^{14}$C]acetate **49**[49], (*S*)-1,2,2-triphenyl-1,2-ethandiol 2-[$^{14}$C]acetate **50**[7c] and a variety of higher alkyl and aralkyl [$^{14}$C]acetates **51**[50].

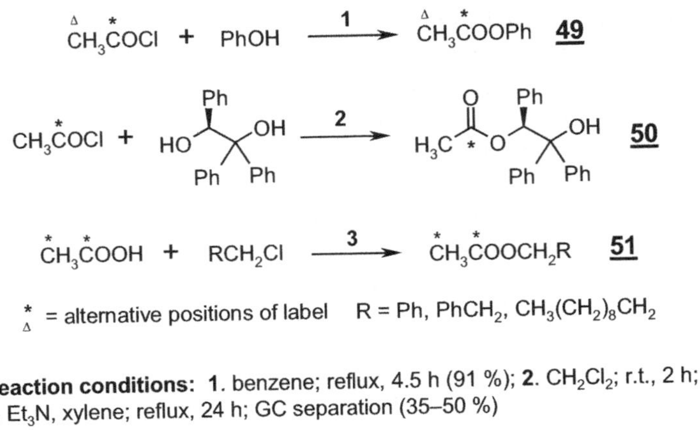


*Figure 6.21* Methods for the preparation of higher alkyl ($n > 4$), phenyl and complex esters of [$^{14}$C]acetic acid

Depending on the respective reaction partner, acetic acid esters can react either as C–H acidic compounds or as acylating agents. Both are illustrated by the self-condensation of ethyl [1-$^{14}$C]acetate in the presence of 0.5 equivalent of sodium ethoxide or triphenymethyl sodium to give ethyl [1,3-$^{14}C_2$]acetoacetate (Claisen condensation). In the first case, however, because of the relatively low radiochemical yields (40–45%) obtained by this procedure, it is of minor importance for the preparation of labeled ethyl acetoacetate[51]. The deprotonation of alkyl [$^{14}$C]acetates with LiHMDS followed by acylation with unlabeled or labeled acyl halides to labeled give $\beta$-keto esters is discussed in Section 6.4. Claisen condensation of alkyl [$^{14}$C]acetates with esters lacking $\alpha$-hydrogens (i.e. ethyl formate, diethyl oxalate, aromatic/heteroaromatic carboxylic acid esters) proceed unidirectionally and are valuable pathways in the synthesis of ethyl [$^{14}$C]formyl acetate (52)[52], diethyl [$^{14}$C]-oxaloacetate (53)[53] and ethyl 3-oxo-3-pyrid-3-yl[2-$^{14}$C]acetate (54)[54]. The last example

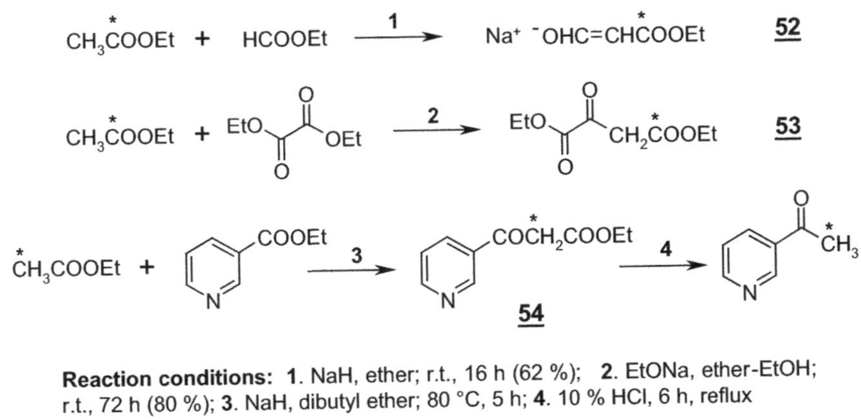


*Figure 6.22* Unidirectional Claisen condensation of alkyl [$^{14}$C]acetates to esters lacking $\alpha$-hydrogens

illustrates the utility of this type of reaction for the synthesis of [$^{14}$C]methyl aryl/heteroaryl ketones, not otherwise accessible by Friedel–Crafts acylation.

Ethyl formyl[1-$^{14}$C]acetate (**52**) and ethyl [3-$^{14}$C]oxaloacetate (**53**) are only stable as esters; the free acids must be used *in situ*. The former is used as a starting material for the standard preparation of [4-$^{14}$C]uracil (**55**)[52]. The latter has been condensed with appropriately substituted 1,3-phenylenediamines, which upon heating of the resulting bis-imine derivative **56** affords the corresponding phenanthrolines **57** in radiochemical yields of around 20%[53].

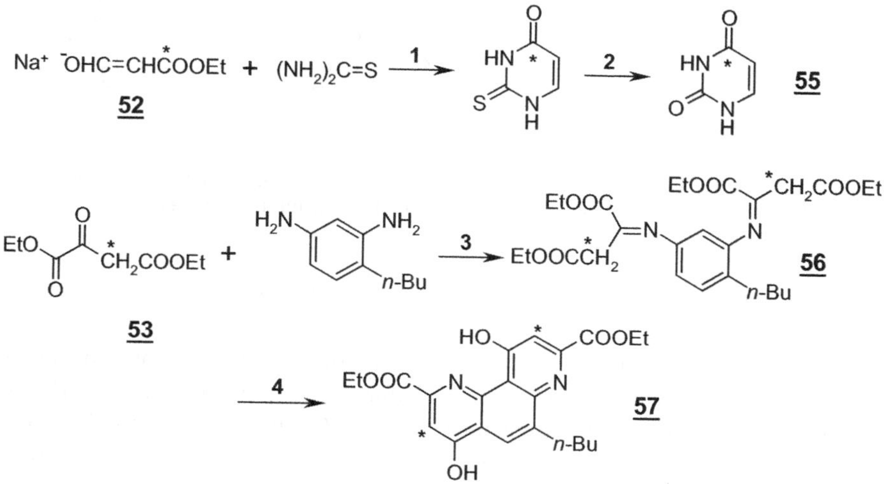


**Reaction conditions: 1.** ethanol; reflux, 3 h; **2.** aq. chloroacetic acid; reflux, 1 h; **3.** benzene; reflux, 16 h; azeotropic removal of water; **4.** dibutyl phthalate; 250 °C, 30 min

***Figure 6.23*** *Use of ethyl formyl[1-$^{14}$]acetate and diethyl [3-$^{14}$]oxaloacetate in the synthesis of labeled heterocycles*

An abnormal unidirectional Claisen reaction has been reported for the addition of deprotonated cyclohexyl [1-$^{14}$C]acetate (**58**) to the carbonyl group of the $\alpha$-*N*-Boc-substituted lactone **59**, providing hemiketal **60** (Figure 6.24). The competitive condensation of the enolate anion of **59** with **58** is probably hindered for steric reasons by the deprotonated *N*-Boc group or because of its negative charge. Subsequent catalytic reduction furnished a mixture of the *N*-Boc-amino diol **61** and its epimer. Deprotection of the desired epimer followed by base-catalyzed cyclization provided lactam **62**, the key intermediate in the synthesis of [3-$^{14}$C]castanospermine **63**[55].

In the presence of a strong base such as LDA, the $\alpha$-carbon of acetic acid esters can react with the carbonyl group of aldehydes or ketones to give $\beta$-hydroxy esters, which may or may not be dehydrated to $\alpha,\beta$-unsaturated esters. This type of reaction has been exploited for the preparation of racemic [2-$^{14}$C]mevalonolactone (**64**)[56] (Figure 6.25) and the benzofuran[1-$^{14}$C]acetamide PD 126,212 (**67**)[48]. In the first case $\alpha$-deprotonated trimethylsilyl [2-$^{14}$C]acetate was treated with 1-trimethylsilyloxy-3-butanone. Acidic work-up resulted in hydrolytic cleavage of the TMS groups and formation of the lactone system

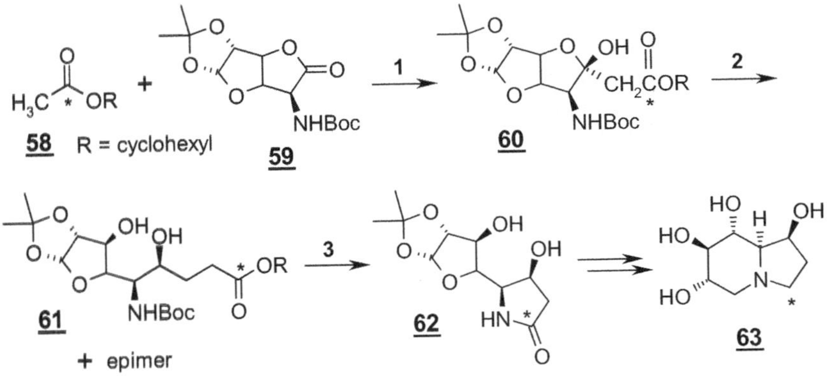

**Reaction conditions:** 1. LDA (3 equiv.), THF; −78 °C, 2 h, −78 °C to −16 °C, 1.5 h; **2a.** $H_2$, $PtO_2$, ethyl acetate; r.t., 24 h, **b.** chromatographic separation; **3a.** HCOOH; 5 °C, 1 h; r.t., 6 h, **b.** $H_2O$; chromatography through Dowex 1X2- 100 (OH-form)

***Figure 6.24*** *Abnormal Claisen condensation of an N-protected α-amino lactone*

in overall 67% radiochemical yield. In the second case deprotonated *tert*-butyl [1-$^{14}$C]acetate reacted with 6,7-dihydrobenzo(b)furan-4(5H)-one to give the α,β-unsaturated ester **65** which could be aromatized to give **66**. Subsequent saponification and coupling of the liberated carboxyl group to the respective amine fragment provided the

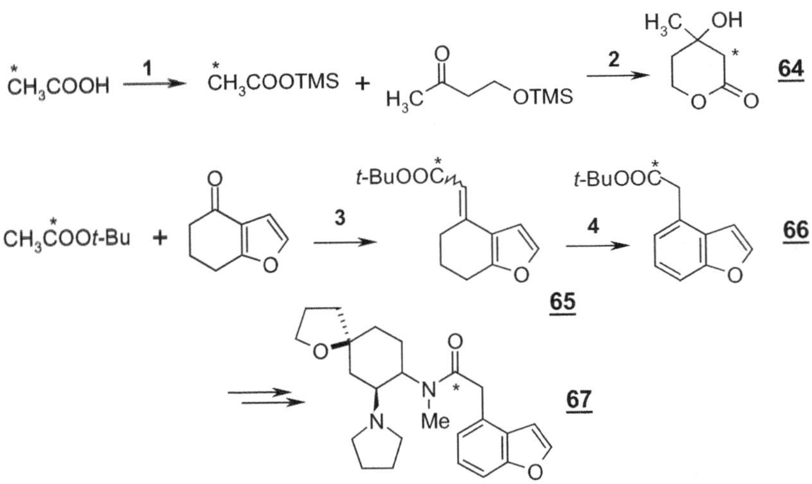

**Reaction conditions:** **1a**. *N*-TMS-imidazole; 5 °C, **b**. fractional distillation (98 %); **2a**. LDA, THF; −78 °C, 1.5 h, **b**. Ketones THF; −78 °C, 2 h, **c**. 6N HCl; **3a**. LDA, THF; −78 °C, 45 min, **b**. 6,7-dihydrobenzo(b)furan-4(5H)one, THF; −70 °C, 16 h, −70°C to 10 °C (40 % from [1-$^{14}$C]acetic acid); **4**. S, 5 % Pd/C, *N*-methylpyrrolidine; 165 °C, 6 h; 200 °C, 30 min

***Figure 6.25*** *Syntheses with deprotonated [$^{14}$C]acetic acid esters*

target compound **67**[48]. This sequence can be regarded as an alternative to the Reformatsky method using $BrZnCH_2{}^{14}COOR$ (Section 6.2.3).

[$^{14}$C]Acetates bearing an appropriate chiral ester function allow for the preparation of chiral β-hydroxy[$^{14}$C]carboxylic esters with high enantiomeric excesses. The diastereoselective addition of doubly deprotonated (*S*)-2-hydroxy-1,2,2-triphenyl[1-$^{14}$C]acetate (*S*)-'[$^{14}$C]HYTRA' or 'chiral [$^{14}$C]acetate', **50**) to the highly substituted 2-(pyrrol-1-yl)-propionaldehyde **68**, for example, afforded the respective chiral β-hydroxy ester **69** in 70% radiochemical yield and an enantiomeric excess exceeding 85%. Subsequent transesterification with MeONa converted **69** into the methyl ester **70**, which served as a key intermediate in the carbon-14 labeling sequence of atorvastatin (**71**), a hypocholesterolemia drug[7c].

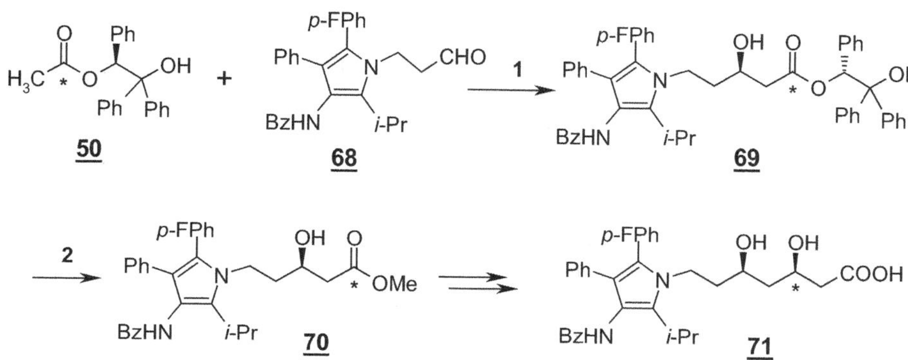

**Reaction conditions: 1a. 50**, LDA (2 equiv.), THF; –78 °C to –10 °C, 1.5 h, **b. 68**, THF; –78 °C, 2 h; **2.** MeONa, THF–MeOH 2:1; 0 to 5 °C, 3 h

***Figure 6.26*** *Preparation and application of a chiral β-hydroxy [$^{14}$C]ester through addition of (S)-[$^{14}$C]HYTRA (chiral [$^{14}$C]acetate) to an aldehyde*

An alternative method using [$^{14}$C]acetic acid esters as mild acylation agents, has been exploited for the carbon-14 labeling of steroids and farnesylacetic acid. Reaction of phenyl [1-$^{14}$C]acetate (**49**) with enol lactone **72** in the presence of NaH followed by treatment of the resulting β-keto lactones **73** with either acid or base provides the respective steroids [$^{14}$C]**74** in radiochemical yields of 20–30%[49] (Figure 6.27). The specific activity of the isolated [3-$^{14}$C]steroids has been found to be 10–20% lower than that of the phenyl ester employed, when the label is located at C1. This is the consequence of a side reaction in which the enolate of the phenyl ester **49**, which is generated in equilibrium with the enolate of **72**, attacks the undeprotonated form of **72** at its carbonyl position to give intermediate **75**. Upon its saponification, de[$^{14}$C]carboxylation occurs to give the unlabeled 1,5-diketone **76**, which then cyclizes to the unlabeled enone **74**. The loss of radioactivity and decrease of the specific activity can be avoided if phenyl [2-$^{14}$C]acetate is employed. In this case the side reaction deposits the label at C4 as documented for [4-$^{14}$C]cholestenone[49] (for further information see Chapter 5, Section 5.5.2).

Under similar reaction conditions ethyl[1-$^{14}$C]acetate adds to γ-butyrolactone to provide β-[1-$^{14}$C]acetyl-γ-butyrolactone (**77**) (Figure 6.28), which upon hydrolysis gave 5-chloro-2-[carbonyl-$^{14}$C]pentanone (**78**). Subsequent reaction with aqueous base converted **78** into

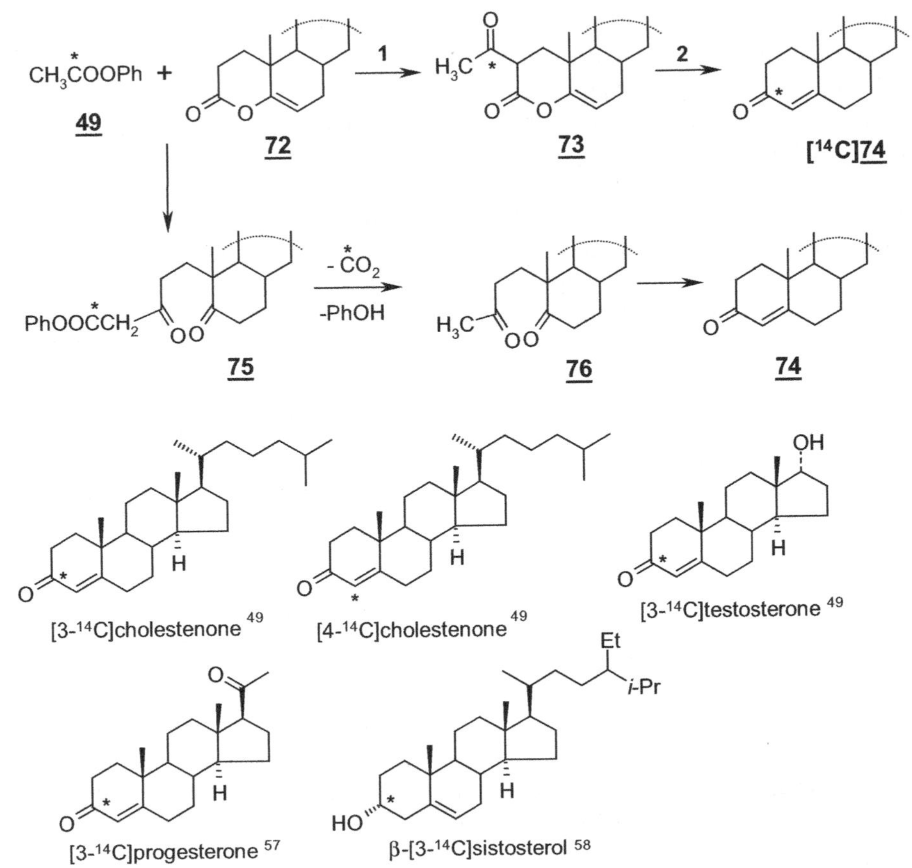


**Reaction conditions: 1.** NaH, benzene; r.t., 42 h; **2.** acetic acid–conc. HCl 8:1; reflux, 27 h

***Figure 6.27*** *Use of phenyl [1-$^{14}$C]acetate for the synthesis of steroids labeled at C3 and examples of steroids labeled at C3 and C4*

methyl cyclopropyl [$^{14}$C]ketone (**79**), which served as a labeled $C_5$ building block in the synthesis of [5-$^{14}$C]farnesylacetic acid (**80**)[59].

As already mentioned, treatment of alkyl acetates with Grignard reagents provides tertiary alcohols. The intermediately generated ketones are generally not isolable but react spontaneously with residual Grignard reagent. This process can be exploited for the preparation of 1-methylcycloalkanols by employing di-Grignard compounds, as has been done several times for the preparation of carbon-13- and carbon-14-labeled 1-methylcyclohexanol (**81**)[60] (Figure 6.29). The latter is obtained in yields in the range of 65–75% when ethyl [1-$^{13/14}$C]acetate is treated with three equivalents of pentamethylenebis (magnesium bromide). It has served as a key intermediate for [1-$^{13/14}$C]toluene, which has been further transformed into [2-$^{13/14}$C]-labeled *o*-nitrotoluene, *o*-toluidine and *o*-toluic acid, and [1-$^{13}$C]benzoic acid and [1-$^{13}$C]benzonitrile. For this purpose **81** is dehydrated by

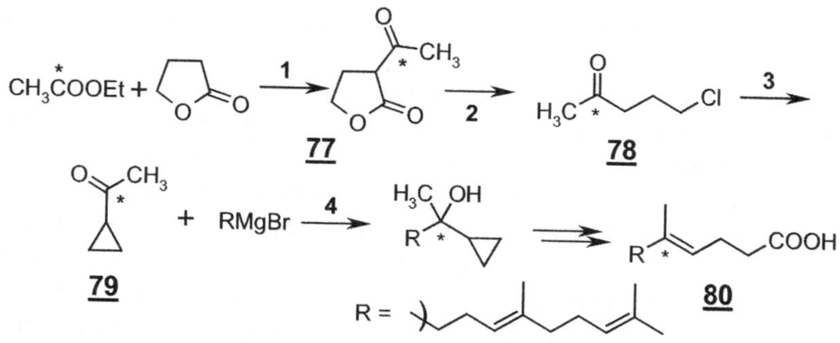

**Reaction conditions:** **1.** Na, benzene, 85–90 °C, 17 h; **2.** conc. HCl–$H_2O$ 2:1, Δ (fractional distillation); **3.** NaOH–$H_2O$, reflux 1 h; **4.** ether; 0–5 °C, 3 h; r.t., 16 h

***Figure 6.28*** *Preparation of methyl cyclopropyl [$^{14}$C]ketone as a carbon-14-labeled $C_5$ building block for chain elongation of halides*

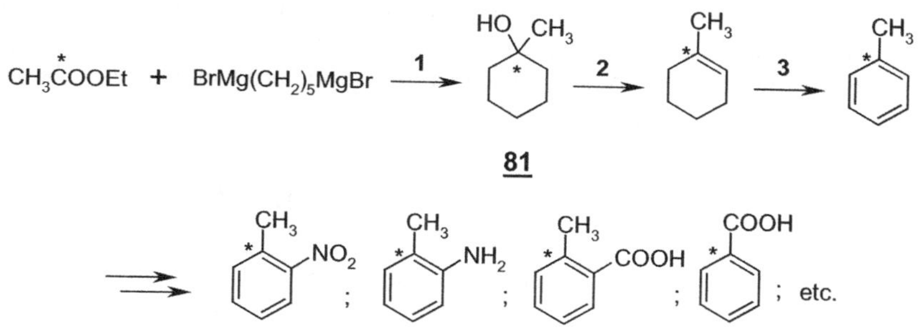

**Reaction conditions:** **1.** ether; 0 °C, 24 h; reflux, 1 h; **2.** 85 % $H_3PO_4$; 140 °C; **3.** vapor phase dehydrogenation over 30 % Pd/C at 400 °C (overall yield 27 %)

***Figure 6.29*** *Preparation of carbon-13/14-labeled toluene from ethyl [1-$^{13/14}$C]acetate*

treatment either with phosphoric acid or iodine at elevated temperatures and aromatized by vapor phase dehydrogenation over Pd/C[61a,b,c] or Pt/$Al_2O_3$[60d].

## 6.2 Halo[$^{14}$C]acetates

Bromo[$^{14}C_n$]acetic acid and its esters, and to a much less extent the corresponding chloro derivatives, are among the most important and frequently used low molecular weight building blocks, especially for the preparation of a broad variety of additional labeled intermediates. Reaction with KCN, for example, provides alkyl cyano[$^{14}$C]acetates, which serve as precursors for mono- or dialkyl [$^{14}$C]malonates and alkyl [$^{14}$C]acetoacetates. With triphenylphosphine, triethyl phosphite or bis(2,2,2-trifluoroethyl) phosphite the corresponding (alkoxy[$^{14}$C]carbonylmethylene)-triphenylphosphoranes (Wittig–Horner reagents), alkyl diethylphosphono[$^{14}$C]acetates (Horner–Wadsworth–Emmons reagents) and alkyl bis(2,2,2-trifluoroethyl) phosphono[$^{14}$C]acetates (Still–Gennari reagents) are obtained. These are indispensable tools for the preparation of labeled (*E*)- and (*Z*)-configured $\alpha,\beta$-unsaturated singly/multiply labeled carboxylic acids/esters and thereby for the extension of the skeleton of halides and carbonyl compounds by two labeled carbon atoms. Reaction with ammonia or *N*-protected derivatives affords [$^{14}$C]glycines, which are essential intermediates in the synthesis of racemic and enantiomerically pure natural and unnatural $\alpha$-[$^{14}$C]amino acids, as well as of alkyl diazo[$^{14}$C]acetates, valuable precursors for the synthesis of labeled [$^{14}$C]cyclopropene- and [$^{14}$C]cyclopropanecarboxylic acid derivatives.

Bromo- and chloro[$^{14}$C]acetic acids are usually prepared by $\alpha$-halogenation of [$^{14}$C]-acetic acid with elemental bromine or chlorine in the presence of catalytic amounts of red phosphorus (Hell–Vollhard–Zelinski reaction)[62] or a phosphorus halide (e.g. $PCl_3$ or $PBr_3$)[61] followed by hydrolysis of the resulting halo[$^{14}$C]acyl halides. A number of slightly modified procedures are available[63]. Free acetic acid is not the reacting species; instead the reaction proceeds via an enolizable derivative such as the acyl halide or anhydride. In the Hell–Vollhard–Zelinski method the reaction is often accelerated by addition of catalytic amounts (0.02–0.06 equivalents) of an acetyl halide or acetic anhydride. There is very good evidence that the halogen of the catalytic acyl halides does not enter the $\alpha$-position, so that a scrambling of the halogens involved is not expected[64]. The most frequently used procedure is illustrated in Figure 6.30. The resulting crude product can easily be purified by fractional distillation and/or sublimation.

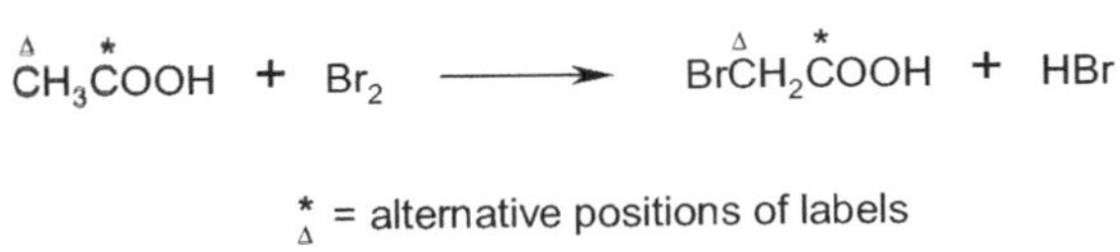


**Reaction conditions:** **1a**. $P_{red}$ (0.01 equiv). $CH_3COCl$, $CH_3COBr$ or $(CH_3CO)_2O$ (0.02–0.06 equiv.), reflux, 1.5 h,**b**. $H_2O$ (85–90 %)

***Figure 6.30*** *Standard procedure for the preparation of bromo[$^{14}$C]acetic acid*

Slightly modified conditions enable the preparation of chloro[$^{14}$C]acetic acid [$Cl_2$, $Ac_2O$ (0.35 equiv.), $P_{red}$ (0.75 equiv.), $PCl_5$ (0.25 equiv.) and iodine (0.005 equiv.); reflux, 2.5 h][62f,65]. Alternative procedures for the synthesis of chloro-, bromo- and iodo[$^{14}$C]acetic acid start with the initial preparation of [$^{14}$C]acetyl chloride/bromide[66] or anhydrides[67] upon treatment of the free acids (Section 6.1), which then react with elementary bromine or iodine[68].

The above methodologies are not applicable to iodo- or fluoro[$^{14}$C]acetic acid. Fluoro-[$^{14}$C]acetic acid is obtained in 40–60% yield through halogen–halogen exchange upon treatment of methyl bromo[$^{14}$C]acetate with KF in a stainless steel microautoclave at 220 °C for 2.5 h and subsequent saponification of the resulting ester[69]. Iodo[$^{14}$C]acetic acid is accessible through nucleophilic displacement of bromoacetic acid with potassium iodide in methanol or acetone[70].

Chloro- and bromo[$^{14}$C]acetic acids/esters have three reactive centers available for further chemical transformations: the carboxyl group, the acidic methylene group and the halogen substituent. These chemistries are discussed in the following sections.

### 6.2.1 Reaction at the Carboxyl Group

The chemical transformations of the carboxyl group include formation of halo[14C]acetyl halides, amides, imides, alcohols, aldehydes and amines. Methods for the preparation of halo[14C]acetyl halides from [14C]acetic acid have already been mentioned. In addition, chloro[14C]acetyl chloride has become available through treatment of chloro[14C]acetic acid with thionyl chloride[71] or phthaloyl chloride, whereas for bromo[14C]acetyl chloride, thionyl chloride[72] or phosphorus pentachloride[73] is recommended. Numerous applications of halo[14C]acetyl halides have been published, out of which the following *C*-, *N*- and *O*-acylations may illustrate their versatility.

Friedel–Crafts acylation of fluorobenzene using bromo[1-14C]acetyl chloride provided exclusively 2-bromo-4′-fluoro[carbonyl-14C]acetophenone (**82**), which was an early intermediate in the synthesis of [14C]fluvastatin (**85**), a HMG-CoA reductase inhibitor[74]. Reaction of **82** with *N*-isopropylaniline followed by $ZnCl_2$-mediated cyclization of the resulting *β*-aminoketone **83** gave 3-(4-fluorophenyl)-1-isopropyl-1*H*-[3-14C]indole (**84**), a key intermediate in the synthetic pathway selected.

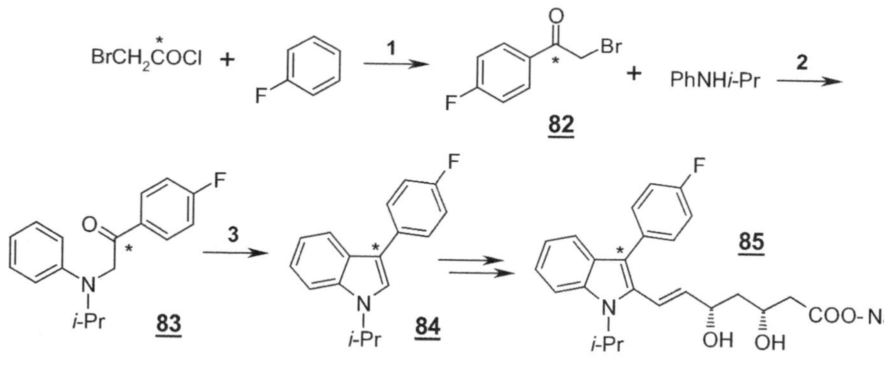


**Reaction conditions: 1.** $AlCl_3$; 80 °C, 1 h; **2.** EtOH; 78 °C, 1 h; **3.** $ZnCl_2$, EtOH; reflux, 5 h (ca. 50 % overall)

***Figure 6.31*** *Use of bromo[1-14C]acetyl bromide in C-acylations applied to the synthesis of 3-phenyl[3-14C]indole derivatives*

Friedel–Crafts acylation of 6-chlorooxindole (**86**) (Figure 6.32) with chloro[2-14C]acetyl chloride afforded the corresponding 5-(chloro[1-14C]acetyl) derivative **87**, which was reduced to give the 2-chloro[1-14C]ethyl analog **88**. Displacement of the 2-chloro function with amine **89** gave the antipsychotic [14C]CP 88-058 (**90**)[68].

The reaction of 7-aminocephem derivative **91** with bromo[1-14C]acetyl chloride in the presence of pyridine illustrates the use of bromo[14C]acetyl halides in *N*-acylation reactions of complex amines[73] (Figure 6.33).

The *N*-acylation of lithiated (2*S*)- or (2*R*)-bornane-10,2-sultam (**92**) (Figure 6.34) with bromo[14C]acetyl bromide illustrates the ready accessibility of bromo[14C]acetyl (2*S*)- or (2*R*)-bornane-10,2-sultam ((2*S*) or (2*R*)-[14C]BABS, **93**). Compound **93** is a versatile building block for the synthesis of e.p. carbon-14-labeled *α*- and *β*-amino acids,

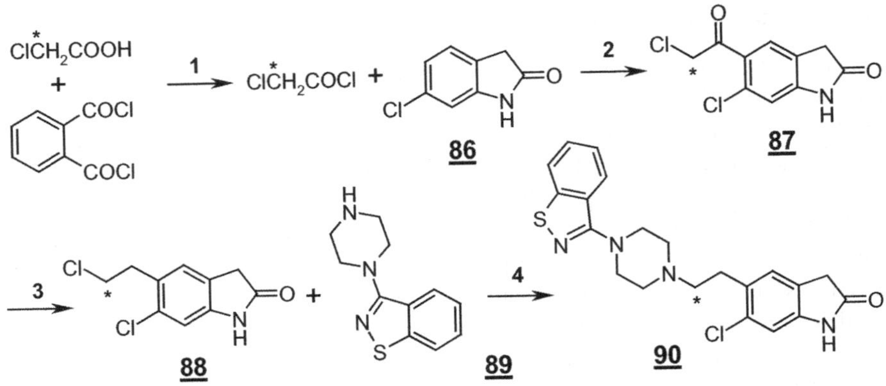


**Reaction conditions:** 1. 80 °C, 3 h; 2. $AlCl_3$, $CS_2$; reflux, 9 h; 3. $Et_3SiH$, $CF_3COOH$; 25 °C, 20 h; 4. $Na_2CO_3$, $H_2O$; 110 °C, 24 h

*Figure 6.32* Friedel–Crafts acylation of 6-chlorooxindole with chloro[2-$^{14}$C]acetyl chloride

α-hydroxy-β-amino acids, α- and β-branched carboxylic acid, chiral acrylates, etc. (see Chapter 11, Sections 11.3.2 and 11.3.9)[75].

The alcoholysis of halo[$^{14}$C]acetyl halides is one of the standard methods for the preparation of low molecular weight alkyl chloro- and bromo[$^{14}$C]acetates[66c–d]. Other methods comprise the esterification of the free haloacids with diazomethane or -ethane (methyl/ethyl halo[$^{14}$C]acetates)[76], with the respective primary alcohol in the presence of HCl gas[77] or with 1,3-dicyclohexylcarbodiimide and 4-pyrrolidinopyridine (*tert*-butyl and benzyl halo[$^{14}$C]acetates)[67a,78].

Reduction of bromo[1-$^{14}$C]acetic acid with $BH_3{\cdot}Me_2S$ in ether[79] or THF[80] or of methyl bromo[1-$^{14}$C]acetate with $AlH_3$ in ether[76b] furnishes 2-bromo[1-$^{14}$C]ethanol (**94**) in

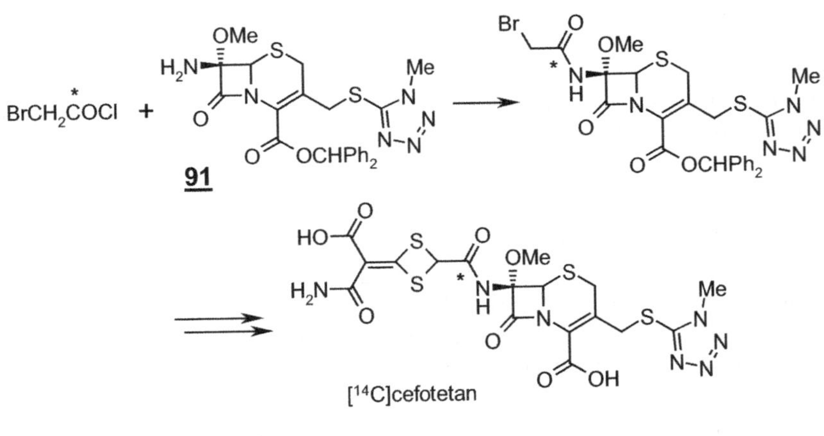


**Reaction conditions:** pyridine, dichloromethane; –30 °C, 20 min

*Figure 6.33* Use of bromo[$^{14}$C]acetyl chloride in *N*-acylation reactions

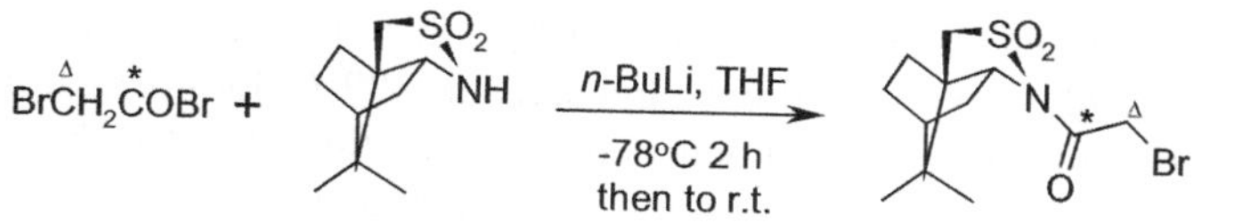


**92** **93**

*/Δ = alternative locations of label

***Figure 6.34*** *Use of bromo[$^{14}$C]acetyl bromide in N-acylation of amides, imides and sulfonamides: preparation of (2S-N-bromo[$^{14}$C]acetyl)bornane-10,2-sultam*

radiochemical yields ranging from 30%[79] up to 80%[80]. 2-Bromoethanol contains both active and proactive electrophilic centers, which allows for sequential substitution reactions. This potential was exploited for the preparation of an *N*-aryl-*N*′-[$^{14}$C]alkylurea **97**. Substitution of the bromo group of **94** with cyanide followed by hydrolysis and chlorodehydroxylation provided 3-chloro[3-$^{14}$C]propionic acid (**95**), which was transformed through standard procedures to the isocyanate **96**. Final reaction with *tert*-butylaniline produced **97**[76b].

If the corresponding nucleophilic partners (or centers) are located in the same molecule, then **94** may be exploited for the construction of heterocyclic systems. This is illustrated in

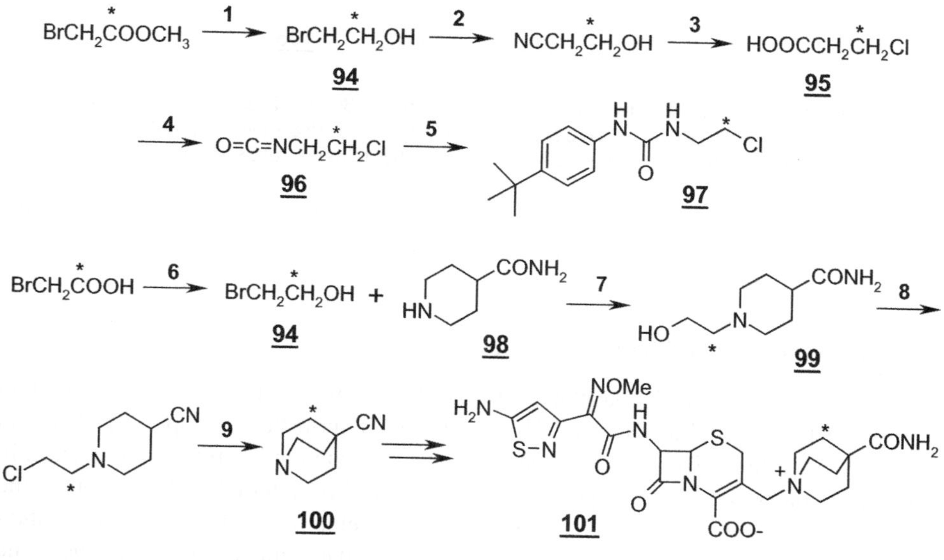


**Reaction conditions: 1a.** LiAlH$_4$, AlCl$_3$, ether; r.t., 5 min, **b.** methyl bromo [1-$^{14}$C]-acetate, ether; 0 °C, 30 min (50 %); **2.** NaCN, ethanol; reflux, 20 h; **3.** conc. HCl; 130 °C, 4 h; **4a.** phthaloyl dichloride; 140 °C, 3 h; **b.** NaN$_3$, (Bu$_4$N)(HSO$_4$),CH$_2$Cl$_2$; r.t., 2 h, **c.** CH$_2$Cl$_2$; reflux, 20 h; **5.** 4-*tert*-butylaniline, CH$_2$Cl$_2$–H$_2$O 40 °C, 20 h; **6a.** BH$_3$·Me$_2$S, ether; reflux, 4 h;**b.** aq. HCl; **7.** KI, K$_2$CO$_3$, *i*-PrOH reflux, 12 h; **8.** SOCl2, CH$_3$CN; reflux, 5 h; **9.** LDA, THF; –78 °C, 30 min; 0 °C, 30 min

***Figure 6.35*** *Preparation and use of bromo[1-$^{14}$C]ethanol in sequential 1,2-substitution reaction sequences*

the synthesis of 4-cyano[3-$^{14}$C]quinuclidine (**100**), which began with the *N*-alkylation of isonipecotamide (**98**) to give *N*-hydroxyethyl derivative **99**. Chloro-dehydroxylation upon treatment of **99** with $SOCl_2$ and simultaneous dehydration afforded the corresponding chloronitrile. Subsequent $\alpha$-deprotonation with LDA and intramolecular *C*-alkylation converted the chloronitrile into (**100**). This compound was a late intermediate in the synthesis of [$^{14}$C]cefclidine (**101**)[79].

The reduction of ethyl bromo[$^{14}$C]acetate can be stopped at the aldehyde level when the reaction is performed with DIBAL-H at low temperature. This has been demonstrated for doubly carbon-13-labeled material. The resulting bromo[$^{13}C_2$]acetaldehyde (**102**) was not isolated but trapped with $Ph_3P$ to give the corresponding Wittig reagent **103**, which was used for two sequential $C_2$ extensions of the aldehyde **104** in the synthesis of [8, 9, 10, 11-$^{13}C_4$]leucotriene $C_4$ (**105**)[81].

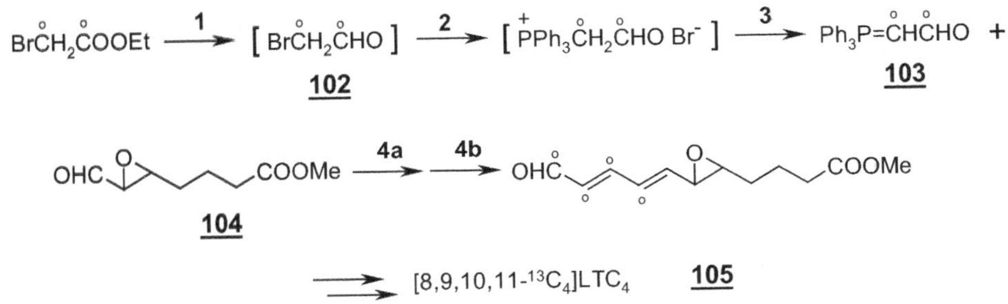


**Reaction conditions: 1a.** DIBAL–H, $CH_2Cl_2$;–78 °C, 30 min, **b.** aq, $NH_4Cl$;–78 °C to r.t.; **2.** $PPh_3$, $CH_3CN$; reflux, 30 min;**3.** $Et_3N$, $CH_2Cl_2$; r.t. (52 %overall); **4a.**1 equiv **103**, benzene; reflux, 30 min (46 %), **b.**1 equiv **103**, toluene; reflux, 30 min (41 %)

***Figure 6.36*** *Preparation and application of carbon-13-labeled bromoacetaldehyde for the extension of aldehydes by two labeled carbon atoms*

### 6.2.2 Reactions at the Methylene Group

Treatment of alkyl $\alpha$-halo[$^{14}$C]acetates with aldehydes and ketones in the presence of an appropriate base (e.g. MeONa, $NaNH_2$, *tert*-BuONa) furnishes $\alpha,\beta$-epoxy [$^{14}$C]esters ([$^{14}$C]glycidic esters). This so-called classical Darzens reaction is a two step sequence, in which an initial Knoevenagel-type reaction is followed by internal $S_N2$ reaction of the resulting $\alpha$-halo-$\beta$-hydroxy [$^{14}$C]ester *in situ*. It is preferentially performed with alkyl chloro[$^{14}$C]acetates, which gives good results with aromatic aldehydes and ketones. The generally poor yields achieved with aliphatic aldehydes can be significantly increased (up to 80%) when the $\alpha$-chloro [$^{14}$C]ester is initially deprotonated with LiHMDS in THF at −78 °C, prior to the addition of aldehydes to the lithium enolate thus generated. With unsymmetrical ketones and aldehydes, the favored reaction product is usually the stereoisomer with the ester function *trans* to the larger group at the $\beta$-carbon atom. The resulting glycidic esters can either be converted into the corresponding homologous aldehydes (through decarboxylation of the free acids) or into $\alpha,\beta$-functionalized carboxylic acid derivatives via nucleophilic opening of the epoxide ring. The latter pathway has been followed for the carbon-14 labeling of SK&F 104353 (**110**). As illustrated in Figure 6.37, methyl chloro[2-$^{14}$C]acetate was treated with the aromatic aldehyde **106** in the presence of sodium methoxide to give the corresponding racemic *trans*-epoxy[2-$^{14}$C]ester **107**. Ring opening with methyl 3-mercaptopropionate converted **107** into a 55:45-mixture of the two regioisomers **108** and **109**. The undesired isomer **109** was destroyed by subjecting the mixture to retroaldol conditions, and **108** was carried on through optical resolution and final hydrolysis to give the target compound **110** in an overall radiochemical yield of 5%[82].

The direct preparation of e.p. [$^{14}$C]glycidic acids/esters has been made possible by the development of a closely related diastereoselective procedure[75,83]. According to this

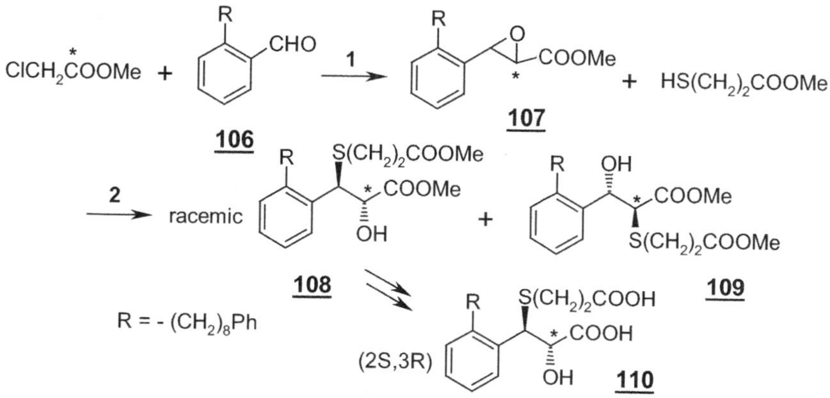


**Reaction conditions:** **1**. NaOMe, $CH_2Cl_2$–MeOH 1:2; –80 °C to r.t., 1 h; r.t.,16 h **2**. $Et_3N$, MeOH; r.t., 16 h

***Figure 6.37*** *Application of the Darzens condensation of methyl chloro[2-$^{14}$C]acetate*

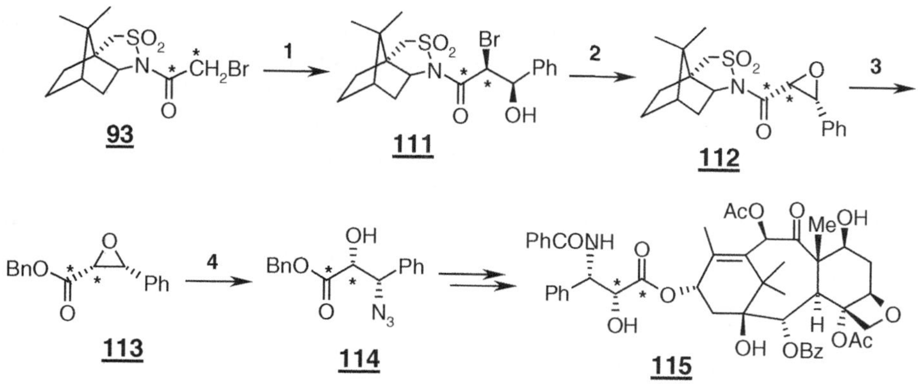


**Reaction conditions:** **1a**. $TiCl_4$, *N*-ethylpiperidine, $CH_2Cl_2$; –78 °C, 5 min, **b**. PhCHO; –78 °C, 10 min; **2**. $K_2CO_3$, DMF, $H_2O$ (traces); r.t., 16 h; **3**. LiOBn,THF; –10 °C, 2.5 h; **4**. $NaN_3$, HCOOMe, MeOH–$H_2O$ 8:1; 50 °C, 24 h

***Figure 6.38*** *Diastereoselective preparation and application of an e.p. [$^{14}$C]glycidic ester*

method the trichlorotitanium enolate, generated from (2*R*)- or (2*S*)-[$^{14}$C]BABS (**93**), *N*-ethylpiperidine and $TiCl_4$, reacts *in situ* with aliphatic and aromatic aldehydes to provide *syn* products with (2*S*,3*R*) or (2*R*,3*S*) configurations, respectively. In the example illustrated in Figure 6.38, addition of benzaldehyde to the enolate provides (2*S*,3*R*)-2-bromo-3-hydroxy-[$^{14}$C]acyl (2*S*)-bornan-10.2-sultam (**111**) in >98% d.e. Unlike the classical Darzens method this reaction stops at the $\alpha$-bromo aldol stage. Epoxy ring closure to provide **112** can, however, be readily effected through treatment of **111** with $K_2CO_3$. In the given example, subsequent cleavage of the auxiliary with lithium benzoxide followed by regioselective opening of the epoxide ring of **113** with $NaN_3$ afforded e.p. benzyl (2*R*,3*S*)-3-azido-2-hydroxy-3-phenyl-[1,2-$^{14}C_2$]propionate (**114**) in 50% overall radiochemical yield. These intermediates were key to the synthesis of the carbon-14-labeled anticancer drug paclitaxel (**115**).

### 6.2.3 Reactions at the Halogen Atom

The halo function of haloacetates, as a primary halide and activated by the neighboring carboxyl function, undergoes easy displacement by nucleophiles of many types. This propensity opens the way for a wide variety of chemistries, which are discussed in the following sections.

*Replacement of Halogen by Carbon Nucleophiles*

With one exception, there is no principal difference between the use of alkyl haloacetates in alkylation reactions of carbon nucleophiles in radiosynthesis and in normal preparative organic chemistry. The exception is that carbon-14 synthesis alkyl halo[$^{14}$C]acetates are additionally used to extend the carbon skeleton of various substrates by incorporating a singly or doubly labeled two-carbon unit, which is then followed by elimination of an unlabeled one-carbon unit. As can be seen in Figure 6.39, dialkyl 2-acyl [4-$^{14}$C]succinate **116**, prepared according to the strategy depicted, decarboxylated upon acid catalyzed hydrolysis, thereby converting an unlabeled $\beta$-ketoester **115** into a one-carbon extended labeled $\gamma$-ketoester **117**. The latter was converted via the dibromoketone into the quorum sensing inhibitor **118**[77]. In the second example, the different functionalizations of the succinate carboxyl groups facilitated the selective hydrolysis and transformation of one without affecting the other. Thus, alkylation of unlabeled imide **119** and subsequent selective alkaline hydrolytic cleavage of the imide function provided succinic acid monoester **120**, which upon Hofmann rearrangement of the free carboxyl group and acid catalyzed ester cleavage afforded the one-carbon extended $\beta$-amino acid ester **121**[84]. If chiral imides or esters are employed as in the second example, then singly or doubly labeled $\beta$-amino acids become available in enantiomerically pure form[75,84].

*Replacement of Halogen by Nitrogen Nucleophiles*

By far the most common type of chemistry associated with the nitrogen–halogen replacement of carbon-14-labeled haloacetates is that forming [$^{14}$C]glycine derivatives (Figure 6.40). For example, treatment of chloro[$^{14}$C]acetic acid with aqueous ammonia[85] or of ethyl bromo[$^{14}$C]acetate with potassium phthalimide followed by saponification of the ester and cleavage of the protecting group[66c,86] provides [$^{14}$C]glycine in 60–80% radiochemical yields.

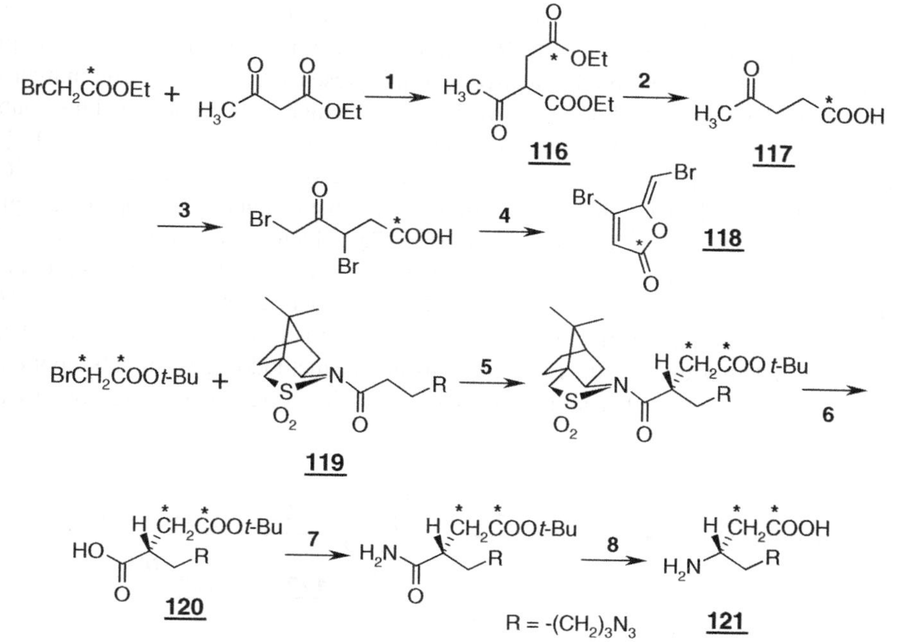

**Reaction conditions :** **1**. EtONa, EtOH; 60–65 °C (microwaves), 16 min; **2**. 2NHCl; 130 °C (microwaves), 10 min (70 % overall); **3**. $Br_2$, HBr, HOAC, $CH_2Cl_2$; reflux, 10 h; **4**. $H_2SO_4$; 100 °C, 10 min; **5**. NaHMDS, THF/toluene/HMPA 1:1:1; −78 °C, 3 h; **6**. LiOH, $H_2O_2$, THF/$H_2O$ 3:1; 0 °C, 4.5 h; **7a**. DCC, HOSU, THF; r.t., 4 h, **b**. conc. $NH_3$.aq; r.t., 16 h; **8a**. $PhI(OCOCF_3)_2$, DMF/H2O 4:1; r.t., 4 h, **b**. $CF_3COOH/H_2O$ 20:1; r.t., 40 min (20 % overall)

***Figure 6.39*** *Use of alkyl bromo[$^{14}C_n$]acetates in the α-alkylation of ester and imide enolates for the extension of carbon chains by a labeled one carbon unit*

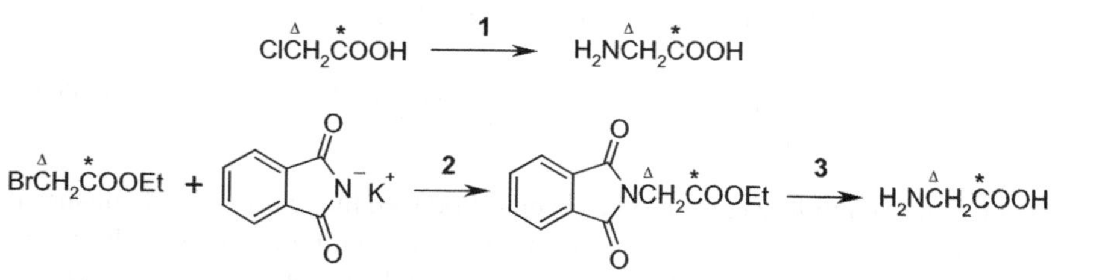

* Δ = alternative positions of label

**Reaction conditions:** **1**. conc. aq. $NH_3$, $(NH_4)_2CO_3$, $H_2O$; 60 °C, 8–12 h; **2**. DMF; r.t., 1–3 d; **3**. conc. HCl/HOAc 1:1; reflux, 16 h

***Figure 6.40*** *Preparation of [$^{14}C$]glycine*

[$^{14}$C]Glycine is a key intermediate for the preparation of racemic as well as enantiomerically pure labeled $\alpha$-amino acids. Reaction of *N*-acetyl [1-$^{14}$C]glycine, for example, with aromatic and heteroaromatic aldehydes in the presence of excess acetic anhydride affords 4-arylidene/heteroarylidene 5-oxazolones **122**, which upon reduction of the double bond ($H_2$, Pd/C or HI, $P_{red}$) and saponification give racemic $\alpha$-[1-$^{14}$C]amino acids **123** in radiochemical yield of 30–40% (Erlenmeyer's azlactone synthesis)[87]. Treatment of [$^{14}$C]-azlactone **124** with MeOH in the presence of triethylamine resulted in ring opening to give *N*-acetyl $\alpha,\beta$-unsaturated [$^{14}$C]amino acid ester **125**, which upon hydrogenation in the presence of the chiral catalyst (*S,S*)-[Et-DuPHOS-Rh]$^+$ followed by hydrolysis produced the nonproteinogenic (*S*)-[1-$^{14}$C]amino acid **126** in excellent enantioselectivity of ≥98% e.e. and 40% overall radiochemical yield[88] (similar chemistry is discussed in more detail in Chapter 11). Unfortunately, the Erlenmeyer azlactone methodology cannot be applied to aliphatic carbon-14-labeled $\alpha$-amino acids because the corresponding oxazolones are obtained only in very low yields.

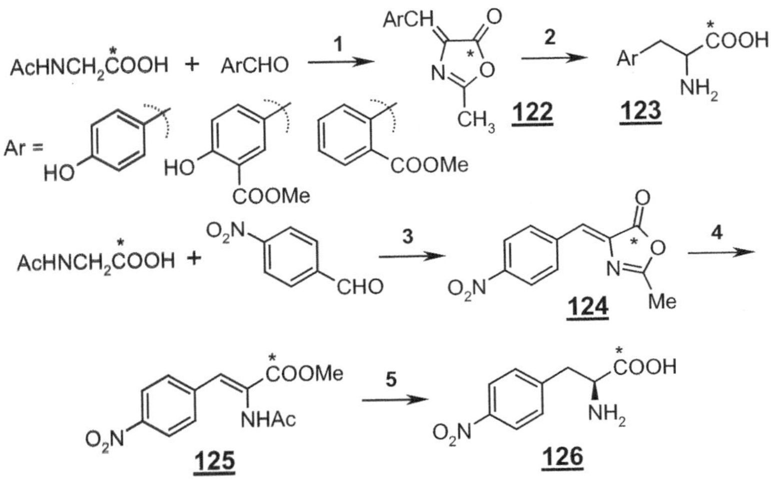

**Reaction conditions: 1.** $Ac_2O$, KOAc; 90–120 °C, 2.5–5 h; **2a.** HI, $P_{red}$; Δ, 4 h, **b.** 4NHCl; Δ, 10 h; **3.** $Ac_2O$; NaOAc; 100–110 °C; **4.** $Et_3N$, MeOH; 45–50 °C; **5a.** 40 psi $H_2$, (*S,S*)-Et-DuPHOS-Rh, MeOH; r.t., **b.** 5N HCl; Δ

***Figure 6.41*** *Preparation of carbon-14-labeled α-amino acids from aromatic aldehydes via Erlenmeyer's azlactone synthesis*

Recently, the Schmidt reagent **128** has been introduced into carbon-13 chemistry. It is a highly versatile building block for the synthesis of $\Delta^{2,3}$-amino acids through reaction of aromatic and aliphatic aldehydes and ketones in excellent yields and high *Z*-selectivity. It is readily accessible by halogenation of *N*-Boc-[$^{13}$C]glycine methyl ester and Arbusov reaction of the resultant $\alpha$-bromo derivative **127** with trimethylphosphite. Reaction of **128** with aldehydes and unsymmetrical ketones yields $\Delta^{2,3}$-amino acids as mixtures of *E*- and *Z*-isomers. Formation of the *Z*-isomer is favored when aldehydes are used. Subsequent (*S,S*)-[Rh-Et-DuPHOS]$^+$-catalyzed enatioselective hydrogenation was shown to provide the corresponding (*S*)-$\alpha$-[1-$^{13}$C]amino acids in 97% chemical yields and 92–99% e.e.

(for more information see Chapter 11, Section 11.2.1)[89]. Though this reaction was applied to carbon-13-labeled materials, there is little doubt that it is also applicable to synthesis with carbon-14.

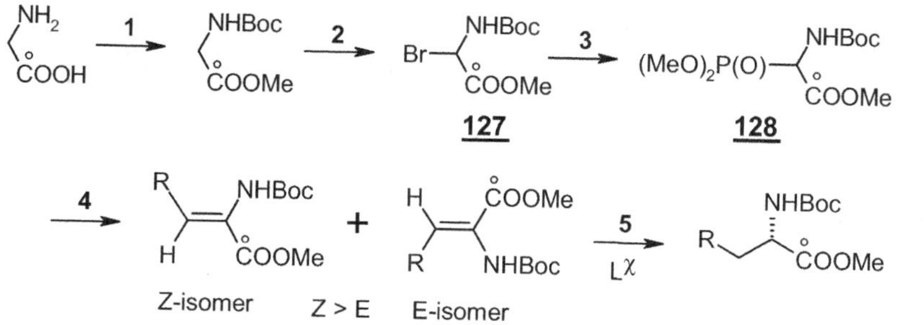


R = H, Ph, *p*-MeO-Ph, indolyl, *i*-Pr; BnOCO; BocNHCO; $Boc_2N(CH_2)_3$ : e.e. = 92-95%
Lχ = [(*S*,*S*)-Et-DuPhos-Rh(COD)]$^+$OTf$^-$

**Reaction conditions: 1a**. $SOCl_2$, MeOH, **b**. $(Boc)_2O$, $Et_3N$; **2**. NBS, $CCl_4$; h.v,r.t.; **3**. $P(OMe)_3$, $CH_2Cl_2$; reflux; **4**. RCHO; DBU or $Me_2NC(=NH)NMe_2$, $CH_2Cl_2$;reflux; **5**. $H_2$ (1–5 atm), RhLχ; MeOH; 50 °C

***Figure 6.42*** *Synthesis and use of Schmidt reagent in the synthesis of e.p. α-[$^{13}$C]amino acids*

The use of linear chiral labeled glycinates (e.g. (2*S*) or (2*R*)-[$^{14}$C]DPMGBS, **129**) offers another alternative. Synthons **129** are readily accessible through treatment of *N*-bromo [$^{14}$C]acetyl (2*S*) or (2*R*)-bornane10,2-sultam (**93**) with benzophenone imine. For example deprotonation of **129** with *n*-BuLi in the presence of 30% HMPA or DMPU followed by alkylation with 2-iodopropane gave the isopropyl derivative **130**. Changing of *N*-protecting groups and hydrolytic removal of the auxiliary provided (*S*)-[1-$^{14}$C]valine (**131**) in 46% overall yield and e.e. > 99%[90]. This procedure was shown to have wide applicability to other α-amino acids in excellent enantiomeric purity (ee ≥ 98%) and overall yields of 30–50%[75] (for additional information see Chapter 11, Section 11.3.6).

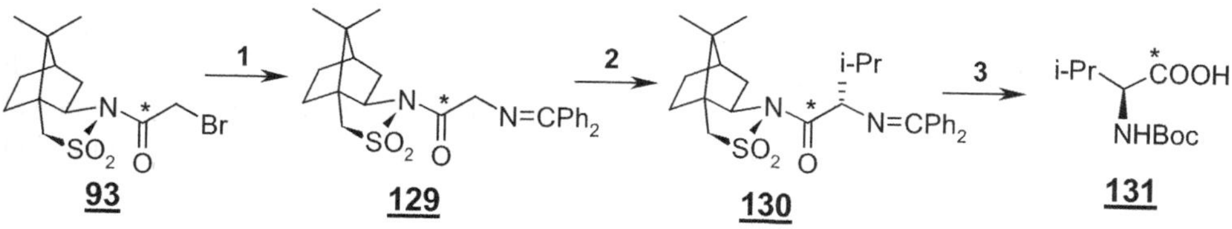


**Reaction conditions: 1**. $Ph_2C{=}NH$, Et*i*-$Pr_2N$, molecular sieves 3 A, acetonitrile; 70 °C, 3.5 h; **2a**. BuLi, THF; –78 °C, 60 min, **b**. *i*-PrI, HMPA; –78 °C, 30 min; 5 °C, 16 h; **3a**. 1 N HCl, THF; r.t., 1 h, **b**. $(Boc)_2O$, THF; r.t., overnight, **c**. LiOH, aq. THF; 0 °C, 2 h

***Figure 6.43*** *Example of the preparation and application of chiral linear [$^{14}$C]glycinates for the synthesis of e.p. (S)-[1-$^{14}$C]valine*

Linear [$^{14}$C]glycinates may be replaced by cyclic analogs, which are obtained when the free amino and carboxyl ends of [$^{14}$C]glycine are bridged by a one- or two-carbon unit containing a chiral auxiliary center, as illustrated in Figure 6.44. Treatment of phenyl bromo [$^{14}$C]acetate with (*S*)-phenylglycinol and blocking of the NH function with $Boc_2O$, for example, afforded the respective (5*S*)-[3-$^{13/14}$C]oxazinone building block (Dellaria's oxazinone, **132**). Alkylation of the deprotonated reagent followed by hydrolytic cleavage of the oxazinone system and hydrogenolytic removal of the chiral auxiliary unit provided α-[$^{13/14}$C]amino acids in excellent enantiomeric purity and yields of 40–50%[91,92] as demonstrated by the synthesis of (2*S*)-[2-$^{13}$C]aspartic acid (**133**)[92] (for further examples and a more detailed discussion see Chapter 11, Section 11.3.7).

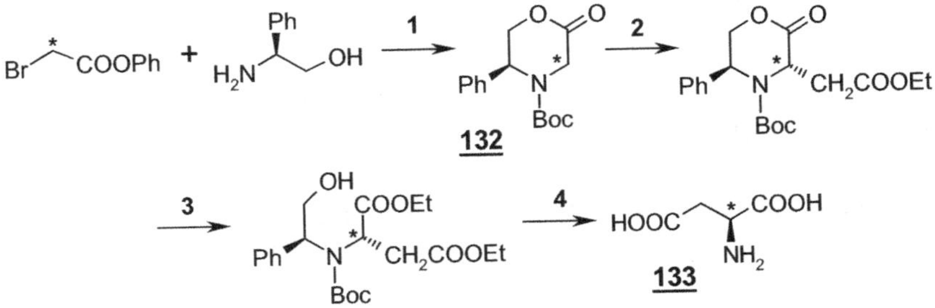


**Reaction conditions: 1a.** *i*-$Pr_2$EtN, $CH_3CN$; r.t., 5 h, **b.** $Boc_2O$, $CH_3CN$; r.t., 5 h; **2a.** LiHMDS, THF; –78 °C, 1 h, **b.** $BrCH_2COOEt$, r.t., 1 h; **3.** $HCl_{gas}$, EtOH, reflux, 5 h; **4.** $H_2$, 10 % Pd/C; r.t., 1.5 d

***Figure 6.44*** *Preparation and application of chiral cyclic [$^{13}$C]glycinate (**132**) in the synthesis of e.p. labeled α-[$^{13}$C]amino acids*

Incorporation of a [$^{14}$C]glycine unit into the ring system of various benzodiazepines has been exploited for bioavailability and biotransformation studies in spite of the risk of metabolic lability of the label[93]. One route starts from the reaction of [1-$^{14}$C]glycine and *N*-carbethoxyphthalimide to give phthaloyl[1-$^{14}$C]glycine (**134**) (Figure 6.45). Preparation of the acyl halide and amide formation with the anilines **135** provided amides **136**. Subsequent removal of the phthaloyl group and concluding cyclization furnished [2-$^{14}$C]diazepam (**137**, R = Me) and its 1-desmethyl metabolite (R = H)[94]. In an alternative procedure the unlabeled aminobenzophenone **138** was treated with *N*-carbobenzyloxy[1-$^{14}$C]glycyl chloride, which is readily accessible through reaction of the *N*-protected [1-$^{14}$C]glycine with $PCl_5$. Removal of the *N*-protecting group from the resulting amide **139** and base-mediated cyclization provided benzazepines **140**[71,95].

Labeled heterocyclic systems of the imidazopyridine type are accessible from halo[$^{14}$C]-acetates. For example, the reaction between 2-aminopyridine and chloro[2-$^{14}$C]acetic acid produced (2-imino-1,2-dihydropyridin-1-yl)[2-$^{14}$C]acetic acid (**141**) in 76% yield (Figure 6.46). Treatment of **141** with $POCl_3$ induced cyclization and chloro-dehydration to furnish 2-chloro[3-$^{14}$C]imidazo[1,2-a]pyridine (**142**) in 99% yield. The latter was a key intermediate in the synthesis of the carbon-14-labeled herbicide MON 37500 (**143**)[96].

*Alkyl diazo[$^{14}$C]acetates* Treatment of alkyl [$^{14}$C]glycinates with sodium nitrite in weakly acidic medium provides alkyl diazo[$^{14}$C]acetates **144** in about 90% yield (Figure 6.47).

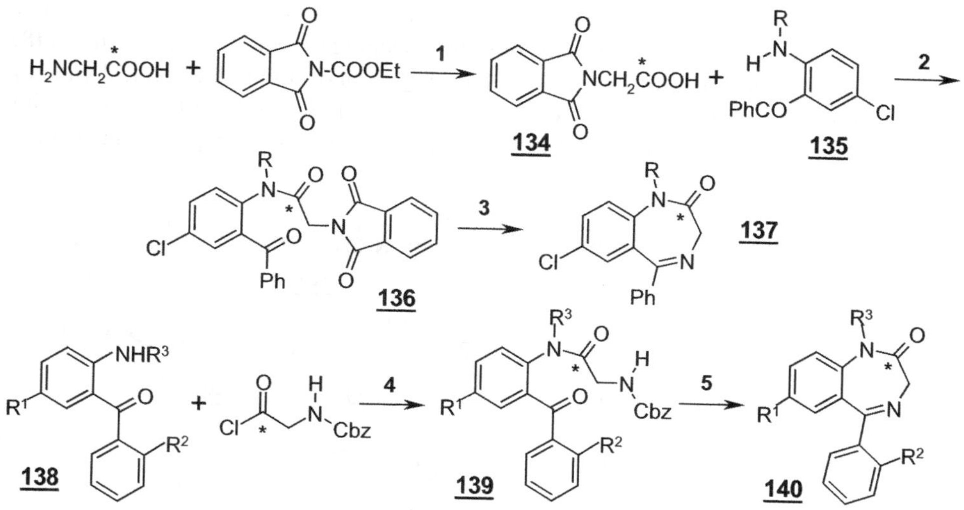

$R^1$ = H, Cl or Br; $R^2$ = H or Cl; $R^3$ = H or Me

**Reaction conditions:** **1**. aq. $Na_2CO_3$; r.t., 30 min; **2a**. **134** $SOCl_2$; reflux, 1 h, **b**. **135** toluene; r.t., 16 h; **3**. hydrazine hydrate, EtOH; reflux, 21 h; **4a**. ether; 10 °C, 40 min, **b**. 2 N NaOH, r.t., 2 h; **5a**. 30 % HBr, HOAc; r.t., 1 h, **b**. 10 % NaOH; 0 °C, 10–20 min

***Figure 6.45*** *Use of [1-$^{14}$C]glycine for the synthesis of [2-$^{14}$C]benzazepines*

These compounds are indispensable building blocks for the preparation of labeled cyclopropane- and cyclopropenecarboxylic acids, as illustrated by the following three examples. In the first case 2,5-dimethylhexa-2,4-diene reacted with ethyl diazo[2-$^{14}$C]acetate at 120–140 °C in the presence of copper powder as a catalyst[97] or at room temperature in the presence of rhodium(II)pivalate[98] to provide a 1:1 *cis/trans* mixture of ethyl [2-$^{14}$C]chrysanthemates **145** in 30–70% yield. Treatment of this mixture

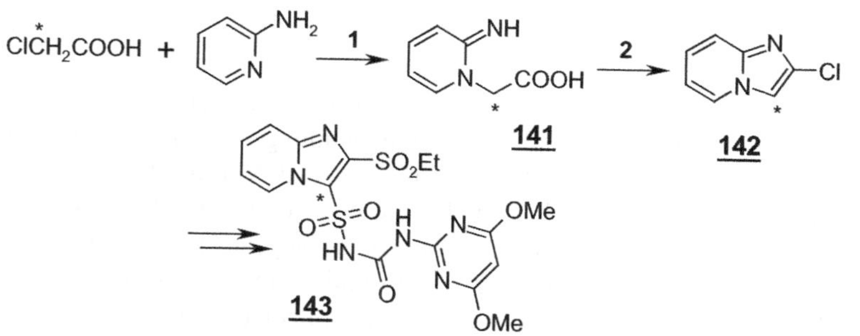

**Reaction conditions:** **1**. $Et_3N$, $H_2O$; 90 °C, 5 h; **2**. $POCl_3$, toluene; reflux, 16 h

***Figure 6.46*** *Use of chloro[$^{14}$C]-acetic acid in the preparation of a carbon-14-labeled heterocycle: imidazo[1,2-a]pyridine*

with $PdCl_2(C_6H_5CN)_2$ altered the ratio to 9:1 in favor of the desired *trans*-isomer. After separation and saponification *trans*-acid **146**, the insecticidal ester [$^{14}$C]pyrethrin I (**148**) was prepared by esterification with pyrethrolone **147** in an overall radiochemical yield of 60%[97]. Optical resolution of *trans*-**146** as a quinine salt followed by esterification with α-cyano-3-phenoxybenzyl bromide (**149**) furnished (1*R*)-*trans*-[$^{14}$C]cyphenotrin (**150**)[97].

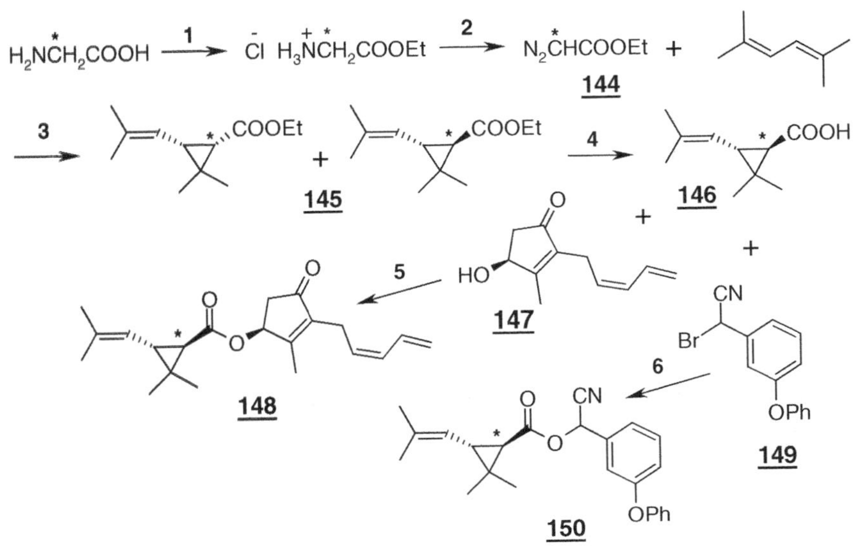


**Reaction conditions: 1**. EtOH, $HCl_{gas}$; 70–80 °C, 2 h; **2**. $NaNO_2$, NaOAc, $H_2SO_4$, $H_2O$; r.t.; **3**. $Rh_2(piv)_4$; $CH_2Cl_2$; r.t., 1 h; **4a**. $PdCl_2(C_6H_5CN)_2$, $CHCl_3$; r.t., 4 h, **b**. LC-separation (silica gel; hexane–ethyl acetate), **c**. NaOH, EtOH; reflux, 2 h; **5**. $(CF_3CO)_2O$, $CH_2Cl_2$; r.t., 20 min; **6a**. opt. resolution (quinine), **b**. $Bu_4N^+Br^-$, $K_2CO_3$, $CH_3CCl_3$; 70 °C, 5 h

***Figure 6.47*** *Synthesis and application of ethyl [2-$^{14}$C]diazoacetate in the synthesis of [1-$^{14}$C] cyclopropanecarboxylic acid derivatives*

Figure 6.48 illustrates an interesting example in which the diazo[$^{14}$C]acetate **152** was generated intramolecularly through sequential coupling of *N*-Boc[2-$^{14}$C]glycine with allylic alcohol **151**, acid-catalyzed *N*-deprotection and *N*-nitrosation. Cyclization was accomplished upon heating of **152** in the presence of the Cu(II) complex depicted. Cleavage of one chlorine and hydrolytic opening of the bicyclic lactone system was followed by removal of the undesired *E*-isomer by crystallization to provide (*Z*)-cyclopropane-[1-$^{14}$C]carboxylate **153**, which served as a key intermediate in the synthesis of γ-[$^{14}$C]-cyphalothrin (**154**), a further pyrethroid insecticide[99].

Reaction of ethyl diazo[2-$^{14}$C]acetate with methyl stearolate (**155**) produced the cycloproprenoid diacid diester **156**. Removal of the cyclopropene carboxyl group was accomplished by saponification to the diacid, dehydration and decarbonylation upon treatment with $HClO_4/Ac_2O$ and subsequent $NaBH_4$ reduction of the resulting cyclopropenium perchlorate, to yield the target compound, [$^{14}$C]sterculic acid (**157**)[100].

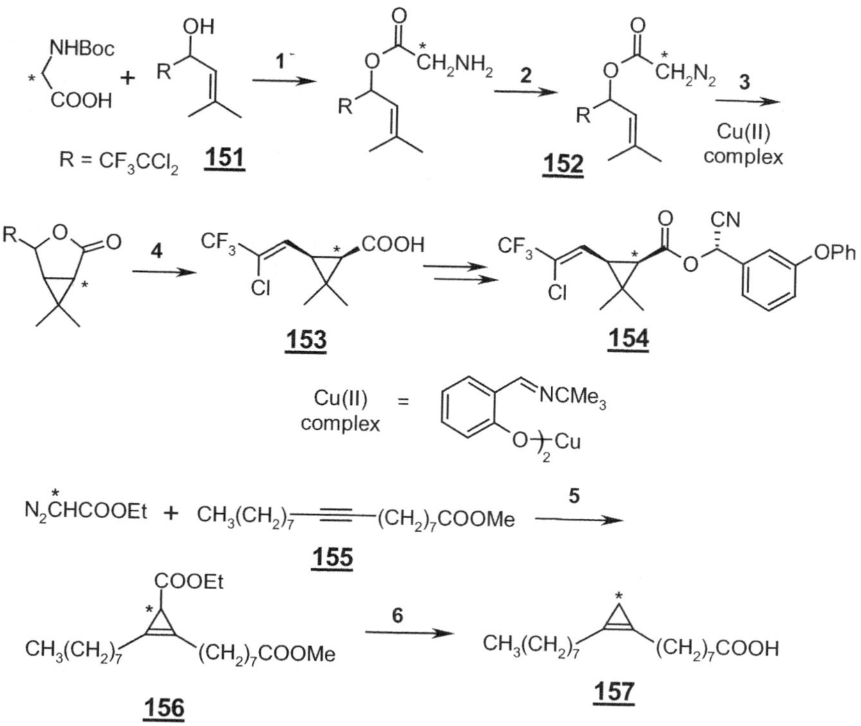


**Reaction conditions: 1a**. *i*-PrN=C=N*i*-Pr, DMAP, toluene/$CH_2Cl_2$; r.t. 1 h,**b**. 4 N HCl, dioxane; r.t., 90 min; **2**. $NaNO_2$, Na-dihydrogencitrate, $H_2O$, $CH_2Br_2$; 0–5 °C, 60 min; **3**. Cu(II) complex, dichloroethane; reflux, 5.5 h; **4a**. Zn, MeOH; reflux, 2.5 h, **b**. 0.5N HCl, r.t., **c**. recrystallization from hexane; **5**. Cu powder; 120 °C; **2a**. KOH, propanol; 105 °C, 2 h, **b**. 70 % $HClO_4$, $Ac_2O$; 0–15 °C, 15 min,**c**. $NaBH_4$, DMSO–pyridine 3:1; 5 °C

***Figure 6.48*** *Intramolecular generation and application of an diazo[$^{14}$C]acetate unit*

### *Replacement of Halogen with Phosphorus Nucleophiles*

By far the main interest in the reaction of halo[$^{14}$C]acetates with phosphorus nucleophiles is for the preparation of phosphoryl-stabilized carbanions for use in Wittig and related reactions. The presence of the additional electron-withdrawing ester group provides additional stabilization, significantly modifying the reactivity of the ylide species and the stereochemical course of its reactions. The two phosphorus reagents discussed here include the triphenylphosphonium salt type **158**, precursors of Wittig methylenetriphenylphosphoranes, and the trialkylphosphonoacetate type **159**, applied in the Horner–Wadsworth–Emmons family of reactions[101].

**158** $Ph_3\overset{+}{P}\overset{\Delta}{C}H_2\overset{*}{C}OR$ $X^-$ **159** $(R''O)_2P(O)\overset{\Delta}{C}H_2\overset{*}{C}OR$

R = OR$^1$ or NR$^1$R$^2$

$\overset{*}{\Delta}$ = alternative positions of label

***Figure 6.49*** *Carbon-14-labeled phosphorus reagents*

*Carbon-14-labeled (alkoxycarbonylmethylene)triphenylphosphoranes ($Ph_3P={}^*CH^{\Delta}CO$ OR) and (aminocarbonylmethylene)triphenylphosphoranes ($Ph_3P={}^*CH^{\Delta}CONRR'$)* The presence of the electron-withdrawing ester group in the $\alpha$-position of the ylides prepared by deprotonation of phosphonium salts **158**, and (alkoxycarbonylmethylene)triphenylphosphoranes **160**, means that these species are less nucleophilic than unsubstituted alkylidenetriphenylphosphoranes. Accordingly, they react readily with aldehydes but more slowly or not at all with ketones. The $\alpha,\beta$-unsaturated esters formed with aldehydes may be (*E*)- or (*Z*)-configured, depending on the solvent and substrate, though (*E*)-selectivity usually predominates[101]. Ylides such as **160** can be readily prepared by alkylation of triphenylphosphine with alkyl bromo[$^{14}$C]acetates in benzene followed by deprotonation of the resulting precipitated phosphonium salt with 0.1 N NaOH; yields are usually >85%[76a,78a,102,103].

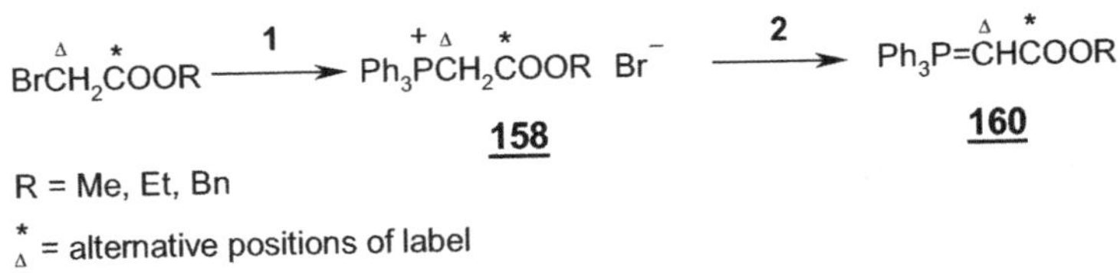

**Reaction conditions:** **1**. $Ph_3P$, benzene; r.t., 16 h; **2**. 0.1 N NaOH, $H_2O$; (pH ≥ 9), 0 °C

***Figure 6.50*** *Standard procedure for the preparation of $Ph_3P = {}^{14}CHCOOR$*

Ylide **160** has found wide application for the extension of the carbon skeleton of terpenes, steroids and polyacetylenes by an isotopically labeled $C_2$ unit[76a,102,104]. An example is the side-chain construction leading to 24-methylidene[24-$^{14}$C]cholesterol (**165**) for use in biosynthetic studies (Figure 6.51). Olefination of (20*S*)-3-oxopregn-4-ene-20-carbaldehyde (**161**) with [1-$^{14}$C]**160**, for example, afforded the (22*E*)-ester **162** in 80% radiochemical yield. Intermediate steps to hydrolyze the ester group and reduce the $\alpha,\beta$-double bond, as well as to modify the A/B ring functionality, produced **163**. Treatment of the corresponding acyl chloride with *i*-PrMgCl/$CdCl_2$ provided 24-oxo[24-$^{14}$C]cholest-5-en-3$\beta$-yl acetate (**164**), which upon methylenation completed the synthesis of **165**[102]. Similar chemistry was used to prepare carbon-13-labeled bile acids[103].

The single recently published application of the amide form was in the synthesis of the histone deacetylase inhibitor vorinostat (**169**) (Figure 6.52). Phosphonium salt **166** was prepared by the reaction of triphenylphosphine with chloro[1-$^{14}$C]acetanilide, itself prepared via chloro [1-$^{14}$C]acetyl chloride. Olefination of aldehyde **167** with the ylide produced from **166** furnished the $\alpha,\beta$-unsaturated amide **168** in 85% yield[105]. The tributylphosphonium analog was employed as a starting material for the synthesis of [$^{14}$C]GV 1550526, a glutamate co-agonist[105b].

*Trialkyl phosphono[$^{14}$C]acetates ($(EtO)_2P(O)^*CH_2{}^{\Delta}COOR$)* The use of trialkyl phosphono[$^{14}$C]acetates **159** in olefination reactions (Horner–Wadsworth–Emmons or HWE reactions) is generally superior to the use of the related Wittig reagents. The carbanions, generated by a suitable base, are more nucleophilic than the latter, so that they react even with ketones. Furthermore, they often show an improved stereoselectivity. They provide 100% *E*-configured acrylates when they react with linear aldehydes, and (*E*)-alkenes

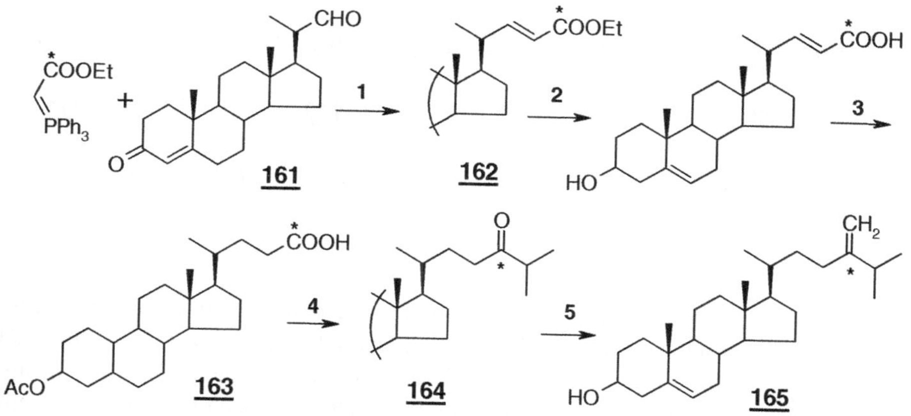


**Reaction conditions:** **1**. MeCN; r.t., 72 h; **2a**. AcCl, $Ac_2O$, reflux, 16 h, **b**. $NaBH_4$, EtOH –$H_2O$; 5 °C, 2.5 h,**c**. NaOH, EtOH–$H_2O$; reflux, 15 h; **3a**. $Ac_2O$, pyridine; r.t.,16 h, **b**. $H_2$, $PtO_2$, dioxane/AcOH; r.t., 24 h; **4a**. oxalyl chloride, benzene; r.t., 5h, **b**. *i*-PrMg/$CdCl_2$, ether; 0 °C, 20 min; r.t., 10 h; **5a**. $CH_2Br_2/TiCl_4/Zn$, $CH_2Cl_2$, **b**. NaOMe, MeOH

***Figure 6.51*** *Use of $Ph_3P = CH^{14}COOEt$ in the synthesis of 24-methylene[24-$^{14}C$]cholesterol*

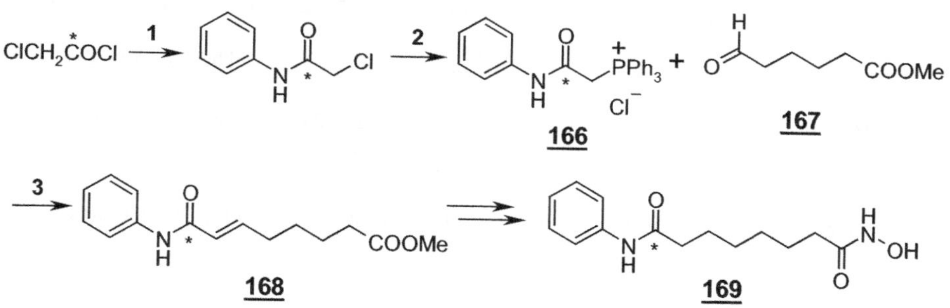


**Reaction conditions:** **1**. aniline, DMAP, THF; 0 °C, 1 h; **2**. $PPh_3$, toluene; 150°C, 5 min (microwaves); **3**. **166**, LDA, THF; 0 °C, 1 h then **167**; r.t., 18 h

***Figure 6.52*** *Preparation and use of $Ph_3P = CH^{14}CONHPh$ in the synthesis of [$^{14}C$]vorinostat*

generally predominate with $\alpha$-branched aldehydes and ketones[101,106]. Reaction of triethyl phosphono[1, 2-$^{14}C_2$]acetate with 2-pentanone, for example, gave a 85:14 *E*/*Z*-mixture of ethyl 3-methyl-2-[1, 2-$^{14}C_2$]hexenoate (**170**) and (**171**) in 70–80% yield[106] (Figure 6.53).

HWE reagents **159** are easily prepared from trialkyl phosphites and alkyl bromo[$^{14}C$]-acetates using conditions of the Michaelis–Arbusov reaction[101,104]. With very few exceptions the corresponding methyl and ethyl derivatives have been employed so far. Different procedures have been recommended, depending on the amount of labeled starting material. Larger quantities (>2 mmol) are obtained in 90–93% yields when equimolar amounts of the neat components are heated together for 4–6 h at 150–165 °C[66d,106,107]. Smaller amounts are made available in 80% to nearly quantitative yield through treatment of the

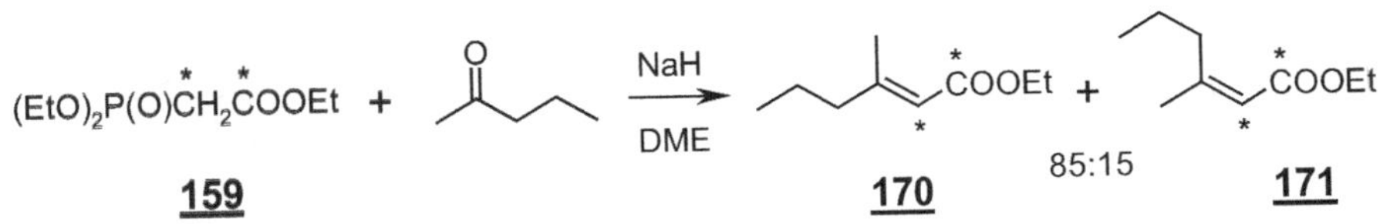

**Figure 6.53** *Stereoselectivity of the olefination reaction of triethyl phosphono[1, 2-$^{14}C_2$] acetate with 2-pentanone*

alkyl bromo[$^{14}$C]acetates either with excess trialkylphosphite at 120 °C for 4 h[108] or with equivalent amounts of the reactants refluxed in ethereal solution for 16–32 h[109] followed by fractional distillation. An alternative procedure, which avoids the preparation of alkyl bromo[$^{14}$C]acetates, starts with the alkylation of diethyl phosphite with [$^{14}$C]methyl iodide. The resulting diethyl methylphosphonate (**172**) is deprotonated with MeLi/LiBr, carboxylated, then esterified with diazomethane or diazoethane[110].

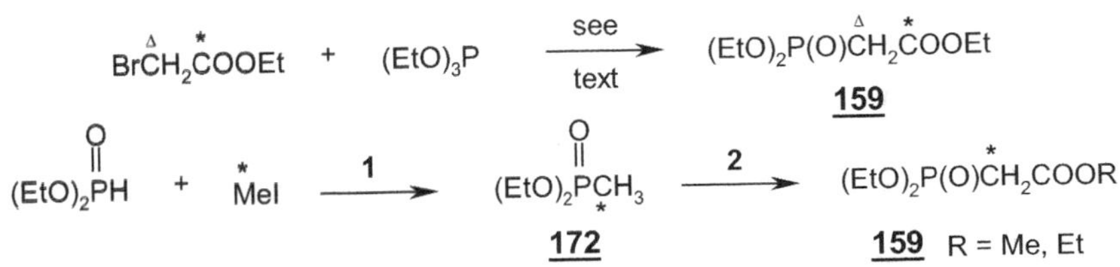

**Reaction conditions:** 1. NaOEt, benzene; 25 °C, 12 h; **2a**. MeLi/LiBr, THF, 65°C, 1 h, **b**. $CO_2$;–65 °C, 2 h,**c**. $RCHN_2$, ether; r.t. (25 % overall)

**Figure 6.54** *Procedures for preparation of triethyl phosphono[$^{14}$C]acetates*

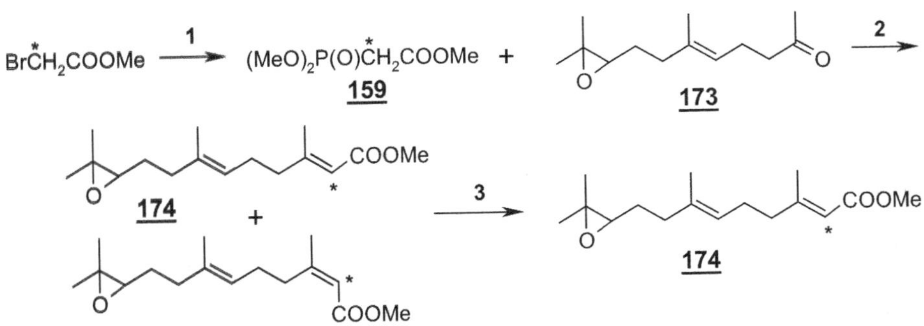

**Reaction conditions:** 1. $(MeO)_3P$,120 °C, 4 h; **2a**. **159**, NaH, ether; reflux, 4 h, **b**. **173**, reflux, 20 h; **3**. preparative GC separation

**Figure 6.55** *Use of trimethyl phosphono[2-$^{14}$C]acetate in the synthesis of racemic insect [2-$^{14}$C]juvenile hormone III*

Labeled phosphonoacetates have found wide application in the carbon-13 and carbon-14 syntheses of terpenoids such as [2-$^{14}$C]geranylgeraniol[107], (13*E*)-[$^{13}$C] and [$^{14}$C]retinoic acid[109b,110,111], [$^{14}$C]insect juvenile hormone III[108,109a] and [$^{14}$C]pseudomonic acid[112], as well as diverse structures like (*R*)-(−)-[1, 7-$^{13}C_2$]shikimic acid[67d], [$^{13}$C]coniferyl alcohol[113] and [$^{13}$C]valine[66d]. The carbon-14-labeled insect juvenile hormone III (**174**), for example was obtained in 67% yield as a 4:1 *E*/*Z*-mixture, when 9,10-epoxy geranyl acetone (**173**) reacted with deprotonated trimethyl phosphono[2-$^{14}$]acetate. The desired *E*-isomer was separated by preparative GC[109a].

Further applications of the HWE reagent **159** comprise the development of low molecular weight carbon-14-labeled building blocks for the three- to five-carbon extension of the skeleton of halides and carbonyl compounds. In the simplest case, alkylation of deprotonated **159** with methyl iodide furnishes the homologous propionate reagent **175**, which has been employed as a starting material in the synthesis of [$^{14}$C]E-5090 (**179**), an orally active IL-1 inhibitor. Reaction of **175** with the substituted 1-naphthalene-

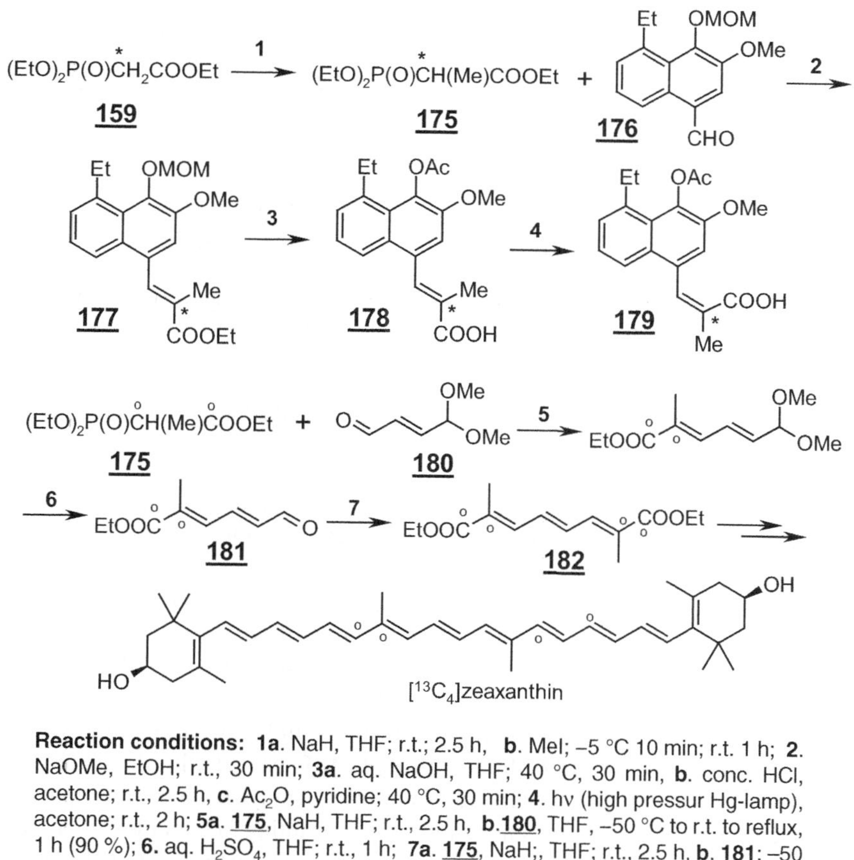


**Reaction conditions:** **1a**. NaH, THF; r.t.; 2.5 h, **b**. MeI; −5 °C 10 min; r.t. 1 h; **2**. NaOMe, EtOH; r.t., 30 min; **3a**. aq. NaOH, THF; 40 °C, 30 min, **b**. conc. HCl, acetone; r.t., 2.5 h, **c**. $Ac_2O$, pyridine; 40 °C, 30 min; **4**. hν (high pressur Hg-lamp), acetone; r.t., 2 h; **5a**. **175**, NaH, THF; r.t., 2.5 h, **b**.**180**, THF, −50 °C to r.t. to reflux, 1 h (90 %); **6**. aq. $H_2SO_4$, THF; r.t., 1 h; **7a**. **175**, NaH;, THF; r.t., 2.5 h, **b**. **181**; −50 °C to r.t. to reflux, 1 h (90 %)

*Figure 6.56 Use of carbon-13/14-labeled triethyl phosphonopropionate in the extension of aldehydes by an isotopically labeled $C_3$ unit*

carbaldehyde **176** afforded ethyl (2*E*)-3-(1-naphthalenyl)-2-methyl-2-[2-$^{14}$C]propenoate (**177**). Subsequent saponification of the ester group, cleavage of the MOM-ether and acylation of the released phenolic OH-group converted **177** into the *E*-configured [$^{14}$C]-E-5090 isomer **178**. Concluding photoisomerization provided an *E*/*Z*-mixture, from which the desired (2*Z*)-isomer **179** was separated by preparative HPLC. The overall radiochemical yield was 13%[114]. The [1, 2-$^{13}C_2$] isotopomer of **175** was used in two sequential three-carbon extensions (each ca. 90% yield) of aldehyde **180** as shown to give [$^{13}C_4$]diester **181**, the key intermediate in the synthesis of the carotenoid [12,12′,13,13′-$^{13}C_4$]zeaxanthin (**182**)[115].

HWE reaction with [2-$^{14}$C]acetone converts **159** into methyl 3-methyl-2-[2, 3-$^{14}C_2$]-butenoate (**183**), a doubly labeled $C_5$-building block. It adds as a vinologous ester in an aldol-like condensation reaction, for example, to $\beta$-ionylideneacetaldehyde (**184**) to give a 2:1-mixture of ethyl (13*E*)- and (13*Z*)-retinoates. Upon ester hydrolysis followed by fractional crystallization from methanol, (all-*E*)-[13, 14-$^{14}C_2$]retinoic acid (**185**) was isolated in 25% radiochemical yield[110].

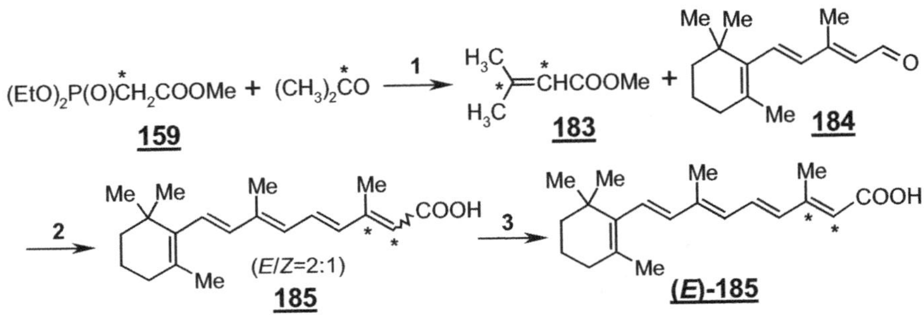


**Reaction conditions: 1a**. **159**, NaH, ether; r.t., 1 h, **b**. [2-$^{14}$C]acetone; r.t., 16 h; **2a**. $KNH_2$, ether; –65 °C, 1 h ; r.t., 17 h, **b**. isomer separation by fractional crystallization from MeOH

***Figure 6.57*** *Preparation and use of the labeled $C_5$-building block **183** in the synthesis of (all-E)-[13, 14-$^{14}C_2$]retinoic acid*

Another labeled $C_5$-building block has been made available through HWE reaction of **159** with pyruvic aldehyde dimethylacetal (**186**)[109b] (Figure 6.58). The resulting 5-hydroxy-4-methyl-2(*H*)-[3-$^{14}$C]furanone (**187**) proved to be radiochemically stable, so that it could be stored for some time. It may be regarded as a masked 2-methylmaleic semialdehyde, which upon reaction with Wittig reagents affords homologous 3-methyl-2,4-pentadienoic acids. Since Wittig reagents are readily accessible from halides and triphenylphosphine, these partial sequences can be combined to a procedure that allows halides to be extended by five carbon atoms. This is illustrated in Figure 6.58 by the synthesis of (13*Z*)-[14-$^{14}$C]retinoic acid (**189**). In this example **187** reacted with the ylide generated from '$C_{15}$ Wittig reagent' **188**, generated *in situ* from 1-vinyl-$\beta$-ionol and triphenylphospine hydrochloride. Subsequent isomerization of the resulting *E*/*Z*-mixture upon treatment with iodine in the presence of light and fractional crystallization afforded isomerically pure **189** in 17% overall radiochemical yield[109b].

The Horner–Wadsworth–Emmons methodology has also been successfully applied to the synthesis of a broad spectrum of enantiomerically pure compounds singly or multiply labeled

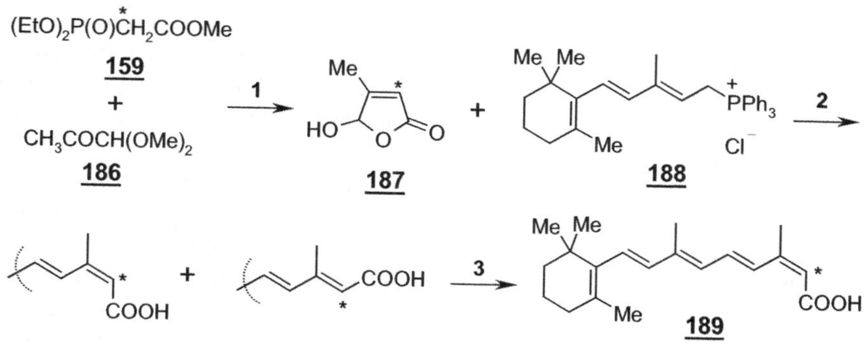


**Reaction conditions: 1a**. MeONa, DMF/MeOH 3:2; r.t., 4 h, **b**. 6 N HCl; reflux, 4 h; **2a**. KOH, isopropanol; –5 °C, 30 min; 0 °C, 10 min; **3**. $I_2$, ether; light, r.t., 1.5 min

***Figure 6.58*** *Preparation and use of the labeled $C_5$-building block* ***187*** *in the synthesis of (13E)-[14-$^{14}$C]retinoic acid*

with isotopic carbon[75]. Triethyl phosphono[$^{14}$C]acetyl (2*S*) or (2*R*)-bornane-10.2-sultam ((2*S*) or-(2*R*)-[$^{14}$C]PABS, **190**), which is readily prepared by the reaction of triethylphosphite with bromo[$^{14}$C]-acetyl (2*S*) or (2*R*)-bornane-10.2-sultam ((2*S*)- or-(2*R*)-[$^{14}$C]-BABS, **93**), reacts with aldehydes (but not with ketones) to give chiral acrylates. Carbon-14-labeled chiral acrylates have been used as key intermediates in the synthesis of $\alpha$-, $\beta$-, $\alpha$, $\beta$-functionalized or branched labeled e.p. carboxylic acid derivatives (via Michael addition) as well as of labeled e.p. cyclohexen-3-ene carboxylic acids (via [2 + 4]-cycloadditions).

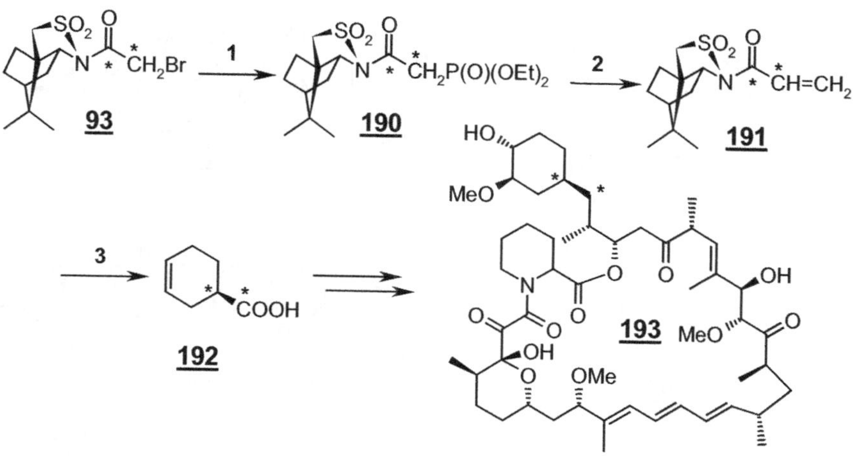


**Reaction conditions: 1**. $P(OEt)_3$; 100 °C, 3 h; **2**. $(HCHO)_2$, galvinoxyl, $K_2CO_3$, THF; 70 °C, 30 min; **3a**. 1,3-butadiene, $EtAlCl_2$, $CH_2Cl_2$; –78 °C, 4 h, **b**. LiOH, aq. THF; r.t., 3 h

***Figure 6.59*** *Preparation and use of (+)-[$^{14}C_2$]PABS (**190**) for the preparation of an e.p. [$^{14}C_2$] cyclohexenecarboxylic acid*

For example, Diels–Alder reaction between butadiene and [1, 2-$^{14}C_2$]acryloyl (2*S*)-sultam **191**, obtained from (2*S*)-[$^{14}C_2$]PABS and paraformaldehyde, followed by hydrolytic removal of the auxiliary, afforded (1*R*)-3-[1, 7-$^{14}C_2$]cyclohex-3-ene carboxylic acid **192**, a precursor for the biological preparation of [$^{14}C_2$]rapamycin (**193**)[116] (for further information see Chapter 11, Section 11.3.4).

Replacement of the ethoxy groups with 2,2,2-trifluoroethoxy groups inverts the *E*-selectivity of the HWE reagent in olefination reactions, thereby providing predominantly *Z*-acrylates (Still–Gennari reactions)[104]. The corresponding building block (Still–Gennari reagent) **194** is conveniently accessible in around 62% yield through alkylation of deprotonated bis(2,2,2-trifluoroethyl)phosphite with alkyl bromo[1, 2-$^{14}C_2$]acetate[117].

The utility and potential of this reagent have been impressively documented with the carbon-14 labeling of (+)-discodermolide, a polyketide natural product showing promising antitumor activities[115]. Reaction of deprotonated **194**, which is best effected with KHMDS in the presence of 18-crown-6, with the aldehyde fragment **195** afforded the (Z)-[$^{14}C_2$]acrylate **196** in 90% yield. Blocking the free hydroxyl group as a carbamate, transformation of the ester group into an aldehyde function, and stereoselective aldol reaction of the resulting intermediate **197** with the chiral boron enolate of fragment **198** gave **199**, a late intermediate to the target.

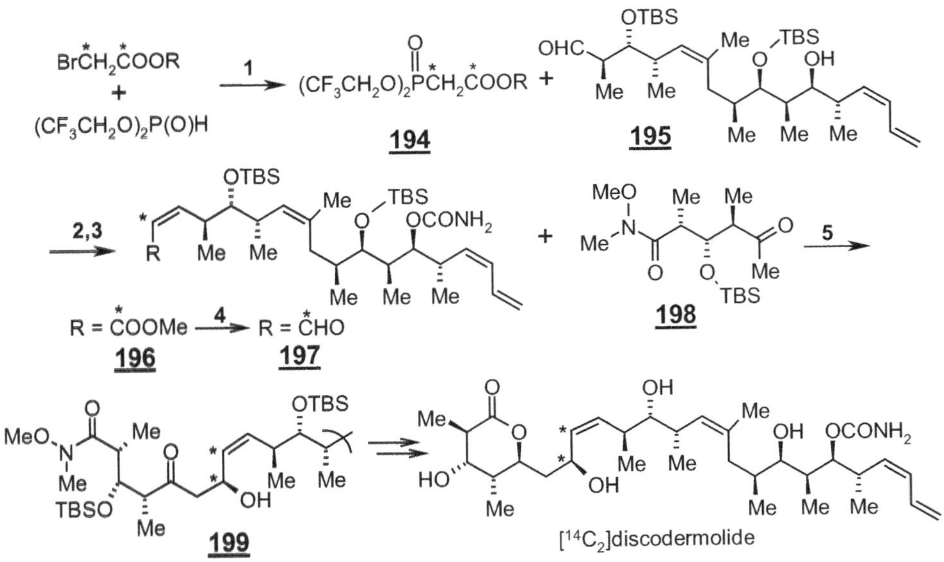


**Reaction conditions: 1**. NaH, *n*-$Bu_4NI$, ether; –40 °C, 5 min; 0 °C, 4 h; **2**. KHMDS, 18-crown-6, THF; –20 °C, 2 h; **3a**. $CCl_3C(O)NCO$, $CH_2Cl_2$; 20 °C, 1 h, **b**. $K_2CO_3$, MeOH; 20 °C, 1 h, **4a**. **196**, DIBAL-H, $CH_2Cl_2$; –78 °C, 1.5 h, **b**. Dess–Martin periodinane, $CH_2Cl_2$; 25 °C, 1 h; **5a**. DIPCl. $Et_3N$, $Et_2O$; r.t., 1.5 h, **b**. $CH_2Cl_2$; –78 °C to –27 °C, 16 h, **c**. 30 % aq. $H_2O_2$, MeOH; 0 °C, 1 h

***Figure 6.60*** *Preparation and application of the Still–Gennari reagent* ***194*** *in the synthesis of Z-[$^{14}C_2$]acrylate, key intermediate in the synthesis of (+)-[$^{14}C_2$]discodermolide*

*Replacement of Halogen by Silicon Functionalities*

The Wittig–Horner and Horner–Wadsworth–Emmons reagents discussed above can be replaced by alkyl (or benzyl) trimethylsilyl[$^{14}$C]acetates (TMS acetates, e.g., **200**). The corresponding carbanions add to carbonyl compounds to give $\alpha$-silylated $\beta$-hydroxy esters, which upon work-up eliminate $Me_3SiOH$ to produce mixtures of labeled (*E*)- and (*Z*)-$\alpha$, $\beta$-unsaturated esters with a slight preference toward (*Z*)-isomers (Petersen olefination). Anions of TMS acetates are more reactive than the analogous phosphonate anions, and this gives them advantages for some applications. For example, readily enolizable ketones may be inert to phosphonate anions or undergo an undesired enolate condensation reaction because the ketone $\alpha$-proton transfer reaction is faster than carbonyl addition; this is less likely with the analogous TMS acetate anions. The latter are also less prone than their phosphonate counterparts to attack groups susceptible to nucleophilic substitution, so that they may be more tolerant of ancillary substrate functionality[118].

Carbon-14-labeled TMS acetates are most directly prepared from the corresponding haloacetates, as preparation by the reaction of lithiated acetate esters with trimethylsilyl chloride (TMSCl) would result in a mixture of *C*- and *O*-silylated products. The preferred procedure involves reaction of TMSCl with the organozinc reagent of the haloacetate in a modified Reformatsky reaction. An application of this Peterson-olefination sequence is illustrated by the preparation of the antipsoriatic vitamin D derivative **204**[119]. Reaction of the anion of benzyl trimethylsilyl[1, 2-$^{14}C_2$]acetate (**200**) with TBS-protected cyclohexanone-3,5-diol afforded the expected benzyl cyclohexylidene[1, 2-$^{14}C_2$]acetate (**201**) in moderate yield. The subsequent transformation of **201** to the phosphine oxide **202** followed

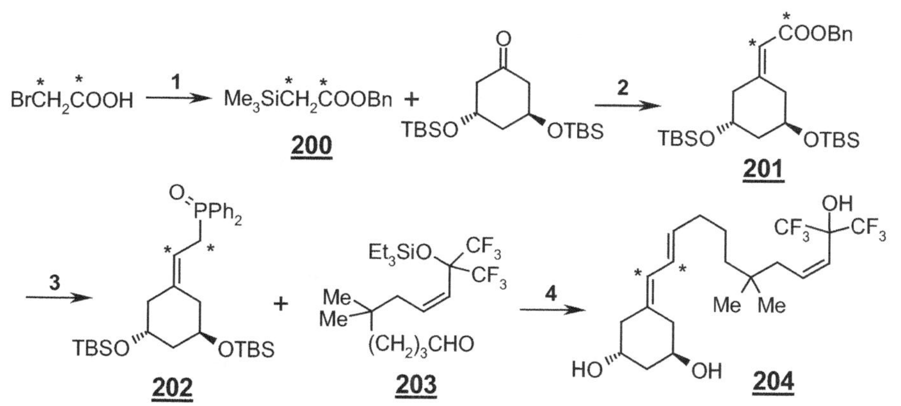


**Reaction conditions: 1a**. BnOH, DCC, DMAP, dichloromethane; 2 °C, 4 h (97%), **b**. Zn, TMSCl, ether; 22 °C, 4 h, **c**. fractional micro-scale distillation (74 %); **2a**. **200**, LDA, THF; –78 °C, **b**. ketone, THF; –78 °C to –20 °C; 3.5 h; **3a**. DIBAL-H, toluene; –78 °C, 2.5 h, **b**. NCS, DMS, dichloromethane; –20 °C to 0 °C, 2 h, **c**. Ph2PH, BuLi, THF; –78 °C, 4 h, **d**. $H_2O_2$; 20 °C, 2 h; **4a**. **202**, BuLi, THF;–78 °C, **b**. **203**, THF; –78 °C to 20 °C, 4 h, **c**. TBAF, THF; 40 °C, 15 h, **d**. prep. HPLC-separation (4 % overall)

***Figure 6.61*** *Preparation and use of benzyl trimethylsilyl[1, 2-$^{14}C_2$]acetate for the Peterson olefination step in the synthesis of [$^{14}C_2$]RO 65-2299*

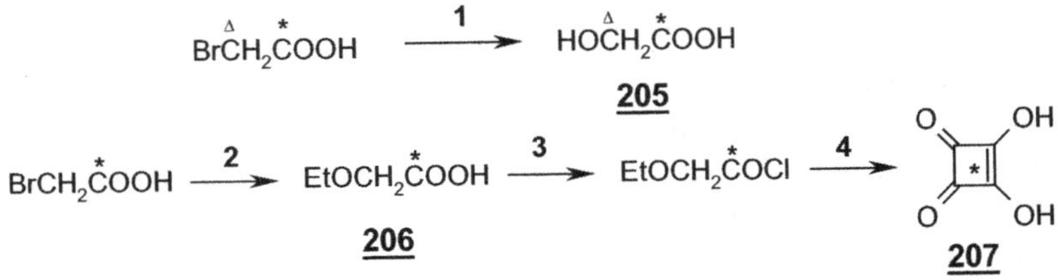

**Reaction conditions:** **1**. 3 N NaOH, 50 °C, 3 h; **2**. NaOEt, EtOH, reflux 3 h; **3**. phthaloyl dichloride; 80 °C, 1 h; 120 °C, 1.5 h; **4a**. $Et_3N$, $(EtO)_2C{=}C(OEt)_2$, hexane; reflux 4 h; r.t., 16 h, **b**. 6 N HCl; 100 °C, 30 min

***Figure 6.62*** *Preparations of [$^{14}$C]glycolic acid and ethoxy[1-$^{14}$C]acetic acid and application of the latter*

classical procedures. Intermediate **202** functioned as a coupling component to the aldehyde fragment **203**, which resulted in doubly protected [$^{14}C_2$]RO 65-2299 (**204**, R = TBS). Cleavage of the protecting groups and HPLC separation of the undesired Z-isomer afforded the labeled drug substance **204** (R = H).

*Replacement of Halogen by Oxygen Nucleophiles*

[$^{14}$C]Glycolic acid and its derivatives are much less important in carbon-14 synthesis than [$^{14}$C]glycines and phosphono[$^{14}$C]acetates. [$^{14}$C]Glycolic acid (**205**) is conveniently

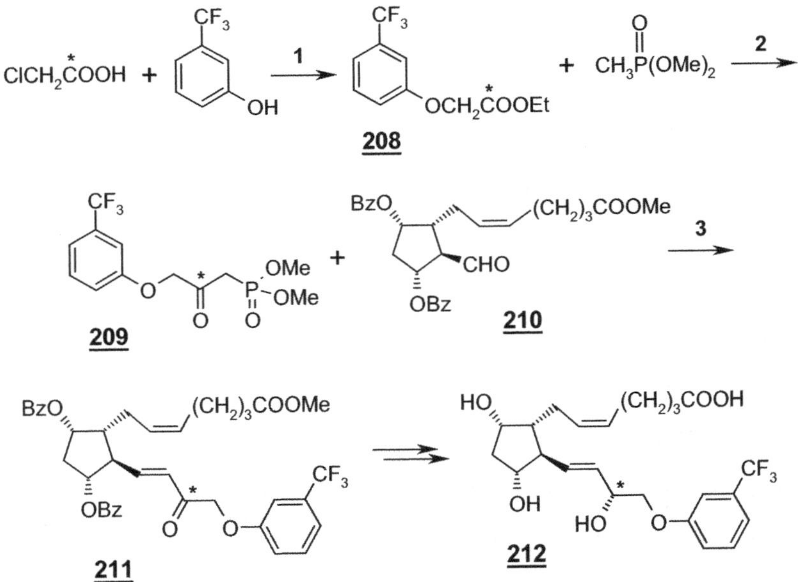

**Reaction conditions:** **1a**. 3-trifluoromethylphenol, 2 N NaOH, reflux, 16 h, **b**. EtOH, conc. $H_2SO_4$, reflux, 16 h; **2a**. BuLi, THF; –78 °C, 30 min, **b**. **208**, THF; 78°C, 2 h; –78 °C to r.t., 16 h; **3**. 1 N NaOH, toluene; r.t., 16 h

***Figure 6.63*** *Preparation and application of an aryloxy[$^{14}$C]acetate ester*

prepared in nearly quantitative yields by hydrolysis of bromo[$^{14}$C]acetic acid with aqueous sodium hydroxide at 50 °C for 3 h, acidification and extraction into ethyl acetate or as an aqueous solution by heating overnight at 120 °C[120]. The corresponding *O*-alkyl and aryl ethers are readily accessible through halogen replacement with alkoxides or phenolates. For example, bromo[1-$^{14}$C]acetic acid reacted with sodium ethoxide to give ethoxy[1-$^{14}$C]acetic acid (**206**). Its conversion into the acyl chloride followed by dehydrohalogenation upon heating with triethylamine generated ethoxy[$^{14}$C]ketene; the latter reacted *in situ* by a [2 + 2]cycloaddition with tetraethoxyethene to furnish [$^{14}$C]squaric acid (**207**)[121].

An example of the preparation and use of an aryloxy[$^{14}$C]acetate is in the synthesis of a labeled prostaglandin F2$\alpha$ analog **212**[122] (Figure 6.63). Ethyl 2-(3-trifluoromethylphenoxy)[1-$^{14}$C]acetate (**208**) was obtained in 44% overall yield through etherification of sodium 3-trifluoromethylphenolate with chloro[1-$^{14}$C]acetic acid followed by esterification. Aryloxy ester **208** was submitted to reaction with lithiated dimethyl methylphosphonate to give phosphonate **209**, which underwent Wittig–Horner coupling with 2,3,5-trisubstituted cyclopentylaldehyde **210** to provide the *O*-protected intermediate **211**, a late intermediate to the target molecule **212**.

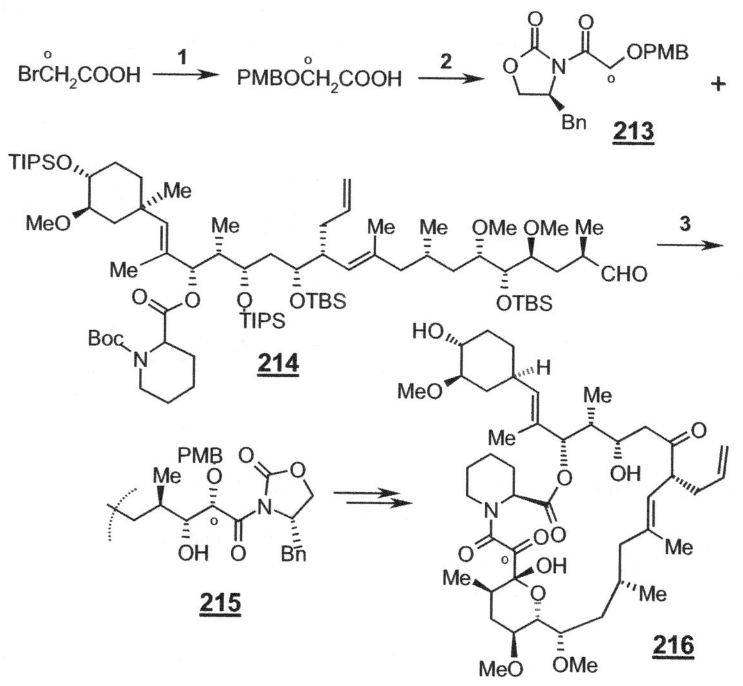


**Reaction conditions: 1**. 4-MeO-BnOH, NaH, toluene; 90 °C, 4 h; **2a**. pivaloyl chloride, $Et_3N$, THF; 0 °C, **b**. (*S*)-4-benzyl-2-oxazolidinone, *n*-BuLi, THF; −78 °C to 0°C; **3**. *n*-$Bu_2BOTf$, Et3N, toluene; −50 °C, 4 h

***Figure 6.64*** *Synthesis and application of carbon-13-labeled chiral glycolate* **213** *and its application in the synthesis of [9-$^{13}$C]FK 506*

Unlike chiral [$^{14}$C]glycinates, chiral [$^{13/14}$C]glycolates have rarely been used in isotope chemistry. The most striking exception is the use of (*S*)-4-benzyl-3-(4-methoxybenzyl)-[2-$^{13}$C]acetyl-1,3-oxazolidin-2-one (**213**) in the highly diastereoselective aldol reaction with the aldehyde fragment **214** to give **215**, key intermediate in the synthesis of [9-$^{13}$C]FK 506 (**216**) (Figure 6.64). The oxazolidinone reagent was prepared through etherification of 4-methoxybenzyl alcohol with bromo[2-$^{13}$C]acetic acid, activation of the carboxyl group with pivaloyl chloride and coupling of the resulting mixed anhydride to deprotonated (*S*)-4-benzyl-2-oxazolidinone (Evans's auxiliary). Its boron enolate, generated *in situ* from di-*n*-butylboron triflate and triethylamine, reacted with **214** to give the $\alpha,\beta$-dihydroxy derivative **215** in 90% yield and high diastereomeric purity[123].

A further class of building blocks available from halo[$^{14}$C]acetate is [$^{14}$C]glyoxylic acid esters. For example, benzyl bromo[2-$^{14}$C]acetate reacts with silver nitrate to give benzyl nitrooxy[2-$^{14}$C]acetate (**217**), which upon treatment with tetrabutylammonium fluoride provided benzyl [2-$^{14}$C]glyoxylate (**218**) in 60% overall radiochemical yield. An application of this building block was as a starting material for the synthesis of carbon-14-labeled antibacterial tricyclic lactams ('tribactams')[124]. Reaction with the anion of the $\beta$-lactam **219** gave the hemiaminal **220**. Its conversion to a phosphonium salt opened the way for the required ring closure via an intramolecular Wittig–Horner olefination to produce the tribactam ring system. Hydrogenolysis of the benzyl group in the presence of $NaHCO_3$ gave sodium salt **221** in 28% overall radiochemical yield from benzyl [2-$^{14}$C]glyoxylate[124].

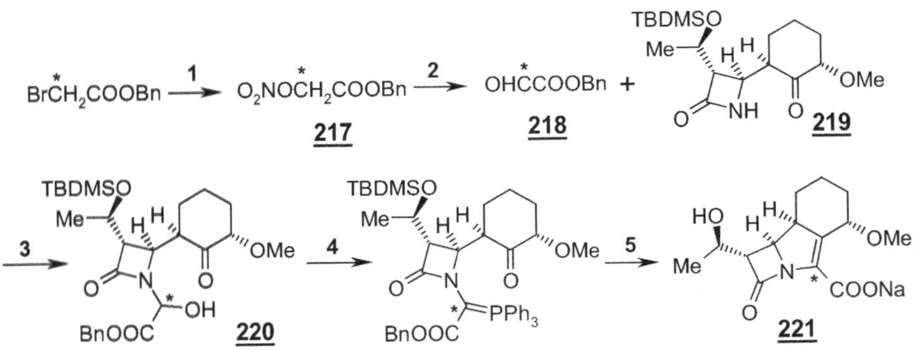


**Reaction conditions: 1.** $AgNO_3$, acetonitrile (dark); r.t., 72 h (95 %); **2.** $Bu_4NF$, acetonitrile; r.t., 1 h (60 %); **3.** $Et_3N$, 3A mol. sieves, THF; $\Delta$; **4a.** $SOCl_2$, 2.6-lutidine, THF, **b.** $PPh_3$, NaBr, 2,6-lutidine, ethyl acetate; **5a.** 1 N HCl, EtOAc/cyclohexane, **b.** aq. $K_2CO_3$, EtOAc, **c.** propanol; reflux, **d.** $H_2$, 10 % Pd/C, ethanol/water, $NaHCO_3$, **e.** freeze drying

***Figure 6.65*** *Synthesis of benzyl [2-$^{14}$C]glyoxylate and its use in the synthesis of a labeled tribactam*

## *Replacement of Halogen by Sulfur Nucleophiles*

Sulfur compounds are generally more nucleophilic than their oxygen analogs, so that the halogen replacement of chloro- or bromo[$^{14}C_n$]acetates by thiols takes place even more smoothly. Accordingly, for example, thiol **222** and ethyl bromo[2-$^{14}$C]acetate reacted even at −78 °C. [2-$^{14}$C]Benzothiophene (**223**) was formed rapidly in the consecutive step in the presence of sodium ethoxide at −78 °C. A two step conversion of the ester into the aldehyde species **224**, Wittig olefination with $Ph_3P{=}CH_2$ and concluding catalytic hydrogenation converted the primary product into the 3,4,5,6-tetrahydro-4-methyl[2-$^{14}$C]thieno[4,3,2-ef]-benzazepine derivative **225**[125].

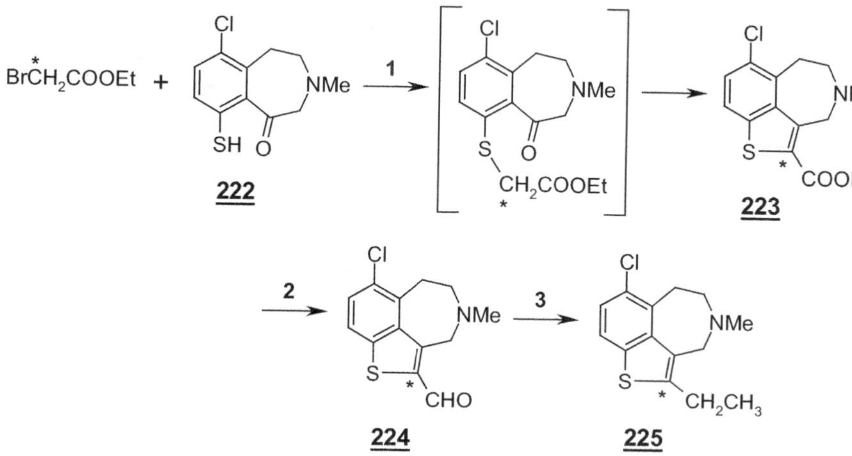


**Reaction conditions: 1**. EtONa, EtOH; −78 °C, 20 min; −78 °C to 0 °C, 90 min; **2a**. $LiAlH_4$, THF; 0 °C to r.t., 90 min, **b**. $MnO_2$, $CH_2Cl_2$, r.t., 1 h; **3a**. $Ph_3PCH_3Br$, BuLi;−10 °C, 90 min, **b**. $H_2$, $PtO_2$, EtOH; r.t., 3.5 h

***Figure 6.66*** *In situ generation of a [2-$^{14}$C]thioglycolate and its use in a Knoevenagel condensation mediated ring closure to labeled [2-$^{14}$C]benzthiophene*

*Alkyl (dimethylsulfanylidene)[$^{13}$C]acetates (S-ylides)* Ethyl bromoacetate reacts with dimethylsulfide to give dimethyl(ethoxycarbonylmethylene) sulfonium bromide. The corresponding ylide, generated upon treatment with an appropriate base (e.g. DBU) has been found to add to $\alpha,\beta$-unsaturated carbonyl compounds to furnish 2-acylcyclopropanecarboxylic esters. This chemistry has been exploited for the preparation of stable isotope-labeled compounds but should be applicable to carbon-14 (Figure 6.67). Treatment of the ylide **226**, accessible from ethyl bromo[1,2-$^{13}C_2$]acetate and dimethylsulfide, with 2-cyclopenten-1-one yielded ethyl 2-oxobicyclo[3.1.0][6-$^{13}$C]hexane-6-[$^{13}$C]carboxylate (**227**). Bucherer–Berg reaction with K$^{13}$CN and $^{15}NH_4Cl$ converted **227** into the corresponding [$^{13}C_3$, $^{15}$N]hydantoin **228**, which upon hydrolysis furnished racemic 2-[$^{15}$N]-amino[6-$^{13}$C]bicyclo[3.1.0]hexane-2.6-di[$^{13}C_2$]carboxylate (**229**)[126]. This reactivity is similar to that of diazoacetates (previous section), except sulfur ylides require activated double bonds, whereas diazoacetates can also add to isolated double bonds.

*Alkyl 2-(phenylsulfonyl)[$^{13}$C]acrylates* Treatment of alkyl bromo[$^{13}$C]acetates with sodium benzenesulfinate provides alkyl 2-(phenylsulfonyl][$^{13}$C]acetates **230** (Figure 6.68). Sequential methylation, selenylation and oxidative elimination of PhSeOH coverts **230** into alkyl 2-(phenylsulfonyl)[$^{13}$C]acrylates **231**, powerful and versatile Michael

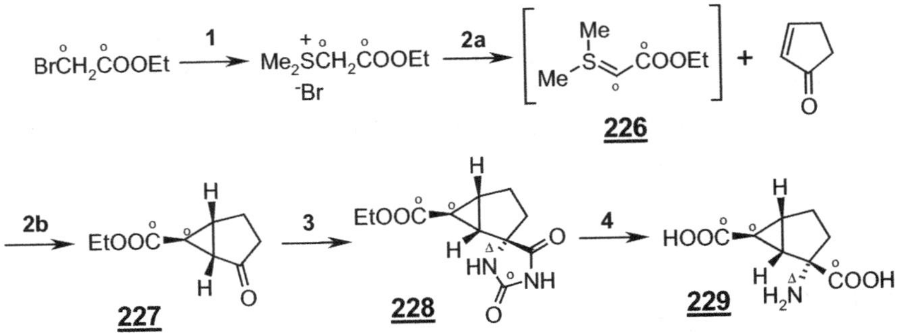

**Reaction conditions:** **1**. $Me_2S$, acetone; r.t., 44 h; **2a**. DBU, acetonitrile, 15 min, **b**. r.t., 44 h; **3**. $K^{13}CN$, $^{15}NH_4Cl$, $Na_2CO_3$, EtOH; 40 °C, 15 h; **4**. 48 % HBr, reflux, 24 h

***Figure 6.67*** *Preparation and use of ethyl (dimethylsulfanylidene)[$^{13}C_2$]acetate* **226** *in the synthesis of a 2-acyl[$^{13}C_2$]cyclopropanencarboxylic acid derivative*

acceptors, especially since the phenylsulfonyl group can be removed reductively with $SmI_2$. This sequence applied with the ethyl ester to the addition of deprotonated Dellaria's chiral oxazinone **232** and the concluding routine removal of the auxiliary provided (*S*)-[4-$^{13}$C]-glutamic acid[127]. Though this reaction has been exploited so far only for carbon-13 synthesis, any concern about stability of the phenylsulfonylacrylate in case of carbon-14 should be unfounded, considering the mild reaction conditions of its generation and the possibility to trap it *in situ*.

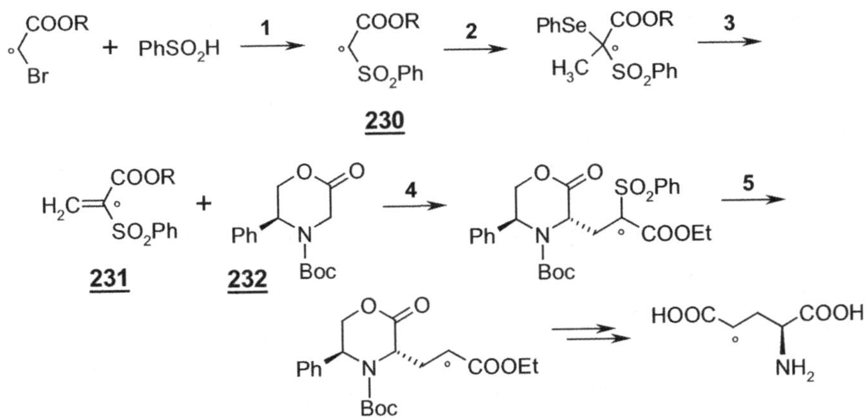

**Reaction conditions:** **1**. NaH, DMF; r.t., 3 d; **2a**. NaH, THF, then MeI, **b**. NaH, THF, HMPA then PhSeBr; **3a**. $H_2O_2$, $CH_2Cl_2$; 0 °C, 15 min, **b**. $NaHCO_3$, 0 °C; **4**. **232**, NaHMDS, THF; –78 °C, 1 h, **b**. **231**, –78 °C, 1 h; **5**. $SmI_2$,THF/MeOH; 0 °C, 1.5 h

***Figure 6.68*** *Synthesis and use of ethyl 2-(phenylsulfonyl)[2-$^{13}$C]acrylate (**231**) a powerful Michael acceptor*

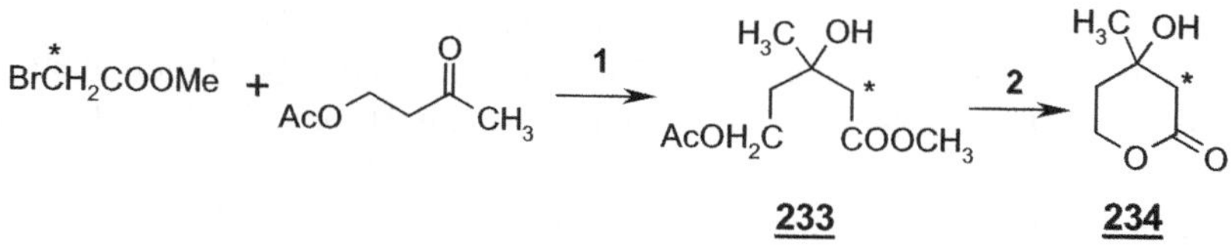


**Reaction conditions: 1**. activated zinc filings, ether; reflux 4 h; **2a**. 1N KOH, MeOH; r.t., 40 h, **b**. distillation (10 °C, 0.1 mm) (53 %)

***Figure 6.69*** *Reformatsky approach to the synthesis of racemic [2-$^{14}$C]mevalonolactone from methyl bromo[2-$^{14}$C]acetate*

### *Alkoxycarbonyl[$^{14}$C]methylenezinc Bromide (Reformatsky Reagents)*

Treatment of alkyl bromo[$^{14}$C]acetate with activated zinc in ether or benzene provides organozinc compounds, which are less reactive than the analogous magnesium compounds in that they react chemoselectively with aldehydes and ketones to give $\beta$-hydroxy[$^{14}$C]-

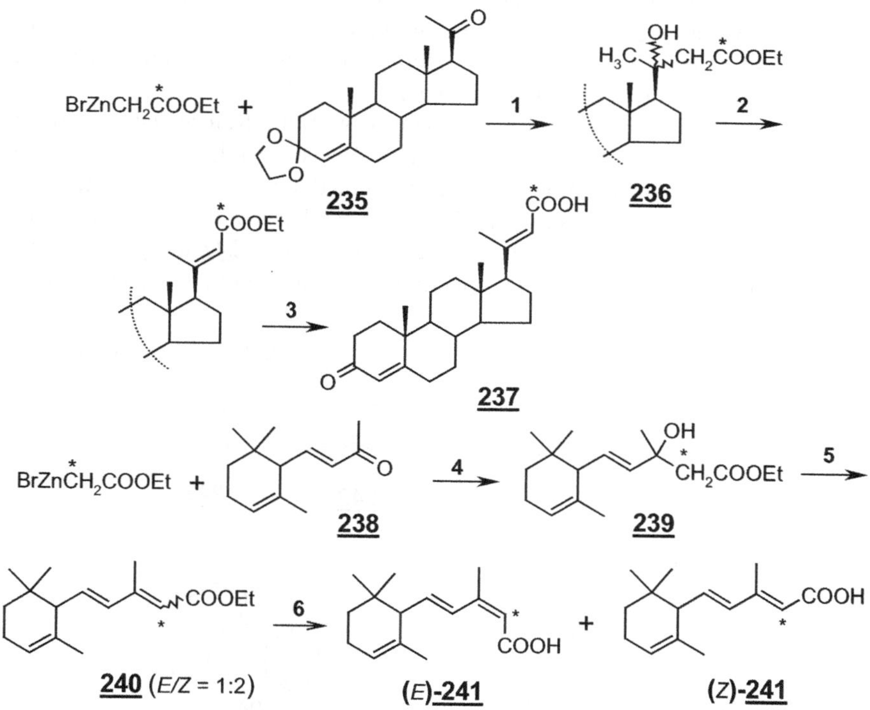


**Reaction conditions: 1**. $I_2$(cat.), benzene; reflux, 30 min; **2**. $POCl_3$, pyridine; r.t., 22 h; **3a**. *p*-TsOH, acetone; r.t., 48 h, **b**. 50 % aq. NaOH/MeOH; reflux, 3 h; **4**. benzene; reflux, 2 h; **5**. *p*-TsOH, benzene; refux, 1 h, **6a**. Prep TLC separation, **b**. KOH, EtOH; reflux, 1 h

***Figure 6.70*** *Applications of ethyl ethoxy[$^{14}$C]carbonylmethylenezinc bromide (Reformatsky reagent) for the preparation of $\alpha,\beta$-unsaturated carboxylic acids*

esters, but they do not react with esters (Reformatsky reaction). This selectivity has been exploited in the synthesis of racemic [2-$^{14}$C]mevalonolactone (**234**) (Figure 6.69). In this example BrZn$^{14}CH_2COOMe$, generated *in situ* from methyl bromo[2-$^{14}$C]acetate and zinc, reacted with 3-oxobutyl acetate to give racemic 3-hydroxy-3-methyl-5-acetoxy-[2-$^{14}$C]pentanoate (**233**). Subsequent saponification of both ester functions, acidification and lactonization upon distillation converted **233** into **234** in 52% overall yield. The compound has been used for investigations into the biosynthesis of cholesterol[128].

Dehydration of the $\beta$-hydroxy ester Reformatsky products to the corresponding $\alpha,\beta$-unsaturated derivatives can be effected by heating with $POCl_3$-pyridine, *p*-TsOH or molten $KHSO_4$ under acidic conditions. By this means the Reformatsky reaction has been used as an alternative to the Wittig–Horner method, especially in the synthesis of carbon-14-labeled steroids and terpenoids.

Reaction of progesterone-3-ketal **235**, for example, with ethoxy[$^{14}$C]carbonylmethylzinc bromide afforded the $\beta$-hydroxy ester **236** (Figure 6.70). Dehydration followed by acid catalyzed cleavage of the ketal function furnished 3-keto-23-nor[24-$^{14}$C]chola-4,20(22)-(*E*)-dienic acid (**237**) in 14% overall yield[129]. Under similar conditions $\alpha$-ionone (**238**) and ethoxycarbonyl[$^{14}$C]methylzinc bromide afforded the corresponding $\beta$-hydroxy ester **239**, which upon dehydration gave a 1:2-mixture of ethyl 2,4-(*Z,E*)- and (*E,E*)-[2-$^{14}$C]ionylideneacetetes (**240**). Concluding chromatographic separation and saponification gave the (*E*)- and (*Z*)-free acids in 17% overall radiochemical yield[130].

The results of the olefination reactions obtained with the Reformatsky approach are similar to those of the Wittig–Horner and the Horner–Wadsworth–Emmons methodologies. But in contrast to these, the position of the newly formed double bond in the Reformatsky approach is not always certain, as exemplified in the synthesis of the kappa opiate agonist [$^{14}$C]PD-17,302 (**246**), where a mixture of double bond isomers **242** and **243** was obtained. In this case, fortunately, the lack of selectivity was unimportant because both were

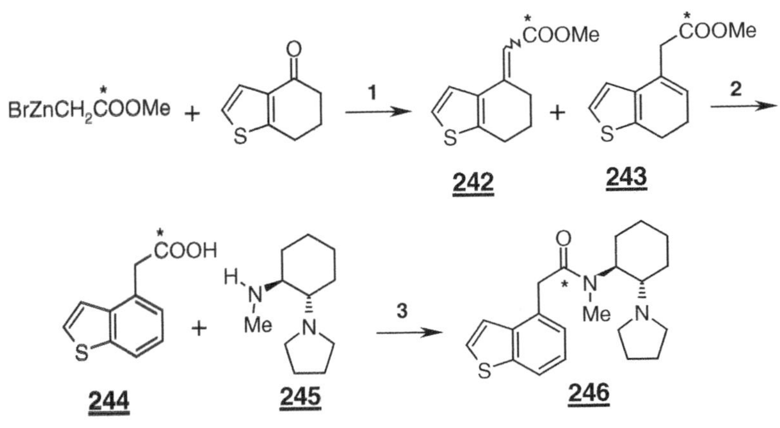


**Reaction conditions: 1a**. reflux, 5 min, **b**. *P*-TSOH, benzene; **2a**. 5 % Pd/C, S, *N*-methylpyrrolidone; reflux, 2.5 h; **b**. KOH, MeOH; r.t.; 6 h, **c**. $SOCl_2$, reflux; **3a**. $SOCl_2$, reflux, **b. 245**, $CH_2Cl_2/Et_2O$ (20 % overall)

***Figure 6.71*** *Reformatsky reaction and dehydration sequence in the synthesis of [$^{14}$C] PD-17,302*

converted equally well to the desired benzothiophene system **244** under the aromatization conditions utilized. Subsequent routine transformations and amide formation with the amine **245** gave the target drug substance **246** in an overall radiochemical yield of 20%[48].

## 6.3 [$^{14}$C]Acetone

Like [$^{14}$C]acetylene (see Chapter 8), [$^{14}$C]acetone has been an important precursor for the synthesis of multiply substituted [ring-$^{14}$C]arenes, as well as for the carbon-14 labeling of the side chains of a broad variety of $\beta$-blocking agents. Differently labeled acetone isotopomers are prepared either by pyrolysis of lithium or barium [1-$^{14}$C]- or [2-$^{14}$C]acetate at 450–460 °C (Figure 6.72, Procedure A)[131,132], by reaction of [$^{14}$C]acetyl chloride with dimethylcadmium (Procedure B)[16], reaction of [$^{14}$C]methylmagnesium iodide with acetaldehyde followed by oxidation of the resulting 2-[1-$^{14}$C]propanol (Procedure C)[133] or by [$^{14}$C]carboxylation of methyllithium and subsequent hydrolysis of the resulting dilithiated ketal (Procedure D)[134].

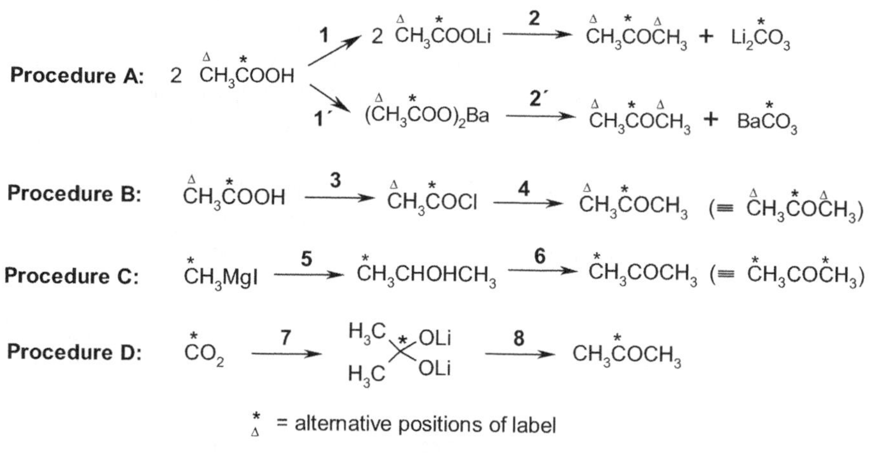


**Reaction conditions:** **1a** .LiOH, $H_2O$,**b**. 110 °C, 2 h; **2**. 490 °C, 20 min (80 %); **1a´**. $Ba(OH)_2$, $H_2O$, **b**. 120 °C, 2 h; **2´**. 450–460 °C, 1.5–2 h (95–99 %/47–49 %); **3**. o-$C_6H_4(COCl)_2$; 140–200 °C, 30 min; **4a**. $Me_2Cd$, di(ethylene glycol) diethyl ether; 50 °C, 15 min; 70 °C, 30 min, **b**. fractional distillation (65 %); **5**. $CH_3CHO$, ether reflux, 1 h; **6**. chromic acid, $HClO_4$, $H_2O$; 40 °C, 48 h; ractional distillation (60–75 %); **7a**. MeLi, THF; r.t., **b**. $Ph_2NH$, THF, **8a**. 1N $H_2SO_4$, **b**. fractional distillation (40–50 %)

*Figure 6.72 Standard procedures for the preparation of [$^{14}$C]acetone isotopomers*

Since Procedures B, C and D require fractional distillation to purify the product, as well as additional chemical manipulations, Procedure A has become the method of choice, even though half of the radioactivity originally contained in C1 of the acetate is diverted back to lithium or barium [$^{14}$C]carbonate. The use of barium [$^{14}$C]acetate is favored, since the yields are generally by 10–15% higher. Procedure B, although it produces [$^{14}$C]acetone labeled (for most practical purposes owing to its symmetrical structure) in all three positions, is especially suited for the preparation of unsymmetrical methyl [$^{14}$C]ketones. For the abovementioned reasons it is less frequently used, though the overall radiochemical yields (50–60%) are comparable to Procedure A (45–49% with barium [$^{14}$C]carbonate). Procedure C is mostly of historical interest; the overall radiochemical yields are 60–70% (based on [$^{14}$C]methanol). Procedure D has been exploited so far only for the preparation of carbon-11-labeled acetone. However, because of the shortness of the synthesis, it might be of interest also for the carbon-14 isotopomer. Trapping of excess or unreacted MeLi by addition of diphenylamine in the carbon-11 case was essential, in order to suppress formation of *tert*-[1-$^{11}$C]butanol through reaction with [1-$^{11}$C]acetone released during the hydrolysis of the intermediately formed di-lithiated ketal. Radiochemical yields up to 50% have been reported[134c].

### 6.3.1 Reaction at the Carbonyl Group

The spectrum of chemical transformations of the carbonyl functions includes the reaction of [$^{14}$C]acetone with organometallic reagents (e.g. RLi, RMgX, $BrZnCH_2COOR'$, $RTi(OR')_3$, Wittig or Horner–Wadsworth–Emmons reagents (olefinations), sulfonium reagents (epoxidations), and C–H acidic compounds (Knoevenagel reactions) as well as reductions with complex metal hydrides, which in the latter case give [$^{14}$C]isopropanol and follow-up building blocks such as [$^{14}$C]isopropyl iodide and [$^{14}$C]isopropylamine.

The reaction with organolithium and Grignard compounds have found widest application in the preparation of tertiary [1-$^{14}$C]alcohols[135,136] and tertiary [1-$^{14}$C]amines[135,136a]. With MeLi or *p*-chlorobenzylmagnesium chloride, for example, the respective *tert*-[1-$^{14}$C]-alcohols <u>247</u> and <u>248</u> were obtained, which upon treatment with benzyl or sodium cyanide gave the amides <u>249</u> and <u>250</u> in an acid-catalyzed Ritter reaction. Subsequent hydrolytic

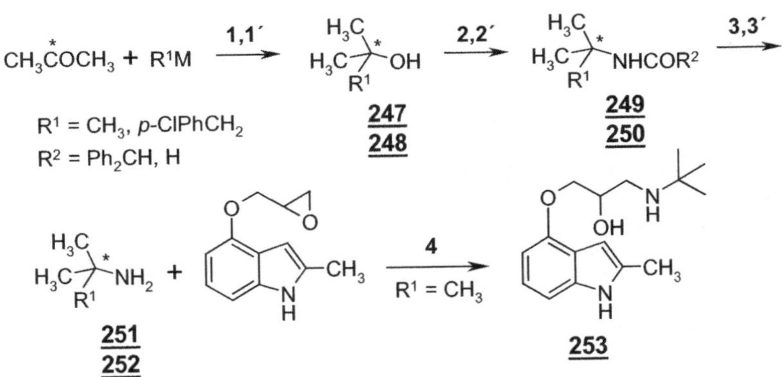


**Reaction conditions:** **1a**. MeLi, ether; 20 °C, 4 h, **b**. phenoxyethanol, He, T-solvent; 100 °C, 2.5 h (70–90 %); **2**. $Ph_2CHCN$, $H_2SO_4$/AcOH; 20 °C, 12 h; **3**. NaOH, triethylene glycol; 190 °C, 4 h (45 % from [$^{14}$C]acetone); **4**. HMPA, 60 °C, 20 h; **1´**. *p*-$ClPhCH_2MgCl$, ether; 40 °C, 1 h (80 %); **2´**. NaCN, $H_2SO_4$/AcOH; 70 °C, 2 h; **3´**. aq. HCl/EtOH; reflux, 4 h; r.t., 16 h (30 % from [$^{14}$C]acetone)

**Figure 6.73** *Use of [$^{14}$C]acetone for the preparation of tert-[$^{14}$C]alcohols and amines*

cleavage released the respective *tert*-[1-$^{14}$C]amines **251** and **252** in overall radiochemical yields of 30–45%. Amine **252** has been used as a key intermediated for the synthesis of [$^{14}$C] bopindolol (**253**), an antihypertensive $\beta$-blocking agent[135].

The *tert*-[1-$^{14}$C]alcohols **247** and **248** have also served as key intermediates in the synthesis of carbon-14-labeled terbinafine (**255**)[137], an orally active antimycotic drug, and of alaproclate[136b], an antidepressant drug candidate. In the first case **247** was prepared by reaction of [$^{14}$C]acetone and the reagent tri(nonyloxy)methyltitanium. This suppressed competing enolate formation that often occurs with organolithium or magnesium reagents and increased the yield of **247** to 90%[137]. Subsequent treatment with HCl and coupling of the resulting *tert*-[1-$^{14}$C]butyl chloride to the acetylenic precursor **254** provided **255** in an overall radiochemical yield of 30%. The latter reaction appears to provide convenient, generally applicable access to *tert*-[$^{14}$C]alkyl-substituted acetylenes bearing further functionalities. The synthesis of [$^{14}$C]alaproclate (**256**) was accomplished in classical fashion by *O*-acylation of **248** with 2-bromopropionyl bromide followed by amino-de-halogenation of the resulting $\alpha$-bromo ester with $NH_3$ in ethanol.

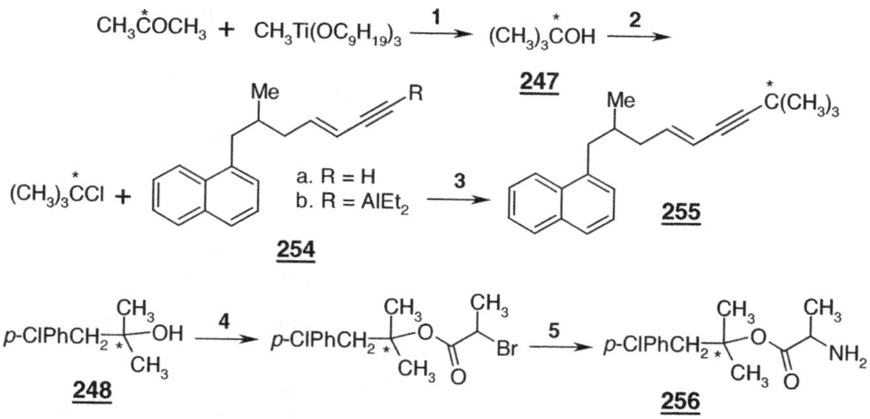


**Reaction conditions: 1a**. r.t., 16 h, **b**. oleic acid; 110 °C, 4.5 h (90 %); **2**. conc. HCl; r.t., 1 h; **3a**. **8a**, *n*-BuLi, hexane; –20 °C, 5 min, b. $Et_2AlCl$, $Me_3{}^{14}CCl$, dichloroethane, –196 °C, c. $Et_2AlCl$; –20 °C, 30 min; r.t., 30 min ; **4a**. $CH_3CHBrCOBr$, N, N-dimethylaniline; r.t., 2 h; 40 °C, 6.5 h, **b**. liq. $NH_3$, EtOH; r.t., 14 h; 30 °C, 4 h

***Figure 6.74*** *Applications of tertiary [1-$^{14}$C]alcohols*

Reaction of [2-$^{14}$C]acetone with vinylmagnesium bromide (Figure 6.75, Procedure A) provided 2-methyl-3-[2-$^{14}$C]buten-2-ol (**257**), which upon treatment with $PBr_3$ rearranged to give 3,3-dimethyl[3-$^{14}$C]allyl bromide (**258**). The latter was used for the alkylation of ethyl acetoacetate, thereby extending the carbon skeleton of the $\beta$-keto ester by a labeled five-carbon (isoprenyl) unit[138]. The initially formed alkylated intermediate **259** was not isolated but immediately saponified and decarboxylated to give ketone **260**. Subsequent Horner–Wadsworth–Emmons olefination and reduction of the separated *trans*-ester **261** converted **260** into [7-$^{14}$C]geraniol (**262**).

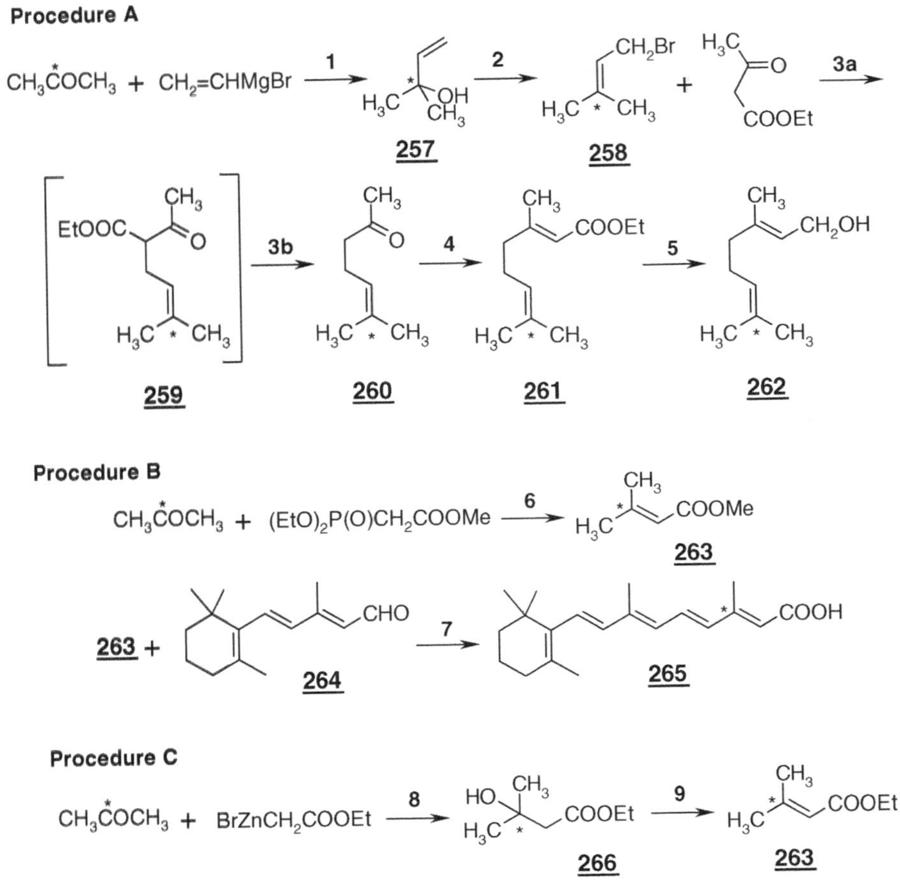


**Reaction conditions:** **1**. THF; r.t., 6 h; **2**. $PBr_3$, pyridine, THF; –10 °C, 2 h; **3a**. EtONa, EtOH; reflux, 12 h, **b**. aqueous KOH; reflux, 2.5 h; **4a**. $(EtO)_2P(O)CH_2COOEt$, NaH, DME; 55–60 °C, 24 h, **b**. preparative TLC separation; **5a**. $LiAlH_4$, ether; –78 °C, 1 h, **b**. r.t., 2 h; **6a**. $(EtO)_2P(O)CH_2COOEt$, NaH, ether; 5 °C, 1 h, **b**. [$^{14}$C]acetone; r.t.,18 h; **7**. $KNH_2$, ether; –65 °C, 1 h, r.t., 17 h; **8**. $BrCH_2COOEt$, Zn, benzene/toluene, reflux, 1 h; **9**. $KHSO_4$, 160 °C, 3 h

***Figure 6.75*** *Use of [$^{14}$C]acetone for the extension of carbon chains by labeled five-carbon units*

The synthesis of *trans*-[13-$^{14}$C]retinoic acid (**265**) (Procedure B) may serve as an additional example of the use of an [$^{14}$C]acetone-based building block. Horner–Wadsworth–Emmons olefination of [2-$^{14}$C]acetone with methyl diethyl-phosphonoacetate afforded methyl 3-methyl-2-[3-$^{14}$C]butenoate (**263**). Upon deprotonation, this coupled with $\beta$-ionylideneacetaldehyde (**264**) to provide **265**, thereby extending the carbon skeleton of **264** again by a labeled five-carbon unit[110]. Building block **263** was also shown to be accessible (Procedure C) through Reformatsky synthesis of [2-$^{14}$C]acetone with zinc and

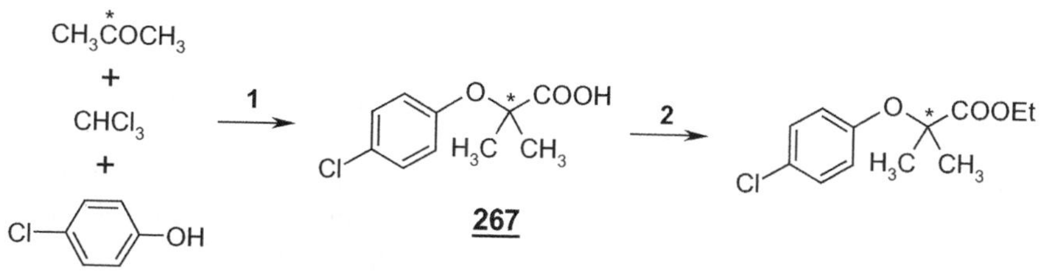

**Reaction conditions:** **1**. KOH, DMF; 0 °C, 3 h; r.t., 18 h; **2**. EtOH, conc. $H_2SO_4$; reflux, 18 h

*Figure 6.76* *Synthesis of [2-$^{14}$C]clofibrate from [2-$^{14}$C]acetone*

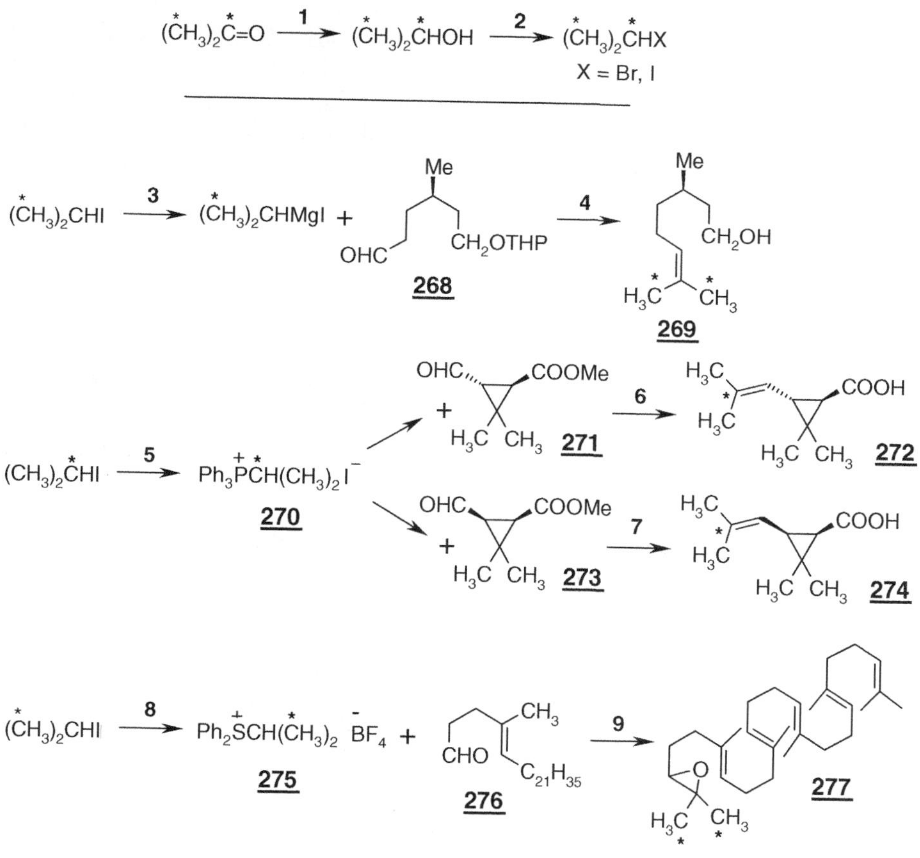

**Reaction conditions:** **1a**. $LiAlH_4$, di(ethylene glycol) diethyl ether; r.t., 16 h, **b**. 2-phenoxyethanol; r.t. to 130 °C; **2**. $PBr_3$; reflux, 3 h **or** $I_2$, $P_{red}$, reflux, 2 h; **3**. ether; r.t.; **4**. HOAc; reflux; **5**. $Ph_3P$, toluene; 105–115 °C, 18 h; **6a**. **270**, *n*-BuLi, THF; r.t., 15 h, **b**. **271**, –20 °C, 18 h, **c**. NaOH, aq. EtOH; reflux, 2 h; **7a**. NaH, DMSO; 80–90°C, 30 min, **b**. **270**; r.t., 30 min, **c**. **273**, r.t., 23 h, **d**. NaOH, aq. EtOH; reflux, 2 h; ; **8**. $Ph_2S$, $AgBF_4$, $CH_2Cl_2$; r.t., 2 h; **9a**. PhLi, THF; -78 °C, 15 min; **b**. **276**, THF; –78°C, 15 min; r.t., 1–2 h

*Figure 6.77* *Synthesis and use of 2-halo[$^{14}$C]propanes*

ethyl bromoacetate followed by dehydration of the intermediately formed β-hydroxy ester[139].

Under basic conditions chloroform reacts with [$^{14}$C]acetone to give 1,1-dichloro-2,2-dimethyl[2-$^{14}$C]oxirane (Figure 6.76). This product could be trapped *in situ* with *p*-chlorophenol and the resulting adduct hydrolyzed to give 2-(*p*-chlorophenoxy)-2-methyl[2-$^{14}$C]propionic acid ([2-$^{14}$C]clofibric acid, **267**), whose ethyl ester is used as an antihyperlipidemic drug[140].

$LiAlH_4$ reduction of [$^{14}$C]acetone followed by treatment of the resulting [$^{14}$C]isopropanol with phosphorus tribromide or red phosphorus–iodine converts it into [$^{14}$C]isopropyl bromide or the corresponding iodide[141]. These have served as key building blocks for the preparation of, for example, [1, 3-$^{14}$C]isopropylmagnesium iodide[142], [2-$^{14}$C]isopropyltriphenylphosphonium iodide (**270**)[143] and [1, 3-$^{14}$C]isopropylsulfonium tetrafluroborate (**275**)[144], which reacted with the carbonyl precursors **268**, **271**, **273** and **276** to give β-[8-$^{14}$C]citronellol (**269**)[142], (1*R*)-*cis*- and (1*R*)-*trans*-[propenyl-1-$^{14}$C]chrysanthemic acids (**272** and **274**)[143a,c] and 2,3-epoxy[1-$^{14}$C]squalene (**277**)[144] as depicted in Figure 6.77.

Nitrogen nucleophiles such as hydroxylamine and isopropylamine react with [$^{14}$C]-acetone to give the corresponding oximes or imines, which can be readily reduced with

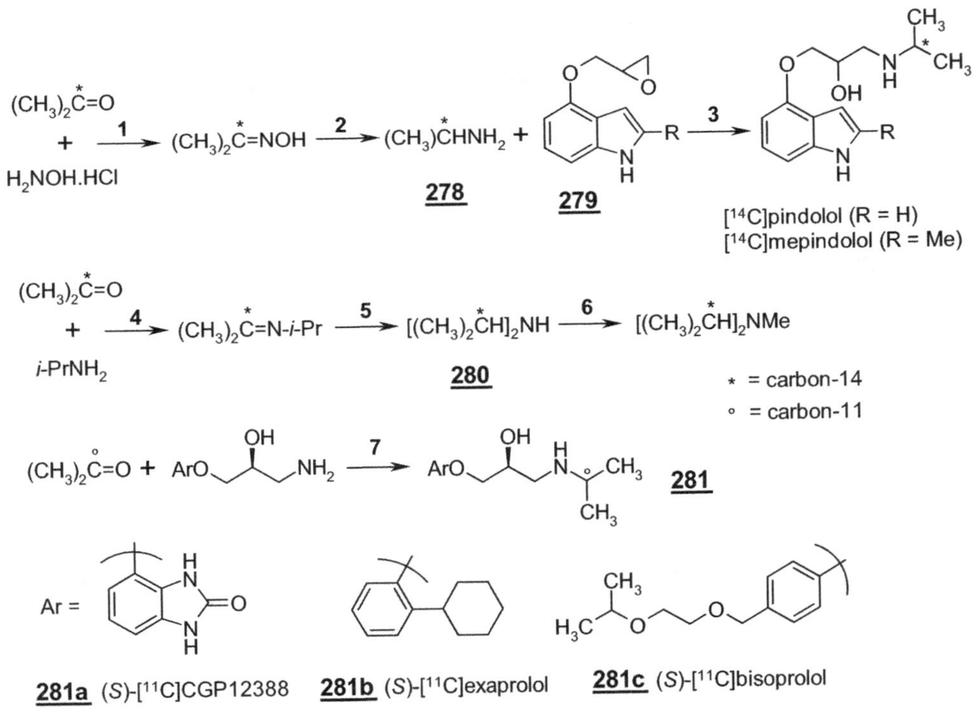


**Reaction conditions:** **1**. NaOH, $H_2O$; r.t., 24 h or $BaCO_3$, ether; reflux, 20 h; **2**. $LiAlH_4$, ether; 0 °C, 30 min; reflux, 3 h; **3**. toluene; 80 °C; 20 h; **4**. conc. HCl, $MgSO_4$; r.t.; 16 h; **5**. $LiAlH_4$, ether; reflux, 5 h; **6**. aq. HCHO, HCOOH; 80 °C, 45 min; reflux, 4 h; **7**. $NaBH_3CN$, $CH_3COOH$, methanol; 120 °C

**Figure 6.78** *Synthesis and applications of [$^{11/14}$C]isopropylamines from [$^{11/14}$C]acetone*

complex metal hydrides to furnish the corresponding [$^{14}$C]amines, e.g. [2-$^{14}$C]isopropylamine (**<u>278</u>**)[145] and di[2-$^{14}$C]isopropylamine (**<u>280</u>**)[145a]. Amine **<u>278</u>** has frequently been used for carbon-14 labeling of the side chain of $\beta$-blocking agents through reaction with the appropriate epoxy derivative **<u>279</u>**[135]. A more elegant and shorter route avoids the intermediate preparation of [2-$^{14}$C]isopropylamine and instead uses the reductive amination pathway, which allows the direct coupling of [2-$^{14}$C]acetone to the selective 2-hydroxypropylamine precursor in the presence of sodium cyanoborohydride[135]. This alternative has been exploited[146] for the carbon-11 labeling of numerous $\beta$-blockers as ligands for PET investigations. Since these unlabeled precursors are readily accessible with high enantiomeric excesses from (2*S*)- or (2*R*)-glycidyl tosylate, the compounds **<u>281a–c</u>** were obtained in excellent optical purity.

Treatment of [1, 3-$^{14}$C]acetone with semicarbazide affords the respective semicarbazone, which was decomposed following the Huang–Minlon protocol to give [1, 3-$^{14}$C] propane[147].

### 6.3.2 Reaction at the Methyl Group

The reaction of [$^{14}$C]acetone with nitromalondialdehyde to provide *p*-nitrophenol labeled with carbon-14 at C1 and/or C2 and C6 is the most important representative of aldol-type reactions on [$^{14}$C]acetone (Figure 6.79)[148]. It is also the second widely applied route to substituted singly or multiply labeled benzenes besides the [$^{14}$C]acetylene approach (see Chapter 8). Its significance is increased by the fact that replacement of [$^{14}$C]acetone by labeled methyl alkyl ketones or homologous dialkyl ketones gives rise to 1,2,4-tri- or 1,2,4,6-tetrasubstituted derivatives[148a]. Reaction of alkyl [$^{14}$C]acetoacetates with nitromalondialdehyde formed 5-nitro[ring-$^{14}$C]salicylic acid in nearly quantitative yield[148a,149].

Figure 6.79 shows a summary of those group interconversions published so far for *p*-nitrophenol labeled with isotopic carbon. Obviously they represent only a small spectrum of potential alternatives. For example, it has recently been pointed out[150] that Pd-catalyzed cross-coupling transformations on aryl triflates (Route 5) would significantly increase the spectrum of valuable building blocks.

Protection of the phenolic hydroxy function through etherification[131a,d,151] followed by reduction of the nitro group[148d,149b,151,152], diazotization and transformation of the diazonium group has been widely exploited for the preparation of *p*-halo[$^{14}$C]anisoles[153], *p*-methoxy substituted [$^{14}$C]benzaldehyde[153], benzonitrile[148c,d,152], and benzoic acid[148c,152] following Route 2. Route 3 demonstrates that reduction of the nitro group[148f,154,155] followed by replacement of the amino group with cyanide can also be accomplished without protection of the phenolic hydroxyl group[154]. The tetraphenoxyphosphonium bromide-mediated replacement of the phenolic hydroxyl group with bromine and the subsequent reduction of the nitro function[148d] opens an access to 2,4-disubstituted [ring-$^{14}$C]anilines (Route 4). The [$^{13}$C]cyano-dehydroxylation is the only application of *p*-nitrophenyl triflate in isotope chemistry known so far despite its synthesis potential[156]. 2,4-Dinitro[1-$^{14}$C]phenol, obtained through treatment of 4-nitro[1-$^{14}$C]phenol with nitrating acid (Route 6), has been used as an analytical reagent for the determination of free amino functions in peptides and proteins[157]. It is also a highly interesting candidate for multiply functionalized derivatives. *p*-Nitro[$^{14}$C]aniline, prepared through replacement of the methoxy group upon treatment of 4-nitro[$^{14}$C]anisole with conc. ammonia at elevated temperature pressure[148a,b], is a highly interesting key intermediate for the preparation of *p*-substituted nitro[$^{14}$C]benzenes (Route 7).

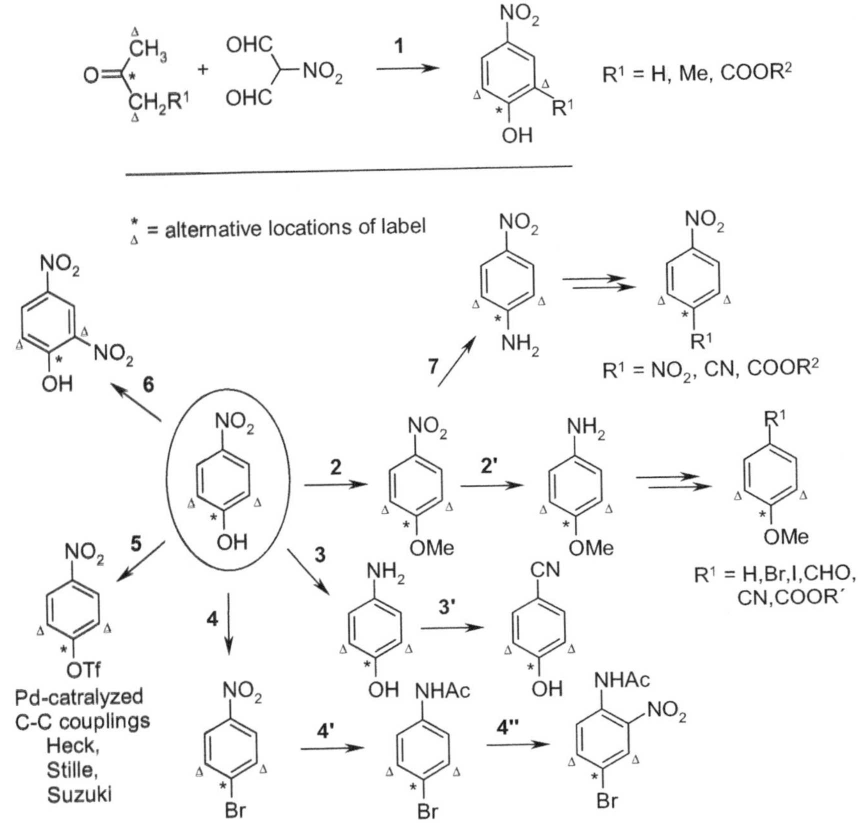


**Reaction conditions:** **1**. aq. NaOH; 0 °C, 2 h; 20 °C, 10 h (50–70 %); **2**. MeI, KOH, DMSO; r.t., 2 h (99 %) **or** $CH_2N_2$, ether; r.t., 20 h (90 %) **or** $Me_2SO_4$, $K_2CO_3$, DMF; r.t., 1 h (93 %); **2'**. $H_2NNH_2.H_2O$, Pd/C, EtOH; reflux, 2 h (94 %); **3**. $H_2$, Pd/C, MeOH; r.t., 6 h (97 %) **or** Sn, HCl; r.t., 30 min; 100 °C (75–80 %); **3'a**. $NaNO_2$, aq. HCl, 0–5 °C, 2 h, **b**. $K_2Ni(CN)_4$, $NH_4Cl$, aq. $NH_3$; 50–60 °C, 15 min; **4a**. $(PhO)_3PCl_2$, 100 °C, 3 h, **b**. EtBr; reflux, 2.5 h, **c**. 200 °C, 15 min (56 %) ; **4'a**. Fe, conc. HCl, toluene; reflux, 2 h (82 %), **b**. $Ac_2O$, py ; **4''**. $HNO_3$; **5**. $CF_3SO_2Cl$, pyridine ; 5 °C to r.t.; r.t, 30 min ; **6**. $H_2SO_4/HNO_3$ 1:1; HOAc; 35– 40 °C, 4 h; **7**. conc. $NH_3$, 170 °C, 20 h (80 %)

***Figure 6.79*** *Preparation and application of p-nitro[$^{14}$C]phenol to the synthesis of multi-substituted [$^{14}$C]benzenes*

Acylation of [2-$^{14}$C]acetone with acetic anhydride under Lewis acid conditions affords [2, 4-$^{14}$C]pentane-2,4-dione ([2, 4-$^{14}C_1$]acetylacetone, **282**; the carbon-14 of the acetone reactant is either at C2 or C4 but not both in the same molecule) as illustrated with the first example in Figure 6.80[158]. Under basic conditions acylation of [2-$^{14}$C]acetone

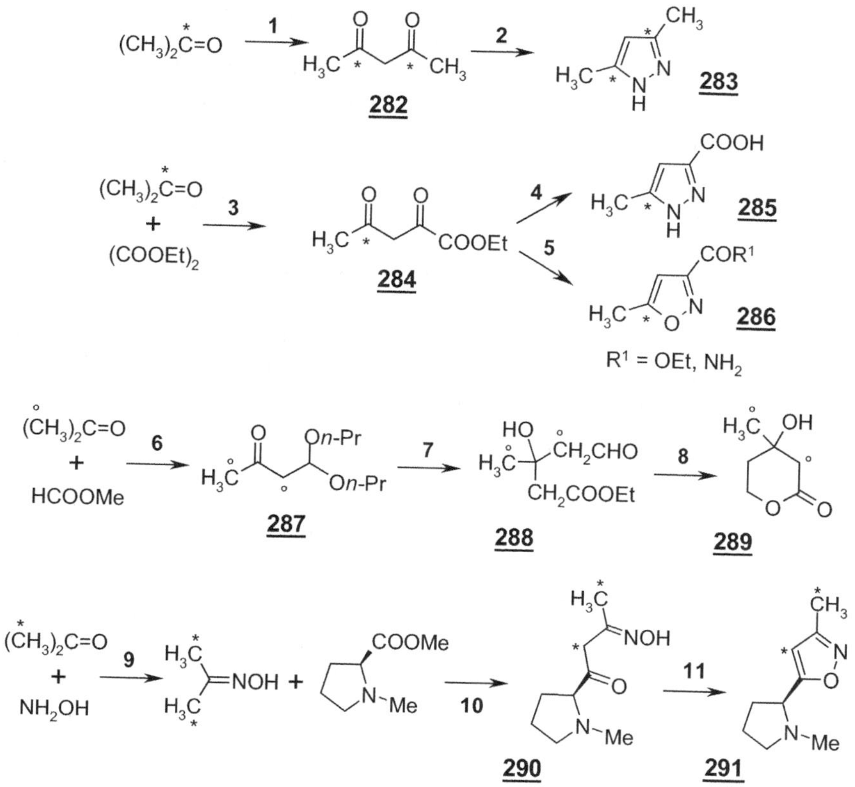


**Reaction conditions: 1.** $Ac_2O$, $BF_3$;–10 °C, 45 min (85 %); **2.** $H_2NNH_2.H_2SO_4$, 10% aq. NaOH; 0–5 °C, 1 h; **3.** EtONa, EtOH/benzene; r.t., 16 h (76 %); **4.** $H_2NNH_2.H_2SO_4$, AcONa, $H_2O$; 50 °C, 4 h; **5a.** $(H_2NOH)_2.H_2SO_4$, EtOH; reflux, 4 h, **b.** conc. aq. $NH_3$; r.t., 1.5 h; **6a.** MeONa, ether; reflux, 4 h, **b.** *n*-PrOH, TMSCl; r.t., 2 h (66 %); **7a.** EtOAc, LDA, THF; –78 °C, 30 min, **b.** **287**, THF; –78 °C, 30 min, **c.** 10-camphorsulfonic acid, acetone; r.t., 10 min; **8.** $NaBH_4$, MeOH; 0 °C, 20 min; r.t., 1 h; **9.** $H_2O$; r.t.; 20 h; **10a.** *n*-BuLi (2 equiv.), THF; 0 °C, 1.5 h, **b.** *N*-methyl-(*S*)-proline methyl ester, THF; 0 °C, 6.5 h; **11.** MsCl, $Et_3N$; 0 °C to r.t., 24 h

***Figure 6.80*** *Acylation reactions at acetone labeled with isotopic carbon*

with diethyl oxalate to furnish ethyl [4-$^{14}$C]acetopyruvate (2,4-dioxo[4-$^{14}$C]pentanoic acid ethyl ester, **284**)[158a,159], and [1, 3-$^{13}C_2$]acetone with methyl formate to give [1, 3-$^{13}C_2$]-acetoacetaldehyde (trapped *in situ* as the acetal **287** to increase its stability)[160]. Aldol-type reaction of **287** with deprotonated ethyl acetate, acid-catalyzed deacetalization, selective reduction of the liberated carbonyl function of **288** and cyclization in acidic methanol gave [4, 6-$^{13}$C]mevalonolactone (**289**), which was used for biosynthesis studies on trichothecene mycotoxins[160]. The $\beta$-dicarbonyl compounds **282** and **284** are valuable building blocks for the preparation of labeled 3,5-substituted pyrazoles and isoxazoles, as verified by the synthesis of **283**, **285** and **286**. The same holds for [$^{14}$C]ABT-418 (**291**), a cholinergic channel activator of the isoxazole type, which was obtained in two steps by of doubly deprotonated [1, 3-$^{14}$C]acetone oxime with *N*-methyl proline methyl ester followed by cyclization of the resulting ketooxime **290**[161].

Oxidation of [1, 3-$^{14}$C]acetone with selenium dioxide provided [1, 3-$^{14}$C]methylglyoxal (**292**) in 25–40% radiochemical yield[132b,162]. Aldol reaction with doubly deprotonated $\beta$-keto acid **293** and simultaneous decarboxylation converted **292** to the 3-hydroxy-2,5-[1,3-$^{14}$]diketone **294**. This cyclized upon treatment with base to give doubly labeled allethrolone (**295**), the alcohol component of allethrine, a potent synthetic insecticide[132b,162a].

Though half of the label gets lost as [2-$^{14}$C]acetate when [1, 3-$^{14}C_2$]acetone is treated with iodine, this reaction is still a standard preparation procedure for [$^{14}$C]iodoform. This compound is an indispensable precursor for [$^{14}$C]methylene iodide[163], which is obtained by reduction of $^{14}CHI_3$ with sodium arsenite in alkaline medium. Diiodo[$^{14}$C]methane in turn is used as a reagent of choice for the Zn–Cu-mediated cyclopropanation of double bonds (Simmons–Smith reaction).

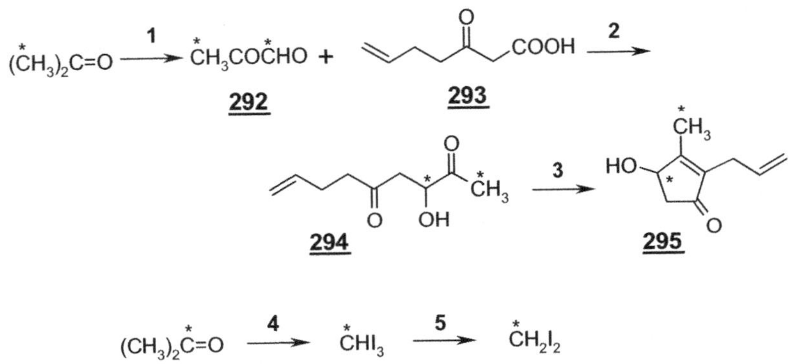


**Reaction conditions: 1**. $SeO_2$, $H_2O$; reflux, 4 h (40 %); **2**. $H_2O$ at pH 8; 35 °C, 5–6 h; **3**. aq. NaOH; r.t., 1.5 h; **4**. KI, NaOCl, $H_2O$; 0 °C, 16 h (r.c.y. 96 %); **5**. $As_2O_3$, NaOH, $H_2O$; 55–60 °C, 1.5 h (89 %)

***Figure 6.81*** *Oxidation reactions at [$^{14}$C]acetone*

## 6.4 Alkyl [$^{14}$C]Acetoacetate

There are few carbon-14-labeled building blocks for which so many preparative procedures are available than for alkyl [$^{14}$C]acetoacetates (see Figure 6.82). Most of them start from unlabeled/labeled malonic acid derivatives (mixed dialkyl malonates, monoalkyl malonates, Meldrum's acid), which are acylated with labeled/unlabeled acetyl chloride and subsequently converted through selective hydrolysis (or alcoholysis) and decarboxylation. Routes 1 and 2, which have been widely applied, use alkyl *tert*-butyl malonate[9b,164] or monoalkyl malonate[165] as coupling components, which upon treatment with [1-$^{14}$C]acetyl chloride and ester cleavage with *p*-toluenesulfonic acid (Route 1) or by simple acidic work-up (Route 2) decarboxylate to give ethyl [3-$^{14}$C]acetoacetate in radiochemical yields of 70–90%. A slightly modified alternative (Route 3) utilizes [1-$^{14}$C]acetyl imidazolide in place of [1-$^{14}$C]acetyl chloride. The imidazolide is generated either *in situ* from [1-$^{14}$C]acetic acid and 1,1′-carbonyldiimidazole, or from [1-$^{14}$C]acetyl chloride and imidazole, and is treated with magnesium ethyl malonate, readily accessible from commercially available potassium ethyl malonate and magnesium ethoxide[165b,166]. Routes 1–3 are most often used to prepare C3- and/or C4-labeled [$^{14}$C]acetoacetates because it is not so easy to prepare the mixed dialkyl and monoalkyl [$^{14}$C]malonates needed for preparation of C1- and/or C2-labeled [$^{14}$C]acetoacetates. Acylation of Meldrum's acid (Route 4) provides a facile route to [2-$^{14}$C]acetoacetates (starting from [2-$^{14}$C]malonic acid; see also Section 6.5 below) in addition to acetoacetates labeled at C3 and C4, and also easy access to a variety of esters through the choice of alcohol in the final alcoholysis/decarboxylation step[19,167]. The remaining routes have been used less frequently. Route 5 resembles Route 1, but has the advantage that it is well-suited to the preparation of [1-$^{14}$C]acetoacetates if benzyl [$^{14}$C]cyanoacetate is used. On the other hand, the Pd-catalyzed hydrogenolytic cleavage of the benzyl ester group may raise concerns about the risk of radioactive contamination, depending on the equipment used for this step[168]. Route 6 involves the acylation of deprotonated labeled or unlabeled alkyl acetates at low temperatures. In this case two equivalents of base (LiHMDS) have to be employed in order to suppress re-protonation of the ester enolate by the alkyl [$^{14}$C]acetoacetate formed. Yields obtained are in the range of 70–75%[169]. Routes 7 and 8 represent the classical base-mediated ($Ph_3CNa$, EtONa) Claisen condensation of two equivalents of labeled/unlabeled alkyl acetates, but these routes are of minor importance in carbon-14 synthesis because the radiochemical yields are usually only about 40–55%[51,168b,170]. However, they may be of some interest for the synthesis of carbon-13-labeled materials, since they can easily be used to prepare multiply labeled species with up to four carbon-13 atoms in one simple reaction step. Methods involving the carboxylation of deprotonated acetone using $^{14}CO_2$ or diethyl [$^{14}$C]carbonate[171b] have found some applications in carbon-11 chemistry[171a], but have not become routine procedures in carbon-14 synthesis so far.

Routes 1–6 can also be used for the preparation of a broad range of carbon-14-labeled $\beta$-keto esters and acids by replacing [$^{14}$C]acetyl chloride with other acyl chlorides. Alternatively, higher homologs can be prepared by alkylation of doubly deprotonated alkyl acetoacetates (see Section 6.4.1)

The methylene group of alkyl acetoacetates is even more acidic ($pK_a \sim 11$) than that of dialkyl malonates ($pK_a$ 13–14). Thus, it can be easily deprotonated, so that the resulting enolate can readily undergo alkylation, acylation, Knoevenagel and Michael reactions etc. Subsequent hydrolysis of the ester function of the resulting $\alpha$-substituted derivative under acidic or slightly basic conditions provides the free $\beta$-keto acids, which readily decarboxylate to give the corresponding methyl ketones (ketonic cleavage). Strong bases, however, attack the $\alpha$-substituted acetoacetates at the carbonyl function and cleave them into alkali acetate and the respective carboxylic ester. Accordingly, alkyl acetoacetates may react as $^{-}CH_2COCH_3$ (or $^{2-}CHCOCH_3$) or as $^{-}CH_2COOR$ (or $^{2-}CHCOOR$) synthons. These two follow-up transformations determine also the position of label. Methyl ketones as target compounds require introduction of the label into C4 and/or C3 of the acetoacetate. However, if the acetoacetate methodology is exploited for the synthesis of labeled carboxylic esters/acids, the label is limited to C1 and C2. In contrast to the reactivity of acetoacetate anions, acetoacetate dianions (i.e., $^{-}CH_2CO^{-}CHCOOR$) preferentially react with electrophiles from the C4 position, as examples in the following subsections will show.

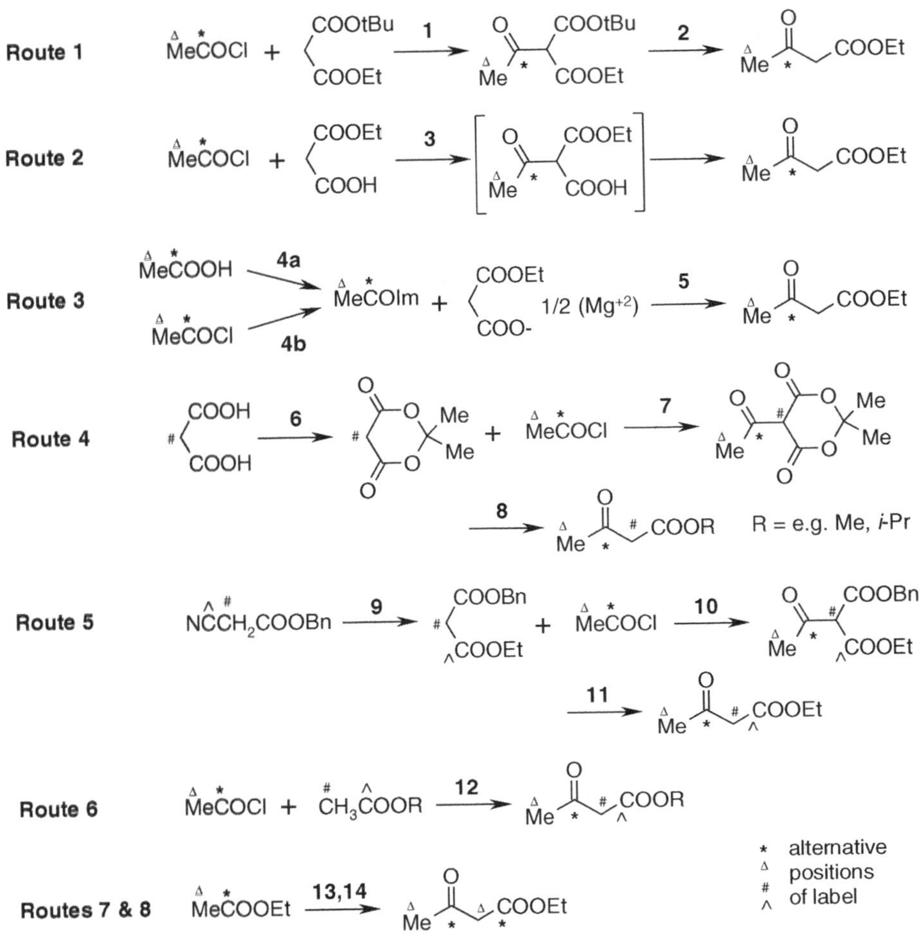

**Reaction conditions:** **1a**. malonate, $Mg(OEt)_2$, ether; reflux, 6 h; **b**. acid chloride, ether; reflux, 30 min; **2**. *p*-TsOH, benzene, reflux, 1.5 h (71–75 %); **3a**. malonate, *n*-BuLi (2 equiv.), THF; –70 °C, 1 h, **b**. acid chloride, THF; –60°C, 10 min, **c**. acidic work-up (88 %); **4a**. CDI, THF; 50 °C, 1 h, **b**. imidazole, THF; 40 °C, 6 h; r.t., 12 h; **5**. THF, 45°C, 2 h; r.t., 12 h; **6**. acetone, $Ac_2O$, $H_2SO_4$ (cat.); r.t., 2 h (65 %); **7**. pyridine, $CH_2Cl_2$; 2 °C, 40 min; r.t., 40 min; **8**. ROH; reflux, 60 min (55–85 %); **9**. EtOH, HCl(g); –5 °C, 36 h; **10**. $Mg(OEt)_2$, ether; reflux, 30 min; **11**. $H_2$, 10 % Pd/C; EtOAc; r.t., 35 min; **12**. LiHMDS, THF; –80 °C, 2 h (72 %); **13**. KH, cyclohexane; 50 to 80 °C, 1 h; reflux 3 h; **14**. $Ph_3CNa$, ether; r.t., 3 min (45%)

**Figure 6.82** *Syntheses of alkyl [$^{14}$C]acetoacetates*

## 6.4.1 Alkylation Reactions

The alkylation of ethyl [3-$^{14}$C]acetoacetates to give labeled methyl ketones has been extensively used for the preparation of labeled isoprenoids, such as [10, 15-$^{14}C_2$]squalene (**300**)[164a] (Figure 6.83) and [3′-$^{14}$C]coenzyme Q10 (**306**)[172] (Figure 6.84). Access to [carbonyl-$^{14}$C]geranylacetone (**298**) was achieved when deprotonated ethyl [3-$^{14}$C]-

acetoacetate was treated with geranyl chloride (**296**) followed by hydrolysis and decarboxylation of the ester function of the initially formed $\beta$-keto ester **297**. [10, 15-$^{14}C_2$]-Squalene was obtained when **298** reacted with 1,4-dibromobutane and magnesium (Barbier reaction), and the resulting 1,6-diol **299** was dehydrated via a two-step sequence.

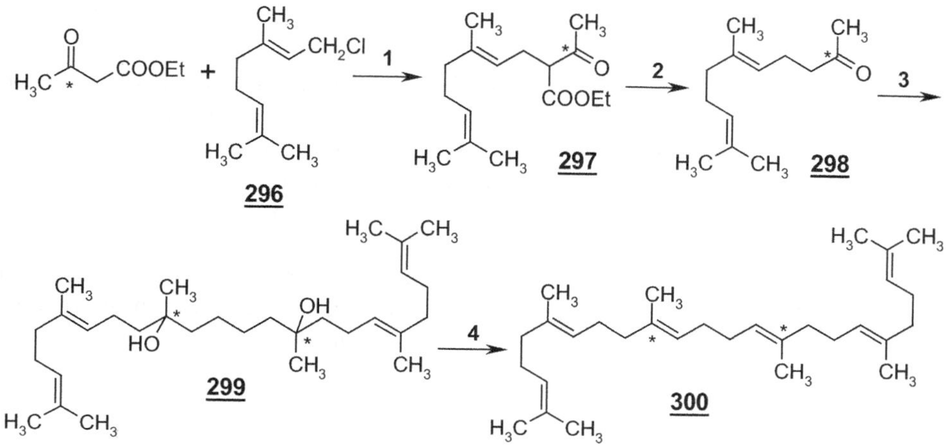


**Reaction conditions:** **1**. EtONa, EtOH; reflux, 25 h; **2**. NaOH, EtOH/$H_2O$; reflux, 48 h; **3**. $Br(CH_2)_4Br$, Mg, ether; reflux, 16 h; **4a**. $PBr_3$, benzene; 90–100 °C, 12 h, **b**.collidine, reflux, 3 h

***Figure 6.83*** *Use of ethyl [3-$^{14}$C]acetoacetate in the synthesis of [10, 15-$^{14}C_2$]squalene*

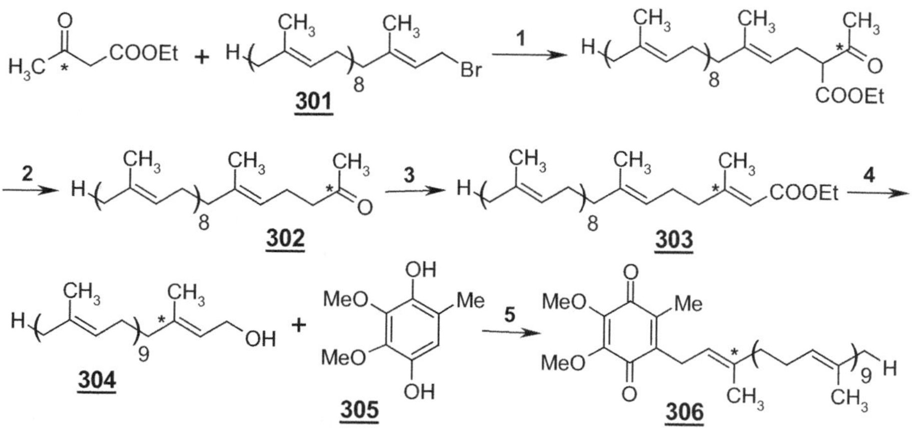


**Reaction conditions:** **1**. EtONa, dioxane; 10–15 °C, 2 h; **2**. aq. NaOH, EtOH dioxane; 80°C, 4 h; **3**. $(EtO)_2P(O)CH_2COOEt$, EtONa, benzene; 5–10 °C, 4 h; **4**. $LiAlH_4$, ether; reflux, 2 h; **5a**. **305**, $H_3BO_3$, toluene; reflux, **b**. **304**, toluene–hexane 1:1; 45 °C, 2 h, **c**. $PbO_2$, HOAc; reflux

***Figure 6.84*** *Use of ethyl [3-$^{14}$C]acetoacetate in the synthesis [3′-$^{14}$C]coenzyme $Q_{10}$*

The first steps of the [$^{14}$C]squalene synthesis resemble the pathway selected for the synthesis on [3′-$^{14}$C]coenzyme Q10 (**306**) (Figure 6.84). In this case alkylation with solanesyl bromide (**301**) and subsequent hydrolysis and decarboxylation of the ester function converted ethyl [3-$^{14}$C]acetoacetate into [2-$^{14}$C]solanesylacetone **302**. Chain extension of **302** in a Horner–Wadsworth–Emmons reaction, reduction of the resulting $\alpha,\beta$-unsaturated ester **303** and coupling of the resulting alcohol ([3-$^{14}$C]decaprenol, **304**) to 2-methyl-5,6-dimethoxy-1,4-hydroquinone (**305**) provided the coenzyme in an overall radiochemical yield of 8% after oxidation. Methodologically comparable pathways have also been published for the synthesis of [5-$^{14}$C]farnesylacetic acid[173] and [3′-$^{14}$C]menaquinone[173b].

Reactions with functionalized alkyl halides significantly enhance the range of application of alkyl [$^{14}$C]acetoacetates. Alkylation of ethyl [3-$^{14}$C]acetoacetate with $\omega$-halo esters, for example, followed by acid-mediated ketonic cleavage opens access to labeled ($\omega$-1)-keto acids/esters, as demonstrated by the synthesis of the [4-$^{14}$C]levulinic acid (**307**) a $\gamma$-keto acid. The latter has served as a key intermediate in the synthesis of the 2,3-substituted indole **308**[19b]. Step 2 in Figure 6.85, the ketonic cleavage, has recently been shown to be accelerated by a factor of 20 when the reaction is submitted to microwave heating[77].

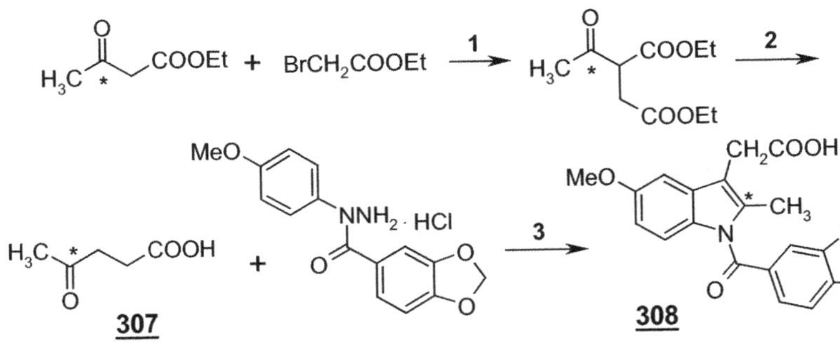


**Reaction conditions: 1.** Na, toluene; 90 °C, 4 h; **2.** HCl 7 %, 90 °C, 4 h; **3.** acetic acid; 75–80 °C, 3 h

***Figure 6.85*** *Use of ethyl [3-$^{14}$C]acetoacetate in the synthesis of $\gamma$-[4-$^{14}$C]keto acids[4-$^{14}$C] (e.g., levulinic acid) and its application to the synthesis of a substituted [3-$^{14}$C]indole*

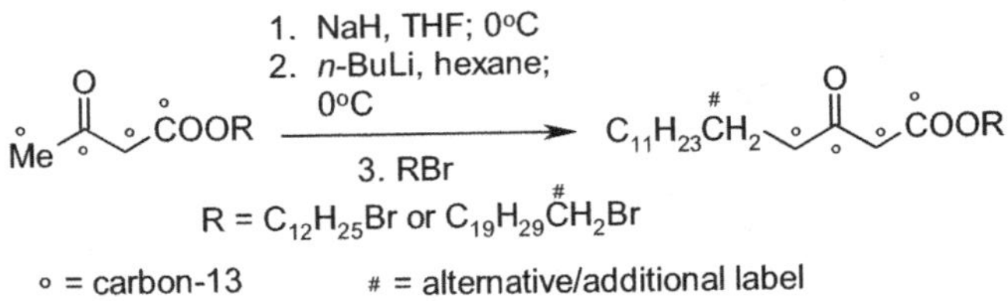


***Figure 6.86*** *Synthesis of homologous $\beta$-keto esters through alkylation of doubly deprotonated ethyl acetate labeled either in C1–C4 or in C5*

Whereas alkylations at C2 are facilitated selectively by deprotonation by weak bases (e.g. alkoxides) or one equivalent of a stronger base, a second deprotonation by a strong base gives a dianion that adds preferentially at C4. As shown in Figure 6.86, higher homologs of labeled acetoacetate can be prepared by alkylation of doubly deprotonated alkyl acetoacetates generated by sequential treatment with equimolar amounts of sodium hydride and *n*-butyllithium. This alternative has been exploited so far only for the synthesis of multiply carbon-13 labeled *β*-keto esters, and of material whose label was introduced into the remote position C5 by the use of an [1-$^{13}$C]alkyl halide[175].

### 6.4.2 Acylation Reactions

Acylation of ethyl [$^{14}$C]acetoacetate plays only a minor role in carbon-14 chemistry of *β*-keto esters. Some interesting examples are documented in Figure 6.87. In the first case, ethyl [1-$^{14}$C]acetatoacetate was treated with myristoyl chloride and the resulting 1,3-dicarbonyl[$^{14}$C]carboxylic ester submitted to mild basic hydrolysis, providing selectively 3-oxo[1-$^{14}$C]hexadecadecanoic acid[176]. In the second example ethyl [3-$^{14}$C]acetatoacetate reacted with 2,5-dichlorobenzohydroxamoyl chloride (**309**) to give ethyl 5-(2,6-dichlorophenyl)-3-methyl-4-[3-$^{14}$C]isoxazolylcarboxylate (**310**). Saponification of the ester function, activation of the carboxyl group with thionyl chloride and coupling of the resulting acyl chloride with 6-aminopenicillanic acid (**311**) provided [$^{14}$C]dichloxacillin (**312**) in overall 30% radiochemical yield[177].

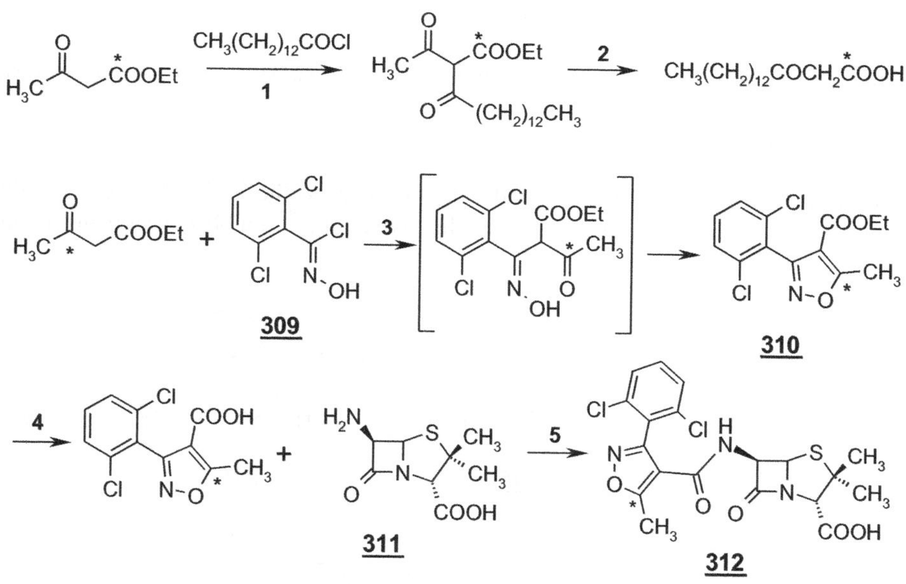

**Reaction conditions**: NaOEt, ether; r.t.; **2.** $NaHCO_3$; EtOH–$H_2O$, 10 °C, 30 min; **3.** NaH, DMF; 0 °C, 1.5 h; r.t., 16 h; **4.** KOH, EtOH/$H_2O$ 1:1; reflux, 1 h; **5a.** $SOCl_2$; reflux, 1 h, **b.** **311**, acetone/$H_2O$ (pH 7.2) 1:3; 0 °C, 1 h

***Figure 6.87*** *Use of ethyl [3-$^{14}$C]acetoacetate in an acylation reaction to produce a homologous β-[$^{14}$C]keto acid (synthesis of [$^{14}$C]dichoxacillin)*

For historical reasons the acylation of unlabeled sodium ethyl acetoacetate with [2-$^{14}$C]acetyl chloride is mentioned. The product of this reaction, the symmetrical 3-ethoxycarbonyl[2,4-$^{14}C_1$]pentan-2,4-dione (**313**), was used for the synthesis of

(*S*)-[4-$^{14}$C]threonine (**315**). For this purpose it was treated with benzenediazonium chloride, producing the azo-intermediate **314** with loss of one acetyl group. Subsequent classical chemical steps including an optical resolution procedure transformed **314** into (2*S*,3*R*)-[4-$^{14}$C]threonine (**315**)[178]. An obvious disadvantage to this route is the loss of 50% of the radioactivity in step 2.

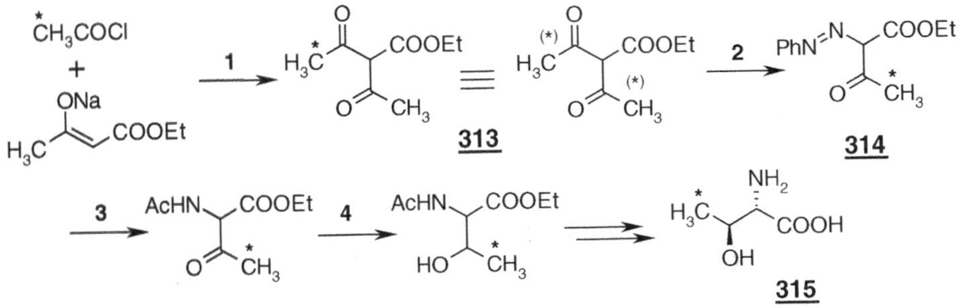


**Reaction conditions:** 1. ether; −7 °C, 0.5–1.5 h; 2. Ph–N≡N$^+$ Cl$^-$, $NaHCO_3$, $EtOH/H_2O$, 10 °C, 20 min; 3. Zn, $Ac_2O$/HOAc; 10 °C 1.5 h, r.t., 3 h; 4. $H_2$, $PtO_2$, 1N NaOH (1.5 equiv.), $H_2O$; 25 °C, 4 h

***Figure 6.88*** *Synthesis of* (2S,3R)-*[4-$^{14}$C]threonine via [$^{14}$C]acetylation of ethyl acetoacetate*

### 6.4.3 Aldol Reactions

There is apparently only one published example of a synthesis that includes an aldol reaction of a carbon-14-labeled acetoacetate. In this example[74] doubly deprotonated methyl

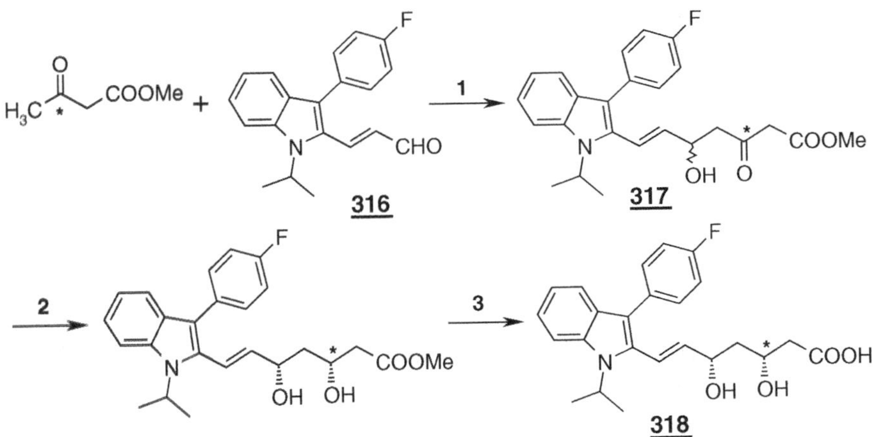


**Reaction conditions:** 1a. NaH, THF; −15 °C, 1 h, b. BuLi, THF; −15 °C, 15 min, c. **316**, THF; −10 °C, 30 min; 2. $NaBH_4$, $Et_2BOMe$; THF–MeOH 4:1; −78 °C, 3 h; 3. NaOH, MeOH; r.t., 2 h

***Figure 6.89*** *Use of an aldol reaction of methyl [3-$^{14}$C]acetoacetate in the preparation of [$^{14}$C]fluvastatin*

[3-$^{14}$C]acetoacetate reacted with the $\alpha,\beta$-unsaturated aldehyde **316** to give the methyl 5-hydroxy-3-oxo-6-[3-$^{14}$C]heptenoate derivative **317**. Its reduction with $NaBH_4$ in the presence of diethylmethoxyborane[179] and saponification provided the corresponding *syn*-3,5-dihydroxy-6-heptenoic acid ([$^{14}$C]fluvastatin, **318**) in 40% overall radiochemical yield.

### 6.4.4 Knoevenagel–Michael Reactions

In carbon-14 chemistry, Knoevenagel–Michael sequences with alkyl acetoacetates have found widest application in the synthesis of labeled 1,4-dihydropyridines (Hantzsch dihydropyridine synthesis)[180]. 4-Aryl-1,4-dihydro[$^{14}$C]pyridine derivatives constitute a class of calcium antagonists and calcium channel blockers that play an important role in the treatment of cardiovascular diseases. The basic sequence, depicted in Figure 6.90, involves first the Knoevenagel condensation of an aromatic aldehyde with an alkyl

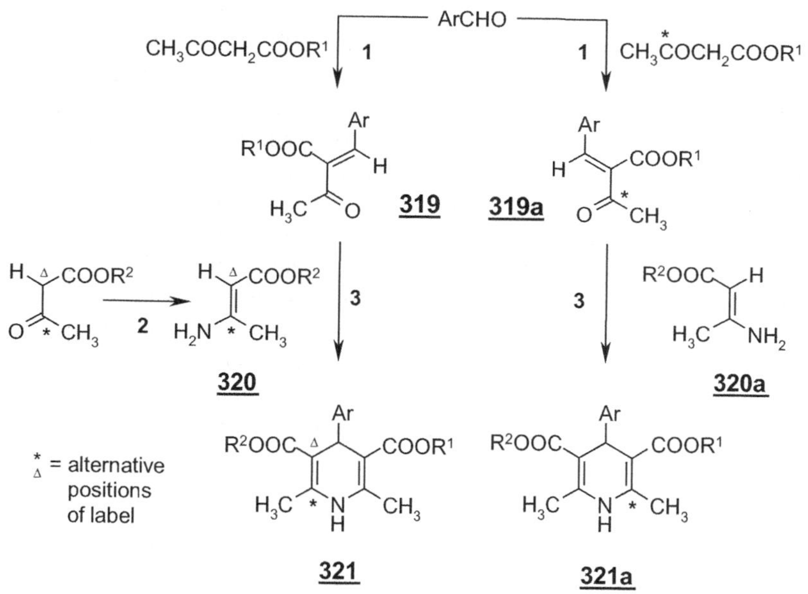


[2-$^{14}$C]nimopidine (ref 167e) : $R^1$ = *i*-Pr, $R^2$ = $CH_2CH_2OMe$, Ar = 3-$NO_2$Ph
[6-$^{14}$C]nisopidine (ref 167f) : $R^1$ = Me, $R^2$ = *i*-Bu, Ar = 2-$NO_2$Ph
[6-$^{14}$C]nitrendipine (ref 167d) : $R^1$ = Me, $R^2$ = Et, Ar = 3-$NO_2$Ph
[6-$^{14}$C]nicardipine (ref 182) : $R^1$ = Me, $R^2$ = $CH_2CH_2N(Me)Ph$, Ar = 3-$NO_2$Ph
[6-$^{14}$C]CV 4094 (ref 181a) : $R^1$ = Me, $R^2$ = -$CH_2CH_2$-(C$_6$H$_4$)-$NCHPh_2$, Ar = 3-$NO_2$Ph
[5-$^{14}$C]YM-09730-5 (ref 167a) : $R^1$ = Me, $R^2$ = (pyrrolidin-3-yl)NBn , Ar = 3-$NO_2$Ph

**Reaction conditions:** **1**. piperidine/HOAc 1:1, isopropanol, r.t., 5 h **2.** $NH_3$ (g); toluene; reflux, 4 h or methanolic $NH_3$; 0 °C, 16 h; **3**. ethanol or propanol; reflux 4–20 h

**Figure 6.90** *Use of alkyl [$^{14}$C]acetoacetates in the synthesis of 1,4-dihydro[$^{14}$C]pyridines*

acetoacetate to give an $\alpha,\beta$-unsaturated esters **319** or **319a**, followed by Michael-addition of an alkyl 3-aminocrotonate (**320** or **320a**) and subsequent ring closure by an intramolecular keto enamine condensation. The 4-aryl-1,4-dihydro[$^{14}$C]pyridine product (**321**, **321a**) may contain isotopic carbon at any of several positions as indicated, depending upon which building block is labeled and in what position. The overall methodology allows considerable structural variation in the aldehyde as well as in the $\beta$-keto ester component. The sequence of the individual steps depends on the reaction conditions applied and the nature of the ester functions. Usually, the unlabeled/labeled $\alpha,\beta$-unsaturated ester and the labeled/unlabeled 3-aminocrotonate are prepared separately and subsequently employed in the Michael-addition to give dihydropyridines labeled at C2 or C6 in overall radiochemical yields of 20–70%. In nearly all cases so far the label has been introduced by the respective 3-amino[3-$^{14}$C]crotonates (route 1), which are readily obtained by treatment of the corresponding alkyl [3-$^{14}$C]acetoacetates either with gaseous ammonia in refluxing toluene[67d–f] or with a methanolic solution of ammonia at 0 °C[181a,b]. Only in one case (i.e. [5-$^{14}$C]FPL 6129) was 3-amino[2-$^{14}$C]crotonate employed[167a]; this component was prepared from [5-$^{14}$C]Meldrum's acid. Thereby, 1,4-dihydropyridines become labeled at C3 or C5. Although all derivatives depicted in Figure 6.90 are chiral, only [6-$^{14}$C]YM-09730-5 was separated into its enantiomers (LiChroprep®, hexane/ethyl acetate)[181a].

For symmetrically substituted 1,4-dihydropyridines (i.e. identical alkyl and ester functions), the aldehyde component may be directly combined in a tandem Knoevenagel–Michael sequence with ammonia and two equivalents of the alkyl [$^{14}$C]acetoacetate[182]. Under these conditions the 3-amino[$^{14}$C]crotonate is generated *in situ*. [2, 6-$^{14}$C]nifedipine (**322**) has been prepared this way[183] (Figure 6.91). In a further modification, the aldehyde is first allowed to react with two equivalents of the alkyl acetoacetate to give the corresponding 1,5-diketone in a tandem Knoevenagel–Michael sequence. Ring closure is then achieved in a two-step sequence by sequential treatment with ammonia and acetic anhydride; this variation was used in the synthesis of [4-$^{14}$C]dithiopyr[184].

With formaldehyde or acetaldehyde as reaction component the corresponding 4-unsubstituted or 4-methyl derivatives were obtained, which in the presence of sodium

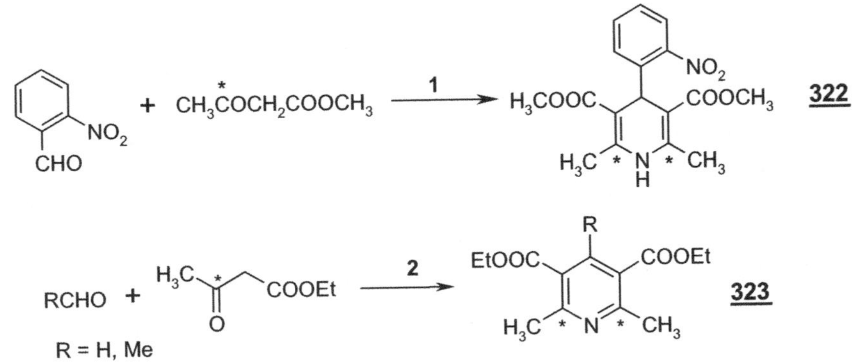


**Reaction conditions: 1**. $NH_3$ (g), EtOH; **2a**. EtOH, $NH_3$, ethanol; reflux, 3 h **2b**. $NaNO_2$, HOAc ; r.t., 2.5 h

**Figure 6.91** *Use of alkyl [$^{14}$C]acetoacetates in a tandem Knoevenagel–Michael reaction sequence*

nitrite immediately underwent further oxidation to give the corresponding diethyl 2,6-dimethyl[2, 6-$^{14}$C]pyridine-3,5-dicarboxylates **323**[185].

The mechanism most often discussed for the preceding chemistries resembles that of the Knorr synthesis of substituted pyrroles, which involves the condensation of alkyl acetoacetates with alkyl 2-hydroxyimino-3-oxobutanoates **324** in the presence of a reducing agent such as zinc or sodium amalgam[168b]. In this case the hydroxyimino component is first reduced, and the resulting alkyl $\alpha$-aminoacetoacetates **325** react *in situ* with the alkyl acetoacetates in a tandem Knoevenagel/keto enamine condensation to give the corresponding dialkyl 3,5-dialkylpyrrole-2,4-dicarboxylates **326**. This sequence has been exploited so far only for the synthesis of material labeled with carbon-13 in different well-defined positions. The requisite labeled 2-hydroxyimino derivatives are conveniently accessible through reaction of the corresponding alkyl acetoacetates with nitrous acid[168b,186].

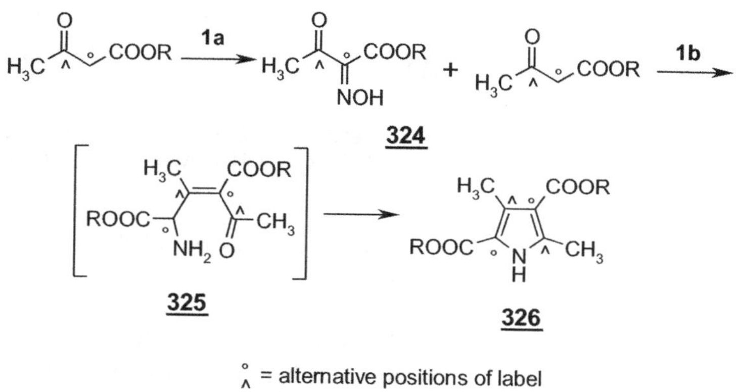


**Reaction conditions**: **1a**. $NaNO_2$, HOAc; r.t., 5 h; **b**. Zn, NaOAc, HOAc; 95-105 °C, 15 min

***Figure 6.92*** *Use of alkyl [$^{13}$C]acetoacetates in the synthesis of substituted [$^{13}$C]pyrroles*

Michael addition of ethyl [3-$^{14}$C]acetoacetate to ethyl crotonate followed by an intramolecular tandem enolate acylation opens an alternative approach to highly substituted labeled resorcinols (e.g., **329**) (see also Section 6.5.1). Aromatization of the initially formed ethyl 6-methyl-4-hydroxy-2-oxo[1-$^{14}$C]cyclohex-3-ene-1-carboxylate (**327**) is readily

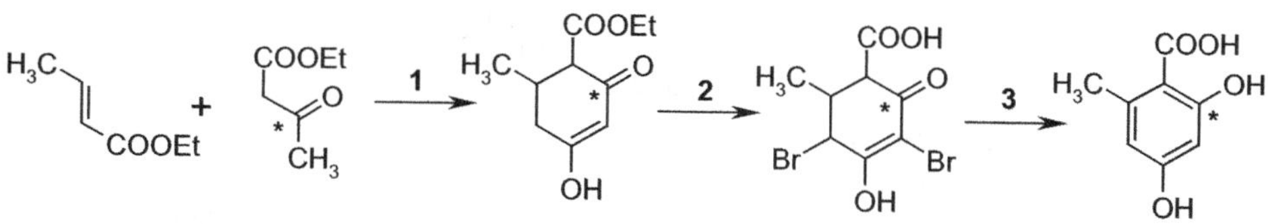


**327** **328** **329**

**Reaction conditions: 1**. EtONa, EtOH; 80 °C, 2 h; **2a**. $Br_2$, HOAc, r.t., 24 h, **b**. conc. $H_2SO_4$; r.t., 40 min; **3**. $H_2$, 2 % Pd–$CaCO_3$, NaOH, MeOH–$H_2O$; r.t., 3 h (37% overall)

***Figure 6.93*** *Application of ethyl [3-$^{14}$C]acetoacetate to the synthesis of substituted [$^{14}$C]resorcinols*

achieved by bromination and subsequent simultaneous hydro-dehalogenation and HBr elimination of the resulting dibromo intermediate **328**[187].

### 6.4.5 Reactions at the Functional Groups

Knoevenagel condensation of cyanoacetamide to ethyl [3-$^{14}$C]acetoacetate and cyclization afforded 3-cyano-4-methyl-2,6-dihydroxy[4-$^{14}$C]pyridine (**330**). Chloro-dehydroxylation upon treatment with $POCl_3$, Pd-catalyzed hydro-dehalogenation and saponification of the cyano function converted **330** into 4-methyl[3-$^{13}$C]nicotinic acid (**331**)[188].

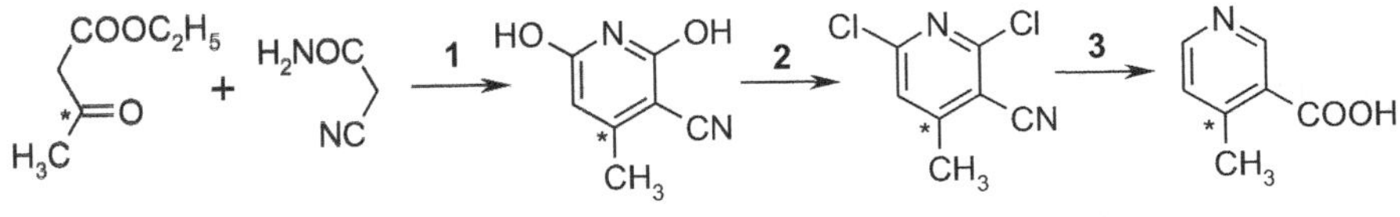


**330** **331**

**Reaction conditions: 1.** KOH, MeOH; reflux, 5 h; **2.** $POCl_3$; 180 °C, 5 h; **3a.** $H_2$ (3 atm), $PdCl_2$, NaOAc, MeOH; r.t., 18 h, **b.** NaOH, ethylene glycol; 175 °C, 18 h

***Figure 6.94*** *Knoevenagel reaction of bifunctional agents on labeled ethyl acetoacetate: Application to the synthesis of 4-substituted [4-$^{14}$C]nicotinic acid*

Condensation of acetoacetates with hydrazines has been widely exploited in medicinal chemistry for the production of pharmacologically highly effective pyrazolin-5-ones, such as antipyrine, propyphenazone and phenylbutazone, showing either pronounced analgesic or anti-inflammatory properties. In carbon-14 synthesis this methodology has been applied to the labeling of derivatives belonging to the same structural type. Reaction of ethyl [3-$^{14}$C]acetoacetate with methylhydrazine, for example, gave 1,3-dimethyl[3-$^{14}$C]-pyrazolin-5-one (**332**), a key intermediate in the synthesis of 4-(2,4-dichlorobenzoyl)-1,3-dimethyl-5-hydroxy[3-$^{14}$C]pyrazole (**333**). The latter is available in two steps through

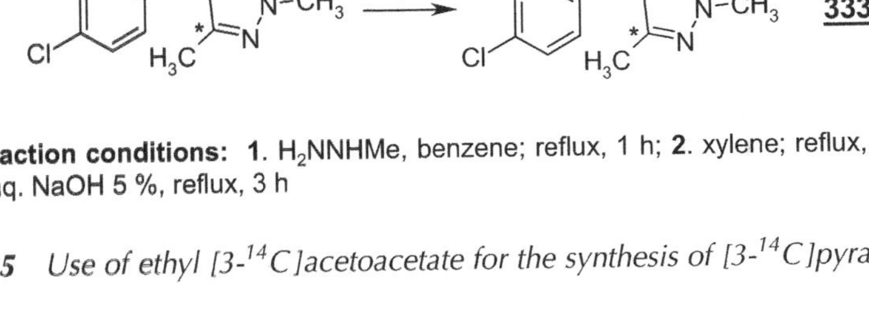


**332** **333**

**Reaction conditions: 1.** $H_2NNHMe$, benzene; reflux, 1 h; **2.** xylene; reflux, 2 h; **3.** aq. NaOH 5 %, reflux, 3 h

***Figure 6.95*** *Use of ethyl [3-$^{14}$C]acetoacetate for the synthesis of [3-$^{14}$C]pyrazolin-5-ones*

reaction of **332** with 2,4-dichlorobenzamidine followed by hydrolysis of the imide function of the intermediate[189].

With nitrous acid, ethyl [3-$^{14}$C]acetoacetate forms the corresponding isonitroso derivative, which is readily hydrolyzed to give ethyl 2,3-dioxo[3-$^{14}$C]butyrate (**334**). This compound served as a key intermediate in one of the classical preparation procedures of [2-$^{14}$C]lactic acid, which was accessible in two steps through treatment with aqueous sodium hydroxide and thermolytic decarboxylation[190].

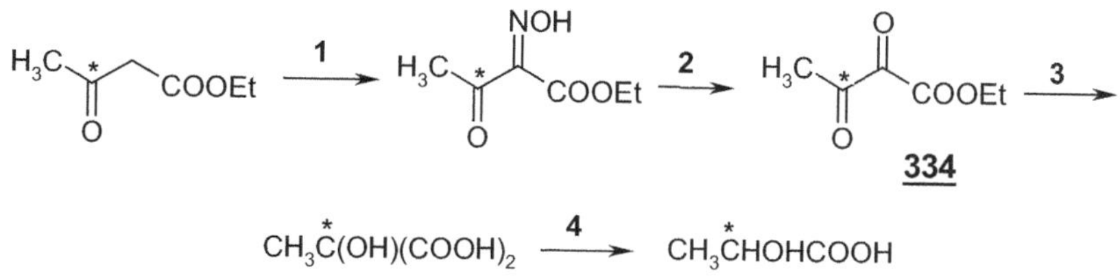


**Reaction conditions**: **1**. $NaNO_2$, $H_2SO_4$, $H_2O$; 0 °C, 2 h; **2**. $HNO_2$, $As_2O_3$; ether; 10°C, 3 h; **3**. NaOH, $H_2O$; r.t., 2 d; **4**. $H_2O$; 80 °C, 48 h

***Figure 6.96*** *Use of ethyl [3-$^{14}$C]acetoacetate in the synthesis of (R,S)-[2-$^{14}$C]lactic acid*

## 6.5 [$^{14}$C]Malonates

Malonic acid labeled with carbon-14 at C2 or at C1,3 and the corresponding mono- and diesters have become important and highly versatile building blocks for the synthesis of a broad range of labeled compounds, such $\alpha,\beta$-unsaturated acids and esters, cycloalkanes, $\beta$-keto esters, $\alpha$-amino acids, phenols, dihydropyridines/pyridines, pyrimidines and quinolines. In addition, they are widely used as acetic acid synthons ($^{-}CH_2COOR$; $^{2-}CHCOOR$) for the extension of carbon chains of alkyl, allyl and benzyl halides as well as of saturated and $\alpha,\beta$-unsaturated carbonyl compounds. This Section discusses all malonates, wherever labeled, and unlabeled malonates used with carbon-14-labeled reaction partners. However, it is the [2-$^{14}$C]malonates that have been most often used in radiochemical synthesis.

Classical preparations of [$^{14}$C]malonates start from labeled/unlabeled chloro- or bromoacetic acid. $S_N2$ reaction with unlabeled/labeled KCN in aqueous solution at pH 8, followed by acidic or basic hydrolysis of the resulting cyanoacetate and acidification provides [$^{14}$C]malonic acid. This can either first be purified through precipitation as the calcium salt (Figure 6.97, Step 3), from which it is released with aqueous HCl[191], or directly extracted into ether and, if necessary, further purified by sublimation at 100 °C and <1 Torr[192]. Esterification can be accomplished either by treatment of the silver salt with the desired alkyl halide (Step 6)[191,192], by reaction with diazoalkanes ($CH_2N_2$, $CH_3CHN_2$)[192] or directly from the sodium salt upon heating with $EtOH/H_2SO_4$ (Step 5)[193]. Modern, more elegant procedures start from labeled/unlabeled acetic acid esters, thereby obviating the need to prepare halo[$^{14}$C]acetic or [$^{14}$C]cyanoacetic acid. Deprotonation of the respective labeled/unlabeled ester with LDA[194] or LiHMDS[195] and quenching of the enolate with unlabeled/unlabeled carbon dioxide[194] or alkyl chloroformate[195] affords directly the corresponding labeled mono- or diester, respectively. The latter procedure is of particular interest. $\alpha$-Alkylation of an unlabeled base-sensitive ester with a chloro[$^{14}$C]formate whose ester function is sensitive to acid allows replacement of the unlabeled ester group with a labeled one by subsequent selective saponification and decarboxylation. This is discussed in more detail in Chapter 10.

Monoesters, valuable intermediates for the preparation of $\beta$-keto acid esters, are also accessible through selective saponification of the corresponding diesters. Also, alkylation of the mono-salt of [$^{14}$C]malonic acid with reactive primary and secondary alkyl halides in a dipolar solvent (acetonitrile, HMPA) has been reported[196].

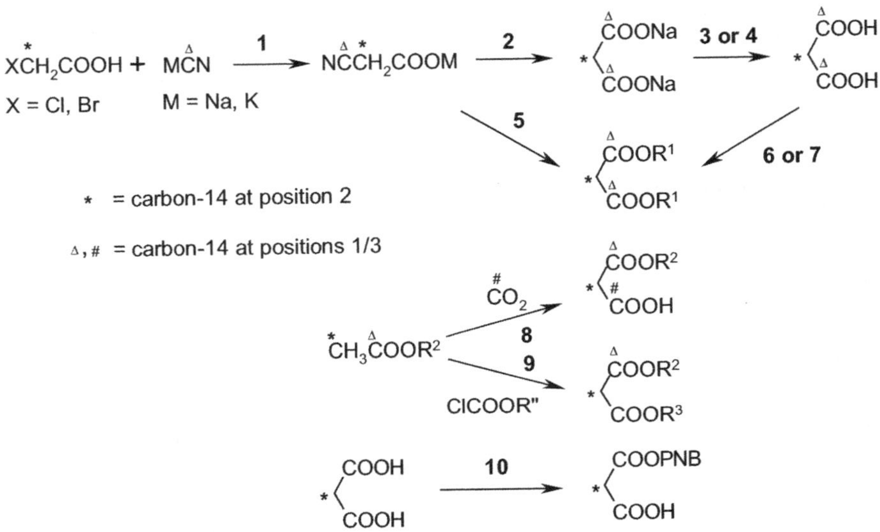


**Reaction conditions: 1**. $H_2O$ (pH 8); 90–100 °C, 30–60 min; **2**. 12.5 N NaOH, 100 °C, 1.5 –2 h; **3a**. $CaCl_2$, $H_2O$, r.t., 15–20 h, **b**.12 N HCl (75–85 %); **4**.conc. HCl, ether extraction, sublimation at 100 °C, <1mmHg (75–80 %); **5**. conc. $H_2SO_4$, EtOH/benzene; reflux, 4.5 h; **6a**. aq. $AgNO_3$; r.t., 4–15 h, **b**. RI; reflux, 3 h (94 %); **7**. $RCHN_2$ (R= H,$CH_3$), ether (>95 %); **7**. conc. $H_2SO_4$, EtOH/benzene; reflux, 15 h (70%); **8a**. LDA, ether; –80 °C, 15 min, **b**.$^{14}CO_2/CO_2$, –80 °C, 30 min (67 %); **9a**. LHMDS, THF; –75 °C, 30 min, **b**. ClCOOR, THF; –75 °C, 2 h (85 %); **10**. *p*-$NO_2PhCH_2Br$, $Et_3N$, acetonitrile; reflux, 3 h (47 %)

***Figure 6.97*** *Routes for the preparation of [$^{14}$C]malonic acid and the corresponding mono- and diesters*

The acidity of the methylene group of dialkyl malonates is strongly enhanced (p$K_a$ 13.7) by the two electron-withdrawing ester functions, allowing ready deprotonation. The resulting enolates react with electrophiles such as primary and secondary alkyl and benzyl halides or sulfonates, aldehydes, reactive ketones, orthoesters, $\alpha$,$\beta$-unsaturated carbonyl or carboxyl systems, nitric acid, nitrous acid and bromine, to afford a broad spectrum of unlabeled and labeled advanced intermediates. Tertiary halides fail to react because of competing dehydrohalogenation. Both ester functions can also undergo several transformations, such as saponification, aminolysis, reduction etc., thereby increasing the number of valuable labeled products[197].

### 6.5.1 Reactions at the Methylene Group

The most common applications of [$^{14}$C]malonates by far are those involving reactions at the methylene group. The various categories of reactions are discussed in the following eight subsections. The first six involve the formation of carbon–carbon bonds via alkylations, arylations, different types of methylenations, etc. The remaining subsections cover formation of nitrogen–carbons with the malonate methylene group, and oxidations at C2.

*α-Alkylation of [$^{14}$C]Malonates*

Deprotonation of malonic esters is routinely effected with EtONa or *tert*-BuOK in the respective alcohols as solvents. The subsequent reaction with equimolar amounts of alkylation agents generally provides the corresponding monoalkylated species in good to excellent yields (see Figure 6.98). A difficulty often encountered is the formation of dialkylated byproducts. In alcoholic solution, formation of these side products does not occur to an appreciable extent, since the acidity of the alcoholic hydroxy group is sufficient to preferentially protonate the monoalkylated enolate ion rather than that of the more acidic diethyl malonate[198]. The use of stronger bases such as NaH in aprotic solvents (DMF, DMSO, HMPA), however, facilitates dialkylation as well as competing alkylation at the oxygen; for steric reasons this effect may become more serious with low molecular weight alkylating agents[199]. For example, in the alkylation of diethyl [2-$^{14}$C]malonate with $CH_3I$, the amount of dialkylated byproduct was reported to be in the range of traces up to 30%[200–203]. However, as can be seen in Figure 6.98, also findings to the contrary have been published claiming that a broad spectrum of alkylating agents ranging from methyl

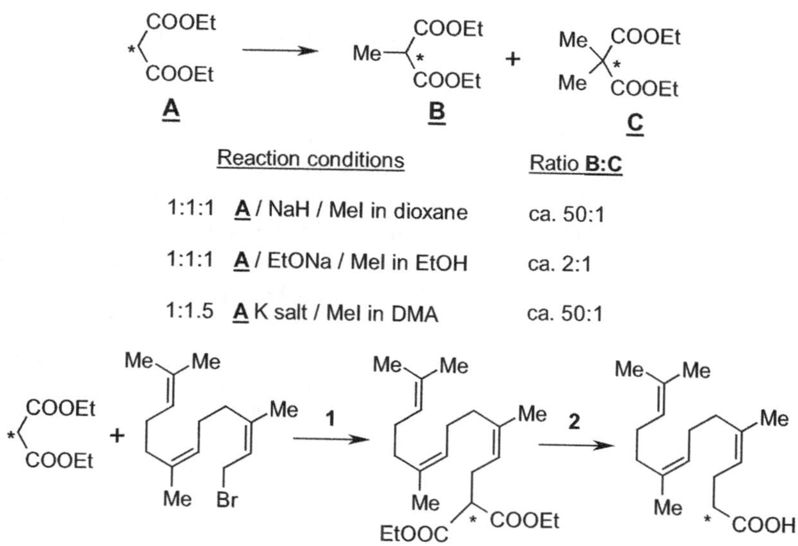


**Reaction conditions: 1**. EtONa (0.94 equiv.), RBr (1 equiv.), EtOH; reflux, 2 h; **2a.** KOH (85 %) EtOH; reflux, 3 h; r.t. 18 h, **b**. acidic workup, **c**. 150 °C, 5 h

***Figure 6.98*** *Synthesis of monoalkylated [2-$^{14}$C]malonates*

iodide[200,203] to farnesyl bromide[204] in stoichiometric ratios have given good results. Accordingly optimal conditions in view of base, solvent and reaction temperature will have to be searched in pilot studies in any case in order to minimize the amount of dialkylated byproduct. Because of the high cost of diethyl [2-$^{14}$C]malonate, it is undesirable to use excessive amounts in order to suppress dialkylation.

The alkylation reaction mixtures are usually separated either by chromatographic methods or fractional distillation, which may prove difficult when boiling points are close. A recently introduced separation technique (Figure 6.99)[205] exploits the fact that, presumably owing to steric reasons, the saponification rates of labeled mono- and dialkylated malonic esters differ significantly. For example, it was shown that refluxing with aqueous 2 M KOH completely saponified the monoalkylated species without affecting the dialkylated by-product.

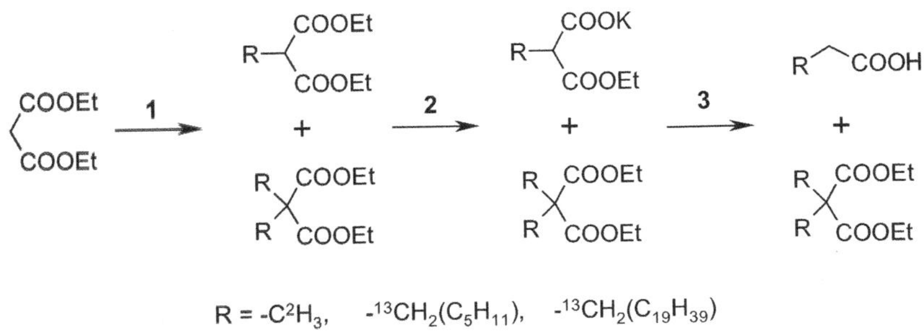


**Reaction conditions:** **1**. NaH, RX, DMF; 60 °C, 1 h (99 %); **2**. 2.8*M* KOH; reflux, 3 h; **3**. 12 N HCl/dioxane; reflux, 24 h (45–70 % overall)

***Figure 6.99*** *Separation of mono- and dialkylated isotopically labeled malonates through selective saponification*

An alternative strategy involves the use of an allyl group as a protective group for one of the acidic methylene hydrogens (Figure 6.100). In isotope chemistry this has been verified so far only for $C^2H_3I$. According to this strategy the malonate is first allylated then subjected to the desired alkylation. De-allylation of the resulting intermediate is achieved upon treatment with ($\eta^2$-propene)Ti(O-*i*-Pr)$_2$ generated *in situ* from *i*-PrMgBr and Ti(O-*i*-Pr)$_4$. For the example shown an overall yield of 75% was achieved[206].

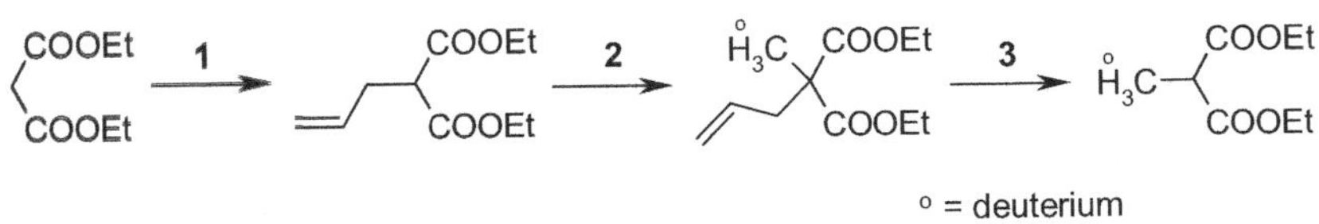


**Reaction conditions:** **1a**. NaH, THF, 20 °C, 1 h, **b**. $CH_2$=$CHCH_2Br$, THF; 20°C,15 h; **2**. NaH, THF; r.t. 1 h, **b**. $C_2H_3I$, THF; r.t., 15 h; **3**. *i*-PrMgBr$_2$, Ti(OiPr)$_4$, ether; –45 °C, 1 h

***Figure 6.100*** *Synthesis of isotopically labeled monoalkylated malonates following a protection/deprotection approach*

These alkylation procedures can benefit greatly from the use of microwave activation, which can reduce reaction times by a factor of 60. Moreover, the often tedious separation of high boiling solvents (DMF, DMSO, HMPA) can be omitted, if reactions are conducted under solvent-free conditions in the presence of *tert*-BuOK as a base and a phase transfer catalyst such as Aliquat 336. Chemical yields are comparable to those obtained using classical procedures and only traces of dialkylated byproducts have been detected. Moreover, microwave treatment of the diester product with LiF or LiCl (Krapcho procedure) under solvent-free conditions provides the corresponding mono-ester by selective decarbalkoxylation[207].

Dialkylation can be effected by using an excess of electrophile (Figure 6.101, Step 1)[208], or carrying out two successive operations, which permits the use of two different alkylating agents (Steps 4, 5)[209]. Alkylation of [$^{14}$C]malonate with one equivalent of an $\alpha,\omega$-dihalide in the presence of two equivalents of base provides a useful route to three- to

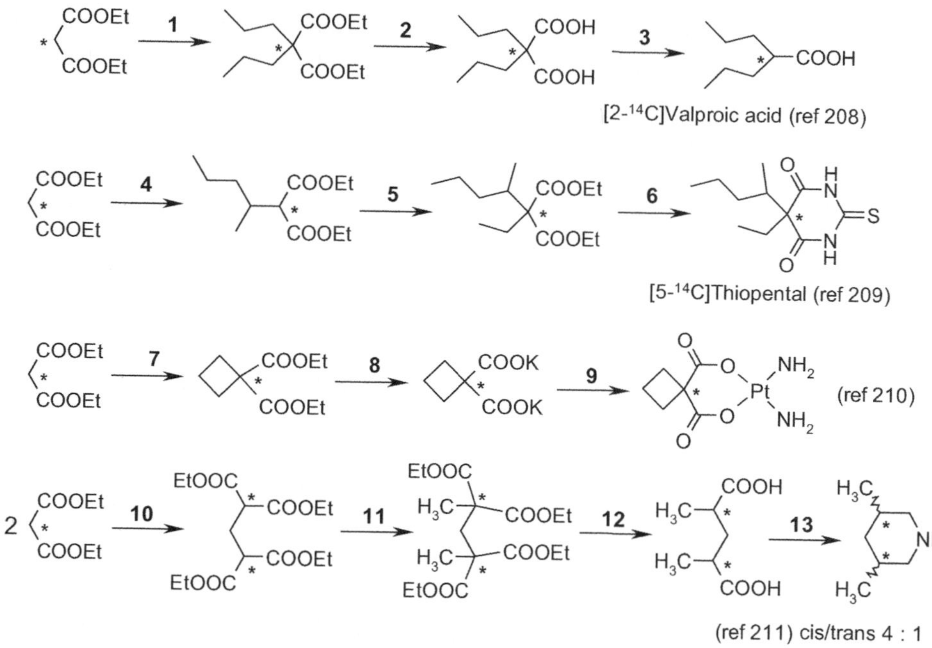


**Reaction conditions:** **1**. NaH, *n*-$C_3H_7Br$ (2.5 eq.); HMPA; r.t., 6 h; **2**. NaOH (10 eq.), MeOH/$H_2O$; reflux, 6 h; **3**.155 °C (73 % overall); **4**. EtONa (2.5 eq.), $ICH_2CH(CH_3)C_3H_7$ (2.1 eq.), EtOH; 90 °C, 1.5 h; **5**. EtONa, EtI (excess); 85°C, 2 h; **6**. $(H_2N)CS$, MeONa, MeOH; 60 °C, 5 h; **7**. EtONa (1.75 eq.), $Br(CH_2)_3Br$ (0.85 eq), EtOH; reflux, 2 h; **8**. KOH, EtOH/$H_2O$; 60 °C, 15 min; **9**. $(NH_3)_2PtI_2$,$AgNO_3$, $H_2O$; 60 °C, 3 h, 5 °C, 64 h; **10**. EtONa (1 eq.), $ICH_2I$ (0.5 eq.), reflux 3 h; **11**. EtNa (2 eq.), MeI ( 2.2 eq.), EtOH; reflux 6 h; **12a**. KOH, EtOH/$H_2O$, reflux, 20 h; **b**. 165–170 °C, 30 min. **13a.** urea, 150–155 °C, 24h **b.** $LiAlH_4$, THF; reflux, 24 h

**Figure 6.101** *Dialkylation of [2-$^{14}$C]malonates using of $\alpha,\omega$-dihalogen alkanes*

seven-membered rings (Step 7)[210], as does the intramolecular alkylation of ω-haloalkyl malonates. This method is also applicable to medium rings (10–14 members) without use of high dilution techniques[199]. With two equivalents of base and two equivalents of [$^{14}$C]-malonate, α,ω-dihalides react to afford α,ω-tetracarboxylic acid derivatives (Step 9)[211].

Saponification of 2-substituted malonate diesters followed by careful acidification provides the corresponding dicarboxylic acids, which on heating decarboxylate to mono-carboxylic acids (e.g., Figure 6.102, example 1). Applied to monoalkylated [2-$^{14}$C]malonic esters, which can either be prepared according to procedures of Figure 6.97 or by selective hydrolysis, the above sequence leads directly to esters of [2-$^{14}$C]carboxylic acids thereby enabling two-carbon extension of the carbon chains of alkyl halides or sulfates. Analogously, dialkylated [2-$^{14}$C]malonic esters are versatile building blocks for labeled α-branched [$^{14}$C]carboxylic acids.

The range of valuable labeled intermediates accessible by these methods can be significantly broadened when functionalized alkylating agents are employed. Thus, α-haloketones provide γ-[$^{14}$C]keto acids (e.g., **335**)[212], and 2-benzyloxyethyl bromide afforded [$^{14}$C]butyrolactone (**336**), which served as a key intermediate for the preparation of carbon-14-labeled 1,4-butandiol, furan, pentanedioic acid, and α-tetralone[213].

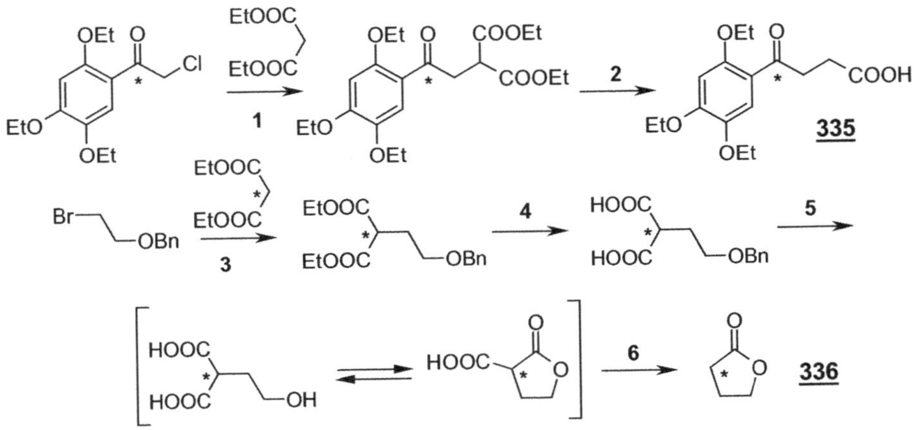


**Reaction conditions:** **1**. sodium diethyl malonate, benzene; 60–65 °C, 2 h; **2**. 4NHCl/AcOH, dioxane; reflux, 7.5 h; **3**. NaH, diethyl[2-$^{14}$C]malonate, DMF; **4**.1 N NaOH/EtOH; **5**. $H_2$, 5 % Pd–C; **6**. 120–140 °C (57 % overall)

***Figure 6.102*** *Use of functionalized labeled/unlabeled alkylating agents in the synthesis of more complex labeled intermediates*

### *α-Arylation of [$^{14}$C]Malonates*

The synthesis of aryl/heteroaryl acetic acids through replacement of aryl/heteroaryl methyl halides with potassium cyanide has proved to be nontrivial and therefore of only limited value in carbon-14 synthesis. The benzylic carbon of the nitrile formed *in statu nascendi* is sufficiently acidic to be alkylated by as-yet unreacted halide to form significant amounts of *bis*-alkylated byproduct, sometimes the major product. Procedures based on transition

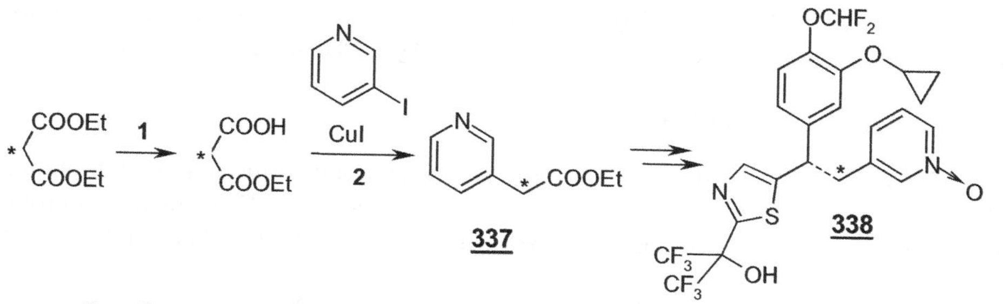


**Reaction conditions**: 1. KOH (1 equiv.), MeOH **2.** CuBr, $MgCl_2$, EtOH; 100 °C, 4 h

***Figure 6.103*** *CuBr-catalyzed α-arylation of [2-$^{14}$C]malonates with iodopyridines*

metal-catalyzed α-arylation of malonates would provide valuable alternatives, as would the Willgerodt–McKillop rearrangement of aryl methyl ketones. Although several broadly applicable methods have been published[214], only the Cu(I)-catalyzed arylation of [$^{14}$C]-malonates using aryl bromides bearing a carboxyl group in the *ortho-* or *peri*-position has been used so far in isotope chemistry. This method, which has been postulated to be dependant on the formation of a Cu–chelate complex with the halide and the carboxyl function, has recently been shown to work also with *o-, m- and p*-iodopyridines to give the respective ethyl pyridineacetic acids in 70–90% yields[215]. When applied to 3-iodopyridine,

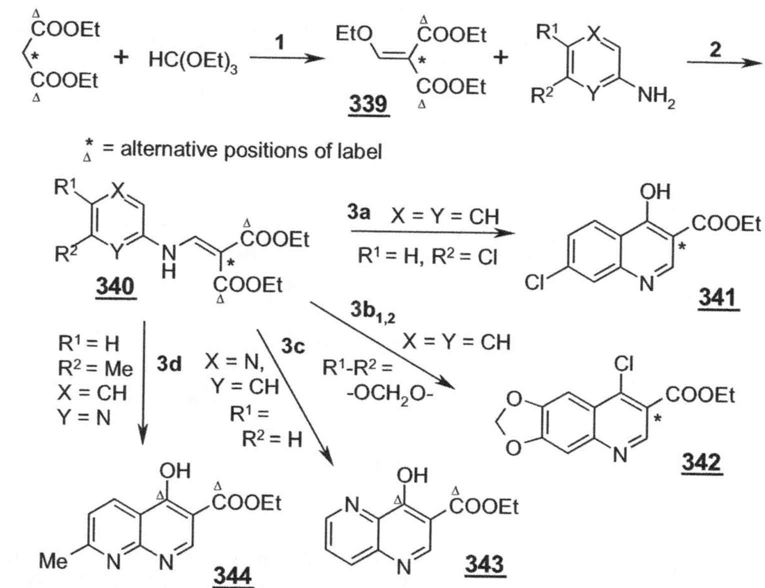


**Reaction conditions 1.** $Ac_2O$, $ZnCl_2$ (cat); 105–130 °C, 5–9 h; 140–180 °C, 4 h; **2.**100–150 °C; 0.5–1.5 h; **3a,b1,c,d.** high boiling solvent; 250–260 °C, **3b$_2$**. $POCl_3$, toluene, reflux, 8 h

***Figure 6.104*** *Preparation and use of dialkyl 2-ethoxymethylene[$^{14}$C]malonates for the synthesis of labeled binuclear heteroaromatic compounds*

ethyl pyrid-3-yl[2-$^{14}$C]acetate (**337**) was obtained in 88% radiochemical yield; it was used as a key intermediate in the synthesis of labeled **338**, a PDE-IV drug candidate.

*α-Alkoxymethylenation of [$^{14}$C]Malonates*

Orthoformates such as $HC(OEt)_3$, in the presence of acetic anhydride and catalytic amounts of an appropriate Lewis acid (e.g. $ZnCl_2$), react with diethyl [$^{14}$C]malonates to provide dialkyl ethoxymethylene[$^{14}$C]malonate (**339**) which serve as highly valuable intermediates

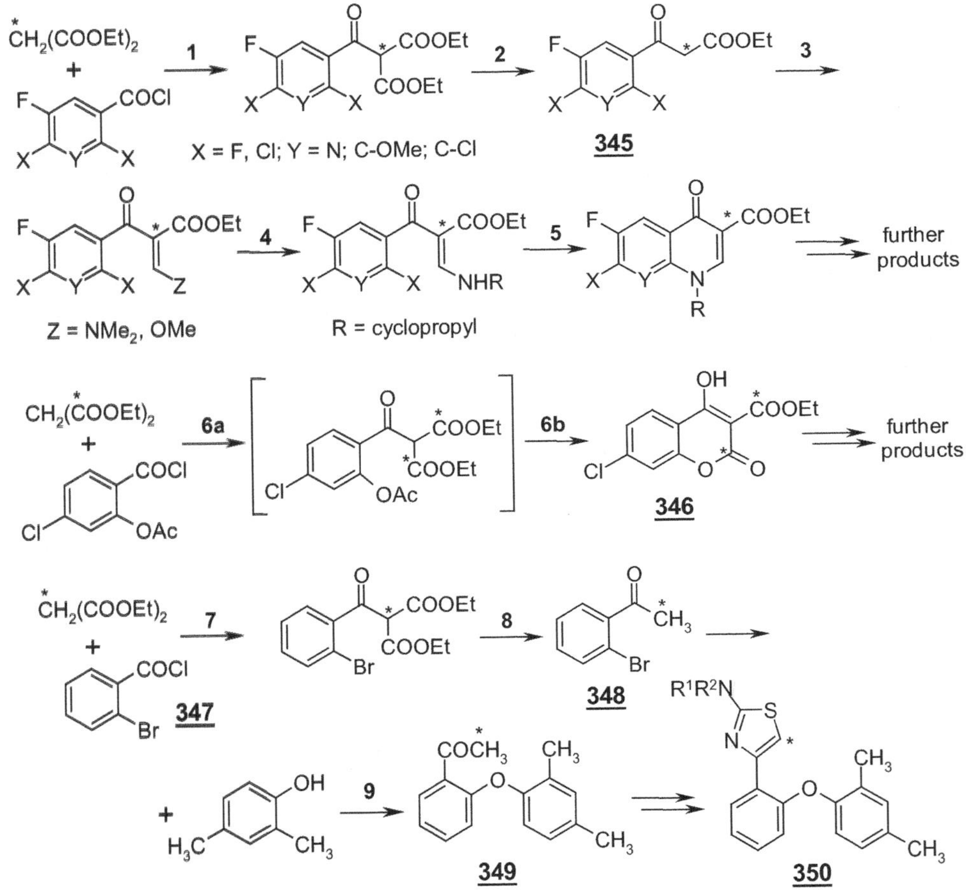


**Reaction conditions: 1a**. $Mg(OEt)_2$, ether; r.t., 2 h, **b**. acid chloride; **2**. *p*-TsOH, $H_2O$; 100–120 °C; 2 h; **3**. $Me_2NCH(OMe)_2$, $Ac_2O$, $CH_2Cl_2$; r.t.; 2 h **or** $HC(OEt)_3$, $Ac_2O$; 140 °C, 4 h; **4**. $R'NH_2$, EtOH; r.t., 1 to 16 h; **5**. $K_2CO_3$, DMF; 90–100 °C, 2–3 h or *t*-BuOK, DMSO, reflux, 1 h; **6a**. NaH, THF; 0°C, 30 min, then acid chloride, **b**. NaOMe; reflux, 30 min; **7a**. $MgCl_2$; $CH_3CN$; 0 °C, 30 min, **b**. RCOCl, $Et_3N$; 0 °C, 2 h; r.t., 15 h; **8**. $HOAc/H_2SO_4$; reflux, 7 h; **9**. CuBr, pyridine; reflux, 7 h

***Figure 6.105*** *α-Acylation of [$^{14}$C]malonates; synthesis and applications of β-keto [2-$^{14}$C] esters*

for the preparation of various binuclear heteroaromatic compounds. Reaction with aromatic or heteroaromatic amines converts these intermediates into alkyl *N*-aryl/heteroaryl-substituted 2-alkoxycarbonyl-3-aminoacrylates (**340**). Subsequent heating of intermediates **340** in a high boiling inert solvent (e.g., diphenyl ether, Dowtherm A) or refluxing with $POCl_3$ in toluene afford 3,4-substituted quinolines **341**[216] and **342**[193], the 1,5-naphthyridine **343**[217] and the 1,8-naphthyridine **344**[218], labeled either at C3 or C4 and the carboxyl position, depending on the labeling pattern of the dialkyl [$^{14}$C]malonate employed.

As the C4 hydroxyl group can be readily replaced with halogen (Cl, Br) and the ester can be split off by saponification and decarboxylation, this approach opens access to a variety of multiply functionalized labeled heteroaromatic compounds.

### *α-Acylation of [$^{14}$C]Malonates*

The reaction of acyl halides or imidazolides with dialkyl [$^{14}$C]malonates, or vice versa of labeled acyl halides with unlabeled malonates, followed by selective ester hydrolysis and decarboxylation has become standard methodology for the preparation of carbon-14-labeled *β*-keto esters (see Section 6.4)[166]. Besides its application to the preparation of various low molecular weight [$^{14}$C]building blocks and intermediates, it has been widely exploited for the synthesis of a broad range of binuclear heterocyclic compounds such as [$^{14}$C]naphthyridines (**345**, Y = N)[220], [$^{14}$C]quinolones (**345**, Y = CH)[221] and [$^{14}$C]coumarins (**346**)[222], as well as for the synthesis of [$^{14}$C]methyl aryl ketones, as illustrated by the examples in Figure 6.105[223]. In the first example, ring closure was effected by treatment of the *β*-keto ester with triethyl orthoformate and *N,N*-dimethylformamide dimethyl acetal, respectively, replacement of the resultant vinylic OMe or $NMe_2$ group by the respective amine, and nucleophilic displacement of the *o*-halogen substituent. This method, which is a reversal of the synthesis sequence presented in the previous Section, has been exploited for the synthesis of various carbon-14-labeled antibacterial drug candidates such as gemifloxacin[220a], moxifloxacin[220b] and cinafloxin[220c]. In the second case, treatment of the *β*-keto diester *in situ* with sodium methoxide and intramolecular cyclization of the resulting phenolate anion provided the coumarin derivative **346**. Finally, saponification and decarboxylation converted *o*-bromobenzoyl chloride (**347**) directly into 2′-bromo[methyl-$^{14}$C]acetophenone (**348**). Subsequent CuBr-catalyzed Ullmann coupling gave diaryl ether **349**, which served as a key intermediate for thiazolyl-substituted derivative **350**, an antimycoticum.

### *Knoevenagel Reaction*

The reactivity of the methylene group of alkyl malonates is sufficient to permit successful Knoevenagel condensations with aldehydes and reactive ketones. These reactions are usually run under weakly basic or neutral conditions in the presence of catalytic amounts of secondary amines (e.g. piperidine), or ammonium or piperidinium acetate[224]. In most cases the *β*-hydroxy esters initially formed cannot be isolated, but dehydrate to give the respective *α,β*-unsaturated diesters. These decarboxylate upon saponification and heating under acidic conditions, leading to *α,β*-unsaturated acids, which may isomerize to the corresponding *β,γ*-unsaturated derivatives. Less reactive ketones may provide good results, if either alkyl cyanoacetates (see Chapter 7) or Meldrum's acid (see Section  below) are employed as reaction partners, or the reaction is performed in the presence of $TiCl_4$ and pyridine (Lehnert procedure[225,226] Figure 106).

Even the methylene group of malonic acid is sufficiently active to couple with aldehydes, if the reaction is performed in pyridine solution in the presence of catalytic amounts of piperidine (Knoevenagel–Doebner modification). Under these conditions only $\alpha,\beta$-unsaturated acids can be isolated as reaction products, with the larger substituent *trans*-oriented to the carboxyl function. This methodology has found wide application in carbon-14 synthesis[227] using either the malonate component or the respective aldehyde for the introduction of the label. In the first example the free carboxyl function of the Knoevenagel product, the *p*-substituted cinnamic acids **351**, obtained in >80% yield, were coupled to substituted pyrrolidines using standard EDAC/HOBt-conditions to afford three potent histamine $H_3$ receptor antagonists **352** in labeled form[228]. In the second example halodecarboxylation of the carboxyl function of intermediate **354**, obtained from [2-$^{14}$C]malonic acid and 1-$\beta$-D-arabinofuranosyl-5-formyluracil (**353**), gave a 27% overall yield of the (*E*)-5-(2-bromo[2-$^{14}$C]vinyl)uracil derivative **355**, a potent antiviral agent[229]. Small heterocyclic aldehydes were condensed with [$^{14}$C]malonic acid to produce in one step $\beta$-heteroaryl[2-$^{14}$C]acrylic acids for use in DNA photolabeling experiments. Yields were fair to good, with some of the products (such as **356**) being thermally labile[227a].

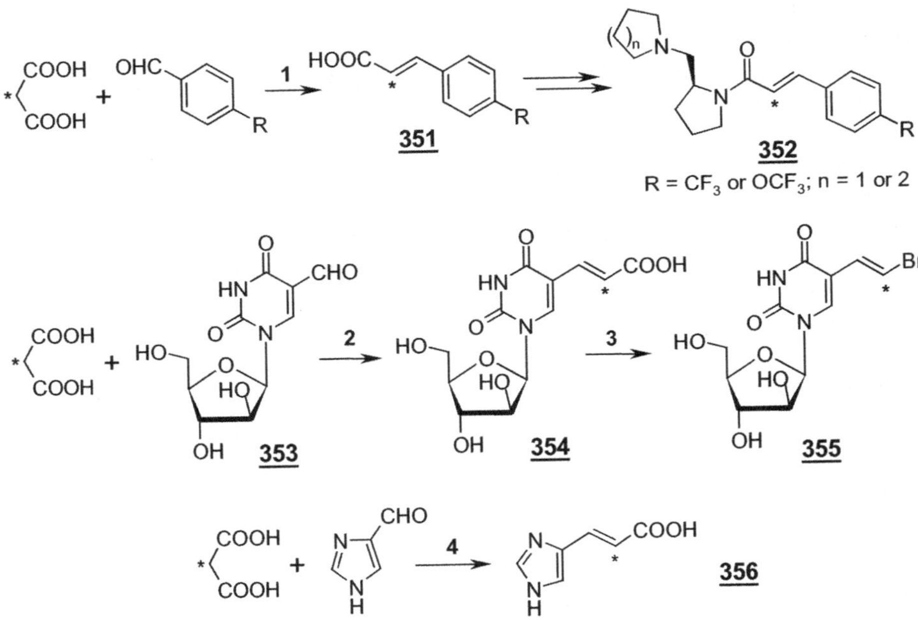


**Reaction conditions:** **1**. piperidine (cat.), pyridine; 90 °C, 3 h; **2**. piperidine (cat.), pyridine; 100 °C, 20 min; **3**. NBS, DMF, $H_2O$; r.t., 10 min; **4**. piperidine (cat.), pyridine; reflux, 2 h

***Figure 6.106*** *Applications of Knoevenagel condensations to [$^{14}$C]malonic acid and aryl/heteroaryl aldehydes*

Furthermore, catalytic hydrogenation of the $\alpha,\beta$-unsaturated acids has been introduced as a convenient alternative to the two carbon chain extension sequence effected by alkylation of dialkyl malonates with primary alkyl and aryl alkyl halides, as exemplified by the synthesis of [$^{14}$C]E2020 (<u>357</u>), a compound with acetylcholinesterase inhibitor activity[227f,230]. Microwave experiments on the Knoevenagel condensation of benzaldehyde and malonic acid in the presence of ammonium acetate under solvent-free conditions have been promising in giving yields of 67–73%[231]. Unfortunately, only carbon-13 labels were employed in these experiments. As many carbon-14-labeled acrylic acids are prone to radiation-induced degradation, the success of this procedure with radioactive components and the stability of the product cannot be certain.

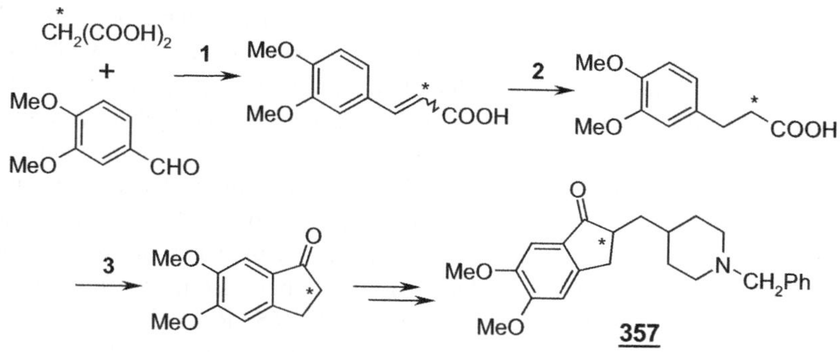


**Reaction conditions**: **1**. piperidine, pyridine; 55 °C, 1 h, 85 °C, 2 h; **2**. $H_2$, 5 % Pd/C, MeOH; r.t.; **3a**. oxalyl chloride, benzene; r.t., 16 h, **b**. $AlCl_3$, benzene; 0 °C, 45 min; r.t., 5 h

***Figure 6.107*** *Use of the Knoevenagel approach for chain extension by two carbon atoms*

Condensations of [$^{14}$C]malonic esters with formaldehyde to give labeled $\alpha,\beta$-unsaturated esters and derivatives are equivalent to those with malonic esters and [$^{14}$C]formaldehyde[232,233]; see Chapter 5, Section 5.4.1.

One example of the condensation of malonate with a carbon-14-labeled ketone is presented in Figure 6.108. In this case, the ketone <u>358</u> was prepared in carbon-14-labeled

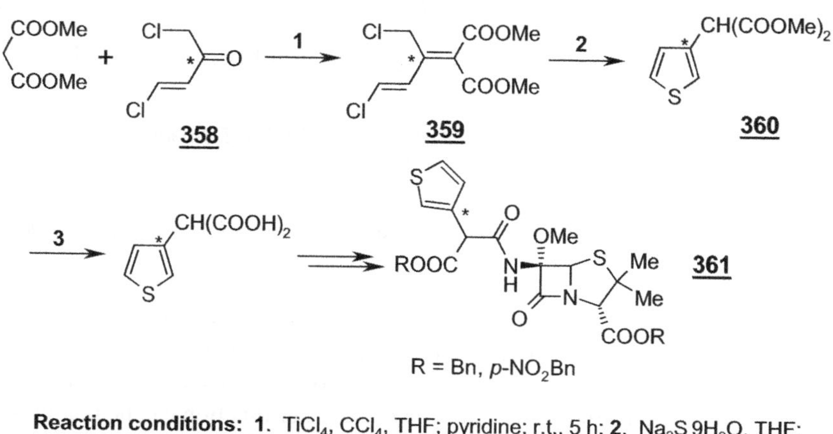


**Reaction conditions:** **1**. $TiCl_4$, $CCl_4$, THF; pyridine; r.t., 5 h; **2**. $Na_2S.9H_2O$, THF; r.t., 20 h; **3**. aq. NaOH; r.t., 1.5 h

***Figure 6.108*** *Knoevenagel condensation of malonate with a [$^{14}$C]ketone*

form by the Lewis acid-catalyzed addition of chloro[1-$^{14}$C]acetyl chloride to acetylene. Condensation of **358** with dimethyl malonate using the Lehnert procedure provided a 64% yield of [$^{14}$C]pentadienoate **359**. Subsequent treatment with sodium sulfide furnished the [3-$^{14}$C]thiophene derivative **360**, the key intermediate in the preparation of the penicillin analog [$^{14}$C]temocillin (**361**)[226].

*Michael Additions*

As in nonisotopic organic chemistry, dialkyl [2-$^{14}$C]malonates have been used for the extension of the carbon skeleton of $\alpha,\beta$-unsaturated carbonyl systems by two-carbon units through Michael addition. This is documented by the preparation of racemic [2-$^{14}$C]jasmonic acid (**365**, R = (*Z*)-3-pentenyl)[234] and its 9,10-dihydro derivative (**365**, R = pentyl)[235]. In both series, reaction of deprotonated diethyl [2-$^{14}$C]malonate with the 2-substituted 2-cyclopenten-1-ones **362** provided chromatographically separable mixtures of the respective *cis*- and *trans*-isomers **363** and **364**. Submitted to saponification and decarboxylation under neutral conditions by heating with water under high pressure (in order to suppresses the retro-Michael addition), both isomers afforded identical 9:1-mixtures of racemic *trans*-[2-$^{14}$C]jasmonic acid (**365**, R = Z-3-pentenyl) and iso-[2-$^{14}$C]jasmonic acid (**366**, R = (*Z*)-3-pentenyl) and the corresponding dihydro derivatives in 40–80% overall yields.

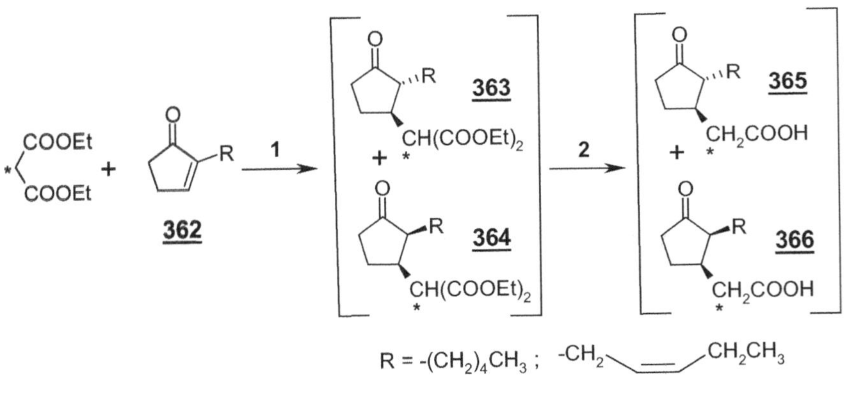


**Reaction conditions: 1.** MeONa, MeOH, r.t., 1.5 h (for R = Z-3-pentenyl); 35 °C, 20 h (for R = pentyl); **2.** $H_2O$, 225 °C, 18–20 h

***Figure 6.109*** *Preparation of racemic [2-$^{14}$C]jasmonic acid and its 9,10-dihydro derivative*

The addition of dialkyl [2-$^{14}$C]malonates to acyclic $\alpha,\beta$-unsaturated ketones has been used more frequently in carbon-14 synthesis. The resulting primary adducts spontaneously cyclize in a tandem Michael–Claisen reaction to give 6-substituted 4-hydroxy-2-oxo[1-$^{14}$C]cyclohex-l-encarboxylate esters **367**. The ester function can be readily split off by saponification and decarboxylation to give the 5-alkyl-3-hydroxy-2-[4, 6-$^{14}C_1$]cyclohexen-1-ones (5-alkyl dihydro[4, 6-$^{14}C_1$]resorcinols) **368**. Both key intermediates can either be further substituted at C3 or, for R$^1$ = H, oxidized to give highly functionalized [1-$^{14}$C]benzoic acid **369** or 1,3-dihydroxy[4, 6-$^{14}C_1$]benzene **370** derivatives.

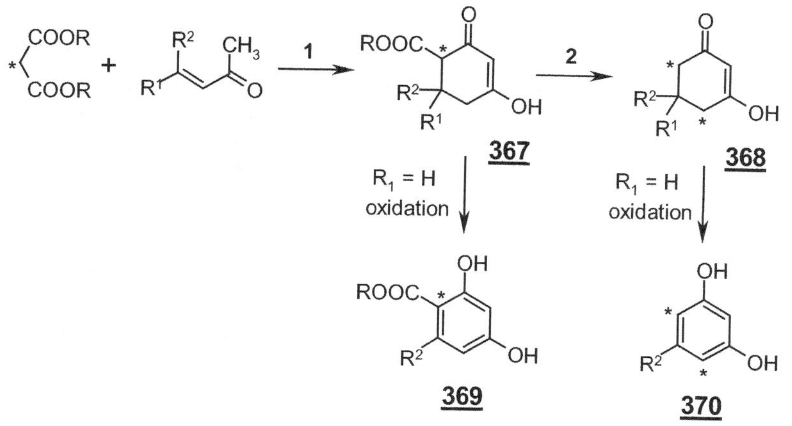


**Reaction conditions:** **1.** MeONa, MeOH; reflux, 3–5 h; **2a.** 10–20 % NaOH, 80–90 °C, 2.5 h, **b.** conc. HCl, 90 °C, 1 h

***Figure 6.110*** *Generic depiction of Michael–Claisen reactions with [2-$^{14}$C]malonates*

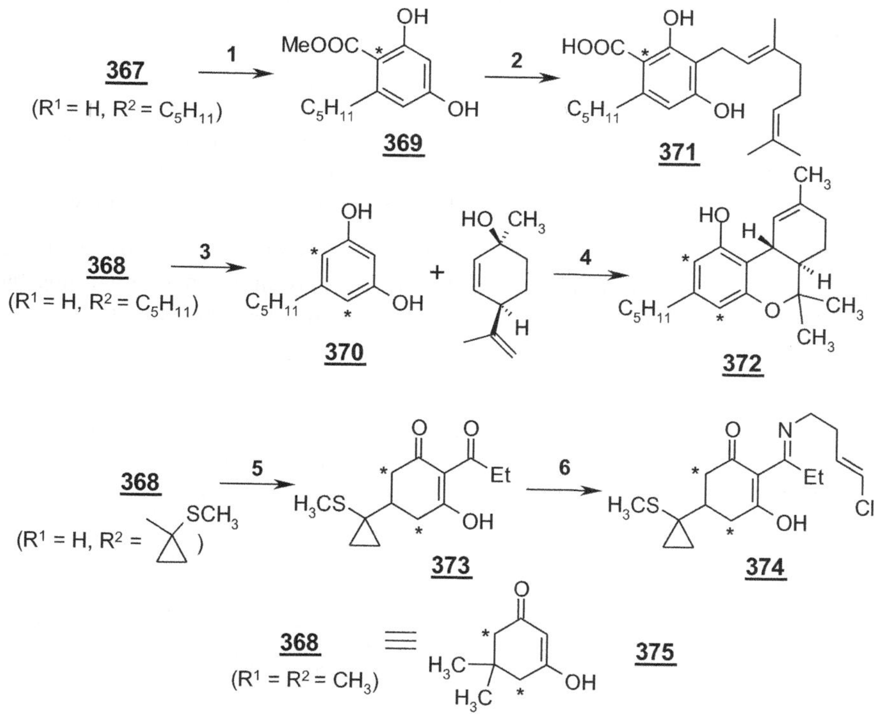


**Reaction conditions:** **1a.** $Br_2$, AcOH; r.t., 3 h, **b.** Raney-Ni; 2N NaOH; 0 °C, 30 min; **2a.** *n*-BuLi, benzene; r.t., 1 h, **b.** geranyl bromide, benzene; reflux, 4 h, **c.** PrSLi, HMPA; r.t., 2.5 h; **3.** $Hg(OAc)_2$, AcOH; reflux 3 h; **4a.** TsOH, benzene; reflux, 2.25 h, **b.** $ZnCl_2$, HCl (g); r.t., 16 h; **5a.** EtCOCl, $Et_3N$, AcOEt, **b.** 4-DMAP, toluene; 110 °C, **c.** $H_2NOCH_2CH{=}CHCl$ . HCl, $K_2CO_3$, MeOH

***Figure 6.111*** *Examples of [2-$^{14}$C]malonate Michael–Claisen reactions in synthesis*

The latter alternative was applied to the syntheses of [1-$^{14}$C]cannabigerolic acid (**371**)[236] and $\Delta^9$-6a,10a-*trans*-tetrahydro[2, 4-$^{14}C_1$]cannabinol (**372**)[237] (Figure 6.111). In the first case, oxidation of **367** ($R^1 = H, R^2 = C_5H_{11}$) was achieved in a two-step process by treatment with bromine followed by dehydrogenation with Raney nickel to give methyl [$^{14}$C]olivetolate (**369**, $R = C_5H_{11}$). Deprotonation, alkylation with geranyl bromide, and cleavage of the ester function converted the methyl [$^{14}$C]olivetolate into **371** in 50–60% radiochemical yield.

Intermediate **368** ($R^1 = H, R^2 = C_5H_{11}$) could be oxidized with $Hg(OAc)_2$ to give olivetol (5-pentyl[2, 4-$^{14}C_1$]resorcinol, **370**). Acid-catalyzed reaction with (+)-*trans-p*-mentha-2,8-dien-1-ol provided (−)-$\Delta^8$-[2, 4-$^{14}C_1$]THC, which upon sequential treatment with $ZnCl_2$-HCl and potassium *tert*-amylate isomerized to the requisite **372**[237]. For the preparation of [4, 6-$^{14}$C]CGA-215684 (**374**), the first alternative (substitution of **368** at C6) was followed[238]. Reaction with propionyl chloride in the presence of triethylamine provided the respective enolester, which on treatment with DMAP rearranged to give the corresponding *C*-acylated product **373**, the penultimate precursor to the target compound **374**. With $R^1$, $R^2 \neq H$, aromatization is not possible. Accordingly, Michael reaction of diethyl [2-$^{14}$C]malonate with mesityl oxide ($R^1$, $R^2 = CH_3$) stops at the dihydroresorcinol level. Saponification and subsequent decarboxylation provided 5,5-[6-$^{14}$C]dimethyl-1,3-cyclohexandione ([6-$^{14}$C]dimedone, **375**), which is commonly used in pharmacology for the radiometric detection of small concentrations of aldehydes (Knoevenagel condensation) in blood and urine[239].

*Reaction with Nitrogen Electrophiles*

There are several reagents with which C2 of a malonate can react as a nucleophile to form C–N bonds. The 'active' nitrogen atom of these reagents must therefore be electrophilic. The following sections describe those that have been used in carbon-14 synthesis.

*Reaction with nitric acid* Diethyl [2-$^{14}$C]malonate reacts with fuming nitric acid to give diethyl 2-nitro[2-$^{14}$C]malonate (**376**), which may serve as an alternative building block to *N*-acylamino[2-$^{14}$C]malonates, standard reagents for the preparation of carbon-

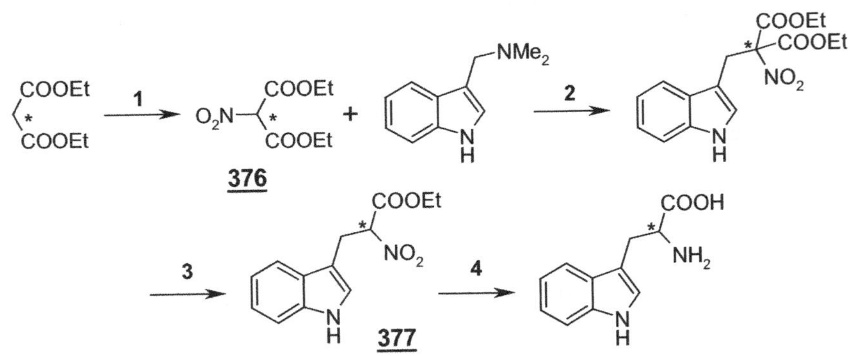


**Reaction conditions :** **1.** $HNO_3$; 15 °C, 4 h ; **2.** toluene; reflux, 4 h ; **3.** EtONa, EtOH-ether; r.t.,16 h; **4.** 10 % HCl; Δ; **5a.** $H_2NNH_2H_2O$, Raney-Ni, EtOH;reflux, 1 h , **b.** 1 N NaOH, reflux (24 % overall)

**Figure 6.112** *Synthesis of racemic [2-$^{14}$C]tryptophan via diethyl 2-nitro[2-$^{14}$C]malonate*

14-labeled amino acids. Alkylation of **376** with gramine, for example, followed by selective saponification of one of the ester functions and decarboxylation furnished ethyl $\alpha$-nitro-$\beta$-(3-indolyl)[2-$^{14}$C]propionate (**377**), which upon reduction of the nitro group and saponification provided racemic [2-$^{14}$C]tryptophan in an overall yield of 25%[240].

*Reaction with nitrous acid* Alkylations of dialkyl 2-acylamidomalonates **379**, **381** and **383**, and the reactions of the respective unlabeled building blocks with labeled alkylation agents, are widely used for the synthesis of $\alpha$-amino acids[241]. Carbon-14-labeled **379**, **381** and **383** contain both the latent $\alpha$-amino and carboxyl functions, as well as the isotopic label. More modern synthons, which in addition possess the required chiral information, are discussed in Chapter 11, Section 11.2.2 (see also ref. 26). Diethyl 2-acetamido[2-$^{14}$C]malonate (**379**), the most widely used building block, is prepared in two steps by treatment of diethyl [2-$^{14}$C]malonate with nitrous acid in glacial acetic acid followed by reduction of the 2-hydroxyimino intermediate **378** with zinc in acetic anhydride/acetic acid[242]. In some cases **379** has been replaced by dialkyl 2-formamido[2-$^{14}$C]malonate (**381**) or 2-phthalimido[2-$^{14}$C]malonate (**383**). Compound **381** is readily accessible by catalytic reduction of the hydroxyimino intermediate **378** and acylation of the free amino group of **380** with formic acetic anhydride[243]. The preparation procedure of compound **383** differs from that of **379** and **381**. In this case diethyl [2-$^{14}$C]malonate is firstly treated with bromine in carbon tetrachloride and the resulting dialkyl 2-bromo[$^{14}$C]malonate **382** further reacted with potassium phthalimide to give the building block in excellent yield[244].

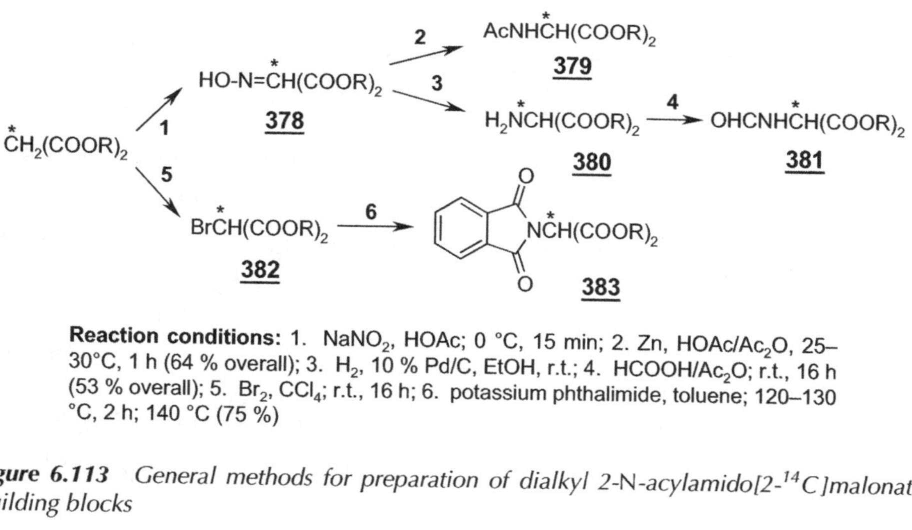


**Reaction conditions:** 1. $NaNO_2$, HOAc; 0 °C, 15 min; 2. Zn, HOAc/$Ac_2O$, 25–30°C, 1 h (64 % overall); 3. $H_2$, 10 % Pd/C, EtOH, r.t.; 4. HCOOH/$Ac_2O$; r.t., 16 h (53 % overall); 5. $Br_2$, $CCl_4$; r.t., 16 h; 6. potassium phthalimide, toluene; 120–130 °C, 2 h; 140 °C (75 %)

***Figure 6.113*** *General methods for preparation of dialkyl 2-N-acylamido[2-$^{14}$C]malonate building blocks*

Deprotonation of **379**, **381** and **383** can be accomplished with EtONa in ethanol, NaH in toluene or DMF, or with tetramethylguanidine in dichloromethane. Alkyl/benzyl bromides are usually employed as alkylation agents; chlorides and iodides are less commonly used. In a few cases Mannich bases have been used[247]. Also, Michael additions using $\alpha$,$\beta$-unsaturated carbonyl compounds and esters are feasible, as documented with the synthesis of (*R*/*S*)-[2-$^{14}$C]homoserine (from acrolein) and [2-$^{13}$C]- and [2-$^{14}$C]glutamic acid (from methyl acrylate)[248]. Typical reaction conditions are given in Figure 6.114. Depending on the reaction partners, reaction temperatures vary from ambient temperature to 130 °C, the reaction times from 30 min to 72 h and the overall radiochemical yields range from 40% to 90%. In spite of the fact that 379 does not always give the highest yields, its commercially availability and the simplicity of the two step sequence make its use still highly attractive.

In some cases the racemic products of this general approach have been separated enzymatically into their individual enantiomers via treatment of the respective *N*-acyl esters with $\alpha$-chymotrypsin (see Chapter 12).

*Reaction with methanesulfonyl azide* In a diazo transfer reaction, dimethyl [2-$^{14}$C]malonate reacts with methanesulfonyl azide and triethylamine to give the corresponding dimethyl

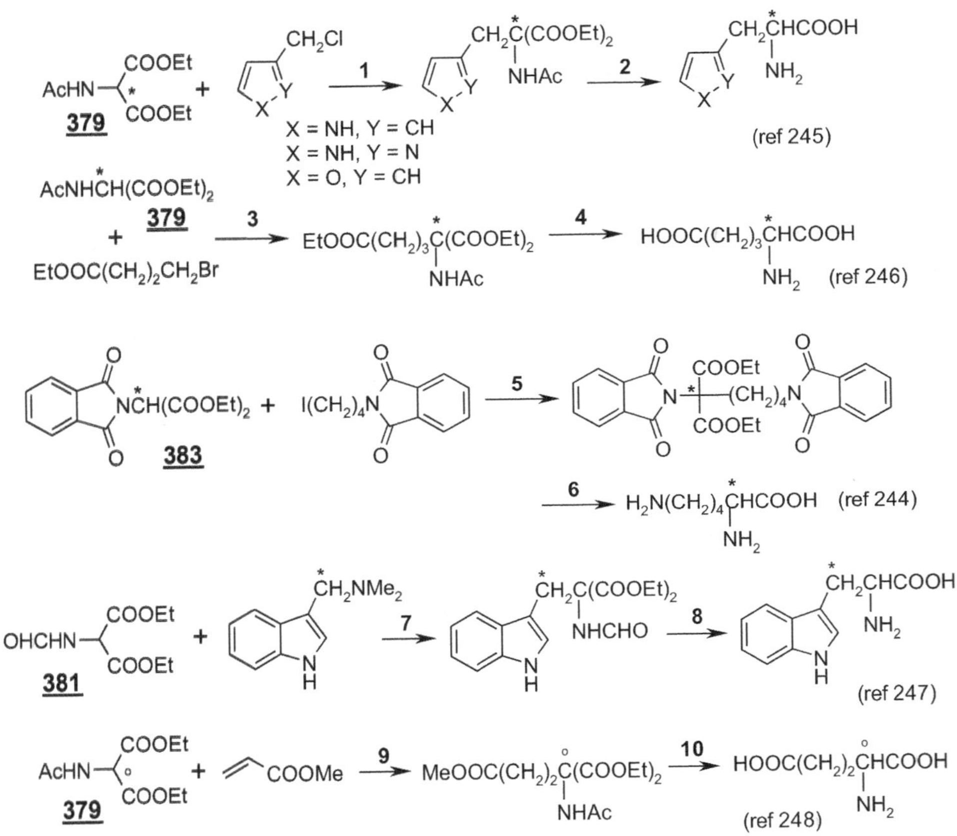


**Reaction conditions:** 1. NaOEt (2 equiv.), EtOH; r.t., 24 h; **2.** HOAc/conc. HCl; reflux, 5 h (95 %); **3.** NaH, toluene; 110 °C, 40 h; **4.** 6N HCl; reflux, 16 h (45 %) **5a.** NaOEt, EtOH, **b.** iodobutylphthalimide, 150–155 °C (neat); 3 h then toluene; reflux, 20 min; **6.** conc. HCl/HOAc; reflux, 20 h ; **7.** NaOH, xylene; reflux, 6 h; **8a.** 2N NaOH; reflux 6 h, **b.** HOAc, reflux, 4 h (44 %) **9.** tetrmethylguanidine, $CH_2Cl_2$; r.t., 72 h; **10.** 6 N HCl; reflux, 5 h (95 %)

**Figure 6.114** *Applications of N-acylamido[2-$^{13/14}$C]malonates and their unlabeled counterparts in the synthesis of racemic carbon-14-labeled $\alpha$-amino acids*

2-diazo[2-$^{14}$C]malonate (**384**). Heating in the presence of rhodium(II) diacetate dimer and 3-bromo-1-propanol transformed **384** into a bimolecular rhodium carbenoid, which inserted into the alcoholic O–H bond. Upon treatment with triethylamine, the resulting 3-bromopropyl ether **385** cyclized to give dimethyl [2-$^{14}$C]tetrahydrofuran-2,2-dicarboxylate (**386**), which served as a key intermediate in the synthesis of [$^{14}$C]SDZ 62-384 (**387**), an antitumor agent[249].

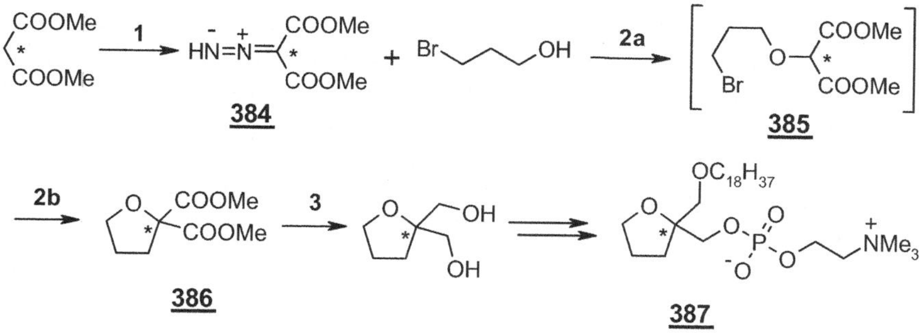


**Reaction conditions: 1**. $MsN_3$, $Et_3N$, THF; r.t., 3 d; **2a**. $Rh_2(OAc)_4$, $CH_2Cl_2$; reflux 18 h, **b**. $Et_3N$; r.t., 48 h; **3**. $LiAlH_4$, ether; 5 °C, 5 h

***Figure 6.115*** *Preparation and application of dimethyl 2-diazo[2-$^{14}$C]malonate*

*Reaction with diazonium salts* The C–H acidity of the methylene group of dialkyl [2-$^{14}$C]malonate is sufficient to couple with aryl diazonium salts using aqueous sodium acetate as a base. In the presence of a tautomerizable hydrogen, the azo compound initially formed (e.g., **388**) isomerizes to the corresponding hydrazone (e.g., **389**). Derivatives of this structural class have been used for the preparation of binuclear heteroaromatic systems. Saponification of the ester groups of hydrazone **389**, for example, generated

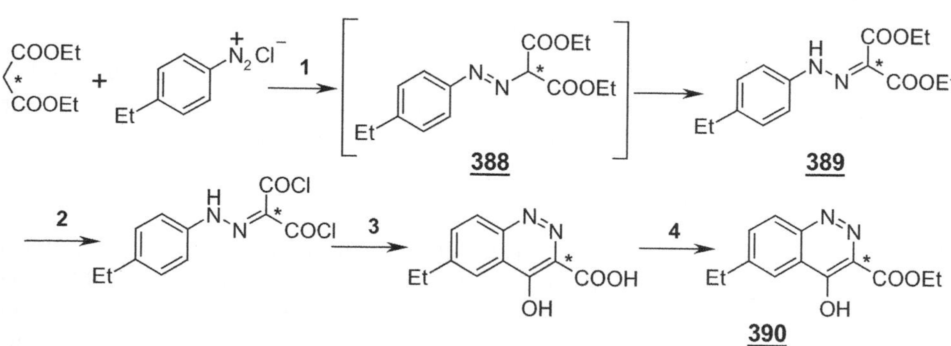


**Reaction conditions: 1**. aq. NaOAc, EtOH; –5 to 0 °C, 1,5 h; r.t. 4 h; **2a**. aq. NaOH/ethanol; r.t., 16 h, **b**. $SOCl_2$, toluene; r.t., 30 min; 70–75 °C, 1 h; **3**. $TiCl_4$, chlorobenzene; 95 to 100 °C, 6 h; **4**. $HCl_{gas}$, EtOH; r.t., 45 min; reflux 5 h

***Figure 6.116*** *Use of diazoniation of [2-$^{14}$C]malonate and its use in the synthesis of a 3,4,6-substituted [3-$^{14}$C]cinnoline*

from *p*-ethylbenzenediazonium chloride and diethyl [2-$^{14}$C]malonate, transformation of the free carboxyl groups into the di-acid chloride and its Lewis acid-catalyzed cyclization provided ethyl 6-ethyl-4-hydroxy-[3-$^{14}$C]cinnol-3-yl carboxylate ([$^{14}$C]ICI 75,186, **390**) in an overall radiochemical yield of 10%[250].

*Oxidation of the Methylene Group*

This type of reaction differs from those discussed so far, in that it does not belong to the class of electrophilic functionalization. Treatment of labeled dialkyl malonates with lead tetraacetate in acetic acid provided dialkyl 2-acetoxy[$^{14}$C]malonates (e.g., **391**). Reduction of **391** with $LiAlH_4$ converted it into [$^{14}$C]glycerol (**392**)[251]. This building block has been widely used in pharmacokinetic and pharmacological investigations of triglycerides. In carbon-14 synthesis it is a versatile precursor for various advanced compounds such as [$^{14}$C]quinolines (Skraup syntheses)[252], [$^{14}$C]triglycerides[253] and some additional small building blocks such as chloro[1-$^{14}$C]acetaldehyde[254].

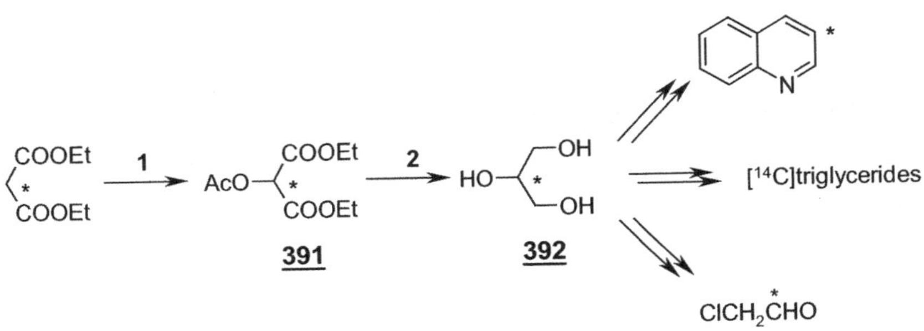


**Reaction conditions: 1.** $Pb(OAc)_4$, HOAc; 90–100 °C, 90 min; **2.** $LiAlH_4$, ether; –20 °C, 45 min; reflux, 1 h

***Figure 6.117*** *Preparation and applications of [2-$^{14}$C]glycerol*

## 6.5.2 Reactions at the Carboxyl Functions

The focus so far on reactions of the malonate methylene moiety does not diminish the utility of synthesis routes that exploit the reactivity on the ester (or acid) groups. In spite of their less frequent use, such strategies have important roles in appropriate cases.

*Reaction with N-nucleophiles*

The coupling of 3-*tert*-butyl [1-$^{14}$C]malonate (**393**) with the $\beta$-amino acid derivative **395** to give amide **396** was developed as an alternative route to the 1,4 addition of **395** to the corresponding [1-$^{14}$C]acrylate **394**, which might have failed due to radiation-induced polymerization. Subsequent treatment with 2,4-bis(4-methoxyphenyl)-1,3,2,4-dithiadiphosphetane-2,4-disulfide (Lawesson's reagent) followed by Raney nickel-mediated desulfurization of the thioamide function formed converted **396** into the target $\beta$-amino acid ester **397**[255]. Intramolecular ester condensation upon treatment of **397** with sodium methoxide and concluding saponification and decarboxylation of the *t*-butyl ester function afforded the trisubstituted 4-[2-$^{14}$C]piperidone derivative **398**.

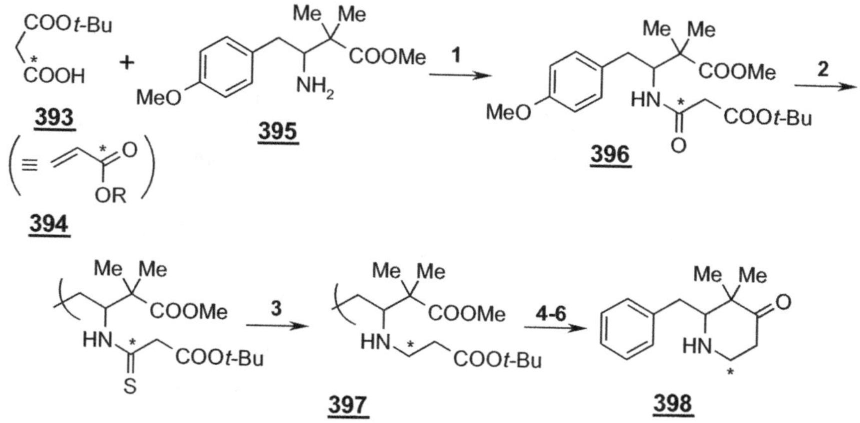


**Reaction conditions:1**. DCC, THF; r.t., 12 h; **2**. Lawesson's reagent, toluene, 100 °C, 4 h; **3**. Raney-Ni, aq. EtOH; **4**. NaOMe, toluene/DMF; 105 °C; **5**. aq. NaOH; reflux; **6**. HCl

***Figure 6.118*** *Use of monoalkyl [$^{14}$C]malonates as synthetic equivalents for [$^{14}$C]acrylates in 1,4-addition reactions*

Treatment of diethyl [2-$^{14}$C]malonate with two equivalents of amines (e.g. the mono-protected diamine **399**) and heating together neat gave [6-$^{14}$C]diamide **400**, which after reduction with borane and hydrolytic cleavage of the protecting group afforded *N'N''*-diethylnor[6-$^{14}$C]spermine (**401**), an anticancer compound[256] (Figure 6.119).

Treatment of diethyl [2-$^{14}$C]malonate with two equivalents of formamide and sodium ethoxide at elevated temperature provided 4,6-dihydro[5-$^{14}$C]pyrimidine (**402**), which served as a key intermediate in the synthesis of 5-amino-4-iodo[5-$^{14}$C]pyrimidine and the corresponding dehalogenated species, both important pharmacophoric substructures in the design of new pharmacologically active drug candidates[257].

However, the reactions with compounds bearing two $NH_2$- or NH-functionalities such as hydrazines, ureas, thioureas, guanidines and acetamidines afford further versatile building blocks for syntheses that are of more strategic relevance (Figure 6.120). In the first case, 4-substituted 5-hydroxy-3(1*H*)-[4-$^{14}$C]pyrazolones (or in the tautomeric form, 4-substituted 3,5-[4-$^{14}$C]pyrazolidinones) **403** are accessible. Phenylbutazone (**403**, $R^2$,$R^3$ = Ph; $R^1$ = Bu) is the most prominent representative, possessing pronounced analgesic and anti-inflammatory activities[258]. Reaction with the more complex benzo-fused dihydrotriazine **404** was also successful, affording [$^{14}$C]azapropazone (**405**)[259].

Sodium alkoxide-mediated reaction of 2,2-disubstituted [2-$^{14}$C]malonates with urea and thiourea produces (Figure 6.121), respectively, 5,5-disubstituted (1*H*,3*H*,5*H*)-[5-$^{14}$C]pyrimidine-2,4,6-triones **406** (or 5,5-disubstituted [5-$^{14}$C]barbiturates) and the corresponding 2-thioxo-(1*H*,5*H*)-[5-$^{14}$C]pyrimidine-4,6-diones **407**, some of which have been widely used as anesthetics (barbital, **406**: $R^1$,$R^2$ = Et, $R^3$,$R^4$ = H, and thiopental, **407**: $R^1$ = Et; $R^2$ = $CHMeCH_2CH_2CH_3$) or as sedative and hypnotic drugs (phenobarbital, **406**, $R^1$ = Et, $R^2$ = Ph, $R^3$,$R^4$ = H; hexobarbital, **406**, $R^1$ = Me; $R^2$ = cyclohexen-1-yl, $R^3$ = Me, $R^4$ = H)[260]. [5-$^{14}$C]Barbituric acid (**408**), the simplest member of this class and routinely prepared by two alternative procedures from urea and [2-$^{14}$C]malonic acid or its diethyl

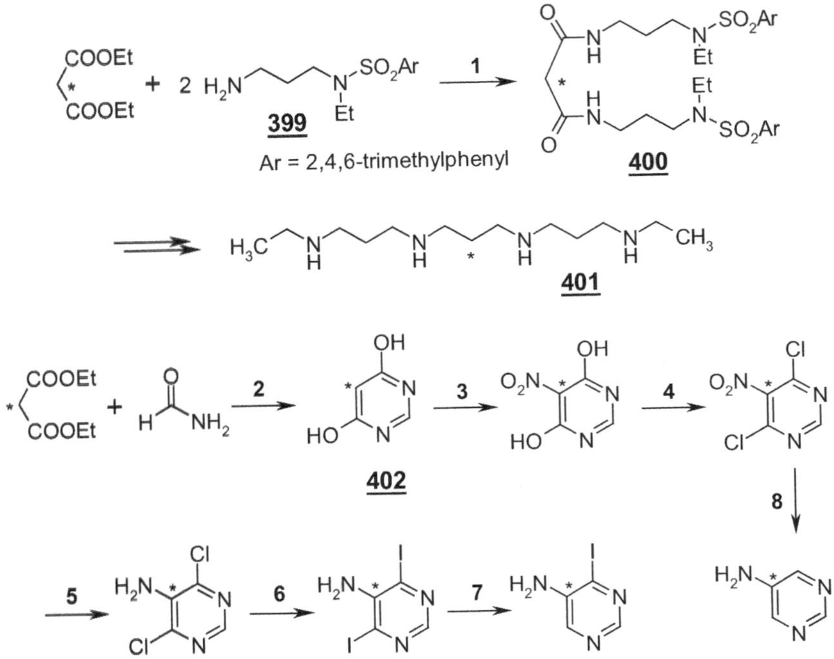

**Reaction conditions: 1**. neat components; 180 °C, 5 h; **2**. EtONa, EtOH; 110°C; 14 h; **3**. $HNO_3$; 0 °C, 2.5 h; r.t., 2 h; **4**. $POCl_3$; *N,N*-dimethylaniline; reflux, 2h; **5**. $SnCl_2$.2 $H_2O$, EtOH; reflux, 1 h; **6**. HI, NaI, $H_2O$; r.t., 48 h; **7**. *i*-PrMgCl, THF; –25°C, 45 min; **8**. 5 % Pd/C, $HCOONH_4$, MeOH; 0 °C, 30 min, r.t., 14 h

***Figure 6.119*** *Coupling of dialkyl [2-$^{14}$C]malonates with two equivalents of amine or formamide*

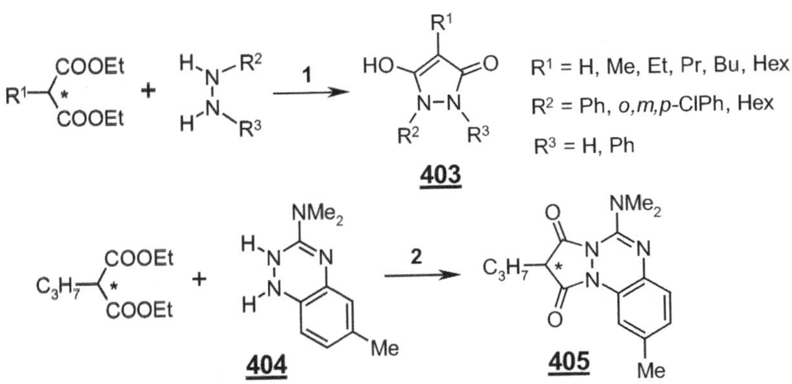

**Reaction conditions: 1**. neat components; 165 °C, 3 h (for $R_1$ = H, Me, Et, Pr) **or** EtONa, EtOH; reflux, 18–20 h (for $R_1$ = Bu, Hex); **2**. EtONa, xylene; 120 °C, 3 h

***Figure 6.120*** *Reactions of [$^{14}$C]malonates with substituted hydrazines*

ester[261], has served as a key intermediate for the preparation of (−)-[4a-$^{14}$C]riboflavin (vitamin B2, **410**), which had been obtained through heating with *N*-(1′-D-ribityl)-3,4-dimethyl(6-carboxyphenylazo)aniline (**409**)[261a].

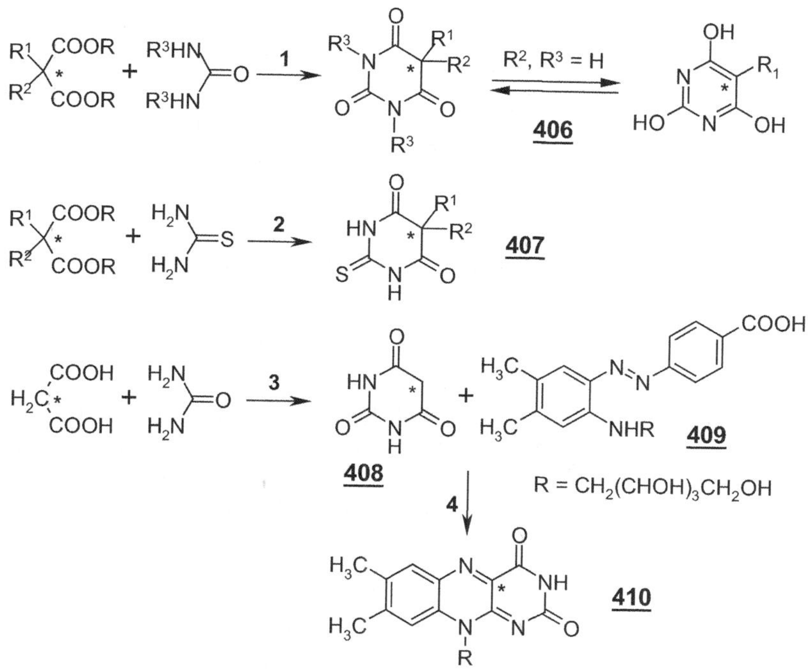


**Reaction conditions: 1**. NaOMe, MeOH; 60 °C, 5 h; **2**. NaOMe, MeOH; 60 °C, 5 h; **3**. $Ac_2O$, AcOH; 65 °C, 16 h; 90 °C, 3 h **or** NaOEt, EtOH; reflux, 15 h; **4**. diethyl malonate, AcOH; 110–120 °C, 24 h

***Figure 6.121*** *Syntheses utilizing the reaction of [$^{14}$C]malonates with ureas and thioureas*

Depending on the substitution pattern and the reagent applied, the hydroxyl functions of barbiturates can be readily replaced with halogen (Figure 6.122). Chlorodehydroxylation of 1,3-dimethylbarbiturate (**411**, accessible from diethyl [2-$^{14}$C]malonate and *N,N*-dimethylurea) with $POCl_3$, for example, provided 1,3-dimethyl-4-chloro[5-$^{14}$C]uracil, whose halogen was replaced with the 4-aryl substituted-1-(γ-aminopropyl)piperazine to give [$^{14}$C]urapidil (**412**), an antihypertensive agent[262].

Guanidines and acetamidines react analogously (Figure 6.123). With guanidine (R = H) and its isopropyl derivative (R = *i*-Pr), 2-amino-4,6-dihydroxy-pyrimidines **413** are formed, labeled either at C5 or at the C4 and C6. In these cases also, the hydroxyl group can be readily replaced with halogen through treatment with $POX_3$ (X = Br,Cl) or $PhPOCl_2$, which subsequently can be split off hydrogenolytically (e.g., to obtain **414**[194]), or substituted with appropriate nucleophiles to produce, for example, 2-amino-4,6-dimethoxy[5-$^{14}$C]pyrimidine **415**, an intermediate in the synthesis of [$^{14}$C]MON 37500[96] (**416**), a sulfonylurea herbicide.

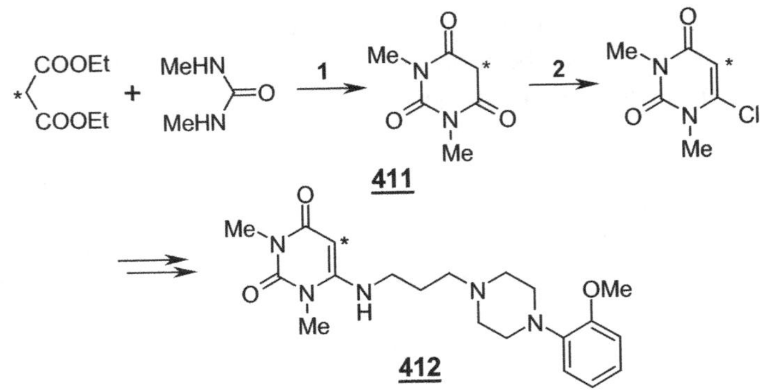


**Reaction conditions:** **1**. EtONa, EtOH; reflux; **2**. $POCl_3$, $H_3PO_4$; reflux

***Figure 6.122*** *Chlorodehydroxylation of [$^{14}C$]barbiturates and application*

The same type of reaction applied to diethyl [1, 3-$^{14}C_2$]malonate and [1-$^{14}C$]acetamidine gave 4,6-dihydroxy-2-methyl[2, 4, 6-$^{14}C_3$]pyrimidine (**417**). Subsequent nitration and replacement of the hydroxyl group with chlorine converted **417** into the requisite coupling component for the preparation of [$^{14}C_3$]moxonidine (**418**), an antihypertensive drug[263].

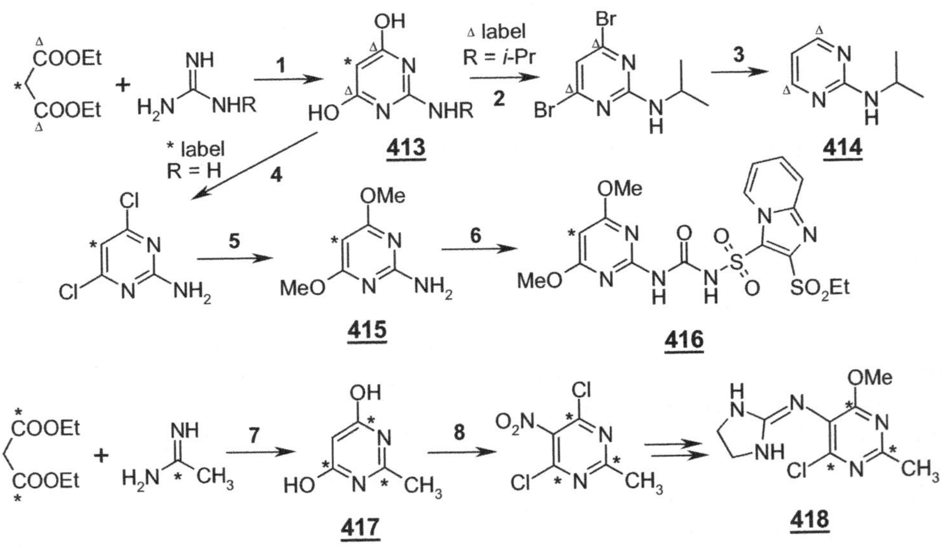


**Reaction conditions:** **1**. 25 % NaOMe, MeOH; reflux, 3.5 h (70–99 %); **2**. $POBr_3$, toluene; reflux, 1.5 h; **3**. $H_2$, Pd/C, NaOH, EtOH; r.t.; **4**. $POCl_3$, $Et_3N$, MeCN; reflux, 1 h; **5**. 25 % NaOMe, MeOH, reflux, 4 h; **6a**. $ArSO_2NH_2$, ClCOOPh, $Et_3N$, MeCN; r.t., 1 h, **b. 415**; 50–55 °C, 45 min (74 %); **7**. NaOEt, EtOH; reflux, 3 h (69%); **8a**. $HNO_3$, AcOH; r.t., 3 h (73 %), **b**. $POCl_3$, $PhNEt_2$; reflux, 2.5 h (66 %)

***Figure 6.123*** *Reactions of [$^{14}C$]malonates with guanidines and acetamidines*

*Meldrum's acid*

Treatment of [2-$^{14}$C]malonic acid with acetone in the presence of acetic anhydride and catalytic amounts of concentrated sulfuric acids provides [5-$^{14}$C]Meldrum's acid (**419**), one of the most versatile low molecular weight building blocks[19,167a,c,264]. As summarized in Figure 6.124, its reactivity at C2 is analogous with that of the malonate diesters already discussed. However, the reactivity of the initial adducts differs. For example, hydrolysis or alcoholysis (including *tert*-BuOH) of the initial adducts with alkyl halides[264b], aldehydes[264b] and acyl chlorides[19,167a,264a] gives directly the [2-$^{14}$C]carboxylic acids/esters, $\alpha,\beta$-unsaturated acids/esters and $\beta$-keto acids/esters, respectively, with simultaneous elimination of acetone and $CO_2$[264b]. In contrast, the 2-(alkoxymethylene)[2-$^{14}$C]malonate derivatives that are obtained[167c] when **419** reacts with ortho esters can undergo aminodealkoxylation to give enamides[19,167a,264a].

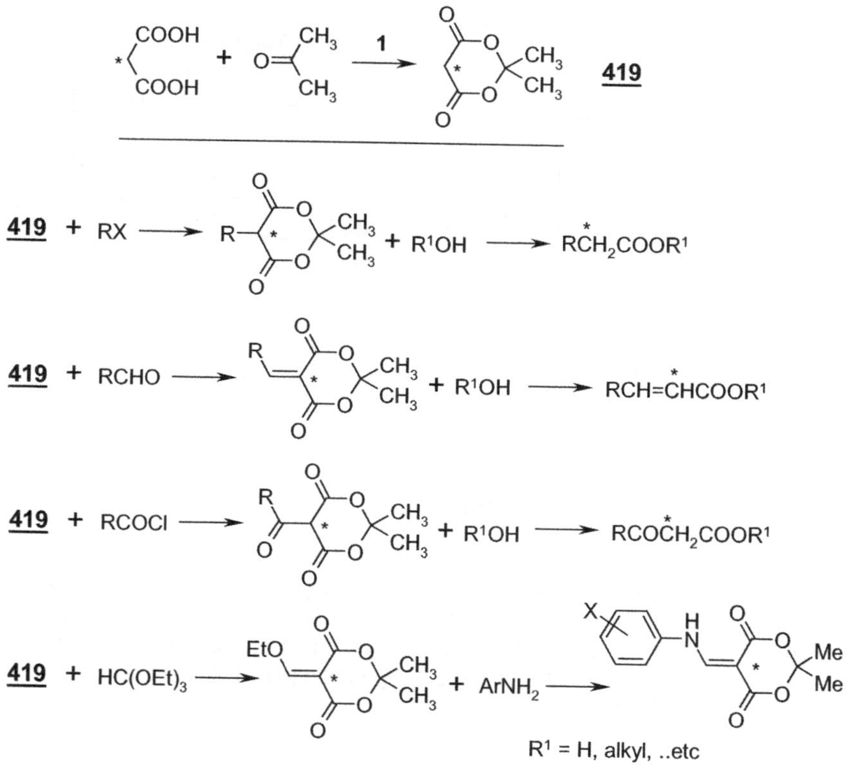


**Reaction conditions: 1.** $Ac_2O$, conc. $H_2SO_4$; 0 °C, 2 h

***Figure 6.124*** *Synthesis and summary of reactions of [5-$^{14}$C]Meldrum's acid*

Examples of the use of [$^{14}$C]Meldrum's acid are shown in Figure 6.125. In the first case, isopropyl 3-amino[2-$^{14}$C]crotonate (**420**), formed upon treatment of the $\beta$-ketoester underwent ring closure by sequential Michael addition to the $\alpha$-methoxycarbonyl substituted α,β-unsaturated carbonyl compound **421** and keto-enamine condensation to give [$^{14}$C]FPL 62129XX (**422**), an antihypertensive calcium channel antagonist[167a,264a]. Secondly, thermolysis of enamide **423** gave a mixture of isomeric [3-$^{14}$C]quinolin-4-ones **424** and **425**. Subsequent treatment of the desired isomer **424** with phosphorus oxychloride provided 4-chloro[3-$^{14}$C]quinoline (**426**), which opens access to ferroquine, an antimalaria drug substance[167c], as well as additional 4-substitued or unsubstituted (via hydrodehalogenation) derivatives.

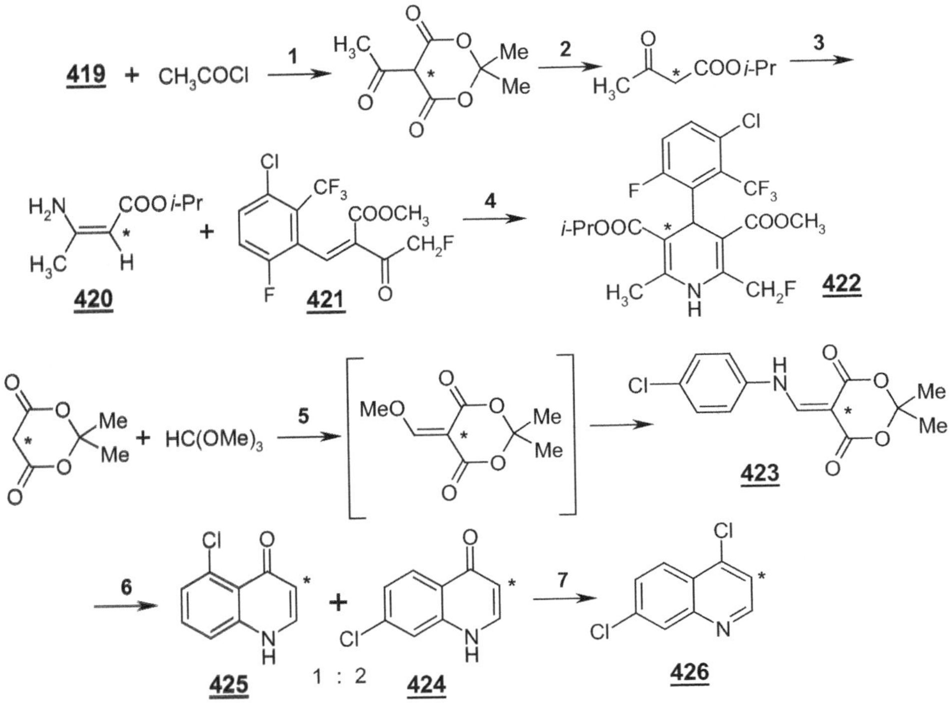


**Reaction conditions:** **1**. pyridine, $CH_2Cl_2$; 2 °C, 30 min; r.t., 40 min; **2**. isopropanol; reflux, 60 min; **3**. $NH_4OAc$, *tert*-BuOH; reflux, 4 h; **4**. *tert*-BuOH; 60 °C, 16 h; **5a**. 80 °C, 3 h, **b**. 3-chloroaniline; r.t., 1 h; **6a**. diphenyl ether, 260°C; reflux, 8 min; r.t.; 16 h; **b**. HPLC separation (MeOH/toluene/$Et_3N$); **7**. $POCl_3$; reflux, 50 min

**Figure 6.125** *Applications of [5-$^{14}$C]Meldrum's acid*

### *Hydride Reduction of the Ester Functions*

$LiAlH_4$ reduction of carbon-14-labeled diethyl malonate, followed by halogenation of the resulting 1,3-diol produces 1,3-dibromo[$^{14}$C]propane (Figure 6.126), a valuable precursor for the synthesis of [$^{14}C_n$]glutaronitrile, [$^{14}C_n$]glutaraldehyde and [$^{14}C_n$]glutaric acid[163a].

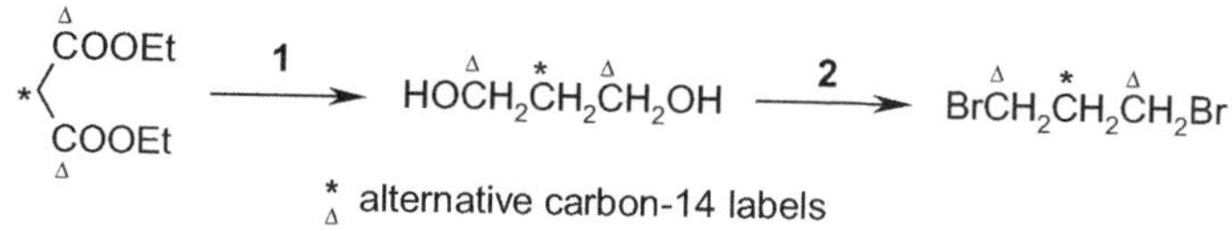


**Reaction conditions :** **1**. $LiAlH_4$, ether; reflux, 1.5 h; **2a**. $Ac_2O$; reflux, 5 h, **b**. 48 % HBr, reflux, 10 h

***Figure 6.126*** *Carbon-14 building blocks via reduction of [$^{14}$C]diethyl malonate*

### *Preparation and Applications of [$^{14}$C]Malonyl Dichloride*

Treatment of [$^{14}$C]malonic acid with oxalyl chloride in the presence of catalytic amounts of DMF affords [$^{14}$C]malonyl chloride in nearly quantitative yield. The structure of the reagent was confirmed by acylation of 2-methoxypropene, which provided a 3:1 mixture of phloroglucinol monomethyl ether (**428**) and 4-hydroxy-6-methyl-2*H*-[2, 4-$^{14}$C]pyran-2-one (**427**). Etherification of the latter with dimethyl sulfate followed by ring opening upon treatment with sodium methoxide and recyclization afforded **429**, which was combined with the original portion of **428** and etherified to give 1,3,5-trimethoxy-[1, 3, 5-$^{14}$C]benzene, which was used as a key intermediate in the synthesis of several carbon-14 labeled cholesterol acyltransferase inhibitors[265].

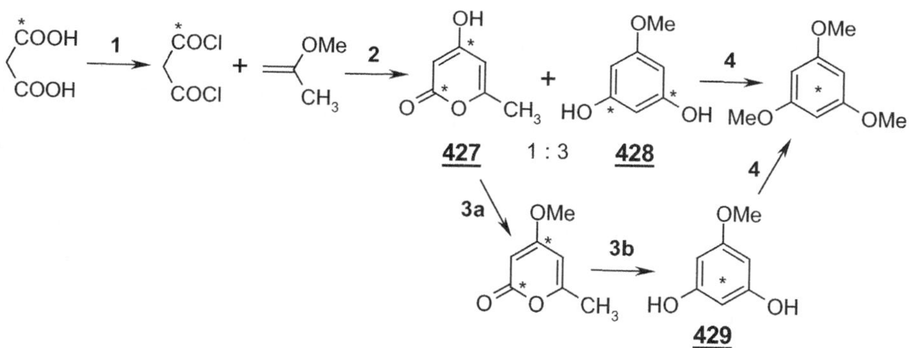


**Reaction conditions:** **1**. oxalyl chloride, cat. DMF; ether; reflux, 65 h; **2a**. ether; –20°C, 3 h; **b**. KOH, $Et_3NBn^+Cl^-$, $H_2O$; r.t., 4 h; **3a**. $(MeO)_2SO_2$, $Na_2CO_3$, acetone; reflux 4 h, **b**. NaOMe, toluene; reflux, 4 h; **4**. $(MeO)_2SO_2$, $K_2CO_3$, acetone; reflux, 44 h

***Figure 6.127*** *Preparation and applications of [$^{14}$C]malonyl dichloride*

## References

1. a. Murray, A. and Williams, D.L., *Organic Syntheses with Isotopes*; Interscience: New York, ***1958***, pp. 34–35; b. Bubner, M. and Schmidt, L.H., *Die Synthese Kohlenstoff-14 markierter organischer Verbindungen*; Verlag Georg Thieme: Leipzig, ***1966***, pp. 107–110.

2. a. Murray, A. and Williams, D.L., *Syntheses with Isotopes*; Interscience: New York, ***1958***, pp. 35–36; b. Cox, J.D.; Turner, H.S. and Warne, R.J., *J. Chem. Soc.*, ***1950***, 3167–3176; c. Bubner, M. and Schmidt, L.H., *Die Synthese Kohlenstoff-14 markierter organischer Verbindungen*; Verlag Georg Thieme: Leipzig, ***1966***, pp. 148–149.
3. Murray, A. and Williams, D.L., *Organic Syntheses with Isotopes*; Interscience: New York, ***1958***, pp. 580–582; b. Heidelberger, C.; Brewer, P. and Dauben, W.G., *J. Am. Chem. Soc.*, ***1947***, *69*, 1389–1390; c. Zwiebel, N.; Turkrvich, J. and Miller, W.W., *J. Am. Chem. Soc.*, ***1949***, *71*, 376–377.
4. Murray, A. and Williams, D.L., *Organic Syntheses with Isotopes*; Interscience: New York, ***1958***, pp. 40–41; b. Tolbert, B.M., *J. Biol. Chem.*, ***1948***, *173*, 205–206.
5. a. Negishi, E.; Swanson, D.R. and Rousset, C.J., *J. Org. Chem.*, ***1990***, *55*, 5409–5412; b. Bailey, W.F. and Punzala, E.R., *J. Org. Chem.*, ***1990***, *55*, 5409–5412; c. Wakefield, B.J., *Organolithium Methods, Best Synthetic Methods*; Academic Press: London, ***1988***, pp. 27–33.
6. a. Elbert, T. and Filip, J., *Radiochem. Radioanal. Letters*, ***1983***, *57*, 211–218; b. Abrams, R., *Experientia*, ***1947***, *3*, 488–489.
7. a. Murray, A. and Williams, D.L., *Organic Syntheses with Isotopes*; Interscience: New York, ***1958***, p. 376; b. Bubner, M. and Schmidt, L.H., *Die Synthese Kohlenstoff-14 markierter organischer Verbindungen*; Verlag Georg Thieme: Leipzig, ***1966***, p. 123; c. Lee, H.T. and Woo, P.W.K., *J. Label. Compd. Radiopharm.*, ***1999***, *42*, 129–133; d. Koltai, E.; Esses-Reiter, K. and Bánfi, D., *J. Label. Compd. Radiopharm.*, ***1982***, *19*, 13–18.
8. a. Murray, A. and Williams, D.L., *Organic Syntheses with Isotopes*; Interscience: New York, ***1958***, p. 375; b. Anker, H.S., *J. Biol. Chem.*, ***1948***, *176*, 1333–1335.
9. a. Murray, A. and Williams, D.L., *Organic Syntheses with Isotopes*; Interscience: New York, ***1958***, p. 380; b. Nakatsuka, I.; Hazue, M.; Makari, Y.; Kawahara, A.; Endo, M. and Yoshitaki, A., *J. Label. Compd. Radiopharm.*, ***1976***, *12*, 395–400.
10. Parnes, H. and de Keczer, S., *J. Label. Compd. Radiopharm.*, ***1985***, *22*, 287–292.
11. Kenner, G.W.; Rimmer, J.; Smith, K.M. and Unsworth, J.F., *J. Chem. Soc. Perkin Trans. I*, ***1977***, 332–340.
12. Shimada, T.; Yanagisawa, T.; Tomiyama, T. and Okazaki, M., *J. Label. Compd. Radiopharm.*, ***1994***, *34*, 79–83.
13. a. Murray, A. and Williams, D.L., *Organic Syntheses with Isotopes*; Interscience: New York, ***1958***, pp. 905–906; b. Cox, J.D. and Turner, H.S., *J. Chem. Soc.*, ***1950***, 3176–3180; c. Bubner, M. and Schmidt, L.H., *Die Synthese Kohlenstoff-14 markierter organischer Verbindungen*; Verlag Georg Thieme: Leipzig, ***1966***, p. 137.
14. a. Murray, A. and Williams, D.L., *Organic Syntheses with Isotopes*; Interscience: New York, ***1958***, pp. 861–863; b. Murray, A. and Ronzio, A.R., *J. Am. Chem. Soc.*, ***1952***, *74*, 2408–2409; c. Bubner, M. and Schmidt, L.H., *Die Synthese Kohlenstoff-14 markierter organischer Verbindungen*; Verlag Georg Thieme: Leipzig, ***1966***, p. 142.
15. Barbry, D. and Couturier, D., *J. Label. Compd. Radiopharm.*, ***1987***, *24*, 603–606.
16. Schweer, K.-H. and Schwarz, A., *Chem. Ber.*, ***1962***, *95*, 620–624.
17. Cohen, S.G.; Salih, E.; Solomon, M.; Howard, S.; Chishti, B.C. and Cohen, J.B., *Biochim. Biophys. Acta*, ***1989***, *997*, 167–175.
18. Minato, H.; Nagasaki, T.; Katsuyama, Y.; Yokoshima, T.; Suga, K. and Ueda, T., *J. Label. Compd. Radiopharm.*, ***1977***, *13*, 103–111.
19. Scherling, D., *J. Label. Compd. Radiopharm.*, ***1989***, *27*, 599–603.
20. Hicks, J.L.; Huang, C.C.; Goel, O.P. and Butler, D.E., *J. Label. Compd. Radiopharm.*, ***1984***, *21*, 937–944.
21. Ferdinandi, E.S., *J. Label. Compd. Radiopharm.*, ***1976***, *12*, 357–364.
22. Duelfer, T.; Duffin, R.J. and Youngstrom, R.E., *J. Label. Compd. Radiopharm.*, ***1988***, *25*, 855–863.
23. Karibe, N.; Nakamura, T.; Mishima, M. and Sugimoto, H., *J. Label. Compd. Radiopharm.*, ***1989***, *27*, 417–421.
24. a. Wheeler, W.J., *J. Label. Compd. Radiopharm.*, ***1992***, *31*, 477–487; b. Dhar, R.K., *Aldrichimica Acta*, ***1994***, *27*, 43–51; c. Brown, H.C. and Ramachandran, P.V., *Acc. Chem. Res.*, ***1992***, *25*, 16–24.
25. a. Moenius, T.; Burtscher, P.; Egger, H.; Bovermann, G. and Oberer, L., *J. Label. Compd. Radiopharm.*, ***1999***, *42*, 1053–1060; b. Corey, E.J.; Shibata, S. and Bakshi, R., *J. Org. Chem.*, ***1988***, *53*, 2861–2863.

26. Voges, R., *Synthesis and Applications of Isotopically Labeled Compounds 1994*; Allen, J. and Voges, R., Eds.; John Wiley & Sons, Ltd: Chichester, ***1995***, pp. 1–26.
27. Ciszewska, G.; Pfefferkorn, H.; Tang, Y.S.; Jones, L.; Tarapata, R. and Sunay, U.B., *J. Label. Compd. Radiopharm.*, ***1997***, *39*, 651–668.
28. a. Kenner, G.W., Rimmer, J., Smith, K.M. and Unworth, J.F., *J. Chem. Soc.,Perkin I*, ***1977***, 332–340; b. Kenner, G.W.; Smith, K.M. and Unsworth, J.F., *J. Chem. Soc. Chem. Commun.*, ***1973***, 43–44; c. McKillop, A.; Swann, B.P. and Taylor, E.C., *J. Am. Chem. Soc.*, ***1971***, *93*, 4919–4920.
29. a. Seltzer, S., *J. Label. Compd.*, ***1973***, *9*, 643–650; b. Brewitt, T.J.; Filer, C.N.; Lacy, J.M. and Peng, C.T., *J. Label. Compd. Radiopharm.*, ***2001***, *44*, 413–416.
30. Testa, B. and Jenner, P., *Drug Metabolism, Chemical and Biochemical Aspects*; Marcel Dekker: New York, ***1976***, pp. 143–151.
31. Pfeifer, S. and Zimmer, C., *Die Pharmazie*, ***1975***, *30*, 625–633.
32. Pilichowski, J.-F. and Godeneche, D., *J. Label. Compd. Radiopharm.*, ***1979***, *16*, 861–875.
33. Allen, J., The role of isotopes in the development of $\beta$-adrenergic blocking agents, in *Isotopes in the Physical and Biomedical Sciences*, Buncel, E. and Jones, J.R., Eds.; Elsevier: Amsterdam, ***1987***, pp. 426–505.
34. Illi, V.O.; Heindi, J.; Balzer, J. and Schwartz, W., *J. Label. Compd. Radiopharm.*, ***1981***, *18*, 707–711.
35. Barnes, R.D., *J. Label. Compd.*, ***1974***, *10*, 207–212.
36. Poullain, D. and Noel, J.-P., *J. Label. Compd. Radiopharm.*, ***1998***, *41*, 387–398.
37. Montelaro, R.C. and Rueckert, R.R., *J. Biol.Chem.*, ***1975***, *250*, 1413–1421.
38. Murray, A. and Williams, D.L., *Organic Syntheses with Isotopes*; Interscience: New York, ***1958***, pp. 412–413.
39. Rao, Ch.S.; Chu, J.-J.; Liu, R.-S. and Lai, Y.-K., *Biorg. Med. Chem.*, ***1998***, *6*, 2193–2204.
40. Anjaneyulu, B., *J. Label. Compd. Radiopharm.*, ***1985***, *22*, 745–750.
41. Brown, R.F.C. and Smith, R.J., *Austr. J. Chem.*, ***1972***, *25*, 607–624.
42. Banks, W.R.; Hawi, A.A. and Digenis, G.A., *J. Label. Compd. Radiopharm.*, ***1989***, *27*, 407–416.
43. a. Bringmann, G.; Pokorny, F.; Wenzel, M.; Wurm, K. and Schneider, C., *J. Label. Compd. Radiopharm.*, ***1997***, *39*, 29–38; b. Horton, W.J.; Adams, S.P. and Niermeyer, M.W., *J. Label. Compd. Radiopharm.*, ***1977***, *13*, 611–617; c. Le Sann, C.; Simpson, T.J.; Smith, D.I.; Watts, P. and Willis, C.L., *Tetrahedron Lett.*, ***1999***, *40*, 4093–4096; d. Clare, J.E.; Lawrie, K.W.M.; Willis, C.L. and Yuen J., *Synthesis and Applications of Isotopically Labeled Compounds*; Vol. 8, Dean, D.C.; Filer, C.N. and McCarthy, K.E., Eds.; John Wiley & Sons, Ltd: Chichester, ***2004***, pp. 1–4; e. Simpson, T., in *Synthesis and Applications of Isotopically Labeled Compounds 1994*; Allen, J. and Voges, R., Eds.; John Wiley & Sons, Ltd: Chichester, ***1995***, pp. 855–862.
44. a. Rajgarhia, V.B.; Priestly, N.D. and Strohl, W.R., *Anal. Biochem.*, ***1995***, *224*, 159–162; b. Clough, R.C.; Barnum, S.R. and Jaworski, J.C., *Anal. Biochem.*, ***1989***, *176*, 82–84; c. Hsi, R.S. P., *J. Label. Compd.*, ***1973***, *9*, 435–442.
45. a. Riordan, J.F. and Vallee, B.L., *Methods Enzymol.*, ***1967***, *11*, 565–570; b. Kas, J. and Rauch, P., *Top. Curr. Chem.*, ***1983***, *112*, 163–220.
46. a. Murray, A. and Williams, D.L., *Organic Syntheses with Isotopes*; Interscience: New York, ***1958***, pp. 416–417; b. Ropp, G.A., *J. Am. Chem. Soc.*, ***1950***, *72*, 2299–2301.
47. Tolbert, B.M.; Christenson, F.A.; Chang, F.N.-H. and Sah, P.P.T. *J. Org. Chem.*, ***1949***, *14*, 525–529.
48. Hays, S.J.; Hicks, J.L.; Butler, D.E. and Huang, C.C., *J. Label. Compd. Radiopharm.*, ***1990***, *28*, 15–24.
49. Turner, R.B., *J. Am. Chem. Soc.*, ***1950***, *72*, 579–585.
50. Comes, R.A. and Jenkins, R.W. Jr., *J. Label. Compd.*, ***1972***, *8*, 575–579.
51. Figge, K. and Voss, H.P., *J. Label. Compd.*, ***1973***, *9*, 23–42.
52. a. Grossman L. and Wisser, D.W., *J. Biol. Chem.* ***1954***, *209*, 447–450; b. Jezdic, V.; Razumenic, N.; Skakun, M.; Albahari, S. and Odavic-Josic, J., *J. Label. Compd.*, ***1970***, *6*, 88–98.
53. Bigg, D.C.H., *J. Label. Compd. Radiopharm.*, ***1976***, *12*, 571–575.
54. Beher, W.T.; Baker, G.D. and Madoff, M., *J. Biol. Chem.*, ***1959***, *234*, 2388–2390.
55. Gill, H.S., *J. Label. Compd. Radiopharm.*, ***1998***, *41*, 201–210.

56. Banerji, A. and Kalena, G.P., *J. Label. Compd. Radiopharm.*, ***1983***, *20*, 163–166.
57. Gut, M., *Helv. Chim. Acta*, ***1943***, *36*, 906–909.
58. Otto, P.Ph.H.L.; Besemer, A.C. and Wientjens, W.H.J.M., *J. Label. Compd.*, ***1970***, *6*, 111–123.
59. Hazue, M.; Miyake, K. and Endo, M., *J. Label. Compd.*, ***1969***, *5*, 220–229.
60. a. Bloom, A.J., *J. Label. Compd. Radiopharm.*, ***2004***, *47*, 266–267; b. Ott, D.G., *Syntheses with Stable Isotopes of Carbon, Nitrogen and Oxygen*; John Wiley & Sons Inc: New York, ***1981***, pp. 116, 134; c. Braden, G.A. and Hollstein, U., *J. Label. Compd. Radiopharm.*, ***1976***, *12*, 507–516; d. Rieker, A.; Scheffler, K. and Müller, E., *Liebigs Ann. Chem.*, ***1963***, *670*, 23–30; e. Bak, B.; Christiansen, J.J.; Lipschitz, L. and Nielsen, J.T., *Acta Chem. Scand.*, ***1962***, *16*, 2318–2320.
61. Climie, I.J.G. and Evans, D.A., *J. Label. Compd. Radiopharm.*, ***1977***, *13*, 311–313.
62. a. Fields, M.; Rothchild, S. and Leaffer, M.A., *J. Am. Chem. Soc.*, ***1952***, *74*, 2435–2436; b. Ropp, G.A., *J. Am. Chem. Soc.*, ***1950***, *72*, 4459–4461; c. Schepartz, B. and Gurin, S., *J. Biol. Chem.*, ***1949***, *180*, 663–673; d. Sakami, W.; Evans, W.E. and Gurin, S., *J. Am. Chem. Soc.*, ***1947***, *69*, 1110–1112; e. Bubner, M. and Schmidt, L.H., *Die Synthese Kohlenstoff-14 markierter organische Verbindungen*; Verlag Georg Thieme: Leipzig, ***1966***, pp. 133–134; f. Ott, D.G., *Syntheses with Stable Isotopes of Carbon, Nitrogen and Oxygen*; Wiley-Interscience: New York, ***1981***, pp. 33–34; g. Weygand, F. and Swoboda, O.P., *Chem. Ber.*, ***1956***, *89*, 18–21.
63. Murray, A. and Williams, D.L., *Organic Syntheses with Isotopes*; Interscience: New York, ***1958***, pp. 326–330.
64. Smith, M.B. and March, J., *March's Advanced Organic Chemistry*; 5th Edn, John Wiley & Sons, Inc: New York, ***2001***, pp. 777–778.
65. a. Ostwald, R., *J. Biol. Chem.*, ***1948***, *173*, 207–209; b. Murray, A. and Williams, D.L., *Organic Syntheses with Isotopes*; Interscience: New York, ***1958***, pp. 330–331; c. Bubner, M. and Schmidt, L.H., *Die Synthese Kohlenstoff-14 markierter organische Verbindungen*; Verlag Georg Thieme: Leipzig, ***1966***, pp. 132–133; d. Lee, C.C.; Kung, F.L.; Hahn, B.-S. and Robson, A.J., *J. Label. Compd*, ***1971***, *7*, 46–51.
66. a. Mezó, I. and Teplán, I., *J. Label. Compd.*, ***1972***, *8*, 359–363; b. Bloch, K., *J. Biol. Chem.*, ***1949***, *179*, 1245–1254; c. Kurumaya, K.; Okazaki, T.; Seido, N.; Akasaka, Y.; Kawajiri, Y. and Kajiwara, M., *J. Label. Compd. Radiopharm.*, ***1989***, *27*, 217–235; d. Yuan, S.-S. and Foos, J., *J. Label. Compd Radiopharm.*, ***1981***, *18*, 563–569.
67. a. Pozzi, O.R. and Arciprete, C.P., *J. Label. Compd. Radiopharm.*, ***1993***, *33*, 259–262; b. Roberts, J.L. and Poulter, C.D., *J. Org. Chem.*, ***1978***, *43*, 1547–1550; c. Cho, H.; Heide, L. and Floss, H.G., *J. Label. Compd. Radiopharm.*, ***1992***, *31*, 589–594.
68. Howard, H.R.; Shenk, K.D.; Smolarek, T.A.; Marx, M.H.; Windels, J.H. and Roth, R.W., *J. Label. Compd Radiopharm.*, ***1994***, *34*, 117–125.
69. Saunders, B.C. and Worthy, T.S., *J. Chem. Soc.*, ***1953***, 1929–1934.
70. Gillespie, J.M. and Springell, P.H., *Biochem. J.*, ***1961***, *79*, 280–286.
71. Dain, J.G., *J. Label. Compd. Radiopharm.*, ***1987***, *24*, 499–504.
72. Hartman, J.D.; Huang, C.C. and Butler, D.E., *J. Label. Compd. Radiopharm.*, ***1984***, *21*, 347–352.
73. Tamazawa, K. and Arima, H., *J. Label. Compd. Radiopharm.*, ***1982***, *19*, 681–688.
74. Tang, Y.S.; Jones, L. and Sunay, U.B., *J. Label. Compd. Radiopharm.*, ***1998***, *41*, 1–7.
75. Voges, R., *J. Label. Compd. Radiopharm.*, ***2002***, *45*, 867–897.
76. a. Bonnafous, J.C. and Mousseron-Canet, M., *Bull. Soc. Chim. Fr.*, ***1971***, 4551–4552; b. Azim, E.; Dupuy, J.-M.; Maurizis, J.-M.; Gaudreault, R.C.; Veyre, A. and Madelmont, J.-C. *J. Label. Compd. Radiopharm.*, ***1997***, *39*, 559–566.
77. Persson, T.; Johansen, S.K.; Martiny, L.; Givskov, M. and Nielsen, J., *J. Label. Compd. Radiopharm.*, ***2004***, *47*, 627–634.
78. Huang, C.C., *J. Label. Compd. Radiopharm.*, ***1987***, *24*, 675–681.
79. Sugiyama, I.; Woolley, G.T.; Mizuo, H. and Yamauchi, H., *J. Label. Compd. Radiopharm.*, ***1992***, *31*, 663–669.
80. Fidder, A.; Noort, D. and Benshop, H.P., *J. Label. Compd. Radiopharm.*, ***1999***, *42*, 269–274.
81. Raferty, M.J. and Gaskell, S.J., *J. Label. Compd. Radiopharm.*, ***1991***, *29*, 313–318.
82. Shu, A.Y.L.; Villani, A.J. and Heys, J.R., *J. Label. Compd. Radiopharm.*, ***1990***, *28*, 143–153.

83. a. Rodriguez, I. and Voges, R., *J. Label. Compd. Radiopharm.*, ***2000***, *43*, 169–176; b. Oppolzer, W. and Zhou, G. *PhD Thesis*, Université Geneve, ***1993***.
84. Voges, R. and Metz Y., in *Synthesis and Applications of Isotopically Labeled Compounds*; Vol. 7, Pleiss, U. and Voges, R., Eds.; John Wiley & Sons, Ltd: Chichester, ***2001***, pp. 307–310.
85. a. Murray, A. and Williams, D.L., *Organic Syntheses with Isotopes*; Interscience: New York, ***1958***, pp. 159–160; b. Ott, D.G., *Syntheses with Stable Isotopes of Carbon, Nitrogen and Oxygen*; Wiley-Interscience: New York, ***1981***, pp. 39–40; c. Bubner, M. and Schmidt, L.H., *Die Synthese Kohlenstoff-14 markierter organische Verbindungen*; Verlag Georg Thieme: Leipzig, ***1966***, pp. 154–155; d. Graul, E.H. and Steiner, B., *Arzneim.-Forsch.*, ***1958***, *8*, 94–98.
86. a. Watts, R.W.E. and Crawhall, J.C., *Biochem. J.*, ***1959***, *73*, 277–286; b. Panigot, M.J.; Fesik, S.W. and Curly, R.W., *J. Label. Compd. Radiopharm.*, ***1995***, *36*, 439–444; c. Ott, D.G., *Syntheses with Stable Isotopes of Carbon, Nitrogen and Oxygen*; Wiley-Interscience: New York, ***1981***, pp. 40–41
87. a. Adriaens, P. and Vanderhaeghe, H., Synthesis of some non-proteinogenic $\alpha$-amino acids, in *Isotopes in the Physical and Biomedical Sciences*, Vol. 1, Part A., Buncel, E. and Jones, J.R., Eds.; Elsevier: Amsterdam, ***1987***, pp. 330–402; b. Larsen, P.O. and Wieczorkowska, E., *J. Label. Compd. Radiopharm.*, ***1979***, *10*, 287–296; c. Larsen, P.O. and Wieczorkowska, E., *Acta Chim. Scand.*, ***1977***, *B31*, 109–113; d. Krasna, A.I.; Peyser, P. and Sprinson, D.B., *J. Biol. Chem.*, ***1952***, *198*, 421–426.
88. Ciszewska, G. and Ray T., in *Synthesis and Applications of Isotopically Labeled Compounds*; Vol. 7, Pleiss, U. and Voges, R., Eds.; John Wiley & Sons, Ltd: Chichester, ***2001***, pp. 299–301.
89. a. Nakajima, T.; Nakayama, K. and Shimizu, I., *J. Label. Compd. Radiopharm.*, ***2007***, *50*, 622–623; b. Schmidt, U.; Griesser, H.; Leitenberger, V.; Lieberknecht, A.; Mangold, R.; Meyer, R. and Riedl, B., *Synthesis*, ***1992***, 487–490; c. Schmidt, U.; Lieberknecht, A. and Wild, J., *Synthesis*, ***1984***, 53–60; d. Schmidt, U.; *Synthesis*, ***1988***, 159–172.
90. a. Burtscher, P.; Kohler, B.; Metz, Y.; Voges, R. and Wenger, R., *J. Label. Compd. Radiopharm.*, ***2000***, *43*, 205–216; b. Martin, A.; Chassaing, G. and Vanhove A., in *Synthesis and Applications of Isotopically Labeled Compounds 1994*, Allen, J. and Voges, R., Eds.; John Wiley & Sons, Ltd: Chichester, ***1995***, pp. 761–766.
91. Takatori, K.; Nishihari, M. and Kajiwara, M., *J. Label. Compd. Radiopharm.*, ***1999***, *42*, 701–708.
92. a. Dellaria, J.F. and Santasiero, B.D., *J. Org. Chem.*, ***1989***, *54*, 3916–3926; b. Dellaria, J.F. and Santasiero, B.D., *Tetrahedron Lett.*, ***1988***, *29*, 6079–6082.
93. Liebman, A.A., Isotopically Labeled Benzodiazepines in *Isotopes in the Physical and Biomedical Sciences*, Vol. 1, Part A., Buncel, E. and Jones, J.R., Eds.; Elsevier: Amsterdam, ***1987***, pp. 194–210.
94. a. Hsi, R.S.P., *J. Label. Compd.*, ***1974***, *10*, 389–398; b. Hsi, R.S.P. and Johnson, T.D., *J. Label. Compd. Radiopharm.*, ***1976***, *12*, 613–619; c. Hsi, R.S.P., *J. Label. Compd.*, ***1973***, *9*, 435–441.
95. a. Tegyey, Zs.; Maksay, G. and Otvös, L., *J. Label. Compd. Radiopharm.*, ***1979***, *16*, 377–385.
96. Maxwell, B.D.; Boyé, O.G. and Ohta, K., *J. Label. Compd. Radiopharm.*, ***2005***, *48*, 397–406.
97. a. Kanamaru, H.; Kamada, T.; Yoshitake, A. and Nakatsuka, I., *Radioisotopes*, ***1986***, *35*, 103–108; b. Kanamaru, H.; Kamada, T.; Yoshitake, A. and Nakatsuka, I., *Radioisotopes*, ***1986***, *25*, 169–173.
98. Bloom, A.J. and Winston, P.M., *J. Label. Compd. Radiopharm.*, ***1999***, *42*, 1004–1005.
99. Johnson, P.L., *J. Label. Compd. Radiopharm.*, ***2007***, *50*, 47–53.
100. Pawlowski, N.E.; Nixon, J.E.; Lee, D.J. and Sinnhuber, R.O., *J. Label. Compd.*, ***1974***, *10*, 45–48.
101. Maryanoff, B.E. and Reitz, A.B., *Chem. Rev.*, ***1989***, *89*, 863–927.
102. Monteagudo, E.S.; Burton, G. and Gros, E.G., *Helv. Chim. Acta*, ***1990***, *73*, 2097–2100.
103. Lai, C.K.; Byon, C.Y.; Gut, M.; Mostowicz, D. and Anderson, W.G., *J. Label. Compd. Radiopharm.*, ***1984***, *21*, 627–637.
104. a. Bohlmann, F., von Kapp-Herr, W., Rybak, Ch. and Repplinger, J., *Chem. Ber.*, ***1965***, *98*, 1736–1743; b. Brill, T.B. and Landon, S.J., *Chem. Rev.*, ***1984***, *84*, 577; c. Brückner, R., *Reaktionsmechanismen, 3. Auflage*; Elsevier: Amsterdam, ***2004***, pp. 458–482; d. Wadsworth, W.S., *Org. Reactions*, ***1977***, *25*, 73–253.

105. a. Soli, E.D. and Braun, M.P., *J. Label. Compd. Radiopharm.*, ***2006***, *49*, 437–443.; b. Almi, M. and Guelfi, S., *J. Label. Compd. Radiopharm.*, ***1998***, *41*, 1073–1074; c. Minato, H.; Katsuyama, Y.; Nagasaki, T.; Irisawa, J. and Igarashi, K., *J. Label. Compd. Radiopharm.*, ***1978***, *14*, 455–460.
106. Rabinowitz, J.L. and Zanger, M., *J. Label. Compd.*, ***1972***, *8*, 657–664.
107. Walter, W.M. Jr., *J. Label. Compd.*, ***1967***, *3*, 54–57.
108. Hafferl, W.; Zurflüh, R. and Dunham, L., *J. Label. Compd.*, ***1971***, *7*, 331–339.
109. a. Adams, P.H., *J. Label. Compd. Radiopharm.*, ***1988***, *25*, 395–402; b. Liebman, A.A.; Burger, W. and Malarek, D.H., *J. Label. Compd. Radiopharm.*, ***1990***, *28*, 525–541.
110. Kaegi, H.; Chew, E. and Chien, P.-L., *J. Label. Compd. Radiopharm.*, ***1980***, *17*, 745–751.
111. Tanumihardjo, S.A., *J. Label. Compd. Radiopharm.*, ***2001***, *44*, 365–372.
112. Freer, R.; Mahoney, A.; Mitchell, M.B.; Morecombe, D.J. and Patrick G.L., in *Synthesis and Applications of Isotopically Labeled Compounds 1988*; Baillie, T.A. and Jones, J.R., Eds.; Elsevier: Amsterdam, ***1989***, pp. 295–298.
113. Newman, J.; Rej, R.N.; Just, G. and Lewis, N.G., *Holzforschung*, ***1986***, *40*, 369–373.
114. Tanaka, M.; Sakurai, H. and Yoshimura, T., *J. Label. Compd. Radiopharm.*, ***1991***, *29*, 719–723.
115. Khachik, F.; Beecher, G.R.; Li, B.W. and Englert, G., *J. Label. Compd. Radiopharm.*, ***1995***, *36*, 1157–1172.
116. Moenius, T.; Voges, R.; Mahnke, M.; Burtscher, P.; Metz, Y. and Guenat, Ch., *J. Label. Compd. Radiopharm.*, ***1999***, *42*, 29–41.
117. Rodriguez, I.; Kohler, B. and Metz, Y., Carbon-14 labeling of discodermolide, a natural anti-cancer product, Internal Report, ***2001***, Novartis Pharma AG.
118. a. Smith, M.B. and March, J., *March's Advanced Organic Chemistry*; 5th Edn; John Wiley & Sons, Inc.: New York, ***2001***, pp. 1228–1229; b. Fuhrhop, J. and Penzlin, G., *Organic Synthesis*; 2nd Edn.; Verlag Chemie: Weinheim, ***1994***, p. 33; c. Ager, D.J., *Synthesis*, ***1984***, 384–398.
119. a. Hartung, T.; Hilpert, H. and Mohr P., in *Synthesis and Applications of Isotopically Labeled Compounds*; Vol. 7, Pleiss, U. and Voges R., Eds.; John Wiley & Sons, Ltd: Chichester, ***2001***, pp. 189–192; b. Shimoji, J.; Taguchi, H.; Oshima, K.; Yamamoto, H. and Nozaki, H., *J. Am. Chem. Soc.*, ***1974***, *96*, 1620–1621; c. Perlman, K.L.; Swenson, R.E.; Paaren, H.E.; Schnoes, H.K. and DeLuca, H.F., *Tetrahedron Lett.*, ***1991***, *32*, 7663–7666.
120. O'Connor, E.J.; Tomita, Y. and McDermott, A.E., *J. Label. Compd. Radiopharm.*, ***1994***, *34*, 735–740.
121. Heys, J.R. and Chew, E.H., *J. Label. Compd. Radiopharm.*, ***1984***, *21*, 409–412.
122. White, D.F., *J. Label. Compd. Radiopharm.*, ***1977***, *13*, 23–39.
123. Askin, D.; Reamer, R.A.; Joe, D.; Volante, R.P. and Shinkai, I., *Tetrahedron Lett.*, ***1989***, *30*, 6121–6124.
124. Almi, M.; Guilfi, S.; Tamburini, B.; Vigelli, G.; Cable, K.M.; Waterhouse, I. and Wells G.N., in *Synthesis and Applications of Isotopically Labeled Compounds 1994*, Allen, J. and Voges, R., Eds.; John Wiley & Sons, Ltd: Chichester, ***1995***, pp. 667–670.
125. a. Landvatter, S.W., *Heterocycles*, ***1996***, *43*, 1189–1199; b. Landvatter S.W.; in *Synthesis and Applications of Isotopically Labeled Compounds 1994*; Allen, J. and Voges, R., Eds.; John Wiley & Sons, Ltd: Chichester, ***1995***, pp. 683–688.
126. Wheeler, W.J.; O'Bannon, D.D.; Kennedy, J.H.; Mon, J.A.; Tharp-Taylor, R.W.; Valli, M.J. and Kuo, F., *J. Label. Compd. Radiopharm.*, ***2005***, *48*, 605–620.
127. Takatori, K.; Sakamoto, T. and Kajiwara, M., *J. Label. Compd. Radiopharm.*, ***2006***, *49*, 445–453.
128. a. Cornforth, J.W.; Cornforth, R.H.; Popjak, G. and Gore, I.Y., *Biochem. J.*, ***1958***, *69*, 145–155; b. Isler, O.; Rüegg, R.; Gey, K.F. and Pletscher, A., *Helv. Chim. Acta*, ***1957***, *40*, 2369–2373.
129. Maier, M.S.; Seldes, A.M. and Gros, E.G., *J. Label. Compd. Radiopharm.*, ***1985***, *22*, 467–474 and 833–836.
130. a. Lehmann, H.; Gros, D. and Schütte, H.R., *Z. Chem.*, ***1972***, *12*, 416; b. Engel, B.G.; Würsch, J. and Zimmermann, M., *Helv. Chim. Acta*, ***1953***, *36*, 1771–1776; c. Wolf, G.; Kahn, S.G. and Johnson, B.C., *J. Am. Chem. Soc.*, ***1957***, *79*, 1208–1212; d. Garbers, C.F., *J. Chem. Soc.*, ***1956***, 3234–3236.
131. a. Bubner, M. and Schmidt, L.H., *Die Synthese Kohlenstoff-14 markierter organischer Verbindungen*; Verlag Georg Thieme: Leipzig, ***1966***, pp. 134–135; b. Jablonskai, I. and Dutka, F., *J. Radioanal. Nucl. Chem. Lett.*, ***1990***, *114*, 173–177.

132. a. Roe, A. and Finley, J.B., *J. Am. Chem. Soc.*, ***1952***, *74*, 2442–2443; b. Murray, A. and Williams, D.L., *Organic Syntheses with Isotopes*; Interscience: New York, ***1958***, pp. 634–636.
133. a. Roe, A. and Abenesius E. L., *J. Am. Chem. Soc.*, ***1952***, *74*, 2402–2403; b. Murray, A. and Williams, D.L., *Organic Syntheses with Isotopes*; Interscience: New York, ***1958***, pp. 632–634.
134. a. Berger, C.; Maziere, M.; Prenant, C. and Comar, D., *Int. J. Appl. Rad. Isotop.*, ***1980***, *31*, 577–578; b. Prenant, C. and Crouzel, C., *J. Label. Compd. Radiopharm.*, ***1990***, *28*, 645–651; c. Soloviev, D.V.; Matarrese, M.; Moresco, R.M.; Todda, S.; Bonasera, T.A.; Sudati, F.; Simonelli, P.; Magni, F.; Colombo, D.; Carpinelli, A.; Kienle, M.G. and Faszio, F., *Neurochem. Int.*, ***2001***, *38*, 169–180.
135. Voges, R.; Griesser, R. and Schreier E., in *Synthesis and Applications of Isotopically Labeled Compounds*; Duncan, W.P. and Susan, A.B., Eds.; Elsevier: Amsterdam, ***1983***, pp. 209–214.
136. Engler, J. and Pallos, L., *J. Label Compd*, ***1973***, *9*, 81–85; b. Gawell, L., *J. Label. Compd. Radiopharm.*, ***1986***, *23*, 947–949.
137. Andres, H., *J. Label. Compd. Radiopharm.*, ***1989***, *27*, 1307–1315.
138. Durst, H.D. and Leek, E., *J. Label. Compd.*, ***1971***, *7*, 52–55.
139. a. Bloch, K.; Clark, L.C. and Harary, I., *J. Biol. Chem.*, ***1954***, *211*, 687–699; b. Murray, A. and Williams, D.L., *Organic Syntheses with Isotopes*; Interscience: New York, ***1958***, pp. 58–59.
140. Ferdinandi, E., *J. Label. Compd.*, ***1975***, *11*, 287–289.
141. a. Bubner, M. and Schmidt, L.H., *Die Synthese Kohlenstoff-14 markierter organischer Verbindungen*; Verlag Georg Thieme: Leipzig, ***1966***, pp. 139–143; b. Murray, A. and Williams, D.L., *Organic Syntheses with Isotopes*; Interscience: New York, ***1958***, pp. 868–870.
142. a. Dauben, W.G. and Bradlow, L.H., *J. Am. Chem. Soc.*, ***1952***, *74*, 559–562; b. Banthorpe, D.V. and Boullier, P.A., *J. Chem. Soc. Perkin Trans I*, ***1977***, 114–119.
143. a. Kanamaru, H.; Yoshitake, A. and Nakatsuka, I., *Radioisotopes*, ***1986***, *35*, 242–248; b. Danieli, B. and Grasso, G., *J. Label. Compd.*, ***1965***, *1*, 275–281; c. Crombie, L.; Doherty, C.F. and Pattenden G., *J. Chem. Soc. (C)*, ***1970***, 1076–1080.
144. Bascoul, J.; Nicolaidis D.; Crastes de Paulet, A. and Pichat, L., *Bull. Soc. Chim. France*, ***1973***, 2318–2320.
145. a. Colombini, C.; Terbojevich, M. and Peggion, E., *J. Label. Compd.*, ***1965***, *1*, 195–200; b. Kiechel, J.R.; Niklaus, P.; Schreier, E. and Wagner, H., *Xenobiotica*, ***1975***, *5*, 741–748.
146. a. Elsinga P.H.; van Waarde, A.; Jaeggi, K.A.; Schreiber, G.; Heldoorn, M. and Vaalburg, W., *J. Med. Chem.*, ***1997***, *40*, 3829–3835; b. van Waarde, A.; Doorduin J; de Jong, J.R.; Dierckx, R.A. and Elsinga, P.H., *Neurochem. Intl.*, ***2008***, *52*, 729–733.
147. Dass, D.V.; Martin, R.W.; Odell, A.L. and Quinn, G., *J. Label. Compd. Radiopharm.*, ***1988***, *25*, 225–227.
148. a. Bubner, M. and Schmidt, L.H., *Die Synthese Kohlenstoff-14 markierter organischer Verbindungen*; Verlag Georg Thieme: Leipzig, *1966*, pp. 172–176; b. Korte, F. and Barkenmeyer, H., *Chem. Ber.*, ***1957***, *90*, 2739–2745; c. Swartz, G.L. and Gulick, W.M., *J. Label. Compd.*, ***1975***, *11*, 525–534; d. Viswanatha, V. and Hruby, V.J., *J. Org. Chem.*, ***1979***, *44*, 2892–2896; e. Geletneky, C.; Balzer, H.; Bock, W. and Berger, S., *J. Label. Compd Radiopharm.*, ***1995***, *36*, 503–507; f. Akira, K.; Hasegawa, H. and Baba, S., *J. Label. Compd. Radiopharm.*, ***1995***, *36*, 845–853.
149. a. Swan, G.A. and Wright, D., *J. Am. Chem. Soc.*, ***1956***, *78*, 1549–1557; b. Kratzl, K. and Vierhapper, F.W., *Monatshefte Chem.*, ***1971***, *102*, 224–232.
150. Herbert, J.M.; Kohler, A.D.; Le Strat, F. and Whitehead, D., *J. Label. Compd. Radiopharm.*, ***2007***, *50*, 441–442.
151. a. Hales, N.J.; Heaney, H.; Hollinshead, J.H. and Ley, S.V., *Tetrahedron*, ***1995***, *51*, 7741–7754; b. Hales, N.J. and Heaney, H., *Tetrahedron Lett.*, ***1975***, *46*, 4075–4078.
152. Baldwin, J.E.; Bansal, H.S.; Chondrogianni, J.; Gallagher, P.E.; Taha, A.A.; Taylor, A. and Thaller, V., *J. Chem. Res. (S)*, ***1984***, 176–177.
153. Liebman, A.A.; Burger, W. and Cupano J., in *Synthesis and Applications of Isotopically Labeled Compounds 1988*; Baillie, T.A. and Jones, J.R., Eds.; Elsevier: Amsterdam, ***1988***, pp. 605–608.
154. Edgar, B.; Latter, A.; Rustidge, D. and Willcocks, K., *Synthesis and Applications of Isotopically Labeled Compounds 1991*; Buncel, E. and Kabalka, G.W., Eds.; Elsevier: Amsterdam, ***1992***, pp. 232–234.

155. Bettens, B. and Gonze, J.M., *J. Label. Compd. Radiopharm.*, ***1970***, *7*, 23–32.
156. Andres, H. and Schnelli, P., Internal Report ***2001***, Novartis Pharma AG, CH-4002, Basel.
157. Heise, K.H. and Bubner, M., *Isotopenpraxis*, ***1976***, *12*, 125–128.
158. a. Thomas, R.C., *J. Label. Compd. Radiopharm.*, ***1978***, *15*, 461–468; b. Doorenbos, H.E.; Muelder, W.W. and Waas, M.N., *J. Label. Compd.*, ***1972***, *8*, 701–703.
159. Minato, H.; Nagasaki, T. and Katsuyama, Y., *J. Label. Compd.*, ***1975***, *11*, 275–279.
160. Zamir, L.O.; Lin, M. and Nguyen, C.D., *J. Label. Compd. Radiopharm.*, ***1988***, *25*, 1081–1086.
161. Surber, B.W.; Elliott, R.L.; Kopecka, H. and Thomas, S.B., *J. Label. Compd. Radiopharm.*, ***1996***, *38*, 857–861.
162. a. Winteringham, E.P.W.; Harrison, A. and Bridges, P.M., *Biochem. J.*, ***1955***, *61*, 359–365; b. Murray, A. and Williams, D.L., *Organic Syntheses with Isotopes*; Interscience: New York, ***1958***, pp. 467–649; c. Clelland, J.D. and Thornalley, P.J., *J. Label. Compd. Radiopharm.*, ***1990***, *28*, 1455–1464.
163. a. Jones, A.R., *J. Label. Compd.*, ***1975***, *11*, 77–80; b. Kupczyk-Subotkowska, L. and Shine, H.J., *J. Label. Compd. Radiopharm.*, ***1993***, *33*, 301–304.
164. a. Dauben, W.G. and Bradlow, H.L., *J. Am. Chem. Soc.*, ***1952***, *74*, 5204–5206; b. Murray, A. and Williams, D.L., *Organic Syntheses with Isotopes*; Interscience: New York, ***1958***, pp. 446–447.
165. a. Wierenga, W. and Skulnick, H.J., *J. Org. Chem.*, ***1979***, *44*, 310–314; b. Bram, G. and Vilka, M., *Bull. Soc. Chim. Fr.*, ***1964***, 945–951; c. Andres, H. and Bühler, K., Isradipine, labeling with carbon-14 in the dihydropyridine moiety, document; Sandoz Ltd: ***1982***, pp. 1–12; d. Maul, W., Scherling, D. and Seng, F, *J. Label. Compd. Radiopharm.*, ***1981***, *18*, 453–463.
166. Lee, H.; Woo, P.; Santay, L.; Travalent, A.; Roark, H.; Capriris, T. and Huang, C.C., *Synthesis and Applications of Isotopically Labeled Compounds 1988*; Baillie, T. and Jones, J.R., Eds.; Elsevier: Amsterdam, ***1989***, pp. 595–598.
167. a. Wilkinson, D.J., *Synthesis and Applications of Isotopically Labeled Compounds 1988*; Baillie, T. and Jones, J.R., Eds.; Elsevier: Amsterdam, ***1989***, pp. 427–430; b. Oikawa, Y.; Sugano, K. and Yonemitsu, O., *J. Org. Chem.*, ***1978***, *43*, 2087–2088; c. McNeill, A.; Murrel, V. and Taylor, K., *J. Label. Compd. Radiopharm.*, ***2004***, *47*, 1019–1027; d. Maul, W. and Scherling, D., *J. Label. Compd. Radiopharm.*, ***1989***, *27*, 457–463; e. Scherling, D., *J. Label. Compd. Radiopharm.*, ***1989***, *27*, 599–603; f. Scherling, D. and Pleiss, U., *J. Label. Compd. Radiopharm.*, ***1988***, *25*, 1393–1400.
168. a. Dahn, H. and Hauth, H., *Helv. Chim. Acta*, ***1959***, *42*, 1214–1224; b. D'Allesandro, G. and Sleiter, G., *J. Label. Compd. Radiopharm.*, ***1980***, *17*, 814–824.
169. Hütter, P. and Zeller, K.P., *Synthesis*, ***1985***, 334–335.
170. a. DeGraw, I.J. and Christie, P.H., *J. Label. Compd. Radiopharm.*, ***1982***, *19*, 945–952; b. Sakami, W.; Evans, W.E. and Gurin, S., *J. Am. Chem. Soc.*, *69*, 1110–1113; ***1947***, c. Murray, A. and Williams, D.L., *Organic Syntheses with Isotopes*; Interscience: New York, ***1958***, pp. 348–350.
171. a. Straatmann, M.G.; Hortmann, A.G. and Welch, M.J., *J. Label. Compd. Radiopharm.*, ***1974***, *10*, 175–179; b. Parter, W.R.; Spitznagle, L.A. and Trager, W.F., *J. Label. Compd.*, ***1976***, *12*, 577–582.
172. Hamamura, K.; Yamatsu, I.; Minami, N.; Yamagishi, Y.; Inai, Y.; Kijima, S. and Nakamura, T., *J. Label. Compd Radiopharm.*, ***2002***, *45*, 823–829.
173. Nishioka, K.; Nakatsuka, I. and Kanamaru, H., *Radioisotopes*, ***1988***, *37*, 133–139.
174. Shimada, K.; Tadano, K.; Satoh, T.; Hashimoto, K.; Tanaka, S. and Yuzuriha, T., *J. Label. Compd. Radiopharm.*, ***1989***, *27*, 1293–1298.
175. Yuan, S.S. and Ajami, A.M., *J. Label. Compd. Radiopharm.*, ***1984***, *21*, 525–532.
176. Jones, J.A. and Blecher, M., *J. Lipid. Res.*, ***1966***, *7*, 422.
177. Ferdinandi, E.S., *J. Label. Compd. Radiopharm.*, ***1978***, *14*, 751–755.
178. a. Meltzer, H.L. and Sprinson, D.B., *J. Biol. Chem.*, ***1952***, *197*, 461–464; b. Murray, A. and Williams, D.L., *Organic Syntheses with Isotopes*; Interscience: New York, ***1958***, pp. 181–186.
179. a. Chen, K.M.; Hardtman, G.E.; Prasad, K.; Repic, O. and Shapiro, M.J., *Tetrahedron Lett.*, ***1998***, *28*, 155–158; b. Chen, K.M.; Gunderson, K.G.; Hardtman, G.E.; Prasad, K.; Repic, O. and Shapiro, M.J., *Chem. Lett.*, ***1987***, 1923.
180. Pleiss, U., *J. Label. Compd. Radiopharm.*, ***2007***, *50*, 818–830.

181. a. Arima, H.; Tamazawa, K. and Takeuchi, M., *J. Label. Compd. Radiopharm.*, ***1988***, *25*, 161–170; b. Tada, N.; Kajino, M.; Watanaba, M. and Hayashi, N., *J. Label. Compd. Radiopharm.*, ***1989***, *27*, 1345–1348.
182. Iwanami, M.; Shibanuma, T.; Fujimoto, M.; Kawai, P.; Tamazawa, K.; Takenaka, T.; Takahashi, K. and Murakami, M., *Chem. Pharm. Bull.*, ***1979***, *27*, 1426–1440.
183. Duhm, B.; Maul, W.; Medenwald, H.; Patschke, K. and Wegner, L.A., *Arzneim. Forsch.*, ***1972***, *22*, 42–43.
184. Mehrsheikh, M.E.; Clizbe, L.A.; Singh, H. and Purdum, W.R., *J. Label. Compd. Radiopharm.*, ***1990***, *28*, 437–443.
185. Racz, W.J. and Marks, G.S., *Biochem. Pharmacol.*, ***1972***, *21*, 143–151; b. Racz, W.J. and Marks, G.S., *Biochem. Pharmacol.*, ***1972***, *21*, 1511–1514.
186. Murray, A. and Williams, D.L., *Organic Syntheses with Isotopes*; Interscience: New York, ***1958***, pp. 139–140.
187. Unverricht, A.; Pfützner, P.; Stecher, B. and Schütte, H.R., *Z. Chem.*, ***1972***, *12*, 289–290.
188. Leete, E. and Leete, S.A.S., *J. Org. Chem.*, ***1978***, *43*, 2122–2125.
189. Yanami, T., *Ann. Rep. Sankyo Res. Lab.*, ***1984***, *36*, 174–177.
190. Davis, H.W.; Grovenstein, E. and Neville, O.K., *J. Am. Chem. Soc.*, ***1953***, *75*, 3304–3305.
191. a. Murray, A. and Williams, D.L., *Organic Syntheses with Isotopes*; Interscience: New York, ***1958***, pp. 117–121 and 438–439; b. Bánfi, D.; Volford, J. and Mészáros, Z., *J. Label. Compd.*, ***1975***, *11*, 409–417.
192. Bubner, M. and Schmidt, L.H., *Die Synthese Kohlenstoff-14 markierter organischer Verbindungen*; Georg Thieme Verlag: Leipzig, ***1966***, pp. 158–159.
193. Yoshitake, A.; Makari, Y. and Endo, M., *J. Label. Compd.*, ***1974***, *10*, 589–598.
194. Pabiot, J.M.; Pichat, L.; Esanu, A.; Pommier, J. and Laurent, J.P., *J. Label. Compd. Radiopharm.*, ***1982***, *19*, 195–201.
195. Mueller, M.E. and Leete, E., *J. Org. Chem.*, ***1981***, *46*, 3151–3152.
196. Nishioka, K. and Kanamaru, H., *J. Label. Compd. Radiopharm.*, ***1991***, *29*, 1051–1060.
197. a. Cope, A.C.; Holmes, H.L. and House, H.O., *Org. React.*, ***1957***, *9*, 107; b. Cox, R.F.B. and McElvain, S.M., *Org. Synth. Coll. Vol 2; Wiley: New York*, ***1943***, *2*, 279–282.
198. Carruthers, W., *Some Modern Methods of Organic Chemistry*; 3rd Edn.; Cambridge University Press: Cambridge, ***1986***, pp. 4–8.
199. Smith, M.B. and March, J., *March's Advanced Organic Chemistry*; 5th Edn.; John Wiley & Sons Inc: New York, ***2001***, pp. 548–551.
200. Friedman, S.M.; Kaneda, T. and Corcoran, J.W., *J. Biol. Chem.*, ***1964***, *239*, 2386–2391.
201. a. Acker, T.E.; Brenneisen, P.E. and Tanenbaum, S.W., *J. Am. Chem. Soc.*, ***1966***, *88*, 834–837; b. Zaugg, H.E.; Dunnigan, D.A.; Michaels, R.J.; Swett, L.R.; Wang, T.S.; Sommers, A.H. and DeNet, R.W., *J. Org. Chem.*, ***1961***, *26*, 644–651.
202. a. Benes, J.; Dobiasova, M. and Veres, K., *Coll. Czech. Chem. Commun.*, ***1964***, *29*, 3178–3181; b. Janak, J.; Novak, J. and Sutovski, J., *Coll. Czech. Chem. Commun.*, ***1962***, *27*, 834–837.
203. Ogawa, K.; Niisawa, K. and Nozaki, K., *Synthesis and Applications of Isotopically Labeled Compounds 1988*; Baillie, T.A. and Jones, J.R., Eds.; Elsevier: Amsterdam, ***1989***, pp. 633–636
204. Coppi, G. and Bernardi, G., *Arzneim.-Forsch.*, ***1969***, *19*, 1518–1519.
205. Kasumov, T. and Brunengraber, H., *J. Label. Compd. Radiopharm.*, ***2006***, *49*, 171–176.
206. a. Allevi, P.; Longo, A. and Anastasia, M., *J. Label. Compd. Radiopharm.*, ***1999***, *42*, 1085–1091; b. Yamazaki, T.; Kasatkin, A.; Kawanaka, Y. and Sato, F., *J. Org. Chem.*, ***1996***, *61*, 2266–2267.
207. Hernandez, L.; Casanova, E.; Loupy, A. and Petit, A., *J. Label. Compd. Radiopharm.*, ***2003***, *46*, 151–158.
208. Ranada, V.V. and Alter, A., *J. Label. Compd. Radiopharm.*, ***1980***, *17*, 733–743.
209. Block, W. and Ebigt, I., *Arzneim -Forsch.*, ***1957***, *7*, 572–576.
210. Haynes, U.J.; Swigor, J.E. and Pittman, K.A., *J. Label. Compd. Radiopharm.*, ***1987***, *24*, 535–539.
211. Crouse, G.D. and Terando, N.H., *J. Label. Compd. Radiopharm.*, ***1989***, *27*, 465–471.
212. Hayashi, N.; Toga, T. and Murata, T., *J. Label. Compd.*, ***1974***, *10*, 609–615.
213. Landvatter, S.W.; Garnes, K.T. and Heys J.R., in *Synthesis and Applications of Isotopically Labeled Compounds 1994*; Allen, J. and Voges, R., Eds.; John Wiley & Sons, Ltd: Chichester, ***1995***, pp. 631–634.

214. a. Beare, N.A. and Hartwig, J.F., *J. Org. Chem.*, ***2002***, *67*, 541–555; b. Henessy, E.J. and Buchwald, S.L., *Organic Letters*, ***2002***, *4*, 269–272.
215. a. Ho, J.Z. and Braun, M.P., *J. Label. Compd. Radiopharm.*, ***2007***, *50*, 277–280; b. Bruggink, A. and McKillop, A., *Tetrahedron*, ***1975***, *31*, 2607–2619.
216. Wang, T.S.T.; Fawwaz, R.A. and Van Heertum, R.L., *J. Label. Compd. Radiopharm.*, ***1995***, *36*, 313–320.
217. Yoshitake, A.; Kamada, T.; Gomi, H. and Nakatsuka, I., *J. Label. Compd. Radiopharm.*, ***1981***, *18*, 1415–1421.
218. Volford, J.; Mészáros, Z. and Kovács, G., *J. Label. Compd.*, ***1973***, *9*, 231–236.
219. Shin, H.I.; Rim, J.G.; Lee, K.S.; Kim, Y.S.; Nam, D.H.; Shin, H.I.; Chang, J.H.; Oh, C.Y. and Ham, W.H., *J. Label. Compd. Radiopharm.*, ***2004***, *47*, 779–786.
220. Ekhato, I.V. and Huang, C.C., *J. Label. Compd. Radiopharm.*, ***1993***, *33*, 869–880.
221. Seidel, D.; Conrad, M.; Brehmer, P.; Mohrs, K. and Petersen, U., *J. Label. Compd. Radiopharm.*, ***2000***, *43*, 795–805.
222. Lee, B.H. and Clothier, M.F., *J. Label. Compd. Radiopharm.*, ***1993***, *33*, 823–826.
223. a. Pleiss, U.; Conrad, M. and Baasner, B., *J. Label. Compd. Radiopharm.*, ***1995***, *36*, 275–279; b. Saus, A.; Klinger, K.H. and Bickel, E., *Arzneim. Forsch./Drug Res.*, ***1977***, pp. 2735–36; c. Ranade, V.V.; Lee, B.H. and Clothier, M.F., *J. Label. Compd. Radiopharm.*, ***1984***, *21*, 381–389.
224. Brückner, R., *Reaktionsmechanismen, Organische Reaktionen, Stereochemie, Moderne Synthesemethoden,*; Elsevier: Spektrum Akademischer Verlag, ***2004***, *3*, pp. 566–569.
225. a. Lehnert, W., *Tetrahedron*, ***1972***, *28*, 663–666; b. Lehnert, W., *Tetrahedron*, ***1973***, *29*, 635–638; c. Campaigne, E. and Beckman, J.C., *Synthesis*, ***1978***, 385–388.
226. Freer, R.; Hatt, B.W.; Mitchell, M.B. and Morecombe, D.J., *J. Label. Compd. Radiopharm.*, ***1987***, *24*, 377–388.
227. a. Mohammad, T. and Morrison, H., *J. Label. Compd. Radiopharm.*, ***1991***, *29*, 1019–1026; b. Robbins, R.J. and Schmidt, W.F., *J. Label. Compd. Radiopharm.*, ***2004***, *47*, 797–806; c. Balba, H.M. and Still, G.G., *J. Label. Compd. Radiopharm.*, ***1978***, *15*, 309–319; d. DeBardeleben, J.F. and Teng, L.C., *J. Label. Compd.*, ***1970***, *6*, 34–39; e. De Pooter, H.; Pé, I. and Van Sumere, C.F., *J. Label. Compd*, ***1974***, *10*, 135–141; f. Larsen, P.O. and Wieczorkowska, E., *J. Label. Compd.*, ***1974***, *10*, 287–296; g. Herbert, M.; Rochas, G. and Pichat, L., *J. Label. Compd.*, ***1968***, *4*, 240–242; h. Haider, K., *J. Label. Compd.*, ***1966***, *2*, 174–183; i. Chen, B.-C.; Bednarz, M.S.; Zhang, H.; Guo, P.; Jemal, M.; Robl, J.A.; Biller, S.A.; Sundeen, J.E.; Balasubramanian, B. and Barrish, J.C., *J. Label. Compd. Radiopharm.*, ***2006***, *49*, 311–319; j. Beejmohun, V.; Grand, E.; Lesur, D.; Mesnard, F.; Fliniaux, M.-A. and Kovensky, J., *J. Label. Compd. Radiopharm.*, ***2006***, *49*, 463–470; k. Murray, A. and Williams, D.L., *Organic Syntheses with Isotopes*; Interscience: New York, ***1958***, pp. 1032–1035.
228. Johansen, S.K., *J. Label. Compd. Radiopharm.*, ***2005***, *48*, 1025–1029.
229. Kumagai, M.; Sakata, S.; Yamaguchi, T.; Ikeda, K. and Machida, H., *J. Label. Compd. Radiopharm.*, ***1989***, *27*, 503–507.
230. a. Iimura, Y.; Mishima, M. and Sugimoto, H., *J. Label. Compd. Radiopharm.*, ***1989***, *27*, 835–839; b. Kuo, F.; Clodfelter, D.K. and Priest, T.R., *J. Label. Compd. Radiopharm.*, ***2007***, *50*, 706–710.
231. Ma, G. and Hayes, S.E., *J. Label. Compd. Radiopharm.*, ***2004***, *47*, 895–901.
232. Schlewer, T.; Stampf, J.-L. and Benezra, C., *J. Label. Compd. Radiopharm.*, ***1980***, *17*, 297–302.
233. De Keyser, J.-L.; De Cock, C.J.C.; Poupaert, J.H. and Dumont, P., *J. Label. Compd. Radiopharm.*, ***1989***, *27*, 909–916.
234. Knöfel, H.D. and Gross, D., *Z. Naturforsch. C: Bioscience*, ***1988***, *43*, 29–31.
235. Unverricht, A. and Gross, D., *J. Label. Compd. Radiopharm.*, ***1986***, *23*, 515–518.
236. Porwoll, J.P. and Leete, E., *J. Label. Compd. Radiopharm.*, ***1985***, *22*, 257–271.
237. Liebman, A.A.; Malarek, D.H.; Dorsky, A.M. and Kaegi, H.H., *J. Label. Compd.*, ***1971***, *7*, 241–246.
238. Key, J.A.; McKillican, B.P. and Norris G.A.in *Synthesis and Applications of Isotopically Labeled Compounds 1994*; Allen, J. and Voges, R., Eds.; John Wiley & Sons, Ltd: Chichester, ***1995***, pp. 621–626.
239. Volford, J.; Szarvas, T.; Tyihak, E. and Trezl, J., *Org. Soedin, Mechennye Radioact. Izot. Mater. Simp.*, 2nd, Stran-Chlenov SEV ***1982***, *1*, 82–87.

240. Lehmann, H.; Gross, D. and Schütte, H.R., *Z. Chem.*, ***1971***, *11*, 426–427.
241. a. Adriaens, P. and Vanderhaege, H., Synthesis of some labeled non-proteinogenic amino acids, in *Isotopes in the Physical and Biomedical Sciences*, Vol. *1*, Part A., Buncel, E. and Jones, J.R., Eds.; Elsevier: Amsterdam, ***1987***, pp. 330–401; b. Murray, A. and Williams, D.L., *Organic Syntheses with Isotopes*; Interscience: New York, ***1958***, pp. 196–197 and 216–217; c. Pichat, L. and Beaucourt, J.P., *J. Label. Compd.*, ***1974***, *10*, 103–112; d. Andersen, L. and Jaroszewski, J.W., in *Synthesis and Applications of Isotopically Labeled Compounds 1997*; Heys, J.R. and Melillo, D.G., Eds.; John Wiley & Sons, Ltd: Chichester, ***1998***, pp. 403–406.
242. a. Murray, A. and Williams, D.L., *Organic Syntheses with Isotopes*; Interscience: New York, ***1958***, pp. 400–401; b. Salter, R. and Schnelli, P., Novartis Pharma Ltd., personal communication, ***2005***.
243. Lee, C.-J. and Serif, G.S., *Biochemistry*, ***1970***, *9*, 2068–2071.
244. Arnstein, H.R.V.; Hunter, G.D.; Muir, H.M. and Neuberger, A., *J. Chem. Soc.*, ***1952***, 1329–1334.
245. Tolman, V.; Hanus, J. and Veres, K., *J. Label. Compd.*, ***1968***, *4*, 243–247.
246. a. Pichat, L. and Beaucourt, J.P. *J. Label. Compd.*, ***1974***, *10*, 103–112; b. Andersen, L. and Jaroszewski, J, in *Synthesis and Applications of Isotopically Labeled Compounds 1997*, Heys, J. R. and Melillo, D.G., Eds.; John Wiley & Sons, Ltd: Chichester, ***1998***, pp. 403–405; c.Trown, P. W.; Smith, B. and Abraham, E.P., *Biochem. J.*, ***1963***, *86*, 284–288.
247. a. Schreier, E., *Helv. Chim. Acta*, ***1976***, *59*, 585–606; b. Murray, A. and Williams, D.L., *Organic Syntheses with Isotopes*; Interscience: New York, ***1958***, pp. 247–252.
248. Potrzebowski, M.J.; Stolowich, N.J. and Scott, A.I., *J. Label. Compd. Radiopharm.*, ***1990***, *28*, 355–361.
249. Kinder, F.R.; Tang, Y.S. and Sunay, U.B., *J. Label. Compd. Radiopharm.*, ***1992***, *31*, 829–835.
250. Longworth, S.W.; Bigg, D.C.H.; White, D.F. and Burns, J., *J. Label. Compd.*, ***1974***, *10*, 423–430.
251. a. Gidez, L.I. and Karnovsky, M.L., *J. Am. Chem. Soc.*, ***1952***, *74*, 2413–2415; b. Bubner, M. and Schmidt, L.H., *Die Synthese Kohlenstoff-14 markierter organische Verbindungen*; Georg Thieme Verlag: Leipzig, ***1966***, pp. 159–160; c. Schweer, K.H., *Chem. Ber.*, ***1962***, *95*, 1799–1804; d. Greenfield, J.C.; Leonhard, N.J. and Nystrom, R.F., *J. Label. Compd. Radiopharm.*, ***1976***, *12*, 545–550.
252. a. Pastan, I.; Tsai, L. and Stadtman, E.R., *J. Biol. Chem.*, ***1964***, *239*, 902–906; b. Gross, D.; Feige, A. and Schütte, H.R., *Z. Chem.*, ***1965***, *5*, 21.
253. a. Mattson, F.H. and Volpenhein, R.A., *J. Biol. Chem.*, ***1962***, *237*, 53–55; b. Guermont, J.P. and Pichat, L., *Bull. Soc. Chim. Fr.*, ***1958***, 1338–1339; c. Weiss, S.B. and Kennedy, E.P., *J. Am. Chem. Soc.*, ***1956***, *78*, 3550; d. Schmidt, L.H. and Bubner, M., *Kernenergie*, ***1963***, *6*, 82–83.
254. Greenfield, J.C.; Leonard, N.J. and Nystrom, R.F., *J. Label. Compd. Radiopharm.*, ***1976***, *12*, 545–550.
255. Andres, H.; Burtscher, P.; Metz, Y. and Moenius, Th., *J. Label. Compd. Radiopharm.*, ***2004***, *47*, 270–271.
256. Hicks, J.L. and Huang, Y., *J. Label. Compd. Radiopharm.*, ***1997***, *39*, 275–284.
257. Latli, B.; Jones, P.J.; Krishnamurthy, D. and Senanayake, C.H., *J. Label. Compd. Radiopharm.*, ***2008***, *51*, 54–58.
258. Gagnon, P.E.; Boivin, J.L.; MacDonald, R. and Yaffe, L., *Can. J. Chem.*, ***1954***, *32*, 823–837.
259. Klatt, L. and Koss, F.W., *Arzneim.-Forsch.*, ***1973***, *23*, 913–921.
260. Block, W. and Ebigt, I., *Arzneim.-Forsch.*, ***1957***, *7*, 572–576.
261. a. Korte, F. and Ludwig, G., *Liebigs Ann. Chem.*, ***1958***, *615*, 94–99; b. Stolle, W.T.; Easter, J.A.; Chew, E.H.; McGrath, J.P.; Palmer, J.R. and Hsi, R.S.P., *J. Label. Compd. Radiopharm.*, ***1994***, *34*, 1187–1199.
262. Ludwig, G.; Vergin, H. and Zech, K., *Arzneim.-Forsch.*, ***1977***, *27*, 2077–2086.
263. Czeskis, B.A., *J. Label. Compd. Radiopharm.*, ***2004***, *47*, 699–704.
264. a. Wilkinson, D.J., *Appl. Radiat. Isot.*, ***1988***, *39*, 554; b. Schmook, F.P. and Polansky, O.E., *Monatshefte Chem.*, ***1969***, *100*, 1640–1653.
265. Lee, H.T.; Travalent, A.M. and Woo, P.W.K., *J. Label. Compd. Radiopharm.*, ***1990***, *28*, 1143–1148.

# 7

# Preparation of Carbon-14-Labeled Compounds via the [$^{14}$C]Cyanide Tree

## 7.1 Metal [$^{14}$C]Cyanides

Metal [$^{14}$C]cyanides (M = Na, K, Zn, Cu) are some of the most versatile reagents in carbon-14 synthesis. They are available commercially at low cost or can be prepared with relative ease (and often used *in situ*). They are good starting materials for the preparation of additional carbon-14 building blocks, and they can be readily incorporated into synthetic intermediates because of their strong nucleophilicity. This reactivity can be tuned by changing the metal counterion (in either a stoichiometric or a catalytic sense), varying the solvent and modifying the leaving group it replaces. Finally, the organic [$^{14}$C]nitrile functions thus introduced are subject to a variety of chemical modifications, allowing the preparation of a wide array of carbon-14-labeled compounds.

### 7.1.1 Preparation

The most common salt of trade and utility is $K^{14}CN$. A convenient and safe method of preparation on a scale of 1–10 mmol is reduction of barium [$^{14}$C]carbonate with potassium azide at elevated temperatures; the procedure is highly reproducible and gives radiochemical yields in the range of 90–98%[1,2]. In this procedure an intimate mixture of barium [$^{14}$C]-carbonate, potassium azide and carefully dried sea sand is heated at temperatures slowly increasing from 450 to 700 °C. The resulting crude product is acidified with 85% phosphoric acid, and $H^{14}CN$ released is expelled with helium into a methanolic solution of potassium methoxide, from which the $K^{14}CN$ is isolated in solid form by evaporation of the solvent. (*Caution: HCN is volatile and extremely toxic.*) Sodium [$^{14}$C]cyanide can be prepared the same way, except using methanolic sodium methoxide. Potassium [$^{14}$C]-cyanide is indefinitely stable as a dry solid at ambient temperature; the sodium salt is somewhat less so.

*Preparation of Compounds Labeled with Tritium and Carbon-14* Rolf Voges, J. Richard Heys and Thomas Moenius

Larger quantities of alkali metal [$^{14}$C]cyanides are produced in a four-step sequence by reduction of $^{14}CO_2$ with hydrogen at 350 °C over a nickel catalyst to form [$^{14}$C]methane. This intermediate reacts with ammonia at 1150 °C over a platinum catalyst, and the resulting mixture of $H^{14}CN$ and $NH_4{}^{14}CN$ is trapped in an aqueous alkali hydroxide solution. Radiochemical yields of 98–100% have been reported[3].

Two other frequently used metal salts are $Cu^{14}CN$ and $Zn(^{14}CN)_2$. The former is readily accessible in crystalline form in 80–90% yield by treatment of a weakly basic aqueous solution of sodium or potassium cyanide with excess $CuSO_4 \cdot 5H_2O$ in the presence of sodium hydrogensulfite at 20–60°C[4] in radiochemical yields of 80–90%. The latter is conveniently obtained as a white precipitate in 90% yield upon treatment of an aqueous solution of $K^{14}CN$ with $ZnCl_2$[5].

### 7.1.2 Introduction of [$^{14}$C]Cyanide into Organic Substrates

Cyanide ion, because of its powerful nucleophilic character, forms C–CN bonds with reaction partners either by displacement of a leaving group or by nucleophilic addition to appropriate acceptors. Reactions of an electrophilic character (i.e., attack of nucleophilic partners on the cyano carbon) are achievable when cyanide is converted into derivatives such as cyanogen bromide (Section 7.2.2), cyanate (Section 7.2.3), thiocyanate (Section 7.2.4) and orthoformate (Section 7.2.5).

*[$^{14}$C]Cyanide Introduction by Displacement Reactions*

Alkali metal (K, Na) [$^{14}$C]cyanides have been extensively employed in the formation of alkyl [$^{14}$C]nitriles via nucleophilic displacement of leaving groups such as primary halides, sulfonates and trialkylammonium salts in examples too numerous to mention. In special cases (e.g., **1**) other leaving groups have been used, such as benzotriazol-1-yl[6]. The most important considerations in the choice of leaving group are often the ease of substrate preparation and the compatibility of the displacement reaction with substrate ancillary functional groups.

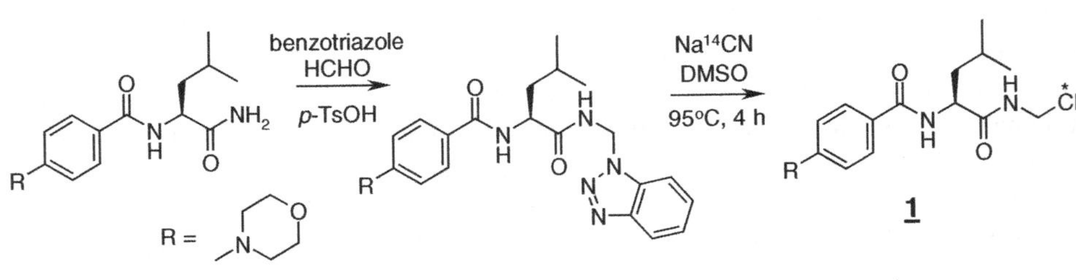


*Figure 7.1* Alkyl [$^{14}$C]cyanide via displacement of a benzotriazol-1-yl group

With secondary halides or sulfonates only moderate yields are obtained, while tertiary substrates fail due to side reactions involving elimination. The most effective solvents are usually polar ones such as DMF, DMSO, aqueous methanol or ethanol and 2-methoxyethanol. Tertiary nitriles can be prepared by $SnCl_4$-catalyzed reaction of the corresponding halides with trimethylsilyl[$^{14}$C]cyanide in the presence of a Lewis acid or base (Section 7.2.1), or by $\alpha$-alkylation of [$^{14}$C]nitriles. In the presence of appropriate bases, primary and secondary [$^{14}$C]nitriles can be deprotonated and alkylated to give secondary or tertiary derivatives[7]. Aliphatic nitriles usually require strong bases such as LDA or lithium *N*-isopropyl- *N*-cyclohexylamide, whereas benzyl cyanides can be deprotonated with NaOH or tetraalkylammonium hydroxides. A number of examples of $\alpha$-alkylation of [$^{14}$C]cyanoacetate are discussed (Section 7.2.6); additional information is available from conventional organic chemistry sources[1].

In contrast to [$^{14}$C]carboxylations commonly used for one-[$^{14}$C]carbon homologations, [$^{14}$C]cyanations can be performed on precursors containing unprotected hydroxy, amino and ester substituents. The resulting functionalized [$^{14}$C]nitriles play an important role in the preparation of numerous multifunctional reagents and intermediates. One of the simplest of these, [$^{14}$C]cyanoacetate, is of sufficient importance to warrant its own detailed discussion (Section 7.2.6). Even $\omega$-haloalkyl [$^{14}$C]nitriles are accessible, if the respective $\alpha,\omega$-dihalides are employed in sufficient excess[8,9]. Various carbon-11 and carbon-14-labeled $\omega$-haloalkyl nitriles thus obtained have served as key intermediates in the preparation of two classes of Wittig reagents, $[Ph_3P(CH_2)_n{}^{14}CN]^+X^-$ and $[Ph_3P(CH_2)_n{}^{14}COOH]^+X^-$ (X = Br or I), widely used for the preparation of labeled unsaturated nitriles[8] and 5-*cis*-[1-$^{14}$C]prostadienoic acids[9]. Di[$^{14}$C]nitriles resulting from the reaction of dihalides with two equivalents of alkali [$^{14}$C]cyanide are common intermediates for the preparation of a broad spectrum of labeled cyclic $\beta$-keto esters, cyclic ketones and *N*-heterocycles.

Aromatic and heteroaromatic halides fail to react with alkali metal cyanides under the conditions normally used for aliphatic nucleophilic displacement reactions. Traditionally, the preparation of aryl [$^{14}$C]nitriles has been accomplished either by the Sandmeyer reaction or the Rosenmund–von Braun reaction. In the former, cyano-dediazoniation of aryl diazonium salts is accomplished with Cu$^{14}$CN or a mixture of K$^{14}$CN and CuCN (or CuCl) under neutral conditions and mild temperatures[10]; arylamines containing acid labile functions can be diazotized with commercially available nitrosonium tetrafluoroborate in organic solvents[10a].

The Rosenmund–von Braun reaction involves the treatment of aryl halides with Cu$^{14}$CN or a 2:1-mixture of K$^{14}$CN/CuI in pyridine, quinoline or an appropriate aprotic polar solvent such as *N*-methylpyrrolidine (NMP), DMF, DMAC, HMPA at temperatures ranging from 150–250 °C[11]. This reaction has been postulated to proceed via organocopper intermediates. The reactivity of aryl–halogen bonds is in the order I > Br > Cl ≫ F. The difference in reactivity between Ar–Br and Ar–Cl is normally sufficient to permit selective cyanation of the Ar–Br bonds (for example, compounds **2** and **3**[12]) (Figure 7.2). Discrimination between Ar–Br and Ar–I bonds, however, is more difficult to accomplish, and therefore mixtures of iodo- and bromo-substituted aryl [$^{14}$C]nitriles can be expected whenever iodo, bromo aryl dihalides are employed[11,13]. The selective replacement of iodine in compound **4**[13] has to be regarded as an exception. Ar–F bonds are generally unreactive. Various substituents, such as $NO_2$, CHO, COR, COOH, COOR, $CONR^1R^2$, O–alkyl, OH and $NH_2$ are tolerated. The main disadvantage of this versatile procedure is the requirement for high reaction temperatures, so that molecules containing thermolabile substituents or substructures may not survive. For example, [$^{14}$C]cyano-deiodination of **5** gave low yields under the usual conditions with either K$^{14}$CN/CuI and Cu$^{14}$CN; in contrast, the same reaction on the truncated molecule **6** proceeded smoothly[14].

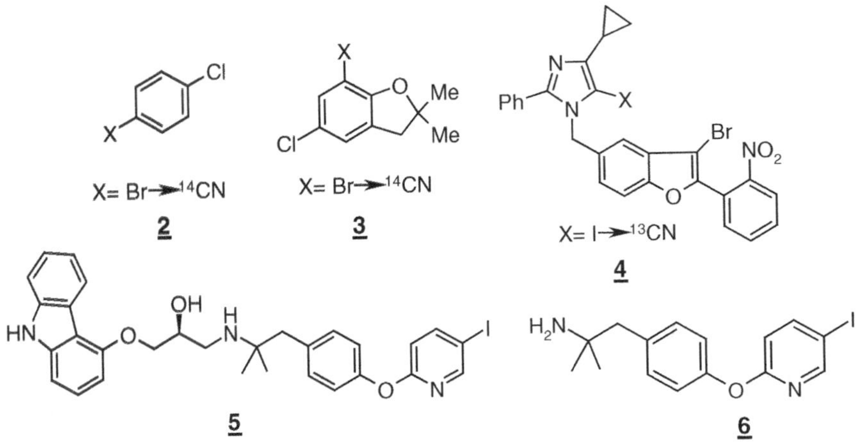


*Figure 7.2* Rosenmund–von Braun [$^{14}$C]cyano-dehalogenation of aryl halides

Palladium-catalyzed cross-coupling of aryl iodides and triflates (ArI > ArOTf ≫ ArBr) with KCN (the Sekiya–Ishikawa procedure[15]) was introduced into carbon isotope chemistry as an alternative possessing a number of advantages over the Rosenmund–von Braun procedure[16]. These include milder reaction conditions, the ability in most cases to use K$^{14}$CN, better utility in heteroarene substrates and even greater tolerance of ancillary functional groups. Pd(PPh$_3$)$_4$ and Pd(OAc)$_2$ have been the most routinely employed catalysts, although [Pd$_2$(dba)$_3$]/DPPF has been recommended as being more reactive[17]. The procedure works under much milder conditions (THF, DMF, NMP, 60–80 °C) and gives excellent results in the presence of electron-deficient substituents. Yields, however, drop when electron-rich substituents are present or the substrate possesses multiple nitrogen and oxygen functionalities. This method permits more discrimination between aryl bromides and aryl iodides, and allows the cyano-deoxygenation of aryl triflates. For example, cyanation of 4-bromoiodobenzene with K$^{14}$CN, Pd(PPh$_3$)$_4$ in refluxing THF furnished exclusively 4-bromobenzo[$^{14}$C]nitrile (**7**) in 85% yield[16b] (Figure 7.3), whereas the uncatalyzed use of Cu$^{14}$CN afforded a mixture of the 4-bromo- and 4-iodo[$^{14}$C]-nitriles. More vigorous reaction conditions (Pd(OAc)$_2$, HMPA, 130–140 °C) were required to effect complete [$^{14}$C]cyano-debromination of compound **8**[18]. An alternative, nickel-based catalytic system was used for [$^{14}$C]cyano-debromination of **9**, after the failure of other methods (KCN/Pd(PPh$_3$)$_4$ in THF; CuCN/Pd(OAc)$_2$ in DMF or NMP). In contrast, a 51% yield of [$^{14}$C]nitrile was obtained by the use of NiBr$_2$(PPh$_3$)$_2$/Zn/PPh$_3$[19]. Although phenol functions suppress the cyanation reaction unless protected with the bulky *tert*-butyldimethylsilyl group, the reactivity of the corresponding aryl triflates has been clearly demonstrated in model studies[17], so that phenols can be regarded as useful precursors for the preparation of aryl [$^{14}$C]nitriles. An illustration of the applicability of [$^{14}$C]cyano-deoxygenation in a complex molecule in good yield is the case of compound **10**[20], which was prepared under mild conditions (NMP, 80°C) through reaction with K$^{14}$CN in the presence of 3:1 Pd$_2$(dba)$_3$/DPPF.

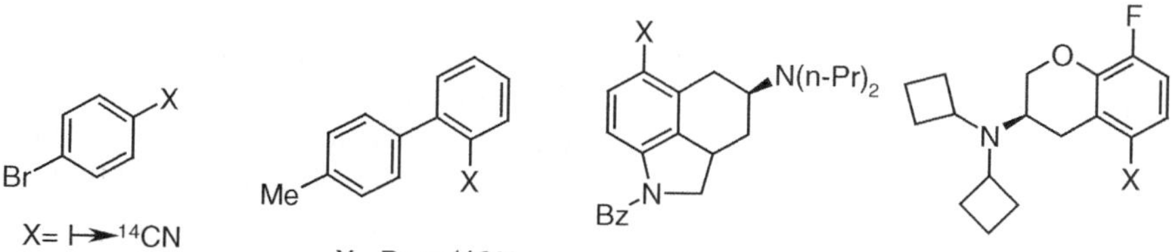


X= I→$^{14}$CN **7** | X= Br→$^{14}$CN **8** | X= Br→$^{14}$CN **9** | X= OTf→$^{14}$CN **10**

*Figure 7.3 Palladium-catalyzed cross-couplings to install aryl [$^{14}$C]cyano groups*

The reaction can sometimes be enhanced by adding catalytic amounts of CuI; for example, yields for the cyanation of substrates containing electron-rich substituents increase by a factor of five to ten. Example **11** demonstrates that even free amino and hydroxy groups seem to be at least partially tolerated. In this reaction also, Ar–Br bonds are reported to react more slowly, which would allow discrimination between those and Ar–I bonds[21].

Replacement of KCN with the highly covalent $Zn(CN)_2$ has emerged as an additional powerful alternative because the concentration of cyanide ions, which are believed to be detrimental to the catalytic cycle, is significantly lowered. In the presence of $Pd(PPh_3)_4$, high yields of nitriles are obtained from aryl bromides at 80 °C in a few hours and from iodoanilines in 30 min.[22]. Under $Pd_2(dba)_3$ catalysis in the presence of P(*tert*-Bu)$_3$ or P(*o*-tolyl)$_3$, even aryl bromides possessing a variety of substituents including unprotected aryl hydroxy and aryl amine functions react at mild temperatures (DMF or NMP, 80°C)[23]. Examples **12**[24], **13**[25], and **14**[26] show the application of $Zn(^{14}CN)_2$ in carbon-14 syntheses.

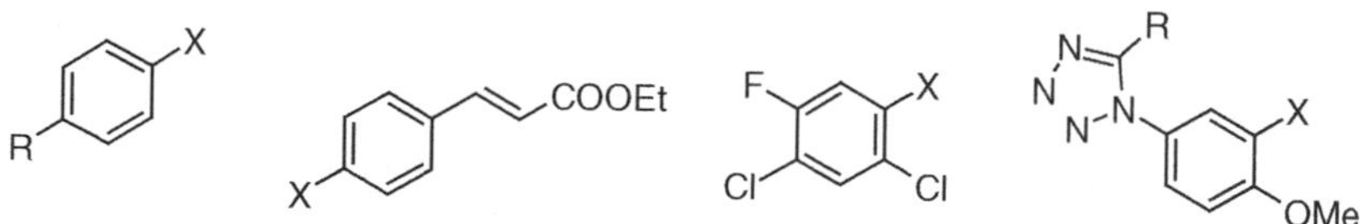


R = OH, NH$_2$; X = I→$^{14}$CN **11** | X = Br→$^{14}$CN **12** | X = Br→$^{14}$CN **13** | R = H,CF$_3$; X = I→$^{14}$CN **14**

**Reaction conditions:** **11**. K$^{14}$CN, Pd(PPh$_3$)$_4$, 10 mol% CuI, THF, reflux (20–50 %, R=OH or 4–25 %, R=NH$_2$); **12**. Zn($^{14}$CN)$_2$, Pd$_2$(dba)$_3$, P(*o*-tolyl)$_3$, NMP, 80 °C (87 % from K$^{14}$CN); **13**. Zn($^{14}$CN)$_2$, Pd(PPh$_3$)$_4$, NMP, 120 °C (59 %); **14**. Zn($^{14}$CN)$_2$, Pd(PPh$_3$)$_4$, DMF, 80 °C (76 % R = H; >90 % R = CF$_3$)

*Figure 7.4 Zn($^{14}$CN)$_2$ in aryl [$^{14}$C]cyano-dehalogenations*

### *[$^{14}$C]Cyanide Introduction by Nucleophilic Additions*

The 1,4-addition of [$^{14}$C]cyanide to appropriate Michael acceptors, such as $\alpha,\beta$-unsaturated carboxylates **15**[27] and **16**[28a], alkyl 3,3-disubstituted 2-cyanoacrylates **17**[29] and dialkyl 2-arylidenemalonates **18**[30], give rise to [$^{14}$C]succino-nitriles and *β*-[$^{14}$C]-cyanocarboxylic acids, which are interesting synthetic intermediates owing to the different reactivity of the two functional groups. Addition of crown ethers (e.g., 18-crown-6 or kryptofix 2.2.2) has been recommended to accelerate the reactions and improve yields[28,31].

The alternative approach, namely the addition of *α*-deprotonated [$^{14}$C]nitriles to appropriate Michael acceptors, can also be effective, as illustrated with the reaction of **19**[32]. [$^{14}$C]Nitriles for this approach are most readily accessible by nucleophilic displacement (previous section). If the Michael acceptor is methyl acrylate, even *α*-substituted benzyl [$^{14}$C]cyanides react, giving tertiary [$^{14}$C]nitriles[7a,29,32], a class of compounds not accessible by nucleophilic displacement with [$^{14}$C]cyanide ion.

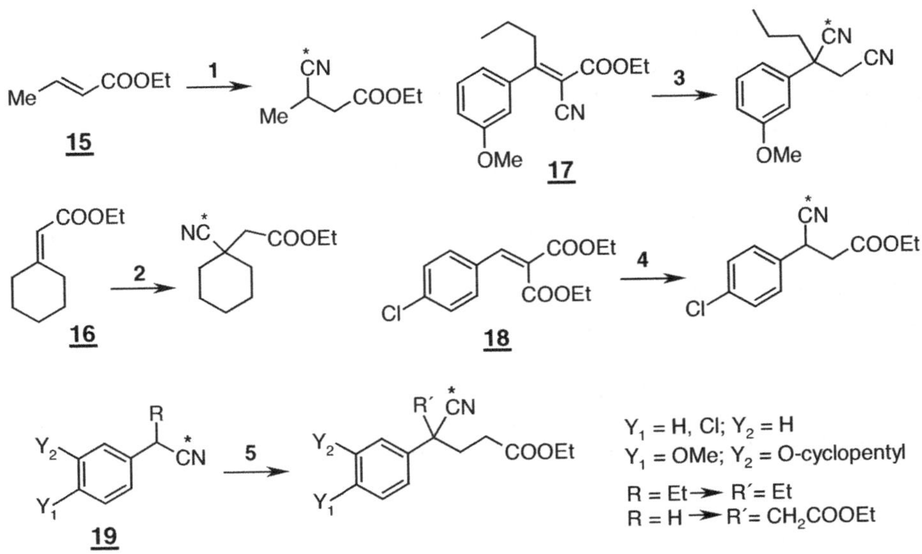


**Reaction conditions: 1.** K$^{14}$CN, aq. EtOH; **2.** K$^{14}$CN, 18-crown-6, MeCN; **3.** K$^{14}$CN, EtOH; 70 °C, 16 h (72 %), **4.** K$^{14}$CN, aq. EtOH; reflux, 16 h (74 %); **5.** $CH_2$=CHCOOEt, $BnMe^3N^+$ $OH^-$ (92 %)

***Figure 7.5*** *Preparation of aliphatic [$^{14}$C]nitriles via 1,4-addition reactions*

The [$^{14}$C]cyanide ion readily undergoes nucleophilic addition to aldehydes and ketones, giving *α*-hydroxy[$^{14}$C]nitriles ([$^{14}$C]cyanohydrins), and to imines, furnishing *α*-amino-[$^{14}$C]nitriles. *α*-Amino[$^{14}$C]nitriles are also formed when aldehydes and ketones are treated with [$^{14}$C]cyanide in the presence of ammonia or ammonium salts. Aliphatic aldehydes and ketones give satisfactory results in this reaction, whereas aryl alkyl ketones give low yields, and diaryl ketones do not react at all. The latter two classes of compounds can, however, be converted into cyanohydrins by treatment with $Me_3Si^{14}CN$ in the presence of a Lewis acid or base (Section 6.2.1). Upon treatment with $POCl_3$/pyridine, [$^{14}$C]cyanohydrins eliminate water to give *α*,*β*-unsaturated [$^{14}$C]nitriles, which have been exploited for the synthesis of labeled steroids[33] such as [20-$^{14}$C]- and [21, 21-$^{13}C_2$]pregn-5-en-3*β*-ol-20-one (**20**)[33a,b].

*N*-Activation in quinolines (through acylation) and pyridines (through etherification of the respective *N*-oxides) allows the nucleophilic addition of [$^{14}$C]cyanide to the position *alpha* to nitrogen, resulting in the formation of 2-[$^{14}$C]cyanoquinolines and [2-$^{14}$C]-cyanopyridines (Reissert–Henze reaction). This has been exploited for the preparation of the 2-[$^{14}$C]cyano derivative **21** in good yield[34].

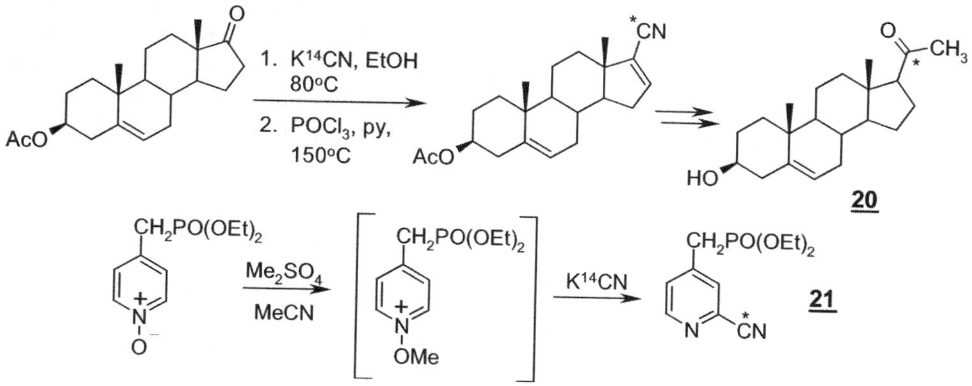


***Figure 7.6*** *Labeled steroids via [$^{14}$C]cyano-dehydration of a steroidal ketone and [$^{14}$C] cyanation of a pyridine through N-activation*

### 7.1.3 Synthetic Transformations of Organic [$^{14}$C]Nitriles

Although nitrile groups within molecules are subject to a limited number of reactions, mainly through nucleophilic attack on the carbon atom, a broad range of chemistry is nevertheless enabled. The nitrile function's reactivity is of four types. These are *reductions* (catalytic hydrogenation, nucleophilic additions of hydride) through which nitriles are converted to imines or aminomethyl groups; *hydrolyses* (nucleophilic additions of hydroxide), which result in primary amides or carboxylic acids; *nucleophilic additions* to the carbon center by agents other than hydrogen, hydride or hydroxide. These nucleophiles can include alkoxy, amino and thio groups, and carbanions; and, finally, [3 + 2]cycloadditions to 1,3-dipolar compounds that give five-membered heterocyclic reaction products.

These direct reactions on nitrile functions often produce simple transformations. It is the potential for further transformation of the immediate products, especially in association with nearby functionality, that gives the chemistry its ability to access a wide range of synthetic targets.

*Reductive Processes*

Reduction of [$^{14}$C]nitriles with metal hydrides such as DIBAL and LiH[Al(OEt)$_3$] provides labeled imines, which can be readily hydrolyzed to [1-$^{14}$C]aldehydes. With $SnCl_2$ and HCl[26], imine products hydrolyze to [1-$^{14}$C]aldehydes during workup, whereas with Raney nickel and $NaH_2PO_2$ in aqueous acetic acid[26,35] or Raney nickel and formic acid[36] [1-$^{14}$C]-aldehydes are formed directly. $LiAlH_4$ and $BH_3$·THF reduce [$^{14}$C]nitriles to [1-$^{14}$C]amines.

The ease of preparing [$^{14}$C]aldehydes through partial reduction of [$^{14}$C]nitriles and hydrolytic workup of the intermediate imines, and the broad variety of possible transformations is illustrated in reductive amination of **22** to provide the carbon-14-labeled NK-1 receptor antagonist **23**[26]; the aldol reaction of **24** with acetone followed by acid catalyzed cyclization to give $\beta$-[$^{14}$C]ionone **25**[37], the key intermediate for the synthesis of a wide variety of carbon-14-labeled retinoids and carotenoids; the Hantzsch synthesis of substituted [4-$^{14}$C]dihydropyridine **27** through reaction of **26** with ethyl acetoacetate in the presence of ammonia[38]; reaction of **28** with substituted 2-oxo-2*H*-cyclohepta[b]furan-3-carboxylate to afford [2-$^{14}$C]azulene **29**[39]; and by conversion of aldehyde **30** into the corresponding hydroxylamine, followed by $\alpha$-chlorination and thioether formation to furnish the naturally occurring glycoside 2-desulfophenethylglucosinolate (**31**)[40].

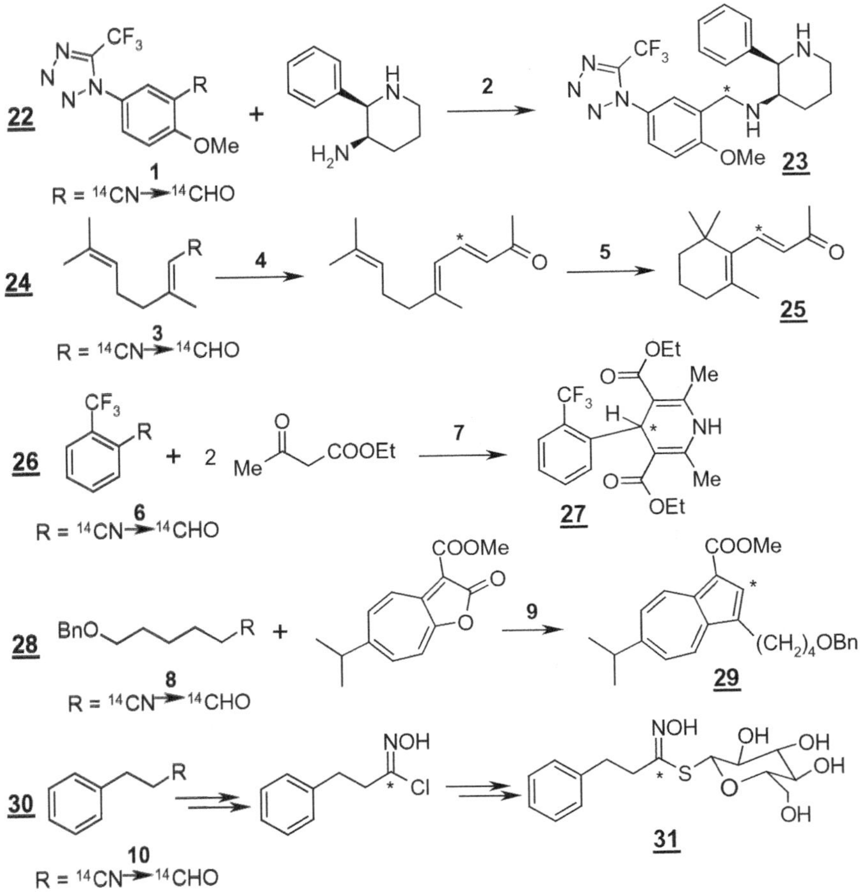


**Reaction conditions: 1**. Raney-Ni, aq. $NaHPO_4$, AcOH, pyridine; **2**. $NaBH(OAc)_3$, $CH_2Cl_2$; **3**. DIBAL, hexane; **4**. 2N NaOH, acetone; **5**. $H_2SO_4$, $MeNO_2$; **6**. DIBAL, toluene; **7**. aq. $NH_3$; **8**. DIBAL, toluene; **9**. morpholine, EtOH; reflux; **10**. DIBAL, hexane; **11a**. $H_2NOH{\cdot}HCl$, $Na_2CO_3$, $H_2O$,**b**. NCS, py, $CH_2Cl_2$

***Figure 7.7*** *Preparation and applications of [1-$^{14}$C]aldehydes from [$^{14}$C]nitriles*

The palladium-catalyzed hemihydrogenation of protected 2-deoxy-D-aldonic acid [$^{14}$C]-nitriles has become an established procedure for the preparation of 2-deoxy-D-[1-$^{14}$C]-glucose (**32**), 2-deoxy-D-[1-$^{14}$C]ribose and various 5-deoxy-[1-$^{14}$C]pentoses[41].

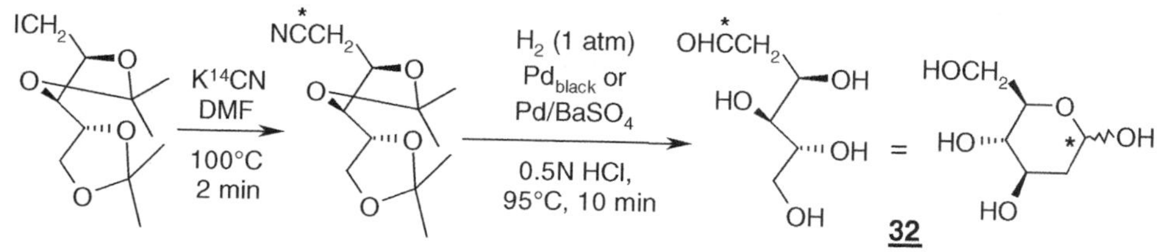


**32**

*Figure 7.8* Standard procedure for preparation of 2-deoxy-D- or L-[1-$^{14}$C]hexoses and pentoses

[$^{14}$C]Imines are not only essential precursors for [$^{14}$C]aldehydes, but are themselves useful building blocks, especially for the synthesis of labeled heterocyclic compounds. Treatment of imine **33**, for example, with diazomethane furnished triazoline **34**[36]. Hydrogenation of nitrile **35** in the presence of aniline **36** gave the reductive amination product **37** in 36% yield, undoubtedly by way of an imine[42]. The synthesis of 4-hydroxy[3-$^{14}$C]indole (**38**) through hydrogenation of 2-benzyloxy-6-nitrobenzyl [$^{14}$C]cyanide may be regarded as a process in which an *in situ* generated imine is trapped intramolecularly by an appropriate reaction partner ($NH_2$ group) nearby[43].

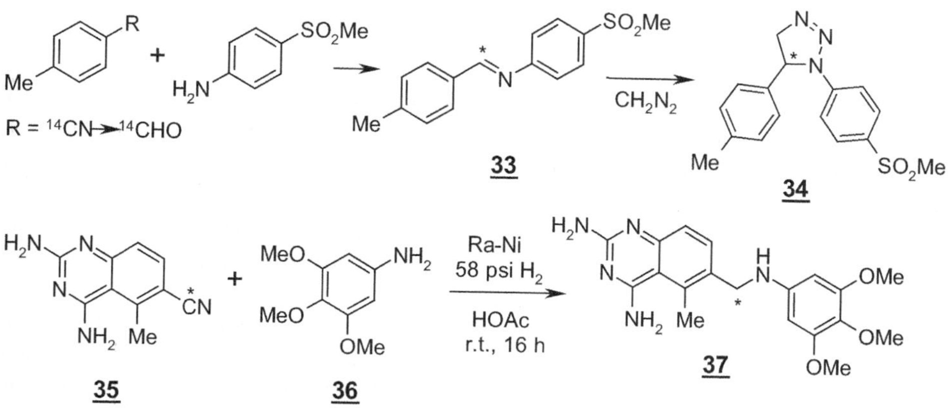


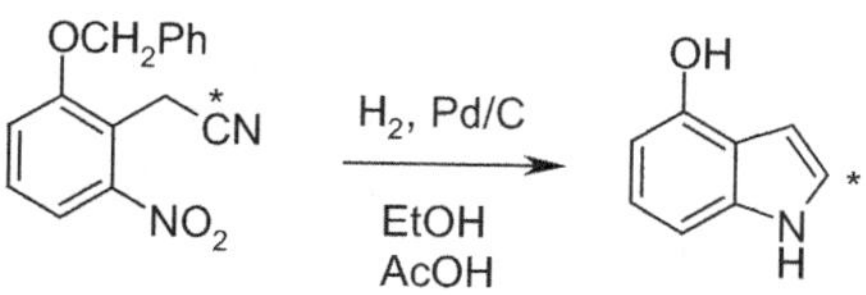


**38**

*Figure 7.9* Preparation and applications of [$^{14}$C]imines

Reductions of [$^{14}$C]nitriles to amino[$^{14}$C]methylene compounds as final synthetic targets is straightforward and useful[44], but can be even more valuable as setups for further transformations, such as in the preparation of [$^{14}$C]piperidine derivative **39**[45], the [$^{14}$C]-tetrahydro-$\beta$-carboline **40**, an endogenously formed neurotoxin resulting from exposure to chloral hydrate[46], and for application of the Tiffeneau–Demjanov rearrangement to give [2-$^{14}$C]cycloalkanones, such as **41**[47].

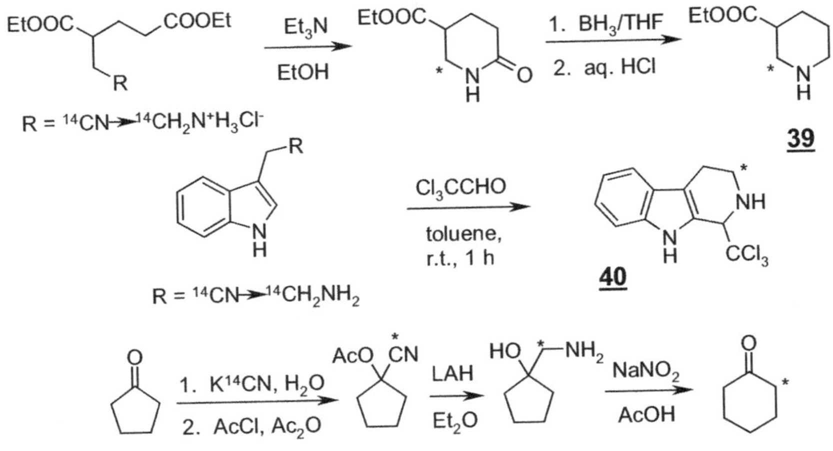


***Figure 7.10*** *Preparation and applications of [1-$^{14}$C]amines*

*Hydrolytic Processes (O-Nucleophiles)*

Alkyl and aryl [$^{14}$C]nitriles can be hydrolyzed by chemical or chemoenzymatic[48] methods to labeled amides and carboxylic acids following routine protocols. Since alkali metal [$^{14}$C]-cyanides are stable, nonvolatile and easy to handle reagents, the latter transformation can be regarded as an alternative to [$^{14}$C]carboxylation of organolithium or organomagnesium precursors (Chapter 5, Section 5.1), in cases where the requisite organometallic species are unviable or the use of gaseous $^{14}CO_2$ has to be avoided for experimental or radiation protection reasons.

Hydrolysis methods for mono[$^{14}$C]nitriles and the spectrum of chemical transformations of the resulting [1-$^{14}$C]carboxylic acids differ little from those for their unlabeled counterparts, so that the synthesis of [1-$^{14}$C]tetralones **42** and [1-$^{14}$C]indanones **46** from $\omega$-arylalkyl [$^{14}$C]nitriles will suffice as examples (Figure 7.11). Acid-catalyzed hydrolysis of 4-phenylbutyro[$^{14}$C]nitrile followed by polyphosphoric acid-mediated cyclization of the resulting phenyl[1-$^{14}$C]butanoic acid converted it into **42** in 89% radiochemical yield. Sequential bromination and oxime formation provided the $\alpha$-bromooxime **43**, which upon treatment with polyphosphoric acid underwent a Beckman rearrangement to provide the ring-expanded tetrahydro-2*H*-1-[2-$^{14}$C]benzazepine-2-one derivative **44**. This was a key intermediate for the preparation of a series of labeled ACE inhibitors[49]. The acid-catalyzed cyclization of a mixture of 3-(2-hydroxyphenyl[1-$^{14}$C]propionic acid and its ethyl ester, prepared by treatment of the propio[$^{14}$C]nitrile with ethanolic HCl proved to be a two-step sequence. In the first step, 2-[2-$^{14}$C]chromanone (**45**) was formed and had to be submitted to

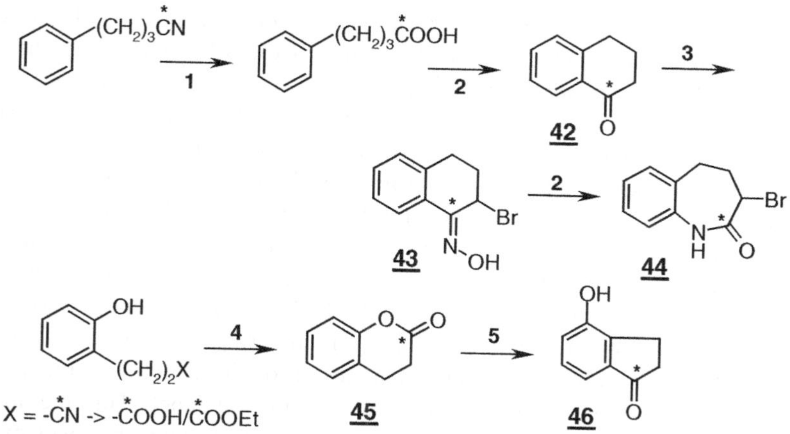


**Reaction conditions:** **1**. HOAc, $H_2SO_4$; reflux; **2**. PPA, heat; **3a**. $Br_2$, **b**. $NH_2OH$; **4a**. HCl, EtOH, **b**.*p*-TsOH, toluene; **5**. $AlCl_3$, NaCl; 190 °C

***Figure 7.11*** *Acid-catalyzed cyclizations via aralkyl [$^{14}$C]nitriles*

a fusion with $AlCl_3$/NaCl in order to obtain the desired hydroxy[1-$^{14}$C]indanone **46** in 60% overall yield[50].

Di[1, *n*-$^{14}C_2$]carboxylic acids and their esters resulting from hydrolysis of the corresponding $\alpha,\omega$-di[$^{14}$C]nitriles are highly valuable and common intermediates for a broad spectrum of building blocks such as 1,3-[1, 4-$^{14}C_2$]butadiene and [$^{14}$C]cycloalkanones. In the first case reduction of [1, 4-$^{14}C_2$]succinic acid followed by a routine three-step conversion of the resulting 1,4-[1, 4-$^{14}C_2$]butanediol into [1, 4-$^{14}C_2$]tetramethylene-1,4-bis(trimethylammonium) hydroxide and thermal decomposition afforded the target compound (Figure 7.12), which is a useful building block in Diels–Alder reactions[51].

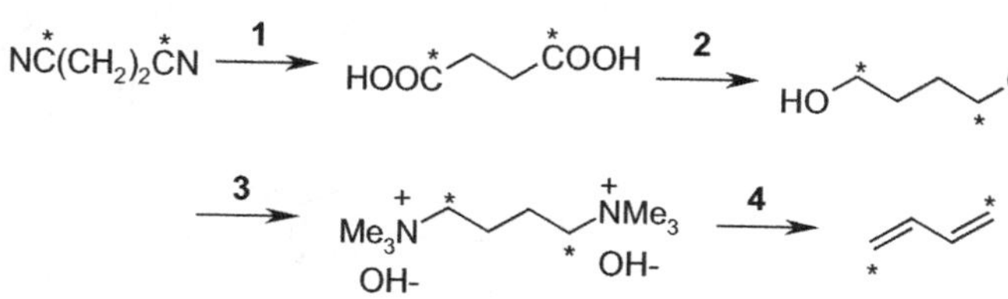


**Reaction conditions:** **1**. conc. HCl; **2**. $BH_3 \cdot$THF, 0 °C; **3a**. 48 % HBr, **b**. $Me_3N$, EtOH, **c**. $Ag_2O$, $H_2O$; **4**. 250 °C (30–57 % overall, 1–15 mCi/mmol

***Figure 7.12*** *Synthesis of [1,4-$^{14}C_2$]butadiene via an $\alpha,\omega$-di[$^{14}$C]nitrile*

Pyrolysis of the barium salts of $HOO^{14}CCH_2(CH_2)_nCH_2{}^{14}COOH$ ($n = 4$–5) furnishes [carbonyl-$^{14}$C]cycloalkanones with simultaneous loss of one carbon unit, which is eliminated as $^{14}CO_2$ (Figure 7.13, steps 1, 2). In spite of this disadvantage the procedure has been exploited several times for the preparation of [1-$^{14}$C]cyclohexanone (**47**) because of its simplicity and the potential to recover the eliminated $^{14}CO_2$ as $Ba^{14}CO_3$[28a,52]. Dieckmann condensation of the corresponding dialkyl esters mediated by base (e.g. NaH, NaOEt, KO*t*-Bu[28b], colloidal K[53]) affords cyclic β-keto esters (steps 3, 4). High yields and easy handling favor the colloidal potassium variant. This type of reaction, which is most successful for five-, six-, and seven-membered rings, also suffers from inevitable loss of half of the carbon-14 label, if the β-keto ester is converted into the cyclic ketone through hydrolysis and decarboxylation. So far, only a few applications, such as *allo*-berberane[28b] and **48**[53], have been reported. However, the intramolecular reductive ester condensation promoted with sodium in aprotic solvents in the presence of TMSCl (acyloin condensation) produces cyclic α-hydroxy ketones bearing both labels in the ring system[54,55] (steps 5, 6). Accordingly, their conversion to cycloalkanones through reductive removal of the hydroxy group with Zn/HCl proceeds without any loss of label. This methodology, which works especially well for medium and large size ring systems, has been exploited for the preparation of carbon-14-labeled 6- and 10-membered rings, as demonstrated by the synthesis of *trans*-1,2-diamino[1,2-$^{14}C_2$]cyclohexane (**49**)[54] and [1,2-$^{14}C_2$]cyclodecanone (**50**)[55]. An alternative to the Dieckmann methodology is treatment of the dinitrile with a lithium amide to induce cyclization, followed by hydrolysis and decarboxylation of the intermediately formed cyclic β-keto nitriles (Thorpe–Ziegler cyclization)[56] (Section 7.1.3.3). Preparation of [1-$^{14}$C]cycloalkanones in one step without loss of radioactivity is achieved by reaction of $^{14}CO_2$ with α,ω-dimagnesio- or dilithioalkanes (Chapter 5, Section 5.1).

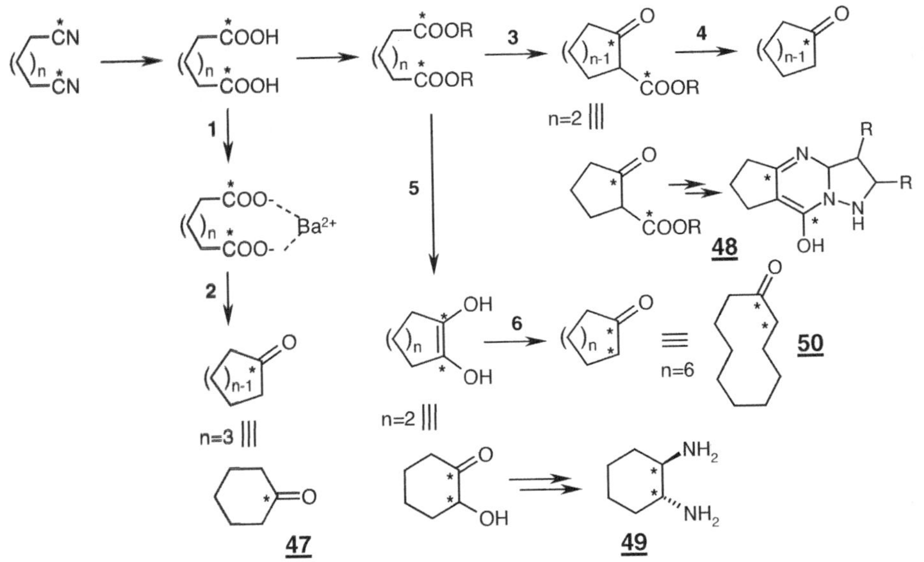


**Reaction conditions: 1.** $BaCO_3$ or $CaCl_2$; **2.** Δ, $-^{14}(CO_2)$; **3.** *tert*-BuOK, benzene; reflux. 40 min or K(colloidal), toluene; 50 °C, 1h; **4.** NaOEt, xylene; r.t, 1 h; **5.** Na. xylene; 130 °C, 30 h or Na, TMSCl, toluene, reflux; **6.** Zn, HCl-HOAc; r.t.

***Figure 7.13*** *Preparation of carbon-14-labeled alicyclic compounds from α,ω-di[$^{14}$C]nitriles*

Partial hydrolysis–cyclization of di[$^{14}$C]nitriles in contrast produces dilabeled amine heterocycles. [2,6-$^{14}C_2$]Glutarimides are the most common products because of their utility as intermediates to pyridines; they have been prepared by several different methods in yields of up to 80% from dinitrile[57,58]. Illustrations of the utility of [2,6-$^{14}C_2$]glutarimide through its transformation into [2,6-$^{14}C_2$]pyridines include the preparations of compounds **51**[57b], **52**[58c] and **53**[59].

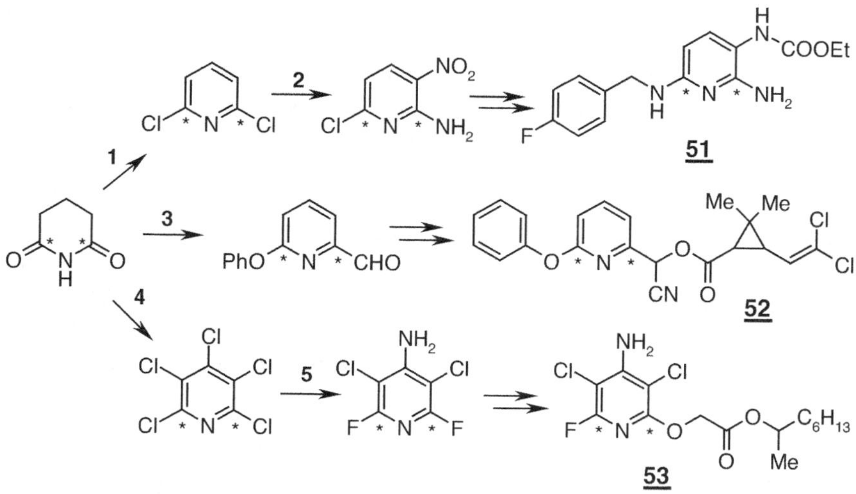

**Reaction conditions:1**. $PCl_5$, $PCl_3$, r.t., 48 h (60–75 %); **2a**. $HNO_3$, $H_2SO_4$,108 °C, 6 h (55 %), **b**. $NH_3$ (g), *i*-PrOH, r.t., 22 h (78 %); **3a**. AcOH, HBr, $PBr_3$,128 °C, 22 h (96 %), **b**. phenol, NaH, DMSO, 80 °C, 30 min, **c**.*n*-BuLi, $Et_2O$, −78 °C to −30 °C then DMF, −78 °C to 0 °C, acid workup (74 %); **4**. $Cl_2$, $PCl_3$, $FeCl_3$, $I_2$, 250 °C, 28 h (91 %); **5a**. CsF, MeCN, reflux, 24 h, **b**. $NH_4OH$, $H_2O$,MeCN, 5 °C to r.t., 30 min (82 %)

***Figure 7.14*** *Preparation of carbon-14-labeled pyridine derivatives from [2,6-$^{14}$C]glutarimide*

The reaction also worked well with substituted succinonitriles and glutaronitriles, giving rise to substituted [2-$^{14}$C]succinimide **54**[27] and [2-$^{14}$C]glutarimide **55**[60], as shown in Figure 7.15.

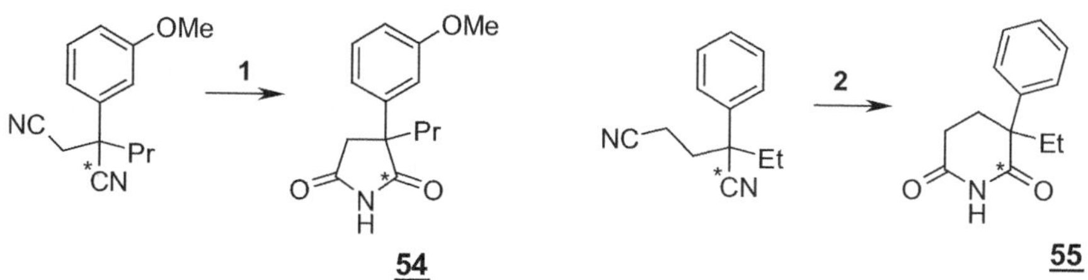

**Reaction conditions: 1**. KOH, $H_2O$, EtOH, ethylene glycol; 110 °C, 20 h; **2**. KOH, $H_2O$, EtOH, ethylene glycol; heat

***Figure 7.15*** *Preparation of substituted [$^{14}$C]succinimides and [$^{14}$C]glutarimides from [$^{14}$C] dinitriles*

The classical and still very commonly used method for the preparation of amino acids labeled with carbon-14 at the carboxyl group is the amino[$^{14}$C]cyanation of aldehydes or ketones and subsequent acid hydrolysis (Strecker reaction)[61] (Figure 7.16). The Bucherer modification of the Strecker reaction involves the reaction of aldehydes or ketones with [$^{14}$C]cyanide and ammonium carbonate to form a hydantoin, which is then hydrolyzed by acid[62] or base[63] to the amino [1-$^{14}$C]acid[62]. If achiral aldehydes are employed and a homochiral product is desired, resolution of the resulting racemic product is required. To improve on this radiochemically wasteful approach, various stereoselective syntheses using chiral auxiliaries have been developed, as exemplified by the preparation of (*S*,*S*)-2,7-diamino[1,8-$^{14}C_2$]suberic acid (**56**)[64]. (Detailed discussion of various stereoselective options can be found in Chapter 11, Section 11.3.2). Use of chiral [$^{14}$C]carbonyl compounds leads to the formation of diastereomeric products, which are often easily separable by simple chemical separation methods, such as crystallization in the case of **57**[62]. Other syntheses of $\alpha$-amino [1-$^{14}$C]acids include routes proceeding via [$^{14}$C]cyanodehalogenation of $\alpha$-haloalkyl phthalimides, which are accessible through Hundsdiecker degradation of *N*-phthalylamino acids, followed by hydrolysis and removal of the phthalyl protection group from the amino function[65]. $\beta$-Amino [1-$^{14}$C]acids can be prepared analogously by [$^{14}$C]cyanide displacement of leaving groups vicinal to (protected) amines[66].

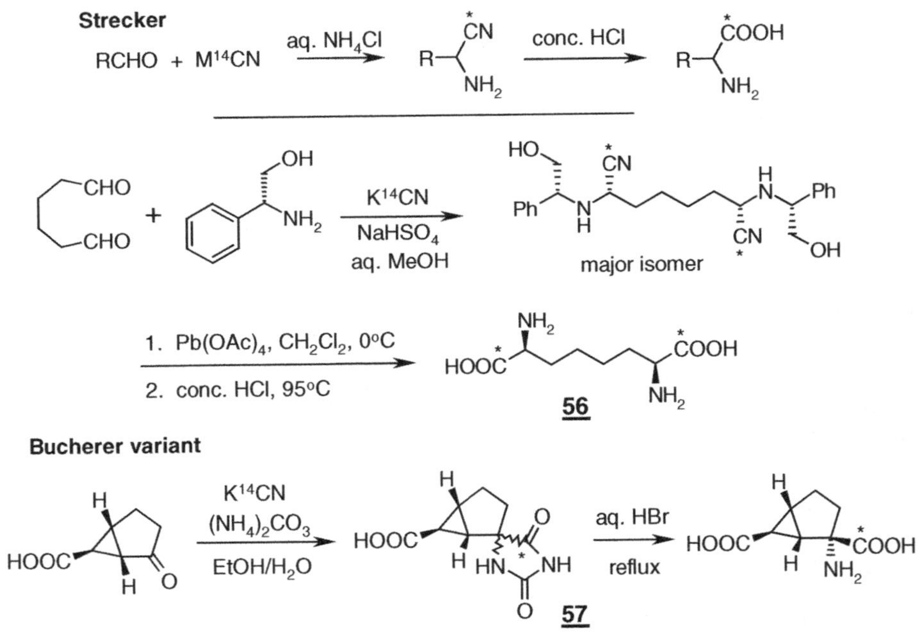


*Figure 7.16* *Strecker and Bucherer methods for synthesis of $\alpha$-amino [1-$^{14}$C]acids*

The range of transformations achievable through hydrolysis of [$^{14}$C]cyanohydrins includes the formation of $\alpha$-hydroxy or $\alpha$,$\beta$-unsaturated [1-$^{14}$C]carboxylic acids and the Kiliani–Fischer synthesis of [1-$^{14}$C]aldoses. In the former, simultaneous acid-catalyzed partial hydrolysis of the nitrile function and elimination of water converted acetone [$^{14}$C]-cyanohydrin into [1-$^{14}$C]methacrylic acid amide[67]. Subsequent acid-catalyzed hydrolysis or methanolysis in the presence of hydroquinone as a radical scavenger (to inhibit radical-induced polymerization) afforded [1-$^{14}$C]methacrylic acid (**58**) as the free acid or its methyl ester in good yield.

Following the classical Kiliani protocol, homologation and labeling of aldoses at C1 is accomplished through treatment of the next lower sugar (e.g., **59**) with [$^{14}$C]cyanide. Hydrolysis of the resulting epimeric [$^{14}$C]cyanohydrins, followed by separation of the mixture of C2-epimeric acids (for example through fractional crystallization of their barium salts), lactone formation (to **60**) and concluding reduction with Na–Hg provides the respective e.p. D-[1-$^{14}$C]aldoses (**61**); e.g., arabinose, ribose, glucose and their respective C2 epimers) in radiochemical yields of 10–55%[68]. In a more recent variant, the [$^{14}$C]-cyanohydrins are converted directly by partial hydrogenation over Pd/$BaSO_4$[69].

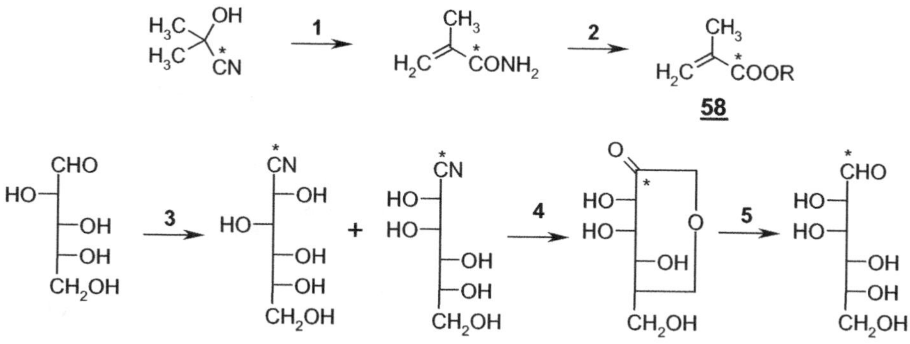

**58**

**59** **60** **61**

**Reaction conditions:** **1**. 98 % $H_2SO_4$, Cu powder; 130 °C; **2**. $H^+$, $H_2O$, hydroquinone (69 %) or $H^+$, MeOH, hydroquinone (75 %); **3a**. $Na^{14}CN$, NaOH; aq. $NaHCO_3$, r.t. 48 h, **b**. 60 °C, 18 h, **c**. $Ba(OH)_2$, 100 °C, 1 h; r.t. 2 to 5 d; **4a**. Ion exchange-H-form, **b**. lyophilization, crystallization from methyl cellosolve; **5**. Na–Hg, $H_2C_2O_4$, $Na_2C_2O_4$; $H_2O$, r.t.

***Figure 7.17*** *Reactions of [$^{14}$C]cyanohydrins*

### *Further Nucleophilic Processes*

Additions of carbon, nitrogen, oxygen and sulfur nucleophiles to [$^{14}$C]nitriles, in most cases aryl [$^{14}$C]nitriles, provide an additional array of options for synthetic transformations, many of them especially well-suited to the preparation of heterocyclic compounds.

The simplest of carbon nucleophilic additions to [$^{14}$C]nitriles are those of Grignard reagents and alkyl- or aryllithiums. Such a reaction was used, for example, in the preparation of [20-$^{14}$C]pregnenolone (**63**) (Figure 7.18), whose labeled [$^{14}$C]acetyl side chain was produced by addition of methylmagnesium bromide to the $\alpha,\beta$-unsaturated [$^{14}$C]nitrile **62**, followed by catalytic hydrogenation of the vinylic double bond[70]. Similarly, addition of methylmagnesium bromide to 2-propoxybenzo[$^{14}$C]nitrile furnished the corresponding acetophenone which, upon condensation with ethyl formate and reaction of the resulting β-ketoaldehyde enolate with cyanoacetamide, provided the [6-$^{14}$C]pyridone **64**, a phosphodiesterase inhibitor[71]. A special case of this methodology is the intramolecular Thorpe–Ziegler condensation of dinitriles, in which a carbanion *alpha* to one of the nitrile groups is generated with LDA or lithium methylanilide and adds *in situ* to the second nitrile function to give a cyclic β-ketonitrile, such as **65**.[56a] This reaction, which is closely related to the Dieckman condensation discussed above, gives good results for five- to eight- and > 14-membered rings. In the radioactive case it suffers, however, from the same disadvantage that, upon removal of the β-nitrile group through hydrolysis and decarboxylation, half of the label is lost.

In a totally different Lewis acid catalyzed reaction, benzo[$^{14}$C]nitrile reacted with an anilinoborane to give specifically the *ortho*-amino substituted [carbonyl-$^{14}$C]benzophenone **66** in good yield[72].

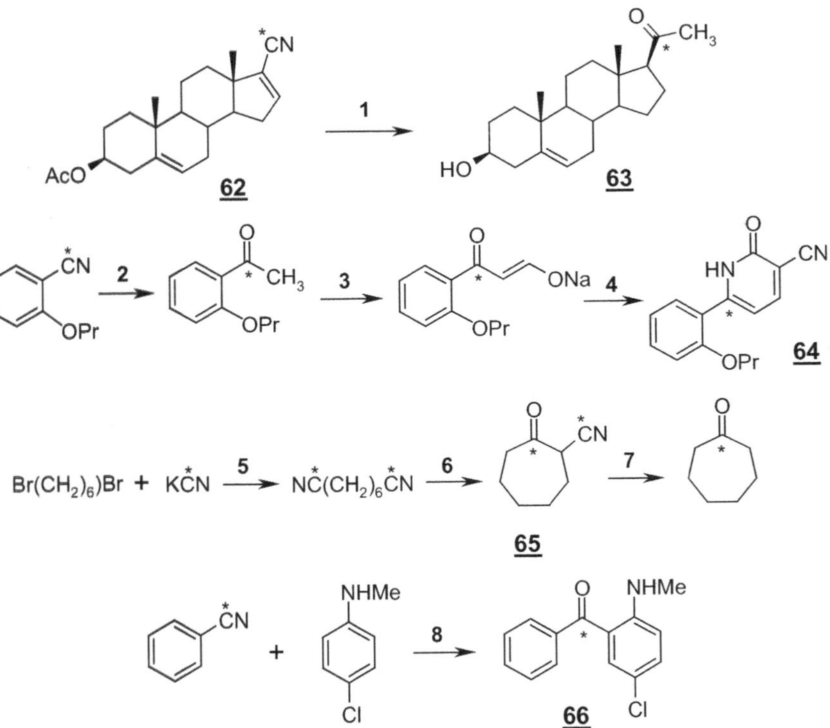


**Reaction conditions: 1a**. $CH_3MgBr$, benzene; reflux, 46 h, **b**. $H_2$, Ra-Ni, EtOH/aq. NaOH; r.t., 3 h; **2**. $CH_3MgBr$, $Et_2O$, benzene; **3**. HCOOEt, NaOMe, $Et_2O$, r.t., 2.5 h; **4**. $NCCH_2CONH_2$, piperidine acetate, $H_2O$, reflux, 20 h; **5**. EtOH, $H_2O$; **6**. Ph(Me)NLi, toluene; **7**. $H_3O^+$, heat ($-^{14}CO_2$); **8**. $BCl_3$, $AlCl_3$, tetrachloroethane, reflux, 5.5 h

***Figure 7.18*** *Additions of carbon nucleophiles to [$^{14}$C]nitriles and Lewis acid arylation of an aryl [$^{14}$C]nitrile*

Addition of *N*-nucleophiles converts nitriles into amidoximes, amidines, amidrazones, etc., which are valuable intermediates, especially for the preparation of heterocycles. For example amidoxime **67**, prepared from *p*-hydroxybenzo[$^{14}$C]nitrile and hydroxylamine, reacted with nicotinoyl chloride to give 3,5-disubstituted 1,2,4-[3-$^{14}$C]oxadiazole **68**, the penultimate precursor to a hypocholesterolemic investigational drug[73]. The *O*-alkylation of nicotin[$^{14}$C]amidoxime (**69**), when treated with *N*-(3-chloro-2-hydroxy)propylpiperidine, shows an alternative transformation pathway for the amidoxime function. Subsequent chloro-diazoniation furnished *N*-alkoxyimidoyl chloride **70**[74].

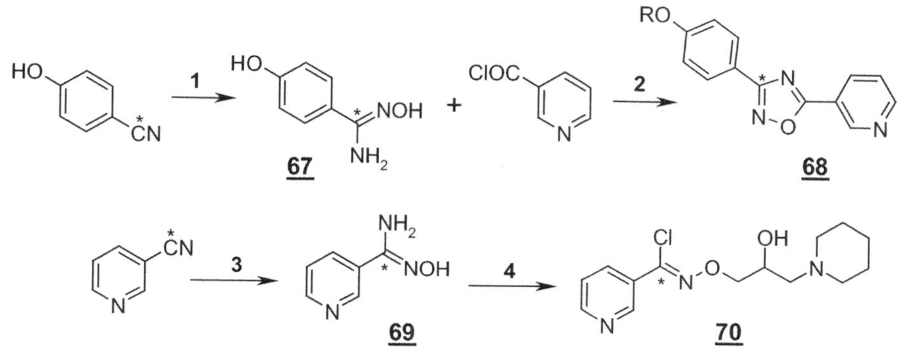

**Reaction conditions: 1**. $HONH_2$, aq. EtOH; 70 °C, 5 h; **2**. pyridine; r.t. to 110 °C; **3**. aq. $HONH_2$; reflux, 3 h; **4a**. $ClCH_2CH(OH)CH_2NR_2$, KOH, *i*-PrOH, **b**. $NaNO_2$, aq. HCl

***Figure 7.19*** *Addition of hydroxylamine to [$^{14}$C]nitriles*

The addition of ammonia and primary amines to nitriles to give amidines requires either drastic reaction conditions or some manner of nitrile activation. The latter may be accomplished either by the use of an appropriate Lewis acid catalyst (e.g., $AlMe_3$, $MeAl(Cl)NMe_2$) or partial hydrolysis (e.g., $H_2O_2$, NaOH) followed by *O*-alkylation of the resulting amide with $Et_3O^+BF_4^-$ to give an iminoester fluoroborate, whose ethoxy group can undergo facile *in situ* displacement by a nitrogen nucleophile. All three approaches have been utilized with [$^{14}$C]nitriles. The first is exemplified by the formation of 3-deaza[2-$^{14}$C]-guanine (**72**) through addition of ammonia to 3-carbomethoxypyrrol-2-aceto[$^{14}$C]nitrile (**71**) at high temperature and pressure[75] (Figure 7.20). [$^{14}$C]Amidine **73** was obtained by $AlMe_3$-catalyzed amination of 2-ethoxybenzo[$^{14}$C]nitrile. Treatment of **73** with hydrazine produced the corresponding [$^{14}$C]amidrazone, which reacted *in situ* with enol ester **74** to afford the [3-$^{14}$C]triazinone **75**, a key intermediate in the synthesis of [$^{14}$C]vardenafil[76].

Finally, amidine **76** was prepared by the iminoester fluoroborate sequence (Figure 7.21). The utility of the amidine for generation of 1,3-diazines is demonstrated by the reaction of **76** with ethyl 4-chloro-2-methylthio-5-pyrimidinecarboxylate to afford the [2-$^{14}$C]-pyrimido-[4,5]pyrimidone **77**, and with ethyl cyanoacetate-2-oxime to furnish the [2-$^{14}$C]-pyrimidine-6-one **79**. Compounds **77** and **79** were precursors for the synthesis of carbon-14-labeled phosphodiesterase inhibitors **78** and **80**, respectively[71]. Another option for

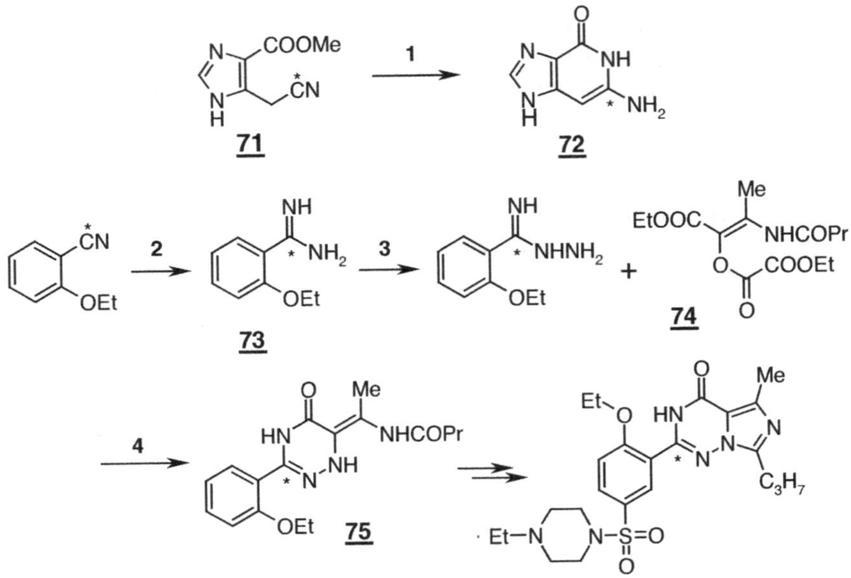


**Reaction conditions: 1**. aq. $NH_3$; 130 °C (under pressure), 7 d; **2**. $NH_4Cl$, $AlMe_3$, toluene; reflux, 9 h **3**. $H_2NNH_2$, EtOH; 0–5 °C, 15 min **4**. **74** reflux, 5 h

***Figure 7.20*** *Direct additions of ammonia and hydrazine to [$^{14}$C]nitriles*

diazine synthesis, by condensation of an amidine with a 1,3-diester, is illustrated by the reaction of [1-$^{14}$C]acetamidine with diethyl [1,3-$^{14}C_2$]malonate to give 4,6-dihydroxy-2-methyl[2,4,6-$^{14}C_3$]pyrimidine (**81**) in 69%[77]. Intermediate **81** was converted to the antihypertensive moxonidine **82**.

Iminoesters are also accessible in good to excellent yields through addition of alcohol to nitriles, either under acid catalysis (e.g., HCl, EtOH, 10–20 °C) or base catalysis (NaOMe, MeOH, r.t.). In most cases where this reaction has been used on [$^{14}$C]nitriles, the alkoxy group has served as a temporary function to be displaced by a nucleophile appropriately substituted for formation of five- or six-membered heterocycles. For example, iminoester **83** (Figure 7.22) reacts with diaminopyrimidinedione **84** to give the carbon-14-labeled xanthine **85**, an advanced intermediate in the synthesis of an endothelial cell adhesion molecule (ECAM) inhibitor[24]. In two other examples, the imino alkoxy group underwent displacement by an acylhydrazine to give an initial adduct which subsequently underwent cyclization to produce a triazolone, in the case of **86**[78], or a triazole, in the case of **87**[4c].

Hydrogen sulfide and aliphatic and aromatic thiols react readily with [$^{14}$C]nitriles to give [$^{14}$C]thioamides in yields of 45–70%. Upon treatment with $\alpha$-halocarbonyl compounds they form [$^{14}$C]thiazoles (Hantzsch thiazole synthesis). This was exploited for the synthesis of 2-(4-chlorophenyl)[2-$^{14}$C]thiazol-4-acetic acid (**89**, X = COOH)[79] (Figure 7.23) and the 2-(2-phthalimidoethyl)-[2-$^{14}$C]thiazole derivative **91**[80], which were obtained when the corresponding thioamides **88** and **90** were treated with 1,3-dichloroacetone and bromoacetaldehyde diethylacetal, respectively. As illustrated by the third example in Figure 7.23,

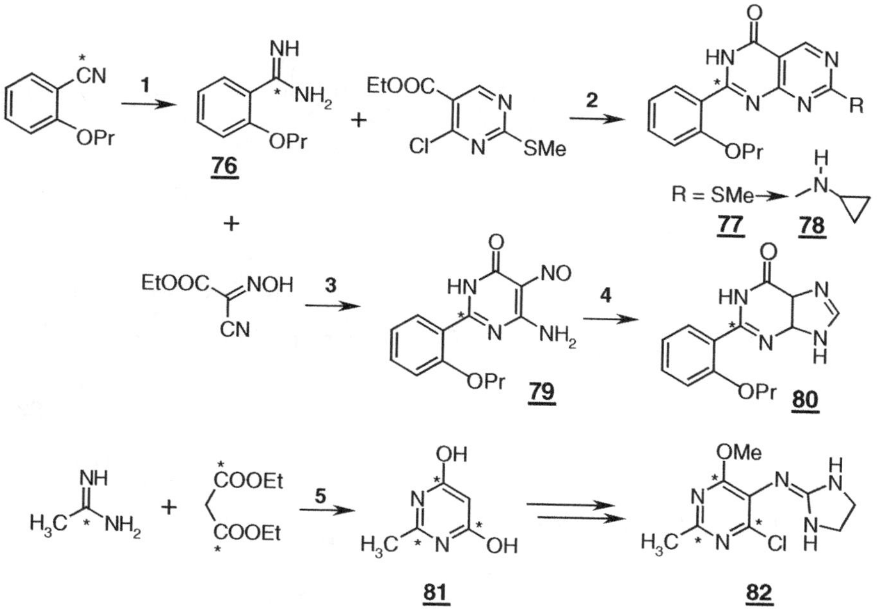

**Reaction conditions: 1a**. 30 % $H_2O_2$, NaOH, MeOH/$H_2O$, 60 °C, 18 h **b**. ($Et_3O$)($BF_4$), $CH_2Cl_2$, r.t., 42 h then $NH_3$, EtOH, r.t., 96 h **2a**. $Et_3N$, MeCN; r.t., 18 h; **3**. NaOEt, EtOH; reflux, 1 h; **4a**. $Na_2S_2O_4$, $NaHCO_3$, aq. MeCN, **b**. HCOOH; reflux, 1.5 h; **5**. NaOEt, EtOH; 85 °C, 3 h

*Figure 7.21* *Nitrogen nucleophile additions to [$^{14}$C]nitriles via iminoester fluoroborates*

[$^{14}$C]nitriles (e.g., 92) and 2-aminothiophenols react to afford [$^{14}$C]benzothiazoles (e.g., 93) in a tandem reaction starting with the nucleophilic addition of the thiol group to the nitrile function[81].

*Cycloadditions at the Nitrile Group*

[$^{14}$C]Tetrazoles are readily accessible in excellent yields through [3 + 2]cycloaddition of $Me_3SnN_3$ to nitriles as illustrated by the synthesis the of [$^{14}$C]KT3-671 (95), an effective AII-receptor agonist (Figure 7.24). Refluxing a mixture of 94 and $Me_3SnN_3$ in toluene for 72 h gave 95 in 85% yield[82].

## 7.2 Preparation of Other Building Blocks from [$^{14}$C]Cyanide

The synthesis of additional one- to three-carbon building blocks is the fourth major area of application of alkali metal [$^{14}$C]cyanides (Figure 7.25). These include trimethylsilyl [$^{14}$C]-cyanide, [$^{14}$C]cyanogen bromide, alkali metal [$^{14}$C]cyanates and [$^{14}$C]thiocyanates, triethyl [$^{14}$C]orthoformate, alkyl [$^{14}$C]cyanoacetates, and [$^{14}$C]diazomethane (via *N*-[$^{14}$C] methyl-*N*-nitroso-*p*-toluenesulfonamide). These building blocks are discussed in the sections below.

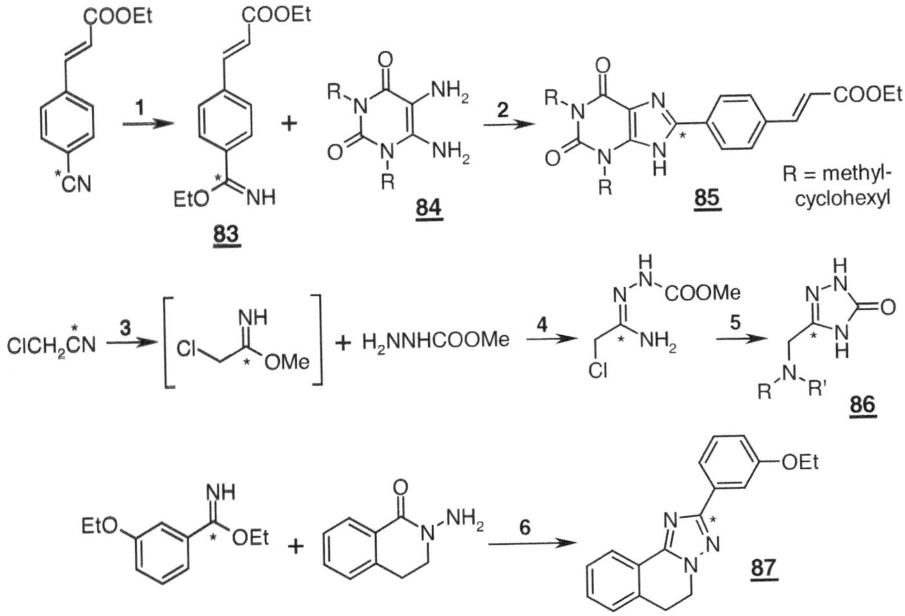


**Reaction conditions:** **1**. HCl(g), EtOH; **2**. AcOH, dioxane; reflux **3a.** NaOMe, MeOH, $Et_2O$; r.t.; **b**. AcOH; **4**. RR'NH, $K_2CO_3$, DMSO; r.t., 2h; **5**. DIEA, xylene; 150 °C, 3h **6a**. 90–130 °C, $Et_2O$; r.t.; **b**. NaOEt, EtOH; reflux

***Figure 7.22*** *Formation of [$^{14}$C]iminoesters by addition of alcohols to [$^{14}$C]nitriles*

[2-$^{14}$C]Glycine should be mentioned as a building block accessible from [$^{14}$C]cyanide in good overall yield[83]. This building block is an alternative to [1-$^{14}$C]glycine, which is usually produced through a [$^{14}$C]carboxylation step (Chapter 5, Section 7.1), and to [2-$^{14}$C]- and [1,2-$^{14}C_2$]glycine, which are accessible through nitro[$^{14}$C]methane (Chapter 5, Section 6).

### 7.2.1 Trimethylsilyl[$^{14}$C]cyanide (TMS$^{14}$CN)

Trimethylsilyl [$^{14}$C]cyanide is conveniently accessible from trimethylsilyl chloride, Na$^{14}$CN and NaI in the presence of lutidine or pyridine[84]. In carbon-14 synthesis it is normally generated *in situ* and used immediately in subsequent Lewis acid catalyzed chemical transformations. It is useful for some reactions in which alkali metal [$^{14}$C]cyanides normally fail, such as cyanide addition to acetals and aromatic and heteroaromatic aldehydes and ketones[1]. Examples include the $ZnI_2$-catalyzed addition of TMS$^{14}$CN to 2-furaldehyde followed by hydride reduction of the resultant silylated cyanohydrin to the aminoalcohol __96__[84a], and the $SnCl_4$-catalyzed formation of D-[1-$^{14}$C]allonitrile __98__, obtained from reaction of TMS$^{14}$CN to the *O*-acetylated hemiacetal __97__ (Figure 7.26). Subsequent addition of $H_2Se$ to the nitrile group, cyclocondensation of the intermediate selenoamide *in situ* with ethyl bromopyruvate and deprotection yielded __99__, precursor to [2-$^{14}$C]CI-935, a compound with antitumor and antiviral activity[84b].

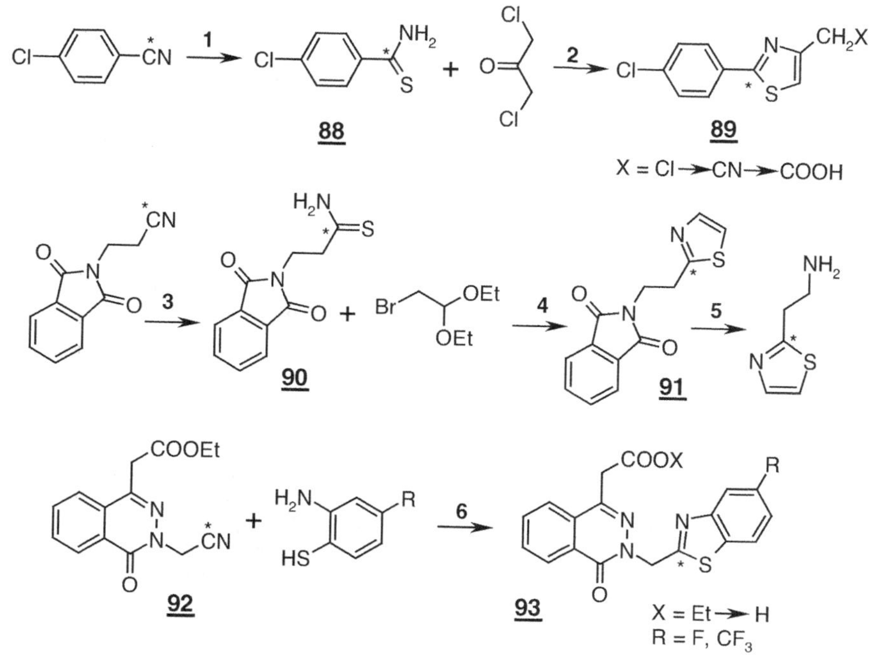

**Reaction conditions: 1**. $H_2S$, $Et_3N$, MeOH; **2a**. acetone; reflux, **b**. $H_2SO_4$; **3**.$H_2S$, $Et_3N$, DMF; **4**. acetone; reflux, 2 h; **5**. 5 N HCl; reflux; **6**. EtOH; reflux, 20 h

***Figure 7.23*** *S-Additions to [$^{14}$C]nitriles*

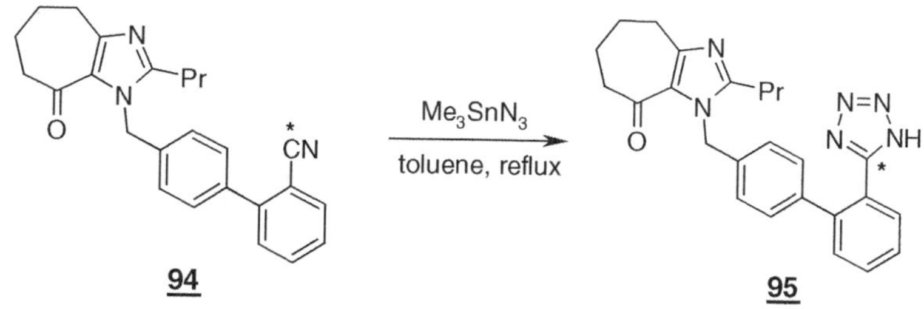

***Figure 7.24*** *Formation of a 1-phenyl[1-$^{14}$C]tetrazole from an aryl[$^{14}$C]nitrile*

### 7.2.2 [$^{14}$C]Cyanogen Bromide (Br$^{14}$CN)

Halogenation of alkali metal [$^{14}$C]cyanides with elemental bromine or chlorine gives the corresponding [$^{14}$C]cyanogen halides, which can be used as reagents for electrophilic cyanations[85]. In contrast to its importance in normal organic synthesis (cleavage of *tert*-amines, peptides) few applications have been reported so far in carbon-14 synthesis. Reaction with one equivalent of ammonia or a primary or secondary amine affords unsubstituted or substituted [$^{14}$C]cyanamides **100** (Figure 7.27), which can be valuable intermediates. In the

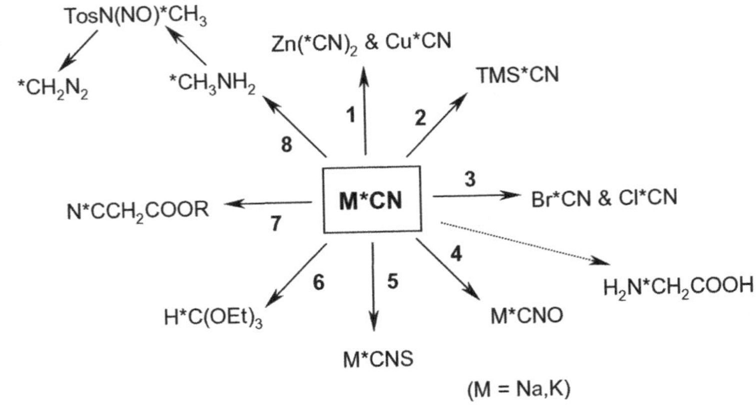


**Reaction conditions: 1a**. $ZnCl_2$, aq. EtOH; r.t; **1b**. $CuSO_4$-$5H_2O$, $NaHSO_3$, $H_2O$; r.t., 1–2 h; **2**. TMSCl, NaI, lutidine, MeCN; r.t., 4d; **3a**. $Cl_2$, $CHCl_3$; –40 °C to r.t., 16 h; **3b**. $Br_2$, $H_2O$; 0–5 °C, 2 h; **4**. $KMnO_4$, cat $Cu(OH)_2$, $H_2O$; 50 °C, 5 min; **5**. S, acetone, reflux, 45 min; **6a**. $H_3PO_4$, **b**. EtOH/ether, HCl(g), –12 °C, 72 h, **c**. $EtOH/CHCl_3$, reflux; 1 h; **7**. $BrCH_2COOH$, $H_2O$ (pH 7–8); 90°C, 1 h; **8a**. $H_2$, $PtO_2$, HCl/HOAc; r.t., **b**. *p*-$CH_3C_6H_4SO_2Cl$, 2N NaOH/dioxane; or aq.NaoH $CH_2Cl_2$, **c**. $NaNO_2$, $HOAc/H_2O$; 0 °C, 1 h, **d**. aq. KOH, ether/EtOH, 65 °C

***Figure 7.25*** *Conversion of alkali metal [$^{14}$C]cyanides into additional [$^{14}$C]building blocks*

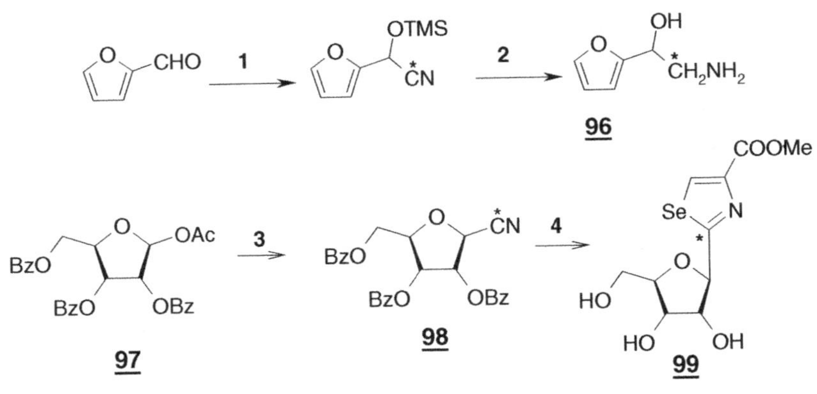


**Reaction conditions: 1**. $K^{14}CN$, TMSCl, NaI, pyridine; $CH_3CN$; reflux, 44 h; **2**. $LiAlH_4$, ether; r.t., 2 h (76 % over two steps); **3**. $TMS^{14}CN$, $SnCl_4$, $CH_3CN$, r.t., 3.5 h; **4a**. $H_2Se$, DMAP, EtOH; 25 °C, 40 min, **b**. $BrCH_2C(O)COOEt$, r.t.

***Figure 7.26*** *Lewis-catalyzed formation of [$^{14}$C]cyanohydrins with $TMS^{14}CN$*

presence of an excess of amine, symmetrical [$^{14}$C]guanidines, such as **101**[86], are produced. Reaction of a different amine with **100** gives an unsymmetrically disubstituted [$^{14}$C]-guanidine. With reaction partners bearing two NH functions in a vicinal or 1,3-relationship, [$^{14}$C]cyanogen bromide can react intramolecularly to give 5- or 6-membered *N*-heterocyclic compounds including fused imidazoles, triazoles and pyrimidines, as illustrated by the syntheses of such varied carbon-14-labeled compounds as the antibronchoconstrictor and antianaphylactic compound **102**[85a], the food mutagen MeIQx (**103**)[87] and CGS 15943 (**104**)[88].

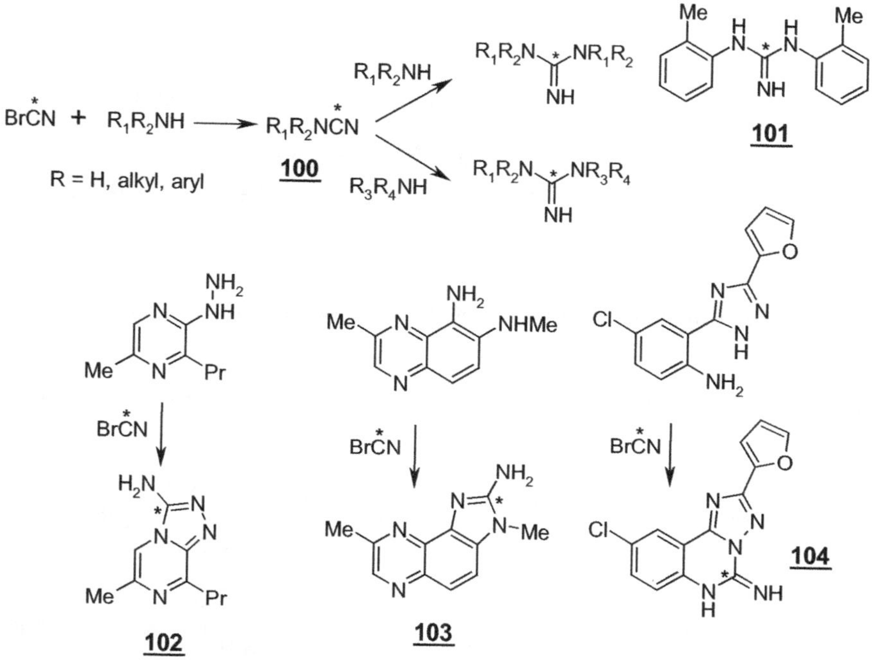


*Figure 7.27* Br$^{14}$CN in the synthesis of [$^{14}$C]guanidines and [$^{14}$C]triazenes

In slightly basic aqueous solution (pH 8–9.5) unsubstituted [$^{14}$C]cyanamide dimerizes to [$^{14}C_2$]dicyandiamide (**105**) (Figure 7.28). This compound can react with a second substituted cyanamide (e.g., *N,N*-diallylcyanamide) to give 6-*N*-substituted 2,4-diamino-*s*-[2,4-$^{14}C_2$]-triazine such as **106**[89]. *N,N*-dialkyl [$^{14}$C]cyanamides can be converted into 6-chloro-2,4-di(*N,N*-dialkyl)amino-*s*-triazines **107**[90] when treated with phosgene and *in situ* generated $NH_4Cl$. The chlorine can be readily replaced with a further amine, giving rise to [2,6-$^{14}C_2$]-triazines substituted with amino groups at C2, C4 and C6. See also Chapter 9 for reactions of [$^{14}$C]cyanamide and [$^{14}$C]guanidine.

### 7.2.3 Alkali Metal [$^{14}$C]Cyanates (M$^{14}$CNO; M = Na, K)

Potassium [$^{14}$C]cyanate is most easily prepared by oxidation of K$^{14}$CN with $KMnO_4$[91]. Addition of ammonia and amines to K$^{14}$CNO is its main application in carbon-14 synthesis. The ammonolysis of K$^{14}$CNO upon heating with aqueous ammonia is one of the standard procedures for preparing [$^{14}$C]urea[1,92]. Upon its reaction with primary or secondary amines, unsymmetrical [$^{14}$C]ureas such as *N*-[$^{14}$C]carbamoyl aspartic acid **108** are obtained[93] (Figure 7.29). Reaction of two equivalents of K$^{14}$CNO with hydrazine produced

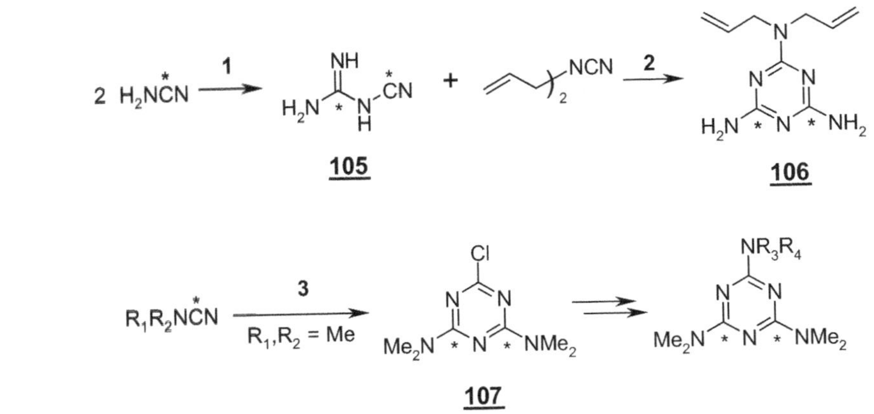

**Reaction conditions:** **1**. $NH_4OH$; **2**. KOH, Cellosolve; 155 °C, 15 min; **3**.$COCl_2$, HCl

***Figure 7.28*** *[$^{14}$C]Cyanamide in the synthesis of fused [$^{14}$C]pyrimidines and [$^{14}$C]imidazoles*

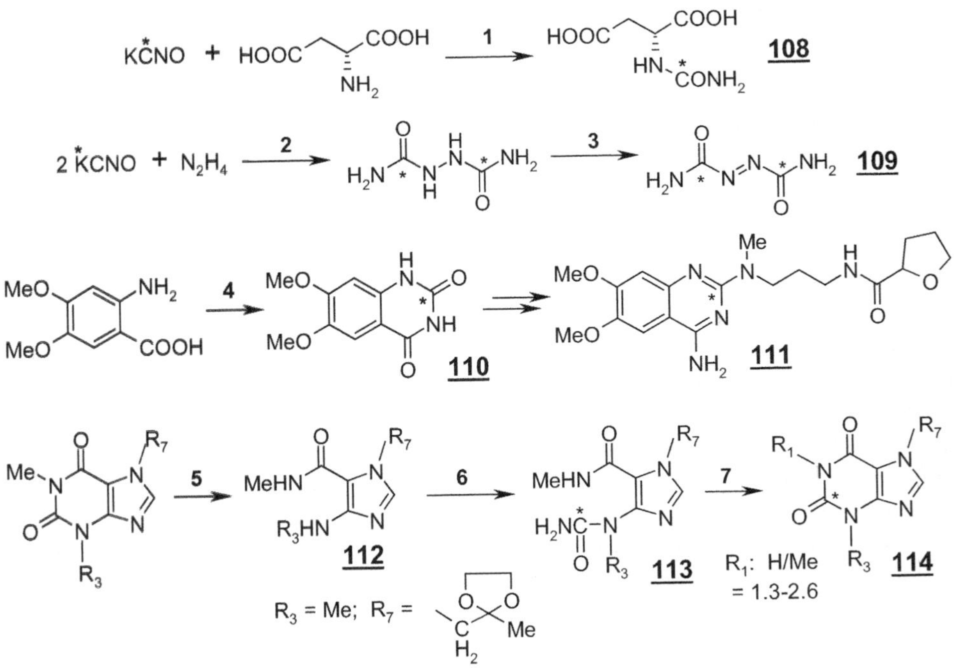

**Reaction conditions:** **1a**. KOH, $H_2O$; r.t., **b**. HCl, $H_2O$; r.t.; **2a**. $H_2O$; r.t., 4h **b**. aq. HOAc; r.t.; **3**. $NH_4NO_3$, cat. $Cu(OAc)_2$, HOAc, $H_2O$; reflux, 15 min; **4a**. $K^{14}CNO$, aq. HOAc; r.t., 2–5 h, **b**. NaOH, 90 °C, 30 min (60 %); **5**. aq. NaOH; reflux, 2 h; **6**. K*CNO, aq. HOAc; r.t., 24 h; **7**. 100 °C, 16 h

***Figure 7.29*** *Use of $K^{14}CNO$ in the synthesis of linear and cyclic [$^{14}$C]urea derivatives*

[$^{14}C_2$]biurea, which upon oxidation gave [$^{14}C_2$]azidodicarbonamide (**109**), an investigational inhibitor of HIV-1[94]. An additional functional group in the amine component (COOH, $NH_2$) may react with the initially formed urea and effect ring closure. This has been exploited for the synthesis of various heterocyclic compounds such as 6,7-dimethoxy[2-$^{14}$C]quinazoline-2,4-dione (**110**), an intermediate in the synthesis of the antihypertensive compound **111**[95] and 3-methyl-7-(2-oxopropyl)[2-$^{14}$C]xanthine (**114**)[96]. In both cases, the respective precursors were first stirred with aqueous K$^{14}$CNO at room temperature, then the initially generated urea intermediates were heated to induce ring closure. Intermediate **113** had been shown before to cyclize in two directions to give the $N_1$-demethylated derivative ($R_1 = H$) as the major product. This procedure is also applicable to other xanthines, although the overall yields are low due to concomitant hydrolytic processes that generate $^{14}CO_2$ and $NH_4{}^{14}CO_3$. The precursors corresponding to **112** are conveniently accessible by alkaline hydrolysis of the respective unlabeled target molecules. Overall, the disadvantages of the procedure may be offset by the low cost of K$^{14}$CNO and the rapidity of the short one-pot process. The use of unlabeled target compound as the starting material for the sequence is a special case of a reconstitution strategy, which is discussed in Chapter 9.

### 7.2.4 Alkali Metal Thio[$^{14}$C]cyanate (M$^{14}$CNS; M = Na, K)

Reaction of alkali metal [$^{14}$C]cyanides with elemental sulfur gives the corresponding [$^{14}$C]thiocyanates[97]. Ammonium [$^{14}$C]thiocyanate has also been prepared from [$^{14}$C]-thiourea[98]. The readily accessible [$^{14}$C]thiocyanates can be alkylated at sulfur, acylated at nitrogen, and undergo nucleophilic addition at carbon. They have been used for the carbon-14 labeling of various unsymmetrical thioureas, thiosemicarbazones, guanidines, as well as of substituted imidazoles, thiazoles, guanosine and deazaguanosine nucleotides.

*S*-Alkylation of [$^{14}$C]thiocyanate prepares the sulfur atom for displacement by a nucleophile, as in the selective *ortho*-cyanation of phenol by its reaction with methyl [$^{14}$C]thiocyanate in the presence of $BCl_3$ and $AlCl_3$[99]. Furthermore, *S*-alkylated [$^{14}$C]-thiocyanates isomerize upon heating to give the corresponding [$^{14}$C]isothiocyanates, which upon reaction with *N*-nucleophiles (amines, hydrazines, etc.) provide the respective *N,N′*-disubstituted [$^{14}$C]thioureas, [$^{14}$C]thiosemicarbazones, etc. In the preparation of [$^{14}$C]-GYKI 21 683 (**116**)[100] for example (Figure 7.30), the *S*-benzyl product, formed by the reaction of *p*-chlorobenzyl chloride with [$^{14}$C]thiocyanate, was isomerized by thermal rearrangement to the corresponding *N*-(*p*-chlorobenzyl)[$^{14}$C]isothiocyanate **115**. Subsequent reaction with 2-(methylamino)ethanol afforded **116**.

A different approach is required to prepare *N*-aryl [$^{14}$C]isothiocyanates and thence unsymmetrical aryl alkyl [$^{14}$C]thioureas; this route takes advantage of a second mode of thiocyanate ion reactivity. As exemplified in the preparation of CGP 20308 (**119**)[101], initial acylation of [$^{14}$C]thiocyanate generated the *N*-acyl derivative, whose carbon center is made more electrophilic by the electron-withdrawing acyl group. Addition of an arylamine therefore produced the *N*-acyl-*N′*-aryl[$^{14}$C]thiourea **117**. Subsequent mild hydrolytic cleavage of the benzoyl group and pyrolysis of the resulting *N′*-unsubstituted aryl[$^{14}$C]-thiourea afforded the corresponding aryl isothiocyanate **118** in 80% yield. Its reaction with 3-mercaptopropionic acid gave the dithiocarbamate **119**[101]. In contrast, the pyrolysis of an

*N*-methyl-*N'*-aryl[$^{14}$C]thiourea, obtained by reaction of methyl-[$^{14}$C]isothiocyanate with an aniline, gave only 13% radiochemical yield of aryl[$^{14}$C]isothiocyanate[102]. Analogous to the formation of **117**, the addition of amine **120** to benzoyl [$^{14}$C]isothiocyanate gave the unsymmetrical *N*-acylated [$^{14}$C]thiourea **121**. *S*-Methylation and subsequent ring closure upon treatment with ammonia at high temperature afforded the 9-[2-$^{14}$C]deazaguanine derivative **122** in moderate overall yield[103].

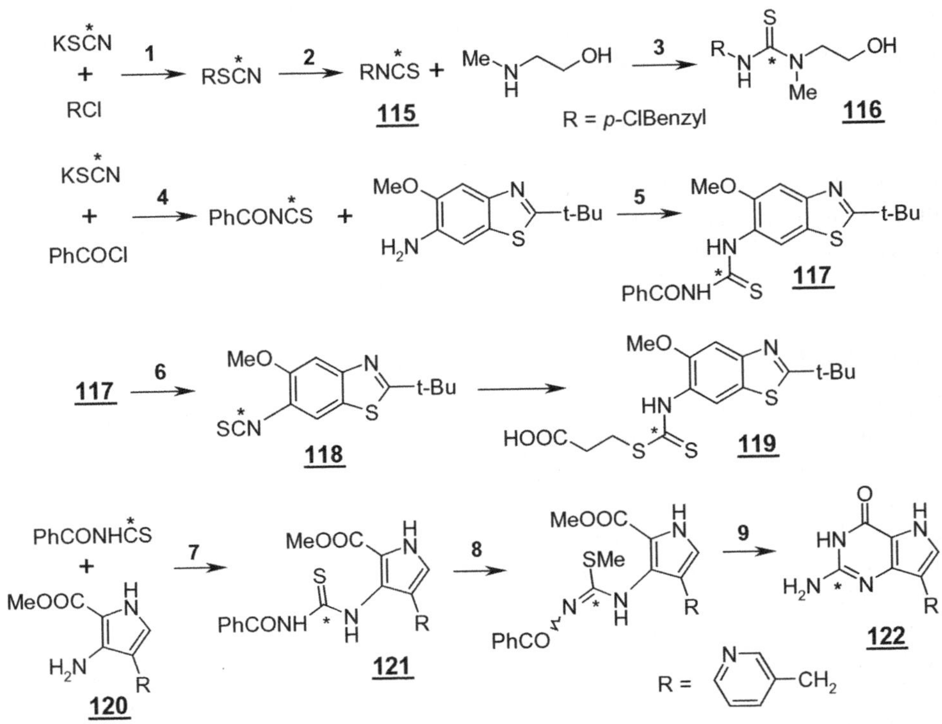


**Reaction conditions: 1**. MeCN; reflux; **2**. 210 °C, 3 h; **3**. MeCN; reflux; **4**. acetone; reflux; 0.5 h **5**. acetone; reflux; 1 h **6a**. aq. NaOH; reflux, **b**. $(NH_4)_2SO_4$ (s), xylene; reflux; **7**. $CH_2Cl_2$; 15 min, r.t.; **8**. MeI, DBN, $CH_2Cl_2$; 0–5 °C, 30 min; **9**. $NH_3$,MeOH; 100 °C (under pressure), 10 h

***Figure 7.30*** *Applications of KS$^{14}$CN in the synthesis of [$^{14}$C]isothiocyanate, [$^{14}$C]thiourea, and [$^{14}$C]deazaguanine*

Nitrogen nucleophiles add to the carbon of thiocyanate to produce thioureas and thiosemicarbazides, and this was exploited to prepare [$^{14}$C]thiosemicarbazide **123**, an intermediate to the carbon-14-labeled aminoguanidine **124**[104]. Activation of **123** by *S*-methylation, reaction with *n*-pentylamine and condensation with the aldehyde component completed the synthesis of **124**.

The same mechanisms are at work in the preparation of several cyclic systems, by applying an intramolecular reaction strategy involving the reactions of [$^{14}$C]thiocyanate with appropriate bifunctional partners. For example, condensation with ethyl 2-amino-acetoacetate gave the 4,5-difunctionalized 2-mercapto[2-$^{14}$C]imidazole **125**[97a]. With aminoacetaldehyde dimethylacetal the unsubstituted analog **126** was obtained, and could be desulfurated (Raney-Ni, formic acid) to provide [2-$^{14}$C]imidazole; further functionalized derivatives were similarly prepared[105]. A three-component condensation provided the 3,4-disubstituted 2-mercapto[2-$^{14}$C]imidazole **127**[106]. Finally, condensation of [$^{14}$C]thiocyanate with several complex *N*-chloroamidines **128** provided labeled aminothiadiazoles **129**, which are key intermediates in the preparation of [$^{14}$C]cephalosporins **130**[107].

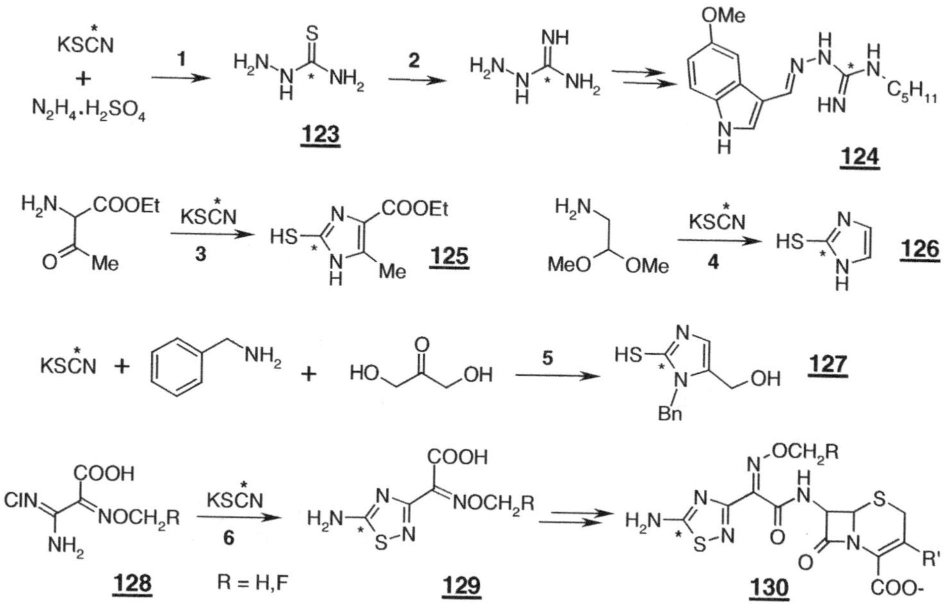

**Reaction conditions: 1.** 100 °C, 2 h; **2a.** MeI,MeOH;reflux, **b.** $C_5H_{11}NH_2$; reflux, 4 h; **3.** HCl, $H_2O$, EtOH; reflux, 2 h (84%); **4.** aq. HCl; 90 °C, 7 h; **5.** *n*-BuOH, AcoH; 80-90°C; **6.** $Et_3N$, MeOH; r.t., 16 h

***Figure 7.31*** *Syntheses initiated through addition of N-nucleophiles to [$^{14}$C]thiocyanate*

### 7.2.5 Triethyl [$^{14}$C]Orthoformate [$H^{14}C(OEt)_3$]

The complete ethanolysis of $K^{14}CN$ in the presence of dry HCl gas and conc. $H_2SO_4$ furnishes triethyl [$^{14}$C]orthoformate[108]. In normal synthetic organic chemistry triethyl orthoformate is used for the conversion of aldehydes and ketones into acetals, enolethers and ketals, respectively, as well as for the formylation of *C*- and *N*-nucleophiles. In carbon-14 labeling the most frequent use has been for the preparation of purines and related compounds labeled at a methine position, through reaction of triethyl [$^{14}$C]orthoformate with appropriate $\omega,\omega'$-diamine precursors. One class of precursor is a substituted pyrimidine such as **131** (Figure 7.32 ), which reacts with the labeled reagent (in 10–25% excess) in the presence of catalytic amounts of a strong acid (*p*-TosOH, MsOH) under strictly anhydrous conditions to afford the [8-$^{14}$C]purine derivative, examples being **132**[109], **133**[108] and **134**[110]; radiochemical yields were 55–86%. The second precursor class includes diamino imidazoles, which have been used under similar reaction conditions to produce [2-$^{14}$C]purines such as [$^{14}$C]griseolic acid diester **135**[111] in about 29% yield and, in 59% yield the [$^{14}$C]imidazodiazepinone derivative **136**, an intermediate in the synthesis of [$^{14}$C]-pentostatin[112]. Lastly, one example has been reported demonstrating that this type of reaction may be extended to other, appropriately 1,2-, 1,3- or 1,4-functionalized substrates; thus, condensation of triethyl [$^{14}$C]orthoformate with a highly substituted aryl hydrazide furnished 2-aryl substituted [5-$^{14}$C]1,3,4-oxadiazole **137** in a low, but useful radiochemical yield of 10%[113].

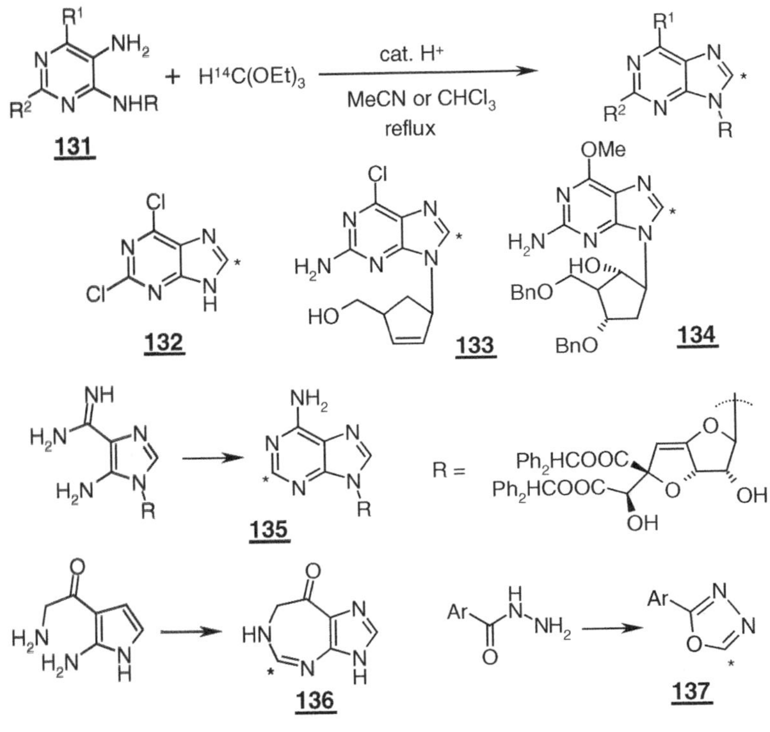


*Figure 7.32* $H^{14}C(OEt)_3$-effected synthesis of [$^{14}$C]purines and related 5,7-bicyclic heterocycles

### 7.2.6 [$^{14}$C]Cyanoacetic Acid [$^{14}CNCH_2COOH$]

Cyanoacetic acid, labeled with carbon-14 at any of the positions, is readily accessible through treatment of labeled/unlabeled chloro- or bromoacetic acid (see Chapter 6, Section 6.2.3) with unlabeled/labeled sodium or potassium cyanide in aqueous solution at pH 8 (Figure 7.33, Route A)[114a–c]; use of these conditions prevents liberation of $H^{14}CN$ and formation of glycolate. Its ethyl ester, which is the derivative most frequently employed, is obtained either by esterification of the free acid with ethanol under acid catalysis[114a], by use of ethereal diazoethane[114d,e] or by treatment of the alkali salt with diethyl sulfate in dioxane

(Route B)[114d]. It can also be prepared by cyano-dehalogenation of labeled/unlabeled ethyl bromoacetate with unlabeled/labeled potassium cyanide in refluxing ethanol (Route C)[115]. Deprotonation of labeled acetonitrile with LDA or LiHMDS followed by quenching of the anion with diethylcarbonate or ethyl chloroformate has to be regarded as a highly promising alternative, since it avoids preparation and isolation of [$^{14}$C]acetic acid and bromo[$^{14}$C]-acetic acid intermediates. However, it has been so far only applied to the synthesis of carbon-13-labeled material (Route D)[116].

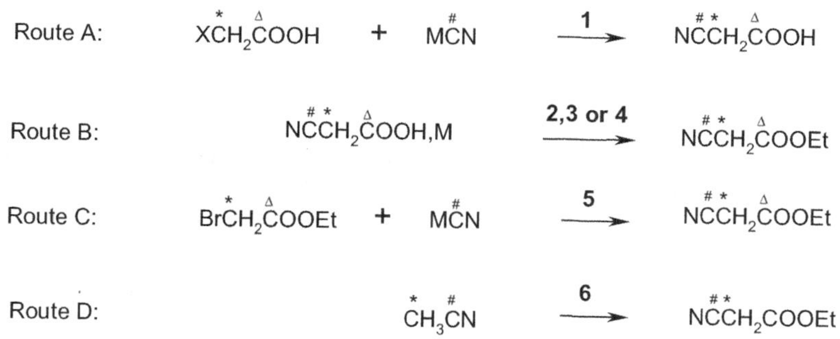


X = Cl, Br; M = Na, K     * Δ # = alternative locations of carbon-14

**Reaction conditions:** **1**. $H_2O$ (pH 8); r.t., 30 min; 85–90 °C, 0.1–1.5 h (80–90 %); **2**. EtOH, conc. $H_2SO_4$; reflux 2 h; **3**. diazoethane, ether; r.t.; **4**. diethyl sulfate, dioxane; 95–100 °C, 60 min; **5**. EtOH; reflux, 5 h (48 %); **6a**. LDA, THF;−60 °C, 30 min; r.t., 30 min, **b**.$(EtO)_2CO$, THF; −60 °C, 2 h (76 %)

***Figure** 7.33 Synthesis of carbon-14-labeled cyanoacetic acid and its ethyl ester from appropriate halogenated precursors or labeled acetonitrile*

[$^{14}$C]Cyanoacetate is a versatile building block because it possesses three reactive centers in as many carbons, in addition to its potential to be labeled in any or all of the positions. The following discussion is organized from a practical perspective, according to the initial synthetic transformations of [$^{14}$C]cyanoacetate, even though similar final synthetic targets may have resulted from different routes.

### *Reactions at the Methylene Group*

As with malonate, the two strong electron-withdrawing functional groups of cyanoacetate significantly enhance the acidity of the methylene group. Accordingly, it can be readily deprotonated and the anion reacts with a broad spectrum of electrophiles, such as alkyl and activated aryl halides, carbonyl compounds, $\alpha,\beta$-unsaturated carbonyl or carboxyl derivatives, and nitrous acid.

*Alkylations and arylations* Simple alkylations of [$^{14}$C]cyanoacetates are of lesser importance than of dialkyl [$^{14}$C]malonates because the products, $\alpha$-[$^{14}$C]cyanoalkanoic esters, are usually more readily available via [$^{14}$C]cyano-dehalogenation of $\alpha$-bromo esters, or by alkoxy-carbonylation of alkyl [$^{14}$C]nitriles with chloroformates or dialkylcarbonates (analogous to Routes C and D in Figure 7.33). Both of these alternatives avoid problems associated with the mixture of di-alkylated, mono-alkylated and unreacted species often obtained in alkylations of cyanoacetates. While 2-[$^{14}$C]cyanopropionic acid (**138**), for example, a starting material for the synthesis of [6-$^{14}$C]thymine and [6-$^{14}$C]thymidine, was prepared by cyano-dehalogenation of α-bromopropionic acid with sodium [$^{14}$C]cyanide in 84% yield[117], the corresponding methyl 2-[$^{13}$C]cyanopropionate (**139**), prepared from methyl [$^{13}$C]cyanoacetate and methyl iodide, had to be separated by careful fractional distillation from the dialkylated byproduct methyl 2-[$^{13}$C]cyanoisobutyrate (5%) and the starting material (40%), and was obtained in only 33% yield[118].

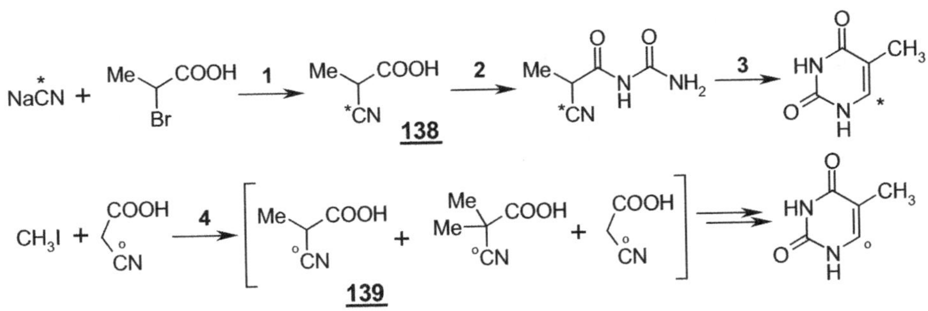


**Reaction conditions:** **1**. NaOH, MeCN–$H_2O$; 50 °C, 2.5 h; **2**. urea, $Ac_2O$; 90 °C, 2 h; **3**. $H_2$, $PtO_2$, AcOH/$H_2O$ 1:1; 70 °C (overall yield 59 %); **4a**. $CH_3I$, NaH, THF; 0 °C, 30 min, **b**. separation by fractional distillation

***Figure 7.34*** *Alternative syntheses of 2-([$^{13/14}$C]cyano)acetate homologs*

Alkyl [$^{14}$C]cyanoacetates can react, as do dialkyl malonates, with activated aryl halides to produce labeled 2-aryl[$^{14}$C]cyanoacetates in generally good yields. Treatment of 1-chloro-4-nitrobenzene with methyl [$^{14}$C]cyanoacetate in the presence of a base and a polar solvent at elevated temperature provided methyl 2-[$^{14}$C]cyano-2-(4-nitrophenyl)-acetate (**140**) in nearly 90% yield (Figure 7.35). Subsequent acid hydrolysis and concomitant decarboxylation converted this intermediate into 4-nitrophenyl[1-$^{14}$C]acetic acid **141**. The equivalence of the specific activities of starting material and product in this reaction demonstrates that hydrolysis and decarboxylation of the ester function occurred prior to the hydrolysis of the cyano group. Subsequent reactions converted **141** into the photoaffinity ligand *N*-(2-*p*-azidophenyl[1-$^{14}$C]ethyl)norlevorphanol (**142**)[119]. *o*-Fluoronitrobenzene reacted with [$^{14}$C]cyanoacetate under similar conditions to give the arylated product **143**, but in this case the ethoxycarboxyl group was removed under Krapcho conditions, giving the [$^{14}$C]nitrile **144** which could undergo different chemistry compared with **141**. Nevertheless, the [$^{14}$C]nitrile group was hydrolyzed in a separate reaction and taken on to 1*H*-indazole[$^{14}$C]carboxylic acid (**145**)[120]. It is notable that in both of these syntheses [$^{14}$C]cyanoacetate is utilized as a 2-carbon extension on an aryl foundation.

α-Arylation of alkyl cyanoacetates (and dialkyl malonates) with unactivated aryl halides are also possible, but the reactions have to be conducted in the presence of copper halide[121] or palladium catalyst systems such as Pd(dba)$_2$, P(*tert*-Bu)$_3$[122].

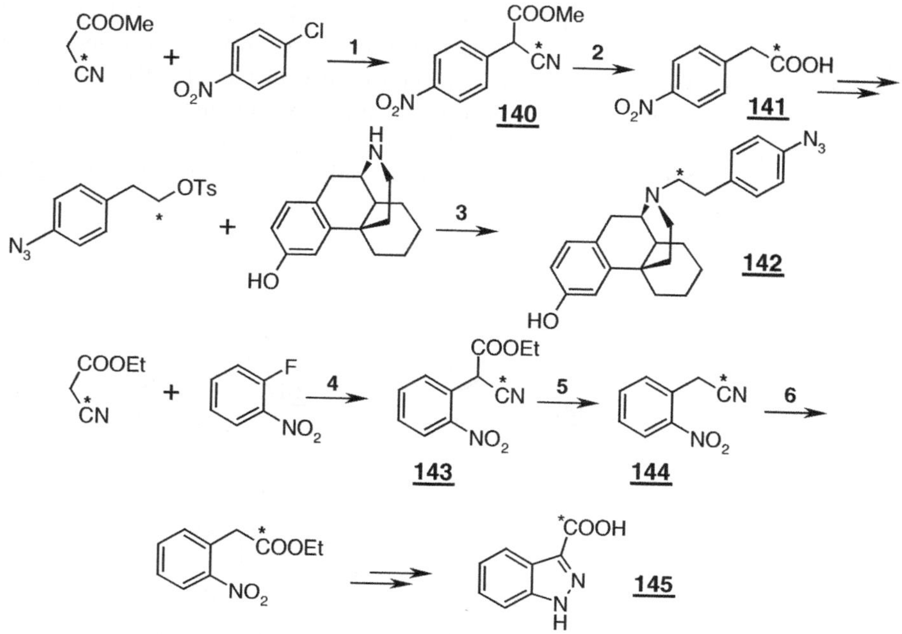

**Reaction conditions: 1**. NaH, HMPA; 95–100 °C, 15 h; **2**. 50 % HOAc/conc. $H_2SO_4$ 20:1, reflux, 60 h; **3**. $Na_2CO_3$, MeOH; 41–43 °C, 60 h; **4**. $K_2CO_3$, DMSO 100 °C, 4 h (98 %); **5**. NaCl, $H_2O$, DMSO; 140 °C, 6 h; **6a**. aq. HCl-HOAc; reflux, 6h, **b**. EtOH, $H_2SO_4$; reflux, 36h

***Figure** 7.35* *Synthesis of aryl-C-$^{14}$C frameworks by arylation of [$^{14}$C]cyanoacetate esters*

*Reactions with aldehydes and ketones (Knoevenagel reactions)* Ketones do not readily undergo Knoevenagel condensations with dialkyl malonates, but they will react readily in the presence of primary or secondary amines with the more reactive alkyl cyanoacetates to give $\alpha$-cyanoacrylates. In contrast to Knoevenagel condensations with aldehydes, the reactions with ketones are not stereoselective and frequently create *E/Z*-stereoisomeric mixtures. Nevertheless, cyanoacetates are valuable building blocks for extending the carbon skeleton of aldehydes and ketones by two carbons because the ester function can be selectively saponified and decarboxylated under mild conditions without affecting the [$^{14}$C]cyano group and, if desired, the double bond can be reduced either catalytically or with complex hydrides.

For example, condensation of the $\delta$-cyano ketone **146** with methyl [$^{14}$C]cyanoacetate gave (Figure 7.36) $\alpha$-[$^{14}$C]cyanobenzo[$\alpha$]quinolizin-2-ylideneacetate derivative **147** in 52% yield. Reduction of the double bond followed by selective de-methoxycarbonylation provided the dinitrile, a key intermediate in the synthesis of carbon-14-labeled 4$\alpha$-hydroxy-*allo*-berbane (**148**)[28b]. Despite loss of half of the label during formation of the final ring by Thorpe–Ziegler condensation, this route was considered superior to the alternative longer route.

Under analogous conditions to the first example, ethyl [$^{14}$C]cyanoacetate reacted with quinuclidin-3-one to afford the corresponding $\alpha$-cyanoacrylate. Catalytic reduction followed by mild saponification and decarboxylation of the ester function converted this intermediate into the 2-substituted [cyano-$^{14}$C]acetonitrile **149**. The value of this sequence, compared with one involving a [$^{14}$C]cyano-dehalogenation, was in avoidance of the intermediacy of an alkyl halide, which is incompatible with the strongly nucleophilic quinuclidine nitrogen. Nitrosation of **149** to produce the oximinonitrile, followed by its treatment with $SO_2Cl_2$ provided 3-chloro-3-(3-chloro-1,2,5-[3-$^{14}$C]thiadiazol-4-yl)-1-aza-bicyclo[2,2,2]octane (**150**), a late intermediate in the synthesis of [$^{14}$C]LY 297 802 (**151**), an analgesic drug candidate[123].

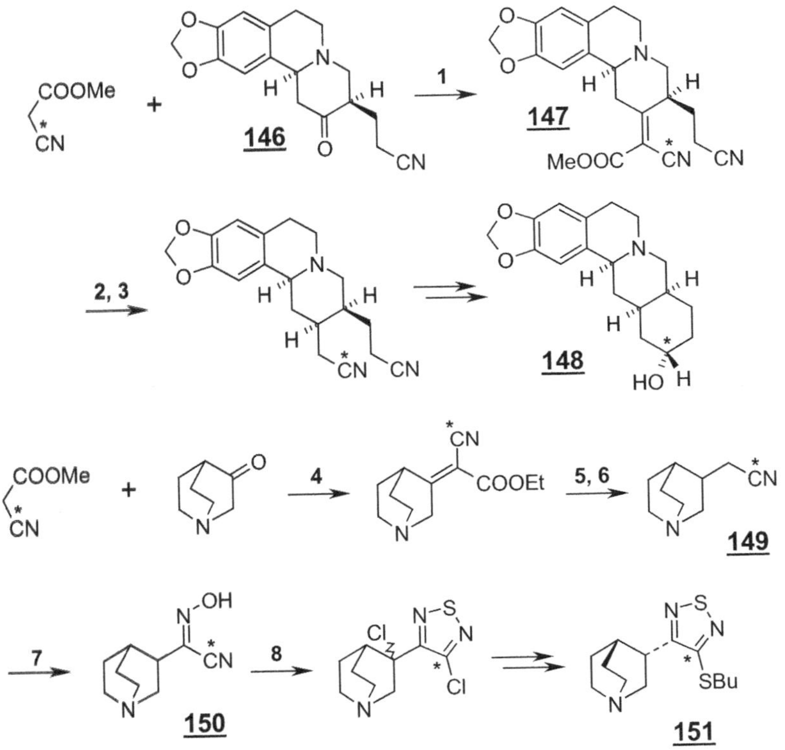


**Reaction conditions:** **1**. $NH_4OAc$, HOAc, benzene; reflux, 2 h; **2**. $NaBH_4$, $MeOH/CH_2Cl_2$ 1:1; 0 °C; **3**. $NaCl/DMSO/H_2O$, 170 °C; **4**. $NH_4OAc$, HOAc, toluene; reflux; **5**. $H_2$, 10 % Pd/C, EtOH; r.t.; **6**. EtONa, EtOH; Δ; **7**. *i*-amylONO; 40–50 °C; **8**. $S_2Cl_2$, DMF

***Figure 7.36*** *Two-carbon homologations of ketones via Knoevenagel condensations with [$^{14}$C]cyanoacetates*

In cases where the nitrogen atom of the [$^{14}$C]cyano function is needed for further chemical transformations, [$^{14}$C]cyanoacetates are used instead of dialkyl [$^{14}$C]malonates in Knoevenagel reactions on aldehydes, as documented in Figure 7.37. In a simple example, condensation of [$^{14}$C]cyanoacetic acid with benzaldehyde provided the α-[$^{14}$C]cyanoacrylic acid **152**. Pyrolytic decarboxylation provided 3-phenylacrylo[$^{14}$C]nitrile, which was subjected to $AlCl_3$-promoted arylation to give 3,3-diphenylpropio[$^{14}$C]nitrile. Transformation of this intermediate into amine products **153** was accomplished by standard methods[124]. Base-catalyzed condensation of methyl [$^{14}$C]cyanoacetate with 3-thiophenecarboxaldehyde provided the analogous $\alpha$-[$^{14}$C]cyanoacrylate, which was reduced by catalytic hydrogenation to give the 2-substituted [$^{14}$C]cyanoacetate **154**. Heteroarylation of **154** with the pyrimidine derivative **155** provided intermediate **156**. Subsequent saponification and decarboxylation of the ester group followed by ring closure upon reduction of the nitro group converted **156** into the target 9-deazapurine **157**[125].

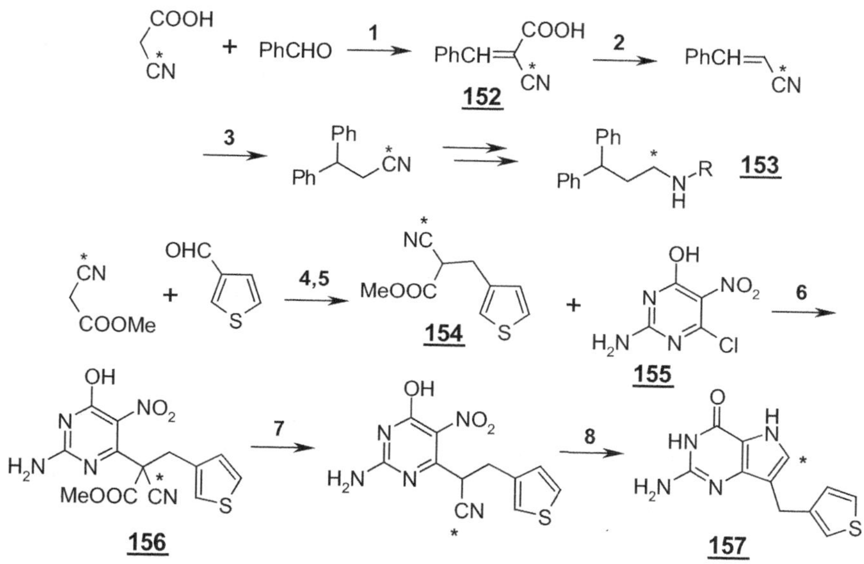

**Reaction conditions:** **1**. NaOH, $H_2O$; 70 °C, 30 min; **2**. Cu powder; 240 °C, 10min; **3**. $AlCl_3$, PhH; reflux, 3 h; **4**. *i*-$Pr_2NH$, *i*-PrOH; r.t., 30 min; **5**. $H_2$, 5 % Pd/C; MeOH/THF 4:1; r.t., 15 h; **6**. $K_2CO_3$, DMSO; 70 °C, 5.5 h; **7**. aq. NaOH; 0 °C, 5 min; r.t. 5 h; **8**. Raney-Ni, $NaH_2PO_2$, pyridine/HOAc/$H_2O$ 2:1:1; 45–50 °C, 18 h

***Figure 7.37*** *[$^{14}$C]Cyanoacetates in Knoevenagel condensations with aldehydes*

*Reactions with other electrophilic reagents* A number of other reagents have been used in condensations with [$^{14}$C]cyanoacetate to give intermediates useful for the elaboration of more complex structures. Shown below are two examples of the use of formyl synthons (amidine and orthoformate) as part of a synthesis resulting in the three-carbon homologation/labeling of substrates. *tert*-Butyl [$^{14}$C]cyanoacetate reacted with the amidine **158** to give the *tert*-butyl 3-amino-2-[$^{14}$C]cyanoacrylate **159**. Sequential one-pot treatment with strong acid to cause ester hydrolysis and decarboxylation, with base to induce cyclization and oxalyl chloride to promote chloro-dehydration gave the [$^{14}$C]cyanoquinoline **160**, an intermediate for the preparation of multiple target compounds[126]. Reaction of ethyl [$^{14}$C] cyanoacetate with triethyl orthoformate provided the ethoxymethylene derivative **161**. Nucleophilic displacement of ethoxide by the *ortho*-lithiated dimethoxynaphthalene provided the 3-aryl-2-[$^{14}$C]cyanoacrylate **162**. Subsequent hydrogenation, ester hydrolysis and decarboxylation afforded the 3-arylpropio[$^{14}$C]nitrile **163**, an intermediate in the synthesis of ring systems exemplified by **164**[127]. In both of these examples, the sequence resulted in addition of a three-carbon unit to the unlabeled partner.

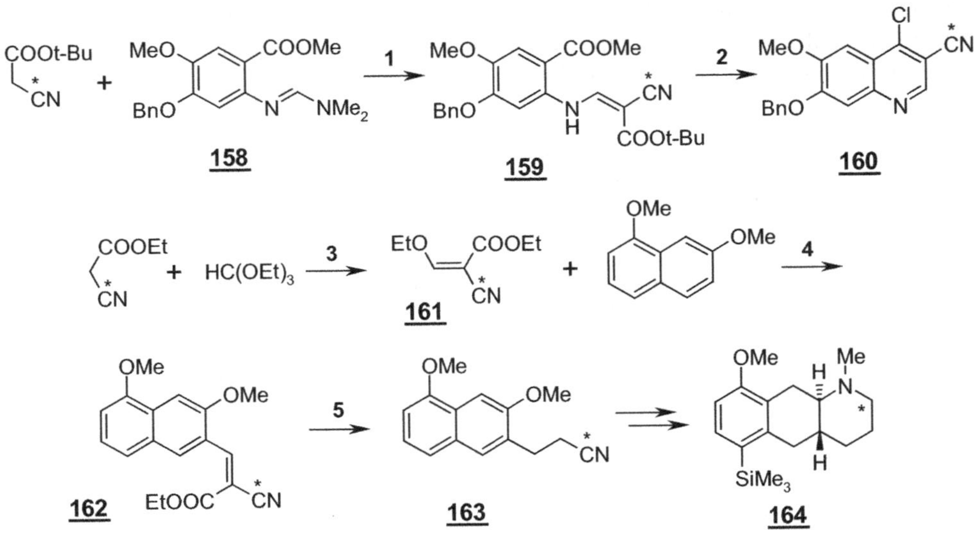

**Reaction conditions:** **1**. *i*-PrOH; r.t.; **2a**. $F_3CSO_3H$, **b**. DBU, **c**. $(COCl)_2$; **3**. $(AcO)_2O$; 145 °C, 5 h; **4**. *n*-BuLi, THF; −20 °C to −70 °C, 1 h to −20 °C; **5a**. $H_2$, 5 % Pd/C, THF/EtOH; 25 °C, 22 h, **b**. NaOH, EtOH; 90 °C, 3 h, **c**. NaCl, DMA, $H_2O$; 150 °C, 2.5 h

***Figure 7.38*** *Syntheses via condensations of formyl synthons with [$^{14}$C]cyanoacetates*

Like diethyl 2-acetamido[2-$^{14}$C]malonate (see Chapter 6, Section 6.5), commercially available carbon-14-labeled ethyl 2-acetamido-2-cyanoacetate (**165**) (Figure 7.39) is a versatile building block for the preparation of racemic neutral, acidic and basic [$^{14}$C]amino acids[128]. In contrast to the corresponding malonate reagent, the nitrile can be used as an masked carboxyl group and be retained through the decarboxylation stage, thereby permitting double labeling (i.e., CN and C2) for achievement of higher product specific activities. The synthesis of **165** parallels that of the malonate: nitrosation of ethyl [$^{14}$C]-cyanoacetate followed by catalytic reduction of the resulting 2-[$^{14}$C]cyano-2-hydroxyimino ester with hydrogen in the presence of acetic anhydride[129]. Reaction of ethyl [$^{14}$C]-cyanoacetate with carbon disulfide in the presence of base, followed by *in situ* *S*-methylation gave the [$^{14}$C]cyanoacrylate bis-thioether **166**, which condensed with acetamidine in the presence of base to provide the [$^{14}$C]cyanopyrimidone **167**, an intermediate in the preparation of the inotropic agent pelrinone[130].

[2-$^{14}$C]**165** has utility in preparation of racemic [2-$^{14}$C]amino acids, as the anion, usually prepared with EtONa in ethanol, reacts with alkylating agents such as **168**, **169** and **170** (Figure 7.40). Heating of the product in acid induces hydrolysis–decarboxylation of the

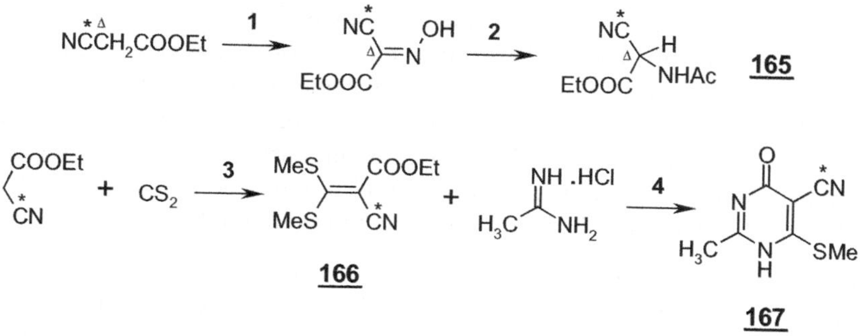

**Reaction conditions:** **1**. $NaNO_2$, AcOH–$H_2O$; 10–15 °C; 3–5 h; **2**. $H_2$, 5 % Pt/C; $Ac_2O$; r.t. (85 %); **3a**. KOH, dioxane; r.t., 20 min, **b**. $Me_2SO_4$, EtOH, 0 °C; **4**. NaH, DMF; r.t., 4 h

***Figure 7.39*** *Use of [$^{14}$C]cyanoacetate in the preparation of labeled α-amino acids and pyrimidines*

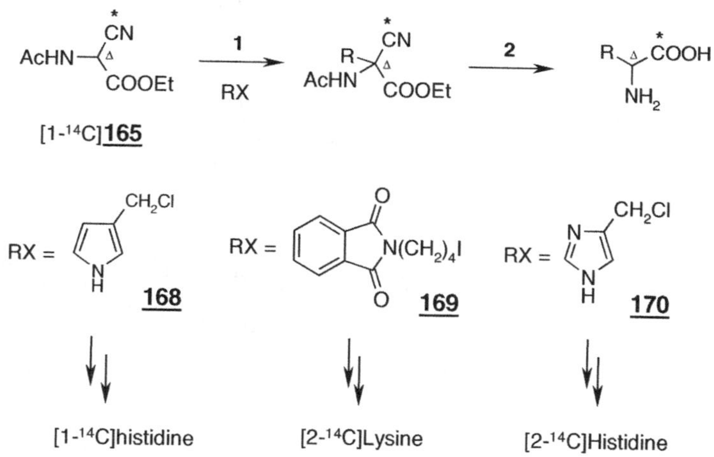

**Reaction conditions:** **1**. EtONa, EtOH; r.t, 2 h (**168**), reflux, 15 h (**169**), r.t., 6 h (**170**); **2**. 48 % HBr, reflux 4 h (**168**, 87 %), conc. HCl; reflux, 15 h (**169**, 70 %), 15% $H_2SO_4$ ; reflux, 12 h (**170**, 65 %)

***Figure 7.40*** *Synthesis of racemic [1/2-$^{14}$C]amino acids from ethyl 2-acetamido-2-cyano-[1/2-$^{14}$C]acetate*

ester faster than hydrolysis of the nitrile group, thereby furnishing the corresponding racemic amino acid without reduction in specific activity[131].

*Reactions at the Functional Groups*

Classically, [$^{14}$C]cyanoacetates have been used in the synthesis of pyrimidine derivatives (e.g. [6-$^{14}$C]uracil, 5-fluoro[$^{14}$C]uracil) and purines (e.g. [4-$^{14}$C]guanine, (8-[4-$^{14}$C]-

azaguanine, [4-$^{14}$C]xanthine, [4,6-$^{14}$C]adenine, 8-[4,6-$^{14}$C]azaadenine and [4-$^{14}$C]uric acid) labeled with carbon-14 at metabolically stable positions. However, more recently this building block has been used in the synthesis of complex drug candidates.

The functional groups, as well as the methylene, undergo a number of transformations, either sequentially or simultaneously in which either the carboxyl and cyano groups, or the methylene and cyano groups, participate in cyclocondensation reactions. These two synthetic approaches are distinct tactically, but can both be utilized to produce cyclic products.

*Sequential functional group transformations and cyclocondensations* Reduction of the nitrile group of [$^{14}$C]cyanoacetates gives $\beta$-[3-$^{14}$C]alanine, a useful building block in its own right. For example, it has been used in the preparation of [6-$^{14}$C]uracil (**171**) and its 5-fluoro derivative **172**, in which the initial catalytic reduction of [$^{14}$C]cyanoacetic acid was followed by treatment with KCNO under acidic conditions at elevated temperatures to produce 5,6-dihydro[6-$^{14}$C]uracil. Introduction of the 5,6-double bond was accomplished by bromination and subsequent thermolytic dehydrobromination. Fluorination of **171** with trifluoromethyl hypofluorite, followed by elimination of $CF_3OH$ upon sublimation, provided 5-fluoro[6-$^{14}$C]uracil (**172**), a potent, widely used antineoplastic drug, in 63% overall radiochemical yield[132]. It is apparent that the corresponding [2-$^{14}$C]-

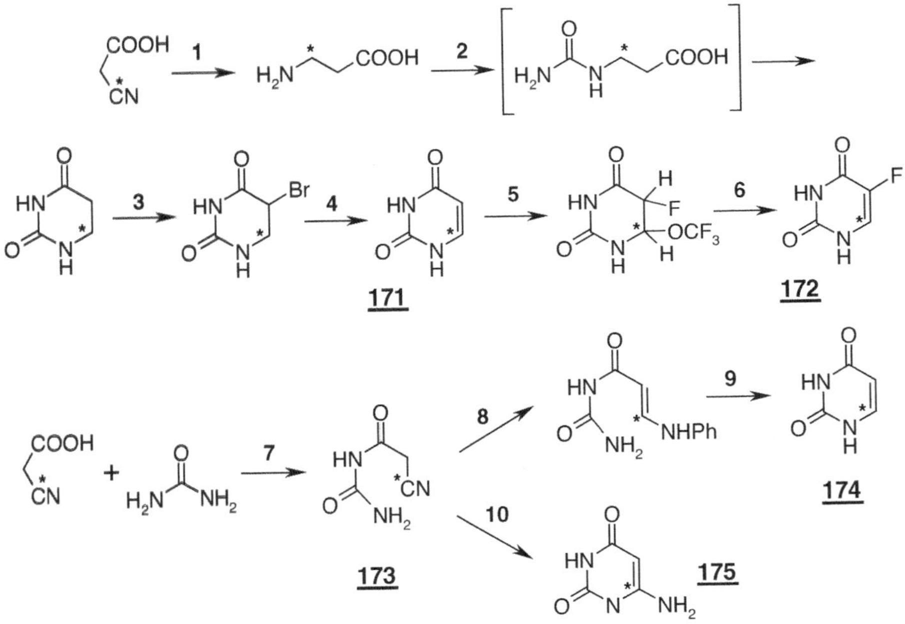


**Reaction conditions: 1.** $H_2$, 10 % Pd/C, 1N HCl; r.t., 2.5 h; **2.** KCNO, 1N HCl; 170 °C, 0.5 h; **3.** $Br_2$, AcOH; 105 °C; **4.** 200 °C, 30 min; **5.** $CF_3OF$, $CF_3COOH$, $H_2O$, $CFCl_3$; –30 °C to r.t, 3 h; r.t., 1 h; **6.** high vacuum sublimation; **7.** urea, $Ac_2O$; 100°C, 30 min; **8.** $H_2$, Raney-Ni, $PhNH_3^+ Cl^-$, $H_2O$; r.t. 10 h; **9.** HCl (g), EtOH; 4 °C, 4d; **10.** EtONa, EtOH; 80 °C, 15 min

**Figure 7.41** *Synthesis of [6-$^{14}$C]pyrimidine-2,4-diones from [$^{14}$C]cyanoacetic acid*

isotopomers could be prepared analogously from $KN^{14}CO$ and unlabeled $\beta$-alanine (Section 7.2.3).

An alternative procedure, which, however, has only been exploited so far for the synthesis of carbon-13-labeled material, avoids the drastic reaction conditions of the above sequence. It begins with the condensation of ethyl [$^{13}$C]cyanoacetate with urea to afford [$^{13}$C]cyanoacetylurea **173** (Figure 7.41). Partial reduction of **173** and treatment of the intermediately generated 3-phenylamino[3-$^{14}$C]acryloylurea with ethanolic HCl gave [4-$^{13}$C]uracil **174** in 17% yield. Under basic conditions **173** cyclizes in nearly quantitative yield to 6-amino[6-$^{13}$C]uracil **175**, which serves as a key intermediate for the preparation of uric acid labeled with isotopic carbon[133].

The preparation of [$^{14}$C]aminopyrimidines and thence purine derivatives has been accomplished beginning with ammonolysis of [$^{14}$C]cyanoacetate ester to give the corresponding [$^{14}$C]cyanoacetamide. Dehydration then gave the symmetrical [1,3-$^{14}C_1$]malondinitrile **176**. Electrophilic functionalization at the methylene group with benzenediazonium chloride and cyclization with formamidine, into 4,6-diamino-5-phenylazo-6(1H)-[4,6-$^{14}C_1$]pyrimidine (**177**). Note that **177** was a mixture of [4-$^{14}$C] and [6-$^{14}$C] isotopomers. Reductive cleavage of the diazo function of **177** provided the triaminopyrimidine, which served as a common intermediate to [4,6-$^{14}C_1$]adenine (**178**)[134], by condensation with formic acid formamide, and to [4,6-$^{14}C_1$]azaadenine (**179**) via diazotization and spontaneous ring closure[135].

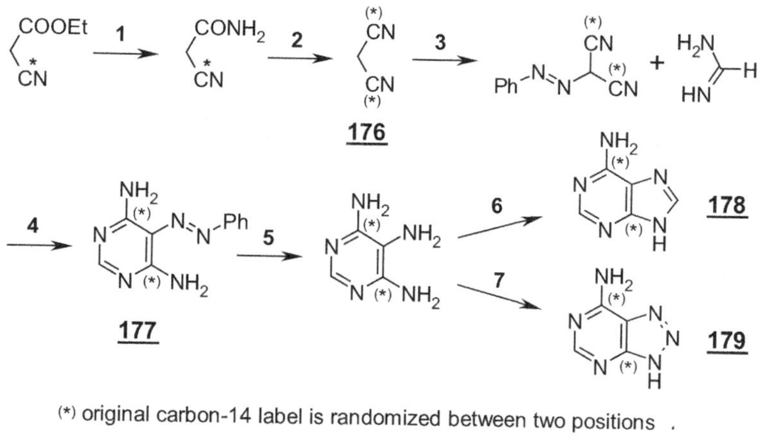


**Reaction conditions: 1.** aq. $NH_3$; r.t., 30 min; **2.** $PCl_5$, 95–100 °C; **3.** Ph–N≡N$^+$ Cl$^-$, 2N NaOAc/MeOH 4:1; 0 °C; **4.** BuONa, BuOH; reflux, 4 h; **5.** Zn, HOAc; r.t., 15 min; **6.** HCOOH, $HCONH_2$; 160 °C, 2.5 h; **7.** $NaNO_2$, NaOAc/HOAc; 5 °C

***Figure 7.42*** *Synthesis of [4,6-$^{14}C_1$]purines via [1,3-$^{14}C_1$]malondinitrile*

*Cyclocondensations with simultaneous involvement of two functional groups* 6-Amino [6-$^{13}$C]uracil (**175**, see above) is also readily accessible through cyclocondensation of ethyl [$^{14}$C]cyanoacetate with urea. Subsequent two-step amination through nitrosation and reduction of the nitroso group converted **175** into 5,6-diamino[6-$^{14}$C]uracil. Heating of this intermediate with urea provided [4-$^{14}$C]uric acid (**180**) in an overall radiochemical yield of 48% (Traube synthesis)[136]. When formamide was employed in the last step, [4-$^{14}$C]-

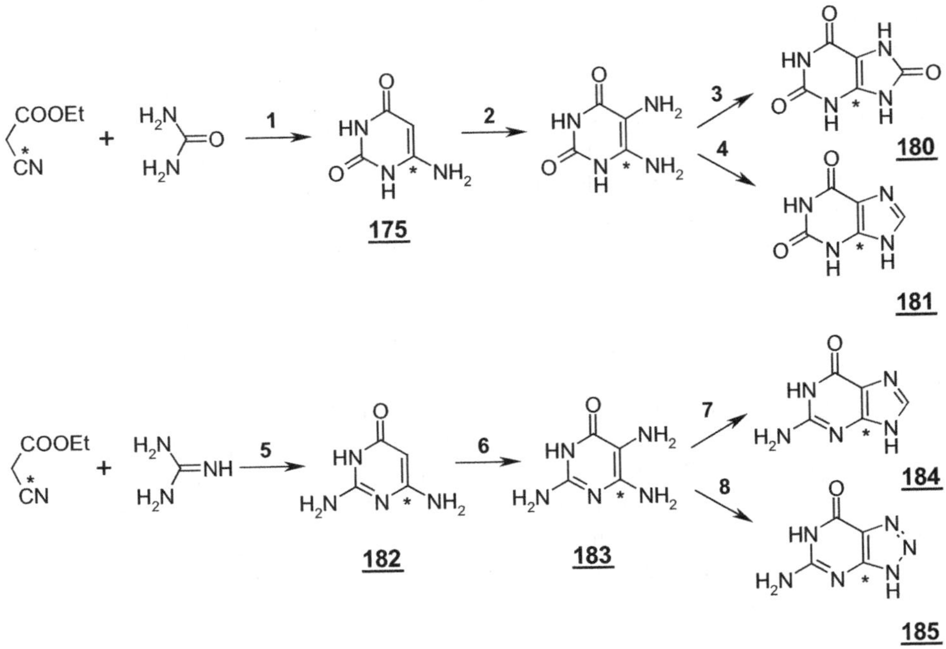


**Reaction conditions:** **1**. EtONa, EtOH; reflux, 2 h; **2a**. $NaNO_2$, HOAc; 0 °C, **b**. $Na_2S_2O_4$, $H_2O$; reflux; **3**. urea; 160 °C, 1 h; **4**. formamide; reflux, 30 min; **5**. MeONa, MeOH; reflux, 8 h; **6a**. $NaNO_2$, NaOAc/HOAc; 0 °C, 3–5 h, **b**. $Na_2S_2O_4$, aq. $H_2SO_4$; 50 °C, 15 min; **7**. HCOONa, HCOOH; reflux, 15 h; **8**. aq. $NaNO_2$, NaOAc; reflux, 1.5 h

***Figure 7.43*** *Synthesis of [4-$^{14}$C]purines via cyclocondensations with ethyl [$^{14}$C]cyanoacetate*

xanthine (**181**) was formed[137]. Replacement of urea by guanidine in the cyclocondensation gave 2,4-diamino-6(1*H*)-[6-$^{14}$C]pyrimidinone (**182**), and the two-step amination provided 2,4,5-triamino-6(1*H*)-[6-$^{14}$C]pyrimidinone (**183**). This compound served as a common intermediate to [4-$^{14}$C]guanine (**184**, 45%)[138] and [4-$^{14}$C]azaguanine (**185**)[134b].

Cyclocondensations of ethyl [$^{14}$C]cyanoacetate to produce other product types are exemplified (Figure 7.44) by the reaction with benzopyranone nitrile **186** to give in 85% yield the fused [$^{14}$C]pyridyl amino ester **187**[139], a reaction which apparently proceeds via a pyranone ring opening and reclosure sequence as indicated. The tetrasubstituted [2-$^{14}$C]thiophene **188** was produced in regiospecific fashion by the reaction of ethyl [$^{14}$C]-cyanoacetate with 3-mercapto-2-butanone catalyzed by an amine base[140]. The 2-amino and 3-ethoxy-carbonyl groups of **188** were subjected to further transformations; it is likely that a range of other $\alpha$-mercaptoketones could be used in this reaction to vary the C4 and C5 substituents in the product.

Cyano[2-$^{14}$C]acetamide is also a highly valuable building block. Its reaction with vinylogous amides followed by elimination of dimethylamine and cyclization offers a convenient route to ring-labeled 2(1*H*)-[3-$^{14}$C]pyridinones, cardiotonic agents, as illustrated by the two examples **189**[141] and **190**[142] (Figure 7.45). In the case of **190** subsequent

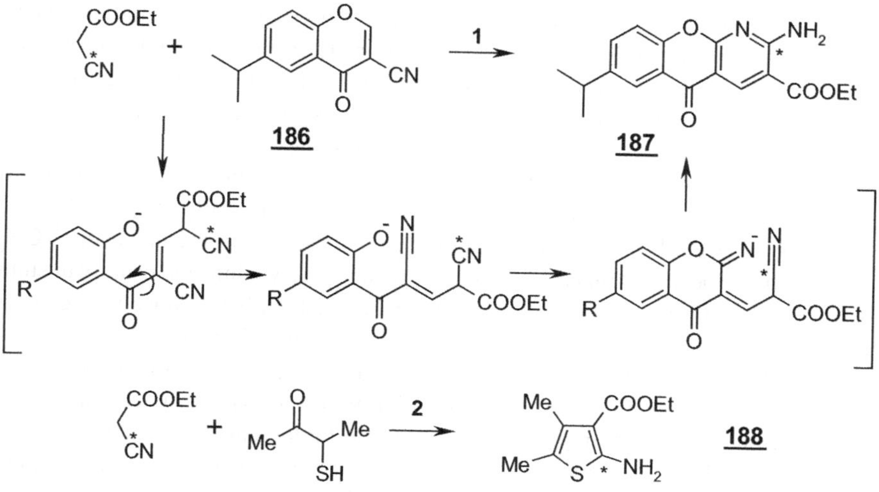


**Reaction conditions:** **1**. piperidine, EtOH; reflux, 1.5 h (85 %); **2**. pyrrolidine, EtOH (76 %)

***Figure 7.44*** *Other cyclocondensations with [$^{14}$C]cyanoacetate*

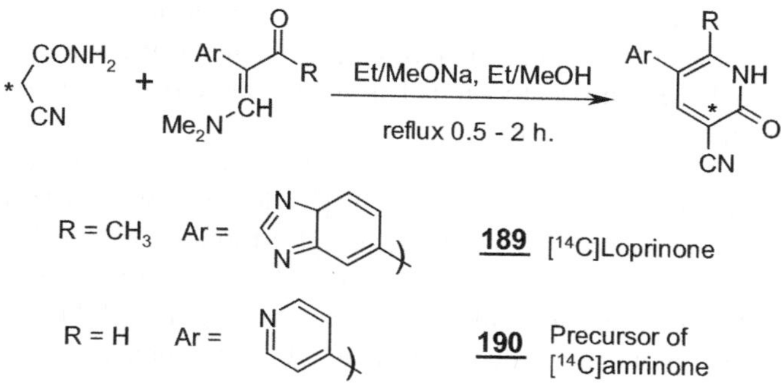


***Figure 7.45*** *Utilization of cyano[2-$^{14}$C]acetamide in synthesis*

partial hydrolysis of the cyano function followed by Hofmann rearrangement of the resulting amide function provided [$^{14}$C]amrinone (**<u>190</u>**, CN → $NH_2$).

### 7.2.7 [$^{14}$C]Diazomethane ($^{14}CH_2N_2$)

[$^{14}$C]Diazomethane is prepared by base-mediated decomposition of *N*-[$^{14}$C]methyl-*N*-nitroso-*p*-toluenesulfonamide with potassium hydroxide in ether–aqueous ethanol at 60–65 °C[143]. The resulting ethereal [$^{14}$C]diazomethane solution is distilled off, trapped in a separate receiver and dried over KOH pellets. For stability reasons it should be stored at −70 °C no longer than overnight. The requisite precursor is accessible via reaction of [$^{14}$C]-methylamine with *p*-toluenesulfonyl chloride and subsequent nitrosation of the resulting

*N*-[$^{14}$C]methyl-*p*-toluenesulfonamide. [$^{14}$C]Methylamine can be prepared by catalytic reduction of alkali metal [$^{14}$C]cyanide[145a] or by reaction of [$^{14}$C]methyl iodide with potassium phthalimide[143b,144] (Chapter 5, Section 5.5). 1-[$^{14}$C]Methyl-3-nitro-1-nitrosoguanidine and *N*-nitroso-*N*-[$^{14}$C]methylurea have been recommended as alternative $^{14}CH_2N_2$ sources[145].

Diazomethane is a valuable reagent for one-carbon extension of acyl halides and anhydrides, as well as for ring expansion reactions of cyclic ketones[146]. It is also widely used in small-scale organic synthesis for the esterification of carboxylic acids and the etherification of phenols, enols and alcohols[147]. In these reactions, however, $^{14}CH_2N_2$ installs the label in positions that are potentially metabolically labile and usually unsuitable for use in metabolism studies. For *in vitro* studies where metabolism is not an issue, tritiated diazomethane is usually preferred because of its higher specific activity.

However, the use of $CH_2N_2$ in the homologation of aldehydes, ketones and activated carboxylic acids, and in ring formation with olefinic or acetylenic precursors, presents unique opportunities for labeling target molecules at biologically more stable positions. Aldehydes and cyclic ketones give fairly good yields of methyl ketones and ring-expanded homologs, respectively. Acyl halides and anhydrides form diazoketones, which upon treatment with HCl or HBr furnish $\alpha$-haloketones, whereas HI reduces them to methyl ketones[148]. Upon heating in the presence of catalytic amounts of silver oxide or benzoate, diazoketones rearrange to ketenes, which can be trapped with water, alcohols, phenols or amines to give carboxylic acid derivatives with the carbon chain extended by one labeled atom (Arndt–Eistert synthesis). Unfortunately, due regard has to be paid to the highly toxic character of this reagent and the risk of explosion even in dilute solution. These disadvantages, and the fact that it must often be employed in considerable excess to achieve satisfactory yields, have significantly limited its use. It is therefore not surprising that only a few applications have been reported so far.

The conversion of aldehydes into [$^{14}$C]methyl ketones has found several applications in the labeling of anthracycline antibiotics such as [$^{14}$C]daunarubicin (**<u>192</u>**, R = H), [$^{14}$C]-doxorubicin (**<u>193</u>**) and analogs[145b,149]. 9-Desacetyl-9-formyl-*N*-trifluoroacetyl-daunorubicin (**<u>191</u>**), the requisite aldehyde precursor, was made available by periodate cleavage of the respective 1,2-diol. Its reaction with 2.5 equivalents of $^{14}CH_2N_2$, followed by hydrolysis of the trifluoroacetyl protecting group, furnished [$^{14}$C]daunorubicin (**<u>192</u>**, R = H) in 15%

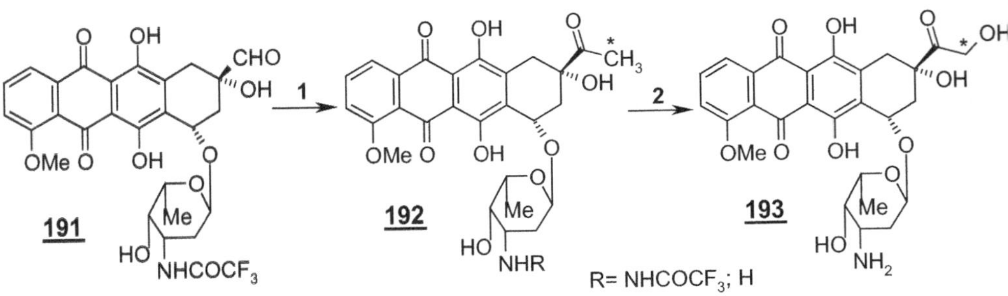


**Reaction conditions:** **1a.** $^{14}CH_2N_2$, $Et_2O/CH_2Cl_2$ 1:1; r.t., 3 h, **b.** 0.1N NaOH, 0 °C, 60 min; **2a.** $Br_2$, MeOH/dioxane, r.t., **b.** aq. HCOONa, r.t., 100 h

***Figure 7.46*** *Carbon-14 labeling of anthracycline using [$^{14}$C]diazomethane*

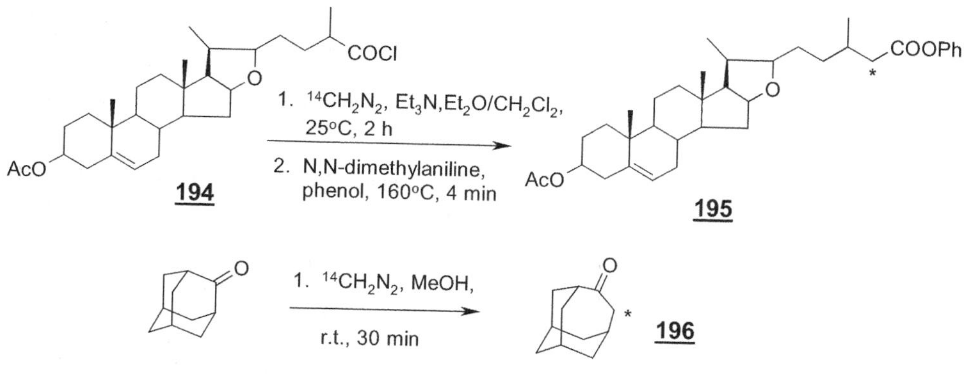


*Figure 7.47* Carbon-14 labeling through $^{14}CH_2N_2$ mediated homologations of acid chlorides and ring expansions

overall radiochemical yield. Bromination of the labeled methyl group and subsequent hydrolysis converted **192** into [$^{14}$C]doxorubicin[149]. Some related compounds were prepared by similar procedures[149c].

There are a few cases documenting the utility of $^{14}CH_2N_2$ for the labeling of carbon skeletons at metabolically stable positions through homologation of carboxylic acids or through ring enlargement of cyclic ketones. In an example of the former, $\Delta^5$-furostene-3$\beta$-acetoxy-26-oic acid chloride (**194**), generated *in situ* from the free acid and oxalyl chloride, was treated with two equivalents of $^{14}CH_2N_2$ to produce the diazoketone. Rearrangement of the diazoketone by heating with an amine base in the presence of phenol produced phenyl ester **195**, which was hydrolyzed to give the corresponding [$\alpha$-$^{14}$C]carboxylic acid in 7% overall radiochemical yield[150]. In an example of ring expansion, adamantanone was treated with two equivalents of $^{14}CH_2N_2$ in methanol at ambient temperature for 30 min, giving [5-$^{14}$C]-homoadamantan-4-one (**196**) in 50% radiochemical yield[151].

## References

1. a. Pichat, L., Synthesis and uses of isotopically labeled cyanides, in *The Chemistry of the Cyano Group*, Rappoport, Z., Ed.; *The Chemistry of Functional Groups*, Vol. X, Patai, S., Ed.; Interscience, John Wiley & Sons, Ltd: London, ***1971***, pp. 744–790; b. Rappoport, Z., *The Chemistry of the Cyano Group*; Interscience: New York, ***1970***.
2. Perry, C.W.; Burger, W. and Delaney, C.M., *J. Label. Compd. Radiopharm.*, ***1979***, *16*, 645–649; b. Murray, III, A. and Williams, L., *Organic Syntheses with Isotopes*; Interscience: New York, ***1958***, pp. 565.
3. Bánfi, D. and Mlinko, W., in *Synthesis and Applications of Isotopically Labeled Compounds 1985*, Muccino, R.R., Ed.; Elsevier: Amsterdam, ***1986***, pp. 319–324.
4. a. Murray, III, A. and Williams, L., *Organic Syntheses with Isotopes*; Interscience: New York, ***1958***, pp. 571–572; b. Crowe, A.M.; Johnston, C.E.A.; Lawrie, K.W.M. and Saunders, D., *J. Label. Compd. Radiopharm.*, ***1992***, *31*, 409–416; c. Sartori, G.; Consonni, P. and Omodei-Salé, A., *J. Label. Compd. Radiopharm.*, ***1981***, *18*, 545–553; d. White, D.F. and Burns, J., *J. Label. Compd. Radiopharm.*, ***1977***, *13*, 393–400.
5. Murray, A. III, and Williams, L., *Organic Syntheses with Isotopes;* Interscience: New York, ***1958***, pp. 38–39.

6. Falgueyret, J.-P.; Desmarais, S.; Oballa, R.; Black, W.C.; Cromlish, W.; Khougaz, K.; Lamontagne, S.; Massé, F.; Riendeau, D.; Toulmond, S. and Percival, M.D., *J. Med. Chem.*, ***2005***, *48*, 7535–7543.
7. a. Chaudhuri, N.K.; Servando, O.; Sung, M.-S. and Baumann, H., *J. Label. Compd. Radiopharm.*, ***1984***, *21*, 193–201; b. Yoshitake, A.; Kanamaru, H.; Shono, F. and Nakatsuka, I., *J. Label. Compd. Radiopharm.*, ***1981***, *18*, 391–402.
8. a. Hörnfeldt, K. and Långström, B., *J. Label. Compd. Radiopharm.*, ***1994***, *34*, 707–715; b. Schultz, W.E.; Machulla, H.J. and Feinendegen, F.E., *J. Radioanal. Nucl. Chem. Lett.*, ***1989***, *135*, 199–205.
9. White, D.F., *J. Label. Compd. Radiopharm.*, ***1977***, *13*, 23–39.
10. a. Hsi, R.S.P.; Stolle, W.T. and Easter, J.A., in *Synthesis and Applications of Isotopically Labeled Compounds 1991*, Buncel, E. and Kabalka, G.W., Eds.; Elsevier: Amsterdam, ***1992***, pp. 196–201; b. Moreau, M.F.; Parry, D.; Bayle, M.; Papon, J.; Labarre, P. and Veyre, A., *J. Label. Compd. Radiopharm.*, ***1995***, *36*, 805–813; c. Sisenwine, S.F.; Tio, C.O. and Ahern, J., *J. Label. Compd. Radiopharm.*, ***1976***, *12*, 501–505; d. Stiver, J.F.; Data, J.B. and Christian, J.B., *J. Label. Compd.*, ***1970***, *6*, 150–154; e. Johnson, H.E. and Waller, G.R., *J. Label. Compd.*, ***1968***, *4*, 295–298.
11. Ellis, G.P. and Romney-Alexander, T.M., *Chem. Rev.*, ***1987***, *87*, 779–794.
12. O'Bannon, D.D. and Wheeler, W.J., *J. Label. Compd. Radiopharm.*, ***1991***, *29*, 625–631.
13. Carr, R.M.; Cable, K.M. and Wells, G.N., *J. Label. Compd. Radiopharm.*, ***1994***, *34*, 887–897.
14. Czeskis, B.A.; Wheeler, W.J. and Clodfelter, D.K., *J. Label. Compd. Radiopharm.*, ***2006***, *49*, 663–673.
15. Sekiya, A. and Ishikawa, N., *Chem. Lett.*, ***1975***, 277–278.
16. a. Allen, J.; Parent, G. and Rivron, L., in *Synthesis and Applications of Isotopically Labeled Compounds 1991*, Buncel, E. and Kabalka, G.W., Eds.; Elsevier: Amsterdam, ***1992***, pp. 202–206; b. Allentoff, A.; Ciszewska, G.; Markus, B.; Pfefferkorn, H.; Ray, T.; Jones, L. and Duelfer, T., in *Synthesis and Applications of Isotopically Labeled Compounds 1997*, Heys, J.R. and Melillo, D.G., Eds.; John Wiley & Sons, Ltd: Chichester, ***1998***, pp. 185–188.
17. Takagi, K.; Sasaki, K. and Sakakibara, Y., *Bull. Chem. Soc. Jpn.*, ***1991***, *64*, 1118–1121.
18. Ueyama, N.; Yanagisawa, T. and Tomiyama, T., *J. Label. Compd. Radiopharm.*, ***1995***, *36*, 915–919.
19. a. Wheeler, W.J.; Swanson, S.P.; Varie, D.L. and O'Bannon, D.D., *J. Label. Compd. Radiopharm.*, ***2005***, *48*, 139–148; b. Sakakibara, Y.; Okuda, F.; Shimobayashi, A.; Kirino, K.; Sakai, M.; Uchino, N. and Takagi, K., *Bull. Chem. Soc. Jpn.*, ***1988***, *61*, 1985–1990.
20. Werner, T.F.; Sohn, D. and Johansson, R., *J. Label. Compd. Radiopharm.*, ***2000***, *43*, 437–447.
21. a. Allentoff, A.J.; Markus, B.; Duelfer, T.; Wu, A.; Jones, L.; Ciszewska, G. and Ray, T., *J. Label. Compd. Radiopharm.*, ***2000***, *43*, 1075–1085; b. Allentoff, A.; Markus, B.; Wu, A. and Ray, T., in *Synthesis and Applications of Isotopically Labeled Compounds*, Vol. 7, Pleiss, U. and Voges, R., Eds.; John Wiley & Sons, Ltd: Chichester, ***2001***, pp. 209–212.
22. Tschaen, D.M.; Desmond, R.; King, A.O.; Fortin, M.C.; Pipik, B.; King, S. and Verhoeven, T.R., *Synth. Commun.*, ***1994***, *24*, 887–890.
23. Ramnauth, J.; Bhardwaj, N.; Renton, P.; Rakhit, S. and Maddaford, S.P., *Synlett.*, ***2003***, 2237–2239.
24. Cable, K.M., *J. Label. Compd. Radiopharm.*, ***2007***, *50*, 509–510.
25. Koermeling, D., *J. Label. Compd. Radiopharm.*, ***2002***, *45*, 269–270.
26. Cable, K.M. and Wells, G.N., *J. Label. Compd. Radiopharm.*, ***2000***, *43*, 29–45.
27. Pichat, L.; Beaucourt, J.P.; Krausz, F. and Moulineau, C., *J. Label. Compd. Radiopharm.*, ***1976***, *12*, 347–355.
28. a. Huang, C.C.; Liedtke, B.; Hoffmann, G. and Goel, O., in *Synthesis and Applications of Isotopically Labeled Compounds 1994*, Allen, J., Ed.; John Wiley & Sons, Ltd: Chichester, ***1995***, pp. 617–620; b. Szammer, J.; Dörnyei, G. and Szántay, C., *J. Label. Compd. Radiopharm.*, ***1994***, *34*, 1041–1045.
29. Blackburn, C.E., *J. Label. Compd.*, ***1970***, *6*, 21–33.
30. Küng, W.; Faigle, J.W.; Kocher, E. and Wirz, B., *J. Label. Compd. Radiopharm.*, ***1983***, *20*, 213–231.

31. Antoni, G. and Långström, B., *J. Label. Compd. Radiopharm.*, ***1989***, *27*, 571–576.
32. Saunders, D. and Mack, J.F., in *Synthesis and Applications of Isotopically Labeled Compounds*, Vol. 7, Pleiss, U. and Voges, R., Eds.; John Wiley & Sons, Ltd: Chichester, ***2001***, pp. 213–216.
33. a. Caballero, G.M. and Gros, E.G., *J. Label. Compd. Radiopharm.*, ***1991***, *29*, 805–812; b. Caballero, G.M. and Gros, E.G., *J. Label. Compd. Radiopharm.*, ***1993***, *33*, 907–911.
34. Allentoff, A.J.; Desai, M.; Markus, B. and Duelfer, T., *J. Label. Compd. Radiopharm.*, ***1996***, *38*, 989–998.
35. Billek, G.; Kindl, H.; Schimpl, A. and Schmook, F.P., *J. Label. Compd.*, ***1969***, *5*, 3–7.
36. a. Matloubi, H.; Shafiee, A.; Saemian, N.; Shirvani, G. and Daha, F.J., *J. Label. Compd. Radiopharm.*, ***2004***, *47*, 31–36; b. Matloubi, H.; Shafiee, A.; Saemian, N.; Shirvani, G. and Daha, F.J., *Appl. Radiat. Isotopes*, ***2004***, *60*, 665–668.
37. Azim, E.; Auzeloux, P.; Maurizis, J.-C.; Braesco, V.; Grolier, P.; Veyre, A. and Madelmont, J.-C., *J. Label. Compd. Radiopharm.*, ***1996***, *38*, 441–451.
38. Walkenstein, S.S.; Intoccia, A.P.; Flanagan, T.L.; Hwang, B.; Flint, D.; Weinstock, J.; Villani, A.J.; Blackburn, D. and Green, H., *J. Pharm. Sci.*, ***1973***, *62*, 580–584.
39. Yanagisawa, T.; Yokota, M.; Tomiyama, T. and Ikegami, S., *J. Label. Compd. Radiopharm.*, ***1994***, *34*, 205–212.
40. Svanem, P.-J.; Bones, A.M. and Rossiter, J.T., *Phytochemistry*, ***1997***, *44*, 1251–1255.
41. a. Mestelan, G.; Aubert, F.; Beaucourt, J.-P.; Comar, D. and Pichat, L., *J. Label. Compd. Radiopharm.*, ***1979***, *16*, 661–668; b. Bayly, R.J. and Turner, J.C., *J. Chem. Soc. (C)*, ***1966***, 704–708; c. Snyder, J.R. and Serianni, A.S., *Carbohydrate Res.*, ***1987***, *163*, 169–188; d. Walker, T.E.; Ehler, D.S. and Unkefer, C.J., *Carbohydrate Res.*, ***1988***, *181*, 125–134.
42. Hicks, J.L.; Hartman, J.D.; Skeean, R.V.; Huang, C.C.; Kepler, J.A. and Liang, C.-H., *J. Label. Compd. Radiopharm.*, ***1991***, *29*, 415–429.
43. Czeskis, B.; Clodfelter, D.K. and Wheeler, W.J., *J. Label. Compd. Radiopharm.*, ***2002***, *45*, 1143–1152.
44. Wheeler, W.J., *J. Label. Compd. Radiopharm.*, ***1992***, *31*, 119–124.
45. Ullas, G. and Kelly, T., in *Synthesis and Applications of Isotopically Labeled Compounds*, Vol. 8, Dean, D.C.; Filer, C.N. and McCarthy, K.E., Eds.; John Wiley & Sons, Ltd: Chichester, ***2004***, pp. 349–352.
46. Bringmann, G.; Feineis, D.; Brückner, R.; God, R.; Grote, C. and Wesemann, W., *Eur. J. Pharm. Sci.*, ***2006***, *28*, 412–422.
47. Yoshitaki, A.; Makari, Y.; Kawahara, K. and Doi, T., *J. Label. Compd. Radiopharm.*, ***1976***, *12*, 247–257.
48. a. Deigner, H.P.; Freyberg, C. and Heck, R., *J. Label. Compd. Radiopharm.*, ***1993***, *33*, 171–180; b. Allen, J.; Denoux, M., Philippe, N.; Rivron, L. and Roy, S.N., *J. Label. Compd. Radiopharm.*, ***2007***, *50*, 624–626.
49. a. Chaudhuri, N.K.; Patera, R.; Markus, B. and Sung, M.-S., *J. Label. Compd. Radiopharm.*, ***1987***, *24*, 1177–1184; b. Chaudhuri, N.K.; Markus, B. and Sung, M.-S., *J. Label. Compd. Radiopharm.*, ***1988***, *25*, 739–748.
50. Arima, H. and Tamazawa, K., *J. Label. Compd. Radiopharm.*, ***1985***, *22*, 1217–1226.
51. a. Diel, B.N.; Han, M. and White, J.M., *J. Label. Compd. Radiopharm.*, ***2007***, *50*, 407–409; b. Susán, A.B. and Duncan, W.P., *J. Label. Compd. Radiopharm.*, ***1981***, *18*, 1227–1234.
52. Speer, R.J.; Humphries, M.L. and Roberts, A., *J. Am. Chem. Soc.*, ***1952***, *74*, 2444; b. Lotfield, R.B., *J. Am. Chem. Soc.*, ***1951***, *73*, 4707–4709.
53. a. Luche, J.L.; Petrier, C. and Dupuy, C., *Tetrahedron Lett.*, ***1984***, *25*, 753–756; b. Crowley, J.I. and Rapoport, H., *J. Am. Chem. Soc.*, ***1970***, *92*, 6363–6365; c. Hartung, T., Synthesis of C-14 Labeled 5-$HT_6$ Receptor Antagonists, presented at the *10th IIS–CED Conference, Bad Soden, Germany*, June 6–7, ***2002***
54. Wooley, G.T.; Jordan, S.I. and Campbell, J.B., *J. Label. Compd. Radiopharm.*, ***1999***, *42*, 893–894.
55. a. Prelog, V.; Urech, H.; Bothner-By, A.A. and Würsch, J., *Helv. Chim. Acta*, ***1955***, *38*, 1095–1113; b. Murray III, A. and Williams, L., *Organic Synthesis with Isotopes*; Interscience: New York, ***1958***, pp. 534–539
56. a. Bottini, A.T.; Frost II, K.A.; Anderson, B.R. and Dev, V., *Tetrahedron*, ***1973***, *29*, 1975–1981; b. Allinger, N.L.; Nakazaki, M. and Zalkow, V., *J. Am. Chem. Soc.*, ***1959***, *81*, 4074–4080.

57. a. Bonacorsi, Jr., S.J.; Burrell, R.C.; Luke, G.M.; DePue, J.S.; Rinehart, J.K.; Balasubramanian, B.; Christopher, L.J. and Iyer, R.A., *J. Label. Compd. Radiopharm.*, ***2007***, *50*, 65–71; b. Choi, Y.M.; Kucharczyk, N. and Sofia, R.D., *J. Label. Compd. Radiopharm.*, ***1987***, *24*, 1–14; c. Noel, J.P. and Pichat, L., *J. Label. Compd. Radiopharm.*, ***1983***, *20*, 1243–1256; d. Kelly, T.P.; Rodgers, T.R. and Wright, C., *J. Label. Compd. Radiopharm.*, ***2001***, *44*, 965–966.
58. a. McKendry, L.H., *J. Label. Compd. Radiopharm.*, ***1981***, *18*, 629–614; b. McKendry, L.H., *J. Label. Compd. Radiopharm.*, ***1983***, *20*, 25–35; c. McKendry, L.H., *J. Label. Compd. Radiopharm.*, ***1984***, *21*, 401–408.
59. McKendry, L.H., *J. Label. Compd. Radiopharm.*, ***1990***, *28*, 405–410.
60. Askam, V.; Eweiss, N. and Nicholls, P.J., *J. Label. Compd. Radiopharm.*, ***1983***, *20*, 1331–1338.
61. a. Billings, R. and Sullivan, H.R., *J. Label. Compd.*, ***1967***, *3*, 17–19; b. Pichat, L.; Guermont, J.P. and Phung, N.L., *J. Label. Compd.*, ***1968***, *4*, 251–255; c. Valcavi, U., *J. Label. Compd.*, ***1972***, *8*, 687–693; d. Aberhart, D.J. and Weiller, B.H., *J. Label. Compd. Radiopharm.*, ***1983***, *20*, 663–668; e. Adriaens, P. and Vanderhaeghe, H., Synthesis of some labeled non-proteinogenic amino acids, in *Isotopes in the Physical and Biomedical Sciences*, Vol. 1, Part A, Buncel, E. and Jones, J.R., Eds.; Elsevier: Amsterdam, ***1987***, pp. 330–402.
62. a. Wheeler, W.J.; O'Bannon, D.D.; Kennedy, J.H.; Monn, J.A.; Tharp-Taylor, R.W.; Valli, M.J. and Kuo, F., *J. Label. Compd. Radiopharm.*, ***2005***, *48*, 605–620; b. Arendt, A. and Kolodziejczyk, A., *J. Label. Compd.*, ***1973***, *9*, 457–461.
63. Miura, Y.; Hayashida, K.; Chishima, S.; Yoshikawa, M. and Takeyama, S., *J. Label. Compd. Radiopharm.*, ***1988***, *25*, 1061–1072.
64. Villani, A.J.; Saunders, D.; Shu, A.Y.L. and Heys, J.R., *J. Label. Compd. Radiopharm.*, ***2002***, *45*, 49–57.
65. Fromm, H.J., *J. Org. Chem.*, ***1957***, *22*, 611–612.
66. Wheeler, W.J. and O'Bannon, D.D., in *Synthesis and Applications of Isotopically Labeled Compounds 1991*, Buncel, E. and Kabalka, G.W., Eds.; Elsevier: Amsterdam, ***1992***, pp. 181–186.
67. Urban, J., *Coll. Czech. Chem. Commun.*, ***1959***, *24*, 4050–4052.
68. a. Murray III, A. and Williams, L., *Oraganic Syntheses with Isotopes*; Interscience: New York, ***1958***, pp. 973–1023; b. Blazer, R.M. and Whaley, T.M., *J. Am. Chem. Soc.*, ***1980***, *102*, 5082–5085; c. Isbell, H.S.; Karabinos, J.V.; Frush, H.L.; Holt, N.B.; Schwebel, A. and Galkowaki, T.T., *J. Res. Natl. Bur. Stand.*, ***1952***, *48*, 163–171.
69. King-Morris, M.J.; Bondo, P.B.; Mrowca, R.A. and Serianni, A.S., *Carbohydrate Res.*, ***1988***, *175*, 49–58.
70. Porto, A.M. and Gros, E.G., *J. Label. Compd.*, ***1968***, *4*, 276–278.
71. Lawrie, K.W.M.; Novelli, C.E.A.; Saunders, D. and Coates, W.J., *J. Label. Compd. Radiopharm.*, ***1995***, *36*, 891–898.
72. a. Harrison, J.M., *J. Label. Compd. Radiopharm.*, ***1983***, *20*, 1011–1015; b. Sugasawa, T.; Toyoda, T.; Adachi, M. and Sasakura, K., *J. Am. Chem. Soc.*, ***1978***, *100*, 4842–4852.
73. Hayashi, N.; Toga, T. and Miyake, A., *J. Label. Compd. Radiopharm.*, ***1978***, *14*, 185–189.
74. Rutkai, Gy.; Koltai, E. and Urögdy, L., *Synthesis and Applications of Isotopically Labeled Compounds 1994*, Allen, J., Ed.; John Wiley & Sons, Ltd: Chichester, ***1995***, pp. 655–657.
75. Hartman, J.D.; Alks, V.; Huang, C.C. and Cook, P.D., *J. Label. Compd. Radiopharm.*, ***1986***, *23*, 35–44.
76. Seidel, D.; Brehmer, P.; Schoof, Y.; Weinberg, U. and Nowakowski, M., *J. Label. Compd. Radiopharm.*, ***2003***, *46*, 1019–1032.
77. Czeskis, B.A., *J. Label. Compd. Radiopharm.*, ***2004***, *47*, 699–704.
78. Elmore, C.S.; Dean, D.C.; Zhang, Y. and Melillo, D.G., *J. Label. Compd. Radiopharm.*, ***2004***, *47*, 837–846.
79. White, D.F. and Burns, J., *J. Label. Compd. Radiopharm.*, ***1977***, *13*, 393–400.
80. Armitage, A.M. and Saunders, D., *J. Label. Compd. Radiopharm.*, ***1985***, *22*, 1075–1080.
81. Mylari, B.L. and Zembrowski, W.J., *J. Label. Compd. Radiopharm.*, ***1991***, *29*, 143–148.
82. Ueyama, N.; Yanigasawa, T. and Tomiyama, T., *J. Label. Compd. Radiopharm.*, ***1995***, *36*, 915–919.
83. a. Wishart, D.S.; Sykes, B.D. and Richards, F.M., *J. Label. Compd. Radiopharm.*, ***1992***, *31*, 1019–1028; b. Ehrensvard, G. and Stjernholm, R., *Acta Chim. Scand.*, ***1949***, *3*, 971–974.

84. a. Mehrsheikh, A., *J. Label. Compd. Radiopharm.*, ***1996***, *38*, 631–636; b. Woo, P.W.K., *J. Label. Compd. Radiopharm.*, ***1988***, *25*, 1149–1155.
85. a. White, D.F. and Burns, J., *J. Label. Compd. Radiopharm.*, ***1975***, *12*, 171–175; b. ($Br^{14}CN$) Murray, III, A. and Williams, L., *Organic Syntheses with Isotopes;* Interscience: New York, ***1958***, pp. 589–590; c. Graff, S.; Engelman, M.; Gillespie, H.B. and Graff, A.M., *Cancer Res.*, ***1951***, *11*, 388–392; d. ($Cl^{14}CN$) Murray, III, A. and Williams, L., *Organic Syntheses with Isotopes;* Interscience: New York, ***1958***, p. 794–795.
86. Westerberg, G.; Kärger, W.; Onoe, H. and Långström, B., *J. Label. Compd. Radiopharm.*, ***1994***, *34*, 691–696.
87. Grivas, S. and Olsson, K., *Acta Chem. Scand., Ser. B*, ***1985***, *39*, 31–34.
88. Chaudhuri, N.K.; Markus, B.; Sung, M.-S. and Desai, M., in *Synthesis and Application of Isotopically Labeled Compounds 1991*, Buncel, E. and Kabalka, G.W., Eds.; Elsevier: Amsterdam, ***1992***, pp. 226–231.
89. Thomas, R.C., *J. Label. Compd. Radiopharm.*, ***1978***, *14*, 807–811.
90. Do, C.T.; Nguyen, H.N.; Hoellinger, H. and Pichat, L., *J. Label. Compd. Radiopharm.*, ***1981***, *18*, 1009–1014.
91. Murray, III, A. and Williams, L., *Organic Syntheses with Isotopes;* Interscience: New York, ***1958***, pp. 572–573.
92. a. Murray, III, A. and Williams, L., *Organic Syntheses with Isotopes;* Interscience: New York, ***1958***, pp. 593–594; b. Haley, E.E. and Lambooy, J.P., *J. Am. Chem. Soc.*, ***1954***, *76*, 2926–2929.
93. Murray, III, A. and Williams, L., *Organic Syntheses with Isotopes;* Interscience: New York, ***1958***, pp. 305–306.
94. Hristova-Kazmierski, M.K. and Kepler, J.A., *J. Label. Compd. Radiopharm.*, ***1999***, *42*, 203–206.
95. Allen, J., *J. Label. Compd. Radiopharm.*, ***1983***, *20*, 1283–1286.
96. Gebert, U.; Phull, G. and Troke, J.A., in *Synthesis and Applications of Isotopically Labeled Compounds 1988*, Baillie, T.A. and Jones, J.R., Eds.; Elsevier: Amsterdam, ***1989***, pp. 289–294.
97. a. Villani, A.J.; Petka, L.; Blackburn, D.W.; Saunders, D.; White, G.R. and Winster, J., *J. Label. Compd. Radiopharm.*, ***1989***, *27*, 1395–1402; b. Zolyomi, G.; Toldy, L. and Bánfi, D., *J. Label. Compd. Radiopharm.*, ***1973***, *9*, 243–251.
98. Patil, S.P.; Kalgutkar, D.B.; Jaychandran, N. and Unny, V.K.P., *J. Label. Compd. Radiopharm.*, ***2007***, *50*, 420–421.
99. Matloubi, H.; Ghandi, M. and Saemian, N., *Appl. Radiat. Isot.*, ***2002***, *57*, 501–504.
100. Esses-Reiter, K. and Reiter, J., *J. Label. Compd. Radiopharm.*, ***1981***, *18*, 1731–1735.
101. Anjaneyulu, B., *J. Label. Compd. Radiopharm.*, ***1987***, *24*, 957–969.
102. Anjaneyulu, B.; Maller, R.K.; Nagarajan, K.; Kueng, W. and Wirz, B., *J. Label. Compd. Radiopharm.*, ***1985***, *22*, 313–327.
103. Pathak, V.P. and Montgomery, J.A., *J. Label. Compd. Radiopharm.*, ***1993***, *33*, 81–85.
104. Andres, H., *J. Label. Compd. Radiopharm.*, ***1999***, *42*, 1008–1009.
105. Baker, M.W.; Mewse, M.V.; Shutler, S.W.; Taylor, L.M. and Titman, R.B., *J. Label. Compd. Radiopharm.*, ***1997***, *39*, 685–693.
106. Saemian, N.; Shirvani, G. and Matloubi, H., *J. Label. Compd. Radiopharm.*, ***2007***, *50*, 576–577.
107. a. Kai, Y. and Chiku, S., *J. Label. Compd. Radiopharm.*, ***1994***, *34*, 1069–1073; b. Sugiyama, I.; Nomoto, S.; Mizuo, H. and Yamauchi, H., *J. Label. Compd. Radiopharm.*, ***1991***, *29*, 1309–1313.
108. Gopinathan, M.B. and Kepler, J.A., *J. Label. Compd. Radiopharm.*, ***1991***, *29*, 645–649.
109. Valsborg, J.S.; Knutsen, L.J.S.; Lundt, I. and Foged, C., *J. Label. Compd. Radiopharm.*, ***1995***, *36*, 457–464.
110. Ogan, M.D.; Kucera, D.J.; Pendri, Y.R. and Rinehart, J.K., *J. Label. Compd. Radiopharm.*, ***2005***, *48*, 645–655.
111. Sato, S.; Mikoshiba, I.; Kawai, K. and Hirai, K., *J. Label. Compd. Radiopharm.*, ***1993***, *33*, 195–203.
112. Woo, P.W.K. and Lee, H.T., *J. Label. Compd. Radiopharm.*, ***1990***, *28*, 445–454.
113. Bergin, J.A.; Booth, H.; Bragg, R.A.; Bushby, N.; Harding, J.R.; Jordan, A. and King, C.D., *J. Label. Compd. Radiopharm.*, ***2007***, *50*, 431–432.
114. a. Murray, III, A. and Williams, L., *Organic Syntheses with Isotopes;* Interscience: New York, ***1958***, pp. 117–118; b. Murray, III, A. and Williams, L., *Organic Syntheses with Isotopes;*

Interscience: New York, ***1958***, pp. 119–121; c. Murray, III, A. and Williams, L., *Organic Syntheses with Isotopes;* Interscience: New York, ***1958***, pp. 167–168; d. Bubner, M. and Schmidt, L.H., *Die Synthese Kohlenstoff-14 markierter organischer Verbindungen;* Georg Thieme: Leipzig, ***1966***, pp. 157–158.

115. Tolman, V.; Cabak, J. and Benes, J., *J. Label. Compd. Radiopharm.*, ***1974***, *10*, 39–43.
116. Van den Berg, E.M.M.; Richardson, E.E.; Lugtenburg, J. and Jenneskens, L.W., *Synthetic Commun.*, ***1987***, *17*, 1189–1198.
117. a. Balaban, A.T. and Bally, I., Isotopically Labeled Pyrimidines and Purines, in *Isotopes in the Physical and Biomedical Sciences*, Vol. 1, Part A, Buncel, E. and Jones, J.R., Eds.; Elsevier: Amsterdam, ***1987***, pp. 288–329; b. Redwine, C.M. and Whaley, T.W., *J. Label. Compd. Radiopharm.*, ***1979***, *16*, 315–320; c. Lukas, J.H. and Gerber, G.B., *J. Label. Compd.*, ***1965***, *1*, 229–230.
118. Lawson, J.A.; DeGraw, J.I. and Anbar, M., *J. Label. Compd.*, ***1975***, *11*, 489–499.
119. DeGraw, J.I. and Engstrom, J.S., *J. Label. Compd.*, ***1975***, *11*, 233–239.
120. Coelho, Jr., R.V. and Schildknecht, K., *J. Label. Compd. Radiopharm.*, ***2007***, *50*, 675–678.
121. a. Osuka, A.; Kobayashi, T. and Suzuki, H., *Synthesis*, ***1983***, 67–68; b. Suzuki, H.; Kobayashi, T.; Yoshida, Y. and Osuka, H., *Chem. Lett.*, ***1983***, 193–194.
122. Beare, N.A. and Hartwig, J.F., *J. Org. Chem.*, ***2002***, *67*, 541–555.
123. Wheeler, W.J.; O'Bannon, D.D.; Chay, S.H. and Herman, J.L., in *Synthesis and Applications of Isotopically Labeled Compounds*, Vol. 8, Dean, D.C.; Filer, C.N. and McCarthy, K.E., Eds.; John Wiley & Sons, Ltd: Chichester, ***2004***, pp. 361–364.
124. Volford, J. and Harsányi, K., *J. Label. Compd.*, ***1973***, *9*, 219–230.
125. Hicks, J.L., *J. Label. Compd. Radiopharm.*, ***1995***, *36*, 1029–1035.
126. Olszewski, J.D.; Berger, D.M. and May, M.K., *J. Label. Compd. Radiopharm.*, ***2007***, *50*, 578–579.
127. Moenius, T.; Andres, H.; Nozulak, J.; Salter, R.; Ray, T.; Burtscher, P.; Schnelli, P.; Züger, C. and Voges, R., *J. Label. Compd. Radiopharm.*, ***2007***, *50*, 616–619.
128. Bouthillier, L.P. and D'Iorio, *Rev. Can. Biol.*, ***1950***, *9*, 382–385.
129. Murray, III, A. and Williams, L., *Organic Syntheses with Isotopes;* Interscience: New York, ***1958***, pp. 442–444.
130. Hicks, D.R. and Hangeland, J.J., *J. Label. Compd. Radiopharm.*, ***1987***, *24*, 747–751.
131. Murray, III, A. and Williams, L., *Organic Syntheses with Isotopes*; Intersciences: New York, ***1958***, pp. 225–226, 279–281, 292.
132. a. Perry, C.W.; Burger, W.; Bader, G.J. and Liebman, A.A., *J. Label. Compd.*, ***1975***, *11*, 583–589; b. Fritzson, P., *Acta Chem. Scand.*, ***1955***, *9*, 1239–1240.
133. Triplett, J.W.; Mack, S.W.; Smith, S.L. and Digenis, G.A., *J. Label. Compd. Radiopharm.*, ***1978***, *14*, 35–41.
134. Murray, III, A. and Williams, L., *Organic Syntheses with Isotopes;* Interscience: New York, ***1958***, pp. 748–752.
135. Bennett, E.L., *J. Am. Chem. Soc.*, ***1952***, *74*, 2420–2422.
136. a. Murray, III, A. and Williams, L., *Organic Syntheses with Isotopes;* Interscience: New York, ***1958***, pp. 769–774; b. Bentley, R. and Neuberger, A., *Biochem. J.*, ***1952***, *52*, 694–699.
137. Ayrey, G. and Yeomans, M.A., *J. Label. Compd. Radiopharm.*, ***1976***, *12*, 323–345.
138. Murray, III, A. and Williams, L., *Organic Syntheses with Isotopes;* Interscience: New York, ***1958***, pp. 762–764.
139. Hayashi, N. and Imanishi, M., *J. Label. Compd. Radiopharm.*, ***1986***, *23*, 339–341.
140. Maxwell, B., in *Synthesis and Applications of Isotopically Labeled Compounds*, Vol. 8, Dean, D.C.; Filer, C.N. and McCarthy, K.E., Eds.; John Wiley & Sons, Ltd: Chichester, ***2004***, pp. 161–164.
141. Yamanaka, M.; Suda, S. and Tadano, K., *J. Label. Compd. Radiopharm.*, ***1992***, *31*, 125–129.
142. Gubitz, F.W., *J. Label. Compd. Radiopharm.*, ***1981***, *18*, 755–758.
143. a. Fittkau, S., *Pharmazie*, ***1980***, *35*, 646–647; b. Rhee, S.W.; Ryan, K.J.; Tracy, M.; Kelson, A.B.; Clizbe, L.A.; Chang, M.-H.; Park, J.-S.; Roh, J.-K.; Kong, J.-Y.; Yang, J.; Kim, W.-B. and Ok, K.D., *J. Label. Compd. Radiopharm.*, ***1997***, *39*, 773–785.

144. Murray, III, A. and Williams, L., *Organic Syntheses with Isotopes;* Interscience: New York, ***1958***, p. 492.
145. a. Bubner, M. and Schmidt, L., *Die Synthese Kohlenstoff-14 markierter organischer Verbindungen;* VEB Georg Thieme: Leipzig, ***1966***, pp. 166–167; b. Vishnuvajjala, B.R.; Kataoka, T.; Cazer, F.D.; Witiak, D.T. and Malspeis, L., *J. Label. Compd Radiopharm.*, ***1978***, *14*, 77–82.
146. Pizey, J.S., *N-Bromosuccinimide, Diazomethane, Manganese Dioxide, Raney Nickel in Synthetic Reagents* Vol. 2; Wiley: New York, ***1974***, pp. 1–311.
147. Smith, M.B. and March, J., *Advanced Organic Chemistry,* 5th Edn.; John Wiley & Sons, Inc.: New York, ***2001***, p. 479.
148. Pussemier, L.; Vande Velde, V. and Piraux, M., *J. Label. Compd. Radiopharm.*, ***1979***, *16*, 651–657.
149. a. Vicario, G.P.; Penco, S. and Arcamone, F., U.S. Patent 4,211,864, ***1980***, July 8; b. Vicario, G.P. and Arcamone, F., German Offen. DE 2806454, ***1977***; c. Felicini, C.; Giribone, D.; Pignatti, A. and Fontana, E., *J. Label. Compd. Radiopharm.*, ***2002***, *45*, 1184–1186; d. Fontana, E.; Giribone, D; Bargiotti, D.; Faiardi, G. and Suarato, A., in *Synthesis and Applications of Isotopically Labeled Compounds 1994*, Allen, J. and Voges, R., Eds.; Elsevier: Amsterdam, ***1995***, pp. 719–722.
150. Bennett, R.D.; Sauer, H.H. and Heftmann, E., *J. Label. Compd.*, ***1969***, *5*, 160–167.
151. Schleyer, P.v.R.; Funke, E. and Liggero, S.H., *J. Am. Chem. Soc.*, ***1969***, *91*, 3965–3967.

# 8

# Preparation of Carbon-14-Labeled Compounds via the [$^{14}C_2$]Acetylene Tree

[$^{14}C_2$]Acetylene, which can be prepared in high yield from barium [$^{14}C$]carbonate, is most often used for the production of [U-$^{14}C$]benzene, and thence a vast array of labeled aromatic compounds. The preparation and handling of [$^{14}C_2$]acetylene is not difficult to achieve in the average radiochemistry laboratory, and it has been used as a two-carbon building block in a variety of syntheses. In addition, it is the source of a number of other two- to four-carbon building blocks, making its utility relatively broad, even aside from its uses as the source of [U-$^{14}C$]aromatic derivatives.

## 8.1 [$^{14}C_2$]Acetylene ($H^{14}C\equiv{}^{14}CH$)

Quantities of [$^{14}C_2$]acetylene in the range of 1–20 mmol are most efficiently prepared by heating a mixture of barium [$^{14}C$] carbonate with excess barium filings under an atmosphere of helium or hydrogen to red heat, which indicates the onset of the exothermic formation of barium [$^{14}C_2$]carbide (Figure 8.1). The resulting black mass, mainly consisting of barium [$^{14}C_2$]carbide, barium oxide and unreacted barium, is cautiously hydrolyzed and the liberated [$^{14}C$]acetylene is expelled with a stream of helium[1]. Larger quantities are better produced by reduction of [$^{14}C$]carbon dioxide with molten lithium and subsequent hydrolysis of the lithium carbide thus formed[2]. In both cases radiochemical yields of about 95% are obtained.

Unlike its enormous importance for the preparation of labeled secondary building blocks, in particular of labeled aromatic compounds, its use as a primary building block has been limited so far. Its monoanion, generated *in situ* upon deprotonation with Li, $NaNH_2$ or $KNH_2$ in liquid ammonia, *n*-BuLi in THF or with EtMgBr in ether, reacts with carbonyl compounds,

*Preparation of Compounds Labeled with Tritium and Carbon-14* Rolf Voges, J. Richard Heys and Thomas Moenius

$$2\,Ba^{14}CO_3 + 5\,Ba \xrightarrow{\Delta} 6\,BaO + Ba^{14}C_2 \xrightarrow{H_2O} H^{14}C{\equiv}^{14}CH$$

$$2\,^{14}CO_2 + 10\,Li \xrightarrow{\Delta} Li_2{}^{14}C_2 + 4\,Li_2O \xrightarrow{H_2O} H^{14}C{\equiv}^{14}CH$$

***Figure 8.1*** *Preparative methods for [$^{14}C_2$]acetylene*

$CO_2$, or alkyl halides to [$^{14}C_2$]alkynols, [$^{14}C_2$]propiolic acid or [$^{14}C_2$]alkynes. Excess base generates the dianion, which furnishes the corresponding bis-funtionalized alkynes.

The reaction of monodeprotonated [$^{14}C_2$]acetylene with carbonyl compounds has been exploited as a means of extension of the carbon chain of various terpenes and steroids by two [$^{14}$C]carbon atoms. In the simplest case, reaction of potassium [$^{14}C_2$]acetylide with steroid ketone **1** and subsequent acid catalyzed cleavage of the enol ether protecting group gave 17$\alpha$-[$^{14}C_2$]ethynyltestosterone (**2**)[3]. The sequential addition of deprotonated [$^{14}C_2$]acetylene to carbonyl compounds opens access to symmetrical or unsymmetrical [2,3-$^{14}C_2$]alkyn-1,4-diols is exemplified in the synthesis of all-*trans*-$\beta$-[15,15′-$^{14}C_2$]-carotene ([$^{14}C_2$]provitamin A). Thus, treatment of lithium [$^{14}C_2$]acetylide with terpene aldehyde **3** followed by double deprotonation of the resultant alkynol **4** and reaction with a second equivalent of **3** provided alkyne-1,4-diol **5** the requisite key intermediate. Subsequent acid-catalyzed dehydration of **5** followed by Lindlar's catalyst-mediated partial hydrogenation and photoisomerization afforded the final product[4].

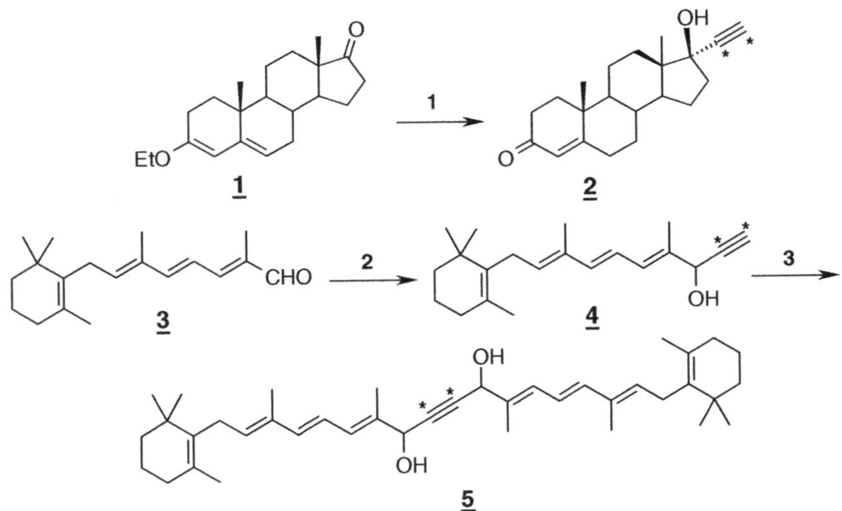


**Reaction conditions: 1a**. K, liq. $NH_3$, −78 °C, **b**. $H^{14}C{\equiv}^{14}CH$; −78 °C to −50 °C, **c**. **1**, benzene/ether 1:1, −78 °C, 3 h; −78 °C to r.t., 16 h, **d**. aq. HCl, EtOH; r.t., 16 h; **2a**. Li, liq. $NH_3$, −78 °C, **b**. $H^{14}C{\equiv}^{14}CH$; −78 °C to −50 °C, **c**. **3**, ether; −40 °C, 2 h; r.t., 1 h; **3**. PhLi, ether; reflux, 30 min then **4**, ether; reflux, 1 h

***Figure 8.2*** *Uses of [$^{14}C_2$]acetylene in [$^{14}C_2$]alkynylations*

Further versatility of [$^{14}C_2$]alkynols as synthetic intermediates is demonstrated in synthesis of β-[6,7-$^{14}C_2$]ionone (**8**)[5], a key intermediate for the synthesis of labeled terpenes. The reaction sequence started with the addition of monolithium [$^{14}C_2$]acetylide to 6-methyl-5-hepten-2-one and $Ag^+$-catalyzed rearrangement of the resultant alkynol **6** upon treatment with acetic anhydride in the presence of catalytic amounts of $H_3PO_4$ to afford [1,2-$^{14}C_2$]citral (**7**). Chain extension through aldol condensation with acetone and acid-catalyzed cyclization converted **7** to β-[6,7-$^{14}C_2$]ionone. Further transformations provided 13-*cis*-[6,7-$^{14}C_2$]retinoic acid[5]. Bis-hydroboration of [$^{14}C_2$]alkynol **9** with 9-borabicyclo[3.3.1]nonane, oxidative work up and hydrolytic cleavage of the carboxylic acid protecting group provided racemic [4,5-$^{14}C_2$]mevalonolactone[6] (**10**). Compound **9** was previously prepared from lithium [$^{14}C_2$]acetylide and masked acetoacetic acid[6].

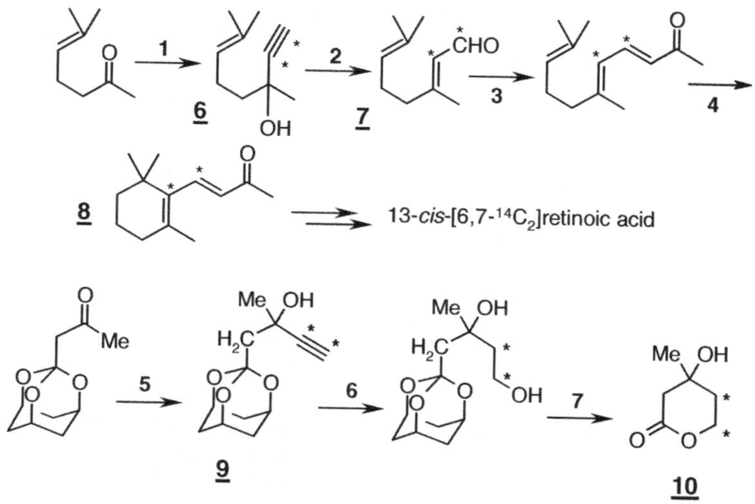


**Reaction conditions**: **1a**. $HC^{14}{\equiv}^{14}CH$, *n*-BuLi, THF; –60 °C, 1 h, **b**. ketone, 0 °C, 1 h; r.t., 16 h; **2a**. $Ac_2O$, $H_3PO_4$ (cat.), r.t., 16 h, **b**. $Ag_2CO_3$(cat), NaOH, HOAc; 90 °C, 1.5 h; **3**. acetone, NaOH, water; 40 °C, 3 h; **4**. aq. $H_2SO_4$ –20°C, 4 min; 7 °C, 10 min; **5a**. $HC^{14}{\equiv}^{14}CH$, *n*-BuLi, THF/hexane; –78 °C,15 min, **b**. ketone, –78°C, 15 min; r.t., 75 min; **6a**. 9-BBN, THF/hexane, r.t., 17 h, **b**. $H_2O_2$, 6N aq. $Na_2CO_3$; 0 °C, 10 min; r.t., 1.5 h; **7a**. $CF_3COOH/H_2O$ 5:1, r.t., 1 h, **b**. aq. $Na_2CO_3$, pH 8; 60 °C, 30 min

***Figure 8.3*** *Preparation and utilization of [$^{14}C_2$]alkynols*

Although unlabeled (or $^{13}C$-labeled) acetylene has been employed in various two-carbon extension reactions of [$^{14}C$]alkyl halides and sulfonates[7], only one application has been reported so far for [$^{14}C_2$]acetylene. In this case the monoanion reacted with $Br(CH_2)_3Cl$ to afford 1-chloro[4,5-$^{14}C_2$]pent-4-yne (**11**) (Figure 8.4). Subsequent cyano-dechlorination, hydrolysis of the cyano function and a second alkylation with 1-chloro-2,5-tetradecadiyne gave the multiply unsaturated carboxylic acid **12**[8].

Acid-catalyzed hydrocarboxylation of acetylene with $Ni(CO)_4$ in methanol furnishes methyl acrylate **13**, a frequently employed dienophile in [2 + 4]cycloadditions. This efficient and convenient access to cyclic derivatives has been only sparsely exploited in carbon-14 syntheses, probably due to the risk of radiation-induced polymerization of carbon-14-labeled acrylates when the specific activitiy exceeds 10 mCi/mmol (concerning radiolytically stable [$^{14}C_2$]acrylates at a specific activity of 100 mCi/mmol and their application in diastereoselective syntheses[9] see also Chapter 11, Section 11.3.1). The abovementioned procedure applied to [$^{14}$C]acetylene provided methyl [2,3-$^{14}$C]acrylate (**13**) in 50% radiochemical yield. Reaction with *trans,trans*-1,4-diacetoxy-1,3-butadiene furnished methyl 2β,5β-diacetoxy[1,6-$^{14}$C]cyclohex-3-enecarboxylate (**14**), the key intermediate in the synthesis of racemic [1,2-$^{14}$C]shikimic acid (**15**)[10].

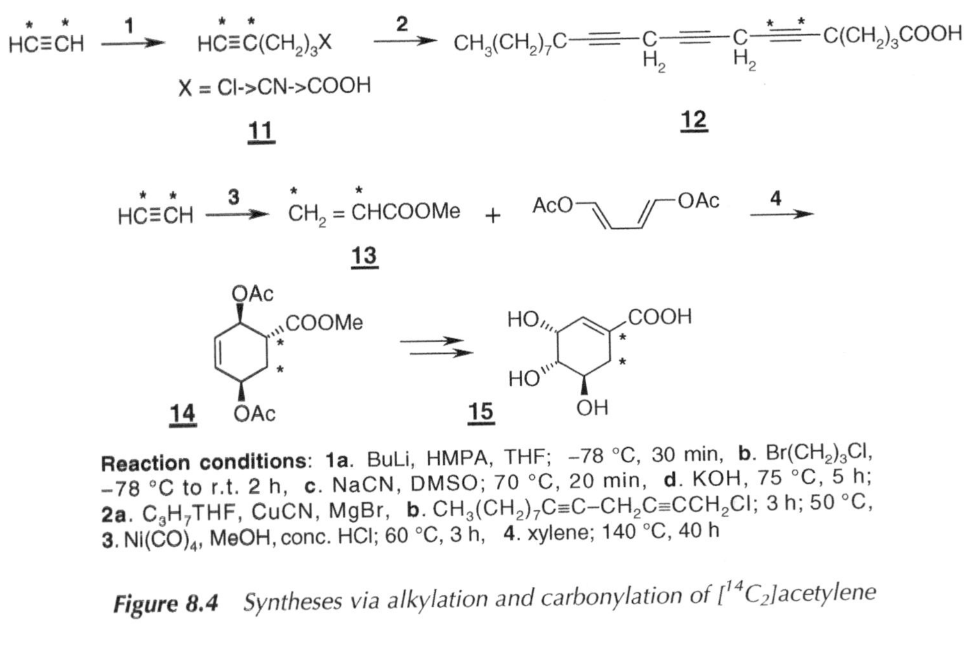


**Reaction conditions**: **1a**. BuLi, HMPA, THF; −78 °C, 30 min, **b**. Br($CH_2$)$_3$Cl, −78 °C to r.t. 2 h, **c**. NaCN, DMSO; 70 °C, 20 min, **d**. KOH, 75 °C, 5 h; **2a**. $C_3H_7$THF, CuCN, MgBr, **b**. $CH_3(CH_2)_7C{\equiv}C{-}CH_2C{\equiv}CCH_2Cl$; 3 h; 50 °C, **3**. $Ni(CO)_4$, MeOH, conc. HCl; 60 °C, 3 h, **4**. xylene; 140 °C, 40 h

***Figure 8.4*** *Syntheses via alkylation and carbonylation of [$^{14}C_2$]acetylene*

Despite its own valuable synthetic potential, the use of [$^{14}C_2$]acetylene as a starting material for various building blocks is of much higher relevance. Mercury(II)-catalyzed hydration, for example, gives [$^{14}C_2$]acetaldehyde (Figure 8.5, Route 1)[11]. The same reaction carried out in the presence of ammonium persulfate furnishes [$^{14}C_2$]acetic acid (Route 2)[12]. Trapping of its mono- or dianion with formaldehyde or carbon dioxide affords [2,3-$^{14}C_2$]propynol[13], [2,3-$^{14}C_2$]butyne-1,4-diol[13], [2,3-$^{14}C_2$]propiolic acid[14] and [2,3-$^{14}C_2$]acetylenedicarboxylic acid[15], respectively (Routes 3–6). UV irradiation of a mixture of HBr and [$^{14}C_2$]acetylene produces 1,2-dibromo[$^{14}C_2$]ethane (Route 8)[16]. Reduction with chromium(II) chloride followed by a two-step epoxidation of the initially formed [$^{14}C_2$]ethylene converts [$^{14}C_2$]acetylene into [$^{14}C_2$]ethylene oxide (Route 7)[1d,1e]. Finally, catalytic homotrimerization[1b] or co-trimerization with other alkynes[17] provides [$^{14}C_n$]benzene or substituted [$^{14}C_n$]benzenes, respectively, the central starting materials for the vast majority of substituted benzenoid aromatic compounds (Route 9).

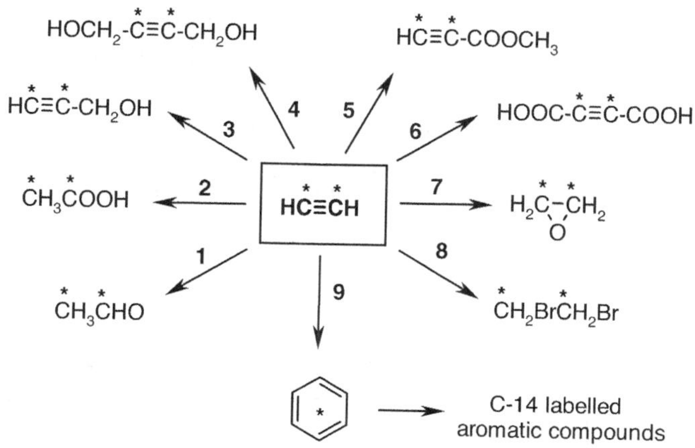


**Reaction conditions: 1.** $HgSO_4$, 10 % $H_2SO_4$; 95 °C, 25 min **or** $^{14}CH_3{}^{14}COCl$, $Bu_3SnH$, $Pd(PPh_3)_4$; 1-methylnaphthalene, 0 °C, 30 min; **2a.** HgO, $(NH_4)_2S_2O_8$, 44 % $H_2SO_4$; r.t., 16 h, **b.** $Cu_2O$, 40 °C, 1 h; **3a.** *n*-BuLi, THF; −78 °C, 30 min, **b.** HCHO; r.t., 60 min; **4a.** MeLi, ether, r.t., 1 h, **b.** HCHO; r.t., 64 h; **5a.** *n*-BuLi, THF; −78 °C, 15 min, **b.** $CO_2$, −78 °C, 1 h, **c.** 2,6-di-*tert*-butyl-*p*-cresol, tetraglyme; −78 °C to r.t., 2 h, **d.** $CH_3OTos$, $Bu_4NHSO_4$, 50 °C, 16 h; **6a.** *tert*-BuLi, THF; −78 °C, 15 min, **b.** $CO_2$;−78 °C, 2 h; −78 to r.t.; **7a.** $CrCl_2$ , 25 % HCl; r.t., 6–48 h, **b.** 1.8 N HOCl; r.t., 2 h, **c.** KOH; −78 °C to −50 °C **8.** $HBr_{gas}$, air (cat.); hν, r.t., 45 min; **9.** KC-Perlkatalysator PKN ex Kali-Chemie; 100 °C, 1 h

***Figure 8.5*** *Syntheses of [$^{14}C_2$]building blocks and [$^{14}C_2$]intermediates from [$^{14}C_2$]acetylene*

## 8.2 [$^{14}C_2$]Acetaldehyde ($^{14}CH_3{}^{14}CHO$)

Three processes are available for the preparation of [$^{14}C_2$]acetaldehyde from [$^{14}C_2$]-acetylene; all give 80–90% radiochemical yields. The most direct procedure involves Hg(II)-catalyzed hydration of [$^{14}C_2$]acetylene[11a,c]. Utilizing $HgSO_4$ in 10% $H_2SO_4$ at 95 °C followed by distillation of the reaction mixture gives [$^{14}C_2$]acetaldehyde as a ~0.1 M aqueous solution[11a]. The second process makes use of the $Pd(PPh_3)_4$-catalyzed reduction of [$^{14}C_2$]acetyl chloride with $Bu_3SnH$[11b]. In spite of providing the compound in anhydrous form, it suffers from the need for two additional steps, preparation of [$^{14}C_2$]-acetic acid and its conversion into [$^{14}C_2$]acetyl chloride. Similarly, the third method provides [$^{14}C_2$]acetaldehyde in anhydrous form but proceeds via $LiAlH_4$ reduction of [$^{14}C_2$]acetic acid in a high-boiling solvent, distillation of the resultant [$^{14}C_2$]ethanol, and finally its oxidation[11e].

The utility of aqueous [$^{14}C_2$]acetaldehyde is illustrated by its Horner–Wadsworth–Emmons olefination with diethyl (cyanomethyl)phosphonate under phase transfer conditions to give [3,4-$^{14}C_2$]crotonitrile (**16**) in 83% radiochemical yield. Subsequent Michael addition of α-aminonitrile **17** produced the masked γ-ketoacid derivative **18** (Figure 8.6), which served as a key intermediate in the synthesis of [$^{14}C_2$]CI 930 (**19**)[11a]. For the $TiCl_4$-promoted aldol reaction with the propiophenone derivative **20**, however, the use of anhydrous [$^{14}C_2$]acetaldehyde was essential. The reaction proceeded with excellent stereoselectivity, providing exclusively the racemic *syn*-diastereomers **21**. Subsequent lipase-mediated resolution, Mitsunobu cyclization and stereoselective $NaBH_4/CeCl_3$ reduction converted **21** into (+)-[$^{14}C_2$]calanolide (**22**)[18]. In a third example, reaction of [$^{14}C_2$]acetaldehyde with two equivalents of acetoacetate followed by acid-induced aldol ring closure and decarboxylation provided enone **23**, which was efficiently converted to the symmetrical phenol **24** labeled at one methyl group and its adjacent ring carbon. It was used as an intermediate in the synthesis of a carbamate insecticidal agent for plants[11c].

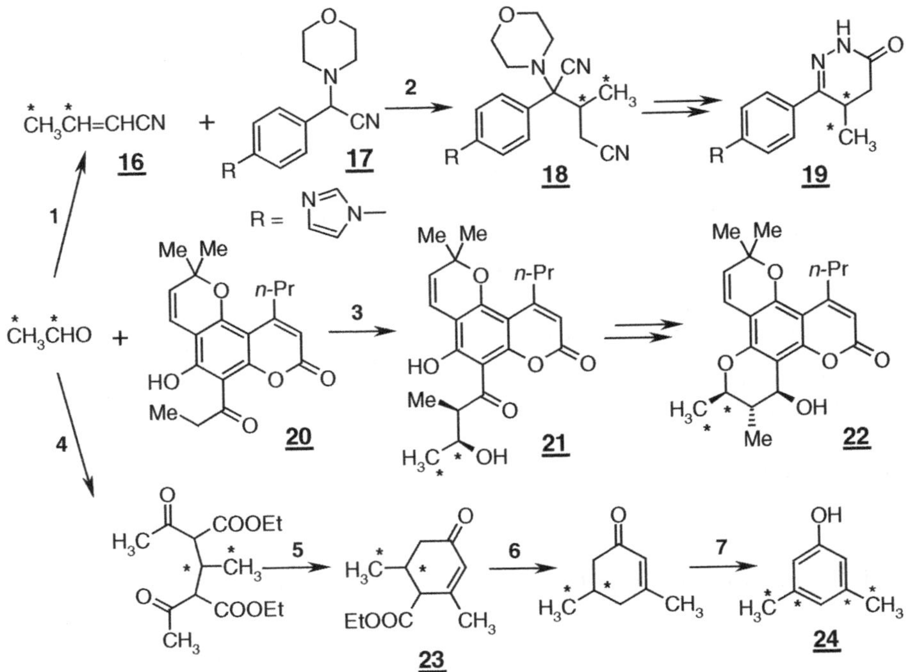

**Reaction conditions:** **1**. $(EtO)_2P(O)CH_2CN$, 1,19-phenanthroline (cat.), $Bu_4N^+I^-$, $CH_2Cl_2$/50 % NaOH 1:1; r.t., 30 min; **2**. methanolic KOH, THF; reflux 24 h; **3a**. $TiCl_4$, $CH_2Cl_2$; –40 °C, **b**. LDA, THF/heptane/ethylbenzene; –78 °C, 30 min, **c**. $^{14}CH_3{}^{14}CHO$; –78 °C, 2 h; **4**. excess ethyl acetoacetate, piperidine, 0 °C, 48 h; **5**. cat. $H_2SO_4$, glac. AcOH; 100 °C, 1 h; **6**. aq. NaOH/EtOH; reflux 15 min; **7**. NiO, Dicalite; 320 °C in flow-through tube.

***Figure 8.6*** *Use of [$^{14}C_2$]acetaldehyde in olefination and aldol reactions*

## 8.3 [1,2-$^{14}C_2$]Acetic Acid ($^{14}CH_3{}^{14}COOH$)

The procedure used for the preparation of [1,2-$^{14}C_2$]acetic acid is closely related to that described above for the preparation of [$^{14}C_2$]acetaldehyde. In this case, however, the Hg(II)-catalyzed hydration of [$^{14}C_2$]acetylene is followed by the addition of ammonium persulfate to oxidize the [$^{14}C_2$]acetaldehyde *in situ*, and decomposition of excess oxidant with $Cu_2O$[12]. [1,2-$^{14}C_2$]Acetic acid replaces its monolabeled isotopomers when the specific activity of the latter is not sufficient, for example in cases of low-dose or high molecular weight drug substances. Its use as an essential starting material for the biological labeling of rapamycin in the metabolically stable positions 1,2,36,37 is a striking example and is discussed in Chapter 11, Section 11.3.6[9]. The utility of [$^{14}C_2$]acetic acid parallels that of its monolabeled isotopomers, discussed in Chapter 6, Section 6.1.1.

## 8.4 2-[2,3-$^{14}C_2$]Propyne-1-ol ([2,3-$^{14}C_2$]Propargyl Alcohol) and 2-[2,3-$^{14}C_2$]Butyne-1,4-diol

The mono- and dianions of [$^{14}C_2$]acetylene, readily accessible by deprotonation with *n*-BuLi (1 equivalent) and MeLi (2.5 equivalents), respectively, can be trapped with monomeric formaldehyde to give [2,3-$^{14}C_2$]propargyl alcohol in 40% yield and 2-[2,3-$^{14}C_2$]butyne-1,4-diol in 72% yield. Both compounds may serve as valuable intermediates for the preparation of additional low-molecular weight building blocks and intermediates. [2,3-$^{14}C_2$]propargyl alcohol, for example, can be reduced with aqueous $CrCl_2$ to [2,3-$^{14}C_2$]allyl alcohol, which was shown to be activated through tosylate formation toward nucleophilic displacement of the hydroxy function to produce radiolabeled allyl derivatives[13]. Catalytic reduction of [2,3-$^{14}C_2$]but-2-yne-1,4-diol using Wilkinson's catalyst gave a 1:1 mixture of [2,3-$^{14}C_2$]but-2-ene-1,4-diol and [2,3-$^{14}C_2$]-butane-1,4-diol, which was not separated but directly submitted to chromic acid oxidation to give [3,4-$^{14}C_2$]furan, a valuable diene in cycloadditions, in moderate yield[13].

## 8.5 Methyl [2,3-$^{14}C_2$]Propiolate ($H^{14}C{\equiv}^{14}CCOOMe$) and Dimethyl [2,3-$^{14}C_2$]Acetylenedicarboxylate ($HOOC^{14}C{\equiv}^{14}CCOOH$)

Both compounds are synthetically accessible through carboxylation of the corresponding anionic species. The propiolate obtained in the mono-carboxylation was not directly isolated, but esterified *in situ* with methyl tosylate. The radical scavenger 2,6-di-*tert*-butyl-*p*-cresol was added to minimize the risk of radiation-induced polymerization reactions[14]. Literature reports of reactions on isolated carbon-14 labeled propiolates or acetylenedicarboxylates are few, an exception being that of the catalytic hydrogenation of dimethyl [2,3-$^{14}C_2$]acetylenedicarboxylate[15c], but the suitability and versatility of their unlabeled and carbon-13 labeled counterparts have been confirmed in numerous publications. Their reaction with $\alpha$-pyrones, for example, could furnish doubly labeled aromatic systems with well defined substitution patterns, as demonstrated by the synthesis of methyl [1,3,$\alpha$-$^{13}C_3$]-salicylate (25), an intermediate in the preparation of [$^{13}C_3$]pindolol[19].

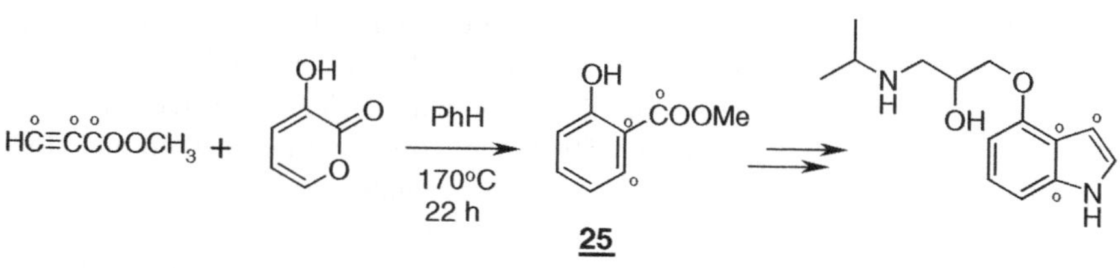


*Figure 8.7* Synthesis of multiply carbon-13-labeled aromatic compounds through [2 + 4]cycloaddition of a labeled propiolic ester with an $\alpha$-pyrone

## 8.6 1,2-[$^{14}C_2$]Dibromoethane ($Br^{14}CH_2{}^{14}CH_2Br$)

1,2-Dibromo[$^{14}C_2$]ethane is conveniently accessible in nearly quantitative yield through UV irradiation of a mixture of gaseous HBr (2 equivalents), [$^{14}C_2$]acetylene and catalytic amounts of dry air[16]. It serves as a starting material for further low molecular weight intermediates. Its treatment with KCN in refluxing aqueous ethanol, for example, provides [2,3-$^{14}C_2$]succinonitrile[20], which is key to the preparation of [2,3-$^{14}C_2$]succinic acid and [2,3-$^{14}C_2$]succinic anhydride, itself an intermediate in the synthesis of [$^{14}C_2$]naphthalene and [$^{14}C_2$]benzo-fused compounds more generally[21]. Reaction of 1,2-dibromo[$^{14}C_2$]ethane with KOAc in acetic acid at 190 °C furnishes 1,2-diacetoxy[$^{14}C_2$]ethane; subsequent alcoholysis with NaOMe gives [$^{14}C_2$]ethylene glycol[22], oxidative cleavage of which with $Pb(OAc)_4$ is a classical preparation of [$^{14}C$]formaldehyde on the 1–5 mmol range[16]. [$^{14}C_2$]-Ethylenediamine has been prepared in high yield by reaction with potassium phthalimide followed by hydrazinolysis[23]. Upon treatment of 1,2-dibromo[$^{14}C_2$]ethane with 1,2-bis (dibenzylamino)ethane, 1,4-dibenzyl[2,3-$^{14}C_2$]piperidine is obtained, which when subjected to hydrogenolytic debenzylation gives correspondingly labeled piperazine, an essential constituent of a number of drug substances[24].

## 8.7 [$^{14}C_2$]Ethylene Oxide

[$^{14}C_2$]Acetylene can be reduced with aqueous chromium(II) chloride to give [$^{14}C_2$]-ethylene in nearly quantitative yield. Subsequent sequential treatment with 1.8 M hypochlorous acid and solid KOH provides [$^{14}C_2$]ethylene oxide in an overall 85% radiochemical yield[1d,e].

[$^{14}C_2$]Ethylene oxide has been used for the preparation of unsymmetrically *N,N'*-substituted [2,3-$^{14}C_2$]piperidines and [$^{14}C_2$]aziridinium salts, as demonstrated with the examples in Figure 8.8. Treatment with *N*-substituted 2-aminoethanol **26** provided the bis-(2-hydroxyethyl)amine **27**. Activation of both hydroxy groups through mesylation and subsequent ring formation by reaction with a second amine provided the doubly labeled target compound **28**[25]. In contrast, differential protection of the two hydroxyl groups of the initial adduct, as in **29**, can be exploited to generate an intermediate that is set up for intramolecular nucleophilic attack, leading to the aziridium salt **30**[26].

## 8.8 [$^{14}C_n$]Benzene and the Synthesis of Ring-Labeled Aromatic Compounds

As previously mentioned, [U-$^{14}C$]benzene can be conveniently prepared in almost 90% yield by KC-Perlcatalysator-catalyzed trimerization of [$^{14}C_n$]acetylene[1b,27]. Also known as KC-Perkator, the catalyst is a family of silica–alumina matrices activated with $Cr_2O_3$[17]. [$^{14}C$]Benzene of specific activity up to 150–200 mCi/mol can be handled and stored in relative safety, but material of higher specific activity tends to polymerize spontaneously. The co-trimerization of [$^{14}C_2$]acetylene with higher alkynes has occasionally been conducted. For example, *o*-[3,4,5,6-$^{14}C_4$]xylene (**31**) (Figure 8.9) was prepared in 50% yield by trimerization of a 1:1 molar mixture of [$^{14}C_2$]acetylene and 2-butyne[17]. Steric effects evidently mitigate against statistical combination of the two substrates and favor the xylene

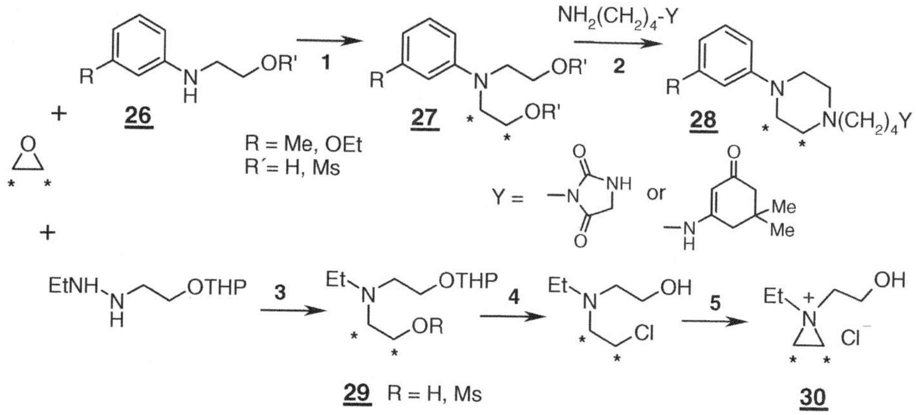

**Reaction conditions:** **1a**. 105 °C, 4 h; 25 °C, 18 h, **b**. MsCl, $Et_3N$, $CH_2Cl_2$; 25 °C, 2 h; **2**. LiBr, $CH_3CN$; reflux, 30–40 h; **3a**. 95 % EtOH; 4 °C, 30 min; r.t., 8 h, **b**. *n*-BuLi, MsCl, ether; r.t., 15 h; **4**. 12 % HCl, r.t., 30 h; **5**. 0.1N NaOH(> pH 7.4–7.5); r.t., 30 min

***Figure 8.8*** *Synthesis of [2,3-$^{14}C_2$]piperidines and [$^{14}C_2$]aziridinium salts from [$^{14}C_2$]ethylene oxide*

product over more highly substituted products. Subsequent oxidation with 30% nitric acid converts **31** into *o*-[3,4.5,6-$^{14}C_4$]toluic acid or [3,4,5,6-$^{14}C_4$]phthalic acid in 50% or 98% yield, respectively, depending on the reaction conditions applied.

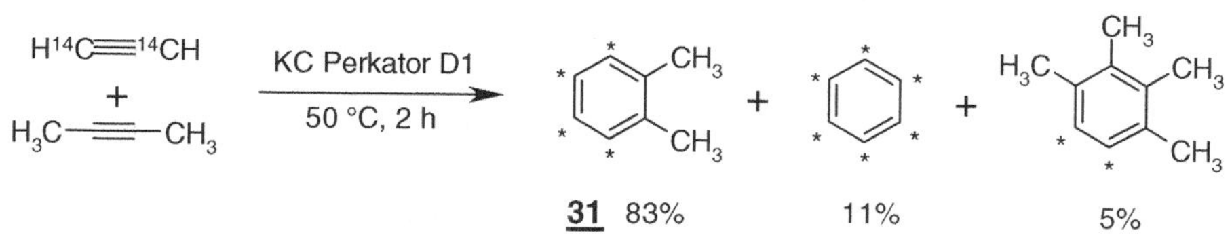

***Figure 8.9*** *Cotrimerization of [$^{14}C_2$]acetylene with 2-butyne*

Mono- or difunctionalizations of [U-$^{14}$C]benzene followed by various functional group interconversions and additional substitutions enable the preparation of a vast range of differently functionalized ring-labeled species; this is by far the most frequently used general route. The alternative approach, discussed in Chapter 6, Sections 6.3.2 and 6.5, can furnish [$^{14}$C]benzene derivatives with up to four substituents in one step by condensation of low molecular weight ketones or $\beta$-keto esters with sodium nitromalondialdehyde, either or both of which may be isotopically labeled. Surprisingly, very few examples have been reported so far utilizing [2 + 4]cycloaddition (Diels–Alder) reactions, although these techniques have found wide application in the synthesis of substituted unlabeled aromatic compounds and some labeled with stable isotopes. With the exception of a few more recent techniques (e.g. directed *ortho* metalations (DoM; Snieckus[28]), transition metal catalyzed reactions (e.g. Heck[29], Stille[30], Suzuki[31], Buchwald[32] reactions) or the use of highly functionalized organomagnesium reagents (Knochel[33]) most procedures are based on classical organic methods such as electrophilic and nucleophilic aromatic substitutions (see Figure 8.10). Interestingly, transition metal-catalyzed reactions have made [ring-$^{14}$C]aryl halides highly valuable precursors, since they permit the introduction of label at a very late synthesis step in contrast to classical pathways (see Sections 7.8.3 and 7.8.4 below). Subsections below correlate with the routes indicated in Figure 8.10.

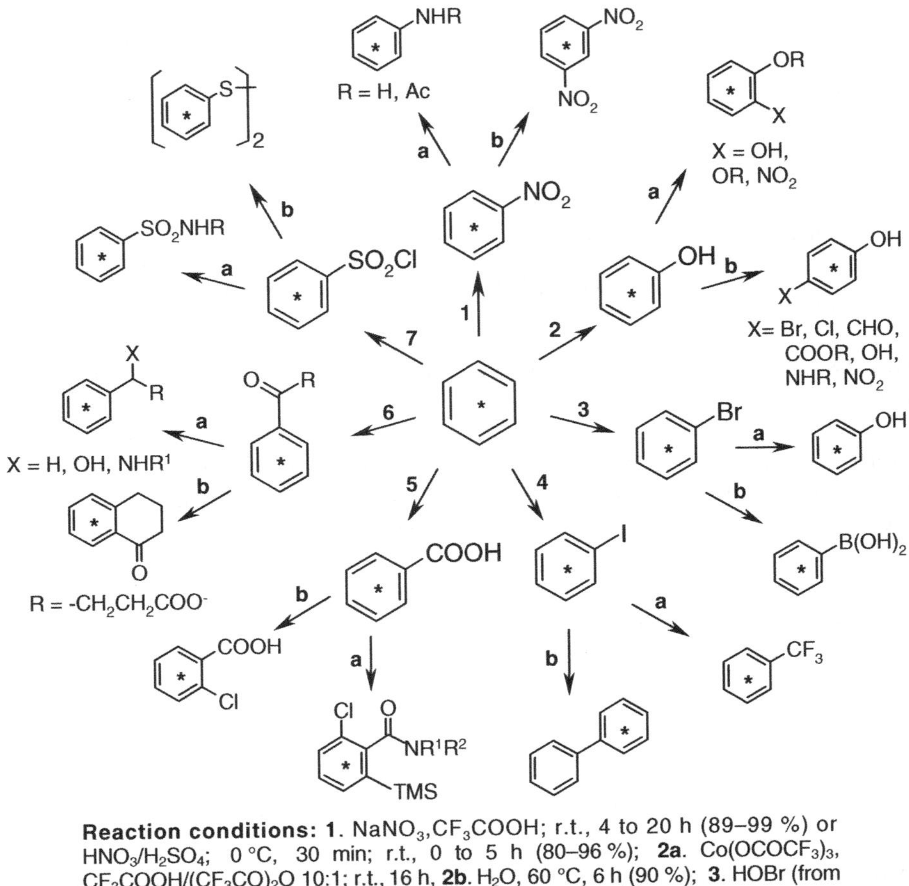


**Reaction conditions:** **1**. $NaNO_3$,$CF_3COOH$; r.t., 4 to 20 h (89–99 %) or $HNO_3/H_2SO_4$; 0 °C, 30 min; r.t., 0 to 5 h (80–96 %); **2a**. $Co(OCOCF_3)_3$, $CF_3COOH/(CF_3CO)_2O$ 10:1; r.t., 16 h, **2b**. $H_2O$, 60 °C, 6 h (90 %); **3**. HOBr (from $Br_2$, HgO, $H_2O$); r.t., 14 h (67 %) or 48 % HBr, $Bu_4N^+Br^-$, 30 % $H_2O_2$; 62–64 °C, 2.7 h (55–71 %) or $Br_2$, aq. $HNO_3$, $AgNO_3$; 0 °C, 5 h (67 %); **4**. $I_2$, 45% $HNO_3$; 85–90 °C, 1.5 h (75–80 %); **5a**. $(COCl)_2$, $AlCl_3$, $CS_2$; reflux, 3 h, **b**. 2.5N NaOH, 80 °C, 16 h (66 %) or **a**. 1,2-(dichloromethylendioxy)benzene, $AlCl_3$, $CH_2Cl_2$; r.t., 5 min, 100 °C, 5 min, **b**. 20 % KOH, reflux, 30 min (76 %); **6a**. RCOCl, $AlCl_3$, $ClCH_2CH_2Cl$; r.t., 15 h or neat components, 55°C, 2 h, **b**. succinic anhydride or acetic anhydride $AlCl_3$, $CHCl_2CHCl_2$; r.t., 16 h, 50°C, 5 h, **c**. X=H: Zn-Hg, conc. HCl/HOAc 16:1, toluene; reflux, 48 h or $H_2$, 10% Pd/C, HOAc, 75°C, 3 h (65%) or $NH_2NH_2.H_2O$, ethylene glycol, 95°C, 17 h; X=NH: $H_2NOMe{\cdot}HCl$, pyridine, EtOH; r.t., 20 h; $BH_3$.THF; (1*S*,2*R*)-1-aminoindano-2-ol, THF; -20°C, 15 h; r.t., 16h; 70°C, 4 h; **7**. $ClSO_3H$, $CCl_4$, r.t. or fuming $H_2SO_4$, 60 °C, 2.5 h, $SOCl_2$, 60 °C (>95 %)

***Figure 8.10*** *Basic procedures for the synthesis of substituted [ring-$^{14}$C]aromatic compounds from [U-$^{14}$C]benzene*

### 8.8.1 Nitrobenzene Branch

Nitration of [U-$^{14}$C]benzene with $NaNO_3$ in trifluoroacetic acid[27a,34c–e] or $HNO_3$/$H_2SO_4$[1b,28b,36] has for many years been the main approach to a broad variety of substituted ring-labeled aromatic compounds. Subsequent catalytic reduction with Pd/C in trifluoroacetic acid, methanol, ethanol, 5% HCl/ethanol or acetic anhydride furnishes [U-$^{14}$C]-aniline[1b,27a,34d,35] and [ring-$^{14}$C]acetanilide[34c], respectively, in yields ranging from 78–98% (Route 1a in Figure 8.10). A broad spectrum of classical functional group interconversions and substitution reactions has been introduced that permit conversion of both key intermediates into numerous mono-, di-, tri-, and tetra-substituted ring-labeled derivatives in satisfactory yields. Nitration under more aggressive reaction conditions (fuming nitric acid, elevated reaction temperatures) provides 1,3-dinitro[U-$^{14}$C]benzene[36] which can be either selectively or completely reduced to 3-nitro[U-$^{14}$C]aniline[37a,b] and 1,3-[U-$^{14}$C]phenylenediamine[36], respectively, both of which have been used for the preparation of *meta*-disubstituted derivatives.

Classical diazonium salt formation by reaction of [U-$^{14}$C]aniline with $NaNO_2$ in acidic medium and subsequent replacement of the diazonium group by various functional groups (OH, F, Cl, Br, I, CN) have been routinely applied in the preparation of monosubstituted [U-$^{14}$C]benzene derivatives. Controlled thermal decomposition of benzenediazonium tetrafluoroborate, for example, was exploited to prepare fluoro[U-$^{14}$C]benzene[1b], while heating of the corresponding diazonium sulfate or iodide in aqueous acidic medium provides [U-$^{14}$C]phenol[34a,b,c] and iodo[U-$^{14}$C]benzene[38], respectively, in good yields. The classical Sandmeyer reaction, in which the replacement of the diazonium group by halogen or cyanide is accomplished in the presence of the corresponding copper(I) salt, has been used for the preparation of chloro[U-$^{14}$C]benzene[39a], and [ring-U-$^{13}$C]benzonitrile[40]. Hydrolysis of the nitrile has been introduced as one of the standard routes to [ring-U-$^{14}$C]benzoic acid. Reduction of the diazonium group with $SnCl_2$ provides [U-$^{14}$C]phenylhydrazine, which serve as a key intermediate in the synthesis of substituted [$^{14}C_n$]indoles[41a]. However, in the meantime most of these reactions have been replaced by methods that allow direct mono-functionalization of [U-$^{14}$C]benzene (Branches 2–7 in Figure 8.10). These and further functionalizations will be discussed in subsequent sections. Treatment of anilines with $BCl_3$ and nitriles in the presence of an appropriate second Lewis acid (e.g. $AlCl_3$, $ZnCl_2$) gives rise to the regioselective formation of *o*-amino aryl alkyl/aryl ketones. With chloroacetonitile, *o*-chloroacetanilide is obtained, which upon reductive cyclization and elimination of water produces 2,3-unsubstituted indoles[41b,c]. Applied to [U-$^{14}$C]aniline, this procedure furnishes [phenyl ring-U-$^{14}$C]indole in an overall radiochemical yield of 40%[42] (Figure 8.11).

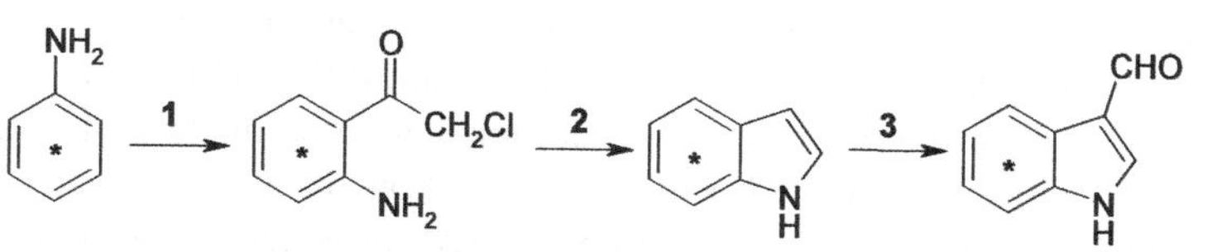


**Reaction conditions: 1a.** $ClCH_2CN$, $BCl_3$, $ZnCl_2$, $ClCH_2CH_2Cl$; 75 °C, 2h, **b.** 1N HCl; 75 °C, 1 h; **2.** $NaBH_4$, dioxane–$H_2O$ 1:1; reflux, 1 h; **3.** DMF, $POCl_3$; 35 °C, 1 h

***Figure 8.11*** *Use of [U-$^{14}$C]aniline in the preparation of [phenyl ring-U-$^{14}$C]indole*

The free amino group strongly activates the aromatic ring towards electrophilic attack, thereby making such reactions difficult to stop at the monosubstitution stage. Accordingly, reaction of [U-$^{14}$C]aniline with *N*-chlorosuccinimide in dichloromethane, or with bromine in acetic acid, results in the formation of 2,4,6-trihalogenated species[43]. Nevertheless, halogenation with a limited amount of *N*-bromosuccinimide in benzene at ambient temperature gives a 3:5 mixture of *p*- and *o*-bromo[U-$^{14}$C]aniline, which can be easily separated by column chromatographic methods. Reductive de-diazoniation of the *o,p* mixture with hypophosphorous acid has been recommended as a good alternative route to [U-$^{14}$C]bromobenzene, giving higher overall yields (75%) and purity than the direct Sandmeyer reaction of diazotized [U-$^{14}$C]aniline with CuBr (35%)[39b]. *N*-Acetylation of [U-$^{14}$C]aniline with acetic anhydride in $CCl_4$, benzene or acetic acid not only improves its radiolytic stability, but also lowers the activation power of the free amine group. Accordingly, chlorination of acet[U-$^{14}$C]anilide with $Bu_4N^+(ICl_4)^-$[44] in dichloromethane or with one equivalent of NCS in acetic acid[43a] furnishes a mixture of *p*- and *o*-chloroacet-[U-$^{14}$C]anilide in ratios of 2.5:1 to 4:1. Employment of a larger excess of NCS (3.3 equivalents) leads to the formation of 2,4-dichloroacet[U-$^{14}$C]anilide in 96% yield[43a]. Halogenation with bromine in acetic acid[39a] or NBS in $CCl_4$[45] affords *p*-brominated product in yields ranging from 88% to 94% and in more than 99% isomeric purity. Reaction of a solution of acet[U-$^{14}$C]anilide in acetic acid/$H_2SO_4$ 1:2 with conc. $HNO_3/H_2SO_4$ 3:2 affords preferentially *p*-nitroacet[U-$^{14}$C]anilide in 60% yield[46], while $Cu(NO_3)_2 \cdot 3H_2O$ in HOAc produces a 3:1 mixture of *o*- and *p*-nitroacet[U-$^{14}$C]anilide in 85% yield, which has to be separated by fractional crystallization[47]. Deprotection with refluxing HCl or by heating with 5% aqueous NaOH liberates the amine function, allowing for further functional group interconversions. Sandmeyer reaction applied to 2-nitro[U-$^{14}$C]-aniline or 2-bromo- and 4-bromo[U-$^{14}$C]aniline, for example, facilitates the synthesis of 1,2-dinitro-[U-$^{14}$C]benzene[48], 1,2-dibromo[U-$^{14}$C]benzene[39b], 2-bromo[ring-U-$^{14}$C]-benzonitrile[39b], 1,4-dibromo[U-$^{14}$C]benzene[45], *p*-chlorobromo[U-$^{14}$C]benzene[39a], 4-chloro[ring-$^{14}$C]-benzoic acid[39a], and 4-chloro[U-$^{14}$C]phenylboronic acid[49]. The latter was coupled in a Suzuki reaction with 2-chloronitro[U-$^{14}$C]benzene to give 4-chloro-5′-nitro[U-$^{14}$C]-biphenyl, a key intermediate in the synthesis of [$^{14}$C]BAS 510F (**32**), a novel fungicide (Figure 8.12). 1,2-Dinitro[U-$^{14}$C]benzene was converted in three steps to 4-amino-3-fluoro [U-$^{14}$C]phenol by nucleophilic displacement of one nitro group with $Bu_4N^+F^-$, reduction of the resulting 2-fluoronitro[U-$^{14}$C]benzene and acid-catalyzed rearrangement[49a].

Since the diazonium group can be removed reductively with aqueous $H_3PO_2$ or with $NaBH_4$ in nonaqueous solution, 2,4-disubstituted[U-$^{14}$C]anilines can be readily converted into the corresponding *m*-substituted [U-$^{14}$C]benzenes. This facilitated, for example, the synthesis of 1,3-dichloro[U-$^{14}$C]benzene from 2,4-dichloro[U-$^{14}$C]aniline in a 60% yield[50].

Functional group interconversions on 3-nitro[U-$^{14}$C]aniline or [U-$^{14}$C]phenylene-diamine, both readily available by reduction of 3,5-dinitro[U-$^{14}$C]benzene (Route 1b in Figure 8.10), open up an alternative route to additional *m*-substituted ring-labeled building blocks. Cuprous chloride-mediated Sandmeyer reaction on 3-nitro[U-$^{14}$C]aniline, followed by reduction of the nitro group with $SnCl_2$/HCl provides 3-chloro[U-$^{14}$C]aniline. Halogenation of the corresponding *N*-acetyl derivative with NCS in acetic acid gives a 2:1 mixture of 3,4- and 2,5-dichloroacet[U-$^{14}$C]anilide in 90% overall yield. HPLC separation and liberation of the amino function of the 3,4 -isomer with refluxing ethanolic HCl facilitates further transformations[51]. Diazonium salt formation and subsequent iodo-dediazoniation furnishes 1,2-dichloro-4-iodo[U-$^{14}$C]benzene, which was used, for example,

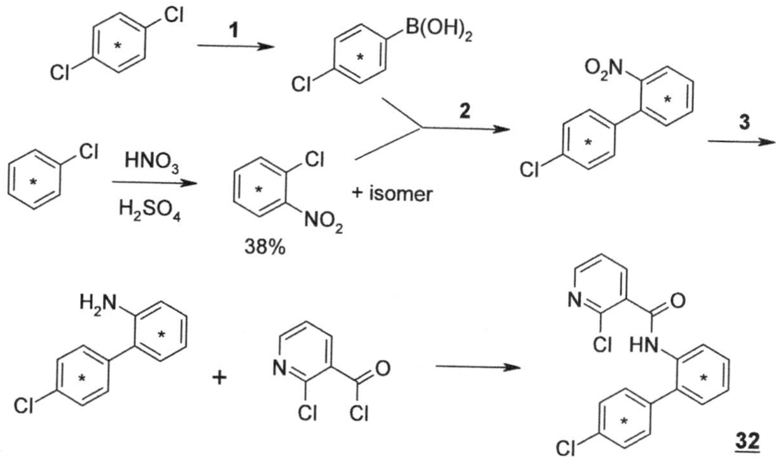


**Reaction conditions:** **1a**. Mg, ether; reflux, **b**. $B(OMe)_3$; –70 °C, 30 min; 0°C, 30 min, **c**. aq. HCl; r.t.; **2**. $(Ph_3)_4Pd$, 2N aq. $Na_2CO_3$, toluene-ethanol; 80°C, 16 h; **3**. $MoO_3$,$Na_2SeO_3$, $NaBH_4$

***Figure 8.12*** *Preparation of a functionalized [U-$^{14}$C]biphenyl through Suzuki coupling of substituted [U-$^{14}$C]phenylboronic acids*

as a key intermediate in the Cu powder-mediated Ullman synthesis of 3,3′,4,4′,5-pentachloro-[1′,2′,3′,4′,5′,6′-$^{14}C_6$]biphenyl[52].

Subjection of 3-chloro[U-$^{14}$C]aniline to a second CuCl-mediated Sandmeyer reaction furnishes 1,3-dichloro[U-$^{14}$C]benzene[37a], which is also directly available from [U-$^{14}$C]-phenylendiamine[36]. Deprotonation with *n*-butyllithium in THF at −78 °C followed by carboxylation affords 2,6-dichloro[U-$^{14}$C]benzoic acid in approximately quantitative yield. Subsequent nitration with $HNO_3$/$H_2SO_4$ gives 2,6-dichloro-3-nitro[ring-U-$^{14}$C]-benzoic acid, a key intermediate in the synthesis of [$^{14}$C]CI-958 (**33**), an anticancer agent[36] (Figure 8.13).

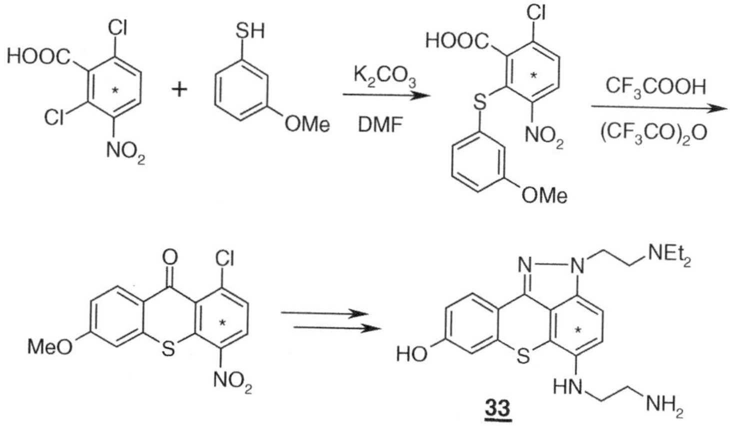


***Figure 8.13*** *Preparation and utilization of 3-nitro-2,6-dichloro[ring-U-$^{14}$C]benzoic acid*

3-Nitro[U-$^{14}$C]acetanilide may also serve as a starting material for 2,3-dichloro[U$^{14}$C]-aniline. Treatment with $HNO_3$ at ambient temperature affords 2,3-dinitroacet[U$^{14}$C]-anilide. Liberation of the amine function and subsequent CuCl-mediated Sandmeyer reaction occurs with simultaneous nucleophilic replacement of the nitro group at C2. The sequence is completed by reduction of the resulting 2,3-dichloro-nitro[U-$^{14}$C]benzene with Sn/HCl[37b].

### 8.8.2 Phenol Branch

The acyloxylation of [U-$^{14}$C]benzene using cobalt(III) trifluoroacetate in trifluoroacetic acid/trifluoroacetic anhydride is one of the more modern synthetic alternatives to the classical route to monosubstituted labeled benzenes[53a]. Addition of water hydrolyzes the resulting trifluoroacetoxy derivative to give [U-$^{14}$C]phenol in 90% radiochemical yield. Etherification with methyl iodide/$K_2CO_3$ or dimethoxymethane in the presence of *p*-TosOH[54] makes possible directed *ortho* metallation with *n*-BuLi and subsequent 1,2-functionalization by transmetallation with *n*-BuMgCl and reaction with oxygen (in the first instance) and reaction of the aryllithium intermediate with methyl iodide (in the second). Concluding deprotection of the 2-substituted phenol ethers with 2N HCl or trimethylsilyl iodide results in the formation of free [U-$^{14}$C]catechol and [U-$^{14}$C]cresol, respectively. Futher *ortho* functionalizations using D*o*M techniques have not been reported so far, although a broad spectrum of 1,2-disubstituted ring-labeled benzenes should become available in excellent regioselectivity and chemical yields[28,55]. A more recent and higher yielding [U-$^{14}$C]catechol synthesis involves *ortho* hydroxylation of (2-benzoyl-5-nitro)-phenyl [U-$^{14}$C]phenyl ether (**34**); subsequent cleavage with piperidine provided the labeled catechol in about 80% yield from [U-$^{14}$C]phenol[56].

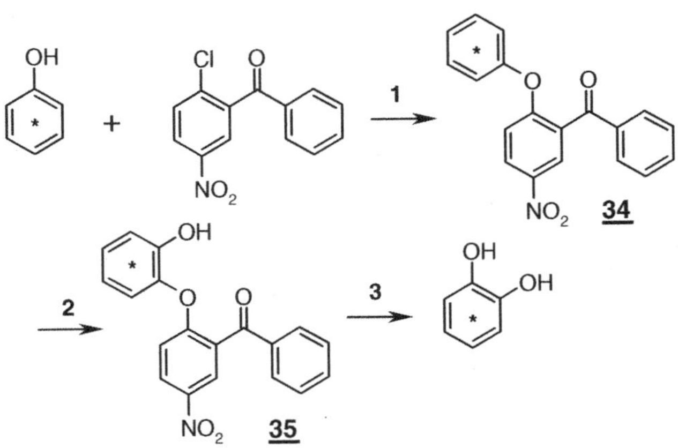


**Reaction conditions: 1**. NaH; hexane/THF 1:3; 70 °C, 8.5 h; **2**. 35 % $H_2O_2$, conc. $H_2SO_4$–HOAc 1:3; r.t., 60 min; **3**. piperidine; 110 °C, 90 min

***Figure 8.14*** *Ortho-hydroxylation of [U-$^{14}$C]phenol*

Alternatively, etherification of 35 with diazomethane enables the preparation of [ring-U-$^{14}$C]guaiacol. Repetition of the hydroxylation, *O*-methylation and release sequence allows the conversion of [U-$^{14}$C]phenol into [U-$^{14}$C]pyrogallol dimethyl ether (2,6-dimethoxy[U-$^{14}$C]phenol) in 5–10% overall radiochemical yield[57].

Nitration of [ring-U-$^{14}$C]anisole with $HNO_3$ in acetic anhydride at 5–10 °C affords preferentially the *ortho* derivative in isomeric ratios of 2.5:1 to 7.5:1 and 60–80% yield[27c,56b]. Reduction of this product with $SnCl_2$/HCl, hydrazine-Raney Ni or $H_2$-Pd/C, diazonium salt formation, and subsequent functional group interconversions on the diazonium group opens up access to various *o*-substituted [U-$^{14}$C]phenols and anisoles. In contrast, treatment of [U-$^{14}$C]phenol with dilute $HNO_3$ at ambient temperature furnishes mixtures of labeled *o*- and *p*-nitrophenol in poor regioselectivity, although ratios can be altered somewhat by choice of nitration conditions[58]. Under more vigorous conditions 2,4-dinitro[U-$^{14}$C]phenol is formed. With $Cu(NO_3)_2$ in acetic acid as a nitration reagent, the ratio of the isomers becomes reversed (*o*/*p* ratio from 2:3 to 1:2)[34c]. The regioselective preparation of *p*-nitro-[U-$^{14}$C]phenol following this approach has not been reported so far. Isomerically pure *p*-nitro[$^{14}C_n$]phenol ($n = 1$–3), however, is formed by condensation of [$^{14}C_n$]acetone with sodium nitromalondialdehyde (Chapter 6, Section 6.3.2).

The range of procedures offering regioselective functionalizations of [U-$^{14}$C]phenol is only complete with the inclusion of methods that furnish preferentially or exclusively *p*-substituted derivatives. Treatment of [U-$^{14}$C]phenol with $CCl_4$ and 20% aqueous NaOH in the presence of $\beta$-cyclodextrin and copper powder furnishes, for example, exclusively *p*-hydroxy[ring-U-$^{14}$C]benzoic acid in radiochemical yields of 45–75%[59]. Subsequent esterification with diazomethane and treatment with sulfuryl chloride affords either the 2-chloro or the 2,6-dichloro derivatives depending on the reaction temperature and the excess of sufuryl chloride employed[59a].

$SnCl_4$-catalyzed Friedel–Crafts alkylation of [4-$^{14}$C]phenol, [U-$^{14}$C]guaiacol, and [U-$^{14}$C]catechol with dichloromethyl methyl sulfide provides the respective hydroxylated [ring-U-$^{14}$C]benzaldehydes in radiochemical yields in the range of 50–60%, contaminated only with traces of their *ortho* isomers in the case of 4-hydroxy[ring-U-$^{14}$C]benzaldehyde and [ring-U-$^{14}$C]protocatechualdehyde, but with 15% [ring-U-$^{14}$C]-isovanillin in the case of [ring-U-$^{14}$C]vanillin, which, however, could be easily separated by routine column chromatography[27c]. This outcome contrasts significantly with the low regioselectivity of the Gattermann formylation of [ring-U-$^{14}$C]anisole with $Zn(CN)_2$/HCl in the presence of $AlCl_3$, for which a *p*/*o* ratio of only 1.5:1 has been reported[60]. The latter findings disagree with results of other authors who observed a much higher regioselectivity (*p*/*o* ratio of 8:1)[61].

*p*-Halogenated [U-$^{14}$C]phenols can also be prepared in high regioselectivity. [ring-U-$^{14}$C]Anisole, for example, reacted with HCl and *m*-chloroperbenzoic acid in DMF to give its 4-chloro derivative in 90% regioselectivity. The corresponding *ortho* isomer was formed in only 9%. For free [U-$^{14}$C]phenol, however, the regioselectivity of this reaction was much less pronounced, the *p*/*o* ratio being 2.3:1. Increasing amounts of HCl and *m*-chloroperbenzoic acid produced 2,4-dichloro[U-$^{14}$C]phenol in 84% regioselectivity and 61% radiochemical yield, based on [U-$^{14}$C]phenol[34a,62].

A much higher regioselectivity was observed when [U-$^{14}$C]phenol reacted with equimolar amounts of bromine in carbon disulfide. The crude reaction mixture contained only 3% of the *ortho* isomer, which could be easily removed by column chromatography to give the *para* derivative in 93% yield. It was used to direct the *O*-acyl migration of the respective [U-$^{14}$C]phenyl acetate exclusively into one of the *ortho* positions (Fries rearrangement) and thus to avoid the formation of a mixture of *o*- and *p*-acetophenones as had been observed in a previous experiment[34b].

Iodination of [ring-U-$^{14}$C]anisole, as well as of the analogous benzyl ether, with equimolar amounts of iodine and HgO occurs selectively at C4 in yields exceeding 90%. No trace of the *ortho* isomer could be detected by $^{1}$H-NMR in these cases[63]. This pronounced *para* selectivity has also been reported for a series of differently substituted unlabeled phenyl ethers[64]. The $PdCl_2[P(o\text{-tolyl})_3]_2$-catalyzed cross coupling of *p*-iodo-[U$^{14}$C]phenyl benzyl ether with a zinc intermediate **36** derived from *N*-protected (*S*)-*β*-iodoalanine methyl ester was recently exploited for a short elegant synthesis of (*S*)-[ring-U-$^{14}$C]tyrosine (Figure 8.15)[63].

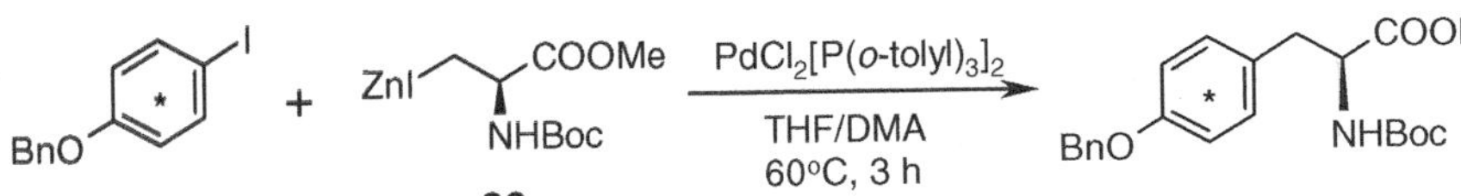

**36**

***Figure 8.15*** *Palladium-catalyzed cross coupling with 1-benzyloxy-4-iodo[U-$^{14}$C]benzene*

### 8.8.3 Bromobenzene Branch

While there has as yet been no report of a simple, high-yield method for monochlorination of [U-$^{14}$C]benzene, three different procedures for the efficient preparation of bromo-[U-$^{14}$C]benzene exist. The first procedure involves the reaction of [U-$^{14}$C]benzene with a aqueous hypobromite solution freshly prepared from bromine and yellow HgO at ambient temperature for 14 h. The crude product, isolated from the reaction mixture in almost quantitative yield, consisted of 85% mono- and 15% *p*-dibromo[U-$^{14}$C]benzene. This mixture was used without separation for the CuO-catalyzed hydrolysis with 4N $Na_2CO_3$ at 250 °C to give [U-$^{14}$C]phenol in 80% yield[61]. The second procedure used elemental bromine, generated *in situ* from 48% HBr and $H_2O_2$. After addition of $Bu_4N^+Br^-$ the reaction mixture was kept at 65 °C for 3 h and treated with additional amounts of bromine. The crude monobromo product contained 17% 1,4- and 2% 1,2-dibrominated materials, which were removed by fractional distillation to furnish pure bromo[U-$^{14}$C]benzene in radiochemical yields of 55–71%[65]. In the third procedure [U-$^{14}$C]benzene was kept in aqueous $HNO_3$ for 5 h at 0 °C with a slight excess of 1:1 bromine and silver nitrate. Steam distillation furnished pure monobromo product in 63% yield[66]. In an illustration of the potential of transition metal-catalyzed reactions in the synthesis of carbon-14-labeled compounds from [$^{14}$C]bromobenzene, Figure 8.16 shows the conversion of bromo[U-$^{14}$C]-benzene to [U-$^{14}$C]phenylboronic acid, followed by $Pd(Ph_3)_4$-catalyzed cross coupling with the diprotected L-phenylalanine derivative **37** to give (*S*)-3-phosphonomethyl-5-[phenyl-U-$^{14}$C]phenylalanine (**38**)[67].

The palladium-catalyzed amination of bromo[U-$^{14}$C]benzene with 2,6-dichloroaniline to afford 2,6-dichloro-*N*-[U-$^{14}$C]phenylaniline (**39**), an intermediate in the synthesis of [$^{14}$C]diclofenac (Figure 8.17) is one of the very few applications in carbon-14 synthesis of Buchwald's reaction[68], which enables the preparation of [$^{14}$C]arylamines from [$^{14}$C]aryl

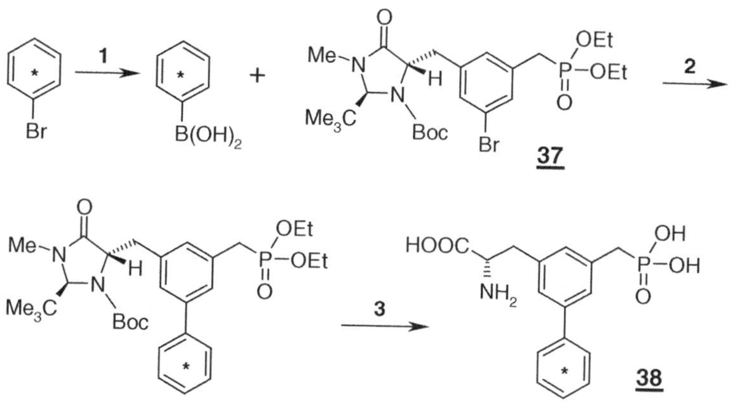


**Reaction conditions:** **1a**. Mg, THF, reflux, 1 h, **b**. $B(OMe)_3$; −70 °C, 30 min, 0 °C, 30 min; **c**. HCl, $H_2O$; r.t. **2**. $(Ph_3)_4Pd$, 2N aq. $Na_2CO_3$, toluene/ethanol 6:1; 85 °C, 24 h; **3a**. $CF_3COOH$, $CH_2Cl_2$ r.t., 16 h, **b**. $Me_3SiBr$, $CH_2Cl_2$ r.t., 16 h, **c**. 0.75 N HCl, Dowex 50, 115–120°C, 48 h

***Figure 8.16*** *Preparation of a [U-$^{14}$C]phenyl-substituted phenylalanine derivative via Suzuki coupling*

halides in excellent yields while tolerating $NO_2$, CN, and COOR substituents. The amination of 1,2-dichloro-3-iodo[U-$^{14}$C]benzene with piperidine to give **40** may emphasize the potential of this methodology for radiosynthetic use[69]. Zn(HMDS)$_2$, which would afford *N*-unsubstituted [$^{14}$C]anilines, has not been tested so far as coupling reagent in carbon-14 labelling[32a,70]. Expansion of the Buchwald reaction to the arylation of amides fails. An analogous coupling can, however, be achieved by replacing Pd-catalysts by copper (Gomberg variant), as demonstrated with the synthesis of 1-[U-$^{14}$C]phenylindole (**41**).

### 8.8.4 Iodobenzene Branch

In a procedure recommended for the preparation of iodo[U-$^{14}$C]benzene, a mixture of [U-$^{14}$C]benzene and 1.8 equivalents of iodine in 45% nitric acid was heated at 85 °C for 1.5 h. Vacuum distillation of the reaction product furnished a 4:1 mixture of iodo[U-$^{14}$C]-benzene and [U-$^{14}$C]benzene in almost quantitative radiochemical yield. Without separation into its components, the mixture was subjected to a copper catalyzed Ullman reaction at 288–290 °C to give [U-$^{14}$C]biphenyl in 65% overall radiochemical yield[71].

Iodo[U-$^{14}$C]benzene was used as a starting material for the labeling of trifusal, an antithrombotic agent. For this purpose it was coupled with trifluoromethyl iodide in DMF in the presence of copper at 130 °C for 20 h to give trifluoromethyl[U-$^{14}$C]benzene in 64% yield. Subsequent nitration ($NaNO_2$, $CF_3SO_3H$), reduction of the resulting *m*-nitro group (Fe, conc. HCl) followed by diazonium salt formation and copper(II) nitrate/copper-(I) oxide-assisted hydroxy-dediazoniation afforded 3-trifluoromethyl[U-$^{14}$C]phenol in 70% yield. Kolbe reaction with $CO_2/K_2CO_3$ and acetylation of the phenolic hydroxyl group completed the synthesis sequence[38].

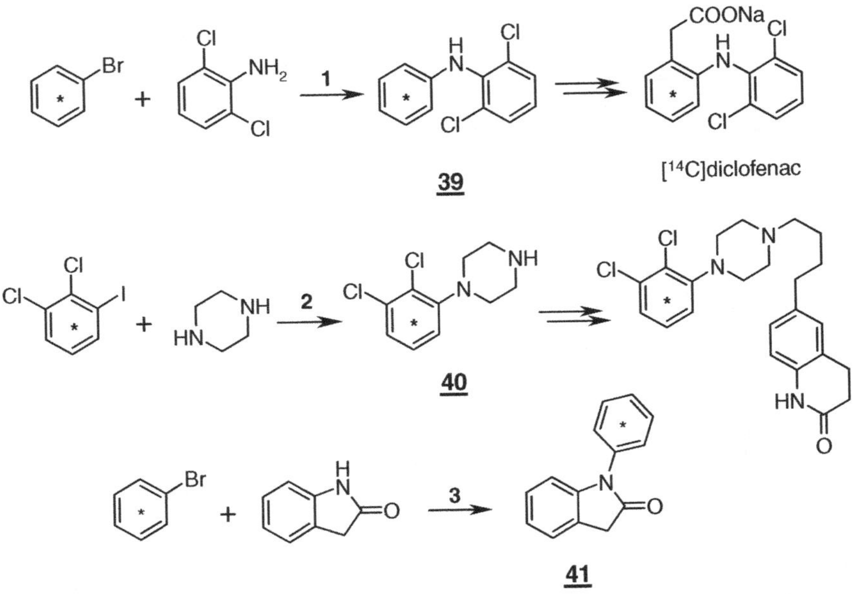


**Reaction conditions:** **1**. $Pd(OAc)_2$, xantphos, *tert*-BuONa, DMF; 130 °C (microwave, 120 W), 13 min; **2**. $Pd_2(dba)_3$,(*S*)-BINAP,*tert*-BuONa, toluene; 120 °C, 45 h; **3**. CuI, $MeNH(CH_2)_2NHMe$, $K_3PO_4$, dioxane; reflux, 16 h

***Figure 8.17*** *Palladium-catalyzed amination of [$^{14}$C]aryl halides (Buchwald reaction)*

### 8.8.5 Benzoic Acid Branch

Two standard procedures have been developed to directly produce ring-labeled benzoic acid in high yields. In the first case [U-$^{14}$C]benzene is treated with a suspension of oxalyl chloride and $AlCl_3$ in $CS_2$ in a Friedel–Crafts acylation. The resulting $\alpha$-keto acid decarbonylates upon heating in 2.5 N NaOH to give [ring-U-$^{14}$C]benzoic acid in 66% overall yield[72]. Comparable yields of 70–80% were achieved when [U-$^{14}$C]benzene was treated with 1,2-(dichloromethylenedioxy)benzene in dichloromethane in the presence of $AlCl_3$, followed by saponification of the resulting catechol mono[ring-U-$^{14}$C]benzoate with 20% KOH[27c,73].

*Ortho*-Thallation of [ring-U-$^{14}$C]benzoic acid with thallium(III) trifluoroacetate in $CF_3COOH$ and subsequent halo-dethallation with KI in water[74] or $Cu_2Cl_2{\cdot}2H_2O$ in acetonitrile[75] furnishes *ortho*-halogenated [ring-$^{14}$C]benzoic acid derivatives in fair to good yields, 25% to 75%. The chloro-dethallation sequence was exploited in the synthesis of [$^{14}$C]SK&F 86466[72].

Directed *ortho* metallation processes offer an alternative route to 2-substituted [ring-U-$^{14}$C]benzoic acid derivatives[28], as demonstrated by the synthesis of *N*-ethyl-2-chloro-6-trimethylsilyl[ring-U-$^{14}$C]benzamide (**43**)[76] (Figure 8.18). The carboxyl group of [ring-U-$^{14}$C]benzoic acid was first converted to a tertiary amide (**42**), a directing metallation group. Sequential *ortho* deprotonation with *s*-BuLi in the presence of TMEDA, trapping of the resulting lithiated species with trimethlysilyl chloride, repeated deprotonation with *sec*-BuLi/TMEDA, followed by halo-delithiation with hexachloroethane and concluding acid-catalyzed hydrolysis of the masked hemiaminal furnished the target 2,6-disubstituted [U-$^{14}$C]benzamide derivative **43**.

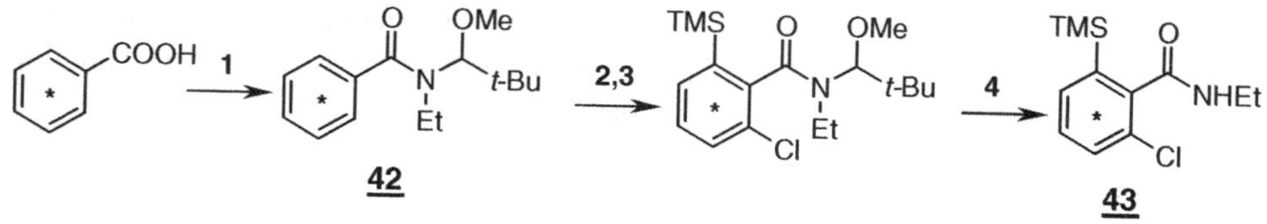

**Reaction conditions: 1a.** $(COCl)_2$, cat. DMF; $CH_2Cl_2$; r.t., 30 min, **b.** *tert*-BuCH=NEt, $CH_2Cl_2$, r.t., 1.25 h, **c.** MeOH, $Et_3N$; r.t., 45 min; **2a.** *s*-BuLi, TMEDA, THF; –78 °C, 90 min, **b.** TMSCl, –78 °C, 15 min; **3a.** *s*-BuLi, TMEDA, ether; –78 °C, 15 min, –25 °C, 1 h, **b.** $CCl_3CCl_3$, ether; –78 °C, 10 min; **4.** HCl, HOAc; r.t., 45 min

***Figure 8.18*** *Directed* ortho *metallation in the preparation of functionalized [ring-U-$^{14}$C] benzoic acid derivatives*

### 8.8.6 Alkyl Phenyl Ketone Branch

Friedel-Crafts acylation of [U-$^{14}$C]benzene with acyl halides or anhydrides provides alkyl [U-$^{14}$C]phenyl ketones which have found wide application in carbon-14 synthesis. Huang-Milon reduction of the carbonyl function with hydrazine, for example[77], converted undecyl-[U-$^{14}$C]phenyl ketone into *n*-dodecyl[U-$^{14}$C]benzene, which upon sulfonation gave *n*-dodecyl [U-$^{14}$C]benzene-4-sulfonic acid, a labeled anionic surfactant[77a]. Enantioselective reduction of the oxime methyl ether, obtained from [ring-U-$^{14}$C]acetophenone and *O*-methylhydroxylamine, yielded (*S*)-(-)-1-[ring-U-$^{14}$C]phenylethylamine of 97% enantiomeric purity. Subsequent purification and coupling to the carboxyl group of *N*-substituted (*R*)-$\alpha$-methyltryptophan afforded [$^{14}$C]CI-1021, an antiemetic drug candidate (see also Chapter 11, Section 11.2.2)[78]. The $\alpha$-aminoketone accessible through $\alpha$-bromination of [ring-U-$^{14}$C]acetophenone and alkylation of the resultant [ring-U-$^{14}$C]phenacyl bromide with 2-aminothiazoline served as a key intermediate to labeled levamisole, an adjuvant agent with 5-fluorouracil in the treatment of human colon cancer[79]. With excess halogen (e.g. chlorine) trichloro[ring-U-$^{14}$C]acetophenone could be generated[80].

4-[ring-U-$^{14}$C]Phenylbutyric acid has been a frequent intermediate in the synthesis of [ring-U-$^{14}$C]phthalic acid, 1-[phenyl-U-$^{14}$C]tetralone and [1,1a,2,3,4,4a-$^{14}$C]naphthalene and related derivatives. It is available in two steps by $AlCl_3$-catalyzed acylation of [U-$^{14}$C]-benzene with succinic anhydride, followed by reduction of the carbonyl function, either with Zn–Hg and HCl or catalytically with 10% Pd/C in HOAc; overall radiochemical yields are about 65%. Cyclization to 1-[phenyl-U-$^{14}$C]tetralone is mediated by trifluoroacetic anhydride or polyphosphoric acid. Oxidation of the latter with alkaline $KMnO_4$ furnishes [ring-U-$^{14}$C]phthalic acid; subsequent treatment with refluxing thionyl chloride converts it into in its anhydride[34d,81]. Alternatively, 1-[phenyl-U-$^{14}$C]tetralone can be reduced ($LiAlH_4$ in ether or $NaBH_4$ in EtOH) to 1,2,3,4-tetrahydro[1a,4a,5,6,7,8-$^{14}$C]naphth-1-ol. Dehydration upon heating in dry dimethylsulfoxide and subsequent dehydrogenation with 10% Pd/C furnishes [1a,1,2,3,4,4a-$^{14}$C]naphthalene[82].

### 8.8.7 Sulfonylbenzene Branch

Chlorosulfonation of [U-$^{14}$C]benzene with chlorosulfonic acid in $CCl_4$ or in a two-step one-pot reaction with fuming sulfuric acid and subsequent treatment wth thionyl chloride furnishes [U-$^{14}$C]benzenesulfonyl chloride in nearly quantitative yield[79]. This was employed in the preparation of various labeled sulfonamides[83a,b]. Furthermore, reduction with $LiAlH_4$ affords di[U-$^{14}$C]phenyl disulfide which was used for the labelling of an insecticide of the alkylphosphonodithioate type[83c].

## References


1. a. Hyman, M.R. and Arp, D.J., *Analytical Biochem.*, ***1990***, *190*, 348–353; b. Anjaneyulu, B.; Maller, R.K.; Nagarajen, K.; Kueng, W. and Wirz, B., *J. Label. Compd. Radiopharm.*, ***1985***, *22*, 313–327; c. Tamers, M.A., *Int. J. Applied Radiat. Isotopes*, ***1975***, *26*, 676–680; d. Cox, J.D. and Warne, R.J., *J. Chem. Soc.*, ***1951***, 1893–1898; e. Murray, A. and Williams, D.L., *Organic Syntheses with Isotopes;* Interscience: New York, ***1958***, pp. 912–916; f. Bubner, M. and Schmidt, L., *Die Synthese Kohlenstoff-14 markierter Verbindungen;* Georg Thieme Verlag: Leipzig, ***1966***, pp. 103–104.
2. a. Würsch, J., *Atompraxis*, ***1961***, *7*, 463–465; b. Pichat, L. and Baret, C., *Tetrahedron*, ***1957***, *1*, 269–272; c. Whaley, T.W. and Ott, D.G., *J. Label. Compd. Radiopharm.*, ***1974***, *22*, 461–468; d. Evans, E.A., in *Isotopes: Essential Chemistry & Applications*, Elvidge, J.A. and Jones, J.R., Eds.; The Chemical Society: London, ***1979***, p. 44.
3. a. Riegel, B. and Liu, Y.C., *J. Org. Chem.*, ***1951***, *16*, 1610–1614; b. Murray, A. and Williams, D.L., *Organic Syntheses with Isotopes;* Interscience: New York, ***1958***, pp. 1087–1089.
4. Inhoffen, H.H.; Schwieter, U.; Chichester, C.O. and Mackinney, G., *J. Am. Chem. Soc.*, ***1955***, *77*, 1053–1057; b. Murray, A. and Williams, D.L., *Organic Syntheses with Isotopes*; Interscience: New York, ***1958***, pp. 1111–1114.
5. Muccino, R.R. and Wasiowich, C.A., *J. Label. Compd. Radiopharm.*, ***1980***, *17*, 463–469.
6. Rousseau, B.; Beaucourt, J.P. and Pichat, L., *J. Label. Compd. Radiopharm.*, ***1983***, *20*, 537–548; b. Rousseau, B.; Beaucourt, J.P. and Pichat, L., *Tetrahedron Lett.*, ***1982***, *23*, 2183–2186.
7. Porwoll, J.P. and Leete, E., *J. Label. Compd. Radiopharm.*, ***1985***, *22*, 257–271.
8. Pilgrim, W.R.; Nedoncelle, P.; Shroot, B.; Maignan, J. and Restlé, S., *J. Label. Compd. Radiopharm.*, ***1991***, *29*, 769–775.
9. Moenius, Th.; Voges, R.; Mahnke, M.; Burtscher, P.; Metz, Y. and Guenat, Ch., *J. Label. Compd. Radiopharm.*, ***1999***, *42*, 29–41.
10. Chabannes, B.; Pichat, L.; Herbert, M. and Pacheco, H., *J. Label. Compd.*, ***1965***, *1*, 102–108.
11. a. Hays, S.J.; Hicks, J.L. and Huang, C.C., *J. Label. Compd. Radiopharm.*, ***1987***, *24*, 1185–1193; b. Moenius, T.; Burtscher, P. and Voges, R., Internal Report, ***1982***, Sandoz Pharma AG, CH-4002, Basel; c. Williams, E.A.; Meikle, R.W. and Redemann, C.T., *J. Agr. Food Chem.*, ***1965***, *13*, 210–213; d. Murray, A. and Williams, D.L., *Organic Syntheses with Isotopes;* Interscience: New York, ***1958***, pp. 618–620; e. Ehrensvärd, G.; Reio, L.; Saluste, E. and Stjernholm, R., *J. Biol. Chem.*, ***1951***, *189*, 93–108; f. Cramer, R.D. and Kistiakowsky, G.B., *J. Biol. Chem.*, ***1941***, *137*, 549–555; g. Susan, A.B. and Duncan, W.P., *J. Label. Compd. Radiopharm.*, ***1981***, *18*, 1755–1758.
12. Elbert, T. and Filip, J., *Radiochem. Radioanal. Lett.*, ***1983***, *57*, 211–218.
13. Heys, J.R.; Rotert, G.A.; Hicks, J.L. and Chew, E.H., *J. Label. Compd. Radiopharm.*, ***1983***, *20*, 1171–1181.
14. a. Dawson, W.H. and Dunlap, R.B., *J. Label. Compd. Radiopharm.*, ***1979***, *16*, 335–339; b. Liebermann, D.F. and Yuan, Y.S., *J. Label. Compd. Radiopharm.*, ***1982***, *19*, 605–609; c. Wiggert, D. and Ackermann, P., Internal Report, ***1982***, CIBA-Geigy AG, CH-4002, Basel; d. Midland, M.M., *J. Org. Chem.*, ***1975***, *40*, 250–2252.
15. a. Topper, Y.S., *J. Biol. Chem.*, ***1949***, *177*, 303–307; b. Somayaji, V.V.; Hall, T.W.; Wiebe, L.I. and Knaus, E.E., *J. Label. Compd. Radiopharm.*, ***1988***, *25*, 1325–1331; c. Nystrom, R.F.; Loo, Y.H.; Mann, K.M. and Allen, J.R., *Nucleonics*, ***1950***, *7*, 57–60.
16. a. Leitch, L.C. and Morse, A.T., *Can. J. Res.*, ***1952***, *303*, 924–927; b. Schreier, E. and Bühler, K., Internal Report, ***1970***, Sandoz Pharma AG, CH-4002 Basel; c. Murray, A. and Williams, D.L., *Organic Syntheses with Isotopes;* Interscience: New York, ***1958***, pp. 1480–1482.
17. de Suzzoni, S. and Noel, J.P., *J. Label. Compd. Radiopharm.*, ***1994***, *34*, 1201–1206.

18. Gaddam, S.; Khilevich, A.; Filer, C.; Rizzo, J.D.; Giltner, R.; Flavin, M.T. and Xu, Z.-Q., *J. Label. Compd. Radiopharm.*, ***1997***, *39*, 901–906.
19. Andres, H. and Voges, R., *Therapie*, ***1987***, *42*, 417–422.
20. Murray, A. and Williams, D.L., *Organic Syntheses with Isotopes;* Interscience: New York, ***1958***, pp. 129–131.
21. Smith, M.B. and March, J., *March's Advanced Organic Chemistry*, 5th Edn.; John Wiley & Sons, Inc.: New York, ***2001***, pp. 712–714.
22. Baldwin, J.E. and Cianciosi, S.J., *J. Am. Chem. Soc.*, ***1992***, *114*, 9401–9408.
23. Popplewell, D.S. and Wilkins, R.G., *J. Chem. Soc.*, ***1955***, 2521–2522.
24. a. Kagemoto, A.; Kishi, M. and Minaki, Y., *J. Label. Compd.*, ***1973***, *9*, 489–496; b. Stock, M. and Schütte, H.R., *Isotopenpraxis*, ***1990***, *26*, 189–190.
25. Hays; S.J., *J. Label. Compd. Radiopharm.*, ***1987***, *24*, 351–360.
26. Mistry, J.S.; Abraham, D.J. and Hanin, I., *J. Label. Compd. Radiopharm.*, ***1989***, *27*, 1105–1114.
27. a. Parnes, H. and deKeczer, S., *J. Label. Compd. Radiopharm.*, ***1986***, *23*, 777–784; b. Noel, J.-P. and Pichat, L., *J. Label. Compd. Radiopharm.*, ***1977***, *8*, 77–96; c. Haider, K., *J. Label. Compd. Radiopharm.*, ***1966***, *2*, 174–183.
28. Snieckus, V., *Chem. Rev.*, ***1990***, *90*, 879–933.
29. a. Heck, R.F., *Palladium Reagents in Organic Synthesis;* Academic Press: New York, ***1985***; b. Heck, R.F., Palladium Catalyzed Vinylation of Organic Halides, in Org. Reactions, ***1982***, *27*, 345–390; c. de Meijere, A. and Meyer, F.E., *Angew. Chem. Int. Edn Engl.*, ***1995***, *33*, 2379–2411.
30. a. Farina, V.; Krishnamurthy, V. and Scott, W.J., *Org. Reactions*, ***1997***, *50*, 1–87; b. Stille, J.K., *Angew. Chem. Int. Edn Engl.*, ***1986***, *6*, 508–524.
31. a. Suzuki, A., The Suzuki reaction with arylboron compounds in arene chemistry, in *Modern Arene Chemistry*, Astruc, D., Ed.; Wiley-VCH: Weinheim, ***2002***, pp. 53–106; b. Alonso, F.; Beletskaya, I.P. and Yus, M., *Tetrahedron*, ***2008***, *64*, 3047–3101; c. Yin, L. and Liebscher, J., *Chem. Rev.*, ***2007***, *107*, 133–173.
32. a. Buchwald, S.L., *Angew. Chem. Int. Edn Engl.*, ***1995***, *34*, 1348–1350; b. Hartwig, J.F., *Angew. Chem. Int. Edn Engl.*, ***1998***, *37*, 2046–2067.
33. Knochel, P.; Dohle, W.; Gommermann, M.; Kneisel, F.F.; Kopp, F.; Korn, T.; Sapountzis, J. and Ahn Vu, V., *Angew. Chem. Int. Edn Engl.*, ***2003***, *42*, 4302–4320.
34. a. Georgin, D.; de Suzzoni-Dezard, S. and Noel, J.P., *J. Label. Compd. Radiopharm.*, ***1996***, *38*, 645–649; b. Immoos, J.E.; Brattesani, A.G. and Sieving, P.F., *J. Label. Compd. Radiopharm.*, ***1996***, *38*, 511–522; c. Chew, E.H. and Heys, J.R., *J. Label. Compd. Radiopharm.*, ***1981***, *18*, 525–533; d. Susàn, A.B.; Ebert, D.A. and Duncan, W.P., *J. Label. Compd. Radiopharm.*, ***1979***, *16*, 579–589; e. Kaegi, H.H., *J. Label. Compd. Radiopharm.*, ***1981***, *18*, 445–447.
35. Chaudhuri, N.K. and Markus, B., *J. Label. Compd. Radiopharm.*, ***1990***, *28*, 322–327.
36. Huang, C.C.; Hicks, J.L. and Showalter, H.D.H., in *Synthesis and Applications of Isotopically Labeled Compounds 1991*, Buncel, E. and Kabalka, G.W., Eds.; Elsevier: Amsterdam, ***1992***, pp. 187–192.
37. a. Koltai, E.; Kling, F. and Rutkai, G., *J. Label. Compd. Radiopharm.*, ***1995***, *36*, 903–908; b. Kimura, I.; Yokoyama, N.; Kasahara, K.; Ueda, T.; Atsumi, K.; Filer, C. and Wright, C., *J. Label. Compd Radiopharm.*, ***1993***, *33*, 87–95.
38. Noel, J.P.; Pichat, L.; Rimbau, V. and Forn, J., *J. Label. Compd. Radiopharm.*, ***1982***, *19*, 373–384.
39. a. Noel, J.P. and Pichat, L., *J. Label. Compd. Radiopharm.*, ***1982***, *19*, 821–836; b. Weller, P.E. and Hanzlik, R.P., *J. Label. Compd. Radiopharm.*, ***1988***, *25*, 991–998; c. Fields, M.; Gibbs, J. and Walz, D.E., *Science*, ***1950***, *112*, 591–593.
40. Akira, K.; Hasegawa, H. and Baba, S., *J. Label. Compd. Radiopharm.*, ***1995***, *36*, 845–853.
41. a. Czeskis, B.A. and Wheeler, W.J., *J. Label. Compd. Radiopharm.*, ***2005***, *48*, 407–419; b. Sugasawa, T.; Adachi, M.; Sasakura, K. and Kitagawa, A., *J. Org. Chem.*, ***1979***, *44*, 578–586; c. Sugasawa, T.; Toyada, T.; Adachi, M. and Sasakura, K., *J. Am. Chem. Soc.*, ***1978***, *100*, 4842–4852.
42. Wager, C.A.B. and Miller, S.A., *J. Label. Compd. Radiopharm.*, ***2006***, *49*, 615–622.
43. a. Sandrock, K.; Attar, A.; Bieniek, D.; Klein, W. and Korte, F., *J. Label. Compd. Radiopharm.*, ***1978***, *14*, 197–204; b. Smith, M.B. and March, J., *March's Advanced Organic Chemistry*, 5th Edn.; John Wiley & Sons, Inc.: New York, ***2001***, p. 705.

44. Burgos, A. and Ellames, G.J., *J. Label. Compd. Radiopharm.*, ***1995***, *36*, 1071–1076.
45. Hafferl, W. and Juppe, G., *J. Label. Compd.*, ***1968***, *4*, 80–89.
46. Sloane, N.H.; Untch, K.G. and Johnson, A.W., *Biochim. Biophys. Acta*, ***1963***, *78*, 588–593.
47. Maul, W.; Scherling, D. and Seng, F., *J. Label. Compd. Radiopharm.*, ***1981***, *18*, 453–462.
48. a. Gillam, M.G., *J. Label. Compd. Radiopharm.*, ***1989***, *27*, 1409–1413; b. *Vogel's Textbook of Practical Organic Chemistry*, 5th Edn., Furniss, B.; Hannoford, A.J.; Smith, P.W.G. and Tatchell, A.R., Eds.: Longman: London, ***1991***, pp. 941–942.
49. Schlecker, R., *J. Label. Compd. Radiopharm.*, ***2002***, *45*, 1178–1179.
50. Madegard, G.; Mestre, P.; Raimond, P. and Noel, J.-P., *J. Label. Compd. Radiopharm.*, ***1995***, *36*, 1123–1132.
51. Volford, J.; Koltai, E.; Kling, F. and Volford, A., *J. Label. Compd. Radiopharm.*, ***1996***, *38*, 661–665.
52. Bergman, A.; Nilson, A.; Riego, J. and Örn, U., *Acta Chem. Scand.*, ***1990***, *44*, 1071–1076.
53. a. Heys, J.R.; Chew, E.H. and Saugier, J.H., *J. Label. Compd. Radiopharm.*, ***1982***, *19*, 801–805; b. Yatsu, M.; Atsumi, K.; Fukasaku, N. and Yokoshima, T., in *Synthesis and Applications of Isotopically Labeled Compounds 1991*, Buncel, E. and Kabalka, G.W., Eds.; Elsevier: Amsterdam, ***1992***, pp. 568–569.
54. Noël, J.P. and Pichat, L., *J. Label. Compd. Radiopharm.*, ***1980***, *17*, 833–840.
55. Townsend, C.A. and Bloom, L.M., *Tetrahedron Lett.*, ***1981***, 22, 3923–3924.
56. a. Ji, R. and Schäffer, A., *J. Label. Compd. Radiopharm.*, ***2003***, *46*, 551–558 and 781; b. Kratzl, K. and Vierhapper, F.W., *Monatsh. Chem.*, ***1971***, *102*, 2224–2322.
57. a. Miller, J.J.; Olavesen, A.H. and Curtis, C.G., *J. Label. Compd. Radiopharm.*, ***1974***, *10*, 151–160; b. Loudon, J.D. and Scott, J.A., *J. Chem. Soc.*, ***1953***, 265–268.
58. Schofield, K., *Aromatic Nitration;* Cambridge University Press: Cambridge, ***1980***, pp. 243–246.
59. a. Ellames, G.J. and Kunec, E.K., *J. Label. Compd. Radiopharm.*, ***1991***, *29*, 29–34; b. Elbert, T.; Havlicek, L.; Wimmer, Z. and Bubner, M., *J. Label. Compd. Radiopharm.*, ***1994***, *34*, 377–382.
60. Kim, J.H.; Creger, C.R. and Couch, J.R., *J. Label. Compd.*, ***1969***, *5*, 35–41.
61. Herbert, M.; Pichat, L. and Langourieux, Y., *J. Label. Compd.*, ***1974***, *10*, 89–102.
62. a. Chung, K.H.; Kim, H.J.; Kim, H.R. and Ryu, E.K., *Synth. Commun.*, ***1990***, *20*, 2991–2997; b. Chung, K.H.; Kim, K.M.; Kim, J.M. and Ryu, E.K., *Synth. Commun.*, ***1991***, *21*, 1917–1922.
63. Kendall, J.T., *J. Label. Compd. Radiopharm.*, ***2000***, *43*, 505–514.
64. Orito, K.; Hatakeyama, T.; Takeo, M. and Suginome, H., *Synthesis*, ***1995***, 1273–1277.
65. Woo, P.W.K. and Lee, H.T., *J. Label. Compd. Radiopharm.*, ***1994***, *34*, 1–10.
66. Collins, C.J. and Bonner, W.A., *J. Am. Chem. Soc.*, ***1955***, *77*, 77–99.
67. Andres, H. and Schnelli, P., Synthesis of [$^{14}$C]SDZ EAB 515, ***1992***, unpublished results
68. Kaegi, F.; Metz, Y.; Rodriguez, I. and Ruetsch, R., *J. Label Comp. Radiopharm.*, ***2004***, *47*, 272–273.
69. Bonarcosi, S.J.; Waller, S.C. and Rinehart, J.K., *J. Label. Comp. Radiopharm.*, ***2006***, *49*, 1–9.
70. Lee, D.Y. and Hartwig, J.F., *Org. Lett.*, ***2005***, *7*, 169–1172.
71. Hoizumi, K. and Moriya, T., *J. Label. Compd.*, ***1974***, *10*, 499–508.
72. Villani, A.J.; Etzkorn, F.; Rotert, G.A. and Heys, J.R., *J. Label. Compd. Radiopharm.*, ***1988***, *25*, 1339–1347.
73. a. Nguyen, H.N.; Herbert, M.; Nguyen, D.X. and Pichat, L., *J. Label. Compd.*, ***1971***, *7*, 299–304; b. Gross, H.; Rusche, J. and Mirsch, M., *Chem. Ber.*, ***1963***, *96*, 1382–1386.
74. a. McKillop, A.; Hunt, J.D.; Zelesko, M.J.; Fowler, J.S.; Taylor, E.C.; McGillivray, G. and Kienzle, F., *J. Am. Chem. Soc.*, ***1971***, *93*, 4841–4844; b. Taylor, E.C.; Kienzle, F.; Robey, R.L.; McKillop, A. and Hunt, J.D., *J. Am. Chem. Soc.*, ***1971***, *93*, 4844–4850.
75. Uemura, S.; Ikeda, Y. and Ichikawa, K., *Tetrahedron*, ***1972***, *28*, 5499–5504.
76. Maxwell, B.D., *J. Label. Compd. Radiopharm.*, ***1997***, *39*, 267–273.
77. a. Chien, D.H.T. and Duncan, W.P., *J. Label. Compd. Radiopharm.*, ***1983***, *20*, 131–134; b. Streitwieser, A. and Reif, L., *J. Am. Chem. Soc.*, ***1984***, *86*, 1988–1993.
78. Zhang, Y., *J. Label. Compd. Radiopharm.*, ***2000***, *43*, 1087–1093.
79. Feil, V.J., *J. Label. Compd. Radiopharm.*, ***1996***, *38*, 1071–1076.
80. Gatto, G.J. and Mertel, H.E., *J. Label. Compd. Radiopharm.*, ***1990***, *28*, 943–953.

81. Wright, C. and Ullas, G.V., *J. Label. Compd. Radiopharm.*, ***2002***, *45*, 1265–1268.
82. a. Auger, P.; Malaiyandi, M.; Wightman, R.H. and Williams, D.T., *J. Label. Compd. Radiopharm.*, ***1993***, *33*, 263–276; b. Walker, G.H. and Hathway, D.E., *J. Label. Compd. Radiopharm.*, ***1976***, *12*, 199–206.
83. a. Gutman, A.D. and Brokke, M.E., *J. Label. Compd.*, ***1966***, *2*, 317–319; b. Nagasaki, T.; Watanabe, F.; Katsuyama, Y.; Hamada, Y.; Ohtani, M. and Narisada, M., *J. Label. Compd. Radiopharm.*, ***1992***, *31*, 23–38; c. Kalbfeld, J.; Pitt, H.M. and Hermann, D.A., *J. Label. Compd.*, ***1969***, *5*, 351–354.

# 9

# Preparation of Carbon-14-Labeled Compounds via the [$^{14}$C]Cyanamide Tree

[$^{14}$C]Cyanamide is prepared in 70–80% radiochemical yield by heating barium [$^{14}$C]-carbonate in a stream of ammonia, according to the procedure of Figure 9.1, the final product being isolated by extraction into ether[1,2].

Its barium salt is the source of a number of additional carbon-14 building blocks, such as [$^{14}$C]guanidine, [$^{14}$C]urea and [$^{14}$C]thiourea. [$^{14}$C]Cyanamide is itself also a versatile building block. The utility of each of these for the synthesis of carbon-14-labeled compounds is discussed in the Sections below.

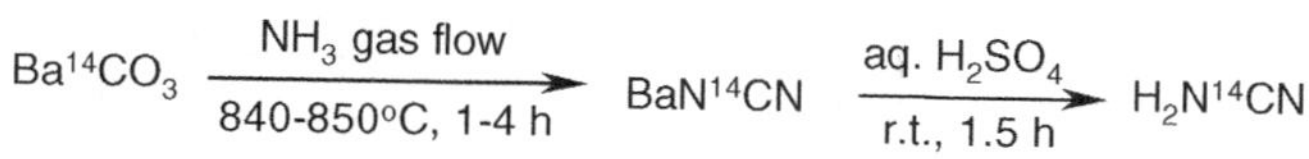


*Figure 9.1* Preparation of [$^{14}$C]cyanamide

## 9.1 [$^{14}$C]Cyanamide ($H_2N^{14}C{\equiv}N$)

[$^{14}$C]Cyanamide undergoes nucleophilic attack at the central carbon to furnish [$^{14}$C]-dicyanamide, [$^{14}$C]guanidines, *O*-alkyl [$^{14}$C]isoureas, [$^{14}$C]urea, [$^{14}$C]thiourea etc. Reaction with aldehydes forms [$^{14}$C]aminals, which are suitable for further transformations. Despite the versatility of [$^{14}$C]cyanamide, it has so far found only limited use as a building block in carbon-14 syntheses. The examples below provide a general overview of these uses.

Monosubstituted [$^{14}$C]guanidines are straightforwardly available through treatment of [$^{14}$C]cyanamide with alkyl or arylamine hydrochlorides in *n*-butanol at elevated temperature under anhydrous conditions, as in the preparation of 1-(2-methyl-5-nitrophe-

*Preparation of Compounds Labeled with Tritium and Carbon-14* Rolf Voges, J. Richard Heys and Thomas Moenius

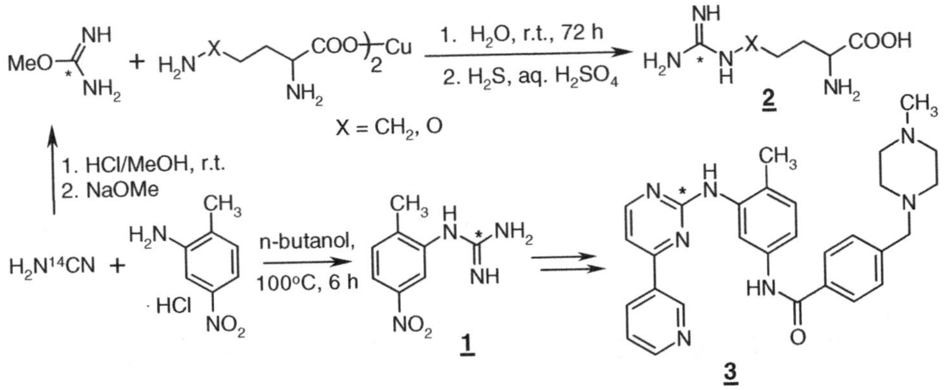

**Figure 9.2** *Monosubstituted [$^{14}$C]guanidines from [$^{14}$C]cyanamide*

nyl)[$^{14}$C]guanidine (**1**)[3], or via aminolysis of *O*-methyl[$^{14}$C]isourea, as illustrated by the synthesis of [guanidino-$^{14}$C]arginine (**2**, X = $CH_2$)[4] or [guanidino-$^{14}$C]canavanine (**2**, X = O)[5] (Figure 9.2).

Compound **1** served as a key intermediate in the synthesis of [$^{14}$C]imatinib (**3**), a drug with excellent efficacy for the treatment of chronic myelogenous leukemia[3]. Other antileukemia agents were also labeled utilizing [$^{14}$C]cyanamide as a starting material (Figure 9.3). 1*β*-D-arabinofuranosyl[2-$^{14}$C]cytosine (**5**), for example, was prepared by

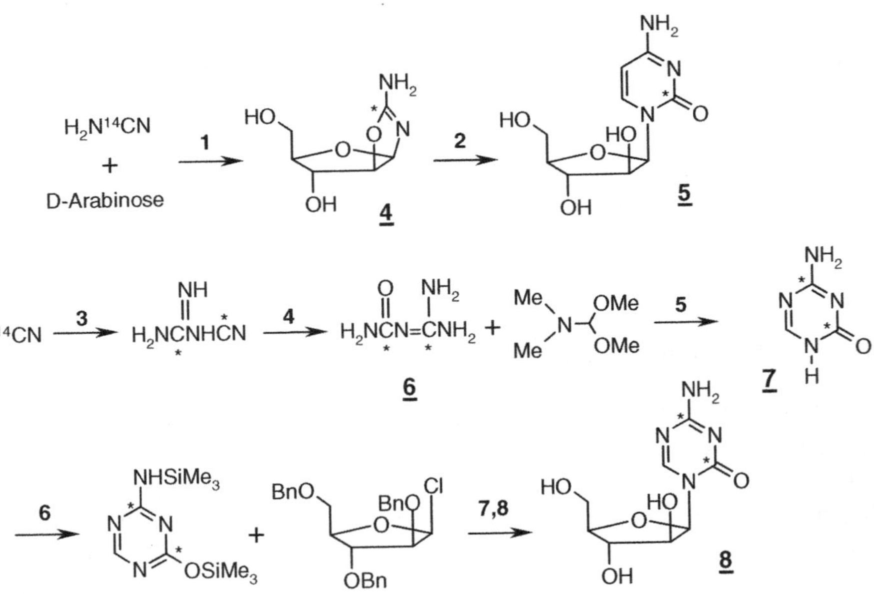

**Reaction conditions:** 1. MeOH, aq. $NH_3$; 2. HC≡CCN, DMA; 3. aq. $NH_3$; reflux; 4. HCl, MeOH; 5. MeOH; 6. $(Me_3Si)_2$, aq. $(NH_4)_2SO_4$; 7. $ClCH_2CH_2Cl$; reflux; 8. $H_2$, 10 % Pd/C, methanolic HCl

**Figure 9.3** *Nucleosides from [$^{14}$C]cyanamide*

treatment of [$^{14}$C]cyanamide with D-arabinose followed by reaction of the resultant oxazoline (**4**) with cyanoacetylene[2]. The synthesis route for the corresponding aza-analog **8** proceeded via [$^{14}C_2$]guanylurea (**6**), which was prepared in two steps through dimerization of [$^{14}$C]cyanamide and partial hydrolysis of the resulting [$^{14}C_2$]dicyandiamide ([$^{14}C_2$]cyanoguanidine). Reaction with dimethylformamide dimethylacetal gave 5-aza [2,4-$^{14}C_2$]cytosine (**7**). *N,O*-Protection by treatment with HMDS followed by coupling to 2,3,5-tri-*O*-benzyl-D-ribofuranosyl chloride and concluding 10% Pd/C-catalyzed hydrogenolytic cleavage of the benzyl groups converted **7** into **8**[6].

The major importance of [$^{14}$C]cyanamide, however, lies in its convenient conversion to [$^{14}$C]guanidine (**9**), [$^{14}$C]urea (**10**) and thiourea (**11**), all essential building blocks for the carbon-14 labeling of numerous five- and six-membered heterocyclic compounds such as pyrimidines, purines, triazines, imidazoles and thiazoles.

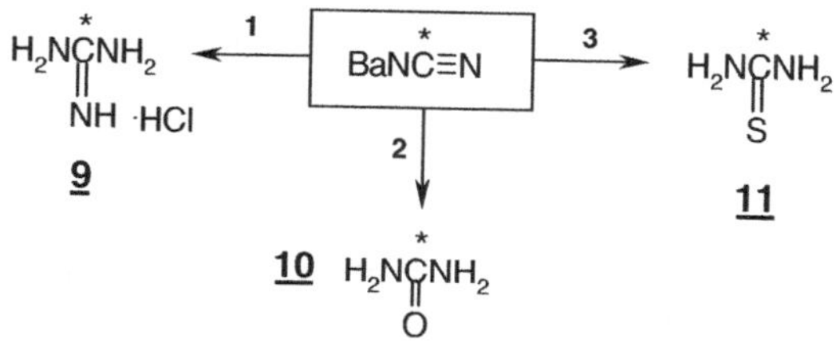


**Reaction conditions: 1a**. $NH_4NO_3$, 165 °C, 20 min, **b**. ammonium picrate, $H_2O$; r.t., 16 h, **c**. HCl/ether; **2a**. $CO_2$, $H_2O$, r.t., **b**. aq. $H_2SO_4$; 40–60 °C, 2–3 d; **3**.$H_2S$, $Ba(OH)_2$, $H_2O$; 85–100 °C, 20 h

***Figure 9.4*** *Synthesis of [$^{14}$C]]building blocks from barium [$^{14}$C]cyanamide*

[$^{14}$C]Guanidine is prepared by heating a mixture of barium [$^{14}$C]cyanamide and ammonium nitrate. The resulting crude [$^{14}$C]guanidinium nitrate is purified by precipitation as picrate and the reagent isolated through treatment with ethereal HCl as indicated in Figure 9.4 provides pure [$^{14}$C]guanidine hydrochloride (**9**), a good form for storage, in 75% overall yield[7]. Acidic hydrolysis of [$^{14}$C]cyanamide after liberation from the barium salt furnishes [$^{14}$C]urea (**10**) in nearly quantitative yield[8]. [$^{14}$C]Thiourea (**11**) is obtained in nearly quantitative yield by treatment of barium [$^{14}$C]cyanamide with hydrogen sulfide and barium hydroxide and heating. Barium ions are removed by saturation of the reaction mixture with $CO_2$ and filtration of the precipitated carbonate, and [$^{14}$C]thiourea is isolated by evaporation of the filtrate and sublimation of the residue[9].

## 9.2 [$^{14}$C]Guanidine ($H_2N^{14}C(=NH)NH_2$)

[$^{14}$C]Guanidine has emerged as a valuable building block for the preparation of various carbon-14-labeled pyrimidines and purines. Treatment with ethyl cyanoacetate in the presence of sodium ethoxide, for example, afforded 2,4-diamino-6-oxo[2-$^{14}$C]pyrimidine (**12**). Subsequent nitrosation at C5 and reduction of the nitroso group with sodium hydrogen sulfite converted **12** into 2,4,5-triamino-6-oxo[2-$^{14}$C]pyrimidine (**13**). Ring closure upon heating of **13** with formic acid furnished [2-$^{14}$C]guanine (**14**) in an overall radiochemical yield of 68–74%[10, 14b]. Substitution of malondinitrile for ethyl cyanoacetate gives 2,4,6-triamino[2-$^{14}$C]pyrimidine, which can be readily converted into 2,6-diamino[2$^{14}$C]purine following the same synthetic pathway described before[7, 14b]. 2-Amino[2$^{14}$C]pyrimidine derivatives **15** are obtained when [$^{14}$C]guanidine is treated with 1,3-dicarbonyl compounds such as acetylacetone and malondialdehyde bis(diethyl acetal)[10–12]. Finally, 5,6-disubstituted 2,4-diamino[2-$^{14}$C]pyrido[2, 3-d] pyrimidines **17** become available when [$^{14}$C]guanidine reacts with substituted 2-bromo-3-cyanopyridines **16**[13].

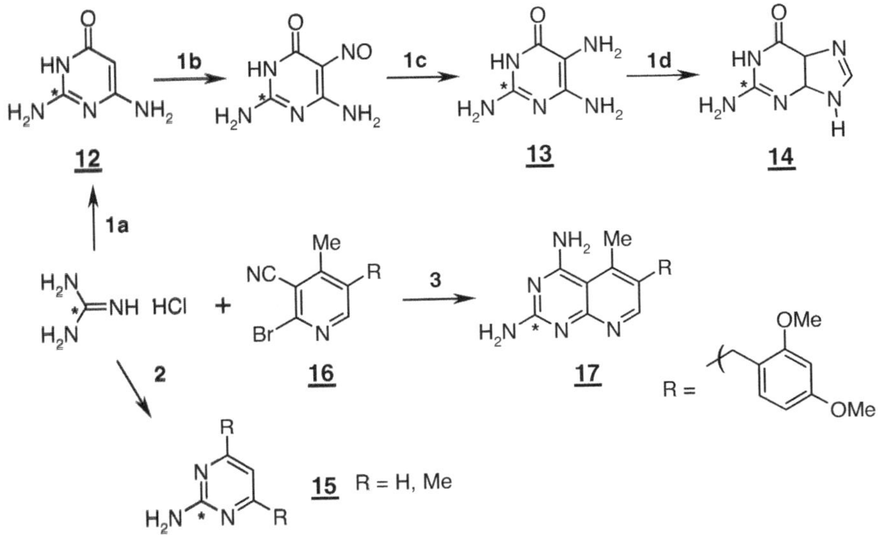


**Reaction conditions: 1a**. NCCH$_2$COOEt, EtONa, EtOH; reflux, 3 h, **b**. NaNO$_2$, HCl, EtOH; reflux, **c**. NaHSO$_3$, aq. NaOH; 70–80 °C, 45 min, **d**. HCOOH; reflux, 15 h; **2**. R=H: (EtO)$_2$CHCHCH$_2$(OEt)$_2$, EtONa, EtOH; refux, 2 h **or** R=Me: CH$_3$COCH$_2$COCH$_3$, Na$_2$CO$_3$, H$_2$O; 100 °C, 12 h; **3** . **16**, NaH, *tert*-BuOH; r.t., 35 min, 65 °C, 50 min, reflux, 4 h

***Figure 9.5*** *Applications of [$^{14}$C]guanidine to the synthesis of [2-$^{14}$C]purines and [2-$^{14}$C] pyrimidines*

## 9.3 [$^{14}$C]Urea, $H_2N^{14}CONH_2$

[$^{14}$C]Urea has been extensively used for the carbon-14 labeling of a broad spectrum of pyrimidine and purine derivatives[14]. Reaction with ethyl $\beta$-ethoxyacrylate or the free acid (formylacetic acid) generated *in situ* upon treatment of L-malic acid or esters **18** (R = H) in fuming sulfuric acid affords [2-$^{14}$C]uracil (**19**, R = H) in 70% yield[15]. Cyclocondensation with diethyl $\beta$-methylmalate (**18**, R = Me) following a slightly modified procedure allows for the preparation of [2-$^{14}$C]thymine (**19**, R = Me) in 60% yield (Figure 9.6). Subsequent incubation of [2-$^{14}$C]thymine with 2′-deoxyribose-1-phosphate in buffer solution at pH 7.4 in the presence of thymidine phosphorylase offers a convenient access to [2-$^{14}$C]thymidine in overall 40% overall radiochemical yield from [$^{14}$C]urea[16]. Sodium butoxide-mediated

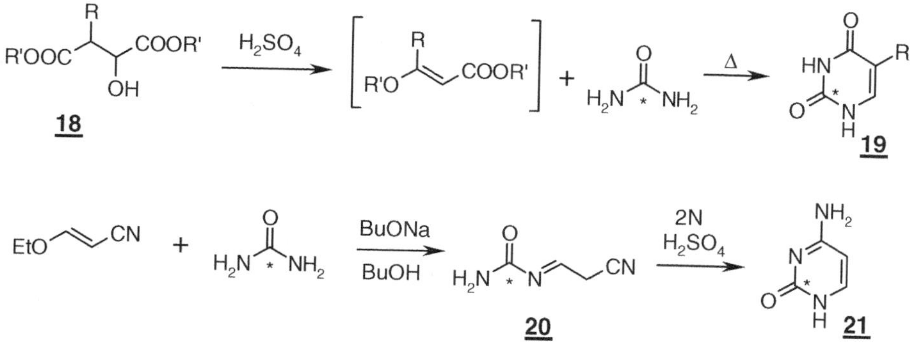

**Figure 9.6** *Application of [$^{14}$C]urea in the synthesis of [2-$^{14}$C]pyrimidines*

condensation of $\beta$-ethoxyacrylonitrile with [$^{14}$C]urea followed by acid-catalyzed cyclization of the initially formed intermediate **20** provided [2-$^{14}$C]cytosine (**21**) in 40% radiochemical yield[17].

Further labeled pyrimidines such as [2-$^{14}$C]barbituric acid (**22**, $R^1$,$R^2$ = H) and its derivatives such as [2-$^{14}$C]barbital (**22**: $R^1$,$R^2$ = Et), and [2-$^{14}$C]pentobarbital (**22**: $R^1$ = Et, $R^2$ = 1-methylbutyl) are readily available from cyclocondensation of [$^{14}$C]urea and the respective diethyl malonates in the presence of sodium ethoxide or butoxide[18]. For [2-$^{14}$C]-phenobarbital ($R^1$ = Et, $R^2$ = Phe), decarboxylation of the malonate component was observed as a main side reaction, so that the respective malonyl dichloride had to be employed in order to achieve satisfactory yields[19]. Analogous syntheses using unlabeled urea and labeled malonate derivatives are described in Chapter 6, Section 6.5.

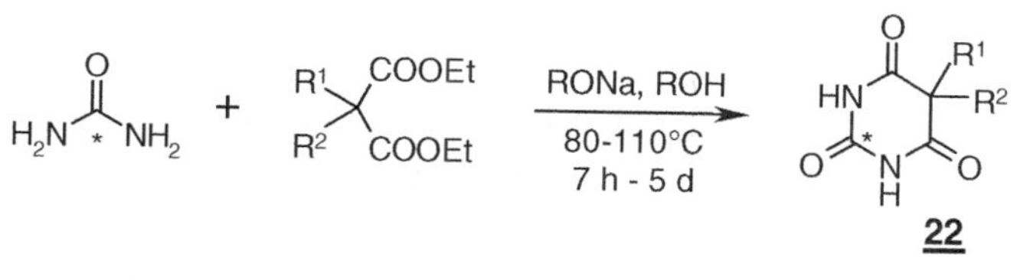

**Figure 9.7** *[2-$^{14}$C]Barbituric acid derivatives from [$^{14}$C]urea*

[2-$^{14}$C]Barbituric acid has been exploited as a key intermediate in the synthesis of various pyrrolo[2,3-d][2-$^{14}$C]pyrimidines[20]. Chloro-dehydroxylation of the former with phenylphosphonic dichloride furnished 2,4,6-trichloropyrimidine (**23**), a building block useful for a regioselective substitution of chlorine by different amines (Figure 9.8). Amino-dechlorination proceeded sequentially first at C2. In the presence of excess pyrrolidine under mild conditions, the 2,4-dipyrrolidinyl derivative **24** was obtained. Heating of **24** with various primary and secondary amines at 155–170 °C produced the 6-alkylamino derivatives **25**. These cyclize with acylic or cyclic 2-bromoketones to give di- and tricyclic products **26** and **27**[20,21]. Reaction of **24** with piperazine yielded **28**, an intermediate for the preparation of [$^{14}$C]trilazad (**29**), a drug for the treatment of head and spinal cord injury[21].

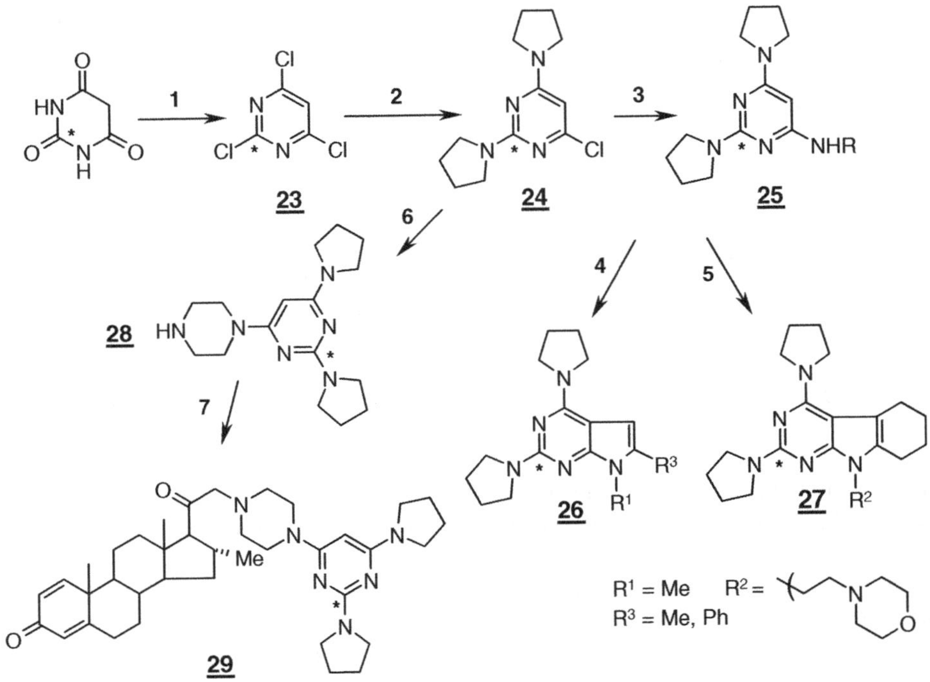

**Reaction conditions:** **1**. $PhPOCl_2$; 180 °C; **2**. pyrrolidine, THF; r.t.; **3**. aq. $MeNH_2$, pyridine; 170 °C, 26 h **or** aminoethylmorpholine, 155 °C, 16 h; **4**. $RCOCH_2Br$ (R= Me, Ph); *i*-$Pr_2NEt$, acetonitrile; r.t., 16 h; **5**. 2-bromocyclohexanone, *i*-$Pr_2NEt$, acetonitrile; 95 °C, 34 h; **6**. piperazine, pyridine, Δ; **7**. steroid α-chloroketone, NaI, $K_2CO_3$, acetone

***Figure 9.8*** *Synthesis of pyrrolo[2,3-d][2-$^{14}$C]pyrimidines from [2-$^{14}$C]barbituric acid*

Replacement of dialkyl malonate with ethyl cyanoacetate in the above described general method provided 6-amino[2-$^{14}$C]uracil (<u>**30**</u>) (Figure 9.9). Chloro-dehydroxylation was accomplished with $POCl_3$/2-methylpyridine to afford <u>**31**</u>, which upon reaction with thiourea and hydrolysis gave [2-$^{14}$C]pyrimidine <u>**32**</u> possessing three different substituents[21]. With malonaldehyde diethyl acetal in the presence of acid instead of dialkyl malonate, 2-hydroxy[2-$^{14}$C]pyrimidine (<u>**33**</u>) is accessible. Subsequent chloro-dehydroxylation provided 2-chloro[2-$^{14}$C]pyrimidine (<u>**34**</u>), a useful building block whose substituent readily undergoes nucleophilic displacement. For example, treatment of <u>**34**</u> with the 4-substituted piperidine <u>**35**</u> provided a carbon-14 isotopomer of the anxiolytic agent buspirone (<u>**36**</u>) in 42% yield[22].

At temperatures above 140 °C, [$^{14}$C]urea decomposes to give $NH_3$ and [$^{14}$C]isocyanic acid, which can trimerize to give [$^{14}C_3$]cyanuric acid (<u>**37**</u>). A practical preparation of <u>**37**</u> consists in heating a mixture of [$^{14}$C]urea in the presence of $NH_4Cl$[23]; the yield is about 60%. In analogy to the preparation of <u>**23**</u> and <u>**31**</u>, chloro-dehydroxylation of <u>**37**</u> provides [U-$^{14}$C]cyanuric chloride (<u>**38**</u>)[24] (Figure 9.10).

*In situ* generated [$^{14}$C]isocyanic acid is probably the active agent when [$^{14}$C]urea is used to ring-close free or monosubstituted 1,2- or 1,3-diamines or amino amides at

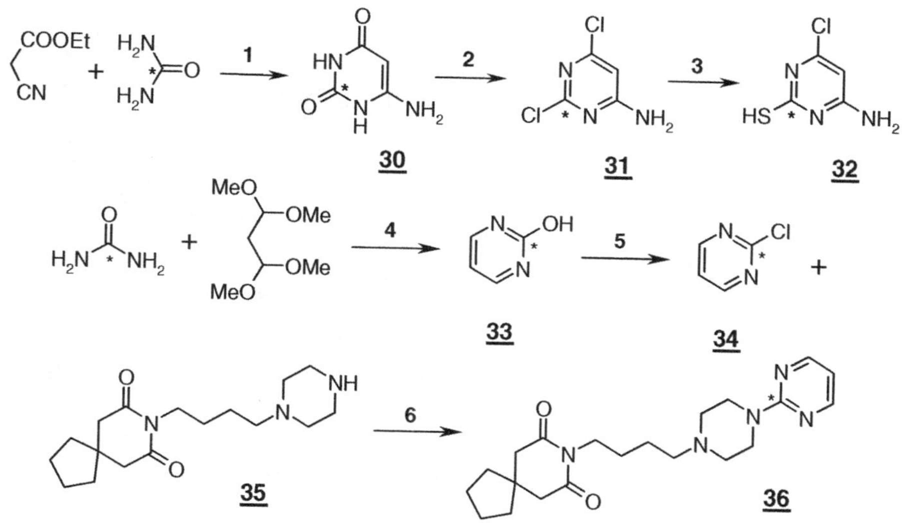


**Reaction conditions:** **1**. NaOEt, EtOH; **2**. $POCl_3$, 2-methylpicoline; **3a**. thiourea, **b**. NaOH; **4**. conc. HCl, EtOH; 85 °C; **5**. $POCl_3$; reflux; **6**. $Et_3N$, EtOH; 125 °C, 24 h

***Figure 9.9*** *[2-$^{14}$C]Pyrimidines from [$^{14}$C]urea and cyanoacetates*

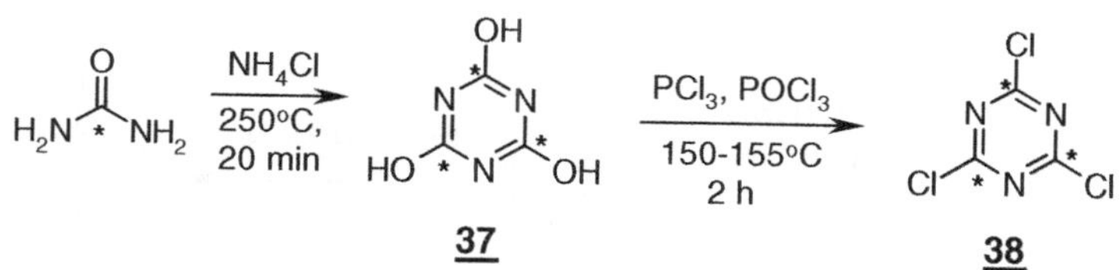


***Figure 9.10*** *[$^{14}C_3$]Cyanuric acid and [$^{14}C_3$]cyanuric chloride from [$^{14}$C]urea*

elevated temperatures. For example, heating of monoalkylated phenylenediamine derivatives **39** with [$^{14}$C]urea furnished the respective [2-$^{14}$C]benzimidazole-2-one derivatives **40**, key intermediates in the labeling of astemizole (**41**)[25] and KB 2413[26], two antihistaminic agents (Figure 9.11). The cyclocondensation of the *o*-aminobenzamide precursor **42** with [$^{14}$C]urea to give [$^{14}$C]ketanserin (**43**), an antihypertensive drug, belongs to the same reaction type[27].

6-Amino[2-$^{14}$C]uracil (**30**), accessible from ethyl cyanoacetate and [2-$^{14}$C]urea as previously mentioned, is one of the key intermediates in the Traube-type synthesis of labeled uric acid and xanthine, whose methyl homologues are important biologically active natural products. Nitrosation at C5 with nitrous acid and reduction of the nitroso group with sodium dithionate installed the 5-amino group. Heating of the resulting 5,6-[2-$^{14}$C]diaminouracil (**44**) in formamide or formic acid resulted in the initial formation of the *N*-formamide at C5, which cyclized with elimination of water to afford [2-$^{14}$C]xanthine (**45**)[14a]. *N*-methylation with dimethyl sulfate in basic medium under adequate reaction conditions enabled the separate preparation of [2-$^{14}$C]theobromine[28]

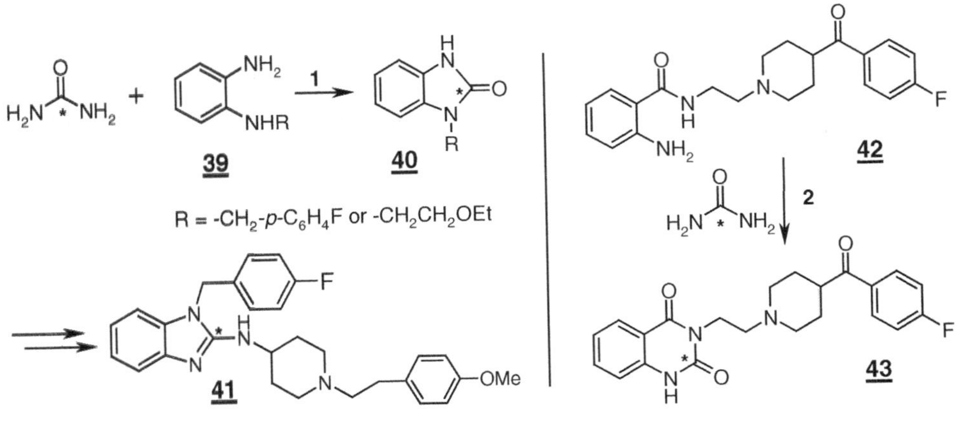


**Reaction conditions: 1**. (R = *p*-$C_6H_4F$) *n*-amyl alcohol; 150 °C, 24 h; (R = $CH_2CH_2OEt$) xylene; 140 °C, 24 h; **2**. xylene; 135 °C, 40 h.

***Figure 9.11*** *Benzo-fused [$^{14}$C]heterocycles from [$^{14}$C]urea*

(44%), [2-$^{14}$C]theophylline[14a] (61%) and [2-$^{14}$C]caffeine[29] (72%). Replacing formic acid or formamide for urea in the cyclization reaction converted **44** into [2-$^{14}$C]-uric acid (**46**)[30].

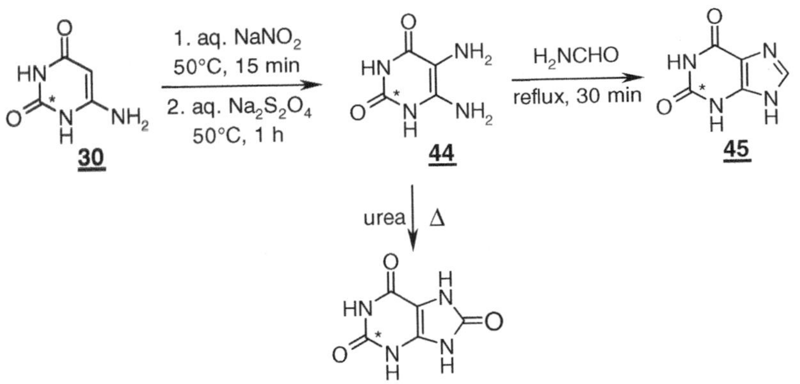


***Figure 9.12*** *Synthesis of [2-$^{14}$C]xanthine and [2-$^{14}$C]uric acid from [2-$^{14}$C]urea*

## 9.4 [$^{14}$C]Thiourea, $H_2N^{14}CSNH_2$

[$^{14}$C]Thiourea has been utilized to a greater extent than [$^{14}$C]urea because of its enhanced reactivity and the versatility of its reaction products for further elaboration. Cyclization reactions upon treatment of [$^{14}$C]thiourea with 1,3-dicarbonyl compounds to give 2-thio-

[2-$^{14}$C]pyrimidines (route A, Figures 9.13–9.15) and with 2-halocarbonyl derivatives to form 2-amino[2-$^{14}$C]thiazoles (route B, Figure 9.16) are the most common applications. Other difunctional reaction partners, e.g. vicinal haloamines, result in 2-imino[2-$^{14}$C]thiazolidines (Figure 9.17). *S*-activation through methylation allows for its nucleophilic displacement for appropriate nucleophiles, e.g. amines, providing acyclic or cyclic guanidines (Figures 9.18 and 9.19). Further transformations of the sulfur function of the resultant products include replacement by the corresponding oxygen functions upon treatment with chloroacetic acid or $H_2O_2$, catalytic desulfurization using Raney nickel, or alkylation and oxidation to alkylthiones, which are displaceable by nucleophiles under appropriate reaction conditions.

The substitution pattern of pyrimidines prepared according to route A depends on the oxidation state of the dicarbonyl partner. Condensation of [$^{14}$C]thiourea with diethyl malonate, for example (Figure 9.13), gave 4,6-dihydroxy-2-mercapto[2-$^{14}$C]pyrimidine (**47**)[31] in 80% yield. Subsequent *S*-methylation (with concomitant *O*-methylation) followed by *S*-oxidation converted the –SH group of **47** into the nucleofuge $-SO_2CH_3$ function of **48**, which is sufficiently activated for a base-promoted nucleophilic displacement by the alcohol **49** to give **50**[31] and by the thiol **51** to give **52**[32], respectively.

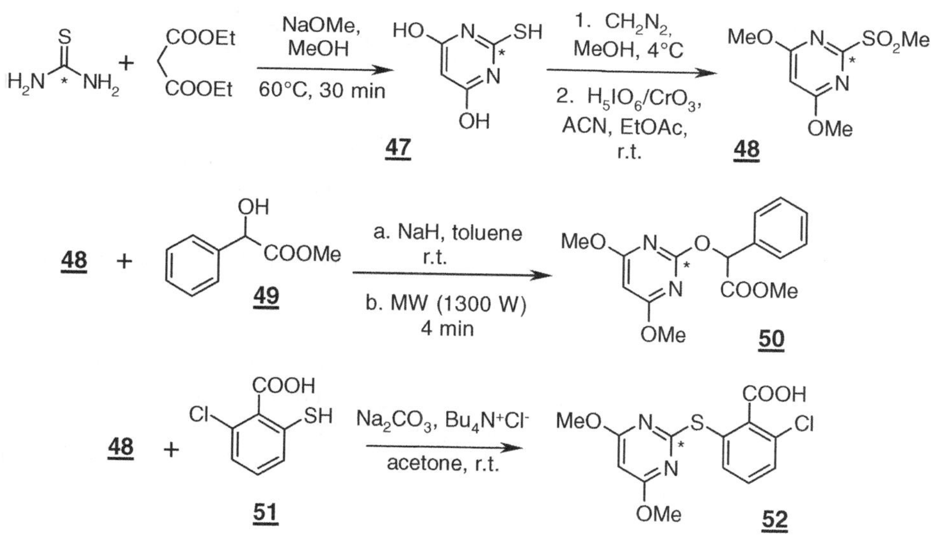


***Figure 9.13*** *Preparation of a [2-$^{14}$C]pyrimidines from [$^{14}$C]thiourea and a 1,3-diester*

Condensation of [$^{14}$C]thiourea with ethyl 2-formylacetates or their diethyl acetals provides 6-hydroxy-2-mercapto[2-$^{14}$C]pyrimidines **53** (Figure 9.14). Subsequent acid-catalyzed oxo-desulfurization produced [2-$^{14}$C]uracil[33]. Substituted 2-formylacetates give the corresponding 5-substituted uracils **54**, ranging from [2-$^{14}$C]thymine (R = Me) to those possessing larger alkyl[34] and functionalized C5-substituents[33]. Execution of the sequence starting with diethyl ethoxymethylenemalonate through oxodesulfurization and final transformation of the C5-carboxyl group upon treatment with $SF_4$ in THF gave 5-trifluoromethyl[2-$^{14}$C]uracil (**55**)[35]. Thiation of 5-alkyl[2-$^{14}$C]uracils with phosphorus pentasulfide furnishes 4-thiono derivatives, which upon reaction with ammonia in methanol converts [2-$^{14}$C]uracils into the corresponding [2-$^{14}$C]cytosines[36]. Finally, 6-substituted [2-$^{14}$C]thiouracils **56** are obtained when $\beta$-keto esters are employed as the 1,3-dicarbonyl reaction partners in the cyclocondensation reaction[33, 37], and upon oxo-desulfurization they afford the corresponding 6-substituted [2-$^{14}$C]uracils **57**.

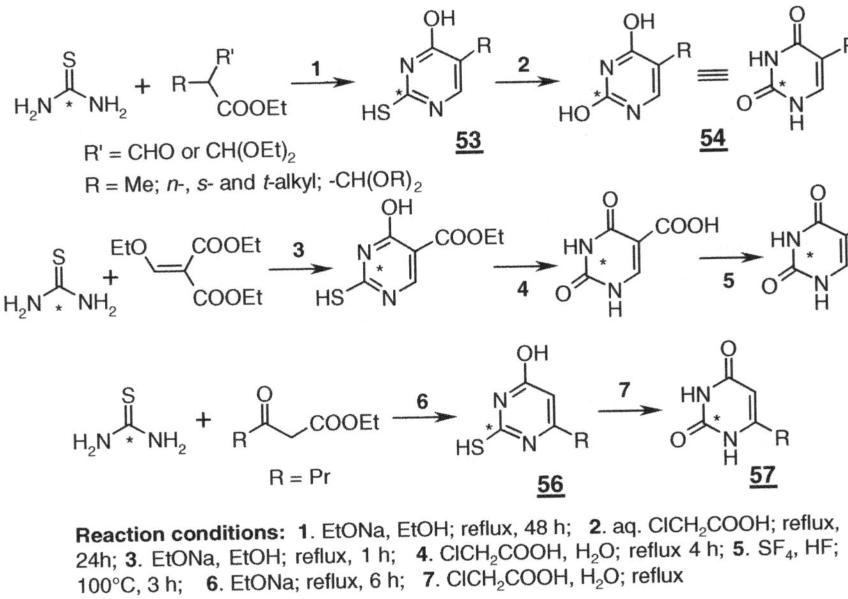


***Figure 9.14*** *Preparation of [2-$^{14}$C]pyrimidines from [$^{14}$C]thiourea and 3-acyl ester equivalents*

The susceptibility of the sulfur function of thiouracils to nucleophilic displacement was exploited in the synthesis of [$^{14}$C]SK&F 93944 (**59**)[9a]. For this purpose 5,6-methyl-3-pyridylmethyl[$^{14}$C]thiouracil (**58**) was first alkylated with methyl iodide, then the resulting 2-methylthio intermediate was treated with the $\omega$-substituted butylamine to give **59** in an overall radiochemical yield of 38% from [$^{14}$C]thiourea.

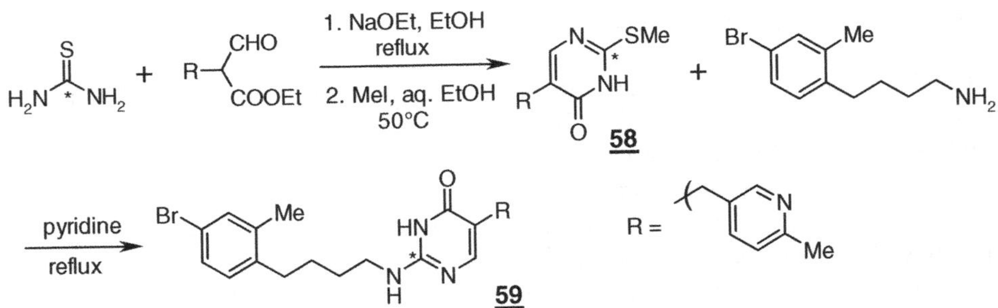


***Figure 9.15*** *Preparation of a [2-$^{14}$C]pyrimidine from [$^{14}$C]thiourea and a 2-formyl ester*

The reaction of α-halocarbonyl compounds or their equivalents with [$^{14}$C]thiourea leads to [$^{14}$C]isothiouronium salts that spontaneously cyclize to 2-amino[2-$^{14}$C]thiazoles (route B). The simplest member of the class, **60**, was an intermediate in the preparation of [$^{14}$C]-niridazole (**61**)[38]. This reaction has been widely exploited for the preparation of more complex 2-amino[2-$^{14}$C]thiazole derivatives as building blocks for the preparation of compounds such as the cephalosporin [$^{14}$C]E1101 (**62**)[39] and [$^{14}$C]cefepime (**63**)[40], and the β-lactam antibiotic [$^{14}$C]AMA-1080 (**64**)[41].

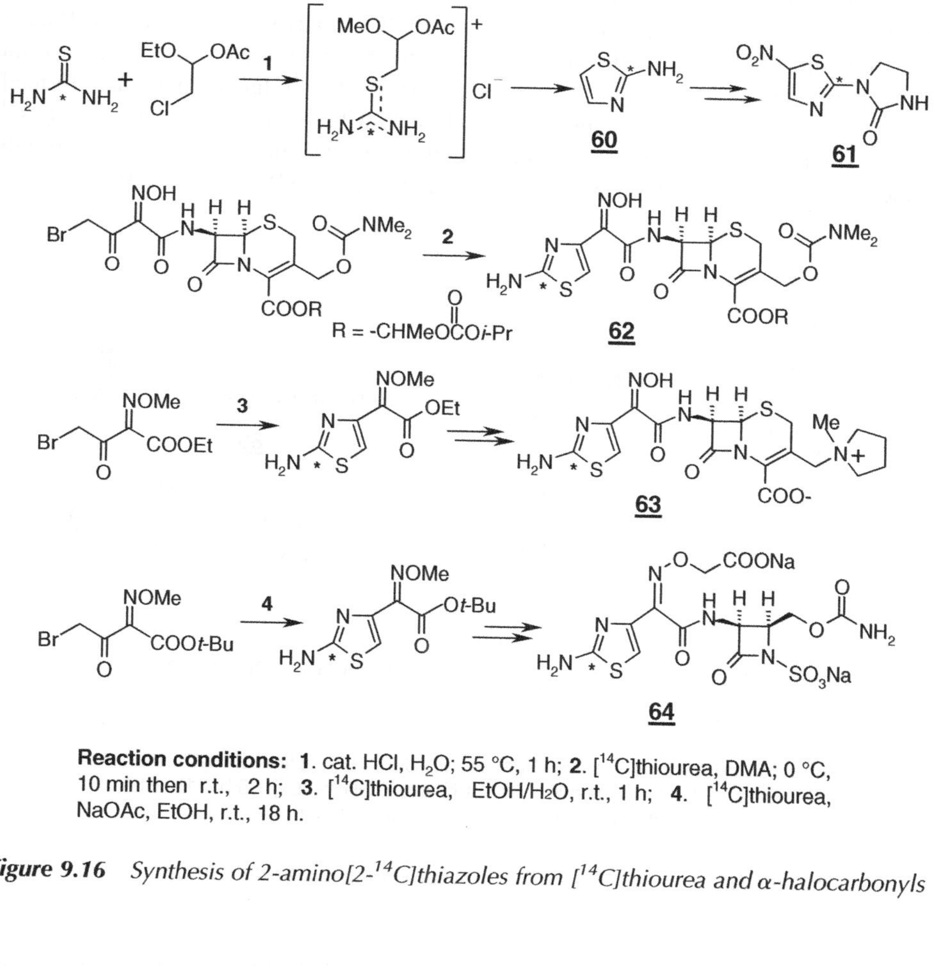

**Reaction conditions:** **1**. cat. HCl, $H_2O$; 55 °C, 1 h; **2**. [$^{14}$C]thiourea, DMA; 0 °C, 10 min then r.t., 2 h; **3**. [$^{14}$C]thiourea, EtOH/$H_2O$, r.t., 1 h; **4**. [$^{14}$C]thiourea, NaOAc, EtOH, r.t., 18 h.

***Figure 9.16*** *Synthesis of 2-amino[2-$^{14}$C]thiazoles from [$^{14}$C]thiourea and α-halocarbonyls*

Condensation of [$^{14}$C]thiourea with the vicinal haloamine **65** produced the 2-imino-[2-$^{14}$C]thiazolidine **66** in 77% yield[42] (Figure 9.17). In this case the substrate was designed so that further treatment with strong acid led to an efficient cyclization to carbon-14-labeled (*R*,*S*)-tetramisole (**67**), a broad spectrum anthelmintic drug substance of the imidazo[1,2-b] thiazole type (89% yield).

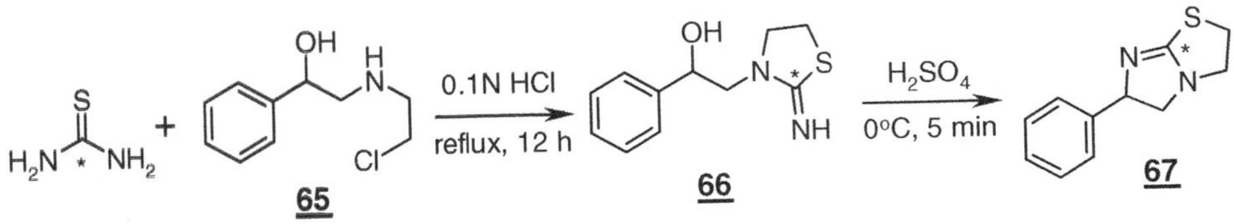


*Figure 9.17* Synthesis of an 2-amino[2-$^{14}$C]thiazoline from [$^{14}$C]thiourea and an α-haloamine

The sulfur replacement strategy can also be applied to [$^{14}$C]thiourea itself. Formation of [$^{14}$C]isothiouronium salts upon alkylation of [$^{14}$C]thiourea with dimethyl sulfate or ethyl bromide sets the stage for thiol displacement by appropriate nucleophiles. In the case of primary and secondary amines, substituted [$^{14}$C]guanidines are formed, as illustrated by the examples **68**[43] and $N_\alpha$-Boc,$N_\omega$-Cbz-[guanido-$^{14}$C]arginine (**69**)[44]. When α-aminoacetaldehyde diethyl acetal was used in the displacement step, the initially formed guanidine cyclized intramolecularly to 2-amino[2-$^{14}$C]imidazole (**70**)[45].

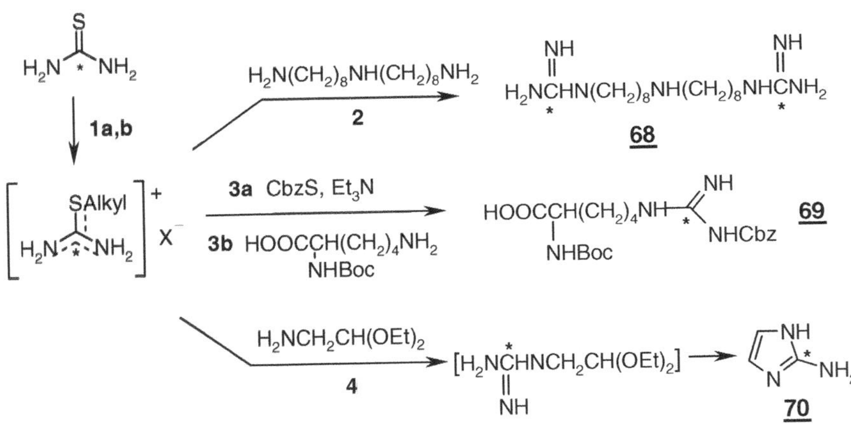


**Reaction conditions: 1a**. MeI, MeOH; $Me_2SO_4$, $H_2O$, reflux; **2**. $H_2O$, reflux, 2 h; **3a**. aq. EtOH; r.t., 20 min; **3b**. $Et_3N$, aq. EtOH; reflux, 1.5 h; **1b**. EtBr, EtOH; 60°C, 2 h; **4a**. 90°C, 1.5 h; **b**. HCl, reflux, 15 min

*Figure 9.18* S-Activation in [$^{14}$C]thiourea for nucleophilic desulfurization

Desulfurization of [$^{14}$C]thiourea can also be accomplished by reduction with Raney-Ni in the presence of $NH_4OAc$, providing [$^{14}$C]formamidine acetate in nearly quantitative yield[46]. This was used in a reaction with 5-amino-2-fluoropyridine-4-carboxylic acid (**71**) to produce the carbon-14-labeled pyrido[3,4-d]pyrimidine **72**, an intermediate in the synthesis of [$^{14}$C]PD0205520 (**73**), an EGFR tyrosine kinase inhibitor.

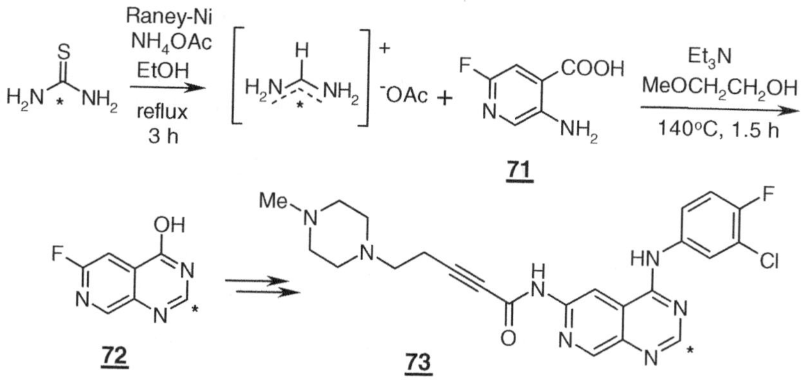


***Figure 9.19*** *Preparation and utilization of [$^{14}$C]formamidine*

# References


1. Murray, A. and Williams, D.L., *Organic Synthesis with Isotopes;* Interscience: New York, ***1958***, pp. 588–589.
2. Hsi, R.S.P., *J. Label. Compd. Radiopharm.*, ***1972***, *8*, 407–413.
3. Salter, R.; Bordeaux, K.; Burtscher, P.; Metz, Y.; Moenius, Th.; Rodriguez, I.; Voges, R. and Zueger, C., *J. Label. Compd. Radiopharm.*, ***2006***, *49*, 208–210.
4. Murray, A. and Williams, D.L., *Organic Synthesis with Isotopes;* Interscience: New York, ***1958***, pp. 272–273.
5. Rosenthal, G.A.; Downum, K.R. and Mattler, J.E., *Analytical Biochem.*, ***1983***, *133*, 277–282.
6. Taylor, G.F.; Zamani, K. and Kepler, J.A., *J. Label. Compd. Radiopharm.*, ***1988***, *25*, 1073–1080.
7. a. Murray, A. and Williams, D.L., *Organic Synthesis with Isotopes;* Interscience: New York, ***1958***, pp. 591–592; b. Bennett, L.L., *J. Am. Chem. Soc.*, ***1952***, *74*, 2432–2438.
8. a. Murray, A. and Ronzio, A.R., *J. Am. Chem Soc.*, ***1949***, *71*, 2245–2247; b. Murray, A. and Williams, D.L., *Organic Synthesis with Isotopes;* Interscience: New York, ***1958***, pp. 592–593.
9. a. Cashyap, M.M.; Mitchell, M.B.; Osborne, D.C. and Saunders, D., *J. Label. Compd. Radiopharm.*, ***1985***, *22*, 1239–1250; b. Murray, A. and Williams, D.L., *Organic Synthesis with Isotopes;* Interscience: New York, ***1958***, pp. 596–597.
10. a. Murray, A. and Williams, D.L., *Organic Synthesis with Isotopes*; Interscience: New York, ***1958***, pp. 760–763; b. Nguyen, H.N.; Herbert, M.; Nguyen, D.X. and Pichat, L., *J. Label. Compd.*, ***1971***, *7*, 299–305.
11. Hoellinger, H.; Nguyen, H.N. and Pichat, L., *J. Label. Compd.*, ***1973***, *9*, 161–166.
12. Bastide, J. and Badon, R., *J. Label. Compd. Radiopharm.*, ***1989***, *27*, 715–720.
13. Hill, J.A.; Witsowaty, C. and Darnofall, M.E., *J. Label. Compd. Radiopharm.*, ***1993***, *33*, 1119–1130.
14. a. Ayrey, G. and Yeomans, M.A., *J. Label. Compd. Radiopharm.*, ***1979***, *12*, 323–345; b. Balaban, A.T. and Bally, I., in *Isotopes in the Physical and Biomedical Sciences*, Vol 1, Part A, Buncel, E. and Jones, J.R., Eds.; Elsevier: Amsterdam, ***1987***, pp. 288–329.
15. a. Herbert, R.B. and Reichard, P., *Acta Chem. Scand.*, ***1955***, *9*, 251–254; b. Rutman, J.; Cantorow, A. and Paschkis, K.E., *Cancer Res.*, ***1954***, *14*, 119–123; c. Bendich, A.; Getler, H. and Brown, C.G., *J. Biol. Chem.*, ***1949***, *177*, 565–569.
16. a. Vander Borght, T.; Pauwels, S.; Lambotte, L.; De Saeger, C. and Beckers, C., *J. Label. Compd. Radiopharm.*, ***1990***, *28*, 819–822; b. Vander Borght, T.; Pauwels, S. and Lambotte, L., *Appl. Rad. Isotop.*, ***1991***, *42*, 103–104.

17. Codington, J.F.; Fecher, R.; Maguire, M.H.; Thomson, R.Y. and Brown, G.D., *J. Am. Chem. Soc.*, ***1958***, *80*, 5164–5168.
18. a. Murray, A. and Williams, D.L., *Organic Synthesis with Isotopes;* Interscience: New York, ***1958***, pp. 711–712. and 1120–1122; b. Haley, E.E. and Lambooy, J.P., *J. Am. Chem. Soc.*, ***1954***, *76*, 2926–2929.
19. Dugger, H.A. and Oppenheimer, E., *J. Label. Compd.*, ***1974***, *10*, 171–173.
20. a. Easter, J.A. and Stolle, W.T., *J. Label. Compd. Radiopharm.*, ***2001***, *44*, 797–810; b. Stolle, W.T.; Easter, J.A.; Chew, E.H.; McGrath, J.P.; Palmer, J.R. and Hsi, R.S.P., *J. Label. Compd. Radiopharm.*, ***1994***, *34*, 1187–1199.
21. Stolle, W.T.; Hsi, R.S.P. and Easter, J.A., in *Synthesis and Applications of Isotopically Labeled Compounds*, Vol. 7, Pleiss, U. and Voges, R., Eds.; John Wiley & Sons, Ltd: Chichester, England, ***2001***, pp 272–275.
22. Bonacorsi, Jr., S.A.; Burrell, R.C.; Luke, G.M.; Depue, J.S.; Rinehart, J.K.; Balasubramanian, B.; Christopher, L.J. and Iyer, R.A., *J. Label. Compd. Radiopharm.*, ***2007***, *50*, 65–71.
23. Patil, S.P. and Padmanabhan, D., *J. Label. Compd. Radiopharm.*, ***2002***, *45*, 539–542.
24. Murray, A. and Williams, D.L., *Organic Synthesis with Isotopes;* Interscience: New York, ***1958***, pp. 795–798.
25. Thijssen, J.B.A.; Knaeps, A.G. and Heykants, J.J.P., *J. Label. Compd. Radiopharm.*, ***1983***, *20*, 861–868.
26. Awata, N. and Satomi, O., *J. Label. Compd. Radiopharm.*, ***1987***, *24*, 331–338.
27. Janssen, C.G.M.; Lenoir, H.A.C.; Thijssen, J.B.A.; Knaeps, A.G.; Verluyten, W.L.M. and Heykants, J.J.P., *J. Label. Compd. Radiopharm.*, ***1988***, *25*, 783–792.
28. Frischknecht, P.M. and Baumann, T.W., *J. Label. Compd. Radiopharm.*, ***1979***, *16*, 669–672.
29. Heftmann, E., *J. Label. Compd.*, ***1971***, *7*, 463–465.
30. Canellakis, E.S. and Cohen, P.P., *J. Biol. Chem.*, ***1955***, *213*, 379–398.
31. Mathew, K.M.; Ravi, S.; Unny, V.K.P. and Savaprasad, N., *J. Label. Compd. Radiopharm.*, ***2006***, *49*, 699–705.
32. Ravi, S.; Mathew, K.M.; Padmanabham, D.; Unny, V.K.P. and Sivaprasad, N., *J. Label. Compd. Radiopharm.*, ***2006***, *49*, 339–343.
33. Jezdić, V.; Razumenić, N.; Skakun, M.; Albahari, S. and Odavić-Josić, J., *J. Label. Compd.*, ***1970***, *6*, 88–98.
34. Szabolcs, A.; Kruppa, G.; Sági, J. and Ötvös, L., *J. Label. Compd. Radiopharm.*, ***1978***, *14*, 713–725.
35. Zambito, A.J. and Chamberlin, E.M., *J. Label. Compd.*, ***1971***, *7*, 495–499.
36. Skakun-Todorovic, M and Jezdic, V. J., *J. Label. Compd.*, ***1973***, *9*, 475–482.
37. Thornhill, D.P. and Sitar, D.S., *J. Label. Compd.*, ***1971***, *7*, 145–148.
38. Faigle, J.W. and Keberle, H., *J. Label. Compd.*, ***1969***, *5*, 173–180.
39. Negi, S.; Komatsu, Y.; Tsuruoka, A. and Uemura, Y., *J. Label. Compd. Radiopharm.*, ***1995***, *38*, 1–11.
40. Standridge, R.T. and Swigor, J.E., *J. Label. Compd. Radiopharm.*, ***1993***, *33*, 759–765.
41. Imanishi, M.; Tomimoto, M.; Watanabe, M. and Hayashi, N., *J. Label. Compd. Radiopharm.*, ***1986***, *23*, 951–956.
42. Janssen, C.G.M.; Thijssen, J.B.A. and Verluyten, W.L.M., *J. Label. Compd. Radiopharm.*, ***2002***, *45*, 591–600.
43. Hudson, H.R.; Pianka, M. and Powroznik, L., *J. Label. Compd. Radiopharm.*, ***1980***, *17*, 383–387.
44. Koltai, E.; Bajusz, S.; Széll, E. and Zólyomi, G., *J. Label. Compd. Radiopharm.*, ***1987***, *24*, 659–665.
45. Fong, M.T. and Leaffer, M.A., *J. Label. Compd. Radiopharm.*, ***1986***, *23*, 981–985.
46. Zhang, Y.; Huang, Y. and Huang, C.C., *J. Label. Compd. Radiopharm.*, ***2005***, *48*, 485–496.

# 10

# Reconstitution Strategies

As discussed in Chapter 2, the vast majority of carbon-14 syntheses, and some tritium syntheses, utilize labeled low molecular weight building blocks, either commercially available or readily accessible from barium [$^{14}$C]carbonate or tritium gas. Target molecules are built up by utilizing one or more labeled building blocks in a series of consecutive synthetic steps (*construction* strategy). An alternative approach, *reconstitution*, starts with unlabeled target compounds and subjects them to procedures that conclude with the target in isotopically labeled form. The reconstitution strategy is often advantageous when the target is structurally complex or contains several stereogenic centers, and would pose a substantial synthetic challenge. There are two kinds of reconstitution approaches. In one, called *replacement* (Section 10.1), labels are introduced into unlabeled targets either by replacing individual atoms (e.g. hydrogen) or units (i.e. [$^{12}$C]methylene or [$^{12}$C]carboxylate) with their labeled counterparts without affecting the structural framework of the target. In the other, *disconnection–reconnection* (Section 10.2), a part of the structural framework is excised by a controlled degradative process, then built back in using an appropriate isotopically labeled building block.

## 10.1 Replacement Strategies

This category of strategies comprises one-step $^1H/^3H$, $^{12}CH_2/^{14}CH_2$ and $^{12}CO_2/^{14}CO_2$ exchange reactions, as well as multistep processes such as halogenations/triodehalogenations and consecutive oxidations/tritioreductions including the formation of C=O and C=C intermediates.

### 10.1.1 $^1H/^3H$ Replacement Strategies

Most tritium-labeled compounds are prepared by one of several hydrogen/tritium replacement processes. This is made possible by the facility with which carbon–hydrogen bonds can be broken and re-formed, by the simplicity of many such reactions, and by the different

*Preparation of Compounds Labeled with Tritium and Carbon-14* Rolf Voges, J. Richard Heys and Thomas Moenius

reactivity of hydrogens within molecules. The most direct method of tritium labeling without affecting the framework of target compounds is also the most trivial as a strategy, namely the exchange of $^1$H by $^3$H. The many variations of this method are discussed in Chapter 3. The most popular indirect method, halogenation followed by tritiodehalogenation, is discussed in Chapter 4, Section 4.2.

*Two-Step Oxidation–Reduction Strategies of Hydroxyl Functions*

Among the strategies not involving cleavage of any part of the target framework are a variety of oxidation–reduction procedures. The most common is used for preparing tritium-labeled targets possessing a primary or secondary hydroxyl function, and it involves oxidation to the aldehyde or ketone then reduction with a tritiated reducing agent. Two representative examples of primary alcohols are the preparation of [$^3$H]1[1] and [$^3$H]2[2].

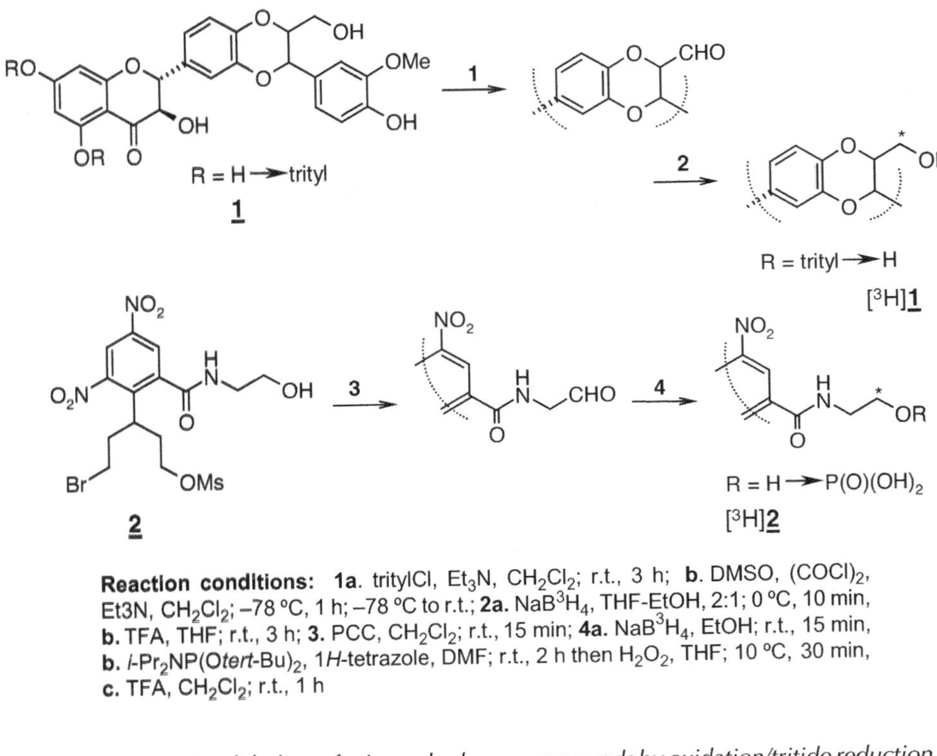


**Reaction conditions: 1a**. tritylCl, $Et_3N$, $CH_2Cl_2$; r.t., 3 h; **b**. DMSO, $(COCl)_2$, Et3N, $CH_2Cl_2$; –78 °C, 1 h; –78 °C to r.t.; **2a**. $NaB^3H_4$, THF-EtOH, 2:1; 0 °C, 10 min, **b**. TFA, THF; r.t., 3 h; **3**. PCC, $CH_2Cl_2$; r.t., 15 min; **4a**. $NaB^3H_4$, EtOH; r.t., 15 min, **b**. *i*-$Pr_2NP$(O*tert*-Bu)$_2$, 1*H*-tetrazole, DMF; r.t., 2 h then $H_2O_2$, THF; 10 °C, 30 min, **c**. TFA, $CH_2Cl_2$; r.t., 1 h

**Figure 10.1** *Tritium labeling of primary hydroxy compounds by oxidation/tritide reduction (1)*

The use of $RuO_4$ generated *in situ* from $RuCl_3$ and $NaIO_4$ converts primary alcohols into the corresponding carboxylic acids. Subsequent activation of the carboxyl function through formation of mixed anhydrides or active esters followed by tritide reduction allows for the incorporation of two tritium atoms, thereby producing compounds of enhanced specific activities. This sequence was exploited for the tritium labeling of the substituted cyclopropylglycine derivative 6. The requisite precursor 4 was obtained from $RuO_4$ oxidation of 3, a protected version of the target compound. Activation of the carboxyl group as

*N*-hydroxysuccinimide ester, followed by $NaB^3H_4$ reduction and cleavage of the protecting groups, provided [$^3$**H**]**6** at a specific activity of 14 Ci/mmol[3].

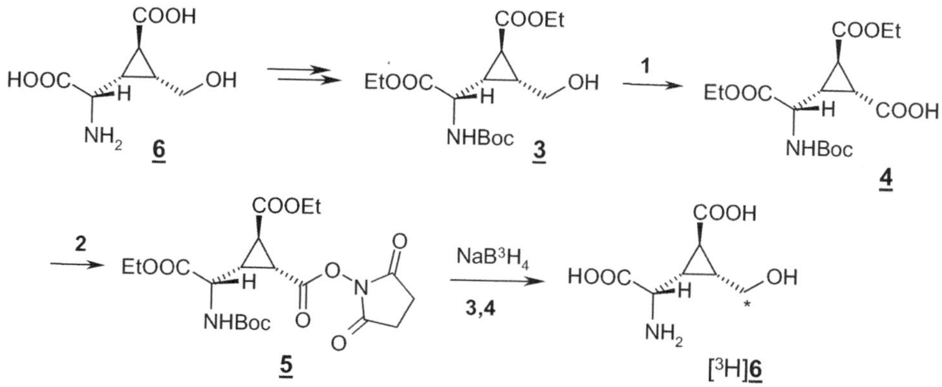

**Reaction conditions: 1**. $RuCl_3$, $NaIO_4$, $CCl_4$, $H_2O$; **2**. *N*-hydroxysuccinimide, EDCI, $CH_2Cl_2$; **3**. $NaB^3H_4$, THF; **4a**. TFA, $CH_2Cl_2$, **b**. NaOH

***Figure 10.2*** *Tritium labeling of primary hydroxy compounds by oxidation/tritide reduction (2)*

Tritium labeling of secondary alcohols by the oxidation–reduction approach involves tritioreductions of ketones, and when stereogenic centers are present in the molecule two diastereomeric tritiated alcohols can result. In the examples shown in Figure 10.3, conditions were successfully developed for the reductions of the ketones **8** and **10**, derived from the influenza virus hemagglutinin protein inhibitor **7**[4] and baccatin-III (**9**)[5], that in each case gave the desired diastereomer with high diastereoselectivity. The key intermediate **8** was prepared by coupling of the corresponding aroyl and cyclohexanone fragments, the latter made available by Swern oxidation of the cyclohexanol derivative **7**.

In the case of [1-$\beta$-$^3$H]cyclosporin A (**11**) (Figure 10.4) the keto precursor **12**, which is readily accessible through oxidation of **11** with NCS–dimethyl sulfide (Kim–Corey procedure) or DMSO, DCCI, and dichloroacetic acid (Pfitzer–Moffat variant), exists as a keto–enol mixture, eliminating one of the adjacent stereogenic centers. Accordingly, the subsequent $NaB^3H_4$ reduction generated a set of four diastereomers [$^3$**H**]**11** and **13**–**15**. The required diastereomer, however, was formed in 95% stereoselectivity and the residual diastereomers could be separated by two consecutive preparative HPLC operations[6].

The presence of additional oxidizable functionalities (e.g. OH, $NH_2$) may complicate these two-step strategies by requiring protection–deprotection steps, as illustrated by the example in Figure 10.5. In addition, it demonstrates that the stereoselectivity of the reductive step can be significantly improved by the use of a bulky tritide reagent such as Li(*sec*-Bu)$_3$B$^3$H (Li-Selectride, see Chapter 4, Section 4.3.7)[7]. In this example, the C23-carbonyl group and the C15,C17,C31,C61-hydroxy groups of **16** (Figure 10.5) were initially blocked by acid catayzed internal ketalization (step 1) and selective stepwise acylation (steps 2,3). Oxidation of the C35 hydroxy group and controlled hydrolytic cleavage of the reactive phenolic acetyl group under mildly basic conditions afforded key

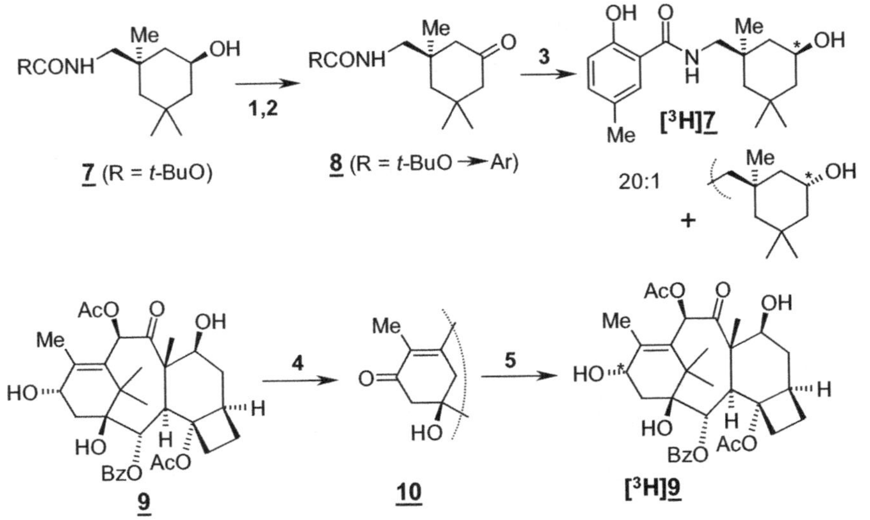


**Reaction conditions:** **1**. oxalyl chloride, DMSO, $Et_3N$, $CH_2Cl_2$; –50°C, 1 h; **2a**. 4N HCl-dioxane, r.t. ; 1 h, **b**. 2-acetoxy-5-methylbenzoyl chloride, $Et_3N$, DMF- $CH_2Cl_2$; 0°C, 1 h; **3a**. $LiB^3H_4$, THF; 0 to 25°C., 55 min, **b**. HPLC separation; **4a**. $MnO_2$, $CHCl_3$; r.t., 20 h, **b**. $Et_3SiCl$, pyridine; r.t., 24 h; **5a**. $NaB^3H_4$, $BF_3.OEt$, THF; r.t., 1 h, **b**. **10**; THF; r.t., 1 h, **c**. 48% HF, pyridine; r.t., 48 h

*Figure 10.3* *Tritium labeling of secondary hydroxy compounds by oxidation/tritide reduction in the presence of stereogenic centers (1)*

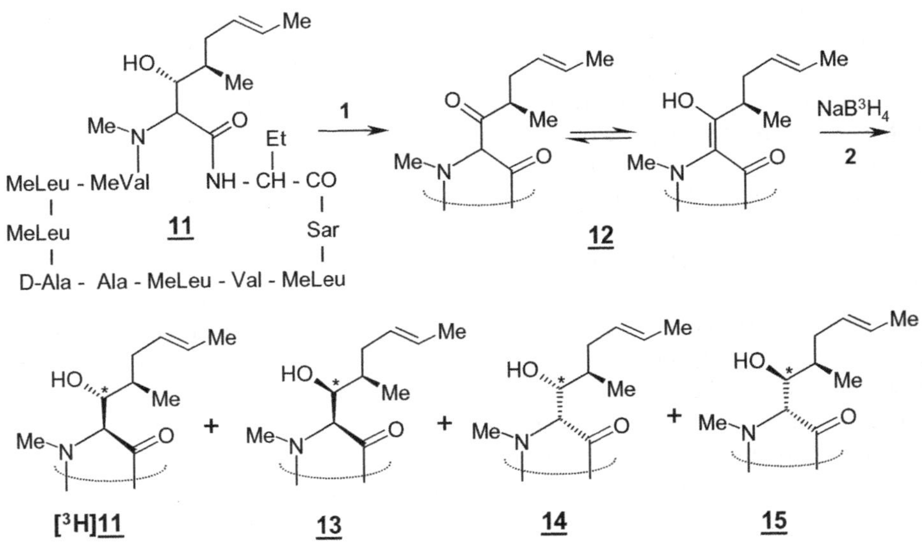


**Reaction conditions: 1**. NCS, $Me_2S$, THF **or** DMSO, DCCI, dichloroacetic acid, *tert*-BuOMe; 15 °C, 20 min; r.t., 4 h; **2**. $NaB^3H_4$, THF; 50 °C, 30 min; **3**. HPLC-separation

*Figure 10.4* *Tritium labeling of secondary hydroxy compounds by oxidation tritide reduction in the presence of stereogenic centers (2)*

intermediate **19** (steps 4,5). The subsequent tritide reduction and cleavage of the ketal function and the C13 trichloroaetyl protecting group (steps 6, 7) provided [$^3$H]**16** in diastereomerically pure form.

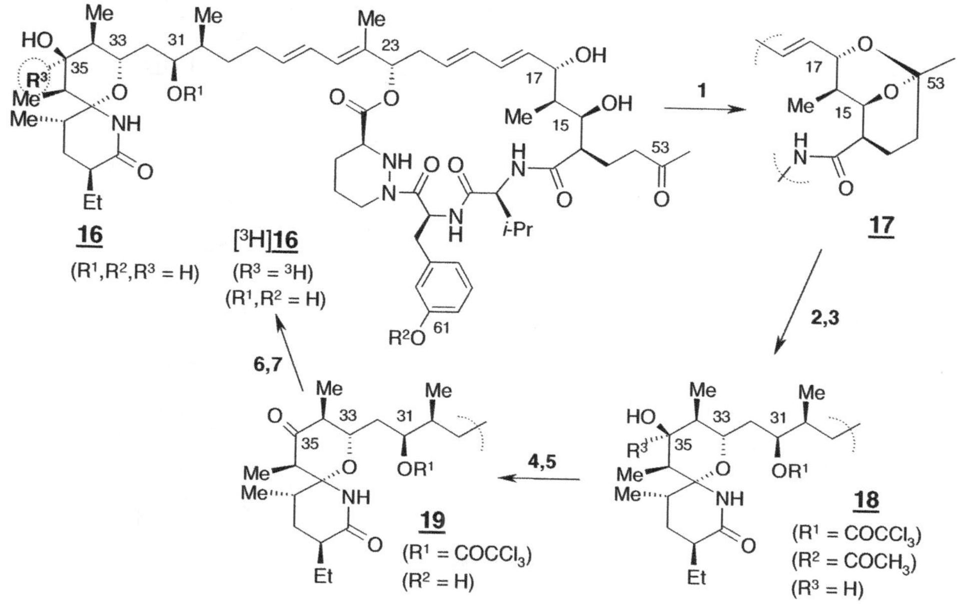


**Reaction conditions:** **1**. TFA (cat.), acetonitrile; **2**. $Ac_2O$, DMAP (cat), $CH_2Cl_2$–pyridine 27:1; **3**. $(Cl_3CO)_2O$, DMAP (cat.), $CH_2Cl_2$–pyridine 31:1; **4**. periodinane (Dess–Martin reagent), $CH_2Cl_2$; **5**. 1 M aq.$NaHCO_3$–MeOH 10:1 **6**. Li(*s*-Bu)$_3$B$^3$H, THF; –15 °C, 1 h; **7**. 1 N HCl, THF–$H_2O$ 1:1

***Figure 10.5*** *Tritium labeling of secondary hydroxy compounds by tritide reduction in the presence of stereogenic centers (3)*

A larger number of examples exists in which this oxidation–tritioreduction sequence has been used on intermediates in the syntheses of tritium-labeled target molecules[8].

The inverse sequence, namely tritioreduction of aldehydes followed by reoxidation of the resulting [$^3$H]alcohols, has been reported less often (see Figure 10.45). By simple statistics, the specific activity of the [$^3$H]aldehyde resulting from this procedure would be expected to be half that of the [$^3$H]alcohol. However, the existence of a kinetic isotope effect[9] in the oxidation and steric influences in certain compounds[10] can significantly influence the outcome, usually towards higher-than-statistical retention of tritium.

*Two-Step α,β-Dehydrogenation/Tritioreduction of Carbonyl Compounds*

Tritioreduction of $\alpha,\beta$-unsaturated carbonyl or carboxyl systems is the basis for another strategy for tritium labeling of target compounds (Figure 10.6). In a first step a double bond is introduced into the $\alpha,\beta$-position of a saturated carbonyl (or carboxyl) compound by an appropriate oxidizing reagent (e.g. DDQ, $(PhSeO)_2O$, or PheSeX (X=Br,Cl), or PhSeSePh followed by oxidation with $H_2O_2$, MCPBA or $NaIO_4$), then in a second step it is subjected to tritioreduction. This protocol has found frequent application in steroid labeling[11].

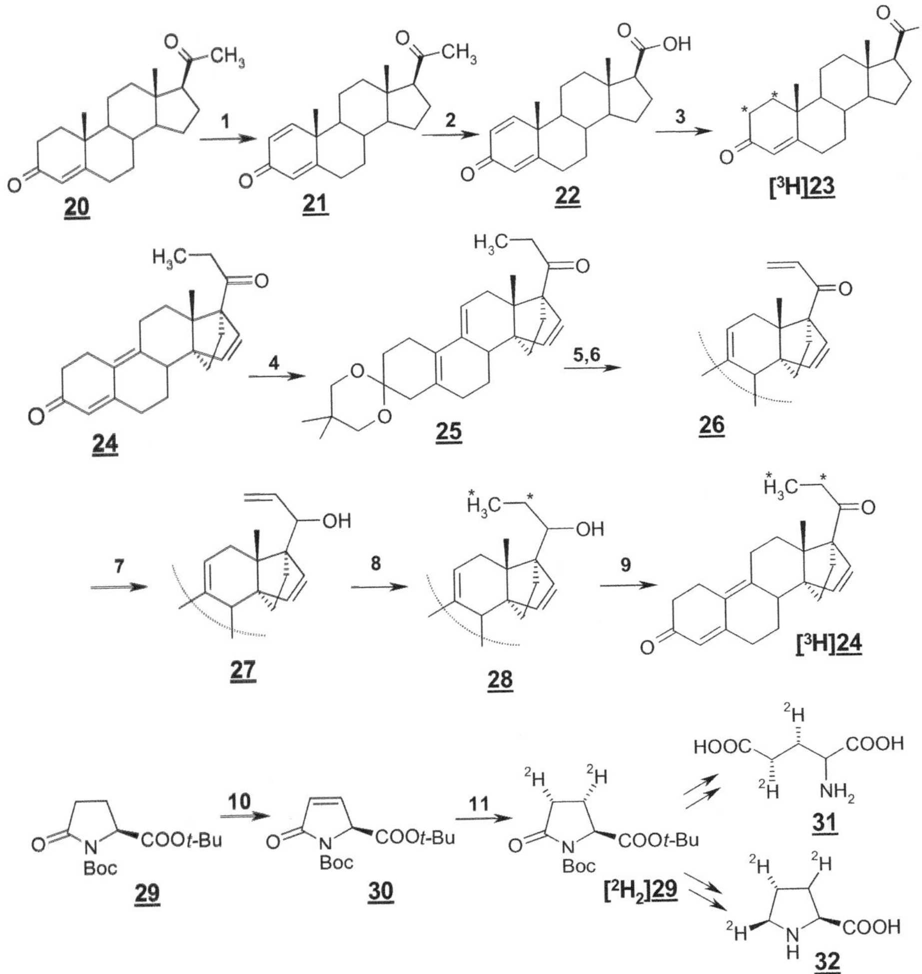

**Reaction conditions: 1.** DDQ, benzene; reflux, 18 h; **2a.** $Br_2$, NaOH, $H_2O$–*tert*-BuOH 5:1; 0–7 °C, 6 h, **b.** 21, *tert*-butanol, 0-7°C, 7 h, **c.** $Na_2S_2O_3$, 15 min; **3.** $^3H_2$,$(PPh_3)_3RhCl$, benzene–EtOH 3:1; r.t., 3 h; **4.** 2,2-dimethyl-1,3-propanediol, $HC(OMe)_3$, *p*-TsOH, $CH_2Cl_2$; r.t., 3 h; **5.** LDA, THF; −78 °C, 20 min; PhSeCl, −78 °C; 20 min; −78 °C → r.t.; **6a.** MCPBA, $CH_2Cl_2$; −35 °C, 30 min, **b.** pyridine, THF; −35 °C → r.t.; **7.** $NaBH_4$, $CeCl_3$; THF; r.t.; **8.** $^3H_2$, $(PPh_3)_3RhCl$, toluene, r.t., 24 h; **9a.** PCC, $CH_2Cl_2$; r.t., **b.** HOAc–$HClO_4$; **10a.** NaHMDS, DMPU, THF; −78 °C, 30 min, **b.** PhSeCl, THF; −78 °C, 2 h, **c.** $H_2O_2$, THF; r.t., 2 h; **11.** $^2H_2$, PdO, $MeO^2H$; r.t., 30 min

***Figure 10.6*** *Tritium labeling of carbonyl compounds through oxidative α,β-dehydrogenation/tritioreduction*

*α,β*-Dehydrogenation of progesterone (20) with DDQ, for example, furnished dienone 21, which upon oxidation of the methyl ketone side chain and reduction of the C1,C2-double bond with tritium gas in the presence of Wilkinson's catalyst gave 3-oxo[1,2-$^3$H] androst-4-ene-17*β*-carboxylic acid (23), key intermediate in the synthesis of 21-diazo-

[1,2-$^3H_2$]progesterone[12]. Analysis by $^3$H-NMR showed comparable incorporation of tritium into C1 (57%) and C2 (43%), but much more tritium in the less hindered $\alpha$-face (79%) compared with the $\beta$-face (21%). In the case of the gestagen precursor **25**, the introduction of the $\alpha$,$\beta$-double bond into the ketal-protected starting material was accomplished by treatment of the deprotonated material with PhSeCl followed by MCPBA oxidation and elimination of phenylselenic acid. Sequential hydride reduction of the carbonyl function, reduction of the resulting allylic alcohol side chain with tritium gas and Wilkinson's catalyst, reoxidation of the alcohol function and acid-catalyzed cleavage of the ketal converted **24** ($\Delta^{5:10,9:11}$) into the ketone **[$^3$H$_2$]24** ($\Delta^{4,9}$)[13]. The third example demonstrates that this strategy is not restricted to steroid derivatives. In this case the requisite $\alpha$,$\beta$-unsaturated precursor **30** was prepared from *tert*-butyl (2*S*)-*N*-BOC-pyroglutamate following a protocol similar to that for the previous example: Reduction of the double bond with deuterium gas in deuteromethanol in the presence of PdO, pretreated with excess deuterium gas in order to avoid $^1H/^2H$ scrambling, followed by deprotection and ring opening upon heating of the resultant [3,4-$^2H_2$]pyroglutamate (**[$^2$H$_2$]29**) afforded (2*S*,3*S*,4*R*)-[3,4-$^2H_2$]glutamic acid (**31**) in 60% e.e. owing to a partial racemization at the $\alpha$-position in the course of the reaction sequence. Stepwise reduction of the lactam carbonyl group with $LiEt_3BH$ and $Bu_3Sn^2H$ in the presence of $BF_3{\cdot}OEt_2$ opens access to (2*S*,3*S*,4*R*,5*S*)-[3,4,5-$^2H_3$]proline (**32**)[14].

The inverse procedure, double bond tritiation followed by re-oxidation, allows for the tritium labeling of $\alpha$,$\beta$-unsaturated carbonyl compounds (Figure 10.7). Applied to the complex steroid **33** it provided **[$^3$H]35**, in which 95% of the tritium was located at C2[15]. If most tritium in the intermediate **34** had been on the $\alpha$-face and equally distributed between C1 and C2, in analogy with the preparation of **[$^{14}$C]24** under very similar conditions, then the benzeneselenic anhydride-mediated dehydrogenation must have constituted a net 1$\alpha$,2$\beta$ elimination. This selectivity is consistent with the DDQ dehydrogenation of [1$\alpha$,2$\alpha$-$^3$H]androst-4-ene-3,17-dione (**36**) to give [2-$^3$H]androsta-1,4-dien-3,17-dione (**37**) with little tritium at C1, and DDQ dehydrogenation of [1$\beta$,2$\beta$-$^3$H]testosterone (**38**) to provide mainly 17$\beta$-hydroxy[1-$^3$H]androsta-1,4-dien-3-one (**39**)[16].

### 10.1.2 $^{12}C/^{14}C$ Replacement Strategies

There are two strategies allowing for a $^{12}C/^{14}C$ replacement without any previous or subsequent cleavage of a fragment of the target molecule: (a) cross-metathesis between a target molecule bearing a terminal ethylidene function and [$^{14}$C]ethylene, catalyzed by a Ru–carbene complex (2nd generation Grubbs' catalyst) and (b) $^{12}CO_2/^{14}CO_2$ exchange upon heating sodium or potassium carboxylates in an atmosphere of $^{14}CO_2$ at elevated temperatures. Despite being of only limited use, both offer quick one-step access to labeled target compounds.

#### *Cross-Metathesis-Effected $^{12}C/^{14}C$ Exchange*

Cross-metathesis is an interconversion of two olefinic substrates to produce two new olefinic species with swapped atoms or groups ($A{=}B + C{=}D \rightarrow A{=}C + B{=}D$). The interconversion is catalyzed by transition metal alkylidene complexes and occurs through alternating [2 + 2]cycloaddition and cycloreversion steps, via a metallacyclobutane

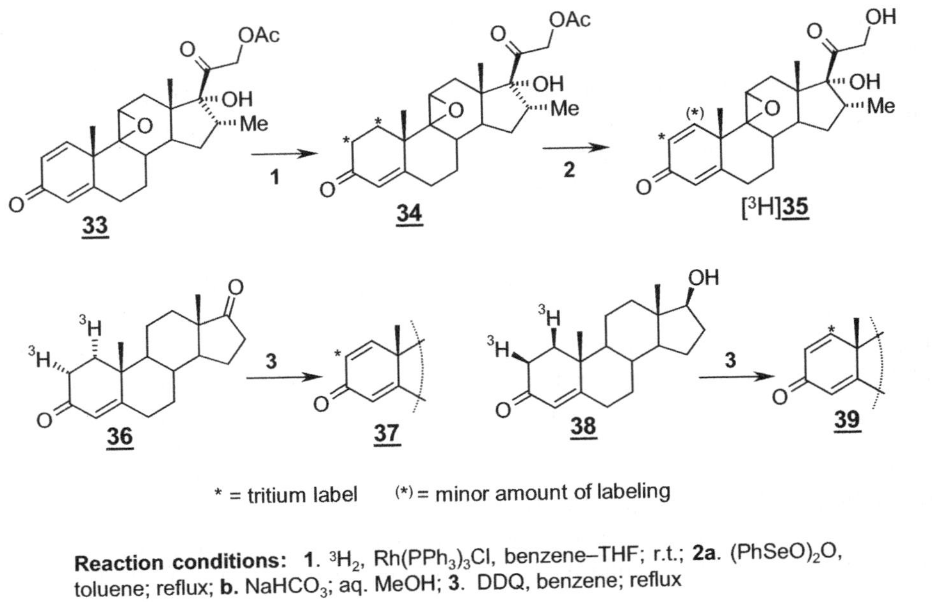

Figure 10.7 Tritium labeling of α,ß-unsaturated carbonyl compounds through tritioreduction/ α,β-dehydrogenation

intermediate[17]. Applied with [$^{14}C_2$]ethylene and a substrate possessing a terminal olefinic bond, it should constitute an example of a replacement approach by producing the structurally identical but terminal-labeled olefinic compound and [$^{14}C_1$]ethylene. This strategy was applied to the carbon-14 labeling of sagopilone (**40**), a synthetic epothilone derivative (Figure 10.8). Treatment of **40** (0.28 mmol) with a large excess of [$^{14}C_2$]ethylene (75 mCi, 1.3 mmol at 57 mCi/mmol) in the presence of **41**, Grubbs' 2nd generation Ru catalyst, afforded 2.54 mCi of [$^{14}$C]**40** at a specific activity of 26.3 mCi/mmol. The modest radiochemical yield of 3.4% is offset by the brevity of the procedure compared with alternative synthetic sequences[18].

### $^{12}CO_2/^{14}CO_2$ *Exchange Procedures*

In selected cases, the exchange of –COONa for –$^{14}$COONa by reacting a carboxylate with $^{14}CO_2$ is the quickest way to incorporate carbon-14 in a position of acceptable metabolic stability. In contrast to the broad spectrum of $^{1}H/^{3}H$ exchange methods, however, such transcarboxylation processes have been seldom used. Only aliphatic sodium or potassium carboxylates, containing at least one $\alpha$-hydrogen but lacking –$NH_2$, –OH and halogen substituents, undergo exchange to incorporate carbon-14 when heated in $^{14}CO_2$ at 1 atm for 1–2 h at temperatures ranging from 300 to 450 °C (Figure 10.9, Equation 1). Depending on the excess of $^{14}CO_2$ employed, radiochemical yields normally vary between 50 and 90% of the theoretical value and chemical yields generally exceed 80%[19]. Aromatic carboxylates tend to disproportionate under these reaction conditions to the thermodynamically most stable dicarboxylates and the parent hydrocarbon[20].

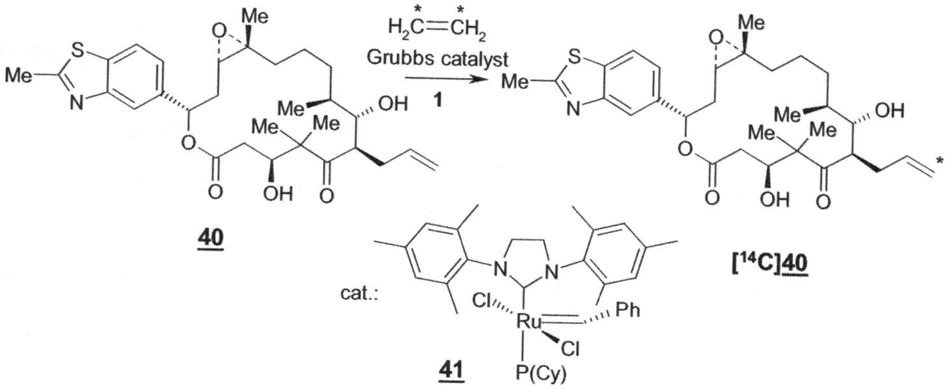


**Reaction conditions:** [$^{14}C_2$]ethylene, 6.5 mol% **41**, $CH_2Cl_2$; r.t., 16 h; 3.4 % radiochemical yield

***Figure 10.8*** *$^{12}C^{14}C$ replacement via cross-metathesis of a terminal olefinic bond*

The successful synthesis of racemic [$^{14}$C]ibuprofen (**[$^{14}$C]42**, 42.9 mCi, 17.6 mCi/mmol) by reaction of the sodium salt of the unlabeled drug substance with a deficiency of $^{14}CO_2$ (1.15 mmol, 52 mCi/mmol)[21] demonstrates that, whenever possible, transcarboxylation procedures should be taken into consideration.

$$R_1R_2CHCOONa,K + \overset{*}{C}O_2 \rightleftharpoons R_1R_2CH\overset{*}{C}OONa,K + CO_2 \qquad \text{equation 1}$$

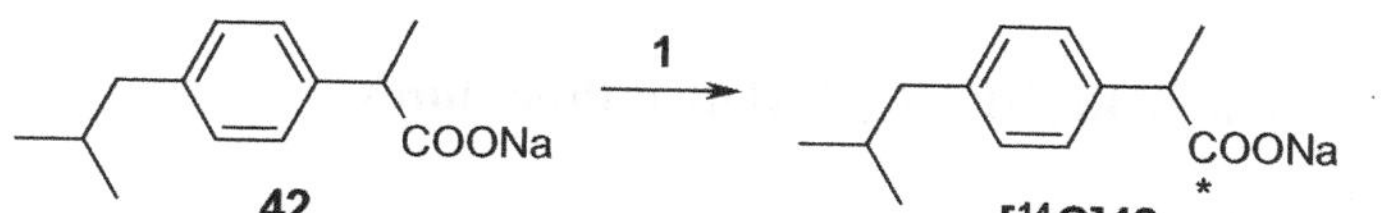


**Reaction conditions:** 1. $^{14}CO_2$ (0.36 equiv., 52 mCi/mmol); 350°C, 8 h

***Figure 10.9*** *Thermal $CO_2/^{14}CO_2$ replacement; application to the synthesis of racemic [carboxyl-$^{14}$C]ibuprofen*

Thermal transcarboxylation reactions are speculated to proceed via intermediate carbanions, which are trapped by $^{14}CO_2$ to give the corresponding [$^{14}$C]malonates, which then decarboxylate to labeled monoacids. This sequence can also be performed in two separate steps as illustrated by the synthesis of 6,11-dihydro[b,e]thiepin-11-one-3-yl[1-$^{14}$C]acetic acid (**[$^{14}$C]43**) (Figure 10.10). Reaction of the unlabeled drug substance **43** with two equivalents of LDA followed by carboxylation of the resulting dianion furnished malonate **44**, which spontaneously decarboxylated during workup to give **[$^{14}$C]43** at a specific activity of 9.1 mCi/mmol[22].

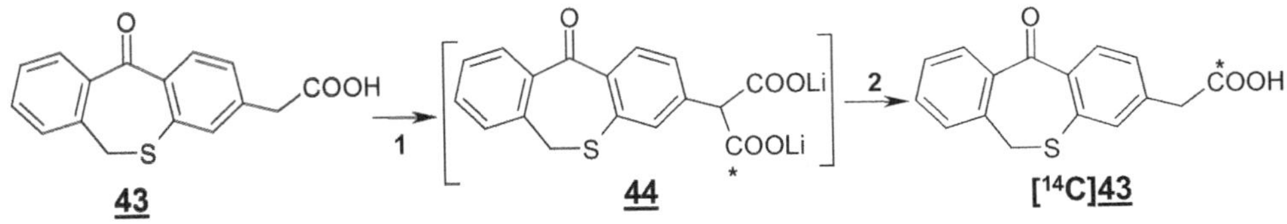

**Reaction conditions: 1a.** *n*-BuLi, *i*-$Pr_2NH$, THF, 0 °C, 15 min; **b.** HMPA; –78 °C, 30 min; **c. 43**, $^{14}CO_2$, –78 °C; **2.** 2 N HCl–THF

***Figure 10.10*** *Base-mediated $CO_2$/$^{14}CO_2$ replacement; application to the synthesis of 6,11-dihydro[b,e]thiepin-11-one-3-yl [1-$^{14}$C]acetic acid*

## 10.2 Disconnection–Reconnection Strategies

Aside from the $^1H/^3H$ exchange strategies discussed elsewhere in this text, those involving disconnection and replacement of molecular fragments by their labeled counterparts or analogs thereof, are of more preparative relevance, particularly for the synthesis of target compounds labeled with carbon-14. Among those methodologies available, the facile cleavage of heteroatom–carbon bonds opens access to a broad spectrum of compounds, especially those required for *in vitro* investigations where the heteroatom–[$^{14}$C]/[$^3$H]alkyl bond is certain to remain intact. As most of the required transformations are routine in normal synthetic chemistry, only special processes particularly suited to isotopic chemistry are discussed in the following sections. However, replacement of moieties that involve *disconnection* and *reconnection* of carbon–carbon single or double fragments requires a broad range of synthetic operations ranging from simple nucleophilic substitutions up to more complex Wittig–Horner–Emmons olefinations using enatiomerically pure substrates, previously prepared by sophisticated stereoselective methods. Therefore, this area of chemistry is discussed in more detail.

### 10.2.1 Dealkylation–Re[$^3$H/$^{14}$C]alkylation Procedures

*N-[$^3$H/$^{14}$C]Alkylamines*

Among alkylamines, it is most frequently the methyl homologs that have been converted to their isotopomers. Straightforward application of the *disconnection–reconnection* strategy requires a *N*-demethylation in the initial step; this is routinely accomplished by one of three methods: the classic von Braun reaction using cyanogen bromide followed by reductive cleavage of the resulting cyanamide (Zn–AcOH or $H_2$, Raney nickel)[23]; *N*-acylation using ethyl, $\alpha$-chloroethyl, phenyl or vinyl chloroformate and subsequent $H^+$-catalyzed solvolysis of the resultant carbamates[24]; or the Fe(II)-catalyzed rearrangement of amine oxides to secondary amines and aldehydes[25]. The last method does not require protection of functional groups such as hydroxy and primary and secondary amino. The carbon-14 labeling of methysergide (**45**)[26], *O*-protected morphine-6-$\beta$-D-glucuronide (**46**)[27] and [methyl-$^{14}$C]galantamine (**47**)[28] (i.e., changing $R=CH_3$ to $R={}^{14}CH_3$ in each case) (Figure 10.11) serve as illustrative examples.

However, as discussed in Chapter 5, Section 5.5.1, the remethylation steps can often be problematic. The reactivity of amines increases as primary < secondary < tertiary, so

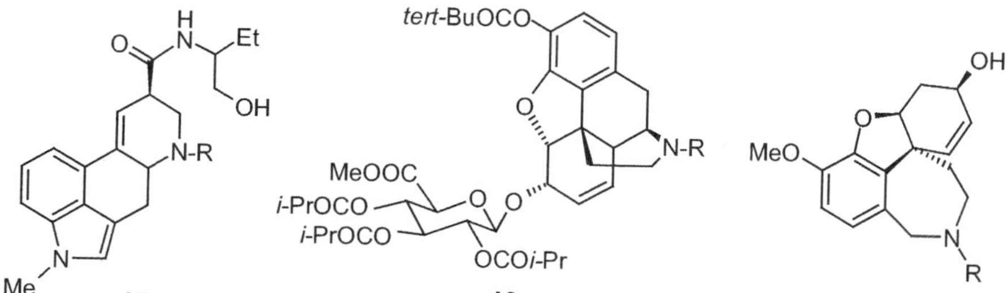


**45** R = $CH_3$ → CN → H → $^{14}CH_3$

**46** R = $CH_3$ → $COOCHClCH_3$ → H → $^{14}CH_3$

**47** R = $CH_3$ → $CH_3$,*N*-oxide → H → $^{14}CH_3$

**Reaction conditions:** **45**: **1**. BrCN, $CH_2Cl_2$; r.t., 6 h; **2**. $H_2$, Raney nickel, DMF; r.t., 24 h; **3**. $^{14}CH_3I$, *i*-$Pr_2NEt$, $CH_3CN$; r.t., 24 h; **46**: **1**. $ClCOOCHClCH_3$, $CH_2Cl_2$; reflux, 6 h; **2**. methanolic HCl; 45 °C, 2 h; **3**. $^{14}CH_3I$, $K_2CO_3$, DMF; r.t., 3 h; **47**: **1**. MCPBA, $CH_2Cl_2$; r.t., 75 min; **2**. $FeSO_4$, MeOH; 0 °C, 90 min; r.t., 1 h; **3**. $^{14}CH_3I$, *i*-$Pr_2NEt$, MeOH; r.t., 18 h

***Figure 10.11*** *Labeling of complex tertiary amines through N-demethylation–N-re[$^{14}$C]methylation procedures*

overalkylation frequently occurs, resulting in waste of [$^3$H]- or [$^{14}$C]methyl iodide. On the other hand, if a large excess of amine is used to mitigate this problem, it is sometimes difficult to separate the desired product from the excess amine reactant. There are two solutions to this problem, at least for tertiary amine target compounds. One consists of *N*-quaternization with benzyl or $\alpha$-methylbenzyl bromide, selective demethylation, and remethylation with labeled methyl iodide followed by hydrogenolytic debenzylation, as shown for [$^{13}$C, $^{13}H_3$]naratriptan (**[$^{13}$C, $^2H_3$]48**)[29] (Figure 10.12). The other involves installation of the [$^3$H]methyl group first, to give a [$^3$H]quaternary salt, and subsequent removal of one methyl group, as demonstrated by the preparation of [$^3$H]GR151004 (**[$^3$H]49**)[29]. The drawback inherent in the latter procedure is the lack of selectivity in the cleavage step between labeled and unlabeled methyl groups (theory would predict a small secondary isotope effect favoring cleavage of the former[30]). In this case the product specific activity was even lower than expected (39 Ci/mmol vs 2/3 $\times$ 80 Ci/mmol = 53.6 Ci/mmol).

*O-Dealkylation–Re[$^{14}$C]alkylation Procedures*

The methods used for *O*-dealkylations (ether cleavages) and re[$^{14}$C]alkylations (etherifications) are numerous and correspond to those used in normal synthetic organic chemistry. Reaction with 48% HBr[31], 48% HBr–HOAc 1:1, $AlCl_3$ in toluene (example **50**), $BBr_3$ in dichloromethane[32] or lithium butanethiolate (in the presence of acid labile functions, example **51**)[28] are the most frequently procedures used in radiosynthesis (Figure 10.13). Re[$^{14}$C]alkylation is usually accomplished by means of the Williamson reaction, involving treatment of alkoxides or aroxides with alkyl bromides or iodides in a polar solvent such as DMSO, DMF or $H_2O$. It should be pointed out that the presence of ancillary tertiary amino functions may cause serious problems, owing to the difficulty of discriminating between *O*-alkylation and *N*-quaternization. An excellent solution is given for [*O*-methyl-$^{14}$C]**52** in

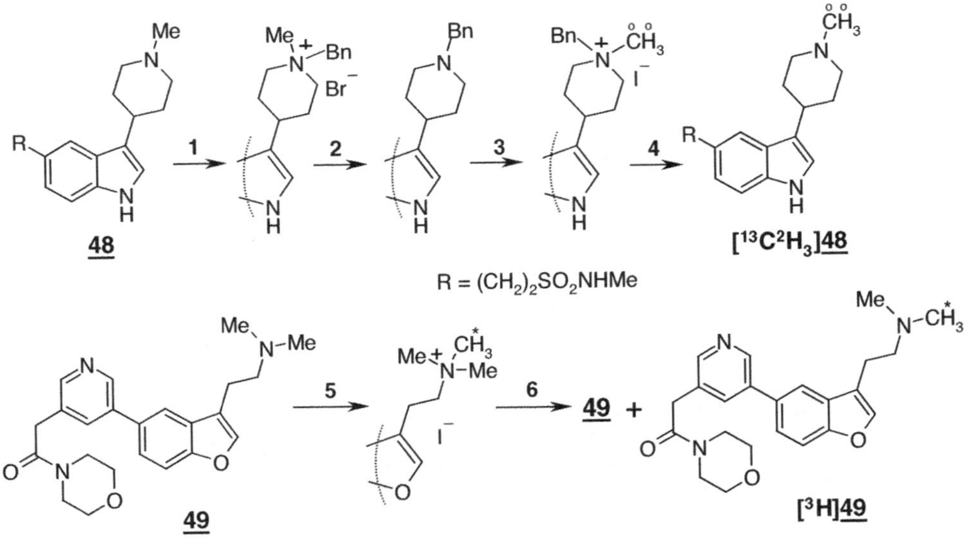


**Reaction conditions: 1** . $PhCH_2Br$, acetone; r.t., 1–3 d (74–99 %); **2.** lithium trisiamylborohydride, THF; reflux, 26 h (44 %); **3.**$^{13}C^2H_3I$, acetone; 20 °C, 72 h (87 %); **4.** IRA 400 ($Cl^-$), $H_2$, Pd/C, aq. EtOH (82 %); **5.** $C^3H_3I$ (80 Ci/mmol), toluene/$CH_2Cl_2$; 20 °C, 24 h (49 %); **6.** DABCO, NMP; 200 °C, 4 h (5 %)

***Figure 10.12*** *$^3H/^{14}C$ labeling of complex tertiary amines through N-[$^3H/^{14}C$]quaternization/selective N-dealkylation*

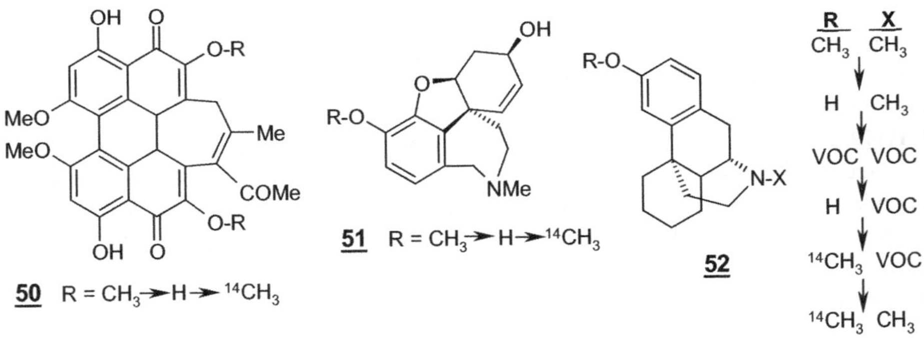


**Reaction conditions:** **50: 1**. $AlCl_3$, benzene; reflux, 1 h; **2.** $^{14}CH_3I$, CsF, THF; r.t. 24 h; **51: 1a.** $CH_3(CH_2)_3SH$, *n*-BuLi, HMPA; –10 °C→ r.t., **b. 51**, r.t.→ 92°C; **2a**. KOH, 1,2-dimethyl-2-imidazolidinone; r.t. 4 h, **b.** $^{14}CH_3I$; r.t., 18 h; **52: 1.** 48 % HBr; reflux, 18 h; **2.** vinyl chloroformate, proton sponge, 1,2-dichloroethane; 60 °C, 16 h; **3.** NaOH, dioxane–water 3:11, 50 °C, 4 h; **4a.** NaH, DMF; r.t.,1 h, **b.**$^{14}CH_3I$, DMF; r.t., 16 h; **5.** $LiAlH_4$, THF; 0 °C, 30 min; 0 °C → r.t., 16 h

***Figure 10.13*** *Labeling of complex ethers through O-demethylation–O-re[$^{14}C$]methylation procedures*

Chapter 5, Section 5.5.1. It involves *O*-dealkylation of dextromethophan (<u>52</u>) with 48% HBr, *N,O*-acylation with vinyl chloroformate, selective hydrolytic cleavage of the vinyl ester group and subsequent *O*-alkylation. Finally, treatment of the carbamate with $LiAlH_4$ regenerated the *N*-methyl group[32].

*S-Dealkylations–Re[$^3H/^{14}C$]alkylations*

The replacement of *S*-methyl groups with [$^3$H]- or [$^{14}$C]methyl groups in thioethers such as methionine resembles one of the above described procedures for tertiary *N*-methylamines. It is routinely used for labeling peptides with tritium or carbon-14 in their methionine residues[33], including in one case a peptide containing a cysteine–cysteine disulfide bond. The procedure involves (Figure 10.14) the reaction between isotopic methyl iodide and the respective peptide <u>53</u> in buffer to give the dimethylsulfonium derivatives <u>54</u>, followed by *S*-demethylation of <u>54</u> by incubation with excess mercaptoethanol. The specific activities of peptides **[$^3$H/$^{14}$C]<u>53</u>** were generally half that of the labeled methyl iodide employed, consistent with the expected nonselective removal of one methyl group of <u>54</u>.

A more elegant approach that avoids waste of labeled methyl iodide starts with oxidation of the thioether to the corresponding sulfoxide, which upon reaction with trifluoroacetic anhydride rearranges to give the trifluoroacetoxymethylsulfide <u>56</u> (Pummerer rearrangement) (Figure 10.15). One-pot hydrolytic cleavage of the trifluoroacetate residue and elimination of HCHO upon treatment with NaOH, and trapping of the resulting thiolate *in situ* with [$^{14}$C]methyl iodide, affords the labeled target compound. This sequence was successfully exploited for the preparation of **[$^{14}$C]<u>55</u>** and **[$^{14}$C]<u>57</u>**, which served as penultimate precursors for two carbon-14-labeled Cox-II inhibitors[34].

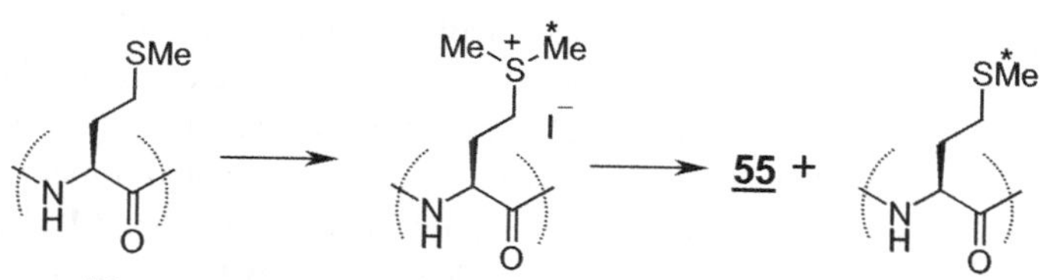


53 54 [$^3$H; $^{14}$C]53

***Figure 10.14*** *$^3H/^{14}C$ labeling of complex thioethers through S-[$^3H/^{14}C$]alkylation–selective dealkylation*

*C-Methyl/C-[$^3H/^{14}C$]Methyl Replacement*

The replacement of C-alkyl groups with their labeled counterparts is not as simple as with alkyl groups attached to heteroatoms. One potential general strategy, that of removing aromatic side chains through side chain oxidation (e.g. $KMnO_4$, $K_2Cr_2O_7$ etc.[35]) followed by halo-decarboxylation, then reconstituting them with isotopic side chains by transition metal-catalyzed cross-couplings (Stille, Suzuki, Kumada), has not been published so far. For aliphatic compounds no general solutions are available, and only substrate-specific solutions are possible. The carbon-14 and tritium labeling of mitomycin C (<u>63</u>), a potent anticancer drug against a variety of solid tumors, at a metabolically stable position may

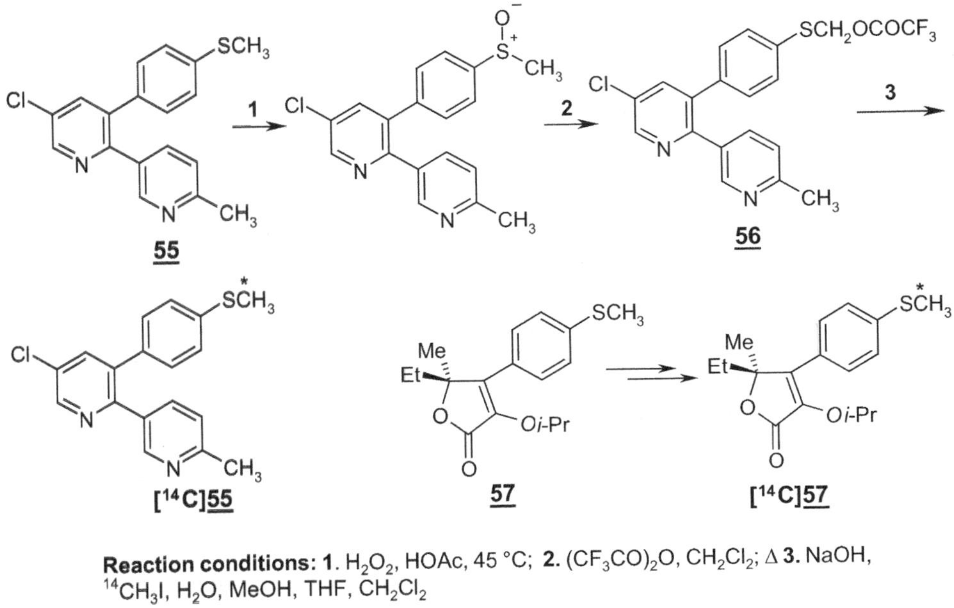


*Figure 10.15 S-methyl/[$^{14}$C]methyl replacement via Pummerer rearrangement; application to Cox inhibitor precursors*

serve as an example (Figure 10.16). The requisite key intermediate **59** was accessible from mitomycin A (**58**) in a four-step sequence involving liberation and ketalization of the carbonyl function at C7, $N_{1a}$-acetylation and introduction of the conjugated C=C bond at C6 through $\alpha$-selenylation and oxidative $\beta$-elimination. 1,2-addition of *N*-(phenylseleno) morpholine gave intermediate **60**, which spontaneously eliminated a methyleneiminium unit in a *retro*-Mannich reaction to give phenylseleno derivative **61**. Base-mediated alkylation using either [$^{14}$C]- or [$^{3}$H]methyl iodide followed by $Bu_3SnH$-initiated reductive deselenylation converted **61** into **62**. Concluding hydrolysis of the *N*-acetyl group and introduction of the amino group at C7 to give labeled [$^{3}$**H**/$^{14}$**C**]**63** was accomplished through treatment of **62** with methanolic ammonia[36].

## 10.2.2 $CO_2$/$^{14}CO_2$ Replacement Strategies

*Malonate Approach*

The one-pot transcarboxylation approach to [$^{14}$C]carboxylic acids, described in Section 10.1.2, can also be performed in two separate steps comprising [$^{14}$C]carboxylation of the carbanion generated *in situ* followed by isolation and thermal decomposition of the resultant [$^{14}$C]malonic acid species. This strategy was followed in the synthesis of 17$\beta$-[16-$^{14}$C]estradiol (**67**) (Figure 10.17). Dimethyl marrianolate methyl ether (**65**) was obtained through base-catalyzed ring cleavage of cyclic ketone **64** in a haloform reaction, followed by esterification of the newly generated carboxylic acids. Deprotonation of **65**, trapping of the ester enolate with $^{14}CO_2$ and subsequent saponification provided tricarboxylic acid **66**.

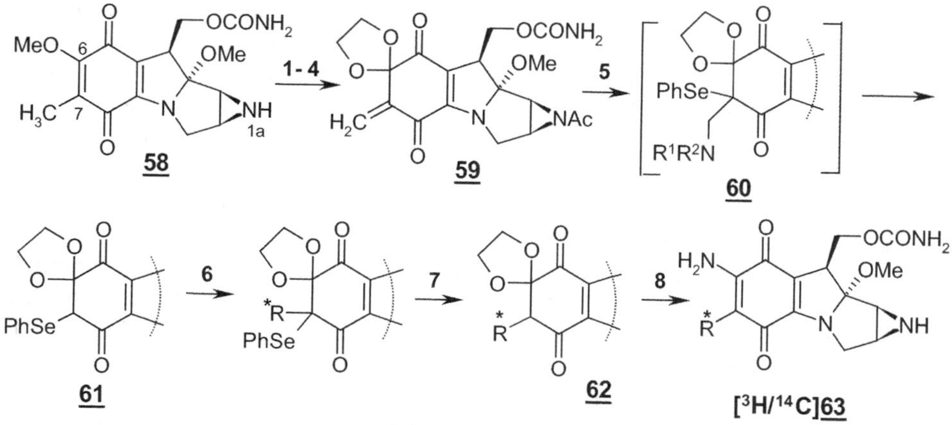


$R^1R^2NH$ = morpholine
R* = [$^3$H]methyl or [$^{14}$C]methyl

**Reaction conditions:** **1**. KOH, ethylene glycol, THF; r.t.; **2**. $Ac_2O$–pyridine, $CHCl_3$; r.t.; **3**. PhSeBr, $Et_3N$, acetonitrile; r.t.; **4**. MCPBA, $K_2CO_3$, $CHCl_3$; –40°C → r.t.; **5**. *N*-(phenylseleno)morpholine, $CHCl_3$; r.t.; **6**. [$^{14}$C]- or [$^3$H] methylodide, $K_2CO_3$, acetone; r.t, 48 h; 35 °C, 20 h; **7**. *n*-$Bu_3SnH$, $Et_3B$, benzene; r.t., 4 h; **8**. $NH_3$, MeOH; r.t., 16 h

***Figure 10.16*** *Methyl/[$^3$H/$^{14}$C]Methyl replacement at an α-methyl-α,β-unsaturated carbonyl compound via a retro-Mannich reaction: application to [$^{14}$C]mitomycin*

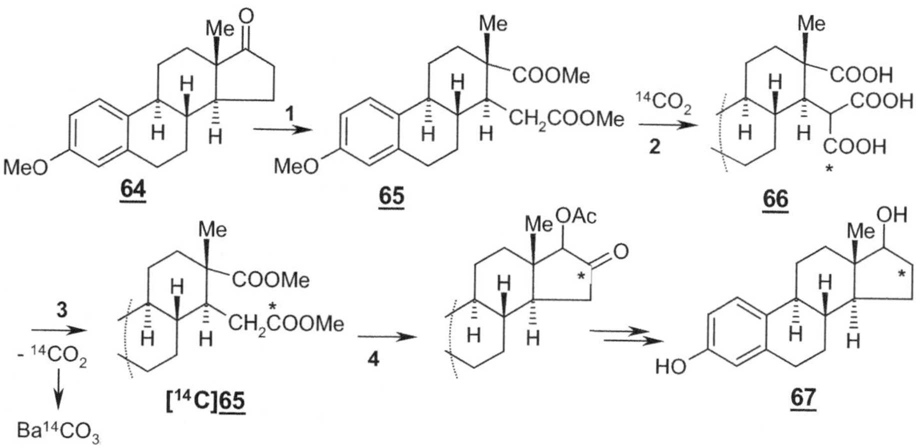


**Reaction conditions**: **1a**. $I_2$, KOH; MeOH–$H_2O$; r.t., 1.5 h, **b.** $H^+$, $CH_2N_2$, ether; r.t.; **2a**. $Ph_3CNa$, ether; r.t., 5 min, **b** .$^{14}CO_2$, ether; 0 °C → r.t.; **c**. aq. KOH-MeOH, reflux, 2 h then acidify.; **3a**. 180 °C; 1 h, **b**. $CH_2N_2$, ether; **4a**. Na, liquid ammonia–2:3: 1 h b. $Ac_2O$,py

***Figure 10.17*** *$CO_2$/$^{14}CO_2$ replacement via monolabeled malonic acid*

Thermal decarboxylation followed by esterification with diazomethane converted **66** into [$^{14}$C]**65** in 83% chemical yield (42% of $^{14}CO_2$ employed). The [$^{14}$C]carbon dioxide evolved during the decarboxylation stage was trapped and regenerated as barium [$^{14}$C]carbonate (44% of $^{14}CO_2$ employed)[37]. Concluding acyloin condensation, *O*-acylation removal of the carbonyl group at C16 through desulfurization of the corresponding thioketal and saponification of the OAc group at C17 provided **67** in an overall radiochemical yield of 13%.

Clearly, a major disadvantage of this procedure is loss of half of the radioactivity in the decarboxylation stage. The method can be improved by conducting the decarboxylation on an appropriate half-ester rather than on a di-acid species. If, for example, an appropriate alkyl chloro[$^{14}$C]formate is utilized as the electrophile, a carbon-14-labeled mixed malonate can be created, capable of being transformed in a more controlled way. This strategy was followed in the synthesis of [$^{14}$C]SB 209670 (**70**, R = H, R$^1$ = –$CH_2CH_2OH$) and [$^{14}$C]SB 217242 (**70**, R = H, R$^1$ = –$CH_2COOH$)[38] and of [$^{14}$C]LY 300502 ((–)-[$^{14}$C]**71**)[39].

Methyl esters **68** and **72** (Figure 10.18) were deprotonated and the resultant ester enolates trapped with isopropyl and ethyl chloro[$^{14}$C]formate, respectively, to give the mixed malonates **69** and **73**. Selective cleavage of the methyl ester function in each case with NaCN in DMSO or DMPU and concomitant decarboxylation converted **69** and **73** into the labeled esters **70** and **74**. In the first case cleavage of the OMEM-group, re-etherification of the resultant phenol with ethylene carbonate and methyl bromoacetate, respectively, and saponification of the ester functions afforded the required labeled drug substances. In the latter case *N*-deprotection and re-lactamization upon heating in toluene afforded labeled LY300502 ((–)-[$^{14}$C]**71**) in an excellent overall radiochemical yield of 48%.

### *Halo-Decarboxylation Procedures*

A more frequently used procedure for carbon-14 labeling involves initial conversion of carboxylic acid derivatives to halides decreased by one carbon. The halides can be converted into the corresponding [$^{14}$C]carboxylic acids either by reaction with alkali [$^{14}$C]cyanide and hydrolysis of the resulting [$^{14}$C]nitriles, or by metallation (Li, Mg) and carboxylation of the resulting organometallic species with $^{14}CO_2$. In the original version (Hunsdiecker reaction), a silver salt of the carboxylic acid undergoes reaction with molecular halogen ($Br_2 > I_2$ or $Cl_2$) (Figure 10.19, Route 1).

This reaction gives good results for linear and branched alkyl carboxylic acids, producing primary, secondary and tertiary halides. Many substituents are tolerated, as long as they are not located at the $\alpha$-position. Aromatic acids bearing electron-withdrawing groups also undergo the reaction with acceptable yields in most cases. Olefinic and electron-rich aromatic acids, however, are subject to side reactions (e.g. electrophilic substitution) and are, therefore, unsuitable. Silver carboxylates selected for Hunsdiecker degradation have to be of high chemical purity, which is often difficult to achieve. Therefore, two more convenient alternatives have been developed: use of a mixture of HgO, the respective halogen and the free acid, (Route 2a), and reaction of the free acid with $Pb(OAc)_4$ in the presence of halide ions or *N*-chlorosuccinimide (Route 2b). All these methods have been extensively applied to the synthesis of labeled target molecules.

Reaction of dry silver perfluorodecanoate (**76**) with iodine at 120 °C for 8 h, for example, furnished perfluorononyl iodide (**77**) in 75% chemical yield (Figure 10.20). Iodine/lithium

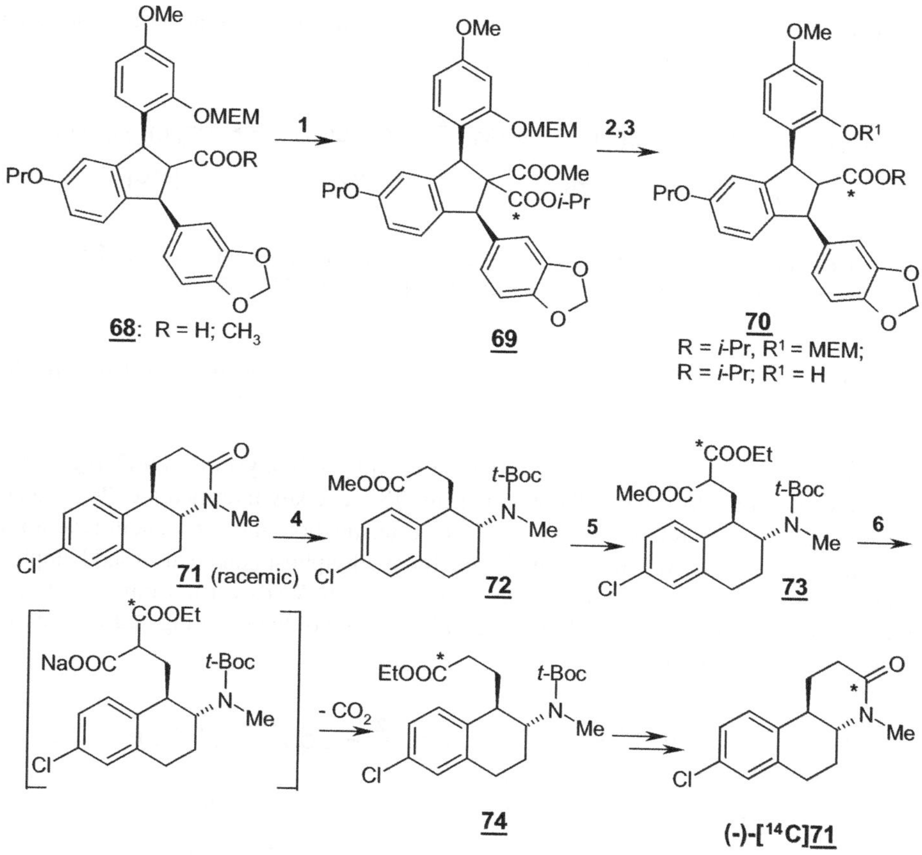


**Reaction conditions: 1a.** 68, LDA, THF; −78 °C; **b.** $Cl^{14}COO$*i*-Pr; **2.** NaCN, aq. DMSO; 165 °C; **3.** methanolic HCl; r.t.; **4a.** MeOH, $H_2SO_4$; Δ, **b.** (−)-di-*p*-toluoyl-L-tartaric acid, **c.** $Boc_2O$, aq. $NaHCO_3$; r.t., 2 h; **5.** LICA, THF; −78 °C, **b.** $Cl^{14}COOEt$, THF; −78 °C, 0.5 h; **6.** NaCN, DMPU; 85 °C, 8 h

***Figure 10.18*** *$CO_2/^{14}CO_2$ replacement via mixed dialkyl malonates: application to the synthesis of [$^{14}$C]217242, [$^{14}$C]SB 209670 and [$^{14}$C]LY 300502*

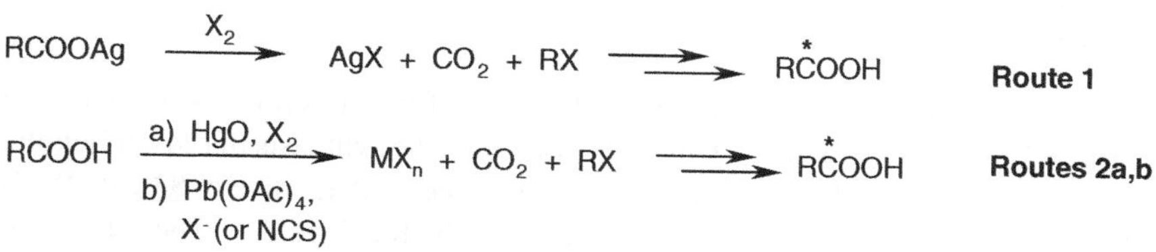


***Figure 10.19*** *Variants of the classical Hunsdiecker halo-decarboxylation procedure*

exchange followed by carboxylation with $^{14}CO_2$ at low temperatures regenerated the parent acid [$^{14}$C]75 in carbon-14-labeled form[40].

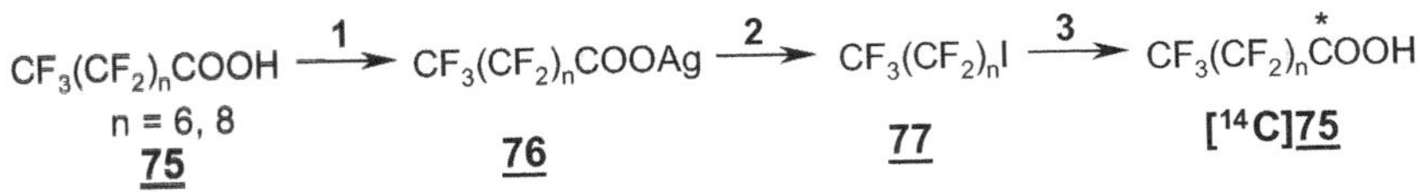


**Reaction conditions:** **1**. $Ag_2O$, $H_2O$; 50–60 °C, 3.5 h; **2**. $I_2$; 120 °C, 8 h; **3a**. MeLi, ether; −105 °C, **b**. $^{14}CO_2$; −196 °C; −100 °C → −75 °C, 20 min, **c**. 2N $H_2SO_4$

***Figure 10.20*** *$CO_2/^{14}CO_2$ replacement via Hunsdieker halo-decarboxylation–re[$^{14}$C] carboxylation*

For the preparation of prostacyclin derivative ([3-$^{14}$C]nonyl)SM-10902 (**82**), halo-decarboxylation of (3*S*)-3-methylheptanoic acid (**78**) gave key intermediate **79**. Grignard reagent preparation, carboxylation with $^{14}CO_2$, esterification and subsequent reaction of the resulting ethyl ester **80** with lithiated dimethyl methylphosphonate afforded ketophosphonate **81** in 70% overall radiochemical yield. Horner–Wadsworth–Emmons reaction with the corresponding aldehyde derivative and reduction of the resulting $\alpha,\beta$-unsaturated ketone converted **81** into **82**[41].

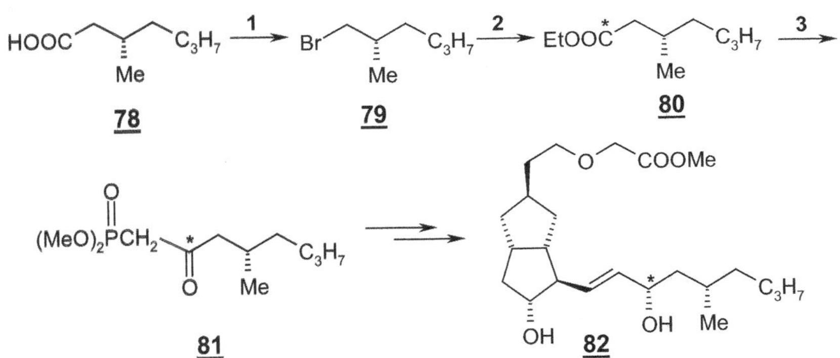


**Reaction conditions:** **1**. HgO, $Br_2$, $CCl_4$; reflux, 1 h; **2a**. Mg, $I_2$ (cat), THF; reflux, 1h, **b**.$^{14}CO_2$, THF, -20°C, 2h, **c**. EtOH, conc. $H_2SO_4$; 70°C, 4.5 h; **3a**. $(MeO)_2PMe$, *n*-BuLi, THF; -60°C, 1.5 h, **b**. 80°C, THF; -60°C, 4 h

***Figure 10.21*** *$CO_2/^{14}CO_2$ replacement via modified Hunsdieker halo-decarboxylation/ re[$^{14}$C]carboxylation*

The use of $Pb(OAc)_4$ in the presence of LiCl has been found in some cases[42] to be superior to the classic Hunsdiecker method. This is illustrated by the synthesis of [24-$^{14}$C]cholic acid ([$^{14}$C]**83**) (Figure 10.22). Heating of the *O*-formyl protected derivative of **83** with $Pb(OAc)_4$ and LiCl furnished chloro intermediate **84** in 62–68% chemical yield[42a]. The carboxyl group was re-introduced by cyano-dechlorination of **84**. Subsequent simultaneous hydrolytic cleavage of the [$^{14}$C]nitrile and formyl functions gave [$^{14}$C]**83** in 41% radiochemical yield over two steps[42c]. The product was used to prepare the bis-amide [$^{14}$C]SCH 209702 (**85**), under investigation as an agent to facilitate entry of therapeutic agents into cells. The procedure for the preparation of [24-$^{14}$C]cavicholic acid ([**$^{14}$C]86**)[42c] and of [24-$^{13}$C]-chenodeoxycholic acid[42b] paralleled that of the synthesis of [**$^{14}$C]83**.

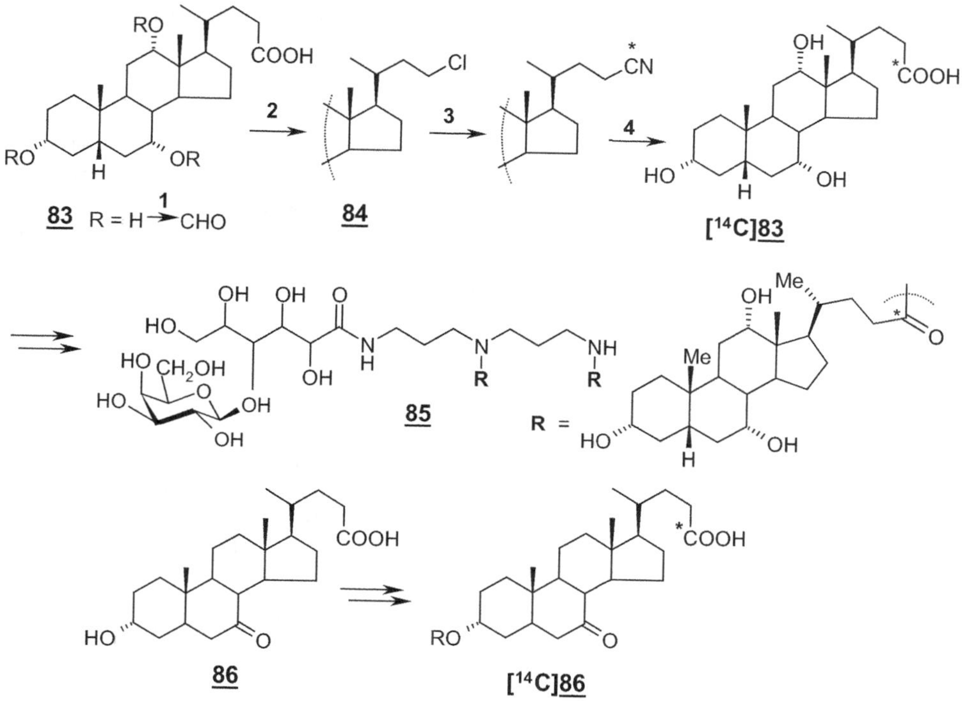


**Reaction conditions: 1.** 88 % formic acid, $Ac_2O$; 55 °C, 1.5 h; **2.** $Pb(OAc)_4$, LiCl, benzene; reflux; **3.** $K^{14}CN$, DMF; 100 °C, 3 h; **4.** NaOH, EtOH; reflux, 48 h

***Figure 10.22*** *$CO_2/^{14}CO_2$ replacement via modified Hunsdiecker halo-decarboxylation/re [$^{14}$C]carboxylation (2)*

The difficulties inherent in the original Hunsdiecker reaction and its modifications stimulated the development of an additional halo-decarboxylation method that involves treatment of thiohydroxamic esters of carboxylic acids with $BrCCl_3$, $ICH_3$ or $CH_2I_2$ in the presence of a radical initiator (Route 3, Barton reaction, Figure 10.23).

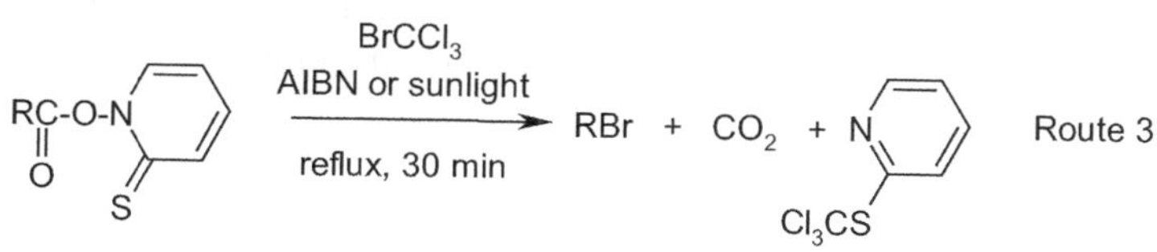


***Figure 10.23*** *Alternative halo-decarboxylation following the Barton procedure*

This method works also for electron-rich aromatic and $\alpha,\beta$-unsaturated carboxylic acids and requires neither heavy metals nor strong oxidants. Among the various thiohydroxamic esters investigated, *N*-hydroxypyridine-2-thione derivatives are exceptionally efficient. They are conveniently available by reaction of the respective acyl chloride with the sodium salt of *N*-hydroxypyridine-2-thione or by DCC-mediated coupling of the free acid. The Barton reaction gives the best yields of bromo derivatives; the corresponding iodo derivatives are produced in much lower yields. Chlorine donors do not react[43].

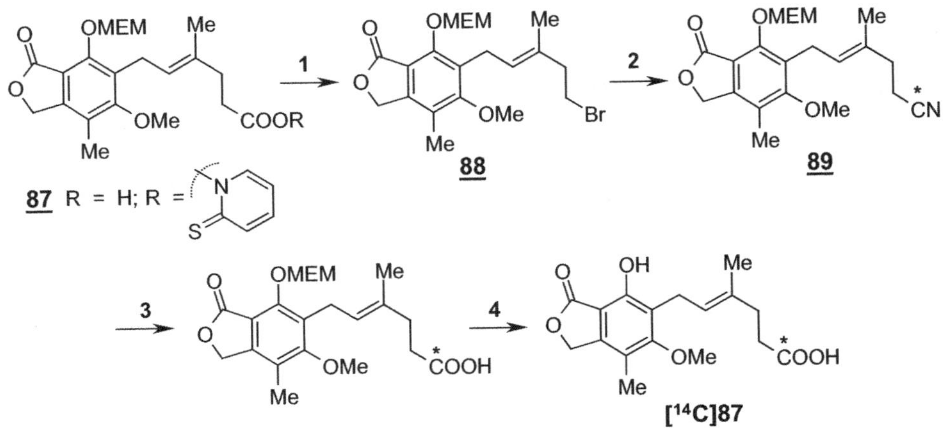


**Reaction conditions: 1a.** hydroxypyridine-2-thiol, DCC, $CH_2Cl_2$; r.t., 45 min, **1b.** $BrCCl_3$; hν; **2.** $Na^{14}CN$, DMSO, 100 °C, 3.5 h; **3.** 20 % KOH, $MeOCH_2CH_2OH$,; reflux, 16 h; **4.** 15 % HCl, THF; reflux, 75 min

***Figure 10.24*** *$CO_2/^{14}CO_2$ replacement via Barton halo-decarboxylation–re[$^{14}$C]carboxylation*

This methodology, applied to the MEM-protected thiohydroxamic ester of mycophenolic acid (**87**), gave the requisite bromide **88** in 43% chemical yield. Treatment with $K^{14}CN$, followed by hydrolysis of the resulting [$^{14}$C]nitrile **89** and acid-catalyzed deprotection converted **88** into **[$^{14}$C]87**, precursor to the immunosuppressive drug mofetil[44].

A similar strategy was followed in the synthesis of e.p. (1*S*,2*S*)- and (1*R*,2*R*)-[carboxy-$^{14}$C]jasmonic acid (**[$^{14}$C]90**) (Figure 10.25). Halo-decarboxylation of the racemic unlabeled target **90** following Barton's procedure gave the corresponding bromide in 67% yield. Protection of the carbonyl function through ketalization, [$^{14}$C]cyano-dehalogenation, and de-ketalization afforded racemic **91**. Treatment of **91** with e.p. semioxamazide **92**, separation of the resultant mixture of diastereomers **93** by preparative HPLC, acid-catalyzed cleavage of the hydrazone and concluding hydrolysis of the resultant e.p. [$^{14}$C]nitrile **94** proceeded in overall 3% radiochemical yield (for both enantiomers)[45].

*Hydroxy-Decarboxylation Approach*

Replacement of carboxyl groups via hydroxy-decarboxylation, although formally involving the oxidative cleavage of a carbon–carbon bond (Section 10.2.4), is discussed here because of

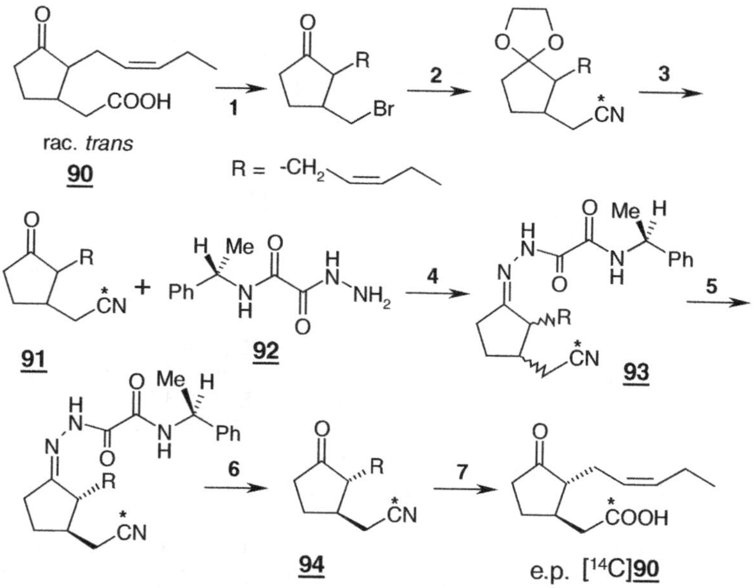

Reaction conditions: 1a. $(COCl)_2$, toluene; r.t., 3 h, b. 2-mercaptopyridine-1-oxide sodium, DMAP, $BrCCl_3$; reflux, 10 min; 2a. $HOCH_2CH_2OH$, TsOH, benzene; reflux, 4.5 h, b. $K^{14}CN$, DMSO; 80 °C, 5 h; 3. HOAc–3N HCl 1:1, reflux, 4 h; THF; r.t., 2 h; 4. TsOH (cat.), benzene reflux, 4 h; 5. prep. HPLC; 6.17 % $H_2SO_4$, benzene; reflux, 16 h; 7. 30 % NaOH, aq. EtOH; reflux, 48 h

*Figure 10.25* *$CO_2/^{14}CO_2$ replacement via Barton halo-decarboxylation–re[$^{14}$C]carboxylation (2)*

its analogy to the preceding chemistries. As applied to the preparation of [$^{14}$C]**95** (Figure 10.26), it starts with the α-hydroxylation of the derived ester **96** using (10-camphorsulfonyl) oxaziridine (Davis' reagent). Reduction of the resultant α-hydroxy ester **97** provided the 1,2-diol **98**, periodate cleavage removed the terminal carbon, and reduction of the resultant aldehyde afforded alcohol **99**. Activation of the hydroxyl group with mesyl chloride, followed by [$^{14}$C]cyanation of the resultant mesylate and concluding hydrolysis of the [$^{14}$C]-nitrile returned the target compound [$^{14}$C]**95** in carbon-14-labeled form[46].

*Amino-Decarboxylation Approach*

When halo-decarboxylations fail, the classic Curtius acyl azide degradation may offer a useful, but lengthy alternative[47] (Figure 10.27). An acyl azide such as **101**, conveniently formed from the carboxylic acid (**100**) either by hydrazinolysis of an ester and subsequent treatment of the hydrazide with $HNO_2$ or $C_5H_{11}ONO$, or by reaction of an acyl halide with sodium azide, loses $N_2$ upon heating and rearranges to an isocyanate. In the presence of water, the isocyanate hydrolyzes to the primary amine (e.g. **102**). Treatment of the amine with nitrous acid gives a diazonium salt which spontaneously decomposes to the corresponding alcohol. Activation of the hydroxy group through mesylation, replacement of the

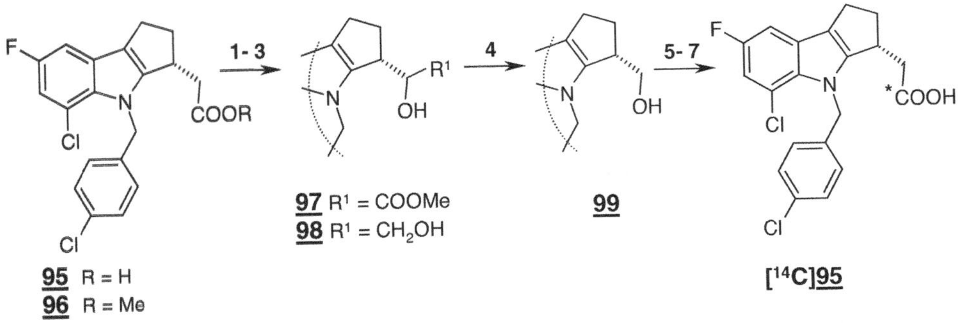


**Reaction conditions: 1.** $CH_2N_2$, ether, r.t.; **2.** KHMDS, Davis reagent, THF; –78 °C, 1 h; **3.** DIBAL, THF; –40 °C, 1 h; **4a.** $NaIO_4$, THF–$H_2O$ 3:1; r.t., 15 min, **b.** DIBAL, THF–toluene 1:1; –40 °C, 1 h; **5.** MsCl, $Et_3N$, $CH_2Cl_2$; –30 °C, 30 min; r.t., 30 min; **6.** $Na^{14}CN$, DMSO; 60 °C, 18 h; **7.** 5 N NaOH, EtOH; reflux, 18 h

***Figure 10.26*** *$CO_2$/$^{14}CO_2$ replacement via hydroxyl-decarboxylation–re[$^{14}$C]carboxylation*

mesylate function with alkali [$^{14}$C]cyanides and subsequent hydrolysis of the resulting [$^{14}$C]nitrile (e.g. **103**) completes the transformation from –COOH (**100**) to –$^{14}$COOH ([$^{14}$C]-**100**). This approach has been used in the synthesis of several ergoline-8-[carbonyl-$^{14}$C]-carboxamides (e.g. [$^{14}$C]LY281067, **104**)[47a].

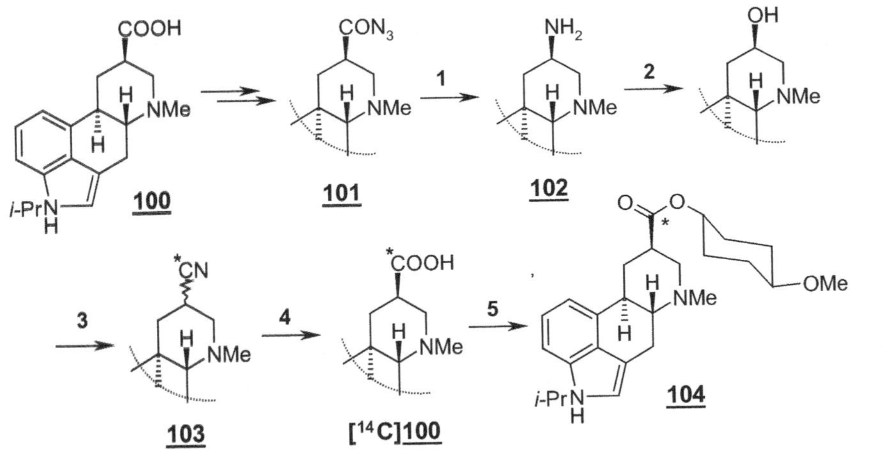


**Reaction conditions: 1a.** MeOH, $H_2SO_4$; r.t., 20 h, **b.** $H_2NNH_2$; reflux, 2,5 h, **c.** $NaNO_2$, 1N HCl, <–10 °C, 30 min; 75 °C, 45 min; **2.** $NaNO_2$, HOAc; 10 °C, 1 h; **3a.** MesCl, pyridine; 10–20 °C, 2 h, **b.** $Na^{14}CN$, DMF; 60 °C, 4 h; **4.** 5N NaOH, MeOH; reflux, 24 h; **5.** *cis*-4-methoxycyclohexyl tosylate, $K_2CO_3$, DMF; 65–70 °C, 22 h

***Figure 10.27*** *$CO_2$/$^{14}CO_2$ replacement via amino-decarboxylation–re[$^{14}$C]carboxylation*

Aromatic diazonium salts need no further transformations but can be directly cyanated with $Cu^{14}CN$, as demonstrated in the preparation of [$^{14}$C]**105**[48]. Alternatively, the use of diphenylphosphoryl azide permits a one-step amino-decarboxylation, without isolation of the intermediately formed acyl azides[49].

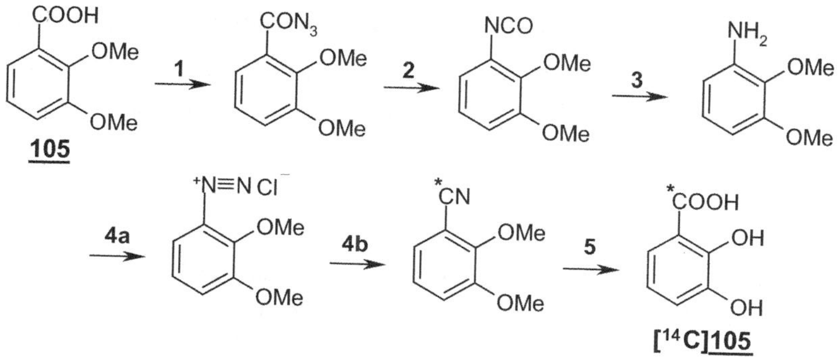

**Reaction conditions: 1.** oxalyl chloride, benzene; reflux, 1 h; **2a** . $NaN_3$, aq. acetone; 10–15 °C; → r.t., 1 h,**b.** 60 °C, benzene; **3.** 20 % aq. HCl; reflux 3 h; **4a.** $NaNO_2$, aq. HCl; 10 °C, 30 min, **b.** $K^{14}CN$, CuCN; $H_2O$; 40 °C, 1 h;**5.** HBr–HOAc 8:3; reflux, 5 h

***Figure 10.28*** *$CO_2$/$^{14}CO_2$ replacement via amino-decarboxylation–re[$^{14}$C]carboxylation (2)*

### 10.2.3 CO/$^{14}$CO Replacement Strategy

The method for replacing carbonyl functions with their labeled counterparts is closely related to the halo-decarboxylation approach because it uses the respective carboxylic acid as intermediate. For example, oxidation of the aldehyde **106** and iodo-decarboxylation of the resulting carboxyl group provided iodoarene **107**. Iodine/lithium replacement followed by reaction of the resultant aryllithium intermediate with dimethyl[$^{14}$C]formamide converted **107** into the corresponding aryl [$^{14}$C]aldehyde [$^{14}$C]**106**, which served as a precursor for the carbon-14 labeling of arylideneaminoguanidinium derivative **108**[50].

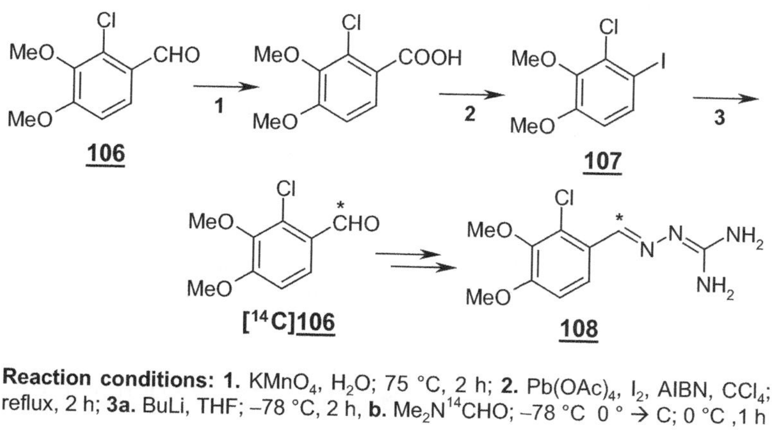

**Reaction conditions: 1.** $KMnO_4$, $H_2O$; 75 °C, 2 h; **2.** $Pb(OAc)_4$, $I_2$, AIBN, $CCl_4$; reflux, 2 h; **3a.** BuLi, THF; –78 °C, 2 h, **b.** $Me_2N^{14}CHO$; –78 °C 0 ° → C; 0 °C ,1 h

***Figure 10.29*** *CO/$^{14}$CO replacement through oxidative halo-decarboxylation–re[$^{14}$C] carbonylation*

### 10.2.4 Oxidative Cleavage of C=C Bonds in the Reconstitution Approach

Ozone and $NaIO_4$ used with catalytic amounts of $OsO_4$, $RuO_4$ or $KMnO_4$ have been frequently used for the cleavage of olefinic bonds for the reconstitutive preparation of isotopically labeled compounds.

*Ozonolytic Procedures*

Compounds containing double bonds can be oxidatively cleaved when treated with ozone at low temperatures. The initially formed ozonides **109** are unstable – sometimes even explosive – and are therefore immediately converted. The spectrum of resultant reaction products strongly depends upon the substitution pattern of the olefin employed and the workup conditions used. Carbonyl products **111a,b** are obtained from hydrolytic workups under slightly reductive conditions (e.g. aq. $NH_4Cl$ containing dimethyl sulfide, trialkylphosphines or phosphites to destroy the ozonides and concomitantly formed $H_2O_2$), alcohols **110a,b** from reductive workups ($NaBH_4$, diborane, or catalytic hydrogenation), and acids or esters **112a,b** from oxidative workups (excess $O_3$, $H_2O_2$, Jones reagent). Functional groups such as –OH, –$NH_2$, and C=O are frequently protected as –OAc, –O*tert*-Bu, –OTMS, –OTHP, –NHAc derivatives or ketals, although such derivatization does not appear to be necessary in all cases. MeOH, $CH_2Cl_2$, $CCl_4$, ethyl acetate and ethyl acetate–acetic acid mixtures are recommended solvents.

Ozone, although inconvenient to handle, has been frequently used for C=C bond cleavages because of its pronounced selectivity. When employed in appropriate amounts it may discriminate between C=C bonds of different reactivity. Alkenes in which the double bond is adjacent to electron-donating groups react many times faster than those in which the double bond is connected to electron-withdrawing groups. Addition of pyridine and use of low temperatures can enhance this selectivity, so that (for example) exocyclic double bonds are attacked rather than endocyclic ones. Conversely, protection of exocyclic C=C bonds, for example by dihalogenation ($Br_2$ in acetic acid), allows selective ozonolytic cleavage of endocyclic bonds. The exocyclic bond may then be regenerated with zinc–acetic acid in a subsequent step[51].

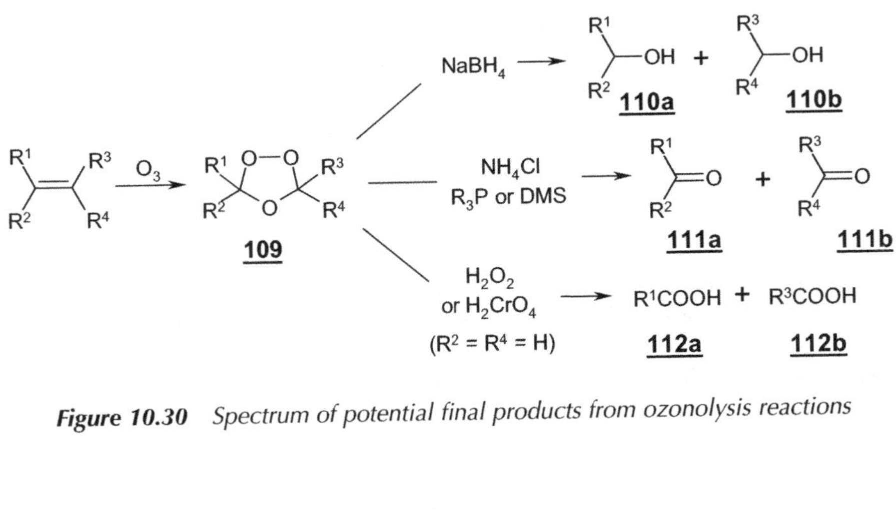


**Figure 10.30** *Spectrum of potential final products from ozonolysis reactions*

Ozonolytic procedures have found frequent application in the synthesis of labeled steroids[11]. In most cases, the resulting carbonyl fragments of interest are treated with singly or multiply labeled Wittig or Horner–Wadsworth–Emmons reagents to regenerate the corresponding labeled olefinic final products or late intermediates (Figure 10.31). The syntheses of [25-$^{14}$C]desmosterol ([$^{14}$C]**113**) and [25-$^{14}$C]lanosterol ([$^{14}$C]**114**)[52] are representative of this approach. Other examples include [23,24-$^{13}C_2$]cholic, -chenodeoxycholic and -ursodeoxycholic acids[53] and the nonsteroidal (+) and (−)-$\alpha$- and $\beta$-[10-$^{14}$C]-pinenes[54], [3-$^{13}$C]Ibuprofen[55], [12-$^{13}$C]$\alpha$-tocopherol[56] and cobalt [$^{14}C_2$]mesoporphyrin[57].

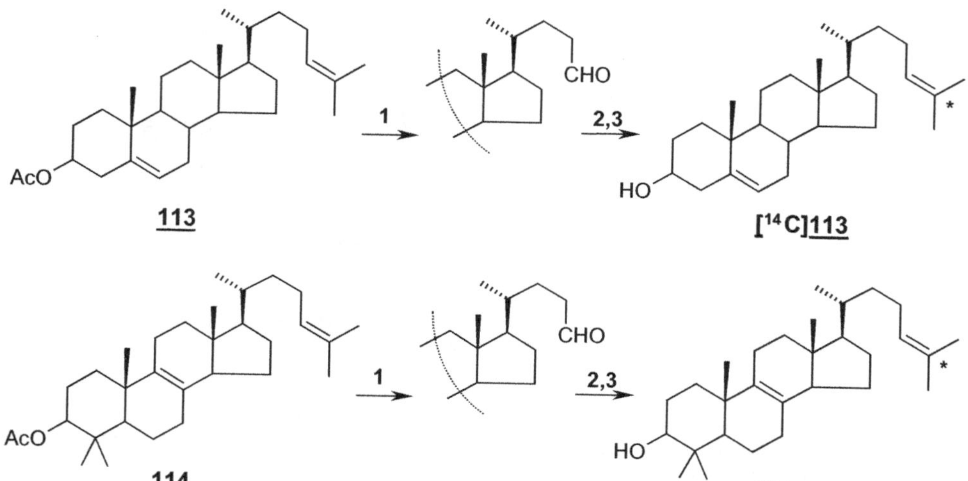


**113** [$^{14}$C]**113**

**114** [$^{14}$C]**114**

**Reaction conditions:** **1.** $O_3$, $CHCl_3$; −20 °C, 15 min, b. aq. $FeSO_4$; **2a.** NaH, DMSO; 70–80 °C, 45 min, **b.** $Ph_3P{=}^{14}CHMe_2$, DMSO; 50 °C, 24 h; **3.** KOH, MeOH; reflux, 2 h

***Figure 10.31*** *$^{12}C/^{14}C$ replacement via sequential ozonolyltic cleavage of C=C bonds and reconstitution*

Cyclic $\alpha$,$\beta$-unsaturated carbonyl compounds containing at least one $\gamma$-hydrogen are known to be degraded to $\delta$-keto acids with loss of one carbon unit. Therefore, subjecting 3-oxo steroids **115** to ozonolysis and oxidative workup afforded $\delta$-keto acids **116**, which in turn could be cyclized to give enol lactones **117** (Figure 10.32). Reaction of **117** with [$^{14}$C]-methylmagnesium iodide generated intermediate 1,5-diketones **118**, which underwent spontaneous cyclization to the bridged tertiary alcohols **119**. Treatment with base (ethanolic NaOH) reverted **119** to 1,5-diketones **118** in a retro-aldol reaction, which then cyclized to afford [$^{14}$C]**115** labeled at the biologically stable C4 (Fujimito–Belleau reaction)[11,58].

This sequence has emerged as a general method for the carbon-14 labeling of a broad spectrum of steroids[11] particularly since [$^{14}$C]**115** may be conveniently reduced to either the corresponding alcohols, deconjugated to $\Delta^{5,6}$-derivatives[11,59], aromatized in ring A to female sex hormones[11], or the 3-carbonyl function can be replaced by a carboxyl group in a $Ph_3Pd(OAc)_2$-catalyzed carbonylation reaction of the corresponding dienol triflate[58b].

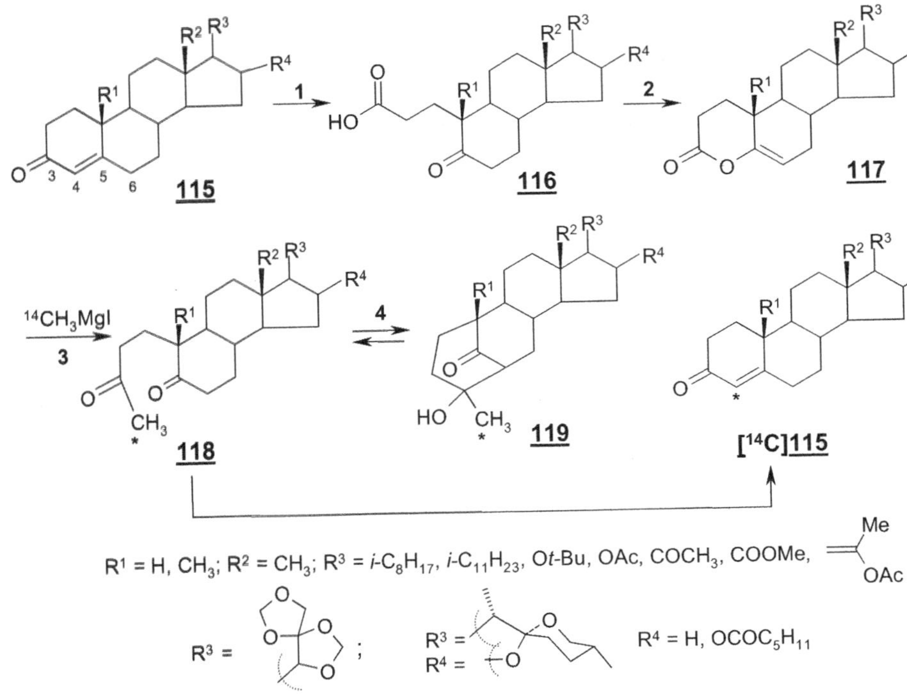


**Reaction conditions:** **1a.** $O_3$, EtOAc–HOAc 3:1; –20 °C, 6 h, **b.** 5 % aq. $H_2O_2$; r.t., 16 h; **2.** NaOAc, $Ac_2O$, 140 °C, 3 h; **3a.** [$^{14}$C]$CH_3MgI$, $Et_2O$; 0–10 °C, 1 h, **b.** 1 N HCl; **4.** 30 % NaOH, EtOH 1:7, r.t., 16 h

***Figure 10.32*** *$^{12}C/^{14}C$ replacement via ozonolyltic cleavage of C=C bonds/reconstitution (Fujimito–Belleau procedure)*

In addition, opening of ring A allows transformation of [$^{14}$C]steroids into corresponding [$^{14}$C]azasteroids[60]; Baeyer–Villiger oxidation of 5-oxo-derivatives results in the formation of [$^{14}$C]brassinosteroids[61].

[$^{14}$C]Methylmagnesium iodide can be replaced in this sequence by phenyl [$^{14}C_n$]-acetate[11,62], since it reacts with deprotonated enol lactones **117** to give the corresponding 2-[$^{14}C_n$]acetyl derivatives **120** (Figure 10.33). Hydrolysis of the enol lactone system with methanolic KOH and decarboxylation of the intermediate $\beta$-diketoacids **121** provides the desired 3-oxo steroids **[$^{14}$C]115**. Since [$^{14}C_2$]acetic acid is easily accessible, this procedure facilitates the simultaneous incorporation of carbon-14 atoms at both C3 and C4. Unfortunately, a minor amount of phenyl[$^{14}C_n$]acetate is also deprotonated in the reaction mixture and attacks the carbonyl function of the enol lactone to give the ring-opened $\beta$-keto ester **122**. Thus, in the case of phenyl [1-$^{14}$C]acetate, the subsequent hydrolysis results in the elimination of some $^{14}CO_2$ and reduction of the specific activity of the resulting material. For $\beta$-[3-$^{14}$C]sitosterol[62] the ratio **[$^{14}$C]115**: **115** amounts to 82:18, corresponding to a 20% diminution of the specific activity.

Cyclic $\alpha$,$\beta$-unsaturated carbonyl compounds lacking a $\gamma$-hydrogen can be ozonized to 1,5-dicarbonyl intermediates, but cyclization to enol lactones is not possible. Reversal

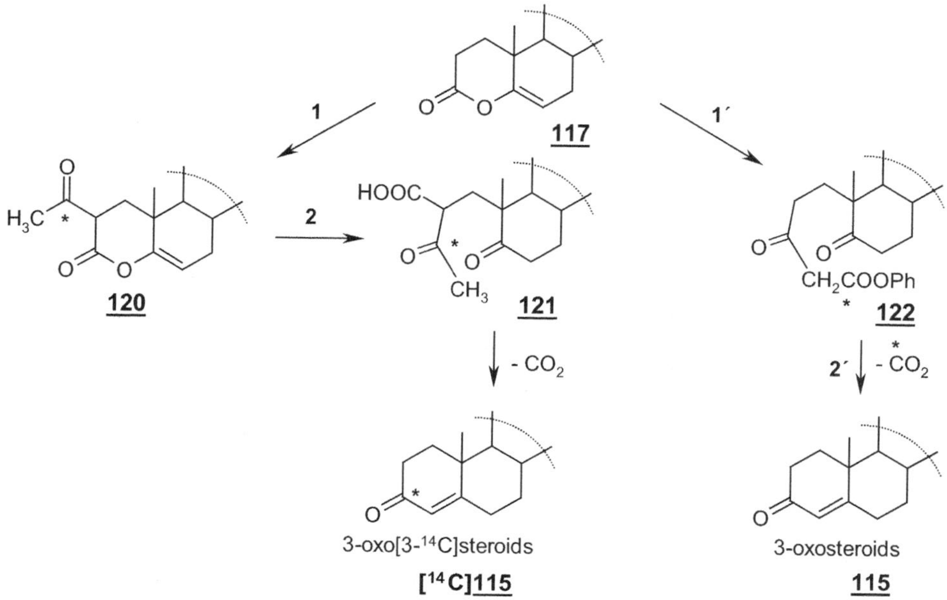


**Reaction conditions: 1/1'** . phenyl [1-$^{14}$C]acetate, NaH, benzene; r.t., 70 h; **2/2'**. KOH, MeOH; reflux, 18 h

***Figure 10.33*** *Use of phenyl [1-$^{14}$C]acetate in the Fujimito–Belleau synthesis of 3-oxo-[3-$^{14}$C] steroids*

of the Fujimoto–Belleau sequence has been shown, however, to offer a solution[63] (Figure 10.34). Initial 1,2-addition of the Grignard reagent ($^{14}CH_3MgI$ in this case) to the ketone **<u>123</u>** was followed by ozonolysis of the resulting allyl alcohol **<u>124</u>** to give **<u>125</u>**. Subsequent base-catalyzed expulsion of the formyl function furnished the keto-aldehyde, which spontaneously cyclized to give [$^{14}$C]**<u>123</u>** in 25–40% yield. Steroid [$^{14}$C]**<u>123</u>** served as an intermediate in the preparation of [$^{14}$*C*]CP-148,623 (**<u>126</u>**) and the closely related [$^{14}$C]CP-242,184, both cholesterol absorption inhibitors[63a,b].

Figure 10.35 illustrates an additional application of the *disconnection–reconnection* approach for the carbon-14 labeling of complex molecules by use of ozonolytic procedures. In the case of [$^{14}C_2$]protoporphyrin IX dimethyl ester (**[$^{14}C_2$]<u>127</u>**), the exocyclic double bonds of unlabeled substrate **<u>127</u>** cleaved ozonolytically to give dialdehyde **<u>128</u>a**, whose electron-rich heterocyclic system remained intact. Wittig olefination of the zinc complex **<u>128</u>b** with $^{14}CH_2{=}PPh_3$ and concluding removal of zinc converted **<u>128</u>b** into the doubly labeled target[57].

The same strategy was exploited for the preparation of 16,17-dihydro[$^{14}$C]GA$_5$ (**[$^{14}$C]<u>129</u>**), a gibberellin plant growth inhibitor labeled at C3 of the ring system (Figure 10.36). Although introduction of the label into C17 (lactone carbonyl) would have been more straightforward, a label at that position was considered to be at risk of metabolic instability. The synthesis sequence started with the incorporation of a labeled $C_1$ fragment at C2 of intermediate **<u>130</u>** followed by excision of a $C_1$ fragment from C3 of the carbon skeleton, restoring the original

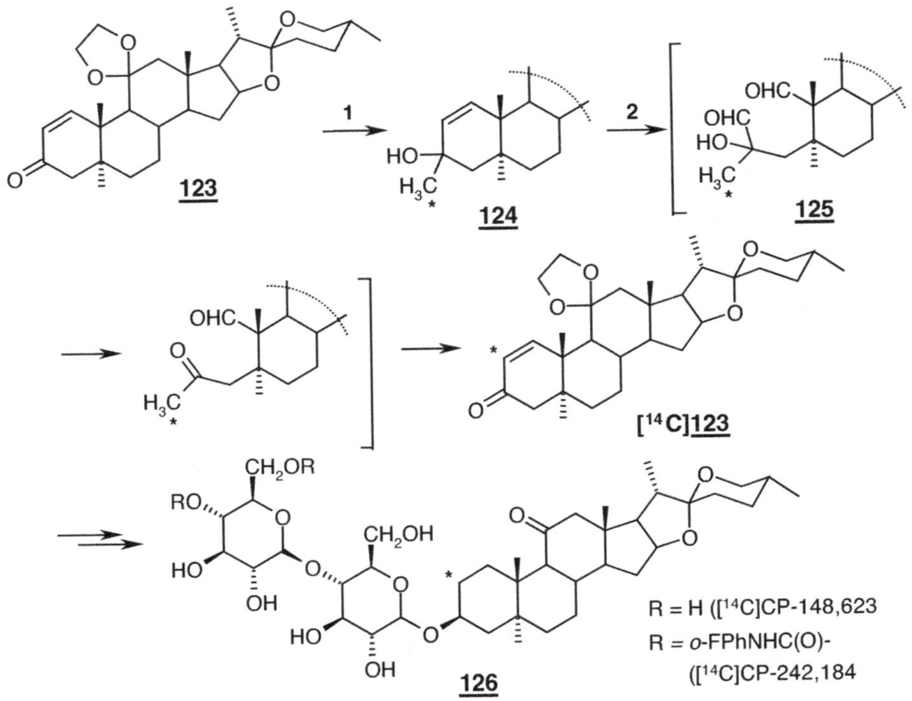


**Reaction conditions**: **1.** [$^{14}$C]MeMgI, ether or THF; –78 °C to r.t.; **2.** $O_3$, EtOAc or $CH_2Cl_2$-MeOH; –78 °C to r.t.; **3.** aq. NaOH, EtOH, r.t.

***Figure 10.34*** *Fujimoto–Belleau procedure on α,β-unsaturated carbonyl compounds lacking γ-hydrogen*

gibberellin system. The requisite precursor **130** was accessed through ozonolysis of the 2,3-double bond followed by a five-step sequence including reduction of the resulting dialdehyde, manipulation of protecting groups and Dess–Martin re-oxidation. Wittig olefination of **130** with $^{14}CH_2{=}PPh_3$, deprotection of the C3-hydroxy group and Dess–Martin oxidation afforded *β*-ketoester **131**, which set up the requisite $C_1$-scission to **132** by means of a *retro*-Claisen reaction. *Anti*-Markovnikov hydration via hydroboration and oxidative workup, followed by Dess–Martin oxidation yielded aldehyde **133**. This intermediate underwent a base-catalyzed intramolecular aldol reaction to furnish the epimeric alcohols **134**, thereby reassembling the original gibberellin skeleton. Target [$^{14}$C]**129**, finally, was obtained in a two-step dehydration of **134**[64].

Though the ozonolysis-based removal and replacement strategy has been exploited predominantly for the preparation of complex carbon-14-labeled compounds, the synthesis of (–)-$\alpha$-[vinylidene-$^3H_2$]kainic acid ([$^3$H]**135**) demonstrates that it can also be usefully applied to tritium-labeled target molecules, even at high specific activities[65] (Figure 10.37). Ozonolytic cleavage of the olefinic bond of the side chain of doubly protected kainic acid, a nonproteinogenic amino acid displaying neuroexitatory activities, furnished the corresponding keto derivative **136**. Reconstitution was accomplished by installing a tritiated methylene group using a Wittig reagent prepared from

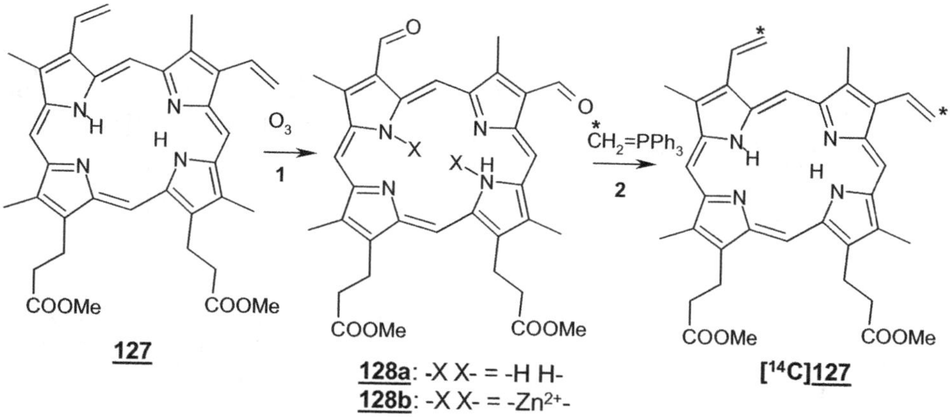


**Reaction conditions:** **1a.** $O_3$, $CHCl_3$–MeOH–$H_2SO_4$ 9:1:0.2; –60 °C, 30 min, **b.** aq.TFA, r.t., **c.** $Zn(OAc)_2$, $CHCl_3$–MeOH 85:15; reflux, 30 min; **2a.** $^{14}CH_2$=$PPh_3$, THF; r.t., 3 h, **b.** TFA; 0 °C, 1 h

***Figure 10.35*** *Application of the disconnection–reconnection strategy to the carbon-14 labeling of complex molecules: [$^{14}C_2$]protoporphyrin IX dimethyl ester*

[$^3H_3$]methyl iodide. Final deprotections provided [$^3$**H**]**135** of specific activity of 58 Ci/mmol in an overall radiochemical yield of 5%.

The strategy outlined here has also been successfully exploited for the preparation of carbon-14-labeled heterocyclic systems containing imidazole, indole, or thiophene subunits. Ozonolysis of (+)-pilocarpine (**137**), for example (Figure 10.38), degraded the imidazole ring while leaving the butyrolactone system and the *cis*-stereochemistry of the C3 and C4 substituents unaffected. Elongation of the side chain at C4 by one carbon atom through hydrolysis of the amide group of **138**, formation of the acyl chloride, reaction with diazomethane and routine transformation of the resultant diazoketone converted **139** to aminomethyl ketone **140**. Imidazole ring formation upon heating of **140** with potassium [$^{14}$C]thiocyanate, followed by oxidative desulfurization and selective *N*-methylation at C1 completed the synthesis of [$^{14}$C]**137**[66].

In the next example (Figure 10.39), Z-protected *N*-desmethyl sumatriptan (**141**) was ozonized in DMF to give ketoformanilide **142**, which reacted with potassium [$^{14}$C]cyanide to form the 2-amino-3-hydroxy[2-$^{14}$C]indoline derivative **143**. Reduction of **143** with $NaBH_4$ in acetic acid produced the respective 3-substituted [2-$^{14}$C]indole [$^{14}$**C**]**141**. Concluding $LiAlH_4$ reduction of the carbamate side chain converted [$^{14}$**C**]**141** to [$^{14}$C]sumatriptan (**144**), an effective agent for the acute treatment of migraine, in 59% overall radiochemical yield from $K^{14}CN$[67].

A similar strategy was applied to the synthesis of the tricyclic benzazepine derivative [$^{14}$**C**]**145** (R = –$CH_2OH$) (Figure 10.40). Oxidative degradation of the trifluoroacetate salt of **145** with ozone gave disulfide **146**. Selective reduction of **146** with $Bu_3P$, immediate *S*-alkylation of the resultant thiol **147** with ethyl bromo[2-$^{14}$C]acetate and NaOEt-mediated cyclization restored the original tricyclic framework in a one-pot procedure. The sequence

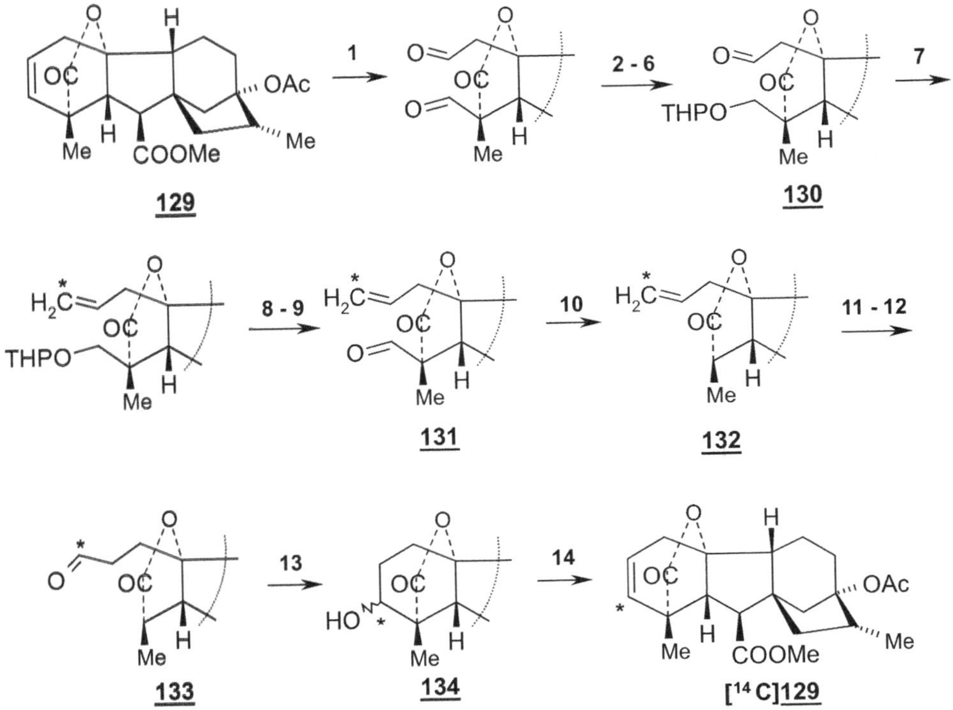


**Reaction conditions: 1a**. $O_3$, $CH_2Cl_2$; –78 °C, **b**. $Me_2S$; –78 °C → r.t., 16 h; **2**. $Zn(BH_4)_2$, DME; 0 °C, 1.5 h; **3**. *tert*-$BuPPh_2SiCl$; imidazole, $Et_3N$, DMF; r.t., 16 h; **4**. DHP, pyridine *p*-TsOH, $CH_2Cl_2$; r.t., 16 h; **5**. $Bu_4NF$, THF; r.t., 5 h; **6**. Dess–Martin periodinane, $CH_2Cl_2$; r.t., 1 h; **7**. $^{14}CH_2{=}PPh_3$, THF; r.t., 1 h; **8**. *p*–TsOH, MeOH; r.t., 16 h; **9**. Dess–Martin periodinane, $CH_2Cl_2$; r.t., 1.5 h; **10**. $K_2CO_3$, MeOH; r.t., 3 min; **11a**. thexylborane, THF; 0 °C; 1.5 h, **b**. $H_2O_2$, aq. $Na_2HPO_4$; r.t., 16 h; **12**. Dess–Martin periodinane, $CH_2Cl_2$; r.t., 1 h; **13**. aq. $NaHCO_3$–MeOH 1:4; r.t., 3 h; **14a**. MsCl, pyridine; r.t., 24 h, **b**. KI, DMF; 130 °C, 48 h

***Figure 10.36*** *Application of the disconnection–reconnection strategy to the carbon-14 labeling of complex molecules: synthesis of 16,17-dihydro[$^{14}C$]$GA_5$*

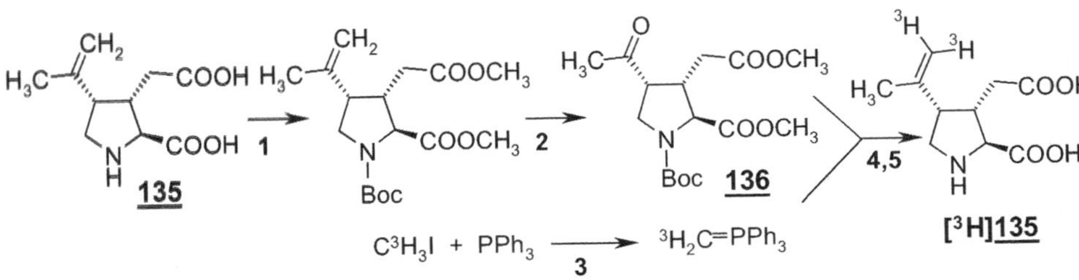


**Reaction conditions: 1a**. MeOH, $SOCl_2$; 60 °C, 16 h;, **b**. $Boc_2O$, DMAP, $Et_3N$; $CH_2Cl_2$; r.t., 4 h; **2**. $O_3$, MeOH–$CH_2Cl_2$ 2:1; –78 °C, 30 min, then $PPh_3$; r.t., 1.5 h; **3a**. THF; r.t., 16 h, **b**. *n*-BuLi, THF; r.t.; **4**. combine $^3H_2C{=}PPh_3$, 136; r.t., 2 h; **5a**. *p*-TsOH, MeOH; r.t., 16 h, **b**. 2N NaOH; r.t, 16 h

***Figure 10.37*** *Use of the ozonolytic cleavage of C=C bonds in the disconnection–reconnection synthesis of (–)-α-[vinylidene-$^3H_2$]-kainic acid*

was completed by reduction of the ester group at C2 of **148**. Surprisingly, ozonolysis of **145** failed when R was ethyl, ethoxycabonyl or halogen[68].

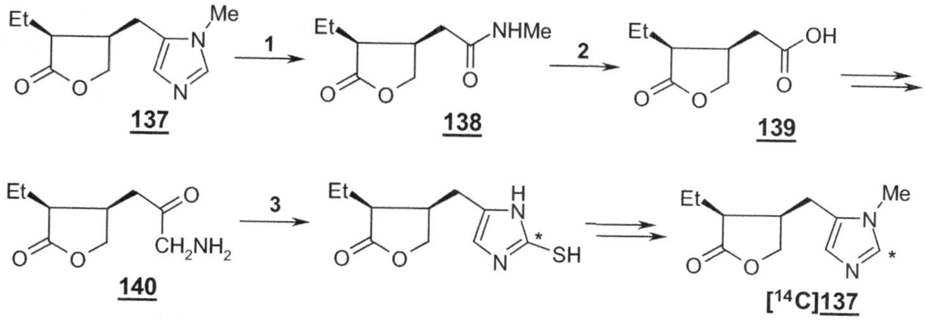


**Reaction conditions: 1.** $O_3$, DMF; **2.** HCl–acetic acid, reflux; **3.** $K^{14}CNS$, DMF

***Figure 10.38*** *Use of the ozonolytic cleavage of C=C bonds in the reconstitution synthesis of heterocyclic substrates: application to the synthesis of (+)-[14C]pilocarpine*

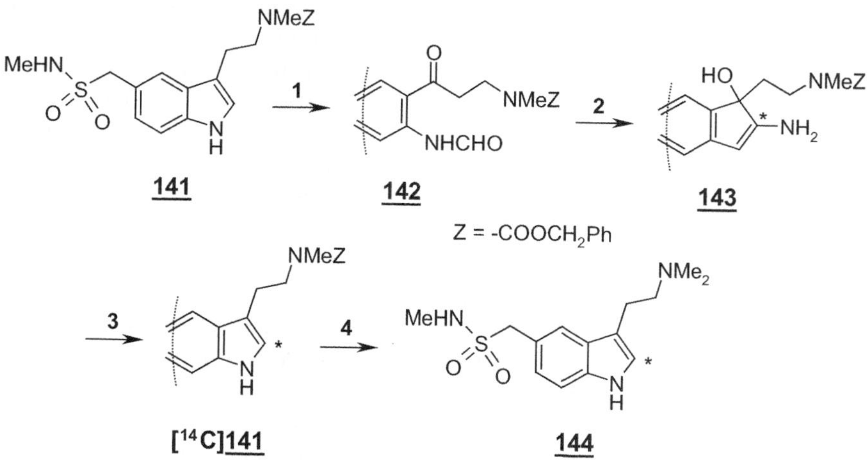


**Reaction conditions: 1.** $O_3$, DMF; −10 °C, 5 min; **2.** $K^{14}CN$, 95 % EtOH; 20 °C, 4 h; **3.** $NaBH_4$, HOAc; 5 °C, 40 min; **4.** $LiAlH_4$, THF; reflux, 1.5 h

***Figure 10.39*** *Use of the ozonolytic cleavage of C=C bonds in the reconstitution synthesis of heterocyclic substrates: application to the synthesis of [14C]sumatriptan*

*Oxidative C=C Cleavage via 1,2-Glycol Intermediates*

Since handling of ozone is inconvenient, it is often replaced by sodium periodate and a catalytic amount of osmium tetroxide[51] for the cleavage of C=C bonds to give the

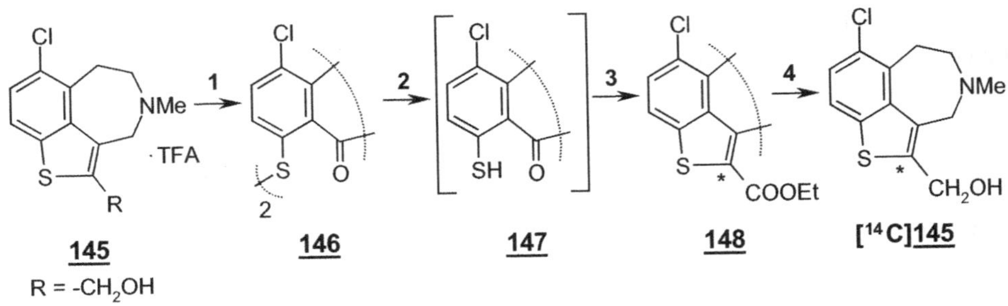


**Reaction conditions: 1.** $O_3$, DMF; **2.** $Bu_3P$, aq. EtOH; **3a.** $Br^{14}CH_2COOEt$, EtOH, –78 °C, **b.** EtONa, EtOH; –78 °C; **4.** $LiAlH_4$, THF

***Figure 10.40*** *Use of the ozonolytic cleavage of C=C bonds in the reconstitution synthesis of heterocyclic substrates: application to the synthesis of 7-chloro-2-hydroxyethyl-3,4,5,6-tetrahydro-4-methyl[2-$^{14}$C]thieno-[4,3,2-e][3]-benzazepine*

respective carbonyl compounds as final products in yields ranging from 50 to 90%. Simultaneously, excess $NaIO_4$ regenerates the osmium tetroxide employed, thus permitting the use of catalytic quantities of this highly toxic reagent. Because of the insolubility of $NaIO_4$ in organic solvents, the reaction is usually carried out in aqueous dioxane, THF, acetone or acetic acid. In anhydrous organic media cleavage reactions can be run by using $Pb(OAc)_4$ as the oxidant. The $NaIO_4/OsO_4$ procedure has been successfully exploited for the carbon-14 labeling of a broad variety of compounds, including [$^{14}$C]oleic acid[69], [$^{14}$C]gibberellin ([$^{14}$**C]149**)[70], [$^{14}$C]permethrin ([$^{14}$C]AY 30,068, a synthetic pyrethroid) (**153**)[71] (Figure 10.41) and the 5-HT$_{1F}$ selective agonist ([$^{14}$**C] 154**)[72] (Figure 10.42).

The target compound selected for a C=C bond cleavage must not necessarily be the final product but it may be a late intermediate, as demonstrated by the synthesis of **153** (second example, Figure 10.41)[71]. In this case the labeling sequence started with cleavage of the double bond of key intermediate **151**. Protection of the newly formed carbonyl function formed through selective reduction and *O*-acetylation allowed for a Fischer indole synthesis using 2-ethylphenylhydrazine to generate the carbazole skeleton of **152**. Subsequent liberation of the carbonyl function through saponification and re-oxidation facilitated the introduction of label through a Wittig olefination using $^{14}CH_2{=}PPh_3$.

The carbon-14 labeling the sumatriptan-derived 5-HT$_1$-receptor agonist **154** depicted in Figure 10.42 corresponds to that starting with ozonolytic cleavage of the indole system. It illustrates that the $OsO_4$–$NaIO_4$ methodology is also applicable to heterocyclic systems[72].

$OsO_4$ has been reported to be replaceable by $RuO_4$, a more vigorous oxidant[51]. Under neutral conditions the $NaIO_4/RuO_4$ system degrades $R^1CH{=}CHR^2$ to aldehydes, whereas under basic or acidic conditions the corresponding carboxylic acids are formed. A $CCl_4$–$CH_3CN$–$H_2O$ mixture is the recommended solvent for this type of reaction.

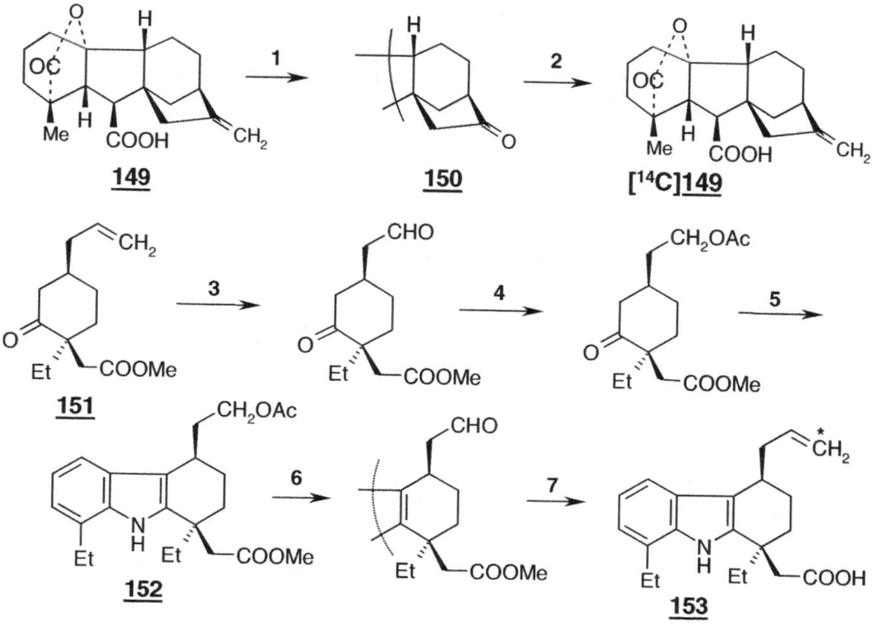


**Reaction conditions: 1**. $OsO_4$ (cat.), $NaIO_4$; aq. THF; r.t., 2d; **2a.** $^{14}CH_3I$,$Ph_3P$, THF; r.t., 2 h, **b**. *tert*-BuOK, toluene–*tert*-BuOH 5:1; r.t., **c**. $Ph_3P{=}^{14}CH_2$, **150**, toluene; r.t., 16 h; **3**. $OsO_4$–$NaIO_4$ 1:150, THF–$H_2O$–*tert*-BuOH 30:10:1; r.t., 1 h; **4a**. $LiAlH(OCEt_3)_3$, THF; –78 °C, 25 min, **b**. $Ac_2O$, $Et_3N$, DMAP; $CH_2Cl_2$; r.t., 20 min; **5a**. 2-ethylphenylhydrazine, toluene; reflux, 2 h, **b**. $BF_3.OEt_2$, HOAc; reflux, 20 min; **c**. $K_2CO_3$, MeOH; r.t., 1.3 h; **6**. Dess-Martin periodinane, $CH_2Cl_2$; 0°C, 10 min; **7**. $Ph_3P{=}^{14}CH_2$, THF; r.t., 5 min

***Figure 10.41*** *Use of the $OsO_4$–$NaIO_4$ cleavage of C=C bonds in disconnection/reconnection syntheses of terminal olefins*

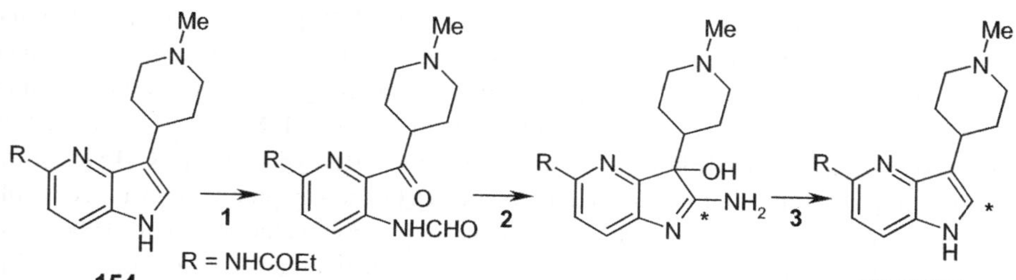


**Reaction conditions**: **1**. $OsO_4$, $NaIO_4$, MeOH; $H_2O$; r.t., 63 h; **2**. $K^{14}CN$, 95 % EtOH; 3 h; **3**. $NaBH_4$, THF–HOAc 1:1; 0 → 5 °C, 2 h

***Figure 10.42*** *Use of the $OsO_4$–$NaIO_4$ cleavage of C=C bonds in the disconnection/reconnection synthesis of labeled heterocyclic compounds*

In cases where ancillary substituents are less susceptible to oxidation, $OsO_4$ can be replaced with alkaline $KMnO_4$, as exploited in the steroid series for examples such as carbon-14 labeled 7$\alpha$-chloro-16$\alpha$-methylprednisolone[73], medroxyprogesterone[74], triamcinolone[75] and dexamethasone (**155**)[76].

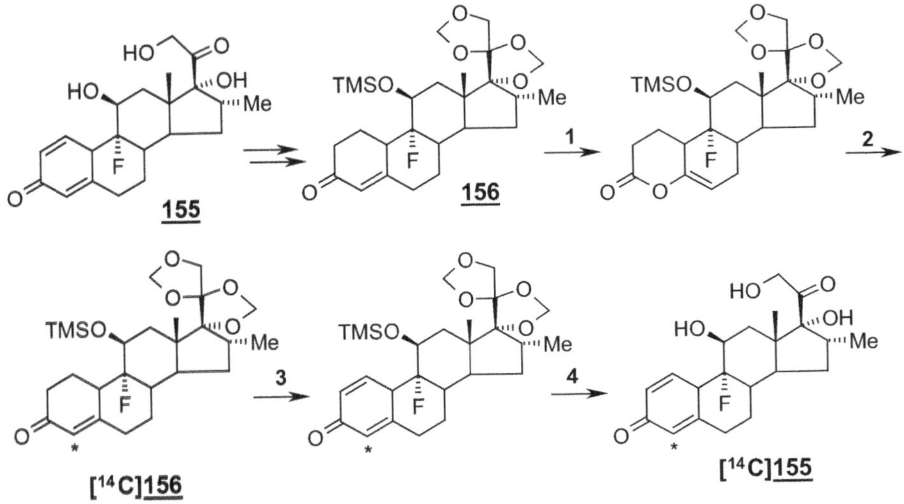


**Reaction conditions: 1a**. $KMnO_4$, $NaIO_4$, $K_2CO_3$, $CH_2Cl_2$–*tert*-BuOH–$H_2O$; r.t., 3.5 h, **b**. $Ac_2O$, NaOAc, $Et_3N$; reflux, 45 min; **2a**. $^{14}CH_3MgI$, ether–benzene 1:1; r.t., 1 h, **b**. lithium 2,6-di-*tert*-butylphenoxide, dioxane; reflux, 4 h; **3**. *m*-iodylbenzoicacid, PhSeSePh, toluene; reflux, 3 h; **4**. $CF_3COOH$–2N HCl; r.t., 3 h

***Figure 10.43*** *Use of the $KMnO_4$–$NaIO_4$-mediated C=C bond cleavage in the reconstitution synthesis of [4-$^{14}$C]dexamethasone*

The use of $Pb(OAc)_4$ in anhydrous organic media is exemplified with the synthesis of [$^{14}$C]safrole and some of closely related derivatives ([$^{14}$**C**]**157**)[77] and of deuterated arachidonic acid [$^2$**H**$_3$]**160a**[78] (Figure 10.44). In the first case the synthesis sequence began with the *cis*-dihydroxylation of the parent compounds **157** using equimolar amounts of $OsO_4$. Subsequent $Pb(OAc)_4$-mediated cleavage of the diols **158** and Wittig olefination of the resultant aldehydes **159** using $Ph_3P$=$^{14}CH_2$ converted **158** into ([$^{14}$**C**]**157**). In the second example methyl 14,15-epoxyarachidonate (**161**), the key precursor, was accessible in an unusual reaction through rearrangement of peroxyarachidonic acid **160c** upon standing at 20 °C, followed by esterification with diazomethane. Acid-catalyzed ring opening of the epoxide **161** and oxidative cleavage of the 1,2-diol formed with $Pb(OAc)_4$ afforded the unstable aldehyde **162**. This was converted into the *cis*-vinyl bromide **164** by Corey–Fuchs olefination using $CBr_4$–$PPh_3$ and selective $Pd(Ph_3)_4$-catalyzed hydrodebromination of the intermediately formed dibromo derivative **163** with $Bu_3SnH$. Concluding Suzuki cross-coupling of **164** to labeled lithium borate reagent **165** regenerated isotopomeric all-*cis* [$^2$**H**$_3$]**160a**.

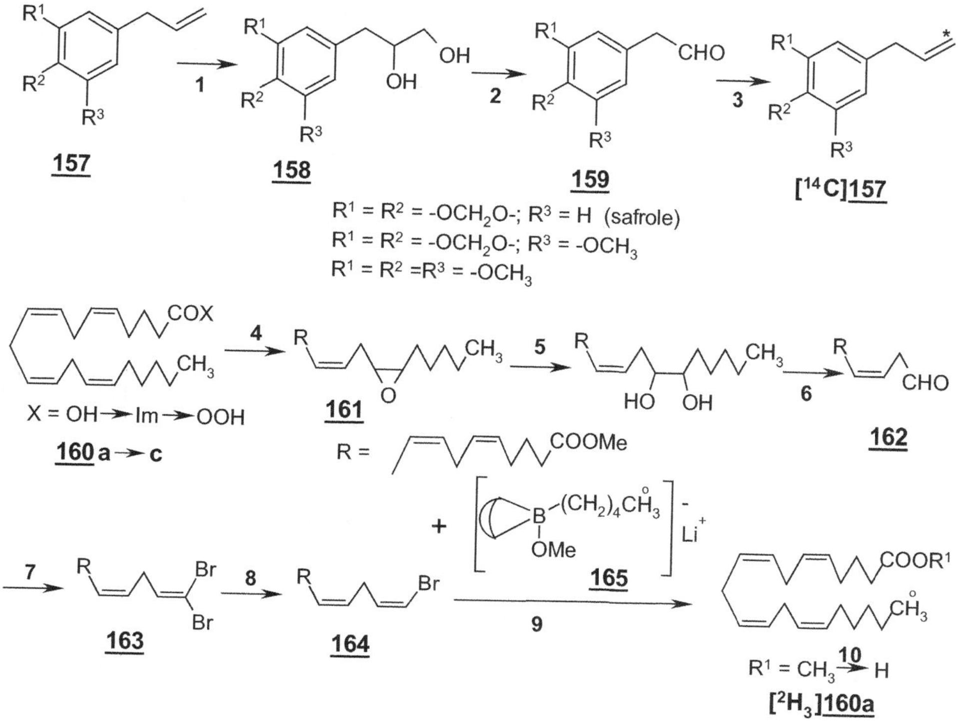


**Reaction conditions: 1.** $OsO_4$, pyridine, ether; r.t.; 2 h; **2.** $Pb(OAc)_4$, benzene; r.t., 3-4 min; **3.** $Ph_3P{=}^{14}CH_2$, THF; -76°C → 10°C; **4.** CDI, $CH_2Cl_2$; 25°C, 20 min, **b.** $H_2O_2$, cat. Li-imidazolide, ether; 0°C, 3 min, **c.** $KHSO_4$, $CH_2Cl_2$, r.t., **d.** $CH_2N_2$, $CH_2Cl_2$; r.t.; **5.** 10% $HClO_4$-THF 3:8; 0°C, 24 h; **6.** $Pb(OAc)_4$, $CH_2Cl_2$; r.t., 30 min; **7.** $CBr_4$, $Ph_3P$, $CH_2Cl_2$; 0°C, 2 h; **8.** $Bu_3SnH$, $(Ph_3P)_4Pd$, benzene; r.t., 16 h; **9.** **165**, $Pd(dppf)Cl_2$, $AsPh_3$, DMF; r.t., 16h; **10.** LiOH, aq. THF; r.t., 12 h

***Figure 10.44*** *Reconstitution via $Pb(OAc)_4$-mediated cleavage of diols in anhydrous organic media*

$H_2O_2$–formic acid, one of the standard methods for preparing 1,2-*anti*-diols, was utilized for the *bis*-hydroxylation and subsequent cleavage of the C=C bond of erucic acid (**166**), facilitating introduction of a carbon-14 label into one of the former olefinic positions through transformation of the resultant 13-oxotridecanoic acid (**167**) into the Wittig reagent **168** and subsequent coupling to [1-$^{14}$C]nonyl aldehyde (**169**)[79] (Figure 10.45). A related approach was used to achieve selective cleavage of one of the terminal isopropylidene units of squalene (**170**), affording the tris-norsqualene aldehyde **171**[80]. In this case *bis*-hydroxylation was accomplished through hydroxy-bromo addition upon treatment of **170** with an equimolar amount of NBS in the presence of water, base-mediated epoxide formation and acid-catalyzed ring opening. Diol cleavage, reduction of the resulting aldehyde **171** with $NaB^3H_4$, re-oxidation of the resultant [1-$^3$H]alcohol according to the strategy described in Section 10.1.1 of this chapter, and concluding Wittig isopropylidenation produced [3-$^3$H]squalene ([3-$^3$**H]170**) with a specific activity of 1.9 Ci/mmol (from $NaB^3H_4$ of 10 Ci/mmol).

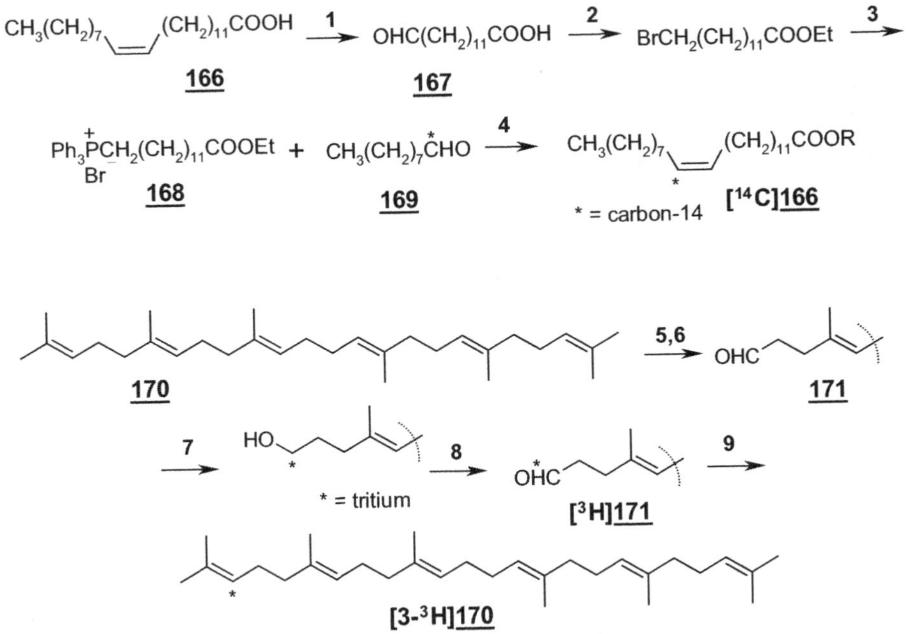


**Reaction conditions:** **1a.** 30 % $H_2O_2$, HCOOH; 55 °C, 1.5 h; r.t. 16 h, **b.** 50 % aq. NaOH; r.t.; 2 h, **c.** $NaIO_4$, $CHCl_3$–$H_2O$ 1:3; r.t., 6 h; **2a.** $NaBH_4$, EtOH; r.t., 2h, **b.** EtOH, $H_2SO_4$ (cat); reflux, 4 h, **c.** NBS, $Ph_3P$, THF; r.t., 30 min; **3.** $Ph_3P$, toluene; reflux, 72 h; **4a.** *n*-BuLi, THF; r.t., **b.** [$^{14}$C] **169**, THF; r.t.; 1.5 h; 50 °C, 0.5 h, **c.** 5% aq. MeOH; reflux; **5a.** NBS, aq. THF; 0°CC, 30 min, **b.** $K_2CO_3$, MeOH; r.t., 3 h, **c.** aq. $HClO_4$, DME; r.t., 2 h; **6.** $NaIO_4$, aq. THF; r.t., 1 h; **7.** $\mathbf{NaB^3H_4}$ (deficiency), MeOH; r.t., 1 h then $NaBH_4$ (xs); r.t., 1 h, **8.** PCC, $CH_2Cl_2$; r.t., 90 min; **9.** $(CH_3)_2CHPPh_3Br$, $NaNH_2$, *n*-BuLi, THF; reflux, 15 min

***Figure 10.45*** *Alternative bishydroxylation methods for the use in the oxidative cleavage of diols in the reconstitution strategy*

### *Oxidative Cleavage of C–C Bonds*

The cleavage of carbon–carbon bonds in disconnection–reconnection approaches is not restricted to double bonds. Appropriately substituted C–C moieties can also be the source of oxidatively cleavable intermediates, such as 1,2-diols, $\alpha$-hydroxy amines, $\alpha$-hydroxy carbonyl compounds and 1,2-diketones. The three-step sequence consisting of $\alpha$-hydroxylation of an ester, reduction of the ester function and $NaIO_4$-mediated cleavage of the resultant diol, is presented as a basis of the hydroxy-decarboxylation strategy in Section 10.2.2. Further prominent examples are the reconstitution synthesis of [$^{14}$C]adriamycinone (**[$^{14}$C]172**), aglycone of [$^{14}$C]adriamycin (**175**), a highly potent anthracycline antibiotic[81] (Figure 10.46), of [$^{14}$C]avermectin B1a (**183**)[82] (Figure 10.47) and 2-desoxo[1′-methyl-$^2H_3$]paraherquamide (**[$^2H_3$]188**) (Figure 10.43), both antiparasitic antibiotics[83].

In the first case, treatment of adriamycinone (**172**) with a 15-fold excess of [$^{14}$C]-methylmagnesium iodide afforded 1,2-diol **173**. $NaIO_4$-mediated cleavage of the diol followed by $\alpha$-bromination of the [$^{14}$C]methyl ketone **174** and hydroxy-debromination converted **173** into the [$^{14}$C]aglycone [$^{14}$**C]172**. Concluding coupling of [$^{14}$**C]172** to the daunosamine moiety provided the intact labeled antibiotic [$^{14}$C]adriamycin **175**.

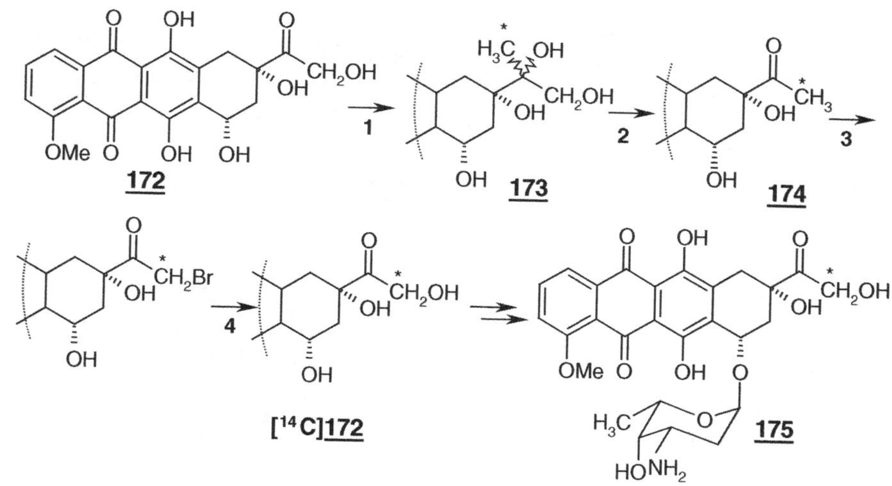


**Reaction conditions: 1.** $^{14}$**CH$_3$MgI** (15-fold excess), THF–ether 10:1; r.t., 16 h; **2.** NaIO$_4$, MeOH–H$_2$O 2:1; 0 °C, 4 h; **3.** Br$_2$, CHCl$_3$; r.t., 5 h; **4.** 0.1N NaOH, acetone–H$_2$O 4:1; 100 °C, 5 min;

***Figure 10.46*** *Preparation of a 1,2-diol for the cleavage of a C–C bond for use in the reconstitution approach*

The synthesis of [23-$^{14}$C]avermectin B1a (**183**) is a highly instructive example demonstrating the utility of the reconstitution strategy for the carbon-14 labeling of complex molecules. In this case the structurally closely related avermectin B$_{2a}$ (**176**, R$^1$, R$^2$ = H) was selected as a starting material because a carbonyl group at C23 was expected to facilitate cleavage of the C22–C23 bond. For this purpose, the hydroxyl groups of **176** at C5 and C4″ (second sugar moiety attached at C13) were selectively protected, then the C23 hydroxyl was oxidized under Swern conditions. α-Hydroxylation of the resulting ketone to give **177** was accomplished in three steps through formation of the silyl enol ether, epoxidation and acid-catalyzed epoxide ring opening. Pb(OAc)$_4$-mediated cleavage of the C22–C23 bond of **177** and transketalization of the product **178** produced the epimeric mixture of methoxy ketal-aldehydes **179**. Wittig olefination of **179** with labeled phosphonate **180**, followed by spiroketalization of the resulting mixture **181** and removal of the protecting groups afforded [23-$^{14}$C]avermectin B$_{1a}$ (**182**)[82] with the normal configuration at C21. The requisite labeled fragment **180** was made available by diastereoselective aldol coupling of the enolate of [1-$^{14}$C]propionyl (2*S*)-bornane-10,2-sultam and (2*S*)-2-methylbutyraldehyde followed by reductive removal of the chiral auxiliary, bromo-dehydroxylation and phosphonium salt formation (Chapter 11, Section 11.3.2).

Finally, paraherquamide A (**184**), a tertiary alcohol, provided exo-olefin **185** when treated with DAST [(diethylamino)sulfur trifluoride]. Subsequent reduction of the oxindole system, followed by dihydroxylation of the exocyclic double bond and cleavage of the resultant 1,2-diol **186**, converted **185** into ketone **187**. Concluding reaction of **187** with

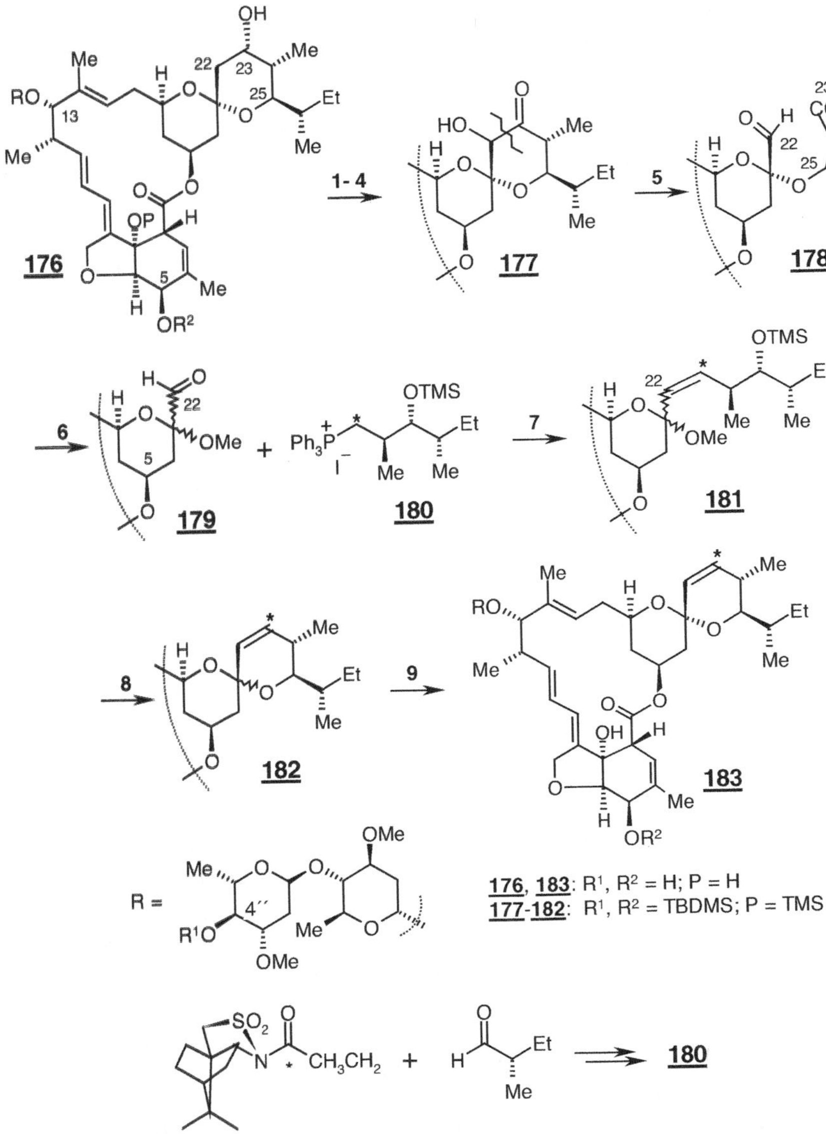


**Reaction conditions: 1a**. TBDMSCl, $Et_3N$, DMF–$CH_2Cl_2$; 20 °C, 16 h, **b**. oxalyl chloride–DMSO; $CH_2Cl_2$; **2**. TMSCl, LiHMDS, THF; –78 °C, 1 h; **3**. MCPBA, $CH_2Cl_2$; 20 °C, 20 min; **4**. 1 % HOAc–MeOH; 20 °C, 1 h; **5**. $Pb(OAc)_4$, benzene; 20 °C, 30 min; **6**. pyridinium tosylate (PPTS) , MeOH; 20 °C, 1.5 h; **7a** . **180**, KHMDS, toluene; r.t., 30 min, **b** . **179**, toluene; –75 °C, 5 min; r.t.; 1.75 h; **8**. PPTS, MeOH; r.t., 2 h; **9**. HF–pyridine; acetonitrile; 20 °C, 2 d

***Figure 10.47*** *Preparation of α-hydroxy carbonyl subunits for the cleavage of C–C bonds for use in the reconstitution approach: synthesis of [23-$^{14}$C]avermectin B1a*

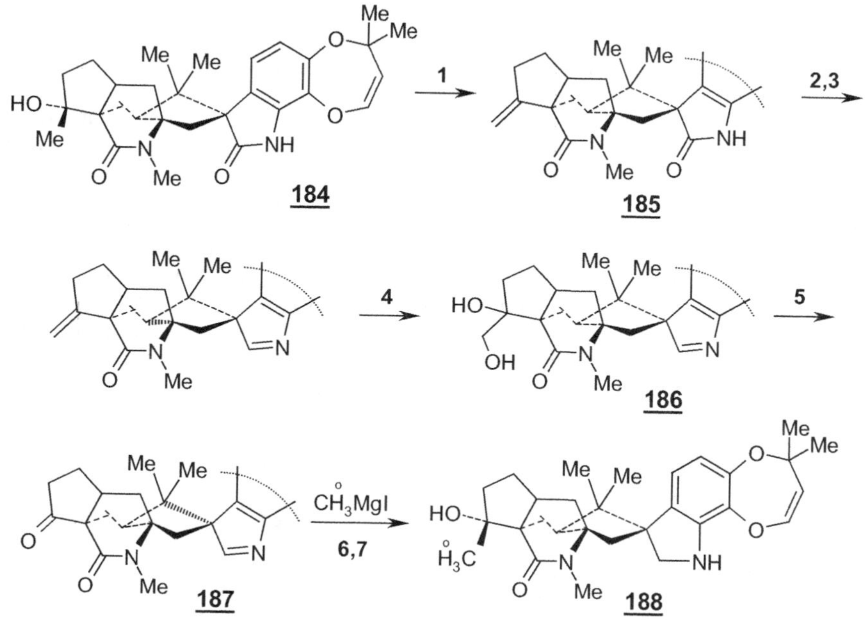


**Reaction conditions:** **1**. DAST, DMAP, $CH_2Cl_2$; r.t., 2 h; **2a** . NaH, THF; r.t., 30 min, **b**. Fmoc-Cl, THF; 0 °C, 10 min; **3a**. $NaBH_4$, MeOH; r.t., 15 min; **b**. piperidine; r.t., 2 h; **4**. $NaIO_4$, THF–$H_2O$ 2:1; 0 °C, 30 min; 5 °C, 16 h; **5**. $NaIO_4$, aq. THF; 5 °C, 16 h; **6a**. $C^2H_3MgI$ (3 equiv.), THF; −78 °C, 30 min; 0 °C, 30 min, **b**. $NaBH_3CN$, MeOH; 0 °C, 30 min

***Figure 10.48*** *Generation of an olefinic compound from an alcohol for the cleavage of a C-C bond for use in the reconstitution approach: synthesis of 2-desoxo[1′-methyl-$^2H_3$] paraherquamide*

$C^2H_3MgI$ and reduction of the indol system afforded the anthelmintic drug substance **188** ([$^2$H]PNU-141962, 2-desoxo[1′-methyl-$^2H_3$]paraherquamide), thereby reconstituting the skeleton of **184** in a slightly modified form. The synthesis depicted was designed as a pilot study for the preparation of the respective tritium and carbon-14-labeled isotopomers. It illustrates a more general C-dealkylation–realkylation approach that is otherwise difficult to accomplish (Section 10.1.2)[83].

## References


1. Hill, J.A. and Bankston, D.D., *J. Label. Compd. Radiopharm.*, ***1995***, *36*, 713–718.
2. Demin, P.M.; Vasiljeva, L.L.; Kochev, D.M.; Pivnitsky, K.K. and Pace-Asciak, C.R., *J. Label. Compd. Radiopharm.*, ***1999***, *42*, 773–779.
3. Wheeler, W.J.; Clodtfelter, D.K.; Kulanthaivel, P.; Pedregal, C.; Stoddard, E.A.; Wright, R.A. and Schoepp, D.D.; *Bioorg. Med. Chem. Lett.*, ***2005***, *15*, 349–351.

4. Dischino, D.D.; Meanwell, N.A.; Morimoto, H.; Pearce, B.C.; Than, C.; Torri, A.F.; Williams, P. G. and Yu, K.-L., *J. Label. Compd. Radiopharm.*, ***1999***, *42*, 1061–1068.
5. Taylor, G.F.; Thornton, S.S.; Tallent, C.R. and Kepler, J.A., *J. Label. Compd. Radiopharm.*, ***1993***, *33*, 501–515.
6. Moenius, Th.; Voges, R.; Susan, A.B.; Jones, L.; Tarapata, R.; Shapiro, M. and Jarema, M.A.C., in *Synthesis and Applications of Isotopically Labelled Compounds 1991*, Buncel, E. and Kabalka, G.W., Eds.; Elsevier: Amsterdam, ***1992***, pp. 149–154.
7. Wagner, J.; Andres, H.; Rohrbach, S.; Wagner, D.; Oberer, L. and France, J., *J. Org. Chem.*, ***2005***, *70*, 9588–9590.
8. a. Atwell, G.J. and Denny, W.A., *J. Label. Compd. Radiopharm.*, ***2007***, *50*, 7–12; b. Fukase, K.; Kinoshita, I.; Suda, Y.; Aoki, Y.; Liu, W.-C.; Oikawa, M.; Lee, S.; Tornus, I.; Dong, H. and Gröger, S., *J. Label. Compd. Radiopharm.*, ***1999***, *42*, 361–372; c. Kurosawa, M.; Zähringer, U.; Seydel, U.; Reitschel, E.Th. and Kusumoto, S., *Synlett*, ***1996***, 252–254; d. Saljoughian, M.; Morimoto, H. and Williams, P.G., *Fusion Technol.*, ***1992***, *21*, 318–324; e. Singh, A.K. and Klein, R.S., *J. Label. Compd. Radiopharm.*, ***1988***, *24*, 1219–1228; f. Rupar, C.A. and Carroll, K.K., *Chem. Phys. Lipids*, ***1976***, *17*, 193–206.
9. Pasto, D.G. and Lepeska, B., *J. Am. Chem. Soc.*, ***1976***, *98*, 1091–1095.
10. See for example Numazawa, M.; Sohtome, N. and Nagaoka, M., *Chem. Pharm. Bull.*, ***2004***, *52*, 722–726.
11. Chauduri, N.K. and Gut, M., Synthesis of isotopically labeled steroids, in *Methods Enzymol.*, ***1969***, *15*, 305–346.
12. Fagart, J.; Sobrio, F. and Marquet, A., *J. Label. Compd. Radiopharm.*, ***1997***, *39*, 791–795.
13. Gay, J. and Rohlfs, G., *J. Label. Compd. Radiopharm.*, ***1999***, *42*, 994–995.
14. Oba, M.; Ohkuma, K.; Hitokawa, H.; Shirai, A. and Nishiyama, K., *J. Label. Compd. Radiopharm.*, ***2006***, *49*, 229–235.
15. Hesk, D.; Delduca, P.; Koharski, D.; McNamara, P.; Magatti, S.; Saluja, S. and Thomas, L., *J. Label. Compd. Radiopharm.*, ***1993***, *33*, 439–442.
16. Al-Rawi, J.M.A.; Elvidge, J.A.; Thomas, R. and Wright, B.J., *J. Chem. Soc. Chem. Commun.*, ***1974***, 1031–1032.
17. a. Trnka, T.M. and Grubbs, R.H., *Acc. Chem. Res.*, ***2001***, *34*, 18–29; b. Grubbs, R.H., *Handbook of Metathesis*; Wiley-VCH: Weinheim, ***2003***, pp. 112–131.
18. a. Maier, F.; Bayer Schering Pharma AG, Berlin; personal communication, May ***2008***; b. Klar, U.; Buchmann, B.; Schwede, W.; Skuballa, W.; Hoffmann, J. and Lichtner, R.B., *Angew. Chem. Int. Edn Engl.*, ***2006***, *45*, 7942–7948.
19. a. Szabolcs, A.; Szammer, J. and Noszko, L., *Tetrahedron*, ***1974***, *40*, 3647–3648; b. Szabolcs, A. and Szammer, J. *J. Label. Compd*, ***1974***, *10*, 113–119; c. Szammer, J. and Szabols, A., *J. Label. Compd. Radiopharm.*, ***1976***, *12*, 189–197; d. McChesney, J.D. and Fraher, P., *J. Label. Compd. Radiopharm.*, ***1979***, *12*, 519–522.
20. Ratusky, J. and Tyka, R., *J. Label. Compd*, ***1969***, *5*, 211–219.
21. Andres, H. and Bühler, K.; Sandoz Ltd, CH-4002, Basel, Internal Report, ***11.09.85*** (unpublished).
22. Parnes, H; *J. Label. Compd. Radiopharm.*, ***1979***, *16*, 771–775.
23. a. Hageman, H.A.; *Org. React.*, ***1953***, *7*, 198; b. Wheeler, W. J.; kau, D.L.K. and Bach, N., *J. Label., Compd., Radiopharm.*, ***1990***, *28*, 273–295.
24. Cooley, J.H. and Evain, E.J., *Synthesis*, ***1989***, 1–7.
25. Monkovic, I.; Wong, H. and Bachand, C., *Synthesis*, ***1985***, 770–773.
26. a. Meier, J. and Schreier, E., *Headache*, ***1976***, 96–104; b. Wheeler, W.J.; Kau, D.L.K. and Bach, N., *J. Label. Compd. Radiopharm.*, ***1990***, *28*, 107–113.
27. Ferguson, J.R.; Hollis, S.J.; Johnston, G.A.; Lumbard, K.W. and Stachulski, A.V., *J. Label. Compd. Radiopharm.*, ***2002***, *45*, 107–13; b. Yost, Y. and Holtzman, J.L., *J. Label. Compd. Radiopharm.*, ***1981***, *18*, 1793–1796; c. Vandersteene, I.V.; Audenaert, K.; Slegers; G. and Dierckx, R.A., *J. Label. Compd. Radiopharm.*, ***1998***, *41*, 171–180.
28. Janssen, C.G.M.; Thijssen, J.B.A. and Verluyten, W.L.M.; *J. Label. Compd. Radiopharm.*, ***2002***, *45*, 841–855. For [*N*-methyl-$^{2}H_3$]galantamine see Rouleau, J. and Guillou, C., *J. Label. Compd. Radiopharm.*, ***2008***, *51*, 236–238.

29. Fellows, I.; Carr, R.M.; deBoek, N.; Montgomery, S. and Waterhouse, I., *J. Label. Compd. Radiopharm.*, ***1998***, *41*, 1127–1143.
30. Melander, L. and Saunders, Jr., W.H., *Reaction Rates of Isotopic Molecules*; Robert K. Krieger: Malabar, FL, ***1987*** (ISBN 0-90974-940-9), pp. 172–174.
31. Liu, J.; Miller, G.G.; Huang, L.; Diwu, Z.; Lown, J.W.; Brown, K.; Moore, R.B.; Tulip, J. and McPhee, M.S., *J. Label. Compd. Radiopharm.*, ***1995***, *36*, 815–823.
32. Senderoff, S.G.; Landvatter, S.W. and Heys, J.R., *J. Label. Compd. Radiopharm.*, ***2000***, *43*, 1283–1288.
33. Barling, P.M.; Preston, J.R.; Bibby, N.J. and Wilson, T., *Anal. Biochem.*, ***1985***, *144*, 542–552.
34. Elmore, C.J.; Dean, D.C.; Marks, T.M.; Braun, M.P.; Egan, M.A. and Melillo, D.G. in *Synthesis and Applications of Isotopically Labeled Compounds, Vol. 8*, Dean, D.C.; Filer, C.N. and McCarthy, K.E., Eds.; John Wiley & Sons, Ltd.: Chichester, ***2004***, pp. 15–18.
35. Smith, M.B. and March, J., *March's Advanced Organic Chemistry*; John Wiley & Sons, Inc.: New York, ***2001***, pp. 1527–1528.
36. a. Kasai, M.; Arai, H. and Kanda, Y., *Synthesis and Applications of Isotopically Labeled Compounds 1991*, Buncel, E. and Kabalka, G.W., Eds.; Elsevier: Amsterdam, ***1992***, pp. 556–559; b. Arai, H. and Kasai, M.; *J. Label. Compd. Radiopharm.*, ***1991***, *29*, 903–908; c. Kasai, M.; Arai, H. and Kanda, Y.; *J. Chem. Soc. Chem. Commun.*, ***1991***, 600–601; d. Kanda, Y., *J. Label. Compd Radiopharm.*, ***1990***, *28*, 1033–1036; e. Kanda, Y. and Kasai, M., *J. Org. Chem.*, ***1990***, *55*, 2515–2518; f. Hase, T.A. and Kukkola, P., *Synth. Commun.*, ***1980***, *10*, 451–456.
37. a. Levitz, M., *J. Am. Chem. Soc.*, ***1953***, *75*, 5352–5355; b. Heer, J. and Miescher, K., *Helv. Chim. Acta*, ***1945***, *28*, 156–165.
38. Mokhallalati, M.K.; Landvatter, S.W. and Baine, N.H., *Synthesis and Applications of Isotopically Labeled Compounds 1997*, Heys, J.R. and Mellilo, D.G., Eds.; John Wiley & Sons, Ltd: Chichester, ***1998***, pp. 157–160.
39. Kuo, F. and Wheeler, W.J., *J. Label. Compd. Radiopharm.*, ***1994***, *34*, 915–920.
40. a. Reich, I.L.; Reich, H.J.; Menahan, L.A. and Peterson, R.E., *J. Label. Compd. Radiopharm.*, ***1987***, *24*, 1235–1244; b. Susàn, A.B. and Nystrom, R.F., *J. Label. Compd*, ***1971***, *7*, 275–282; c. Susàn, A.B. and Nystrom, R.F., *J. Label. Compd*, ***1971***, *7*, 269–274.
41. a. Kurosowa, M.; Kanamaru, H. and Nishioka, K., *J. Label. Compd. Radiopharm.*, ***1997***, *39*, 129–138; b. Masjedizadeh, M.R. and Parnes, H., *J. Label. Compd. Radiopharm.*, ***1993***, *33*, 687–696; c. Tanaka, F.S.; Wien, R.G. and Hoffer, B.L., *J. Label. Compd. Radiopharm.*, ***1984***, *21*, 549–561.
42. a. Hachey, D.L.; Szczepanik, P.A.; Berngruber, O.W. and Klein, P.D., *J. Label. Compd*, ***1973***, *9*, 703–719; b. Carlson, G.L. and Fromm, H., *J. Label. Compd. Radiopharm.*, ***1979***, *16*, 421–434; c. Ren, S.; McNamara, P. and Koharski, D., *J. Label. Compd. Radiopharm.*, ***2008***, *51*, 231–235; d. Tserng, K.Y. and Klein, P.D., *J. Lipid, Res.*, ***1977***, *18*, 400–403; e. Carr, R.M., *J. Label. Compd. Radiopharm.*, ***1994***, *34*, 961–971.
43. a. Barton, D.H.R.; Lacher, B. and Zard, S.Z., *Tetrahedron*, ***1987***, *41*, 4321–4328; b. Barton, D.H. R. and Zard, S.Z., *Pure Appl. Chem.*, ***1986***, *58*, 675; c. Barton, D.H.R.; Crich, D. and Motherwell, W.B., *J. Chem. Soc. Chem. Commun.*, ***1983***, 939–941.
44. Huang, G.T. and Parnes, H., *J. Label. Compd. Radiopharm.*, ***1995***, *36*, 449–456.
45. Loreau, O.; Boullais, C.; Noel, J.P. and Mioskowski, C., *J. Label. Compd. Radiopharm.*, ***1998***, *39*, 159–170.
46. Berthelette, C. and Wang, Z., *J. Label. Compd. Radiopharm.*, ***2007***, *50*, 1–5.
47. a. Marzoni, G.; Wheeler, W.J. and Garbrecht, W.L., *J. Label. Compd. Radiopharm.*, ***1988***, *25*, 429–438; b. Wheeler, W.J.; Kau, D.L.K. and Bach, N.J., *J. Label. Compd. Radiopharm.*, ***1990***, *28*, 273–295; c. Mantegani, S.; Brambilla, E.; Ermoli, A.; Fontana, E.; Angiuli, P. and Vicario, G.P., *J. Label. Compd. Radiopharm.*, ***1991***, *29*, 519–533; d. Hicks, D.R., *J. Label. Compd. Radiopharm.*, ***1984***, *21*, 229–235.
48. Sundaram, M.G., *J. Label. Compd. Radiopharm.*, ***1981***, *18*, 489–494.
49. Brown, C.; Eustache, J.; Frideling, J.P. and Shroot, B., *J. Label. Compd. Radiopharm.*, ***1984***, *31*, 973–983.
50. Almeida, M.; Boman, A. and Lundstedt, T., *J. Label. Compd. Radiopharm.*, ***2002***, *45*, 371–377.

51. Lee, D.G. and Chen, T., *Cleavage reactions, in Comprehensive Organic Synthesis, Vol. 7 (Oxidations)*, Trost, B.M.; Fleming, I. and Ley, S.S., Eds., Pergamon: Oxford, ***1992***, pp. 541–591.
52. Danieli, B. and Russo, G., *J. Label. Compd*, ***1966***, *1*, 275–281.
53. Lai, C.K.; Byon, C.Y.; Gut, M.; Mostowicz, D. and Anderson, W.G., *J. Label. Compd. Radiopharm.*, ***1984***, *21*, 627–637.
54. Njarr, V.C.O. and Banthrope, D.V., *J. Label. Compd. Radiopharm.*, ***1985***, 22, 615–621.
55. Hsi, R.S.P.; Stelzer, L.S. and Stolle, W.T., *J. Label. Compd. Radiopharm.*, ***1989***, *27*, 1115–1125.
56. Urano, S.; Nakano, S. and Matsuo, M., *J. Label. Compd. Radiopharm.*, ***1984***, *21*, 313–321.
57. Woo, P.W.K.; Hayes, R.; Beylin, V.G.; Johnson, S.; Lovdahl, M.; Huang, Y. and Huang, C.C., in *Synthesis and Applications of Isotopically Labeled Compounds 1997*, Heys, J.R. and Melillo, D.G., Eds.; John Wiley & Sons, Ltd: Chichester, ***1998***, pp. 429–432.
58. a. Han, M.; Hayes, B.A.; Prendergast, P.T. and Gupta, S.; *J. Label. Compd. Radiopharm.*, ***2000***, *43*, 1149–1157; b. Shu, A.Y.L. and Heys, J.R., *J. Label. Compd. Radiopharm.*, ***1994***, *34*, 587–596; c. Ferdinandi, E.S., *J. Label. Compd. Radiopharm.*, ***1979***, *16*, 345–350; d. Hoellinger, H.; Nguyen, H.-N. and Pichat, L.; *J. Label. Compd*, ***1974***, *10*, 617–626; e. Schulze, P.E., *J. Label. Compd*, ***1969***, *5*, 312–319; f. Thompson, L.M.; Yates, C.H. and Odell, A.D., *J. Am. Chem. Soc.*, ***1947***, *69*, 726–727; g. Gut, M., *Helv. Chim. Acta.*, ***1953***, *36*, 906–909; h. Belleau, B. and Gallagher, T.F., *J. Am. Chem. Soc.*, ***1951***, *73*, 4458–4460; i. Fujimoto, G.I., *J. Am. Chem. Soc.*, ***1951***, *73*, 1856; j. Turner, R.B., *J. Am. Chem. Soc.*, ***1950***, *72*, 579–585; k. Turner, R.B., *J. Am. Chem. Soc.*, ***1947***, *69*, 726–727; l. Weill-Raynal, J., *Synthesis*, ***1969***, *1*, 49–56.
59. Covey, D.F.; McMillan, P.C.; Wixler, L.L. and Cabell, M., *Biochem. Biophys. Res. Comm.*, ***1988***, *157*, 82–86.
60. Prakash, S.R.; Ellworth, R.L.; Mertel, H.E. and Rasmusson, H. in *Synthesis and Applications of Isotopically Labeled Compounds 1988*, Baillie, T.A. and Jones, J.R., Eds.; Elsevier: Amsterdam, ***1989***, pp. 619–622.
61. Seo, S.; Nagasaki, T; Katsuyama, Y.; Matsubara, F.; Sakata, T.; Yoshioka, M. and Makisumi, Y., *J. Label. Compd. Radiopharm.*, ***1989***, *27*, 1383–1393.
62. Otto, P.Ph.H.L.; Besemer, A.C. and Wientjens, W.H.J.M., *J. Label. Compd*, ***1970***, *6*, 111–122.
63. a. McCarthy, K.E.; Schilknegt, K. and Zandi, K.S. in *Synthesis and Applications of Isotopically Labeled Compounds 1997*, Heys, J.R. and Melillo, D.G., Eds.; J. Wiley & Sons: Chichester, ***1998***, pp. 165–169; b. DeNinno, M.P. and McCarthy, K.E., *Tetrahedron*, ***1997***, *53*, 11007–11020; c. DeNinno, M.P., *J. Am. Chem. Soc.*, ***1995***, *117*, 9927–9928.
64. Beck, E.J.; Twitchin, B. and Mander, L.N., *Can. J. Chem.*, ***2004***, *82*, 293–300.
65. Ahern, D.G.; Seguin, R.J. and Filer, C.N., *J. Label. Compd. Radiopharm.*, ***2002***, *45*, 401–405.
66. Chaudhuri, N.K.; Sung, M.S.; Markus, B. and Desai, M., in *Synthesis and Applications of Isotopically Labeled Compounds 1989*, Baillie, T.A. and Jones, J.R., Eds.; Elsevier: Amsterdam, ***1989***, pp. 587–590.
67. Waterhouse, I.; Cable, K.M.; Fellows, I. and Wipperman, M.D., *J. Labeled Compd. Radiopharm.*, ***1996***, *38*, 1021–1030.
68. Landvatter, S.W., in *Synthesis and Applications of Isotopically Labeled Compounds 1994*, Allen, J. and Voges, R., Eds.; John Wiley & Sons, Ltd: Chichester, ***1995***, pp. 683–688.
69. Campbell, J.B.; Knagg, E.; Newby, M.; Reid, R.G. and Wilde, P.D., in *Synthesis and Applications of Isotopically Labeled Compounds 1991*, Buncel, E. and Kabalka, G.W., Eds.; Elsevier: Amsterdam, ***1992***, pp. 207–213.
70. Lin, J.T. and Heftmann, E., *J. Label. Compd. Radiopharm.*, ***1979***, *16*, 537–545.
71. Hicks, D.R.; Hangeland, J.J.; Mobilio, D. and DeLange, B., *J. Label. Compd. Radiopharm.*, ***1988***, *25*, 89–94.
72. a. Wheeler, W.J.; Chay, S.H.; Herman, J.L. and O'Bannon, D.D., *J. Label. Compd. Radiopharm.*, ***2005***, *48*, 669–681; b. Wheeler, W.J.; Douglas, D.M. and O'Bannon, D.D., in *Synthesis and Applications of Isotopically Labeled Compounds 1997*; Heys, J.R. and D. Melillo, D.G., Eds.; J. Wiley & Sons: Chichester, ***1998***, pp. 491–494.

73. Terasawa, T.; Atsumi, K.; Yokoshima, T.; Ohyabu, S. and Fukasaku, N., *J. Label. Compd. Radiopharm.*, ***1986***, *13*, 1005–1017.
74. Rao, P.N. and Damodaran, K.M., *J. Label. Compd. Radiopharm.*, ***1982***, *19*, 415–424.
75. Rao, P.N.; Cessac, J.W. and Hill, K.A., *J. Label. Compd. Radiopharm.*, ***1982***, *19*, 1115–1124.
76. a. Rao, P.N.; Cessac, J.W.; Hill, K.A., in *Synthesis and Applications of Isotopically Labeled Compounds 1982*, Duncan, W.P. and Susan, A.B., Eds., Elsevier: Amsterdam, ***1982***, pp. 215–216; b. Rao, P.N.; Cessac, J.W.; Hill, K.A., *J. Label. Compd. Radiopharm.*, ***1982***, *19*, 1161–1170.
77. Walker, M.; McKinney, J. and Oswald, E., *J. Label. Compd*, ***1974***, *10*, 405–411.
78. a. Falck, J.R.; Anjaiah, S.; Tuniki, V.R.; Gopal, V.R. and Capdevila, J.H., *J. Label. Compd. Radiopharm.*, ***2006***, *49*, 245–252; b. Manna, S.; Falck, J.R.; Chacos, N. and Capdevila, J., *Tetrahedron Lett.*, ***1983***, *24*, 33–36.
79. Bieniarz, C.; Ramakrishna, K.V. and Kelly, T.P., *J. Label. Compd. Radiopharm.*, ***2000***, *43*, 243–250.
80. a. Nadeau, R.G. and Hanzlik, R.P., *Methods Enzymol.*, ***1969***, *15*, 346–351; b. Ceruti, M.; Grosa, G.; Rocco, F.; Dosio, F and Cattel, L., *J. Label. Compd. Radiopharm.*, ***1994***, *34*, 577–585.
81. Chen, C.R.; Fong, M.T.; Fujiwara, A.N.; Henry, D.W.; Leaffer, M.A.; Lee, W.W. and Smith, T.H., *J. Label. Compd. Radiopharm.*, ***1978***, *14*, 111–117.
82. a. McKillican, B.P.; Perine, J.W.; Simmons, D.P. and Wilson, D.E., *J. Label. Compd. Radiopharm.*, ***2005***, *48*, 25–30; b. McKillican, B.P.; Simmons, D.P. and Wilson, T.E., in *Synthesis and Applications of Isotopically Labeled Compounds*, Vol. 7, Pleiss, U. and Voges, R., Eds.; John Wiley & Sons, Ltd: Chichester, ***2001***, pp. 193–196; c. Shih, T.; Mrozik, H.; Holmes, M. and Fisher, M., *Tetrahedron Lett.*, ***1990***, *31*, 3525–3528; d. Shih, T.; Mrozik, H.; Holmes, H. and Fisher, M., *Tetrahedron Lett.*, ***1990***, *31*, 3529–3532.
83. Lee, B.H., *J. Label. Compd. Radiopharm.*, ***2002***, *45*, 1209–1218.

# 11

# Preparation of Enantiomerically Pure Compounds Labeled with Isotopes of Hydrogen and Carbon

Enantiomers have essentially identical chemical and physical properties except in a chiral environment. Biological systems, providing chiral environments, therefore often distinguish between the enantiomers of pharmacologically active compounds, eliciting different pharmacological and pharmacokinetic effects. (*R*)-Thalidomide, a powerful tranquilizer, for example, shows no teratogenicity, whereas its (*S*)-enantiomer produces profound birth defects. (*R*,*R*)-Chloramphenicol is well known to exhibit excellent antibacterial activities, whereas the (*S*,*S*)-enantiomer is completely inactive. In racemic drugs, enantiomers that are pharmacologically inactive do not contribute to positive medicinal properties but may contribute to undesirable side effects. Therefore, regulatory authorities strongly recommend that either the pharmacological activities of each individual enantiomer be characterized before promoting a racemic drug candidate into clinical Phase 1 or, preferably, that only the active enantiomer be advanced through development. Consequently, the individual enantiomers of chiral investigational drugs should be available in carbon-14- or tritium-labeled form early on, beginning no later than for use in ADME studies in animals.

As with their unlabeled counterparts, e.p. tritiated or carbon-14-labeled compounds can be prepared in three ways: resolution of racemates, stereoselective syntheses and biochemical methods. Each of these methods has its own strengths and limitations, according to the structural complexity of the compound, the availability of nonlabeled reference material(s) and the kinds of in-house experience and resources available. The resolution of racemates and stereoselective syntheses are the topics of this chapter. Biochemical methods are the subject of Chapter 12.

*Preparation of Compounds Labeled with Tritium and Carbon-14* Rolf Voges, J. Richard Heys and Thomas Moenius

## 11.1 Resolution of Racemates[1]

Despite the availability of a steadily increasing number of stereoselective procedures and the inevitable loss of at least 50% of radioactivity, the resolution of racemates is still one of the most widely used strategies for the preparation of e.p. (enantiomerically pure) labeled compounds. Unlike techniques applied on an industrial scale, and as an alternative to classical techniques such as crystallization of diastereomeric salts or chromatographic separation of diastereomeric derivatives, resolution of racemates by preparative chiral HPLC has evolved to be the most powerful method, largely because a variety of high-quality packed columns (e.g. Chiracel®, Chiralpak®, Chiraspher® etc.) have become commercially available. Methods developed in-house for analytical purity controls, or obtained from commercial vendors of chromatographic equipment and supplies, may be used as starting points for development of preparative methods. Illustrative examples (among the many available) include the resolution of racemic [2,3-$^{14}$C]MaxiPost™ (**1**) on Chiracel OD (20 mm × 250 mm, 10 μm; 9 : 1 hexane/isopropanol)[2], the racemic 2-hydroxy[1,2-$^{14}$C]-arylacetic acid derivative **2** on Chiralpak AD (10 mm × 250 mm, 10 μm; 4 : 1 hexane/isopropanol)[3], racemic [6a-$^{14}$C]apomorphine dimethyl ether (**3**) on Chiracel OD (4.6 mm 250 mm, 10 μm; 950 : 50 : 5 hexane/isopropanol/dimethylamine)[4] and methyl (*R,S*)-11-tosyloxy[11-$^{3}$H]stearate (**4**) on Chiralpak AD (4.6 mm × 250 mm, 10 μm; 100:1 heptane/ethanol; 6 mg scale)[5] to give the enantiomers in yields of 36–47% (i.e. 72–94% of the theoretical value) and >98% e.e.s (enantiomeric excesses).

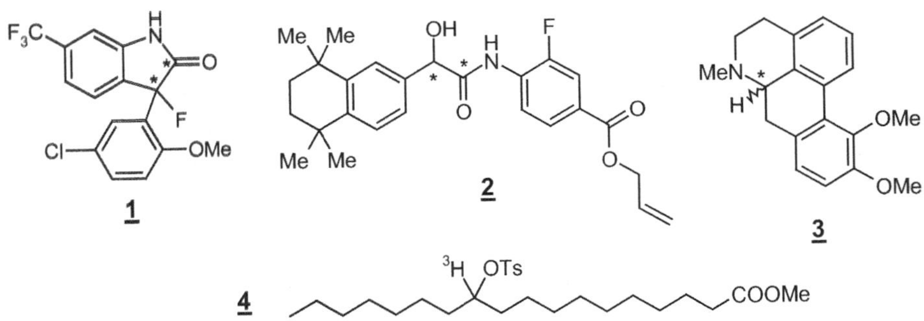


*Figure 11.1 Racemic tritiated and carbon-14-labeled final products and intermediates resolved by preparative chiral HPLC*

Chiral HPLC separation techniques have some significant advantages over classical resolution methods:

- They are not dependent on the presence of functional groups, such as COOH, $NR_2$ etc., required for the formation of diastereomers or diastereomeric salts through covalent or ionic binding to an e.p. chiral reagent;
- Small quantities can be resolved without considerable losses of material (essential for candidates of high specific activity, for example **4**);
- Radiochemical yields are often significantly higher (35–47%) than for the more classical methods (25–35%);

- Resolution can often be conducted at a very late synthesis step, which is more efficient for preparative reasons.

Nevertheless, resolution of acidic or basic radiolabeled racemates is still frequently carried out through differential crystallization of diastereomeric salts with unlabeled e.p. bases (e.g. (−)-brucine, (−)-strychnine, (+)-cinchonine, (−)-quinine, (−)-ephedrine) or acids (e.g. (1*R*)-(−)-10-camphorsulfonic acid, (1*S*)-(−)- and (1*R*)-(+)-3-bromocamphor-8-sulfonic acid, (*S*,*S*)-(−)- and (*R*,*R*)-(+)-tartaric acid, (*S*,*S*)- and (*R*,*R*)-*O*,*O*′-dibenzoyl acids). Neutral candidates, lacking appropriate basic or acidic functionalities but containing for example a hydroxy group can firstly be converted into the respective hydrogen phthalate through reaction with phthalic anhydride. Examples demonstrating the utility of this classic methodology are compound **5**, the penultimate precursor for D-[1-$^{14}$C]chloramphenicol[6]; [$^{14}C_2$]EM652 (**6**), needed for the preparation of e.p. [$^{14}C_2$]acolbifene, a selective estrogen receptor modulator[7]; compound **7**, key intermediate for e.p. (*S*)-[4-$^{14}$C]threonine[8]; and **8**, the penultimate precursor for e.p. [$^{14}$C]paroxetine[9].

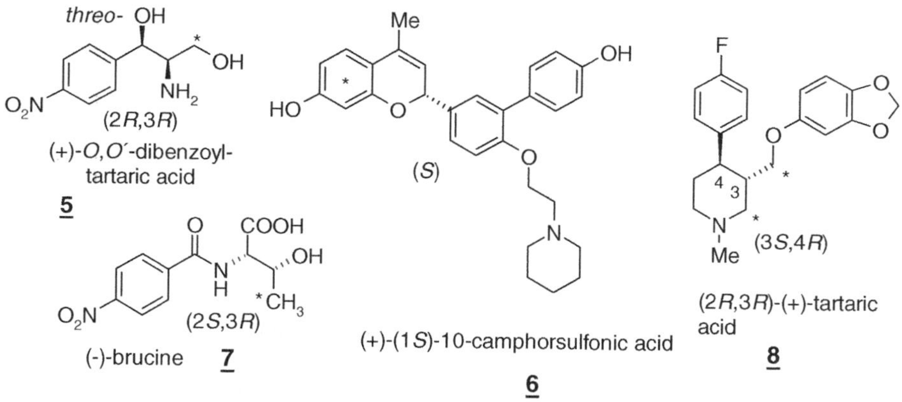


***Figure 11.2*** *Enantiomeric [$^{14}$C]intermediates separated through formation of diastereomeric salts*

Noncrystallinity or insufficient differences in solubilities of diastereomeric salts may prevent resolution by this classical method, but diastereomers prepared by covalent attachment of the labeled racemate to an enantiopure nonlabeled reagent may be separated by normal achiral chromatographic methods. Subsequent cleavage of the nonlabeled reagent from each diastereomer gives the respective labeled enantiomer. Critical to success is that the cleavage procedure does not cause any racemization of the labeled enantiomer. An example of the application of such a strategy is the preparation of [$^{14}$C]SKF 104353 (**11**)[10] (Figure 11.3). Racemic **9** was first coupled to *N*-(2,2,2-trichloroethoxycarbonyl)-(*S*)-proline following the acyl chloride method. Subsequent achiral HPLC separation of the resultant diastereomers **10a,b** and ester cleavage provided the desired 2*S*,3*R*-enantiomer in 36% radiochemical yield.

This methodology is of particular interest when the unlabeled e.p. coupling component is part of the molecular structure being built up and therefore does not need to be removed, as is frequently the case in the synthesis of labeled peptides, such as **12**[11] and **13**[12] (Figure 11.4).

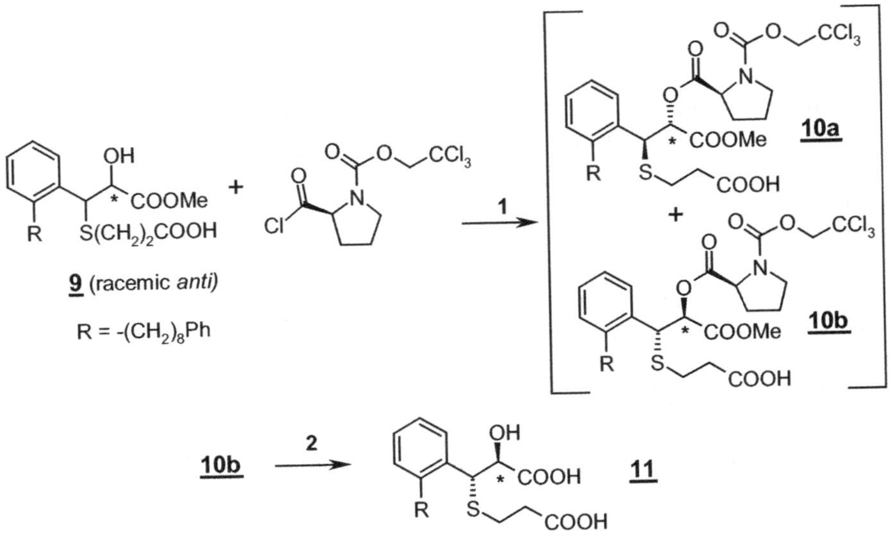


**Reaction conditions: 1**. pyridine; r.t., overnight; **2a** . HPLC on silica gel, 98:2 dichloroethane/ethyl acetate, **b**. LiOH, $H_2O$, dimethoxyethane; 5 °C, 4 h

***Figure 11.3*** *Resolution by HPLC through covalent attachment and subsequent hydrolytic cleavage of a chiral reagent*

Z-(*R,S*)-[3-$^{14}$C]PheOH —**1**→ Z-(*R,S*)-[3-$^{14}$C]Phe-Pro-OH —**2**→

H-(*S*)-[3-$^{14}$C]Phe-PrOH ⟶⟶ H-Tyr-(*R*)-Met-Gly-[3-$^{14}$C]Phe-Pro-OH

**12**

Boc-5,7-Br$_2$-(*R,S*)-Trp-OH ⟶⟶ His-5,7-Br$_2$-(*R,S*)-Trp-Ala-Trp-(*R*)-Phe-Lys —**3**→

90% His-5,7-Br$_2$-(*R*)-Trp-Ala-Trp-(*R*)-Phe-Lys —**4**→ 90% His-(*R*)-[$^3$H]Trp-Ala-Trp-(*R*)-Phe-Lys

10% His-5,7-Br$_2$-(*S*)-Trp-Ala-Trp-(*R*)-Phe-Lys 10% His-(*S*)-[$^3$H]Trp-Ala-Trp-(*R*)-Phe-Lys

—**5**→ His-(*R*)-[$^3$H]Trp-Ala-Trp-(*R*)-Phe-Lys

**13**

**Reaction conditions: 1a**. 2,4,5-trichlorophenol, DCC, THF; r.t., 16 h, **b**. (*S*)-Pro-OH, $Et_3N$, pyridine; **2a**. $H_2$, Pd/C, **b**. HPLC separation; **3**. partial HPLC separation; **4**. $^3H_2$, Pd/C, Pd/$CaCO_3$, DMF; **5**. HPLC separation

***Figure 11.4*** *e.p./d.p. [$^{14}$C]compounds through coupling of a racemic [$^{14}$C]target to a chiral subunit of the target substance followed by HPLC separation of the resulting diasteromers*

Mechanical separation of enantiomers, which crystallize as conglomerates, is of no practical relevance in isotope chemistry. However, formation of true racemates, in which each unit cell contains equal numbers of both enantiomers, is of fundamental importance for the purification of enantiomerically enriched candidates to e.p. materials. Compounds that result from stereoselective synthesis frequently contain, besides the desired enantiomer (e.g. (+)-A), the true racemate (e.g. 1 : 1 (+A)/(−A)) in minor quantities. These two components usually have different solubilities, making it possible to separate them through fractional crystallization. Furthermore, when the scale of the separation, the amount of radioactive material and its specific activity are sufficiently high even racemic product can be resolved by means of this approach. In such cases, portions of unlabeled e.p. target material are added to the labeled racemate and the material is recrystallized until the desired level of enantiomeric purity is achieved. The specific activity of the resulting e.p. labeled compound is reduced by an amount proportional to the amount of added diluent. For example, dilution of racemic [$^{14}$C]CP101,606 at 57 mCi/mmol with e.p. unlabeled (+)-CP101,606 to give a 93 : 7 ratio of enantiomers and subsequent crystallization provided (+)-[$^{14}$C]CP101,606 (**14**) in 23% radiochemical yield with an enantiomeric excess >99% and a specific activity of 5.3 mCi/mmol[13].

**14**

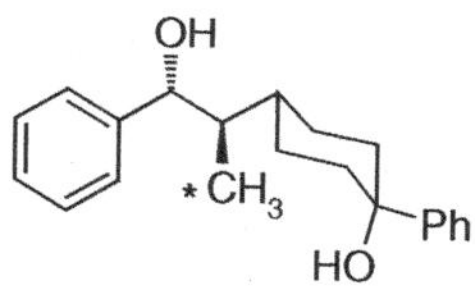


***Figure 11.5*** *(+)-[$^{14}$C]CP101,606 resolved through fractional crystallization*

Chemistry-based kinetic resolution methods, which make use of the preferential reaction of one enantiomer with a chiral reagent (e.g. hydroboration of racemic alkenes with diisopinocampheylborane) or an achiral reagent in the presence of an appropriate chiral catalyst (e.g. Sharpless epoxidation of racemic allylic alcohols with *t*-BuOOH in the presence of (2*R*,3*R*)- or (2*S*,3*S*)-diisopropyl tartrate and $Ti(O\textit{i}\text{-}Pr)_4$) have not been exploited so far for the isolation of e.p. labeled substances. In contrast, biochemical methods have been widely used, particularly for the resolution of racemic $\alpha$-[$^{14}$C]amino acids and various [$^{14}$C]carboxylic acids. Such methods, including:

- Selective cleavage of *N*-acyl groups of racemic *N*-acylated $\alpha$-[$^{14}$C]amino acids by lipases or proteases (method A);
- Selective ester hydrolysis of racemic *N*-protected $\alpha$-[$^{14}$C]amino acid and [$^{14}$C]carboxylic acid esters by proteolytic enzymes (method B);
- Selective *N*-acetylation of racemic $\alpha$-[$^{14}$C]amino acids by acetyl transferases in the presence of vinyl acetate (method C);
- Selective oxidative deamination of (*R*)-enantiomer components within $\alpha$-[$^{14}$C]amino acid racemates by (*R*)-amino acid oxidases (method D);

are mentioned here but are discussed in more detail in Chapter 12. Illustrative examples include the isolation of (*S*)-[1-$^{14}$C]alanine, (*S*)-[1-$^{14}$C]valine, (*S*)-[1-$^{14}$C]norvaline, (*S*)-[1-$^{14}$C]leucine, (*S*)-[1-$^{14}$C]isoleucine, and (*S*)-amino[1-$^{14}$C]isobutyric acid through hog kidney acylase catalyzed *N*-deacetylation[14]; of (*S*)-[1-$^{14}$C]phenylalanine, *p*-fluoro-(*S*)-[1-$^{14}$C]-phenylalanine, (*S*)-[1-$^{14}$C]tyrosine and (*S*)-[2,3-$^{3}$H]phenylalanine through carboxypeptidase A mediated *N*-de-chloroacetylation[14,15]; of (*R*)-[3-$^{14}$C]tryptophan[16], (*S*,*S*)-$\alpha$-[$^{14}$C]-methyltryptophan[17] and (*S*)-*ortho*-[2-$^{14}$C]tyrosine[18] using chymotrypsin; through lipase AK-promoted enantioselective acetylation of the undesired (−)-enantiomer of **15** to provide (+)-[13,14-$^{14}C_2$]pre-CaA, a synthetic precursor to (+)-[$^{14}$C]calanolide A[19]; and finally of (*S*)-[3-$^{11}$C]alanine and (*S*)-[3-$^{11}$C]phenylalanine through immobilized (*R*)-amino oxidase-catalyzed conversion of the (*R*)-enantiomers into the respective $\alpha$-keto acids[20].

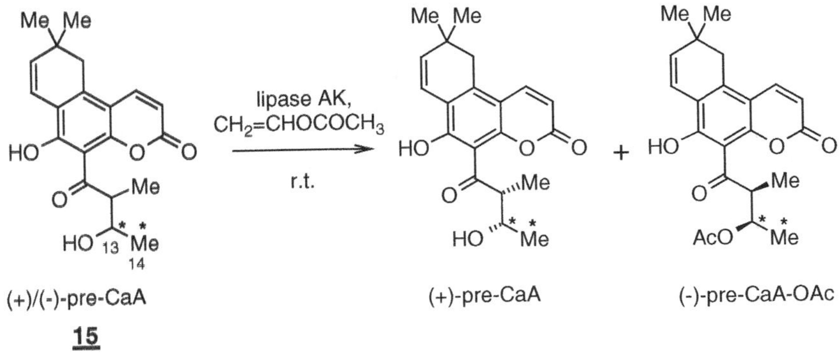


***Figure 11.6*** *Resolution of (+)-[13,14-$^{14}C_2$]pre-CaA, penultimate precursor to (+)-[10,18-$^{14}C_2$]calanolide through lipase AK-catalyzed acetylation of the undesired enantiomer*

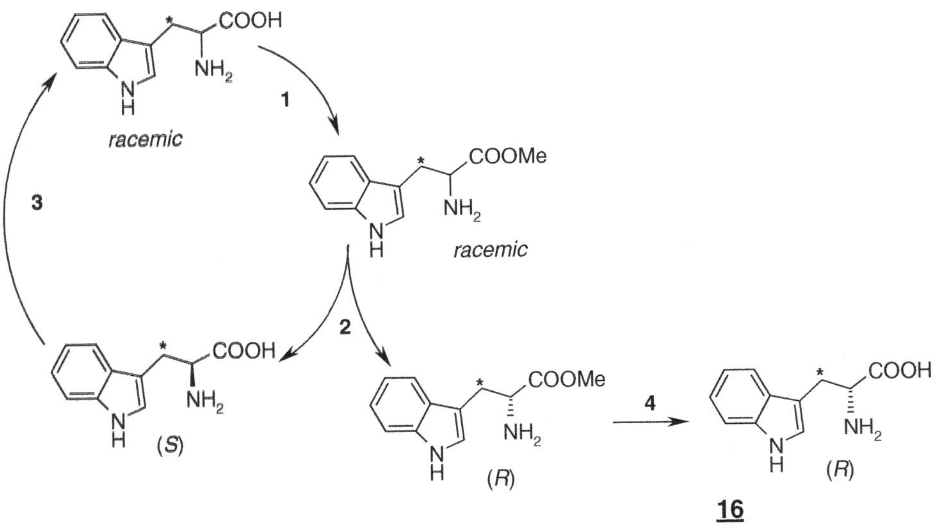


**Reaction conditions:** **1**. HCl/MeOH; 35° C, 20 h; **2**. α-chymotrypsin, 1 N aq. NaCl, pH 6.0–6.3; r.t., 4 h; **3**. $H_2O$; 200 ° C, 8 h; **4**. THF/aq. NaOH, pH 12; r.t., 2 h

***Figure 11.7*** *Preparation of (2R)-[3-$^{14}$C]tryptophan by iterative resolution*

In the ideal case, methods A, B and C provide equal amounts of (*S*)- and (*R*)-α-[$^{3}H/^{14}C$]-amino acids (or derivatives) upon separation. If complete transformation of racemic material is required, re-racemization of the undesired enantiomer can be accomplished chemically by acid catalysis in the presence of catalytic amounts of salicylaldehyde[21], thermal treatment[16] or enzymatically with the aid of transferases[22] and subjection of the undesired product to repeated resolution cycles, as illustrated for (2*R*)-[3-$^{14}$C]tryptophan (**16**).

Whenever enantiomers can be smoothly racemized and one of them is capable of forming an appropriate insoluble salt through reaction with an e.p. chiral reagent, a racemic mixture can be directly converted to the desired enantiomer. This de-racemization procedure (second order asymmetric transformation) has been routinely exploited for the preparation of e.p. arylglycines and cyclic amino acids. It was recently applied to the preparation of (*R*)-phenyl[1-$^{14}$C]glycine (**17**) of high enantiomeric purity (97–98% e.e.) through heating of the racemic ester with one equivalent of (*R*,*R*)-tartaric acid in the presence of a small amount of benzaldehyde followed by acid-catalyzed ester cleavage. The de-racemization is promoted through reversible formation of a Schiff base intermediate[23].

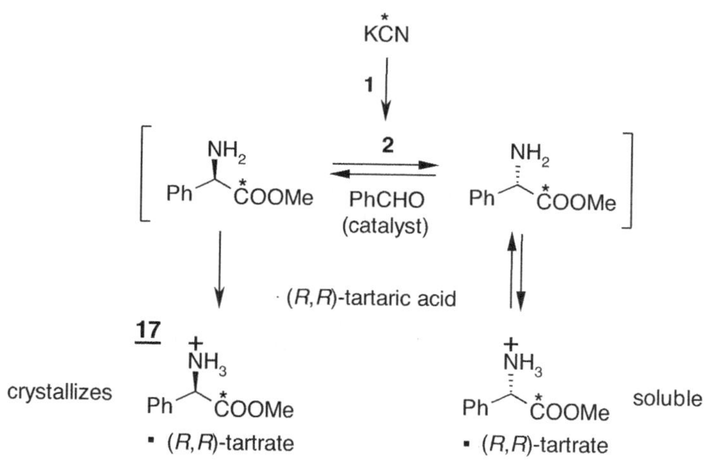


**Reaction conditions: 1a**. PhCHO, $NH_4Cl$, aq. $NH_3$, **b**. HCl, $CH_3OH$; **2**.PhCHO, (*R*,*R*)-tartaric acid, EtOH; r.t.

***Figure 11.8*** *Preparation and resolution of (R,S)-phenyl[1-$^{14}$C]glycine through second order asymmetric transformation*

## 11.2 Enantioselective Synthetic Methods

Three categories of enantioselective procedures are the focus of this section. Two of them can be used with isotopic hydrogen, thereby introducing both deuterium or tritium and chirality into unlabeled substrates. These procedures can also be used with unlabeled reagent to introduce chirality into already labeled compounds. The third procedure is used to transform an achiral or prochiral labeled compound into a chiral one, but cannot introduce isotopic hydrogen or carbon.

### 11.2.1 Hydrogenation/Tritiation of Labeled/Unlabeled $\Delta^{2,3}$-Amino Acid Derivatives

One of the standard methods for preparing enantiomerically pure compounds is the enantioselective hydrogenation of olefins, $\alpha,\beta$-unsaturated amino acids (esters, amides), $\alpha,\beta$-unsaturated carboxylic acid esters, enol esters, enamides, $\beta$- and $\gamma$-keto esters etc. catalyzed by chiral cationic rhodium, ruthenium and iridium complexes[24]. In isotope chemistry, it has only been exploited for the synthesis of e.p. natural and *non*natural $^3$H-, $^{11}$C-, $^{13}$C-, $^{14}$C- and $^{18}$F-labeled $\alpha$-amino acids and small peptides from *N*-protected $\alpha$-(acylamino)acrylates or cinnamates and unsaturated peptides, respectively (Figure 11.9)[25]. This methodology has seen only limited use, perhaps because of perceived radiation safety issues with the use of hydrogenation procedures on radioactive substrates. Also, versatile alternatives are available, including enantioselective metal hydride/tritide reductions, chiral auxiliary-controlled and biochemical procedures (see this chapter, Sections 11.2.2 and 11.3 and Chapter 12).

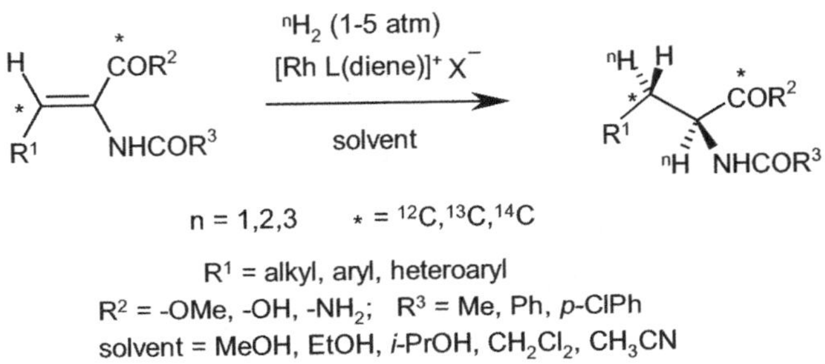


***Figure 11.9*** *Preparation of high enantiomeric purity (R)- or (S)-$\alpha$-amino acids labeled with isotopic hydrogen or carbon through rhodium complex-catalyzed enantioselective hydrogenation of $\Delta^{2,3}$-amino acid precursors*

The requisite precursors, labeled/unlabeled *N*-acylated $\Delta^{2,3}$-amino acids or esters (**20**), are conveniently prepared through reaction of labeled or unlabeled aldehydes with either unlabeled or labeled 5-(4*H*)-oxazolones (**18**) followed by acid- or base-catalyzed hydrolysis or alcoholysis of the resulting 4-arylmethyleneoxazolones **19** (Method 1, Figure 11.10), or with *N*-acylated $\alpha$-phosphonoglycine trimethyl esters **22** in a modified Horner–Wadsworth–Emmons reaction (Method 2, Figure 11.11)[26].

Method 1 only works with aromatic and heteroaromatic aldehydes, and provides predominantly or exclusively Z-olefins **20**, which may be labeled with isotopic carbon at aliphatic positions C1, C2 and/or C3 or in the aromatic/heteroaromatic nucleus. The sequence can be performed in one pot by heating the labeled or unlabeled aldehyde with the respective 5-(4*H*)-oxazolone generated *in situ* from unlabeled or labeled *N*-acylglycine in the presence of acetic anhydride and sodium acetate (classical Erlenmeyer procedure)[27]. For sensitive aldehydes the reaction may have to be carried out at ambient temperatures and reaction times extended up to several days[27g]. In such cases, however, it is highly recommended to first prepare the 4-unsubstituted 5-(4*H*)-oxazolone (variant B, **19**) through cyclodehydration of the *N*-acylglycine with acetic

Method 1, variant A

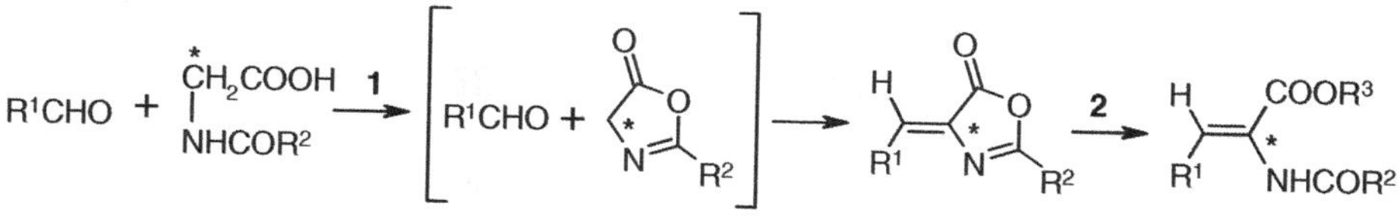

**18** **19** **20**

$R^1$ = Ph; 4-HOPh; 4-AcOPh; 4-$O_2$NPh; 3-MeOPh; 3-MeO,4-AcOPh; 2-naphthyl
$R^2$ = Me; Ph

Method 1, variant B

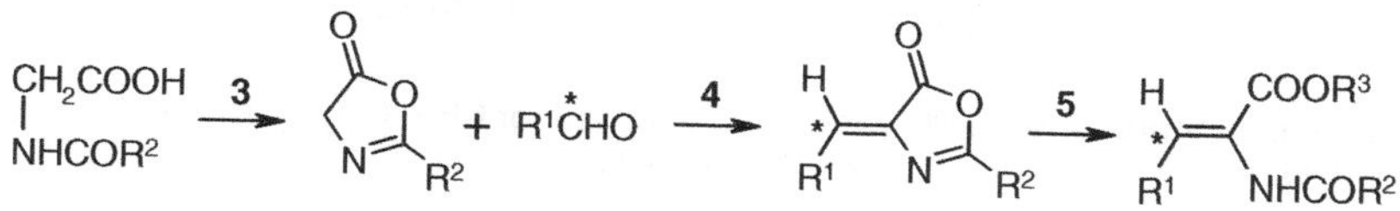

**18** **19** **20**

$R^1$ = Ph; 2-naphthyl; $R^2$ = Me; Ph

**Reaction conditions:** **1**. $Ac_2O$, NaOAc; 100—130 °C, 1—2 h; **2**. $H_2O$/acetone; reflux, 4 h or $Et_3N$, MeOH; 45—50 °C; **3**. *i*-PrN=C=N*i*-Pr, acetonitrile; r.t. or ClCOOEt, $Et_3N$, benzene; r.t.; **4**. DABCO, DBU or $Pb(OAc)_2$, EtOH or $CH_3CN$; 50 to 140 °C; 4 min to 4 h; **5**. acetone/$H_2O$; reflux, 4 h or aq. NaOH; 140 °C, 1 min

***Figure 11.10*** *Synthesis of N-acyl $\Delta^{2,3}$-amino acid through reaction of aldehydes with 5-(4H)-oxazolones (Method 1, variants A and B)*

anhydride, ethyl chloroformate–triethylamine or diisopropyl carbodiimide, then subject it to reaction with the aldehyde component in the presence of an appropriate base (e.g. DABCO, DBU) in a second separate step[28].

Method 2 to *N*-acylated $\Delta^{2,3}$-amino acids involves the use of *N*-protected (Boc, Z) $\alpha$-phosphonoglycine trimethyl ester (**22**), which has been recommended as a universally applicable reagent for Horner–Wadsworth–Emmons reactions with both aliphatic and aromatic aldehydes and ketones to afford the corresponding derivatives in excellent yields (80–99%) and high *Z*-selectivity[29]. The unlabeled reagent is commercially available, but as recently shown with carbon-13 material (Figure 11.11), the corresponding carbon-14-labeled isotopomer should also be readily accessible. For this purpose *N*-protected isotopically labeled glycine methyl ester is halogenated and the resulting **21** subjected to Arbuzov reaction with trimethylphosphite to give **22**. Depending on the reaction partners, the subsequent olefination reaction can be accomplished using either LDA in THF or *tert*-BuOK and $CH_2Cl_2$ at low temperature, or with DBU in refluxing $CH_2Cl_2$[30]. In the latter case a further improvement in yield and *Z*-selectivity can be achieved by replacing DBU with tetramethylguanidine in THF[29a]. This is especially advantageous in cases where the catalyst hydrogenates (*E*)-isomers more slowly and with lower enantiomeric efficiency, so that *E*/*Z*-mixtures would need to be separated chromatographically.

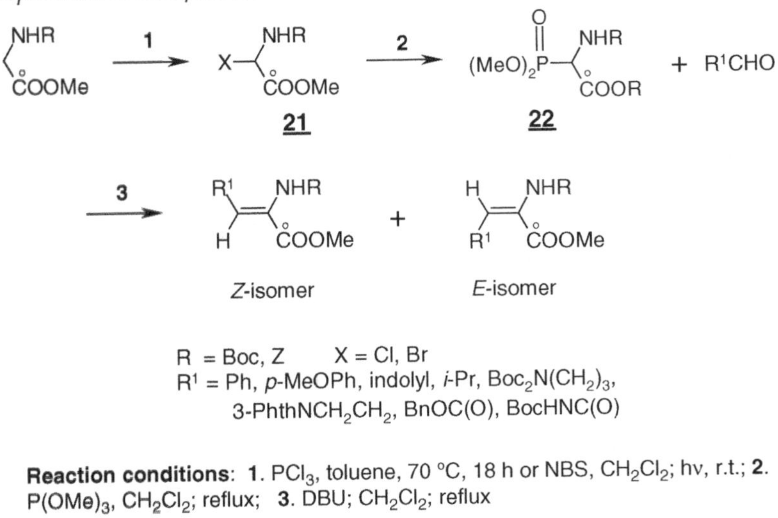


***Figure 11.11*** *Preparation and application of N-Boc α-phosphono[1-$^{13}$C]glycine trimethyl ester to the synthesis of $\Delta^{2,3}$-[1-$^{13}$C]amino acid methyl ester (Method 2)*

Several other methods for the synthesis of *N*-acyl $\Delta^{2,3}$-amino acid derivatives have been described. These include the condensation of aldehydes or ketones with isocyanoacetates followed by transformation of the resultant *N*-formyl derivatives[31], *N*-chlorination of *N*-acyl amino acid esters followed by elimination of HCl using DBU[32], reaction of α-oxo carboxylic acids with amides or methyl carbamates[33] and enzymatic dehydrogenation of *N*-acyl amino acid amides[34]. With the exception of the last, these methods have not been reported so far in isotope chemistry.

Enantioselective hydrogenation of *Z*-*N*-acyl $\Delta^{2,3}$-amino acids, esters or amides mediated by cationic rhodium(I) catalysts complexed to chiral bisphosphine ligands (Figure 11.12) is a long-established method[24]. Prior to 1990, (*R*,*R*)- and (*S*,*S*)-DIPAMP (**23**) were the most effective and versatile ligands, catalyzing the hydrogenation at ambient temperatures and around 1 atm of hydrogen and giving enantiomeric excesses generally exceeding 90%[24d,35]. Accordingly, in isotope chemistry [Rh-(*R*,*R*)-DIPAMP]$^+$ has found application in the synthesis of (*S*)-[2,3-$^3$H$_2$]tryptophan through tritiation of *N*-acetyl $\Delta^{2,3}$-tryptophanamide (methanol, 18 °C, 1.2 atm $^3$H$_2$, 3 h; 94% ee, 40% yield after enzymatic de-amidation)[34a]. Under similar conditions (*S*)-[$^2$H$_7$]tyrosine was obtained in 98% yield and 91% e.e.[27b]. Other ligands besides **23**, for example (*S*,*S*)-Chiraphos (**26**), have been used to prepare (*S*)-[2,3-$^2$H$_2$]Phe and (*S*)-[2,3-$^2$H$_2$]Leu[27f]. Reduction of *N*-acetyl $\Delta^{2,3}$-Ala, -Phe, -Tyr, -Trp and -Dopa with tritium gas in the presence of [Rh-(*R*)-PROPHOS]$^+$**24** has even been used commercially to produce the respective tritiated (2*S*)-α-amino acids with high specific activities and e.e. values of 87–95%[37]. The same catalyst system mediated the reduction of *N*,*N*′-diprotected $\Delta^{2,3}$-ornithine with deuterium gas to provide the corresponding (2*S*,3*S*)-[2,3-$^2$H$_2$]ornithine derivative in 60% chemical yield and 88% e.e.[30b]. However, when **24** was employed in the hydrogenation of *N*-benzoyl- or *N*-*p*-chlorobenzoyl-$\Delta^{2,3}$-[3-$^{11}$C]Phe, the enantiomeric excesses dropped down to 80%, probably because of the high reaction temperature (100 °C) necessary to drive the reaction to completion in 10 min[28d]. With [Rh-(*S*,*S*)-BPPM]$^+$ (25) as a catalyst, 2-*N*-acetylamino-3-(2-naphthyl)[3-$^{14}$C]acrylic acid could be reduced to give the respective (*R*)-[3-$^{14}$C]amino acid in 80% yield and >98% e.e.[28a,b].

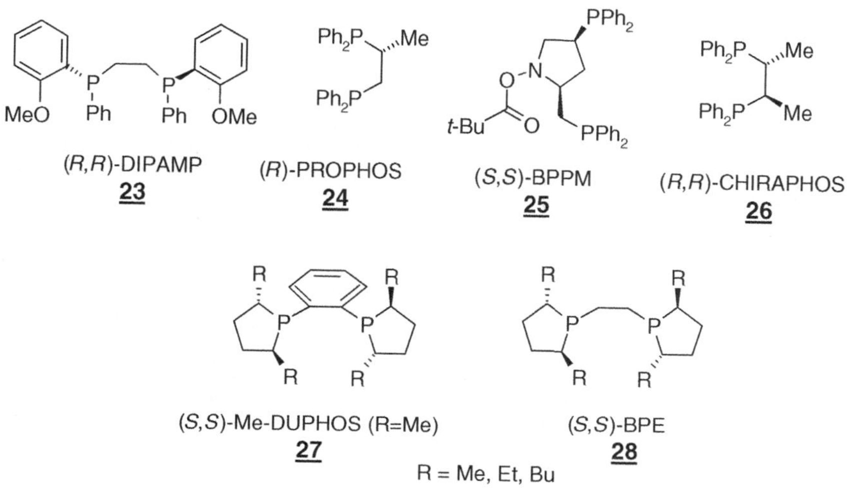


***Figure 11.12*** *Chiral diphospine ligands used in isotope chemistry in the rhodium(I)-catalyzed enantioselective (isotopic) hydrogenation of N-acyl* $\Delta^{2,3}$*-amino acid derivatives*

Systematic investigations into the [Rh-(*S*,*S*)-CHIRAPHOS]$^+$-catalyzed reduction of five *N*-acyl $\Delta^{2,3}$-amino acids ($\Delta^{2,3}$-Ala, -Phe, -Leu, -Tyr, -Dopa) with hydrogen gas have revealed that the enantiomeric excesses of products are dependent on the nature of the solvent used. Reduction of $\Delta^{2,3}$-Phe in ethanol, for example, gave Phe with 89% e.e., but the e.e. using THF was only 74%. Conversely, the same reaction applied to $\Delta^{2,3}$-Leu gave an e.e. of 72% in ethanol and of 87% in THF[27f]. Accordingly, in each special case careful pilot optimization studies will have to precede the hot run.

Unlike the other catalyst systems discussed so far, the electron-rich chiral bisphosphine ligands 27 (DuPHOS) and 28 (BPE) are able not only to reach over 99% e.e.s, but also to catalyze the hydrogenation of (*Z*)- and (*E*)-isomers in equally high enantiomeric excesses to give the same enantiomeric product[38]. They can, therefore, be applied not only in the more difficult hydrogenation of *β*,*β*-disubstituted substrates, but they can also avoid tedious chromatographic separations of *E*/*Z*-isomeric mixtures. Furthermore Z, nitro and bromo functions are well-tolerated, so that in the last case, for example, the resulting halogenated products can be employed in Suzuki cross-couplings to give e.p. amino acids of the bisaryl type. It has been stated that no other of the aforementioned ligand families consistently matches this level of efficiency[24a]. [Rh-Et-DuPHOS]$^+$ (27) was exploited for the synthesis of various (*S*)-*N*-Boc-[1-$^{13}$C]amino acid esters (Phe, *p*-MeOPh, Trp, Ala, Val, *N*-Boc-Asn, Ser, Lys)[30a] and (*S*)-naphth-2-yl[1-$^{14}$C]alanine and (*S*)-4-nitrophenyl[1-$^{14}$C]alanine[27a] through hydrogenation of the respective olefinic precursors in 92–98% e.e.

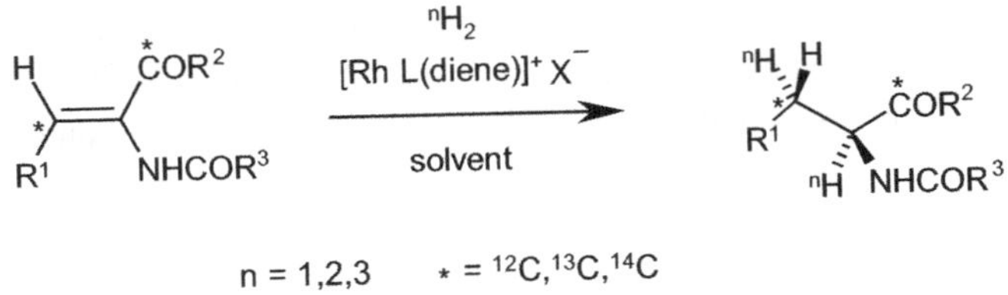


| $R_1$ | $R_2$ | $R_3$ | $^nH_2$ | label | chiral ligand | e.e. | |
|---|---|---|---|---|---|---|---|
| 3-indolyl | $NH_2$ | Me | $^3H_2$ | 2,3-$^3H_2$ | (*R,R*)-DIPAMP | 94 % | 34c |
| 4-AcOPh-$d_4$ | OH | Me | $^2H_2$ | 2,3-$^2H_2$ | (*R,R*)-DIPAMP | 98 % | 27b |
| Me; Ph; 4-HOPh; Ph, 2-naphthyl | OH | Me | $^3H_2$ | 2,3-$^3H_2$ | (*R*)-PROPHOS | 91–98 % | 36 |
| Me, Phe, indolyl, 4-HOPh, 3,4-dihydroxy-Ph | OH | Me | $^3H_2$ | 2,3-$^3H_2$ | (*R*)-PROPHOS | 87–95 % | 37 |
| Phth*N*$CH_2CH_2$ | OMe | Ph | $^2H_2$ | 2,3-$^2H_2$ | (*R*)-PROPHOS | 88 % | 30b |
| Ph | OH | Ph | $^2H_2$ | 2,3-$^2H_2$ | (*S,S*)-CHIRAPHOS | 72–99 % | 27f |
| 4-AcOPh; 3-MeOPh Ph | OH | Ph; 4-ClPh | $^1H_2$ | 3-$^{11}C$ | (*R*)-PROPHOS | 80 % | 28c,d |
| Ph; 2-naphthyl | OH | Me | $^1H_2$ | 3-$^{14}C$ | (*S,S*)-BPPM | 98 % | 28a,b |
| p-$NO_2$Ph; 2-naphthyl | OMe | Me | $^1H_2$ | 1-$^{14}C$ | (*S,S*)-EtDuPHOS | >99 % | 27a |
| Ph; 4-MeOPh; *N*-Boc-indolyl, *N*-Boc-imidazolyl, *i*-Pr; BnOC(O) BocH*N*C(O); $Boc_2N(CH_2)_3$ | OMe | $OCMe_3$ | $^1H_2$ | 1-$^{13}C$ | (*S,S*)-EtDuPHOS | 92–96 % | 30 |

***Figure 11.13*** *Enantioselective reduction of labeled/unlabeled N-acyl $\Delta^{2,3}$-amino acid derivatives with $^1H_2$, $^2H_2$, or $^3H_2$ at 1–5 atm mediated by chiral rhodium(I) catalysts*

The methodology outlined here has been shown to work also with $\Delta^{2,3}$-di- and tripeptides with satisfactory stereoselectivities. [Rh-DIPAMP]$^+$-catalyzed tritiation of *N*-acetyl-$\Delta^{2,3}$-Trp-PhOMe and *N*-acetyl-$\Delta^{2,3}$-Phe-PheOMe have been reported to provide the corresponding (*S,S*)-ditritiated dipeptides with theoretical specific activities and 85% and 90% d.e., respectively[34a,c,39]. In the presence of [Rh-(*R,R*)-BPPM]$^+$, *N*-acetyl-$\Delta^{2,3}$-Phe-PheOMe could be converted into the ditritated (*R,S*)-dipeptide, again with 90% d.e. The resulting tritiated di- and tripeptides may serve as building blocks for larger labeled peptides for *in vivo* investigations into their metabolic and pharmacokinetic profiles[39]. However, catalytic enantioselective hydrogenations of unsaturated larger peptides have given varying results. [Rh-(*R,R*)-DIPAMP]$^+$-catalyzed deuteration of Glu-His-$\Delta^{2,3}$-Trp-Ser-Tyr afforded the expected labeled *N*-terminus of gonadotropin releasing hormone with 99% d.e.. Under the same reaction conditions Boc-*β*-Ala-$\Delta^{2,3}$-Trp-Met-Asp-PheNH$_2$, a pentagastrin precursor, gave the deuterated pentapeptide with only 26% d.e., favoring the (*S*)-[2,3-$^2H_2$]Trp subunit. Under achiral conditions, i.e. with PdO as catalyst, the corresponding diastereomer is formed with 52% d.e. favoring the (*R*)-[2,3-$^2H_2$]Trp-containing diastereomer. This indicates, that for larger peptides, not only the identity and configuration of the catalyst but also the steric and electronic environment of the peptide backbone greatly influence the stereochemical outcome of chiral enantioselective hydrogenations catalyzed by chiral rhodium(I) complexes[34b].

### 11.2.2 Reduction of Labeled Prochiral Carbonyl Compounds and Oximes

The stereoselective reduction of prochiral ketones to enantiomerically highly enriched secondary alcohols[40] has become one of the most important synthetic tools for the preparation of e.p. compounds labeled with isotopic hydrogen and carbon[25]. The convenient availability of appropriate starting materials through routine conversions of labeled carboxylic esters or nitriles with organometallic reagents, and the feasibility of replacing secondary hydroxy groups with azide functionality without racemization ($HN_3$, $Ph_3P$, ROOC–N=N–COOR, Mitsunobu reaction), permits extension of this methodology to the production of e.p. $\alpha$-branched labeled amines. The procedures developed in recent decades involve the use of both stoichiometric reagents and catalytic processes. As the costs for radiolabeled materials significantly exceed those for individual reducing agents, there is no preference of catalytic over stoichiometric methods, as there is in chemical production. Hence, both procedures have been employed in isotopic syntheses, the most important determinant of appropriateness in a particular case being the structure of the substrate. $BH_3{\cdot}THF$ (or $BH_3{\cdot}Me_2S$, catecholborane) in the presence of catalytic amounts of e.p. 1,3,2-oxazaborolidines (**29**, Itsuno reagents), bicyclic oxazaborolidines (**30a–d**, Corey–Bakshi–Shibata or CBS-reagents) or (+)- or (−)-*B*-chlorodiisopinocampheylborane (**31**, DIP-chloride, $Ipc_2BCl$) all efficiently reduce aryl and heteroaryl alkyl ketones, halogen-substituted aryl alkyl ketones, acyclic and cyclic $\alpha,\beta$-enones and $\alpha,\beta$-ynones to give the corresponding alcohols in high yields (>70%) and e.e.s (>85%)[40]. Heteroatoms have been found to have little steric effect on the outcome. In contrast, linear dialkyl ketones show only low stereoselectivities. In these cases much higher e.e.s (65–82%) are achieved when chiral **32** (*B*-iso-2-(2-diethylaminoethyl)apopinocampheyl-9-borabicycl[3.3.1]nonyl hydride) is used as a stoichiometric reducing agent[41].

Itsuno catalyst — (*S*)-**29**

CBS catalysts — (*S*)-**30a**: R = H, Ar = Ph; **b**: R = Me, Ar = Ph; **c**: R = Bu, Ar = Ph; **d**: R = Bu, Ar = Nap

(+)-$Ipc_2BCl$ or (+)-DIP chloride — (+)-**31**

(+)-**32**

***Figure 11.14*** *Chiral catalysts and stoichiometric reagents used in the enantioselective reduction of prochiral ketones*

The use of these reagents in the synthesis of enantiomerically enriched labeled derivatives is illustrated by the following examples. CBS reduction of 3′-bromo- and 3′-methoxy[carbonyl-$^{14}$C]acetophenones (**33**) with $BH_3{\cdot}THF$ in the presence of 0.3 equivalents of (*S*)-**30b** furnished (*R*)-carbinols **34** in chemical yields of 75% and e.e.s of 97%. Subsequent Mitsunobu reaction with hydrazoic acid and reduction of the resulting azides (35) with $Ph_3P$–2N HCl converted 34 into the corresponding e.p. (1*S*)-1-aryl[1-$^{14}$C]-ethylamine derivatives 36 with inversion of the configuration at the chiral center. Both compounds were intermediates in the preparation of such acetylcholinesterase inhibitors such as [$^{14}$C]rivastigmine (37)[42].

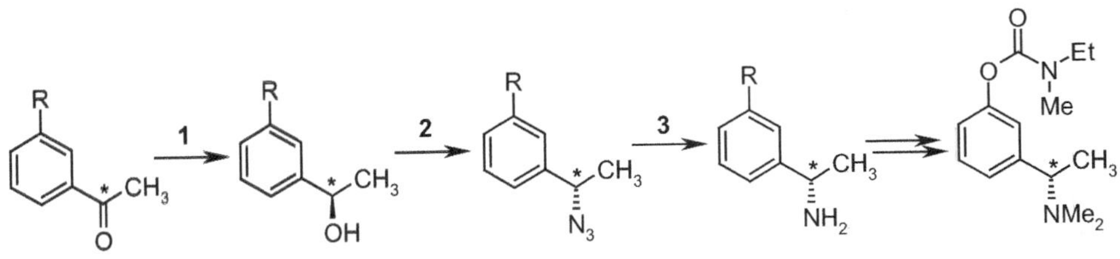

33 34 35 36 37

R = OMe, Br

**Reaction conditions: 1a.** (*S*)-30b, $BH_3 \cdot THF$, THF; 0 °C, 15 min, r.t., 20 min, **b.** 0.2 N aq. HCl; r.t.; **2.** $HN_3$, $Ph_3P$, EtOOC–N=N–COOEt, toluene; 0 °C, 5 min; 20 °C, 10 min; **3.** $Ph_3P$, 2 N HCl, THF; 0 °C, 5 d

***Figure 11.15*** *CBS reduction of prochiral [carbonyl-$^{14}$C]acetophenones and subsequent conversion of carbinols into e.p. inverted chiral 1-aryl[1-$^{14}$C]ethylamines*

The preparation of 39, starting material for a diastereoselective Evans auxiliary-promoted aldol reaction (see Section 11.3.2) and key intermediate in the synthesis of [$^{14}$C]Sch 58235 (40), is an additional example of the compatibility of CBS reductions with a broad variety of substituents. Alcohol 39 was isolated after CBS reduction of ketone 38 with $BH_3 \cdot Me_2S$ in the presence of 0.3 equivalents of (−)-30a in nearly quantitative yield and >99% e.e.[43].

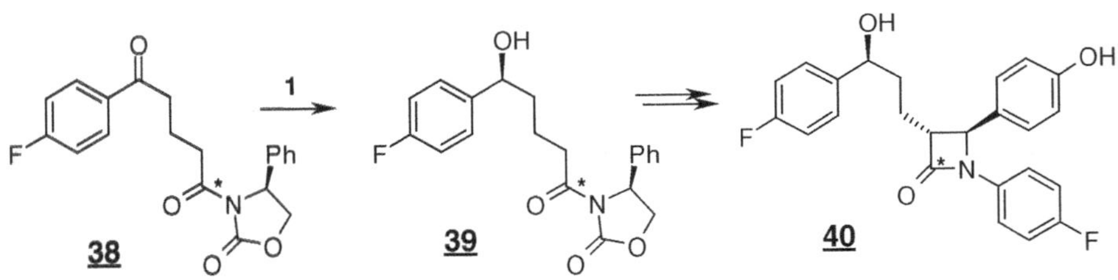

38 39 40

**Reaction conditions: 1.** $BH_3 \cdot Me_2S$, (*R*)-30a, $CH_2Cl_2$; 0°C, 2.5 h

***Figure 11.16*** *CBS reduction in the presence of ancillary function groups*

The utility and equivalence of the alternative stoichiometric approach have been demonstrated with the enantioselective reduction of 41, a 4-aryl-4-oxo[1,4-$^{14}C_2$]butanoic acid derivative, with (+)-$Ipc_2BCl$ to give (4*R*)-hydroxyamide 42 in 80% radiochemical yield and >96% e.e. Further reduction with $LiAlH_4$ converted 42 into (4*R*)-[1,4-$^{14}C_2$]ibutilide (43), an antiarrhythmic agent[44].

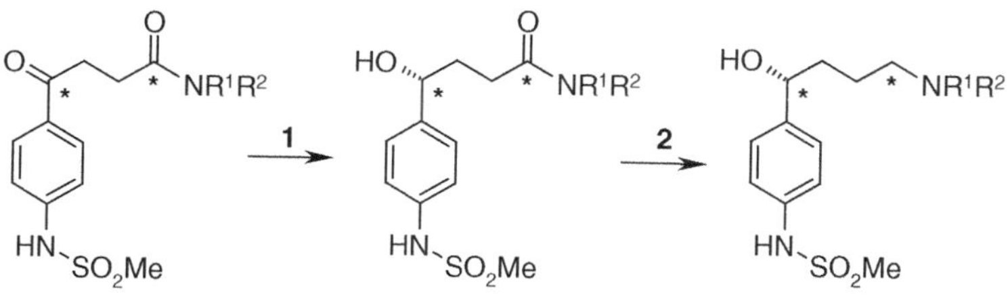

**41** **42** **43**

$R^1 = C_2H_5$; $R^2 = C_7H_{15}$

**Reaction conditions: 1a**. (+)-$Ipc_2BCl$, THF; 40 °C, 1 h, **b.** $(HOCH_2CH_2)_2NH$, r.t., 1.5 h; **2.** $LiAlH_4$, THF; –20 °C, 45 min

***Figure 11.17*** *Enantioselective reduction of an aryl [$^{14}$C]alkyl ketone using stoichiometric amounts of (+)-$Ipc_2BCl$; synthesis of an e.p. substituted γ-hydroxycarboxylic acid derivative*

Both methods are also powerful tools for the enantioselective reduction of prochiral aryl alkyl ketones containing one or more halogen substituents in their alkyl chain. The resulting haloalcohols generally show e.e.s of >90%. They can be cyclized to chiral aryl-substituted [$^{14}$C]epoxides or higher cyclic ethers with retention of configuration at the carbinol stereogenic center, or the halogen can be directly replaced by an appropriate nucleophile.

Reduction of 2-chloro[1-$^{14}$C]acetophenone (**44**) with (–)-$Ipc_2BCl$ or with $BH_3$·THF in the presence of catalytic amounts of (*R*)-oxazaborolidine **30b** furnished (1*R*)-2-chloro-1-phenyl[1-$^{14}$C]ethanol (**45**) in 87–89% chemical yield and with e.e.s of 94–95%. Treatment with 2N NaOH/ether or TBDMS imidazole and subsequently with $Bu_4NF$ converted **45** into (*R*)-(+)-phenyl[1-$^{14}$C]oxirane (**46**) in 40–50% yield. Ring opening with aqueous $NH_3$ provided (*R*)-2-amino-1-phenyl[1-$^{14}$C]ethanol (**47**)[45] and ring opening with TMSCN/CaO gave (*S*)-3-hydroxy-3-phenyl[3-$^{14}$C]propionitrile (**48**)[46]. Derivatives **47** and **48** were converted to e.p. labeled pharmaceutical candidates as shown in Figure 11.18. Since substituted 2-chloro[1-$^{14}$C]acetophenones are readily accessible through reaction of [carbonyl-$^{14}$C]aroyl chlorides with diazomethane in the presence of dry HCl[45] or halogenation of [carbonyl-$^{14}$C]acetophenones with sulfuryl chloride in dry $CCl_4$[46], preparation of aryl[1-$^{14}$C]oxiranes of high enantiomeric purity following the route outlined should be preferable to Sharpless' bishydroxylation of [$^{14}$C]styrenes (Section 11.2.3), since the latter are prone to radiation-induced polymerization.

Analogously, aryl 2-chloro[1-$^{14}$C]ethyl carbinols **49** (Figure 11.19) are accessible by CBS- or $Ipc_2BCl$-mediated reduction of aryl (2-chloro)ethyl[carbonyl-$^{14}$C]ketones; e.e.s are usually >90%. Replacement of chlorine with iodine followed by treatment with a primary amine converted haloalcohol **49** (Ar = thien-2-yl) into the 3-aryl-3-hydroxy[3-$^{14}$C]propylamine derivative **50**, which served as the penultimate intermediate in the preparation of the carbon-14-labeled CNS-active drug duloxetine[47]. Furthermore, Mitsunobu reaction (DEAD or ADDP, $Ph_3P$ or $Bu_3P$) on **49** (Ar = phenyl) in the presence of various phenols provided the respective (*R*)- *β*-haloethers **51** with inversion of the configuration at the stereogenic center. This route was followed for the synthesis of [$^{14}$C]atomoxetine and one of its metabolites[48].

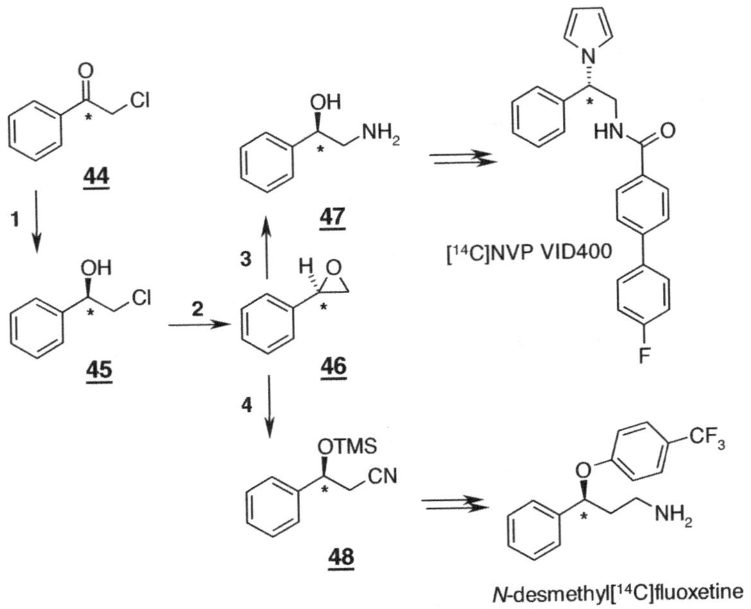


**Reaction conditions:** **1.** (–)-Ipc$_2$BCl, THF; –25 °C, 68 h then $(HOCH_2CH_2)_2NH$, ether; r.t., 2 h or $BH_3 \cdot THF$, (*R*)-**31a,** THF; r.t., 1 h then $H_2O$; **2.** 2 N NaOH, ether; r.t., 1 h or TBDMSCl, imidazole, DMF; r.t., 16 h then $Bu_4NF$, THF; r.t., 3 h; **3.** aq. $NH_3$, MeOH; 40 °C, 3.5 h; **4.** $BH_3THF$, THF; r.t., 16 h then 1 N HCl

***Figure 11.18*** *Synthesis of an aryl aminomethyl[$^{14}$C]carbinol and an aryl 2-aminoethyl [$^{14}$C] carbinol via an aryl[1-$^{14}$C]oxirane of high enantiomeric purity*

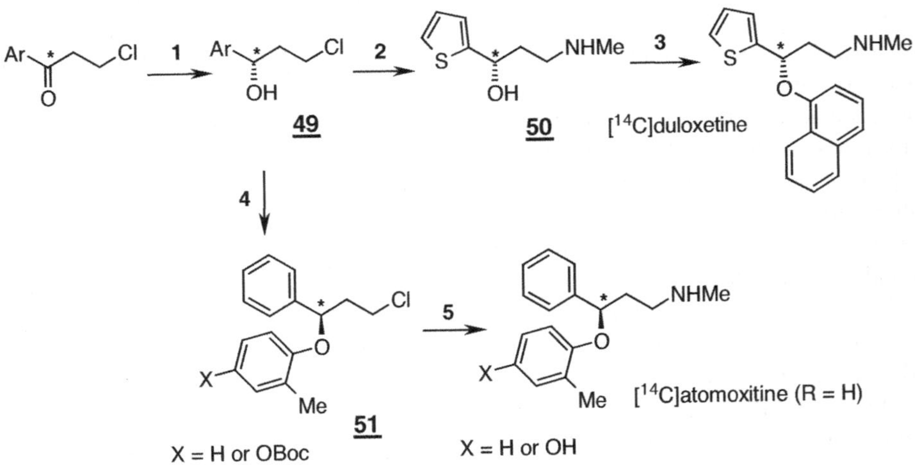


**Reaction conditions:** 1. $BH_3 \cdot THF$, (*R*)-**30b**, THF; 0 °C, 1,5 h; **2a**. (Ar = 2-thienyl) NaI, acetone; reflux, 16 h, **b**. 40 % aq. $MeNH_2$, THF; r.t., 6 h; **3**. 1-fluoronaphthalene, NaH, DMAC; 90 °C, 3 h; **4**. (Ar = phenyl) *o*-cresol, $Ph_3P$, DIAD, THF; 0 °C, 2 h; r.t. 16 h; **5a**. NaI, acetone; reflux, 16 h, **b**. $MeNH_2$, THF; r.t., 6 h

***Figure 11.19*** *Enantioselective synthesis of 3-aryl-3-hydroxy[$^{14}$C]propylamines of high enantiomeric purity*

The need for carbon-14-labeled 1-arylethylamines of high enantiomeric purity for the synthesis of labeled pharmacologically active compounds prompted investigations into alternatives to the reduction of aryl alkyl ketones, in order to circumvent the need for Mitsunobu conversion of the resulting alcohols. Stoichiometric CBS reduction of oxime methyl ethers in the presence of (1*S*,2*R*)-1-aminoindan-2-ol (52) has been found very effective, yielding e.e.s of >90% (variant A, Figure 11.21)[49]. Alternatively, $NaBH_4/ZrCl_4$ in the presence of one equivalent of (*R*)- or (*S*)-1,1-diphenylvalinol (53) as a chiral modifier gives e.e.s of 55–80% (variant B, Figure 11.21)[50].

Application of variant A to the reduction of *anti*-[ring-U-$^{14}$C]acetophenone oxime methyl ether (55) afforded (*S*)-1-[ring-U-$^{14}$C]phenylethylamine (56) in 70% radiochemical yield and 96% e.e.[51]. In contrast to values published in the literature, enantiomeric excesses were only 80% when variant B was employed in the reduction of 3-methoxy[carbonyl-$^{14}$C]-acetophenone oxime methyl ether (57), so that the material had to be further purified through a diastereomeric salt[52].

The chiral complex 54, formed from $LiAlH_4$, (1*S*,2*R*)-(+)- or (1*R*,2*S*)-(−)-*N*-methylephedrine (1 equiv.) and 3,5-dimethylphenol (2 equiv.) and originally introduced for the reduction of ketones with 15–80% e.e.s[53], has also been shown to reduce oximes of $\alpha$-keto esters with good stereoselectivity. Applied to the reduction of diethyl 2-hydroxyimino[4-$^{14}$C]glutarate, it provided [4-$^{14}$C]glutamic acid diethyl ester in >60% yield and 80% e.e.[54].

Chiral oxazaborolidines are also efficient catalysts for the enantioselective reduction of aldehydes labeled with isotopic hydrogen at the carbonyl function. Treatment with catecholborane in the presence of (*R*)- or (*S*)-*B*-*n*-butyloxazaborolidine (30d) affords C1-deuteriated or tritiated primary alcohols with enantiomeric excesses generally exceeding 90%[55]. Use of catecholborane is essential, since for $BH_3{\cdot}THF$ the achiral uncatalyzed process is competitive with the chiral catalyst-mediated pathway, thereby reducing the enantiomeric purity of products. Enantiomerically enriched [1-$^2$H]- and [1-$^3$H]alcohols have been extensively used in the study of enzymatic mechanisms[56] and biosynthetic pathways[57], and as precursors for chiral [$^1$H,$^2$H,$^3$H]acetic acid[58] and [$^1$H,$^2$H]fluoroacetic acid[59]. Equivalent results are obtained when $Bu_3SnH$ is employed in the presence of BITIP (58, prepared *in situ* from (*R*)- or (*S*)-1,1′-bi-2-naphthol and $Ti(O\text{-}i\text{-}Pr)_4$ 2 : 1)[60] (Figure 11.23).

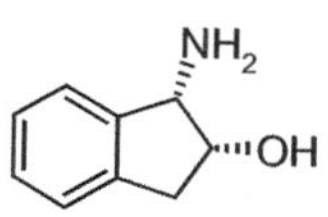


52

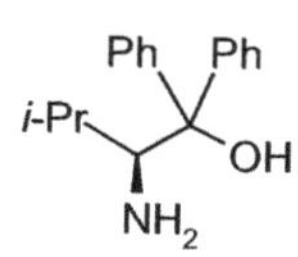


53

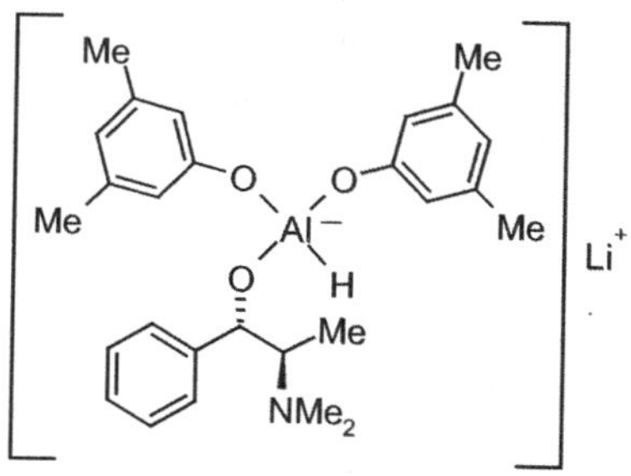


54

***Figure 11.20*** *Chiral ligands/reagents used for the enantioselective reduction of oximes and oxime ethers*

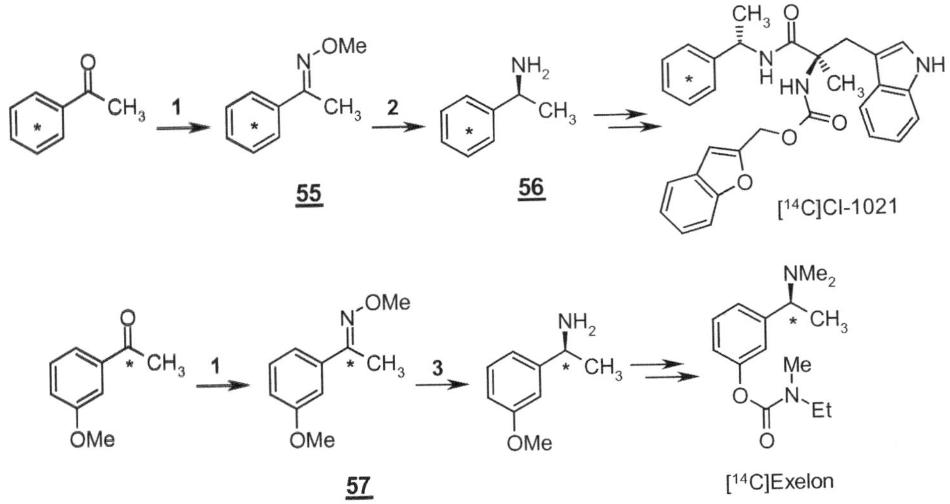


**Reaction conditions:** 1. $MeONH_2\cdot HCl$, pyridine, EtOH; r.t., 20 h; **2.** <u>**52**</u>, $BH_3\cdot THF$, THF; –20 °C, 15 h; **3.** $NaBH_4/ZrCl_4/$<u>**53**</u> 4:1:1, THF; –20 °C,15 h; r.t., 16 h; 70 °C, **b.** 2 N HCl, –10 °C

***Figure 11.21*** *1-[$^{14}C$]arylethylamines of high enantiomeric purity through enantioselective reduction of oxime ethers*

In both latter cases, the protocols outlined start from C1-deuterated/tritiated aldehydes, conveniently accessible either through exchange labeling with $^2H_2$ or $^3H_2$ in the presence of $[Ir(cod)pyr(PCy_3)]PF_6$ in high isotopic abundance[61] (Chapter 3, Section 3.5); through reduction of esters (or aldehydes) with complex deuterides or tritides followed by reoxidation with DMSO/oxalyl chloride (Chapter 4, Section 4.3), periodinane (Dess–Martin oxidation) or $CrO_3$–pyridine[62]; or through quenching of deprotonated dithianes with deuterium or tritium oxide and subsequent cleavage with $HgO–HgCl_2$[58]. Since <u>**30d**</u> and <u>**58**</u> are available in both enantiomeric forms, there is no need to use labeled catecholborane or tributyltin hydride as the reducing agent, as either one would first have to be prepared from lithium deuteride or lithium tritide (Chapter 4, Section 4.3.4).

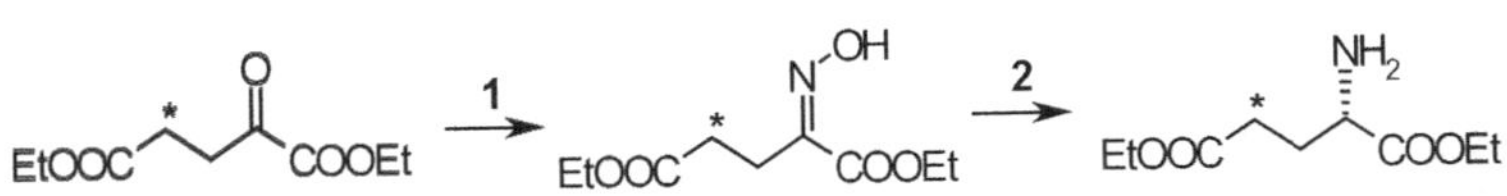


**Reaction conditions:** 1. $H_2NOH\cdot HCl$; **2.** $LiAlH_4$, (1*R*,2*S*)-(–)-*N*-methylephedrine, 3,5-dimethylphenol 1:1:2

***Figure 11.22*** *Complex <u>54</u>-mediated enantioselective reduction of an α-hydroxyimino-alkanoic acid ester to an α-[$^{14}C$]amino acid ester*

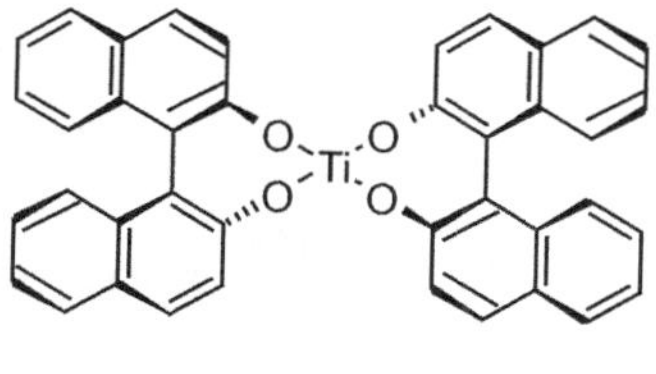

58 (*R*)-BITIP

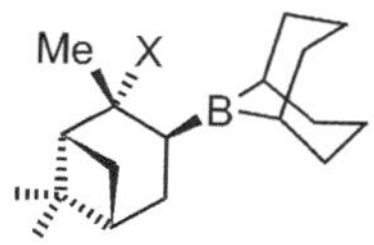

X = $^1$H, $^2$H, $^3$H

(*S*)-[$^n$H]Alpine-Borane

59

***Figure 11.23*** *Chiral ligand and reducing agent used in the enantioselective reduction of aldehyde functions*

Both catalytic versions compare favorably with the stoichiometric methodology using (+)- or (−)-*B*-3α-pinanyl-9-BBN ((+)- or (−)-Alpine Borane, 59) in giving an equivalent stereochemical outcome of 80–100% e.e.[62]. Employment of 1-deutero or 1-tritio aldehydes as substrates is also highly recommended, in order to avoid the inconvenient preparation of labeled reducing agent. The synthesis of (*S*)-[2-$^3$H,$^2$H]acetic acid (60)[58] and (*R*)-[1-$^3$H]-hexadecyloxypropane-1,2-diol (61)[56] may serve as illustrative examples.

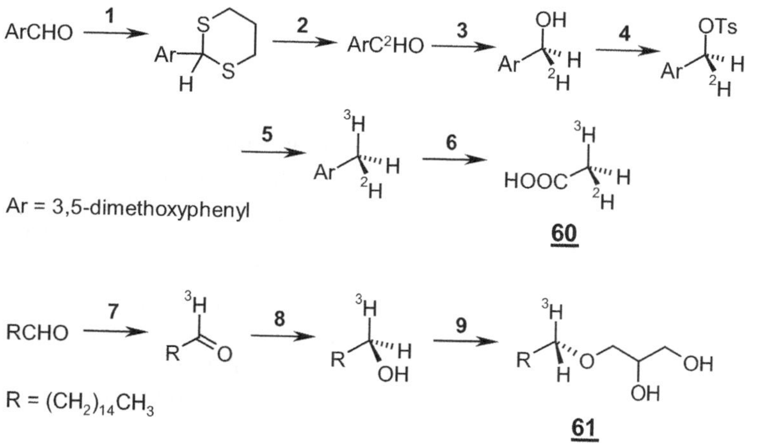


**Reaction conditions**: **1**. HS(CH$_2$)$_3$SH, HCl (g), CHCl$_3$; r.t., 18 h; **2a**. *n*-BuLi, THF; −70 °C, 1 h, **b**. $^2$H$_2$O, THF; r.t., 2 h, **c**. HgO/HgCl$_2$, MeOH, H$_2$O; 80 °C, 2 h; **3**. (+)-58, THF; r.t., 15 h; reflux, 1 h; **4**. NaH, TsCl, ether; −30 °C to −10 °C, 1 h; 4 °C, 2.5 h; **5**. LiEt$_3$B$^3$H, THF; r.t., 1.5 h; **6**. O$_3$, SiO$_2$, pentane; −78 °C, 8 h; 4 °C, 16 h; **7a**. NaB$^3$H$_4$, MeOH; 25 °C, 20 h, **b**. CrO$_3$–pyridine, CH$_2$Cl$_2$; r.t., 18 h; **8**. (+)-59, THF; 25 °C; 46 h; **9a**. TsCl, pyridine; reflux, 30 min, **b**. NaH, *O*-isopropylidene-*syn*-glycerol; 80 °C, 4 h

***Figure 11.24*** *Synthesis of chiral [$^2$H,$^3$H]methyl and [$^2$H,$^3$H]methylene functions*

### 11.2.3 Enantioselective Oxidation of Olefins and Allylic Alcohols

Enantioselective oxidations of olefinic compounds to give e.p. vicinal diols and epoxides have gained enormous importance in synthetic organic chemistry over the last twenty

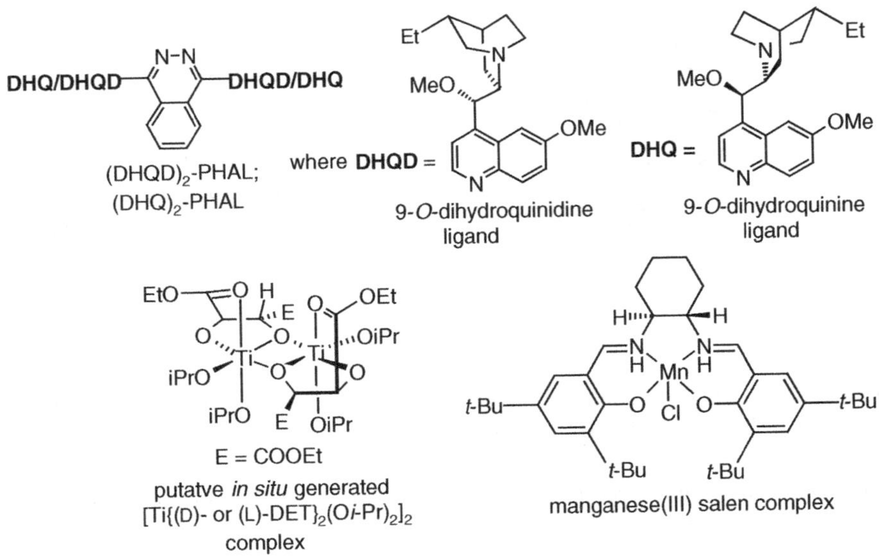


***Figure 11.25*** *Catalysts used in the enantioselective oxidation of isotopically labeled olefinic substrates*

years (Figure 11.25). Nevertheless, only a few applications have been reported in isotope chemistry, most of them dealing with the synthesis of compounds labeled with deuterium or carbon-13. Since they may also find future applications in tritium and carbon-14 syntheses, available isotopic examples are discussed under the following categories: *Sharpless asymmetric bishydroxylation* of olefins oxidized with $OsO_4$ in the presence of a chiral dihydroquinidine or dihydroquinine catalyst[63]; *tert*-butylperoxide-mediated *Sharpless epoxidation* of primary and secondary allylic alcohols in the presence of a chiral titanium catalyst generated *in situ* from titanium tetraisopropylate and diisopropyl D- or L-tartrate[64]; and finally, *Jacobsen epoxidation* of olefinic substrates with chlorine bleach or 4-phenylpyridine-*N*-oxide catalyzed by a chiral manganese(III) salen complex[65].

Application of Sharpless' bishydroxylation method to 6-chloro-3-[1-$^{14}$C]vinylpyridine (**62**), accessible from pyridine-3-[$^{14}$C]aldehyde and methylenetriphenylphosphorane, was reported[66] to provide (*R*)-1-(6-chloropyrid-3-yl)[1-$^{14}$C]ethane-1,2-diol (**63**) in 85% radiochemical yield and 94% e.e.. Such a result may be exceptional, owing to the expected radiolytic instability of the radiolabeled vinylpyridine intermediate. The reaction was executed with commercially available AD-mix-$\beta$, consisting of $(DHQD)_2$-PHAL, $K_2OsO_4$, $K_3Fe(CN)_6$ and $K_2CO_3$, which generates the highly toxic and expensive $OsO_4$ *in situ* (Figure 11.26). Selective monotosylation of the primary hydroxyl function and subsequent treatment with base gave epoxide **64**. Ring opening of the epoxide with the appropriate amine converted **63** into optically active $\beta$-hydroxyamine **65**, an intermediate in the preparation of [$^{14}$C]CP-331,684 (**66**), a $\beta_3$-agonist.

The same procedure, when applied to the $\alpha,\beta$-unsaturated ketones (*E*)-5-benzyloxy-3-[1-$^{13}$C]penten-2-one (**67**) and its [1-$^{13}$C], [2,3-$^{13}C_2$] and the [4-$^{13}$C] isotopomers, provided

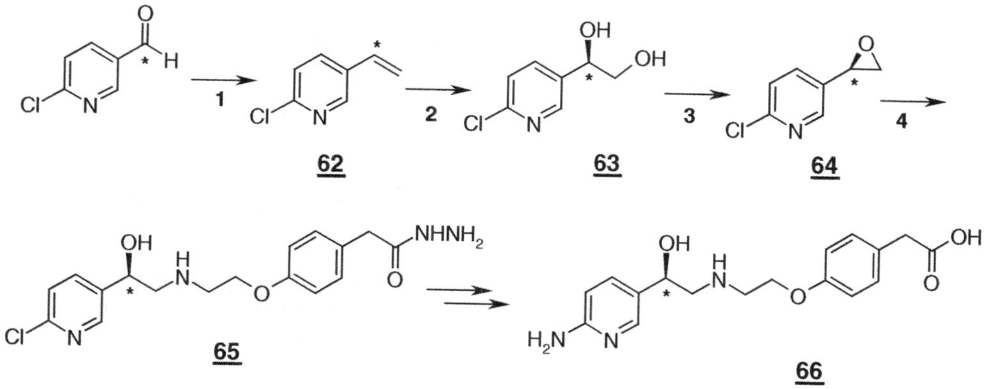


**Reaction conditions**: **1a**. $Ph_3P^{14}CH_3Br$, KHMDS, THF; −10 °C to r.t., 1 h, **b**. aldehyde, −78 °C, 16 h; **2**. AD-mix-β, *t*-BuOH/$H_2O$, 4 °C, 16 h; **3**. TsCl, pyridine; **4a** . *t*-BuOK, DMSO, **b**. $ArOCH_2CH_2NH_2$, toluene/DMSO

***Figure 11.26*** *Application of Sharpless' asymmetric dihydroxylation to the synthesis of a carbon-14-labeled epoxide of high enantiomeric purity*

the (3*S*,4*R*)-dihydroxy derivatives **68**. Concluding hydrogenolytic debenzylation afforded the respectively labeled 1-deoxy-D-xyluloses (**69**), which were used as substrates in investigations in the biosynthesis of moenomycins[67].

The excellent enantioselectivities generally achieved with titanium tartrate-catalyzed epoxidation of allylic alcohols with *tert*-butylhydroperoxide was exploited, for example, for the preparation of (2*S*,3*S*)- and (2*R*,3*R*)-[2,3-$^2H_2$]oxiranes (**73**), two valuable chiral building blocks (Figure 11.28). By using a slightly modified Sharpless procedure, (*E*)-3-triphenylsilyl-2-[2,3-$^2H_2$]propenol (**70**) was converted into the (*S*,*S*)- and (*R*,*R*)-forms, respectively, of 3-triphenylsilyl[2,3-$^2H_2$]glycidol (**71**). The hydroxylmethyl substituent

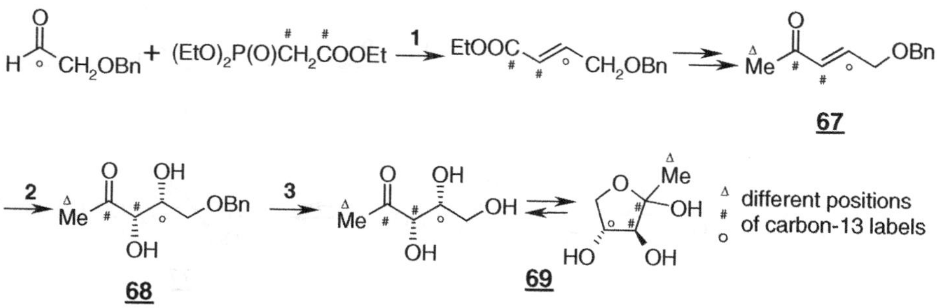


**Reaction conditions: 1**. NaH, benzene; r.t., 3 h; **2**. AD-mix-β, $MeSO_2NH_2$, $NaHCO_3$, aq. *tert*-BuOH; 0 °C, 24 h; **3**. $H_2$, 5 % Pd/C, MeOH; r.t., 16 h

***Figure 11.27*** *Application of Sharpless' asymmetric dihydroxylation to the synthesis of e.p. 1-deoxy-D-xylulose labeled with carbon-13 in different positions*

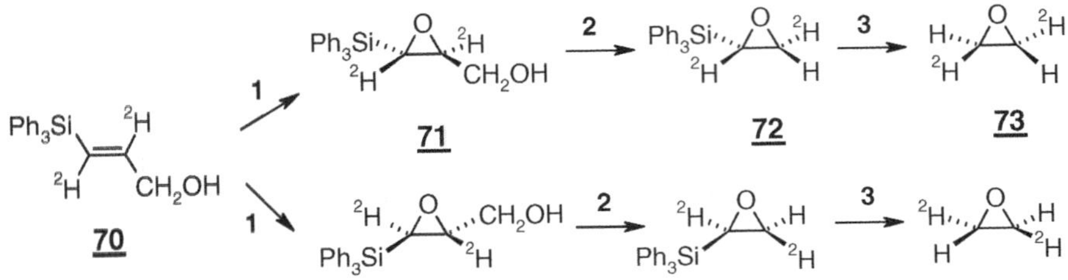

**Reaction conditions**: **1**. *t*-BuOOH, Ti(O*i*-Pr)$_4$, (−)-or (+)-diisopropyltartrate, $CaH_2$, silica gel, $CH_2Cl_2$; −20 °C, 16 h; **2a**. $MnO_2$, benzene; reflux, 16h, **b**. $(Ph_3P)_3RhCl$, benzene; reflux, 4 h; **3**. $Et_4NF$, DMSO; −20 °C, 2 h

***Figure 11.28*** *Synthesis of e.p. (2S,3S)- and (2R,3R)-[2,3-$^2H_2$]oxiranes through Sharpless epoxidation*

was removed through oxidation of the primary hydroxy group to the aldehyde, followed by decarbonylation. Desilylation of the resulting 72 generated the dideutero ethylene oxides in satisfactory yields and enantiomeric excesses of >92%[68].

A successful strategy for extending the Sharpless procedure to other $\alpha,\beta$-unsaturated species is the reduction of $\alpha,\beta$-unsaturated esters to allylic alcohols, followed by enantioselective epoxidation and final regeneration of the carboxyl function through oxidation of the primary hydroxyl group. This sequence was applied to the preparation of [4-$^{13}$C]-EP-475 (77), a potent cysteine protease inhibitor[69]. The requisite allylic alcohol 75 was accessed through reduction of ethyl (*E*)-4-benzyloxy[2-$^{13}$C]crotonate (74). Routine Sharpless epoxidation followed by reoxidation of the hydroxyl group converted 75 into the corresponding (2*S*,3*R*)-2,3-epoxy[2-$^{13}$C]carboxylic acid 76. Coupling of 76 with (*S*)-leucyl-isoamylamide (Leu-Iaa), cleavage of the benzyl group and oxidation of the liberated alcoholic function completed the sequence.

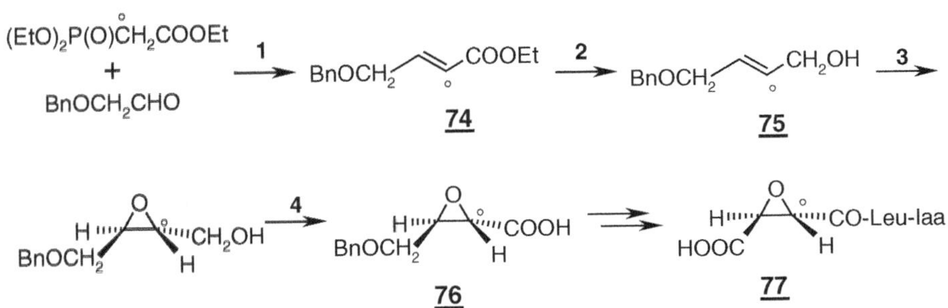

**Reaction conditions**: **1**. NaH, DME; r.t., 3 h; **2**. DIBAL-H, benzene; **3**. *tert*-BuOOH, Ti(O*i*-Pr)$_4$,D-(−)-diethyltartrate, $CH_2Cl_2$; **4**. $O_2$, Pt black, EtOAc/aq. $NaHCO_3$; r.t., 5 d

***Figure 11.29*** *Stereoselective synthesis of (2S,3S)-2,3-epoxy[2-$^{13}$C]succinyl-(S)-leucyl-isoamylamide*

Enantiomerically pure 2,3-epoxy alcohols are valuable synthetic intermediates because the consecutive oxygen functions can be readily exploited for regio- and stereoselective nucleophilic substitutions. Blocking the free hydroxyl group, for example, with a bulky protecting group suppresses the attack of nucleophiles at the adjacent position of the epoxide ring and directs it to the opposite site to give rise to a chiral 3-substituted 1,2-diol[70]. On the other hand, connecting to the hydroxyl group a substituent bearing a (latent) nucleophilic function may facilitate an attack at the adjacent epoxide position to afford a chiral 2-substituted 1,3-diol[71]. Both strategies have found application in the synthesis of e.p. multifunctionalized carbon-13-labeled substrates.

Sharpless epoxidation of (*E*)-2-[3-$^2$H]buten-1-ol (**78**) followed by *in situ* blocking of the free hydroxyl group with the large trityl group afforded the 2,3-epoxy ether **79** in 52% chemical yield and >95% e.e. Lithium di[$^{13}$C]methylcuprate attacked the epoxide **79** at C3 to give the corresponding monoprotected 1,2-diol **80**. Further transformations as depicted converted **80** into highly enriched *N*-Boc-(2*S*,3*S*)-[3-$^2$H;4-$^{13}$C]valine (**81**)[72].

The stereoselective synthesis of D-*ribo*-[1,1-$^2H_2$;1,2-$^{13}C_2$]phytosphingosine (**87**)[73] (Figure 11.31) illustrates the utility of both Sharpless' methods for the preparation of singly or multiply labeled complex molecules with multiple stereogenic centers. Treatment of terminal olefin **82** with AD-mix-$\beta$ using standard conditions afforded the 1,2-diol **83** in 80% e.e., thereby generating a first stereogenic center. Transacetalization followed by reductive ring opening of the resulting 1,3-dioxolane provided the selectively protected derivative **84**. Dess–Martin oxidation, Horner–Wadsworth–Emmons coupling of triethyl phosphono[1,2-$^{13}C_2$]acetate to the resulting aldehyde and reduction of the ester group converted **84** into the unsaturated intermediate **85**. This introduced the desired labeling pattern as well as the requisite allylic alcoholic functionality. Subsequent Ti(O*i*-Pr)$_4$/(*S*,*S*)-(−)-DIPT-catalyzed epoxidation followed by stereospecific transformation of epoxyalcohol **86** through benzoylcarbamate cyclization

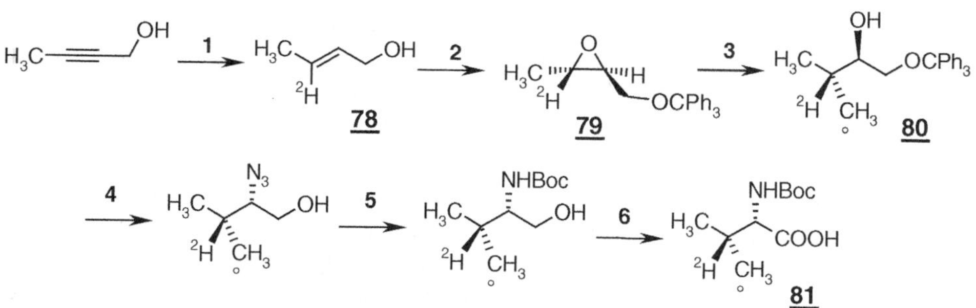


**Reaction conditions**: **1a**. LiAlH$_4$, THF, **b**.$^2H_2O$; **2a**. *tert*-BuOOH, Ti(O*i*-Pr)$_4$,(*R*,*R*)-(+)-DIPT, molecular sieves 3 Å, $CH_2Cl_2$; 8 °C, 16 h, **b**. Ph$_3$CCl, Et$_3$N, DMAP; r.t., 16 h; **3**. ($^{13}CH_3$)$_2$CuLi, BF$_3$·Et$_2$O, THF; −78 °C; **4a**. MsCl, Et$_3$N, **b**.*p*-TosOH, MeOH, **c**. Me$_2$NC(=NH$_2^+$)NMe$_2$ N$_3^-$, $CH_2Cl_2$/cyclohexane; reflux, 16 h; **5a**. H$_2$, 10 % Pd/C, TFA, MeOH; r.t., **b**. Boc-ON, Et$_3$N, aq. dioxane; **6**. RuCl$_3$, NaIO$_4$, CCl$_4$/MeCN/H$_2$O

***Figure 11.30*** *Application of Sharpless' asymmetric epoxidation in the stereoselective synthesis of N-Boc-(2S,3S)-[3-$^2$H;4-$^{13}$C]valine*

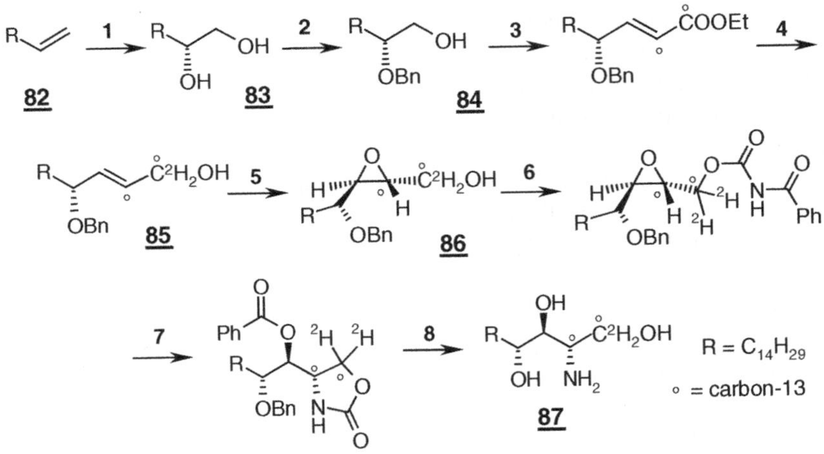


**Reaction conditions:** **1**. AD-mix-β, *t*-BuOH/$H_2O$; 4 °C, 40 h; **2a**. $PhCH(OMe)_2$, *p*-TsOH∘$H_2O$, benzene; r.t., 18 h, **b**. *i*-$Bu_2AlH$, toluene; −78 °C, 30 min; **3a**. periodinane, pyridine, $CH_2Cl_2$; r.t., 3 h, **b**. $(EtO)_2P(O)^{13}CH_2{}^{13}COOEt$, NaH, THF; 0 °C to r.t.; **4**. $LiAl\,^2H_4$–$AlCl_3$, $Et_2O$; 0 °C, 40 min; **5**. *t*-BuOOH, Ti(O*i*-Pr)$_4$, (*S,S*)-(−)-DIPT, molecular sieves 3A °, $CH_2Cl_2$; −30 °C, 5 min; −20 °C, 3 d; **6**. PhCOCl, AgNCO, $CCl_4$; r.t., 4 h; **7**. NaH, THF; reflux, 6 h; **8a**. LiOH, THF/$H_2O$; r.t., 12 h, **b**. $H_2$, 10 % Pd/C, THF; r.t., 12 h, **c**. 2 N NaOH/EtOH; reflux, 12 h

***Figure 11.31*** *Use of the Sharpless' asymmetric dihydroxylation and asymmetric epoxidation in the synthesis of D-ribo-[1,1-$^2H_2$;1,2-$^{13}C_2$]phytosphingosine containing three chiral centers*

and sequential removal of the *N*- and *O*-protecting groups afforded **87** with the desired (2*S*,3*S*,4*R*)-stereochemistry in overall 9% chemical yield.

So far, there is only one report on the application in carbon-14 chemistry of Jacobsen's chiral (salen)Mn-mediated epoxidation of disubstituted *cis*- and trisubstituted olefins with bleach[74]. In this case the methodology was exploited for the preparation of a precursor for the labeling operation (Figure 11.32). $N_1$-protected 1,2-dihydro-8-nitro-quinoline was treated with bleach in the presence of Mn(III)-(*S,S*)-salen chloride and 4-phenylpyridine-*N*-oxide to give the required epoxide **88** in 67% chemical yield and 96% e. e. Replacement of the epoxide ring by the corresponding aziridine (**89**) with inversion at both stereogenic centers was followed by regioselective Lewis acid-catalyzed reductive ring opening to give (*R*)-amine **90** at 99% e.e. Selective blocking of the 3-methylamino group and catalytic reduction of the nitro group converted **90** into **91**, the key intermediate for introduction of label, which was accomplished using [$^{14}$C]phosgene. The target compound, **92**, a dopamine agonistic drug candidate, was intended for drug disposition studies.

## 11.3 Diastereoselective Synthetic Procedures

Chiral auxiliary-mediated diastereoselective synthesis is a widely applied alternative to enantioselective synthesis (Section 11.2), biosynthetic methods (Chapter 12) and resolution

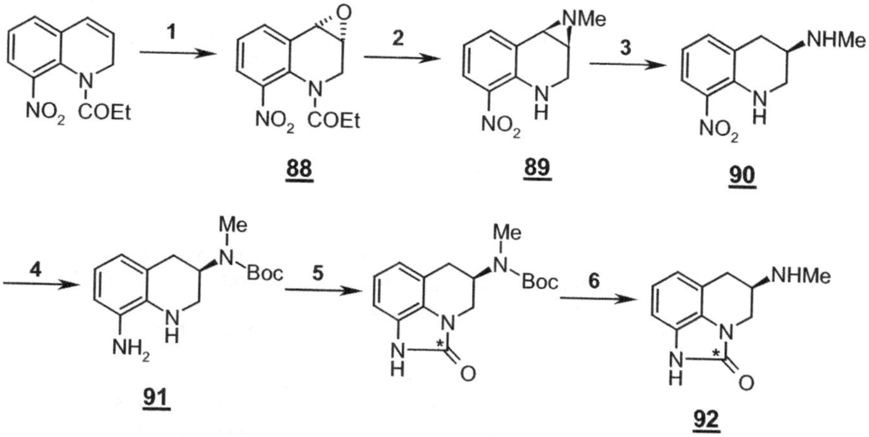

**Reaction conditions:** **1**. aq. NaOCl (Clorox) at pH 10.5, Mn(III)–(*S*,*S*)-salen chloride, 4-phenylpyridine-*N*-oxide, $CH_2Cl_2$; r.t., 2 h; **2a**. aq. $MeNH_2$, MeCN; 90 °C, **b**. NaOMe, MeOH, **c**. $PPh_3$, DEAD; **3**. $NaCNBH_3$, $BF_3{\cdot}Et_2O$, THF; r.t., 30 min; **4a**. $Boc_2O$, THF; r.t., 20 min, **b**. $H_2$, 10 % Pd/C, EtOH; r.t., 1 h; **5**. $^{14}COCl_2$, $Et_3N$, toluene; r.t., 3.5 h; **6**. HCl, MeOH; reflux, 3.5 h

***Figure 11.32*** *Use of the Jacobsen–Katsuki asymmetric epoxidation for the preparation of an e.p. intermediate for preparation of structure* ***92***

of racemates (Section 11.1), for the preparation of e.p. compounds labeled with isotopic hydrogen and carbon[25,75]. Chiral auxiliaries are homochiral molecules covalently attached to substrate molecules for the purpose of controlling the configuration of the newly generated stereogenic centers in a subsequent reaction on the target molecule, and which are removed after they have served their purpose[76]. The overall process is depicted in the Figure 11.33.

There are some practical aspects that favor diastereoselective procedures in comparison with enantioselective alternatives. First, by proper choice of the auxiliary, low molecular weight substrates can be rendered nonvolatile, permitting the subsequent chemical transformations to be performed without any special precautions. Second, since the primary reaction products **P*–Xc** and $\mathbf{P^*}_{anti}$**–Xc** are diastereomers and therefore separable by conventional chromatography, the level of stereoselectivity of Step 2 is less critical than it is for enantioselective processes. In some cases even diastereomeric ratios of 4:1 (60% d.e.) may be acceptable. Accordingly, many more diastereoselective applications in isotope chemistry have been published than enantioselective ones.

Since the vast majority of the substrates employed in diastereoselective syntheses are attached via a carboxyl function to the respective auxiliary, this chapter deals with the diastereoselective $\alpha$-functionalization of chiral carboximides and esters, as well as with the homologation and aldolization alternatives of chiral glycinates, glycolates and haloacetates. Also discussed is the use of chiral $\alpha$,$\beta$-unsaturated representatives in the synthesis of e.p. linear $\beta$- and $\alpha$,$\beta$-substituted carboxylic acid derivatives and cyclic analogs (Diels–Alder reactions). Other methods, in which substrates are coupled, for example, via a carbonyl function in order to facilitate the diastereoselective $\alpha$-alkylation of aldehydes or

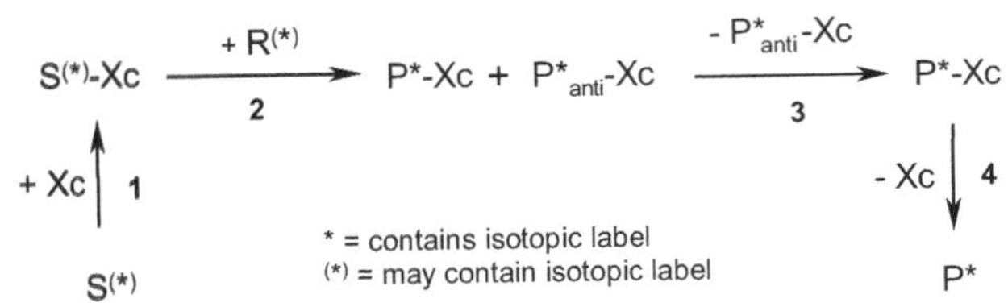


**Reaction steps:**

**1** Covalent coupling of the chiral auxiliary (**Xc**) to the labeled/unlabeled substrate (**S**$^{(*)}$) nearby the stereogenic center to be generated. The coupling should proceed by simple methods in high yields, the auxiliary should be stable under the conditions of the subsequent reaction and readily removed without endangering the newly created stereogenic units of the desired product **P***, and it should facilitate a high degree of stereoselectivity. The influence of the auxiliary should significantly override that of any ancillary stereogenic centers in the substrate and the reagent (auxiliary controlled stereoselective reactions)

**2** Reaction of **S**$^{(*)}$**-Xc** with the unlabeled/labeled reagent **R**$^{(*)}$ produces two diastereomers **P*–Xc** and **P***$_{anti}$**–Xc** in significantly different quantities. In the optimal case only traces of **P***$_{anti}$**–Xc**, the undesired diastereomer, should be formed.

**3** Removal of **P***$_{anti}$**–Xc** and impurities by classical achiral separation procedures.

**4** Cleavage of the auxiliary without racemization of the target molecule **P***. Since those reaction components labeled with isotopic carbon are much more expensive than the auxiliaries employed, there is no need to recycle the latter as is often the case in classical process chemistry. They may even be destroyed in the process (e.g. Dellaria's chiral glycine synthon, see Section 11.3.6.2)

***Figure 11.33*** *Outline of the chiral auxiliary-mediated diastereoselective process*

ketones (Enders' SAMP/RAMP-procedures[77]), are only rarely used. Those auxiliaries that have been extensively employed in the synthesis of isotopically labeled compounds are depicted in Figure 11.34. It also shows the commonly used auxiliary–substrate complexes or *chiral building blocks* derived from them.

The second kind of building blocks, those whose chiral information is not generated by one of the above-mentioned auxiliaries and which are exploited for special applications, are depicted in Figure 11.35.

The reaction (Figure 11.33) of the auxiliary–substrate complex S$^{(*)}$–Xc with the reagent R$^{(*)}$ to form the major diastereomer P*–Xc can be of two types. In the first, S$^{(*)}$–Xc lacks the substituent at the $\alpha$ position (or the $\alpha$- and $\beta$-positions, if they are part of an olefinic system) required to define the class of compound to be created (i.e., $\alpha$-amino acid, $\alpha$-hydroxyacid, $\alpha$-bromo-$\beta$-hydroxyacid, etc.). It is the reagent R$^{(*)}$ that brings this information. This concept is illustrated for an amino acid in Figure 11.36. In other words, the requisite substitution pattern at these positions is generated by an appropriate diastereoselective $\alpha$- or *$\alpha$,$\beta$-functionalization* reaction (e.g. an $\alpha$-alkylation, $\alpha$-aldolizatiom, $\alpha$-amination, $\alpha$-hydroxylation, conjugate addition, Diels–Alder reaction, bishydroxylation etc.).

In the second type, S$^{(*)}$–Xc is already *functionalized*, in that it contains the substituent (e.g. an $\alpha$-amino, $\alpha$-hydroxy substituent) that defines the compound class. The reaction

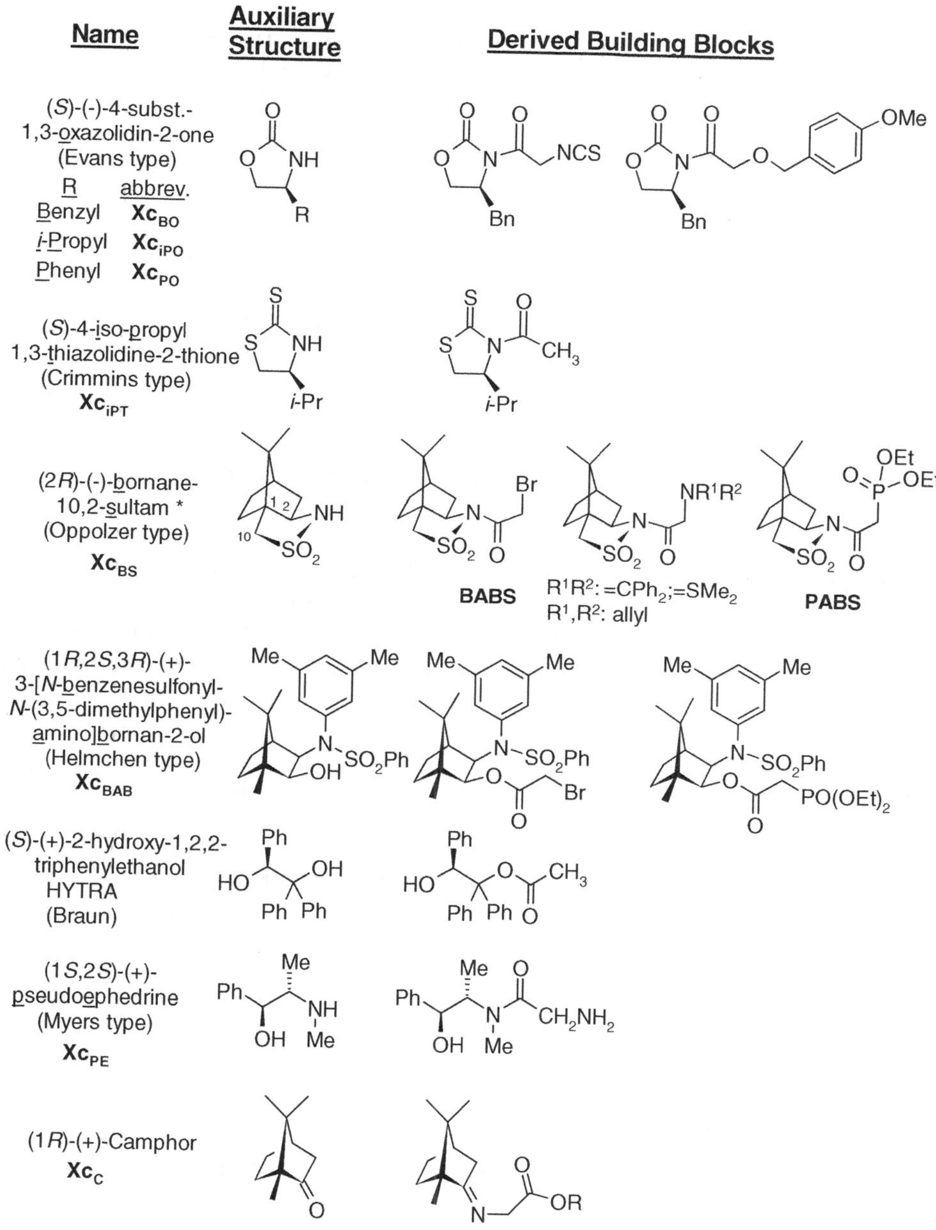


* (1S)-(-)-2,10-camphorsultam is used synonymously

***Figure 11.34*** *Chiral auxiliaries and their derived building blocks used in carbon-14 and tritium syntheses*

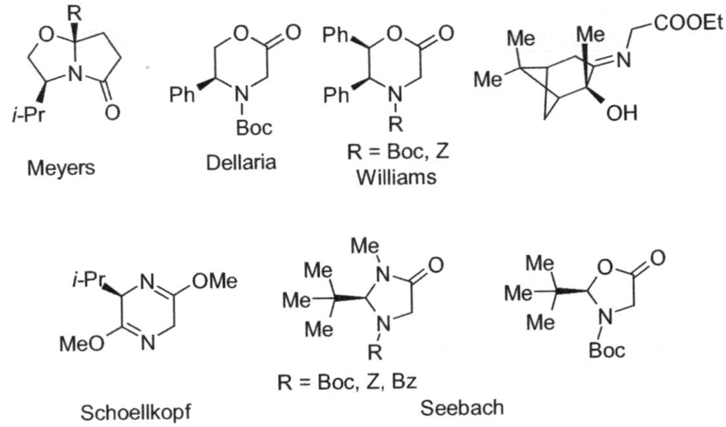


*Figure 11.35* Additional chiral building blocks used in tritium and carbon-14 synthesis

with $R^{(*)}$ (e.g. an alkylation or an aldolization) simply elaborates upon this core. This type of $S^{(*)}$–Xc may therefore be regarded as an elementary unit of the target compound.

These alternative pathways are reflected in the structure of this chapter. Sections 11.3.1–11.3.5 and Section 11.3.10 discuss Type I diastereoselective reactions, and

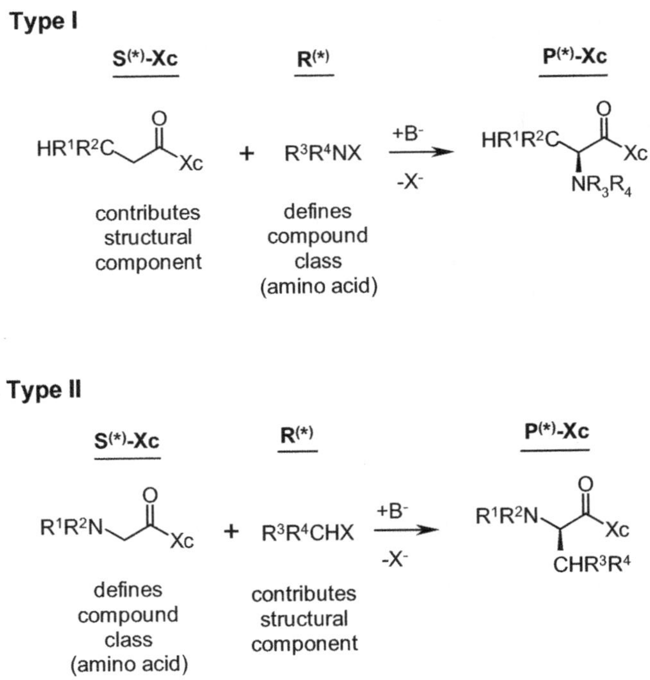


*Figure 11.36* The two types of reactions between $S^{(*)}$–Xc and $R^{(*)}$

Sections 11.3.6–11.3.9 cover Type II reactions. Section 11.3.10, 'Reactions on Chiral α,β-Unsaturated Imides and Esters', appears at the end because, although it is formally a Type I reaction, it utilizes as starting materials chiral haloacetates, which are for practical purposes Type II intermediates.

Type I reactions:

- α-Alkylations of chiral imide enolates;
- Aldol reactions of chiral imide and ester enolates;
- 1,4-Additions of chiral imide enolates to Michael acceptors;
- α-Aminations of chiral imide enolates;
- α-Oxygenation of chiral imide enolates.

Type II reactions:

- α-Alkylation of chiral glycinates;
- Aldol reaction of chiral glycinates;
- Aldol reactions of chiral glycolates;
- Aldol reactions of chiral haloacetates;
- Reactions on chiral α,β-unsaturated imides and esters.

### 11.3.1 α-Alkylation of Chiral Imide Enolates

Carboxylic acids can be alkylated to produce α-branched carboxylic acid derivatives in diastereoselectivities generally exceeding 85% via the process depicted in Figure 11.33: linkage to the auxiliaries (*R*)- or (*S*)-$\mathbf{Xc_{BO}}$, $\mathbf{Xc_{\mathit{i}PO}}$ (Evans) or $\mathbf{Xc_{BS}}$ (Oppolzer) to give the respective chiral imides, deprotonation of the imides to imide enolates with appropriate lithium or sodium bases (e.g. *n*-BuLi, LDA, LiHMDS, NaHMDS), and trapping the enolates with nonactivated primary alkyl iodides or allylic, propargylic and arylmethyl bromides or iodides[78,79]. Secondary alkyl iodides fail or give unsatisfactory results, but the corresponding triflates in the presence of HMPA appears to hold promise[80]. Electrophiles with a predisposition toward $S_N1$-reactivity (e.g. $BnOCH_2X$, acetals, orthoesters) have been shown to α-alkylate chiral *N*-acyl-1,3-oxazolidin-2-ones (Evans' auxiliaries) with >95% d.e. and chemical yields of 90–99%[81]. The observed stereoselectivities imply the formation of Li- or Na-chelated *Z*-enolates in which one of the faces is heavily encumbered, in the *N*-acyl-1,3-oxazolidin-2-ones ($\mathbf{Xc_{BO}}$ or $\mathbf{Xc_{\mathit{i}PO}}$) by the substituent at C4, and in the *N*-acylsultams ($\mathbf{Xc_{BS}}$) by the rigid bridgehead system. Electrophiles have to attack the Cα-position of the substrates predominantly from the opposite face; for the sultam system this is opposite to the lone electron pair on the nitrogen atom.

Evans' and Oppolzer's auxiliaries are of comparable efficacy with regard to chemical yields and the stereoselectivities they induce in the α-alkylation process. In some cases Oppolzer's imide enolates are superior because they give slightly higher stereoselectivities, the resultant α-branched *N*-acylsultams are often more easily purified by simple crystallization, and the auxiliary is easier and cleaner to remove. For both auxiliaries, the *N*-acylated derivatives required for alkylation are readily accessible by treatment of the deprotonated auxiliaries with the respective acyl halides[79a,82,83] or mixed anhydrides[84], generated *in situ* from the free acid (or the sodium salt) and pivaloyl chloride in the presence of LiCl and triethylamine. Alternative methods include the bis(2-oxo-3-oxazolidinyl)phosphine (BOPCl) promoted coupling of the free acid to the auxiliary in the presence of catalytic amounts of DMAP (in the Evans-Crimmins series)[85] and the $AlBr_3$-mediated *N*-acylation of Oppolzer's auxiliary with the respective methyl esters[79a,86]. Owing to their sensitivity to radiation-induced polymerization, radiolabelled acryloyl and crotonyl imides of the Evans-Oppolzer type are preferably prepared by Horner-Wadsworth-Emmons coupling of the corresponding chiral phoshono[$^{14}$C]acetyl precursors to formaldehyde or acetaldehyde as discussed in Section 11.3.10. Higher homologues are accessible either by use of the aforementioned general methods or through reaction of the auxiliary with the acyl chloride in the presence of triethylamine-DMAP, copper(I) chloride and copper powder[88].

Removal of the Evans and Oppolzer auxiliaries after the $\alpha$-alkylation step can be accomplished with little or no racemization in three ways: *hydrolytically* in aqueous THF using LiOH[89] or, more efficiently, with $H_2O_2$ and LiOH[79a,83,90] or *tert*-$Bu_4NOH$[91]; *by transesterification* with lithium alkoxides such as $Li(OBn)_4$[78a] or MeMg(OMe)[76f,92] or under neutral conditions with $Ti(OBn)_4$[89a] or Ti(O*i*-Pr)$_4$-allylic alcohol[90b]; or *reductively* with $LiAlH_4$ or $LiBH_4$ in ether or THF[78a,84,93,94]. These treatments release the substrate in the form of the carboxylic acid, ester, and primary alcohol, respectively. Aldehydes are usually prepared in two steps through reductive cleavage of the sultam or oxazolidinone auxiliary followed by Swern oxidation of the corresponding alcohols with DMSO-oxalyl chloride[94], though for *N*-acylsultams DIBAL-mediated reductive cleavage of the auxiliary can be accomplished[95]. Mercaptolysis with $BnSAlMe_3Li$[95] or transamidation with *N*,*O*-dimethylhydroxylamine[95] convert the Evans-Opplozer imides into thioester and *N*-methoxy-*N*-methylamides, respectively, which serve as intermediates for the preparation of ketones[96].

In isotope chemistry the label can be introduced with the acyl group or with the electrophile, or both. Applications of these options are exemplified in Figures 11.37 and 11.38. In most cases only singly or doubly labeled products are needed, but it is obvious that two or more labels can be incorporated via each reaction partner, leading to products containing larger numbers of labels. Examples of multiply labeled intermediates prepared via Horner–Wadsworth–Emmons chemistry[87] are discussed in Section 11.3.10.

In the first example, the sodium enolate of (*S*)-4-benzyl-3-[1-$^{14}$C]butyryl-1,3-oxazolidin-2-one (**93**) was treated with a tenfold excess of methyl iodide to afford a 95:5 mixture of the two $\alpha$-methylated diastereomers **94** and **95**. Preparative HPLC separation of **94** followed by cleavage of the auxiliary provided the free (*S*)-2-methyl[1-$^{14}$C]butyric acid (**96**) in overall 58% yield from sodium [1-$^{14}$C]butyrate. Acid **96** was a key intermediate in the carbon-14 labeling of pravastatin, an antihyperlipoproteinemic drug, as well as for investigation of the biosynthesis and the biological labeling of avermectin $B_1$a, currently marketed as a miticide/insecticide[97].

The example in Figure 11.38 demonstrates that diastereoselective $\alpha$-alkyations are not limited to simple unsubstituted chiral carboxylates, and that there is no need for the use of large excesses of the alkylating agents. Employment of equimolar amounts of the sodium enolate of (*R*)-4-benzyl-3-(4-nitrophenylacetyl)-1,3-oxazolidin-2-one (**97**) with [$^{14}$C]-methyl iodide gave diastereomerically pure (*R*)-2-(4-nitrophenyl)[3-$^{14}$C]propionyl derivative **98** in 34% radiochemical yield. In this case the auxiliary was removed by reduction with $NaBH_4$ to give (*R*)-2-(4-nitrophenyl)-1-[3-$^{14}$C]propanol (**99**). Standard procedures

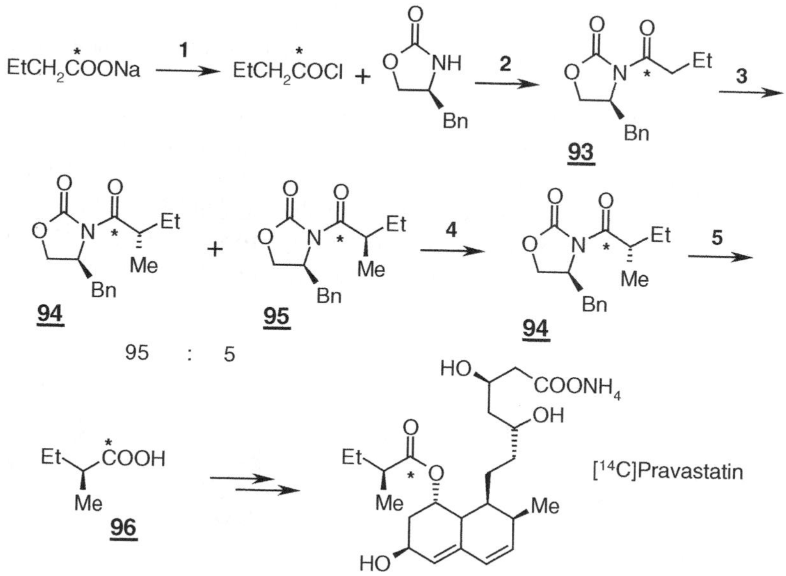


**Reaction conditions:** **1**. $(COCl)_2$, DMF (trace), $CH_2Cl_2$; r.t., 2 h; **2a**. (*S*)–**Xc**$_{BO}$, *n*-BuLi, THF; –78 °C, 1 h, **b**. [1-$^{14}$C]butyryl chloride, $CH_2Cl_2$/THF; –78 °C, 2.5 h; **3a**. NaHMDS, THF; –78 °C, 1 h, **b**. MeI, THF; –78 °C, 1 h; **4**. prep. HPLC; **5**. LiOH, $H_2O_2$, aq. THF; r.t., 1 h

***Figure 11.37*** *Diastereoselective α-alkylation of a (S)-4-benzyl-3-[1-$^{14}$C]acyl-1,3-oxazolidin-2-one using an unlabeled primary alkylating agent*

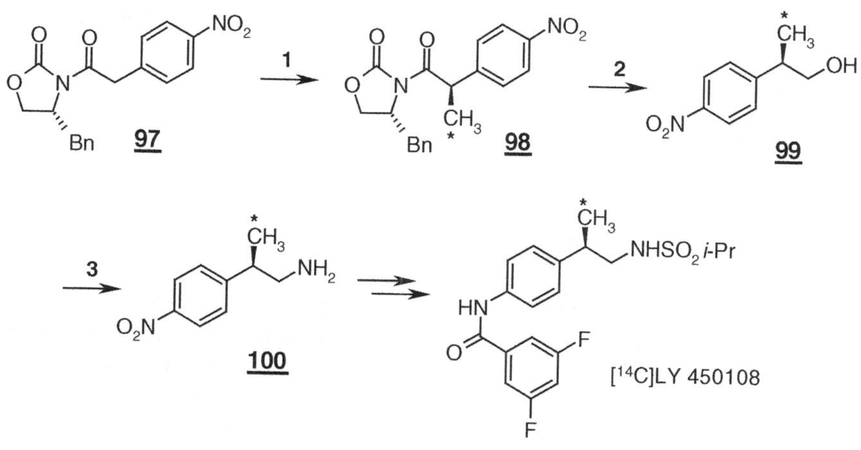


**Reaction conditions:** **1**. NaHMDS, [$^{14}$C]MeI, THF/toluene; –70 °C, 15 min to r.t.; **2**. $NaBH_4$, aq. THF, 3 h; **3a**. phthalimide, $PPh_3$, DEAD, THF; r.t., 18 h, **b**. $H_2NNH_2$, toluene; 80–90 °C, 1 h

***Figure 11.38*** *Diastereoselective α-alkylation of a chiral functionalized (R)-benzyl-3-acyl-1,3-oxazolidin-2-one with a carbon-14-labeled alkylating agent*

converted the alcohol into the corresponding amine **100**, which has been further transformed to [$^{14}$C]LY50108, a drug substance developed for the treatment of cognitive disorders[98].

More sterically demanding electrophiles can also be used effectively, as illustrated in Figure 11.39. Treatment of deprotonated (2*R*)-*N*-([1-$^{14}$C]propionyl)sultam **101** with an excess of (6-methoxypyridin-3-yl)methyl bromide in the presence of $Bu_4NI$ and HMPA provided the $\alpha$-heteroarylmethyl substituted intermediate **102** in 62% yield and >98% d.e. Reductive cleavage of the auxiliary gave (2*S*)-5-(3-hydroxy-2-methyl-[3-$^{14}$C]propyl)-2-methoxypyridine (**103**), a key intermediate in the preparation of [$^{14}$C]NVP-ACQ090, a potent and selective SRIF-antagonist[99].

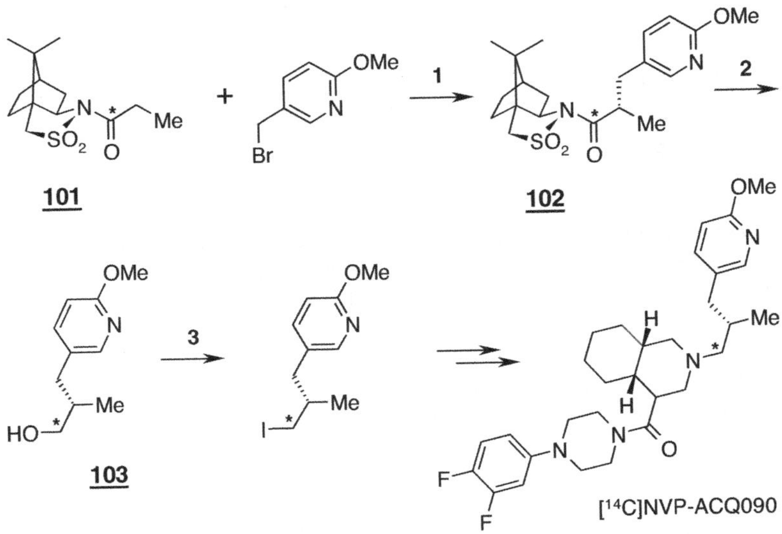


**Reaction conditions: 1a.** NaHMDS, THF; –78 °C, **b.** Bromide, *n*-$Bu_4NI$, HMPA, THF; –78 °C, 5 min; **2.** $LiAlH_4$, THF/$Et_2O$; 0 °C, 3 h; **3.** $MeP(OC_6H_5)_3I$, DMF; 0 °C, 3 h

***Figure 11.39*** *Diastereoselective $\alpha$-alkylation of a N-([1-$^{14}$C]acyl)bornane-10,2-sultam using an unlabeled functionalized alkylating agent*

The use of appropriately functionalized alkylating agents enables extension of the Evans–Oppolzer methodology to the preparation of compounds having more complex structures and labeling patterns. Trapping of the sodium enolates **104** (R = $-(CH_2)_3N_3$ or Ph) with equimolar amounts of labeled bromoacetate afforded the $\alpha$-branched succinates **105** in 62–68% yield and >98% d.e. In both primary intermediates the carboxyl groups were differentiated, one as an amide and one as a *tert*-butyl ester resistant to basic hydrolysis. Saponification with $LiOH/H_2O_2$ cleaved the imide functions to give e.p. $\alpha$-branched [3,4-$^{14}C_2$]succinic acid mono-*tert*-butyl esters **106**, opening the way for Hofmann or Curtius degradation of the free carboxylic acids to afford the corresponding

(*S*)-3-amino[1,2-$^{14}C_2$]carboxylic acid esters **107** in 85–95% yield. In the case of **107** (R = –$(CH_2)_3N_3$), *N*-acylation with isovaleryl chloride, catalytic reduction of the ω-azide function of the side chain and coupling of the newly generated amino group to 4-(*N*-Bocamino)benzoic acid followed by acid catalyzed deprotection led to [$^{14}$C]NVP GPI562 (**108**), a potent inhibitor of platelet aggregation[100a]. On the other hand, **107** (R = Ph) was subjected to transesterification to give **109**, a key intermediate to the doubly labeled protected tripeptide **110**, a building block for the solid phase synthesis of a β-nonapeptide for investigations into its pharmacokinetic and metabolic behavior[100b].

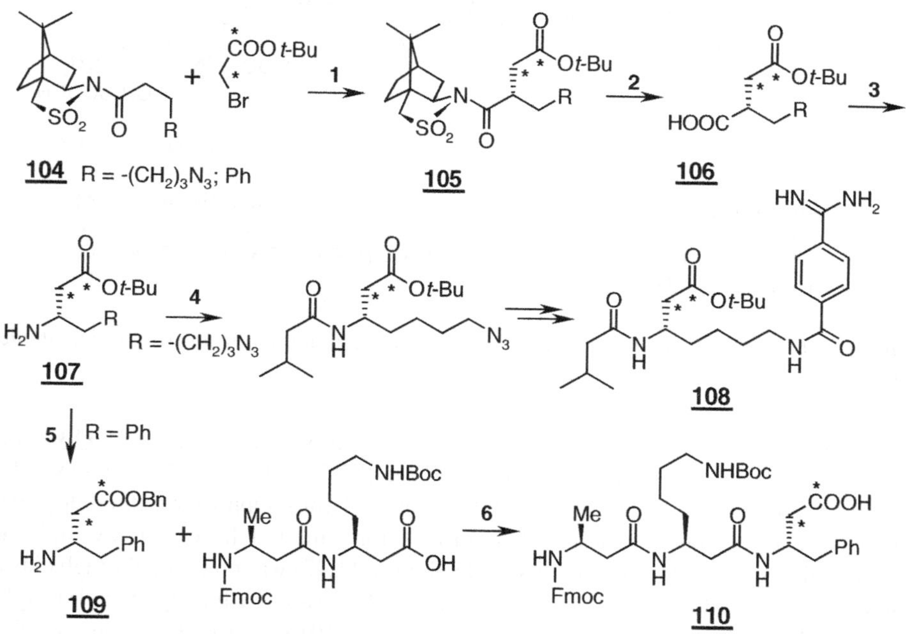

**Reaction conditions: 1a**. NaHMDS, THF; –78 °C, 1 h, **b**. *tert*-butyl bromo-[1,2-$^{14}C_2$]acetate, THF/toluene/DMAP; –78 °C, 2–3 h; **2**. LiOH, $H_2O_2$, aq. THF; 0 °C, 4 h; **3**. R = $CH_2CH_2N_3$: **a**. HOSu, DCC, THF; r.t.; 4 h, **b**. aq. $NH_3$, r.t.; 16 h, **c**. $PhI(OCOCF_3)_2$, $DMF/H_2O$; r.t., 4 h; R = Ph: $(C_6H_5O)_2P(O)N_3$, $Et_3N$, toluene; reflux, 2 h;. **4**. *i*-$PrCH_2COCl$, $K_2CO_3$, *i*-$PrOAc/H_2O$; r.t. 30 min; **5a.** 6 N HCl, dioxane; reflux, 1 h, **b**. $Me_3SiCl$, BnOH; r.t., 16 h; **6a**. BtOH, EDC, HOBt, $CH_2Cl_2$; r.t., 16 h, **b**. $H_2$, 10 % Pd/C, $CF_3CH_2OH$; r.t., 20 h

***Figure 11.40*** *Diastereoselective α-alkylation of chiral (1S)-N-acyl sultams using doubly carbon-14-labeled functionalized alkylating agents*

So far only one example for the successful use of acetals in diastereoselective α-functionalization of chiral isotopically labeled imides has been published. Though this reaction was performed with carbon-13 material, there is little doubt that it is also applicable in carbon-14 synthesis. The titanium enolate prepared from (4*S*)-4-benzyl-3-[(2*R*)-2-[$^{13}$C,$^{2}H_3$]methylpropionyl]-1,3-oxazolidin-2-one (**111**) reacted with trioxane to afford the corresponding α-hydroxymethyl derivative **112** in 83% yield and >96%

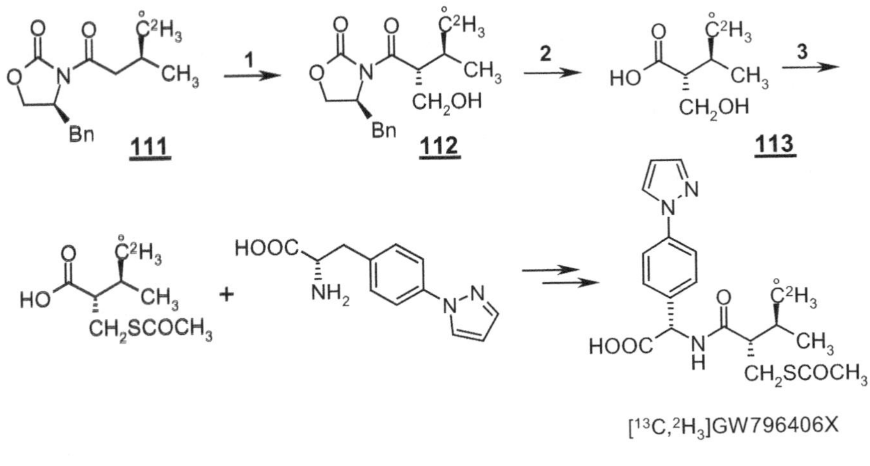


**Reaction conditions:** **1a**. $TiCl_4$, $CH_2Cl_2$; 0–5 °C, 5 min, **b**. *i*-$Pr_2EtN$; 0–5 °C, 1 h, **c**. trioxane, $CH_2Cl_2$; 0–5 °C; 15 min, **d**. $TiCl_4$, 0–5 °C, 1 h; **2a**. LiOH, $H_2O_2$, aq. THF; r.t.; 1 h, **b**. aq. $Na_2SO_3$; 0–5 °C; **3a**. DIAD, $Ph_3P$, $CH_3COSH$, THF; 20°C, 2 h; **b**. $SOCl_2$, $CH_2Cl_2$; 20°C, 1 h

***Figure 11.41*** *Use of an acetal in the diastereoselective α-functionalization of a chiral titanium imide enolate*

d.e. Cleavage of the auxiliary furnished labeled 2-hydroxymethylated acid **113**, which served as an intermediate in the synthesis of [$^{13}C$,$^{2}H_3$]GW796406X in 60% overall yield[101].

Preparations of e.p. tritiated compounds by means of Evans–Oppolzer methodology have not been reported, but syntheses of substrates for tritiation have. For example, the unsaturated (*R*)-[$^{3}H$]ibuprofen precursor **115** (Figure 11.42) was prepared through diastereoselective α-methylation of the sodium enolate of (4*R*,5*S*)-3-(4-*i*-butenylphenyl)acetyl-4-methyl-5-phenyl-1,3-oxazolidin-2-one (**114**) and LiOH/$H_2O_2$ mediated cleavage of the auxiliary. The label was introduced through homogeneous tritiation of **115** using Wilkinson's catalyst to give material of a specific activity of 54 Ci/mmol. This has been made available for fatty acid CoA lipases inhibition studies[102]. In this case, the modified Evans auxiliary was selected since it has been reported[103] to give higher diastereomeric excesses than the original 4-benzyl and 4-isopropyl substituted representatives. The same procedure was applied to the synthesis of [$^{14}C$]ibuprofen, starting, however, from unlabeled saturated **114** and [$^{14}C$]methyl iodide.

The Evans–Oppolzer methodology is not suitable for the creation of quaternary stereogenic centers. Meyer's bicyclic lactam approach[104] is one alternative, and the only one that has been used so far in isotope chemistry. In this case[105] (Figure 11.43), the Meyers lactam derived from (*S*)-valinol and 3-benzoylpropionic acid (**116**) was alkylated successively with [$^{13}C$]methyl iodide and [$^{2}H_3$]methyl iodide to give a mixture of diastereomers with **117** predominating. Hydrolysis and oxidative cleavage of the benzene ring of **117** provided the labeled dimethylsuccinic monoester **118**. Subsequent Barton decarboxylation through treatment of the *N*-hydroxypyridine-2-thione ester with $Bu_3SnH$ and ester saponification converted **118** to (*S*)-[$^{2}H_3$,$^{13}C$]pivalic acid (**119**). Compound **119** was utilized in the investigation of an unusual biomethylation reaction.

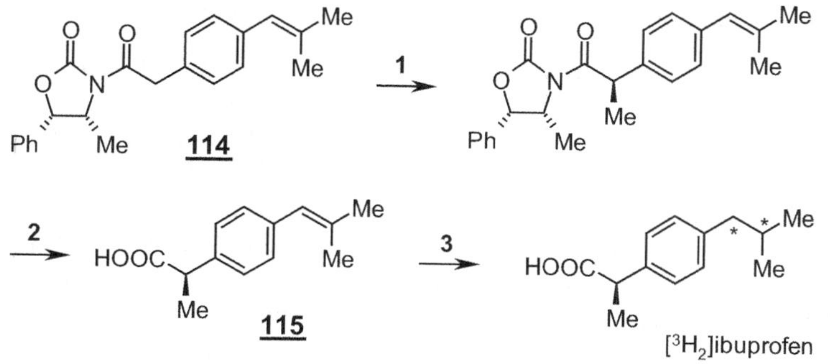


**Reaction conditions:** **1a**. NaHMDS, THF; −78 °C, 1 h, **b**. MeI; −78 °C, 1.5 h; −40 °C, 2 h; **2**. LiOH, $H_2O_2$, aq. THF; 0 °C, 3 h; **3**. $^3H_2$, $(PPh_3)_3RhCl$, EtOH; r.t., 4 h

***Figure 11.42*** *Diastereoselective α-methylation of an unlabeled (4R,5S)-3-acyl-4-methyl-5-phenyloxazolidin-2-one containing a double bond for subsequent introduction of label through catalytic tritiation*

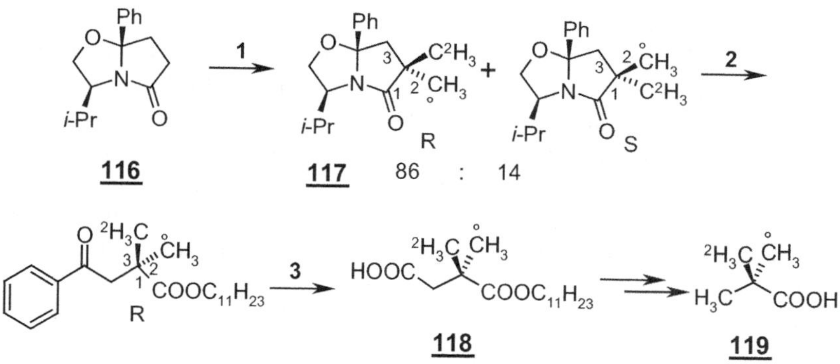


**Reaction conditions:** **1a**. LDA, THF; −78 °C, 2 h, **b**. $\mathbf{C^2H_3I}$, THF; −78 °C, 1–2h, **c**. LDA, THF; −78 °C, 2 h, **d**. $\mathbf{^{13}CH_3I}$, THF; −78 °C, 2 h; **2a**. HPLC separation, **b**. $H_2SO_4$, undecanol; reflux; **3**. $RuCl_3$, $KIO_4$

***Figure 11.43*** *Application of Meyers' bislactam methodology to the synthesis of labeled α,α-dialkylated carboxylic acid derivatives of high enantiomeric purity*

Naturally, carbon-14- and tritium-labeled electrophiles could be used to prepare analogous radiolabeled analogs of various types using this methodology. Furthermore, lactams analogous to **116** possessing these and other axial substituents can readily be prepared by condensation of an e.p. (*R*)- or (*S*)-amino alcohol (e.g. valinol, *tert*-leucinol, phenylalaninol) with succinic anhydride followed by treatment of the resulting imide with a Grignard reagent and acid-catalyzed ring closure. Since various isotopically labeled succinic anhydrides are commercially available, labeled lactams would also be accessible for a yet wider range of labeling options.

## 11.3.2 Aldol Reactions of Chiral Imides and Ester Enolates

Enolates of chiral carboximides linked to Evans' or Oppolzer's auxiliaries (+)- or (−)-**Xc$_{BO}$**, -**Xc$_{iPO}$** and -**Xc$_{BS}$** react with aldehydes, giving rise to *syn* $\alpha$-branched $\beta$-hydroxy carboxylic acid derivatives (*syn*-aldols) with excellent stereoselectivities, generally exceeding 96% and in chemical yields above 70%[76,106]. Both boron enolates and chlorotitanium enolates have been used in isotopic synthesis.

Boron imide enolates of the Evans 3-acyl-1,3-oxazolidin-2-one type (**Xc$_{BO}$**, **Xc$_{iPO}$**) are generated upon treatment with $Bu_2BOTf$ in the presence of *i*-$Pr_2NEt$ in dichloromethane[82a,93,107], whereas for the corresponding Oppolzer *N*-acylsultams (**Xc$_{BS}$**), the superior reagent is $Et_2BOTf$, prepared *in situ* from $Et_3B$ and $CF_3SO_3H$[83]. Chlorotitanium enolates are generated *in situ* with $TiCl_4$ in the presence of an appropriate base such as *i*-$Pr_2NEt$, TMEDA, or 1-ethylpiperidine. Chlorotitanium enolates show similar (sometimes slightly lower) stereoselectivities[108], but often higher yields, compared to boron enolates.

Both procedures normally give (*Z*)-enolates in greater than 99% selectivity. These react with aldehydes via a cyclic six-membered transition state of the Zimmerman–Traxler type, with one face of the imide enolates hindered by the auxiliary, forcing attack of the aldehyde at the opposite face. As with diastereoselective $\alpha$-alkylations (previous section), the stereodirecting influence of the auxiliaries generally overrides that imposed by ancillary stereogenic centers in the aldehyde substrates[76].

An example of the use of chlorotitanium enolates in isotope chemistry is the synthesis of [$^{14}$C]SCH 48461 (**122**), a drug candidate of the $\beta$-lactam type[109] (Figure 11.44). Sequential treatment of carbon-14-labeled imide **120** with $TiCl_4$ and TMEDA and reaction of the resultant chlorotitanium enolate with *p*-anisaldehyde provided the corresponding *syn*-[$^{14}$C] aldol **121** with the desired (*R*,*R*)-configuration at the newly formed stereogenic centers in 90% radiochemical yield and >99% d.e. Hydrolytic cleavage of the auxiliary, followed by coupling of the resultant acid to *p*-anisidine and $\beta$-lactam formation using Mitsunobu conditions afforded the target molecule **122** in 19% overall radiochemical yield and >99% e.e.

A boron enolate was used in the preparation of [$^2H_5$]DPH 140662 (**127**), an HIV-protease inhibitor (Figure 11.45). Accessed from (*S*)-4-benzyl-3-(3-[$^2H_5$]phenylpropionyl)-1,3-oxazolidin-2-one (**123**) through treatment with $Bu_2BOTf$ and *i*-$Pr_2NEt$, the enolate reacted with (benzyloxy)acetaldehyde to give the (*S*,*S*)-*syn*-aldol **124** in high diastereomeric excess. Hydrolytic removal of the auxiliary gave the carboxylic acid **125**, which was transformed through several steps into (*S*,*S*)-1-*N*-Bocamino-1-[$^2H_5$]benzylpropane-2,3-diol (**126**), an advanced intermediate to **127**. The latter was intended for use as an internal standard for LC-MS evaluations of pre-clinical studies[110].

The normal *syn*-selectivity of (*Z*)-boron and chlorotitanium enolates can be overridden by diverting the cyclic transition state of the aldolization processes to an open one[76a,106]. This can be accomplished either by using (*Z*)-*O*-silyl-*N*,*O*-ketene acetals as starting materials[89b,111] or by treating the boron enolate solution with an aldehyde pre-complexed with a second Lewis acid[76a,106,112,113]. In the latter case, enantiopure 2-oxazolidinone-derived boron enolates have been shown to react with aldehydes in the presence of $Et_2AlCl$ to give predominantly *anti*-aldols with *anti*/*syn* ratios of 85:15–95:5 for aliphatic and 75:25 for aromatic aldehydes[106,112]. The stereoselectivity of boron enolates of *N*-acylsultams is not affected by the presence of $Et_2AlCl$. However, when the reaction is carried out in the

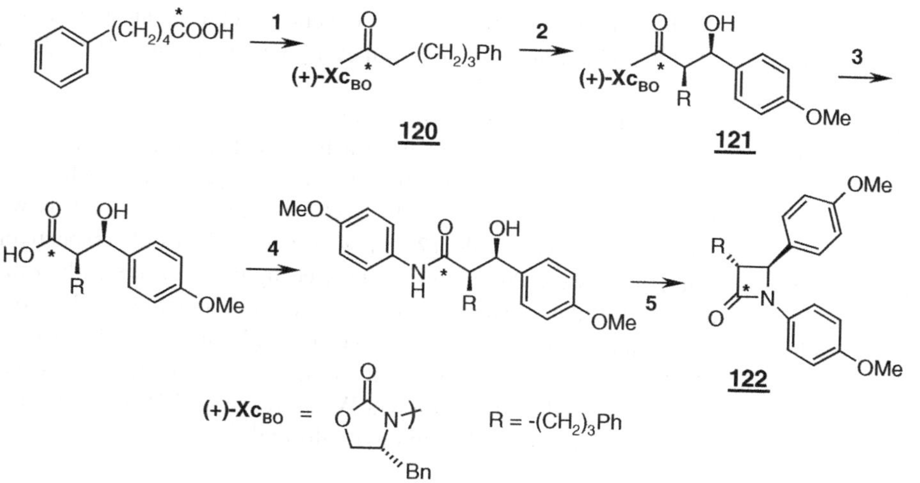

**Reaction conditions: 1a**. $(COCl)_2$, DMF (trace), benzene; 55 °C, 2 h, **b**. (+)-$Xc_{BO}$–H, *n*-BuLi, THF; –78°C, 1 h; then add acid chloride; –78°C to 0°C, 1 h; **2a**. TiCl4, TMEDA, $CH_2Cl_2$; –20 °C, 1 h, **b**. *p*-anisaldehyde, $CH_2Cl_2$; –20 °C, 1 h; –20 °C to 15 °C; **3**. LiOH, $H_2O_2$, aq. THF; 0 °C, 3 h; **4**. *p*-anisidine, DCC, HOBt, $CH_2Cl_2$; r.t., 3 h; **5**. DEAD, $Bu_3P$, THF; –78 °C, 4 h

***Figure 11.44*** *Application of a chiral chlorotitanium imide enolate to the synthesis of an enantiomerically enriched carbon-14-labeled syn-aldol and its conversion to an e.p. β-[$^{14}$C] lactam*

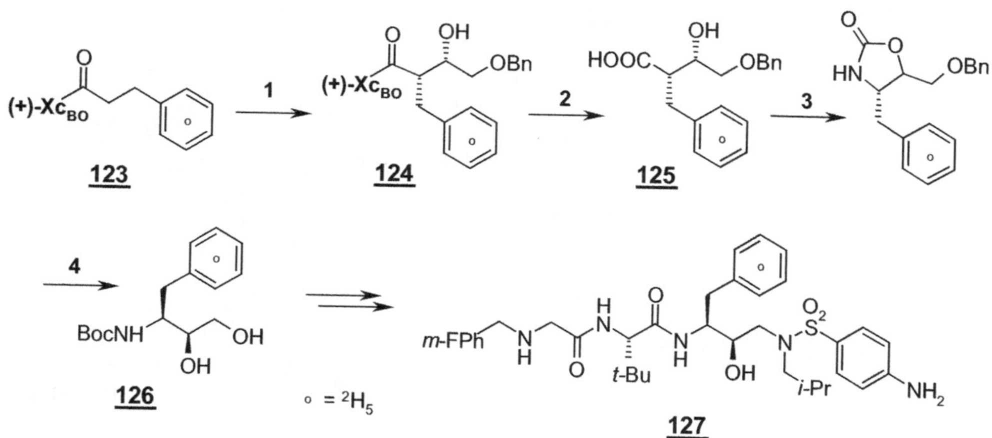

**Reaction conditions: 1a**. $Bu_2BOTf$, *i*-$Pr_2EtN$, $CH_2Cl_2$; r.t., 1 h, **b**. $BnOCH_2CHO$, $CH_2Cl_2$; –78 °C, 30 min, –78 °C to r.t., 2 h; **2**. LiOH, $H_2O_2$, aq. THF; 0 °C, 6 h; **3**. $(PhO)_2P(O)N_3$, $Et_3N$, toluene; reflux, 4 h; **4a**. aq. KOH; 70 °C, 3 h, **b**. $Boc_2O$, $CH_2Cl_2$; r.t., 2 h, **c**. $H_2$, 5 % Pd/C, EtOH; r.t., 18 h

***Figure 11.45*** *Use of a chiral boron enolate in the diastereoselective synthesis of an enantiomerically pure syn-aldol*

presence of $TiCl_4$ *anti*-aldol adducts are formed with excellent stereoselectivities at *anti*/*syn*-ratios of 94 : 6–99 : 1[113], which in most cases can easily be purified to >99% d.e. by simple crystallization.

This tactic was exploited for the synthesis of carbon-14-labeled avermectin, a potent insecticide consisting of a 9 : 1 mixture of avermectin $B_{1a}$ and $B_{1b}$. The boron enolate, generated from (2*S*)-*N*-([1-$^{14}$C]propionyl)bornane-10,2-sultam (**128**) and diethylboron triflate in the presence of *i*-$Pr_2EtN$, reacted with (*S*)-2-methylbutanal as well as with *i*-butanal and $TiCl_4$ to give the *anti*-aldols **129** in 90% and 95% yield, respectively, with 99 : 1 diastereomeric excesses. Protection of the free hydroxy group with benzyl 2,2,2-trichloroacetimidate, in order to suppress a *retro* aldol side reaction, followed by reductive cleavage of the auxiliary with $LiAlH_4$ converted **129** into the (1*S*,2*R*)- and (1*S*,2*R*,3*S*)-alcohols **130**, respectively (for $R^1$ = Me or Et). These intermediates were progressed as shown to the desired avermectins via Wittig reagents **131**. Overall yields were 16% ($B_{1a}$) and 5% ($B_{1b}$) from [1-$^{14}$C]propionyl chloride[114].

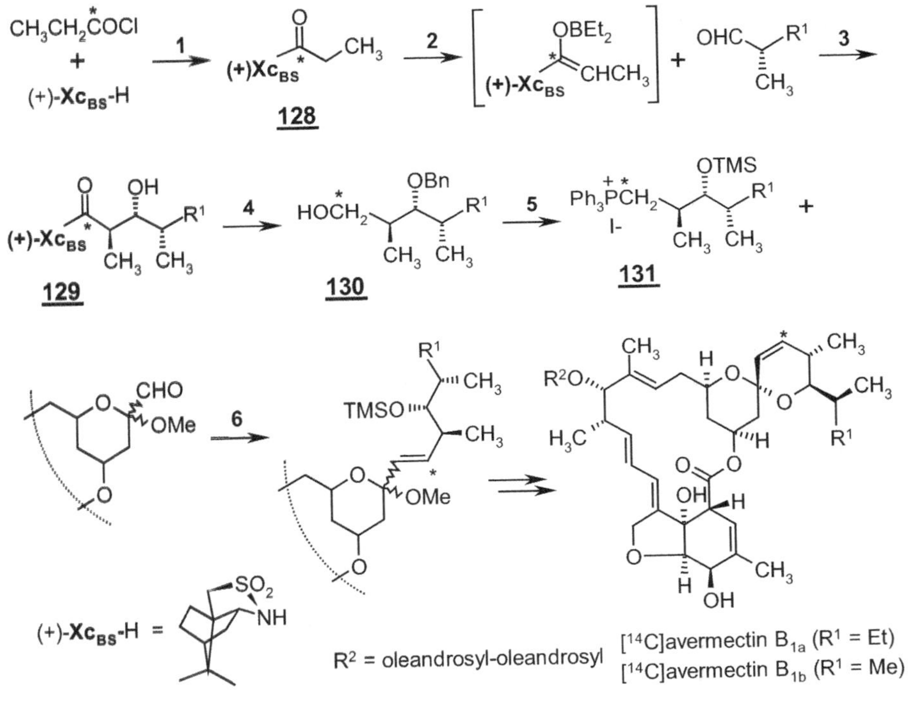


**Reaction conditions:** **1**. NaH, toluene; r.t., 2 h; **2a**. $Et_3B$, $CF_3SO_3H$, hexane; r.t., 40 min, **b**. **128**; *i*-$Pr_2NEt$, $CH_2Cl_2$; −10 °C, 30 min; **3**. $R^1CH(CH_3)CHO$, $TiCl_4$, $CH_2Cl_2$; −78 °C; **4a**. $CCl_3C(OBn)$=NH, $CH_2Cl_2$/cyclohexane, **b**. $LiAlH_4$, THF; **5a**. TsCl, pyridine, **b**. TMSI, **c**. NaI, acetone, **d**. $PPh_3$, $CaCO_3$, acetonitrile; **6**. TMSI, 2,6-lutidine

**Figure 11.46** *Aldol reaction of a chiral carbon-14-labeled boron imide enolate of the sultam type with $TiCl_4$ pre-complexed achiral/chiral aldehydes to give anti-aldols and application to [$^{14}$C]avermectin $B_{1a}$ and $B_{1b}$*

An atypical *anti* product (**134**) was also produced in the reaction of Schiff base **133** with (4*S*)-4-phenyl-3-[(5*S*)-5-(4-fluorophenyl)-5-hydroxypentanoyl]-1,3-oxazolidin-2-one (**132**) in the presence of *i*-$Pr_2EtN$ and TMSCl followed by addition of $TiCl_4$. The resultant compound was obtained in 29% radiochemical yield after crystallization. This result may be explained by intermediate formation of an *O*-silyl-*N*,*O*-ketene acetal. Adduct **134** was used as an intermediate in the preparation of [$^{14}$C]SCH 58235 (**135**), a cholesterol absorption inhibitor[115].

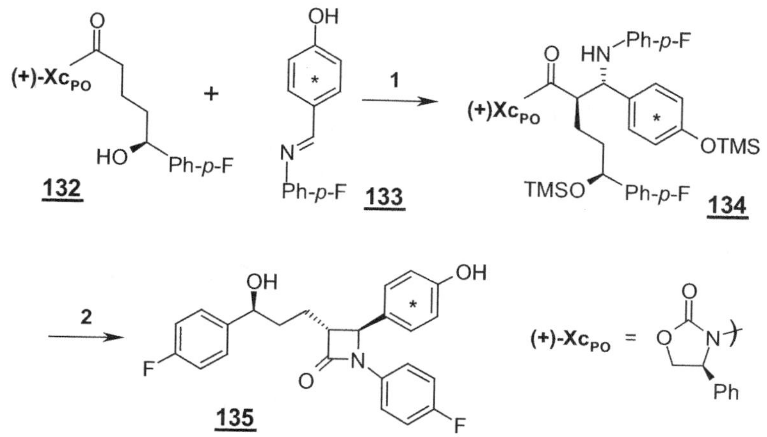


**Reaction conditions:** **1a**. TMSCl, *i*-$Pr_2EtN$, $CH_2Cl_2$; 0–5 °C, 75 min, **b**. $TiCl_4$; −30 °C to −35 °C, 4 h, **c**. BSA, $CH_2Cl_2$; reflux, 30 min; **2a**. BSA, $CH_2Cl_2$; r.t., 15 min, **b**. $Bu_4NF{\cdot}3H_2O$; 0 °C, 2 h

***Figure 11.47*** *Diastereoselective addition of a TMS-protected chiral imide enolate of the 1,3-oxazolidin-2-one type to a Schiff base providing an anti configured adduct*

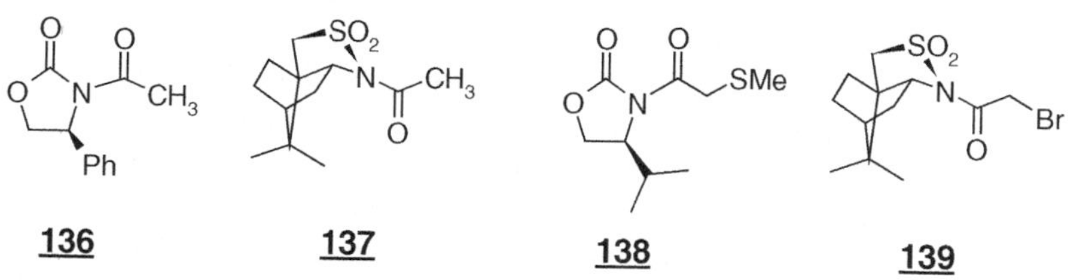


***Figure 11.48*** *Chiral acetyl building blocks*

In contrast to the examples presented so far, chiral building blocks possessing acetyl groups, such as **136** and **137**, show poor stereoselectivities due to a lack of steric interactions in the transition state of the aldolization process.

This limitation can be overcome by employing either the corresponding thioglycolate (e.g., **138**)[116] or bromoacetyl (e.g., **139**)[87] derivative, both of whose heteroatomic substituents improve the stereoselectivity of aldol reactions and can be removed by reductive cleavage afterwards. This strategy was followed for the carbon-14 labeling of Lipid X, a Lipid A-analogue, in the *N*-(*R*)-3-hydroxymyristoyl side chain. Reaction of the chlorotitanium enolate of (2*S*)-(bromo[1-$^{14}$C]acetyl)bornane-10,2-sultam (**140**) with lauraldehyde gave (2*S*,3*R*)-2-bromo-3-hydroxy[1-$^{14}$C]myristoyl sultam (**141**) in 72% yield and diastereomeric ratio of 98:2. Purification by crystallization followed by dehalogenation with Zn/$NH_4Cl$, protection of the hydroxy function and hydrolytic cleavage of the auxiliary provided *O*-benzyl (*R*)-3-hydroxy[1-$^{14}$C]myristic acid (**142**), which was employed as a key intermediate for the subsequent sequence to Lipid X[87].

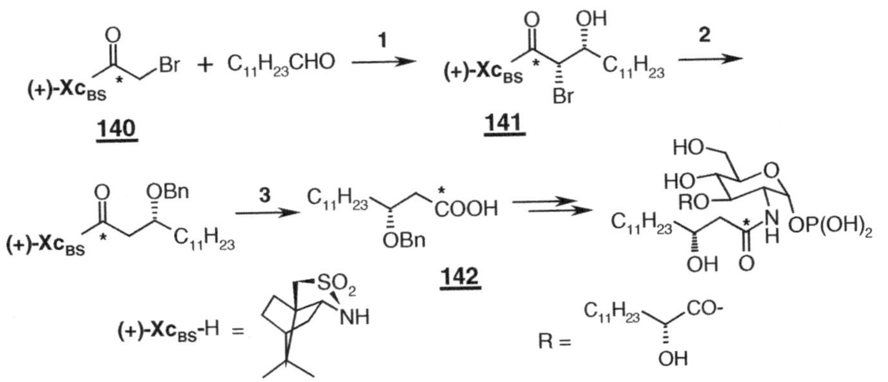


**Reaction conditions: 1a**. $TiCl_4$, 1-ethylpiperidine, $CH_2Cl_2$; –78 °C, 15 min, **b**. aldehyde, $CH_2Cl_2$ ;–78 °C, 1 h ; –50 °C, 3 h; **2**. Zn, $NH_4Cl$, MeOH ; r.t., 16 h; **3a**. $CCl_3C(OBn)=NH$,$CF_3SO_3H$, cyclohexane; r.t., 2 h, **b**. LiOH, $H_2O_2$; 0 °C, 3 h

***Figure 11.49*** *Use of a chiral haloacetimide as a substitute for a chiral acetimide to improve stereoselectivity aldol reactions*

An alternative to **138** and **139** is (*R*)- or (*S*)-2-hydroxy-1,2,2-triphenylethyl acetate ((*R*)- or (*S*)-HYTRA) (**143**), a 'chiral acetate' which does not need an auxiliary heteroatomic substituent. Compound **143** is readily available through acylation of 1,2,2-triphenyl-1,2-ethandiol with acetyl chloride or acetic anhydride. Its dianion, generated by double deprotonation with LDA and transmetallation with $MgBr_2$ or $MgI_2$, reacts with aliphatic and aromatic aldehydes to give $\beta$-hydroxy acid derivatives with diastereomeric ratios ranging from 92 : 8 to 98 : 2[117]. The crude adducts can be readily purified by crystallization. Removal of the auxiliary can be accomplished either hydrolytically (KOH, aq. MeOH) or transesterification (MeONa, MeOH/THF) to provide the respective $\beta$-hydroxy acid/ester in enantiomerically pure form.

This alternative was successfully applied to the carbon-14 labeling of atorvastatin, a potent antihyperlipoproteinemic drug. (*S*)-[acetyl-1-$^{14}$C]HYTRA (**143**) needed as a starting material has been made from [1-$^{14}$C]acetyl chloride following routine procedures (Figure 11.50). Reaction of its dianion with substituted 3-(pyrrol-1-yl)propionaldehyde (**144**) provided $\beta$-hydroxy ester **145** in 68% yield. Transesterification, two-carbon chain elongation through ester condensation with lithiated *tert*-butyl acetate and stereoselective

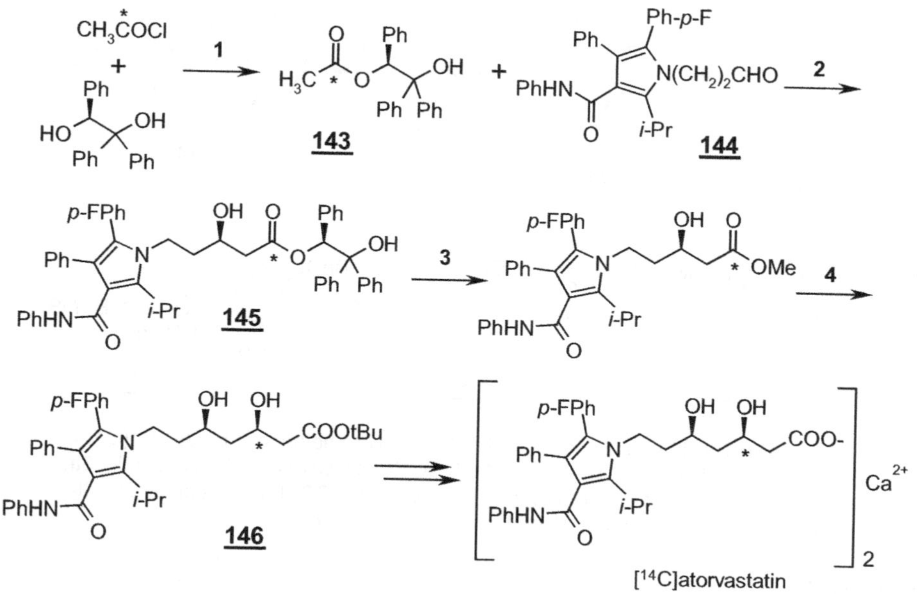

**Reaction conditions: 1**. $CH_2Cl_2$; r.t., 2 h; **2a**. LDA, THF; −78 °C to −10 °C, 1.5 h, **b**. **144**, THF; −78 °C, 2 h; **3**. MeONa, MeOH/THF; 0–5 °C, 3 h; **4a**. $CH_3COO$-*t*-Bu, LDA, hexane; −78 °C, 1 h, **b**. ester; −78 °C to −40 °C, 3 h; −10 °C, 2 h, **c**. $Me_3CCOOH$, $Et_3B$, THF; r.t., 1 h, followed by addition of $NaBH_4$,MeOH; −70 °C, 2.5 h

***Figure 11.50*** *Use of (S)-[acetyl-1-$^{14}$C]HYTRA 'chiral [$^{14}$C]acetate' in the synthesis of an e.p. labeled β-hydroxy acid derivative*

reduction of the β-carbonyl function with $Et_3B/NaBH_4$ afforded the corresponding e.p. dihydroxy ester **146**, the penultimate precursor, in 35% yield (based on **145**)[118].

A second alternative exploits the excellent stereoselectivities achievable with *N*-acyl derivatives of (*R*)/(*S*)-4-substituted 1,3-thiazolidin-2-ones (**147**) in *syn*-aldol reactions (Figure 11.51). This is one of the most promising second-generation auxiliaries recently introduced. Due to the higher affinity of sulfur to titanium, the corresponding transition state is expected to be more rigid, which should lead to improved selectivities. This explains why, in contrast to their classical analogous oxazolidinones, even *N*-acetyl derivatives (analogous to **136** and **137** above) provide β-hydroxy acid adducts with diastereomeric excesses >90% when their tin enolates react with aliphatic aldehydes[106,119]. This was evidenced by the synthesis of carbon-13-labeled 8,9-dihydroisocoumarins. (*R*)-3-[1,2-$^{13}C_2$]acetyl-4-isopropyl-1,3-thiazolidine-2-thione (**148**), the required starting material, was obtained in nearly quantitative yield through reaction of the auxiliary **147** with [$^{13}C_2$]acetic acid and BOP-Cl in the presence of catalytic amounts of DMAP. Its tin enolate reacted with butanal as well as with *i*-butanal to afford the β-hydroxy acyl derivatives **149** as single diastereomers. Replacement of the auxiliary with *N*,*O*-dimethylhydroxylamine and *O*-silylation converted **149** into the corresponding *O*-protected Weinreb amides **150**, which upon treatment with arylmethyllithium **151** gave the enatiomerically enriched *O*-protected dihydroisocoumarin precursor **152**[85,120].

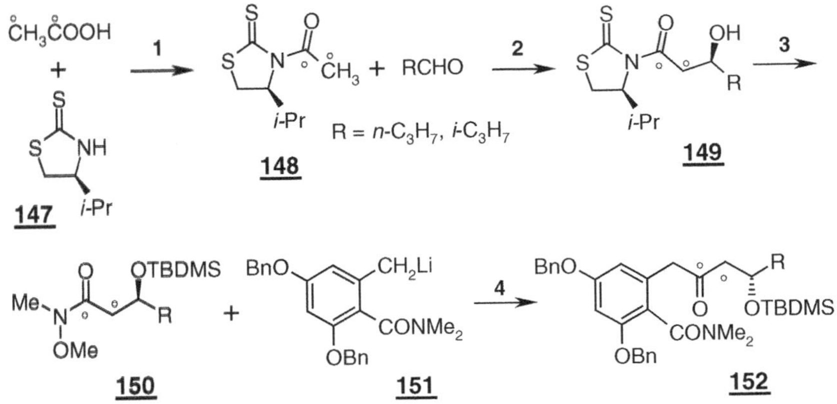


**Reaction conditions:** **1**. BOP-Cl, DMAP, dichloromethane; r.t.; **2a**. $Sn(OTf)_2$, 1-ethylpiperidine, $CH_2Cl_2$; −40 °C, 4.5 h, **b**. RCHO, $CH_2Cl_2$; −78 °C, 20 min; **3a**. MeNHOMe·HCl, $Et_3N$, $CH_2Cl_2$; r.t., 72 h, **b**. TBDMSOTf, 2,6-lutidine, $CH_2Cl_2$; −25 °C, 1 h; −25 °C to r.t., 12 h; **4**. THF; −78 °C, 1.3 h

***Figure 11.51*** *Diastereoselective aldol reactions with (S)-3-[$^{13}C_2$]acetyl-4-isopropyl-1,3-thiazolidine-2-thione*

### 11.3.3 1,4-Additions of Chiral Imide Enolates to Michael Acceptors

The diastereoselective addition of chlorotitanium enolates of chiral carboximides to appropriate Michael acceptors is a strategy capable of simultaneously generating two stereogenic centers (at the C$\alpha$ and C$\beta$ positions), giving rise to $\alpha,\beta$-substituted/branched $\gamma$-functionalized carboxylic acid derivatives[76a,121–124]. In isotope chemistry the scope of this strategy has so far been exploited only for the addition to nitroolefins[124]. The resulting $\beta$- or $\alpha,\beta$-branched $\gamma$-nitrocarboxylic acids have been proposed as versatile building blocks for the synthesis of a series of singly or doubly labeled enantiomerically pure substituted nitrogen heterocycles[124]. In carbon-14 synthesis, low-molecular-weight electrophilic olefins are often prone to radiation-induced polymerization; therefore the label is generally introduced with the carboximide component, as illustrated by the example in Figure 11.52.

Michael addition of the chlorotitanium enolate of (*R*)-3-[1-$^{14}$C]acetyl-4-isopropyl-5,5-diphenyl-1,3-oxazolidin-2-one (**153**) to (*E*)-2-methoxy-$\beta$-nitrostyrene gave a 4:1-mixture of the diastereomers **154** and **155**[125] (Figure 11.52). This modified Evans' building block had shown promise of better diastereoselectivity in previous unlabeled experiments[123]. Nonetheless, chromatographic separation of the desired **154** allowed its isolation in 42% radiochemical yield. Catalytic hydrogenation of the nitro group triggered spontaneous aminolytic cleavage of the auxiliary and cyclization to 4-(2-methoxyphenyl)-2-[2-$^{14}$C] pyrrolidinone (**156**). Hydride reduction provided 3-aryl[5-$^{14}$C]pyrrolidine **157**, a common intermediate in the syntheses of the $\alpha_{2c}$-antagonistic, antischizophrenic drug substances [$^{14}$C]NVP ABE687 (**158**) and [$^{14}$C]NVP ABE651 (**159**)[125].

The use of homologous acyl derivatives of **153** (i.e., **Xc**–C(O)Et, **Xc**–C(O)$CH_2CH_2$Ph) would be expected to give higher diastereoselectivities in such Michael additions, as would the use of the Oppolzer sultam analog of **153**, namely **160**. In cold experiments reportedly

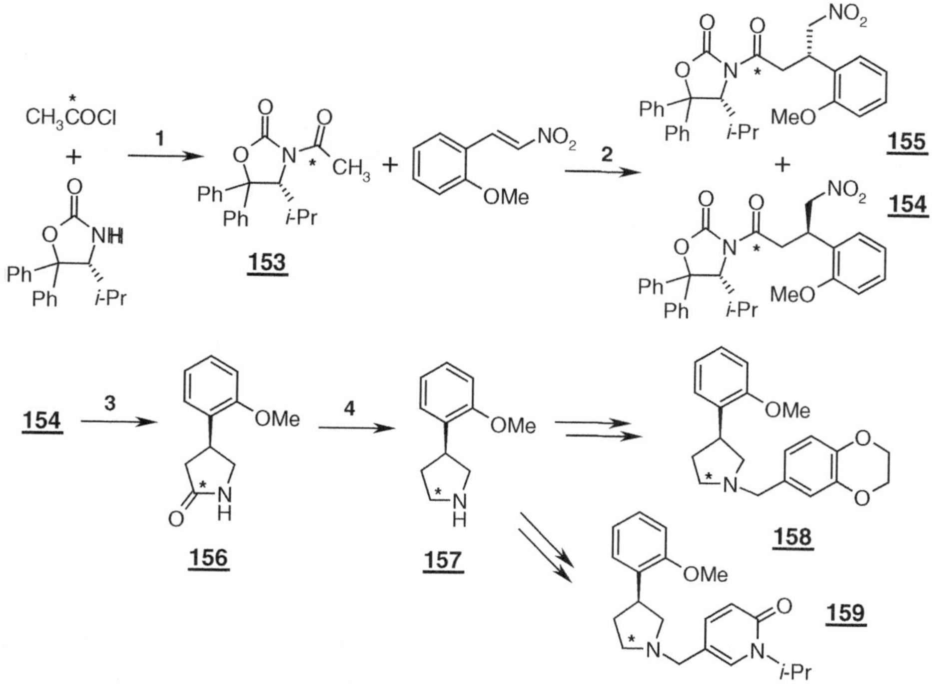


**Reaction conditions:** **1**. *n*-BuLi, THF; 0 °, 30 min, 0 °C to r.t., 3 h; **2a.** 153, $TiCl_4$, *i*-$Pr_2EtN$, $CH_2Cl_2$;–78 °C to 0 °C, 30 min, **b.** (*E*)-2-methoxy-β-nitrostyrene $CH_2Cl_2$; –78 °C, 7 h, **c.** HPLC separation, crystallization; **3.** $H_2$, Raney-Ni, EtOH/$H_2O$/EtOAc; r.t., 17 h, 100 °C; **4.** $LiAlH_4$, THF; 90 °C, 2 h

***Figure 11.52*** *Diastereoselective Michael addition of the chlorotitanium enolate of an Evans-type N-[1-$^{14}$C]acetyloxazolidin-2-one to a β-nitrostyrene*

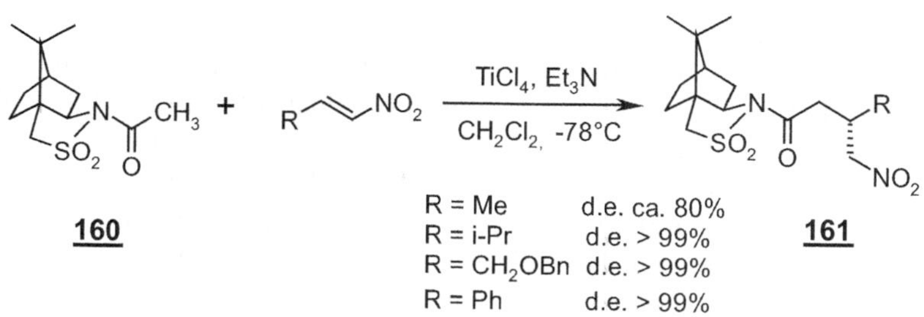


***Figure 11.53*** *Diastereoselectivities achieved in labeling method development using the N-acetylsultam synthon*

developed for use with labeled components[124], *N*-acetylsultam **160** was shown to react with several nitroolefins to give Michael adducts **161** in generally very high d.e.

**161** has been proposed as a valuable intermediate for the preparation of carbon-13/14-labeled substituted prolines and 2-hydroxy-piperidones.

### 11.3.4 α-Amination of Chiral Imide Enolates

The α-amination of chiral imide enolates of the Evans and Oppolzer types with nitrogen electrophiles is a highly attractive and direct route to e.p. proteinogenic and *non*-proteinogenic (*R*)- and (*S*)-amino acids[25,126–129]. Three reagents have emerged as useful *N*-electrophiles: di-*tert*-butyl azodicarboxylate (DBAC)[82c,127], 2,4,6-triisopropylbenzene-sulfonyl azide (trisyl azide)[128] and 1-chloro-1-nitrosocyclohexane (CNC)[129]. Each of these reagents gives an intermediate that needs to be reductively converted in an additional step to give the free or *N*-protected e.p. α-amino acid.

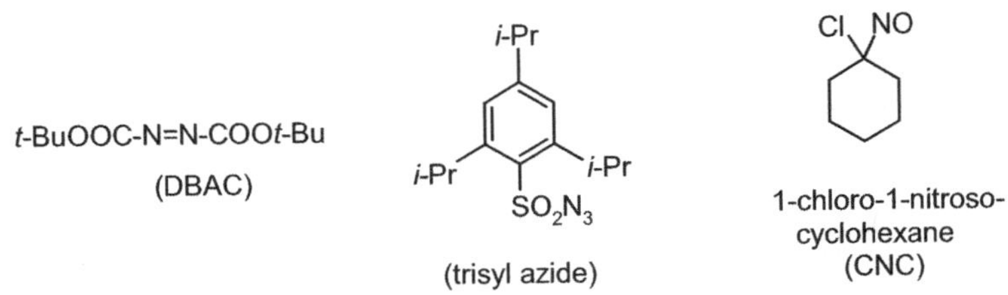


***Figure 11.54*** *N-Electrophiles for amination of chiral imide enolates*

DBAC was the first reagent to be investigated. Generally, addition of DBAC to the lithium imide enolate generated from (*R*) or (*S*)-4-substituted 3-acyl-1,3-oxazolidin-2-ones affords di-Boc-α-hydrazino adducts in >90% yields and >97% d.e. The minor diastereomers and impurities are generally separated by flash chromatography. Subsequent removal of the auxiliary is routinely accomplished either with LiOH (or in sterically demanding cases with $LiOH/H_2O_2$) or via transesterification with lithium benzoxide. Cleavage of *tert*-butyl ester groups with trifluoroacetic acid then provides α-hydrazino acids or esters with little if any racemization[82c,127]. The disadvantage of this method as it is described in classic literature is that conversion of the α-hydrazino acids to α-amino acids requires harsh reaction conditions (Raney Ni, $H_2$ at 500 psi) that are unattractive for use with radioactive materials.

This shortcoming has been overcome, however, as illustrated in a synthesis of (*S*)-[1-$^{14}$C]-phenylalanine (**165**)[130] (Figure 11.55). Treatment of lithiated (*S*)-4-benzyl-3-(3-phenyl[1-$^{14}$C]propionyl)-1,3-oxazolidin-2-one (**162**) with DBAC followed by chromatographic purification afforded the hydrazinedicarboxylate **163** in 46% yield (yields up to 96% have been reported in the literature). Standard procedures converted **163** to the hydrazine **164**. Hydrogenolytic N–N cleavage to the labeled amino acid was accomplished in this case under a more acceptable 50 psi $H_2$ by using aqueous isopropanol as solvent instead of trifluoroacetic acid. The overall radiochemical yield of (*S*)-[1-$^{14}$C]phenylalanine was 11%. It was used as shown for the synthesis of [$^{14}$C]PD 132002 (**166**), a renin inhibitor[123].

In contrast to α-hydrazino acids, catalytic hydrogenolytic cleavage of N–N bonds in α-azido derivatives can be accomplished in a variety of solvents under hydrogen pressures of 15–75 psi[128]. α-Azido derivatives can be formed by the treatment of potassium enolates of (*R*)- or (*S*)-4-substituted-3-acyl-1,3-oxazolidin-2-ones (e.g., **162**, **168** in Figure 11.56) with trisyl azide, followed by a fragmentation of the primary adduct (**167**, **169**) upon quenching of the reaction mixture with AcOH and warming up to ambient temperature[128a].

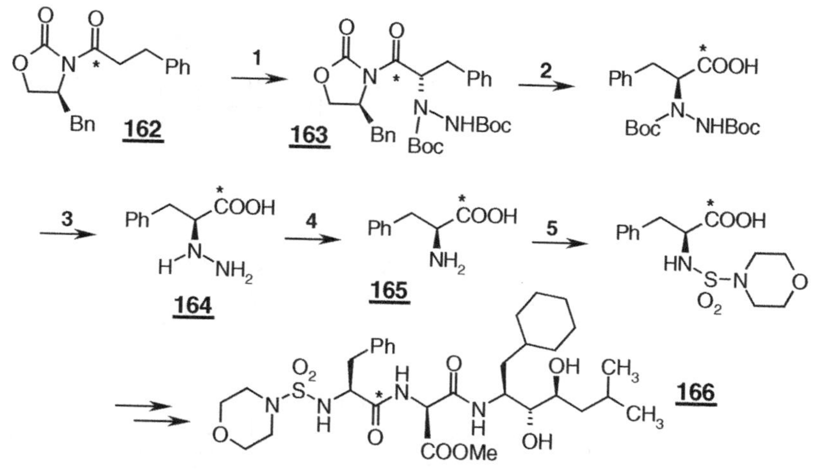

**Reaction conditions: 1a.** LDA, THF; –78 °C, 1.5 h, **b.** DBAC, THF; –78 °C, 30 min; **2.** LiOH, aq. THF; 0 °C, 20 h; **3.** TFA, $CH_2Cl_2$; r.t., 4 h; **4.** $H_2$ (50 psi), Raney-Ni, aq. *i*-PrOH; r.t., 4 h; **5a.** $Me_3SiSiMe_3$, MeCN; reflux, 3 h, **b.** *N*-morpholinosulfonyl chloride, MeCN; r.t.; 17 h

***Figure 11.55*** *Synthesis of (S)-[1-$^{14}$C]phenylalanine via diastereoselective α-hydrazination of (S)-4-benzyl-3-(3-phenyl[1-$^{14}$C]propionyl)-1,3-oxazolidin-2-one*

Nevertheless, despite this advantage a direct comparison of both methods by the same author found that for e.p. (*S*)-[1-$^{14}$C]phenylalanine (**165**) the DBAC variant gave a higher overall yield[130]. This outcome may have to do with the need for careful control of reaction conditions in the α-functionalization step in order to achieve optimal yields (i.e., use of reagents precooled to −78 °C and very short reaction times).

Though the utility of the diastereoselective α-azide formation has been confirmed by several examples in synthetic organic chemistry[128], only one additional application has been reported in radiosynthesis, namely in the preparation of (2*S*,5*S*)-[5-$^3$H]ornithine (**170**). Reaction of the potassium enolate of (*S*,*S*)-3-(5-azido[5-$^3$H]pentanoyl)-4-benzyl-1,3-oxazolidin-2-one (**168**) with trisyl azide proceeded in 45% yield. Subsequent LiOH-promoted hydrolytic removal of the auxiliary and hydrogenolytic cleavage of the two azido functions provided (2*S*,5*S*)-[5-$^3$H]ornithine in 28% overall yield[131].

CNC is the most user-friendly nitrogen electrophile of the three reagents discussed, and has been used for the preparation of a broad variety of carbon-13, carbon-14- and nitrogen-15-labeled (*R*)- and (*S*)-α-amino acids of high enantiomeric purity[25,132]. It reacts with the sodium enolates of (2*S*)- or (2*R*)-*N*-acylbornane-10.2-sultams (e.g., **171**) to give nitrones **172**, which decompose upon hydrolysis with dilute HCl to furnish the corresponding α-hydroxylamino derivatives **173**, generally with diastereomeric excesses >99% and chemical yields >70% (Figure 11.57). Purification of **173**, usually by crystallization, and N–O cleavage with Zn-acetic acid affords the corresponding α-aminoacyl sultams **174**. Removal of the auxiliary with LiOH and chromatographic separation provides the free e.p. isotopically labeled (*R*)- and (*S*)-α-amino acids in overall yields of 40–60%, based on $^{14}CO_2$. β-Substituents (e.g. R = *i*-Pr, (*S*)-*sec*-Bu, *tert*-Bu) seem to have only marginal

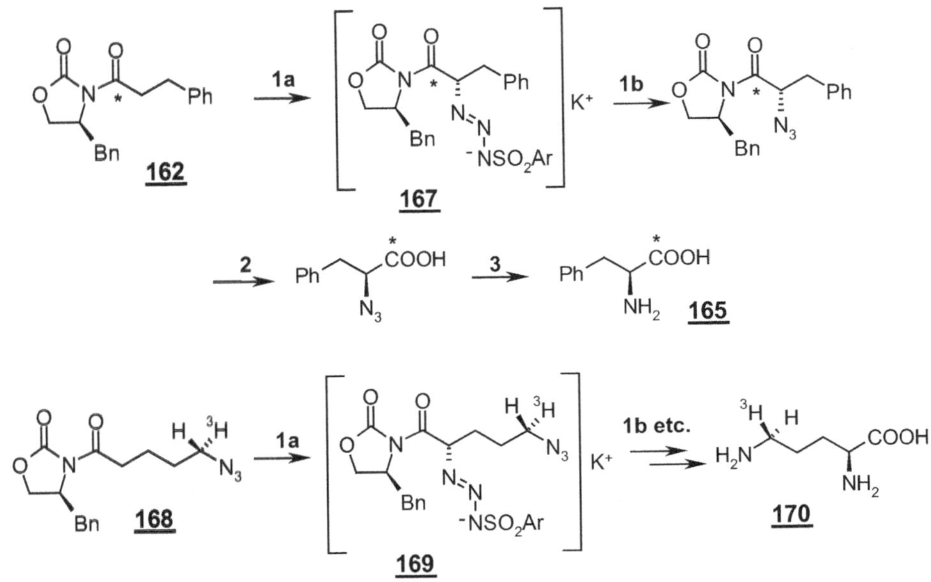

**Reaction conditions: 1a**. KHMDS, THF; −78 °C, 30 min, then trisyl azide, THF; −78 °C, 1–2 min, **b**. AcOH; THF; −78 °C to r.t., 16 h; **2**. LiOH, aq. THF; 0 °C, 30 min; **3**. $H_2$, 10 % Pd/C, EtOH; r.t., 12 h

***Figure 11.56*** *Preparation of (S)-[1-$^{14}$C]phenylalanine and (2S,5S)-[5-$^{3}$H]ornithine following the trisyl azide α-amination protocol*

influence on selectivity and yields of the α-hydroxylamination reaction. The overall nonoptimized radiochemical yields ranging from 25–45%, as observed for (*S*)-*tert*-[1-$^{14}$C]leucine, (*S*)-[1-$^{14}$C]isoleucine and (*S*)-[1-$^{14}$C]valine, are only slightly lower than for the corresponding linear derivatives[87]. In preparing *N*-protected e.p. representatives for immediate use in peptide synthesis, it is advantageous to treat the α-aminoacyl sultams **174** with $Boc_2O$. The resulting *N*-Boc derivatives can easily be repurified by flash chromatography and/or crystallization to guarantee d.e. >99%. Hydrolytic cleavage of the auxiliary provides the respective *N*-Boc-α-[$^{13,14}$C]amino acids in enantiomerically pure form.

The α-[$^{14}$C]amino acids thus obtained are monolabeled at the carboxyl position, and have a maximum specific activity of 62 mCi/mmol. If higher specific activities or different positions of the label are required, the *N*-acylsultam starting materials must be made available by different synthetic routes. As discussed below in more detail (Section 11.3.10) one generally applicable alternative makes use of the Horner–Wadsworth–Emmons olefination reactions between (2*S*)- or (2*R*)-*N*-(phosphono[$^{14}C_n$]acetyl)bornane-sultam ([$^{14}C_n$]PABS, **175**) and unlabeled or C1-labeled aldehydes followed by catalytic hydrogenation of the resulting α,β-unsaturated *N*-acyl derivatives **176**[87] (Figure 11.58). Since [$^{14}C_n$]PABS is readily obtained from selectively or doubly labeled acetate, and carbonyl-$^{14}$C-labeled aldehydes are routinely available by standard procedures, up to three different labeling positions should become available and the achievable specific activities may be as high as 185 mCi/mmol.

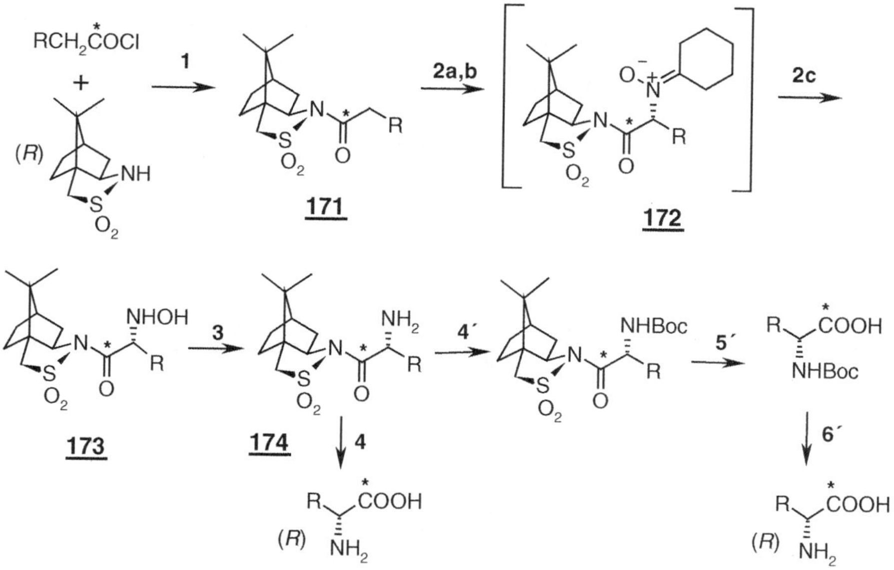

R = $CH_3$, $CH_2SiMe_3$, *n*-$C_3H_7$, *i*-$C_3H_7$, CH(Me)Et, *t*-Bu, $CH_2$-naphth

**Reaction conditions: 1a**. NaH, toluene; r.t., 30 min, **b**. [1-$^{14}$C]acyl halide, −120 °C → r.t; r.t., 75 min; **2a**. NaHMDS, THF; −78 °C, 45 min, **b**. CNC, THF; −78 °C, 1 h, **c**. 1 N HCl; −78 °C → r.t.; **3**. Zn, HOAc/1 N HCl; 0 °C, 60 min; **4a**. LiOH, aq. THF; r.t., 1 h, **b**. Amberlite IR-120, $H_2O$ (pH 4–5); **4´**. $Boc_2O$, THF; r.t., 1.5 h; **5´**. LiOH, aq. THF; 0 °C, 2 h; **6´**. TFA; r.t.

***Figure 11.57*** *Generic synthesis of (R)- and (S)-[$^{14}$C]amino acids of high enantiomeric purity through diastereoselctive α-amination of (2R)- or (2S)-N-[$^{14}$C]acylsultams using CNC*

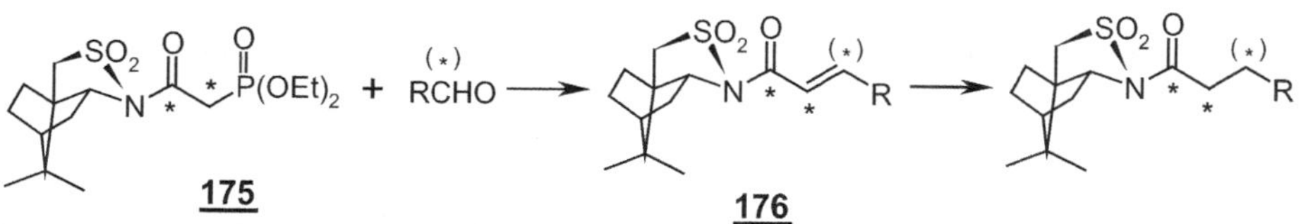

(*) : potential position of carbon-14

***Figure 11.58*** *Preparation of differently/multiply labeled N-acylsultams needed as starting materials for selectively/multiply labeled α-amino acids of high enantiomeric purity*

The need for additional labeling positions may be illustrated by (*S*)-[4-$^{14}$C]SDZ ISQ844 (**179**)[87], whose label was to be located at the benzylic position for metabolic reasons. The synthesis was accomplished starting from 3,4-dimethoxybenz[$^{14}$C]aldehyde. Reaction with unlabeled PABS followed by catalytic hydrogenation of the resultant cinnamoyl intermediate gave *N*-[3-(3,4-dimethoxybenzyl)[3-$^{14}$C]propionyl]sultam (**177**). Application of the CNC methodology to **177** followed by reductive cleavage of the auxiliary provided (*S*)-3-(3,4-dimethoxyphenyl)-2-amino-[3-$^{14}$C]propanol (**178**). Subsequent *N*/*O*-*bis*-acylation, Bishler–Napieralski amide cyclization then cleavage of the ester function converted **178** into the carbon-14-labeled dihydroisoquinoline drug candidate.

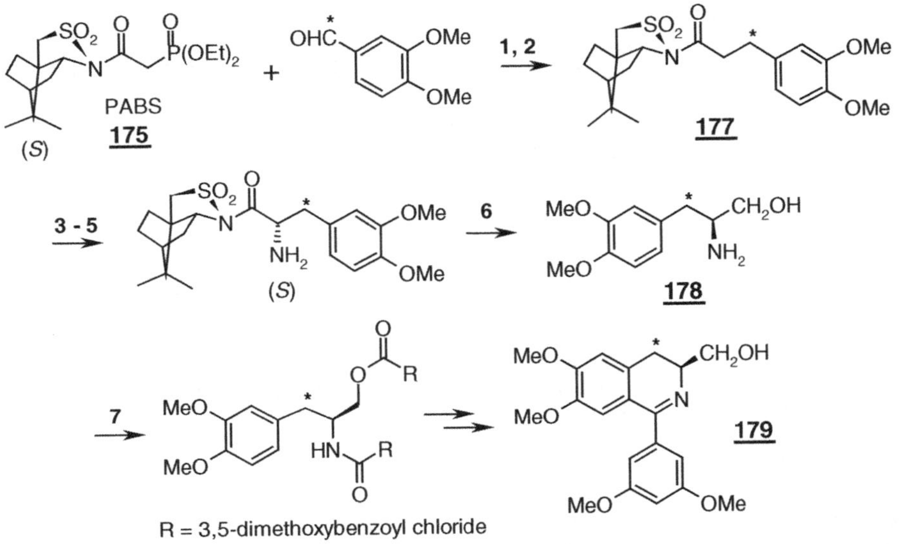


**Reaction conditions:** **1**. DBU, THF; 50 °C, 50 h; **2**. $H_2$, 5 % Pd/$Al_2O_3$, MeOH/THF; r.t., 3 h; **3 – 5**. see general scheme, Figure 11.57; **6**. $NaBH_4$, $I_2$, EtOH; 0 °C, 90 min; r.t., 3 h; **7**. 3,5-dimethoxybenzoyl chloride, $Et_3N$, DMAP, $CH_2Cl_2$; r.t., 3 h

***Figure 11.59*** *Synthesis of a (S)-3-aryl-2-amino[3-α$^{14}$C]propanol and its application to the synthesis of an e.p. 3,4-dihydro[4-$^{14}$C]isoquinoline*

*N*-Boc-protected derivatives of aminoalcohols such as **180** are useful for further elaboration (Figure 11.60). For example, in the synthesis of [$^{14}C_2$]SDZ PRI053 (**184**)[87], Swern oxidation of **180** and trapping of the resulting aldehydes *in situ* with $Ph_3PCH_2COOEt$ provided the key intermediate *N*-Boc-$\alpha,\beta$-unsaturated-$\gamma$-amino acid esters **181**. Epoxidation of **181** with *m*-chloroperbenzoic acid was guided by the ancillary stereogenic center to give a 9:1 mixture of the diasteromeric epoxides with **182** predominating. Stereo- and regioselective opening of the epoxide rings upon treatment with 4-methoxybenzylamine furnished the requisite e.p. $N_4$-Boc-$N_2$-*p*-MeOBn-(2*R*,3*R*,4*S*)-2,4-diamino-3-hydroxy-5-phenyl[3,4-$^{14}C_2$]pentanoic acid ethyl ester fragments (**183**) in about 10% overall nonoptimized yield. Compounds **183** were submitted to further transformations to afford the target compounds **184**.

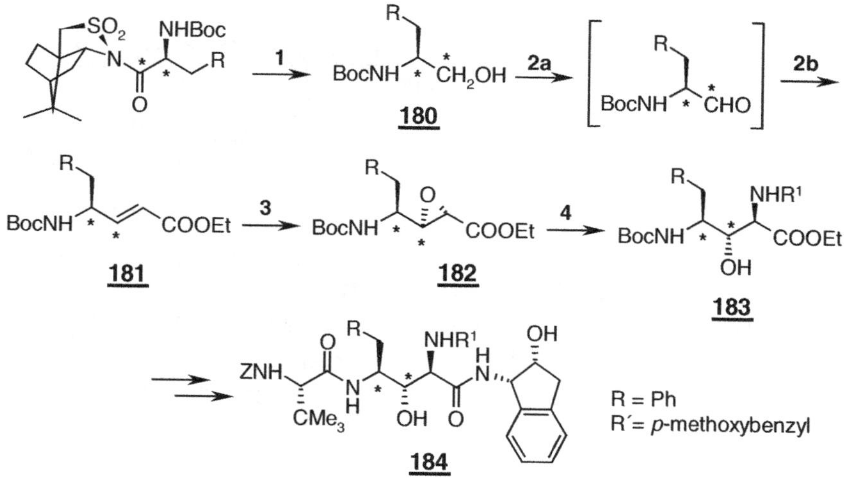


**Reaction conditions:** **1**. $NaBH_4$–AcOH 1:1, dioxane, 80 °C, 3 h; **2a**. $(COCl)_2$, DMSO, $Et_3N$, THF; −60 °C, 30 min; −60 °C to 0 °C, **b**. $Ph_3P{=}CHCOOEt$, $Et_3N$; 0 °C to r.t., 2 h; **3**. MCPBA, $CH_2Cl_2$; r.t., 8 d;**4**. $PMBnNH_2$, EtOH; 70 °C, 20 h

***Figure 11.60*** *Preparation of multiply labeled complex compounds containing mutiple stereogenic centers from N-protected amino aldehydes by a two-carbon elongation*

### 11.3.5 α-Hydroxylation of Chiral Imide Enolates

2-(Phenylsulfonyl)-3-phenyloxaziridine (**186**, Davis reagent, Figure 11.61) has been shown to react with sodium enolates of chiral carboximides of the Evans–Oppolzer type, providing α-hydroxyacyl derivatives with diastereoselectivities of 90–98% d. e.[92,133]. Subsequent titanium(IV) isopropoxide-mediated transesterification with benzyl alcohol and hydrogenolytic debenzylation releases the free α-hydroxy acids in enantiomerically pure form and in yields of 65–75%. This procedure was applied to the synthesis of α-hydroxy[1-$^{14}$C]isovaleric acid[25]. In this case treatment of the (2*S*)-[1-$^{14}$C]isovalerylsultam **185** with Davis reagent **186** afforded the respective α-hydroxyacylsultam **187** in 81% yield without any trace of the undesired diastereomer. In a variation of the routine procedure, the transesterification was followed by protection of the hydroxy group with TBDMSOTf-lutidine prior to the reductive cleavage of the benzyl ester. The resulting *O*-protected (*S*)-2-hydroxy[1-$^{14}$C]isovaleric acid **188** was a key intermediate in the synthesis of [$^{14}$C]NVP 214-103 (**189**), a cyclosporin A-derived cyclopeptolide.

### 11.3.6 α-Alkylation of Chiral Glycinates

The synthesis of e.p. α-[$^{14}C_n$]amino acids through diastereoselective homologation of protected acyclic or cyclic homochiral glycine building blocks (Figure 11.62) by alkylation of their enolates is a valuable alternative to the electrophilic α-amination of chiral α-unsubstituted imide enolates[25,86,87,134–137].

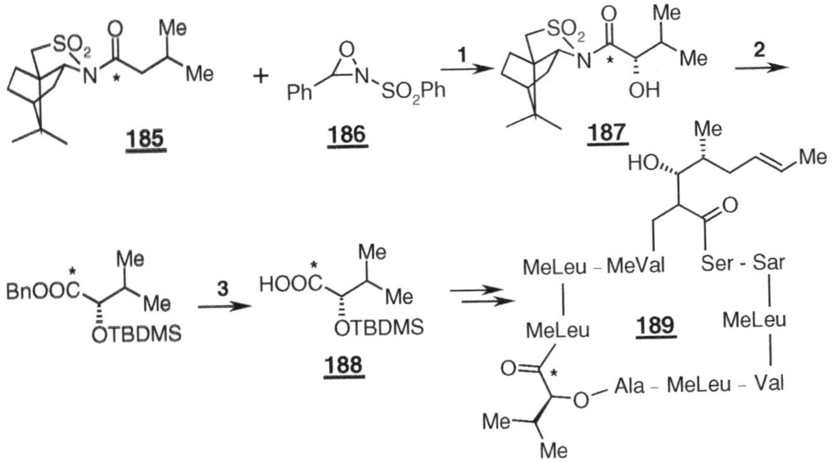


**Reaction conditions: 1a**. NaHMDS, THF; –78 °C, 45 min, **b**. THF; –78 °C, 60 min; **2**. Ti(Oi-Pr)$_4$, BnOH, molecular sieves 4Å; 120 °C, 4 h; **3**. TBDMSOTf, 2,6-lutidine, $CH_2Cl_2$; 0 °C → r.t., 1.5 h; **4**. $H_2$, 5 % Pd/C, MeOH; r.t. 20 h

***Figure 11.61*** *2-(Phenylsulfonyl)-3-phenyloxaziridine-mediated diastereoselective α-hydroxylation of a chiral carboximide*

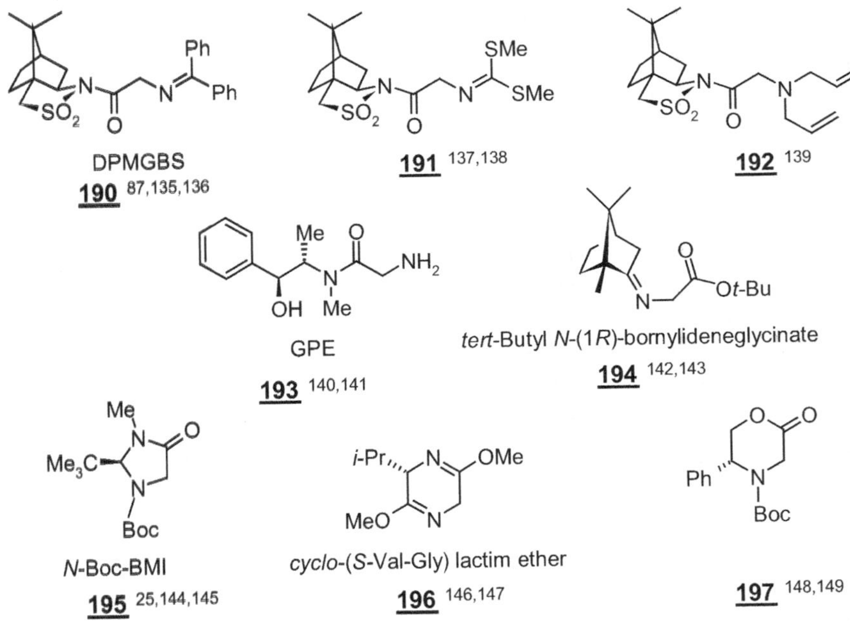


***Figure 11.62*** *Glycine building blocks used for the synthesis of carbon-13/14-labeled e.p. α-amino acids*

*α-Alkylations of Linear Chiral Glycinates*

The building blocks most widely applied to the synthesis of labeled e.p. α-amino acids are the analogously labeled enantiopure *N*-(*N′*-diphenylmethylideneglycyl)bornane-10,2-sultam (DPMGBS, **190**)[25,134,135] and *N*-(*N′*-*bis*(methylthio)methylideneglycyl)-bornane-10,2-sultam (**191**)[86,136,137]. (2*R*)- and (2*S*)-[$^{14}C_n$]**190** are readily available (Figure 11.63) from singly or doubly labeled acetic acid through conversion into bromo [$^{14}C_n$]acetyl bromide, followed by reaction with lithiated (2*R*)- or (2*S*)-bornane-10,2-sultam and replacement of bromine with benzophenone imine in 60–70% yield by the method described in Figure 11.63[87,135]. The synthesis of [$^{13}C_n$]**191** starts with the esterification of [$^{13}C_n$]glycine. The ester reacts with carbon disulfide–methyl iodide to generate *N*-bis(methylthio)methylideneglycine, which is then coupled to (2*S*)- or (2*R*)-bornane-10,2-sultam to give [$^{13}C_n$]**191** in an overall yield of 69%[136]. Synthons **190** and **191** seem to be equipotent with regard to diastereoselectivities and yields. For application to carbon-14 synthesis, **190** shows some advantages because of economic and practical reasons, so that it is used for both isotopes, whereas **191** has been only used so far for the preparation of carbon-13-labeled materials.

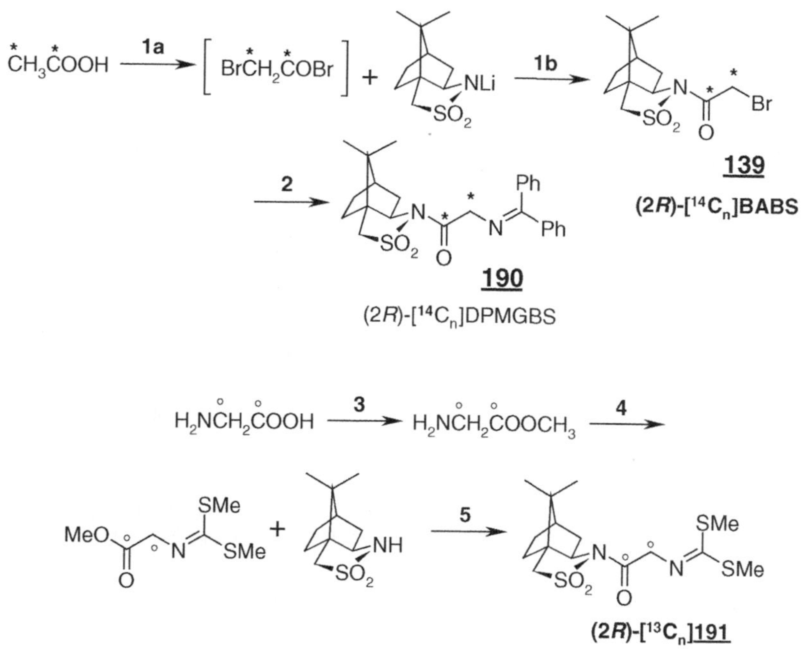


**Reaction conditions:** **1a**. $Br_2$, $PBr_3$, TFAA (cat.); 100–120 °C, 2 h, **b**. THF; −78 °C, 2 h; −78 °C → r.t.; **2**. $Ph_2C$=NH, molecular sieves 3Å, acetonitrile; 70 °C, 3.5 h; **3**. $SOCl_2$, MeOH; reflux; **4**. $CS_2$, MeI; **5a**. $Me_3Al$, toluene; r.t., 15 min, **b**. 50–60 °C, 24 h

***Figure 11.63*** *Procedures for the preparation of (2R)- or (2S)-[$^{14}C_n$]**190** from [$^{14}C_n$]acetic acid and of (2R)- or (2S)-[$^{13}C_n$]**191** from [$^{13}C_n$]glycine*

Deprotonation of (2*S*)- or (2*R*)-[$^{14}C_n$]**190** and [$^{13}C_n$]**191** with *n*-BuLi followed by treatment of the enolates with electrophiles in the presence of DMPU or HMPA gives *N*-substituted $\alpha$-aminoacylsultams in excellent diastereoselectivity (90–98% d.e. crude, ≥99% d.e. after crystallization)[86,134–137]. Suitable electrophiles include nonactivated primary and secondary alkyl iodides, and activated alkenyl and arylmethyl bromides and iodides. Even highly bulky agents such as bromodiphenylmethane and 9-bromofluorene have been shown to react with DPMGBS in 68–72% yields and 95% d.e., when the reactions are performed at ambient temperature[134b]. Enantiopure [$^{13/14}C_n$]-amino acid products are obtained in 50–70% overall yields by hydrolysis of the imine and removal of the auxiliary with LiOH. The synthesis of *N*-Boc-(*S*)-[1,2-$^{14}C_2$]valine (Figure 11.64) at a specific activity of 100 mCi/mmol exemplifies the procedure. Alkylation of (2*R*)-[1,2-$^{14}C_2$]DPMGBS (**190**) with isopropyl iodide gave the protected (2*R*)-[(*S*)-[1,2-$^{14}C_2$])valinyl]sultam **198** in 71% yield and ≥99% d.e. after HPLC purification. Subsequent hydrolytic cleavage of the imine function, blocking of the freed amino group and removal of the auxiliary afforded *N*-Boc-(*S*)-[1,2-$^{14}C_2$]valine (**199**) in about 60% overall yield. This served as a key intermediate for the synthesis of [$^{14}C_2$]valsartan (**200**), a highly efficacious and widely applied antihypertensive drug[135b].

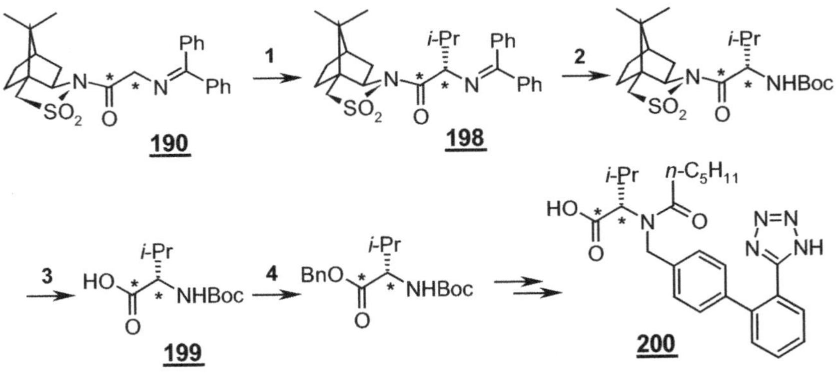


**Reaction conditions: 1a**. *n*-BuLi, THF; −78 °C, 1 h, **b**. DMPU, *i*-PrI, THF; −78 °C, 30 min; 5 °C, 16 h; **2a**. 1 N HCl/THF; r.t., 1 h, **b**. $Boc_2O$, THF; r.t., 16 h; **3**. LiOH, THF; 0 °C, 2 h; **4**. BnOH, EDCI, DMAP (cat), $CH_2Cl_2$; 0 °C, 90 min

***Figure 11.64*** *Use of (2R)-[$^{14}C_2$]DPMGBS for the synthesis of e.p. (S)-[1,2-$^{14}C_2$]valine*

Use of substituted alkyl halides (e.g., $Br(CH_2)_nCOOR$, $I(CH_2)_nCl$) permits the preparation of linear and cyclic neutral, acidic and basic isotopically labeled $\alpha$-amino acids of high enantiomeric purity, as demonstrated by the synthesis of e.p. (*S*)-[2,7-$^{14}C_2$]-pipecolic acid, (*S*)-[2,6-$^{13}C_2$]proline and (*S*)-[1,2-$^{13}C_2$]lysine in excellent yields. Alkylation of (2*R*)-[1,2-$^{14}C_2$]**190** with 1-chloro-4-iodobutane followed by hydrolytic cleavage of the diphenylmethylene group gave 2-amino-6-chlorohexanoyl sultam **201**, which upon treatment with $Bu_4NI$ and *i*-$Pr_2NEt$ cyclized to the pipecolic acid imide **202**. Subsequent cleavage of the auxiliary afforded the free (*S*)-amino acid **203** at a specific activity of 114 mCi/mmol in 45% overall radiochemical yield. It served as a precursor for the biological labeling of rapamycin (**204**) at a specific activity of 60 mCi/mmol[135c].

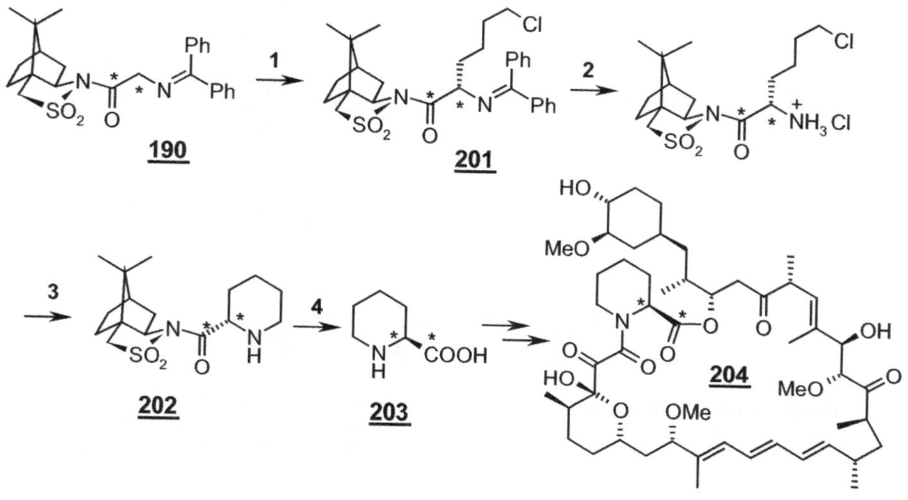

**Reaction conditions:** **1a**. *n*-BuLi, THF; −78 °C, 1 h, **b**. DMPU, **c**. $I(CH_2)_4Cl$, THF; −78 °C → r.t.; r.t., 2 h; **2**. 2 N HCl/THF, r.t., 12 h; **3**. $NBu_4^+ I^-$, *i*-$Pr_2NEt$, $CH_3CN$; 80 °C, 2 h; **4**. LiOH, aq. THF; r.t., 1 h

***Figure 11.65*** *Synthesis of e.p. (S)-2,7-[$^{14}C_2$]pipecolic acid and its use for the biological labeling of [1,2-$^{14}C_2$]rapamycin*

The same route has been applied to synthon (2*R*)-[$^{13}C_2$]**190** and 1-chloro-3-iodopropane, proceeding via **205**, to provide (*S*)-[2,6-$^{13}C_2$]proline in an overall yield of 62%[136a] (Figure 11.66). The preparation of (*S*)-[1,2-$^{13}C_2$]lysine (**207**) diverged from this sequence at **205** by displacement of its chloro substituent with potassium phthalimide, producing **206**. Acid-catalyzed removal of its $=C(SMe)_2$ moiety, cleavage of the auxiliary and saponification of the phthalimide residue furnished **207** in 32% yield[136b].

Additional proteinogenic and nonproteinogenic e.p. labeled $\alpha$-amino acids have been made using building block **192**, which was prepared in $^{13}C_2$-labeled form as illustrated in Figure 11.67. Diastereoselective allylation of **192** followed by replacement of the *N*-protecting groups furnished **208**, an intermediate common to the preparation of (*S*)-[1,2-$^{13}C_2$]aspartic acid (**209**)[138a] and (*S*)-[1,2-$^{13}C_2$]homoserine (**211**)[138a,b]. The former was accessed via oxidative cleavage of the double bond of **208** with ruthenium dioxide–sodium periodate and removal of auxiliary and protecting group. On the other hand, removal of the auxiliary from **208** by transesterification gave protected (*S*)-[1,2-$^{13}C_2$]allylyglycine **210**, from which **211** was prepared in 93% yield by ozonolysis with reductive workup and deprotection. The key intermediate **208** is also available by alkylation of (2*R*)-[$^{13}C_2$]**190** or

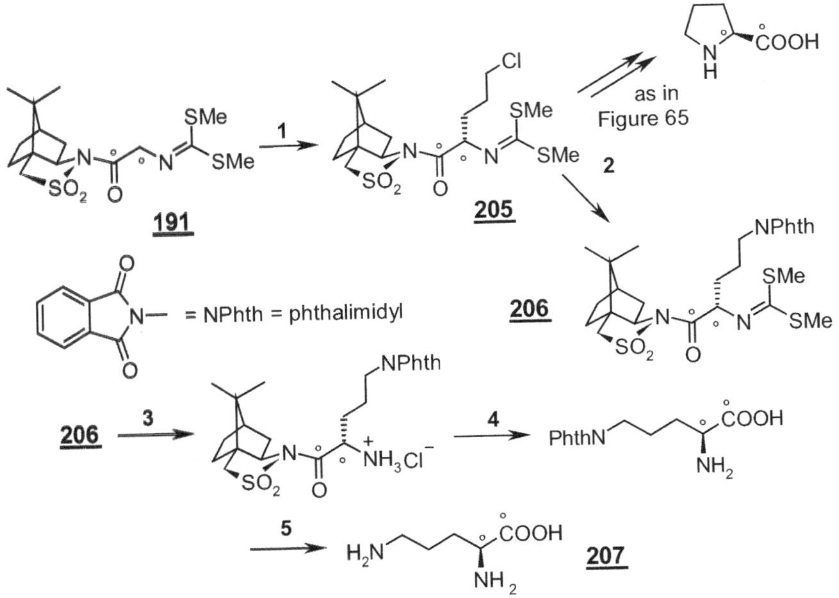


**Reaction conditions: 1a.** *n*-BuLi; THF; −78 °C, 25 min, **b.** HMPA; −78 °C, 15 min, **c.** $I(CH_2)_3Cl$, THF; −78 °C, 2 h; −78 °C → r.t.; **2.** potassium phthalimide, DMF; 80 °C, 12 h; **3.** 1 N HCl, THF; r.t., 25 h; **4.** LiOH, THF; r.t., 1 h; **5.** 3 N HCl; reflux, 3 h

***Figure 11.66*** *Use of (2R)-[$^{13}C_2$]**190** for the preparation of doubly labeled cyclic and basic (S)-α-amino acids of high enantiomeric purity*

(2*S*)-[$^{13}C_n$]**191** with allyl bromide as demonstrated with unlabeled material; these alternatives may be advantageous in the radioactive case, in order to avoid radiation-induced polymerization side reactions.

Since it is usually unnecessary to use a large excess of the alkylating agent to obtain good yields, labeled electrophiles may be employed in the reaction. This results in incorporation of labels at C3 (or more distant positions), as demonstrated with the synthesis of e.p. (*R*)-[3-$^{14}$C]phenylalanine[136a] (Figure 11.68). Also, utilization of two labeled reaction partners furnishes carbon-14-labeled compounds of high specific activity or carbon-13-labeled compounds suitable for use as internal standards in GC/LC-MS assays.

Although bulky alkylating reagents generally give satisfactory results, $Me_3SiCH_2Br$ did not react either with DPMGBS (**190**) or with **191**. The enantiopure reagent glycyl-pseudoephedrine, (+)- or (−)-[$^{14}C_n$]GPE ([$^{14}C_n$]**193**), offered a positive alternative[139,140]. Labeled **193** was readily prepared in at least 40–50% yield (Figure 11.69) by coupling bromo[$^{14}C_n$]acetyl bromide with (+) or (−)-pseudoephedrine, displacement of bromine with azide ion, and catalytic reduction of the azide function. Deprotonation of (+)-[1-$^{14}$C]GPE at −60 °C with slightly less than two equivalents of base (LDA or *n*-BuLi) in the presence of LiCl generated the *O*,*N*-dianion, which upon warming up to −5 °C equilibrated to the thermodynamically more stable enolate. This enolate reacts

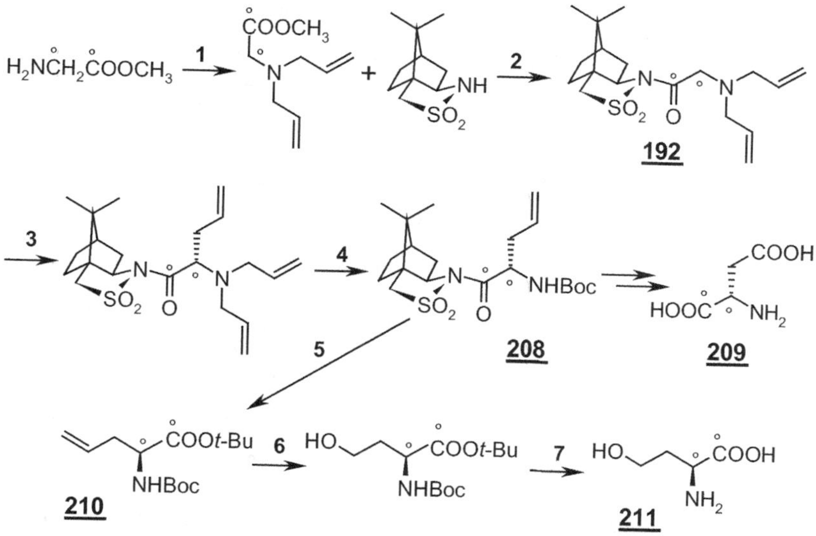

Reaction conditions: 1. $CH_2$=$CHCH_2Br$, *i*-$Pr_2EtN$, toluene; reflux; 2. $AlBr_3$, toluene; 50 °C; 3. $CH_2$=$CHCH_2Br$, $Bu_4NI$, DBU, $K_2CO_3$, DMF; 40 °C; 4a. $Pd(PPh_3)_4$, *N,N*-dimethylbarbituric acid, $CH_2Cl_2$, 35 °C, b. $Boc_2O$, $Et_3N$, 5a. LiOH, $H_2O_2$, aq. THF; r.t., b. *tert*-BuOH, DCC, DMAP, $CH_2Cl_2$;6a. $O_3$, $CH_2Cl_2$,b. DMS, c. $BH_3$–DMS, THF; 7. 5 N HCl, reflux

*Figure 11.67* *Synthesis of (2R)-N-(N′,N′-diallyl[1,2-$^{13}C_2$]glycinyl)bornane-10,2-sultam and application to the synthesis of multiply labeled proteinogenic and nonproteinogenic α-amino acids of high enantiomeric purity*

with alkylating agents to give C-alkylated products exclusively. In the case of $Me_3SiCH_2Br$, yields of 65% and d.e.s >99% were achieved. Removal of the auxiliary with aqueous NaOH and *N*-protection afforded (*S*)-*N*-Boc-trimethylsilyl[1-$^{14}$C]alanine (**212**) in an overall 43% radiochemical yield. It was used as a key intermediate for the

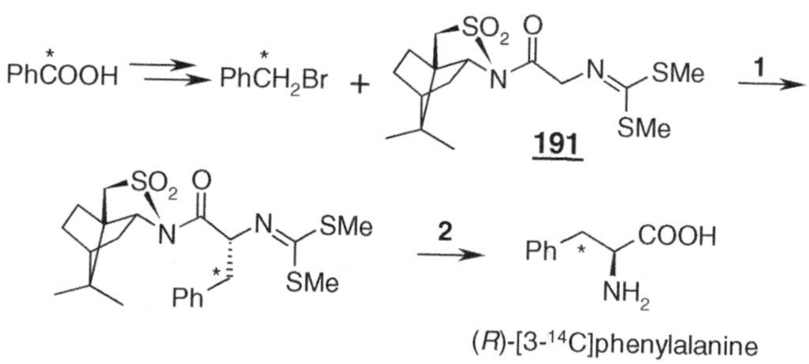

Reaction conditions: 1a. *n*-BuLi, THF; –70 °C, 4 h, b. HMPA, THF; –70 °C, 30 min; –70 °C to r.t.; 2a. 0.75 N HCl, THF; r.t. 16 h, b. LiOH, aq. THF; 0 °C 2.5 h

*Figure 11.68* *Synthesis of (R)-[3-$^{14}$C]phenylalanine by use of a labeled alkylating agent and an unlabeled (2S)-glycinate*

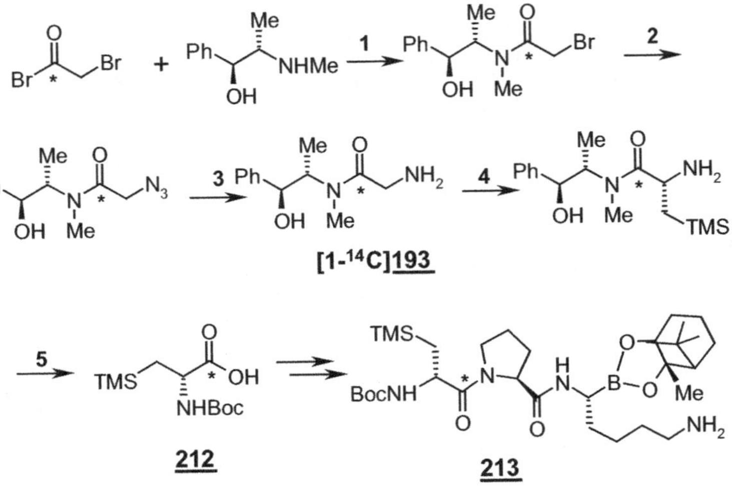


**Reaction conditions:** **1**. $Et_3N$, $CH_2Cl_2$; 0 °C, 1.5 h; **2**. $(Me_2N)_2C{=}NH_2^+\ N_3^-$, $CH_2Cl_2$; 0 °C, 45 min; **3**. $H_2$, 10 % Pd/C, EtOH; r.t., 4 h; **4a** . LDA, LiCl, THF; −60 °C, 15min; −5 °C, 15 min, **b**. $TMSCH_2Br$, THF; r.t., 16 h; **5a**. 1 N NaOH, EtOH; reflux, 4.5 h, **b**. $Boc_2O$, aq. THF; 60 °C, 4 h

***Figure 11.69*** *Synthesis and application of [1-$^{14}$C]glycyl (+)-pseudoephedrine to the synthesis of (S)-N-Boc-trimethylsilyl[1-$^{14}$C]alanine*

preparation of [$^{14}$C]NVP MTH 958 (**213**), a peptidometic antithrombotic drug substance[140].

An alternative linear chiral glycinate, particularly efficacious with allylic and benzylic alkylating agents, is a *N*-bornylideneglycinate (e.g., **194**) derived from camphor. Primary alkylating agents give low d.e.s with this reagent, but secondary alkyl electrophiles also react with good stereoselectivity to give products in reasonable yields and d.e.s[141]. In a recent application[142], deprotonation of *tert*-butyl *N*-(1*R*)-(+)-bornylideneglycinate (**194**) with LDA and reaction with 2-bromo[$^{14}$C]methylnaphthalene gave *tert*-butyl *N*-(1*R*)-(+)-bornylidene-3-naphth-2-yl-(*R*)-[3-$^{14}$C]alanate (**214**) in 57% yield and 58 mCi/mmol after separation from the 9:1 diastereomeric ratio of products (Figure 11.70). The *E/Z* ratio of the starting imine (**194**) is immaterial, as *N,O*-chelation of the enolate with $Li^+$ produces a homochiral complex, so that the diastereoselectivity of the alkylation is solely the result of the facial selectivity of the incoming electrophile. Hydrolysis of the imine with mild acid, *N*-acetylation and ester hydrolysis furnished (*R*)-*N*-acetyl-3-naphth-2-yl[3-$^{14}$C]alanine (**215a**), the key intermediate for the solid-phase synthesis of a carbon-14-labeled decapeptide with utility in shortening drug treatment during *in vitro* fertilization procedures. A triple carbon-14-labeled isotopomer was prepared via reaction under similar conditions of [$^{14}C_2$]-**194** with the labeled bromomethylnaphthalene to afford *tert*-butyl (*R*)-*N*-Fmoc-3-(naphth-2-yl)-[1,2,3-$^{14}C_3$]alanate in 82% yield and thence, with a different *N*-protection, to **215b** having a specific activity of 161 mCi/mmol[142].

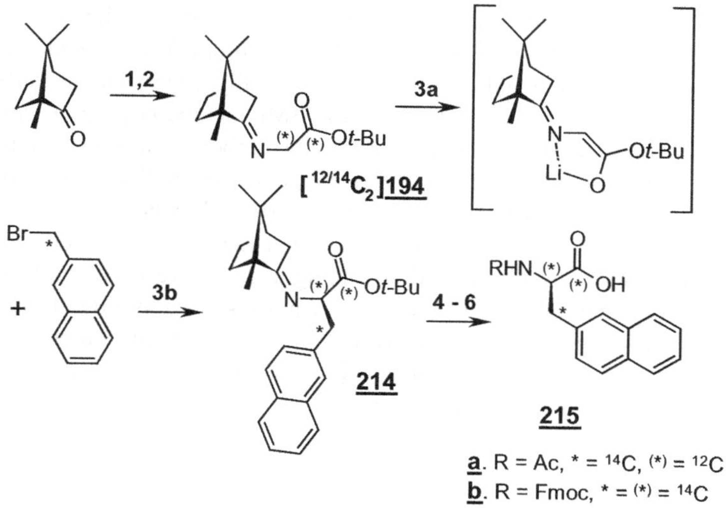


**Reaction conditions:** **1**. $P_2S_5$, $NaHCO_3$, glyme; 90 °C, 1.5 h; **2**. *tert*-butyl glycinate or *tert*-butyl $[^{14}C_2]$glycinate, toluene; reflux, 16 h; **3a**. LDA, THF, HMPA; −78 °C, 15 min, **b**. $Br^{14}CH_2$naphth-2-yl; −78 °C, 3 h; **4**. aq citric acid, THF; 80 °C, 4 h; **5**. $Ac_2O$, pyridine; r.t., 16 h or Fmoc-OSu, $Na_2CO_3$, aq. THF; r.t., 15 min; **6**. TFA, $CH_2Cl_2$; r.t., 2 h

***Figure 11.70*** *Use of the norbornyl auxiliary in the preparation of singly- and multiply-labeled (R)-3-(naphth-2-yl)alanine*

## *α-Alkylation of Cyclic Chiral Glycinates*

Cyclic chiral glycinates such as **195**, **196** and **197** (Figure 11.62), in which both glycine ends are bridged by a one- or two-carbon unit containing the stereogenic center, are powerful alternatives to linear glycinates. The enolates generated by deprotonation of these cyclic reagents are rigid, forming planar or slightly bent conformations in which one of the faces is shielded by the bulky substituent at the stereogenic center, thereby forcing approach of the alkylating agent to the opposite face. Trapping of these enolates with primary nonactivated alkyl iodides and activated arylmethyl bromides or iodides generally affords adducts in chemical yields ranging from 60 to 80% and diastereomeric excesses of 80–98%[143,144]. Problems arise especially for the alkylation of (*R*)- or (*S*)-Boc-BMI (**195**) with secondary alkyl iodides, with which the chemical yields drop down to 27%[145]. In such situations yields can be raised up to 92% by replacing **195** with the corresponding lactim ether, which can readily be prepared by treating **195** with trimethyloxonium tetrafluoroborate[144d].

(*R*)- or (*S*)-Boc-BMI (**195**) is prepared from achiral starting materials (glycine methyl ester, *N*-methylamine, pivalaldehyde). In order to obtain the reagent in enantiomerically pure form, an optical resolution via mandelic acid or chiral HPLC is required, which in practice reduces the yield of each enantiomer to 20% (vs 50% theoretical). Furthermore, harsh reaction conditions are necessary to cleave the imidazolidone system after it has served its purpose. These disadvantages have restricted the applications of **195** to alkylations with primary alkyl iodides and arylmethyl bromides, or iodides that do not possess acid-sensitive functional groups and are readily accessible in carbon-13/14-labeled form. Within these limitations, the Boc-BMI methodology is a valuable one, which has even been used for the preparation of carbon-11-labeled $\alpha$-amino acids, including (*R*)- and (*S*)-[$\beta$-$^{11}$C]alanine (>98% e.e., corrected radiochemical yield 20%), (*R*)- and (*S*)-[$\beta$-$^{11}$C]phenylalanine (>98% e.e., corrected radiochemical yield 75%), *p*-chloro[$\beta$-$^{11}$C]phenylalanine (>98% e.e., corrected radiochemical yield 20%). In all cases published so far, only C1-labeled alkylating agents have been employed[146]. The use of the Boc-BMI methodology in carbon-14 synthesis is exemplified in Figure 11.71 by the preparation of e.p. (*R*)-[3-$^{14}$C]phenylalanine, which served as a precursor for the solid phase synthesis of [$^{14}$C]octreotride (**216**), a cyclic octapeptide used for the treatment of gastrointestinal endocrine tumors[143].

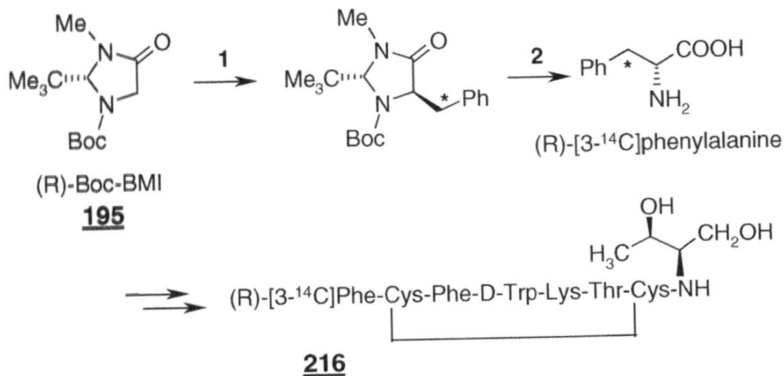


**Reaction conditions: 1a**. LDA, THF; −78 °C, 30 min, **b**. **Ph$^{14}$CH$_2$Br**, THF; −78 °C, 2 h; −78 °C to r.t., 16 h; **2a**. TFA, $CH_2Cl_2$; r.t., 3 h, **b**. Dowex *50Wx8*, H$^+$-form; 105 °C, 24 h

***Figure 11.71*** *Use of (R)-Boc-BMI for the synthesis of (S)-[3-$^{14}$C]phenylalanine*

In contrast to their acyclic counterparts, Boc-BMI (**195**) and Schöllkopf's lactim ether (**196**) are useful for the preparation of labeled $\alpha,\alpha$-dialkylated amino acids of high enantiomeric purity. This is exemplified by the preparation of (*S*)-$\alpha$-methyl-*p*-chloro-[$\beta$-$^{11}$C]phenylalanine of >98% e.e. (Figure 11.72)[146b].

Schöllkopf's cyclo-[(*S*)- or (*R*)-ValGly] bislactim ether (**196**) is prepared in four steps from glycine ethyl ester. The latter reacts with Z-(*R*) or (*S*)-valine to give Z-(*R*)/(*S*)-Val-Gly-OMe (**217**), which upon cleavage of the Z-group and heating cyclizes to form the corresponding diketopiperazine **218**. Subsequent treatment of **218** with trimethyloxonium tetrafluoroborate affords **196**. Since yield and number of steps for preparation of **196** are often similar to those generally needed for the preparation of labeled alkylating agents, the bislactim ether is often employed as a labeled reaction partner. This is illustrated for (*S*)-[1,2-$^{13}C_2$]leucine in Figure 11.73[147b].

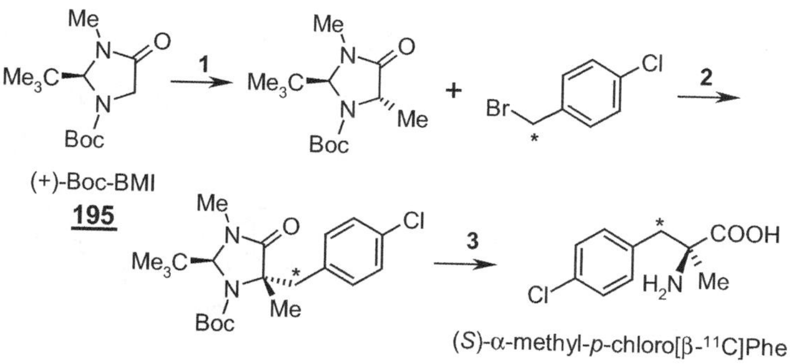


**Reaction conditions: 1a**. LDA, THF; −78 °C, 30 min, **b**. MeI, THF; −78 °C, 1 h; −78 °C → r.t.; **2a**. LDA, THF; −78 °C, 20 min, **b**. ***p*-ClPh$^{11}$CH$_2$Br**, THF; −78 °C, 3 min; **3**. 57 % HI; 200 °C, 7 min

*Figure 11.72* *Use of Boc-BMI for the synthesis of an e.p α,α-dialkyl-α-[$^{14}$C]amino acid*

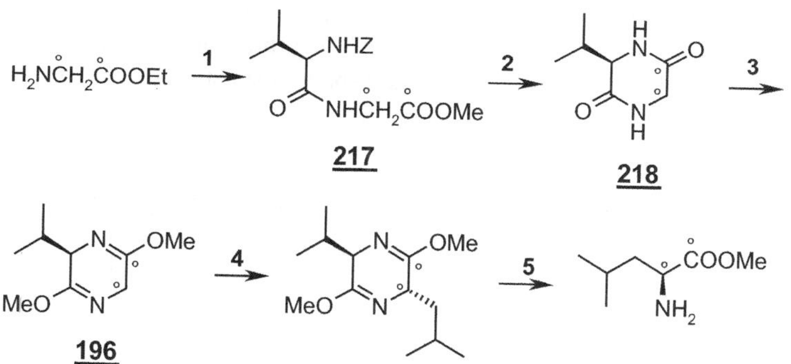


**Reaction conditions: 1**. Z-(*R*)-Val, TBTU, Et$_3$N, dioxane, 0 °C, 1 h; r.t., 13 h; **2a**. H$_2$, 5 % Pd/C, EtOH, **b**. toluene; reflux, 7 d; **3**. Me$_3$OBF$_4$, CH$_2$Cl$_2$; 40 °C, 72 h; **4a**. BuLi, THF; 78 °C, 15 min, **b**. (CH$_3$)$_2$CHCH$_2$I; THF; −78 °C, 5 h; **6**. 0.25 N HCl; r.t., 12 h

*Figure 11.73* *Synthesis and application of Schöllkopf's cyclo-(R)-Val-[1,2-$^{13}$C$_2$]Gly) bislactim ether to the synthesis of (S)-[1,2-$^{13}$C$_2$]leucine of high enantiomeric purity*

In carbon-14 synthesis, Schöllkopf's methodology has been exploited for the synthesis of e.p. (*S*)-*p*-methoxy[*β*-$^{14}$C]tyrosine (**220**) (Figure 11.74). Alkylation of (*R*)-**196** with 4-methoxy[*α*-$^{14}$C]benzyl bromide furnished **219** in 83% yield and >95% d.e. Sequential acid catalyzed cleavage of the ring of **219** and of the resultant methyl ester gave e.p. **220** in 28% overall radiochemical yield. This amino acid was used in the synthesis of [$^{14}$C]PD 0326234 (**221**)[147a].

Isotopic syntheses using Dellaria's glycinate (**197**) have been reported only for a few carbon-13-labeled *α*-amino acids, despite the relative ease of preparing enantiopure isotopically labeled **197** and the excellent diastereoselectivities usually achieved, even with low-molecular-weight alkylating agents (e.g., MeI gives >98% d.e.)[148,149].

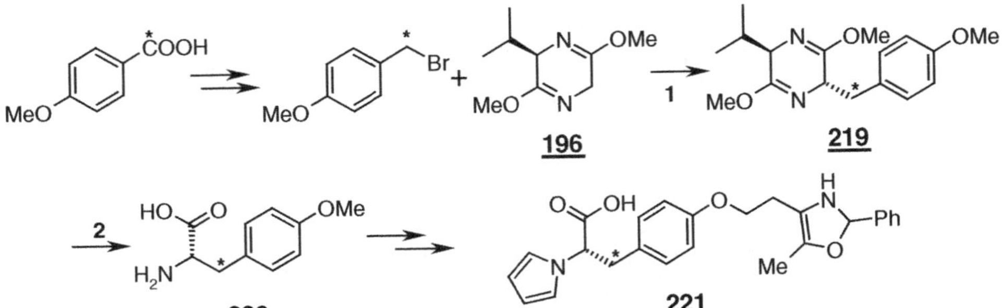


**Reaction conditions: 1a**. *n*-BuLi, THF; −78 °C, 15 min, **b. *p*-MeOPh$^{14}$CH$_2$Br**, THF; −78 °C, 5 h; **2a**. 0.25 N HCl; r.t., 12 h, **b**. 40 % HBr; reflux, 2 h

***Figure 11.74*** *Application of Schöllkopf's methodology to the preparation of e.p. (R)-p-methoxy[β-$^{14}$C]tyrosine of > 98% e.e*

The requirement for 75–90 psi $H_2$ pressures for the ring cleavage may have contributed to this lack of use. As for Seebach's and Schöllkopf's synthons, alkylating agents attack the enolate from the less hindered face *anti* to the 5-phenyl substituent.

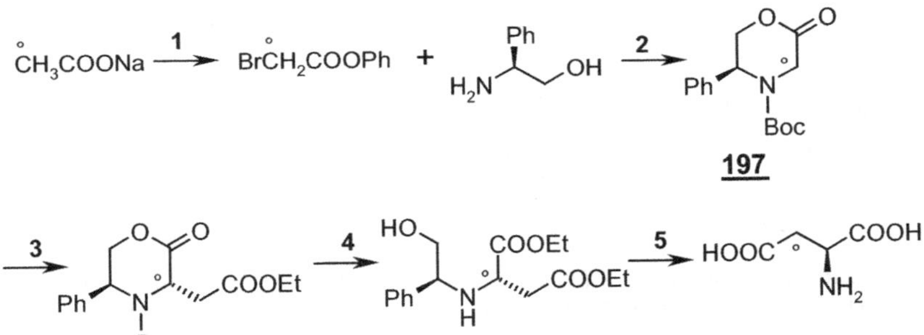


**Reaction conditions: 1a**. PhCOOH, PhCOBr, 120 °C, 16 h, **b**. Br$_2$; r.t., 5 min; reflux, 5 h, **c**. PhOH, CH$_2$Cl$_2$; r.t., 1.5 d; **2a**. (*S*)-2-phenylgycinol, *i*-Pr$_2$EtN, acetonitrile; r.t., 5 h, **b**. Boc$_2$O, acetonitrile; r.t., 5 h; **3a**. LiHMDS, THF; −78 °C, 1.5 h, **b**. BrCH$_2$COOEt, THF; −78 °C, 1.5 h; **4**. HCl, EtOH; reflux 5 h; **5a**. H$_2$ (75 psi), 10 % Pd/C, EtOH; r.t.; 1.5 d, **b**. 6 N HCl, 85 °C, 5 h

***Figure 11.75*** *Synthesis of Dellaria's synthon labeled in the backbone and application to the synthesis of e.p. (S)-[3-$^{13}$C]aspartic acid*

The preparation of (*S*)-[3-$^{13}$C]aspartic acid illustrates the preparation and use of this cyclic glycinate. (*S*)-[$^{13}$C]<u>**197**</u> was prepared by alkylation of 2-phenylglycinol with phenyl bromo[2-$^{13}$C]acetate, readily available in two steps from sodium [2-$^{13}$C]acetate, and heating of the resultant reaction mixture with Boc$_2$O. The overall yield from sodium [$^{13}$C]acetate was 35–40%[148c]. Deprotonation of (*S*)-[$^{13}$C]<u>**197**</u> with LiHMDS and trapping of the enolate with ethyl bromoacetate gave <u>**222**</u> in 87% yield. In general, nonactivated alkyl iodides and activated allyl and arylmethyl bromides provide primary products in 65–90% yields and diastereomeric excesses of >95%[148a,b,149]. Release of the free e.p. $\alpha$-amino acids is accomplished by acidic hydrolysis and catalytic hydrogenolytic cleavage of 2-phenylethanol. As is evident in the above reaction scheme, Dellaria's methodology opens a convenient pathway to triply labeled $\alpha$-amino acids, since doubly labeled sodium acetate and C1-labeled alkylating agents labeled are routinely accessible and in many cases commercially available.

### 11.3.7 Aldol Reactions of Chiral Glycinates

The stannous enolates of Evans' (*R*)- and (*S*)-4-benzyl-3-(isothiocyanoacetyl)oxazolidin-2-ones **223** (Figure 11.76) react with aliphatic and aromatic aldehydes to afford 5-substituted 2-thioxo-oxazolidine-4-carboxylates, which are readily hydrolyzed to give *syn*-(2*S*,3*R*)- and (2*R*,3*S*)-$\beta$-hydroxy-$\alpha$-amino acids, respectively, in fairly good yields (50–70%) and excellent stereoselectivities (90–99% d.e.)[94,150]. Although **223** is readily accessible from the respective chloroacetyl precursor through treatment with sodium azide and transformation of the resultant azide with $Ph_3P$ and $CS_2$, the less commonly used ethyl (1*R*,2*R*,5*R*)-((2-hydroxypinan-3-ylene)amino)acetate (**224**) and its (+)-antipode are the only linear chiral glycinate exploited so far in aldol reactions in isotope chemistry[151].

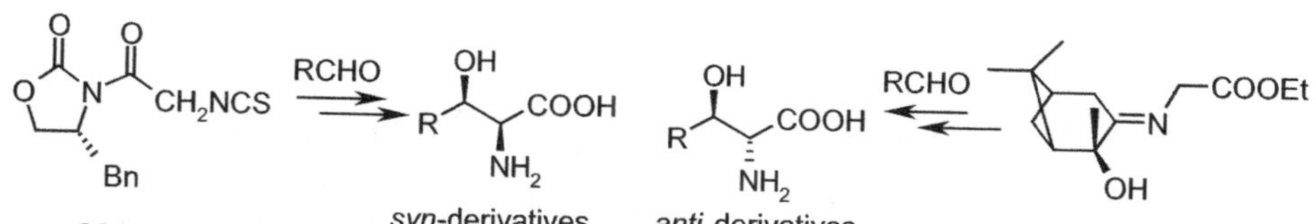


*Figure 11.76* *Acyclic chiral glycinates* **223,224** *appropriate for the preparation of e.p. $\beta$-hydroxy-$\alpha$-amino acids*

Reagent **224** is readily prepared by refluxing ethyl glycinate and the desired antipodal 2-hydroxypinan-3-one in benzene in the presence of $BF_3$-etherate. The titanium enolate, generated by treatment with $ClTi(OEt)_3$ in the presence of triethylamine, reacts with aldehydes to give (2*R*,3*R*)- and (2*S*,3*S*)-$\beta$-hydroxy-$\alpha$-amino acyl derivatives, which upon hydrolytic cleavage of the auxiliary furnish the free acids of the *allo*-threonine type in excellent stereoselectivities (>98% e.e.)[151]. As illustrated in Figure 11.77, application of this strategy to (2*E*)-2-[1-$^2$H]hexadecenal (**225**) furnished the (2*R*,3*R*)-adduct **226** in 70% yield after MPLC separation. Hydrolytic cleavage of the auxiliary, followed by reduction of the resulting ester **227** with $NaB^2H_4$ converted the initial adduct into D-erythro-[1,1,3-$^2H_3$]-sphingosine (**228**). Use of $NaB^3H_4$ as a reducing agent afforded the respective [1,1-$^3H_2$,3-$^2$H]analog[152].

Alternative routes to labeled *anti*-$\beta$-hydroxy-$\alpha$-amino acids of the *allo*-threonine type are described in Section 11.3.9. They involve aldol reactions of haloacetyl-Evans and -Oppolzer auxiliaries to give *syn*-$\beta$-hydroxy-$\alpha$-halo derivatives, whose $\alpha$-halo groups are then inverted by nucleophilic displacement with azide ion[153].

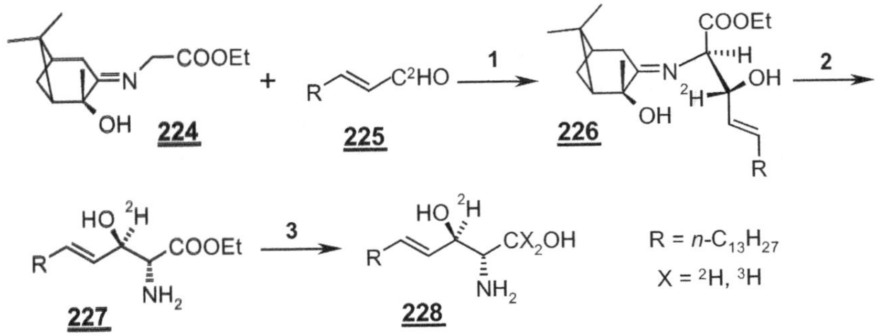


**Reaction conditions:** **1**. ClTi(OEt)$_3$, Et$_3$N, CH$_2$Cl$_2$; 0 °C, 4 h; **2**. 1 N HCl, THF; r.t.; **3**. NaBX$_4$ (X= $^2$H, $^3$H) , aq. EtOH; 0 °C, 3 d

*Figure 11.77* *Use of the pinanyl chiral glycinate for the synthesis of e.p. labeled β-hydroxy-α-amino acids of the allo-threonine type and its application to the synthesis of labeled sphingosine*

In contrast to the linear glycinates discussed above, the utility of lithium or titanium enolates of the cyclic glycinates **229**, **230** and **231** for the preparation of e.p. isotopically labeled β-hydroxy-α-amino acids of the threonine type has been documented by several examples[126,154,155].

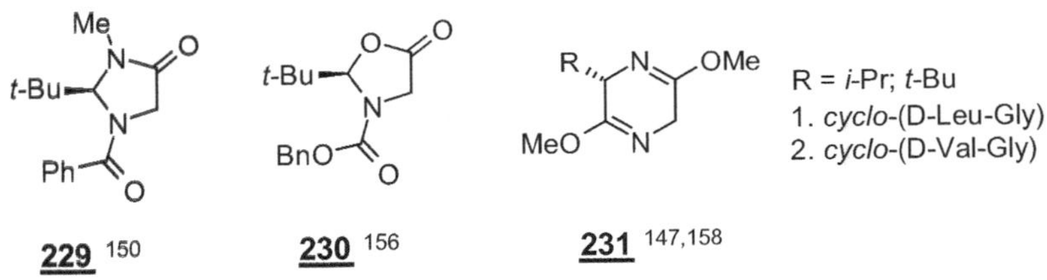


*Figure 11.78* *Homochiral cyclic glycinates for the preparation of labeled β-hydroxy-α-amino acids of high enantiomeric purity*

The lithium enolate of (*R*)- or (*S*)-1-benzoyl-2-*tert*-butyl-3-methylimidazolidin-4-one (Seebach's reagent, **229**), for example, furnishes the respective primary aldol adducts in excellent diastereoselectivities (d.e. >90%). Their hydrolysis to the free acids, however, requires such drastic reaction conditions (6N HCl, 150 °C, sealed tube) that these building blocks are of very limited use. For this reason the corresponding (*R*)- and (*S*)-1,3-oxazolidin-5-ones **230** are recommended as more useful alternatives. In their case hydrolysis of the primary adducts can be accomplished under very mild reaction conditions, namely hydrogenolytic removal of benzyloxycarbonyl function followed by hydrolytic cleavage of the heterocyclic system with LiOH/aq. THF or 1N HCl at room temperature[155].

This procedure has been successfully applied to the synthesis of [$^{14}$C]SM-11044 (**235**), a β-adrenoceptor agonist[156]. The primary adduct **233**, obtained from reaction of the enolate of **230** and 3,4-methylenedioxybenz[$^{14}$C]aldehyde (**232**), was cleaved following the recommended sequence to furnish (2*S*,3*R*)-2-amino-3-(3,4-methylenedioxyphenyl)-3-hydroxy[3-$^{14}$C]propionic acid (**234**) in 61% radiochemical yield and >99% enantiomeric purity. In a parallel set of experiments, it was found that acidic hydrolysis of the primary adduct obtained from **232** and Seebach's reagent **229** with refluxing conc. HCl proved too destructive, giving only trace amounts of **234**. Reversible *N*-protection, coupling of the carboxyl group to pyrrolidine, re-liberation of the amino function followed by reductive *N*-alkylation through treatment with *β*-aryl substituted propanal–$NaBH_3CN$ and concluding *O*-deprotection provided [$^{14}$C]SM-11044 (**235**) in an overall radiochemical yield of 11% from barium [$^{14}$C]carbonate.

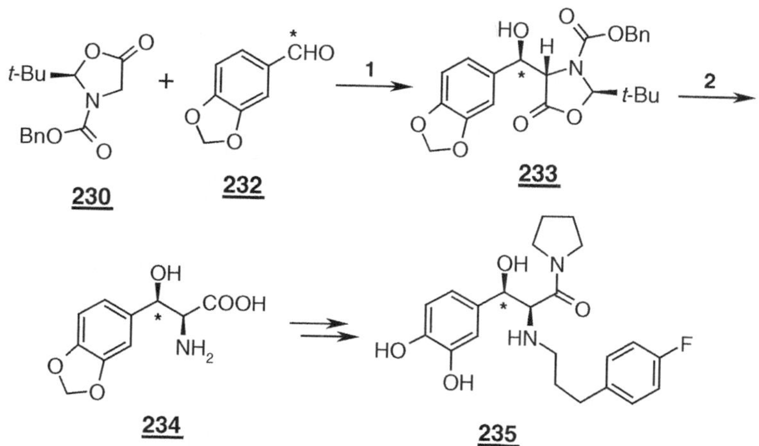


**Reaction conditions: 1a .230**, LiHMDS, THF/hexane; −78 °C, 1 h, **b. 232**, THF; −78 °C, 3.5 h (66 %); **2a.** $H_2$, 10 % Pd/C; 3.5 h, **b.** 1 N HCl; r.t., 1 h (99 %)

***Figure 11.79*** *Use of glycinate **230** for the synthesis of e.p. (2S,3R)-2-amino-3-(3,4-methylenedioxyphenyl)-3-hydroxy[3-$^{14}$C]propionic acid*

For success with this method two issues[155] must be taken into account. First, the lithium enolate of **230** is rather unstable in THF and can only be generated with LiHMDS at −78 °C. Deprotonation with LDA, NaHMDS or KHMDS leads to decomposition. Second, aliphatic aldehydes give mixtures of **233** and bicyclic carbamates. Upon hydrolysis both components afford the same *β*-hydroxy-*α*-amino acid without loss of the enantiomeric purity.

Schöllkopf's bislactim ether of *cyclo*-((*R*)-Val-Gly) (**231**) has been applied to the synthesis of (2*S*,3*R*)-[3,4-$^{13}C_2$]threonine and (2*S*)-[3-$^{13}$C]serine of high enantiomeric purity[147c,157]. In the first case the titanium enolate must be employed in order to achieve sufficient *threo* selectivity (>92%). This was readily generated through transmetallation of the lithium enolate with $ClTi(NEt_2)_3$. Treatment of the titanium enolate with [1,2-$^{13}C_2$]-acetaldehyde and chromatographic removal of traces of the diastereomer furnished the major adduct **236** in 58% yield. Hydrolysis of **236** produced a mixture of (2*S*,3*R*)-[3,4-$^{13}C_2$]-threonine (**237**) and valine. Separation of the desired product using cationic ion exchange chromatography afforded the desired doubly labeled *β*-hydroxy-*α*-amino acid in enantiomerically pure form with no detectable amounts of *allo*-threonine.

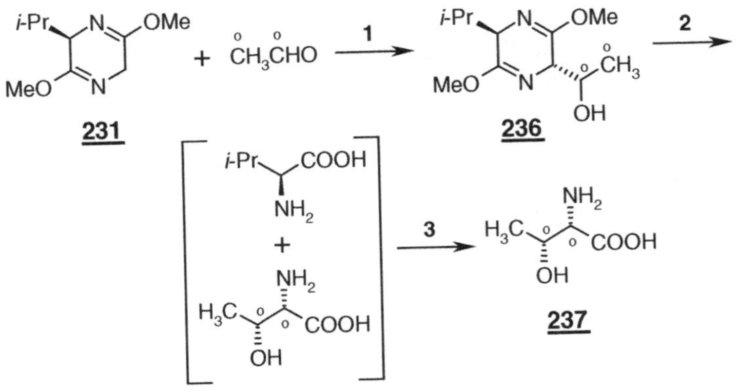


**Reaction conditions: 1a** . *n*-BuLi, THF; –70 °C, 15 min, **b**. ClTi(NEt$_2$)$_3$, hexane; –70 °C, 45 min, **c. [1,2-$^{13}$C$_2$]acetaldehyde**, THF; –70 °C, 16 h; **2a**. 0.25 N HCl; r.t., 1 h, **b**. conc. HCl, reflux, 4.5 h; **3a**. separation via cationic ion exchange (AG-50 W X8-resin), 0.1 N HCl, **b**. 1,2-epoxybutane; reflux, 1 h

***Figure 11.80*** *Use of Schöllkopf's bislactim ether for the synthesis of (2S,3R)-[3,4-$^{13}$C$_2$] threonine of high enantiomeric purity*

### 11.3.8 Aldol Reactions of Chiral Glycolates

The aldol reactions of *O*-protected (*R*)- or (*S*)-hydroxyacetyl 4-substituted 1,3-oxazolidin-2-ones (*O*-protected chiral glycolates) show some peculiarities. The boron enolates of (*S*)- and (*R*)-3-*p*-methoxybenzyloxyacetyl-4-benzyl-1,3-oxazolidin-2-one (238) react with aldehydes to provide the expected *syn* aldols[96a,158,159]. In contrast, *anti* aldols are formed in high stereoselectivities with the corresponding stannous enolates when the reactions are run in the presence of TMEDA[160].

So far, chiral glycolates have not been extensively used in isotope chemistry. Only the reaction of the glycolate 238 with the aldehyde fragment 239 has been reported (Figure 11.81). Glycolate 238 is readily accessible through reaction of sodium bromo-[2-$^{13}$C]acetate with sodium 4-methoxybenzoxide followed by coupling of the resultant *O*-protected [2-$^{13}$C]glycolic acid to the auxiliary by means of the mixed anhydride method (Me$_3$CCOCl, Et$_3$N; see Section 11.3.2). Deprotonation of 238 with triethylamine and di-*n*-butylboron triflate, followed by addition of 239 furnished the expected (2*S*,3*R*)-aldol 240 in 90% yield. The aldol was converted via a previously used sequence to [$^{13}$C]FK 506 (241), a potent immunosuppressive agent, labeled with carbon-13 at C9[161].

### 11.3.9 Aldol Reactions of Chiral Haloacetates

Chiral haloacetates of the Evans and Oppolzer type are highly potent and versatile synthons for the preparation of a broad variety of e.p. intermediates. The boron enolates of Evans' haloacetates and the titanium enolates of Oppolzer's derivatives undergo aldol additions with aliphatic and aromatic aldehydes to provide *syn* 2-halo-3-hydroxy carboxylic acid derivatives with excellent diastereoselectivities, generally exceeding *syn*/*anti* ratios >50:1[153,162,163]. In the Oppolzer series, only with acetaldehyde (diasteromeric ratio

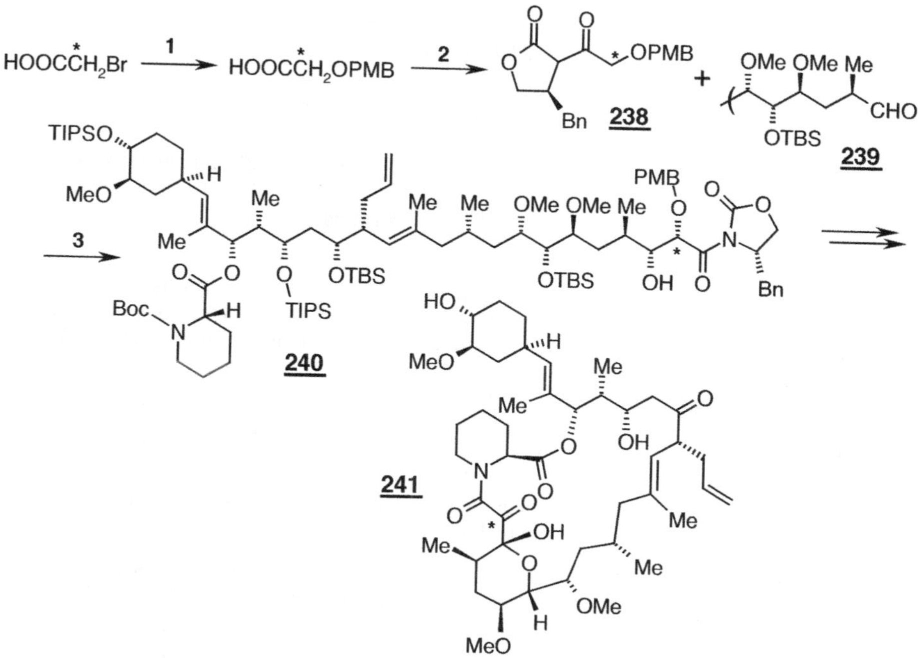


Reaction conditions: 1. *p*-MeOPhCH$_2$OH, NaH, toluene; 90 °C, 4 h; 2a. Me$_3$CCOCl, Et$_3$N, THF; 0 °C, b. (*S*)-4-benzyl-2-oxazolidinone, *n*-BuLi, THF; –78 °C→0 °C; 3. Et$_3$N, *n*-Bu$_2$BOTf, toluene, –50 °C, 4 h

***Figure 11.81*** *Use of a chiral [$^{13}$C]glycolate synthon for the preparation of an isotopically labeled syn aldol*

97:3) and formaldehyde (3:1) have lower values been observed[87,163]. Purification of the reaction mixtures by flash chromatography or (in the Oppolzer series) simply by crystallization provides d.p. materials in 50–60% (in the Evans series) and 70–95% yield (Oppolzer series). Subsequent chemical transformations allow the preparation of e.p. $\beta$-hydroxy acids, *anti* $\alpha$-substituted $\beta$-hydroxy acids or *cis* 2,3-epoxy esters. These options are summarized in Figure 11.82.

The easy accessibility of haloacetates labeled at C1 and/or C2 and aldehydes labeled at the carbonyl group makes it feasible to selectively prepare adducts labeled at any or all of three different positions, all using the same procedures. However, in carbon-14 chemistry, the synthetic potential of isotopically labeled haloacetates and haloaldol adducts has been exploited so far only for Oppolzer's auxiliaries, and only some of the potential options have been executed so far[87]. The preparation of the bromo[$^{14}C_n$]acetyl (+) and (−)-bornane-10,2-sultams ((2*S*)/(2*R*)-[$^{14}C_n$]BABS, $n = 1,2$; **139**) as depicted in Figure 11.83 is also described in Section 11.3.6.

Both antipodes are obtained in >70% yield upon heating sodium [1-$^{14}$C]- and [1,2-$^{14}C_2$]-acetate with a mixture of $Br_2$ and $PBr_3$ followed by vacuum transfer of the resulting bromo[$^{14}C_n$]acetyl bromide into a suspension of the lithiated auxiliary. The requisite

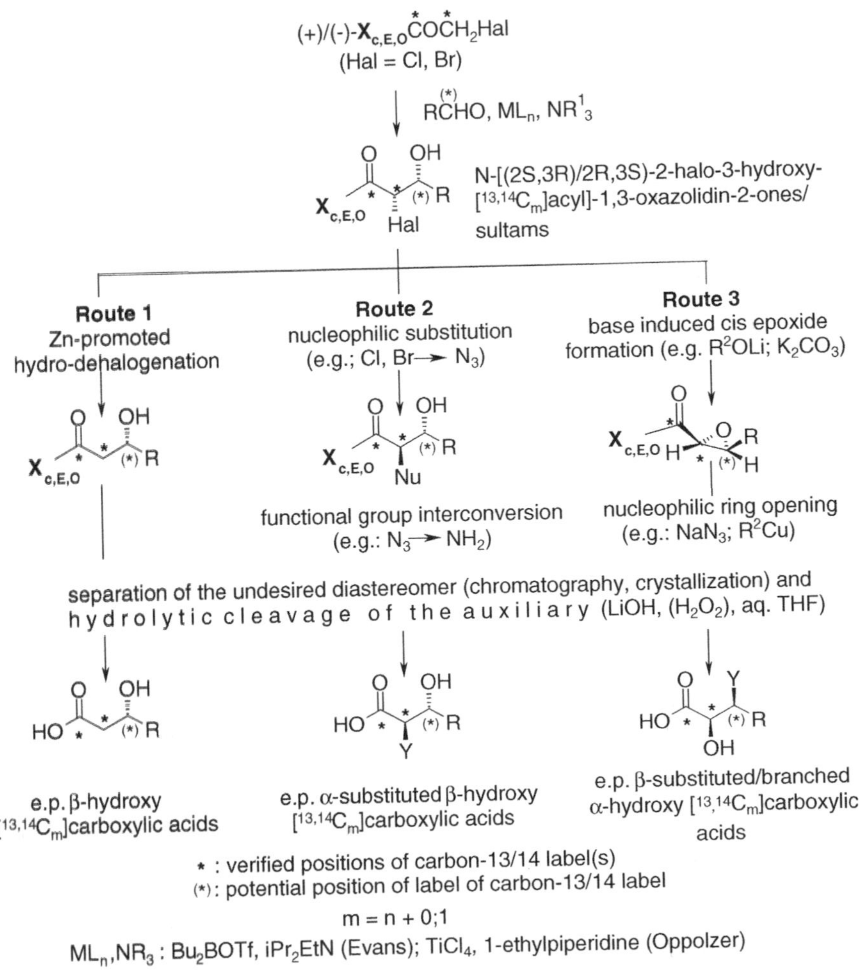


***Figure 11.82*** *Labeling options and potential applications of d.p. syn-2-halo-3-hydroxy[$^{14}C_n$] acyl conjugates of Evans and Oppolzer chiral auxiliaries accessible from [$^{14}C_n$]haloacetates*

[$^{14}$C$_n$]-haloaldol adducts are generated by sequential addition of 1-ethylpiperidine, $TiCl_4$ and the respective aldehyde to the −78 °C cold solution of (+)/(−)-[$^{14}$C$_n$]BABS in dichloromethane[87,163,164] (Figure 11.83).

As elaborated in Sections 11.3.6 and 11.3.8, the spectrum of valuable labeled building blocks, such as α-amino acids, β-hydroxy α-amino acids of the *allo*-threonine type, chiral (*E*)-enoyl imides, is significantly broadened by replacing the halogen in carbon-13/14-labeled (2*S*)/(2*R*)-BABS with various functionalities (e.g. –N=CPh$_2$, –NCS, –PO(OEt)$_2$). In addition, replacement of the bromo group subsequent to aldolization provides further options:

(A) The option to reductively dehalogenate chiral α-halo-β-hydroxy imides, resulting from aldol reactions with chiral bromoacetates, offers an alternative route to chiral

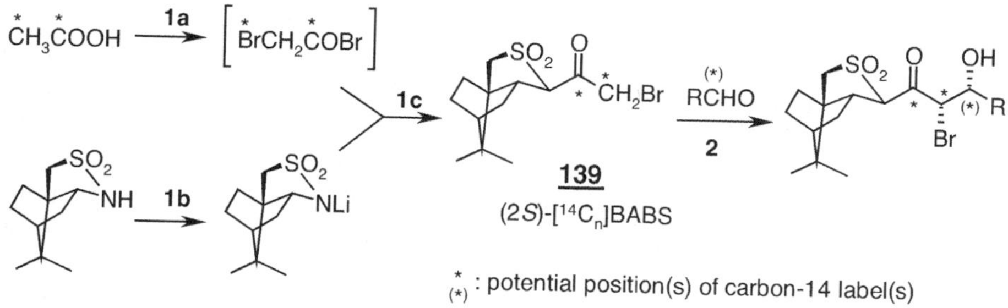


**Reaction conditions:** **1a** $PBr_3$, $Br_2$, TFAA (trace); 100–120 °C, 2 h, **b**. *n*-BuLi, THF; −78 °C, 45min, **c (a+b)**. −78 °C, 2 h, then −78 °C to 0 °C, 1 h; **2a**. $TiCl_4$, 1-ethylpiperidine, dichloromethane,−78 °C, 5 min, **b**. RCHO; −78 °C, 10–15 min

***Figure 11.83*** *Synthesis of (2S)-[1,2-$^{14}C_n$]BABS and its application to the synthesis of d.p. singly/multiply labeled syn bromo[1,2-$^{14}C_2$]aldols*

β-hydroxyacids in cases where poor diastereoselectivities are obtained with the use of unsubstituted chiral acetates. See the synthesis of (*R*)-3-hydroxy[1-$^{14}$C]myristic acid, Section 10.3.2[87]

(B) Displacement of bromo by azido in the bromoaldol adduct (**242**), obtained from (2*S*)-[1,2-$^{14}C_2$]BABS (**139**) and formaldehyde, followed by catalytic reduction of

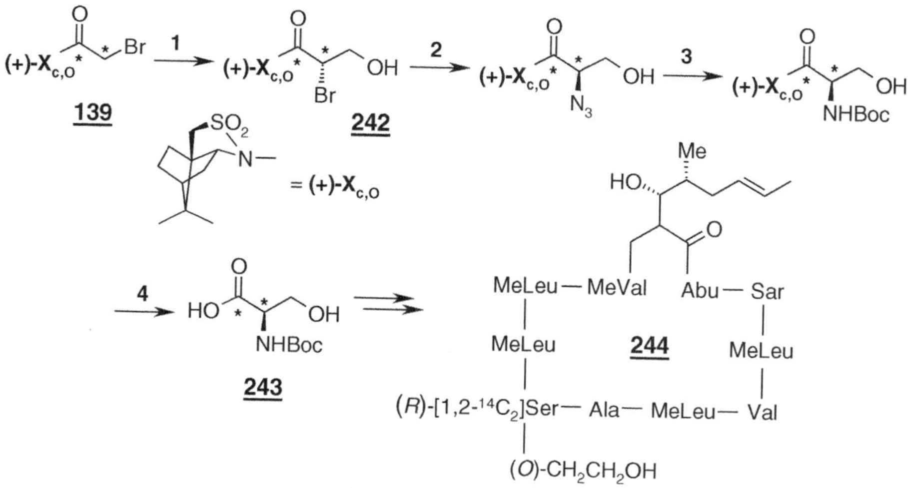


**Reaction conditions:** **1a**. $TiCl_4$, 1-ethylpiperidine, $CH_2Cl_2$; −78 °C, 5 min; **b**. HCHO, −78 °C, 1 h; −50 °C, 0.5 h; **2**. $NaN_3$, DMSO; 40 °C, 4.5 h; **3a**. $H_2$, 10 % Pd/C, MeOH/THF; r.t., 2.5 h, **b**. $Boc_2O$, aq. *tert*-BuOH; r.t., 4.5 h; **4**. LiOH, aq. THF; r.t., 2.5 h

***Figure 11.84*** *Preparation of (R)-N-Boc-[1,2-$^{14}C_2$]serine and its application to the synthesis of [(R)-[1,2-$^{14}C_2$]Ser$^8$]NVP IMM 125*

the azide function, protection of the $NH_2$ group and hydrolytic cleavage of the auxiliary, provided e.p. (*R*)-*N*-Boc-[1,2-$^{14}C_2$]serine (**243**) in 46% overall radiochemical yield. Despite the unsatisfactory stereoselectivity of the aldol reaction in this case (d.e. ~75%), this procedure was superior to all others investigated, especially since the undesired diastereomer could be readily separated by flash chromatography. Compound **243** was used as a building block for the preparation of [$^{14}$C]NVP IMM 125 (**244**), a cyclosporin A derivative showing antiasthmatic activities[87].

(C) This alternative involves the base-promoted conversion of d.p. haloaldol adducts to *cis*-2,3-epoxy carboxylic esters, followed by regio- and stereoselective opening of the epoxide ring to afford e.p. $\beta$-substituted-$\alpha$-hydroxy acid derivatives[162b,163]. In the example depicted in Figure 11.85, (2*S*,3*R*)-2-bromo-3-hydroxy-3-phenyl[1,2-$^{14}C_2$]-propionyl sultam (**245**), accessed from the titanium enolate of (2*S*)-[1,2-$^{14}C_2$]BABS and benzaldehyde in 78% chemical yield and >98% d.e., provided (2*R*,3*R*)-3-phenyl-[1,2-$^{14}C_2$]oxirane-2-carboxylic acid benzyl ester (**246**) when treated with base followed by cleavage of the auxiliary. Regioselective opening of the epoxide ring with $NaN_3$, protection of the hydroxy group and reduction of the azide function converted **246** into benzyl (2*R*,3*S*)-*O*-*tert*-butyldimethylsilyl phenyl[1,2-$^{14}C_2$]isoserinate (**247**). This intermediate was used as indicated for the preparation of the side chain which, when attached to the baccatin III system, produced paclitaxel (**248**), a highly efficacious anti-cancer drug, doubly labeled with carbon-14 in its phenylisoserine side chain[164].

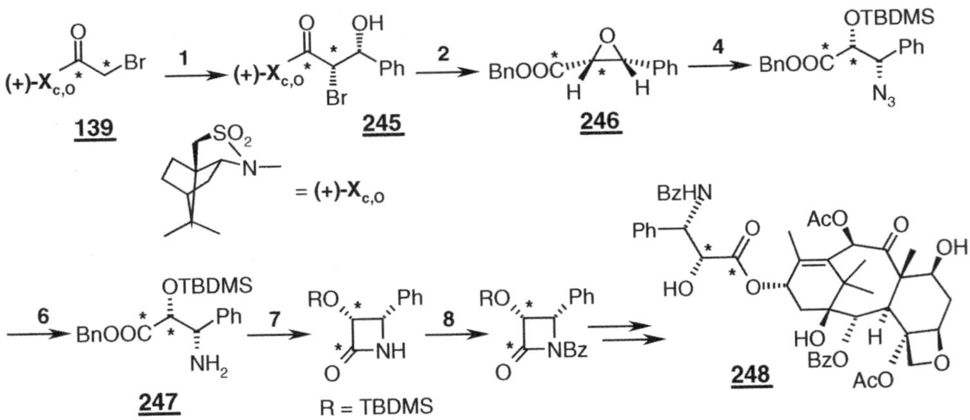


**Reaction conditions: 1a**. $TiCl_4$, 1-ethylpiperidine, $CH_2Cl_2$; –78 °C, 5 min, **b**. PhCHO; –78 °C, 10 min; **2**. $K_2CO_3$, DMF (trace $H_2O$), **3**. BnOLi, THF; –10 °C, 4 h; **4**. $NaN_3$, HCOOMe, aq. MeOH; 50 °C, 24 h, **5**. TBDMSOTf, 2,6-lutidine, $CH_2Cl_2$; 0 °C to r.t., 1 h; **6**. $PPh_3$, aq. THF; 60 °C, 3 h; **7**. *tert*-BuMgCl, ether; –10 °C, 2 h; **8**. BzCl, DMA (cat.), $Et_3N$, $CH_2Cl_2$; –10 °C, 1.5 h

***Figure 11.85*** *Diastereoselective synthesis of an e.p. (2R,3S)-β-hydroxy-α-amino[1,2-$^{14}C_2$]-carboxylic acid from (2S)-[$^{14}C_2$]BABS via a (2R,3R)-oxirane[1,2-$^{14}C_2$]carboxylic acid ester*

### 11.3.10 Reactions on Chiral α,β-Unsaturated Imides and Esters

Auxiliary-controlled diastereoselective reactions on chiral α,β-unsaturated imides and esters linked to Evans', Oppolzer's or Helmchen's auxiliaries (e.g., **249**) significantly expand the pool of e.p. building blocks[121]. For example, 1,4-addition of organocuprates or Grignard reagents (giving **250**) followed either by trapping of the resultant imide or ester enolates with an appropriate electrophile (alkyl iodide, CNC, DBAC, NBS; see Sections 11.3.1 and 11.3.4) furnishes enantiomerically enriched α,β-branched/functionalized carboxylic acid derivatives (**251**) with up to two newly created stereogenic centers[76f,g,121,165–167]. Other transformations suited to chiral α,β-unsaturated imides include Diels–Alder reactions[168,169], 1,3-dipolar cycloadditions[170], $Pd(OAc)_2$-catalyzed cyclopropanations[171] and $OsO_4$- or $RuCl_3/NaIO_4$-mediated bishydroxylations[172]. These provide synthetic approaches to additional targets such as e.p. alicyclic, heterocyclic, α,β-dihydroxylated and cyclopropanated carboxylic acid derivatives.

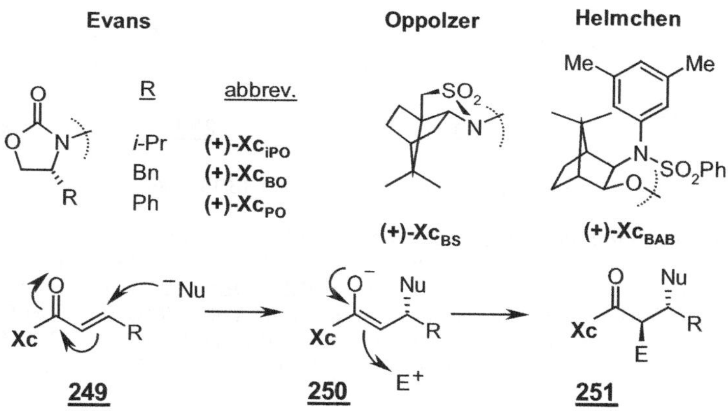


*Figure 11.86* *Auxiliaries used for the preparation of chiral α,β-unsaturated imides and esters labeled with isotopic carbon (only one antipode depicted): Generic scheme of diastereoselective conjugate additions*

Chiral α-unsubstituted α,β-unsaturated, isotopically carbon-labeled imides or esters (such as **249** above) needed as starting materials for such applications are best prepared by Horner–Wadsworth–Emmons reaction of e.p. (*R*)-or-(*S*)-*N*-diethylphosphonoacetyl 4-substituted 1,3-oxazolidin-2-ones or bornane-10,2-sultams[87,135c] or of (1*R*,2*S*,3*R*)/(1*S*,2*R*,3*S*)-3-[*N*-phenylsulfonyl-*N*-(3,5-dimethylphenyl)amino]bornyl diethylphosphonoacetates (e.g., **252**) with aldehydes[167a]. For purposes of isotopic synthesis, labels could be incorporated either via the haloacetyl group or the aldehyde component, or both, as suggested in Figure 11.87. Entities of the **249**-type are also accessible through aldol reaction of the respective homochiral acetyl derivatives **253** with aldehydes, *O*-acetylation of the resultant diastereomeric β-hydroxy intermediates **254** and elimination of HOAc upon treatment with DBU[168b,173b]. Both methods avoid the separate preparation of labeled acrylic acid derivatives (acid, esters), which, at least in the radioactive case and especially for lower homologs, are highly prone to radiation-induced polymerization. They also allow for the introduction of isotopic carbon atoms at three or more different positions, which provides options for the regioselective labeling of different positions for biosynthesis studies, or for achievement of specific activities up to 180 mCi/mmol. $\alpha$-Alkylated derivatives, higher homologous acrylates or materials of low specific activity may be connected to the auxiliaries following routine routes employing acyl chlorides or mixed anhydrides prepared *in situ* (see Section 11.3.1).

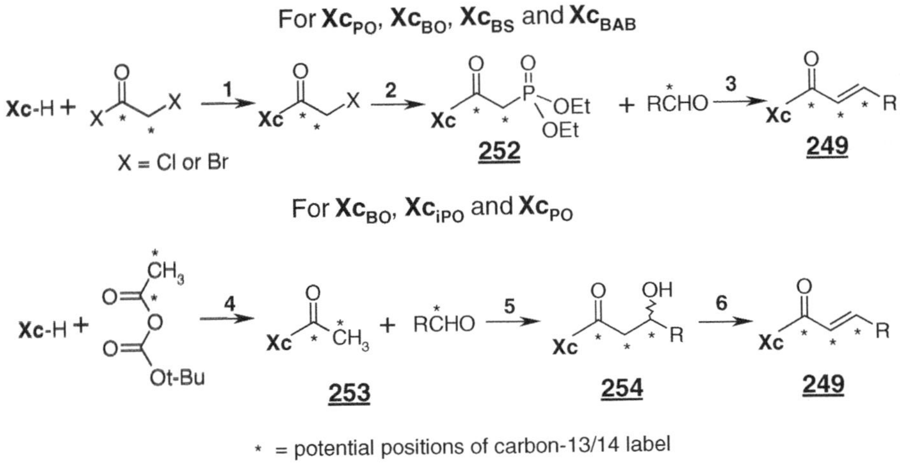


**Reaction conditions: 1a** . (X = Br): **Xc**$_{PO}$-H or **Xc**$_{BS}$-H, *n*-BuLi, THF; −78 °C, 45 min, **b**. $BrCH_2COBr$, −78°C, 2 h then −78 °C → 0 °C, 1 h (75-80%); (X = Cl): **Xc**$_{BAB}$-H, mol sieves 4., $CCl_4$; reflux 16 h; **2**. 252, $P(OEt)_3$; 100-170 °C, 3 h (up to 99%); **3**. DBU, LiBr, acetonitrile; r.t., 3 h; **4**. *n*-BuLi, THF; −78 °C, 0.5 h; $CH_3CO$-OCOt-Bu, 3 h; **5**. $TiCl_4$, *i*-$Pr_2EtN$; $CH_2Cl_2$, −78°C, 30 min then RCHO; −78 °C, 6 h; −78 °C → r.t., 16 h; **6a**. $Ac_2O$, pyridine; $CH_2Cl_2$; r.t., 16 h, **b**. DBU, −$CH_2Cl_2$; r.t., 16 h

*Figure 11.87 Preparation of chiral homologous $\alpha$-unsubstituted (E)-acryloyl imides and esters indicating convenient potential labeling positions*

Out of this arsenal of synthetic procedures, only Michael[87,135a,173] and Diels–Alder additions[87,135c] have been exploited so far for the preparation of e.p. compounds labeled with isotopic carbon or hydrogen. For applications in 1,4-additions, starting materials derived from Oppolzer's and Helmchen's auxiliaries have shown some superiority to the original Evans' derivatives, though most weak points of the latter have been overcome by utilizing the C4-phenyl auxiliary (**Xc**$_{PO}$) instead of the isopropyl (**Xc**$_{iPO}$) or benzyl (**Xc**$_{BO}$) analogs. Nevertheless, the latter two have been more commonly used.

### *1,4-Additions*

Addition of EtMgBr to (2*R*)-*N*-[(*E*)-[1,2-$^{14}C_2$]crotonyl]bornane-10,2-sultam (**256**) and trapping of the resulting magnesium enolate with 1-chloro-1-nitrosocyclohexane afforded (2*R*)-*N*-[(2*S*,3*S*)-(2-hydroxyamino-3-methyl)[1,2-$^{14}C_2$]pentanoyl]bornan-10,2-sultam (**257**) with 85% d.e., as assessed by radio-HPLC. Chromatographic separation of the undesired diastereomers followed by reductive *N*-methylation of the $\alpha$-*N*-hydroxyamino group, N/O-hydrogenolysis and hydrolytic cleavage of the auxiliary converted **257** into e.p. (*S*,*S*)-*N*-methyl[1,2-$^{14}C_2$]isoleucine (**258**) in 34% overall radiochemical yield. This compound served as a key intermediate in the synthesis of [$^{14}C_2$]NVP NIM 811 (**259**), a nonimmunosuppressive cyclosporin derivative[87]. [1,2-$^{14}C_2$]Crotonoyl imide **256** used as a starting material was obtained from (2*R*)-*N*-[phosphono[1,2-$^{14}C_2$]acetyl]bornane-10,2-sultam ((2*R*)-[1,2-$^{14}C_2$]PABS, **255**) and acetaldehyde according to the procedure of Figure 11.88. The 85–88% diastereoselectivity induced at the $\beta$-position lies in the range reported for 1,4-additions of most alkyl Grignard reagents to chiral crotonoyl sultams[121].

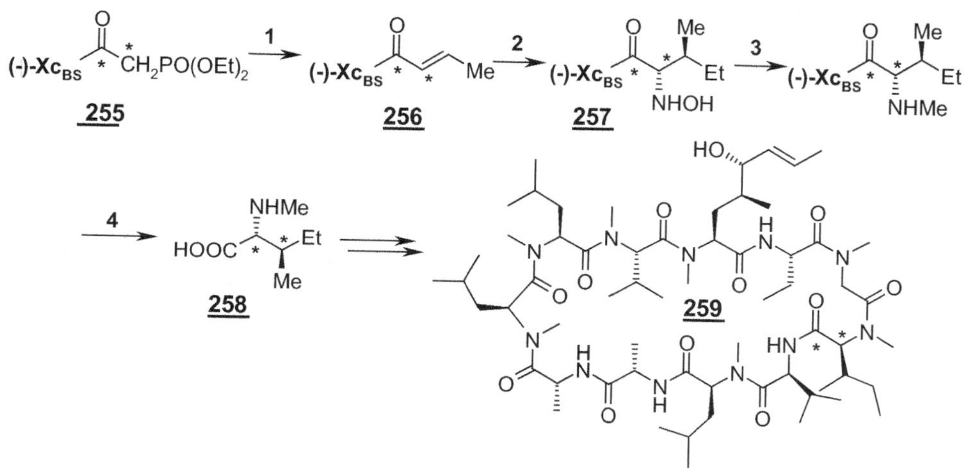

**Reaction conditions:** **1**. MeCHO, DBU, LiBr, acetonitrile; r.t., 3 h; **2a**. EtMgBr, THF; −80 °C, 16 h, **b**. 1-chloro-1-nitrosocyclohexane, THF; −78 °C; 1 h, **c**. 1 N HCl; −78 °C; **3a**. aq. HCHO (pH 4–6), NaBH₃CN, MeOH; r.t., 1.5 h, **b**. Zn, 1 N HOAc/1 N HCl; 0 °C, 30 min; r.t., 2 h; **4**. LiOH, aq. THF; r.t., 16 h

***Figure 11.88*** *Application of the tandem version of diastereoselective Michael addition/ $\alpha$-hydroxyamination to the synthesis of (S,S)-N-methyl[1,2-$^{14}C_2$]isoleucine*

The 1,4-addition in step 2 above does not work with methyl Grignard reagents because they add preferentially in a 1,2-fashion[121,165b,c,171]. In such cases, as well as for the transfer of aryl and vinyl groups, $PBu_3$-stabilized Gilman reagents ($R_2CuLi{\cdot}PBu_3$)[165c] have to be employed. This tactic was used as illustrated by the following two examples (Figure 11.89). Thus, treatment of (2*R*)-*N*-(crotonoyl)bornane-10,2-sultam (**256**) with *in situ* prepared ($^{13}CH_3$)$_2CuLi{\cdot}PBu_3$ then $BF_3{\cdot}OEt_2$, followed by quenching of the resultant enolate provided (2*R*)-*N*-((3*R*)-3-methyl[4-$^{14}$C]butanoyl)sultam ([$^{13}$C]**260**) in 89% yield and 82% d.e. Removal of the auxiliary followed by a one-carbon homologation sequence gave methyl (4*S*)-4-methyl-3-oxo[5-$^{13}$C]pentanoate (**261**), thereby shifting the stereogenic center by one position. Concluding enzymatic reductive amination of the carbonyl function provided (2*S*,4*S*)-[5-$^{13}$C]leucine (**262**), which was used for investigations in studies of secondary metabolic reactions[173a]. Similarly, [$^{14}$C]**260**, prepared by methylation of doubly labeled **256** with $Me_2CuLi{\cdot}PBu_3$, served as a key intermediate for the preparation of e.p. $\beta^2$-[1,2-$^{14}C_2$]HVal (**263**)[87], a $\beta^2$-amino acid required as a building block for labeled $\beta$-peptides[174]. Compound **263** was obtained by stereoselective $\alpha$-alkylation of [$^{14}$C]**260** with $BrCH_2COO$*tert*-Bu, proteolytic cleavage of the *tert*-butyl ester, Curtius degradation of the freed carboxyl group and concluding hydrolytic removal of the auxiliary.

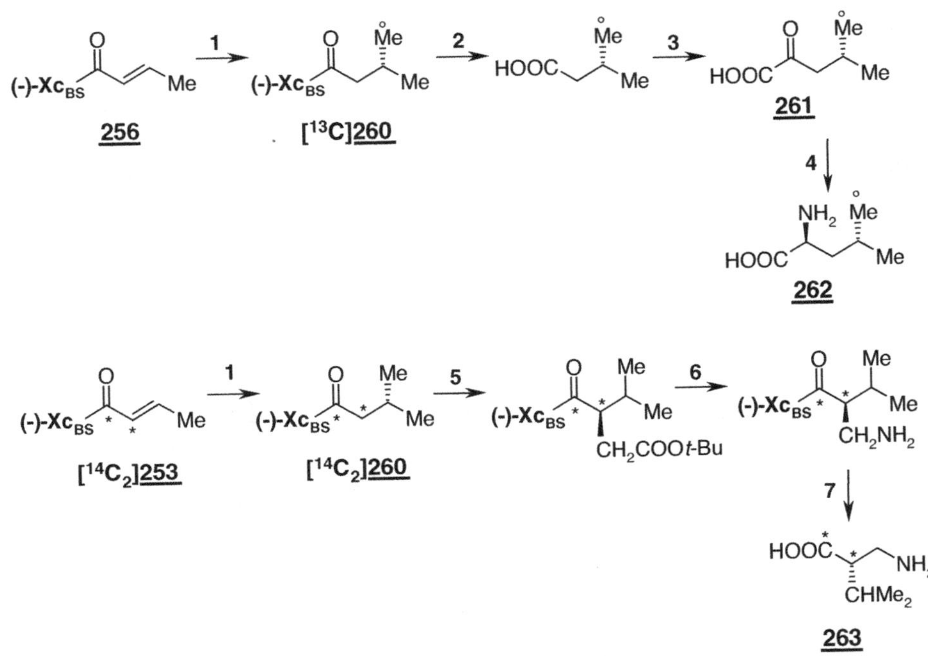


**Reaction conditions: 1a. [$^{13}$C]MeLi** or **MeLi** (2 equiv.), CuI.PBu$_3$, ether/toluene; –5 °C, 30 min, **b**. BF$_3$.OEt$_2$, toluene; –78 °C, 6 h; **2**. LiOH, H$_2$O$_2$, aq. THF; 0 °C; **3a**. PPh$_3$CHCN, EDCI, DMAP; CH$_2$Cl$_2$; 0 °C → r.t., 16 h, **b**. O$_3$, CH$_2$Cl$_2$/MeOH; r.t., 2 h; **4**. leucine dehydrogenase; **5a**. NaHMDS, THF, –78 °C, 1.5 h, **b**. BrCH$_2$COO-*t*-Bu, Bu$_4$NI, HMPA, THF; –78 °C, 10 h; –50 °C, 2 h; **6a**. TFA, CH$_2$Cl$_2$; 0 °C, 2 h, **b**. ClCOOEt, Et$_3$N, THF; 0 °C, 1 h, **c**. NaN$_3$, aq. THF; 0°C, 1 h, **d**. toluene, 100 °C, 2 h, **e**. *t*-BuOH, toluene; 100 °C; **7**. LiOH, H$_2$O$_2$, aq. THF; r.t., 16 h

***Figure 11.89*** *Sequential Michael addition/chain elongation and $\alpha$-functionalization, respectively, applied to the synthesis of (2S,4S)-[5-$^{13}$C]leucine and $\beta^2$-[1,2-$^{14}C_2$]-HVal*

Significantly better yields (80–95%) and diastereoselectivities (to >98% d.e.) in 1,4-additions of organocuprates at the $\beta$-position can be achieved by use of (1*R*,2*S*,3*R*)/(1*S*,2*R*,3*S*)-3-[*N*-phenylsulfonyl-*N*-(3,5-dimethylphenyl)amino]bornan-2-ol (Helmchen's auxiliary) in place of Oppolzer's auxiliary, and of $RCu{\cdot}BF_3$ (Yamamoto reagents) instead of Gilman's reagents ($R_2CuLi{\cdot}PBu_3$)[173a,175]. Yamamoto reagents are prepared *in situ* from RLi and equimolar amounts of CuI and $BF_3{\cdot}OEt_2$ in dry ether. In the example depicted in

Figure 11.90, the chiral crotonate **264** was treated with a seven-fold excess of $^{13}CH_3Cu \cdot BF_3$ to give (3*R*)-3-methyl[4-$^{13}$C]butanoate (**265**) in >98% d.e. and 79% yield[173a]. Cleavage of the auxiliary afforded the e.p. acid **266** in quantitative yield. This was further converted to (2*S*,4*R*)-[5-$^{13}$C]leucine (**267**) following the same pathway described above for e.p. (2*S*,4*S*)-[5-$^{13}$C]leucine. A weak point of this method is that, in contrast to Oppolzer's approach for which a 2–2.5 fold excess of the organometallic reagent is sufficient, a five- to seven-fold excess is essential in order to achieve satisfactory yields[173a]. Therefore, in the radioactive case introduction of the label via the organocuprate would be very inefficient.

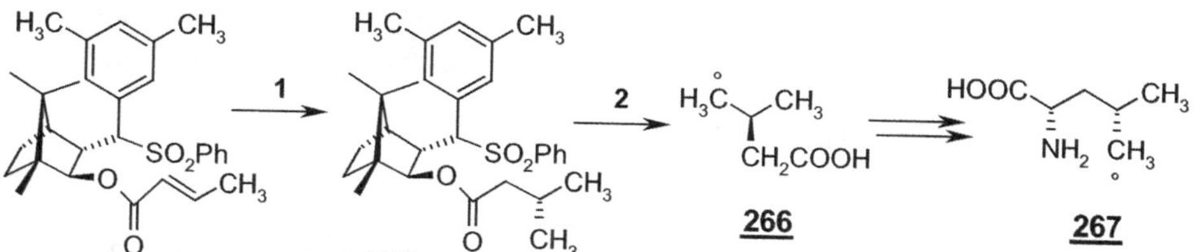


**264** **265** **266** **267**

**Reaction conditions: 1a.** $^{13}CH_3Li$, CuI (1eq.), ether; −10 °C, 20 min, **b.** $BF_3 \cdot OEt_2$, **262**, ether; −78 °C, 1 h; −40 °C, 3 h; −20 °C, 3 h (79 %); **2.** 2 N KOH, MeOH; reflux (100 %)

***Figure 11.90*** *Synthesis of (3R)-3-methyl[4-$^{13}$C]butanoic acid through 1,4-addition of [$^{13}$C] MeCu·BF$_3$ to chiral crotonate linked to Helmchen's auxiliary*

*Diels–Alder Reactions*

Auxiliary-controlled diastereoselective Diels–Alder addition of chiral acrylates, crotonates or fumarates to dienes is one of the fundamental methodologies for simultaneous formation of two bonds to give diastereomerically pure adducts with up to four stereogenic centers and largely predictable stereochemistry[168,169]. Despite its enormous potential, it has found only very limited application to the synthesis of e.p. intermediates labeled with isotopes of carbon or hydrogen. The $EtAlCl_2$-catalyzed addition of (2*S*)-*N*-([1,2-$^{14}C_2$]acryloyl)bornane-10,2-sultam (**268**) to 1,3-butadiene in the presence of galvinoxyl, a radical scavenger to suppress polymerization, is an illustrative example. It has been exploited to prepare (2*S*)-*N*-[(1*R*)-[1,7-$^{14}C_2$]cyclohex-3-enylcarbonyl]sultam (**269**) in 53% radiochemical yield and >98% radiochemical purity after chromatographic purification[87,135c]. Cleavage of the auxiliary, epoxidation of the double bond, and treatment with base initiated intramolecular attack on the epoxide to give lactone **270**. Subsequent hydrolytic opening of the lactone ring provided (1*R*,3*R*,4*R*)-3,4-dihydroxy[1,7-$^{14}C_2$]cyclohexane (**271**). Compound **271** was used together with (*S*)-[1,7-$^{14}C_2$]pipecolic acid as precursors for the biological labeling of rapamycin (**272**) at a specific activity of 100 mCi/mmol, sufficiently high for pharmacokinetic and metabolic studies in animals and man[135c] (see also Section 11.3.6). Acryloyl starting material **268**, was prepared in 52% radiochemical yield through reaction of (2*S*)-[1,2-$^{14}C_2$]PABS (**253**) with paraformaldehyde in the presence of galvinoxyl, following the

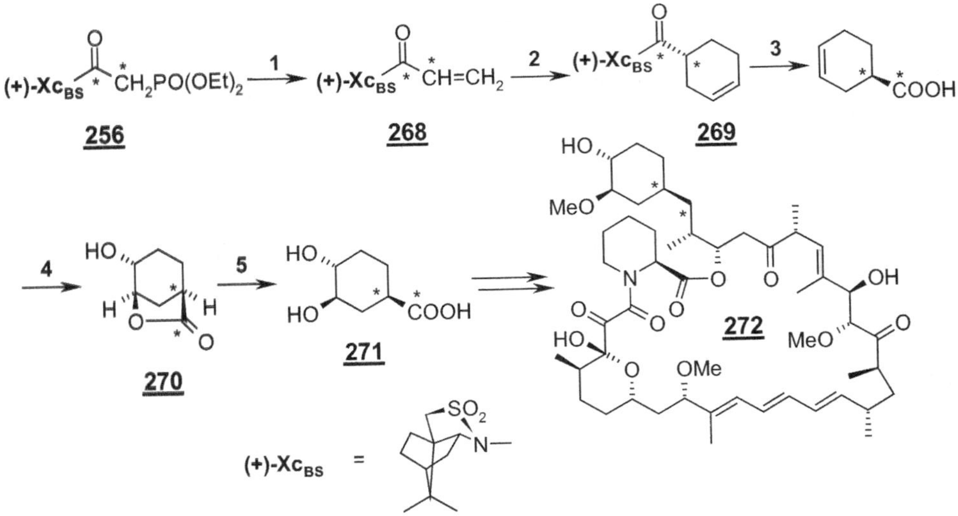


**Reaction conditions:** **1**. $(HCHO)_n$, galvinoxyl, $K_2CO_3$, THF; 70 °C, 30 min; **2**. 1,3-butadiene, $EtAlCl_2$, $CH_2Cl_2$; −78 °C, 4 h; **3**. LiOH, aq. THF; r.t., 7 h; **4a**. MCPBA, $CCl_4$; 0 °C, 10 min; r.t.,16 h, **b**. $Et_3N$, $SiO_2$; 70 °C, 5 h; **5**. 5 N HCl/THF; r.t., 16 h

***Figure 11.91*** *Diels–Alder route for the preparation of (1R,3,4R)-[1,7-$^{14}C_2$]cyclohexane carboxylic acid*

Horner–Wadsworth–Emmons method already described. In contrast to what is known from other carbon-14-labeled acrylates, 256 proved to be unexpectedly stable towards radiation-induced polymerization, probably due to the bulky auxiliary.

## References


1. Eliel, E.L.; Wilen, S.H. and Mander, L.N., *Stereochemistry of Organic Compounds;* Wiley Interscience: New York, ***1994***, p. 297.
2. Dischino, D.D.; Gribkoff, V.K.; Hewawasam, P.; Luke, G.M.; Rinehart, J.K.; Spears, T.L. and Starrett, J.E., Jr., *J. Label. Compd. Radiopharm.*, ***2003***, *46*, 139–149.
3. Dischino, D.D.; Lee, C.-W.; Belema, M. and Zusi, C., *J. Label. Compd. Radiopharm.*, ***2003***, *46*, 159–165.
4. Kitson, S.L. and Knagg, E., *J. Label. Compd. Radiopharm.*, ***2006***, *49*, 517–531.
5. Schneider, C.; Boeglin, W.E.; Lai, S.; Cha, J.K. and Brash, A.R., *Anal. Biochem.*, ***2000***, *284*, 125–135.
6. Murray, A. and Williams, D.L., *Organic Syntheses with Isotopes;* Interscience: New York, ***1958***, pp. 382–386.
7. Sanceau, J.-Y.; Labrie, F. and Gauthier, S., *J. Label. Compd. Radiopharm.*, ***2004***, *47*, 741–752.
8. a. Murray, A. and Williams, D.L., *Organic Syntheses with Isotopes;* Interscience: New York, ***1958***, pp. 181–184; b. Meltzer, H.L. and Sprinson, D.B., *J. Biol. Chem.*, ***1952***, *197*, 461.
9. Willcocks, K.; Barnes, R.D.; Rustidge, D.C. and Tidy, D.J.D., *J. Label. Compd. Radiopharm.*, ***1993***, *33*, 783–794.
10. Shu, A.Y.L.; Villani, A.J. and Heys, J.R., *J. Label. Compd. Radiopharm.*, ***1990***, *28*, 143–153.

11. Koltai, E.; Bajusz, S.; Széll, E. and Zolyomi, G., *J. Label. Compd Radiopharm.*, ***1987***, *24*, 645–657.
12. Shu, A.Y.L. and Heys, J.R., *Int. J. Peptide Protein Res.*, ***1993***, *42*, 432–439.
13. McCarthy, K.E.; Miller, S.A.; Chenard, B.L.; Butler, T.W.; Dumont, M.L. and Stemple, J.Z., *J. Label. Compd. Radiopharm.*, ***1997***, *39*, 973–985.
14. Loftfield, R.B. and Eigner, E.A., *Biochim. Biophys. Acta*, ***1966***, *130*, 449–457.
15. Gerday, Ch. and Verly, W.G., *J. Label. Compd.*, ***1968***, *4*, 335–337.
16. Andres, H., SMS 201-995, Labeling with carbon-14 in the (*R*)-tryptophan moiety, Internal Document, Sandoz Ltd, December ***1986***
17. Mzengeza, S.; Venkatachalam, T.K.; Rajagopal, S. and Diksic, M., *Int. J. Appl. Radiat. Isot.*, ***1993***, *44*, 1167–1172.
18. Petitclerc, C.; D'Iorio, A. and Benoiton, N.L., *J. Label. Compd.*, ***1969***, *5*, 265–269.
19. Gaddam, S.; Khilevich, A.; Filer, C.; Rizzo, J.D.; Giltner, J.; Flavin, M.T. and Xu, Z.-Q., *J. Label. Compd. Radiopharm.*, ***1997***, *39*, 901–906.
20. Antoni, G. and Långström, B., *J. Label. Compd. Radiopharm.*, ***1987***, *24*, 125–243.
21. a. Mitulovi, G.; Lämmerhofer, M.; Maier, N.M. and Lindner, W., *J. Label. Compd. Radiopharm.*, ***2000***, *43*, 449–461; b. Yamada, S.; Hongo, C; Yoshioka, R. and Chibata, I., *J. Org. Chem.*, ***1983***, *48*, 843–846.
22. Schnell, B.; Faber, K.; Kroutil, W., *Adv. Synth. Catal.*, ***2003***, *345*, 653–666.
23. Kuo, F.; Clodfelter, D.K.; Priest, T.R. and Kau, D.L.K., *J. Label. Compd. Radiopharm.*, ***2004***, *47*, 599–608.
24. a. Brown J.M., Hydrogenation of functionalized carbon–carbon double bonds, in *Comprehensive Asymmetric Catalysis*, Vol. 1, Jacobsen, E.N.; Pfaltz, A. and Yamamoto, H., Eds.; Springer: Berlin–Heidelberg–New York, ***1999***, pp. 122–182; b. Pfaltz, A.; Blankenstein, J.; Hilgraf, R.; Hörmann, E.; McIntyre, St.; Menges, F.; Schönleber, M.; Smidt, P.; Wüstenberg, B. and Zimmermann, N., Iridium-catalyzed enantioselective hydrogenatiion of olefins, in *Adv. Synth. Catal.*, ***2003***, *345*, 33–43; c. Nagel, U. and Albrecht, J., Enantioselective hydrogenation of *N*-acyl dehydroamino acids, in *Top. Catal.*, ***1998***, *5*, 3–23; d. Knowles, W.S., Asymmetrische Hydrierungen, in *Angew. Chem.*, ***2002***, *114*, 2096–2107; e. Noyori, R., Asymmetrische Katalyse; Kenntnisstand und Perspektiven, in *Angew. Chem.*, ***2002***, *114*, 2108–2123; f. Kotha, S., Opportunities in asymmetric synthesis: an industrial approach, *Tetrahedron Report No. 352*, ***1994***, 3640–3662; g. Spindler, F.; Häusel, R.; Blaser, H.U., *Synth. Methods Organomet. Inorg. Chem.*, ***2002***, *10*, 78–83.
25. Voges R, Stereoselective procedures in the synthesis of e.p. isotopically labeled compounds, in *Synthesis and Applications of Isotopically Labeled Compounds 1994*, Allen, J. and Voges, R., Eds.; John Wiley & Sons, Ltd: Chichester, ***1995***, pp. 1–26.
26. a. Schmidt, U.; Lieberknecht, A. and Wild, J., *Synthesis*, ***1988***, 159–172; b. Shin, C.G.; Takahashi, N. and Yonezawa, Y., *Chem. Pharm. Bull.*, ***1990***, *38*, 2020–2023; c. Rao, Y.S. and Filler, R., *Synthesis*, ***1975***, 749–764.
27. a. Ciszewska, G. and Ray, T., in *Synthesis and Applications of Isotopically Labeled Compounds 2000*, Pleiss, U. and Voges, R., Eds.; John Wiley & Sons, Ltd: Chichester, ***2001***, pp. 300–302; b. Kendall, J.T., *J. Label. Compd. Radiopharm.*, ***2000***, *43*, 917–924; c. Larsen, P.O. and Wieczorkowska, E., *Acta Chem. Scand.*, ***1977***, *31*, 109–114; d. Larsen, P.O. and Wieczorkowska, E., *J. Label. Compd.*, ***1974***, *10*, 287–296; e. Read, G.; Vining, L.C. and Haskins, R.H., *Can. J. Chem.*, ***1962***, *40*, 2357–2362; f. Fryzuk, M.D. and Bosnich, B., *J. Am. Chem. Soc.*, ***1977***, *99*, 6262–6267; g. Takahashi, K.; Kirk, K.L. and Cohen, L.A., *J. Label. Compd. Radiopharm.*, ***1986***, *23*, 1–8; h. Herbst, R.M. and Shermin, D., *Org. Synth. Coll.* Vol. 2, ***1943***, 1–3.
28. a. Parnes, H. and Shelton, E.J., *Int. J. Peptide Protein Res.*, ***1986***, *27*, 239–244; b. Parnes, H.; Shelton E.J. and Huang, G.T., *Int. J. Peptide Protein Res.*, ***1986***, *28*, 403–410; c. Halldin, C. and Långström, B., *Int. J. Appl. Radiat. Isot.*, ***1984***, *35*, 779–782; d. Halldin, C.; Långström, B., *Int. J. Appl. Radiat. Isot.*, ***1984***, *35*, 945–948; e. Halldin, C.; Långström, B., Medical application of Cyclotron III, *Ann. Univ. Turkuensis*, ***1984***, *D17*, 200–203; f. Buck, S. and Ide, W.S., *Org. Synth. Coll.* Vol. 2, ***1943***, 55–56; g. VandenBerg, G.E.; Harrison, J.B.; Carter, H.E. and Magerlein, B.J., *Org. Synth.* ***1973***, *5*, 946–948; h. Tripathy, P.K. and Mukerjee, A.K., *Synthesis*, ***1985***, 285–288.

29. a. Schmidt, U.; Griesser, H.; Leitenberger, V.; Lieberknecht, A.; Mangold, R.; Meyer, R. and Riedl, B., *Synthesis*, ***1992***, 487–490; b. Schmidt, U.; Lieberknecht, A. and Wild, J., *Synthesis*, ***1984***, 53–60; c. Schmidt, U.; Lieberknecht, A.; Schanbacher, U.; Beuttler, T. and Wild, J., *Angew. Chem. Intl. Edn Engl.*, ***1982***, *21*, 776–777.
30. a. Nakajima, T.; Nakayama, K. and Shimizu, I., *J. Label. Compd. Radiopharm.*, ***2007***, *50*, 622–623; b. Baldwin, J.E., Merritt, K.D. and Schofield, C.J., *Tetrahedron Lett.*, ***1993***, *34*, 3919–3920.
31. a. Wüst, H.H.; Bardenhagen, J. and Schöllkopf, U., *Liebigs Ann. Chem.*, ***1985***, 1825–1837; b. Schöllkopf, U.; Gerhart, F.; Schröder, R. and Hoppe, D., *Liebigs Ann. Chem.*, ***1972***, *766*, 116–129.
32. a. Poisel, H. and Schmidt, U., *Chem. Ber.*, ***1975***, *108*, 2547–2553; b. Poisel, H. and Schmidt, U., *Angew. Chem. Int. Edn Engl.*, ***1976***, *15*, 294–295; c. Poisel, H., *Chem. Ber.*, ***1977***, *110*, 942–947.
33. a. Makowski, M.; Rzeszotarska, B.; Kubica, Z. and Pietrzynski, G., *Liebigs Ann. Chem.*, ***1985***, 893–900; b. Makowski, M.; Rzeszotarska, B.; Kubica, Z.; Wieczorek, G., *Liebigs Ann. Chem.*, ***1984***, 920–928; c. Shin, C.; Yonezawa, Y. and Watanabe, E., *Tetrahedron Lett.*, ***1985***, *26*, 85–88.
34. a. Hammadi, A.; Meunier, G. and Genet, R., in *Synthesis and Applications of Isotopically Labeled Compounds 1994*, Allen, J. and Voges, R., Eds.; John Wiley & Sons, Ltd: Chichester, ***1995***, pp. 131–136; b. Hammadi, A.; Meunier, G.; Tarride, J.L.; Menez, A. and Genet, R., in *Synthesis and Applications of Isotopically Labeled Compounds 1997*, Heys, J.R. and Melillo, D. G., Eds.; John Wiley & Sons, Ltd: Chichester, ***1998***, pp. 197–201; c. Pinto-Alphandary, H.; van Craeynest-Jimonet, C.; Morgat, J.L. and Fromageot, P., *J. Label. Compd. Radiopharm.*, ***1988***, *15*, 1273–1279.
35. a. Knowles, W.S.; Sabacky, M.J. and Vineyard, B.D., U.S. Patent No. 4,005,127, ***1977***; b. Knowles, W.S., *Acc. Chem. Res.*, ***1983***, *16*, 106–112; c. Vineyard, B.D.; Knowles, W.S.; Sabacky, M.J.; Bachman, G.L. and Weinkauff, D.J., *J. Am. Chem. Soc.*, ***1977***, *99*, 5946–5952; d. Knowles, W.S.; Sabacky, M.J.; Vineyard, B.D. and Weinkauff, D.J., *J. Am. Chem. Soc.*, ***1975***, *97*, 2567–2568.
36. a. Parnes, H. and Shelton, E.J., in *Synthesis and Applications of Isotopically Labeled Compounds 1985*, Muccino, R.R., Ed.; Elsevier: Amsterdam, ***1986***, pp. 159–160; b. Griesser, R. and Düblin, R., Sandoz Pharma Ltd., Basle, Switzerland, unpublished results, ***1982***.
37. Kent, B; GE Healthcare, Amersham, Forest Pharm, Cardiff, UK; personal communication to RV
38. a. Burk, M.J.; Gross, M.F.; Harper, T.G.P.; Kalberg, C.S.; Lee, J.R. and Martinez, J.P., *Pure Appl. Chem.*, ***1996***, *68*, 37–44; b. Burk, M.J.; Gross, M.F. and Martinez, J.P., *J. Am. Chem. Soc.*, ***1995***, *117*, 9375–9376; c. Burk, M.J.; Lee, J.R. and Martinez, J.P., *J. Am. Chem. Soc.*, ***1994***, *116*, 10847–10848; d. Burk, M.J.; Feaster, J.E.; Nugent, W.A. and Harlow, R.L., *J. Am. Chem. Soc.*, ***1993***, *115*, 10125–10138; e. Burk, M.J.; Feaster, J.E. and Harlow, R.L., *Tetrahedron Asymmetry*, ***1991***, *2*, 569–592.
39. a. Levine-Pinto, H.; Morgat, J.L.; Fromageot, P.; Meyer, D.; Poulin, J.-C. and Kagan, H.B., *Tetrahedron*, ***1982***, *38*, 119–123; b. Meyer, D.; Poulin, J.-C.; Kagan, H.B.; Levine-Pinto, H.; Morgat, J.L. and Fromageot, P., *J. Org. Chem.* ***1980***, *45*, 4680–4682; c. Ojima, I. and Shimizu, N., *J. Am. Chem. Soc.*, ***1986***, *108*, 3100–3102; d. Ojima, I.; Yoda, N.; Yatabe, M.; Tanaka, T. and Kogure, T., *Tetrahedron*, ***1984***, *40*, 1255–1268.
40. a. Itsuno, S., Enantioselective reduction of ketones, in *Organic Reactions*, Vol. 52, John Wiley & Sons, Inc.: New York, ***1998***, pp. 395–576; b. Itsuno, S., Hydroboration of carbonyl groups, in *Comprehensive Asymmetric Catalysis*, Vol 1, Jacobsen, E.N.; Pfaltz, A. and Yamamoto, Y., Eds.; Springer: Berlin–Heidelberg–New York, ***1999***, pp. 290–315; c. Corey, E.J. and Helal, C.J., *Angew. Chem. Int. Edn Engl.*, ***1998***, *37*, 1986–2012; d. Corey, E.J., *Pure Appl. Chem.*, ***1990***, *62*, 1209–1216; e. Wallbaum, S. and Martens, J., *Tetrahedron: Asymmetry*, ***1992***, *3*, 1475–1504; f. Brown, H.C. and Ramachandran, P.V., *Pure Appl. Chem.*, ***1994***, *66*, 201–212; g. Dhar, R.K., *Aldrichim. Acta*, ***1994***, *27*, 43–50; h. Ramachandran, P.V. and Brown, H.C., in *Reductions in Organic Synthesis*, Abdel-Magid, A., Ed.; Oxford University Press: Oxford, ***1996***, Chapter 5; i. Daverio, P. and Zanda, M., *Tetrahedron: Asymmetry*, ***2001***, *12*, 2225–2259.
41. Weissman, S.A. and Ramachandran, P.V., *Tetrahedron Lett.*, ***1996***, *37*, 3791–3794.

42. Moenius, Th.; Kiechle, P. and Achini, R., SDZ ENX792, Labeling with carbon-14, Internal Document, Sandoz Pharma AG, February ***1993***
43. Hesk, D.; Bignan, G.; Lee, J.; Yang, J.; Voronin, K.; Magatti, C.; McNamara, P.; Koharski, D.; Hendershot, S.; Saluja, S. and Wang, S., *J. Label. Compd. Radiopharm.*, ***2002***, *45*, 145–155.
44. Stolle, W.T.; Stelzer, L.S.; Hester, J.B.; Perricone, S.C. and Hsi, R.S.P., *J. Label. Compd. Radiopharm.*, ***1994***, *34*, 929–942.
45. Moenius, Th.; Burtscher, P.; Egger, H.; Bovermann, G. and Oberer, L., *J. Label. Compd. Radiopharm.*, ***1999***, *42*, 1053–1060.
46. Wheeler, W.J., *J. Label. Compd. Radiopharm.*, ***1992***, *31*, 477–487.
47. Wheeler, W.J. and Kuo, F., *J. Label. Compd. Radiopharm.*, ***1995***, *36*, 213–223.
48. Kuo, F.; Clodfelter, D.K. and Wheeler, W.J., *J. Label. Compd. Radiopharm.*, ***2004***, *47*, 615–625.
49. Didier, E.; Loubinoux, B.; Ramos Tombo, G.M. and Rihs, G., *Tetrahedron*, ***1991***, *47*, 4941–4958.
50. a. Itsuno, S.; Sakurai, Y.; Shimizu, K. and Ito, K., *J. Chem. Soc. Perkin Trans. I*, ***1990***, 1859–1863; b. Itsuno, S.; Sakurai, Y.; Shimizu, K. and Ito, K., *J. Chem. Soc. Perkin Trans. I*, ***1989***, 1548–1549.
51. Zhang, Y., *J. Label. Compd. Radiopharm.*, ***2000***, *43*, 1087–1093.
52. Ciszewska, G.; Pfefferkorn, H.; Jones, L.; Tarapata, R. and Sunay, U.B., in *Synthesis and Applications of Isotopically Labeled Compounds 1997*, Heys, J.R. and Melillo, D.G., Eds.; John Wiley & Sons, Ltd: Chichester, ***1998***, pp. 153–155.
53. a. Vigneron, J.-P. and Bloy, V., *Tetrahedron Lett.*, ***1979***, 2683–2686; b. Vigneron, J.P. and Jacquet, I.; *Tetrahedron*, ***1976***, *32*, 939–944; c. Jacquet, I. and Vigneron, J.P., *Tetrahedron Lett.*, ***1974***, 2065–2068.
54. Wurz, R.E.M.; Kepner, R.E. and Webb, A.D., *Prep. Biochem.*, ***1989***, *19*, 209–215.
55. Corey, E.J. and Link, J.O., *Tetrahedron Lett.*, ***1989***, *30*, 6275–6278.
56. Taguchi, H.; Paal, B. and Armarago, W.L.F., *J. Chem. Soc. Perkin Trans. I*, ***1997***, 303–307.
57. a. Parry, R.J. and Lii, F.-L., *J. Am. Chem. Soc.*, ***1991***, *113*, 4704–4706; b. Leeper, F.J. and Battersby, A.R., in *Synthesis and Applications of Isotopically Labeled Compounds 1988*, Baillie, T.A. and Jones, J.R., Eds.; Elsevier: Amsterdam, ***1989***, pp. 41–46.
58. Kobayashi, K.; Jadhav, P.K.; Zydowsky, T.M. and Floss, H.G., *J. Org. Chem.*, ***1983***, *48*, 3510–3512.
59. Goss, R.J.M.; Meddour, A.; Courtieu, J.; O'Hagan, D., in *Synthesis and Applications of Isotopically Labeled Compounds*, Vol. 7, Pleiss, U. and Voges, R., Eds.; John Wiley & Sons, Ltd: Chichester, ***2001***, pp. 163–166.
60. Keck, G.E. and Krishnamurthy, D., *J. Org. Chem.*, ***1996***, *61*, 7638–7639.
61. Chapelle, M.R. and Morgan, A.D., in *Synthesis and Applications of Isotopically Labeled Compounds*, Vol. 8, Dean, D.C.; Filer, C.N. and McCarthy, K.E., Eds.; John Wiley & Sons, Ltd: Chichester, ***2004***, pp. 55–58.
62. a. Midland, M.M.; Greer, S.; Tramontano, A. and Zderic, S.A., *J. Am. Chem. Soc.*, ***1979***, *101*, 2352–2355; b. Midland, M.M. and Greer, S., *Synthesis*, ***1978***, 845–846; c. Midland, M.M.; Tramontano, A. and Zderic, S.A., *J. Am. Chem. Soc.*, ***1977***, *99*, 5211–5215.
63. a. Kolb, H.C. and Sharpless, K.B., in *Transition Metals for Organic Synthesis*, Vol. 2, 2nd Edn., Beller, M. and Bolm, C., Eds.; Wiley-VCH: Weinheim, ***2004***, pp. 275–297; b. Muniz, K., in *Transition Metals for Organic Synthesis*, Vol. 2, 2nd Edn., Beller, M. and Bolm, C., Eds.; Wiley-VCH: Weinheim, ***2004***, pp. 298–307; c. Sharpless, K.B., *Angew. Chem. Intl. Edn Engl.*, ***2002***, *41*, 2024–2033; d. Marko, I.E. and Svendsen, J.S. in *Comprehensive Asymmetric Catalysis II*, Jacobsen, E.N.; Pfaltz, A. and Yamamoto, H., Eds.; Springer: Berlin, ***1999***, pp. 713–784; e. Kolb., H.C.; VanNieuwenhze, M.S. and Sharpless, K.B., *Chem. Rev.*, ***1994***, *94*, 2483–2547.
64. a. Katsuki, T., in *Comprehensive Asymetric Catalysis II*, Jacobsen, E.N., Pfaltz, A. and Yamamoto, H., Eds.; Springer: Berlin, ***1999***, pp. 621–648; b. Katsuki, T. and Martin, V.S., *Org. React.*, ***1996***, *48*, 1–299; c. Besse, P. and Veschambre, H., *Tetrahedron*, ***1994***, *50*, 8885–8927.
65. a. Brandes, B.D. and Jacobsen, E.N., *J. Org. Chem.*, ***1994***, *59*, 4378–4380; b. Jacobsen, E.J., in *Catalytic Asymmetric Synthesis;* Ojima, I., Ed.; VCH: New York, ***1993***, pp. 159–202;

c. Zhang, W. and Jacobsen, E.N.; *J. Org. Chem.*, ***1991***, *56*, 2296–2298; d. Zhang, W.; Loebach, J. J.; Wilson, S.R. and Jacobsen, E.N., *J. Am. Chem. Soc.*, ***1990***, *112*, 2801–2803.
66. Collins, M.R.; McCarthy, K.E. and Zandi, K.S., in *Synthesis and Applications of Isotopically Labeled Compounds*, Vol. 8, Dean, D.C.; Filer, C.N. and McCarthy, K.E., Eds.; John Wiley & Sons, Ltd: Chichester, ***2004***, pp. 309–312.
67. Schuricht, U.; Endler, K.; Hermig, L.; Findeisen, M. and Welzel, P., *J. Prakt. Chem.*, ***2000***, *342*, 761–772.
68. Schwab, J.M.; Ray, T. and Ho, C.K., *J. Am. Chem. Soc.*, ***1989***, *111*, 1057–1063.
69. Yabe, Y. and Rich, D.H., *Peptide: Chem. Biol.; Proc. Am. Peptide Symp.*, ***1998***, *10*, 289–291.
70. Hoagland, S.; Morita, Y.; Bai, D.L.; Marki, H.-P.; Kees, K.; Brown, L. and Heathcock, C.H., *J. Org. Chem.*, ***1988***, *53*, 4730–4735.
71. Knapp, S.; Kukkola, P.J.; Sharma, S.; Dhar, T.G.M. and Naughton, A.B.J., *J. Org. Chem.*, ***1990***, *55*, 5700–5710.
72. Meara, J.P. and Rich, D.H., *Peptide: Chem. Biol.; Proc. Am. Peptide Symp.*, ***1990***, *11*, 941–943.
73. Li, S.; Pang, J.; Wilson, W.K. and Schroepfer, G.J. Jr., *Tetrahedron: Asymmetry*, ***1999***, *10*, 1697–1707.
74. Heier, R.F.; Moon, M.W.; Stolle, W.T.; Easter, J.A. and Hsi, R.S.P., *J. Label. Compd. Radiopharm.*, ***1996***, *38*, 1087–1098.
75. McCarthy, K.E., Recent advances in the design and synthesis of carbon-14 labeled pharmaceuticals from small molecule precursors, in *Curr. Pharm. Design* **2000**, *5*, 1057–1083.
76. a. Mulzer, J., Basic principles of asymmetric synthesis, in *Comprehensive Asymmetric Catalysis*, Jacobsen, E.N; Pfaltz, A. and Yamamoto, H., Eds.; Springer: Berlin–Heidelberg–New York, ***1999***, Chapter 3, pp. 34–98; b. *Asymmetric Synthesis*, Aitken, R.A. and Kilenyi, S.N.; Blackie: London, ***1992***, Chapter 5, pp. 83–142; c. Ager, D. J.; Prakash, I. and Schaad, D.R., *Aldrichimica Acta*, ***1997***, *30*, 3–12; d. Ager, D. J.; Prakash, I. and Schaad, D.R., *Chem. Rev.*, ***1996***, *96*, 835–875; e. Arya, P. and Qin, H., *Tetrahedron*, ***2000***, *56*, 917–947; f. Oppolzer, W., *Pure Appl. Chem.*, ***1990***, *62*, 1241–1250; g. Oppolzer, W., *Pure Appl. Chem.*, ***1988***, *60*, 39–48; h. Oppolzer, W., Camphor derivatives as chiral auxiliaries in asymmetric synthesis, in *Tetrahedron*, ***1987***, *43*, 1970–2003.
77. a. Job, A.; Janeck, C.F.; Bettray, W.; Peters, R. and Enders, D., The SAMP/RAMP-hydrazone methodology in asymmetric synthesis, *Tetrahedron*, ***2002***, *58*, 2253–2329; b. Rothenbacher, K. P.; Fischer, I.U.; Wodarz, R. and Rettenmeier, A.W., in *Synthesis and Applications of Isotopically Labeled Compounds 1994*, Allen, J. and Voges, R., Eds.; John Wiley & Sons, Ltd: Chichester, ***1995***, pp. 109–112.
78. a. Evans, D.A.; Ennis, M.D. and Mathre, D.J., *J. Am. Chem. Soc.*, ***1982***, *104*, 1737–1739; b. Evans, D.A.; Wu, L.D.; Wiener, J.J.M.; Johnson, J.S.; Ripin, D.H.B. and Tedrow, J.S., *J. Org. Chem.*, ***1999***, *64*, 6411–6417.
79. a. Oppolzer, W.; Moretti, R. and Thomi, S., *Tetrahedron Lett.*, ***1989***, *30*, 5603–5606; b. Oppolzer, W.; Rosset, St. and De Brabander, J., *Tetrahedron Lett.*, ***1997***, *38*, 1539–1540.
80. a. Lehman de Gaeta, L.S.; Czarniecki, M. and Spaltenstein, A., *J. Org. Chem.*, ***1989***, *54*, 4004–4005; b. Decicco, C.P.; Nelson, D.J.; Corbett, R.L. and Dreabit, J.L., *J. Org. Chem.*, ***1995***, *60*, 4782.
81. Evans, D.A.; Urpi, F.; Somers, T.C.; Clark, J.S. and Bilodeau, M.T., *J. Am. Chem. Soc.*, ***1990***, *112*, 8215–8216.
82. a. Evans, D.A.; Bartroli, J. and Shih, T.L., *J. Am. Chem. Soc.*, ***1981***, *103*, 2127; b. Gage, J.R.; Evans, D.A.; *Org. Synth.* ***1990***, *68*, 83–91; c. Evans, D.A.; Britton, T.C.; Dorow, R.L.; Dellaria, J. F.; *Tetrahedron* ***1988***, *44*, 5525–5540; d. Evans, D.A.; Chapman, K.T. and Bisaha, J., *J. Am. Chem. Soc.*, ***1988***, *110*, 1238–1256.
83. Oppolzer, W.; Blagg, J.; Rodriguez, I. and Walther, E., J. *Am. Chem. Soc.*, ***1990***, *112*, 2767–2772.
84. Ho, G.-J. and Mathre, D.J., *J. Org. Chem.*, ***1995***, *60*, 2271–2273.
85. Jordan, A.C.; Axford, L.C.; Harding, J.R.; O'Connell, Y.; Simpson, T.J. and Willis, C.L., *J. Label. Comp. Radiopharm.*, ***2007***, *50*, 338–341.
86. Oppolzer, W.; Moretti, R. and Zhou, C., *Helv. Chim. Acta*, ***1994***, *77*, 2363–2380.
87. Voges, R., *J. Label. Compd. Radiopharm.*, ***2002***, *45*, 867–897.

88. Lee, J.Y.; Chung, Y.J. and Kim, B.H., *Synlett*, ***1994***, 197.
89. a. Evans, D.A.; Ellman, J.A. and Dorow, R., *Tetrahedron Lett.*, ***1987***, *28*, 1123–1126; b. Oppolzer, W.; Starkemann, C.; Rodriguez, I. and Bernadelli, G., *Tetrahedron Lett.*, ***1991***, *32*, 61–64.
90. a. Evans, D.A.; Britton, T.C. and Ellman, J.A., *Tetrahedron Lett.*, ***1987***, *28*, 6141–6144; b. Oppolzer, W., Lienhard, P., *Helv. Chim. Acta*, ***1992***, 75, 2572–2582.
91. Hasegawa, T.; Yamamoto, H., *Synlett.*, ***1998***, 882–883.
92. Evans, D.A., Morrissey, M.M. and Dorow, R.L., *J. Am. Chem. Soc.*, ***1985***, *107*, 4346–4348.
93. Evans. D.A.; Sjogren, E.B.; Bartroli, J. and Dow, R.L., *Tetrahedron Lett.*, ***1986***, *27*, 4957–4960.
94. Evans, D.A. and Weber, A.E., *J. Am. Chem. Soc.*, ***1986***, *108*, 6757–6761.
95. Oppolzer. W.; Darcel, C.; Rochet, P.; Rosset, St and De Brbander, J., *Helv. Chim. Acta*, ***1997***, *80*, 1319–1337.
96. a. Evans, D.A. and Bender, S.L., *Tetrahedron Lett.*, ***1986***, *27*, 799–802; b. Evans. D.A.; Bender, S.L. and Morris, J., *J. Am. Chem. Soc.*, ***1988***, *110*, 2506.
97. a. Wallace, M.A.; Dean, D.C.; Ellsworth, R.L.; Melillo, D.G.; Marks, T. and White, R.F., *J. Label. Compd. Radiopharm.*, ***1993***, *33*, 697–702; b. Wallace, M.A.; Dean, D.C.; Ellsworth, R. L.; Marks, T.; Melillo, D.G. and Masurekar, P., in *Synthesis and Applications of Isotopically Labeled Compounds 1994*, Allen, J. and Voges, R., Eds.; John Wiley & Sons: Chichester, England, ***1995***, pp. 605–612.
98. Czeskis, B.A.; O'Bannon, D.D.; Wheeler, W.J. and Clodtfelter, D.K., *J. Label. Comp. Radiopharm.*, ***2005***, *48*, 85–100.
99. Moenius, Th.; Troxler, Th. and Masero, R., *J. Label. Compd. Radiopharm.*, ***2007***, *50*, 613–615.
100. a. Voges, R. and Metz, Y., in *Synthesis and Applications of Isotopically Labeled Compounds*, Vol. 7, Pleiss, U. and Voges, R., Eds.; John Wiley & Sons: Chichester, England, ***2001***, pp. 307–310; b. Rodriguez, I.; Metz, Y. and Voges, R., *J. Label. Compd. Radiopharm.*, ***2001***, *44*, 963–964.
101. Wadsworth, A.H. and Lawrie, K.W.M., *J. Label. Compd. Radiopharm.*, ***2007***, *50*, 500–501.
102. Zhang, Y.; Hong, Y.; Huang, C.C. and Belmont, D., *J. Label. Compd. Radiopharm.*, ***2006***, *49*, 237–244.
103. Goldstein, S.W.; Overman, L.E. and Rabinowitz, M.H., *J. Org. Chem.*, ***1992***, *57*, 1179–1189.
104. a. Romo, D. and Meyers, A.I., Chiral non-racemic lactams. Vehicles for the construction of natural and unnatural products containing quaternary carbon centers, *Tetrahedron*, ***1991***, *47*, 9503–9569; b. Meyers, A.I.; Wallace, R.H.; Harre, M. and Garland, R., *J. Org. Chem.*, ***1990***, *55*, 3137–3143; c. Meyers, A.I.; Lefker, B.A.; Sowin, T.J. and Westrum, L.J., *J. Org. Chem.*, ***1989***, *54*, 4243–4246.
105. Giner, J.-L.; Hormann, R. and Arigoni, D., in *Synthesis and Applications of Isotopically Labeled Compounds 1994*, Allen, J. and Voges, R. Eds.; John Wiley & Sons: Chichester, England, ***1995***, pp. 723–726.
106. Cowden, J.C. and Paterson, I., in *Organic Reactions*, Paquette, L.A., Ed.; Wiley: Chichester, England, ***1997***, *51*, 3–200.
107. Evans, D.A., Nelson, J.V.; Vogel, E. and Taber, T.R., *J. Am. Chem. Soc.*, ***1981***, *103*, 3099–3111.
108. Evans, D.A.; Rieger, D.L.; Bilodeau, M.T. and Urpi, F., *J. Am. Chem. Soc.*, ***1991***, *113*, 1047–1049.
109. Hesk, D.; Bowlen, C.; Hendershot, S.; Koharski, D.; McNamara, P.; Rettig, D. and Saluja, S., *J. Label. Compd. Radiopharm.*, ***1996***, *38*, 1039–1046.
110. Ekhato, I.V.; Liao, Y. and Plesescu, M.J., *J. Label, Compd. Radiopharm.*, ***2004***, *47*, 821–835.
111. Evans, D.A.; Tedrow, J.S.; Shaw, J.T. and Downey, C.W., *J. Am. Chem. Soc.*, ***2002***, *124*, 392–393.
112. a. Walker, M.A. and Heathcock, C.H., *J. Org. Chem.*, ***1991***, *56*, 5747–5750; b. Guerlavais, V.; Carroll, P.J. and Jouillié, M.M., *Tetrahedron: Asymmetry*, ***2002***, *13*, 675–680.
113. Oppolzer, W. and Lienard, P., *Tetrahedron Lett.*, ***1993***, *34*, 4321–4324.
114. a. McKillican, B.P.; Perine, J.W.; Simmons, D.P. and Wilson, T.E., *J. Label. Compd. Radiopharm.*, ***2005***, *48*, 25–30; b. McKillican, B.P.; Simmons, D.P. and Wilson, T.E.in *Synthesis and*

*Applications of Isotopically Labeled Compounds*, Vol. 7, Pleiss, U. and Voges, R., Eds.; John Wiley & Sons: Chichester, England, ***2001***, pp. 193–196.
115. Hesk, D.; Bignan, G.; Lee, J.; Yang, K.; Voronin, K.; Magatti, C.; McNamara, P.; Koharski, D.; Hendershot, S.; Saluja, S. and Wang. S., *J. Label. Compd. Radiopharm.*, ***2002***, *45*, 145–155.
116. Woo, P.W.K., *Tetrahedron Lett.*, ***1985***, *26*, 2973–2976.
117. Devant, R.; Mahler, U. and Braun, M., *Chem. Ber.*, ***1988***, *121*, 397–406.
118. Lee, H.T. and Woo, P.W.K., *J. Label. Compd. Radiopharm.*, ***1999***, *42*, 129–133.
119. a. Murakami, N.; Sugimoto, M. and Kobayashi, M., *Bioorg. Med. Chem.*, ***2001***, *9*, 57–67; b. Crimmins, M.T. and She, J., *Synlett*, ***2004***, *8*, 1371–1374; c. Crimmins, M.T.; King, B.W.; Tabet, E.A. and Chaudhary, K., *J. Org. Chem.*, ***2001***, *66*, 894–902; d. Crimmins, M.T. and Chaudary, K., *Organic Lett.*, ***2000***, *2*, 775–777; e. Crimmins, M.T.; King, B.W. and Tabet, E.A., *J. Am. Chem. Soc.*, ***1997***, *119*, 7883–7884.
120. Axford, L.C.; Simpson, T.J. and Willis, C.L., *Angew. Chem. Intl. Edn Engl.*, ***2004***, *43*, 727–730.
121. Rossiter, B.E. and Swingle, N.M. *Asymmetric Conjugate Addition* in *Chem. Rev.*, ***1992***, *92*, 771–778.
122. Evans, D.A.; Bilodeau, M.T.; Somers, T.C.; Clardy, J.; Cherry, D. and Kato, Y., *J. Org. Chem.*, ***1991***, *56*, 5750–5752.
123. a. Brener, M. and Seebach, D., *Helv. Chim. Acta*, ***1999***, *82*, 2365–2374; b. Hintermann, T. and Seebach, D., *Helv. Chim. Acta*, ***1998***, *81*, 2093–2126.
124. a. Clare, J.E.; Lawrie, K.W.M.; Willis, C.L. and Yuen, J., in *Synthesis and Applications of Isotopically Labeled Compounds*, Vol. 8, Dean, D.C.; Filer, C.N. and McCarthy, K.E., Eds.; John Wiley & Sons: Chichester, England, ***2004***, pp. 1–4; b. Clare, J.E.; Willis, C.L.; Yuen, J.; Lawrie, K.W.M.; Charmant, J.P.H. and Kantacha, A., *Tetrahedron Lett.*, ***2003***, *44*, 8153–8156; c. Crosby, S.R.; Hateley, M.J. and Willis, C.L., *Tetrahedron Lett.*, ***2000***, *41*, 397–401.
125. Moenius, Th, Seiler, M.P., Nozulak, J. and Burtscher, P., *J. Label. Compd. Radiopharm.*, ***2002***, *45*, 1180–1181.
126. a. Duthaler, R.O., Recent development in the stereoselective synthesis of $\alpha$–amino acids, *Tetrahedron Report Number 349, Tetrahedron*, ***1994***, *50*, 1539–1650; b. Kelly, M.N.; Sutherland, A. and Willis, C.L., *Nat. Prod. Reports*, ***1997***, 205–219.
127. a. Evans, D.A.; Britton, T.C.; Dorow, R.L. and Dellaria, J.F. Jr., *J. Am. Chem. Soc.*, ***1986***, *108*, 6395–6397; b. Trimble, L.A. and Vederas, J.C., *J. Am. Chem. Soc.*, ***1986***, *108*, 6397–6399.
128. a. Evans, D.A.; Britton, T.C.; Ellman, J.A. and Dorow, R.L., *J. Am. Chem. Soc.*, ***1990***, *112*, 4011–4030; b. Evans, D.A. and Ellman, J.A., *J. Am. Chem. Soc.*, ***1989***, *111*, 1063–1072; c. Evans, D.A. and Britton, T.C., *J. Am. Chem. Soc.*, ***1987***, *109*, 6881–6883.
129. a. Oppolzer, W. and Tamura, O., *Tetrahedron Lett.*, ***1990***, *31*, 991–994; b. Oppolzer, W.; Tamura, O. and Deerberg, *Helv. Chim. Acta*, ***1992***, *75*, 1965–1978.
130. a. Lee, H.T.; Hicks, J.L. and Johnson, D.R., *J. Label. Compd. Radiopharm.*, ***1991***, *29*, 1065–1072; b. Herbert, J.M., *J. Label. Compd. Radiopharm.*, ***1998***, *41*, 859–862.
131. Parry, R.L.; Ju, S. and Baker, B.J., *J. Label. Compd. Radiopharm.*, ***1991***, *29*, 633–643.
132. a. Lodwig, S.N. and Unkefer, C.J., *J. Label. Compd. Radiopharm.*, ***1996***, *38*, 239–248; b. Lodwig, S.N.; Silks, L.A. and Unkefer, C.J., *J. Label. Compd. Radiopharm.*, ***1996***, *38*, 161–167.
133. Oppolzer, W., personal communication to RV, October 19, ***1993***
134. Martin, A.; Chassaing, G. and Vanhove, A., in *Synthesis and Applications of Isotopically Labeled Compounds 1994*, Allen, J. and Voges, R., Eds.; John Wiley & Sons: Chichester, England, ***1995***, pp. 761–766; b. Josien, H.; Martin, A. and Chassaing, G., *Tetrahedron Lett.*, ***1991***, *32*, 6547–6550.
135. Burtscher, P.; Kohler, B.; Metz, Y.; Voges, R. and Wenger, R., *J. Label. Compd. Radiopharm.*, ***2000***, *43*, 205–216; b. Moenius, Th.; Voges, R.; Burtscher, P. and Züger, Ch., *J. Label. Compd. Radiopharm.*, ***2000***, *43*, 1245–1252; c. Moenius, Th.; Voges, R.; Mahnke, M.; Burtscher, P.; Metz, Y. and Guenat, Ch., *J. Label. Compd. Radiopharm.*, ***1999***, *42*, 29–41.
136. a. Koltai, E.; Alexin, A.; Rutkai, Gy. Toth-Sarudy, E., *J. Label. Compd. Radiopharm.*, ***1998***, *41*, 977–982; b. Lodwig, S.N. and Unkefer, C.J., *J. Label. Compd. Radiopharm.*, ***1998***, *41*, 983–991; c. Unkefer, C.J. and Lodwig, S.N., in *Synthesis and Applications of Isotopically Labeled Compounds*, Buncel, E. and Kabalka, G.W., Eds.; Elsevier: Amsterdam, ***1992***, pp. 337–342.

137. Oppolzer, W.; Moretti, R. and Thomi, S., *Tetrahedron Lett.*, ***1989***, *30*, 6009–6010.
138. a. Rees, D.O.; Bushby, N.; Harding, J.R.; Song, C. and Willis, C.L., *J. Label. Compd. Radiopharm.*, ***2007***, *50*, 399–401; b. Rees, D.O.; Bushby, N.; Cox, R.J.; Harding, J.R.; Simpson, T.J. and Willis, C.L., *ChemBioChem*, ***2007***, *8*, 46–50; c. Workman, J.A.; Garrido, N.P., Sanson, J.; Roberts, E., Wessel, H.P. and Sweeney, J.B., *J. Am. Chem. Soc.*, ***2005***, *127*, 1066–1067.
139. a. Myers, A.G.; Gleason, J.L.; Yoon, T. and Kung, D.W., *J. Am. Chem. Soc.*, ***1997***, *119*, 656–673; b. Myers, A.G. and Gleason, J.L., *J. Org. Chem.*, ***1996***, *61*, 813–815; c. Myers, A. G.; Yoon, T. and Gleason, J.L., *Tetrahedron Lett.*, ***1995***, *36*, 4555–4558; d. Myers, A.G.; Gleason, J.L. and Yoon, T., *J. Am. Chem. Soc.*, ***1995***, *117*, 8488–8489; e. Myers, A.G.; Yang, B.H., Chen, H. and Gleason, J.L., *J. Amer. Chem. Soc.*, ***1994***, *116*, 9631–9632; f. Myers, A.G.; Yang, B.H.; Chen, H.; McKinstry, L.; Kopecky, D.J. and Gleason, J.L., *J. Am. Chem. Soc.*, ***1997***, *119*, 6496–6511.
140. Voges, R.; Kohler, B.; Metz, Y. and Riss, B., in *Synthesis and Applications of Isotopically Labeled Compounds*, Vol. 7, Pleiss, U. and Voges, R., Eds.; John Wiley & Sons: Chichester, England, ***2001***, pp. 295–299.
141. a. McIntosh, J.M. and Leavitt, R.K., *Tet. Lett.*, ***1986***, *27*, 3839;. b. McIntosh, J.M. and Mishra, P., *Can. J. Chem.*, ***1986***, *64*, 726–734.
142. Wiegerinck, P.H.G.; Hofstede, L.W.H.; Sperling, F.M.G.M. and Vader, J.F., *J. Label. Compd. Radiopharm.*, ***2008***, *51*, 195–201.
143. Andres, H.; Schnelli, P.; Sandoz Pharma AG, CH-Basel, Switzerland; personal communication to RV; lecture ICI meeting, Oxford 3. ***1989***
144. a. Seebach, D.; Dziadulewicz, E.; Behrendt, L.; Cantoreggi, S. and Fitzi, R., *Liebigs Ann. Chem.*, ***1989***, 1215–1232; b. Fitzi R. and Seebach, D., *Tetrahedron*, ***1988***, *44*, 5277–5292; c. Seebach, D. and Fadel, A., *Helv. Chim. Acta*, ***1985***, *68*, 1243–1250; d. Blank, St. and Seebach, D., *Angew. Chem. Int. Edn Engl.*, ***1993***, *32*, 1765–1766; e. Seebach, D.; Gees, T. and Schuler, F., *Liebigs Ann. Chem.*, ***1993***, 785–793; f. Fitzi, R. and Seebach, D., *Angw. Chem.*, ***1986***, *98*, 354–363.
145. Seebach, D.; Juaristi, E.; Miller, D.D; Schickli, C. and Weber, T, *Helv. Chim. Acta*, ***1987***, *70*, 237–261.
146. a. Fasth, K.J.; Malmborg, P. and Langström, B., *J. Label. Compd. Radiopharm.*, ***1991***, *30*, 401–402; b. Plenevaux, A.; Al-Darwich, M.J.; Lemaire, Ch.; Delfiore, G. and Comar, D., *Appl. Radiat. Isot.*, ***1994***, *45*, 361–369.
147. a. Zhang, Y. and Huang, C.C., in *Synthesis and Applications of Isotopically Labeled Compounds*, Vol. 8, Dean, D.C.; Filer, C.N. and McCarthy, K.E., Eds.; John Wiley & Sons: Chichester, England, ***2004***, pp. 21–23; b. Lankiewicz, L.; Nyassé, B.; Fransson, B.; Grehn, L. and Ragnarsson, U., *J. Chem. Soc. Perkin Trans. I*, ***1994***, 2503–2510; c. Karstens, W.F.J.; Berger, H.J.F.F.; van Haren, E.R.; Lugtenburg, J. and Raap, J., *J. Label. Compd. Radiopharm.*, ***1995***, *36*, 1077–1096; d. Raap, J.; van der Wielen, C.M.; Lugtenburg, J., *Recl. Trav. Chim. Pays-Bas*, ***1990***, *109*, 277–286; e. Subramassian, P.K., Woodard, R.W., *Int. J. Peptide Protein Res.*, ***1986***, *28*, 579–585.
148. a. Takatori, K; Hayashi, A.; Kajiwara, M., *J. Label. Compd. Radiopharm.*, ***2004***, *47*, 787–795; b. Takatori, K.; Toyama, S.; Narumi, S.; Fujii, S. and Kajiwara, M., *J. Label. Compd. Radiopharm.*, ***2004***, *47*, 91–94; c. Takatori, K.; Nishihara, M. and Kajiwara, M., *J. Label. Compd. Radiopharm.*, ***1999***, *42*, 701–708.
149. a. Dellaria, J.F. Jr., and Santarsiero, B.D., *J. Org. Chem.*, ***1989***, *54*, 3916–3926; b. Dellaria, J.F. and Santarsiero, B.D., *Tetrahedron Lett.*, ***1988***, *29*, 6079–6082.
150. Evans, D.A. and Weber, A.E., *J. Am. Chem. Soc.*, ***1987***, *109*, 7151–7157.
151. a. El Achqar, A.; Boumzebra, M.; Roumestant, M.L. and Viallefont, P., *Tetrahedron*, ***1988***, *44*, 5319–5332; b. El-Hadrami, M.; Lavergne, J.-P.; Viallefont, Ph.; Chiaroni, A.; Riche, C. and Hasnaoui, A., *Synth. Commun.*, ***1993***, *23*, 157–163; c. Solladié-Cavallo, A. and Simon, M.C., *Tetrahedron Lett.*, ***1989***, *30*, 6011–6014.
152. Li, S.; Pang, J; Wilson, W.K. and Schroepfer, G.J. Jr., *J. Label. Compd. Radiopharm.*, ***1999***, *42*, 815–826.
153. Evans, D.A.; Sjogren, E.B.; Weber, A.E. and Conn, R.E., *Tetrahedron Lett.*, ***1987***, *28*, 39–42.

154. Schöllkopf, U., Enantioselective synthesis of nonproteinogenic amino acids, in *Top. Curr. Chem.*, ***1983***, *109*, 65–84.
155. Blaser, D. and Seebach, D., *Liebigs Ann.Chem.* ***1991***, 1067–1078.
156. Kurosawa, M.; Kanamaru, H. and Nishioka, K., *J. Label. Compd. Radiopharm.*, ***1996***, *38*, 285–297.
157. a. Beulshausen, T.; Groth, U. and Schöllkopf, U., *Liebigs Ann. Chem.*, ***1991***, 1207–1209; b. Groth, U.; Schöllkopf, U. and Tiller, T., *Liebigs Ann. Chem.*, ***1991***, 857–860.
158. a. Jones, T.K.; Mills, S.G.; Reamer, R.A.; Askins D.; Desmond, R.; Volante, R.P. and Shinkai, I., *J. Am. Chem. Soc.*, ***1989***, *111*, 1157–1159; b. Jones, T.K.; Reamer, R.A.; Desmond, R. and Mills, S.G., *J. Am. Chem. Soc.*, ***1990***, *12, 2998.*
159. Evans, D.A.; Kaldor, S.W.; Jones, T.K.; Clardy, J. and Stout, T.J., *J. Am. Chem. Soc.*, ***1990***, *112*, 7001–7031.
160. a. Evans, D.A.; Gage, J.R.; Leighton, J.L. and Kim, A.S., *J. Org. Chem.*, ***1992***, *57*, 1961–1963; b. Evans, D.A.; Gage, J.R. and Leighton, J.L., *J. Am. Chem. Soc.*, ***1992***, *14*, 9434–9453; c. Evans, D.A.; Glorius, F.; Burch, J.D., *Organic Lett.*, ***2005***, *7*, 3331–3333.
161. Askin, D.; Reamer, R.A.; Joe, D.; Volante, R.P. and Shinkai, I., *Tetrahedron Lett.*, ***1989***, *30*, 6121–6124.
162. a. Abdel-Magid, A.; Lantos, I. and Pridgen, L.N., *Tetrahedron Lett.*, ***1984***, *25*, 3273–3276; b. Abdel-Magid, A.; Pridgen, L.N.; Eggleston, D.S. and Lantos, I., *J. Am. Chem. Soc.*, ***1986***, *108*, 4595–4602; c. Pridgen, L.N.; Abdel-Magid, A.F.; Lantos, I.; Shilcrat, S. and Eggleston, D.S., *J. Org. Chem.*, ***1993***, *58*, 5107–5117.
163. Zhou, C., Asymmetric alkylation, Michael addition and aldol reactions of Cα–nitrogen or bromo–substituted *N*–acylsultams; Thesis, University of Geneva, Switzerland, ***1993***, pp. 63–64
164. Rodriguez, I. and Voges, R., *J. Label. Compd. Radiopharm.*, ***2000***, *43*, 169–176.
165. a. Oppolzer, W., *Pure & Appl. Chem.*, ***1990***, *62*, 1241–1250; b. Oppolzer, W., *Pure & Appl. Chem.*, ***1988***, *60*, 39–48; c. Oppolzer, W; Poli, G.; Kingma, A.J.; Starkemann, C. and Bernardinelli, G., *Helv. Chim. Acta*, ***1987***, *70*, 2201–2214; d. Oppolzer, W.; Kingma, A.J. and Poli, G., *Tetrahedron*, ***1989***, *45*, 479–488; e. Oppolzer, W. and Kingma, A.J., *Helv. Chim. Acta*, ***1989***, *72*, 1337–1345.
166. a. Nicolás, E.; Russell, K.C. and Hruby, V.J., *J. Org. Chem.*, ***1993***, *58*, 766–770; b. Han, Y. and Hruby, V.J., *Tetrahedron Lett.*, ***1997***, *38*, 7317–7320; c. Qian, X.; Russel, K.C.; Boteju, L.W. and Hruby, V.J., *Tetrahedron*, ***1995***, *51*, 1033–1054; d. Williams, D.R.; Kissel, W.S. and Li, J.J., *Tetrahedron Lett.*, ***1998***, *39*, 8593–8596; e. Bergdahl, M.; Iliefski, T.; Nilsson, M. and Olsson, Th., *Tetrahedron Lett.*, ***1995***, *36*, 3227–3230; f. Davies, S.G. and Sanganee, H.J.; *Tetrahedron: Asymmetry*, ***1995***, *6*, 671–674.
167. a. Helmchen, G. and Wegner, G., *Tetrahedron Lett.*, ***1985***, *26*, 6051–6054; b. Urban, E.; Riehs, G. and Knühl, G., *Tetrahedron Lett.*, ***1995***, *36*, 4773–4776; c. Urban, E.; Knühl, G. and Helmchen, G., *Tetrahedron*, ***1996***, *52*, 971–986.
168. a. Evans, D.A; Chapman, K.T. and Bisaha, J., *J. Am. Chem. Soc.*, ***1984***, *106*, 4261–4263; b. Evans, D.A.; Chapman, K.T. and Bisaha, J., *Tetrahedron Lett.*, ***1984***, *25*, 4071–4074.
169. a. Oppolzer, W., *Angew. Chem., Int. Edn Engl.*, ***1984***, *23*, 876–889; b. Oppolzer, W.; Dupuis, D.; Poli, G.; Raynham, T.M. and Bernardinelli, G., *Tetrahedron Lett.*, ***1988***, *29*, 5885–5888; c. Oppolzer, W.; Chapuis, C and Bernadelli, G., *Helv. Chim. Acta*, ***1984***, *67*, 1397–1401; d. Vandewalle, M.; Van der Eycken, J.; Oppolzer, W. and Vullioud, C., *Tetrahedron*, ***1986***, *42*, 4035–4043; e. Oppolzer, W.; Rodriguez, I.; Blagg, J. and Bernardinelli, G., *Helv. Chim. Acta*, ***1989***, *72*, 123–130; f. Achmatowicz, M.; Chapuis, C.; Rzepecki, P. and Jurczak, J., *Helv. Chim. Acta*, ***1999***, *82*, 182–190; g. Thom, C.; Kocienski, P. and Jarowicki, K., *Synthesis*, ***1993***, 475–477; h. Helmchen, G.; Karge, R. and Weetman, J., in *Modern Synthetic Methods*, Vol. 4; Scheffold R., Ed.; Springer Verlag: Berlin, ***1986***, pp. 261–306.
170. a. Kim, B.H. and Curran, D.P., *Tetrahedron*, ***1993***, *49*, 293–319; b. Curran, D.P. and Heffner, T. A., *J. Org. Chem.*, ***1990***, *55*, 4585–4595; c. Curran, D.P.; Kim, B.H.; Daugherty, J. and Heffner, T.A., *Tetrahedron Lett.*, ***1988***, *29*, 3555–3558.
171. Vallgarda, J. and Hacksell, U., *Tetrahedron Lett.*, ***1991***, *32*, 5625–5628.

172. a. Song, C.E.; Roh, E.J.; Lee, S. and Kim, I.O., *Tetrahedron: Asymmetry*, ***1995***, *6*, 871;. b. Lee, A. W.M.; Chan, W.H.; Yuen, W.H.; Xia, P.F. and Wong, W.Y., *Tetrahedron: Asymmetry*, ***1999***, *10*, 1421–1424; c. Oppolzer, W. and Barras, J.-P., *Helv. Chim. Acta*, ***1987***, *70*, 1666–1674.
173. a. Fletcher, M.D.; Harding, J.R.; Hughes, R.A.; Kelly, N.M.; Schmalz, H.; Sutherland, A. and Willis, C.L., *J. Chem. Soc. Perkin Trans. 1*, ***2000***, 43–52; b. Harding, J.R.; Hughes, R.A.; Kelly, N.M.; Sutherland, A. and Willis, C.L., *J. Chem. Soc. Perkin Trans. 1*, ***2000***, 3406–3416.
174. Abele, S. and Seebach, D., *Eur. J. Org. Chem.*, ***2000***, 1–15.
175. Yamamoto, Y.; Yamamoto, S.; Yatagai, H.; Ishihara, Y. and Maruyama, K., *J. Org. Chem.*, ***1982***, *47*, 119–126.

# 12

# Biotransformations in the Preparation of Compounds Labeled with Carbon and Hydrogen Isotopes

Biotransformations are selective chemical modifications of molecules by biological entities ranging from isolated enzymes (cell-free systems) to intact organisms[1]. In particular cases they are exploited by the chemical and pharmaceutical industries for production of chemicals; this approach can be especially advantageous for structurally complex compounds[2]. Preparative biotransformations have been used on a laboratory scale by radiochemists to prepare isotopically labeled compounds[3] necessary for drug development and other applications in the life sciences. Their synthetic use for isotope labeling includes:

- selective modifications of individual functional groups;
- synthesis of structurally complex molecules from less complex isotopically labeled precursors;
- synthesis of key intermediates for *reconstitution* approaches to labeled compounds (see Chapter 10).

Enzymes are proteins able to catalyze reactions with very high regio- and/or stereo-selectivities under mild conditions. Those enzymes combining low substrate specificity (broad applicability) with high product selectivity are of optimal suitability for organic synthesis. Further increase of structural variety and stability by protein- and medium-engineering[4] has extended their synthetic applicability significantly and made them important tools in organic synthesis. The broad scope of their synthetic use is documented in numerous papers, reviews and textbooks[5]. Although successful in particular applications, biotransformations are less common in isotopic labeling.

As the complexity of the biological system varies from highly purified enzymes to whole cell-containing systems, so do some of the practical dimensions of the experiments vary. Therefore this chapter is organized accordingly.

*Preparation of Compounds Labeled with Tritium and Carbon-14* Rolf Voges, J. Richard Heys and Thomas Moenius

## 12.1 Application of Isolated Enzymes

In one-step biotransformations particularly, isolated enzymes have distinctive advantages, since they can be used simply, almost like any chemical reagent, and – importantly for isotopic labeling – they do not result in any isotope dilution. Since a large number of enzymes are commercially available, many types of reactions can be carried out.

Examples of the use of isolated enzymes in isotopic labeling include *N*- or *O*-acylations, hydrolyses, redox reactions and group transfer reactions. Most of them aim for applications such as racemate resolution, catalysis of reactions and synthesis of enantiomerically pure molecules.

### 12.1.1 Optical Resolutions via Derivatives

Despite the availability of several chemical strategies for diastereo- and enantioselective syntheses in isotopic labeling (see Chapter 11), kinetic racemate resolution remains a major application of enzymes in the chemistry of isotopically labeled compounds. Depending on the molecular entity, different possibilities have been described for achieving enantioselective differentiation through either derivatization (e.g. *N*-acyl formation and cleavage, *O*-acyl formation and cleavage) or selective degradation (e.g. redox reactions).

***N*-Acyl Formation:** Enzymatically catalyzed *N*-acylation of racemates is used for preparative synthesis of carbon-14-labeled (*S*)-amino acids[6]. (*R*)-Amino acid acetyl transferase (EC group[7]: 2.3.1) isolated from fresh yeast, acylates (*R*)-amino acids selectively in the presence of acetyl-CoA and makes them easily separable from the unacylated (*S*)-amino acids. This enzyme shows low substrate specificity. Out of 43 amino acids investigated, only one (cysteine) showed no stereospecificity and a few (e.g. glutamic acid, proline, isovaline, asparagine) gave acylation rates too low for preparative purposes. (*S*)-Amino acids thus prepared usually contain <0.1% of the corresponding (*R*)-isomers. In contrast, isolated acylated (*R*)-amino acids are contaminated with up to 13% of nonenzymatically acylated (*S*)-amino acids; this limits the utility of the procedure to the synthesis of enantiomerically pure (*S*)-amino acids. Radiochemical yields of up to 80% (relative to the racemate) were obtained by repeated cycling via racemization of the separated, undesired (*R*)-enantiomer (azlactone-pathway: AcOH, $Ac_2O$; 100 °C, 0.5 h; hydrolysis: 2 N HCl; 100 °C, 3 h).

***N*-Acyl Cleavage:** Preparative isolation of labeled (*R*)-amino acids from the racemate has been performed by selective hydrolysis of the *N*-acetyl derivatives, as applied to (*R*)-[$^{14}$C]-serine for example[8]. After selective hydrolysis of *N*-acetyl-(*S*)-[$^{14}$C]serine by natural aminoacylase the unreacted *N*-acetyl-(*R*)-[$^{14}$C]serine was separated and subsequently hydrolyzed chemically to the (*R*)-[$^{14}$C]serine. The same approach (except using porcine kidney acylase) was applied to the synthesis of enantiomerically pure [$^{13}$C]alanine[9]; in this case both enantiomers were obtained in pure form. The commercially available aminoacylases of porcine kidney and *Asperigillus sp.* show distinctively different substrate specificities[10], especially with regard to $\alpha$-methyl-$\alpha$-amino acids. *O*-Acyl-, *N*-acyl-*N*-alkyl amino acids (including *N*-acylproline) are not suitable substrates for aminoacylases.

***O*-Acyl Cleavage:** Carboxypeptidases (EC 3.4) are exopeptidases, which catalyze the release of C-terminal (*S*)-amino acids from peptides. Carboxypeptidase variants include A (for neutral and acidic amino acids), B (for basic amino acids) and Y (accepting a broad range of substrates). Carboxypeptidase A was used for the selective hydrolysis of the trifluoroacetate of (*S*)-[3-$^{14}$C]tyrosine from the racemate to provide its enantiomerically pure (*S*)-isomer on a hundred-mCi scale[8,11].

Selective hydrolysis of esters is a well-established procedure for the resolution of chiral carboxylic acids. Enzymes such as hydrolases, lipases and proteases are utilized. Due to their high selectivity for (*S*)-amino acids, proteases have been widely used in the selective transformations of amino acids and their derivatives. $\alpha$-Chymotrypsin – a serine protease – catalyzes not only the hydrolysis of amide bonds, but also the cleavage of various esters, including $\alpha$-alkyl-$\alpha$-amino acid esters. One application is the synthesis of (*S*)-$\alpha$-[$^{14}$C]-methyltryptophan (**3**)[12]. The anion synthesized by LDA-deprotonation of *N*-benzylidene tryptophan methyl ester (**1**) was alkylated with [$^{14}$C]methyl iodide to obtain the $\alpha$-[$^{14}$C]-methyl *N*-benzylidene derivative **2**, which was hydrolyzed under acidic conditions. Subsequent selective cleavage of the ester group with $\alpha$-chymotrypsin provided the enantiomerically pure (*S*)-amino acid **3** in 33% radiochemical yield.

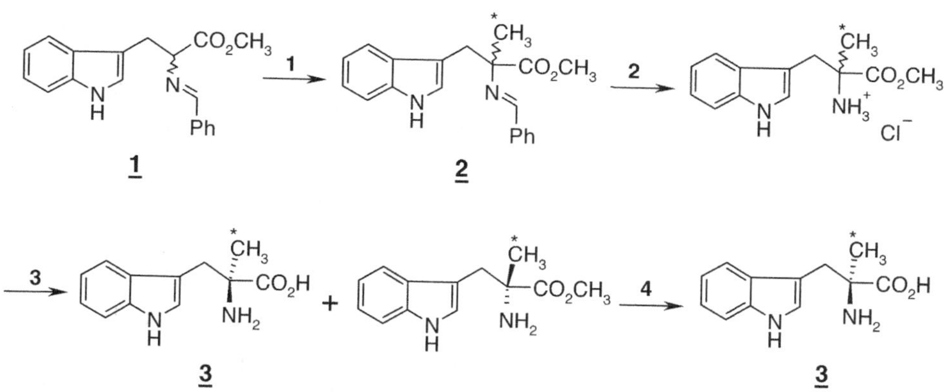

**Reaction conditions: 1a**. LDA, THF; –78 °C, 30 min; **b**. $^{14}CH_3I$, THF; –78 °C, 4 h; 25 °C, 12 h; **2**. HCl, $Et_2O$; 25 °C; **3**. $\alpha$-chymotrypsin, pH 5.0; 37 °C, 24 h, **4**. flash chromatography

***Figure 12.1*** *Synthesis and resolution of (S)-α-[$^{14}$C]methyltryptophan (**3**)*

Preparation of the naturally occurring nonproteinogenic amino acids (2*S*,1′*R*)- and (2*S*,1′*S*)-cyclopentenyl[2-$^{13}$C]glycine **5** and **7**[13] (Figure 12.2) extends this approach to the optical resolution of molecules having more than one chiral center. Chemical synthesis produced the four diastereomers of the compound, which were separated by achiral chromatography into the mixture of the two racemates (2*R*,1′*S*), (2*S*,1′*R*) (**4a,b**) and (2*R*,1′*R*), (2*S*,1′*S*) (**6a,b**). Treatment with $\alpha$-chymotrypsin at pH 6.9 resolved the enantiomers; subsequent separation of the free acids from the unreacted esters, followed by chemical hydrolysis of the *N*-acyl moiety (1 M trifluoroacetic acid; reflux, 6 h) afforded the required (2*S*,1′*R*)- and (2*S*,1′*S*)-cyclopentenyl[2-$^{13}$C]glycine diastereomers **5** and **7** in optical purities of >96.5 and 98% e.e., respectively (determined by NMR spectroscopy). The yields of 30% and 47% are significantly higher than those reported (6%) for a more tedious chemical approach[14].

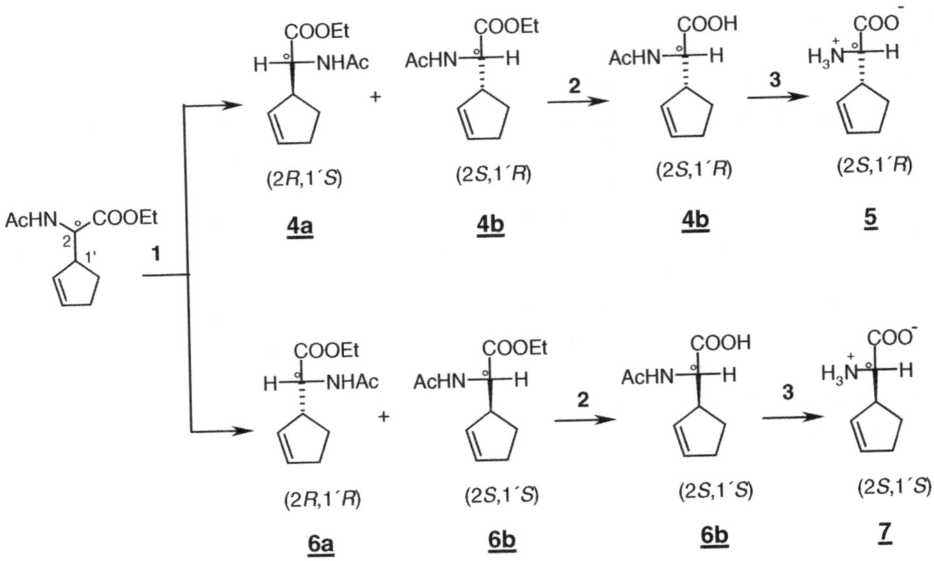


**Reaction conditions: 1**. separation by HPLC; **2**. α-chymotrypsin, MeOH, 0.5 M $NH_4OAc$, pH 6.9; 24 h; **3**. 1 M trifluoroacetic acid; reflux, 6 h

***Figure 12.2*** *Synthesis of (2S,1′R) and (2S,1′S)-cyclopentenyl[2-$^{13}$C]glycine*

Other enzymes widely used in organic syntheses are lipases. Lipases are serine hydrolases that catalyze the hydrolysis of lipids to fatty acids. An example of their use is in the resolution of racemic methyl 4-benzyloxy-2-(hydroxymethyl)[1-$^{13}$C]butyrate (8), an intermediate for the carbon-13 labeling of the antiviral agents penciclovir and famciclovir. From a limited number of candidates *Candida cylindracea* lipase turned out to be the enzyme of choice. At optimal reaction conditions (pH 5–7, 35 °C) no nonenzymatic hydrolysis was detectable. Due to insufficient enzymatic stereospecificity, however, the hydrolysis provided the unreacted dextrorotary ester **9** at an optical purity of 94% e.e.[15] and the hydrolyzed levorotary acid **10** at an optical purity of 70% e.e.

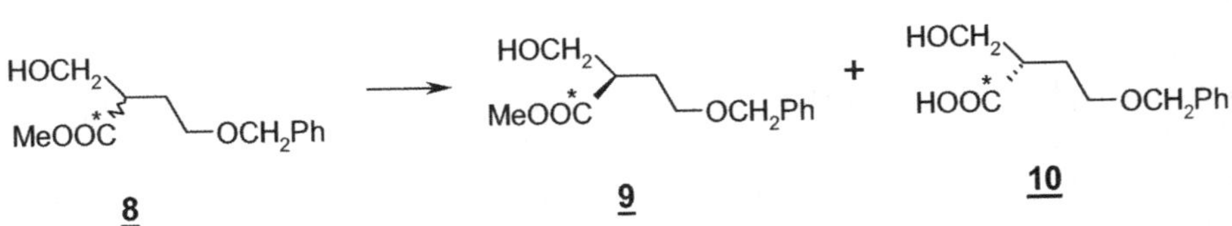


**Reaction conditions:** lipase from *Candida cylindracea*, water, pH 5.0, 35 °C

***Figure 12.3*** *Synthesis of (+)-methyl 4-benzyloxy-2-(hydroxymethyl)[1-$^{13}$C]butyrate by resolution*

***O*-Acyl Formation:** Compared with hydrolysis reactions, esterification and transesterification reactions are much slower and require the use of activated esters to facilitate the reaction and to make it kinetically irreversible. These include trichloroethyl esters, trifluoroethyl esters, enol esters, thioesters and vinylcarbamates. Lipases isolated from *Pseudomonas* species are highly selective for the hydrolysis of esters of secondary alcohols, and therefore also for the corresponding reverse reactions.

An illustrative example is the resolution of the carbon-14-labeled hydroxyindane derivative **11**, an intermediate in the synthesis of the serotonin uptake inhibitor MDL 27777A. Under catalysis by lipase from *Pseudomonas flourescens*, vinyl acetate reacted with the alcohol **11** in *tert*-butyl methyl ether solvent. After flash-chromatography on silica gel the acylated (**12**, 43% yield; 97% e.e.) and nonacylated (**13**, 47% yield, 98% e.e.) compounds were obtained in excellent optical purities[16].

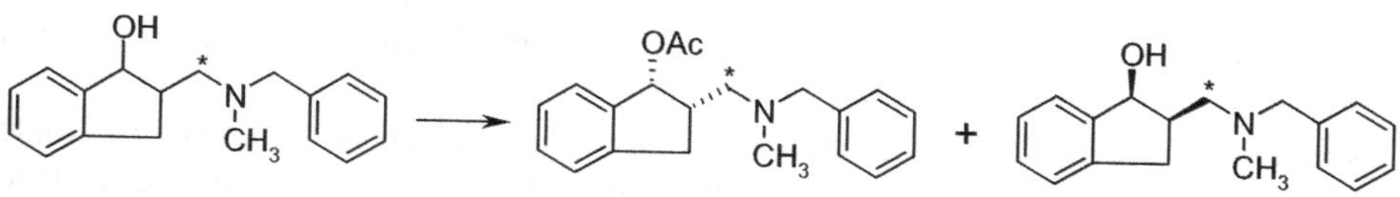


11 12 13

**Reaction conditions:** vinyl acetate (2 equiv.), *tert*-butyl methyl ether, Lipase P; 22 °C, 4 d

***Figure 12.4*** *Kinetic resolution of a hydroxyindane derivative by lipase-catalyzed transesterification*

**Redox reactions:** Another way to resolve racemic mixtures is the enzyme-catalyzed oxidative transformation of the undesired enantiomer. This approach was used[17] for the synthesis of (*S*)-[3-$^{11}$C]valine (**14**). For this purpose chemically synthesized racemic [3-$^{11}$C]valine was treated with immobilized (*R*)-amino acid oxidase (EC 1.4.3.3). This converted the undesired (*R*)-enantiomer to [3-$^{11}$C]pyruvic acid (**15**), which could be easily removed by HPLC. (*S*)-[3-$^{11}$C]Valine was obtained in e.e.s of 90–99%, even though reaction times were held to 17 min or less because of the 20 minute half-life of carbon-11.

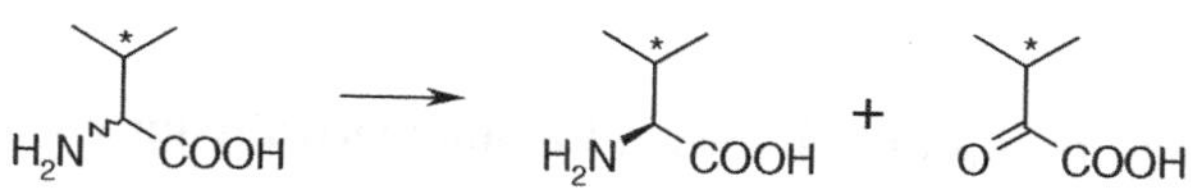


14 15

**Reaction conditions:** R-amino acid oxidase, 0.1 M tetrasodium pyrophosphate buffer, pH 8.3, disodium flavine dinucleotide; 37 °C, 10–17 min

***Figure 12.5*** *Resolution of (S)-[3-$^{11}$C]valine by selective oxidative deamination*

Enzymatic resolution of racemates by selective *N*- or *O*-acyl cleavage (described above) requires completely acylated material. Since chemical acylation is often incomplete (even a significant excess of trifluoroacetic anhydride does not completely convert tyrosine into its trifluoroacetyl derivative) the racemate requires prepurification. One way to circumvent this problem is to pretreat the crude acetylated product with amino acid oxidase. For example, in the preparation of (*S*)-[3-$^{14}$C]tyrosine, the crude racemic trifluoroacetate was treated with (*R*)-amino acid oxidase to quantitatively convert the remaining unacetylated (*R*)-tyrosine into its corresponding pyruvic acid, and subsequent treatment of the mixture with carboxypeptidase A hydrolyzed the (*S*)-trifluoroacetate, resulting in precipitation of the desired (*S*)-[3-$^{14}$C]tyrosine[8].

Racemic 3-hydroxy[3-$^{14}$C]butyric acid was resolved using the stereospecific 3-hydroxybutyrate dehydrogenase enzyme (EC 1.1.1.30) which converted the undesired (*R*)-isomer to [3-$^{14}$C]acetoacetic acid; the latter was decarboxylated *in situ* by addition of perchloric acid and heating. Subsequent purification of the unreacted (*S*)-(+)-enantiomer by preparative TLC on silica gel gave a yield of 32.5% in 98% purity. This procedure was an improvement over a previous method that required the addition of acetaldehyde and alcohol dehydrogenase (EC 1.1.1.1) to the reaction mixture to force the equilibrium redox reaction to completion[18].

**Supplemental use of racemase enzymes:** The main disadvantage of kinetic racemate resolution is the limitation of yield to 50%, unless the undesired isomer is racemized and the resolution repeated. An elegant alternative to this tedious, terative procedure is the dynamic, kinetic racemate resolution (Figure 12.6), which combines a fast racemization equilibrium (*S*)-A⇔ (*R*)-A with an enantioselective transformation in order to accomplish the deracemization process. In addition to the chemical methods (such as thermal-, acid/base-, deprotonation/protonation processes) enzyme-catalyzed methods also find application for this racemization process. The use of racemases (EC. 5.1) has recently been reviewed[19].

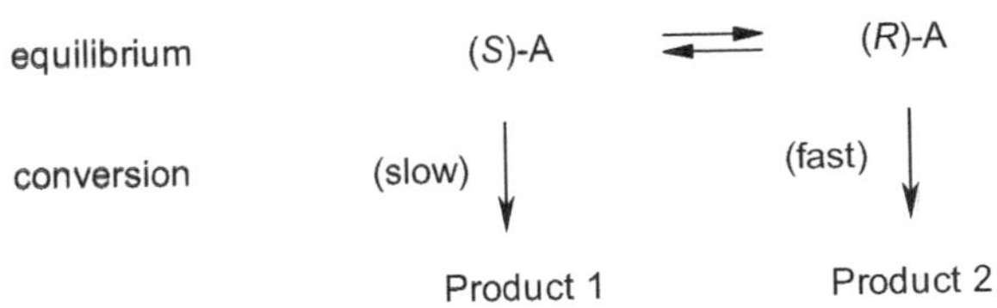


***Figure 12.6*** *Principle of the dynamic, kinetic racemate resolution*

### 12.1.2 Synthesis of Isotopically Labeled, Enantiomerically Pure Compounds

A relatively large number of different enzymes have been used to prepare enantiomerically pure isotopically labeled compounds.

**Use of aminotransferases:** Aminotransferases are pyridoxal 5′-phosphate-dependent enzymes that catalyze the reversible transfer of an amino group from an $\alpha$-amino acid to an $\alpha$-ketoacid in a two-step process. Of the many transaminases, aspartate aminotransferase (EC 2.6.1.1) is synthetically most useful since spontaneous decarboxylation of the generated oxaloacetate to pyruvate shifts the equilibrium to the product side.

Glutamic–pyruvic transaminase from porcine heart (EC 2.6.1.2) was used for the synthesis of (*S*)-[4-$^{13}$C]glutamic acid (**18**) from labeled 2-oxo-glutarate (**17**) and (*S*)-alanine (**16**) as an amino group donor[20].

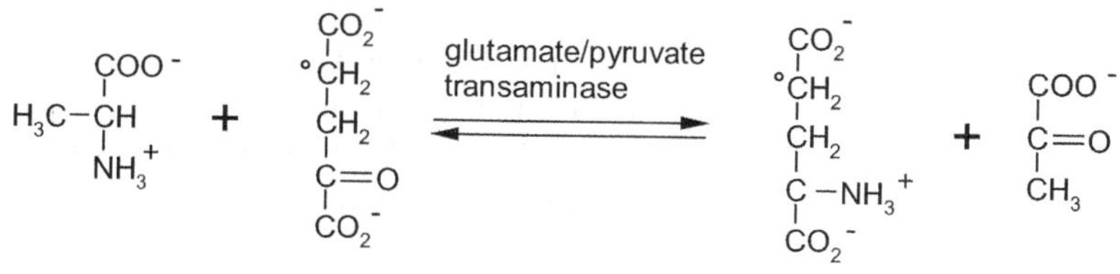


16 17 18

***Figure 12.7*** *Synthesis of (S)-[4-$^{13}$C]glutamic acid by glutamic–pyruvic transaminase (EC 2.6.1.2)*

The same type of conversion – a reductive amination of achiral 2-oxoglutaric to enantiomerically pure (*S*)-glutamic acid – was performed in the presence of ammonium ions, using glutamate dehydrogenase (bovine liver GDH). The necessary coenzyme NADH was regenerated by the simultaneous oxidation of ethanol to acetaldehyde. Performing the reaction under slight vacuum removes the acetaldehyde and shifts the equilibrium to the product side[21]. Chemical and enzymatic reactions to synthesize a variety of differently carbon-13-labeled 2-oxoglutaric acids as well as their conversions into labeled (*S*)-glutamine and (*S*)-proline have been described[21,22].

The combination of transaminases with amino acid oxidases opens opportunities for selectively inverting chiral centers. An example of such an operation is in the synthesis of (*S*)-[1-$^{13}$C]valine (**20**). Racemic [1-$^{13}$C]valine, produced by Strecker synthesis from isobutyraldehyde, K$^{13}$CN and ammonia in 83% yield, was treated with (*R*)-amino oxidase and branched-amino acid amino transferase in order to invert the (*R*)-enantiomer into the (*S*)-enantiomer via the $\alpha$-ketoacid **19**. Addition of a substantial excess of glutamic acid drove the aminotransferase reaction to completion, giving (*S*)-[1-$^{13}$C]valine in 75% overall yield from K$^{13}$CN on a 30 mmol scale. Due to the low substrate specificity of the enzymes, it is anticipated that this approach will also be effective for most standard amino acids[23].

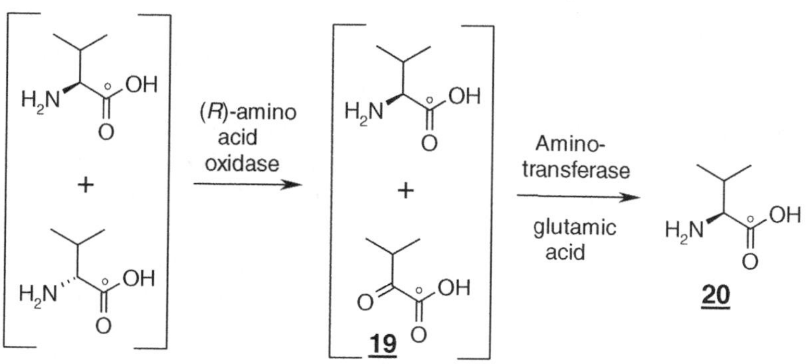


19 20

**Reaction conditions:** 100 mM Tris HCl, pH 8.5, bovine serum, pyridoxal 5′-phosphate, catalase, (R)-amino acid oxidase, branched-amino acid amino transferase

***Figure 12.8*** *Enzymatic transformation of (R,S)-[1-$^{13}$C]valine into its e.p. (S)-enantiomer*

**Applications of ammonia lyases:** Ammonia lyases catalyze the reversible, regio- and stereoselective addition of ammonia to alkenes. The family of lyases includes aspartate, methyl aspartate, histidine, tyrosine and phenylalanine ammonia lyases. The system consisting of phenylalanine ammonia lyase (PAL, EC 4.3.5.1), cinnamic acid, water and ammonium ions was optimized in terms of pH, temperature, reaction time, concentration and buffer for the synthesis of (2*S*,3*S*)- and (2*S*,3*R*)-[3-$^3$H]phenylalanines.

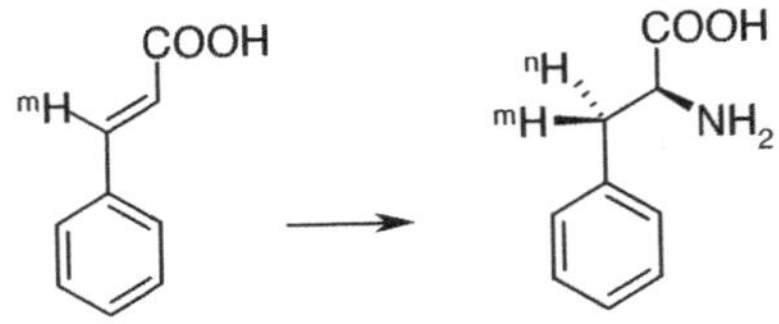


**Reaction conditions:** $m$ = 3/n=1 and $H_2O$ or $m$ = 1/n=3 and $^3$HHO, phenylalanine ammonia lyase, aq. $NH_3$, pH 10; 30 °C, 120 h

***Figure 12.9*** *Synthesis of (2S,3S)- and (2S,3R)-[3-$^3$H]phenylalanine*

This method enables the selective preparation of diastereomerically tritiated (2*S*)-phenylalanines, in which either of the $\beta$-hydrogens is substituted with tritium (Figure 12.9). Use of tritiated water gives (2*S*,3*S*)-[3-$^3$H]phenylalanine ($m = 1$; $n = 3$) whereas use of tritiated cinnamic acid provides (2*S*,3*R*)-[3-$^3$H]phenylalanine ($m = 3$; $n = 1$). Chemical yields were 34–38% on a 0.1 mmol scale, and product specific activities were 25 mCi/mmol and 3 mCi/mmol, respectively[24]. It is likely that increasing the specific activity of [$^3$H]-cinnamic acid would have produced (2*S*,3*R*)-product of significantly higher specific activity. This cannot be expected for the (2*S*,3*S*)-diastereomer, because the enzyme requires a significant volume of aqueous buffer for activity, which would preclude the use of very high specific activity tritiated water. An analogous procedure was used to prepare (*S*)-phenylalanine isotopomers labeled with carbon-14 at C1 and C2[25]. Due to the substrate specificity of PAL, application to the corresponding (*S*)-tyrosine isotopomers failed. Instead, the latter were prepared from the [$^{14}$C]phenylalanines by oxidation with (*S*)-phenylalanine 4′-monooxygenase (EC 1.14.16.1)[25].

**Applications of tryptophan synthetase:** Tryptophan synthetase (EC 4.2.1.20) is a pyridoxal phosphate-dependent enzyme that, in the cell, catalyzes the $\alpha,\beta$-elimination of water from serine to form a pyridoxyl-bound $\alpha$-aminoacrylate, which undergoes Michael addition of indole to form the named amino acid. This type of reaction has been used to prepare (*S*)-tryptophan isotopomers with a variety of labeling patterns by use of different labeled indoles and (*S*)-serines[26] in yields of up to 98% based on indole and 92% based on (*S*)-serine.

The enzyme's ability to form this Michael acceptor intermediate has been exploited for the preparation of other labeled amino acids by using alternative nucleophiles. For example, tryptophan synthetase, isolated from a strain of *E. coli* engineered to overexpress this enzyme, was applied to the synthesis of several carbon-13 and deuterium isotopomers of *S*-benzyl-(*S*)-cysteine from incubations of the corresponding labeled (*S*)-serines in the presence of benzyl mercaptan as the nucleophile instead of indole. The resulting

*S*-benzyl-(*S*)-[$^2$H/$^{13}$C]cysteines were chemically deprotected using sodium/liquid ammonia to obtain the corresponding [$^2$H/$^{13}$C]cysteines[27].

**Applications of lactate dehydrogenase:** L-Lactate dehydrogenases (LDH) prepared from mammalian and bacterial sources, are available commercially. They are known to produce (2*S*)-$\alpha$-hydoxy acids with different hydrocarbon side-chains by reduction of $\alpha$-keto acids in high enantiomeric purities and good chemical yields (Figure 12.10). As NADH-dependent enzymes, LDHs are used in combination with other enzymes, such as formate dehydrogenase (FDH), to regenerate the co-factor *in situ*.

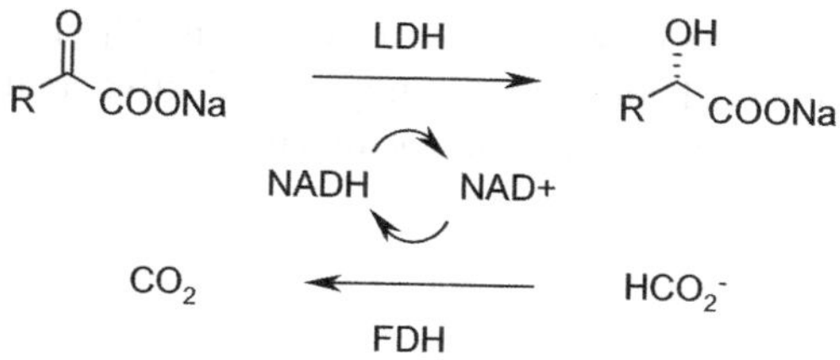


***Figure 12.10*** *(S)-Lactate dehydrogenase-mediated reduction of $\alpha$-keto acids*

LDH isolated from *Bacillus stearothermophilus* (BS-LDH) was applied[28] in this protocol to a range of isotopically labeled novel $\alpha$-keto acids with side chains containing olefinic, ketonic, methyl ether and acetal moieties in scales of 1–100 mmol. The resulting (2*S*)-$\alpha$-hydroxy acids were converted, via their triflate derivatives, to the corresponding (2*R*)-amino acids in >75% chemical yields and >98% e.e.s by two alternative methods:

- $S_{N2}$ displacement with sodium azide in DMF and catalytic reduction of the resultant azide under catalytic hydrogenation conditions; or
- Treatment with the Li-salt of $(Boc)_2NH$ and subsequent acid-catalyzed cleavage of the Boc protecting groups.

The complementary approach to the synthesis of isotopically labeled (2*R*)-hydroxy esters (<u>22</u>) and (2*S*)-amino acids using the corresponding (*R*)-lactate dehydrogenase from *Staphylococcus epidermis* (SE-LDH) suffers from its significantly lower substrate specificity. One way to circumvent this problem is to invert the stereogenic center of <u>21</u> chemically by application of the Mitsunobu reaction to give the (2*R*)-hydroxy esters <u>22</u> (99% e.e.) and thence the corresponding (2*S*)-$\alpha$-amino acids.

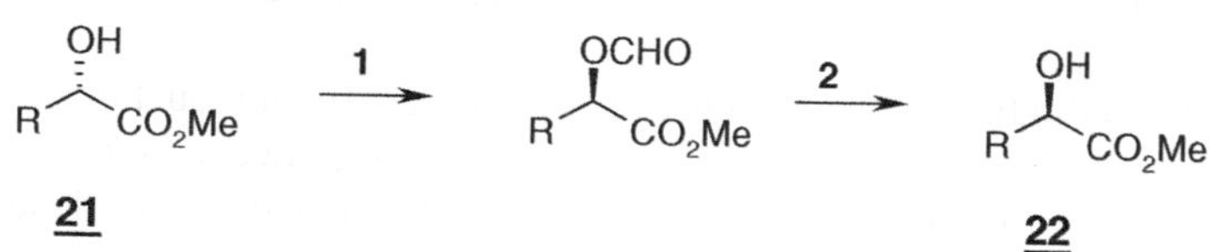


**Reaction conditions: 1.** $HCO_2H$, $PPh_3$, DEAD; **2.** $H^+$, MeOH

***Figure 12.11*** *Stereoinversion of (2S)-hydroxy esters under Mitsunobu conditions*

**Use of amino acid dehydrogenases:** Direct conversion of $\alpha$-keto acids into (*S*)-$\alpha$-amino acids is also accomplished by amino acid dehydrogenases. A series of (*S*)-$\alpha$-amino acids (including Ala, allo-Ileu, Ileu, Leu, Phe, Ser, Val) have been prepared from $\alpha$-keto acids in good yields and excellent enantioselectivities. Through use of [$^{15}$N]ammonium formate as the source of isotope label, this approach has provided access to deuterium- and carbon-13- as well as nitrogen-15-labeled isotopomers[29]. A good example is the preparation of (*S*)-leucine isotopomers **26**, selectively labeled in either diastereotopic methyl group. The requisite keto intermediates **25** were prepared either by diastereoselective alkylations using Evans' *N*-propionyl oxazolidinone **23** or Michael addition of [$^{13}$C]methylcuprate to the *N*-crotonyl bornane-10,2-sultam **24**[30] followed by $C_2$- and $C_1$-chain elongation, respectively. Subsequent leucine dehydrogenase-catalyzed reductive aminations converted the $\alpha$-keto acids into the isotopomeric $\alpha$-amino acids **26**.

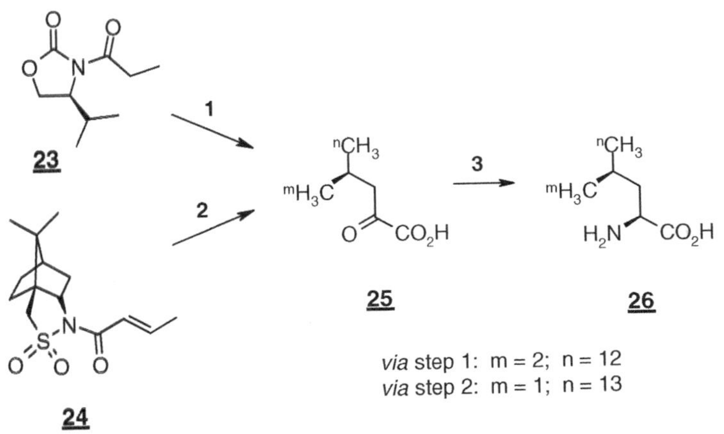


**Reaction conditions: 1a**. NaHMDS, $C^2H_3I$, THF, HMPA, **b**. $LiAlH_4$, THF, **c**. Ph3P, $Br_2$, $PhNO_2$, **d**. Mg, THF, $Et_2O$, then $(CO_2Et)_2$, −78 °C; **2a**. $(^{13}CH_3)_2CuLi{\cdot}PBu_3$, **b**. LiOH, $H_2O_2$, **c**. $Ph_3PCHCN$, EDCI, DMAP, **d**. $O_3$, MeOH, $CH_2Cl_2$; **3a**. NaOH, **b**. leucine dehydrogenase, NADH, $HCO_2NH_4$, FDH

***Figure 12.12*** *Stereoselective synthesis of (2S)-leucine derivatives (**26**) isotopomeric in their methyl groups*

A potential drawback of this approach is the limited substrate specificity of this enzyme superfamily, which prevents application particularly to those substrates having heteroatoms in the side chain. MOM-protection of the hydroxy group, however, allowed extension of this approach to the labeling of $\alpha$-amino-$\beta$-hydroxy acids, such as (2*S*, 3*S*)-[$^{15}$N]allothreonine and (2*S*, 3*R*)-[$^{15}$N]threonine[31].

**Use of oxidoreductases:** Second only to hydrolases, oxidoreductases are the most widely used enzyme classes in organic chemistry. Their significance for isotope labeling derives from their utility not only for the direct introduction of tritium and deuterium labels but also for regioselective oxidations of functional groups.

Enzyme-catalyzed redox processes involve the transfer of two electrons in either one or two steps, each process requiring different coenzymes. Nicotinamide adenine dinucleotide (NAD) and its 2′-phosphate (NADP) coenzymes are involved in many two-electron oxidations catalyzed by dehydrogenases. The nicotinamide ring system is redox active, accepting either a hydride or two electrons and a proton to form the 1,4-dihydronicotinamides NADH or NADPH. Because of the high costs of the isotopic cofactors NAD$^n$H and NADP$^n$H (and even the nonisotopic versions), they are usually used in catalytic quantities that are regenerated *in situ* by a coupled enzymatic reduction using less expensive isotope sources. The complete set of stereospecifically deuterated coenzymes is available through the systems comprising $^2$HCOOH/formate dehydrogenase, $(CH_3)_2C^2HOH$/alcohol dehydrogenase, or [1-$^2$H]glucose/glucose dehydrogenase[32].

An application of this approach is the synthesis of chiral primary alcohols labeled with tritium at one of the enantiotopic $\alpha$-hydrogens[33]. Liver alcohol dehydrogenase, for example, reduces benzaldehyde to (*R*)-[methylene-$^3$H]benzyl alcohol (**27**) in the presence of NAD/NAD$^3$H, which is regenerated by oxidation of (*R*,*S*)-[1-$^3H_1$]ethanol to acetaldehyde. Because the enzyme transfers exclusively the Re-positioned isotope of (*R*,*S*)-[1-$^3H_1$]ethanol, the theoretical radiochemical yield is only 50% (the other 50% remains in acetaldehyde), and the specific activity of the product is only half that of this isotope source. One way to overcome this drawback is to use achiral, secondary alcohols such as [1-$^3$H]-cyclohexanol or [1-$^3$H]cyclopentanol (**28**) as isotope sources. The corresponding (*S*)-antipode of **27** would be accessible by analogous reduction of benz[$^3$H]aldehyde using nonisotopic cofactors.

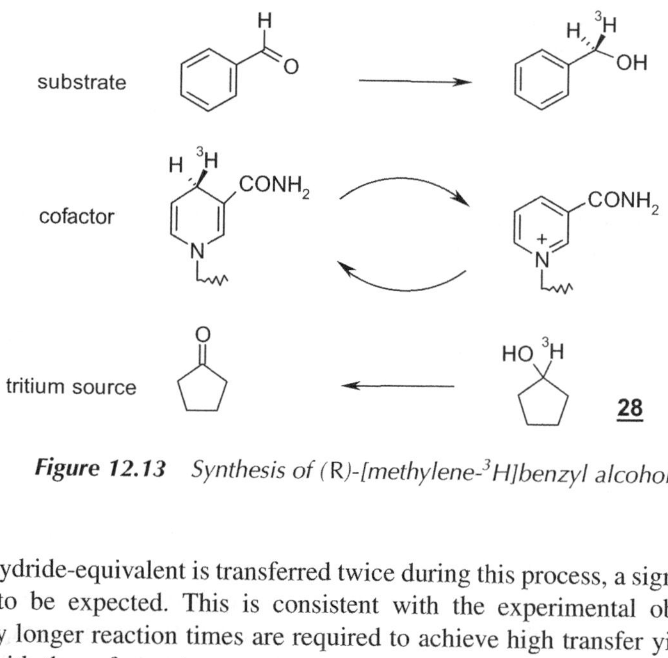


**27**

**28**

***Figure 12.13*** *Synthesis of (*R*)-[methylene-$^3$H]benzyl alcohol*

Since a hydride-equivalent is transferred twice during this process, a significant isotope effect has to be expected. This is consistent with the experimental observation that significantly longer reaction times are required to achieve high transfer yields of tritium compared with those for protons.

Application of the isotopes hydrogen, deuterium and tritium enabled the synthesis of 'chiral acetate', first reported in 1969. Its chemical and enzymatic syntheses as well as its use in investigation of stereochemical details of biochemical reactions are described in a recent paper[34].

### 12.1.3 Conjugation Reactions

Metabolic Phase II transformations include conjugations with glucuronic acid, sulfuric acid, glutathione, glycine and acetic acid. Glucuronide conjugates are polar metabolites that facilitate the renal excretion of drugs; their relevance to organismal biochemistry and their chemical and enzymatic syntheses have been reviewed[35].

The main Phase II metabolite of valproic acid (**29**), a broad spectrum anticonvulsant, is the acyl glucuronide 1-*O*-valproyl-*β*-D-glucopyranuronic acid (**30**). This metabolite was prepared in carbon-14 and tritium-labeled forms enzymatically in one step in around 52% yields using UDP-glucuronosyltransferase[36]. The positions of labels were not specified. In contrast, the chemical synthesis starting from valproic acid and 2,3,4,6-tetrabenzyl-D-glucopyranose required three steps and provided the required *β*-anomer in only 6.8% yield.

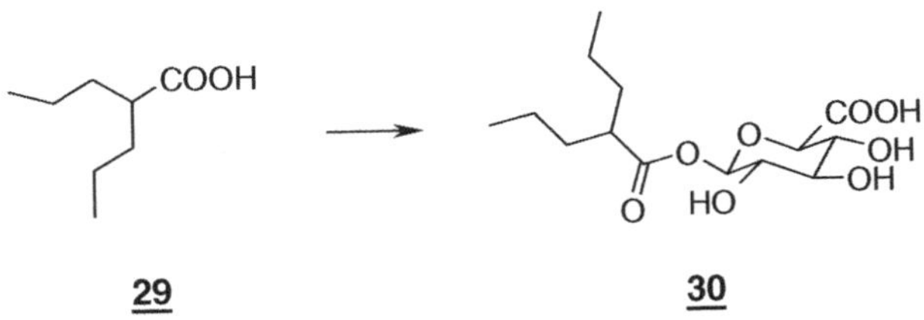


**Reaction conditions:** UDP-glucuronic acid, UDP-glucuronosyltransferase, glycerine–NaOH buffer, pH 8.0; 37 °C, 48 h

***Figure 12.14*** *Synthesis of 1-O-[$^{14}$C/$^{3}$H]valproyl-β-D-glucopyranuronic acid*

**Use of multi-enzyme systems:** Combining two or more purified enzymes in the same pot can facilitate the preparation of more structurally complex molecules. One example of this is the bi-enzymatic system of PG-synthase and PGH-converting enzyme isolated from various mammalian tissues. This mixture was used to converted various tritium-labeled polyunsaturated fatty acids to prostaglandins D, E and F and thromboxane B at high specific activities and yields of 20–80%[37].

## 12.2 Application of Cell-Containing Systems

The use of living cell systems to carry out the preparation of complex isotopically labeled compounds is long established[3b]. They have also been used to achieve functional group transformations in cases where the use of isolated enzymes is not feasible, as when the latter are commercially unavailable, difficult to isolate, unstable and/or require complex cofactor systems to retain their activity. Cell preparations, for example homogenized tissues and partially purified fractions therefrom, are often used for similar reasons, although the applications may be very different. Homogenized liver tissues or microsomal fractions obtained from their centrifugation can be easily prepared for exploitation of their enzyme activities in oxidative metabolism or conjugation. On the other hand, multistep biosyntheses of complex labeled compounds can be conducted using whole-cell systems, such as bacterial or fungal cell fermentations, or living plant tissues or even entire organisms. Biosyntheses using whole-cell systems usually present more complex experimental challenges, ranging from maintenance of good conditions for their growth to ensuring that the radioactive products possess the desired specifications.

### 12.2.1 Transformations of Functional Groups

As for isolated enzymes, cell-containing systems have been used for a number of functional group transformations. Examples range from redox reactions to group-transfer reactions.

**Redox reactions:** In order to mitigate the environmental hazard of 2,4,6-trinitrotoluene (TNT) a procedure was developed for its reduction[38]. Treatment with *Saccharomyces cervisiae* (32 °C, 3–4 d) converted carbon-14-labeled TNT to a mixture of its four mono- and diamino- metabolites in a one-step procedure.

Microbial hydroxylation is a widely used method for the synthesis of mammalian Phase I metabolites. An example is the synthesis of 1′-hydroxy[$^{13}C_4$]midazolam (**31**)[39] using a culture of *Beauveria sulfurescens* ATCC7159. Of particular interest is the stereo- and regioselective hydroxylation of [butyrate-1-$^{14}$C]compactin (**32**) by *Mucor hiemelus* MF5021 (pH 5.3; 28 °C, 48 h) to [$^{14}$C]pravastatin (**33**)[40] of >95% radiochemical purity in 32% yield.

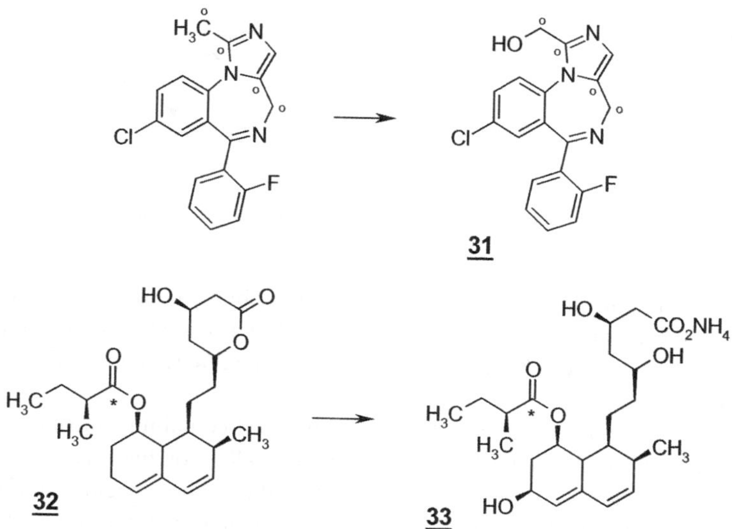


*Figure 12.15* Preparations of [$^{13}C_4$]hydroxymidazolam and [$^{14}$C]pravastatin by microbial hydroxylation

Additional applications make use of microsomes and liver homogenates. An example is the liver microsome-mediated *in vitro* transformation of 2,6-diethyl[U-$^{14}$C]aniline into 4-amino-3,5-diethyl[U-$^{14}$C]phenol, one of the predominant metabolites of the herbicide alachlor in rats[41]. S9 liver homogenate was used to synthesize the main metabolite **35** of fasudil by hydroxylation of the labeled drug substance **34** (66% yield)[42].

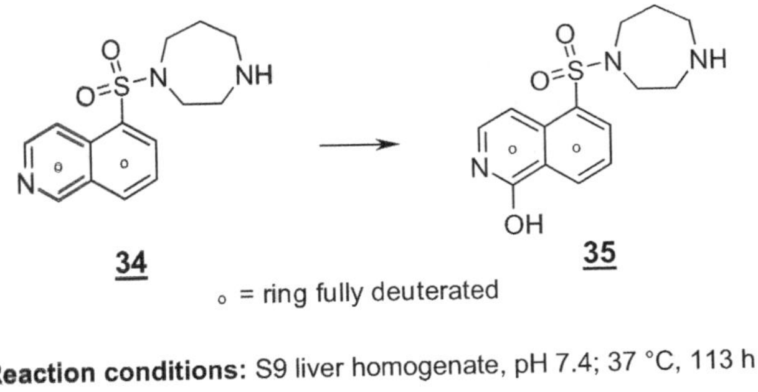


***Figure 12.16*** *Synthesis of the main metabolite of Fasudil™*

**Hydrolysis of nitriles:** Hydrolysis of [$^{14}$C]nitrile functions represents a well-established route for the synthesis of [$^{14}$C]carboxylic acids. For cases in which the harsh reaction conditions of chemical hydrolysis (e.g., 2 N NaOH, reflux) are incompatible with sensitive functionalities, application of enzymes might provide an alternative (Figure 12.17). Enzymatic hydrolysis of nitriles results in amides if catalyzed by *nitrile hydratases* or in the corresponding carboxylic acids if catalyzed by nitrilases[43].

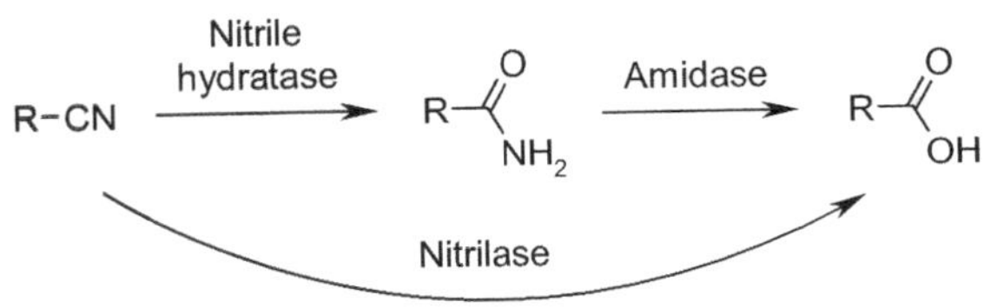


***Figure 12.17*** *Enzyme-catalyzed hydrolysis of nitriles*

Hydrolysis of the nitriles **36a** and **36b**[44] by an immobilized enzyme system (nitrilase/amidase activity) prepared from *Rhodococcus sp.* SP409 provided the corresponding acids **37** in 40–50% yields. However, the more bulky substrate **36c** was unreactive under these conditions (Figure 12.18).

A recent publication[45] describes the use of the cell-containing nitrilase activity of *Agrobacterium* DSM6336 and *Rhodococcus* sp SRL4281 for hydrolysis of the nitrile moiety of **38** to the carboxyl group of **39** in 75–80% radiochemical yield, without affecting the ancillary methyl ester function. Based on experimental data with unlabeled analogs there was evidence that purified enzymes of a commercially available nitrilase kit would have been successful as well. Both cell strains and the enzyme kit, however, failed to hydrolyze sterically more demanding substrates, such as **40**.

**Transfer reactions:** Preparative glucuronidation of labeled compounds has been reported using liver homogenates, hepatic microsomes[46] and intact organs. For example, the acyl glucuronide of racemic 2-phenyl[1-$^{14}$C]propionic acid was prepared using isolated perfused rat liver[47], and a 770 mg quantity (19% yield) of the glucuronide of [$^{2}H_{10}$] cipemastat/Ro 32-6414) was obtained[48] from the urine of female marmosets dosed with the deuterium-labeled parent drug.

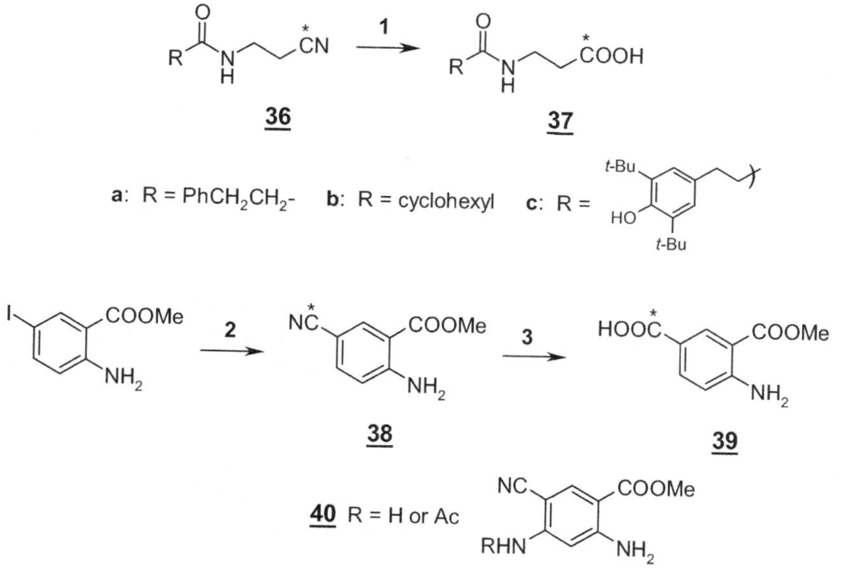


**Reaction conditions:** **1**. nitrilase *Rhodococcus* sp. (SP409), 0.1 μM phosphate buffer, pH 7; **2**. $Cu^{14}CN$, CuI, DMF, 160 °C; **3**. nitrilase, *Rhodoccus* 4281 or *Agrobacterium* DSM1663

***Figure 12.18*** *Microbial hydrolysis of [$^{14}$C]nitrile derivatives*

### 12.2.2 Fermentative Synthesis of Structurally Complex Molecules by Incorporation of Labeled Precursors

Addition of isotopically labeled precursors – simple, readily available compounds – to an appropriate biological system can result in their intact or metabolically modified incorporation into complex products – primary or secondary metabolites – which might be difficult to obtain by chemical synthesis.

**General considerations:** The biosynthetic incorporation of added (labeled) intermediates usually competes with their endogenous nonlabeled counterparts; therefore they are rarely undiluted in their transformation into products. The efficiency of utilization of these exogenous labeled precursors, and the isotopic abundance found in the labeled product is characterized by terms such as incorporation (Inc), dilution factor (Dil) and specific incorporation (Spec Inc)[49,79a]. For cases in which precursor molecules are incorporated into product molecules in a 1:1 ratio, these terms are defined as follows.

$$\text{Inc} = \text{total radioactivity (product)}/\text{total radioactivity (precursor)} \times 100(\%)$$
$$\text{Dil} = \text{molar specific activity (precursor)}/\text{molar specific activity (product)}$$
$$\text{Spec Inc} = \text{molar specific activity (product)}/\text{molar specific activity (precursor)} \times 100(\%)$$

In order to describe the catalytic efficiency of microorganisms, the productivity number PN was introduced[50]. PNs allow comparison of the efficiency of whole cells or crude extracts of cells in bioconversions.

$$PN = \text{product [mmol]/weight of biocatalyst [kg]} \times \text{reaction time [h]}$$

These measures are strongly influenced by a variety of experimental parameters such as the producing organism, the medium, the precursor, the process conditions (temperature, concentration of the precursor, time-points of addition and harvest) and, for radioactive precursors, the radiosensitivity of the organism.

Since the experimental parameters determine not only the quality of the process (i.e. amount of waste, capacity, costs) but also the chemical and physical characteristics (i.e. position of label, chemical and biological stability, isotopic abundance) of the labeled target molecule, they need to be carefully optimized in pilot studies.

**Experimental set-up:** With a few exceptions, microbial labeling techniques use batch-procedures. In case of radioactive isotopes special radiation safety activities might be necessary, especially when volatile activity is applied or generated during the fermentation process. For example, a chamber system is described for the production of isotopically labeled tobacco[51], which provides both for the safe handling of radioactivity and for control of various environmental factors such as relative humidity, temperature, carbon dioxide, nutrients and auxiliary lighting. For aerobic fermentations, simple, air-tight, temperature-controlled shaker-systems[52] are usually adequate, allowing removal of any evolved $^{14}CO_2$ from the air by circulation through aqueous 2 N NaOH. When tritiated precursors are used, work-up of the fermentation broth has to take into consideration the likelihood of byproduct tritium in the aqueous medium and in exchangeable positions of all components.

**Primary metabolites:** Primary metabolites are compounds vital to the growth, development and function of the organism (i.e. its structure, its metabolism and/or its cell regulation); examples are proteins, peptides, poly- and oligosaccharides, nucleic and deoxynucleic acids and lipids. Labeled with isotopes through biotransformation processes, they can be broken down by subsequent chemical or enzymatic processes to furnish their labeled monomeric components (i.e. $\alpha$-amino acids, carboxylic acids, nucleotides, sugars etc.) for a wide variety of uses.

An important example for this approach is the incorporation of labeled $^{13/14}CO_2$ in phototrope biomass. Algae, yeast, bacteria, plant parts (leaf, root) or even intact plants are used, depending on the target molecule. It has the distinct advantage of being inexpensive and relatively straightforward experimentally. In addition, the method can give products having very high Spec Inc and general labeling of all carbon positions.

Selection criteria for the organism include its physiological properties, the yield of the fermentation product, the by-product spectrum and the organism's resistance to radiation. An example is the algae *anacystica nidulans*, which compared with *chlorella* is significantly more radiation resistant[53]. This offers the opportunity for assimilation of isotopically pure $^{14}CO_2$, making possible the preparation of fermentation products of nearly theoretical specific activities (Dil approaching 1; isotopic dilution by inoculation is normally negligible). Plants are generally more radiosensitive than algae and isolated leaves.

Typical of this approach, incubation of *anacystica nidulans* results in a mixture of highly labeled nucleotides, nucleosides, desoxynucleic acids, fatty acids and proteins, which subsequently requires preparative separation. Applications of this technique are described in earlier publications[54]. More recent reports describe the preparation of uniformly labeled α-amino acids and proteins/peptides[55]; saccharides such as amylose, sucrose, glucose, fructose, and starch[56]; fatty acids; menthol[57]; solanesol[58] and shikimic acid[59].

Knowledge of physiological processes opens additional opportunities. For example, inhibition of 5-enolpyruvylshikimate 3-phosphate synthase by addition of the herbicide glyphosphate suppresses metabolic degradation, resulting in the accumulation of shikimic acid and thereby facilitating its preparative synthesis[59].

Exemplifying the potential of this approach is its application to the synthesis of [$^{11}$C]-glucose. Photoassimilation of Swiss chard plants converted 1.5 Ci of $^{11}CO_2$ into a mixture of labeled sucrose, fructose and glucose. The optimized process, including incubation, extraction, and separation produced 24–30 mCi of [$^{11}$C]glucose within 60 min[60].

The use of alternative carbon sources, as well as the incorporation of alternative isotopes (e.g., nitrogen-15, oxygen-17) in primary metabolites is also described. Examples are *Escherichia coli* and *B. subtilis*, which under optimized reaction conditions incorporate [$^{13}$C]acetate and/or [$^{15}$N]ammonium sulfate into correspondingly labeled amino acids[61]. Immobilized bacteria were used to obtain nitrogen-15 enriched (*S*)-aspartic acid, (*S*)-alanine, (*S*)-tyrosine and (*S*)-DOPA in gram quantities[62].

**Secondary metabolites:** Unlike primary metabolites, secondary metabolites are not essential for the survival of cells. In most cases their specific functions are not known, but in cases where their effects have been recognized they are at the organismal level and usually of an ecological nature. Defense against predators, parasites or disease; interspecies competition; or facilitation of reproductive processes[63] are some examples. It is not surprising, therefore, that the pharmacological properties of plants and other organisms are usually inherent in one or more secondary metabolites[64], which are therefore chemicals of distinct interest in drug research. Alkaloids and antibiotics are two of the largest categories of such compounds.

Growing plants or plant tissue cultures have been used to prepare isotopically labeled secondary metabolites; an example is that of three pyrrolizidine alkaloids from *Senicio vulgaris*[65].

Microbiological fermentations have been widely used in the pharmaceutical industry for production of pharmacologically useful compounds. In fact, numerous examples of carbon-14 and tritium-labeled compounds prepared by fermentation have been described in the literature, including paspalic acid[66], aflatoxins[67], cyclosporin A[68], acarbose[69], salinomycin[70], tacrolimus (FK506)[71], avermectin[72], erythromycin A[73], Ziracin®[74], rapamycin[75], geldanamycin[76] and (arguably a secondary metabolite) methane[77]. Most of these publications are strictly production-oriented and lack detailed process optimization; however, some of the knowledge has been utilized in the laboratory-scale preparation of isotopically labeled compounds by fermentation. Reviews on applications[78] and methodological details[69b,70,79] have appeared.

### 12.2.3 Specific Requirements for Fermentations Using Isotopically Labeled Compounds

This section focuses on the optimization of key parameters such as selection of microorganism, precursor, medium and fermentation conditions. Experimental standard procedures such as organism development, procedures for inoculation, processing, work-up and purification are beyond the purview of this book.

**Microorganisms:** Microorganisms available for use vary widely from wild-types to those that have been fully optimized for the production of specific metabolites, and therefore they will have to be evaluated for their suitability to a particular labeling goal[69b]. Optimization of a microorganism for a certain purpose usually involves some sort of genetic modification that enhances the relevant metabolic pathway. One such example is a strain of *Escherichia coli* lacking the ability to metabolically degrade glucose and trehalose[80]. Addition of [$^{14}$C]glucose to the fermentation of this strain produced $\alpha$,$\alpha$-[$^{14}$C]trehalose in 80% isolated radiochemical yield on a small scale, and a 50% yield on a 30 mCi scale. Mutational modifications may also be used to reduce the level of endogenously produced compound, thereby leading to increased product specific activities.

**Addition of inhibitors:** Blocking of specific metabolic pathways with suitable inhibitors can be a useful strategy for increasing production of a desired product. For example, addition of borate to the methylotroph *Methylobacterium extorquens* AM1[26] suppresses the expression of the enzymes formate decarboxylase and glycine decarboxylase, opening the opportunity for stereoselective biosynthesis of [$^2$H/$^{13}$C/$^{15}$N]serine (**44**) from the appropriately labeled glycine (**43**) or methanol (**41**) as precursor (Figure 12.19). Thus, methanol dehydrogenase catalyzes the oxidation of methanol to formaldehyde which, without being degraded by formate decarboxylase, is stereospecifically condensed to glycine via methylenetetrahydrofolate (**42**). Therefore, variation of the labeled precursors ([1-$^{14}$C]- or [2-$^{14}$C]glycine **43** and [$^{14}$C]methanol (**41**) allows the specific labeling of the serine isotopomers **44**. Other cases in which additives increase incorporation, yield and/or purity of labeled materials have been reported[70,81].

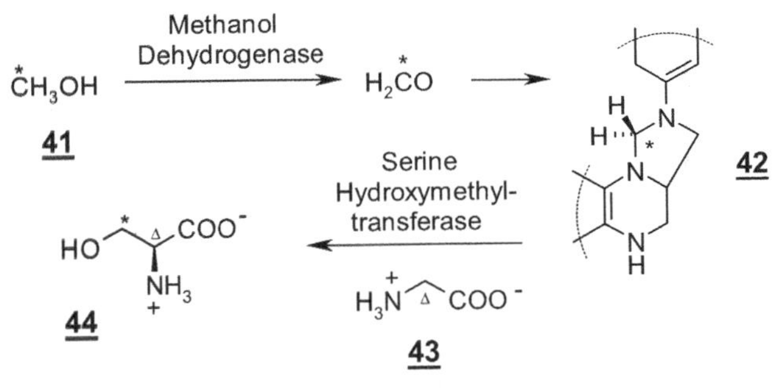


*Figure 12.19 Microbial synthesis of differently labeled serines*

**Precursor selection:** An important step in designing a microbial labeling process is the identification of one or more suitable labeled precursors. Generally, secondary metabolites are biosynthesized via primary metabolites from five metabolic sources. These are amino acids, shikimic acid (shikimic acid pathway), acetate and its homologues (polyketide pathway), mevalonic acid (isoprene pathway) and carbohydrates[64]. Selection of a suitable precursor is primarily influenced by the biosynthetic pathway(s) involved, but also depends on the desired position of label in the product and the availability of labeled precursors.

Even if the appropriate biosynthetic pathways are operative, poor incorporation into the product can occur if the added precursor is

- not sufficiently taken up by the organism;
- incorporated into the molecule structure of the microorganism[82];
- subject to alternative metabolic pathways[52b,73];
- highly diluted by endogenous compound.

Accordingly, the incorporation might vary significantly from precursor to precursor. An example demonstrating this point is the fermentative production of verruculogen (**45**) by the surface culture of *penicillium simplicissimum*[83] (Figure 12.20). In this process the relative incorporation of the five precursors tryptophan, mevalonate, proline, methionine and acetate was 29 : 14 : 11 : 2 : 1. From these results the precursors finally selected for the preparative scale was a mixture of [$^{14}$C]proline and [$^{14}$C]mevalonate.

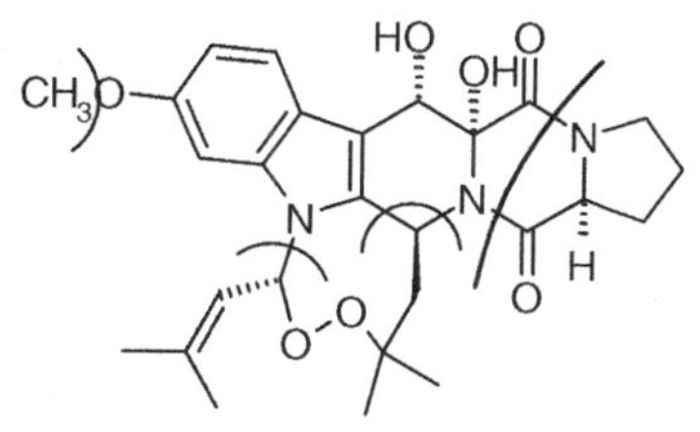


***Figure 12.20*** *Incorporation of different precursors into verruculogen*

Easily accessible labeled precursors such as carbon-14-labeled acetate, propionate and butyrate can be useful for labeling structurally complex compounds. Examples are FK506[71], erythromycin A[73] and geldanamycin[76]. Disadvantages of this approach, however, are high dilution and low incorporation (FK506 Inc = 0.59%, geldanamycin Inc = 0.62%) resulting in relatively low total and specific activities and large amounts of radioactive waste.

More efficient is the labeling of *N*-, *O*-, *S*- and *C*-methyl moieties using (*S*)-[methyl-$^{3}$H/$^{14}$C]methionine, and several applications have been described[68,82,84], despite the risk of metabolic lability of the label at *O*- and *N*-methyl groups. Feeding of *Micromonospora carbonacea* with (*S*)-[methyl-$^{3}$H/$^{14}$C]methionine was used[74] to label ziracin (Sch 27899, **46**), an oligosaccharide of the everninomicin type, with tritium and carbon-14 in the indicated *C*- and *O*-methyl- as well as *O*-methylene moieties. Optimized fermentation conditions gave an incorporation of 7.8% for tritium and 18.7% for carbon-14, resulting in specific activities of 288 mCi/mmol and 44 mCi/mmol, respectively. The difference in specific activities is caused by the addition of different amounts of labeled methionine and therefore reflects the different degrees of endogenous dilution. Dosing of **46** to rats did indeed result in considerable metabolic loss of label, but sufficient radioactivity remained with the molecule to make the studies successful; the carbon-14-labeled material appeared to be superior[85].

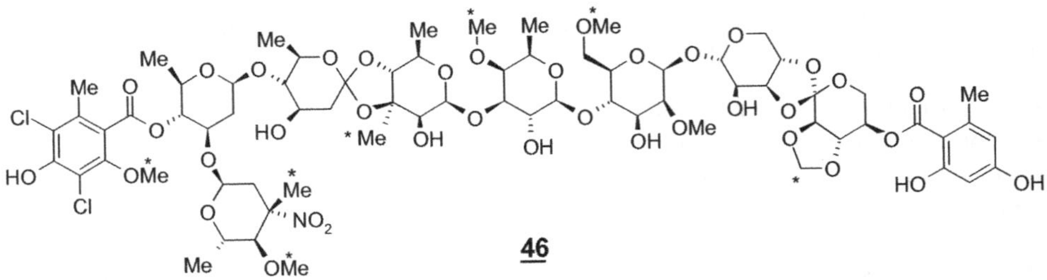

**46**

***Figure 12.21*** *Tritium and carbon-14-labeled Ziracin® (Sch 27899)*

The regioselective methylation of the *meta*-hydroxyl group of caffeic acid (**47a**) and 5-hydroxyferulic acid (**47b**) was accomplished using an extract of poplar and *S*-adenosyl [methyl-$^{14}$C]methionine[86].

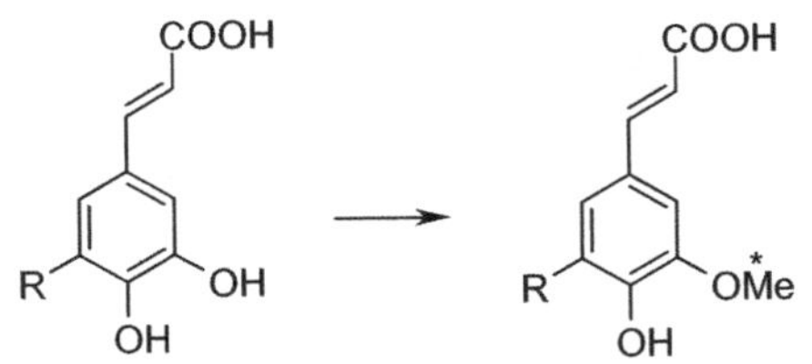

**47a**, R = H: caffeic acid ⟶ sinapic acid

**47b**, R = OMe: 5-hydroxyferulic acid ⟶ ferulic acid

***Figure 12.22*** *Synthesis of [$^{14}$C]sinapic acid and [$^{14}$C]ferulic acid*

Whenever utilization of such basic precursors does not produce satisfactory results, more advanced precursors have to be considered. Even if this involves significantly greater expense, the disadvantage may be offset by easier processing and less radioactive waste. An example is the fermentative (*Streptomyces avermitilis*) carbon-14 labeling of avermectin (**48**), a pentacyclic 16-membered macrolide with antiparasitic activity (Figure 12.23). Changing the precursor from sodium [1-$^{14}$C]propionate[87] to (*S*)-2-methyl-[1-$^{14}$C]-butyric acid[72] not only increased the incorporation from 0.69% to 14.3%, but also simplified the production process significantly. Analysis of the resulting product, however, identified 1.5% of 26-epi-avermectin as a new by-product. Since the latter precursor was applied in enantiomerically pure form, partial racemization of the 2-methyl[1-$^{14}$C]butyryl-CoA adduct was postulated as the reason. Interestingly, addition of the corresponding *N*-acetyl cysteamine thioesters, described as phosphopantotheine mimics, did not improve the incorporation rates, as seems to be the case in a number of other studies[88].

Another example is the fermentative labeling of rapamycin (**51a**) using *Streptomyces hygroscopicus*[75] (Figure 12.24). Addition of structurally complex precursors[89] such as (1*R*,3*R*,4*R*)-3,4-dihydroxy[1-$^{14}$C]cyclohexane [$^{14}$C]carboxylic acid (**49**), DHCCA, 111.7 mCi/mmol) and (*S*)-[$^{14}$C$_2$]pipecolic acid (**50**), PA, 114.2 mCi/mmol) showed high

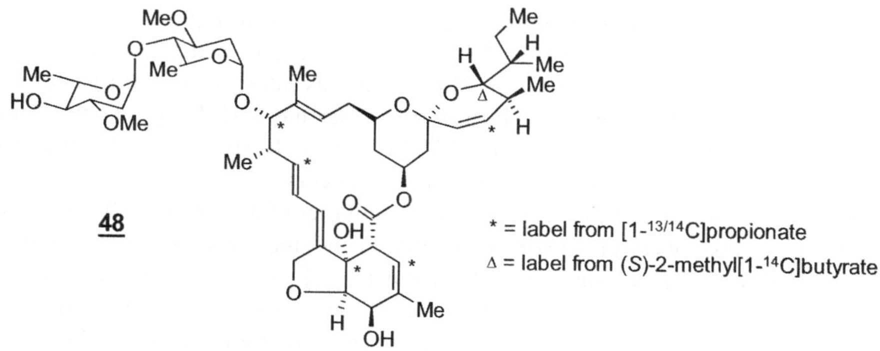


*Figure 12.23* $[^{14}C]$Avermectin $B_1$

incorporation (DHCCA: 29%, PA: 50%) with low dilution, thereby justifying the considerable effort (15 synthetic steps) required for their syntheses.

ADME studies of everolimus (**51b**), a rapamycin derivative with immunosuppressive activity, suggested that it was metabolically cleaved at two sites, indicated by hashed lines in the structure. Selective labeling of both molecular fragments was therefore conducted in order to allow complete elucidation of the metabolic fate of everolimus. Both precursors (DHCCA and PA) were inoculated to the fermentation broth of *Streptomyces hygroscopicus*. To label both parts at comparable specific activities the higher dilution of DHCCA was compensated for by higher concentrations of DHCCA. In the case, addition of a 3:2 mixture of DHCCA and PA produced $[1,2,36,37\text{-}^{14}C_4]$-rapamycin at a specific activity of 100.4 mCi/mmol in 13% radiochemical yield[75]. Final chemical conversion produced $[^{14}C_4]$everolimus.

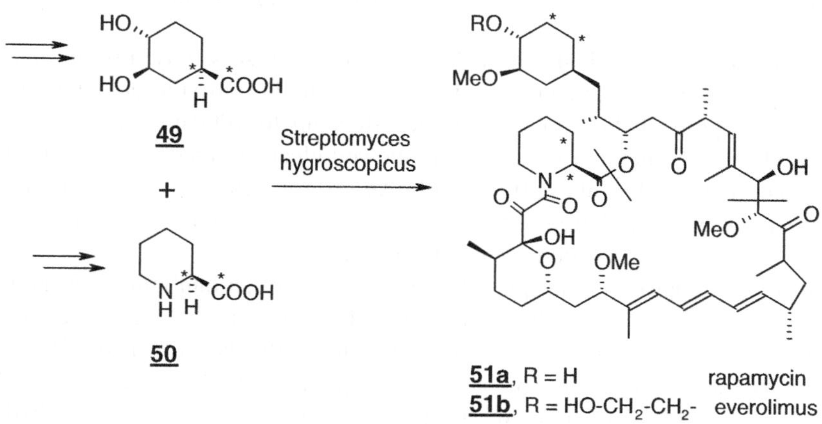


*Figure 12.24* Microbial labeling of $[^{14}C_4]$rapamycin

**Concentration of precursor:** The variation of the concentration may have a significant influence on features of the fermentation process, in particular the on incorporation (Inc), the dilution factor (Dil) and the mass yield of the metabolite.

**Influence on dilution and incorporation:** As mentioned in the rapamycin example, the higher endogenous dilution of [$^{14}C_2$]DHCCA can be counteracted by increasing its concentration in the fermentation broth. This effect was intensively investigated in the production of [$^3H/^{14}C$]paspalic acid (**52**) from racemic [$^3H/^{14}C$]tryptophan using the strain NRRL3080 of *Claviceps paspali*[66]. The results clearly showed that the extent of the dilution strongly depends on the amount of precursor fed, and on the amount of material produced by the organism. High specific activities of the target might be obtained by adding large amounts of precursor. On the other hand, these results also demonstrated that with significantly increasing amounts of precursor its incorporation (Inc) declines, resulting in incomplete utilization of the radioactivity. Therefore, optimization of the precursor concentration requires finding an appropriate compromise between dilution and incorporation.

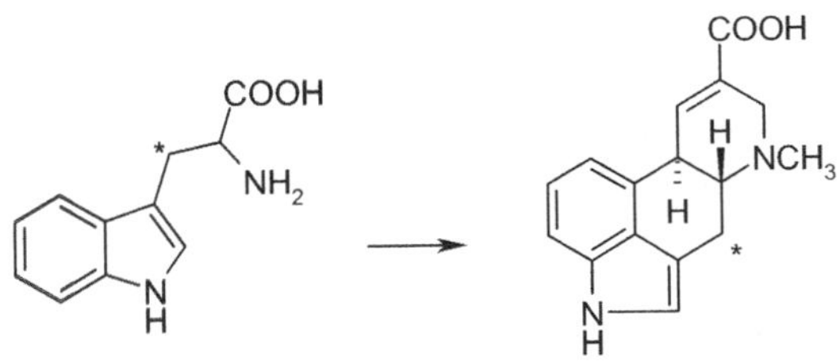


**52**

**Reaction conditions:** Inoculation of the production medium of sorbitol, ammonium succinate and mineral salts with *Claviceps paspali*; after 5 d at 22–24 °C addition of racemic [3-$^{14}$C]tryptophan, fermentation for 10 d

***Figure 12.25*** *Microbial labeling of [$^{14}$C]paspalic acid from racemic [3-$^{14}$C]tryptophan*

**Influence on the chemical yield:** Addition of precursors may also influence the chemical yields of the fermentation. In the case of the paspalic acid fermentation, addition of tryptophan had no measurable effect up to 1000 mg per liter of broth. (*S*)-[methyl-$^3H/^{14}C$]methionine, however, showed clear product inhibition in the fermentation of SCH27899[74] above 0.5 mg/mL. As a consequence, the concentration of the latter precursor was kept below 0.5 mg/mL over the entire incubation period of 5 d by regular feeding of small portions (pulse-feeding).

An alternative approach to keeping the concentration of an inhibiting precursor below a critical level was pursued for the fermentation of nemadectin[90]. Acetate and propionate in higher concentrations were found to inhibit not only the growth of *Streptomyces cyaneogriseus sp. noncyanogenus*, but also the production of the desired metabolite. Therefore, the amide derivatives of the precursors were added, and slow hydrolysis of the amides *in situ* provided continuous, low concentrations of acetate and propionate, thereby minimizing their inhibitory effects.

The microbial labeling of clavulanic acid may serve as an example in which the precursor stimulates the production of the product, but synergistically inhibits the formation of the by-product[91]. Long-term fermentation of *streptomyces clavuligerus* NRRL3585 showed a concentration-dependent stimulatory effect of ornithine, while simultaneously inhibiting the competing biosynthesis of cephamycin. This identified ornithine as the precursor of choice for the fermentation of clavulanic acid.

**Choice of optimal time-point for addition and harvest of products:** The choice of the most appropriate time point for addition of the precursor and harvest of the products depends on the organism's course of development. Chronological development of microorganisms consists of an initial period of constant biomass (adaptation phase), a period of increasing biomass (growth phase), a period of maximal, but constant biomass (stationary phase) and a final period of decreasing biomass (death phase).

If the labeled precursor is added at a time-point earlier than the microorganism can use it for the desired biosynthesis, it may undergo other biochemical transformations and be depleted. Alternatively, harvesting should take place when the maximal amount of the product is present and before degradation occurs. For example, secondary metabolites are produced during the stationary phase; therefore the optimal time point for precursor addition is at the end of the growth phase, and harvesting at a time point just before the death phase begins[70,92].

Since the timing of a microorganism's development phases vary considerably not only by strain, but also according to the variables of the fermentation conditions, they need to be characterized for each individual case. An example of such investigations is in the fermentation of [$^{3}$H]naphthyridinomycin from (*S*)-[methyl-$^{3}$H]methionine[84a].

**Fermentation medium:** Since the objective of fermentative labeling is the efficient utilization of radiolabeled precursor, the presence of unlabeled precursor in the fermentation medium should be minimized. In some cases this might mean that the established fermentation is not suitable for microbial isotope labeling and must be modified to eliminate or reduce the amount of that component of the medium. The necessary tradeoffs among what are usually sub-optimal alternatives aim at achieving a balance of chemical yield, incorporation levels and dilution, are illustrated by the examples of [$^{14}$C]riboflavin[93] and [$^{14}$C]bicyclomycin[94].

An alternative is the use of resting cells, which allows independent optimization of growth- and bioconversion conditions. One medium recipe may be used to grow the cells, from which they are harvested intact, and another medium may then be used for feeding the labeled precursor. This tactic was very effective for the incorporation of [$^{14}C_2$]acetate into aflatoxin B1 in a resting medium containing only glucose[95]. A variation of this technique[69] involved growing *Actinoplanes-Mutante SN1667/47* in a medium containing unlabeled glucose, after it was fully developed, transferring it to a production medium lacking unlabeled glucose but containing D-[U-$^{14}$C]glucose (330 mCi/mmol). Two molecules of labeled glucose were incorporated into the core-unit of the product, [$^{14}$C]acarbose, whose specific activity of about 138 mCi/mmol indicated a dilution of less than 1:5.

**Fermentation temperature:** Enzyme stability and acceleration of chemical reactions show opposite temperature dependencies, resulting in a distinct temperature optimum of the underlying fermentation process. Examples such as the fermentative production of the antibiotic viomycin[92] show that this temperature range for optimal production rates might be quite small (viomycin production: 25 °C – 523 mg/L, 30 °C – 657 mg/L; 37 °C – 0 mg/L). Since microbial degradation of the already-formed product is governed by the same principle, the conditions (temperature, incubation time) for netto formation (= actual isolable amount) needs to be optimized. As an example, for the microbial production of

[$^{14}$C]nucleosides, long-term incubation of *Escherichia coli* at 2 °C is clearly superior to short-term incubation at 37 °C[96]. Thus, the strong impact of the fermentation temperature and the time of incubation on the yeild of product supports the importance of defining them in pilot studies.

**Sensitivity to isotopic substitution:** There are two ways in which isotopes can alter the outcome of fermentation processes: through radiation effects and through heavy-atom effects. Heavy-atom effects are alterations (usually slowing) in the rates of chemical reactions compared with those occurring with lighter isotopes. As an example, several microorganisms show distinctive slower replication rates in $^{2}H_{2}O$[78,79b].

Furthermore, radiation has the potential to damage biological material; therefore fermentations with radioactive substrates might suffer from radiation-induced effects. Attempts to synthesize highly tritiated cyclosporin A by fermentation failed due to the radiation sensitivity of the microorganism *Trichoderma polysporum*[97]. In this case, (*S*)-[methyl-$^{3}$H]-methionine concentrations of 243 μCi/mL significantly reduced the productivity of the organism. Radioactivity concentrations above 1 mCi/mL caused irreversible damage to the mycelium, halting cyclosporin A production. Similar behavior was reported for the fermentation of [2a, 8a, 10a, 14a-$^{3}$H]geldanamycin (53)[76] (Figure 12.26). Since radioactivity concentrations above 2.5 mCi/mL were found to damage the mycelium, the amount of [3-$^{3}$H]propionic acid (19.3 Ci/mmol) added at day four after inoculation was limited in order to avoid producing concentrations of greater than 0.65 mCi/mL. Consequently, the fermentation process produced [2a, 8a, 10a, 14a-$^{3}$H]geldanamycin at a specific activity of only 76 mCi/mmol, highly diluted by endogenous compounds.

53

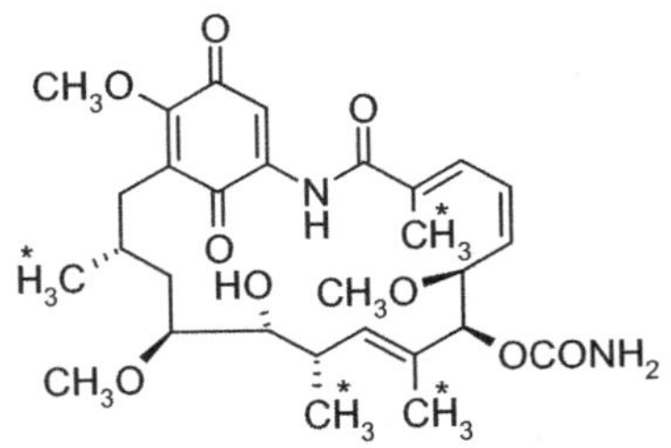


***Figure 12.26*** *[2a, 8a, 10a, 14a-$^{3}$H]Geldanamycin*

Unfortunately, because of wide differences in the radiosensitivity of different organisms, generalizations about the nature and extent of the radiosensitivity have little value. Therefore, fermentation with radioactivity requires pilot studies to determine the range of specific activities that can be used.

## 12.3 Biocatalyzed Synthesis of Key Intermediates for *Reconstitution* Approaches

For highly complex molecules, a combination of biotransformations with chemical synthesis ('*chemo-enzymatic*' approach) can open up opportunities superior to organic synthesis alone[98]. By this approach, an appropriate enzyme-catalyzed or microbial derivatization (or degradation) of the unlabeled target can provide an intermediate suitable for *reconstitution* by chemical synthesis with simultaneous introduction of the label. Simple examples from the literature include the oxidation–reduction and the demethylation–remethylation strategies. Applications of these tactics are described below.

### 12.3.1 Oxidation–Reduction Approach

A common strategy for tritium labeling is the oxidation–reduction approach, which provides fast and direct access to structurally complex molecules. While in the vast majority of cases the introduction of the label can be accomplished by reduction using one of many available selective reagents, regioselective oxidation often remains a problem.

*Hydroxy–Keto Interconversions*

The underlying strategy may be illustrated by the tritium labeling of the ganglioside GM1 ($\beta$-Gal-(1-3)-$\beta$-GalNAc-(1-4)-[$\alpha$-Neu5Ac-(2-3)]-$\beta$-Gal-(1-4)-$\beta$-Glc-(1-1)-Cer) (54). As an alternative to the previous microbial synthesis which produced GM1 of rather low specific activity, a chemo-enzymatic approach was developed[99]. Under optimized conditions, galactose oxidase (EC 1.1.3.9) specifically oxidized the primary hydroxyl group of the terminal sugar moiety of 54 to its corresponding aldehyde (55). Reduction of 55 with $NaB^3H_4$ (followed by treatment with unlabeled $NaBH_4$) produced the specifically labeled [$^3$H]GM1 ([$^3$H]54) at a specific activity of 1.2 Ci/mmol. Another example is the tritium labeling of riboflavin[100]. The vitamin $B_2$-aldehyde-forming enzyme isolated from *Schizophyllium commune* selectively oxidizes the terminal alcohol moiety of riboflavin to the aldehyde, which can then be reduced with sodium borotritide. As reported, use of $NaB^3H_4$ (27 Ci/mmol) resulted in [5′-$^3$H]riboflavin at a specific activity of 12.5 mCi/mmol.

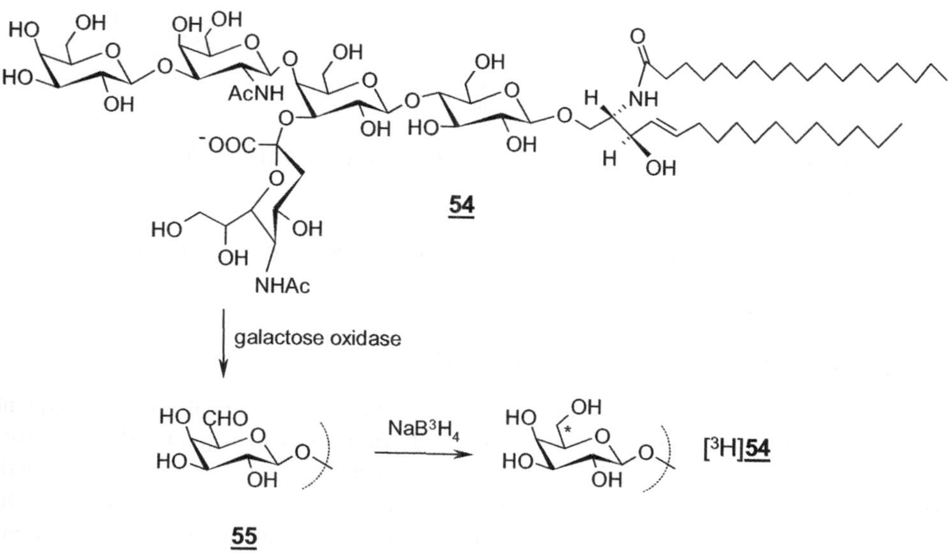

54
55
[$^3$H]54

*Figure 12.27 Tritium labeling of the ganglioside GM1*

*Alkyl–Alkenyl Interconversions*

An enzyme extract of tryptophan side chain oxidase from *Pseudomonas* oxidizes *N*-acetyl tryptophanamide to *N*-acetyl $\Delta^{2,3}$-tryptophanamide. Enantioselective [Rh-DIPAMP]$^{+}$-catalyzed reduction (DIPAMP = (1*R*, 2*R*)-bis[(2-methoxyphenyl)phenylphosphino]ethane) with tritium gas afforded the (*S*)-enantiomer in an optical purity of 94 % and nearly theoretical specific activity[101].

This procedure was applied to the deuterium and tritium labeling of the *N*-terminal fragment (pGlu–His–Trp–Ser–Tyr) of the gonadotropin releasing hormone LH-RH, and to pentagastrin (Boc–$\beta$Ala–Trp–Met–Asp–Phe–$NH_2$)[102]. pGlu–His–$\Delta^{2,3}$-Trp–Ser–Tyr, obtained from oxidation of the former peptide with (*S*)-tryptophan-2′,3′-oxidase (isolated from *Chromobacterium violaceum*) was reduced in the presence of [Rh-(*R,R*)-DIPAMP]$^{+}BF_4^{-}$ ($^2H_2$/$^3H_2$ gas mixture, 14 bar) to obtain the labeled isotopomer in a good yield (68%) with a diastereomeric excess of 99% (Figure 12.28 ). It is not known whether this reduction would work at the 1 atm or lower pressure normally used in laboratory tritium gas-handling equipment.

However, when this approach was applied to dehydro-pentagastrin (Boc–$\beta$Ala–$\Delta^{2,3}$-Trp–Met–Asp–Phe–$NH_2$) it became evident that the structure of the catalyst was not the only factor influencing the stereoselectivity of the reaction. Achiral hydrogenation conditions (PdO) led to a 52% diastereomeric excess in favor of the (*R*)-tryptophan derivative. Asymmetric induction was also very low in the presence of chiral rhodium catalysts: 44% in favor of the (*R*)-diastereomer with [Rh-(*S,S*)-DIOP]$^{+}$ (DIOP = 2,3-*O*-isopropylidene-2,3-dihydroxy-1,4-bis(diphenylphosphino)butane) and only 26% in favor of the (*S*)-diastereomer with [Rh-(*R,R*)-DIPAMP]$^{+}$ $BF_4^{-}$.

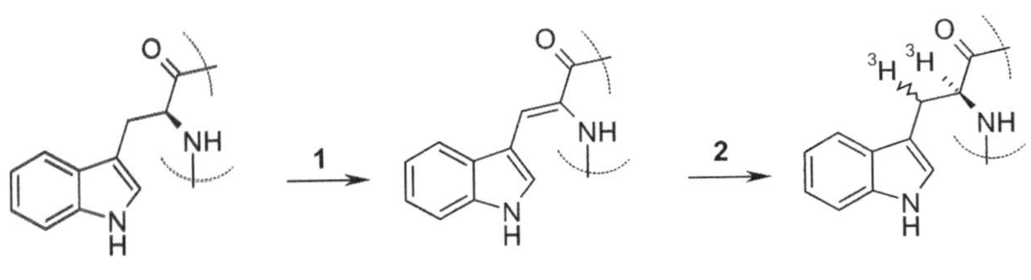


**Reaction conditions**: **1**. L-tryptophan 2′,3′-oxidase; **2**. [Rh(COD)(R, R)-dipamp]$BF_4$,$^2H_2$/$^3H_2$, 5-14 atm, 72 h

***Figure 12.28*** *Tritium labeling of the tryptophanyl moiety of a peptide fragment*

## 12.3.2 Dealkylation–Realkylation Approach

The combination of microbial excision of a structural fragment and subsequent re-incorporation of its labeled analog by chemical synthesis is an elegant and straightforward approach to the preparation of labeled compounds. As recently communicated, SSR126768 (**56**) was demethylated by incubation with *Mortierella isabellina* MMP108 to obtain the required SSR1136660 (**57**). Remethylation under standard conditions demonstrated that **57** was a precursor suitable for carbon-11 (and therefore probably tritium or carbon-14) labeling[39].

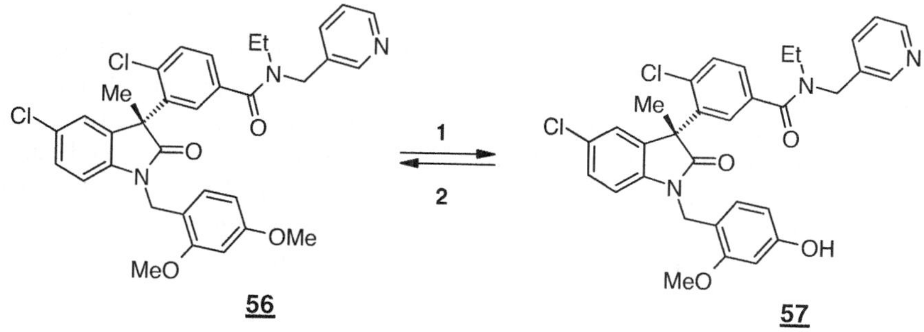

**Reaction conditions: 1**. *Mortierella isabellina MMP108*; **2**. NaH, $CH_3I$, DMF; 10 min

***Figure 12.29*** *Dealkylation–realkylation of SSR126768*

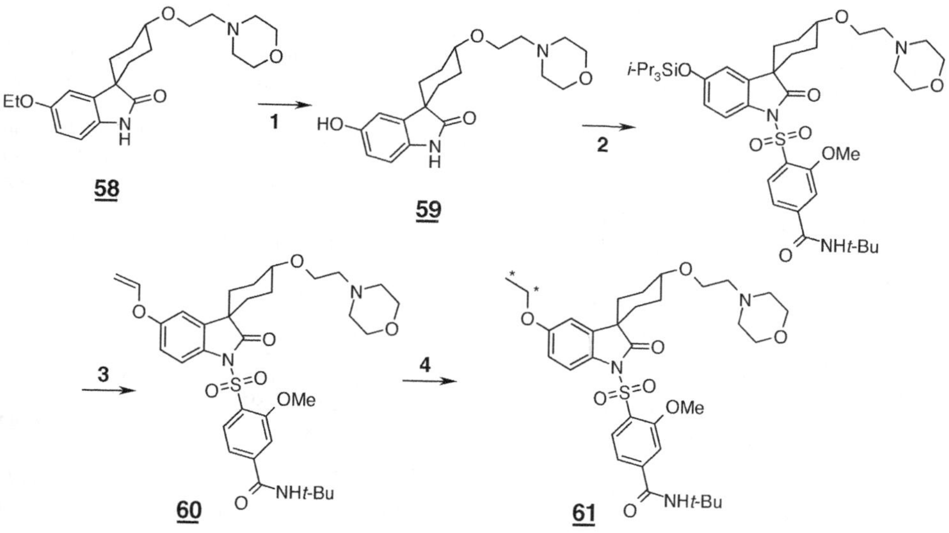

**Reaction conditions: 1**. *Cunninghamella Echinulata* LCP 73.2203 (60 %); **2a**. TIPSCl, $NEt_3$, DCM, cat. DMAP; **b**. NaOH, cat. $BnNEt_3^+$ $Cl^-$, $RSO_0Cl$, DCM (45 %); **3**. TBAF, tetravinyltin, $Cu(OAc)_2/O_2$; 4. $^3H_2$, undefined conditions

***Figure 12.30*** *Tritium labeling of SD121463 (**61**)*

An example[39] of a multistep chemoenzymatic approach is the preparation of [$^3$H]SR121960 (**61**). Attempts to dealkylate **61** itself using a culture of the fungus *Cunninghamella echinulata* LCP73.2203 were unsuccessful, but the microbial deethylation of the synthetic intermediate **58** did work. Sequential protection of the free phenolic hydroxyl group with TIPSCl, *N*-acylation with the appropriate arylsulfonyl chloride, release of the phenolic hydroxyl group upon treatment with $Bu_4NF$ and re-etherification with $(CH_2{=}CH)_4Sn$ converted **59** into the vinyl ether precursor **60**. Successful catalytic tritiation of the double bond afforded [$^3H_2$] SR121463 (**61**) at a specific activity of 56 Ci/mmol.

## References


1. Crout, D.H.G. and Christen, M., *Modern Synthetic Methods*, Vol. 5, Scheffold, R. Ed.; Springer: Berlin, Heidelberg, ***1989***, pp. 1–114.
2. a. Patel, R.N., *Enzyme and Microbial Technology*, ***2002***, *31*, 804–826; b. Straathof, A.J.J.; Panke, S. and Schmid S., *Curr. Opinion Biotechnol.*, ***2002***, *13*, 548–556.
3. a. Mahnke, M. and Moenius, Th., *J. Label. Compd. Radiopharm.*, ***2001***, *44*, 969; b. Evans, E.A., in *Synthesis and Applications of Isotopically Labelled Compounds 1991*, Buncel, E. and Kabalka, G.W., Eds.; Elsevier: Amsterdam, ***1992***, pp. 1–12.
4. a. DeSantis G. and Jones J.B., *Curr. Opinion Biotechnol.*, ***1999***, *10*, 324–330; b. Carrera, G. and Riva, S., *Angew. Chem.*, ***2000***, *113*, 2313–2341.
5. a. Wong, C.H. and Whitesides, G.M., *Tetrahedron Organic Chemistry Series*, Vol. 12; Pergamon: New York, ***1994***; b. Silverman, B., *The Organic Chemistry of Enzyme-Catalyzed Reactions;* Academic Press: San Diego, ***2000***.
6. a. Winkler, F.J.; Kühnl, K.; Medina, R.; Schwarz-Kaske, R. and Schmidt, H-L., *Isotop. Environ. Health Stud.*, ***1995***, *31*, 161–190, b. Schmitt, J. and Zenk, M.H., *Proc. of the Second Int. Conf. Methods Preparation and Storage of Labelled Compounds*, Brussels (Belgium), November 28th–December 3rd 1966, ***1968***, pp. 305–316.
7. Nomenclature of enzymes follows IUPAC- and IUBMB-recommendations (see http://chem.qmul.ac.uk/iubmb/).
8. Lester, D.J., in *Synthesis and Applications of Isotopically Labelled Compounds 1988*, Baille, T. Y. and Jones, J.R., Eds.; Elsevier: Amsterdam, ***1989***, pp. 423–426.
9. Ilida, K. and Kajiwara, M., *J. Label. Compd. Radiopharm.*, ***1991***, *29*, 201–216.
10. Uchida, K. and Kainosho, M., *J. Label. Compd. Radiopharm.*, ***1991***, *29*, 867–874.
11. Viswantha, V. and Hruby, V.J., *J. Org. Chem.*, ***1979***, *44*, 2892–2896.
12. Mezengeza, S.; Ventekatachalam, T.K.; Rajagopal, S. and Diksic, M., *Appl. Rad. Isotop.*, ***1993***, *44*, 1167–1172.
13. a. Andersen, L. and Jaroszewski, J.W., in *Synthesis and Applications of Isotopically Labelled Compounds 1997*, Heys, J.R. and Melillo D.G., Eds.; John Wiley & Sons, Ltd: Chichester, ***1998***, pp. 403–406; b. Andersen, L.; Nielsen, B. and Jaroszewiski, J.W., *Chirality*, ***2000***, *12*, 665–669.
14. Santoso, S.; Kemmer, T. and Trowitsch, W., *Liebigs Annalen der Chemie*, ***1981***, 658–667.
15. Sime, J.T.; Barnes, R.D.; Elson, S.W.; Jarvest, R.L. and O'Toole, K.J., *J. Chem. Soc. Perkin Trans. I*, ***1992***, 1653–1658.
16. Cregee, R.J.; Wagner, E.R.; Freedman, J. and Margolin, A.L., *J. Org. Chem.*, ***1990***, *55*, 4237–4238.
17. Antoni, G. and Långström, B., *Appl. Radiat. Isot.*, ***1987***, *38*, 655–659.
18. Dombrowski, G.; Chao, K.L. and Swiatek, K.R., *J. Label. Compd. Radiopharm.*, ***1979***, *17*, 365–368.
19. Schnell, B.; Faber, K. and Kroutil, W., *Adv. Synth. Catal.*, ***2003***, *345*, 653–666.
20. Goux, W.J.; Rench, L. and Weber, D.S., *J. Label. Compd. Radiopharm.*, ***1993***, *33*, 181–193.

21. Cappon, J.J.; Baart, J.; van der Walle, G.A.M.; Raap, J. and Lugtenburg, J., *J. Recl. Trav. Chim. Pays-Bas.*, ***1991***, *110*, 158–166.
22. Cappon, J.J.; Baart, J.; van der Walle, G.A.M.; Raap, J. and Lugtenburg, J., *Indian J. Chem.*, ***1992***, *31B*, 813–816.
23. Polach, K.J.; Shah, S.A.; Laluppa, J.C. and LeMaster, D.M., *J. Label. Compd. Radiopharm.*, ***1993***, *33*, 809–815.
24. Jemielity, J.; Kanski, R. and Kanska, M., *J. Label. Compd. Radiopharm.*, ***2001***, *44*, 295–304.
25. Augustyniak, W.; Kanski, R. and Kanska, M., *J. Label. Compd. Radiopharm.*, ***2001***, *44*, 553–560.
26. Hanners, J.L.; Gibson, R.; Velarde, K.; Hammer, J.; Alvarez, M.; Griego, J. and Unkefer, C.J., *J. Label. Compd. Radiopharm.*, ***1991***, *29*, 781–790.
27. Unkefer, C.J.; Hanners, J.L. and Ehler, D., *J. Label. Compd. Radiopharm.*, ***1991***, *29*, 1241–1246.
28. Willis, C.L., in *Synthesis and Applications of Isotopically Labelled Compounds 1994*, Allen, J. and Voges R., Eds.; John Wiley & Sons, Ltd: Chichester, ***1995***, pp. 591–596.
29. a. Kelly, N.M.; O'Neill, B.C.; Probert, J.; Reid, G.; Stephan, R.; Wang, T.; Willis, C.L. and Winton, P., *Tetrahedron Lett.*, ***1994***, *35*, 6533–6536; b. Kelly, N.M.; Reid, G.; Willis, C.L. and Winton, P., *Tetrahedron Lett.*, ***1996***, *37*, 1517–1520.
30. Fletcher, M.D.; Kelly, N.M.; Sutherland, A. and Willis, C.L., in *Synthesis and Applications of Isotopically Labelled Compounds 1997*, Heys, J.R. and Melillo D.G., Eds.; John Wiley & Sons, Ltd: Chichester, ***1998***, pp. 161–164.
31. Sutherland, A. and Willis, C.L., *Tetrahedron Lett.*, ***1997***, *38*, 1837–1840.
32. Ottolina, G.; Riva, S.; Carrea, G.; Danieli, B. and Buckmann, A.F., *Biochim. Biophys. Acta*, ***1989***, *998*, 173–178.
33. Battersby, A.R., *CIBA Foundation Symposium*, ***1985***, *111*, 22–30.
34. Ault, A., *J. Chem. Ed.*, ***2003***, *80*, 333–345.
35. Spahn-Langguth, H. and Benet, L.Z., *Drug. Metab. Rev.*, ***1992***, *24*, 5–47.
36. Yamamura, N.; Muratmatsu, S.; Suzuki, K.; Uchiyama, M. and Nakajima, E., *Radioisotopes*, ***1999***, *48*, 383–388.
37. a. Shram, S.I.; Lazurkina, T.Y.; Shevchenko, V.P.; Nagayev, I.Y. and Myasoedov, N.F., *J. Label. Compd. Radiopharm.*, ***1998***, *41*, 741–753; b. Myasoedov, N.F. and Susàn, A., in *Synthesis and Applications of Isotopically Labelled Compounds 1997*, Heys, J.R. and Melillo D.G., Eds.; John Wiley & Sons, Ltd: Chichester, ***1998***, pp. 453–457.
38. Kröger, M. and Fels, G., *J. Label. Compd. Radiopharm.*, ***2002***, *45*, 249–255.
39. Allen, J.; Brasseur, D.; de Bruin, B.; Denoux, M.; Perard, S.; Philippe, N. and Roy, S.N., *J. Label. Compd. Radiopharm.*, ***2007***, *50*, 342–346.
40. Wallace, M.A.; Dean, D.C.; Ellsworth, R.L.; Melillo, D.G.; Marks, T. and White R.F., *J. Label. Compd. Radiopharm.*, ***1993***, *33*, 697–702.
41. Logusch, S.J.; Feng, P.C.C.; Fujiwara, H.; Hutton, W.C. and Wratten, S.J., *J. Agric. Food Chem.*, ***1999***, *47*, 2125–2129.
42. Maier, F. and Zorn, L., Abstracts of the 11th Conference IIS-CED, Bad Soden (Germany), September 25, ***2003***.
43. a. Cohen, M.A.; Sawden, J. and Turner, N.J., *Tetrahedron Lett.*, ***1990***, *31*, 7223–7226; b. Singh, R.; Sharma, R.; Tewari, N.; Geetanjali, and Rawat D.S., *Chemistry and Biodiversity*, ***2006***, *3*, 1279–1287.
44. Deigner, H.P.; Freyberg, C. and Heck, R., *J. Label. Compd. Radiopharm.*, ***1993***, *33*, 171–180.
45. Allen, J.; Denoux, M.; Philippe, N.; Rivron, L. and Roy, S.N., *J. Label. Compd. Radiopharm.*, ***2007***, *50*, 624–626.
46. Tsai, M.C. and Gorrod, J.W., *Drug Metabolism and Drug Interactions*, ***1999***, *15*, 223–237.
47. Shackleford, D.M.; Hayball, P.J.; Reynolds, G.D.; Hamon, D.P.G.; Evans, A.M.; Milne, R.W. and Nation, R.L., *J. Label. Compd. Radiopharm.*, ***2001***, *44*, 225–234.
48. Wiltshire, H.R.; Prior, K.J.; Dhesi, J. and Maile, G., *J. Label. Compd. Radiopharm.*, ***2001***, *44*, 149–164.
49. Campbell, I.M., *Phytochemistry*, ***1975***, *14*, 683–693.

50. Simon, H.; Bader, J.; Günther, H.; Neumann, S. and Thanos, I., *Angew. Chem. Int. Edn Engl.*, ***1985***, *24*, 539–553.
51. a. Gager, F.L.; Core, M.T.; Bass, R. and Jenkins, R.W., *J. Label. Compd.*, ***1969***, *5*, 339–345; b. Bass, R.; Jenkins, R.W.; Newell, G. and Osdene, H., *Int. J. Appl. Rad. Isotop.*, ***1975***, *26*, 753–757.
52. a. Park, D.L.; Mislivec, P.B. and Heath, J., *J. Food Protection*, ***1977***, *40*, 90–93; b. Schenk, T.; Breitschwerdt, A.; Kessels, G.; Schuphan, J. and Schmidt, B., *J. Label. Compd. Radiopharm.*, ***1997***, *39*, 705–710.
53. Oldham, K.G. and Carre, N.G., Ger. Offen. DE72-2212092 19720313, ***1972***, pp. 2–18.
54. Schütte, H.R., *Radioaktive Isotope in der organischen Chemie;* VEB Deutscher Verlag: Berlin, ***1966***, p. 110.
55. Ling, J.R.; Cooper, B.; Parker, S. and Armstead, I., *J. Label. Compd. Radiopharm.*, ***1992***, *31*, 417–426.
56. Gupta, S., *Abstracts of the 1st Conference IIS-CED, Bad Soden (Germany)*, March 14, ***1991***.
57. Haut, S. and Core, M., *J. Label. Compd. Radiopharm.*, ***1982***, *19*, 117–120.
58. Hassam, S., *J. Label. Compd. Radiopharm.*, ***1985***, *22*, 1261–1271.
59. Krüper, A.; Gehrke, P. and Amrhein, N., *J. Label. Compd. Radiopharm.*, ***1990***, *28*, 714–718.
60. Dence, C.; Lechner, K.; Welch, M. and Kilbourn, M., *J. Label. Compd. Radiopharm.*, ***1984***, *21*, 743–750.
61. Myasoedov, N.F.; Kuznetsova, O.B.; Turkiya, M.A.; Basiladze, M.B.; Dierashvili, M.S. and Zhgenti, S., in *Synthesis and Applications of Isotopically Labelled Compounds 1991*, Buncel, E. and Kabalka, G.W., Eds.; Elsevier: Amsterdam, ***1992***, pp. 608–611.
62. a. Kahana, Z.E. and Lapidot, A., in *Stable Isotopes*, Schmidt, H.L.; Förstel, H. and Heinzinger, K., Eds.; Elsevier: Amsterdam, ***1982***, pp. 747; b. Kerby, N.W.; Rowell, P. and Stewart, W.D.P., UK Patent Application GB 2206881 A, ***1989***.
63. a. Crozier, A.; Clifford, M.N. and Ashihara, H., Eds.; *Plant Secondary Metabolites*, Blackwell: Oxford, ***2006***; b. Keller, N.P.; Turner, G. and Bennett, J.W., *Nature Rev. Microbiol.*, ***2005***, *3*, 937–947.
64. Bentley, R., *Crit. Rev. Biotechnol.*, ***1999***, *19*, 1–40.
65. Segall, H.J.; Brown, C.H. and Paige, J., *J. Label Compd. Radiopharm.*, ***1983***, *20*, 671–689.
66. Schreier, E., *Helv. Chimica Acta*, ***1976***, *59*, 585–606.
67. Salhab, A.S. and Edwards, G.S., *J. Label. Compd. Radiopharm.*, ***1978***, *15*, 41–46.
68. Kobel, H.; Loosli, H.R. and Voges, R., *Experientia*, ***1983***, *39*, 873–876.
69. a. Maul, W.; Müller, L.; Pfitzner, J.; Rauenbusch, E. and Schutt, H., *Drug Res.*, ***1989***, *39*, 1251–1253; b. Seidel, D.; Wehlmann, H. and Lenz, J., *J. Label. Compd. Radiopharm.*, ***2007***, *50*, 933–943.
70. Figdor, S.K. and Fried, J.H., in *Synthesis and Applications of Isotopically Labelled Compounds 1991*, Buncel, E. and Kabalka, G., Eds.; Elsevier: Amsterdam, ***1992***, pp. 389–396.
71. O'Connor, S.P.; Ellsworth, R.L.; Omstead, M.N.; Jenkins, R.G. and Kaplan, L., *J. Label. Compd. Radiopharm.*, ***1992***, *31*, 103–108.
72. Wallace, M.A.; Dean, D.C.; Ellsworth, R.L.; Marks, T.; Melillo, D.G. and Masurekar, P., in *Synthesis and Applications of Isotopically Labelled Compounds 1994*, Allen, J. and Voges R., Eds.; John Wiley & Sons, Ltd: Chichester, ***1995***, pp. 605–612.
73. Walker, A.E.; Pothuluri, J.V.; Heinze, T.M.; Volmer, D. and Cerniglia, C.E., *J. Label. Compd. Radiopharm.*, ***1997***, *34*, 59–67.
74. Hesk, D.; Gunnarsson, I.; Hendershot, S.; Koharski, D.; McNamara, P.; Schwartz, J.L.; Thonoor, M. and Wirth, M., *J. Label. Compd. Radiopharm.*, ***1999***, *42*, 159–167.
75. Moenius, Th.; Voges, R.; Mahnke, M.; Burtscher, P.; Metz, Y. and Guenat, Ch., *J. Label. Compd. Radiopharm.*, ***1999***, *42*, 29–41.
76. Taylor, G.F.; Foarde, K.K.; Webber, T.D. and Kepler, J.A., *J. Label. Compd. Radiopharm.*, ***2001***, *44*, 165–168.
77. Daniels, L. and Zeikus, J.G., *J. Label. Compd. Radiopharm.*, ***1983***, *20*, 17–24.
78. Simon, H. and Guenther, H., *Biotechnology 6b*, Rehm, H.J. and Reed, G., Eds.; Wiley-VCH: Weinheim, ***1988***, pp. 177–192.

79. a. Floss, H., *Radiotracer Tech. Appl.*, ***1977***, *2*, 689–732; b. Hoeksema, S., in *Synthesis and Applications of Isotopically Labelled Compounds 1991*, Buncel, E. and Kabalka, G.W., Eds.; Elsevier: Amsterdam, ***1992***, pp. 636–639.
80. Horlacher, R. and Peist, R., in *Synthesis and Applications of Isotopically Labelled Compounds*, Vol. *7*, Pleiss, U. and Voges R., Eds.; John Wiley & Sons, Ltd: Chichester, ***2001***, pp. 315–318.
81. Monjil, N.; Mechlinski, W. and Schaffner, C.P., *J. Antibiotics*, ***1976***, *29*, 438–443.
82. Hatano, K.; Harada, S. and Kishi, T., *J. Antibiotics*, ***1975***, *28*, 15–20.
83. Day, J.B. and Mantle, P.G., *Appl. Environ. Microbiol.*, ***1982***, *43*, 514–516.
84. a. Zmijewski, M.J.; Miller-Hatch, K. and Goebel, M., *Antimicrobial Agents and Chemotherapy*, ***1982***, 787–793; b. Hsi, R.S.P.; Witz, D.F.; Visser, J.; Stolle, W.T. and Ditto, C.L., *J. Label. Compd. Radiopharm.*, ***1989***, *17*, 147–165.
85. Gunnarsson, I.; Hesk, D.; Koharski, D.; Schwartz, J.L.; Thonoor, M.; Wirth, M. and McNamara, P., in *Synthesis and Applications of Isotopically Labelled Compounds 1997*, Heys, J.R. and Melillo, D.G., Eds.; John Wiley & Sons, Ltd: Chichester, ***1998***, pp. 399–401.
86. Marigo, G.; Riviere, D. and Boudet, A.M., *J. Label. Compd. Radiopharm.*, ***1981***, *18*, 695–702.
87. Ku, C.C.; Hwang, S.C.; Kaplan, L.; Nallin, M.K. and Jacob, T.A., *J. Label. Compd. Radiopharm.*, ***1985***, *22*, 451–459.
88. Simpson, T., in *Synthesis and Applications of Isotopically Labelled Compounds 1994*, Allen, J. and Voges, R., Eds.; John Wiley & Sons, Ltd: Chichester, England, ***1995***, pp. 855–862.
89. a. Lowdon, P.A.S.; Böhm, G.A.; Staunton, J. and Leadlay, P.F., *Angew. Chem.*, ***1996***, *108*, 2395–2397; b. Paiva, N.; Domain, A. and Roberts, M., *Enzyme Microb. Technol.*, ***1993***, *15*, 581–585; c. Paiva, N.L. and Demain, A.L., *J. Natural Products*, ***1991***, *54*, 167–177.
90. Ahmed, Z.H.; Fiala, R.R. and Bullock, M.W., *J. Antibiotics*, ***1993***, *46*, 614–622.
91. Romero, J.; Liras, P. and Martin, J.F., *Appl. Environ. Microbiol.*, ***1986***, *52*, 892–897.
92. Tam, A.H.K. and Jordan, D.C., *J. Antibiotics*, ***1972***, *25*, 524–529.
93. Lowry, L.K.; Herman, Y.F. and Sauberlich, H.F., *J. Label. Compd.*, ***1970***, *5*, 363–370.
94. Iseki, M.; Miyoshi, T.; Konomi, T. and Imanaka, H., *J. Antibiotics*, ***1980***, *33*, 488–493.
95. Mabee, M.S.; Chipley, J.R. and Applegate, K.L., *J. Label. Compds.*, ***1972***, *9*, 277–279.
96. Filip, J.; Pritasil, L.; Nejedly, Z.; Vesely, J. and Cihak, A., *J. Label. Compd. Radiopharm.*, ***1978***, *14*, 297–305.
97. Bogner, S.; Kobel, H.; Loosli, H.R.; Masero, R. and Voges, R., OL27-400 Internal Report Sandoz AG, ***1981***.
98. Filer, C., *J. Label. Compd. Radiopharm.*, ***2001***, *44*, 931–945.
99. Sonnino, S.; Chigorno, V. and Tettamanti, G., *Methods Enzymol.*, ***1999***, *311*, 639–653.
100. Kekelidze, T.N.; Edmondson, D.E. and McCormick, D.B., *J. Label. Compd. Radiopharm.*, ***1995***, *34*, 953–960.
101. Pinto-Alphandary, H.; van Craeynest-Jimonet; Morgat, J.L. and Fromageot, P., *J. Label. Compd. Radiopharm.*, ***1988***, *25*, 1273–1279.
102. a. Genet, R.; Hammadi, A.; Fusai, G.; Muriel, G. and Menez, A., in *Proceedings of the European Peptide Symposium, 24th September* ***1996***, pp. 413–414; b. Hammadi, A.; Meunier, G.; Tarride, J.-L.; Ménez, A. and Genet, R., in *Synthesis and Applications of Isotopically Labelled Compounds 1997*, Heys, J.R. and Melillo, D.G., Eds.; John Wiley & Sons, Ltd: Chichester, ***1998***, pp. 197–201.

# Index

Printed and bound by CPI Group (UK) Ltd, Croydon, CR0 4YY
16/09/2026
14963389-0007